REVIEWS in MINERALOGY
Volume 12

FLUID INCLUSIONS

An introduction to studies of all types of fluid inclusions, gas, liquid, or melt, trapped in materials from earth and space, and their application to the understanding of geologic processes.

EDWIN ROEDDER
UNITED STATES GEOLOGICAL SURVEY
RESTON, VIRGINIA 22092

Series Editor: PAUL H. RIBBE
Department of Geological Sciences
Virginia Polytechnic Institute & State University
Blacksburg, Virginia 24061

MINERALOGICAL SOCIETY OF AMERICA

REVIEWS in MINERALOGY

(Formerly: SHORT COURSE NOTES)

ISSN 0275-0279

Volume 12: FLUID INCLUSIONS

ISBN 0-939950-16-2

ADDITIONAL COPIES of this volume as well as
those listed below may be obtained at moderate cost from:

Mineralogical Society of America

2000 FLORIDA AVENUE, N.W. WASHINGTON, D. C. 20009

FOREWORD

"FLUID INCLUSIONS" is the first single-author contribution to be published by the Mineralogical Society of America in its ten-year-old series begun as "Short Course Notes" but since 1980 issued under the title "Reviews in Mineralogy". This is Volume 12, and it is the largest single volume (644 pages) with the most references (2001), diagrams (131) and photographs (387).

Edwin Roedder, who assembled this encyclopedic work after more than 30 years' study of fluid inclusions, is eminently qualified for the undertaking. He has been editor and primary contributor to "Fluid Inclusion Research -- Proceedings of COFFI", since its inception in 1968; these volumes provide citations and English abstracts of the 800-900 items published each year on fluid inclusions. In 1965 Roedder edited an English translation of a massive work by N.P. Yermakov and others [see Ermakov, 1950], and in 1972 he published "Composition of Fluid Inclusions", U.S. Geological Society Professional Paper 440JJ. As retiring President of the Mineralogical Society of America, he delivered his presidential address, "The Fluids in Salt", exactly six months ago at the Society's annual meeting in Indianapolis, Indiana.

The organization of this book is somewhat different from previous volumes in that the Table of Contents lists only chapter titles: detailed contents are given on the first page of each chapter. Additional unique features (made possible by the fact that there was no time pressure imposed by an impending Short Course!) are the Subject Index and the Locality Index at the end of Chapter 19, followed by an extensively cross-referenced bibliography.

Other titles in the "Reviews in Mineralogy" series are listed on the opposite page.

Paul H. Ribbe
Series Editor
Blacksburg, VA

1 / V / 84

CONTENTS

*A chapter outline is given on the initial page of each chapter.

ABBREVIATIONS USED IN THIS BOOK

T   - Temperature
P   - Pressure
V   - Volume
T_h, P_h - Temperature and pressure of homogenization**
T_t, P_t - Temperature and pressure of trapping**
T_d, P_d - Temperature and pressure of decrepitation**
T_m - Temperature of melting**
T_e - Temperature of eutectic**
T_n - Temperature of nucleation**
X_{CO_2} - Mole fraction CO_2 (similarly, X_{CH_4}, X_{H_2O}, etc.)
kbar   - Kilobars pressure
atm   - Atmospheres pressure
δ - In isotopic ratios, the difference, stated in parts per
             thousand (per mil), between the unknown and a standard
             (usually SMOW for H and O, and PDB for carbon).
equiv. - Equivalent
Salinity - Expressed as wt % NaCl equiv.; the concentration of NaCl
             needed to achieve the same depression of the freezing
             point.

**For details on definitions and modifications, see Table 7-3, page 198.

PREFACE and ACKNOWLEDGMENTS

This book has been written mainly to help the newcomer in fluid-inclusion work learn how to use fluid inclusions and to avoid many of the pitfalls and blind alleys that beset anyone starting in a new field of research. Of course, it is impossible to avoid all such diversions. However, too often, writers of scientific papers (and some editors) seem to believe that it is undesirable or even demeaning to report experimental details and the various problems that had to be overcome in the work. I do not agree with this approach. Why should subsequent workers be frustrated and waste much time solving problems that others have already solved? Give them the benefit of previous experience so that they can get on with new work; in so doing, they will encounter enough new problems of their own.

One difficulty in presenting a subject such as fluid inclusions is the surprising degree to which the chapters are interrelated. I have tried to strike an appropriate compromise between repeated referral to other chapters and excessive repetition, because everything cannot be put into logical sequence without redundancy.

Chapters 11-18 attempt to discuss the many applications of fluid inclusions to the study of and understanding of geologic processes and the geologic environments in which they acted. For the reader's convenience, I have categorized all environments from which fluid inclusions have been studied into these eight chapters. The arbitrary dividing lines between such environments are never sharp, nor generally acceptable, particularly if more than one geologist is asked, so I hope the reader will forgive me if my semantics disagree with his or hers; the differences are of no real consequence to the points being made.

Although some of the data and ideas in this book are new, other parts come from earlier papers of my own or from those on which I have been a coauthor. I make no apology for this, as I see no point in using quotation marks or trying to rephrase one's own words. Only about a third of the text is taken more-or-less directly from these earlier works (with modifications). Similarly, many but not all the photomicrographs have been used earlier. In the choice of examples, I have leaned heavily on those from my own experience and papers, mainly because this procedure is less prone to errors from misquotation, and because I have all the negatives of the photomicrographs I made in these studies.

In a petrography class, in 1939, my teacher, Dr. Donald M. Fraser, showed me some inclusions in Precambrian quartzite in which the bubbles were rapidly bouncing around in their tiny cells, as they presumably had been for more than a billion years. This so intrigued me that after completing graduate work (more than 30 years ago) I started studying fluid inclusions. I hope that some aspect of this book may, in the same way, intrigue others.

I have tried to help the reader by including chapter outlines and a detailed index, and in the References I have listed the page(s) where each item is cited, as this also can help the reader to become acquainted with the rather large and scattered literature and some of its applications. The overall organization is somewhat of an adaptation of the news reporter's outline -- "who, what, when, where, and why": what kinds of information inclusions provide, when and where inclusions form, how they change, how to prepare material and make microthermometric measurements[1], how to interpret these data, and then what has been found in applications of fluid-inclusion studies to each of a series of different geologic environments.

----

[1] The use of trade names in this publication is for identification only and does not imply endorsement by the U.S. Geological Survey.

As in most developing areas of science, numerous erroneous concepts, procedures, and statements have been published (including some of my own). I have a file of several hundred of these errors, but most do not merit attention and hence are not mentioned in this volume, except where they may have led to more than occasional confusion or misunderstanding by later workers. Caveat emptor.

I have been helped by many people in the preparation of this volume. Individuals from all over the world have sent data and reprints, and others have supplied original prints of some of the published illustrations that are used here. Most important, over the years I have profited from discussions with many of my colleagues; hence, the acknowledgements for this book could well be a tabulation of all those people. I am particularly indebted to my coauthors where I have used our joint efforts in this book; e.g., R.J. Bodnar was coauthor of a paper used as the major source for part of Chapter 9. In 1981 the Mineralogical Association of Canada held a well-attended short course on Fluid Inclusions: Applications to Petrology, and issued a handbook (Hollister and Crawford, 1981) that covers some of the same material presented here. Although the planning of present volume largely preceded my activity with the short course, I profited greatly from the experience gained in the preparation and presentation of my contributions to the course, and from the many interactions with other inclusionists involved.

Many others, too numerous to list, have helped in a variety of ways. Special mention is needed, however, for H.E. Belkin and E.L. Libelo, who skillfully made many of the photographic prints used here, and, with the cooperation of the USGS library staff, located many obscure references.

Several individuals helped with the typography of some of the first chapters, but special thanks are due N. Teed for most of this work. I am also thankful to A. Sangree for editing, and to my wife Kathleen for frequent editorial consultations -- as well as for her considerable patience.

Parts of the manuscript were reviewed by too many individuals to list completely, but the following have contributed thoughtful reviews of one or more chapters: E.C. Alexander, Jr., A.T. Anderson, Jr., L.D. Ashwal, C.E. Barker, M.D. Barton, P.B. Barton, Jr., H.E. Belkin, S.C. Bergman, P.R.L. Browne, I.-M. Chou, M.L. Crawford, C.G. Cunningham, P.J. Eadington, N.K. Foley, J. Guha, D.M. Harris, I. Haapala, D.O. Hayba, P. Heald-Wetlaufer, R.T. Helz, R. Henley, L.S. Hollister, S.S. Howe, R. Kreulen, S.D. Ludington, P.T. Lyttle, H.O.A. Meyer, S. Morasse, R. Petrovich, N.M. Ratcliff, T.J. Reynolds, W.I. Rose, Jr., L.P. Rowan, R.L. Rudnick, R.O. Rye, E.T.C. Spooner, T.G. Theodore, G.C. Ulmer, and R.W.T. Wilkins. I owe a special debt to J.W. Hedenquist and R.J. Bodnar, who reviewed all of them. These colleagues should not be held accountable for failing to catch any of the many errors and omissions that I am sure remain. I would appreciate having all such shortcomings called to my attention.

I am especially grateful, and the members of the Mineralogical Society of America should be equally grateful, for the continuing hard work and selfless contribution of P.H. Ribbe, Editor of "Reviews in Mineralogy". Only authors in the series really know how large and thankless a task he has.

Edwin Roedder
959 U.S. Geological Survey
Reston, VA 22092

17/II/84

Chapter 1

INTRODUCTION to FLUID INCLUSIONS

CONTENTS

GENERAL NATURE OF FLUID INCLUSIONS

"A Fluid Inclusion Speaks Its Mind"

Do you like me?
Why do you look
At this tiny thing
That I am?

I see your eyes
And you want to sing to me
To discover my soul

Don't tell me
That you want to know me
Just to learn my secrets.
Caught you there
With your soul bare.

You want to know my past.
Was I homogeneous?
Was I heterogeneous?
These are the questions
I see in your eyes.

Can you hear me?
I am trying to say something.
I know you can love me.
And when you do
I will tell it to you all.

Jayanta Guha
Chicoutimi, March 6, 1982

With the notable exception of those crystals in metamorphic samples that have grown in the solid state, all crystals in all terrestrial and extraterrestrial samples have grown from some kind of fluid. The minerals of many meteoritic and lunar samples, and of terrestrial igneous rocks, have grown from fluid silicate melts; some of the minerals from these sample groups have grown in the presence of an additional low-density vapor phase. New crystals in many sedimentary and some metamorphic rocks, and in almost all ore deposits, formed from an aqueous fluid containing various solutes. After crystallization, the minerals of practically all terrestrial sedimentary, metamorphic, and igneous rocks

1

have been fractured one or more times, and the fractures have healed in the presence of liquid or gaseous fluids. During these processes of crystal growth and fracture healing, small quantities of the surrounding fluid medium are commonly trapped as fluid inclusions in the host crystal. If solid materials are also present in the fluid, they may become enclosed by the growing crystal as solid inclusions; often some fluid is also trapped as the host crystal surrounds the solid inclusion.

In much of the inclusion literature, the term fluid inclusion has been used only for those inclusions that trapped a fluid that remains in large part fluid at surface temperature, and the term "melt inclusion" has been used for those that have become essentially solid at surface temperature. As will be seen, however, a continuum exists between these two extremes, and because the processes involved in trapping, the methods of study, and the problems in interpretation of the data are basically the same for all such inclusions, I use the term fluid inclusion for all. The term is thus used here to refer to the state of the trapped material at the time of trapping (i.e., a liquid or a gas) and not to its condition as we observe it now. During cooling to surface temperatures, the fluid in many inclusions has formed crystals, and the fluid in inclusions that have trapped a silicate melt may have cooled to form a glass. The term liquid inclusion (or gaseous inclusion) can be used when it is necessary to differentiate between melt inclusions and those that remain fluid at room temperature. Further ambiguity can be eliminated by the use of compositional terms such as gaseous, aqueous, CO_2, oil, or melt inclusion.

Minerals from certain metamorphic rocks, particularly those recrystallized in a water-deficient environment (Yoder, 1955), may be free of all fluid inclusions (gas, liquid, or melt); if any inclusions are present, they are most apt to consist of CO_2. Most materials believed to have formed deep in the earth, such as diamonds and the minerals of some eclogites, are essentially free of fluid inclusions. The most notable exceptions are the CO_2 inclusions in the minerals in the dunitic and peridotitic xenoliths or "nodules" brought to the surface with many alkalic basalts (Roedder, 1965a). Liquid inclusions have even been reported in certain meteorites (Yasinskaya, 1967; Fieni et al., 1978; Warner et al., 1983).

Inclusions are seldom larger than 1 mm and commonly go unnoticed. On the other hand, museum specimens that have single inclusions containing tens or even hundreds of milliliters of fluid are known (Hidden, 1882; Prikazchikov, 1959; Prikazchikov et al., 1964; Rankin and Greenaway, 1978). The number of inclusions in any given sample is usually related inversely to inclusion size[1], for very small inclusions are much more abundant than large ones. In most samples, inclusions in the range 1 to 10 μm outnumber all inclusions larger than 10 μm by a factor of 10 or even 100, and electron microscopy has revealed large numbers of inclusions as small as ~0.02 μm (2×10^{-6} cm; Green and Radcliffe, 1975). Presumably, a size continuum exists down to single water molecules (~2×10^{-8} cm) trapped along grain boundary dislocations or bound in the structure (Spear and Selverstone, 1983). Ordinary white quartz or calcite is usually white from the presence of perhaps 10^9 inclusions/cm^3. Most studies of fluid inclusions are made on populations in the range 10-100 μm, although some have been made on inclusions averaging <10 μm, particularly in metamorphic terranes.

----

[1] In this book, as in most inclusion literature, the term "size" is used rather loosely, because inclusions may be very irregular in shape. Size generally refers to the largest dimension of the main inclusion body and ignores thin extensions. A more exact definition is seldom warranted in inclusion studies.

Historical Perspectives

Other than brief mention in some of the ancient literature, the first specific description of inclusions is by Abu Reykhan al-Biruni, a central-Asian scholar of the 11th century (Lemmlein, 1950). Robert Boyle's description of a large moving bubble in quartz (Boyle, 1672) is apparently the first reference in English. The early naturalists showed considerable attention and interest in these large inclusions (Dewey, 1818; Dwight, 1820; and others as listed by Smith, 1953). Although several earlier reports were made on the nature of the trapped fluids (e.g., Dolomieu, 1792), the first actual "analytical" work to establish the composition of specific inclusions was that done by Breislak (1818), Davy (1822), Brewster (1823a), and Nicol (1828); such work was used as very strong evidence to support the Neptunist theory of the formation of minerals and rocks from water. In the following 155 years, a large number of investigations of fluid inclusions was made, for a variety of reasons. Many investigations were aimed at using or disproving the thesis proposed by Sorby (1858) that the gas bubbles present in the fluid of most inclusions are the result of differential shrinkage of the liquid and the enclosing mineral during cooling from the higher temperature of trapping (T_t) to the temperature of observation.[2] Sorby showed that the coefficients of expansion on heating (and, conversely, the coefficients of contraction on cooling) for a variety of liquid solutions resembling the fluids in inclusions were one or two orders of magnitude greater than the coefficients for the enclosing or host minerals. Hence, he reasoned that the temperature of trapping can be estimated by heating the sample to the point at which the bubbles disappear, i.e., the temperature of homogenization (T_h). Although most of the early work is descriptive only, occasional qualitative analytical data are found (e.g., Zirkel, 1870), and a few quantitative analyses were made without the benefit of modern techniques and apparatus. In spite of these limitations, the work of such men as H.C. Sorby and Ferdinand Zirkel is surprising; many of their conclusions about the significance of inclusions have stood the test of time, and many of their observations are still valid and useful.

Inclusions were a subject of intense study and debate by many geologists during Sorby's time, and some of the conclusions from their study were so disturbing to certain schools of geologic thought that many efforts were made to discredit them. Phillips (1875) pointed out that there are large variations in the composition and nature of different inclusions in the same sample and that Sorby had obviously oversimplified the problem. One of the most serious questions challenging the validity of fluid-inclusion evidence was that concerning leakage. Skinner (1953) and several other workers in that period published evidence that under certain conditions, inclusions do leak. Fifteen years later, he recanted (Roedder and Skinner, 1968), but even today, some are reluctant to accept inclusion data for two reasons: (1) proof of secondary origin of many inclusions; and (2) the mistaken belief that widespread or universal leakage of fluid inclusions is a proven fact. It is generally agreed that, in nature, some fluid inclusions have leaked, particularly in certain special metamorphic environments, but that such leakage is negligible in many other environments. Actually, these very ambiguities of origin and subsequent

----

[2] Frequently the gas or vapor bubbles in the liquid (which range from high-pressure gas to nearly a vacuum) are termed "libellae" in the older literature, and whole inclusions, regardless of phases present, are termed "vacuoles" or "lacunae," but because of changing usage, none of these terms is free of ambiguity. Probably the term "bubble" is most appropriate for the gas or vapor phase in a surrounding fluid (or glass), but unfortunately some geologists apply this term incorrectly to the whole inclusion of liquid and a gas or vapor bubble.

environment may cause the data to have greater significance, if the techniques used are adequate to resolve the ambiguities. Error can arise when the conclusions drawn are broader than the available facts (and their limitations) justify. Unfortunately, several papers containing such conclusions have been published, some overly commendatory and others derogatory toward inclusion study, and these papers have caused considerable misunderstanding. Regardless of their origin and history, inclusions do represent actual samples of fluids formerly existing at some time in the history of the earth. As such, they are important clues in understanding geologic processes.

Literature Sources and Summaries

Literature originally published in English. The literature on fluid inclusions is very large and very scattered. Smith (1953) presented a comprehensive annotated bibliography of more than 400 papers on the subject published before 1953, but as a result of a considerable increase in interest in fluid inclusions in recent years, the literature has had a very short doubling time since then. Anyone interested in inclusions should read Sorby's classic article (Sorby, 1858), which shows him to be just as remarkably perspicacious in this field as he was in many others. The present volume is being published about 75 years after the death of Sorby and 125 years after his pivotal paper. He was truly a man for all sciences and has been widely recognized as the "Father of Microscopical Petrography" (D.W. Humphries, as quoted by Johnson, 1979); he also was certainly the "Father of Fluid Inclusion Study."

Gübelin (1953) published a 220-page book dealing with the identification of gemstones on the basis of their inclusions, both solid and fluid. A series of short courses on fluid inclusions has been offered to small groups in recent years, at Imperial College, London (1978); Memorial University, St. Johns, Newfoundland (1979); La Trobe University, Australia (1979); by the Mineralogical Association of Canada at Calgary (1981); and by the Indian Institute of Technology in Bombay (1982). These courses have all had notes available for the registrants, but except for those for the 1981 course (Hollister and Crawford, 1981), the notes have not been formally published. An extensive review of data has been published (Roedder, 1972) on the composition of fluid inclusions and on the significance and validity of the many methods used to obtain such data. As that paper is no longer in stock at the U.S. Government Printing Office, some segments from it are given here, mainly in Chapters 4 and 5.

In addition to the short courses, numerous scientific meetings in the West have dealt specifically with fluid-inclusion studies. Some of these meetings have been organized under the aegis of COFFI (an acronym for Commission on Ore-Forming Fluids in Inclusions) at various international meetings, and others have been individually organized (see Roedder, 1968a, for details). None of these meetings even approaches the size of the conferences held in the USSR (see below).

Books from the USSR. G.B. Naumov, Mironova and Naumov (1976) reported that their bibliographic files contained 2900 entries on fluid inclusions as of 1975. More than two-thirds of the publications are in Russian, and 70% of the 2900 were published in the period 1965-1975. As little of this Russian literature was covered by Smith (1953), Lemmlein (1956b) presented a short review of much of the Russian inclusion work as an appendix to a Russian translation of Smith's book. This appendix has been translated into English (Lemmlein, 1956b). Periodic reviews have been published in Russian on the fluid-inclusion work in the West, but not the reverse, even though the bulk of all fluid-inclusion literature is published in Russian. An excellent short review of the various current studies in the USSR presented by Bakumenko and Dolgov (1977) has since been translated into English.

4

Other than a 126-page French book by Deicha (1955), all non-English books on inclusions have come from the Soviet Union. A 169-page book in Ukrainian on inclusion study (Kalyuzhnyi, 1960) emphasized the methods of identification and study of the phases present, particularly as used by Kalyuzhnyi and his coworkers. Numerous books on inclusions have been published in Russian: Lesnyak (1964, 219 pp.); Kostyleva (1964, 98 pp., on decrepitation); D.N. Khitarov (1965a, 189 pp., mainly on composition, 336 references); Ikorskii (1967a, 121 pp., on organic inclusions in the Khibiny alkalic massif); Ermakov and Dolgov (1979, 271 pp., on "Thermobarogeochemistry," a synonym for fluid-inclusion study commonly used in the USSR); Moiseenko and Malakhov (1979, 200 pp., on methods and apparatus for studies of inclusions); Rekharsky (1980, 200 pp., also on methods and apparatus for studies of inclusions), and others.

The first book on the subject in any language was by Ermakov (1950; sometimes the name is transliterated and indexed as "Yermakov.") This 460-page Russian book has been translated into English; it summarizes the extensive Russian work in the field up to 1950. Subsequent Soviet book literature on the subject presents somewhat of a bibliographic tangle, and, in addition, some of the volumes are exceedingly rare. Since the publication of Ermakov's book, a series publication, the "Transactions of the All-Union Research Institute of Piezooptical Mineral Raw Materials" (Trudy VNIIP, Vsesoyuznyy Nauchno-Issledovatel'skiy Institut P'ezioopticheskogo Mineral'nogo Syr'ya), was established, with Ermakov as chief editor. Of this series, volume I, part 2 (177 pp., Moscow, 1957), and volume II, part 2 (134 pp., Moscow, 1958) dealt with inclusions and have been translated and are bound in with the published translation of Ermakov's book. (See headnote to "References.") Additional parts and at least five more volumes of this series are known to have been published but are not available to me.

Symposium volumes. The first Soviet "All-Union Conference on Mineralogical Thermometry and Barometry," dealing mainly with inclusion studies, was held in Moscow about 20 years ago (May 17-24, 1963), where more than 60 papers were presented. A 328-page volume containing 42 papers from the symposium was issued under the title "Mineralogical Thermometry and Barometry" (Smirnov et al., 1965) but has not been translated. One of the papers in this volume gives a review of more than 400 analyses of inclusions taken from the literature (Khodakovskiy, 1965). A 264-page volume containing 38 additional papers stemming from this conference was issued in 1966 by the "Nedra" Press of Moscow (Ermakov, 1966a) as volume 9 of the "Transactions of the All-Union Research Institute for the Synthesis of Mineral Raw Materials."

The second symposium on "geothermobarometry" (mainly by inclusions), held in Novosibirsk in 1965, resulted in 102 papers, issued in two volumes also entitled "Mineralogical Thermometry and Barometry." Volume 1 (Ermakov, 1968a, 368 pp.) presented 47 papers dealing with general physical and chemical problems of ore formation and the use of inclusion studies as an exploration tool. Volume 2 (Ermakov, 1968b, 320 pp.) contained 53 papers dealing with the thermodynamic regime of ore formation, detailed studies of individual minerals from specific stages of mineralization, and methods of inclusion study.

The third symposium, held in the Moscow area but inaccessible to Westerners, resulted in 161 papers and a 280-page volume of abstracts (Entin, 1968). The program of the fourth, in 1973 at Rostov-on-Don, listed only 144 papers presented, but 236 extended abstracts were published in a 351-page volume (Ermakov and Trufanov, 1974). The fifth, in 1976 at Ufa, included 255 papers (Ermakov, 1976). The sixth, in 1978 at Vladivostok, had two volumes of abstracts, volume I - Thermobarogeochemistry in Geology, and II - Thermobarogeochemistry and Ore Genesis (Ermakov, 1978). The seventh was scheduled for L'vov in 1983, but has been rescheduled for 1984.

In addition to the above series of symposia, several major special symposia on inclusions have been held in the USSR, for example, one in 1975 at L'vov, on "Carbon and its compounds in endogenetic processes of mineral formation" (Anonymous, 1975), and numerous other smaller conferences. Most of these symposia have been announced, and, except for a few that were classified and for which the abstracts are unavailable, the abstracts have been translated and indexed (see Roedder, 1968a).

Several smaller symposia have been held in the West, some of which have resulted in special volumes. Of these, three of the most notable are reported in Schweizerische Mineralogische und Petrographische Mitteilungen, v. 50, part 1, 1970 (208 pp); Bulletin de Minéralogie, v. 102, nos. 5-6, 1979 (197 pp); and Chemical Geology, v. 37, no. 1/2, 1982 (Special Issue) (213 pp). The first Chinese conference on fluid inclusions was held in 1977 in Guangxi Province in the People's Republic of China; it resulted in a volume of 335 pp. plus many plates, containing 45 papers and 10 abstracts on fluid inclusions (Chinese Geological Society, 1981). Another smaller conference (48 papers) was held in 1981 (Chi, 1981). Fluid-inclusion studies are now going on in at least 27 countries of the world.

Abstract volumes. Every year since 1968, the world literature on fluid inclusions and closely related work has been abstracted in English in Fluid Inclusion Research--Proceedings of COFFI (Roedder, 1968a). This set of volumes is sometimes referred to simply as "COFFI." Although originally aimed only at studies of the ore fluids, the volumes cover studies of all types of fluid inclusions (aqueous, organic, silicate melt, sulfide melt, gas, etc.).

DATA OBTAINABLE FROM FLUID INCLUSIONS

The geologist works as a detective in attempting to reconstruct the events of the remote past from the evidence of the present. Unlike the detective, however, the geologist works with events that took place millions or even billions of years ago, and the clues he has to piece together are unusually meager in view of the long history. As detailed in Chapters 11-18, fluid inclusions provide a record, albeit complex, fragmentary, and miniscule, of fluids long since gone from the face of the earth; hence, they provide a rich source of small but valuable clues for unravelling past geologic processes. Regardless of their origin and history, inclusions do represent actual samples -- with rare exceptions, the only samples -- of fluids existing at some time in the geologic history of a rock. As such, they are important clues in understanding the geologic modus operandi--the temperature, pressure, density, and composition of the fluids that formed or traversed the rock. In many rocks, they provide the only such data.

One of the earliest and most outstanding examples of the usefulness of inclusions is found in the study of ore deposition. Far too commonly, the ore-forming fluid for a deposit was simply assumed to have been "rich" only in those constituents now present, even in proportion to their abundance in the deposit! However, most ore fluids contained, in addition to the ore elements deposited, large amounts of volatile constituents and soluble salts that passed through the deposit leaving almost no trace -- except the fluid inclusions -- and hence much has been learned about the processes of ore deposition from a study of these inclusions.

Temperature. The use of fluid inclusions for deciphering the temperature of past geologic events was first proposed by H.C. Sorby in 1858. Since then, data from fluid-inclusion geothermometry have been reported in several thousand scientific papers, and inclusions have taken their place as one of our best and

Figure 1-1. Serial photomicrographs of a large primary inclusion in sphalerite (S) from Creede, Colorado (sample ER 57-34), after equilibration at the temperatures indicated. The horizontal bars are oscillatory striae on the cavity walls. On heating, the bubble (v) decreased in volume to the homogenization temperature (or "filling temperature"); this inclusion would be said to "homogenize in the liquid phase at 210°C." From Roedder (1972).

certainly most widely applicable geothermometers. In spite of this common usage, the most astounding aspect of fluid inclusions is seldom mentioned -- the enormous number of such clues to temperature that nature has provided. Each cubic centimeter of white quartz or calcite may contain a billion fluid inclusions, each a self-contained recording geothermometer, preserving for us the temperature of a specific moment in the past--the moment of sealing of that particular inclusion. The fact that most of these billion inclusions may record the same temperature (or several temperatures representing the small number of geologic processes that the sample has experienced) does not detract from the wonder of it for me.

Basically this use of fluid inclusions for geothermometry is a result of differential shrinkage of the host mineral and the inclusion fluid on cooling from the temperature of trapping to that of observation. The fluid shrinks far more than the host, and in the simplest case, the difference shows up as a bubble in the fluid at surface temperature. We need only reverse this process by heating the sample until the bubble disappears (i.e., the fluid inclusion homogenizes), as viewed through the microscope, to obtain the temperature at which the bubble first appeared millions or billions of years ago (Fig. 1-1). Numerous caveats must be observed, and corrections (sometimes major) must be made, as detailed in later chapters, but, in essence, that is the homogenization method for geothermometry. The temperatures obtained range from well over 1000°C for many silicate-melt inclusions (i.e., those in which the fluid was a melt or lava) to room temperature and even below. Microscope studies such as these, and also those studies at very very low temperatures, are sometimes lumped under the term microthermometric studies.

Pressure. Data obtained from fluid inclusions can provide information on the pressure of the environment at the time of trapping. A variety of procedures have been used for this purpose. All are based on experimental data on the thermodynamic properties of similar fluids, and many merely provide constraints on the minimum or maximum pressure of formation, but even this limited information can be valuable. The measured pressures range from near atmospheric to many kilobars.

Density. If the composition and density of each of the phases now present in a fluid inclusion (e.g., liquid, gas, or crystals) can be determined, along with their individual volumes, the total average density of the material in the inclusion can be calculated. Such densities are important in understanding the past circulation of fluids in the earth's crust, as this circulation is frequently driven by density differences. Fluid inclusions provide the only direct data we have on the density of these former fluids.

Composition. Many nondestructive procedures are available for obtaining qualitative, semiquantitative, and even quantitative information on the composition of fluid inclusions from their phase behavior at low and high temperatures. Still others require opening the inclusion for analysis. The nondestructive procedures generally involve the identification and characterization of the individual phases present. Except for magmatic and high-grade metamorphic rocks, by far the most abundant type of inclusion consists of a liquid of low viscosity and a gas or vapor bubble having a volume generally less than that of the liquid when viewed at surface temperatures. The ratio of liquid volume to total volume is sometimes termed the degree of fill. The liquid is generally a water solution with less than 10 wt % solutes, but concentrations range from more than 50 wt % to practically 0%. The solutes consist of major amounts of Na, K, Ca, Mg, Cl, and SO_4, and lesser amounts of many other ions. Many individual ions in this list may predominate, although Na and Cl are generally the most abundant. CO_2, both as liquid and gas, is not uncommon and may be dominant, and CH_4 is present in many inclusions. If the vapor bubble is purely the result of shrinkage, it may consist only of water vapor at the few millimeters of vapor pressure -- practically a vacuum -- characteristic of most water solutions at room temperature. In other inclusions, the vapor bubble is a highly compressed gas (most commonly CO_2) that was originally dissolved in the trapped fluid. Much can be determined about the composition of fluid inclusions by studying their phase behavior at low temperatures, sometimes even to liquid nitrogen temperatures (-196°C).

If, upon cooling from the temperature of trapping to surface temperatures, the fluid becomes supersaturated with a soluble salt such as NaCl, or any other mineral, these minerals may crystallize from the fluid as one or more new crystals, called daughter minerals[3]. These daughter minerals may, in turn, have inclusions of liquid plus gas in them that were trapped during their crystallization on cooling. In addition to the formation of daughter minerals, some further crystallization of the host (enclosing) mineral from the fluid nearly always takes place, but is merely added to the inclusion walls and is usually both insignificant and invisible.

Occasionally during the growth of a crystal, solid crystals of the same or other phases become trapped. It is not uncommon for such trapped solid inclusions to cause simultaneous trapping of some of the surrounding fluid as well.

Not uncommonly, the originally homogeneous fluid splits upon cooling to form two immiscible fluids in addition to a shrinkage bubble. Some authors use the general term "three-phase inclusion," without definition, when referring to such inclusions; others use the same term to mean inclusions containing a crystal, liquid, and vapor. The resulting ambiguity is sometimes unresolvable and always undesirable. Usually, when there are two liquids and a gas bubble in a given inclusion, one is liquid water solution, one is liquid CO_2, and the gas bubble is CO_2 under pressure. Less often, the second liquid, immiscible with water, is a liquid hydrocarbon; such inclusions may be formed by trapping of an originally inhomogeneous suspension of hydrocarbon fluid in water.

Inclusions in igneous rocks and meteorites (sometimes called magmatic inclusions) may consist of a clear to brownish silicate glass and a shrinkage bubble, and sometimes one or more daughter crystals such as pyroxene or feldspar. More rarely, the inclusions consist of sulfide melt, CO_2, or hydrocarbon fluids.

----

[3] In the rather extensive Russian literature on such daughter minerals, they are described by the term "mineraly-uzniki," which may be translated literally as "captive" or "prisoner" mineral; as this may give an incorrect concept of origin, I have introduced the term "daughter mineral" (Roeder, 1963). This change is made particularly necessary by the fact that another term, "zakhvachennye," meaning "trapped," is sometimes used in the Russian literature for solid mineral grains trapped during the formation of the inclusion, i.e., solid inclusions.

Inclusions in rapidly cooled rocks may be predominantly silicate glass. Those in slowly cooled rocks are more apt to have daughter minerals, and silicate-melt inclusions in deep-seated intrusive rocks may be completely crystalline "stony" inclusions, and hence rather difficult to recognize (as they resemble solid inclusions), and to study (as they tend to be opaque). In any given rock, the large inclusions are more apt to be crystalline, and the small, glassy.

As water (plus smaller quantities of CO_2) is the major component of many liquid inclusions, the amount of inclusion fluid present may cause significant errors in chemical determinations of water or (OH) in the host mineral, and actually makes accurate, unambiguous water determinations nearly impossible, regardless of the methods used. Any solutes present in the inclusion fluid are normally reported as part of the total analysis of the sample; not uncommonly, liquid inclusions contribute as much as 100 ppm of each of several nonvolatile constituents to the analysis of separated "pure" minerals. As an extreme example, fluid inclusions containing crystals of NaCl are so common in some granitic feldspar that single-crystal X-ray photographs of the feldspar crystals show powder-diffraction lines of NaCl as well (Roedder and Coombs, 1967).

All the many chemical analyses of fluid inclusions that have been made -- my own included -- are subject to serious limitations. The chemical manipulations may be reasonably straightforward, but the small sample size and the wide variation in composition of both the inclusions and the host mineral make the analyses far from routine and impose large analytic uncertainties. Furthermore, as the volume varies with the cube of the radius, a few relatively large inclusions may carry far more fluid than thousands of smaller ones; this simple fact alone makes it almost impossible to obtain truly <u>duplicate</u> samples for a test of analytical accuracy.

Much more serious, however, are such problems as the possible multiplicity of origin of the inclusions extracted and drastic contamination and/or loss during extraction. Such factors have not been adequately evaluated in many studies. The importance of the nature of the samples used cannot be overemphasized, and the difficulty in obtaining usable material is a major problem in most inclusion studies. The nature of the samples used is the prime consideration in the selection of suitable methods of study and in evaluating the precision, the accuracy, and most important, the significance of any measurements obtained.

APPLICATIONS OF FLUID INCLUSIONS

Because fluid inclusions are almost ubiquitous in geologic samples, their study is applicable to a variety of geologic problems and areas. A few of these applications, discussed in later chapters, include the following:

(1) In the study of ore deposits, fluid inclusions have provided much information that has been used in many ways, both in the immediate problems of mineral exploration, and in the longer-range, but equally important problems of understanding the physical and chemical environment of ore deposition.

(2) In the study of lunar and meteoritic samples, fluid inclusions have helped us reconstruct various extraterrestrial (and possibly early terrestrial) processes.

(3) In gemology, fluid inclusions might be considered as purely negative features, because they are often the imperfections that make the difference between a valuable gemstone and an inexpensive mineral specimen, but they have proven to be valuable defects, in exploration for

gem deposits, in gem identification, in recognition of the original source, and in distinguishing natural from synthetic stones.

(4) In stratigraphy and sedimentation, fluid inclusions have been used as helpful "fingerprints" to identify the nature of the provenance of detrital grains in sandstone, quartzite, and conglomerate.

(5) In complex igneous and metamorphic terranes, fluid inclusions have sometimes clarified the petrogenesis and tectonics, and the pressure and temperature changes during uplift and erosion; they may even be of help in recognizing the precursors to explosive volcanism.

(6) In the search for oil, fluid inclusions containing hydrocarbons and/or water have provided information on the tectonic and pressure-tempera-ture evolution of petroleum basins prior to, during, and after hydro-carbon migration, and may provide information on the contentious issue of the mechanism of primary migration of petroleum.

(7) In some domal salt deposits, high-pressure gas inclusions, presently of unknown origin, have prevented development of parts of the resource, because blowouts into mine openings from natural decrepitation of the inclusions in thousands of tons of such inclusion-rich salt have re-sulted in fatalities. Recognition of such inclusions in core samples may thus prevent mining accidents.

(8) In the study of the evolution of the atmosphere and paleoclimatology, gas inclusions in dated polar ice sheets have permitted reconstruction of paleoconcentrations of CO_2, fluid inclusions in speleothems (cave deposits) have provided data on both paleotemperatures and paleoclima-tology during the last 350,000 years, and inclusions in some samples of bedded salt may provide geochemically useful data on isotopes of the rare gases in the atmosphere (He, Ne, Ar, Xe) back to the Devonian.

(9) In high-pressure, high-temperature solution chemistry, inclusions pro-vide us with elegantly simple, transparent "visual autoclaves" that permit observations of phase behavior at high temperature and pressure that, in turn, can guide the design of the much more difficult labora-tory studies of synthetic systems.

(10) During the drilling of active geothermal systems, fluid inclusion measurements on the cores can guide the drilling by providing valuable data on deep temperatures, and on whether that portion of the system is heating or cooling. Such data can aid both the exploration and the development of geothermal fields.

(11) Study of the fluid inclusions in samples of mantle materials such as the ultramafic nodules in basalt and kimberlite (and possibly even diamond) has provided information on the nature of volatiles in the mantle, and particularly on the somewhat enigmatic genetic interrela-tions of the various forms of carbon derived from the mantle: as a solid solution in silicates, as dense CO_2 inclusions, as carbonatites, and as disseminated carbonate, graphite, or diamond.

(12) In evaluating the safety of sites for both nuclear reactors and atomic-waste repositories, fluid inclusions have played a role. At nuclear reactor sites, they have provided valuable evidence concerning the time since the last movement on faults at the site; at atomic-waste-repository sites, they themselves may present a significant hazard that should be considered in the engineering design of the repository.

Chapter 2

The ORIGIN of INCLUSIONS

CONTENTS

INTRODUCTION

A _correct_ determination of the origin of fluid inclusions is crucial to any useful interpretation of P-V-T-X data derived from them. The many ways by which a homogeneous fluid may become trapped within a crystal are similar whether the fluid is a silicate magma, a water solution, or a dense mixture of gases. Perhaps more than 99% of all fluid inclusions were originally formed by the trapping of a homogeneous fluid, but not all are formed this way. In some environments, most inclusions are formed from heterogeneous systems of two or more fluids; the distinction is crucial to the interpretation. As a consequence, inclusions formed from a homogeneous fluid may be considered "normal," and are discussed first, followed by the "exceptional" inclusions, formed by trapping of heterogeneous fluids. Regardless of the mode of formation, the fluid must be saturated with respect to the host crystal phase in order to become trapped.

TRAPPING OF INCLUSIONS FROM A HOMOGENEOUS FLUID

(i.e., "normal" inclusions)

When crystals grow or recrystallize in a fluid medium of any kind, growth irregularities result in the trapping of small portions of the fluid in the solid crystal. Such irregularities may be sealed off during the growth of the surrounding part of the host crystal, yielding _primary_ fluid inclusions. Healing of fractures formed during crystal growth yields _pseudosecondary_ inclusions, and healing of fractures formed at some later time yields _secondary_ inclusions. The inclusions in any given sample are seldom of only one generation. In some minerals from hydrothermal deposits, the primary inclusions are sparse and randomly distributed and relatively large, but the secondary inclusions in the same sample are small, very numerous, and in planar arrays. Size is not an unambiguous criterion, however; in some samples the only primary inclusions are rare and as small as most secondary inclusions. As these two fluids may have been trapped millions of years apart, the fluids may have grossly different compositions, making the distinction between primary and secondary origin of critical importance. The distinction requires a reconstruction of the events that took place in the original growth and subsequent history of the sample, and, as in most of geology, this reconstruction must be based mainly on examination of the resulting final product. Additional insight may be derived from special naturally "quenched" samples representing stages in the growth process, and from laboratory studies of synthetic crystal growth. Unfortunately, the growth rates for synthetic crystals are generally much faster than those for natural crystals and hence may not be applicable.

These various studies are summarized in the form of a tabulation of the most commonly available criteria for distinguishing the origin (Table 2-1, on page 43), but numerous caveats are needed. Some of these criteria, and a few typical examples, are discussed in this chapter; additional examples will be found in the other chapters, because the applicable criteria are almost as varied as the environments of growth. As the criteria are not absolute, and many are merely suggestive or applicable only to certain materials, they _must be applied with care_ and with awareness of the considerable ambiguity that exists. When in doubt, it is far better to assume that the sample does not contain identifiable primaries. Note, however, that this does _not_ preclude obtaining valuable data. As will be shown below, under "Secondary inclusions," the lack of primaries merely makes it much more difficult to work out the sequence of events. Perhaps the most frustrating aspect of fluid-inclusion study is that many (or even most) inclusions simply do not permit application of _any_ of the criteria. This frustration can become intense when careful search of an important sample yields only a few inclusions of usable size, all of which are indeterminate or ambiguous as to origin. C'est la vie!

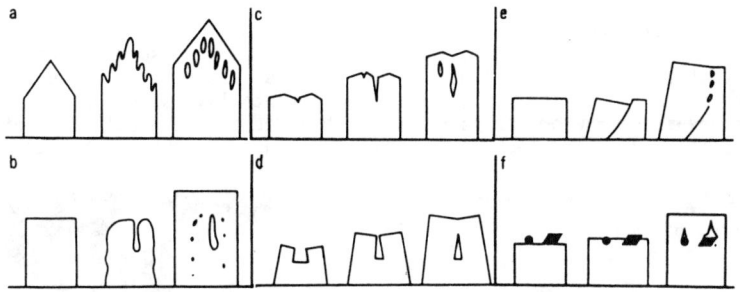

Figure 2-1. Mechanisms of trapping of primary inclusions. (a) Rapid dendritic growth is covered by solid growth. (b) Partial solution of a preexisting crystal yielded a deep reentrant and a curved crystal surface, both of which resulted in trapping inclusions during renewed growth. (c) Inclusions are trapped between individual growth spirals, and sometimes at the centers of growth spirals. (d) Subparallel growth of crystal blocks traps inclusions (e.g., see Roedder, 1972, upper right frontispiece). (e) A fracture in the surface of a growing crystal results in imperfect growth and trapping of inclusions. (f) Any foreign object on the surface of a growing crystal may be enclosed as a solid inclusion but may also enclose some fluid with it.

Primary Inclusions

Intracrystalline inclusions. Most inclusion studies are made on intra-crystalline inclusions (i.e., within single crystals). Any process that inter-feres with the growth of a perfect crystal may cause the trapping of primary inclusions. Some of the more common mechanisms are illustrated in Figure 2-1, but I must hasten to add that we simply do not know the mechanism(s) involved in the formation of many primary inclusions. A crack in the surface of a growing crystal is an imperfection that may result in subsequent irregular growth and trapping of inclusions in the imperfect new growth above the crack (Fig. 2-1e). Many inclusions form as a result of the nonuniform supply of nutri-ents in time or space. Thus, a period of rapid crystal growth can result in porous, dendritic growth (Fig. 2-1a). In a silicate magma, such a growth spurt might be caused by a loss of volatile materials as a result of pressure release (e.g., on eruption), which in turn may increase the degree of supersaturation and cause the growth of a rim of skeletal crystal. Later enclosure by slower, solid crystal growth can trap a zone of inclusions, thus marking the sequence of events (Fig. 2-2). (The bubbles of escaping volatile materials may also be trapped, as will be discussed later.)

Crystal-growth studies have shown that the concentration of even minor constituents in the fluid can grossly affect the perfection of the growing crystals, making them either more or less perfect. If a hollow, shell-like, skeletal, or cavernous crystal later becomes complete, large primary inclusions may be trapped. Some minerals form cavernous or tubular crystals more commonly than others and hence are more apt to be the hosts for relatively large primary inclusions (e.g., olivine and leucite, Figs. 2-3a,b,c). Skeletal or dendritic growth takes place most frequently from silicate magmas. As the temperature drops, a magma may become strongly supersaturated with a given phase if it fails to nucleate. When nucleation finally does take place, growth may be rapid, and hence skeletal or dendritic (Fig. 2-3c), until the degree of super-saturation is reduced. (Skeletal growth is not necessarily caused by strong supercooling, however (Sekerka, 1973).) Succeeding solid growth can cover this skeletal zone, leaving a core containing one large amoeboid inclusion or many smaller ones (Fig. 2-4).

Whether material is supplied to the growing faces by mass flow of the fluid or by diffusion, large inclusions may be trapped as a result of temporary starvation of the centers of faces relative to the faster growing edges. The edges have easier access to the moving nutrient solution, and as pointed out by Wilkins (1979), under steady-state growth conditions, supersaturation is greatest at the corners and least at the centers of the faces. The same result can also be due to easier nucleation of new growth layers on corners or edges. This "starvation" phenomenon is commonly observed in the growth of synthetic crystals and may be the mechanism whereby many natural quartz, halite, and

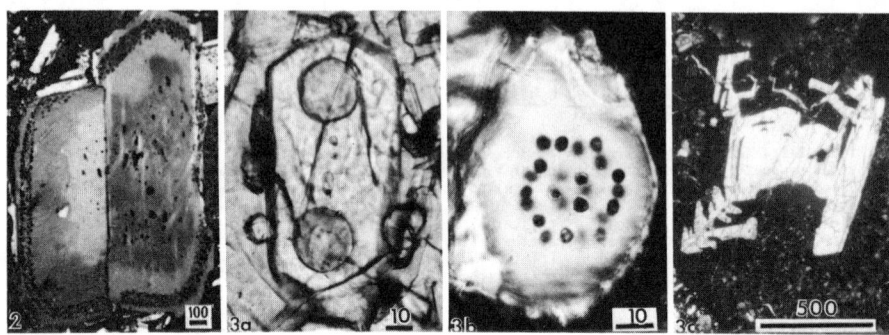

Figure 2-2. Plagioclase phenocryst in basalt from Chile, showing zone of small primary melt inclusions near edge (partially crossed polars). Scale bar in μm. From Roedder (1979a).

Figure 2-3. Primary melt inclusions in phenocrysts. (a) Olivine phenocryst from 1965 Makaopuhi basalt lava lake, Hawaii, USA, (collected at 1,135°C), containing two large primary melt inclusions, trapped by early skeletal growth, from Roedder and Weiblen (1971). (b) Leucite phenocryst in leucite basalt from Capo di Bove, Rome, Italy (USNM rock 929), showing shell of 24 melt inclusions, one for each face of a tetrahexahedron. Three of the six inclusions in the center and several around the outside are deep in the section and hence out of focus. From Roedder and Coombs (1967; see also their plate 4E). (c) Skeletal olivine phenocryst from lunar sample 12010, 30; covering of this dendritic skeleton with solid growth would trap large primary inclusions. Scale bar in μm. From Roedder and Weiblen (1971).

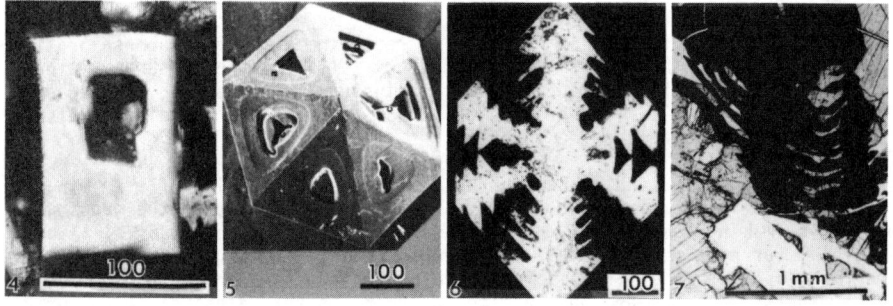

Figure 2-4. Cored plagioclase crystal from lunar basalt sample 10072-43, which presumably started as a skeletal dendritic crystal from rapid growth in a strongly supersaturated melt immediately after nucleation and then grew as solid crystal. Presumably the core area has recrystallized to some degree, yielding a single large multiphase melt inclusion. Scale bar in μm.

Figure 2-5. SEM photograph of a quartz phenocryst from an ignimbrite from the Valley of 10,000 Smokes, Katmai, Alaska, USA, showing reentrants in the centers of the faces, from Clocchiatti and Mervoyer (1976). Covering of such reentrants could yield large melt inclusions (see Fig. 2-6). Scale bar in μm.

Figure 2-6. Section parallel to the c axis through the center of a quartz phenocryst from a liparite, showing skeletal growth. A view of the external surface of this crystal would probably have looked like that seen in Figure 2-5. Scale bar in μm. From Lemmlein (1930).

Figure 2-7. Section of lunar basalt 10047,26, showing skeletal ilmenite crystal with many inclusions. Some of these "inclusions" are reentrants filled with single-crystal silicates that are optically continuous with those outside; others are actual sealed melt inclusions. From Roedder and Weiblen (1970b).

other crystals trap a series of very large inclusions, flattened parallel to a crystal face, with only thin septa of crystal separating each chamber (Figs. 2-5,6). Phenocrysts of ilmenite in the lunar basalts commonly show similar growth (Fig. 2-7). In fluorite, such starvation of the centers of the growing cube faces may result in porous inclusion-rich zones at the center of each cube face, consisting of large numbers of parallel inclusions elongated roughly perpendicular to the cube face (Fig. 2-8).

Even with a uniform supply of nutrients, most crystals grow as a series of almost parallel blocks (mosaic or lineage structure, Buerger, 1934), which may be the major source of primary inclusions. Examples of such blocky structure are plainly visible in most mineral collections. Each of these blocks may represent a specific growth spiral, or they may be bounded by surface flaws in the growing crystal (Zwicky cracks), or perhaps both. Carstens (1968) has shown, by etching of natural quartz crystals, that lineage boundaries originate from regions of high dislocation density. Electron microscopy (e.g., Buseck, 1983) may reveal the origin of some very minute inclusions. Macroscopic examination of synthetic quartz crystals shows that fluid inclusions are frequently trapped between adjacent large growth spirals and sometimes at the centers of spirals. The latter mechanism may explain the common occurrence of thin long tubular inclusions in beryl crystals, arrayed parallel to the c axis (Fig. 2-9), but another process, later etching (see "Pseudosecondary Inclusions" below), may yield apparently identical inclusions. Even more divergent crystal growth, as in curving crystals and divergent crystal bundles, may trap large primary inclusions, but these are sealed poorly and often leak.

If some growth spirals grow faster than others, the surfaces of the crystal can become rough, having many angular reentrants; later growth covering these reentrants produces negative crystal cavities. The trapping of inclusions on the twin plane of twinned crystals (Fig. 2-10) is somewhat analogous. However, similar alignment of inclusions along twin planes is commonly observed in secondary or pseudosecondary inclusions, e.g., in plagioclase and sphalerite (Fig. 2-11). Negative crystal shape to an inclusion, by itself, is frequently but erroneously stated to be proof of primary origin (see Table 2-1). Primary inclusions trapped by such processes are generally large and isolated or randomly arrayed; such characteristics are valid criteria for primary origin. Still other mechanisms of formation may yield primary inclusions that are very regularly arrayed (Table 2-1). The most commonly used criterion for assigning a primary origin is the occurrence of an inclusion in a euhedral crystal projecting into an open vug or vein. Unfortunately, this is simply not a valid criterion (see "Pseudosecondary Inclusions," below).

Any condition that disturbs the growth of an otherwise microscopically perfect crystal face can also cause the trapping of primary inclusions. Thus, a surface crack may cause imperfect growth and trapping of inclusions. By far the most common sources of interference are extraneous solid particles, which hinder the supply of nutrient at the points of contact, while growth of the free part of the face continues. The mechanisms involved in the trapping of such solid inclusions have been treated mathematically and experimentally in terms of fluctuating flow velocities, critical growth rates, particle size, etc. (e.g., Chernov and Temkin, 1977). The object may be pushed along by the force of crystal growth (Becker and Day, 1905), leaving a trail of inclusions behind it (called "rejection paths" by Ermakov, 1950), but generally the crystal grows over it, thus trapping a solid inclusion. If some of the surrounding fluid is also trapped, it is called a composite inclusion. In many composite inclusions, the relative positions of the solid and fluid reflect the growth direction in the crystal (Figs. 2-12a,b). This feature is most commonly observed in composite silicate-melt inclusions, in which the silicate melt will be found on the side of the solid inclusion toward the outside of the host crystal. Such evidence is of more than trivial importance, as it may help to distinguish between solid

15

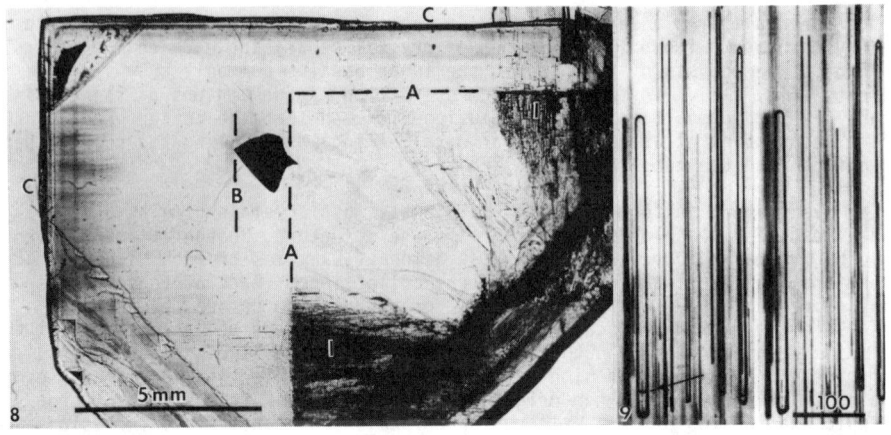

Figure 2-8. Photomicrograph of a section of the corner of a cubic crystal of fluorite, showing the large numbers of subparallel tubular primary inclusions (I), near the centers of the cube faces and perpendicular to them, that sometimes are typical of this mineral. These inclusions presumably indicate "starvation" of the centers of the cube faces, which ceased when the crystal grew to stage A. The opaque object in the center is a crystal of chalcopyrite that nucleated on the fluorite surface just before A, grew somewhat faster than fluorite for a period, expanding over it, and finally was overgrown by fluorite at stage B. Outer edges of cubic crystal are at top and left (C). Sample from Hill mine, Cave-in-Rock, southern Illinois fluorite-Pb-Zn district, Illinois, USA.

Figure 2-9. Photomicrographs of a plate cut parallel to the c axis twin of a beryl crystal, showing long tubular inclusions. Such inclusions have two possible modes of origin (see text). Two photographs taken at different temperatures, to show formation of liquid CO_2 (arrow). Scale bar in μm. Photographs by G.G. Lemmlein (reproduced from Grigoriev et al., 1973, p. 209).

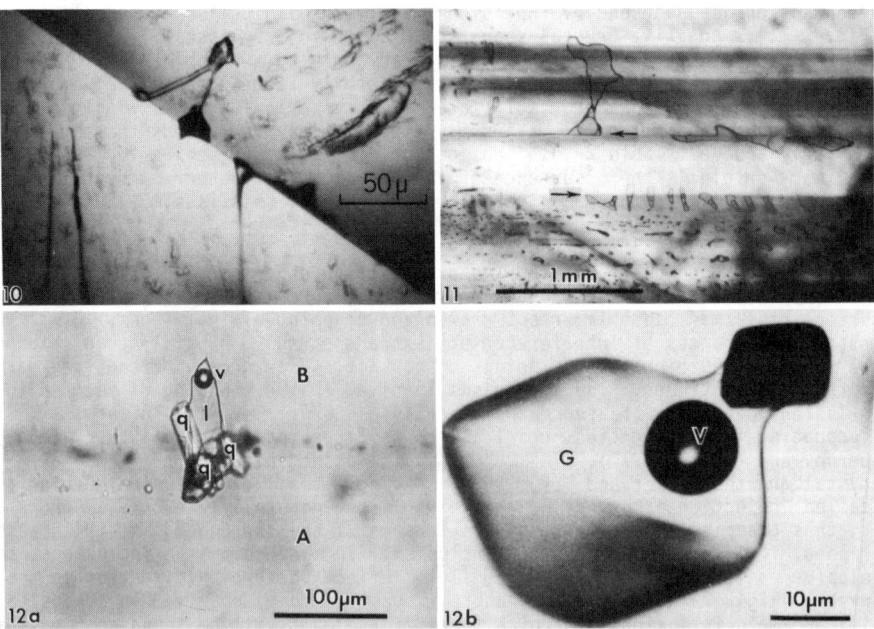

Figure 2-10 to 12 legends on opposite page.

inclusions and daughter crystals, as mentioned below. In aqueous inclusions, recrystallization of the host mineral by means of diffusion through the fluid phase after trapping can eliminate such evidence, but such recrystallization is usually minor in melt inclusions.

Generally, melt inclusions in phenocrysts of olivine are relatively large (Fig. 2-3a,c), but those in pyroxene or plagioclase may be small or rare, and the difference has not been explained. Where inclusions do occur in pyroxene or plagioclase, however, they are commonly in zones and may be very numerous (Fig. 2-2). A large percentage of the work on silicate-melt inclusions has been done in the USSR. A classification of zonal inclusions from an extensive review of Russian melt-inclusion work (Sobolev and Kostyuk, 1975) is based on whether the inclusions started forming, or were all sealed over, at a given growth zone (Fig. 2-13). Sobolev and Kostyuk also described various types of what they call "azonal inclusions" (Fig. 2-14). Their terminology differs, however, from that used in the West. Thus, we would call the inclusions shown in Figures 2-14d, e, and f "zonal."

Partial dissolution of an earlier crystal, forming deep etch pits, as a result of some temporary change in conditions during growth, followed by renewed growth, can result in the trapping of large isolated inclusions in deep reentrants, or in a zone of small inclusions marking the curved solution front (Fig. 2-1b). Wilkins (1979) believed that dislocation etch pits formed during temporary periods of undersaturation of the growth medium are probably a very important source of primary inclusions, and Wilkins and McLaren (1981) have shown that etching of growth dislocations can yield primary fluid inclusions (see "Pseudosecondary Inclusions," below). Lemmlein (1930) showed, however, that such reentrants, particularly in quartz phenocrysts in lavas, are generally growth features and do not result from dissolution (see Figs. 2-5, 6 and also Clocchiatti and Mervoyer, 1976). Leaching followed by renewed growth, trapping large primary inclusions, is not uncommon in hydrothermal ore veins, but a problem exists in distinguishing between inclusions from growth processes and those from dissolution (see "Pseudosecondary Inclusions," below).

The above discussion shows that all primary inclusions are, by definition, irregularities or imperfections in what otherwise might be microscopically perfect crystals. As such, they represent "accidents" in the normal orderly process of crystal growth; hence, their distribution in the crystals may be essentially random. (Secondary inclusions are also imperfections, but generally not random.) Many of the mechanisms discussed so far for the trapping of primary inclusions are readily recognizable in the final product. However, other primary inclusions may show no recognizable evidence as to their origin. Carried one step further, it is not uncommon to find that for no discernible reason, one crystal in a given sample may contain many inclusions, whereas hundreds of other adjacent crystals of the same mineral, presumably formed at the same time, are essentially free of inclusions. The presence of inclusions in the one crystal must be

---

Figure 2-10. Inclusions trapped on a twin plane in a crystal of cassiterite from Araca mine, Bolivia. Scale bar in μm. From Kelley and Turneaure (1970).

Figure 2-11. Plane of pseudosecondary inclusions in sphalerite that cuts across a series of twin planes (horizontal dark bands). Note that even though these inclusions formed after the crystal growth, they are strongly aligned on the twin planes (arrows). Sample USNM R577, Cananea, Mexico.

Figure 2-12a. Primary fluid inclusion in fluorite from Hansonburg, New Mexico, USA, trapped as a result of imperfect enclosure of several quartz crystals (q) which were deposited on the surface of cube of essentially inclusion-free fluorite (A). Further growth of the fluorite (B) enclosed the quartz and trapped a primary fluid inclusion, which now contains liquid (l) and vapor (v). From Roedder et al. (1968).

Figure 2-12b. Melt inclusion, now glass (G) and vapor (V), in olivine trapped as a result of enclosure of a solid inclusion of spinel (black). Crystal grew toward lower left. Kilauea Iki basalt lava lake crust, Hawaii, USA, from a depth of 0.7 feet (21 cm). From Roedder and Weiblen (1971).

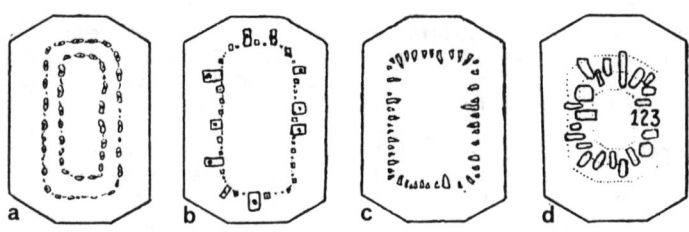

Figure 2-13   Four types of "zonal" primary melt inclusions, in the terminology of Sobolev and Kostyuk (1975; their Fig. 14). See text.

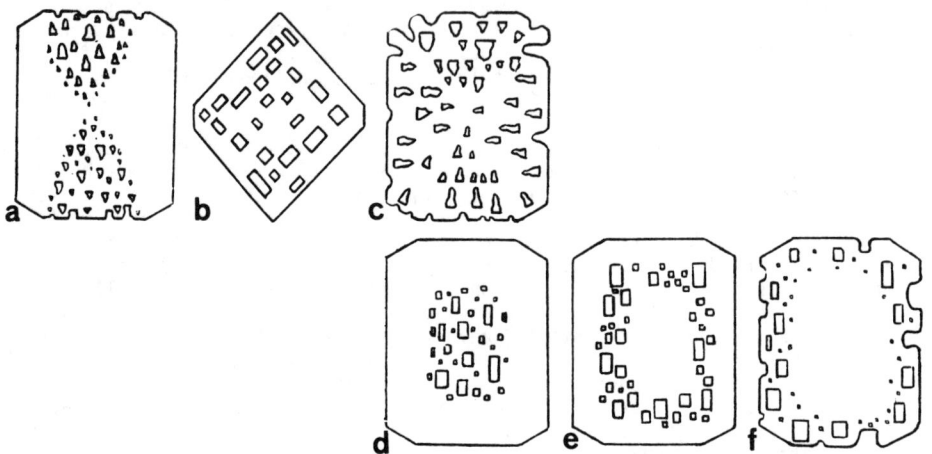

Figure 2-14   Six types of "azonal" primary melt inclusions, in the terminology of Sobolev and Kostyuk (1975; their Fig. 15). See text.

the result of some special local growth phenomenon, but the actual mechanism is seldom recognizable. On the other hand, if many crystals are growing simultaneously in a given environment such as a vein or vug, and the conditions in the whole vein or vug change, an inclusion-rich zone may form in every crystal, marking the change. The cause for the trapping of any individual inclusion in this zone may be unknown, but the initiation or cessation of generally inclusion-forming conditions may be marked very exactly (e.g., Fig. 2-13, 14).

A little confusion may arise in the usage of the term "primary" when reference is made to inclusions in a mineral that has recrystallized. Thus, many salt beds (e.g., Roedder and Belkin, 1979a; Roedder, 1984a) contain some grains that still maintain their original features from crystallization in a Paleozoic basin, and hence are called primary salt crystals, along with other grains of recrystallized salt. Both primary and recrystallized salt crystals may contain primary inclusions, trapped at the two different times of growth of their specific host crystals, and secondary inclusions, trapped at some still later time or times.

Intercrystalline (or intergranular) inclusions. Although practically all major inclusion studies have been of intracrystalline inclusions (i.e., those within a single crystal), a few studies have been made of inclusions on the contact between crystals. Grigoriev (1948) first described the trapping of inclusions on the interface between simultaneously growing crystals (which he called "induction surfaces" and which are called "compromise growth surfaces" in the West; Haynes, 1959). These interfaces may occur between two differently oriented crystals of the same phase, or between different phases. In the first case, the only difference between such intercrystalline inclusions and intra-crystalline inclusions along lineage boundaries (e.g., between blocks which grew on different spirals; Fig. 2-1c,d) is in the degree of structural mismatch of the two crystals.

Although Grigoriev (1948) described only the inclusions forming along the interface between two free-growing crystals, essentially the same type of inclusion is found abundantly on the interfaces between crystals in some metamorphic rocks. Thus Sella and Deicha (1962a,b; 1963) described and showed electron micrographs of shadowed replicas of broken surfaces of various aplitic rocks, quartz from Au veins, etc., in which large numbers of very small negative crystal cavities (presumably former fluid inclusions) were visible. Most were <1 μm in size. Another "metamorphic" rock, recrystallized rock-salt strata, commonly shows such intercrystalline inclusions in such large numbers that even though the inclusions are small individually, they may have some significance in the possible use of the halite bed for the storage of nuclear waste (see Chapter 11).

Because the walls of such intercrystalline inclusions are polycrystalline, and most grain boundaries have some finite permeability, most such inclusions have leaked. Thus the inclusions between salt crystals, although originally full of liquid, are generally full of air when examined in the laboratory unless the samples have been protected carefully after coring. Inclusions in other polycrystalline aggregates, such as agate nodules, may contain liquid, but there is no assurance when this fluid was last flushed out and replaced (see Chapter 11). In the Russian literature (e.g., Ermakov, 1950, p. 37), these intercrys-talline inclusions are considered to be pseudosecondary, but I believe they have the features of true, albeit leaking and hence unreliable, primary inclusions.

The air bubbles in glacial ice represent a special type of intercrystalline inclusions. This air represents a separate fluid phase, from which the crystals grew, which was trapped during the recrystallization of a mass of snow crystals. Except for possible diffusion through the host crystal (see Chapter 3), these air bubbles represent primary inclusions. In a somewhat similar situation, a series of Russian papers has suggested that (presumably intercrystalline) inclu-sions in some types of fine granular quartz represent the water from the dehy-dration of an original silica gel.

Secondary Inclusions

All primary inclusions are surrounded by host mineral deposited at about the same time that the fluid was trapped. Secondary inclusions are those that form by any process after the crystallization of the bulk of the host is essentially complete. Thus, if a crystal is fractured in the presence of a fluid in which it has a finite solubility, the fluid will enter the fracture and start to dis-solve and recrystallize the host crystal, first replacing curved, high-surface-energy surfaces having large and nonrational Miller indices with crystal faces having low Miller indices (Fig. 2-15). The process continues by reducing the amount of such new crystal surface and usually results in the trapping of new secondary inclusions by a process called necking down (see also Chapter 3). The final result can be a plane of sharply faceted negative crystal inclusions (Fig. 2-16). Although the crystal immediately surrounding the inclusion has crystal-lized at the same time that the fluid was trapped, the bulk of the crystal has not, and the meaning of the term "secondary" is apparent. Wilkins and Bird (1980) have shown that the mechanisms of trapping of secondary inclusions in fluorite are different for small (<50 μm) and large (>100 μm) inclusions. The newly recrystallized material is optically continuous and indistinguishable from the major part of the host, but Wilkins et al. (1978, 1981) and Wilkins and Bird (1978, 1980) have shown that in fluorite, a few minutes of irradiation with 2.5 MeV protons or alpha particles on a cleaved surface produces a color banding and

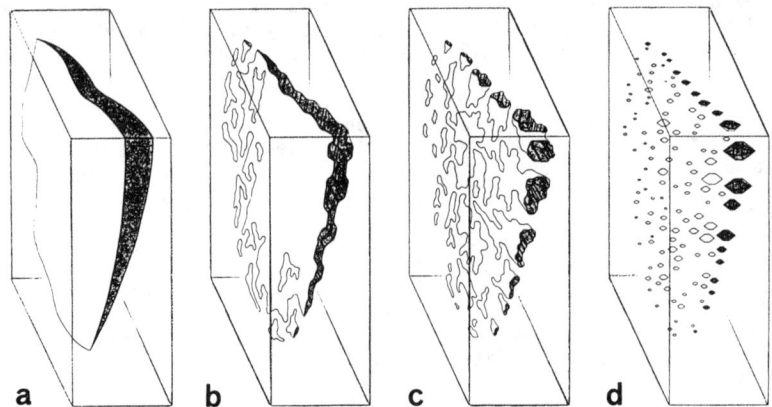

Figure 2-15. Stages in the healing or "necking down" of a crack in a quartz crystal, resulting in secondary inclusions. Solution of the curved surfaces having nonrational indices, and redeposition as dendritic crystal growth on other surfaces, eventually results in the formation of sharply faceted negative crystal inclusions (see Fig. 2-16). If this process occurs after temperature decrease has caused the formation of new phases such as a vapor bubble, the individual inclusions will have a variety of gas/liquid ratios. From Roedder (1962b).

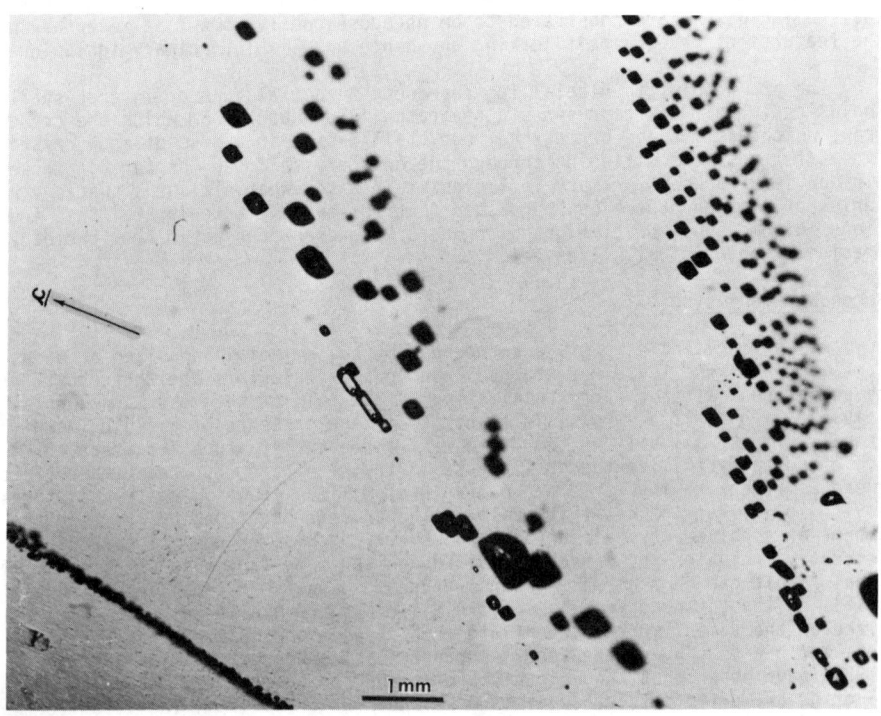

Figure 2-16. Photomicrograph of a plate cut from pegmatitic quartz crystal, Volta Bala, Minas Gerais, Brazil, showing three planes of negative-crystal inclusions. Each is a faceted hexagonal bipyramid, oriented parallel to c axis as shown. Although these look secondary, they are actually pseudosecondary. From Roedder (1965b).

decoration of growth and dislocation features that permits recognition of the newly grown material. Additional corroboration was obtained by etching the matched cleavage surface for a few minutes in 10% NaHSO$_4$ solution at 70°C. Similarly, Sprunt and Nur (1979) showed that differences in the color of catho-doluminescence revealed the zones of new material along healed fractures in quartz. A zone of recrystallized material can sometimes be recognized by a reduction in the normal complement of previously present fluid inclusions. Such a zone is seen in some rock salt, and Wilkins and Barkas (1978) reported it in metamorphic quartz. Further discussion of the process of healing of a fracture is given in Chapter 3, under the heading "Changes in Shape."

Wilkins and Barkas (1978) have divided secondary inclusions into two types, those resulting from brittle deformation (i.e., healed fractures, as described above) and those resulting from ductile (i.e., plastic) deformation. Plastic de-formation features such as deformation lamellae and slipbands, as well as recov-ery features such as subgrain boundaries and deformation-band boundaries may have associated inclusions. Contemporaneous or subsequent recrystallization of the ductilely-deformed material may sweep out all previous fluid inclusions. Once recrystallization has taken place, the origin of any remaining inclusion will be difficult to establish, but planes of very small inclusions (<10 μm) along the boundaries between slightly disoriented parts of a given crystal, as in much sheared or strained quartz, must be postdeformation. The fluids in them may come from outside the crystal or result from processes within the crystal (see later section on "EXSOLUTION INCLUSIONS"). Most of the secondary inclusions reported in the inclusion literature, even the smaller inclusions, probably result from brittle deformation, but those resulting from ductile deformation may be very abundant in some metamorphic samples, and many brittle fractures may have some associated dislocations and plasticity (Wilkins and Barkas, 1978).

Most discussions of secondary inclusions in the following pages deal specif-ically with those resulting from brittle deformation. Thus, a typical sample of white quartz is normally white because of the large numbers of healed shear planes in it, each outlined by thousands of tiny secondary inclusions. Many such samples contain 10^9 inclusions per cubic centimeter (Fig. 2-17). Commonly this healing process results in rows of inclusions having a regular spacing within the plane, from cleavage steps in the original fracture surface (Fig. 2-18). Slight move-ments of the two sides of such a stepped fracture relative to each other can yield tubular secondary inclusions; thus these would represent sealing of the "microtubes" reported by G. Simmons and coworkers (e.g., Shirley et al., 1979). This same group (e.g., Padovani et al., 1982) has shown that "microcracks" are present in many metamorphic rocks, but that these cracks have formed during late stages of retrograde metamorphism and are now open or only partly sealed.

Although the inclusions in some samples are mainly primary (particularly those in the lunar basalts and in some free-growing crystals from vugs), most inclusions in most samples are secondary. As a consequence, the safest presump-tion is that an inclusion is secondary until proven primary. The volume of inclusions increases essentially with the cube of the diameter, so even a very few larger primary inclusions may hold more fluid than thousands of tiny secon-daries in a given sample. The actual volume ratios of primary to secondary in-clusions can only be crudely estimated at best, from measurements on apparently "representative" volumes.

A general misconception among inclusionists is that data from secondary inclusions are meaningless or close to it. Actually, much can be learned about the sequence of events from secondary inclusions, although generally far more observations are required than if primary inclusions from the same events were available. Several procedures for using secondary inclusions are discussed in Chapters 12 and 13 on inclusions in metamorphic rocks (in such rocks, secondar-ies may be all that are available); secondaries have also been used in studies

100 µm
17

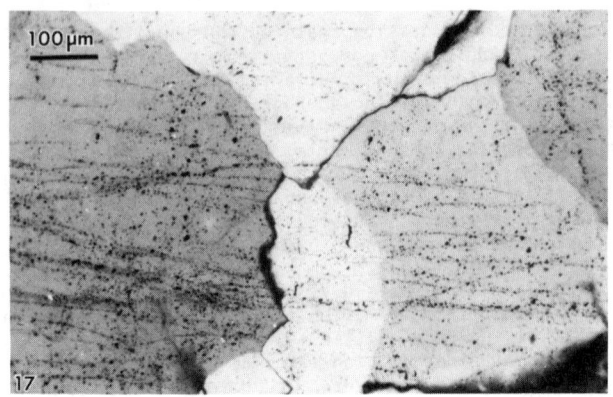

Figure 2-17 to 2-21 legends on opposite page.

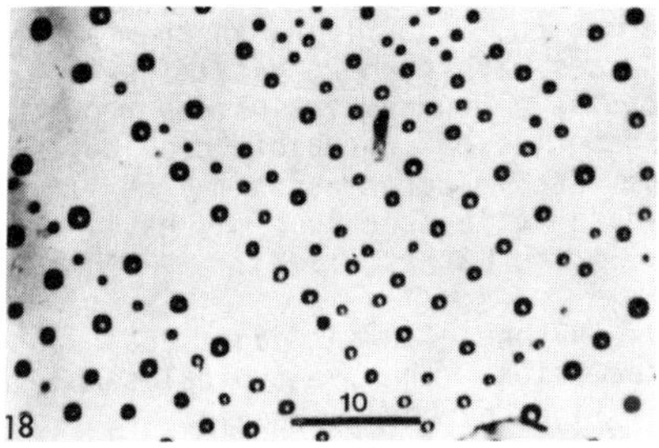

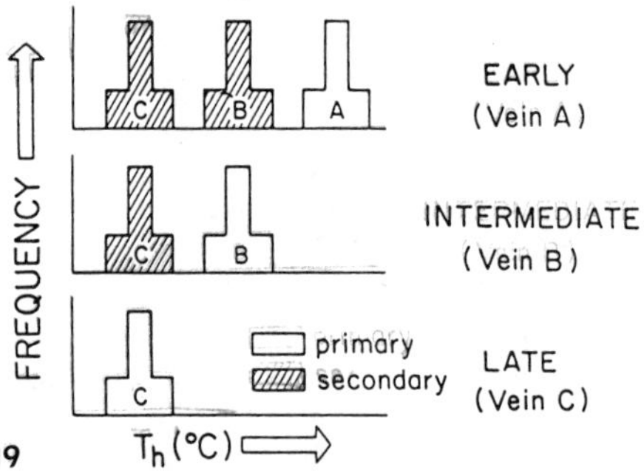

10
18

FREQUENCY
C
B
A
EARLY
(Vein A)
C
B
INTERMEDIATE
(Vein B)
C
primary
secondary
LATE
(Vein C)
T_h(°C)

19

B
P
A
PS
PS
1 mm
20

1mm
21

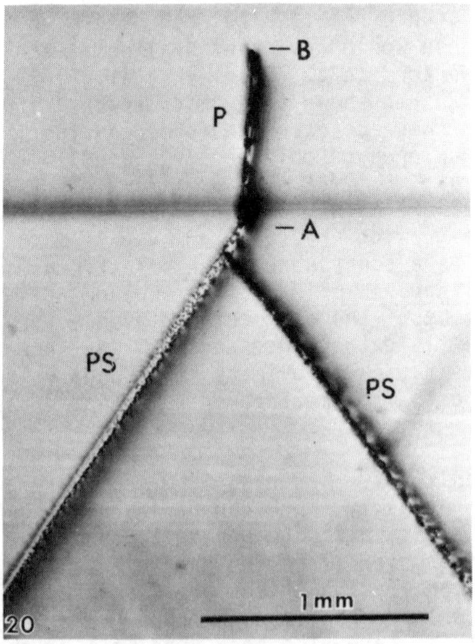

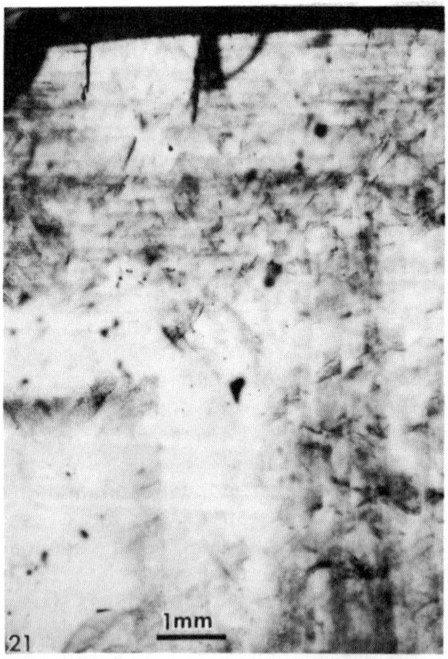

of mineral deposits. Thus Bodnar and Beane (1980) were unable to distinguish primary from secondary inclusions in growth-zoned quartz crystals but could recognize that certain types of inclusions were restricted to certain stages of crystal growth (and to all earlier stages). Preece and Beane (1983) used a similar process-of-elimination approach to determine which fluid-inclusion types were associated with which age of crosscutting veins in a porphyry copper deposit (Fig. 2-19).

Pseudosecondary Inclusions

Ordinarily one thinks of primary inclusions as forming during the growth of the crystal and of secondary inclusions as forming at some later time, from entirely different fluids. There is, however, a zone of overlap between the two types, first recognized by Ermakov (1949) and termed by him "primary-secondary" (also termed "virtual secondary" and "early secondary" in other Soviet papers). If a crystal fractures during its growth, the fluids from which it was growing, and which may be in the process of being trapped as primary inclusions in the rim of the crystal, will enter the fracture and become trapped in the core of the crystal. Because such inclusions look secondary but are formed during the growth of the host crystal, they have been termed pseudosecondary in the West. If there has been a change in the composition of the fluids between the growth of the core and the rim, adjacent inclusions in the core may then have different compositions; in view of such data and the obvious occurrence in a healed fracture, the apparently secondary inclusions frequently will be disregarded. Their true origin can only be recognized if the healed crack can be traced outward to an abrupt ending, at a former growth face, within the crystal (Figs. 2-20, 3-2). Sample material that presents positive proof of pseudosecondary origin is rare, but in my opinion, many of the large healed cleavage or curving fracture planes, now delineated by planes of inclusions, in most euhedral crystals from vugs, are probably pseudosecondary rather than secondary (e.g., Figs. 2-20, 21).

Inclusions in free-growing crystals lining vugs were commonly and erroneously assumed to be primary simply on the basis of this mode of occurrence (e.g., Newhouse, 1932, 1933; Ingerson, 1947; and many others). Most such inclusions are either pseudosecondary or secondary, and the abundance of planes of inclusions in healed fractures in euhedral crystals projecting into open vugs is sometimes surprising. There are many ways in which such a crystal may be fractured, thus trapping pseudosecondary or secondary inclusions, and such

---

Figure 2-17. Photomicrograph of planes of secondary inclusions in quartz from a Au-quartz vein, Grass Valley, California, USA, taken with partly crossed polars. Note that many of the planes cut across grain boundaries, outlining former throughgoing shear fractures. Scale bar in μm.

Figure 2-18. Secondary gas inclusions outlining healed fracture plane in halite, from the Pre-Carpathian depression, Stebnikskoye potash deposit, USSR, of Neogene age. From Petrichenko (1977).

Figure 2-19. Diagram showing how the relative ages of various generations of secondary inclusions can be deduced from their occurrences in veins whose sequence is known. From Preece and Beane (1983).

Figure 2-20. Photomicrograph of a pair of planes of pseudosecondary inclusions in fluorite from Hansonburg, New Mexico, USA. This crystal grew from the bottom upward as a slightly color-banded cube. When the crystal face was at A, the surface of the crystal fractured parallel to the octahedral cleavage directions (here viewed perpendicular to (110)) to trap thousands of pseudosecondary inclusions (PS). Subsequent growth of the fractured crystal resulted in the trapping of primary inclusions (P) until finally at B the irregularity was healed. From Roedder et al. (1968).

Figure 2-21. Another fluorite crystal from the same locality as that shown in Figure 2-20, showing more than 30 growth stages. At the end of each stage the surface of the crystal became covered with cleavage cracks, generally less than 100 μm deep. These subsequently healed, trapping planes (gray patches) of pseudosecondary inclusions, most of which are too small to be resolved at this magnification. Photograph taken in well-collimated lighting; in ordinary light, this plate merely looks cloudy. Each gray patch is a miniature replica of an event as seen in Figure 2-20. From Roedder et al. (1968).

cracking can be of commercial significance, as in the manufacture of synthetic quartz crystals (Laudise, 1979). Whether a natural crystal broke during its growth (yielding pseudosecondary inclusions) or afterward (yielding secondary inclusions) is not always possible to determine. The causes of the stress include: (1) variations in the amount of substitutional or interstitial ions, causing a misfit of the individual zones (many isometric minerals show considerable birefringence in polished plates, because of this mechanism); (2) crystal misfit along lineage growth; (3) solid inclusions whose compressibility or thermal-expansion characteristics differ from those of the host, combined with subsequent changes in P or T; (4) twinning in an anisotropic mineral, combined with subsequent changes in P or T; (5) phase transitions involving volume changes, along with variations in composition causing a small but finite range of inversion temperatures, for example, in quartz (Bambauer, 1961; Young, 1962); (6) thermal shock; (7) mechanical stress from distortion of the base where attached; and (8) transmission of stress by a vug-filling such as calcite or halite, which was later dissolved away. (This last process is seldom considered but may be rather common -- White, 1957, p. 1650; Roedder, 1963, p. 177.)

Pseudosecondary inclusions may also be trapped by the covering of pits formed by partial dissolution, i.e., etching or "hydrothermal leaching." This process is probably more common than generally recognized (Barton et al., 1963) and may yield large fluid inclusions, visibly cutting across growth zones. The etching of a crystal seldom involves simple removal of a uniform layer; for a variety of reasons, certain lines, zones, or areas dissolve at much more rapid rates, yielding etch pits that may be 1000 times as deep as they are wide (Nielsen and Foster, 1960). Similar etch pits were found radiating outward from liquid inclusions (Carstens, 1968) and from buried solid inclusions (Tsinober et al., 1968). These several authors propose that impurities that have diffused to dislocations cause stress, and hence higher solubility, thus yielding the pits. Deep etch pits, presumably following optically invisible dislocations, frequently form in the seed plates used in synthetic quartz-crystal manufacture and are covered by subsequent growth (Fig. 2-22; see also Wilkins and McLaren, 1981). Similar deep pits may be found at the intersections of several growth spirals, or at the tops of surface mounds on the crystals, representing growth on individual spirals; whether these pits are from growth or dissolution processes is not always clear. Still others are generated from the etching of clusters of dislocations resulting from imperfect enclosure of a solid inclusion (Fig. 2-23). Peacor et al. (1983) have suggested that some of the large "primary" inclusions in fluorite form by etching of regions of high dislocation density.

A semantic problem in interpretation arises with respect to inclusions trapped during healing of an etched crystal (Fig. 2-24). Commonly the etching process simply rounds the crystal corners, and when precipitation starts again, these curved crystal surfaces (with nonrational Miller indices) grow rapidly to form crystal faces. Such growth generally traps inclusions, and these would obviously be primary inclusions, because they are surrounded by host crystal that grew simultaneously with the trapping process. If, however, the etching yields deep tubes in the preexisting crystal, which are subsequently covered, or the equivalent occurs as a result of fracturing (Figs. 2-20, 21), the resulting inclusions will essentially be enclosed by a part of the host crystal that grew earlier than the trapping event and so are pseudosecondary.

To make the distinction between obvious primary inclusions, trapped by the enclosure of surface irregularities in an etched surface, and obvious pseudosecondary inclusions from the healing of a deep etch pit or fracture that crosscuts growth zones may be easy in theory but difficult in practice. My own solution to the necessarily arbitrary placement of the dividing line is illustrated in Figure 2-24: If the material making up the entire inclusion wall can be traced continuously into a specific growth zone in the crystal away from the inclusion, the inclusion is a primary inclusion; if it cannot, it is pseudo-

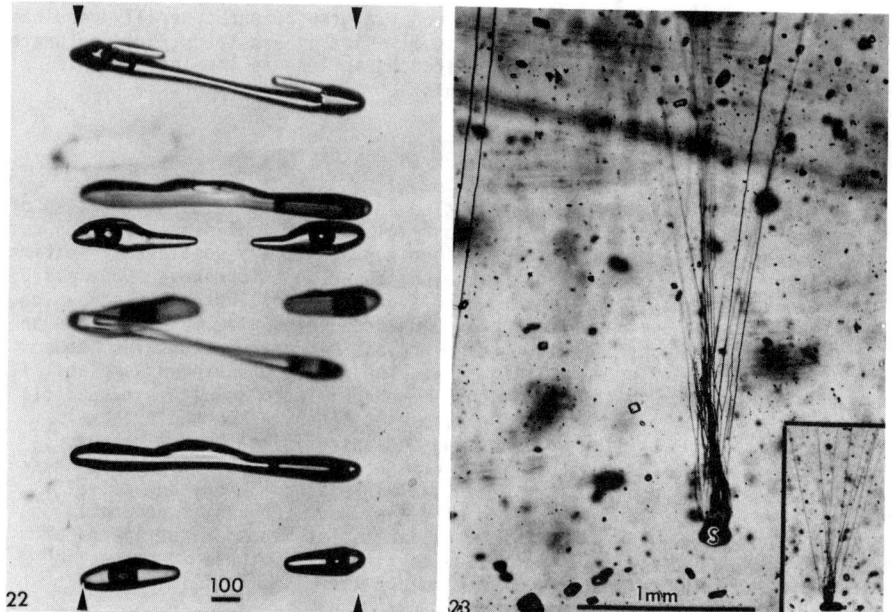

Figure 2-22. Section of synthetic quartz crystal, cut through the region of the seed plate (vertical, indicated by arrows), showing deep etch pits now containing fluid inclusions trapped by later growth. The etch pits are aligned on invisible linear imperfections that cross all the way through the seed plate. Some imperfections were etched all the way through; others are marked by a pair of etch pits from opposite sides. Scale bar in μm. From Roedder (1965b).

Figure 2-23. Topaz, Brazil, showing long array of minute tubes radiating from a trapped solid inclusion (S). The tubes are presumably from the etching of dislocations, and can be traced through many millimeters of the crystal, which grew from the bottom up as viewed here. The two tubes on the left are from another similar array, and still another array in the same crystal is shown in the inset.

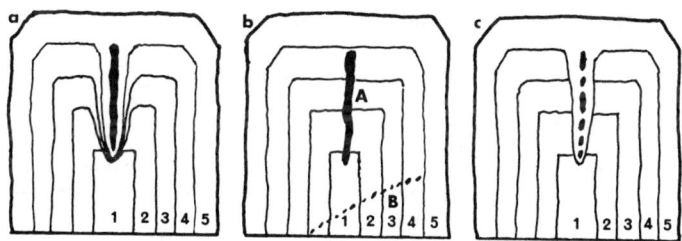

Figure 2-24. Diagrams showing the distinction between primary inclusions in deep reentrants and pseudosecondary inclusions. (a) An obvious primary inclusion (black) has been trapped by the covering during stage 5 of a deep growth reentrant; the walls of the inclusion are lined with stage-5 material. (b) An obviously pseudosecondary inclusion (A) has been trapped by the covering of a deep etch pit formed between stages 4 and 5; a plane of pseudosecondary inclusions (B) has also been trapped by the healing of a fracture formed between stages 4 and 5. None of these inclusions appear to be enclosed by material identifiable as stage 5. (c) Etching occurred between stages 4 and 5, yielding a deep etch pit cutting across all earlier stages. This pit was then partly lined with stage5 material before being filled and covered to trap a series of primary inclusions that may appear to be enclosed in earlier zones but are actually enclosed by stage-5 material. The zoning in most natural crystals simply does not permit such fine distinctions.

secondary. Because the visual recognition of a growth zone generally requires an appreciable thickness of crystal, this division is arbitrary, but fortunately, it is generally not of great significance, so long as the inclusion is recognized to be the result of deep etching.

CLASSIFICATION OF INCLUSIONS

Many attempts have been made to set up an all-inclusive classification of inclusions. Thus, Ermakov (1969) set up 21 classes of inclusions based on composition and origin and earlier established another widely used classification based on the relative proportions of the phases present (Ermakov, 1950, p. 28). The development of classifications can be a scientifically sterile occupation. Classifications are useful only if they elucidate principles by organizing an otherwise confusing mass of data, if they reveal hidden relationships between categories, or if they provide insight into the nature of unknown samples. If you know a lot about an inclusion, you will be able to place it in a specific cubbyhole in most published inclusion classifications, a cubbyhole that is labelled with a name telling you that the inclusion is what you already know it is. Many classifications consist of categories based on arbitrarily chosen numerical ratios of phases present (e.g., divisions at 25, 50, and 75 vol % gas). However, these ratios are part of a continuum and cannot generally be measured accurately. In spite of many explicit statements in the literature to the contrary, particularly in the Soviet literature, there is no real physical, chemical, or genetic significance to the division "line" at 50/50 gas/liquid at room temperature.

By far the most common and useful classification scheme for inclusions is the one I have used here, one that is based on their <u>origin</u>: primary, pseudosecondary, and secondary. Although the genetic information needed to apply this classification is frequently difficult or even impossible to obtain, this categorization represents the most fundamental and important step in interpreting the inclusion data in terms of geologic processes. Additional informative categorization on the basis of bulk composition (silicate, sulfide, aqueous, organic, CO_2) is generally possible. In addition to this categorization, another natural and fundamental division must be made before any inclusion data can be interpreted, a division based on the nature of the fluid at the time of trapping -- was it a homogeneous or a heterogeneous system?

TRAPPING OF INCLUSIONS FROM A HETEROGENEOUS FLUID

(i.e., "exceptional" inclusions)

Liquids Plus Solids

Abundant evidence indicates that at least at certain times in the growth history of many crystals, solid particles of various compositions were present in suspension in the fluids from which they grew. In some deposits, this solid matter was derived from the crushing of coarse vein-filling material or wall rock, or from the dispersion of newly precipitated fine-grained material; in other deposits it may have spontaneously nucleated in the fluid. The coarser particles will fall through the fluid to rest on any available surface, but if there is any vertical component to the laminar fluid movement, the finest material will be winnowed out and carried upward (e.g., see Barton et al., 1971). The textures seen in most veins, however, point to such very slow flow rates that visible evidence of winnowing is seldom found. Thus the "shadows" from the overhang of one crystal over another, which indicate essentially <u>vertical</u> settling of such particles, and which are readily apparent in many specimens from vugs and inclined or even horizontal open veins, provide evidence for a low or near-zero horizontal flow component. Grigoriev (1944) has suggested the study of such "ghosts" to recognize tectonic events involving rotation (see also Poty, 1966).

Many crystals from the Alpine-cleft type of metamorphic veins show multiple zones or "ghosts" of small solid inclusions (e.g., Poty, 1969), representing some sort of episodic events. Some of these solid inclusions have nucleated on the crystal surfaces; others have settled out of the fluid. Commonly, those that settle out act as nuclei for further growth, but either variety may be accidentally trapped as solid inclusions in composite fluid inclusions forming at that time, and hence may be confused with daughter minerals that crystallize out of the fluids trapped in inclusions after sealing (see Chapter 3). The distinction is based mainly on the phase ratios in the inclusions. A mineral that occurs only in some of the fluid inclusions, in highly variable amounts (particularly if it also occurs as simple solid inclusions in the host), is probably an accidental solid inclusion, whereas daughter minerals should be ubiquitous (unless nucleation causes problems; see Chapter 3), and, if present, they should occur in a regular ratio to other phases. An error could be made in interpretation if the fluid that was trapped consisted of a dispersion of very small crystals (or even a colloidal suspension) which later recrystallized to single crystals in each inclusion. If the dispersion were sufficiently fine relative to the size of the inclusions, the phase ratios in the inclusions would be regular. I know of no evidence indicating that this has indeed occurred in nature.

Unambiguously primary fluid inclusions are commonly found in zones of solid inclusions, attached to some of these solid grains. Whether the solid phase is an accidental solid inclusion of the vein-filling material or a daughter mineral, the fluid in the inclusions was presumably saturated with respect to it. This may not be true for crushed wall-rock fragments. Silicate-melt inclusions in particular are commonly found attached to solid inclusions of earlier minerals, enclosed by the host crystal (Fig. 2-12).

Two Immiscible Fluids

General principles. Many different geologic environments have involved two (or even more) immiscible-fluid phases. Sometimes one, the other, or both of these fluids are trapped in inclusions, and sometimes the immiscibility is the cause of the trapping. However, the inclusions from all these types of environment are so similar in trapping mechanisms and behavior that they can be discussed together. In addition, it is important to realize that frequently a continuum exists between liquid-liquid and liquid-gas equilibria. I use the term "immiscibility" in a broad sense, to describe the coexistence of any combination of two or more fluid phases at equilibrium (Roedder and Coombs, 1967, p. 419; see also Chapter 8).

In igneous processes, many types of immiscibility, including both liquid-liquid and liquid-gas, have been recognized from inclusion studies, as summarized in Fig. 2-25, and numerous other types have been recognized in lower temperature environments. By far the most common combinations of immiscible phases in nature that are preserved in inclusions are those consisting of silicate melt/gas (CO_2 or H_2O), aqueous solution/gas (H_2O, CO_2, CH_4, N_2, etc.), and aqueous solution/oil, but other rarer types have been proposed or reported, such as silicate/silicate, silicate/sulfide, silicate/metal, silicate/carbonate, silicate/oxide, saline fluid/hypersaline fluid, hydrocarbon/hydrocarbon, aqueous solution/metallic mercury, aqueous solution/liquid sulfur (in halite), and others. Inclusion evidence of simultaneous occurrence of three immiscible fluids, e.g., silicate/sulfide/CO_2, silicate/saline fluid/gas (H_2O or CO_2), or aqueous solution/oil/gas, is not uncommon, and probably four fluid phases have coexisted in several geologic environments.

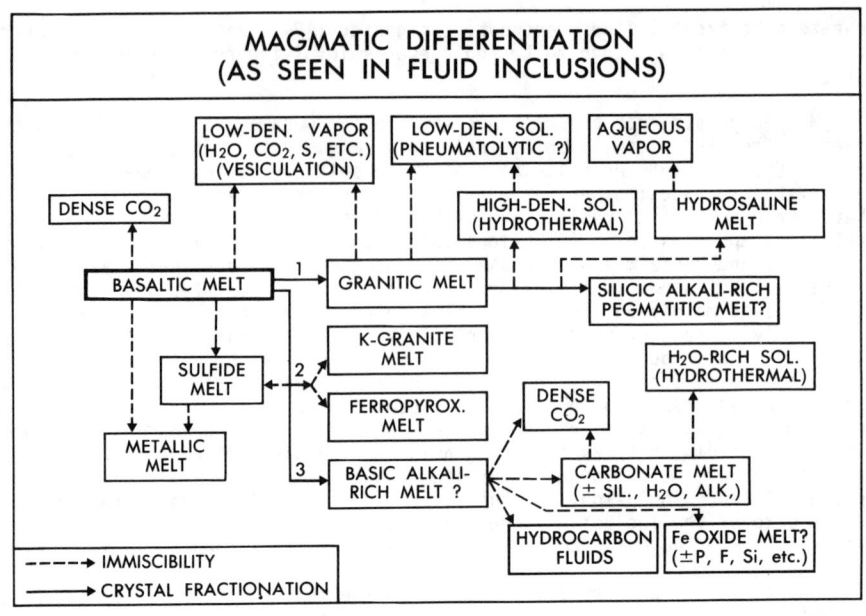

Figure 2-25. Hypothetical suggested outline of magmatic differentiation, as seen in magmatic inclusions of melt, fluid, and vapor, showing the presumed relationship between the several processes of crystal fractionation and immiscibility. From Roedder (1979a).

Preferential nucleation and wetting phenomena. Immiscible bubbles of steam or other gases, and immiscible liquid droplets, appear to nucleate from a previously homogeneous fluid most readily at the interface between the fluid and a solid. They are much less apt to form out in the bulk of the fluid, away from any interface, and even when they do so, the nucleation may well be on a generally invisible suspended speck of solid. (A glass of a carbonated beverage illustrates this as well as several other inclusion phenomena. Thus the beverage manufacturers go to considerable pains to filter their beverages carefully, to remove such suspended nuclei and hence make the carbonation last longer in the glass -- a perfect example of metastable supersaturation.)

Preferential nucleation normally results in globules of the new dispersed phase on the solid surfaces, sometimes arranged in an extremely regular array, although some globules may also nucleate out in the main continuous fluid phase. Various solid phases may even differ in the ease with which nucleation occurs at their surfaces. Thus Bonev (1977) has suggested that steam bubbles in boiling hydrothermal solutions may preferentially nucleate on the normally rather hydrophobic surface of galena, and Petrovskaya et al. (1971) and Petrovskaya (1973) reported evidence of similar phenomena for bubbles of CO_2 in hydrothermal fluids attaching themselves to native gold surfaces and becoming trapped.

In addition to the nucleation phenomena, if a given solid surface is exposed to a heterogeneous system of two fluid phases, differences in the interfacial energies, reflected in the contact angles, normally result in preferential wetting of the surface by one of the two fluids. The nucleation and wetting phenomena may therefore be interrelated. The most obvious example of such preferential wetting is the use of soap or detergents to change a water solution so that it will wet cloth fibers or other surfaces preferentially to their being wet by grease or oil.

The inclusions that are trapped in crystals growing from a heterogeneous system of two fluid phases may reflect both nucleation and wetting phenomena, and very erroneous conclusions can be drawn if these and other features are not considered. As an example, consider bubbles of CO_2 forming as the dispersed phase from a crystallizing basaltic melt. The growing crystals of one (or all) of the phases may become coated with tiny gas bubbles, each of which shields that part of the surface with which it is in contact from further growth, and so it becomes enclosed.[1] Note first that this process may yield a large number of primary gas inclusions, without any melt, in a crystal that actually grew from a melt. This would hardly mislead one into assuming that the silicate crystals grew from a CO_2 fluid.[2] But the identical phenomenon, involving steam bubbles forming on crystals growing from a hydrothermal fluid and trapped without any of the fluid, may easily be misinterpreted as evidence that the crystals grew from a low-density fluid and hence were one more example of "pneumatolysis." The distinction can be of more than just academic importance. Such a nucleation process should be particularly effective if sudden small pressure drops cause periodic effervescence of dissolved gas, or boiling of the fluid, as has been suggested by Barabanov (1958) for some so-called "pneumatolytic" tungsten deposits.

Another useful example of trapping of a dispersed phase, presumably by preferential wetting, is found in the many synthetic gems (ruby, sapphire, spinel, etc.) that are grown as boules, by the crystallization of an adhering film of melt of the same composition, at high temperature. Minute gas bubbles in this melt layer preferentially wet the growing crystal surface and are trapped as rows of tiny gas inclusions. As these gas bubbles outline the curved upper surface of the growing boule, the presence of such curved planes of gas bubbles is useful to distinguish synthetic gems cut from such boules from natural gems (Gübelin, 1953, 1974). The natural stones may also have gas inclusions (high pressure CO_2), but generally in flat arrays, parallel to crystal growth faces, if in planes at all. As the natural stones may be valued at 10^5 times more per carat than boule material, the distinction is important.

The inclusion evidence suggests that no mineral is wet by a CO_2 fluid preferentially over H_2O. There is, however, abundant evidence that a CO_2 fluid wets silicate minerals preferentially over a silicate melt. Most trapped oil/water systems indicate preferential wetting by the water phase, but some fluorite is preferentially wet by the oil phase (Fig. 2-26). Perhaps the best evidence for wetting by oil is found in the southern Illinois lead-zinc-fluorite deposits, where some fluorite is densely packed with primary oil inclusions, each having a flat bottom, marking the spot where a suspended droplet of oil, floating in the water-rich brine from which these crystals grew, adhered to the growing crystal and prevented further growth, and an almost spherical top, where the crystal grew around the droplet (Fig. 2-27). The addition of more droplets of oil to such a globule just before it is completely enclosed by the fluorite yields bottle-shaped inclusions as shown in Figure 2-28 (see also Figs. 11-22, 23, and 24). It should be noted that most oil inclusions in such deposits represent merely an immiscible phase, carried along by the brines; they do not

----

[1] Whether or not the bubble is lifted away by crystal growth is determined by the disjoining pressure as defined by Chernov and Temkin (1977).

[2] Petrichenko (1973) designated two classes of primary inclusions, authigenic, containing the fluid from which the host grew, and xenogenic, containing other materials. Although some immiscible fluid pairs permit this distinction, many do not, so I do not find the terms useful. Thus a CO_2 inclusion enclosed in an olivine crystal in a basalt might immediately be termed "xenogenic," yet the crystallization of the CO_2-saturated basalt was probably different than it would have been if no CO_2 were present.

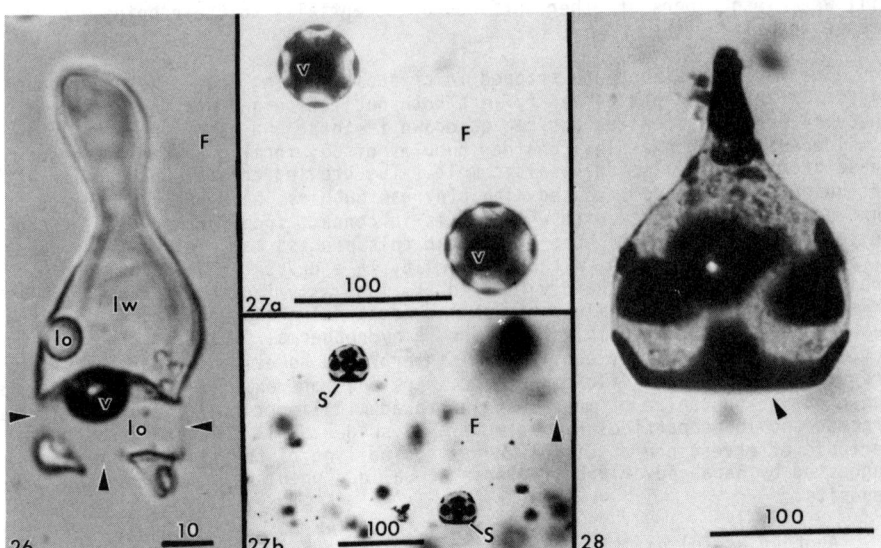

Figure 2-26. Photomicrograph of a large primary inclusion in yellow fluorite (F), showing dark vapor bubble (v), strong brine (lw), and two globules of oil (lo). The oil has an index of refraction almost identical with that of the enclosing fluorite. The oil wets the fluorite host in three zones (arrows), thus isolating two small segments of brine. The oil droplet at the left was not present before freezing in the laboratory; presumably it was dislodged from the main mass by rapid volume changes during the almost instantaneous freezing of the strongly supercooled brine. Sample ER 64-13a1, Koh-i-Maran Range, Kalat Division, Pakistan, provided by Omar B. Raup, U.S. Geological Survey. Scale bar in μm. From Roedder (1972).

Figure 2-27. (a) Photomicrograph of primary oil inclusions in fluorite (F). These originally homogeneous droplets of oil, initially suspended in the saline brines from which the fluorite grew, became attached to the surface of the crystal. After trapping, the oil underwent degradation to form a small amount of a dark immiscible asphaltic phase that preferentially wet certain crystallographic planes of the fluorite walls (the black spots with fourfold symmetry). The more transparent part is enriched in lighter hydrocarbons; in some such inclusions this lighter part vaporizes instantly when the inclusion is opened at room temperature. Subsequent cooling from Tt (~150°C) formed the shrinkage bubbles (v), now containing methane under high pressure. (b) A cross section of a fluorite cube (F), which grew in the direction of the arrow. The two oil droplets first stuck to the growing fluorite crystal along the flat surfaces on the bottom of each inclusion (S), now coated with dark material. Figure 27a is a plan view taken perpendicular to the cube face of two oil inclusions similar to those in Figure 27b. Sample ER 59-57e, West Green mine, southern Illinois fluorite-Pb-Zn district, Illinois, USA. (See also Roedder, 1962b, for a plan view of another similar sample.) Sample courtesy of Dr. James Bradbury, Illinois State Geological Survey. Scale bar in μm. From Roedder (1972).

Figure 2-28. Oil inclusion in yellow fluorite from Ozark-Mahoning's West Green mine, southern Illinois fluorite-Pb-Zn district, Illinois, USA, (sample ER 59-57e; F330 from Illinois State Geological Survey, courtesy R. Grogan). This inclusion originally formed as a spherical immiscible globule of oil, similar to those seen in Figure 2-27, that stuck to the growing surface of the host fluorite cube at the flat bottom surface (see arrow). As the fluorite grew up around it, additional oil droplets in the surrounding brine adhered to the exposed top of the droplet, causing it to develop the neck of the "bottle." Scale bar in μm. From Roedder (1979c).

generally represent exsolution of oil from a formerly homogeneous solution, and should not be confused with this process.

In all the above, I have assumed a state of equilibrium in the system. An additional complication can arise, however, if crystallization is fast, resulting in local disequilibrium in the fluid. Thus Rutherford et al. (1974) have proposed that the crystallization of feldspar from some originally homogeneous basaltic melts in laboratory charges may have resulted in local precipitation of globules of an immiscible, high-iron silicate melt. Fenn and Luth (1973) proposed the same mechanism for the formation of water-rich inclusions in crystals of anhydrous silicates growing from viscous water-bearing silcate melts, both in the laboratory and in nature. This process of enrichment of the fluid in residual gaseous components next to the growing crystal to the point where a new gas phase forms may also explain the report by Zirkel (1873, p. 86) of a hauyne crystal growing in a lava that had an estimated 3.6×10^{11} gas inclusions/cm^3, presumably trapped as bubbles of gas forming on the surface of the hauyne as it grew from the silicate magma. Roedder and Coombs (1967) reported similar phenomena in the trapping of immiscible globules of dense, highly saline aqueous fluids from a granitic melt, and possibly in the trapping of immiscible globules of low-density, carbon dioxide-rich vapor from the subsequent effervescence of the dense saline phase. In all these examples, however, it is difficult to determine whether the equilibria are indeed just local, or represent the stable assemblage for the whole system.

Once a globule of an immiscible fluid adheres to a growing crystal, further exsolution of that dispersed phase from the continuous fluid phase will generally add to it, as growth is easier than nucleation. If the host crystal continues to grow during this addition, long rodlike or toothlike inclusions can result (Fig. 2-29), as a continuation of the process that formed the inclusion in Figure 2-28. [The process has been modeled mathematically (Zhdanov et al., 1980).] Note also that inclusions may trap only one or the other fluid, or both fluids in arbitrary ratios (see next section also); examples of such trapping of two immiscible fluids are common in the lunar basalts (Fig. 2-30). Examples of rodlike inclusions from continuing exsolution of the dispersed phase during crystallization of the host are found in many materials. Thus the crystals of sucrose comprising "rock candy" commonly contain long tubular inclusions of air (Fig. 2-31). Perhaps the most familiar example is the occurrence of tubular air inclusions in ice cubes. Petrichenko (1973) has described similar gas tubes in halite (his Fig. 8) but has called them "secondary." Examples of such tubular inclusions are also common as a result of silicate/silicate immiscibility in the lunar basalts (Fig. 2-32), some silicate/sulfide immiscibility (Fig. 2-33), and some silicate/CO_2 immiscibility (Fig. 2-34; see also Gutmann, 1977). In these examples, the original nucleation sites were apparently in a rather regular array. (Such regularity in nucleation is common and is used in the commercial growth of some special composite materials.)

Phase ratios and their significance. In the general case, boiling (also effervescence) will result in a series of inclusions that have trapped different proportions of liquid and vapor. Apparently under some conditions, the vapor phase is rarely trapped. Pure vapor inclusions (also called "steam" inclusions when the liquid is a water-rich fluid) contain very little liquid and a very large bubble at room temperature; on heating they will homogenize in the vapor phase by evaporation of the liquid, ideally at the same temperature as that at which the liquid inclusions homogenize. (The determination of the evaporation of the liquid will be accurate only when the steam inclusion has a narrow reentrant into which the last bit of fluid phase is concentrated by capillarity; otherwise the specific point of homogenization cannot be recognized). Small amounts of liquid are rather commonly (but not always) trapped with the steam, and more rarely, small amounts of vapor (i.e., a small bubble) are trapped with the liquid. Such types of inclusions will obviously yield homogenization temperatures that are higher than the trapping temperature (the steam inclusion will have too much liquid, and the liquid inclusion too much vapor; Fig. 2-35). In theory, one may use the minimum value determined on a large number of such inclusions as the true trapping temperature, as these are the inclusions that are most likely to have trapped pure liquid or pure vapor. The possibility of mechanisms other than boiling (such as leakage, necking down, or sequential trapping under varying P and T; Chapter 3) being involved in the origin of such inclusions makes the use of such minimum temperatures hazardous at best. Under some circumstances, e.g., at Waiotapu, New

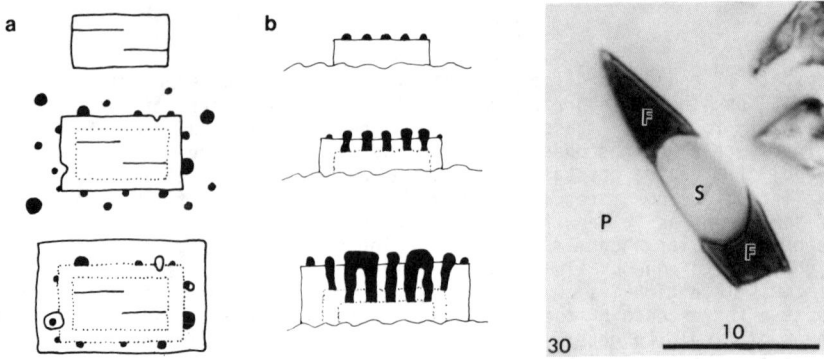

29

30

Figure 2-29. Appearance of inclusions resulting from nucleation of a new dispersed immiscible fluid phase either out in the host melt (a) or on the surface of a host crystal (b). (For examples of the formation of rodlike inclusions, see Figs. 2-31 to 34.) From Roedder (1979a).

Figure 2-30. Primary inclusion of immiscible high-Si (S) and high-Fe (F) melts, now glasses, in plagioclase crystal (P) in lunar sample 12057. Scale bar in μm. From Roedder and Weiblen (1971).

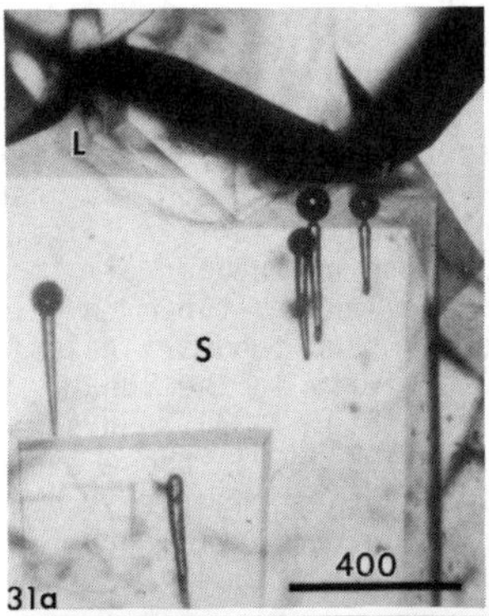

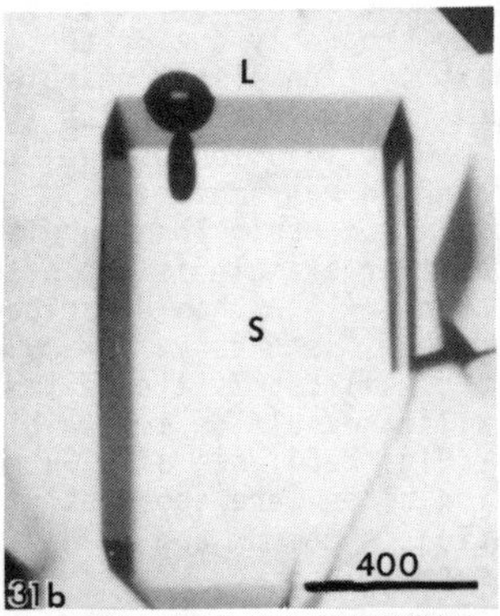

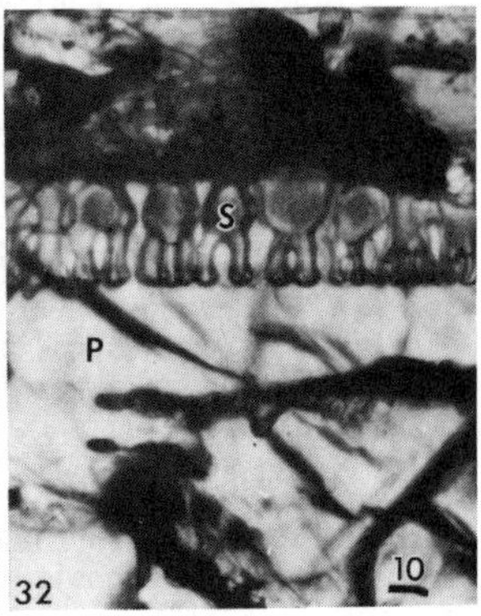

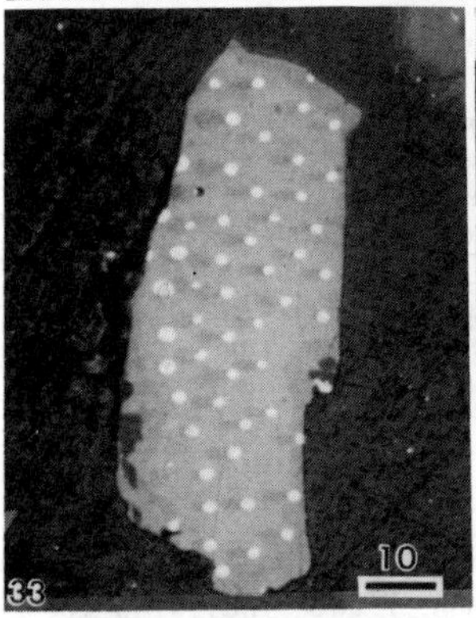

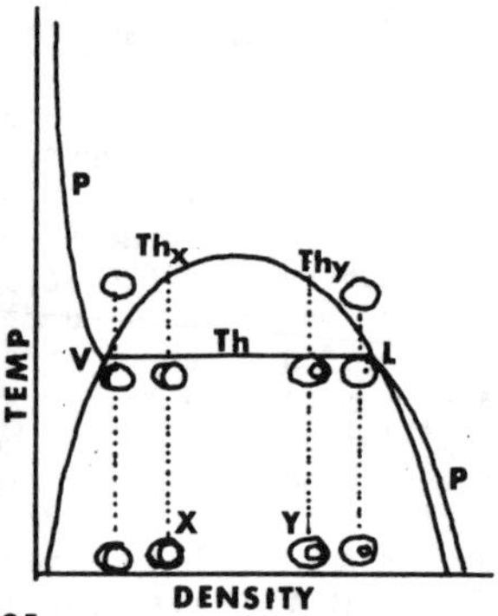

Figure 2-31. Evidence of trapping of fluid inclusions that are not representative of the fluid from which crystal has grown. Synthetic sucrose crystal (S) growing in saturated water solution (L) at room temperature by slow evaporation of water. Exsolution of air, as described by Powers (1958), formed gas bubbles that adhered to the growing crystal surface and formed tubular gas-filled inclusions, all of which (in these photomicrographs) are still connected to the surface. In (a) the bubbles are in a layer of solution above the crystal, except for the inclusion at the bottom, which lost its bubble. In (b), the bubble image is distorted by the primatic crystal termination. Not a single inclusion was found of the syrupy water solution from which the crystals actually formed. Scale bars in μm. From Roedder (1972).

Zealand (J.W. Hedenquist, pers. comm.), vapor-phase inclusions are not trapped, even though a vapor phase was known to be present, so the absence of vapor-phase inclusions cannot be considered to be proof of the absence of boiling.

Not all examples of immiscible fluids necessarily arrived at that condition via simple cooling from a previously homogeneous precursor. In addition to the immiscible oil-water pair described above, in the lunar lavas, an originally homogeneous basalt melt split into immiscible granitic and pyroxenitic fractions (e.g., Fig. 2-30) as a result of crystallization moving the liquid composition into the field of immiscibility. The two melts in such inclusions presumably will homogenize at some high temperature, but this would have no necessary relationship to the natural conditions, unless the appropriate solid phases were present to permit retracing the original path (see Roedder, 1978b, composition 3 in his Fig. 1, and Chapter 18).

"Empty" inclusions, caused by the trapping of a low-density steam phase from the boiling of near-surface fluids, are not uncommon in those epithermal ore deposits where extrapolated P-T conditions indicate that boiling could have occurred. The very small volume of liquid water formed on the inclusion walls by condensation during cooling may form a film that is sometimes even invisible on freezing (Roedder, 1970a, Fig. 7). However, in metamorphic rocks, "empty" inclusions or highly variable gas/liquid ratios are normally the result of processes other than boiling: sequential trapping of fluids under different P-T conditions, leakage of part of the inclusions (either in nature from shearing or decrepitation, or in the laboratory), necking down, or fracturing and later refilling. Such processes can usually be recognized by microscopy (see Chapter 6). "Empty" inclusions may also be filled with dense supercritical gases -- see Chapters 4, 5 and 7.

The trapping of gas or vapor bubbles along with liquid from an originally heterogeneous, two-phase system usually invalidates attempts to use such inclusions for geologic thermometry by the homogenization method, as this method is based on trapping of a single homogeneous fluid phase. On the other hand, proof of the presence of inclusions that <u>simultaneously</u> trapped single fluids, some that have trapped liquid and some that have trapped vapor (Fig. 2-35), gives the best evidence available that the fluids were actually boiling, and the best possible geothermometry.

The trapping of inclusions of even just one of the two phases permits thermometry. The temperature of homogenization (Th) for such inclusions is also the temperature of trapping (Tt), as no "pressure correction" is needed; hence no information is needed on the composition of the fluid. Such inclusions provide an excellent geobarometer as well, but to obtain the pressure of trapping, the compositions of both liquid and vapor are needed, along with appropriate P-V-T-X

Figure 2-32. Planar group of primary glass inclusions of high-Si potash granite composition (S) embedded in a pyroxferroite crystal (P) from lunar sample 12063,9. The bottoms of the "teeth" start at a plane that presumably marks the sudden onset of immiscibility in the melt. The outside of the pyroxferroite crystal is at the top of the row of "teeth." Scale bar in μm. From Roedder and Weiblen (1971).

Figure 2-33. Primary sulfide melt inclusions in a fayalite grain from Soviet Luna-24 soil sample 24077,53, in reflected light. The inclusions are actually in the form of cylindrical rods, nearly perpendicular to this surface. Scale bar in μm. From Roedder and Weiblen (1978).

Figure 2-34. Primary tubular CO_2 inclusions in anorthoclase feldspar megacryst from Lunar Crater Volcanic Field, Nevada, USA, (Bergman, 1982). Outer surface of megacryst at right. Almost all these inclusions have decrepitated during eruption and are now filled with air or lava. Only a few of the smallest still contain CO_2. Scale bar in μm. Sample courtesy of S. Bergman.

Figure 2-35. Temperature-density diagram for a fluid such as H_2O, illustrating the trapping of boiling fluids. Fluid boiling at the pressure of isobar P will consist of liquid of density (L) and vapor of density (V). Inclusions of these two fluids, cooled to 25°C, would appear as shown, and on heating would both homogenize at the same temperature (Th). If any liquid is trapped with the vapor (e.g., inclusion (X)), or if any gas phase is trapped with the liquid (e.g., inclusion Y), the homogenization temperatures will be higher than Th (e.g., Th_x and Th_y).

data (Roedder and Bodnar, 1980). However, to distinguish between small amounts of liquid trapped with the gas and small amounts of liquid condensed from the gas after trapping is sometimes difficult.

Incorrect inferences are frequently drawn in the inclusion literature that the amount of vapor phase trapped, or the ratio of the number of inclusions of vapor to those of liquid phase, give an indication of the relative amounts of these two phases present in the original heterogeneous system. A single tiny inclusion of a gas phase indicates gas saturation just as well as a million such inclusions, but the amount and number of such inclusions in a sample are merely a result of the vagaries of the inclusion-trapping processes discussed above, and do not give any valid indication of the phase ratios in the original two-fluid system at any given time. The trapping of gas inclusions may even give misleading data on the sequence of phase changes with time. Thus if one of a series of zones of primary inclusions in a given crystal is made up of gas-rich inclusions, the common interpretation given in the literature is that gases "played a major role" in the deposition of that zone. All it may really mean, however, is that some minor amounts of gas bubbles formed during the growth of that zone and were preferentially trapped.

Boiling vs effervescence. The trapping of "primary gas" inclusions from an otherwise liquid medium is common. The "gas" bubbles can consist of the vapor of the host liquid (i.e., from true "boiling" of a one-component solution), or of a very minor gaseous constituent in the liquid (i.e., from "effervescing" of a multicomponent solution). Both are examples of immiscibility. Since most natural fluids are multicomponent systems, the term "effervescing" is more appropriate but usually the term "boiling" is used in the literature. Primary gas inclusions in hydrothermal systems tend to be large, but in silicate melts they may be exceedingly minute. Although evolution of a gas by effervescence vs a vapor by boiling of a fluid may have grossly different effects on the composition of the remaining liquid and may occur under very different pressures (e.g., Drummond, 1981; Drummond and Ohmoto, 1979), these two processes actually represent end members of an immiscibility continuum and cannot be easily distinguished. Each volatile species has its own partition coefficient between the vapor and the liquid phase under the prevailing conditions, but most of these partition coefficients have not been measured. Contrary to many statements in the inclusion literature, the ratios of the various volatile species in vapor and in liquid will generally be very different. Thus the "vapor"-phase inclusions found in olivine crystals growing from deep-seated basaltic melt are generally almost pure CO_2 (Roedder, 1965a), yet the basaltic melt may well have contained more H_2O than CO_2 (Chapters 16 and 17).

Immiscibility before or after trapping. If carbon dioxide is present in aqueous fluids, its limited solubility, particularly at low P and T, commonly results in immiscibility. However, an immiscible condition at room temperature does not require immiscibility under the conditions of trapping. The mutual solubilities of CO_2 and H_2O change considerably with temperature and fluid composition (Chapter 8). Homogeneous H_2O-CO_2 fluids trapped in inclusions can, and frequently do, separate on cooling to form two immiscible fluids, CO_2-rich (generally supercritical) and H_2O-rich (subcritical). The constancy of the ratios of volumes of the phases in several inclusions provides the main criterion in distinguishing the trapping of a heterogeneous mixture from the later separation of an originally homogeneous one (Fig. 3-2).

In order to use inclusions trapped from a heterogeneous system to obtain the P and T of trapping in nature from the extent of the field of immiscibility (e.g., in the system H_2O-CO_2-NaCl), it is imperative (but difficult) to prove that the immiscibility seen in the inclusions was actually present at the time of trapping. All too commonly, the mere presence of two inclusion types, CO_2-rich and H_2O-rich, is taken as evidence for the existence of these two fluids

34

as immiscible phases at the time of trapping. Several other processes can yield the same result. For example, immiscibility can occur within the inclusion after trapping, followed by necking down, or two unrelated fluids can be trapped at different times. (Note, however, that necking down may also take place after trapping of composite inclusions from an immiscible liquid pair.) Criteria on which these important distinctions may be made with some assurance are not always available, and this ambiguity may lead to seriously erroneous P and T estimates, as discussed by Roedder and Bodnar (1980).

The enigma of inclusions of fluid having zero solubility for the host. One fascinating but seldom considered paradox in the formation of inclusions from immiscible fluid pairs concerns the exact mechanism whereby fluid inclusions can become sealed even though they apparently contain only a fluid in which the host crystal is essentially insoluble. The usual explanation is that the fluid that is eventually trapped was present as the dispersed phase (i.e., globules) in a continuous phase, and that the crystallization of the host mineral took place only from the continuous phase. In detail, this requires that the host crystal grows around the globule and that this growth extends up against the globule at all times, excluding any trace of the continuous phase. Such a mechanism may seem a little strange but is not too difficult to imagine; e.g., a fluorite crystal grows (from a brine) around an adhering globule of oil, or an olivine crystal grows (from a silicate melt) around a CO_2 globule. The many solid inclusions in such crystals, containing no fluid, present a perfect analogy. However, the extension of this concept to the trapping of secondary inclusions in tiny fractures becomes more difficult.

Many studies of the inclusions in metamorphic quartz (Chapters 12 and 13) reveal the presence of large numbers of tiny secondary inclusions of seemingly pure CO_2 fluid, or even N_2-CH_4 fluids (Tomilenko et al., 1976; Swanenberg, 1980). Fluids of such composition can form by a variety of rock/fluid interactions, but the solubility of quartz in such exotic fluids as supercritical N_2-CH_4 mixtures is presumably very low or even close to zero, so what was the medium through which quartz crystallized (or recrystallized) to provide the sealing and particularly to permit the recrystallization needed to yield negative crystals? Many samples seem to contain no inclusions of an aqueous phase that might be presumed to be responsible for the sealing. Several explanations appear feasible. (1) There may actually be small amounts of a water phase present through which the recrystallization could take place, but as the water will preferentially wet the walls, it will be essentially invisible, particularly in very small inclusions. This appears to be the correct explanation for at least some organic fluid inclusions in quartz (Kvenvolden and Roedder, 1971). (2) There may have been some water phase present in the inclusions, but it has subsequently "dissolved" and dispersed into the host mineral, as suggested by White (1973; see following section). (3) H_2, from the reaction $CH_4 + 2H_2O \rightarrow CO_2 + 4H_2$, may diffuse out of a mixed CO_2-CH_4-H_2O inclusion, resulting in a residue of pure CO_2 (Hollister and Burruss, 1976, p. 173). Such diffusion of H_2 is discussed in Chapters 3 and 15. (4) As suggested by these same authors, immiscibility may be followed by necking down or other separation phenomena, leaving pure CO_2 inclusions. (5) Movement of host material may take place by surface diffusion or through the host crystal itself, rather than through the trapped fluid (see Chapter 3, "Changes in shape"). (6) The solubility of minerals in "exotic" fluids may be larger than is commonly assumed. Thus, Sakhibgariyev and Lashkova (1977) believe that they can prove that petroleum may corrode quartz (although they invoke bacteria and water-oil contacts).

EXSOLUTION INCLUSIONS

Wilkins (1979, and pers. comm., 1978) proposed a new type of inclusion, called exsolution inclusions, that does not fit in any of the three normal categories of inclusion origin, because it forms as a result of internal iso-chemical phenomena within a crystal. The name was first proposed for certain inclusions in metamorphic quartz (e.g., Wilkins and Barkas, 1978), on the basis of earlier work by several investigators.[3/] Various studies of the H content of quartz (see Aines et al., 1983; Aines and Rossman, 1984; and references in Wilkins and Barkas, 1978) show that it varies widely. When these data are reduced to H_2O equivalent (each $H/10^6Si$ is equivalent to 0.15 ppm by weight H_2O), the values are found to range from >500 to <1 ppm H_2O. Bambauer (1961) found a maximum of 180 ppm H_2O in Alpine vein-quartz crystals, and an average of ~5 ppm, but to my knowledge, there are no such analyses of ordinary quartz grains from metamorphic rocks. Luckscheiter and Morteani (1981) reported only 0.1-1.1 ppm H_2O equivalent, plus one sample containing 2.3 ppm, in a series of 29 quartz crystals from veins in a metamorphic terrane (the Penninic series of the Tauern Window, Tyrol, Austria/Italy). White, in a series of papers (e.g., 1973) sug-gested that many of the small inclusions now seen in metamorphic quartz formed along dislocations, perhaps during pressure release, by the migration of water and alkali ions that were formerly part of the quartz structure[4/]. To place this possibility in proper perspective, one should note that just 1 ppm H_2O in quartz is equivalent to ~20,000 ellipsoidal inclusions per cm^3, averaging 5 x 5 x 10 μm.

As migration of H_2O through quartz has been observed in laboratory studies (see Chapter 3), Wilkins and Barkas (1978) suggested that similar migration during metamorphism would explain some of the features of the distribution of inclusions that they have observed. Since then, several studies have shown that the diffusion of H in quartz is surprisingly fast. Kronenberg et al. (1983) showed that after just 2 days at 800°C and 8.9 kb H_2O, 3 mm thick quartz plates had taken in 80-90 $H/10^6Si$, and showed no compositional gradient, i.e., they were homogeneous. Giletti and Yund (1984) showed that the diffusion rates are nonlinear with temperature, and differ greatly with crystallographic direc-tion. The possibility of such partial dissolution of inclusion contents, migra-tion through the crystal, and exsolution to form a new inclusion must always be considered as a viable process in any very slowly cooled terrane (see also Chapter 3). The presence of H_2O in quartz has major effects on strength of the quartz, and Blacic (1981) has made use of these differences to estimate that the diffusivities of H_2O in quartz at high pressure are orders of magnitude greater than those determined at low pressure (see also p. 341).

A major difference should be noted between the fluid in exsolution inclu-sions and those formed by trapping of the normal intergranular fluid phase in metamorphic rocks. The latter has probably come to equilibrium with respect to the mineral assemblage present, whereas the former will only be in equilibrium with the host quartz from which it exsolves. Some of the implications of this difference are discussed by Spear and Selverstone (1983).

[3/] Green and Radcliffe (1975) proposed that CO_2 inclusions precipitated on crystal defects induced by defor-mation or exsolution in olivine. Most of the "fluid precipitates" they described are several orders of magnitude smaller than those described here, but the process may be similar.

[4/] The literature on metallurgy and ceramics contains many examples of the migration of gas atoms and of vacancies in the crystal structure to positions of minimum energy along dislocations and subgrain bound-aries, thus decorating both with minute gas inclusions. Alkali ions are particularly mobile in the quartz structure (see discussion in Roedder, 1958). As the mechanisms of diffusion of H_2O through quartz are not known, the connection, if any, between this and the diffusion discussed in the previous section is also unknown.

RELATION OF INCLUSION COMPOSITION TO THAT OF THE BULK OF THE FLUID

Nature of the Problem

That inclusions can be trapped from either a homogeneous or a heterogeneous fluid system was shown earlier in this chapter. The general assumption is that the material trapped in inclusions from a homogeneous fluid system is representative of the bulk of the fluid present at the time. Most inclusions are in this category. Material trapped in inclusions from a heterogeneous fluid system, however, obviously will _not_ be representative, in that the phase ratio of the material trapped will generally not be the same as that in the bulk of the fluid. The general assumption is, however, that the compositions of the two individual fluid phases are representative of those phases at the time of trapping. Are these two assumptions on the trapping of a representative sample of a given fluid phase valid? I believe that neither of these assumptions is literally true; the fluid trapped in any given fluid inclusion cannot be identical with that of the main mass of the fluid from which it was trapped, because of boundary-layer effects. However, the important point is the magnitude of the differences, which I believe are too small to be of significance, at least to present-day inclusion studies. In this section, the nature of these differences and the evidence as to their magnitudes are reviewed. Because exsolution inclusions do not form by the trapping of a specific fluid phase, they are excluded from this discussion.

Nonrepresentativeness Because of Changes with Time

A rather common derogation of fluid inclusions, from the older literature discussing the validity of fluid inclusions as samples of ore-forming fluids, was that they are not actually samples of the ore fluids but rather are the "last residue" or "the final spent fraction" from the crystallization of the ore (e.g., Ingerson, 1954, 1965). A series of primary inclusions, trapped in zoned crystals of ore or coprecipitated gangue minerals, from zones representing the earliest start of mineralization to the last fraction of a millimeter before mineralization ended with euhedral crystals protruding into the open vein, are certainly not a "final spent fraction." The ore or gangue mineral may continue to crystallize out on the walls of the inclusions after trapping, but the evidence is good (Chapter 3) that the amounts of such precipitation are generally very small and reversible upon reheating.

Regardless of the process whereby the fluid became saturated with respect to the host mineral for the inclusions -- whether by loss of gases, mixing with other fluids, loss of heat, or reaction with wall rocks or earlier vein minerals -- the fluid passing the protruding crystal is the fluid from which that crystal is growing, and if that fluid is trapped in the crystal as a fluid inclusion, that inclusion is a sample of the ore-forming fluid, not a "final spent fraction." After the fluid has passed through the ore deposit and on into a barren vein and has undergone the total of chemical transactions that take place between it and the ore body or its wall rocks, it can be considered a spent residue, but only with respect to that deposit. It may still produce other mineral deposits at higher levels, for example, near-surface or hot-spring manganese deposits (Hewett and Fleischer, 1960).

Nonrepresentativeness Because of Differences in Space

A range of scales of observation must be considered. On a large scale, the composition of the fluid present during a given geologic process will almost certainly differ from point to point. Thus the residual melt between the crystals of a mush at the edge of an intrusive body will have a different composition than the main (central) mass of liquid. Similarly, in metamorphic environments, there is generally a spatial nonuniformity, at any given time, of the intergranular fluid available to become trapped as inclusions. Although the scale may vary, abundant evidence has been obtained on compositionally variable metamorphic rock sequences to show that the mineral assemblage in any given band or bed generally could not have been in equilibrium with that in adjacent bands (e.g., Ferry, 1979). As a consequence, the intergranular fluid through which and with which these mineral assemblages have come to local equilibrium must also have been different, e.g., in such major variables as the H_2O/CO_2 ratio, once mineral/fluid equilibrium was established. Although this local control of the composition has been evaluated in all careful studies of metamorphic inclusions (e.g., Crawford et al., 1979b), it is too frequently ignored.

An ore fluid in the process of reacting with carbonate wall rocks to form skarn will differ from fluid in the main flowing stream. On a still smaller scale, the intergranular fluid within the wall rocks of a vein must certainly differ with location, as shown by the very different mineral assemblages in the various hydrothermal alteration zones around many veins. Regardless of the mechanism(s) by which the intergranular fluid got its dissolved components -- by mass flow outward from the vein, or by diffusion through an essentially static intergranular fluid medium -- inclusions of these intergranular fluids, trapped at various places in these wall rocks, should differ among themselves, and all should differ from the fluid flowing in the nearby vein.

Although the potential and real differences mentioned above seem rather obvious, too often they are forgotten and may cause misunderstanding and misinterpretation. Much more commonly considered in discussions of the degree to which fluid inclusions are representative are those differences that may occur at a still smaller scale -- right at the growing surface of the crystal, i.e., the boundary layer.[5]

Boundary-Layer Effects -- Aqueous Inclusions

We have long known that a liquid adjacent to any interface with another phase differs from the bulk of the liquid. Most studies of mineral-water interfaces (e.g., Adamson, 1976, pp. 196-243) are based on a model involving an electrically charged double layer. Henniker (1949) examined these "vicinal" liquids and found that at equilibrium, all surfaces affected liquids to a depth of many molecular diameters. For low molecular-weight liquids, there is evidence of some effects for several hundred Angstroms (i.e., ~0.03 μm). The observed effects are mainly structural (i.e., polar water molecules are oriented), and as such, may be detected by changes in electrical properties (Adamson, 1976; Roberts and Zundel, 1979), but compositional differences generally are not mentioned.

Under nonequilibrium conditions, as in the steady-state growth of crystals of soluble salts from H_2O solutions, considerable concentration gradients, for the crystallizing material, are set up near the growing face (Humphreys-Owen, 1949). Berg (1938) found the concentration gradients to be largest near the center of the face and detected effects to a distance of about 0.5 mm into the liquid. This early work uncovered several interesting problems (Buckley, 1951) that still have not been adequately explored.

One would expect, intuitively, that concentration gradients would occur near a fast-growing crystal, forming from a strong solution of the same substance. All atoms, ions, or molecular groupings that will become part of the crystal must go through the layer of fluid at the surface of the crystal, and this fluid, in turn, and all other materials in solution in it, must continually

---

[5] The term boundary layer is not used in the hydrodynamic sense but refers to somewhat analogous compositional differences.

be pushed away by the growing crystal. However, to the best of my knowledge, such effects have not been studied under the combination of conditions believed to be appropriate to ore-mineral crystallization: relatively slow crystallization at elevated temperatures, where diffusion is fast, from fluids perhaps high in extraneous dissolved substances, but relatively dilute with respect to the crystallizing phase. In such a situation, surface effects would presumably extend over extremely small distances, on the basis of analogous dissolution studies of Berner (1978), and hence would be of very minor significance to the gross composition of the fluid trapped in inclusions. Also, as very little material is crystallizing out of the fluid at the surface, this crystallization cannot result in significant gradients in the extraneous dissolved substances.

The only discordant notes here are provided by some experimental data reported by Barnes et al. (1969) on relatively large inclusions in synthetic sphalerite crystals. Barnes et al. synthesized inclusion-rich overgrowths of sphalerite on natural seed crystals within sealed Au tubes, at temperatures of 325-375°C and pressures of 0.27-1.0 kbar, by transporting nutrient ZnS about 10 cm down a 15°C gradient along the tube. The solutions used were 0.3-6.0 molal NaOH. The inclusions were typically 0.05-2.0 mm long, and formed only at the interface between the acid-etched seed crystal and the new growth. When a microscope freezing stage was used, the fluids in 106 such inclusions were found to be more dilute than the parent solution (the latter measured by a somewhat different procedure), the difference in molality being nearly constant at 0.2 ± 0.03 for parent-solution concentrations from about 1 to 6 molal NaOH. Barnes et al. stated that "This difference suggests that natural processes forming inclusions are complex and do not necessarily trap samples representative of the parent solution," but to my knowledge, no one has attempted to explain or corroborate these results.

Boundary-Layer Effects -- Silicate-Melt Inclusions

Diffusion rates in viscous fluids such as silicate melts are much lower than in water solutions, so the boundary-layer problems might be expected to be larger and more important. Obviously, there must always be compositional gradients adjacent to any growing crystal face, and diffusion, both toward and away from this zone, is continually leveling out these gradients (Anderson, 1967, 1976). The width of the zone in which significant compositional differences might occur increases with the speed of crystallization, the viscosity of the melt, and the amount of the host phase crystallizing from any given portion of melt; the width decreases with higher diffusion rates and with fluid flow past the crystal (e.g., sinking crystals). Furthermore, different ions have different diffusion rates, so the actual composition of the fluid near the surface that might be trapped in an inclusion is difficult to calculate. In addition to differences in diffusion rates for different ions, Watson (1979) has shown that ion diffusivities in dry granite glass are increased three to four orders of magnitude by the addition of 6.1 to 6.3 wt % H_2O. Inclusions that are small relative to the width of the zone of compositional gradients at the time they are trapped will have a more nonrepresentative composition (i.e., one that differs from that of the bulk melt), whereas large inclusions may be nearly representative of the bulk of the fluid (Fig. 2-36).

To my knowledge, the only report of differences in inclusion composition with size is that of Anderson (1974a), who found that (in olivine) melt inclusions <25 μm in diameter gave discordant results, whereas larger inclusions yielded consistent data. Whether this discordance stems from nonrepresentative trapping, analytical problems, or later changes during cooling is unknown, and further studies are obviously needed. In view of such problems, inclusion size, volume percent vapor bubble, etc., should always be recorded in studies of inclusions.

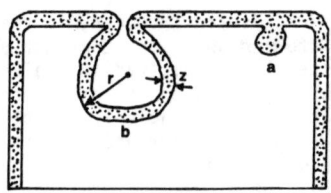

Figure 2-36. Diagram showing the relationship between inclusion radius (r), the width of the compositional gradient around the growing crystal (z), and the relative amount of compositionally modified fluid that will be trapped. Note that the composition of an inclusion with a small radius (inclusion a) should show more effects of any zone z of compositional gradients around a growing crystal than would a large inclusion (b). The width of the compositional gradient is shown by the stippled pattern. When z = 0.1r, the gross composition of the inclusion will contain 27 vol % of fluid from the gradient zone; when z = 0.01 r, the percentage drops to 3. The thickness of the gradient zone within the inclusion reentrants may be greater than that on the outside of the crystal.

The application of these concepts to actual studies of inclusions in magmatic minerals shows that the real effects of boundary-layer problems on melt-inclusion chemistry, as revealed by quenched-in compositional gradients, are generally minor. Bottinga et al. (1966) found a maximum of only several weight-percent differences in major constituents of basaltic glass 3 μm from a growing feldspar crystal, and Donaldson (1975) found differences of similar magnitude adjacent to an olivine crystal. In addition to the evidence from the above quenched-in compositional gradients, there is evidence from "control-line plots" that the real effects of boundary-layer problems on melt-inclusion chemistry are generally minor. The melt inclusions in any given host phase will have lost a variable and unknown amount of that phase from crystallization on the walls after trapping; hence, on a compositional plot, they should fall along the extension of a line connecting the composition of the host with that of the original liquid, i.e., a "control line" (see Chapter 16). The intersection of several control lines, from inclusions in several different but simultaneously crystallizing host phases, should yield the composition of the original melt. The fact that this procedure seems to work (e.g., Watson, 1976) indicates that except for the crystallization of the host phase from the trapped fluid, boundary-layer problems are apparently negligible.

Boundary layers in rhyolitic melts should be larger, in terms of both distance and compositional differences, than those found in basaltic melts, because the viscosities of rhyolites are orders of magnitude higher. Thus Evans and Nash (1979) reported gradients 10-15 μm wide adjacent to various microphenocrysts in natural rhyolitic glass, but again, most such modified compositions will probably fall on control lines for the host phase.

Any ion that is excluded from the growing crystal may be expected to increase in concentration at the growing surfaces. Thus crystallization of anhydrous phases increases the concentration of H^+ adjacent to the crystal, but the diffusion rate for H^+ is probably so much higher than that for all other constituents that the effects on the concentration of H^+ in the trapped inclusions should be minimal. However, if anhydrous phases crystallize rapidly from a melt that is nearly saturated with H_2O, an immiscible H_2O-rich phase may form at the growing surface and become trapped as inclusions (Fenn and Luth, 1973). In the future, when analytical procedures have been improved, the boundary-layer effects not only may be more readily detectable but also may be useful for rate studies, e.g., a comparison of the boundary-layer differences for elements having different diffusion rates. The interpretation of any such gradients will not be simple. Thornber and Huebner (1984), in a study of the dissolution of olivine in basaltic liquids, were able to show conclusively that the compositional gradients in glass adjacent to their olivine crystal interfaces were merely artifacts reflecting efficiency of a quench and are in no way related to equilibrium or dynamic element partitioning during bulk growth of the olivine.

GENERAL PROBLEMS IN THE USE OF INCLUSIONS IN
UNDERSTANDING GEOLOGIC PROCESSES

The application of inclusion data to the understanding of geologic processes is best illustrated by examples from specific environments, as detailed
in later chapters. However, three general problems concerning inclusion origin
are pertinent to many geologic environments and hence are covered briefly in
this section.

Relation of Inclusions to the Process being Investigated

The single most serious problem in all fluid-inclusion research is that
of relating the inclusions to the process being investigated, i.e., determining
the origin of the inclusions. In some samples, the occurrence of obviously
primary inclusions may make this problem seem negligible, but the possibility
of later changes, by leakage, necking down, etc., as discussed in Chapter 4,
should never be dismissed lightly. Many early successes in the application of
the study of fluid inclusions to the solution of geologic problems involved
rather large, recognizably primary inclusions in the minerals of hydrothermal
ore deposits. Many igneous and most metamorphic rocks have had a much more
complex history and hence usually contain many secondary fluid inclusions,
trapped on several different occasions, perhaps millions of years apart, without
the convenience of chronological sequence markers in the form of zoned crystals.
Although large primary inclusions of molten material are trapped in the minerals
of some igneous rocks, yielding "melt" inclusions (now glassy or crystallized),
recognizably primary liquid inclusions are rare or absent in many igneous and
metamorphic rocks. Most inclusions in metamorphic rocks are not only small but
are either demonstrably secondary or of indeterminant origin.

Scarcity of Inclusions in Certain Minerals

Just as in the field geologist's familiar lament that "the critical contacts are always covered," nature seldom puts inclusions in those phases where
we would most like to have them. The trapping of primary inclusions is a
result of interaction between the growth phenomena of the specific mineral and
the local conditions. Many inclusions have been trapped as a result of unknown
causes; they are simply "accidents." Thus, in the lunar basalts, we would like
to delineate the liquid line of descent by the examination of silicate-melt
inclusions in a series of phases that have crystallized in sequence. However,
among the phases present in these basalts, only early olivine and late ilmenite
commonly trap large usable melt inclusions; the major phases, plagioclase and
pyroxene, seldom do, thus frustrating the attempted delineation (Roedder and
Weiblen, 1977b).

Similarly, most studies of fluid inclusions in metamorphic rocks deal with
inclusions in the one phase, quartz, that is least indicative of the specific
metamorphic conditions. This reliance on quartz is in part a result of the
almost ubiquitous occurrence of quartz, but mainly it stems from the fact that
quartz just happens to trap fluid inclusions, both primary and secondary, more
commonly than do many other metamorphic minerals. Primary inclusions would be
very useful in compositionally variable phases such as feldspar, amphibole,
pyroxene, and mica, but they are seldom found. Only a few of the higher grade
metamorphic phases such as garnet may trap good inclusions and preserve them
through subsequent retrograde processes.

Solid and Fluid Compositions -- Which Controlled Which?

Throughout the wide range of natural environments in which fluid-inclusion
studies have been made, interactions between fluid and solid phase(s) have taken
place. Some of these interactions appear to have approached an equilibrium state

before the inclusions were trapped and quenched to the condition in which we now find them, whereas others are obviously far from equilibrium. As most of the ore deposits of the world are the results of these interactions, it is important that we understand what has occurred. At equilibrium, the relative amounts of any two phases are of no consequence, but many of the interactions in nature have been limited because of gross differences in the amounts of fluid and solid phases. Such examples may best be viewed in terms of the composition of the phase present in smallest amount effectively being controlled by that of the other phase. The fluids in the inclusions are always a very minute part of the present system, but at the time of trapping they may have been either the major or the minor phase.

In a magma, the first few crystals that form will have their composition effectively controlled by the very large mass of the magma. Once a small volume of magma is trapped as a melt inclusion within the growing crystal, however, the reverse becomes true. Any subsequent change in melt/crystal equilibria will now be between a very small amount of melt and a relatively much larger amount of crystal. Let us use a melt inclusion in olivine as an example. As trapped, at equilibrium the distribution of Mg and Fe between crystal and liquid (K_D) is effectively controlled by the melt, and given by the expression:

$$K_D = Fe_{xtl} \cdot Mg_{liq} / Mg_{xtl} \cdot Fe_{liq}$$

However, once the inclusion is sealed in an olivine bottle, the mass relationship is reversed. If we assume that on decreasing temperature, at equilibrium, an olivine crystal in contact with that liquid should become more iron rich, a thin layer of the new iron-rich crystal will form on the inside walls of the inclusion. However, this layer is not in equilibrium with the bulk of the crystal, and as the iron diffuses out into the surrounding crystal, more must be extracted from the melt inclusion. Obviously, this process will be limited by the small mass of the available melt and will be quenched at some point during this diffusion.

Immiscibility in a magma can be compared with the first crystallization, as generally only a few small globules of the new immiscible fluid form. The principle is the same, whether these globules be an immiscible sulfide, another silicate melt, a dense CO_2 fluid, or a low-density steam phase. The composition of this small amount of a second phase is effectively controlled by that of the major fluid present.

Hydrothermal vein deposits provide an excellent example of a situation wherein the composition of the solid is effectively controlled by that of the liquid. Most such veins are assumed to have formed as a result of a very small amount of precipitation from very large volumes of fluid, so the crystals of ore minerals were the minor phase. This assumption is reinforced by studies of the daughter phases in the fluid inclusions (Roedder, 1960a). Gangue minerals in the vein (and alteration phases formed on the immediate walls of the vein) also crystallized from, and were effectively controlled by, this fluid.

Farther out in the walls, however, this is no longer true. As fluid migrates out into the walls by mass movement through pores, or as ions diffuse to and from the vein, the composition of the very small volume of interstitial fluid, and hence the composition of any primary or secondary fluid inclusions formed in the wall rocks at this time, will differ from that in the vein. These differences may be quite large, and certainly must be considered in any attempt to reconstruct the chemical variables such as pH or Na/K in the environment of ore deposition from the composition of the inclusions. The presence of a trace of interstitial fluid in the pores of a metamorphic rock, fluid that is available for trapping during fracturing, is an even more extreme example of a fluid whose composition is controlled by the solids.

Table 2-1. Empirical criteria for the origin of fluid inclusions
(revised from Roedder, 1976a; 1979b).
(Does not include exsolution inclusions; see text for caveats;
the order within any group is of no significance.)

Criteria for primary origin

I. Based on occurrence in a single crystal, with or without evidence of direction of growth or growth zonation.

A. Occurrence as a single inclusion (or a small three-dimensional group of inclusions) in an otherwise inclusion-free crystal (Roedder, 1965b, Fig. 10, and 1972, Plate 6).

B. Large size relative to that of the enclosing crystal, e.g., with a diameter ~>0.1 that of crystal, and particularly several such inclusions.

C. Isolated occurrence, away from other inclusions, for a distance of ~>5 times the diameter of the inclusion.

D. Occurrence as part of a random, three-dimensional distribution throughout the crystal (Roedder and Coombs, 1967, Plate 4, Figs. A and B).

E. Disturbance of otherwise regular decorated dislocations surrounding the inclusion, particularly if they appear to radiate from it (Roedder and Weiblen, 1970b, Fig. 9).

F. Occurrence of daughter crystals (or accidental solid inclusions) of the same phase(s) that occur as solid inclusions in the host crystal or as contemporaneous phases.

G. Occurrence along a twin plane (Kelley and Turneaure, 1970), but note that secondary or pseudosecondary inclusions can behave similarly (Fig. 2-11).

II. Based on occurrence in a single crystal showing evidence of direction of growth:

A. Occurrence beyond (in the direction of growth) and sometimes immediately before extraneous solids (the same or other phases) interfering with the growth, where the host crystal fails to close in completely. (Inclusion may be attached to the solid or at some distance beyond, from imperfect growth; Roedder, 1972, Plate 1.)

B. Occurrence beyond a healed crack in an earlier growth stage, where new crystal growth has been imperfect (Roedder, 1965b, Figs. 18 and 19; Roedder et al., 1968, Fig. 15).

C. Occurrence between subparallel units of a composite crystal (Roedder, 1972, Frontispiece upper right).

D. Occurrence at the intersection of several growth spirals, or at the center of a growth spiral visible on the outer surface.

E. Occurrence, particularly as relatively large flat inclusions, parallel to an external crystal face, and near its center (i.e., from "starvation" of the growth at the center of the crystal face), e.g., much "hopper salt."

Table 2-1 (continued)

F. Occurrence in the core of a tubular crystal (e.g., beryl). This may be merely an extreme case of previous item.

G. Occurrence, particularly as a row, along the boundary between two growth sectors.

III. Based on occurrence in a single crystal showing evidence of growth zonation (as determined by color, clarity, composition, X-ray darkening, trapped solid inclusions, etch zones, exsolution phases, etc.).

A. Occurrence in random three-dimensional distribution, with different concentrations in adjacent zones (as from a surge of sudden feathery or dendritic growth).

B. Occurrence as subparallel groups (outlining growth directions), particularly with different concentrations in adjacent zones, as in previous item (Roedder, 1965b, Fig. 11).

C. Multiple occurrence in planar array(s) outlining a growth zone (Roedder and Coombs, 1967, Plate 4E). (Note that if this is also a cleavage direction, there is ambiguity.)

D. Occurrence on a surface from an episode of etching that interrupted normal crystal growth.

IV. Based on growth from a heterogeneous (i.e., two-phase), or a changing fluid.

A. Planar arrays (as in III-C) or other occurrence in growth zones, in which the compositions of inclusions in adjacent zones are different (e.g., gas inclusions in one and liquid in another, or oil and water (Roedder et al., 1968, Fig. 9).

B. Planar arrays (as in III-C) in which trapping of some of the growth medium has occurred at points where the host crystal has overgrown and surrounded adhering globules of the immiscible dispersed phase (e.g., oil droplets or steam bubbles).

C. Otherwise primary-appearing inclusions of a fluid phase that is unlikely to be the mineral-forming fluid, e.g., mercury in calcite, oil in fluorite (Roedder, 1972, Plate 9, Fig. 2) or air in sugar (Roedder, 1972, Plate 9, Fig. 4).

V. Based on occurrence in hosts other than single crystals (i.e., <u>intercrystalline</u> inclusions).

A. Occurrence on a compromise growth surface between two nonparallel crystals. (These inclusions have generally leaked and could also be secondary.)

B. Occurrence within polycrystalline hosts, e.g., as pores in fine-grained dolomite, cavities within chalcedony-lined geodes ("enhydros"), vesicles in basalt, or as crystal-lined vugs in metal deposits or pegmatites. (These last two are among the largest "inclusions," and have almost always leaked.)

C. Occurrence in noncrystalline hosts (e.g., gas bubbles in amber; vesicles in pumice).

Table 2-1 (concluded)

VI.   Based on inclusion shape or size.

    A.   In a given sample, larger size and/or equant shape.

    B.   Negative crystal shape--this is valid only in certain specific sam-
ples; in other samples, both primary and secondary or only the secondary inclusions may have negative crystal shape.

VII.   Based on occurrence in euhedral crystals, projecting into vugs (suggestive, but far from positive--see Roedder, 1967b, p. 523).

Criteria for secondary origin

I.   Occurrence as planar groups outlining healed fractures (cleavage or otherwise) that come to the surface of crystal (note that movement of inclusions during recrystallization can cause dispersion -- Roedder, 1971a,
Fig. 11; see also III-C above).

II.   Very thin and flat; in process of necking down (but note that necking
down may occur either during temperature decline or isothermally, in
primary, secondary, or pseudosecondary inclusions).

III.   Occurrence within a plane that differs compositionally from the rest of
the crystal, e.g., in cathodoluminescence (Sprunt and Nur, 1979; Ebers
and Kopp, 1979).

IV.   Primary inclusions with filling representative of secondary conditions

    A.   Located on secondary healed fracture; hence presumably refilled with
later fluids (Kalyuzhnyi, 1971).

    B.   Decrepitated and rehealed after exposure to higher temperatures or
lower external pressures than at time of trapping; new filling may
have original composition but lower density (Roedder, 1965a, Fig. 18).

V.   Temperature of homogenization (Th) far below that of adjacent presumed
primary inclusions in the same growth zone (on the basis that low-Th early
primaries would be decrepitated by a much hotter later stage of growth;
note, however, that late-stage primaries and pseudosecondaries can be at
any lower temperature, and that barring decrepitation, primary inclusions
can show an increase in Th with stage of formation).

Criteria for pseudosecondary origin

I.   Same occurrence as secondary inclusions, but outer end of the fracture
visibly terminates at a growth surface within crystal (Roedder, 1965b,
Figs. 18 and 19; Roedder et al., 1968, Figs. 12, 14, and 15. See also
III-C under "Primary" above). Frequently tapered in size, the largest
inclusions being near the outer termination.

II.   Generally more apt to be equant and of negative crystal shape than secondary inclusions in same sample (suggestive only).

III.   Occurrence as a result of the covering of etch pits crosscutting growth
zones (Roedder, 1972, Plate I, Fig. 8).

Chapter 3

CHANGES in INCLUSIONS after TRAPPING

"Everything in the world is in the process of change and fluid
inclusions are no exception. From the moment a fluid inclusion
is born, processes are at work which tend to modify it, sometimes beyond recognition of its original character. The ultimate
success of many fluid inclusion investigations depends upon
recognizing where such processes have occurred and to what degree,
thereby avoiding the trap of over-simplistic interpretation"
(Wilkins, 1979, p. 5).

CONTENTS

INTRODUCTION

In this chapter, I discuss the various physical and chemical changes that occur within inclusions after trapping, as well as the significance of these changes in the study of inclusions. Single-phase inclusions, which have not changed significantly since being trapped, are rather frustrating to study. Although they may provide only limited information on the environment of formation, this information can be valuable, as will be detailed in later chapters. Most inclusions are different now from when they were trapped, as new phases have appeared. The phase changes that occurred on cooling in nature can generally be reversed in the laboratory and hence provide valuable data on the pressure, temperature, and composition (P-T-X) conditions of the original trapping. Most of the physical changes, however, are essentially irreversible in the laboratory, and even the fact that they have occurred is sometimes difficult to prove. However, these physical changes may have gross effects on the evaluation of any P-T-X data obtained from the phase assemblage and hence must be carefully considered.

PHASE CHANGES -- DAUGHTER PHASES

After a small volume of fluid has become enclosed within a crystal host, a variety of changes may occur, both in the phase assemblage and in the physical nature of the inclusion. The changes in phase assemblage are the most useful in providing data on the P, T, and X of the fluids from which the crystal grew. Normally, a single homogeneous fluid was originally trapped, but at room temperature, multiple phases are present. All new phases formed (generally isochemically) within the inclusion are termed daughter phases and, if crystalline, daughter crystals.

Crystallization on the Walls

Although usually unseen and frequently ignored, crystallization on the walls of the inclusion leads to the formation of the one universally present "daughter" mineral, the host mineral itself. Most solid substances show an increase in solubility as temperature rises. Thus, one would expect that during the natural cooling of a fluid inclusion trapped at an elevated temperature, crystallization of the host mineral must take place, as the fluid is certainly saturated with respect to it when trapped. Generally, this crystallization takes place on the walls, rather than as a separate crystal. Ermakov (1950) described some special cases in which a visible coating (there described only as "bordure of cognate substance") formed inside the inclusions, lining the walls. He indicated that dissolution of this coating on heating may be remarkably slow, often requiring hours to reach equilibrium, even though the "reaction vessel" is measured in micrometers. In a very few instances, reactions between the trapped fluid and the walls may result in new phases. Thus NaCl-rich inclusions in plagioclase might be expected to form a scapolite daughter phase, but Vanko and Bishop (1982) showed that high NaCl concentrations and temperatures are needed.

Aqueous inclusions. Generally, the amount of crystallization of the walls is immeasurably small, as might be expected in view of the known low solubilities of most minerals at low to moderate temperatures. Many aqueous inclusions show no evidence of dissolution of the walls on heating, or of crystallization on cooling, even though the presence of minute details on the walls of the inclusions can make this qualitative test rather sensitive. Except for inclusions in highly soluble phases such as halite, I cannot recall ever seeing any evidence of dissolution of the walls of aqueous inclusions at <500°C, and only very small amounts are occasionally reported at higher temperatures. Lemmlein and Kliya (1952a) made use of a sensitive interferometric method to measure the

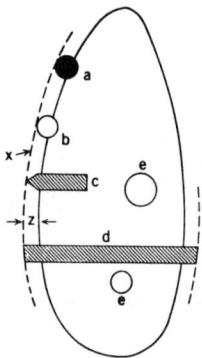

Figure 3-1. Diagram showing evidence of crystallization of a layer of host mineral on the walls of a silicate-melt inclusion after trapping. The original inclusion wall was presumably at (x). A globule of immiscible sulfide melt (a), a vapor bubble (b), and two daughter crystals (c and d), nucleated on the walls. Precipitation of a layer of host crystal of thickness (z) from the trapped melt partly surrounded these daughter phases. This crystallization was simultaneous with part of the the growth of daughter crystal (c), but crystal (d) completed its growth before significant crystallization of host on the walls. The interface (x) is usually invisible. Other vapor bubbles (e) may also nucleate out in the melt. As drawn, the thickness of layer (z) would correspond to the crystallization of one third of the original trapped melt.

amount of deposition of topaz on the walls of large inclusions in pegmatitic topaz on cooling from high temperature (>500°C?); they estimated that as much as 2 wt % had precipitated. Their measurements were such that this 2% represents only the amount of precipitation since cooling to the point where a vapor bubble formed in the inclusion.

Except for special situations (e.g., inclusions that have minute details on the walls), rather major amounts of dissolution are necessary to yield visible changes in the inclusions. Thus, if a cubic inclusion of water 20 μm on an edge in a host of density 3.0 g/cm^3 is enlarged to only 21 μm on an edge by dissolution of the walls on heating (an amount that might easily be overlooked), the volume of the inclusion would now be 15.8% larger, and the solution in it would now contain 32 wt % host mineral in solution.

Silicate-melt inclusions. Significant dissolution of the walls of inclusions during heating is much more common in silicate-melt inclusions, because relatively large volumes of the host crystal may have precipitated on the walls during cooling. In some samples, such precipitation can be recognized by careful examination of the wall of the inclusion where it is in contact with daughter phases. If the daughter phase appears to be embedded in the wall (Fig. 3-1), the most reasonable interpretation is that it nucleated on the wall when the inclusion was larger, and crystallization on the wall since then has tended to surround it. On the other hand, crystallization on the walls can occur without evidence of embedding.

Evidence of crystallization around immiscible sulfide globules or daughter crystals that have nucleated on the wall (Fig. 3-1) is common in inclusions in olivine from basalts, both lunar and terrestrial (see Chapter 18). In contrast to the sulfide blebs, vapor bubbles, although also generally nucleating at the wall, apparently have a contact angle near 180°; therefore, the olivine can continue to grow beneath the bubbles and push them into the remaining melt. As a result, vapor bubbles in olivine, unlike those in some other minerals (e.g., quartz, as shown by Zirkel, 1873, his Fig. 22), rarely appear to be embedded in the walls.

The zone of precipitated olivine, the "bordure of cognate substance" in the Russian literature, is in crystal continuity with the host crystal and may go unnoticed except when a feature such as a buried sulfide globule is recognized for what it is. Minor differences in index of refraction of such zones can sometimes be detected, e.g., in topaz (Voznyak and Kalyuzhnyi, 1974b).

An important consequence of extensive crystallization on the walls of an inclusion is the enlargment of the vapor bubble. The volume change from crystallization will be added to the volume change from differential shrinkage of melt and host on cooling, so such vapor bubbles will appear larger than those from normal shrinkage (see next section).

Shrinkage and Immiscibility (i.e., Fluid Daughter Phases)

The most conspicuous single feature of most fluid inclusions is a vapor or gas bubble that (in liquid inclusions) may move under the influence of gravity or a thermal gradient, and, if small enough (generally only when less than a few micrometers), may be in constant motion (mistakenly called "Brownian" motion; see Chapter 7). Although a small number of inclusions do give evidence of trapping of heterogeneous systems, most minerals from ordinary rocks and ores commonly show millions of inclusions per cubic centimeter, every one a simple two-phase system of gas and liquid with similar G/L ratios, suggesting entrapment of a homogeneous fluid. These two phases result from the fact that we are examining the inclusions at a lower temperature than that at which they were trapped. As the volume coefficient of thermal expansion for most fluids other than silicate melt is one to three orders of magnitude larger than that for most minerals, the container for the inclusions shrinks much less than the fluid it contains on cooling from the temperature of trapping to room temperature. When the pressure in the inclusions drops below the saturation vapor pressure of the contained fluid at that temperature (and hence the equilibrium volume of the fluid becomes less than that of the inclusion), the fluid will split into two phases, liquid and vapor. The relative amounts of these two phases will vary with the original fluid density. Most commonly, a small amount of a vapor phase forms (i.e., a vapor bubble). Sorby (1858), in a long and classic paper on fluid inclusions, proposed that this process could be reversed by heating the inclusions to the temperature of disappearance of the bubble (i.e., the homogenization temperature, Th), and that by adding a pressure correction, he could ascertain the trapping temperature (Tt).

The bubble may be very low pressure vapor (i.e., ~20 mm vapor pressure of H_2O at room temperature) if a water solution of nonvolatile solutes was trapped, or nearly a vacuum if a silicate melt without volatile materials was trapped. Any volatile species present (e.g., H_2O, CO_2, CH_4) will generally partition strongly into this vapor bubble and may make it a dense fluid at high pressures. Whether low or high density, the bubble is a true immiscible fluid daughter phase and must be considered as such.

Fundamental differences exist between aqueous inclusions and silicate-melt inclusions in terms of the amount of shrinkage on cooling. Silicate melts have much smaller thermal-expansion coefficients (see Chapter 16) than aqueous solutions; hence, even though the temperature interval over which melt inclusions have been cooled is generally much larger than that for aqueous inclusions, the shrinkage bubbles in melt inclusions generally occupy a smaller volume percent. Superimposed on these smaller thermal coefficients for melts, however, is another basic difference. The bubble in a melt inclusion can only continue to expand as the inclusion cools if the viscosity of the melt is low enough to permit the expansion in the available time. This viscosity also increases rapidly with temperature decrease, generally by a factor of about 10^3 per 100°C decrease (Fig. 18-19), and soon the melt becomes a glass that is rigid in the normal cooling time frames. If the thermal expansion coefficient for the glass is larger than that for the host, the glass will be put in tension and might even crack free from the host. This is probably the explanation of the fractures commonly seen in the glass of the larger glass inclusions.

In contrast, the bubble in an aqueous inclusion continues to expand as the inclusion cools down to room temperature and even much lower if crystallization does not intervene. If an inclusion contains a fluid of the critical density, the bubble will expand, contract, or remain essentially unchanged after it first forms on cooling, depending on the shape of the solvus (see Chapter 8). In any case, it will become more readily visible with further cooling, as the difference in index of refraction of liquid and vapor increases rapidly. Some fluids have even larger thermal coefficients of expansion than water and hence form even larger bubbles upon a given amount of cooling; this feature provides some crude constraints on the composition of the fluid. Similarly, the much lower thermal coefficients of expansion of strongly saline brines relative to pure water can be used to distinguish brine from fresh water.

Cooling of a silicate melt inclusion can produce another kind of immiscibility if the melt becomes saturated with respect to an iron sulfide melt and precipitates a globule of that fluid, in addition to a vapor bubble. Such immiscible sulfide melts concentrate Ni and Cu from the silicate melt and hence may yield valuable clues to the formation of economically important magmatic sulfide deposits.[1]

"Liquid" vs "gas". The division between "liquid" and "gas" is sometimes considered a semantic problem, but it actually presents a very real scientific problem and has been the source of some misunderstanding in the inclusion literature. I believe that no real distinction exists between the separation of a vapor bubble by shrinkage of the fluid (even if the bubble is almost a vacuum) and the separation of an immiscible fluid; both are simply examples of pairs of immiscible phases (see also discussion in Chapter 2). If the inclusion contains pure water and consists of two phases at 100°C, the liquid will have a density of 0.96 g/cm^3 and the bubble will be steam (water vapor) with a low density (0.0006 g/cm^3), but if the temperature is 370°C, the steam density will be approximately 0.2 g/cm^3, more than 300 times greater. Most ordinary gases in natural fluids, such as CO_2 or CH_4, will preferentially enter the "vapor," or low-density phase so long as two phases are present. The density of this "vapor" may then exceed that of the "liquid," and the "gas bubble" can be seen to sink in the "liquid" of the inclusion, as was first reported by Hartley (1877b). Takenouchi and Kennedy (1964, p. 1062) determined this density inversion point for various isobars in the system H_2O-CO_2, and stated quite appropriately: "The pointlessness of continuing the normal 1-atmosphere distinction between the gas and the liquid on to higher pressures is strikingly evident here."

If the separate CO_2-rich vapor phase is sufficiently dense, on cooling it also may split again into two fluids, representing liquid CO_2 and gaseous CO_2, thus yielding "three-phase inclusions," as illustrated in Figure 3-2.[2] Generally, this splitting occurs at temperatures below the critical point for pure CO_2, 31.0°C, because the inclusion bulk density usually does not equal the critical density, and because most common impurity gases, such as N_2 or CH_4, depress the critical temperature.

Aqueous-brine/aqueous-brine immiscibility. In certain highly saline fluid inclusions, a small amount of a new, dense, solute-rich fluid phase separates from the liquid of the inclusions on heating. This fascinating aspect of the use of highly saline fluid inclusions as "visual autoclaves" was first reported years ago by Kalyuzhnyi (1956; see also Kalyuzhnyi, 1958c; Ermakov et al., 1957), but its compositional significance is still unknown. These workers noted that when certain fluid inclusions are heated above 200°C, a small amount (less than 10 vol %) of a new fluid phase separates from the liquid of the inclusions. The new fluid has an appreciably higher index of refraction than the surrounding fluid and presumably contains a higher concentration of salts.

----

[1] Although silicate/silicate liquid immiscibility occurs in a variety of environments and is a common cause of trapping of such melts as inclusions, such immiscibility rarely is caused within a silicate melt inclusion by cooling after trapping.

[2] The undefined term "three-phase inclusions" has been used by various authors to indicate different phase assemblages (e.g., solid, liquid, and gas, or two liquids and gas). This term must be preceded by a specific definition to avoid ambiguity.

Ermakov et al. (1957, p. 121 in original) reported that the new liquid persisted, with little change in volume, until the disappearance of the gas bubble at 327°C and that the phase changes were reversible on cooling. Presumably still higher temperatures are needed to make the two liquids miscible again. Trufanov et al. (1970), Trufanov (1972) and Ahmad and Rose (1980) reported possibly similar features, and Sobolev et al. (1970) reported that a second liquid phase is present for a narrow range of temperatures near 700-800°C in some 125 inclusions in chkalovite ($Na_2BeSi_2O_6$) from Ilimaussaq, Greenland that homogenize at 860-980°C.

The significance of such immiscibility in terms of specific fluid composition is not known, but in view of the possibly complex phase relations of multiple salt-water-gas systems, such immiscibility with rising temperature under constant volume conditions is perfectly expectable in a manner similar to the "infra-critical" and supra-critical" homogenization in the system CO_2-H_2O (Ypma, 1963). An actual example of such saline immiscibility has recently been found in laboratory studies of the system $KHPO_4-H_2O$ (Marshall et al., 1981). It would be particularly interesting to know if this immiscibility is related to the precipitation (i.e., partial condensation) of a small amount of highly saline fluid from lower-density, lower-salinity fluids in the pure system $NaCl-H_2O$ (p. 333).

"Vapor" inclusions from pure melts. One example of the formation of "vapor" inclusions needs special mention, as it may explain some otherwise puzzling inclusions found in a few lunar samples and meteorites (e.g., Roedder and Weiblen, 1973, p. 1043 and 1048). A gas inclusion within a single crystal can form by two entirely different processes. The most common process is the enclosure of a vapor bubble that adheres to the surface of the growing crystal. However, if the fluid from which the crystal is growing is a pure or essentially pure melt of the host mineral, apparent gas inclusions can form simply by the trapping of this melt. Once the melt is trapped within a rigid crystal, the volume change (decrease) on crystallization must result in the formation of a "vapor" bubble (generally a vacuum). If the melt does not exactly match the composition of the mineral, this bubble will normally be lined with a glass containing the "impurities." All gradations are possible between this and the more usual silicate-melt inclusion, in which some crystallization on the walls, or some crystallization of separate daughter phases, increases the volume percent of the vapor bubble in the remaining inclusion contents.

Surface energy and contact angles. When several different fluids, such as silicate melt, water solution, liquid CO_2, oil, or gas, are in the same inclusion, they usually assume positions of minimum surface energy. This behavior provides some assistance in phase identification. When glass is present, it will be in contact with the walls, i.e., it has a small contact angle or "wets" them, and any other fluids, such as "vapor" (essentially a vacuum bubble) or liquid water or CO_2, occur as bubbles within the glass rather than wetting the walls (Roedder, 1972, p. 11). (Very small bubbles may nucleate on the walls of some silicate-melt inclusions and remain there during cooling.) When a water solution exists together with an immiscible CO_2 fluid (liquid or gaseous), the globule of CO_2 is always within the water solution, which, in turn, wets the walls and fills all minor reentrants in the cavity. If a large volume percentage of liquid (or gaseous) CO_2 is present, it may appear to press against the walls, but freezing the inclusion will show that a film of water is usually present. When a gas bubble occurs with liquid water and liquid CO_2, the bubble is always completely surrounded by the CO_2 liquid, an observation first made by Brewster (1826a) many years before the true nature of liquid CO_2 became known. This simple fact causes considerable difficulty in recognizing the presence of liquid CO_2 when it constitutes $\leftsim 10$ vol % of the sum of L+V, unless the inclusions are strongly flattened (see Chapter 7).

The same relationship is shown by vapor bubbles in inclusions containing water and an immiscible organic fluid. In these, the vapor bubble nearly always occurs in the organic phase, which, in turn, is generally free of the walls.

Only rarely (in some fluorites) have I found an oil phase preferentially wetting the walls of an otherwise brine-filled inclusion (Fig. 2-26).

In a very few examples, the interfacial tensions between three different fluids and their relative volumes are such that the minimum energy configuration is not that of concentric shells but that of "partial engulfing" (e.g., Fig. 3-3; see also Torza and Mason, 1969; Mori, 1978).

Daughter Minerals

On cooling, the originally homogeneous fluids trapped in inclusions generally become saturated with respect to new fluid phases (see above); much less commonly, they become saturated with respect to solid phases other than just that of the walls and nucleate new crystals in the inclusions, called daughter minerals or daughter crystals. On heating in the laboratory, true daughter crystals should redissolve in the fluid, barring kinetic problems, and the temperature at which the solid finally "melts" (i.e., complete dissolution; T_m) provides valuable but somewhat limited constraints on the composition of the original fluid. If an apparent daughter crystal does not redissolve, it may be a result of necking down (see Chapter 2), or of H_2 leakage (discussed in a later section), or the crystal may be an accidentally trapped solid inclusion. By far the most common daughter crystal is a cube of NaCl, but other minerals, such as carbonates, oxides, silicates, sulfates, and sulfides are also found. The identification of daughter minerals is important because it provides compositional data on the fluid that was originally trapped. The methods of identification are discussed in Chapters 4 and 5.

Aqueous inclusions. Most aqueous inclusions containing relatively soluble daughter minerals such as halite contain only a single crystal of each daughter phase, as a result of the combination of two factors. First, most geologic samples have cooled very slowly from the temperature of trapping to surface temperature. In many metamorphic terranes, this cooling period (during uplift) must have lasted for millions of years (Hollister et al., 1979), so plenty of time was available for a single nucleation event, at very slight degrees of supersaturation and even in very small inclusions. Second, even if several nuclei of a given phase formed and grew, the very small distances involved and the presence of an effective low-viscosity solvent (the solution) permit the larger crystal to "eat up" the small one over geologic time, a result of the small but finite differences in surface energy. Such grain growth is important in many industrial processes and is the reason why some very fine chemical precipitates are "aged" before filtration. Two crystals of halite are sometimes seen in a single inclusion (e.g., Guha et al. 1979, Fig. 3; Roedder, 1972, Plate 5, Fig. 2). I have no really satisfactory explanation for such occurrences, but their rarity suggests that the processes described above are generally effective. For every such exception, probably thousands or millions of inclusions each contain a single halite cube (Fig. 3-4). Only very rarely does NaCl form crystals other than cubes (Fig. 13-1B). Large daughter crystals may contain fluid inclusions as well, providing some information on the conditions under which they crystallized during the cooling of the host crystal (Fig. 13-1E).

In aqueous inclusions, the daughter phases that occur as multiple crystals are generally minerals that have a low solubility in the liquid phase, which retards the normal growth of the larger grains at the expense of the smaller ones. Even geologic time is not adequate to preclude multiple crystals of a given daughter phase if the inclusion is large, if the daughter phase has a relatively low solubility in the inclusion fluid, or if the inclusion cooled rapidly after the multiple daughter crystals formed. Thus, small tufts consisting of many acicular crystals of some (generally unidentified) presumed daughter phase are common in some inclusions. Brewster (1845) reported a cloud of tiny crystals of calcite in the fluid of a large inclusion, and I have been told that

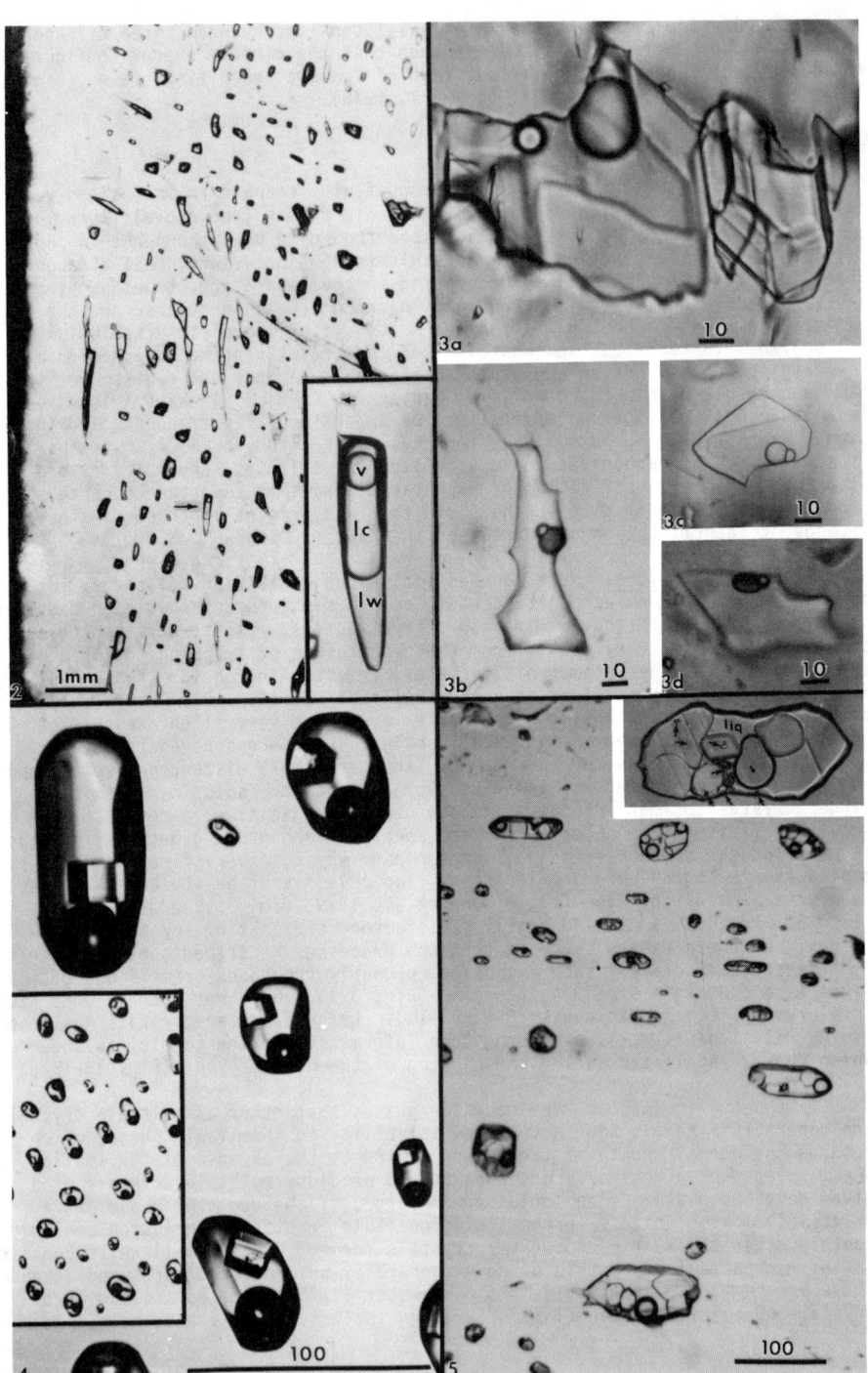

some large inclusions in quartz from a Brazilian pegmatite contain hundreds of tiny loose cubes of pyrite.

The term "uniform phase ratio," without definition, is commonly used as a criterion for a daughter-crystal origin for the crystalline phase. Actually, two very different definitions of the term are involved here that are not always distinguished. When inclusions cool, a daughter phase may nucleate in some but not in others (see next section on Metastability). As a result, the phase ratio in different inclusions will be very different, but the crystals in those that did nucleate will be in uniform ratio and will be valid daughter crystals. In contrast, the trapping of accidental solid inclusions can also result in some inclusions with and others without a crystal. The important difference is that those that contain a crystal will not show a uniform volume percent of that crystal, as there is no necessary correlation between the volume of the inclusion and the size of the solid grain that was trapped. The measurement of phase volumes in inclusions (Chapter 7) is necessarily inexact but important, because it provides the best evidence of the mode of origin of the crystal. Nucleation can be surprisingly consistent, e.g., some highly concentrated solutions will form large numbers of different daughter phases, but each of hundreds of small inclusions will have the same assemblage (Fig. 3-5). The criterion for a daughter-crystal origin should be a uniform phase ratio in those inclusions that contain the phase in question. Note, however, that even this relation does not necessarily hold for daughter crystals in melt inclusions

---

Figure 3-2. Photomicrograph of a plane of pseudosecondary inclusions in pegmatitic quartz and of one inclusion (arrow) from the plane, enlarged (inset). Note the abrupt termination of the plane beneath the outside crystal surface (at left). All inclusions show a uniform ratio (at 24°C) of liquid-water solution (lw; about 0.7 molar NaCl equiv., estimated from the freezing temperature of ~-2.5°C), liquid CO_2 (lc), and gaseous CO_2(v). Homogenization of the two CO_2 phases occurs in the liquid phase at 27.65±0.05°C; at much higher temperatures, probably above 350°C, only one homogeneous "gas" phase is present, containing all the CO_2, H_2O, and salts. A tiny daughter crystal (arrow in inset) is connected by a thin tube. Sample ER 61-28, Volta Bala, Teofilo Otoni, Minas Gerais, Brazil, courtesy of W.D. Johnston, Jr., (his number 52).

Figure 3-3. Photomicrographs showing the behavior of small globules of an immiscible liquid in inclusions. In this sample, each inclusion contains liquid-water solution (lw), a small vapor bubble (v), and a small immiscible globule of a fluid originally thought to be liquid H_2S (ls), because of its physical behavior and the very strong odor of H_2S on breaking the sample. Subsequent Raman and heating-stage studies (Rosasco and Roedder, 1979) have shown that although HS⁻ and $H_2S°$ are present in the water phase, the immiscible droplets (ls) cannot be liquid H_2S. The identity of this phase remains unknown. As the phase ratios of many of these inclusions are rather uniform (v:ls is ~4:1 by volume), the assumption is that a single homogeneous fluid was trapped originally and has separated into three phases on cooling. A few such inclusions show separate spherical (or circular flattened) globules of (v) and (ls) as in (a). This accidental configuration does not have the lowest surface energy. Apparently the system has the lowest surface energy when the (ls) liquid occupies the interface between liquid water and vapor, as in (b), (c), and (d) (Torza and Mason, 1969). In more three-dimensional inclusions, the (ls) globule sticking to the interface may appear, because of foreshortening, to be inside the vapor bubble. In (a), (b), and (d) the plane of focus has been adjusted to show that the globule of (ls) liquid has a considerably higher index of refraction than does the vapor bubble. Sample ER 67-7, yellow quartz, from Xique-Xique, Bahia, Brazil (Johnston and Butler, 1946, p. 615), courtesy of Earl Ingerson. Scale bar in μm. From Roedder (1972).

Figure 3-4. Photomicrograph of a small part of a plane of inclusions in quartz from the Bitsch power plant, Valais, near Brig, Aar massif, Switzerland (near an occurrence of evaporites), consisting of hundreds of inclusions, each containing a single halite cube and a vapor bubble. On heating, the halite dissolves at 196°C, and homogenization (to liquid) is complete at 308°C. Inset shows plane at lower magnification. Scale bar in μm. Photograph and data courtesy of H.A. Stalder and Bernard Poty.

Figure 3-5. Inset: Photomicrograph of a large (272 μm) multiphase inclusion in transparent magnesite crystal from sample ER 62-4, Brumado, Bahia, Brazil (Bodenlos, 1954; see also Rosenberg and Mills, 1966). The phases visible are liquid (liq), vapor bubble (v), and 14 daughter crystals, at least seven of which are different phases. Two of the daughter crystals are isotropic, but six of the small and at least one of the large crystals are anisotropic (see arrows). Such an inclusion could have been formed by a group of loose mineral grains being washed into a reentrant in the surface of a growing crystal and then become trapped as solid inclusions, along with some solution. This inclusion, however, is just one of a plane of many such inclusions (see main Fig.), each containing the same phase assemblage within the limits of available observational techniques. Hence, these must be daughter crystals, and the fluid that was trapped originally must have been a homogeneous solution containing >50 wt % solutes. Scale bar in μm.

(see next section).

Silicate-melt inclusions. The occurrence of daughter minerals in silicate-melt inclusions is much more erratic than it is in aqueous inclusions, and when such daughter minerals occur, they are commonly multiple crystals. These features are a result of two important differences between silicate-melt and aqueous inclusions. First, most silicate-melt-inclusion studies have been made on extrusive rocks, which have generally cooled much faster than have the hosts for most aqueous inclusions (particularly through the important first few hundred degrees, when crystallization is still possible). Second, because the viscosities of silicate melts are very high, both nucleation and growth are severely retarded. Therefore, even though most of the daughter crystals found in silicate-melt inclusions are phases that are "highly soluble" in the melt (i.e., major components such as pyroxene or feldspars), their formation and growth are far less common and regular than are those of the daughter crystals seen in aqueous inclusions. Even if the same phase nucleated in every inclusion, the time of this nucleation in the cooling history would be random. Nuclei that form earlier may grow into much larger daughter crystals than those of the same phase that form later in other inclusions that were trapped at the same time. Note that such behavior directly negates the most important criterion for a daughter-mineral origin -- the uniform ratio.

In view of the problems discussed above, how can one distinguish between solid inclusions and daughter phases, when the standard criterion used for aqueous inclusions is generally not applicable? Although some samples contain numerous silicate-melt inclusions that have uniform phase ratios, many contain inclusions having highly variable phase ratios. Commonly, only a few inclusions will contain any crystal, and this crystal may be a single relatively large one (Fig. 3-6). The only useful criteria for distinguishing between daughter crystals and solid inclusions in such samples are unfortunately rather subjective. Thus, if the crystal is so large that it would be unlikely for the melt with it to have held that much in solution, it is probably a solid inclusion[3]. Similarly, a crystal is probably a solid inclusion if the same phase also occurs as isolated solid inclusions in the host, particularly if it also appears from petrographic evidence to be earlier than the host crystal. Good examples of widely varying crystal/melt ratios are found in the chrome spinel crystals plus silicate melt commonly found embedded in olivine crystals (Fig. 2-12b).

One of the best proofs of daughter-mineral origin is the behavior of the crystal on heating (see Chapters 16 and 18). If it dissolves on heating and recrystallizes on cooling, one can be reasonably certain it is a daughter phase. This is a particularly valid criterion if the same phase first nucleates during heat-up in other originally all-melt inclusions, and then melts at the same temperature.

I suggest another criterion that might be applied that is elegantly simple in concept but not necessarily in practice. If the same melt is trapped in two inclusions in a given host phase, and another phase nucleates and grows as a daughter crystal in only one inclusion, the index of refraction of the remaining glass should be different in the two inclusions. These indices can be obtained by careful crushing in oils. This criterion is merely a logical extension of a finding reported in one of the very early studies of silicate-melt inclusions (Kalyuzhnyi, 1965), which showed differences in the index of refraction of glassy inclusions in different host minerals from a given lava. A difference

----

[3] In considering this aspect, one must be sure to include whatever amount of the host mineral that might be expected to dissolve from the walls. Thus, if the original magma were essentially a solution of pyroxene and feldspar, a melt inclusion trapped in either of these phases could consist of just a crystal of the other phase plus a shrinkage cavity. Conditions approaching this are probably not unusual.

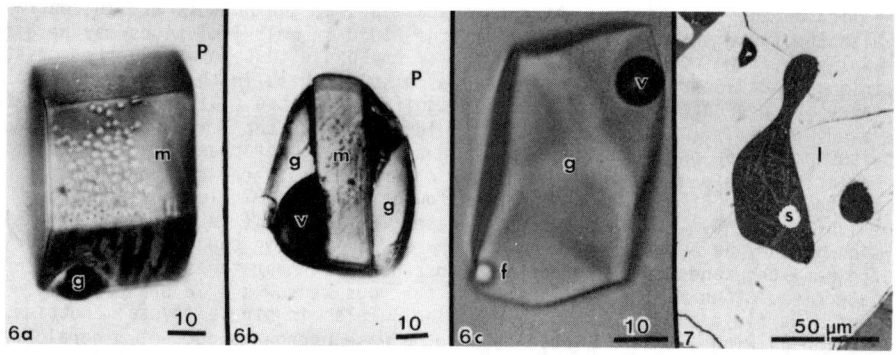

Figure 3-6.   Silicate-melt inclusions containing single crystals; which are daughter phases and which are simply solid inclusions? The crystal in (a) is obviously just a trapped solid that was enclosed along with some silicate melt, now glass (gl), and, in fact, was probably the cause of the trapping of the melt. However, when the enclosed crystal is smaller, as in (b) and (c), the distinction is more difficult and much more commonly ambiguous than it is in aqueous inclusions (see text). Scale bars in μm.

Figure 3-7.   Typical silicate-melt inclusion from the lunar basalts, which has crystallized to yield a mass of small prisms of pyroxene (bright) in a plagioclase-rich glass (dark). (S) is an immiscible sulfide bleb. From Roedder and Weiblen (1972b).

in the index of refraction of the glasses would indicate crystallization of the daughter phase in one. The same difference in index would result if the crystallization of the daughter phase took place on a solid inclusion trapped with the melt. If actual compositions could be obtained, e.g., by microprobe, this ambiguity would be eliminated (Roedder, 1979a, Fig. 39; see also Chapter 16).

The above deals with the common occurrence of glass inclusions containing one or a few large crystals. Perhaps even more commonly, silicate-melt inclusions consist of glass that usually contains a shrinkage bubble and thousands of tiny disseminated crystals of one or more phases (Fig. 3-7). Such crystals are almost certainly daughter crystals from the crystallization of the melt; the large number of crystals of any given phase stems from the high viscosity of the melt, which results in large numbers of nuclei and little growth on each. In many melt inclusions, crystallization is only incipient, giving the glass a faintly granular appearance in strongly collimated lighting. This type of devitrification occurs at temperatures too low to permit appreciable grain growth. Preferential scattering of blue light by these small crystals may cause an otherwise colorless clear glass to become yellow-brown in transmitted light and opalescent bluish in reflected light. Commercial white glasses are white because materials have been added to cause such fine devitrification; they are normally slightly yellowish-brown in transmitted light.

The position of the bubbles in silicate melt inclusions should indicate the direction of gravity at the time of quenching of the melt, i.e., they should be at the top of the inclusions, or at least show a statistical preference for the top. This information could be of value in solving structural problems in old volcanic rocks. Although I have seen qualitative indications of preferential placement of the bubbles in some meteorites and terrestrial volcanic rocks, careful statistical studies are needed to verify this, as the bubbles form wherever a suitable nucleus is present in the wall, and generally do not float free to the top. Viscosities are high also, so presumably only the largest inclusions might reveal an effect from gravity.

Occurrence of daughter-mineral phases outside inclusions. An important difference between silicate-melt and aqueous inclusions is the common occurrence of silicate-melt inclusion daughter phases as part of the host rock. In contrast with the fluid in aqueous inclusions, the fluid in melt inclusions may be essentially the same composition as the rock. Any phase that forms daughter crystals in such inclusions should be expected as a solid phase in the rock as well, because its presence in the inclusions signifies that the fluids had become saturated with it. (The only exception would be a phase that might react chemically with some other phase in the rock.) Thus, the daughter phases in silicate-melt inclusions are usually found among the other phases in the rock outside the host crystal. Many of the daughter phases found in aqueous inclusions are water soluble; if they were ever separate phases in the rock (i.e., outside of inclusions), they have generally been lost by dissolution in the ground water during the many thousands of years the host rock has been near the surface.[4] Such late dissolution may result in some of the vugs we now see in ore deposits. Thus, the fluid inclusions, by preserving samples in single-crystal "bottles," provide us with a unique window to an otherwise unknown aspect of mineralogy. Many rare water-soluble phases have been found in inclusions, and I suspect that many new minerals will eventually be recognized among these daughter crystals, particularly in the inclusions from some pegmatites (see Chapter 14).

Changes in daughter minerals on cooling to surface temperatures. In a few Soviet papers (as referenced in Roedder, 1972), phase changes (i.e., inversions) have been noted in daughter crystals during heating in the laboratory, particularly in alkali-aluminum-fluoride phases. Because such inversions may take place quickly on heating, the reverse change is presumed to have occurred during the original cooling. Recognition of the existence of such inversions can be especially important in the identification of those daughter minerals whose high- and low-temperature forms have a visibly different crystal symmetry or habit.

Two separate and discrete daughter crystals, one of NaCl and one of KCl, are found in many fluid inclusions from various high-temperature geologic environments (e.g., Figs. 15-11,12,13). As NaCl and KCl form a continuous solid solution at high temperatures (minimum at 658°C; Chou, 1982), with a solvus at lower temperatures (solvus maximum at ~480°C; Skeaff et al., 1979), any daughter crystal forming above the solvus will be a mixed crystal. Presumably, these original mixed crystals have exsolved into two separate phases on cooling over geologic time, although during cooling after laboratory homogenization, they may crystallize as a single phase and remain that way (R.J. Bodnar, pers. comm., 1981).

Lack of Phase Change -- Metastability

Significant degrees of metastability (usually stated in terms of temperature) are most commonly encountered in the nucleation of gas bubbles and daughter minerals on cooling inclusions that were previously homogeneous. Some homogeneous, single-phase fluid inclusions, less than ~100 μm in size and generally formed by trapping at less than ~70°C, have persisted for millions of years at rather high negative pressure without forming the "stable" bubble (Roedder, 1967a, 1971a). Sometimes freezing and thawing the inclusion will cause the nucleation of the small bubble that should be present, but when this procedure is used, the bubble may form as a result of permanent deformation of the inclusions (i.e., stretching from expansion on the formation of ice; see below), rather than just nucleation. Once saturation is achieved, salt crystals generally nucleate, even in very tiny inclusions, but other phases having lower degrees of supersaturation are found only in the larger inclusions.

----

[4] A few may be trapped as solid inclusions within single crystals and preserved; e.g., halite is occasionally reported in quartz.

Some fluid inclusions containing "stretched" fluids may well represent stable rather than metastable equilibria, even though the fluid has a density less than it would have in larger volume. As a result of the very rapid increase in internal pressure within a bubble from surface tension as bubble radius decreases, small inclusions that should have a very small bubble under normal equilibrium will not have them. Even if a bubble did nucleate, the surface tension would cause it to "blink out" instantly, thus putting the liquid under tensile stress. As a consequence, very small inclusions will yield significantly incorrect (low) values for Th L-V(L), whereas larger inclusions that trapped the identical fluid will yield almost the correct values. The magnitude of this effect becomes particularly significant when the inclusion diameter is <5 µm (Chapter 10).

These and a variety of other types of metastability have been recognized during laboratory study of the phase changes in inclusions. They are discussed in some detail in Chapter 10, since they can cause serious practical problems in the laboratory work. The most common problems occur during freezing of inclusions, when failure to nucleate newly stable phases has been observed for ice, i.e., metastable supercooling, and even more noticeably for crystals of $NaCl \cdot 2H_2O$ or $CO_2 \cdot 5.75H_2O$ (Roedder, 1963). Inability to freeze small inclusions can actually preclude some investigations. If the gas bubble is eliminated by expansion of the liquid on freezing, the liquid may come under very high negative pressures during melting and show some strange phase behavior, e.g., metastable superheated ice at temperatures at least as high as +6°C (Roedder, 1967a). As a result, determinations of salinity in the absence of a vapor phase may not be valid.

PHYSICAL CHANGES

Changes in Shape

Necking down of homogeneous inclusions. Many, perhaps most, inclusions do not now have the shape they had at the instant of trapping. As originally trapped, many fluid inclusions have relatively large surface areas. For example, healing of narrow cracks may yield thin flat secondary or pseudosecondary inclusions, and long tubes are characteristic of the primary inclusions in some minerals. If the host mineral is at all soluble in the fluid, processes of recrystallization, generally termed necking down, immediately start to reduce the high surface energy of the system. The analogous studies of Berner (1978) indicate that diffusion is not a major rate-limiting factor, but apparently the solubility of the host minerals is. Lemmlein pointed out in a series of papers (1929, and particularly Lemmlein and Kliya, 1952b) that inclusions formed by the healing of cracks introduced into crystals of water-soluble salts immersed in a saturated solution permit these changes to be observed and photographed by time-lapse photography. Dissolution in some areas and deposition in others results in crystal faces forming in place of the original curving fracture surface, and eventually some of the crystal surfaces from opposite sides of the fracture meet (Fig. 2-15). The resulting deep reentrant in the wall rapidly becomes filled with material dissolved from the main part of the walls, so that the whole amoeboid inclusion (e.g., Fig. 2-15c) may coalesce into a single more equant cavity. Elongated inclusions gradually develop bulges separated by thin necks, which eventually become sealed off (Fig. 3-8,9).

The final result of such necking down ("necking off" would be more precise) is the formation of several smaller inclusions that have the same total volume as the original single inclusion but a smaller total amount of surface energy. Nichols (1976) presented a theoretical treatment of the general process. Sometimes the recrystallized host crystal differs slightly from the original host and can be recognized by cathodoluminescence (Sprunt and Nur, 1979) or by phase

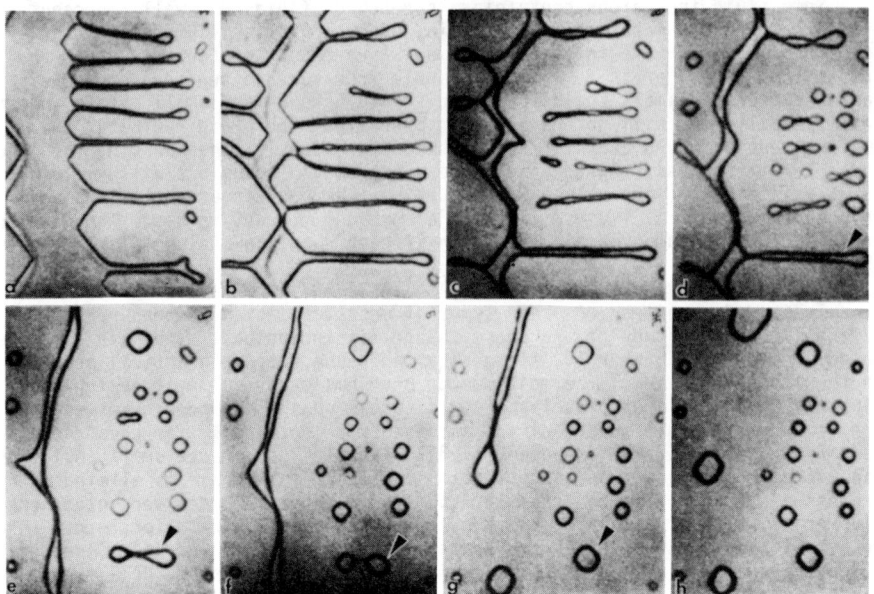

Figure 3-8.   Eight stages in the healing of a crack in a synthetic NaNO₃ crystal, photographed by Lemmlein (1951). Note particularly the group of inclusions forming at the bottom of each picture (arrows); two of these first start to neck down (e and f), but then coalesce as shown in (g).

TIME AND DECREASE IN TEMPERATURE ⟶

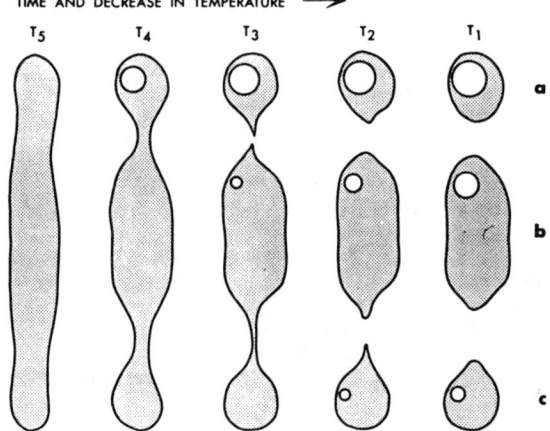

Figure 3-9.   Necking down of a long tubular inclusion. The original inclusion, trapped at temperature T_5, breaks up during slow cooling to form three separate inclusions, (a), (b), and (c). When reheated in the laboratory, inclusion (a) would homogenize <u>above</u> the true trapping temperature T_5, inclusion (b) would homogenize between T_3 and T_4, and inclusion (c) would homogenize between T_2 and T_3. Adapted from Roedder (1962b).

microscopy (Mikhailov, 1981). Presumably, a similar process, on a much smaller scale, yields a string of minute inclusions when solvent and solute molecules are reorganized along dislocations. Practically all secondary and pseudosecondary inclusions are the result of necking down.

The magnitude of the change in surface is significant. Thus, water in a fracture 1 μm wide has a contact surface with the host mineral of 3.6×10^5

$cm^2/mole$. When this total volume of water coalesces to inclusions 10 μm on an edge (i.e., inclusions evenly spaced at 33 μm intervals along the former fracture), this area drops to 10^5 $cm^2/mole$; further coalescence to inclusions 100 μm on an edge would reduce this area by another order of magnitude to 10^4 $cm^2/mole$.

Regardless of the original shape of the inclusion, recrystallization generally occurs if the host mineral has any finite solubility in the fluid of the inclusion under the conditions at which it is held after trapping. Because such recrystallization is probably of very general occurrence, most inclusions now have shapes different from those they had when originally trapped. The differences can be relatively small if cooling has been fast[5]/, if the solubility is very low, or if the original shape was close to the stable shape (as is true of some negative crystals). Thus oil inclusions in fluorite frequently have round shapes that appear to be those of the original round immiscible oil drops, but small inclusions of the immiscible water phase with them have undergone considerable change in shape by recrystallization. On the other hand, differences in the shapes of inclusions having different compositions do not necessarily reflect later changes in shape because the original shapes could also have been different (Fig. 3-10).

The final shape assumed by those fluid inclusions in which time and composition have permitted recrystallization may be either smoothly rounded and globular, or faceted negative crystals. A very weak driving force, slight differences in surface free energy, may drive the system toward the smallest amount of surface (spherical), or the lowest energy surfaces (negative crystal faces). The differences in specific area between such shapes are small. Thus a cavity the shape of a regular octahedron has 18% more surface than a sphere of the same volume. Some secondary or pseudosecondary inclusions in quartz are almost spherical, whereas inclusions of fluids of different composition in other planes, even in the same crystals, may be sharp hexagonal bipyramids that have only a relatively few percent more surface area than the spheres, per unit volume. Presumably such differences result from fluid compositional differences yielding different surface absorption (Adamson, 1976, pp. 248-253).

Calculations have been made to determine which faces on negative crystal cavities represent the lowest energy configuration. McLaren and Phakey (1966) calculated that in α-quartz, the equilibrium form is a polyhedron bounded by six {$10\bar{1}0$} faces, six {$10\bar{1}1$} faces, and six {$0\bar{1}\bar{1}1$} faces. Unfortunately, just as it affects the external faces on a crystal, the composition of the phase with which the crystal is in contact (the inclusion fluid in this case) affects both the crystal forms and the habit of the negative crystals. Thus, in many fluorite samples, I have found hundreds of primary inclusions, all beautifully regular negative crystal cubes, whereas other fluorite samples contain hundreds of inclusions, each a very regular tetrahedron. In some banded crystals, the inclusions in one growth band are tetrahedral, and in the adjacent band, cubic (Roedder, 1977a). Still other inclusions in fluorite are complexly faceted, as discussed below.

Planes of hundreds of minute, almost spherical pseudosecondary inclusions in some fluorite crystals may represent the ultimate in the balance between spherical and faceted shapes; on close inspection these inclusions are seen to be faceted negative crystal shapes, each having 26 faces from the combination of the cube, octahedron, and dodecahedron (Fig. 3-11). Negative crystal shape is frequently but erroneously stated to be a valid criterion of primary origin

----
[5]/ In geologic terms, this cooling apparently must be very fast. Thus Shelton and Orville (1980) showed that new secondary inclusions in quartz were formed and changed to equant forms in 2 to 66 hours at 600°C and 2 kbar H_2O pressure. Similarly, Pecher (1981) noted significant modification of the walls of inclusions in quartz in 1 hour at 500°C and 5 kbar.

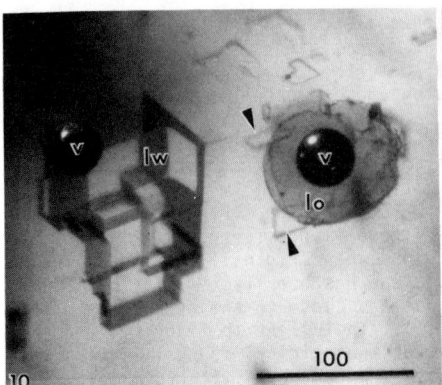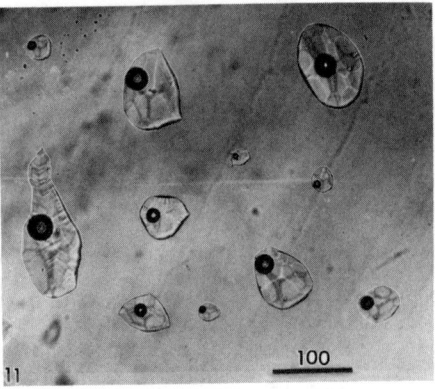

Figure 3-10. Photomicrograph of large primary inclusions of colorless brine (lw) and yellow oil
(lo) in color-zoned purple fluorite. These inclusions are believed to have retained essentially
the shape that they had at the moment of trapping. The brine inclusion (left) results from the
covering of a group of negative crystal reentrants on the surface of the growing cube, as are
commonly seen on the present surface of such crystals. This brine inclusion is considered primary
from its position in the crystal, and not from its shape, as recrystallization can also form such
negative crystal faces, although usually only on <u>much</u> smaller inclusions. The oil was present as
a rounded droplet suspended in these brines; this droplet adhered to the growing fluorite surface
and was enclosed by growth of the fluorite without change in shape. A few small blebs of brine
enclosed with the oil now have a negative crystal shape (arrows). Note the larger volume percentage of vapor bubble (v) in the oil inclusion than in the brine inclusion, although the two inclusions were probably trapped at almost identical temperatures; this difference results from differences in compressibility and thermal expansion characteristics for the two fluids. The two different apparent radii on the bubble in the brine arise from prism effects at the sloping inclusion
walls. Note also that this bubble is actually adhering to the wall (small light-gray oval at
top). This adherence only occurs in very strong brines, where the salinity causes gross changes
in surface wetting characteristics. Photographed in deep purple transmitted light, with added
lateral light for reflection from bubbles. Sample ER 59-3, Hill mine, southern Illinois fluorite-
Pb-Zn district, Illinois, USA. Scale bar in μm. From Roedder (1972).

Figure 3-11. Plane of pseudosecondary inclusions of brine in fluorite. Although some of these
inclusions appear almost globular, they are actually negative crystal cavities, lined by 26 facets.
Only a few of these facets could be made visible in this photograph. Each has a minute birefringent daughter crystal. Sample ER 59-10, southern Illinois fluorite-Pb-Zn district, Illinois, USA.
Scale bar in μm.

for the inclusions, but many obviously secondary inclusions have recrystallized
to form very sharply faceted negative crystals, and etch pits and tubes are
frequently lined with bright crystal faces.

    <u>Relative age or maturity of secondary inclusions</u>. As may be inferred
from Figures 3-8 and 3-9, the shape and size of secondary inclusions may give
at least a <u>rough</u> indication of relative inclusion age, and hence, indirectly,
of inclusion origin. Tuttle (1949) described the steps in the healing of
microfractures in quartz to form planes of inclusions, starting with large,
flat, very thin inclusions and ending with a rather regular array of more-or-
less equant inclusions (Fig. 3-12). Further maturation apparently results in
coalescence of smaller inclusions to yield more widely spaced larger inclusions
but the total volume of inclusions present per unit area is controlled by the
width of the original fracture (Fig. 3-13; see also Fig. 2-16). If several
stages of fracturing have occurred at different times, the inclusions from each
stage may show different degrees of maturity, thus forming a chronological
sequence. Wise (1964) used such features to help in understanding microjointing
in basement rocks. On the other hand, the composition of the inclusions can
affect the final shape. Most descriptions of inclusions of several different
compositions in a given sample indicate some correlation of shape and composition
(e.g., Hollister and Burruss, 1976). Not uncommonly, crystals contain two different generations of secondary inclusions, both old enough to be "mature"
(Swanenberg, 1980), but having two fluids of different composition and two
different shapes (e.g., spherical and faceted).

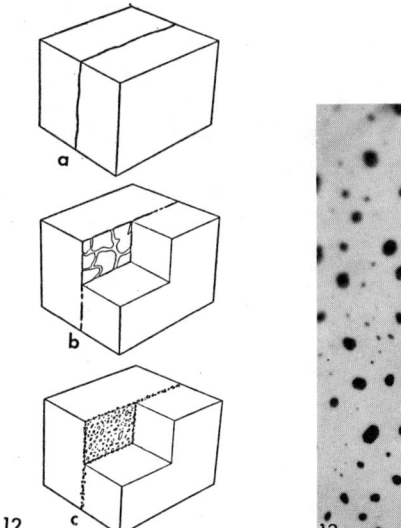
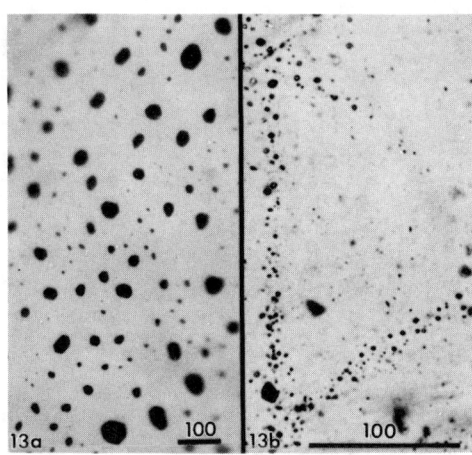

Figure 3-12. Stages in the healing of a microfracture in quartz. From Tuttle (1949).

Figure 3-13. Photomicrographs of planes of secondary inclusions in quartz. (a) Plane in peg-matitic quartz from same sample as Figure 3-2. Plane is parallel to photograph and is mature, in that the inclusions are large, widely spaced, and most small inclusions near to large inclusions are gone. (b) Three nonmature healed fractures in a thin section of low grade metamorphic quartz-ite. Each is strongly inclined to the photograph. The approximate original crack width is esti-mated, from inclusion volumes, to have been ~2 µm in (a) and ~0.01 µm in (b). Scale bars in µm.

Diffusion route for necking down. Nichols and Mullins (1965a,b) studied the process of reduction of surface energy of a rodlike inclusion and pointed out three routes by which the material of the host phase may move to eliminate the instability: (1) diffusion through the inclusion fluid; (2) surface diffu-sion along the fluid/host interface; and (3) diffusion through the host crystal. They also showed that some distinctions are possible between these processes, at least for the coalescence of a tubular inclusion, on the basis of the rela-tive dimensions of the original cylinder and the resulting sphere or series of spheres. In natural materials, diffusion through the fluid seems to be the dominant route. In most examples of the trapping of fluids in which the host phase is essentially insoluble, the inclusions appear to maintain their original shape.

In Chapter 2, the enigma of the trapping of inclusions having zero solubil-ity for the host was mentioned. One of the possible explanations for such trap-ping, and for the later recrystallization and changes in shape of such inclu-sions, is diffusion at the surface or through the host. Thus the CO_2 inclusions in the olivine of peridotite nodules are commonly euhedral, although only some of the inclusions seem to have a film of silicate melt. D.L. Smith et al. (1982) and Wanamaker et al. (1982) have shown that microcracks in olivine heal even at 1000°C in gas, and that surface diffusion was the essential route.

Necking down after phase changes. Recrystallization of the host makes no difference to the total composition of the system, but if any phase change has occurred before the necking down or coalescence, the resulting new inclusions may have grossly different compositions and densities. This possibility is of considerable significance to geologic thermometry and barometry based on inclu-

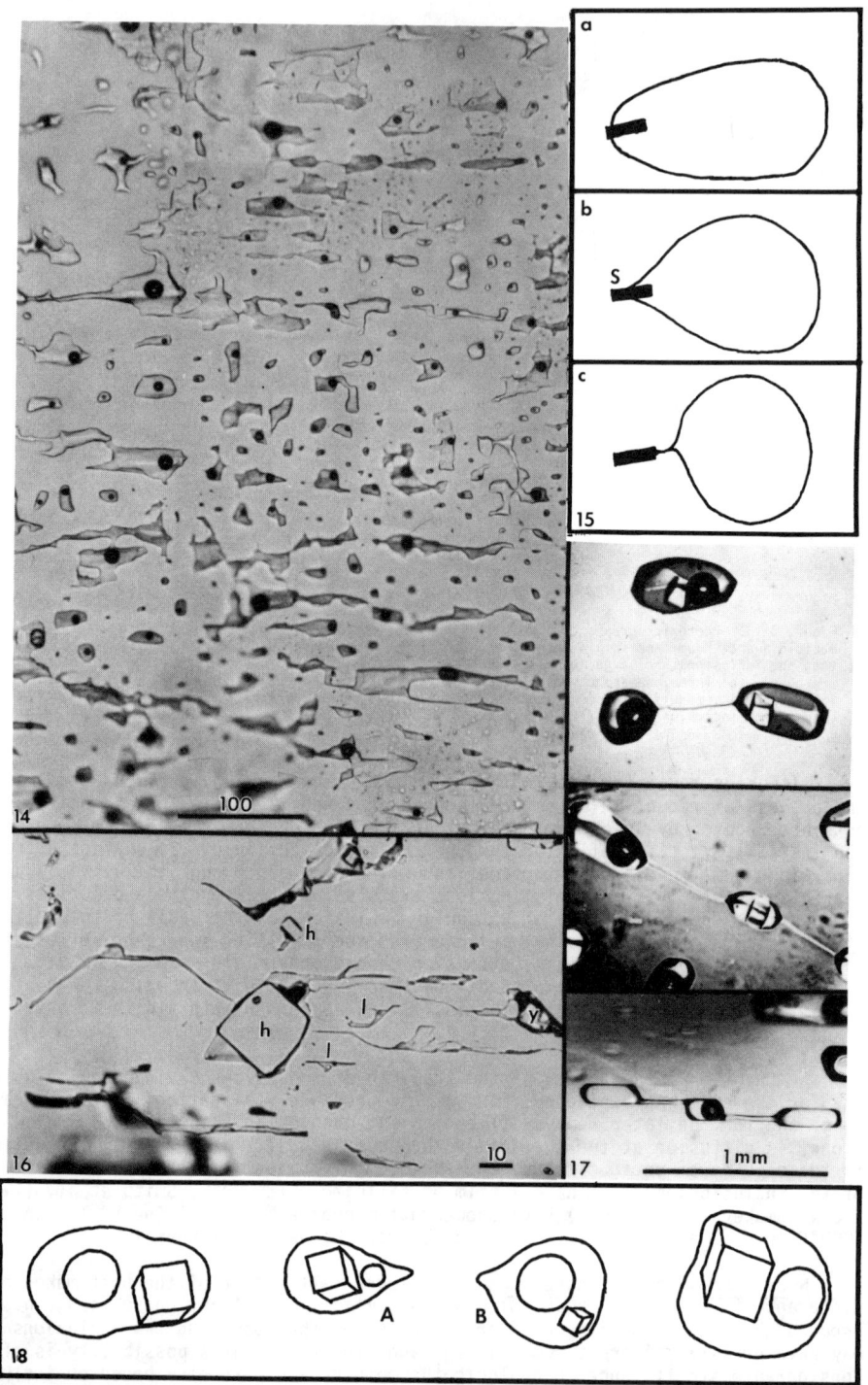

14

15

a

b
S

c

16

h
h
l
l
y
10

17
1mm

100

18
A
B

sion-filling temperatures, because the single gas bubble may be trapped in one of the new inclusions, and the other new inclusions may then nucleate their own gas bubbles with further drop in temperature (Fig. 3-9). Although the evidence of necking down shown in Figure 3-9 is clear, and the effects of such necking on measurements of Th would be large enough to be obvious, unrecognized necking down may be one of the major causes of scatter in the data from some careful geothermometric studies. In many groups of inclusions, most of the actual sealing off probably takes place near the original temperature of formation (Fig. 3-14). If sealing occurs above the temperature of the first phase change on cooling, no problem exists, as a homogeneous phase is still present. However, if only a very small bubble (relative to the final inclusion volumes) forms before the necking-down process, the resulting differences in gas/liquid ratios of the two inclusions might not be large enough to be recognized and hence might not cause these particular inclusions to be excluded in advance.[6]

If a daughter mineral has formed before an inclusion starts to coalesce, it may be left behind, completely surrounded by the host mineral (Fig. 3-15), but a fluid inclusion will be nearby in which that particular daughter mineral is absent (or perhaps is present as a new but much smaller crystal). Quartz crystals from the Colombian emerald mines frequently show sharp cubes of NaCl completely enclosed in the host quartz. A few such cubes show very thin tubes, sometimes almost at the limit of resolution of the microscope, connecting them with a liquid inclusion many micrometers away (Fig. 3-16).

----

[6] The avoidance of such "abnormal" inclusions in the selection of inclusions to be measured, as recommended by Ermakov and Kalyuzhnyi (1957) and others, represents an attempt to eliminate those inclusions whose Th values would not be valid, because of necking down, primary gas, or leakage. However, such inclusions may not be abnormal but may merely be the result of abrupt changes in conditions, which produce inclusions that appear abnormal. The line is difficult to draw for some samples, and the subjectivity involved has caused some doubts about the validity of the method ever since Sorby was accused of introducing bias into his samples by this selection procedure.

Figure 3-14. Plane of pseudosecondary inclusions in fluorite, viewed perpendicular to (111). Although considerable recrystallization and necking down of large flat inclusions has formed multiple smaller inclusions, most (even very tiny ones) now have a vapor bubble, and all have the same Th (144°C) as the adjacent primary inclusions. Most of the apparent one-phase inclusions without bubbles are actually connected with other inclusions by thin tubes. Therefore, the process of necking down and isolation of these individual inclusions from the original layer of fluid filling an open cleavage crack must have taken place above Th and was subsequently arrested completely, as no additional necking took place in the ensuing millions of years at lower temperatures. Because Tt of these inclusions was probably very little above Th, the necking down must have been essentially an isothermal process at about Tt. Sample from Hansonburg Pb mine, New Mexico, USA. From Roedder et al. (1968).

Figure 3-15. Diagram of necking down of an inclusion after the formation of a daughter crystal. The crystal sometimes becomes trapped in the coalescing wall and is left behind. The shape in (b) suggests that the liquid preferentially wets the daughter crystal (S).

Figure 3-16. Plane of pseudosecondary inclusions in quartz, showing an early stage in the process of necking down after daughter minerals have precipitated. Highly birefringent unidentified daughter crystals (y) and a large halite daughter crystal (h, coated with a film of brine) are nearly isolated from each other and have been isolated from the bulk of the fluid of the inclusion (out of this field of view). Necking down has caused individual parts of the original inclusion to be sealed off at various stages in the process, resulting in another liquid-coated isotropic cube (h) and two apparently "low-temperature" inclusions containing very small vapor bubbles (l). Sample ER 63-84Q, Muzo emerald mine, Colombia, courtesy of Banco de la Republica, Bogota. Scale bar in μm. From Roedder (1972).

Figure 3-17. Photomicrographs of necking down in inclusions. Top and middle, in quartz from same locality as that of Figure 3-4, courtesy of J. Touret. (Approximate width of field 0.2 mm.) Bottom, in halite from Salar Grande, Chile, courtesy of G. Erickson. Note that the necking down in these three photographs could have taken place before or after the formation of the daughter phases, but if the connecting tubes were sealed off by further recrystallization at near-surface temperatures, not only the phase ratios, but also the phase assemblage would differ greatly in different inclusions. (Compare with Fig. 3-18.)

Figure 3-18. Diagram of the appearance of inclusions indicating necking down during the formation of daughter phases (compare with Fig. 3-17). Two adjacent inclusions, A and B, have phase ratios (visual estimate) of crystal/liquid/vapor that: (1) differ from the otherwise uniform ratio for inclusions in this plane, and (2) differ in opposite directions (A has too much crystal; B has too much vapor). The projecting points between A and B commonly are not observed, as they are rather transient features that round off soon after the original sealing of the tube that formerly connected inclusions such as these.

In the laboratory, one must be alert to recognize the existence of necking down. When the process has been interrupted, the long (Fig. 3-16) or short (Fig. 3-17) tails between inclusions are easy to see. After sufficient recrystallization, however, the tails disappear, and necking down may only be recognized by the adjacent occurrence of two or more inclusions that differ in phase ratios in opposite directions (Fig. 3-18), and sometimes by abandoned daughter crystals, left surrounded by host crystal. If one believes that no necking down has occurred in his samples, he should remember that all secondary inclusions have formed by this process, but generally isothermally and before any daughter phases formed.

Changes in Position from Gradients

For nearly 100 years, all inclusions were assumed to stay where they first formed; for many inclusions, this assumption is valid. However, if the inclusion walls are more soluble on one side than on the other, material will dissolve, diffuse, and recrystallize, thus effectively moving the inclusion through the crystal. Such solubility differences (i.e., gradients in Gibbs free energy) may arise from gradients in temperature, gravity, stress, or phase composition in the host crystal. As pointed out by R. Petrovich (pers. comm., 1982), dislocations with a screw component are also needed, as these enable the inclusions to dissolve at the front face without requiring very large undersaturations (Petrovic, 1969).

Thermal gradients. The most obvious example of the movement of inclusions is the well-known migration of fluid inclusions in salt crystals up a thermal gradient, first described by Lemmlein (1952). This process has come to prominence recently as a problem in the design of nuclear-waste storage sites in rock salt. Lemmlein claimed that the movement he observed was caused by the thermal gradient from an extremely local (and very small) energy release. Except for some special situations such as geothermal systems, natural geologic environments seldom maintain significant thermal gradients for any considerable length of time, but significant manmade temperature gradients from the radioactivity of nuclear waste can last tens or hundreds of years. The phenomenon has been studied extensively, both in theory and experiment. The literature on both aspects has been summarized by Jenks and Claiborne (1981), but some of their conclusions have been opposed by Roedder and Chou (1982), and not all the parameters can be evaluated at this time.

The rate of inclusion movement in salt in thermal gradients is a function of many variables, but the major controlling parameters are the solubility of the host mineral in the inclusion fluid and its temperature coefficient, the size of the inclusion, the ambient temperature, and the thermal gradient. The rate of movement is a direct function of each of these five parameters. Under some conditions, inclusions can move down the gradient rather than up, and not all the phenomena involved are fully understood (Roedder and Belkin, 1979a, 1980a). The kinetic aspects are the most difficult to model; thus, Olander et al. (1980) have shown that the intersection of a migrating inclusion with one or more dislocations has a major effect on the rate.

Gravitational gradients. Variations in solubility with pressure could, in theory, cause migration in a gravity field. Centrifuge experiments by Anthony and Cline (1970) showed that in very large gravitational gradients, inclusions in KCl will migrate at rates that are measurable. However, these rates are so low that the equivalent rate (upward) resulting from the earth's much lower gravitational field is apparently negligible, or is essentially cancelled by the opposing migration rate (downward) from the geothermal gradient (Roedder, 1984a).

Stress gradients. Static residual strain in crystals apparently can result in significant movement of inclusions. Presumably the host mineral recrystallizes in the wake of the inclusions to a less strained condition. This driving force may seem minuscule, but the available time is long, and I believe that several lines of evidence from the behavior of planes of secondary inclusions in healed fractures support the concept.

In an earlier section, all inclusions in secondary planes were shown to change shape with time. Such changes in shape are only part of the picture. Several processes seem to take place simultaneously during the slow healing of a fracture, but they are not fully understood. On the assumption that different samples present us with different stages in the process, I believe that the steps are as follows: (1) Thin flat amoeboid or semiregular secondary inclusions are formed by sealing off parts of the fracture, generally starting at the thin end (Fig. 2-16). (2) These large but thin planar inclusions break up into many tiny inclusions, frequently faceted but still individually planar and essentially in a plane, in the manner described by Lemmlein and Kliya (1952b). (3) These tiny, apparently sealed inclusions individually recrystallize into more equant forms. (At this stage, the inclusions outlining the plane will vary greatly in size, as a result of the irregularities in the necking process.) (4) The inclusions seem to coalesce, which results in a smaller number of larger inclusions (Fig. 3-13a) with larger spaces between each. (Unless the inclusions themselves move toward coalescence, which I believe is unlikely, this coalescence requires movement of quartz -- see p. 74.) (5) Some of these larger inclusions may migrate away from the original plane, finally yielding randomly disseminated inclusions difficult to distinguish from primary inclusions (Roedder, 1971a).

Steps 1 and 2 are relatively fast; steps 3-5 become increasingly slow but are nonetheless real. Steps 1-3, involving only a change in shape, have been described earlier and seem to be reasonably explained by the obvious reduction in surface free energy. The elimination of the smaller inclusions (step 4) is more difficult to explain. Wise (1964) showed that the smaller inclusions coalesced by some unspecified mechanism, eliminating practically all smaller inclusions in the size distribution. Swanenberg (1980) found a similar disappearance of the small inclusions as planes of inclusions mature. A possible mechanism for coalescence of fluid inclusions under geologic conditions[1] involves migration of the smaller inclusions toward the larger. R. Petrovic (pers. comm., 1982) has suggested very close fluid inclusions could coalesce even in a hydrostatic stress field, if the confining pressure was greater than the internal pressure. Under these conditions, the perturbation of the stress field between two inclusions should cause a migration of the inclusions toward one another (Petrovic, 1969). He estimates, however, that since the rate of migration should be inversely proportional to the 15th power of the distance between the centers of the inclusions (assumed spherical), the migration would not be significant when that distance exceeds the sum of the inclusion radii by more than a few percent. However, if water is soluble in the quartz structure, as discussed in Chapter 2, the small inclusions possibly could dissolve and the larger ones increase in size (see also next section on "Changes in Volume"). Such dissolution would be aided by the internal pressure increase from host/fluid surface-tension forces as the radius decreases.

Step 5 is easy to observe in hand specimen but difficult to document convincingly (Fig. 3-19; see also Stalder, 1980, his Fig. 3 bottom; and Swanenberg, 1980, his Figs. 41, 42, 46). I believe that this migration results from the presence of elongated rodlike zones of high stress in the crystals. These zones are presumably the loci of the dislocations or bundles of dislocations commonly revealed in X-ray topographs of quartz plates. When quartz crystals are etched

----

[1] Weertman (1968) has shown that gas-bubble inclusions in ice grow by coalescence, but that this coalescence is caused by a mechanism involving large amounts of shear, which is not applicable to most rocks.

with acid, deep etch tubes are sometimes produced, indicating the former existence, in an otherwise microscopically perfect crystal, of gross differences in solubility of different parts of the crystal (Nielsen and Foster, 1960). This solubility presumably results from strain along crystal imperfections or dislocations induced during the original crystallization. Swanenberg (1980, p. 81) rejected this concept because he could not find evidence of "crystallographic controls," but as such imperfections are not necessarily perpendicular to growth faces, they do not need to show "crystallographic controls."

The strain causing preferential dissolution can also arise from earlier mechanical stress. Bosworth (1981) described strain-induced preferential dissolution in halite crystals that had been loaded under dry conditions and that subsequently were unloaded and immersed in brine.

If a crystal containing such elongated zones of strain was fractured across the zones, and a plane of secondary fluid inclusions was trapped along the fracture, any fluid inclusion that came in contact with a more soluble line or zone of high stress could be expected to "bore" into the mineral and redeposit more perfect material behind it as it moved (Roedder, 1971a). If the strain resulted from impurities (e.g., along dislocations), the migration would stop unless the impurities were precipitated as solid particles (R. Petrovic, pers. comm.) or swept along in solution in the fluid. The latter process, resembling zone refining, would cause a buildup in the impurities in the liquid. Thus sea ice is purified of its brine inclusions by their migration in thermal gradients (Shreve, 1967); Norman Hubbard (pers. comm., 1981) has suggested that impurities in halite beds might be swept along by the migration of fluid inclusions in thermal gradients. Gerlach and Heller (1966) have proposed that inclusions in halite change shape or move as a result of strain in the crystal.

The actual operation of the "boring" process has not been proved, and minerals other than quartz should be studied as well, but the trains of fluid inclusions going off into the host quartz crystal on either side of some healed fractures are difficult to explain otherwise. Swanenberg (1980) suggested further that as a result of the greater solubility of quartz in H_2O than in CO_2, H_2O inclusions may become separated by movement away from previously associated CO_2 inclusions.

Phase composition instabilities. Wilkins and Sverjensky (1977) have proposed a somewhat similar process, in which secondary fluid inclusions in a plane acted as centers of nucleation and growth of clinopyroxene rods exsolving from the host high-temperature bustamite solid solution. The fluid inclusions moved as much as hundreds of micrometers through the bustamite as the clinopyroxene rods grew. The formation of regularly spaced inclusions in "varved" celestite may also be analogous (Fig. 11-16).

Changes in Volume

Constancy of the volume of the inclusion after trapping is the prime prerequisite for geothermometry on the basis of inclusions, and in most samples I believe that constancy has been maintained. A few of the major exceptions are detailed in the following section.

Reversible changes. Some reversible change can and does occur, by several mechanisms: (1) crystallization on the walls or in the fluid itself; (2) thermal contraction of the host mineral (and precipitated minerals) on cooling; and (3) dilational changes from internal and external pressures. Ermakov (1950) and others placed much emphasis on crystallization of daughter minerals and crystallization on the walls, which reduces the volume of the cavity, and they presumed that this can cause rather gross inaccuracies in Th. Two factors tend to minimize error from this cause: (1) precipitated daughter minerals (and precipitated matter on the walls) will have no effect on the volume if this material redis-

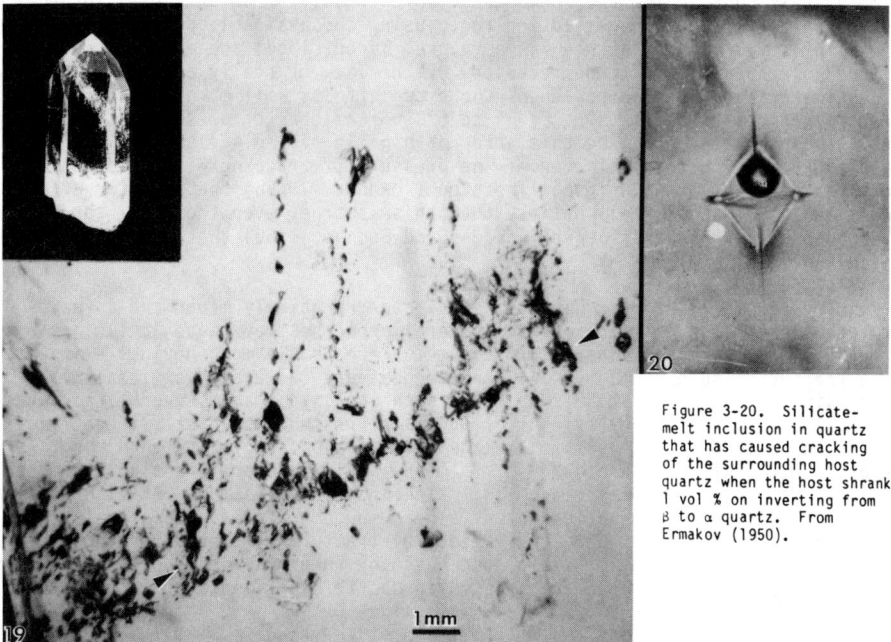

Figure 3-20. Silicate-melt inclusion in quartz that has caused cracking of the surrounding host quartz when the host shrank 1 vol % on inverting from β to α quartz. From Ermakov (1950).

Figure 3-19. Planes of secondary inclusions, in quartz, that have become spread out as much as 9 mm parallel to the c-axis from the original healed fracture (arrows). Sample ER-61-32a, Currul Frio, Brazil. Inset photograph of a 7-cm crystal shows appearance of a similar "plane" in hand specimen. From Roedder (1971a).

solves on reheating, and (2) the volume change upon crystallization or dissolution of a solute is not simply equal to the volume of the solid solute, as has sometimes been assumed. Some phases, notably NaCl, may even have a negative apparent molal volume in certain temperature and concentration ranges. Crystallization under these conditions would increase the apparent volume of the fluid phases.

Ermakov also placed great stress on the thermal-expansion characteristics of the mineral host. This expansion would have an effect on the size of the bubble at room temperature (minerals with larger thermal coefficients having somewhat smaller bubbles), but as the contraction on cooling is exactly reversed on heating, there should be no effect on Th. Petrichenko (1973) stated that the similarity of expansion of brines and host halite invalidates thermometry on such inclusions. I agree that thermometry on halite is meaningless (see Chapter 11), but I disagree with the reason he gives. Even though halite has a large thermal coefficient relative to other minerals, and brines have a low coefficient relative to other fluids, a 25 wt % NaCl solution still expands at almost double the rate of halite (Haas, 1976; Skinner, 1966).

The third mechanism involves two completely separate dilational effects that affect Th but that can probably be neglected in much though not all present-day practice. Although "reversible" in theory, these effects are usually not reversed in laboratory studies. The two effects are: (1) dilation of the entire crystal upon release of the confining pressure, which enlarges the fluid-inclusion cavities within the crystal; and (2) dilation of the host mineral immediately surrounding the inclusions, which results from the local internal pressure during the homogenization run. Using the available data on the compressibility of quartz at room temperature (Birch, 1966) and a series of fairly reasonable assumptions (made necessary by the lack of really applicable data), I estimate that the combination of these two effects will make all measured Th determinations on quartz high by ~4 to 6°C per kilobar of pressure at the time of trapping. As correcting this error at high Th will place the inclusion on a different isochore, estimates of the pressure of trapping will also be seriously affected. I had originally stated (Roedder, 1979b) that the two effects were additive, and I still believe that this is true, even though I stated (erroneously) in another paper (Roedder and Bodnar, 1980) that they would cancel each other.

Irreversible changes. In addition to the relatively minor and in part reversible effects described above, several processes may cause inclusions to undergo major permanent changes in volume. Perhaps the most obvious example is the fracturing so commonly visible in the host crystal around silicate-melt inclusions in quartz (Fig. 3-20). Many such inclusions show several fractures radiating outward for several inclusion diameters into the host. These fractures result from the 1 vol % shrinkage in quartz on cooling through the inversion from the high-temperature beta form to the low-temperature alpha form at ~573°C; this shrinkage places the inclusion under compression. At such temperatures, the glass in the inclusion has already become a rigid, relatively incompressible solid, so the quartz must yield by fracturing. These fractures are commonly parallel to the c axis of the host and sometimes perpendicular to that axis. Similar fractures can be induced in quartz experimentally (Shelton and Orville, 1980).

Still other irreversible volume changes can take place if the host mineral undergoes an irreversible phase change. Wilkins (1979) has pointed out that because the alkali feldspars show large irreversible thermal-expansion effects due to homogenization (Clark, 1966), fluid-inclusion volumes will also be affected.

Similar stresses, and fracturing, can occur if the inclusions contain a fluid under high internal pressure, instead of the host crystal shrinking around a rigid inclusion, as described above. Obviously, if the internal pressure of fluid in an inclusion becomes sufficiently greater than the external confining pressure, the host mineral will fracture and release the pressure. Such decrepitation can occur either in nature or in the laboratory, and the fracture is not always visible under the microscope.[8] The resulting empty inclusions are useful only as evidence that a sufficient pressure differential was achieved. This pressure difference will vary widely, particularly with the strength of the mineral and the size of the inclusion. Thus moderate-sized inclusions (~35 µm) in quartz will generally decrepitate at ~850 bars internal pressure (i.e., pressure difference; Naumov et al., 1966), but smaller inclusions (e.g., ~12 µm) in quartz can withstand ~1200 bars (Leroy, 1979), and 1 µm inclusions can withstand ~6000 bars (Swanenberg, 1980; see Fig. 3-21). Small inclusions of CO_2 in some olivine crystals can withstand 5000-7000 bars internal pressure, even at 1200°C (Roedder, 1965a).

----

[8] Most fractures in transparent minerals are visible only because of total reflection at the interfaces of the cracks. If the width of the crack is small (e.g., <0.5 µm, so that it approaches the wavelength of the light used), and particularly if the crack is filled with a medium of greater index of refraction than air, the reflection becomes less than total. A striking example of the effect of crack width is provided by many clear transparent polycrystalline plastics. When these are bent sharply, they turn white at the bend, but become clear again when the stress is released. Apparently intergranular cracks open up under stress, causing total reflection, and then close up to become essentially invisible. An example of the effects of the medium filling the cracks is seen when clear quartz is cut with a coarse diamond saw. Totally reflecting cracks several millimeters long form out from the cutting point but almost instantly disappear as they become filled with the lubricant water.

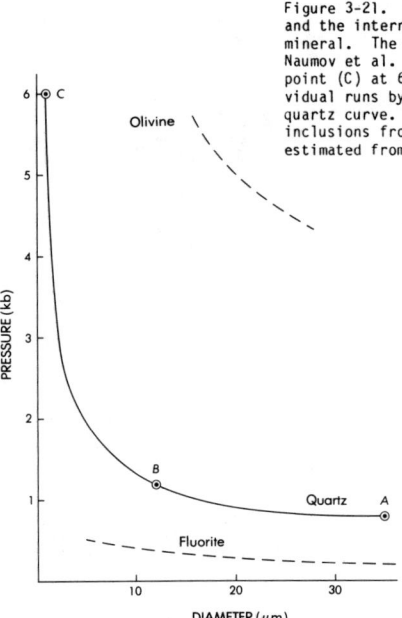

Figure 3-21. Plot showing relationship between the size of inclusions and the internal pressure at which they decrepitate for a given mineral. The curve for quartz is based on point (A) at 850 bars from Naumov et al. (1966), point (B) at 1200 bars from Leroy (1979), and point (C) at 6000 bars from Swanenberg (1980). Numerous other individual runs by Leroy agree reasonably well with the lower part of the quartz curve. The curve for olivine is an estimate, based on CO_2 inclusions from Roedder (1965a and 1981b); that for fluorite is estimated from data from Bodnar and Bethke (1984) and Poland (1982).

A decrepitated, empty inclusion is easily recognized and hence will not cause any problems, but under some conditions, the expansion from internal pressure will cause a fracture to form out into the surrounding crystal that does not extend to the surface. The volume increase represented by this cracking relieves the pressure until the crack can no longer propagate. The crack may subsequently heal to trap numerous inclusions of this expanded fluid in the form of a halo of tiny secondaries around the larger central original inclusion, and all will have the same (lower) density. Recognition of such partial decrepitation, first described by Lemmlein (1956a; Fig. 3-22), is relatively simple, so such inclusions seldom cause errors and can actually provide useful information on the path through P-T space that the host sample followed in nature (Voznyak and Kalyuzhnyi, 1974a). Touret (1977) described such "exploded" inclusions from granulites, and Swanenberg (1980) found numerous examples which he called "decrepitation clusters", in high-grade metamorphic rocks (granulites) of southwest Norway. Sabouraud et al. (1980) suggested that the Th values for quartz and fluorite from a French ore deposit are wrong as a result of natural stretching. Pecher (1981) has produced similar inclusions in quartz experimentally and suggested that recognition of their origin on the basis of petrographic evidence may be difficult.

The internal pressure in the new secondary inclusion formed during partial decrepitation will be greater than the ambient external pressure at the time, as a result of the elastic deformation of the host crystal. Thus, secondary inclusions formed by this process should (at least originally) have a slightly higher density and lower Th than adjacent secondaries formed at the same time by fracturing from external stress. Such differences should be looked for even though they may disappear as a result of slow plastic deformation of the host mineral, as suggested by the work of Gratier (1982) on calcite and quartz.

A similar but much less obvious permanent deformation of the host crystal around an inclusion, generally without visible cracking, may occur in soft

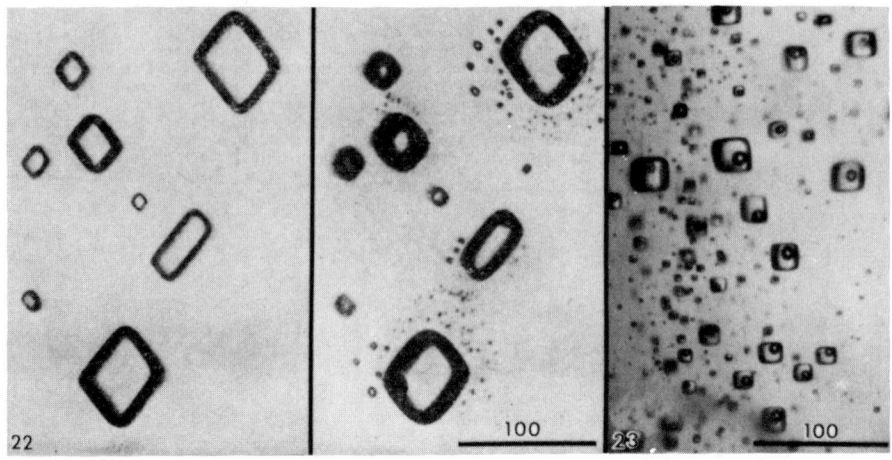

Figure 3-22. Photographs by Lemmlein (1956a) showing a group of inclusions in a synthetic crystal of NaNO₃, before (left) and after (right) overheating and partial decrepitation that has yielded halos of small secondary inclusions. Scale bar in μm.

Figure 3-23. Group of small primary hopper-growth inclusions in halite from Permian Salado beds, New Mexico, USA, after heating to 250°C. These inclusions originally lacked bubbles; now each has one as a result of plastic deformation of the host salt. They now homogenize at temperatures as high as 273°C. Scale bar in μm. From Roedder and Belkin (1979a).

minerals such as fluorite if the internal pressure exceeds a certain finite limit, e.g., by overheating beyond Th (in the laboratory or in nature), or by expansion of ice when an inclusion is frozen that contains such a small bubble that it is eliminated during the freezing (Chapter 10). This deformation or "stretching" may be evidenced only by a rise in Th, and once the specific over- heating limit (i.e., limit of elastic deformation) is exceeded (this will vary with inclusion size as well), the increase in Th is a surprisingly regular function of the amount of overheating and hence the internal pressure (Bodnar and Bethke, 1980, 1984). As a result of such stretching, either in nature or the laboratory, such inclusions can yield reproducible, but erroneously high, values for Th. Because such volume changes put the inclusion on a different isochore, they can have serious effects on estimates of pressure as well.

Bodnar and Bethke (1984) established that the internal pressure needed to initiate stretching (Ps) in fluorite was a function of the inclusion volume: Ps = -147 Log V + 900, where Ps is in bars, and volume, V, is in μm³. Poland (1982; see also Rowan et al., 1983) extended this by examining the stretching of fluorite under external pressures (P_{ex}); she obtained the rela-tionship: Ps = -178.0 Log V + 0.7P_{ex} + 1018.9. As a result, Ps increases with decreasing inclusion volume and increasing external pressure. The pressure range was <1034 bars at <300°C. Once stretching begins, its magnitude is a smooth function of internal pressure. As internal pressure increases, however, the inclusions may suddenly stretch much more, presumably as a result of a new mechanism of deformation. Poland (1982) has called this "super-stretching"; it generally results in a halo of new secondary inclusions.

A particularly instructive case of volume change in inclusions was reported by Bonev (1969, 1977) in galena crystals from a series of Bulgarian lead-zinc deposits. In these crystals, large (1 to 15 mm in width), low-salinity aqueous fluid inclusions containing <10 bars CO_2 pressure (Piperov et al., 1977), are found a few tenths of a millimeter to as much as 2 mm under the outside surfaces of the host crystal. Their presence is made evident by a plastic deformation of the host galena "roof" over the inclusion. Most have deformed inward, some as much as 1 mm; a few are deformed outward. Bonev proved fairly conclusively that this deformation occurred during cooling from the temperature of deposition (280-360°C in the various deposits, from inclusions in associated quartz). Presumably the declining pressure in the inclusions, once vapor bubbles had nucleated (at relatively constant exterior pressure), provided enough pressure differential (estimated to be ~75 bars) to cause the deformation. The exceptions, in which the inclusion wall bulged outward, may have been a result of external pressure release. Similar bulges formed over hidden inclusions on heating at 200°C for 1 hour. The same phenomenon may occur in native gold; Petrovskaya et al. (1971) reported bulges in gold grains over gas inclusions that were found to contain 5-20 bars of a gas consisting of 90% CO_2.

One of the most extreme examples of stretching of inclusions is provided by halite. After slow heating to 250°C, inclusions in halite that originally had no bubbles (but if they had, Th would be ~40°C), developed large bubbles at room temperature and had a new Th near 250°C (Fig. 3-23). "Stretching" of the walls can take place even at low temperatures. For example, slow overheating of an inclusion 20°C above its Th of 20°C caused enough stretching to yield a new Th of 39°C (Roedder and Belkin, 1979a). Such inclusions decrepitate if they are heated rapidly to temperatures above Th. However, salt is so plastic, particularly in the presence of water, that it will not decrepitate if the heating rate is slow; instead, once Th is reached and the internal pressure starts to increase rapidly (~12 bars/°C), the host salt simply expands by permanent deformation. Wilkins et al. (1981), using proton irradiation, have shown that such salt has a set of glide bands at each corner of the expanded inclusion. If inclusions in salt can expand that easily under a few hours of internal pressure in the laboratory, I must assume that they have also expanded (or contracted) in response to natural changes in P and T; hence, the inclusion volume we now see represents some rather arbitrary "quench" conditions after a long period of sequential "anneals." In spite of this easy change in volume, some evidence, e.g., Petrichenko (1973), suggests that adjacent inclusions in halite generally do not leak, either to the surface or by interdiffusion.

In view of the above evidence for relatively rapid volume change in aqueous inclusions in salt in response to pressure differentials, how is it possible for salt to maintain high-pressure gas inclusions, as in the "popping salt" (see Chapter 11)? Two features seem to provide an adequate explanation: (1) the high-pressure gas inclusions are always small, generally only a few micrometers in size; and (2) they generally contain no visible water phase, and water may be essential in weakening the salt.

The most extreme but illustrative example of a mineral that permits changes in the volume of fluid inclusions embedded in it is ice. Ice in the Antarctic sheet contains abundant "fluid inclusions" (air bubbles) in the top 800 m, but they gradually disappear with depth and cannot be detected optically below 1100 m (Gow and Williamson, 1975). This disappearance is not accompanied by any significant loss of air from the ice (as determined by analysis), and all available evidence indicates that the air actually diffuses into the ice in response to increasing overburden pressure. (Gas hydrate formation is also possible.) The existing gas inclusions (air bubbles) are under pressure that increases with depth; those having pressure >~16 bars begin to relax back to this value soon after in-situ pressures are relieved by drilling.

Unfortunately, the significance of the above data may not be limited to galena, halite, and ice. Numerous studies of fluid inclusions in quartz from igneous and metamorphic terranes show that the inclusions result from a complex series of steps involving trapping, deformation and recrystallization (e.g., Wilkins and Barkas, 1978), under a wide range of high pressures and temperatures during geologically long periods of time. Pressure differences between the inclusion and its environment can be large if the P-T path taken by the rock deviates (either way) from the isochore appropriate to the inclusion filling.

Tools have yet to be developed with which to recognize the possibility of volume changes on annealing of inclusions in metamorphic quartz similar to the changes described above in galena, halite, and ice, but the evidence of at least some volume changes seems clear.

The possibility of fluids dissolving and migrating through the host crystal was mentioned above. The crux of the matter is that any changes in volume of an inclusion during these processes requires concomitant movement of host material relative to the bulk of the crystal, either through self diffusion or some mechanical deformation. The increase in volume on partial decrepitation is evidenced by the volume of the crack itself and by the necessary equivalent expansion of the volume of the inclusion as its two halves moved apart. However, in the halite and ice inclusions, no evidence of brittle fracture is seen; in these inclusions, the walls of the inclusion have expanded (or contracted) by ductile (plastic) deformation. Such plastic deformation is perhaps expectable in halite and ice, but can it occur in quartz? Several lines of evidence suggest that it can:

(1) Christie et al. (1964) were the first to show that at high temperature (500-700°C) and confining pressure (20,000 atm), high shear stress could cause plastic deformation in quartz. The specific mechanisms of such deformation were studied by McLaren and Phakey (1965a,b), Paterson and Kekulawala (1979), and Kekulawala et al. (1978); many of the early studies probably do not refer to an equilibrium in the quartz with respect to OH (Peterson and Kekulawala, 1979; Kekulawala et al., 1981). Kirby and McCormick (1979) suggested from their laboratory experiments on creep in quartz that the precipitation of molecular water (i.e., inclusions) results in hardening.

(2) McLaren and Phakey (1966) examined crystals of amethyst and citrine by TEM (transmission electron microscopy) after they had been turned milky by annealing at temperatures above 600°C for a few hours. The milkiness was found to be due to light scattering from nearly spherical "bubbles" whose diameters varied from ~0.02 to 0.1 μm. When annealing times were longer, these "bubbles" became faceted negative crystals. The "bubbles" in amethyst had a tendency to lie in the Brazil twin boundaries. Some evidence was presented for the presence of H_2O molecules in the "bubbles," but regardless of their contents, these "bubbles" require the movement of quartz relative to quartz.

(3) The disappearance of the small inclusions as planes of inclusions mature requires the movement of quartz relative to quartz. If water is soluble in the quartz structure, as discussed in Chapter 2, the small inclusions possibly could dissolve and the larger ones increase in size. Such behavior, which reduces total surface energy, is common in many systems. However, this process would also require transposition of some of the quartz itself, since one cannot eliminate an inclusion in quartz without moving quartz.

(4) Swanenberg (1980, p. 9), in a major study of inclusions in high-grade metamorphic rocks of southwest Norway, noticed a "striking incompatibility between the occurrence of extremely dense [gaseous] inclusions and the high temperature-low pressure conditions of granulite facies metamorphism." He suggested that these dense inclusions formed by partial collapse of preexisting lower density inclusions for which the retrograde P-T path intersected higher density isochores.

(5) Gratier (1982) found that fluid inclusions in quartz show a density decrease of as much as 3% in long-duration (2-month) runs during which the internal pressure was only 10 MPa in excess of the external pressure. These decreases he attributed to host-mineral creep rather than to leakage. Similar studies were reported by Gratier and Jenatton (1983) and by Pecher and Boullier (1983).

(6) Numerous authors (see review by Pecher, 1981) have reported inclusion densities either higher or lower than expected, suggesting later reequilibration.

These various points all indicate that at least under some conditions, fluid inclusions in quartz can change volume by some mechanism, presumably plastic deformation, although some changes may be explained by leakage. (Solid diffusion seems to be ruled out because the temperatures are so far from the melting point.) As a result, these points bring into question the constancy of volume for inclusions in quartz in many metamorphic terranes, where long annealing periods are common. Obviously, volume will not change if the internal pressures in the inclusions exactly match the external pressures. However, the progress of a given metamorphic rock mass through P-T space as it approaches surface conditions does not necessarily follow an isochore (Hollister et al., 1979), so the pressure within an inclusion will, in general, <u>not</u> be equal to that of the surrounding rock (see also p. 341).

If collapse, "migration" of the fluid through the mineral, and reformation of fluid inclusions by dissolution and exsolution does take place on a signifi-cant scale, another important corollary must be considered. This corollary is that the various constituents of a fluid inclusion, such as H_2O, CO_2, and soluble salts, almost certainly will not move at equal rates; hence the composition of both the residual and the new inclusions may be changed by this process. I wonder, particularly, whether the essentially pure CO_2 or CO_2-CH_4-N_2 inclusions in many metamorphic terranes are not the residues from the dissolution of the water component and its migration through the quartz to grain boundaries (and its subsequent loss). This process would be distinct from (but perhaps act in concert with) the reaction of H_2O and CH_4 to yield CO_2 and H, as described by Hollister and Burruss (1976).

Leakage In or Out

The previous section dealt with volume changes, some of which involved loss of constituents. In this section, however, I discuss leakage of material in or out of inclusions by processes involving essentially no volume changes. Although these two phenomena are completely separate in cause, they may have similar effects. The simplest (although not necessarily obvious) example in-volves opening and refilling of previous inclusions by new fluids during defor-mation (Ypma, 1963; Kalyuzhnyi, 1971). Similar replacement has been reported in which primary solid inclusions of anhydrite were replaced by glauconite, thus giving a false impression of primary glauconite inclusions (Sabouraud and Siavochani, 1980).

Most (all?) inclusions are intersected by one or more dislocations, even in relatively perfect crystals (e.g., Scheffen-Lauenroth et al., 1981), and deformed materials have many more. Carstens (1969) showed that arrays of dislo-cations were associated with healed fractures in quartz, and Weathers et al. (1979) found that quartz grains from the Moine thrust zone had a dislocation density of $5 \times 10^8/cm^2$; this density corresponds to an average of five dislocations in every square micrometer. These dislocations represent possible routes for leakage. If pressure gradients are maintained between the inclusion and its external environment, leakage (in or out) is to be expected. Two different routes may be followed by the moving fluid: (1) directly through the host-crystal structure (e.g., the solution and exsolution of H_2O in quartz discussed in Chapter 2), or (2) via various small imperfections in the crystal such as dislocations, or larger imperfections such as fractures. An unambiguous dis-tinction between these routes is seldom possible and, in general, is not of great concern. The most immediate question for inclusion workers is essentially one of the <u>magnitude</u> of leakage, rather than of its existence or routes.

Particularly in past decades, the possibility of movement of fluid into or out of inclusions after they are trapped has been of great concern. Leakage and/or refilling of part of a group or plane of fluid inclusions where inter-sected by another later fracture is so obvious during microscopy that it is seldom missed.[9/] It is particularly common in ore deposits in metamorphic terranes (Ypma, 1963). However, small amounts of leakage, in or out, are much less easily recognized. Several reports of laboratory measurements of leakage of fluid inclusions under pressure gradients have been widely quoted (e.g., Kennedy, 1950a, reporting unpublished work of Grunig; Skinner, 1953; McCulloch, 1959). They showed that within hours to weeks fluid can be moved into or out of fluid inclusions by establishing large pressure gradients (270-1550 atm/<1 mm). Later work and a critical review of all the evidence for leakage (Roedder and Skinner, 1968) have shown that at least some of the earlier experiments indicating leakage may be explained by microfractures introduced during sample preparation, and that most inclusions have not leaked. Roedder and Skinner (1968) focused on the problems of possible leakage of fluid inclusions from vein-type mineral deposits that have generally not undergone the long cooling cycles characteristic of metamorphic rocks. Hence, their conclusions are not necessarily applicable to metamorphic rocks, where both the available time and the possible gradients driving the movement of fluid, as discussed above, are much greater. In addition, Blacic (1981) has shown that diffusivities for "water" in natural quartz are orders of magnitude greater at high pressure than at low pressure (see also Kronenberg et al., 1983).

In addition to the possible movement of H_2O, another notable exception that must be made concerns the possible loss (under special conditions), of hydrogen by diffusion out of the inclusions. The three major possible sources of hydrogen are as follows: (1) It could be an original constituent (many gas equilibria under geologically reasonable conditions have significant concentra-tions of H_2 at equilibrium). (2) H_2 could come from the disproportionation of H_2O. At any temperature, some of the H_2O molecules are dissociated into H_2 and O_2. If the H_2 diffuses away, further disproportionation would effectively cease once the residual O_2 pressure increased, but if this O_2 could be used up in some local reaction, such as oxidation of sulfide to sulfate, or ferrous iron to ferric (Roedder and Skinner, 1968), disproportionation could continue. (3) H_2 could come from reaction of original methane and water ($CH_4 + 2H_2O \rightarrow CO_2 + 4H_2$), as proposed by Hollister and Burruss (1976, p. 173).

Loss of hydrogen may well have occurred in nature, not only because it 'is expectable, so long as a suitable hydrogen "sink" is available, but also because it provides the only logical explanation for certain nondissolvable "daughter minerals" (Roedder and Skinner, 1968, p. 721). The only other explanation, sluggish kinetics, seems unlikely, in view of some very long heating experiments that failed to dissolve even very small crystals of these "daughter minerals." Also, H_2 apparently diffuses within hours (at high temperatures) in some hydrous silicate-melt inclusions (see Anderson and Sans (1975) and Chapter 16). Some experimental evidence from Ypma (1969) has shown that H_2 apparently diffuses out of crushed samples of quartz within minutes at 400-500°C.

In addition to "leakage" of inclusions on decrepitation (either in the laboratory or by overheating from the intrusion of dikes, etc.), deformation of the host rock to the point of shearing will frequently open many or most preexisting inclusions, refill some with new fluids, and cause the trapping of many new inclusions (Ypma, 1963). Because many ore deposits are in faults, shearing is a very common problem. Unfortunately, the tiny fractures through which major leakage (in or out) may occur are frequently so small that they cannot be seen by ordinary light microscopy (see footnote 8 and Chapter 6); however, Kerrich (1976) has shown that visible signs of strain (seen by using

----

[9/] A. Kozlowski (pers. comm.) has noted that cleavage flakes from large tremolite crystals lost most of their visible liquid CO_2 within minutes of sample preparation.

crossed polars) provide a valid warning of possible leakage. He also pointed out that the well-known weakening of quartz to deformation in the presence of structural water (Blacic, 1975) may be involved in the process of loss of inclusion fluids. Aines et al. (1984) reported that the strength of synthetic quartz increases when the H_2O is accumulated into fluid inclusions and cannot react with the quartz framework.

Although the above discussion of leakage may seem disheartening to students of fluid inclusions in some metamorphic terranes, the situation is not hopeless. Certainly not all metamorphic inclusions have leaked, and even those that have leaked can still provide useful information, so long as care is used in the interpretation. In many geologic environments, because of the time, the temperature, the nature of the pathways, and the gradients in pressure and chemical potential, leakage has been undetectable (Roedder and Skinner, 1968). Perhaps the best evidence for such a statement is found in studies of inclusions in minutely zoned crystals (e.g., Woods et al., 1982); the results obtained in such studies would be simply impossible had these inclusions leaked, even slightly.

Chapter 4

NONDESTRUCTIVE METHODS
of DETERMINATION of INCLUSION COMPOSITION

CONTENTS

INTRODUCTION

Most inclusions are extremely small and may average <10 μm in diameter. A single inclusion of this size contains a total of only ~10^{-10} g of material; hence, it cannot be analyzed by ordinary chemical techniques. However, a surprising amount of useful qualitative and semiquantitative compositional data on the nature of solid, liquid, and gaseous phases present in such an inclusion can be obtained by a series of relatively simple nondestructive microscope procedures. These procedures, and their limitations, are discussed in this chapter, along with some of the more illuminating examples of their application from the literature.[1]

Various analytical procedures have shown that although inclusions of organic liquids and gases are common in some localities, most fluid inclusions consist

----

[1] Much of the material in this chapter will be found, in a more extensive form, in an earlier book (Roedder, 1972). Because that book is out of print, the updated material is presented here.

of a liquid-water solution. At room temperature, the gas bubble usually contains only water vapor at ~0.03 atm pressure, but it may consist of CO_2 at pressures as great as 70 atm. Where a second immiscible liquid phase is present, it is generally found to be liquid CO_2.

The daughter minerals that form in some inclusions after trapping are compounds that were sufficiently concentrated in the original fluid to precipitate when saturation was reached on cooling. They have been studied intensively, as they provide compositional data on the original fluids. Small isotropic cubes are by far the most common daughter minerals, as they are found in samples from many environments; they have been identified as halite by several procedures, as described in this chapter. Many other daughter minerals have been recognized, including particularly sylvite, calcite, anhydrite, and hematite, and, less commonly, various other sulfides, sulfates, carbonates, fluorides, and even fluoborates. Some of these species are rare, and many are as yet unidentified. Large amounts of such daughter minerals are found in inclusions from several environments, including certain types of pegmatites and porphyry copper deposits.

Many igneous rocks and meteorites also contain fluid inclusions, but the "fluid" that was trapped was a silicate melt and now consists of silicate crystals or glass; the study and qualitative analysis of such inclusions involves many of the same techniques.

SIGNIFICANCE OF QUALITATIVE AND SEMIQUANTITATIVE DATA

For generations, the geological sciences have suffered under the opprobrium of being qualitative, whereas many other fields have evolved into the intellectually more satisfying concept of quantitative science. Many aspects of geology are rapidly becoming quantitative, so why, in this modern day, should I emphasize methods first applied over a hundred years ago that can only hope to achieve qualitative, or, at best, semiquantitative results? J.H. Mackin, in speaking of empiricism in geology, pointed out in 1971 that the very act of making quantitative measurements may provide a solid sense of accomplishment, but when a number replaces understanding as the objective, real problems lie ahead in the interpretation of geological data. In the following discussions of quantitative methods, which, with a few notable exceptions, are destructive and hence are covered in Chapter 5, it is obvious that I fear that some of the "quantitative" fluid inclusion data fit into Mackin's prophecy.

The problems of determining the composition of fluid inclusions have no panacea. Except for the very large inclusions, whose analyses are relatively easy, no known method or combination of methods will give an accurate unambiguous analysis of any given inclusion in any given mineral. Many different techniques have been used, but none is universally applicable. Because of the extremely wide range in sample material and in the accuracy of the methods used, the available published analyses of fluid inclusions must be examined with considerable care.

The vagaries of inclusion occurrence make it desirable to obtain as much information as possible from those few inclusions that happen to fulfill more than one of the specifications of an ideal sample: optically clear and measurable, large size, and known origin. Not infrequently, good evidence has been presented that the fluids from which a given crystal grew changed very markedly during the growth. As a result, it is almost impossible to obtain truly duplicate samples even of primary inclusions, and the widespread occurrence of secondary inclusions provides additional ambiguity. Thus, all possible nondestructive tests should be applied to any irreplaceable inclusion before a destructive test is used, even though they may yield only qualitative or semiquantitative data. For many inclusions, such data are the only type that can be obtained.

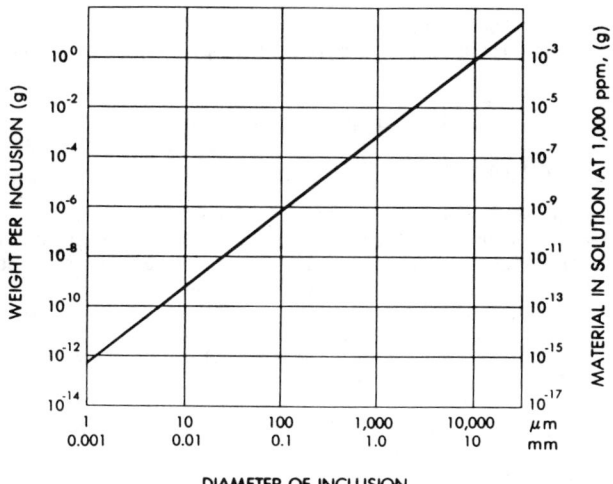

Figure 4-1. Volume and weight of spherical fluid inclusions, assuming a fluid of density 1.0 g/cm³. Modified from Roedder (1958).

As the size of the inclusion to be analyzed decreases, the error and ambiguity in the individual determination increase rather rapidly, simultaneously with an equally rapid decrease in the number of determinable constituents. This is a direct result of the fact that the volume and weight of an inclusion are functions of the cube of its radius (Fig. 4-1). From this, it might appear that only the large inclusions should be analyzed. Unfortunately, large fluid inclusions (greater than a few millimeters) are relatively rare, and very large fluid inclusions (>1 cm), permitting the use of ordinary analytical methods, are known in only a relatively few localities in the world. Even inclusions 1 mm in diameter are too uncommon to be used in most such studies. In addition, several studies have indicated that the probability of leakage into or out of a given inclusion increases rapidly with size.

If a given type of geologic process is to be studied in several places, or if the complexities in a given geologic occurrence are to be unraveled by a study of the fluid inclusions present, suitable inclusions must be found in many samples and parts of samples. Resolution of the problems of analysis of the composition of fluid inclusions must be based on a compromise between the tremendous abundance of the smaller inclusions and the ease and accuracy of analysis of the larger ones. The simple expedient of taking a sample containing a large number of small inclusions, to get the same amount of fluid, can be used where most of the inclusions are presumably of one generation (e.g., Roedder, 1958), but such procedures can introduce many additional problems such as contamination (see Chapter 5), which have caused gross errors in many published analyses.

Obviously, methods of analysis are needed that are applicable to individual very small inclusions, and the petrographic microscope, with various accessories, provides many of them. Most such methods are based on optical identification of the phases present, and they use the rather surprising variety of properties that can be measured with the ordinary petrographic microscope. Most optical identifications are somewhat ambiguous, particularly if only one parameter is used, but additional independent methods are generally available, and they are generally very quick and easy to use. Most significantly, these optical methods are exceedingly sensitive, as they work quite adequately on single inclusions weighing less than 10^{-10}g (i.e., a spherical inclusion <6 µm in diameter).

Only the methods themselves are given here. General information on laboratory procedures for microscopy of inclusion samples will be found in Chapter 6.

NONDESTRUCTIVE PROCEDURES

Composition of Liquid Phases

Fluidity. Even the rough estimates of viscosity possible from observations of bubble movement can be useful to differentiate highly viscous oils from low viscosity water solutions. One common source of trouble in the identification of "liquid" is the differentiation of glass from aqueous liquid (see Chapter 6). Note, however, that highly concentrated water solutions (e.g., $CaCl_2-H_2O$) can be viscous even at room temperature, and I.V. Kulikov (pers. comm., 1983) has noted that the fluid phase in multiphase inclusions from the Tyrnyauz deposit in the USSR is exceptionally viscous.

Color. With one observed exception, the water solutions in inclusions are not colored. Feklichev (1962) reported that in some very large inclusions in beryl, the liquid phase had a greenish-blue color. Some of the larger inclusions in the same samples contained greenish solid matter. Care must be used in determining the color of very small inclusions at very high magnifications (x 1000 to x 1500), as a variety of optical effects yield spurious colors. Many of the various hydrocarbon fluids found in some inclusions are yellow or brown. This color may be used to differentiate oil from water in many samples and is usually assumed to be a positive test. Unfortunately, however, colorless organic liquids also occur in inclusions (Figs. 4-2,4,6; 11-17,18; 12-5; Murray, 1957; Roedder, 1963, p. 201-203).

Wetting characteristics. When several different fluids, such as glass, water solution, liquid CO_2, oil, or gas, occur in the same inclusion, they usually take positions of minimum surface energy. Glass is usually in contact with the walls, i.e., it wets them. Any other fluids such as "vapor" (essentially a vacuum bubble), or liquid water, or CO_2, occur as bubbles within the glass (Roedder, 1965a; Roedder and Coombs, 1967). Ermakov (1950, p. 80) has noted that the bubbles in glass inclusions may appear "pressed against the inclusion wall." This is indeed common, and in some glass inclusions, dozens of tiny gas bubbles will be seen at the walls. Rather than representing the stable, lowest energy position, this probably results from a nucleation phenomenon, as gas bubbles can be seen to nucleate at the inclusion walls in some liquid-water inclusions and then move away from the walls into the liquid, where they coalesce. Zirkel (1873, p. 69) has described glass inclusions in which the bubble appears to be embedded in the wall of the surrounding crystal. Presumably, these result from later recrystallization that changes the shape of the inclusion after nucleation of the bubble. The gas bubbles in glass inclusions that contain daughter minerals frequently are attached to these crystals, and occasionally small daughter minerals in aqueous inclusions will adhere to the liquid-gas interface (Deicha, 1952b).

Where water solution exists together with an immiscible CO_2 fluid (liquid or gaseous), the globule of CO_2 is always within the water solution, which, in turn, wets the walls and fills all minor reentrants in the cavity. Hartley (1876, p. 249) noted that when no water is present, the liquid CO_2 may actually fail to wet the cavity walls. He did not state what criteria he used to verify that no water was present. If a large volume of CO_2 (liquid or vapor) is present, it may press against the walls, but freezing data (Roedder, 1963, p. 192) have shown that a film of water is usually present. When a gas bubble occurs with liquid water and liquid CO_2, the bubble is always completely surrounded by the CO_2 liquid, an observation first made by Brewster (1826a), many years before the true nature of liquid CO_2 became known. This simple fact

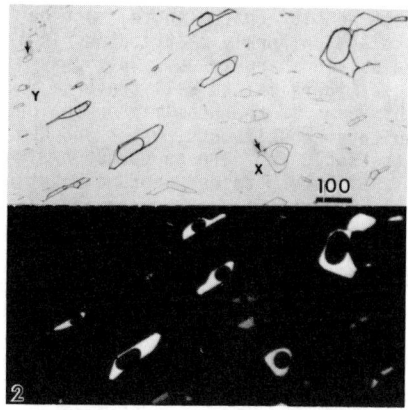

Figure 4-2. Plane of flat secondary organic liquid inclusions in quartz from vein in Precambrian rocks from South-West Africa, in transmitted white light (above) and incident short-wave ultraviolet light (below). The water solution (arrows) in composite inclusions (as at X) and in simple aqueous inclusions (as at Y) does not fluoresce. Scale bar in μm. From Kvenvolden and Roedder (1971).

causes considerable difficulty in recognizing the presence of small amounts of liquid CO_2 phase. For example, in a spherical bubble, a layer that constitutes 10 vol % would make up a rim that would have a thickness of only 3.5% of the bubble radius and hence would be hidden in the normal black border from total reflection at the edge of the bubble. Such films are more visible if the inclusions are strongly flattened. Liquid hydrocarbons, if present with liquid water and vapor, behave like the liquid CO_2 just described.

An extreme example of preferential wetting is shown by certain multiphase oil inclusions in fluorite, in which one of the liquid organic phases wets only certain spots on the walls of the roughly spherical inclusions, corresponding to specific crystallographic planes in the host fluorite structure (Figs. 2-27, 28; 11-22,23).

Infrared (IR) and ultraviolet (UV) absorption; fluorescence. A very few inclusion fluids (organic liquids) fluoresce under ultraviolet illumination, and if the enclosing mineral is transparent to the UV, this simple test is very effective (Fig. 4-2). It was first used by Reese (1898) to identify petroleum inclusions in quartz and has recently had many applications in the oil industry (van Gijzel, 1979). The test must not be considered definitive, as many organic liquids do not fluoresce. Simple diagonally incident UV illumination will often suffice, but the eyes should be adapted to darkness first. If a coverglass is used, it must be of silica glass, as even a very thin coverglass of ordinary glass is opaque to most UV. Burruss et al. (1980) and Burruss (1981b) have shown that the differences in the color of fluorescence of various hydrocarbon phases can provide valuable information on the timing of hydrocarbon migrations, which can be important in studies of the origin of petroleum.

Many of the phases and constituents present in fluid inclusions (particularly H_2O, CO_2, and various organic constituents) show very strong, sharp, and characteristic absorption spectra, particularly in the infrared. The five forms in which CO_2 may occur in inclusions at room temperature -- liquid CO_2, gaseous CO_2, and in solution as H_2CO_3, HCO_3^-, and CO_3^{2-} -- might be identified or even analyzed nondestructively by means of their absorption spectra, obtained by passing an IR beam through a polished plate of the mineral. Although some of the common minerals in which these inclusions occur -- quartz, fluorite, and halite -- are transparent to large segments of the IR spectrum, the absorption that does take place causes some problems, and except in special samples, the method is generally unsatisfactory. The major problems are not in host absorption but in the inherently nonuniform distribution of "sample," that is, inclusion fluid, and the general nature of absorption spectra. Many of the sharp absorption lines for a given constituent are superimposed on a broad absorption band; if they are to be recognized, the sample density (path length) must be controlled. Unless the inclusions are minute and uniformly distributed, it is impossible to control the sample density and obtain sharp lines. Only a very small percentage of the beam passing through a mineral plate will traverse inclusions. The effective path length of this part of the beam may be so long that the broad-band absorption for the ubiquitous constituents, water and CO_2, gives complete absorption (i.e., as though the inclusion were an opaque grain). The rest of the beam traverses no inclusions. I have obtained broad absorption bands, presumably assignable to liquid water, in IR spectra of quartz and fluorite containing known inclusions, but as these inclusions were mainly large enough to be effectively opaque to IR, I conclude that the spectra resulted from a very small amount of water in the form of minute, partly transparent inclusions. Calas et al. (1976) have shown that in spite of these problems, semiquantitative compositional data can be obtained by IR methods on certain samples of fluorite, halite, and sylvite.

Fourier transform infrared spectroscopy (FTIRS) may eventually help in these problems, but is not quite ready at present. Thus, Pasteris et al. (1983) were able to verify the presence of CO_2 and a weak peak for CH_4, presumably from the CO_2 inclusions in mantle xenoliths, using the method. They found quantification difficult. Furthermore, the analysis has to be made of a cylindrical volume of 75 µm diameter through the entire grain.

The IR method has been of greatest use in distinguishing the various possible modes of occurrence of hydrogen (including liquid inclusions) in minerals such as beryl and quartz (Wood and Nassau, 1968; Paterson, 1982). It was used by McLaren and Phakey (1966) to show the change in bonding of the H in quartz on annealing, from presumed OH^- ions in the structure, as formed, to H_2O molecules, presumed to be in newly developed very minute gas inclusions after annealing (see also Bambauer et al., 1969; Luckscheiter and Morteani, 1981). Wilkins and Sabine (1973) have used IR spectroscopy to determine the H_2O content (<100 ppm) of some nominally anhydrous silicates and have used the spectrum shifts on exchange with D_2O (i.e., ^2H_2O) to verify the peak assignments. Aines and Rossman (1980) have shown that the various structural types of protons or water and their bonding in beryl, cordierite, topaz, feldspar, and zircon can be differentiated by single-crystal IR spectroscopy. As long as fluid inclusions are small enough, such procedures should be adequate to distinguish between water in fluid inclusions and that in various structural sites in the host mineral, as in the channels in beryl. Melikhov et al. (1981) used a novel procedure to make such distinctions in synthetic crystals, as an aid to understanding the formation of fluid inclusions. They added a paramagnetic stable radical to the crystal growth medium and studied the resulting crystals using electron paramagnetic resonance (EPR). Aines and Rossman (1984) and Aines et al. (1984) have shown that the distinction between very small inclusions and aggregates of structurally bound molecules of water in quartz is difficult. Their studies indicated that water molecules are the dominant H-containing species in synthetic quartz, but that this water is not in aggregates large enough to form ice when cooled.

Infrared spectroscopy has been used very effectively in determinations of water in silicate glasses. Stolper (1982a, b) has used the method to characterize the molecular species of water in silicate melts and to determine the amounts present. Measurements can be made on samples 100 µm in diameter.

A completely different use of IR is in the study of normally opaque ore minerals. Some of these are transparent in IR. Although H_2O is essentially opaque to large segments of the IR spectrum, it is transparent to IR wavelengths in the range 0.8-1.2 µm, so details of inclusion contents can be seen (Fig. 4-3). Trufanov and Rodzyanko (1973) reported using IR for thermometry of fluid inclusions (presumably in opaque minerals). Plumlee et al. (1983) reported

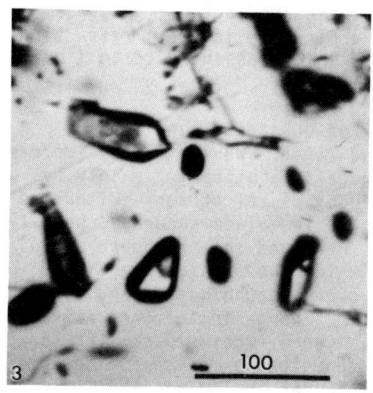

Figure 4-3. Fluid inclusions in an opaque doubly-polished plate of tetrahedrite from Orcopompa, Peru, in transmitted IR illumination (wavelength 0.8-1.2 μm), subsequently converted to visible light for photographic recording. Scale bar in μm. Plate is ~90 μm thick. Photograph by C.J. Hackbarth, courtesy of G.S. Plumlee.

that they were able to see detailed growth banding and to measure Th on inclusions in tetrahedrite and tennantite.

The strong absorption of IR light by liquid CO_2, causing local heating and homogenization, has proven to be useful as one criterion to identify liquid CO_2 inclusions in olivine from the olivine nodules of basalts (Roedder, 1965a, and 1972, pl. 4, Fig. 6). As the IR present in normal microscope illumination is frequently adequate to homogenize such inclusions, it may prevent their recognition unless special efforts are made to obtain illumination free from IR.

One additional nondestructive method for the analysis of liquid phases also involves irradiation of the sample with light. This is Raman spectroscopy. As it can be used to determine certain constituents in liquids, solids, or gases, its use is discussed in a separate section in this chapter.

Index of refraction. The index of refraction of fluids in inclusions may be roughly estimated from the apparent relief they show against the host mineral, or against each other. Thus, glass inclusions (n ~ 1.50) in quartz (n = 1.55) have much narrower black borders (much less total reflection) than do inclusions of water solutions (n ~ 1.35); the same is true for water solutions when compared with liquid CO_2 (n ~ 1.19; Quinn and Jones, 1936). This criterion is so extremely sensitive to inclusion shape in the third dimension, however, that it must be used with considerable caution (see Chapter 6). A close match of the two indices gives very useful information about the inclusion fluids (Fig. 2-26) but may make observations of the inclusions difficult, as in the inclusions of water solutions in cryolite (n = 1.34) in which only the vapor bubble may be visible (Prof. Hans Pauly, pers. comm., 1965). Similarly, Ermakov (1965) reported liquid-water inclusions from chambered pegmatites of Kazakhstan, USSR, so concentrated in salts that they matched or exceeded the index of refraction of the enclosing fluorite (n = 1.434). Some similar inclusions from these deposits, provided through the courtesy of Prof. Ermakov, showed exceedingly strong brines when examined on the freezing stage (Roedder, 1963, p. 180-181), but the index of refraction of the liquid (at room temperature) was less than that of the fluorite in all of them. Perhaps on heating, the solution of the voluminous daughter minerals would raise the index to near that of fluorite.

As the index of refraction of water solutions varies considerably with the salts present (from 1.33 for pure water to about 1.36 for strong brines), quantitative measurements are desirable. These are possible by several methods, the most elegant of which is based on the angle for total reflection at a flat interface on the inclusion wall between the mineral (of known index or indices) and the fluid (of unknown index). This method, using reflected light, was originally described by Brewster (1823a) to determine the index of refraction of a "strange new fluid" he found in inclusions, which was shown to be liquid CO_2 many years later.[2/]

Stegmüller (1952) gave detailed nomograms for obtaining the index of refraction of inclusion fluids using transmitted light. Wahler (1956) improved the technique considerably and made it more generally applicable, mainly by immersing the crystal in a medium of the same index of refraction. With this technique, he measured inclusions to second- and third-place accuracy, permitting estimates of salinity of the water phase. He found the procedure useful also to recognize single-phase gas inclusions, which show an index of ~1.00. Micheelsen (1975) has described a modification of the technique, using oblique pinhole diaphragm illumination and conoscopy. The change in index of refraction of the liquids with temperature might possibly be an effective parameter for their identification; Bokii et al. (1961) have described a method for determining the index of refraction of liquids at temperatures as low as -150°C.

Kalyuzhnyi (1954) described a method, also based on total reflection, for obtaining the index of a liquid or solid inclusion, using a Fedorov (i.e., universal) stage. He has given graphs for conversion of stage inclination to liquid index for inclusions in quartz, topaz, and calcite. Since then, several papers have been published in which this method has been used (Kalyuzhnyi, 1955b; Lisitsyn and Malinko, 1961; Kalyuzhnyi and Shchiritsya, 1962; and Bakumenko, 1964).

I have found that the index of the fluid in the larger inclusions can also be measured by comparing the real thickness with the apparent thickness of an inclusion obtained by focusing on the top and the bottom of it, using a calibrated fine-focus screw (one division = 1 μm actual movement). The real thickness can be obtained from similar measurements of the apparent wall thickness above and below the inclusion, subtracted from the total plate thickness, or by viewing from the side (Roedder et al., 1963, p. 355).

The composition of the fluid in the inclusions will, in general, change as daughter crystals form, and the index should change as well. Although differences in index have not been reported for inclusions with and without nucleated daughter minerals, these should exist and might be useful in estimating composition. The data of Kalyuzhnyi (1965) on the differences in index for glass inclusions in various phenocrysts of a hyalodacite might thus be explained by changes due to crystallization of the host mineral on the inclusion walls.

Thermal expansion and homogenization. Ever since Brewster (1823a) found a "remarkable" fluid in inclusions in topaz, that had a thermal expansion 32 times that of water, this parameter has been used to aid in the identification of the fluids. CO_2 is not the only liquid with a high thermal coefficient of expansion. Many organic fluids have coefficients much larger than those for water solutions. Thus the visual appearance at 25°C of the inclusion shown in Figure 4-4 would suggest that Th would be at least 250°C if this were a water solution. But since the fluid is an organic mixture, with a much greater expansion, it homogenizes at 124.8°C. Sorby (1858) was one of the first to determine

----

[2/] It is important to remember that even though liquid CO_2 will dissolve very little water, many other compounds, particularly organic ones, will dissolve in it (Francis, 1954) and may be present in natural liquid CO_2, affecting its index of refraction and other properties. Brewster measured the index of refraction of two different fluids (1826a, 1826b) which were later given the names "cryptoline" and "Brewstoline" (or "cryptolinite" and "Brewsterlinite"). From the index obtained (1.21, close to the 1.19 index of liquid CO_2) and from other data, it appears that the Brewsterlinite was liquid CO_2; the nature of cryptolinite is obscure, but this fluid is generally presumed to be a water solution. Brewster (1862b) thought both were hydrocarbons.

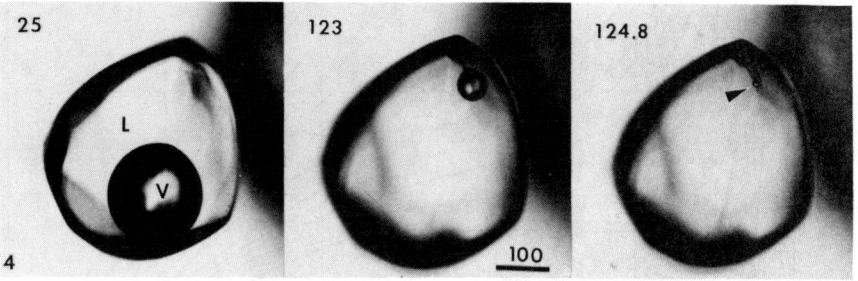

Figure 4-4. Sequential photomicrographs of faceted, negative crystal, primary inclusion in quartz from vein in Precambrian rocks from South-West Africa, containing large vapor bubble (V), taken at the temperatures indicated (°C), showing the large thermal coefficient of expansion of the included organic liquid (L). Homogenization of the tiny bubble (arrow) occurred after an immeasurable rise above 124.8°C. Scale bar in μm. From Kvenvolden and Roedder (1971).

experimentally, with some accuracy, the high-temperature thermal expansions of some salt solutions pertinent to inclusion studies. He sealed fluids into glass capillaries to make "synthetic inclusions." In natural materials, the long straight tubular inclusions are most suitable for this procedure, as they permit relatively accurate phase ratio determinations from the simple linear intercepts of the phase boundaries on a microscope ocular scale. The areas of individual phases in flattened inclusions can be used similarly, but less accurately (Chapter 6).

Although the coefficient of thermal expansion of any liquid is a definitive property, application of the principle to the identification of the fluids in inclusions is not simple. In addition to the problems of measurement of phase ratios mentioned above, the relative phase volumes are involved in determining the apparent expansion. Ostapenko and Khetchikov (1968) have developed a differential equation to calculate these effects. In moderate- to high-density inclusions, and low temperatures, where the vapor phase is of low density and volume, the movement of the meniscus toward the vapor as temperature increases will be a fairly accurate indicator of the thermal expansion of the liquid (Fig. 4-5). At higher temperatures and lower overall inclusion density, however, an increasingly significant part of the total mass of the inclusions will be in the vapor phase, and the mass of the liquid phase may decrease as temperature increases. As a result, inclusions having certain densities will, upon heating, first show an expansion and then a contraction and final disappearance of the liquid phase, although the density of the liquid decreases continuously during heating. Ermakov (1950) has designated this behavior "homogenization with an inversion point" and has shown many graphs of measurements of the "degree of fill" (i.e., volume percent liquid phase) versus temperature for individual inclusions, which he calls "homogenization curves." Kalyuzhnyi (1958c) has calculated the homogenization curves for inclusions of H_2O and of CO_2, and he has shown that in H_2O inclusions, inversions generally take place in those that have only 25 to 35% liquid phase at room temperature.

Llambias (1963) has proposed that the Th can be obtained from measurements at several temperatures, extrapolated on the assumption of constant coefficient of expansion, but Ermakov's experimental curves and Kalyuzhnyi's calculated ones both show that the coefficient of expansion is far from constant.

Weis (1953, p. 676) has recorded a range of 200°C in the experimentally determined filling temperatures of two-phase, liquid-plus-gas inclusions, in a given sample, all having the same apparent degree of fill at room temperature. This might be thought the result of differences in concentration, but the range indicated by his freezing data (-3 to -19°C, corresponding to about 5 to 22 wt

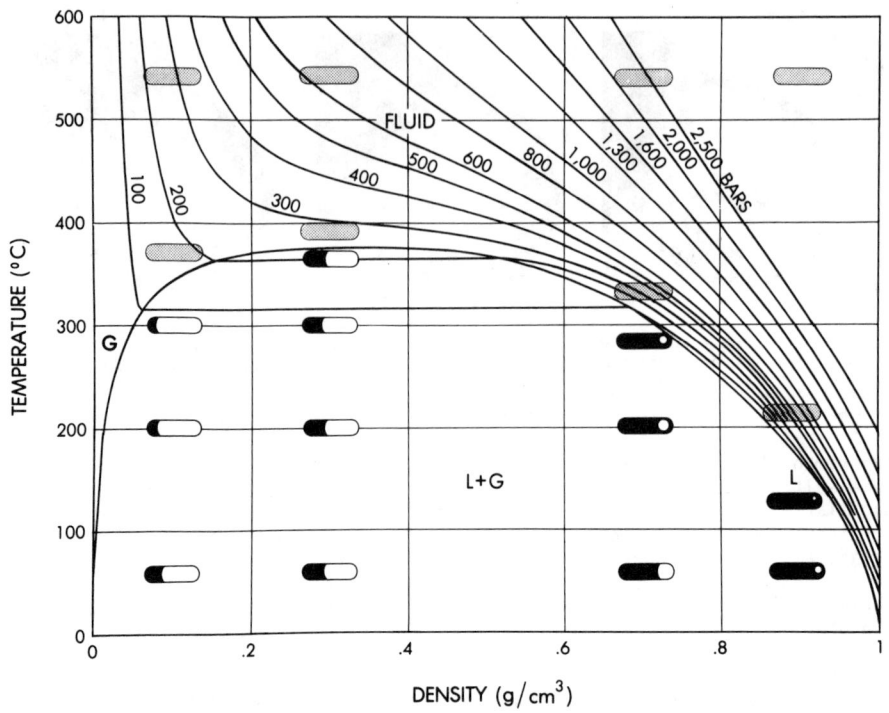

Figure 4-5. Temperature-density diagram for the system H_2O, plotted from the data of Kennedy (1950b) and Maier and Franck (1966). The homogenization behavior of four inclusions, all trapped at 540°C (but at different pressures), is indicated. Liquid - black; gas - colorless; fluid - gray. The inclusions having density 0.9 and 0.7 homogenize in the liquid phase (L); that having density ~0.3 homogenizes at the critical point; that having density 0.1 homogenizes in the gas phase (G). Modified from Roedder (1972).

% salts) is not adequate, assuming that the volumetric properties of the fluids are similar to H_2O-NaCl. The discrepancy may be a result of the inherent inaccuracies in phase volume measurements.

If data were available on density versus temperature for both liquid and gas phases in the multicomponent systems similar to fluid inclusions, such measurements on inclusions of unknown composition could be used to obtain the composition, or at least limit the possible ranges of composition (Ermakov, 1950, p. 116; Klevtsov and Lemmlein, 1959a, 1959b). The composition of fluid inclusions varies widely, and the effects on thermal expansion for two different compositional variables may be equal and opposite. Thus, salts in solution usually reduce the coefficient of expansion of H_2O, and substances such as CO_2 in solution may increase it. As a result, when measured homogenization curves for inclusions are compared with calculated curves from actual data on known simple systems, the deviations may be large. Laboratory data on the more complex systems of interest are just now becoming available (Chapter 8) and should permit interpretation of many of the phenomenological observations of the past.

Th of fluid inclusions varies widely. Aqueous inclusions trapped under near-surface conditions, if they form a vapor bubble at all, will homogenize at near room temperature. Sufficiently low- or high-pressure gas inclusions and certain organic fluids may homogenize at subzero temperatures. CO_2 inclusions that have a density approximately equal to the critical density homogenize at room temperatures (<31°C). If the density is sufficiently high, however, homogenization may occur at subzero temperatures. Thus, Rutherford (1963) reported that some CO_2 inclusions from the Flin Flon massive sulfide deposits in Canada homogenized in the liquid phase on warming to -16°C, and Dolgov et al. (1967) reported similar inclusions in kyanite. Conversely, if the density is sufficiently low (i.e., the CO_2 pressure at trapping was relatively low), homogenization in the vapor phase will occur by evaporation of the liquid phase. Roedder (1963, p. 204) reported this type of homogenization, at <-33°C, for CO_2 inclusions in Colombian emerald. Inclusions of water solutions may homogenize at any temperature up to a maximum of 374°C, the critical point for pure water, and at much higher temperatures for salt solutions (Fig. 8-12; Sourirajan and Kennedy, 1962).

Although most inclusions homogenize below 500°C, the literature of water-rich inclusions shows many that homogenize in the liquid phase at or above 500°C (Roedder, 1972, Table 7). Obviously, these require the presence of high concentrations of salts in solution, as pure-water inclusions cannot have Th above the critical point for water. Some of the highest recorded Th for water-rich inclusions, in the range 840 to 850°C, were reported for otherwise normal-looking, two-phase, liquid-plus-gas inclusions in nepheline (Kerkis and Kostyuk, 1963; Bazarova 1965; Bazarova and Feigin, 1966). At this temperature there was visible evidence of considerable enlargement of the inclusion by solution of the walls, implying an approach to magmatic conditions.

Homogenization of inclusions in alkalic rocks at temperatures as high as 890°, and even >950°C, has been reported by Panina (1966). Although these temperatures seem appropriate for homogenization of silicate glass inclusions, Panina's description of critical behavior in other inclusions in sodalite and cancrinite (Th <630°C) would imply aqueous solutions rather than silicate glass.

The gas bubble or bubbles in glass inclusions generally begin to contract appreciably at about 650°C, although this temperature should vary with glass composition (Clocchiatti, 1975).

New fluid phases formed upon heating. A variety of interesting and eventually informative observations have been made of phase changes occurring in the liquid or fluid phase upon heating. Certain inclusions show an unmixing of the liquid phase to form a small amount (<10%) of a new immiscible liquid, on heating to temperatures above 200°C (see Chapter 3). The significance of this immiscibility, in terms of specific fluid composition, is not known, but in view of the probable configurations of the pressure-temperature-composition (P-T-X) diagrams for water-salt systems, such immiscibility under conditions of rising temperature and constant volume is expectable. In this connection, it might be pertinent to examine the behavior of inclusions in synthetic quartz crystals from those crystal-growth processes where a two-liquid condition in the growth chamber has been proposed (Butuzov and Bryatov, 1956), to see whether such immiscibility can be recognized.

Wahler (1956, p. 113) described an inclusion in quartz from Madagascar that, when heated to about 150°C, developed a "new" gas bubble, which vanished again at 220°C; he believed that this bubble probably was CO_2 and that it resulted from the reversal in the solubility of CO_2 with temperature. The data on the system CO_2-H_2O (see Chapter 8) show that, for considerable range in CO_2-H_2O ratio, immiscibility would occur on heating.

Critical phenomena. Not infrequently, the composition and degree of fill of an inclusion are such that it homogenizes at a critical point. This is recognized by the abrupt and complete fading of the meniscus between liquid and vapor. Many more inclusions will appear to homogenize at a critical point than are actually filled to the critical degree because the liquid and vapor exhibit rather gross changes in density near the critical temperature; i.e., on a temperature-density plot the solvus is almost flat on top (Fig. 4-5). There is only one true critical density (or degree of fill), but as a result of the flat top to the two-phase field, inclusions having densities grossly different from the critical density (on either side) may appear to go through a critical point on rapid heating or cooling. Thus, the slower the heating rate near the critical point, the smaller the number of inclusions that will appear to show critical phenomena. Inclusions having a density greater than the critical density will homogenize in the liquid phase, and show a fading of the meniscus just before the bubble shrinks to disappearance; inclusions having less than the critical density homogenize in a similar manner by the evaporation (disappearance) of the liquid phase (that is, they homogenize in the gas phase), and show a fading of the meniscus just before the bubble expands to fill the inclusion.

The critical temperature is a definitive property for pure substances (Kobe and Lynn, 1953), and one inclusion fluid, CO_2, is sometimes sufficiently pure in nature to exhibit critical phenomena at the critical temperature (+31.1°C) for the pure material. In general, however, inclusion fluids are multicomponent mixtures, and any critical points (C.P.) they show are for such mixtures. Thus, mixtures of methane (C.P., -82.1°C) and ethane (C.P., +32.3°C) have been suggested as the probable major constituents of two monophase inclusions in quartz crystals from Herkimer County, N.Y., showing critical behavior at -5.55°C and -7.85°C (Fig. 11-17). Touray and Sagon (1967) reported CH_4 as a major component of similar monophase inclusions in quartz crystals from a marl in the Pyrenees. Other organic compounds (or mixtures) may have critical temperatures well above 100°C (Fig. 4-6). Inclusions of H_2O solutions may show that the critical temperature for pure H_2O (374.2°C) has been lowered by components such as CO_2; thus, Cameron et al. (1953) found inclusions showing critical phenomena at 311°C. More commonly, the salts in solution raise the critical point from that for pure H_2O. Sourirajan and Kennedy (1962) showed that 25 wt % NaCl raises the critical point for H_2O to about 670°C, and many critical point determinations on fluid inclusions fall in the range of 375 to 450°C. Klevtsov and Lemmlein (1959a) described the homogenization of the gas and liquid phases in inclusions at a critical temperature below that necessary to dissolve all solid halide daughter minerals; such behavior is perfectly expectable in multicomponent systems.

Even where the density of fill is not near the critical density for the specific fluid, it is important to note the rate of change of the volumes of liquid and gas phases with increase in temperature, in addition to the actual Th. This rate of change can provide a crude estimate of the probable critical temperature of fluids trapped at slightly more than the critical density, for as the critical temperature is approached, not only is the thermal expansion of the liquid large, but its rate of increase with temperature becomes progressively greater (note the change in density along the liquid curve in Fig. 4-5), and it becomes effectively infinite at the critical temperature. Thus, the high thermal expansion of the unidentified liquid (CO_2) found by Brewster (1823a) merely reflects a close approach to its critical temperature of +31.1°C. Similarly, for inclusions homogenizing in the gas phase, the rate of increase in density of the gas phase becomes progressively greater as temperature increases (Fig. 4-5). This cannot be measured as conveniently as can the expansion of the liquid, but the rate of change in density can be estimated from the rate of evaporation of the liquid phase into the gas phase. Thus, Roedder (1971b) noted in some samples from Bingham, Utah, that ~12 vol % (at room temperature) of a phase believed to be liquid H_2S evaporated into the vapor bubble at temperatures ranging from 64 to 97°C. These temperatures are well above the critical temperature for liquid CO_2, eliminating this phase as a possibility. Both the temperatures and the density increase of the gas phase are appropriate, however, for liquid H_2S near its critical temperature of 100.4°C. Such data are not definitive but do present strong support to other data which also indicate that this phase is liquid H_2S.

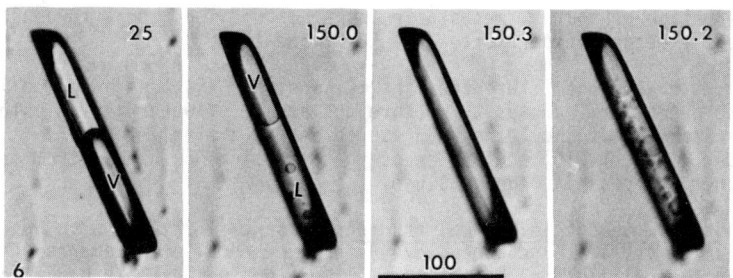

Figure 4-6. Sequential photomicrographs of a secondary inclusion in quartz from a vein in Precambrian rocks from South-West Africa, containing a large vapor bubble (V) in organic liquid (L) taken at the temperatures indicated (°C). The average (filling) density is near the critical density, resulting in homogenization by fading of the meniscus at the critical temperature of 150.3°C, and subsequent violent "boiling" to form two phases on cooling to 150.2°C. Note also the small bubbles of vapor in liquid at 150°C, caused by small thermal gradients in the sample. Scale bar in μm. From Kvenvolden and Roedder (1971).

If the sample is warmed to above 31°C (on the microscope lamp housing, or on the stage with hot air from a handheld hairdryer) and then examined, any inclusion that still contains liquid plus vapor phases cannot consist of CO_2.

Freezing data. Since the pioneering effort of Sir Humphry Davy (1822), various attempts have been made to use the behavior of fluid inclusions at low temperatures to help in the identification of the materials present. Simple cooling with a drop of alcohol, acetone, ether, or spraying with the compressed gases used in cleaning of instruments will frequently prove useful as a qualitative test.

Ethyl chloride (available in handy pressurized tubes as a medical supply) is particularly convenient. A thin copper washer with blotting paper on its upper side, on the end of a thin probe, is used. The blotter is wet with ethyl chloride and the washer is held against the slide and observations made through the central hole. For lower temperatures, a jet of ethyl chloride can be sprayed directly on the top of the slide, or, if an oil immersion objective is being used, the substage condenser can be temporarily swung out and ethyl chloride sprayed up against the underside of the slide.

Microscope freezing stages (see Chapter 7) permit quantitative measures of the temperatures of phase changes. The most commonly measured parameter is, in effect, the depression of the freezing point of the liquid-water solutions (Fig. 8-11a).[3] Within limits discussed in Chapter 7, this depression is a measure of the total solutes in the fluid, and it permits estimation of the salt concentration in inclusions weighing as little as 10^{-10}g. Other phase changes are observable in fluid inclusions at low temperatures, most of which yield at least qualitative compositional data, as discussed in Chapter 8. Among these are the triple point of CO_2 at -56.6°C and the temperature of eutectic melting (Te), representing the first development of visible liquid upon heating a completely frozen inclusion (Fig. 8-11b).

Several metastable phenomena occur commonly during the cooling of inclusions. Although these generally cause some experimental difficulty, they are useful indicators that inclusion fluids are generally very clean and free of spurious solid nuclei and hence have presumably been trapped from exceedingly slow-moving solutions (Roedder, 1962a, 1963). If the expansion on formation of ice eliminates the vapor bubble, sluggishness in its renucleation may lead to

----

[3] Various terms and abbreviations have been used in the past for this point. A consensus has been reached on the use of "temperature of melting" (Tm) or more specifically, "Tm ice," for this point (see Chapter 7).

high negative pressures, metastable "superheated" ice, and erroneously high apparent freezing temperatures (Chapter 7 and 10).

High-precision differential thermal analysis has been proposed as a procedure to detect and perhaps to measure the amount of water in inclusions in a frozen sample. It should be adequate to detect a single inclusion <1mm in size (Roedder, 1972, p. 16), but it seems that practical problems, from metastability and other sources, preclude its use.

Bubble movement in thermal gradients. Sang (1873; see also Hunter and Sang (1873) and Tait and Swan (1874)) reported peculiar movements of the bubbles in some fluid inclusions when a thermal gradient was established across the inclusion.[4] Some bubbles moved up a thermal gradient; i.e., they appeared to be attracted to a warm probe touching the sample and moved through the fluid as far as the inclusion shape would permit. Other bubbles either moved down a thermal gradient, i.e., they appeared to be attracted by a cold probe and repelled by a warm one, or they did not move at all in a thermal gradient even though they were apparently free from the walls. Still others showed very peculiar oscillatory movements, either of the bubble itself or of solid particles in the liquid (Brewster, 1845; Hartley, 1877b; Hawes, 1881).

These phenomena were variously attributed to "unknown cause," "capillarity," or "evaporation and recondensation," until Hartley (1877b) offered a brief explanation (proposed by Stokes) based on changes in surface tension. This seemed to explain some of the apparent conflicts in the data, and the subject was dropped, except for a series of rediscoveries of the phenomenon (Hoagland, 1951; Rush, 1954; Safronov, 1957; Johnson, 1961a,b; Sutton, 1964; Roedder, 1965c).

The basis for all these phenomena appears to be variation in the surface tension from temperature differences over the bubble surface, causing flow of the surface that drags fluid with it. Although all the phenomena are still not fully explained, the motions of the bubbles can be detected very quickly and simply by hot or cold probes (Roedder, 1965c, 1966; see also Fig. 6-4) and sometimes by changes in the illumination system (Fig. 6-3); these motions are useful in inclusion studies in a variety of ways. First, the specific behavior is obviously controlled by the composition of the fluid. Even though the nature of the effects of these compositonal parameters is complex and not understood, differences in the behavior of the bubbles in adjacent inclusions that otherwise seem to be identical indicate differences in composition and presumably origin; these differences frequently correlate with differences in freezing and homogenization temperatures. Second, the phenomena aid in the determination of freezing temperatures in those inclusions in which the last ice crystal sticks to the bubble-liquid interface. In these, the thermal (and compositional?) gradients set up by slight melting of the ice result in a rapid movement of the bubble and ice crystal about the inclusion. Whether the bubble pushes the ice ahead of it or drags it along, the movement is an indication that melting is still taking place, even though the ice may be invisible. Also, many inclusions have only one small clear "window" through which the freezing behavior can be watched; movement of the bubble and ice across this window thus permits more positive identification. Third, the phenomena aid in homogenization experiments. In these, the bubble movement permits an evaluation of the direction of the inevitable thermal gradients in the heating cell, and, by superimposing small thermal gradients on the heated sample near to its homogenization temperature, tiny bubbles may be moved out of dark areas of total reflection. Fourth, as mentioned above under "Fluidity," bubble movement from thermal gradients may be used to discriminate between glass and liquid inclusions. Fifth, thermal gradients are

----

[4] This is presumably not the first description of the phenomenon, as William Nicol (1829) mentioned that the bubble in an inclusion in halite from Cheshire did not move when a hot wire was placed opposite it, thus implying that other bubbles did move under such conditions.

frequently useful to place the bubble in an inclusion in a position more suitable for photography; they were so used in taking a number of the photographs in this book.

Neutron activation. Luckscheiter and Parekh (1979) have shown that neutron activation of unopened inclusions in quartz permitted reasonably accurate determinations of a series of elements. Na and Cl were determined after 0.2 hours irradiation, and Na, K, Mn, As, and Br were determined after a second irradiation of 2 hours. The determination of Mg and Ca required destructive analysis (crushing and leaching). All such analyses are subject to possible contamination from the presence of the same elements in the structure of the host. Thus, Na can be present in the quartz structure in amounts from a few parts per million to a few hundred parts per million (Roedder, 1958; Frondel, 1962). Luckscheiter and Parekh showed, however, that the ratio Na/Cl was reasonably constant in three separate samples, and because the likelihood of appreciable Cl in the quartz structure is small, the Na and Cl found are probably both associated with the fluid in the inclusions.

Composition of Solid Phases

Daughter minerals in fluid inclusions -- solid phases that have crystallized out of the fluid after trapping -- indicate saturation of the fluid with respect to these phases at the conditions of observation. Hence, daughter minerals are extremely useful in determining the composition of the fluid, but only recently have they been given adequate study. Probably the most serious problem in such use lies in the apparent difficulty of identification, but the many methods that can be used provide a combination of identification criteria that is generally unambiguous. A large number of different daughter phases have been recognized (see indices in Roedder, 1968a), but the most common, by far, are isotropic cubes of NaCl (and much more rarely, octahedra, Fig. 13-1b).

General significance and usefulness. The problem of distinguishing between daughter minerals, formed from the fluid of the inclusion, and accidental solid inclusions, trapped along with the liquid, is sometimes difficult. Regularity of phase ratio in several or many inclusions is perhaps the best criterion for a daughter-mineral origin (Fig. 3-5), but recrystallization (i.e., necking down), can isolate already formed daughter minerals from their parent liquid. In addition, small inclusions may not show a daughter mineral that adjacent larger ones have; this is commonly due to failure to nucleate. In NaCl solutions, freezing may cause the nucleation of the stable phase, NaCl crystals, in such supersaturated inclusions (Roedder, 1967a; Touray and Sabouraud, 1970). Any phase with which the solution is supersaturated may be caused to nucleate by freezing, as this withdraws water from the solution and increases the degree of supersaturation.

Fortunately, the distances are so small in fluid inclusions that daughter crystals, at least the water-soluble ones, have generally had time to reach minimum surface energy, that is, one single crystal of each phase (Bienfait and Kern, 1965). Hence, the number of daughter mineral phases is generally equal to the number of crystals. This is not true for very slightly soluble crystals or for those formed during a relatively fast cooling.

Ermakov (1950, p. 24) noted inclusions in which the sequence of formation of several daughter minerals upon cooling is preserved as a record of the sequence of saturations. He also presented what he considered to be evidence (p. 38) of the crystallization of quartz from fluids that were saturated with respect to NaCl, so that NaCl occurs both as solid inclusions and as a daughter mineral.

Every host mineral should be considered as an additional "daughter mineral" phase, as at least small amounts of it must crystallize out from almost every fluid inclusion upon cooling. It is important to remember that each daughter phase, as well as the host phase, represents a material with which the fluid is saturated. Thus, for fluids in the pure system NaCl-H$_2$O, the presence of a daughter crystal of NaCl at room temperature indicates that the fluid contains ~26 wt % NaCl at room temperature and that the original fluid that was trapped contained still more. This estimate of ~26 wt % in solution is valid only if the solution contains no other solutes. If divalent cations are present, such as Ca or Mg, saturation with NaCl can occur at the level of several percent NaCl.

In describing certain synthetic minerals and the crystals formed in inclusions of glass, Sorby (1858, p. 457 and 477) discussed a concept that could be used as an effective aid in the identification of some daughter minerals but that seems to have been forgotten by many later workers. Further crystallization of the host mineral from the trapped fluid will generally occur on the walls of the inclusion, but other phases form separate daughter crystals. Thus, inclusions of a given fluid, saturated with respect to minerals A and B, will show daughter crystals of A in inclusions trapped in B, and vice versa.

Only rarely is the host phase also found as a daughter phase. Kalyuzhnyi et al. (1966) reported daughter crystals of quartz(?) in inclusions in quartz, and Llambias (1963) reported a daughter crystal of borax, presumably identified as such by crystal shape, in a fluid inclusion in a larger borax crystal. Llambias mentioned the presence of organic matter within the host borax crystal. Thus, a film of such material lining the inclusion could have prevented crystallization of the material of the daughter crystal directly on the walls.

Ordinary methods of microscopy. Brewster (1826a, p. 21) was the first to report "squares" (that is, cubes, probably of NaCl) in inclusions. Solid daughter-mineral phases in fluid inclusions are generally so small that only optical methods are adequately sensitive for their nondestructive identification. The volume of each daughter mineral, particularly of phases other than NaCl, is frequently so small, however, that the resolving power of the microscope limits their detection and identification in small inclusions.

Many of the usual petrographic techniques may be used, in modified form, but rather severe limitations are imposed by the nature of the samples. Thus, observations of extinction angles and birefringence are limited by the optical properties of the host mineral. If the host is birefringent, it can be put at extinction, and then many daughter crystals can be checked to obtain maximum values for extinction angles and birefringence. Sorby (1858) pointed out that the inclusions in birefringent plates are best studied with a polarizer in, to eliminate troublesome double images. I have found that for strongly birefringent minerals such as calcite it is important to use only the image of the ordinary ray, which yields much better resolution than that from the extraordinary ray. Because these carbonate minerals are optically negative, this procedure also results in an appreciably greater maximum depth of focus into the plate. The thick plates (0.5-5 mm) normally used for inclusion study usually show so much birefringence (even with normally isotropic host minerals such as fluorite and halite) that both extinction angles and birefringence of solid inclusions and daughter minerals can only be estimated by using many inclusions. In the presence of low birefringence in the host mineral, the presence of low birefringence in daughter crystals can sometimes be recognized by means of the first-order red plate. Low birefringence is particularly hard to verify in high-index minerals because of the masking by reflection polarization at the interface with liquid.

The maximum interference color shown by the daughter-mineral grains can be used in conjunction with their apparent thickness to obtain the birefringence. Equant crystals can be assumed to have a thickness about equal to their diameter, and prismatic crystals can be assumed to have a thickness about equal to the lesser dimension of the prism, but the thickness of thin flat daughter minerals is particularly difficult to estimate. Eppler (1962) has shown that the interference color seen in reflected light from flat inclusion crystals can sometimes be used to obtain such an estimate. Crystal habit, interfacial angles, color, pleochroism, etc. are all useful criteria for identification. Not infrequently, the daughter minerals are sharply faceted crystals that permit identification of their crystal system as well. Generally the inclusions are too deeply buried in the sample, or are too small, for interference figures to be obtained; Kalyuzhnyi (1958a) mentioned grinding and polishing to within 10 or 15 μm of the inclusion to permit the use of such techniques.

One important caveat is in order here. Too frequently the identification of daughter crystals is based on too few parameters. Thus the discovery and verification of dawsonite [NaAl(CO$_3$)(OH)$_2$] by Coveney and Kelly (1971) was followed by many other reports of dawsonite or dawsonite(?), some of which were based mainly on a fibrous radial crystal habit, without recourse to extinction angle, birefringence, or even the general chemical nature of the system.

Opaque daughter minerals, presumably metallic sulfides and oxides, occur in some inclusions. The most common oxide daughter mineral is hematite, but it is frequently in crystals small enough to be bright red (or even yellow in extremely thin plates) and birefringent. The rarity and small size of opaque ore-mineral daughters have been used to place limits on the quantities of metal that may be precipitated from the ore fluids by temperature decrease (Roedder, 1960a; Takenouchi, 1962). The determination of the size, and hence of the volume and weight percent, of opaque grains is usually limited to measurements of the two visible dimensions and a guess as to the third dimension. In some inclusions, the crystal can be moved about for further measurements.

Relatively few parameters exist that can be used for the identification of opaque daughter minerals. In a very few inclusions, the grain is large enough to permit recognition of the color and crystal habit in reflected light (Roedder et al., 1963, p. 367). Not infrequently, the external crystal shapes -- or rather, silhouettes of them -- are visible. Brewster (1853) was able to estimate the relative densities of several different daughter mineral grains by comparing the rates of fall through the liquid when the sample was inverted. One simple but very sensitive test that should be used routinely on opaque grains is for their magnetic properties; if the grains are magnetic, and loose in the cavity, they will move when a small strong magnet is rotated or moved near the microscope objective. As an additional test, Prof. W.C. Kelly (pers. comm., 1967) has been able to determine the Curie temperature of magnetic daughter minerals by observing reversible changes in their properties on the heating stage. [See also Kelly and Turneaure (1970).]

A series of studies of the daughter minerals in large inclusions in pegmatitic quartz, topaz, and fluorite were summarized by Ermakov (1965). In this work, many of the methods described here were used, and measurements by two-circle goniometer, X-ray diffraction, and microhardness determinations were also used on extracted daughter minerals. As many as 30 (presumably all different) daughter minerals were recognized. One-third of these daughter minerals were opaque phases, and many of the other were new minerals. Halite, sylvite, and fluorides made up more than half the total volume, but the following minerals were also identified: hematite, chlorite, albite, quartz, rutile, zircon, muscovite, staurolite, and sphene. Freezing studies on some of these inclusions (Roedder, 1963, p. 181) have shown that even at room temperature, the liquid-water solution contains high concentrations of salts with at least some CaCl$_2$, since Te = -49°C. The samples containing these inclusions came from chambered pegmatites at Bektauata (Kazakhstan) and Volynia, USSR, some of which have central crystal-lined cavities with volumes as great as 200 m^3.

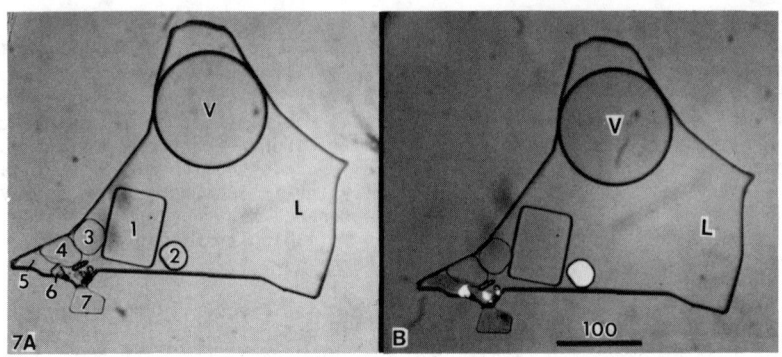

Figure 4-7. Fluid inclusion in topaz from Volynia, USSR, in plain light (left) and partly crossed polars (right). 1 = NaCl; 2 = unnamed Zn-Al chloride; 3 = elpasolite [K$_2$NaAlF$_6$] (index of refraction <liquid); 4 = sylvite; 5, 6, and 7 unknown birefringent phases (plus five smaller unidentified phases); L = liquid; V = vapor. Scale bar in μm. Sample courtesy of Vl. A. Kalyuzhnyi.

Lindgren and Whitehead (1914) described some inclusions in quartz from a hydrothermal ore deposit near Zimapan, Mexico, that have more than 50 wt % NaCl, as solid daughter mineral and solution, plus unknown amounts of other materials in solution. If only NaCl and H$_2$O are assumed to be present, such inclusions would homogenize at about 440°C (Sourirajan and Kennedy, 1962). Similar amounts of daughter minerals are found in fluid inclusions in samples from many other ore deposits.

Estimation of index of refraction. One of the most useful and definitive properties of transparent daughter minerals is the index or indices of refraction, which can be estimated by comparison with whatever known phase is available. Thus, Lemmlein et al. (1962) found crystals of cryolite (n = 1.34 and hence is comparable with or even less than water solutions) in the inclusions of a pegmatitic topaz (Fig. 4-7). Ikorskii (1966) reported inclusions containing rose-colored cubes of villiaumite (NaF, n = 1.336) and organic matter in many of the alkalic rocks of the Khibina massif, USSR. The isotropic cubes (and occasionally octahedrons; see Fig. 13-1b) seen in many inclusions are usually NaCl (n = 1.54), but not infrequently NaCl occurs with KCl (n = 1.49). (The vertical sides of such cubes may make estimates of relative relief rather difficult.) Kalyuzhnyi (1958a) noted that NaCl daughter crystals may be pinkish or yellow, but upon recrystallization after heating and cooling, the color is gone. KCl cubes are sometimes bright yellow. Daughter crystals of NaCl in inclusions in quartz are sometimes in sufficiently close contact with the host that a sensitive comparison can be made between the index of the halite (n = 1.5443) and the ordinary ray of the quartz (n = 1.5443; Fig. 4-8). The universal stage may also be useful in identifying such embedded solid crystals (Bakumenko, 1964), and similar comparison can sometimes be made with closely adjoining daughter minerals as well.

Strong birefringence is easy to recognize even without placing the host crystal at extinction, but if the host crystal has a considerably higher birefringence than the daughter mineral, particularly where the host does not go to complete extinction, as in many thick sections of calcite or quartz, this test is inadequate. In such plates, grains with high birefringence can be recognized by the change in relief relative to the liquid on rotation of the polarizer, without the analyzer. Thus, Slivko (1955, p. 70) showed that one of the many daughter minerals in inclusions in green and multicolored tourmaline has one index of refraction very near that of the solution in which it occurs. It also has a high birefringence and is very soluble, even in refractive index liquids; it is presumably a fluoborate (Slivko, 1958). On very tiny grains, the bright-

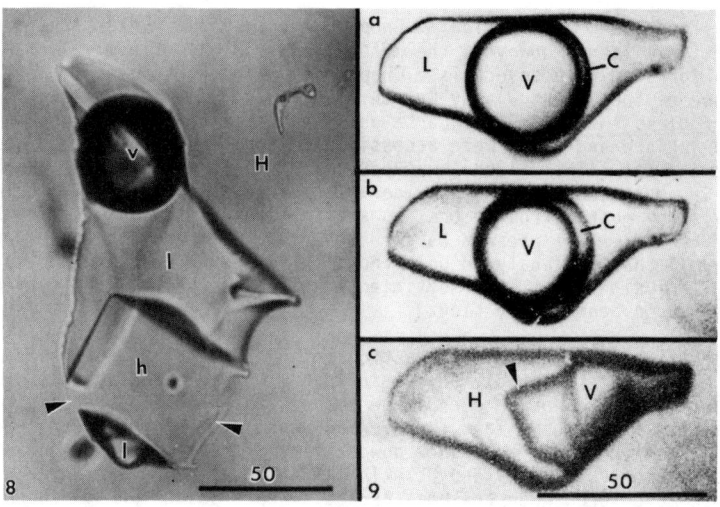

Figure 4-8. Photomicrograph of a large three-phase inclusion in quartz. In addition to the dark vapor bubble (v), a large daughter crystal of halite (h) grew from the liquid (l) in very close contact with the quartz walls, sealing off a small part of the liquid phase (at bottom). The index of refraction of the cube of halite (1.5443) is very close to that of the surrounding quartz host (H) in this orientation, making the contacts almost invisible (see arrows). Scale bar in μm. Sample ER 63-84Q, Muzo emerald mine, Colombia, courtesy of Banco de la Republica, Bogota. From Roedder (1972).

Figure 4-9. Serial photomicrographs of a flattened inclusion taken at different temperatures to show the identification of CO_2 by the formation of invisible crystals of $CO_2 \cdot 5.75H_2O$ on cooling. Figure 4-9a, taken at room temperature, shows liquid-water solution (L), probably about 1 molar NaCl equiv. in salinity, CO_2 vapor (V), and a small film of liquid CO_2 (C). When the inclusion is cooled quickly, more liquid CO_2 condenses from the gas to yield the metastable assemblage shown in Figure 4-9b, taken at about -5°C. When the inclusion is first frozen at -78°C, then equilibrated at -8°C (Fig. 4-9c), the CO_2 slowly reacts with the water solution to form a mass of crystals of $CO_2 \cdot 5.75 H_2O$ (carbon dioxide hydrate, H), eliminating liquid CO_2 as a phase. The "meniscus" between this solid mass and the vapor bubble (v) is jagged (arrow); on heating, the last invisible hydrate crystal melts at 7.60±0.20°C to form liquid CO_2 and water solution again, with a smoothly curved meniscus. During the melting process, the CO_2 hydrate crystals are usually invisible, as they are isotropic, and have an index of refraction almost identical with that of the solution. Scale bar in μm. Sample ER64-187, quartz, from a quartz-anhydrite-calcite vein, Homestake mine, South Dakota, USA. From Roedder (1972).

ening of the centers on slight change of focus can be used for index comparisons in lieu of an actual Becke line.

Use of the electron microscope, electron microprobe, ion microprobe, and X-ray diffraction. As inclusions are generally small, electron-microscope techniques would seem to be very appropriate for their study, but the very shallow penetration of the beam precludes its use on most unopened inclusions. McLaren and Phakey (1965a, b; 1966) have used transmission electron microscopy to study the development of dislocations and associated spherical and faceted negative crystal cavities in very thin flakes of citrine and amethyst, particularly on annealing at temperatures of 600° to 800°C. Akizuki (1967a) used an acetate peel technique to obtain layers of galena thin enough for transmission electron microscopy. The electron microscope has also been used on replicas of fractured surfaces to study the structure of the walls of inclusions and the changes in shape with recrystallization (Akizuki, 1965a,b,c; 1966; 1967b; Kurshev and Trufanov, 1965) and to study the distribution and nature of very minute inclusions on grain boundaries and within grains (Folk and Weaver, 1952; Iwao et al., 1953; Sella and Deicha, 1962a,b, 1963).

Similarly, the electron microprobe would be eminently suitable for determining the composition of daughter minerals, if they could be put into the electron beam. Absorption of both the incident beam and the emitted X-rays generally precludes examination of unopened inclusions. Unless the inclusions are relatively large, however, the mechanical problems of removal of daughter crystals for study are severe (see Chapter 5). The first use of the electron microprobe on inclusions was by Carron (1961), who used it to study the composition of glass inclusions in quartz phenocrysts that were truncated by the polished surface and hence were accessible to the beam. The second was by Dolomanova et al. (1966, 1968), who determined the presence of Fe, Cu, Ti, Ca, Zn, and Cr among the 11 daughter minerals on the walls of a fluid inclusion in an early smoky quartz from an ore deposit in Transbaikal, USSR. They believe that the Fe and Cr are present as chromite. Dr. Dolomanova stated (pers. comm., 1967) that she analyzed only those solid grains on the upper surface of unopened inclusions that were sufficiently close to the polished surface to permit electron-beam penetration.

Identification by X-ray diffraction methods generally requires the extraction of the daughter mineral, unless it is present in large amounts. For example, halite is present in such large amounts in inclusions in some feldspar crystals from Ascension Island (Roedder and Coombs, 1967) that single-crystal X-ray diffraction photographs show powder lines for the three strongest reflections for halite. Vorob'ev and Lozhkin (1976) have shown that certain daughter phases become colored when the host crystals are exposed for 15 minutes to X-rays from a tungsten tube. NaCl becomes an intense yellow; KCl, a deep lilac; NaF, rose; and LiF, brown.

Nondestructive Raman spectroscopy can be used to identify certain solid phases in fluid inclusions, but as this method is not limited to solids and has also been used on liquids and gases, it is discussed in a separate section in this chapter.

Behavior on freezing. When inclusions are frozen, a variety of new solid phases may be formed, and the identification of these frequently yields otherwise unobtainable compositional information (Roedder, 1962a and 1963). The most common new phase is ice, which can be identified by: (1) index of refraction (always appreciably less than that of the solution); (2) very low birefringence; (3) characteristic growth, on cooling, of two parallel flat plates from opposite sides of the round grain obtained by partial melting; (4) parallel extinction and length-fast orientation of these plates; (5) volume increase on freezing and decrease on melting (this registers as an increase in bubble size as melting occurs); and (6) temperature range of stablity. (Care must be taken in using this last criterion on inclusions whose vapor bubble is eliminated by the expansion on freezing. These inclusions may develop high negative pressure upon partial melting, which, in turn, may cause ice to exist, metastably, at temperatures as high as +6°C (Chapter 10).) Many larger inclusions are broken open by the expansion of freezing, as first recorded by Dwight (1820).

Various salts and hydrates may also crystallize on freezing. A particularly significant phase is $NaCl \cdot 2H_2O$ (hydrohalite). This can form directly from the fluid, but if a daughter crystal of halite is present, it can react with the liquid upon cooling to form hydrohalite crystals. As these are strongly birefringent and melt incongruently at temperatures up to a maximum of +0.1°C (in the pure system $NaCl-H_2O$), they provide a very good test for distinguishing a cube of NaCl from KCl, which forms no such hydrate. As little as 10^{-12} g of NaCl can be recognized by this test (Fig. 8-17; Roedder, 1963, p. 178 and 182) but care is needed, as hydrohalite crystals form very slowly. The crystals are commonly very small (~1 μm) but still show first-order white birefringence. In large mass, they may appear opaque. Aqueous inclusions in halite host react slowly with the walls of the host on cooling to form a mass of this phase that is larger than the original inclusion.

Another useful solid phase formed when certain inclusions are cooled is the clathrate compound carbon dioxide hydrate $CO_2 \cdot 5.75H_2O$[5] (Roedder, 1963, p. 188-196), which forms by reaction of H_2O and either gaseous or liquid CO_2. Unfortunately, its index of refraction is very close to that of most water solutions, and it is isotropic, so that it is not always visible even though present in rather large amounts. Sometimes its presence is evidenced only by a jagged interface between gas and "liquid" (Fig. 4-9). W.C. Kelly (pers. comm., 1967) has found that the presence of such invisible crystals can also be detected in those inclusions containing a magnetic daughter mineral. When such a magnetic grain is pulled around in the inclusion by manipulating a magnet on the microscope stage, it bumps into the invisible hydrate crystal. Data on the dissociation pressure of the hydrate (Takenouchi and Kennedy, 1965a; Collins, 1979) may permit estimates of the CO_2 pressures in some inclusions from the freezing data. One serious source of ambiguity in the use of freezing data on this phase is that each of the several clathrate crystal structures permit a wide variety of molecules to be trapped, including H_2S, CH_4, C_2H_6, and many other organic species (Stackelberg and Müller, 1954; Jeffrey, 1963; Poty and Stalder, 1970; see also Fig. 8-22).

Occasionally, freezing will result in the elimination of metastable equilibria within inclusions by causing the formation of new phases that should have been present at surface temperatures. Thus, freezing commonly permits the formation of a gas bubble in single-phase inclusions, formed at temperatures below 100°C, that have persisted for geologic time as metastable stretched fluid, under negative pressure (Roedder, 1963, p. 197; 1967a). Similarly, daughter minerals are not always present in inclusions that should contain them. NaCl crystals usually nucleate, even in very tiny inclusions at the limit of resolution of the light microscope (Roedder, 1967b, p. 533), but Touray and Sabouraud (1970) have described relatively large supersaturated inclusions that nucleated a NaCl daughter crystal only after being frozen and then rewarmed to room temperature. Daughter minerals present in only small amounts are more apt to remain in (supersaturated) solution. Thus, small multiphase inclusions commonly have fewer solid phases than adjacent apparently cogenetic large inclusions, and the missing phases are almost always those present in minor quantity (e.g., see Roedder and Coombs, 1967).

Behavior on heating, in combination with other data. Behavior on heating is a very important aid in the identification of daughter minerals. Brewster (1845) was the first to study this behavior with care and to make use of it to prove the existence of different daughter mineral phases in the same inclusion in topaz. Thus, KCl can be distinguished from NaCl by its much higher temperature coefficient of solubility, as pointed out by Ermakov (1950, p. 57). Below about 200°C, the increase in solubility with temperature in the pure binary systems is approximately eight times greater for KCl than NaCl; however, as a result of the curvature of the three-phase boundary for NaCl, KCl, and solution in the system NaCl-KCl-H_2O (see Fig. 8-25), the difference in solubility is even more noticeable, because NaCl solubility is retrograde along this boundary below about 100°C.

Far too frequently, guesses are made in the literature as to the composition of daughter crystals without considering all possible lines of evidence. Daughter crystals that do not dissolve before 300-400°C are not likely to be highly soluble compounds. Thus, 84 g of $FeCl_3$ will dissolve in 16 g of H_2O at 100°C, and $CaCl_2 \cdot 6H_2O$ will form a similar solution at 20°C, yet these two compositions have been suggested for high-melting daughter phases. Even more to the point, $CaCl_2 \cdot 6H_2O$ melts at 30°C, without any other components present.

----

[5] Although usually referred to by this formula, the theoretical structural formula is $CO_2 \cdot 6H_2O$. Some CO_2 sites are empty, giving the apparent formula $8CO_2 \cdot 46H_2O$ (Stackelberg and Müller, 1954).

Two separate and discrete daughter crystals, one of NaCl and one of KCl, are found in many fluid inclusions from various high-temperature geologic environments (e.g., Fig. 15-11; Roedder, 1972, plates 2, 3, and 7). As NaCl and KCl form a continuous solid solution at high temperatures, with a solvus at lower temperatures (solvus maximum at ~480°C; Skeaff et al., 1979), any daughter crystal forming at temperatures above the solvus will be a mixed crystal. Presumably, these original mixed crystals have exsolved into two separate phases when cooled over geologic time, but when cooled after laboratory homogenization, they may crystallize, and remain, a single phase (R.J. Bodnar, pers. comm.). Although there apparently has been no mention of it in the literature on fluid inclusions, the behavior of NaCl and KCl daughter crystals should be examined in the light of this solvus, as this could yield significant data on the composition of the system.

Yakubova (1952, 1955) used crystal shape and solubility to show the common presence of borax crystals in inclusions in quartz and topaz from a Uralian pegmatite. She verified the identification with microchemical and spectrographic analyses. Ikorskii (1968) described the effects of heating to 360°C and 480°C on bitumens in inclusions in eudialyte from the Khibiny apatite deposits (USSR).

Many of the recorded heating experiments on inclusions having several daughter minerals show that some do not dissolve, even at high temperatures. Thus, Feklichev (1962) found that some of the daughter crystals in pegmatitic beryl did not dissolve even at 750°C, and there are numerous references to daughter minerals, particularly ones that are opaque or that have high indices of refraction, persisting to temperatures above Th for the liquid-plus-gas phases (Ermakov, 1950; Kalyuzhnyi, 1958a). Four possible explanations for such behavior should be considered:

(1) Equilibrium was not attained in the time used. The relatively soluble crystals, such as NaCl and KCl, equilibrate with the solution rapidly when heated, but Ermakov (1950, p. 95) reported that 4 to 5 hours are needed to reach equilibrium, even at high temperatures, for the dissolution of the inclusion walls and of the less soluble daughter minerals.

(2) Equilibrium was attained, but the correct temperature for melting of the last grain of the daughter mineral actually does lie above Th for liquid-plus-gas. There is no a priori reason why a given solution, upon cooling, should not become saturated with a solid phase before it separates into two fluids, liquid plus gas. In such inclusions, Th of the daughter minerals would be closer to Tt than Th of liquid plus gas would be, as pointed out by Smith (1953), Sheftal' (1956) and others.

(3) The "daughter" mineral grain was merely a solid inclusion, trapped accidentally at the time of formation of the inclusion. Perhaps the best criterion for distinguishing between an accidental solid inclusion and a true daughter mineral is the constancy of occurrence, and of volume percent, of each daughter mineral in several or many inclusions. Unfortunately, available sample material does not always permit the application of this criterion. If a given mineral occurs also as solid inclusions in the host mineral, it is difficult to be certain that its occurrence in fluid inclusions is truly as a daughter mineral; certainly the solutions were saturated with respect to it, and hence it should occur as a daughter mineral.

(4) Not infrequently, evidence of necking down of inclusions is found, in which the daughter minerals (and gas bubbles) present at that time are restricted to one of the several smaller inclusions forming at the expense of a larger one; this can cause very diverse phase ratios in different inclusions (Figs. 3-16, 17,18).

Actually, it is difficult to separate the effects of a sluggish approach
to equilibrium with respect to (1) the solution of daughter (and host) minerals,
and (2) changes in the thermal regime in the microscope hot stage at the actual
inclusion site.

Lemmlein et al. (1962) described heating experiments on large primary seven-
phase inclusions in topaz containing 70 vol % solids at room temperature (Fig.
14-2b). Each had quartz, muscovite, cryolite, an undetermined mineral in pseudohexagonal plates of n = 1.51, fluorite, liquid, and gas, in the same volume
ratios. In order to prevent decrepitation of the inclusions from the internal
pressures that built up -- a common and very troublesome problem in much inclusion work -- they heated the samples in a bomb at 3000 kg/cm^2 external pressure.
Their data and photographs (before heating and after quenching) indicate that
not only do the daughter minerals dissolve to form a silicate melt (yielding a
glass on quenching), but the solubility of topaz in these fluids, at 700°C, is
about 10 to 15 vol % (Roedder, 1972, plate 8, Figs. 2-6). The resulting melt
was estimated to contain over 10% H_2O. One interesting point they make is that
in those inclusions that had fractured and lost their volatiles the daughter
minerals did not dissolve. Voznyak (1968) presented considerable evidence that
at least some of these "daughter minerals" are actually solid inclusions,
trapped during the growth of the crystal. However, those phases that occur in
uniform ratios in several inclusions (Figs. 3-4,5) are probably valid daughter
minerals.

Feklichev (1962) eliminated the problem of decrepitation by heating his
samples to the maximum temperature of observation (750°C) before an inclusion
was selected for high-temperature study; by this simple but effective procedure,
he eliminated those inclusions that were not suitable for high-temperature
study. I wish to add that unheated samples should be examined also, because
(1) the preheated samples present a biased sample of inclusions present; (2)
the inclusions may become stretched; and (3) many daughter minerals are very
slow to recrystallize after dissolving.

Lemmlein and Kliya (1952a) used an ingenious interferometric technique on
flat inclusions in topaz to make rough quantitative estimates of the amount of
topaz precipitated on the inclusion walls upon cooling. They showed that during
the original slow cooling of their topaz samples, the nucleation and growth of
the bubble was simultaneous in part with the deposition of topaz on the walls
from the inclusion fluids; a record was left in the form of crescent-shaped
rings, each showing where a former size and shape of flattened bubble had prevented deposition of topaz on the walls against which it was pressed. The
amount of topaz in solution in the inclusions at 200°C was estimated to be
about 2 vol %.

Kalyuzhnyi (1956) described several interesting changes in inclusions in
topaz. Daughter crystals of elpasolite (K_2NaAlF_6) in multiphase inclusions
(e.g., Fig. 4-7) decomposed ("melted incongruently") to form three other phases
of lower index of refraction, apparently also fluorides, in the temperature
range 135 to 170°C. Other inclusions showed decomposition of elpasolite crystals
to still another unknown phase at 250 to 260°C.

Zakharchenko (1955) described some very interesting polyphase inclusions
in quartz from quartz veins in the Pamir, USSR, which contain as many as nine
daughter minerals. They were identified (in part on the basis of destructive
tests on large inclusions), in descending order of volume percent abundance,
as: magnesian calcite (estimated, from Zakharchenko's Fig. 15, to be ~20%);
halite (~10-20%); sylvite (~1%); hematite (~1%); a sericite-like phase (<1%);
clear albite (<1%); and several other unidentified phases.[6/] The fluids present

----
[6/] Spectrographic analyses of the fluid extracts showed small but constant amounts of Al and Fe. The pH of
the diluted fluids (a leachate made after loss of gases) was found to be 8.6.

are H_2O solution, liquid CO_2 (6-8%), and a gas having a very strong odor of H_2S (~35%). During homogenization of the high-temperature inclusions (at temperatures to 360 to 420°C), very significant amounts of the quartz host dissolved. The amount was estimated, from measurements of the inclusion size, to be as much as 160 g/l (Zakharchenko, 1955, p. 45; also stated as 6 vol % or 13 wt %, p. 42). From these data, Zakharchenko calculated that the total volume of mineral matter carried in the solution, at the time of trapping, was 50 to 55 vol %. Apparently even pure CO_2 can dissolve appreciable amounts of unknown solids (e.g., Roedder, 1972, plate 6, Fig. 4).

Glass is a common solid phase in the inclusions in the minerals of volcanic rocks and can be recognized in part by its behavior when heated. Many interesting studies have been made of these inclusions, as some, at least, represent true samples of the magma and all its volatile components (see Chapter 16).

Composition of the Gas Phase

Before the advent of the micro-Raman technique discussed in the next section, relatively little could be determined about the nature of the gas phase by nondestructive tests.[1] Fluid inclusions give us practically the only data we have on the density of the ore fluids (Roedder, 1967b), and the volume percent of gas phase is essential in calculating total density of the fluid that was trapped. Ermakov (1950) categorizes inclusions by their "state of aggregation," based on the manner in which they homogenize (i.e., in the liquid phase, in a supercritical fluid phase, or in the gas phase), under the assumption that these three states are the states in which they were trapped. All that can be definitely determined from most inclusions, however, is the density of the trapped fluid (after corrections for thermal expansion and compressibility of the host mineral) and the minimum temperature of trapping. Depending upon the pressure, all three of Ermakov's "homogenization types" could originate from trapping of supercritical fluids at the same temperature but at different pressures and, hence, different densities (Fig. 4-5).

The above does not mean that the mode of homogenization is of no value. If the composition of the fluid is also known, these two values define the density of the fluid and, hence, the isochore. If the composition is not known, the mode (and temperature) of homogenization place some constraints on composition.

Only rarely can the density be determined directly (Petrichenko and Shaydetskaya, 1968). Generally, the determination of bulk density is based upon the measurement of the volumes of all phases and their identifications as to composition and hence individual density. (In a very few inclusions, corrections must be made for the solution of the inclusion walls upon heating to the trapping temperature.) Simple linear intercepts can be used for long tubular inclusions, and various planimetry methods can be used for flattened inclusions (Fig. 4-10). Sketching the projected image of a flat inclusion on heavy paper, followed by cutting out and weighing the pieces of paper representing the phases, is simple, fast, and surprisingly accurate compared with the use of a planimeter. In thicker inclusions, the spherical bubbles can be measured more accurately than any of the other phase volumes, if the image size is not reduced by a curved upper inclusion wall acting as a negative lens (Roedder, 1972, plate 11, Figs. 7,9). Allman-Ward and Rankin (1979) proposed a shortcut method based on some assumptions as to average inclusion shape. Such assumptions, however, must

----

[1] A novel test, which is applicable only rarely, was used by O'Keefe et al. (1962, 1964) to determine the presence of Ne, He, H, and O in the low-pressure gas in large (~1 cm^3) gas bubbles in tektites. A spectrograph was used to analyze the light produced by an electrodeless gas discharge in the unopened inclusions.

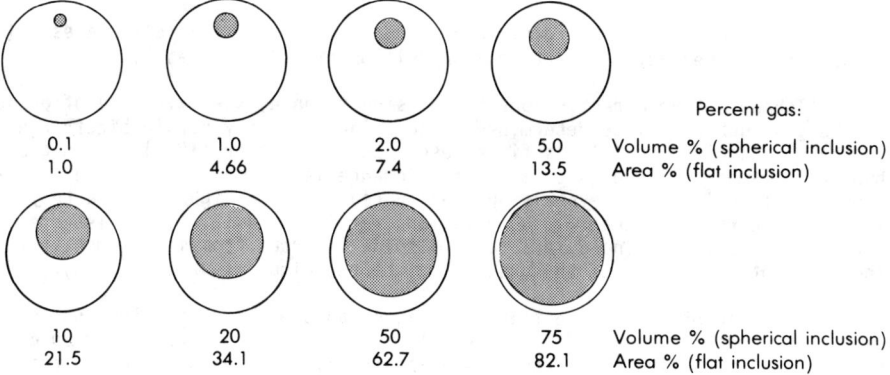

Figure 4-10. The appearance of inclusions that have various volume percentages of gas phase, in which both the gas bubble and the inclusion are assumed to be spherical. The area percentages refer to thin flat inclusions, where both bubble and inclusion are assumed to be circular disks of negligible thickness; in these, area percent equals volume percent. Shaded area is gas phase. From Roedder (1972).

either be extremely limiting in what inclusions will be measured, or yield extremely inaccurate results. Unless care is taken, gross errors may also be introduced in the measurement of volume of a small sphere under the microscope because of the production of a falsely focused image associated with wide-angle illumination (Saylor, 1965). Measurements of partly flattened or distorted bubbles yield only very crude and inconsistent estimates of phase volumes, and visual estimates can be very inaccurate (see Fig. 4-10 and discussion in Chapter 7).

Bodnar (1983) has shown that if the composition of the fluid is known and if adequate P-V-T-X data are available on such solutions, the volume (and density) of the inclusion can be calculated from the volume of the vapor bubble and the Th. Because the bubble is usually spherical, even in very irregular inclusions, this method greatly extends the possibilities for valid density determinations. Furthermore, because the volume of an inclusion is needed in many of the destructive analytical procedures, the procedure should become widely used.

Most tests of the nature of the gas bubble are destructive (see Chapter 5), but two nondestructive phenomena may give qualitative evidence of the presence of carbon dioxide. Hartley (1877a) first made the interesting observation that during heating of some inclusions, the "vapor" bubbles evidently achieve a density greater than that of the "liquid" and sink in it. Work on the system CO_2-H_2O by Tödheide and Franck (1963) and by Takenouchi and Kennedy (1964) has shown that there is indeed a density inversion between "liquid" and "gas" phases in this system. Such density inversions are useful as they impose constraints on the possible range of compositions (see also p. 239).

Roedder (1963, p. 195) showed that at low temperatures dense CO_2 gas will condense to form liquid CO_2, and this liquid, or invisible CO_2 in the gas bubble, will react with water in the surrounding liquid to yield crystals of the clathrate compound carbon dioxide hydrate ($CO_2 \cdot 5.75H_2O$). These crystals are almost invisible, since they are isotropic and their index of refraction matches that of the ordinary water solutions in inclusions. Data on the system H_2O-CO_2-$NaCl$ (Takenouchi and Kennedy, 1965b; Collins, 1979) may permit the estimation of CO_2 pressures in the gas bubble from the temperature of dissociation of such crystals. CO_2 is the most likely compound to form clathrate crystals in cooled natural inclusions, but some ambiguity remains in the identification, in that many other compounds, both organic and inorganic, form similar clathrates. (See "Behavior on freezing," in section "Composition of Solid Phases.")

Although of no direct compositional significance, the existence of pressure in the gas bubbles may be determined in some inclusions by strain birefringence in the surrounding mineral, as first described by Brewster (1820) around gas bubbles in amber. Although strain birefringence is common around solid inclusions in minerals, most crystals apparently require such high pressure to yield visible birefringence that fluid inclusions rarely cause.it. The existence of high pressures in gas inclusions could also be inferred from measurements of the index of refraction of the gas, using the technique of Wahler (1956).

The advent of microscope freezing stages capable of maintaining inclusion samples at temperatures -100°C or lower (Poty et al., 1976) has opened up a new field of investigation of the gas phase in fluid inclusions. Using this equipment, gas inclusions rich in N_2, CH_4, CO_2, and other species have been recognized from their phase behavior at low temperatures (e.g., see Fig. 12-5; Chapter 8; and Burruss, 1981a,b; Guilhaumou et al., 1981; Swanenberg, 1979).

Raman Spectroscopy

The most important recent development in the field of nondestructive analysis of fluid inclusions is the use of Raman spectroscopy. As it can be used for the analysis of certain constituents in solids, liquids, or gases, it is described here, separately from the preceding three sections.

Historical development. Raman spectroscopy is a nondestructive procedure for analysis of certain molecular species in solids, liquids, or gases, based on the principle of Raman scattering, originally described by Raman and Krishnan (1928). When an intense beam of light of a given wavelength is passed through an apparently optically transparent substance, three types of scattered radiation can generally be recognized: Rayleigh scatter, fluorescence, and Raman scatter. Rayleigh-type scattered radiation has the same wavelength as the exciting radiation. If cracks or other imperfections such as fluid inclusions exist, scattered light from these will add to the intensity of the Rayleigh scatter. Fluorescence involves absorbance of the incident radiation and subsequent emission of new radiation of longer wavelength. The new radiation frequently consists of a continuum over a wide band of wavelengths and may be emitted almost instantaneously, or for an appreciable period of time after the excitation, when it is termed "phosphorescence" in the mineralogical literature. (The usage of the terms "fluorescence" and "phosphorescence" varies with the branch of technology involved.) The wavelength of the emitted radiation is a characteristic of the substance and does not shift if the exciting radiation is shifted.

Raman scattered radiation differs from the above two types in several important ways. It emerges at a longer (or shorter) wavelength that is a characteristic of the substance irradiated, but this wavelength is shifted from that of the exciting radiation by a specific amount (expressed in frequency, i.e., wave numbers, in units of cm^{-1}). The most important features are that the magnitude of the shift does not change with a change in the exciting radiation, and the Raman scattered light may be almost perfectly monochromatic. The shift for each Raman line (there may be many for any given substance) is a result of a transfer of energy between the incident radiation and molecular (or crystalline) vibrations in the sample. An additional important difference lies in the low intensity of the Raman scatter, which is 10^{-5} to 10^{-12} less intense than the Rayleigh scatter (Adar et al., 1982a). As a result, rather sensitive detectors are needed, even when relatively large samples are irradiated, but the method has been applied to many fields of research and chemical analysis, including various solutes in water samples (Cunningham et al., 1977).

The advent of the laser resulted in great improvements in Raman spectroscopy (e.g., see Gilson and Hendra, 1970; Tobin, 1971), and most particularly, it made the micro-Raman technique possible. Not only does the laser provide an exceedingly intense monochromatic light source, but its radiation can be focused down to submicrometer dimensions. This permits such high levels of irradiance that signals adequate for qualitative analysis can be obtained from individual particles or sample volumes in the submicrometer range (i.e., 1 picogram, 10^{-12}g), and quantitative data can be obtained on somewhat larger samples.

Two different laboratories worked on the development of apparatus for micro laser Raman spectroscopy (sometimes abbreviated LRS), along different lines: the University of Lille in France (M. Delhaye and coworkers) and in the United States at the National Bureau of Standards (NBS) (G. Rosasco and coworkers). The two laboratories have each published a series of papers, starting in 1975 (Rosasco et al., 1975a,b; Delhaye and Dhamelincourt, 1975; see also reviews by Muggli, 1979; Rosasco, 1980; and Delhaye et al., 1980). The two instruments use similar lasers, spectrometers, and detectors, but the optical paths are different. An important series of papers on the application of the French instrument to fluid inclusions has since been published by a group of French workers in several laboratories (Guilhaumou, Touray, Dubessy, Poty, and others).

A commercial version of the Delhaye instrument has been introduced ("MOLE", for Molecular Optic Laser Examiner, by Instruments SA, Inc., Metuchen, New Jersey) and now a second-generation instrument is available (model U-1000), as well as several other models by other manufacturers. The MOLE permits imaging, on a TV screen, the response of an area of sample at a specific Raman frequency. Although the NBS instrument lacks this imaging capability, it appears to have better signal-to-noise ratio and hence better sensitivity. The original NBS instrument involved a detector optical axis perpendicular to the incident laser beam. For use on fluid inclusions, this required special samples and sample preparation (Rosasco et al., 1975b). A new instrument was subsequently developed, based on back-scattering (Rosasco and Etz, 1977), that was much more adaptable and permitted the use of standard polished sections for fluid-inclusion studies (Rosasco and Roedder, 1979).

Limitations of the Raman method. The four major limitations in Raman microprobe analyses of fluid inclusions are: (1) occasional restriction to the use of probe beams of low irradiance (laser power/area); (2) interference from the fluorescent and Raman scattering of the host; (3) lack of effective Raman peaks for many of the major (monatomic) constituents in fluid inclusions; and (4) difficulty in calibration for quantitative analyses. The first limitation arises from adverse heating effects associated with optical absorption by phases within the inclusion. Physical movement, dissolution, or even destruction of some phases can result if excessive irradiance levels are used. As the reflection from opaque phases comes not only from the outside surface, but also involves atoms beneath the surface, opaque phases can yield Raman spectra, but the irradiance levels must be kept low. Irradiances of a few kilowatt/square centimeter (kW/cm^2) are generally acceptable and transparent phases can be run at 100 kW/cm^2. Apparently the most easily destroyed fluid-inclusion constituents are certain unidentified organic compounds that may polymerize (or decompose) in the beam, depositing dark material on the inner wall of the inclusion, hence causing localized heating (Rosasco and Roedder, 1979). Guilhaumou (1982) reported that liquid aliphatic hydrocarbons are easily destroyed by the laser beam.

The second limitation, due to host-mineral interferences, is essentially a problem of signal-to-noise ratio. In most studies of highly concentrated phases within fluid inclusions (e.g., pure liquid droplets, high-pressure gas bubbles, and daughter minerals), the host interferences have not been too restrictive. By contrast, the analysis for dilute constituents (e.g., atmospheric pressure gas bubbles or low concentration solutes in brines) is difficult or sometimes impossible. For example, the fluorescence of a sphalerite sample was found to be very strong in the vicinity of 2600 cm^{-1} and thus prohibited analysis for HS$^-$ and H$_2$S in the inclusion fluid (Rosasco and Roedder, 1979).

The presence of host interferences implies that the "effective sample volume" is not solely within the desired phase in the inclusion but must include significant parts of the host. In analyses of inclusions, some Raman radiation from the host phase is impossible to avoid; hence, host-mineral lines must be subtracted from the spectra obtained from the inclusion. The host spectrum can be obtained on an inclusion-free area adjacent to the inclusion, and unless there is broad-band interference from fluorescence, the subtraction is simple. Since the lifetime of fluorescence is normally $>10^{-10}$ seconds, whereas Raman radiation occurs in $\sim 10^{-14}$ seconds, time-lapse spectroscopy (Frantz et al., 1982), using very short excitation and recording times, can avoid the fluorescence problem, but increases the instrumental complexity.

The third limitation is basic to the nature of Raman spectra. In general, strong peaks are obtained only from polynuclear species such as SO$_4^{2-}$, HSO$_4^-$, H$_2$S, HS$^-$, HCO$_3^-$, CH$_4$, and N$_2$. The major ions in solution in fluid inclusions, such as K, Na, Mg, Ca, and Cl, give rise to relatively weak spectral effects, or to indirect nonspecific changes in the spectrum of the water or another polynuclear ion; thus, Raman spectra are not expected to provide quantitative measures of these ions in fluid inclusions. Unfortunately, H$_2$O provides rather broad peaks, unless it is studied at very low temperatures.

The fourth limitation refers to calibration for quantitative analyses. Raman spectra of fluid inclusions can provide rapid, highly sensitive nondestructive qualitative data on the polynuclear species present (e.g., Guseva et al., 1983). However, the step from qualitative to quantitative is large. In some cases, the ratios of certain constituents can be calculated from Raman spectra, based on Raman cross section data, but the excitation conditions in fluid inclusion samples are so irregular that quantification is difficult without an internal standard.

The Raman procedure is particularly valuable in fluid-inclusion analyses for the sulfur species present, as SO$_4^{2-}$, HSO$_4^-$, H$_2$S, and HS$^-$ all yield good spectra, and determinations of the sulfur species in fluid inclusions have long been the most difficult and least accurate of all inclusion analyses, even by destructive methods (Roedder et al., 1963). Yet Rosasco and Roedder (1979) found that even after many calibration measurements, the Raman determination of SO$_4^{2-}$ in solution in a fluid inclusion from a porphyry copper deposit could only be stated as 12,000 ± 4000 ppm. They estimated that calibration errors can be reduced to yield an accuracy of $\sim 8\%$ at 10,000 ppm and $\sim 30\%$ at 1000 ppm SO$_4^{2-}$ in fluid inclusions in quartz. Any such calibration is based on the reasonable assumption that no undetected solute is greatly changing the spectra observed (see also Dubessy et al., 1983).

Strengths of the Raman method. Obviously, the major strengths of this method are that it is effective on single small (preferably >5 μm) inclusions and is generally nondestructive. Even if only qualitative data could be obtained, this method should still be considered a major development in fluid-inclusion analysis. In addition, it has potentially very high sensitivity, even on tiny samples. Thus, the sensitivity for SO$_4^{2-}$ could be pushed to <100 ppm routinely, and the reported sensitivity for CO$_3^{2-}$ is ~ 75 ppm. Most important, the method can be used on individual phases of multiphase inclusions. Thus, major CH$_4$ was found and an estimate of its pressure (27±7 atm) was obtained on the "vapor bubble" in a brine inclusion in fluorite from southern Illinois, some N$_2$ and CH$_4$ were found in the liquid CO$_2$ phase in quartz from Brazil, and a separate line was recognized for ^{13}CO$_2$ (Rosasco and Roedder, 1979). The errors involved in

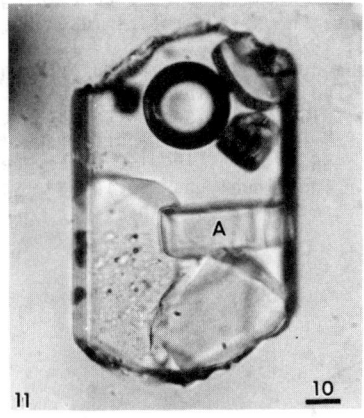

Figure 4-11. Multiphase fluid inclusion in apatite from Cerro de Mercado, Durango, Mexico. Phase (A) was believed to be anhydrite, on the basis of its optical properties; the identification was verified unambiguously by Raman spectroscopy. The other daughter crystals are unknowns. Scale bar in µm. From Rosasco et al. (1975a).

measuring the intensities of these lines are such that, by calibration with an appropriate CO_2 standard, the method could be used to measure a value for $\delta^{13}C$ with a precision of ±20 parts per thousand. Although this precision is very low, the sample used here (~4 x 10^{-9} g) was about five orders of magnitude smaller than the samples used for normal $^{13}C/^{12}C$ determinations by mass spectroscopy (see also Dhamelincourt and Schubnel, 1977), and the method is nondestructive.

Daughter minerals having polynuclear components such as SO_4^{2-} or CO_3^{2-} in their structures also produce strong sharp spectra. Thus, a very large daughter crystal (12 x 40 µm) in an inclusion in apatite that was thought to be anhydrite on the basis of its optical properties was verified as anhydrite by Raman (Fig. 4-11). A much smaller daughter crystal, ~5 µm in size in an inclusion from a porphyry copper deposit (Fig. 15-11), also yielded nine sharp lines in the range 400-1200 cm^{-1}, perfectly matching a reference spectrum of anhydrite. Chalcopyrite also yields a good Raman spectrum. At present, the identification of daughter crystals by Raman spectroscopy is limited by the lack of an adequate library of reference spectra, but even negative data can be useful. Thus, an unknown daughter crystal in inclusions in Colombian emeralds, previously tentatively identified as parisite [(Ce,La)$_2$Ca(CO$_3$)$_3$F$_2$], was shown to have a very different spectrum from that of reference parisite from the same vein and hence is not parisite (Rosasco and Roedder, 1979). Dhamelincourt and Schubnel (1977), and Dele-Dubois et al. (1980) have shown that Raman spectroscopy can be used to identify solid inclusions in diamond, sapphire, and emerald gemstones. Even opaque phases were identified, because reflection of light from a metallic mineral such as pyrite is not from the true surface but involves some penetration of light into the structure. Dele-Dubois et al. (1980) also have shown that the various monoclinic and triclinic pyroxene-group minerals each have rather characteristic spectra. Eventually, it may be possible to analyze qualitatively for ions in solution, such as Ca, by recognition of the Raman spectra of the expected hydrate phases on freezing, such as $CaCl_2 \cdot 6H_2O$, because the water molecules in each such hydrate theoretically should yield different spectra. Dubessy et al. (1982) have given spectra for $NaCl \cdot 2H_2O$, $CaCl_2 \cdot 6H_2O$, $MgCl_2 \cdot 6H_2O$, $MgCl_2 \cdot 12H_2O$, $FeCl_3 \cdot 6H_2O$, and $KCl \cdot MgCl_2 \cdot 6H_2O$ (carnallite). Quantitative analysis would seem to be unlikely by this method, because of the irregular distribution of phases in the frozen inclusion.

By far the most important application of Raman spectroscopy to fluid inclusions has been in the field of gas analyses. Most of these applications have been by a series of French workers and have dealt with inclusions containing CO_2, N_2, H_2S, H_2O, CH_4, and higher hydrocarbons, in various ratios (e.g., Guilhaumou et al., 1978; Dhamelincourt et al., 1979; Bény et al., 1981, 1982; Dubessy et al., 1982; Touray and Guilhaumou, 1984; and particularly an extensive study by Guilhaumou, 1982). These studies have involved quantitative and qualitative analyses of the gases present, plus comparisons with data from freezing studies and the scanty experimental P-V-T data on the appropriate systems, and include some attempts at isotopic analysis of $^{13}C/^{12}C$. The present instruments are such that detection limits for individual gas species are seldom as low as 0.1 mole %, but future improvements are almost certain.

Chapter 5

DESTRUCTIVE METHODS
of DETERMINATION of INCLUSION COMPOSITION

CONTENTS

INTRODUCTION

To be of maximum value, the analysis of any given inclusion should be complete. In order to understand the chemistry of the ore-forming process, for example, we would <u>like</u> to know the concentrations of the major solvents H_2O and sometimes CO_2, the major solute ions Na, K, Ca, Mg, Cl, SO_4 and HCO_3, the minor solute ions Al, Fe, B, Ba, Br, Mn, P, F, and Si, the heavy metals (including very minor constituents such as Au; see Chapter 12), the pH and Eh, the species present for variable valence elements such as S (SO_4^{-2}, H_2S and HS^-) and C (CO_2, CH_4, CO, C_xH_y, H_2CO_3, HCO_3^-), the "organic" compounds if present (methane is usually predominant), the other gaseous species (H_2, He, N_2, O_2, Ar, and even $Hg°$), and the isotopic signatures of the major elements (particularly those of H, C, N, O and S, but also including Ar and He). The common occurrence of several different generations of inclusions in any given sample makes it desirable to be able to determine this composition for <u>single</u> fluid inclusions of <u>known</u> origin, or for groups of contemporaneous inclusions. For most samples, the "shopping list" above is simply wishful thinking in view of the limited capabilities of present-day chemistry, because fluid inclusions are very small. Even for those elements for which an adequately sensitive method is available, extraction of the fluid from the inclusion, without serious loss or contamination, is a major hurdle. The design of truly valid blank runs is also a common source of ambiguity.

Several thousand partial to relatively complete quantitative analyses of aqueous fluid inclusions have been reported in the literature. The different types of geologic environment show grossly different compositional ranges. These inclusion samples have been analyzed by a variety of methods. The concentration of solutes in the inclusion fluid as trapped is generally less than 10 wt %, but may range from more than 50 to practically 0%. The solutes consist of major amounts of the following elements or molecular species: Na, K, Ca, Mg, Cl, and SO_4, smaller amounts of Li, Al, BO_3, PO_4, $HSiO_3$, HCO_3, CO_3, and many others. Many individual elements or ions in this list may predominate, although Na and Cl are generally the most abundant. Free CO_2, both as liquid and gas, is not uncommon and may be dominant. Many of the published analyses have been made for gaseous constituents only.

Many analyses have also been made, by a variety of methods, of the organic gases and liquids that form inclusions in the minerals of some igneous rocks. These show (besides H_2O and CO_2) appreciable amounts of a variety of compounds of high molecular weight, in addition to major amounts of H_2, CO, CH_4 and C_2H_6, but whether the inclusion fluids have an inorganic or an organic origin is still uncertain.

The significance of most of these data is seriously limited by problems of sample selection and extraction procedure. In addition, analytical procedures adequate to provide quantitative data on the very small samples of fluid usually obtained require considerable care to avoid major contamination, and loss, from a variety of sources. Some of the procedures reported (e.g., for pH) cannot fail to give grossly erroneous results except under very limited conditions. Although it might seem extreme, I believe that the possibilities for major errors in inclusion analyses are sufficiently numerous that one should simply discount all analytical reports that do not give details on sample size, and the selection, cleaning, and extraction procedures used, as well as the usual statements of analytical methods, sensitivity, accuracy, precision, blanks, standardization, etc. There is no known panacea for these problems, and because of the many variables involved, it is unlikely that a standardized analytical procedure, suitable for most samples, will be developed.

In much analytical work in geochemical studies, the older methods are completely superceded as new ones become available. This is not true for fluid inclusion analyses. The combination of variability in the physical and chemical properties of host minerals, volumes of actual sample (generally very small), substances of interest in the inclusions, and sources of loss or contamination makes most such studies research projects rather than routine analyses. As a result, the experience of previous workers, even though their qualitative or semiquantitative data were obtained by outmoded analytical instruments or methods, still can provide much food for thought. Far too often, the greatly improved sensitivity and speed of the newer analytical techniques are permitted to obscure the problems of sample selection and preparation which may be much more serious, and which are the prime considerations in evaluating the precision, and most important, the accuracy and significance of any measurements obtained. Most of the recent major breakthroughs in our understanding of geochemical processes through analyses of inclusions are such that improved precision and accuracy in the analyses themselves would not have helped much.

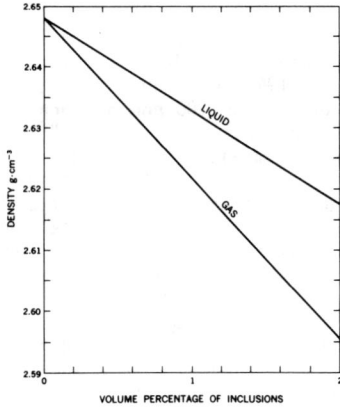

DENSITY g cm⁻³

VOLUME PERCENTAGE OF INCLUSIONS

Figure 5-1. Variation in density of quartz calculated from presence of fluid-filled inclusions containing all liquid solution (density, 1.1 g/cm³) or all gas (density, 0.0 g/cm³). Density of pure quartz assumed to be 2.6484 g/cm³. From Roedder (1958).

SAMPLE SIZE PROBLEM

The amount of inclusion fluid available for quantitative analysis in any given sample is usually disappointingly small, even though the inclusions are exceedingly numerous. Electron microscopy of fractured mineral surfaces has shown the existence of large numbers ($\sim 10^8$/cm²) of very small (>0.02 μm) inclusions on mineral grain boundaries (Sella and Deicha, 1962a,b; 1963). Although very numerous, these inclusions are of such small volume individually that they would amount to only about 0.01 vol % of the rock. Presumably there is no break in the series between these inclusions (~ 0.02 μm in diameter) and single dislocations (~ 0.0002 μm).[1]

In total volume, fluid inclusions seldom comprise more than a few tenths of 1% of the sample, even though they may be exceedingly numerous. Thus, much white quartz and calcite contains $\sim 10^9$ inclusions/cm³, and owe their white color to the inclusions, but as these inclusions average ~ 1 μm in size, the total fluid content is only ~ 0.1 wt % (Fig. 4-1). Samples with >0.1 wt % are relatively rare. Petrichenko (1973) reported some salt samples with as much as 28% fluid, but most minerals contain <0.1% inclusions; 0.01 to 0.001% (i.e., 100 to 10 ppm) is perhaps a good guess for the average range. Zirkel (1873) reported a hauynite with an estimated 3.6×10^{11} inclusions per cubic centimeter, but with an unstated volume percentage. The white appearance of many minerals is due to multiple reflections at the inclusion interfaces, and hence the coefficient of transmission of light can be used as a very rough measure of the inclusion content (Ushakovskii, 1966). "Turbid" feldspars also have large numbers of fluid inclusions (Folk, 1955).

A cubic inclusion 10 μm on an edge contains $\sim 10^{-9}$ g fluid, most of which is generally water; individual major solutes range from 10^{-10} to 10^{-12} g, or less -- far below the sensitivity of most analytical methods even if the entire inclusion is used for a single determination. Quantitative analyses of single, large (~ 1 mm) aqueous inclusions have been made by semimicro-, micro-, or ultra-microchemical methods. More commonly, the fluid from a large number of smaller inclusions is extracted by crushing and leaching to provide enough material for analysis by more conventional methods. In mineral processing operations involving recirculation of processing water, the buildup of salts in solution from inclusions can even affect mineral recovery (Pedan et al., 1978; Kulikov, 1982a).

----

[1] McLaren and Phakey (1966) reported observation of $\sim 10^{14}$ "voids" or "bubbles" per cm³ of white quartz, using transmission electron microscopy. The size ranged from 0.02 to 0.1 μm. At 0.05 μm the quartz would have about 1% voids, the walls of which are presumably wet with liquid water (A. C. McLaren, pers. comm.).

The volatile constituents present in these inclusions, such as H_2O and CO_2, may cause major ambiguities and errors in analytical determinations of these constituents in rocks and minerals (Faber, 1941; Ermakov and Myaz', 1957). Fluid inclusions may also contribute as much as 100 ppm of each of several non-volatile constituents to the total analysis of even "pure" mineral separates and may cause serious errors in the determination of the density of minerals (Piznyur, 1957; Ermakov and Myaz', 1957). Thus Hawes (1881) described a quartz with a high concentration of inclusions that has a density of 2.625 g/cm^3 (vs pure quartz at 2.6484; Fig. 5-1) and Langway (1958) used the density of ice to determine the volume of the numerous gas inclusions present. Density separations have been used to concentrate inclusion-rich grains (Roedder, 1958).

As a result of the irregularities inherent in the processes of trapping of fluid inclusions detailed below, the abundance of inclusions in adjacent samples from any given deposit may vary over many orders of magnitude. In some ore deposits, fluid-inclusion studies have been possible only by the fortuitous discovery of suitable sample material. Furthermore, the sample requirements for individual methods of fluid inclusion study may vary widely, from as little as 10^{-13} g of fluid for a number of the optical methods to 10^{-3} or even 1.0 g for some wet-chemical procedures. Cross-checking one method against another is occasionally possible, but the variations in the nature of the samples, and in the sample requirements for the various methods, may make such comparisons difficult to interpret.

One of the most frustrating aspects of fluid inclusion study is that although large inclusions are much easier to work with, the larger the inclusion, the more likely it is to have leaked, and hence to give spurious data (or none at all). All the largest former fluid inclusions in the world have leaked, because they had polycrystalline rock walls rather than being enclosed within single crystal bottles.

Much of the advertising for new analytical instruments concentrates on the high sensitivity of the method or instrument, usually stated in concentration terms (i.e., ppm or ppb). For fluid inclusions, however, the important parameter is the volume of sample needed to achieve this sensitivity. Knowing the concentration and the volume, one can obtain the actual mass of the element or molecular species that can be detected. When this is calculated, some of the new, highly sensitive methods turn out to be inadequate for the amounts expected in the fluid from many inclusion samples.

QUALITATIVE AND SEMIQUANTITATIVE METHODS

Solids

If a solid daughter mineral can be removed from its inclusion it can be identified by various normal petrographic and chemical techniques, as well as by single-crystal X-ray and electron-microprobe procedures (see section on various microprobes at end of this chapter). Brewster (1823b) presented the first account of loose daughter (?) crystals of calcite extracted from a fluid inclusion in quartz. Minute (~2 μm) single daughter crystals are surprisingly difficult to find in the crushed debris even when they are strongly birefringent and the host is isotropic (Roedder, 1963, p. 175), but Zolensky and Bodnar (1982) described an effective technique (see below).

Once a daughter crystal is isolated from the host, it can be immobilized on a gelatine-coated slide and its indices measured with index liquids (Fairbairn,

1943). If it (or a group of daughter crystals) is large enough, it can be powdered, and a standard x-ray powder-diffraction identification can be obtained, e.g., dawsonite in gold quartz veins (Coveney and Kelly, 1971), ulexite in tin deposits (R. Thomas, 1979) and many others, also in tin deposits (Dolomanova et al., 1976). Ermakov (1965) and Lyakhov (1966) extracted and identified a variety of daughter minerals from Volynian pegmatitic quartz by X-ray diffraction. Lyakhov reported X-ray data on several unknown minerals, and he also verified the presence of halite, sylvite, $FeCl_2 \cdot 2H_2O$ (see also Kalyuzhnyi and Voznyak, 1967), elpasolite [K_2NaAlF_6], cryolite, and a hydrated magnesium carbonate in inclusions in smoky quartz from pegmatites in Volynia.

The Gandolfi X-ray camera (Gandolfi, 1967) permits a powder-diffraction pattern to be obtained from a single very small crystal without crushing. Zolensky and Bodnar (1982) have shown how this can be used to identify a variety of daughter crystals. After completing freezing and heating tests, a tiny "moat" is excavated around the inclusion using a dental diamond drill. The resulting pedestal of host crystal containing the inclusion is broken out and transferred to a crushing stage, cracked open, and the daughter crystal transferred to a fine glass fiber for mounting in the camera. In this way they verified the identification of hematite as the red "daughter mineral" in inclusions from porphyry Cu deposits, and showed the opaque "triangles" to be chalcopyrite. Furthermore, they showed that the larger of the two isotropic cubes in these inclusions (from one locality) was sylvite, rather than halite as is usually determined (or presumed). Graziani (1983) has described several improvements in the extraction procedure. With present X-ray generators, 3- to 6-day exposures are needed to obtain usable patterns, but this can certainly be expected to decrease in the future, as newer generators become available. Thus synchrotron X-radiation has been obtained at 10^6 times the intensity of ordinary X-ray tubes (Weaver and Margaritondo, 1979).

Elaborate and detailed studies of daughter minerals in opened inclusions have been reported by Kalyuzhnyi (1958a, 1960, 1961). A tungsten carbide drill (advanced by the microscope fine focus screw) was used to drill two holes into large inclusions in pegmatitic topaz in a humidified environment, after nondestructive optical tests were complete. The index of refraction of the solution was measured, and the indices of the daughter minerals were obtained by inserting index liquids directly into the drained inclusions. Water and other reagents were added in sequence to determine solubilities and to obtain solutions on which various microchemical and spectrographic tests were performed. The solid crystals which Kalyuzhnyi recognized included halite, sylvite, elpasolite [K_2NaAlF_6], teepleite [$Na_2(BO_2)Cl \cdot 2H_2O$], cryolite [Na_3AlF_6], caracolite [$Na_3Pb_2(SO_4)_3Cl$], quartz, an unnamed new chloride of zinc and aluminum[2]/, and a series of other unidentified phases on which only partial data could be obtained.

In a relatively few inclusions, daughter phases are present as very well-formed crystals, permitting detailed studies. The work of Rankin and Le Bas (1974a) on the identification of nahcolite [$NaHCO_3$] daughter crystals in inclusions in apatite from some East African ijolites and carbonatites, using a combination of microchemical, solubility, crystallographic, and optical determinations is a tour de force in this area.

Infrared absorption has proven very useful in characterizing the various bitumens extracted from the abundant hydrocarbon inclusions in the Khibiny alkalic massif in the USSR (Ikorskii, 1967b, 1968).

2/ This phase is present in small amounts (1-2 vol %) in almost all the inclusions studied (e.g., Fig. 4-7). Its properties are as follows: apparently "rhombic" (orthorhombic?), showing prism, two pyramids, and probably the pinacoid, $\gamma = 1.599\pm0.001$, $\alpha = 1.585\pm0.002$, $+2V = 50°$, colorless, H = 1-2, soluble in water with light-grayish residue, soluble in concentrated H_2SO_4 with effervescence.

Liquids

Sorby (1858, p. 470) was the first to prove without doubt that the liquid in common inclusions was water; he decrepitated inclusions in a closed-end tube, condensed the water vapor with low temperatures and determined the crystal form and melting point of the resulting ice crystals. (He also noted (p. 471) in this same experiment that another substance was given off, which condensed at a higher temperature. This he found to be KCl or NaCl.) Since then many determinations have been made of the H_2O content of fluid inclusions. The quantity of liquid in inclusions can be roughly estimated from weight loss determinations (Ermakov and Myaz', 1957) or from density measurements, as seen in Figure 5-1 (see also Roedder, Ingram, and Hall, 1963, p. 355).

Many papers have been published, particularly in the Soviet Union, in which the amount of inclusion fluid released during the crushing and leaching of a given sample for analysis is estimated from a count of inclusions in a "representative" small sample. However, in addition to the problems of multiplicity of origin of inclusions, in many samples the actual volume, composition, and concentration of the total inclusion fluid in any given portion may be determined almost completely by a small number of erratically distributed larger inclusions.

Some of the earliest studies on the composition of inclusions made use of simple but effective qualitative tests involving evaporation of the inclusion fluids. Thus Nicol (1828) and Newhouse (1932) noted that the evaporation of inclusion fluids (e.g., from Mississippi Valley-type deposits) yielded new crystals, but that the mass stayed wet for days, indicating the presence of a deliquescent substance. I have noted similar behavior for inclusions from Creede, Colorado.[3] Newhouse (1932) made use of the formation of cubes of NaCl from evaporating inclusion fluids on broken mineral surfaces to obtain a crude but good estimate of the salinity of the fluids; he evaporated similar droplets of known salinity to make the test semiquantitative.

Zirkel (1870) described an elegantly simple but extremely sensitive test for Na in fluid inclusions: when a sliver of the mineral decrepitates in a flame, the sudden release of Na-laden steam makes a tiny yellow flash. Perhaps the most commonly obtained qualitative evidence of the composition of inclusion liquids is based on the recognition of solid daughter crystals of NaCl, signifying that the solution is saturated with respect to NaCl. Note, however, that this does not necessarily indicate that the fluid must contain ~26 wt % NaCl at room temperature; the amount of NaCl in solution at saturation is strongly affected by divalent cations such as Ca and Mg, and may be as low as 1-2% NaCl (see Chapter 8).

Zakharchenko (1950) and Skropyshev (1957) have analyzed very large inclusions (up to 1.5 cm^3) for six or seven elements by semiquantitative spectrographic techniques, by simply soaking the porous carbon electrodes in the fluid. Others have made semiquantitative spectrographic measurements on the solids obtained by evaporating leachates from crushed samples. Several investigators have used semi-quantitative spectrographic analyses of samples both with and "without" inclusions to obtain data on the inclusions (for example, Grushkin and Prikhid'ko, 1952; Grushkin, 1958; Skropyshev, 1957). Some authors report crushing samples directly in test reagents for qualitative determinations (Machairas, 1963a; Saitô, 1951). Autoradiography has been used to test for radioactive elements in inclusions (Huntley, 1955; Picciotto, 1950), but with ambiguity from possible

----

[3] The taste test has also been reported (Buerger, 1932; Newhouse, 1932). Its results present data on the chemical characterization of the ore-forming fluids that are only qualitative but vivid and memorable. I found that large (~1.0 mm) primary fluid inclusions in sphalerite from Creede, Colorado, USA, which were accidentally exposed on a new cleavage surface, had a salty but slightly astringent or bitter taste corresponding to an NaCl solution with some Ca or Mg salts.

solid inclusions, contamination from cerium oxide polishing agents, etc.

Loskutov (1962) showed that the water present in inclusions in natrolite could move through the open structure of this zeolite and evaporate during months of museum storage, leaving behind crystals of several salts. He records crystals of hydrous sodium ammonium (?) carbonate, nahcolite ($NaHCO_3$), acid phosphate of sodium and ammonium (stercorite?), dehydrated sodium carbonate (?), and sodium phosphate (?). On occasion, when an aqueous inclusion leaks during a heating run, it will leave the solutes in the cavity. Sawkins (1977) recognized the presence of alkali chloride cubes in leaking inclusions that originally had no daughter crystals, and Roedder and Belkin (1980b) noted birefringent crystals forming in inclusions in salt that leaked, indicating the presence of solutes other than NaCl.

X-ray diffraction can be used to identify the phases in the residues from evaporation of water leaches. Thus Bergman and Blankenburg (1964) reported the identification of the NaCl cubes and solutions in inclusions in Brazilian quartz crystals, and Lamar and Shrode (1953) used X-ray diffraction to identify $CaSO_4$, $CaCO_3$, $MgSO_4$, NaCl, KCl, $MgCl_2$, and basic magnesium carbonate in the solids from evaporation of leachates from ball-milled limestones and dolomites. Dimov et al. (1980) applied electron diffraction to the residues from evaporation of aqueous extracts (see below, under "Transmission electron microscope").

The microscope heating-pressure stage of Ypma (1965) permits crushing an inclusion under external hydrostatic pressure, and hence a determination of the vapor pressure of the fluid in inclusions, (i.e., the pressure in the vapor bubble), at various temperatures. This, in turn, provides an independent evaluation of the salinity. By crushing a series of cogenetic inclusions in a sample at a given temperature but varying external pressure, the pressure at which the vapor bubble neither expands nor contracts on release can be determined.

Above 100°C, the water in inclusions will also vaporize when the inclusions are opened. Deicha (1952a) described an effective visual test for such gas evolution. He used a "visual decrepitometer" with transparent oil as the heating medium so that the tiny bubbles emitted could be seen. Hence, the procedure could be used to determine the temperature at which saline fluid inclusions have vapor pressures >1 atm.

Gases

Methods based on expansion on release. Many different qualitative tests can be used to determine something about the gases in inclusions. The pressure can be estimated by the simple expedient of opening the inclusions while the sample is immersed in a fluid. This was first reported by Davy (1822), and was used to determine the gas pressure in a large vesicle in a tektite (Rost, 1964). The vapor bubbles in many inclusions are almost solely low-pressure water vapor (i.e., ~20 mm pressure at room temperature), as they collapse instantaneously and completely when exposed to atmospheric pressure. Crushing can be done simply and quickly by pressing on the cover glass while watching the grain with the microscope, but the crushing stage provides better control (see Chapter 7).

Although obviously only qualitative or at best semiquantitative, the crushing test is very useful, as it is exceedingly sensitive. As little as 10^{-14} g of a relatively nonsoluble, noncondensable gas (less than a billion molecules) may be detected in this manner, as it will form an easily visible gas bubble, several micrometers in diameter at atmospheric pressure and temperature. (In this context, "noncondensable" means not condensed by 1 atm pressure at room temperature.) Some compositional data can also be obtained by crushing in selected solvents (see Chapter 7). The volume expansion of highly compressed gases, such as liquified CO_2 (at more than 70 bars), provides useful constraints on the composition of the inclusion.

Chemical reactions of the evolved gas with the fluid in which the crushing is performed can be observed with the microscope. Thus Rasumny (1960) was able to identify the CO_2 in very small (5 μm) inclusions by reaction with a $Ba(OH)_2$ solution, and G.R. Helz (pers. comm.) was able to determine the presence of H_2S in inclusions by reaction of the gas with an anhydrous glycerol solution of sodium nitroprusside. A green precipitate formed in 1 minute from a 100 μm bubble containing 20 vol % H_2S.

The gas pressure in the numerous gas inclusions in ice have been determined by opening under a fluid under a controlled, externally applied hydrostatic pressure. Scholander and Nutt (1960) and Nutt (1961) used a solvent for ice (glycerine) as the pressure medium and noted the pressure needed to just balance that in the inclusions at the moment that the slowly advancing solution front first intersected long tubular inclusions. With this method they found that the gas inclusions in Greenland icebergs were under pressures <20 bars. A less precise method was introduced by Hamberg (1895), in which the density of the ice containing inclusions and the volume of air released on melting were used to obtain the air pressure in the inclusions. Similar procedures could be applied to the high pressure gas inclusions in salt.

Ypma (pers. comm., 1965) has shown that several properties of the gas evolved on crushing may be determined with his heating-pressure stage. Thus CO_2, evolved during crushing of inclusions in a Brazilian quartz crystal (part of this sample is illustrated in Fig. 3-2) could be liquefied and revaporized by appropriate adjustments of pressure or temperature. As the bubble gradually became smaller, through solution in the pressure medium (glycerol), the pressure needed for liquefaction increased to more than 80 kg/cm^2, greater than the critical pressure of CO_2 (75.3 kg/cm^2). This is believed to be due to differential solution in the glycerol of the gaseous species present, presumably increasing the concentration of minor constituents.

Still another elegantly simple technique was described and used by Vogelsang and Geissler (1869). They decrepitated a sample in an evacuated tube containing electrodes; the presence of CO_2 was proved by the spectra observed upon excitation of the gas.

Use of odor. If strong-smelling substances are present in inclusions, the odor on crushing can be used as a crude but effective test. Very small quantities of volatile hydrocarbons in some inclusions, such as those in fluorite from the southern Illinois deposits (Fig. 3-10), yield a strong odor of petroleum. A separate hydrocarbon phase is not necessary to provide such odors; Price (1976, 1981b) showed that hydrocarbon solubilities in aqueous phases were surprisingly high.

An exceedingly small amount of H_2S may be detected by its odor. If we assume an average minimum level of detectability of 0.025 ppm (Patty, 1962) and a "sniff volume" of 10 ml, the release of ~10^{-10} g of H_2S should be detectable with the nose. Wright (1881) detected an "unmistakable" odor of H_2S from inclusions in a pegmatitic quartz from Branchville, Connecticut, USA, but the volume of H_2S was too small to measure with his techniques. Some limestones yield a fetid odor on scratching or crushing, but this odor is rather rare in well-crystallized minerals. Notable exceptions are the finding of liquid H_2S in coarse marbles of the Grenville Series (Harrington, 1905), in similar marbles adjacent to a Pb-Zn deposit associated with the Bingham porphyry Cu deposit in Utah (Roedder, 1971b), and an H_2S-bearing phase in a Brazilian quartz crystal (Fig. 3-3).

Dons (1956) noted that an odor of H_2S was evolved from a sedimentary barite crystal when it was rubbed, but not when it was decrepitated, perhaps due to disproportionation of polysulfide ion to form HS^{1-} and SO_4^{2-} at room temperature

(Cloke, 1963). Sphalerite sometimes releases a slight odor of H_2S on breaking, but Hosking and Spry (1955) have shown that sphalerite is the one common sulfide that emits H_2S on being scratched with an iron knife.

Some fluorite that has been subjected to α-particle bombardment emits a strongly pungent but sweetish odor on crushing. This was originally attributed to free fluorine or to ozone liberated by reaction of fluorine with water (Becquerel and Moissan, 1890; Sine, 1925). Heinrich and Anderson (1965) report that the gases evolved on crushing a fetid fluorite-bearing carbonatite in a mass spectrometer were a mixture of C_5 and C_6 hydrocarbons (possibly fluorinated?), F_2, HF, and F_2O, and similarly Kranz (1966) found a variety of fluorinated hydrocarbons in gas inclusions in uraniferous "Stinkspat" fluorite.

QUANTITATIVE METHODS

Release, Extraction and Analysis of Gases

Destructive analysis of the gases in fluid inclusions involves three steps: the release of the gases from the inclusions, the complete extraction of the released gases from the host, and detection of the composition and amount of these gases. The many different combinations of methods that have been used for these three steps make a systematic survey impractical. One generalization is evident -- no single combination of methods can be considered to be optimum. In some methods, the limitations and errors are large, and the resulting data are not always comparable. Many of the analyses for gases from inclusions are reported in the literature as though the gases came only from pure gas inclusions, but most of these probably came from the evaporation of inclusion liquids such as H_2O or CO_2, from gases dissolved in these liquids, or from various sources other than inclusions, including contamination. Freund (1982) and Freund et al. (1981, 1983) have added still another possible complication, "atomic" carbon, in igneous melts and minerals.

Methods of release and extraction of gases. The gases in inclusions (including H_2O) may be released for analysis by various mechanical procedures, or by heating. Karpinskii (1880) proved the presence of high-pressure CO_2 in inclusions by crushing under mercury and absorbing the evolved gas in a barium hydroxide solution. Kalyuzhnyi (1955a) drilled into inclusions under dehydrated glycerol; the evolved bubbles of gas were trapped for measurement as they rose in the fluid. Pfaff (1871) crushed in a stream of inert gas, and Khitarov et al. (1958) drilled into large inclusions in calcite and measured the volume of water by evaporation in a stream of dry air, which was then passed through a weighed absorption tube. A mercury seal around the drill made it possible to open the inclusions directly in the air stream. Maslova (1961) crushed samples in dehydrated glycerol and transferred the evolved gas bubbles for analysis by means of a piston ultramicropipet. Mikhaylova et al. (1973) first used a laser to open inclusions for gas analysis. Kotra and Gibson (1982, 1983) released gases from individual inclusions by a pulsed laser into a stream of He, for subsequent gas chromatography.

Detection by weight, volume, pressure, or absorption. The released gases may be detected and analyzed by any of a variety of methods. Thus sample weight loss, or weight gain of an absorption tube have been widely used. The volumes of gases that are not readily condensed can be measured by standard gas handling procedures. Condensable gas such as H_2O vapor can be converted to H_2 with calcium hydride (Savel'yeva and Naumov, 1979), or to acetylene with calcium carbide (Andrawes and Gibson, 1979). The pressure of the released gases has also been used, sometimes with the aid of cryogenic procedures to separate specific gases (see subsequent discussion of Wahler, 1956, and Harris, 1981a,b,c). A novel ultramicrochemical adaptation of selective absorption procedures described by Dolgov and Shugurova (1965) is discussed below.

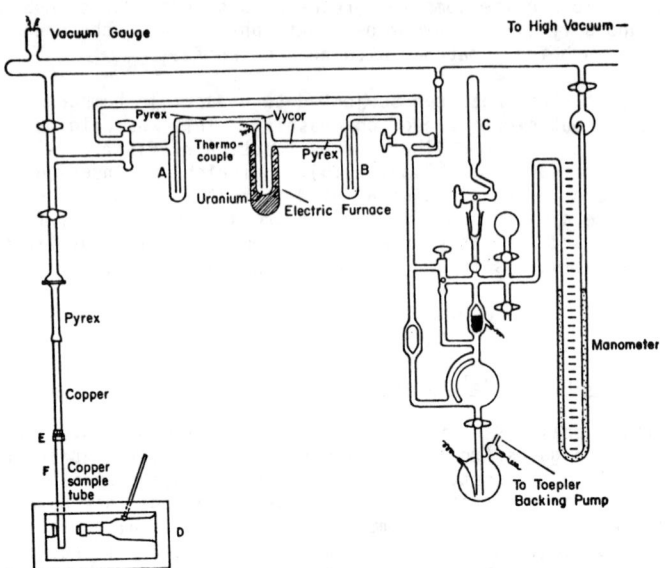

Figure 5-2. Vacuum line for recovery of gases from fluid inclusions, as used by Roedder et al. (1963). See text for details.

Electrical measurements can be very precise, and hence have been adapted to several of these problems. Khitarov and Vovk (1963) and Khitarov (1965b) report using an electrical method, applicable in the submilligram range, for CO_2 evolved from inclusions; they measured the change in the electrical conductivity of a $Ba(OH)_2$ solution as the CO_2 is absorbed and $BaCO_3$ is precipitated. H_2O has also been determined electrically (see subsequent discussion of Cremer et al., 1972).

Kramer (1965) used weight loss to determine the volume of H_2O in inclusions in salt, prior to analysis of the solutes. He weighed selected fragments on a microbalance, after drying to constant weight[4], cracked open the inclusions and evaporated the H_2O, and reweighed the fragments; the entire operation was done in a dry box.

Various vacuum techniques have also been applied to determine the amount of H_2O (and other gases) in inclusions. Barker (1965b) described an effective vacuum crushing device, and Elinson (1968) described a device for crushing in an inert gas stream, and associated gas-handling equipment.

Roedder et al. (1963) and Hall and Friedman (1963) placed samples containing large inclusions (i.e., >~1 mg) in a collapsible metal tube (copper or stainless steel) and squeezed the tube from the outside with a hydraulic press, thus breaking open some of the inclusions (Fig. 5-2; details below, under "Determination of isotopic ratios"). Suess (1951) crushed tektites containing gas vesicles in a sealed tube to determine their gas pressure; he measured the density of the samples before and after crushing to obtain the volume of vesicles. Somewhat similar procedures have been used to obtain the volume of the liquid phase in inclusions (Roedder et al., 1963, p. 389) and the pressure in gas inclusions in ice (Hamberg, 1895).

----

[4] Although the hot plate used was at 125°C (Kramer, 1965, p. 939), the sample temperatures were probably <80°C (pers. comm., 1965).

Vacuum ball milling has been used frequently to open inclusions for gas analyses. Elinson (1949) used a metal ball mill connected to a vacuum line but had considerable difficulty with the rotating seal. Elinson (1956) adopted a simpler technique of grinding in vacuo, with subsequent pumping off and collection of the evolved gases, and made a series of analyses with it (Elinson and Polykovskii, 1961a, 1961b, and 1963). Umova et al. (1957, 1960) used a similar technique. Although not always stated, apparently steel grinding balls were used in all of these studies.

Goguel (1963, 1964) showed that the large quantity of N_2 found in most of the analyses of gases from ball milling was a contaminant evolved from the steel grinding balls as they were abraded in the mill. N is present in igneous rocks mainly in the form of NH_4^+ (Wlotzka, 1961; Stevenson, 1962). It is not known how NH_4^+ will behave on ball milling, but Kranz (1968) obtained many different N compounds, including ammonia, amines, and nitriles on analysis of the gases released on vacuum crushing feldspar. Goguel also showed that on grinding carbonate minerals containing no inclusions, considerable amounts of CO_2 may be formed, particularly if silica is present. To minimize or avoid these difficulties, he devised a vacuum microball-milling technique, using a silica glass mill and tungsten carbide balls on <1 g samples of <2 mm grain size; this was followed by gas-analysis procedures similar to those described by Wahler (1956). Piperov et al. (1977) used a rotatable needle in a vacuum chamber to puncture large inclusions in galena. The water-insoluble gases present in salt samples are very easily extracted by dissolving the sample in water (e.g., the popping salt of Wieliczka, Poland -- see Dumas (1830) and Rose (1839)). Meteorites and rocks have also been dissolved in acids to obtain the gases from them, although much of this gas may not be present as discrete inclusions.

Most analyses have involved release of gas by heating, i.e., decrepitation, in an inert gas stream followed by absorption (Khitarov and Rengarten, 1956; Roedder, 1958, p. 263-266; and Rutherford, 1963), in vacuum (Wright, 1881), or by simple weight loss determinations upon heating. Using this last method, Sorby (1858) found up to 0.4 wt % H_2O in quartz from Cornish granites. Kormushin and Darbadaev (1978) let the gases from decrepitation expand into a rubber bulb in a metal cylinder, then compressed the gases with external pressure on the bulb. Kokubu et al. (1961) heated samples to 1100°C in low-pressure O_2, and then pumped off the gases to obtain water samples for mass-spectrometric determination of the deuterium/hydrogen (D/H) ratios.

Still other methods of extraction of gas inclusions, appropriate to certain samples only, involve complete fusion or solution. Hoy et al. (1962) fused a salt sample containing gas inclusions to obtain a gas sample for analysis. Godbeer and Wilkins (1977) determined total H in quartz by fusing with lithium metaborate flux in pure O_2, followed by adsorption on magnesium perchlorate. Analysis for Ar usually involves fusing of the samples, either with or without a flux, in a vacuum system (Lippolt and Gentner, 1963; Rama et al., 1965).

The interpretation of gas analyses requires considerable care, in terms of what gas was actually sampled, how much it was changed by the sampling, and the accuracy of the analysis itself. Chamberlin (1908, p. 39-40) used a vacuum crushing device to prove that only a very small part of the gases he obtained by heating rock powders in a vacuum came from fluid inclusions. Thus he obtained 0.81 volumes of gas per volume of rock from a quartz sample by heating (his analysis, no. 71) but no measurable gas was evolved upon crushing a portion of the same material (p. 40).

Wahler (1956) investigated the various sources of error in gas analyses from rocks and minerals, and devised an elaborate technique for obtaining and analyzing the gases from selected large inclusions, which were evolved by decrepitation upon rapid heating. Entirely apart from analytical problems such as losses by absorption on the walls of the apparatus, he found many problems inherent in the use of heat to release the gases. For example, diffusion of H_2 from burner gases through the apparatus walls was a serious source of contamination in some experiments reported in the literature. Wahler also presented evidence, mainly from the literature, that much, but not all, of the CO, H_2, CH_4, O_2 and H_2S found by earlier workers by heating rocks and minerals came from various chemical or catalytic reactions of original H_2O and CO_2 (and also possibly organic matter) with each other and with the mineral surfaces present, yielding a new assemblage of gases. A cogent argument in this direction is given by the early work of Travers (1898), who compared the gases evolved upon heating various samples (H_2, CO, and CO_2), with the gases evolved upon solution in acids (CO_2 only). After extensive study of the similar problem of volcanic gas analyses, Shepherd (1938) concluded that it is impossible to relate conclusively the gases found by analysis of volcanic gas, or the gases from heated rocks and lavas, to those originally present. Several studies have shown that the amounts of H_2 (and CO) in the evolved gases increase with the temperature used to extract the gases.

Various investigators using vacuum techniques for inclusion studies have assumed, tacitly or expressly, that the expansion of the fluid inclusion contents into the vacuum upon crushing or decrepitation is so close to "instantaneous" that there is "little or no chance" for changes to occur in the gases, and hence that the species subsequently detected in the vacuum system by one or another method must have been present in the inclusions. The mean velocity of gas molecules is so high that an individual molecule has $\sim 10^8$ collisions per second, so the expansion of the contents of an opened inclusion into a vacuum system is so far from "instantaneous" that any given molecule may have many collisions with the fresh mineral and instrument surfaces (and with other molecules) during the expansion. As a result, absorption on such surfaces can and does take place, and, in addition, at least partial reequilibration toward the species assemblage stable at lower pressures may well occur. In ball milling, particularly, absorption of gases on the large amount of new sample surface can be an important source of error. Thus Kunkel (1950) found that well-cleaned quartz surfaces normally have one or two layers of H_2O molecules that are not released by heating under 500°C in vacuum, and absorption of H_2O on olivine and pyroxene is even less reversible (Nelson and Vey, 1968). Ware and Pirooz (1967) have shown that on breaking open gas bubbles in glass in vacuum, absorption on the new surface causes significant losses of CO_2, SO_2, and H_2O, and even inert' gases such as N_2 and Ar are absorbed and held tenaciously (Khodakov, 1966). Price et al. (1977) showed that whether an obsidian sample is crushed or not exerts a "considerable influence" over the gas composition obtained on vacuum extraction at 1000°C.

Barker and Torkelson (1975) made quantitative studies of the problem of adsorption of gases on new surfaces. They found by mass spectrometric analysis of eight gases, obtained while crushing in vacuum, that adsorption of the various gases differed drastically (Fig. 5-3). Thus both the total amounts and the composition of the gases obtained by vacuum crushing are changed. Similar conclusions were drawn by Barker and Sommer (1973) and by Ikorskii and Evetskaya (1975).

Goguel (1963) also presented good evidence that much of the H_2 and He present in many minerals is lost by diffusion in one year at room temperature, once the mineral is crushed. Ypma (1969) showed that H_2 diffuses out of the quartz structure at 500°C even during the analysis (see also Kovalishin, 1968). Conversely, diffusion into sample surfaces prior to analysis can contaminate samples with gases (for example, Reynolds, 1960).

In addition to the serious problems of extraction of gases detailed above, the analytical procedures for use on small amounts of gas mixtures are notoriously poor. Absorption by solid or liquid reagents, combined with barometry or

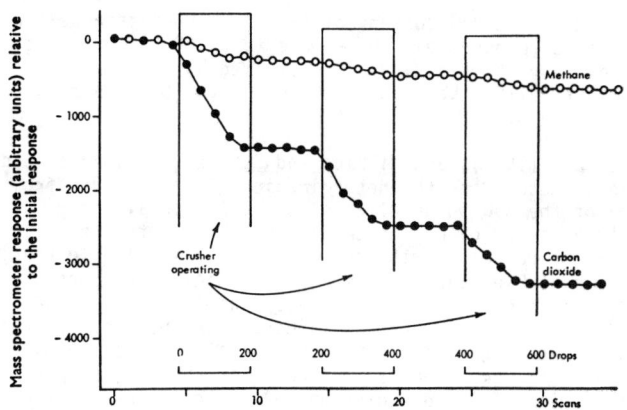

Figure 5-3. Change in partial pressure of CO_2 and CH_4 by adsorption as new surfaces are generated during the crushing quartz. Barker and Torkelson (1975, their Figure 2, p. 214).

volumetry, is the basis of many of the analytical procedures, but it is seldom truly specific.

Dolgov and Shugurova (1965, 1966a,b and pers. comm., 1967) claimed to be able to extract and analyze, to two- or three-place accuracy, for nine constituents (CO_2, NO, O_2, CO, H_2, CH_4, total hydrocarbons, N_2 plus rare gases, and sum of SO_2, NH_3, Cl, H_2S and F) by selective absorption techniques on gases from inclusions as small as 10 μm or even 1 μm. If we assume liquified gas at a density of 1 g/cm^3, this would correspond to a total sample of only 10^{-9} or 10^{-12} g. Dolgov and Shugurova presented experimental data on the analysis of bubbles of known gas mixtures in castor oil or anhydrous glycerol, using absorbers such as KOH and $Cd(C_2H_3O_2)_2 \cdot 2H_2O$ and some ingenious transfer techniques. The bubbles of known gas mixtures that were analyzed, however, were one to six orders of magnitude larger than the unknowns, as they ranged from 1.0 to 0.3 mm in diameter, corresponding to approximately 10^{-6} to 10^{-8} g gas (see also Shugurova, 1968, and Sobolev et al., 1970). Dolgov (1968) reported appreciable fractionation of gases on gradual release from a punctured inclusion.

Dolgov and Shugurova (1968) presented a large number of gas analyses on a wide variety of minerals by these methods, and have also provided gas-analysis data for numerous papers published by other Russian authors. But every gas is soluble to some extent in every absorbent (Grishina, 1979, has tried to minimize this problem); this solubility, combined with the rapidly rising internal pressure in the bubble (and hence rising solubility) as its radius decreases (see Chapter 7 on the use of the crushing stage) would seem to pose an insurmountable problem.

Wahler (1956) gave rather complete experimental details on the method he devised for analysis of inclusion gases. His method for analysis of gas mixtures, based essentially on fractionation by freezing out various constituents at specific temperatures (plus other manipulations), with pressure measurements before and after each step, can yield analyses accurate to 1-2% on as little as 1 mm^3 of gas.

The handling and determination of small quantities of H_2O is particularly difficult, as it adsorbs onto practically all surfaces. For this reason it was converted into H_2 with hot U metal by Roedder et al. (1963) and Knauth and Kumar (1981). Suzuoki et al. (1975) used this method to determine the small amounts of H_2O (0.6 to 4.7 ppm) in peridotite olivine (sum of H_2O in CO_2 inclusions and silicate melt inclusions). Hite et al. (1979) have used a novel method, still under development, involving solution in methanol followed by Karl Fischer titration, to obtain analyses of H_2O in inclusions in salt. Knauth and Kumar

(1981) have studied the H_2O content of salt from Louisiana salt domes, also using another novel procedure. The sample (3 to 50 g) is heated under high vacuum to complete volatilization. The evolved H_2O is converted to H_2 and measured manometrically. Less than 2×10^{-5} g can be measured, corresponding to ~1 ppm in the sample.

Cremer et al. (1972), and Wilkins and Sabine (1973) described a sensitive and effective procedure for the determination of H_2O evolved from samples, based on adsorption of the H_2O in an electrolytic cell containing P_2O_5. The samples were heated in a stream of N_2, which carried the H_2O to the cell. Cremer et al. (1972) were mainly concerned with the analysis of samples in the 10^{-4} g H_2O range, and the possible sources of systematic bias in the method. Wilkins and Sabine (1973) reported that the sensitivity was $1-2 \times 10^{-6}$ g H_2O, or 1-2% of the total H_2O content, whichever was larger. They pointed out that the H_2O content of the carrier gas is far from trivial; it ranged below 7 ppm, but varied from tank to tank. The major finding of this paper, pertinent to many quantitative studies of fluid-inclusion H_2O in rocks, was that a series of nominally anhydrous silicates such as kyanite, andalusite, andradite, pyrope, diopside, rhodonite, andesine, adularia, and olivine, all have OH contents (as weight percent H_2O, but specifically excluding H_2O as liquid inclusions) of at least 0.008, and up to 2.29 for grossular. An H_2O content of 0.008% may seem insignificant, but amounts to 2×10^5 tonnes of H_2O in every cubic kilometer of such rocks. Perhaps even more significant to fluid inclusion studies was that these values, obtained by IR absorption spectra, were verified by observing the changes on replacement of hydrogen with deuterium (D) after treatment with 100 bars D_2O pressure at 750°C for <5 days. Hydrogen diffusion and isotopic exchange through these minerals (except olivine) was surprisingly rapid (see Chapter 12).

An entirely different approach used by Harris (1981a,b,c) for the analysis of gases evolved from samples on heating involves the use of cryopumping and low-temperature vapor-pressure measurements with a capacitance manometer. The method was specifically designed for the analysis of H_2O, CO_2, and SO_2 in silicate glass inclusions. Quantitative analyses of 1 to 5×10^{-8} g H_2O or CO_2 were reported, with blanks (and hence detection limits) of about 10^{-8} g for H_2O and 5×10^{-9} g each for CO_2 and SO_2. Analyses of CO_2 in several basaltic glasses by this method do not agree with previously reported analyses by mass spectrograph (Muenow et al., 1979) in part because the samples run were not true duplicates, and in part for other reasons (see Chapter 16).

Detection by mass spectrometry. The mass spectrometer is used in two different ways in the analysis of gases in inclusions. It is used to analyze the gases present in mixtures (discussed below) and to analyze the isotopic ratios of elements in these gases (discussed in a later section). It is partic- ularly effective in the analysis of the noble gases (Herzog et al., 1962). Its first successful use for inclusions was in the analysis of the gases present in very small bubbles in synthetic glass containing only a fraction of a microliter of a gas mixture (Todd, 1956; Wosinski and Kearney, 1966), but only in recent years has much been done with mass spectrometry of natural fluid inclusions. Mass spectrometry has also been suggested as an exploration tool for mineral deposits, based simply on the analysis of the gases evolved from heating samples (Norman, 1981, 1983; Palin and Norman, 1982; Clifton, 1983a,b). Quantitative determination of the original species present requires accurate knowledge of the instrumental fragmentation patterns, which may vary with the wide range in sam- ple density and composition encountered in such work.

A variety of instruments have been used, mostly conventional arcuate tra- jectory mass spectrometers, but also cycloidal and omegatron trajectory (Bratus' et al., 1968), time-of-flight (TOF; Heinrich and Anderson, 1965) quadrupole (Gibson and Johnson, 1972; Gibson, 1973) and tandem (i.e., MS/MS) mass spectro- meters (Maugh, 1980). Preparation techniques (all in vacuum) include cleavage or crushing at room temperature, high-temperature decrepitation or incremental heating (Zimmermann, 1966; Touray and Lantelme, 1966; Dolomanova and Nosik, 1977), and thermogravimetric analysis (Gibson, 1973). Incremental heating may affect isotopic ratios because of differences in diffusion rates. The gases may be pretreated by cryogenic separation and/or gas chromatography (Kranz, 1966, 1968) to effect separation of individual gases or groups of gases and reduce ambiguity in interpretation of the mass spectra. Mass spectrometry is most appropriate for analysis of volatile compounds in liquid inclusions but has also been used to determine the volatile contents of glass inclusions (Sommer, 1977; see Chapter 16).

Great care should be used in the interpretation of mass-spectrometric analyses of inclusions, particularly for other than noble gases. It is easy to let the extreme sensitivity blind one to the gross inaccuracies that can creep into such results from numerous sources. When working with such small samples, contamination becomes a major problem. Thus H_2 is given off by many metals used in grinding inclusion samples, yet on opening individual inclusions, H_2 is seldom found (Kalyuzhnyi and Svoren, 1978). The most serious limitation other than contamination on all such mass spectrometric gas analyses is that of gas fractionation by absorption (and reaction) on the various surfaces exposed. It is also difficult to design meaningful blank determinations. Such problems become acute in the sample-size range involved in inclusion studies. Unless high-resolution mass spectrometry is used (e.g., see discussion of gases in diamonds, Chapter 17), ambiguity exists due to superposition of isobars (molecules of similar mass) such as $^{12}C^{16}O$ and $^{14}N^{14}N$, and $^{14}N^{16}O^{16}O$ and $^{13}C^{17}O^{16}O$.

Barker (1965a, 1965b, 1966) used both heating to 400°C and crushing in vacuo to release the gases. By using duplicate samples of three fluorites (1965b) he found that all of the CO and most of the H_2 and CH_4 released upon heating to 400°C were apparently formed by reaction during the extraction process. The mass spectra of the hydrocarbon gases evolved on crushing also differed greatly from those evolved on heating to 400°C (Barker, 1966). Chaigneau (1967) also reported considerable differences in the gases emitted from quartz on heating or crushing in a mass spectrometer.

Karasev (1958) found evolution of gas and emission of electrons when he broke quartz crystals in a high vacuum, but he got neither on breaking glass or fused quartz. Vanderslice and Whetten (1962) cleaved natural halite from Baden, and synthetic melt-grown crystals, in a mass spectrometer under high vacuum (10^{-8} mm). They obtained bursts of gas, mainly H_2O, which they attribute to very small amounts of individual H_2O molecules in the structure, rather than to fluid inclusions. Ryan et al. (1968) got similar bursts of gas on cleaving silicates in ultrahigh vacuum, but they reported evidence that the gases were released from the walls of the chamber and the pump, rather than from the samples. Mercer (1967; see also Goldsztaub et al., 1966) found much larger amounts of gas (10^{13} to 10^{14} molecules/cm²) were released on cleaving muscovite in an ultrahigh vacuum. The gas was mainly N_2, with minor H_2, which he believed was trapped between the silicate layers. Heinrich and Anderson (1965) analyzed the gases evolved from crushed samples of a fetid carbonatite and verified the presence of F and several F compounds.

Kranz (1966, 1968) reported the presence of a large number of fluorinated hydrocarbons in the gases released on crushing U-bearing fluorite in vacuum, and gave full details on the extraction and analysis procedures. These studies were extended by Vochten et al. (1977), who used high-resolution mass spectrometry on gases released during crushing similar samples of U-bearing fluorite. Instead of fluorinated hydrocarbons, they found a series of gaseous compounds of S with F, O, or H, including SO, SO_2, SOF, SOF_2, H_2S, HS, S_2F_2, S_2F, S_2, and SF, and present no explanation of the difference.

Mass spectrometry has been widely used in the petroleum industry. In the first use of this technology for the analysis of inclusions, Murray (1957) reported quantitative analyses of fifteen hydrocarbons, plus N_2 and CO_2 but no

H₂O, in hydrocarbon inclusions in quartz. Zimmermann et al. (1979) found more "organic" compounds on heating quartz (to 400°C) than appeared to be present in the fluid inclusions, based on optical examination. They concluded that these compounds were present as submicroscopic inclusions. Many other studies have shown abundant hydrocarbons in the inclusion fluids, either in aqueous solution or as a separate gas and/or liquid phase. CH_4 is most common, and may be a major constituent. Murray (1957) found 64.9 mole % CH_4 in liquid/vapor inclusions (homogenizing in the vapor at 100±5°C), and Touray and Sagon (1967) analyzed CH_4-rich inclusions in quartz that homogenized even higher (~190°C). Preisinger and Huber (1964; see also Arming and Preisinger, 1968) heated very tiny (0.1 mm) grains of feldspars, from several parts of zoned crystals, in an ultra-high-vacuum cycloidal mass spectrometer and determined large differences in the contents of H_2, CH_4, N_2, CO, CO_2, and Ar, between the core and rim. The analyses were made on a total volume of 10^{-8} to 10^{-9} cm^3 of gas, as they report finding 0.8 to 3.3 mm^3 gas/g mineral. Ohmoto (1968) reported extensive studies of the gases in inclusions from the Bluebell mine in British Columbia, made by mass spectrometer on samples released by crushing or decrepitation. He analyzed H_2O, CO_2, H_2, CH_4, N_2, CO, C_2H_6, and C_3H_8, but was unable to detect H_2S. Bratus' et al. (1968) analyzed the gases from individual inclusions (in the mass range 10^{-6} to 10^{-9} g) from pegmatitic quartz. Six constituents, H_2, CH_4, H_2O, N_2, Ar, and CO_2 were determined. Touray (1968) found mass spectrometry of gases released on heating or crushing particularly useful for obtaining the H_2O/CO_2 ratios. Another major recent application of standard mass spectrometry (and an offshoot, the ion microprobe) has been to determine the volatile contents of volcanic samples, in both bulk glass and glass inclusions (discussed in Chapter 16).

Detection by gas chromatography. Gas chromatography is an alternative to mass spectrometry. The first usage on inclusion-type samples was in the analysis of small gas bubbles in glass (Bryan and Neerman, 1962; Helzel, 1969). It is especially appropriate for complex hydrocarbon mixtures (Zimmerman and Poty, 1970; Kvenvolden and Roedder, 1971; Touret, 1976; Ypma, 1979a), and can even be used on normally nonvolatile species (Giddings et al., 1968). However, gas chromatography is more complicated in that two columns are necessary for effective separation of permanent gases (e.g., CO_2, CH_4) and H_2O (Clark and Cable, 1967), and two detectors, (normally thermal conductivity and flame ionization types), may be needed. The detectors are only effective over a limited range of concentrations so careful control of the sample size is required.

Recent improvements in gas-chromatography techniques for small samples, such as fluid inclusions, have been reported by Andrawes and Gibson (1979). The method combines crushing in a stream of He, plus multiple chromatographic columns and the use of the extremely sensitive He ionization detector. The sample is crushed in a relatively simple but effectively designed crusher (Fig. 5-4) that greatly minimizes many of the common problems of high blanks, contamination from air or outgassing of O-rings or the metal of the crusher, adequate transport of evolved gases to the columns, etc. H_2O is converted to C_2H_2 (acetylene) before injection into the multiple chromatographic columns. The reliability of the highly sensitive but somewhat erratic He ionization detector was found to be greatly improved by the addition of 1-5 ppm H_2 to the otherwise very pure He carrier gas with a diffusion chamber (Fig. 5-5) and by other procedures (Andrawes and Gibson, 1978). The detection limits were found to be in the range 0.03 - 0.05 ng for H_2, CH_4, CO_2, C_2H_2, C_2H_4, C_2H_6, and C_3H_8, and 0.4 - 0.7 ng for N_2, O_2, CO, and Ar. Individual inclusions were opened by a pulsed laser and similarly analyzed by gas chromatography (Kotra and Gibson, 1982, 1983), but it seems likely that the extreme (and variable) heating from this procedure will result in major changes in the molecular species present.

When rocks are heated to obtain gases for analysis, much or even most of the organic gases are probably from the pyrolysis of solid organic phases (e.g., Gibson, 1973; Gibson and Johnson, 1972). The nature of the gases obtained is controlled in large part by the pyrolysis conditions (Barker, 1978). Aleksandrova et al. (1980) showed that by pretreating the sample with organic solvents

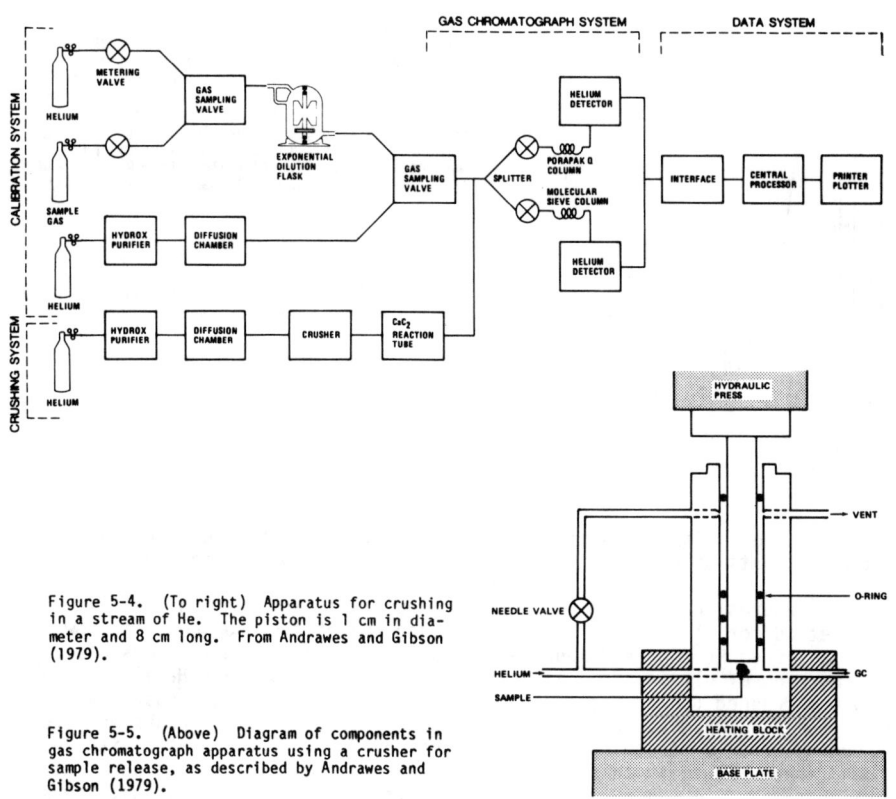

Figure 5-4. (To right) Apparatus for crushing in a stream of He. The piston is 1 cm in diameter and 8 cm long. From Andrawes and Gibson (1979).

Figure 5-5. (Above) Diagram of components in gas chromatograph apparatus using a crusher for sample release, as described by Andrawes and Gibson (1979).

such as acetonitrile, much of the contaminating organic matter is eliminated, and on subsequent heating in the gas chromatograph, the gases obtained are more representative of those of the fluid inclusions alone. Most such contamination and change from reaction on heating can also be avoided by the use of room-temperature crushing of a coarse sample in an evacuated collapsible tube (Roedder et al., 1963); this procedure was used effectively even on 50 g samples of coarse metamorphic rock samples by Kreulen and Schuiling (1982).

Alternative sample preparation has included gas chromatography of isopropyl alcohol into which the inclusion fluid has been released by crushing (Bouberlova, 1977) and fusion of rock or mineral samples (Jeffery and Kipping, 1963).

Apparently the first application of gas chromatography to the study of actual inclusion gases was by Ackermann et al. (1964), who studied "popping salt" from the Werra river district in East Germany. Petersilie and Sørensen (1970) used gas chromatography to analyze organic gases evolved from fluid inclusions in the alkalic rocks of the Ilímaussaq intrusion in Greenland. Similar studies have been made of the gases evolved on stepwise heating of apatite and nepheline from the Khibiny alkalic complex, using both gas chromatography and mass spectrometry (Karzhavin, 1976). Numerous studies have been published in the Soviet Union, (e.g., Malakhov, 1977) and several in France (e.g., Cuney et al., 1976), and the People's Republic of China (e.g., Lu et al., 1982), in which gas chromatography has been used to analyze the gases from liquid inclusions in ore deposits. The analytical choices for both release and detection and their limitations and applicability to geological samples have been reviewed by Ypma (1979a). The method is particularly effective for differ-

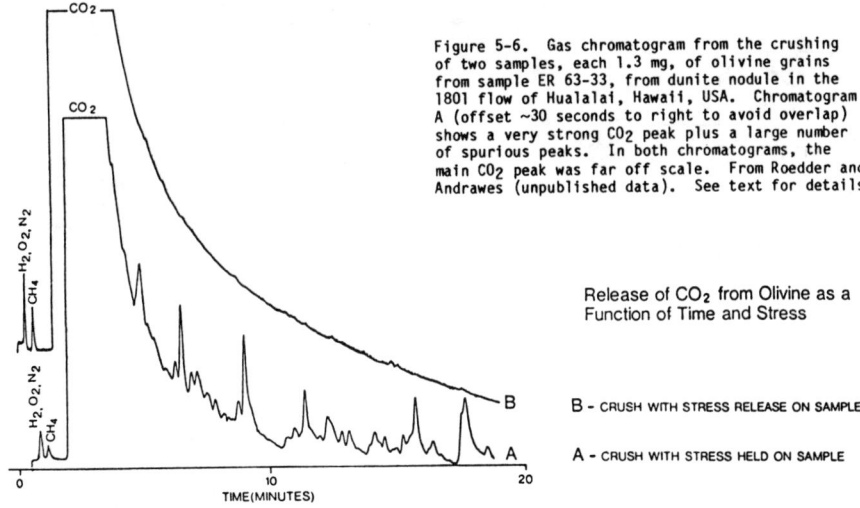

Figure 5-6. Gas chromatogram from the crushing of two samples, each 1.3 mg, of olivine grains from sample ER 63-33, from dunite nodule in the 1801 flow of Hualalai, Hawaii, USA. Chromatogram A (offset ~30 seconds to right to avoid overlap) shows a very strong CO_2 peak plus a large number of spurious peaks. In both chromatograms, the main CO_2 peak was far off scale. From Roedder and Andrawes (unpublished data). See text for details.

Release of CO_2 from Olivine as a Function of Time and Stress

B - CRUSH WITH STRESS RELEASE ON SAMPLE

A - CRUSH WITH STRESS HELD ON SAMPLE

entiating the various S species. Thus Lovell (1979) was able to recognize H_2S, smaller amounts of COS, and traces of CS_2 in a fetid barite sample.

An interesting example of the high sensitivity and hence suitability of this method for fluid inclusion samples, and one of the problems, is shown in Fig. 5-6, from Roedder and Andrawes (unpublished data). In this example, 1.3 mg of grains of olivine from a dunitic nodule brought up by a Hawaiian basalt flow was crushed (chromatogram A). It yielded traces of other gases and a huge peak for CO_2 that went far off scale. A series of other peaks were found throughout the 19-minute run. These peaks could not be correlated with any known gases, and their shape was very uncharacteristic for normal peaks at these elution times. It was then realized that after the initial crushing, the sample was left under pressure, and these many small peaks represented the continued yielding of small parts of the sample, opening additional tiny inclusions. Thus, in effect, new "gas samples" were being added throughout the run. When another similar sample was crushed and the stress immediately relieved, the elution tail from the huge CO_2 peak had no additional peaks (chromatogram B, Fig. 5-6).

Determination of isotopic ratios. In addition to the use of mass spectrometry to analyze gas mixtures from inclusions for their molecular species, there has been a rapidly growing interest in studies of the isotopic ratios of various elements in fluid inclusions, particularly in the field of ore deposition (see Chapter 15). The last decade has seen a great surge in such studies, involving H, O, S, and C in both inclusions and in host minerals. D/H is the most commonly determined ratio on inclusion fluids, and less commonly, $^{18}O/^{16}O$, $^{13}C/^{12}C$, and $^{15}N/^{14}N$. $^{34}S/^{32}S$ would be particularly desirable (especially if the original molecular form of the S in the inclusion (SO_4^-, HS^-, S^{2-} etc.) could also be determined), but the small quantities of S available generally preclude such determinations with presently available techniques.

D/H ratios have been measured on H_2O from many inclusion samples. Thus Kokubu et al. (1961) measured the H_2O which filled large amygdaloidal cavities in basalts and Roedder et al. (1963) measured 15 inclusion samples, mainly from ore deposits, using the vacuum extraction method described above (Fig. 5-2). On crushing in vacuo, the evaporated H_2O and CO_2 are condensed in a cold trap at -196°C; the pressure of noncondensable gases is measured (i.e., noncondensable at these conditions); the H_2O is converted to H_2 by reaction with hot U metal, and it is separated from the CO_2 by another cold trap. After volume measurements, the H_2 is measured isotopically by mass spectrometry and the solutes from the evaporated solution fluids in the crushed debris are leached and analyzed.

Hall and Friedman (1963) reported 33 additional measurements with this method, all on Mississippi Valley-type deposits. Rye (1965) reported D/H ratios for the inclusion waters in a series of samples from Providencia, Mexico. He also has examined the $^{13}C/^{12}C$ and $^{18}O/^{16}O$ ratios in various generations of calcite from this deposit. Similar measurements were made on the C and O in the fluid inclusions in calcite, quartz, and sphalerite by Rye and O'Neil (1968), who showed that the fluid in inclusions in calcite and quartz exchanged O, and those in calcite exchanged C with the host phase during cooling (see also Cole et. al., 1983). Cole (1983) suggested that isotopic disequilibrium could be used to obtain rock/water interaction times.

The combination of several isotopic ratios on the fluid from inclusions (most commonly D/H and $^{18}O/^{16}O$; e.g., Fig. 11-13) along with other data on mineral assemblages and compositions of fluid inclusions, has proved to be of immense value in understanding the source and history of the fluids in a variety of geologic environments (Taylor, 1979a,b; Ohmoto and Rye, 1979; Gat and Gonfiantini, 1981; Perry and Montgomery, 1982). But it is important to remember that one of the major obstacles in such work has been the sample requirement of one to several milligrams of water for each determination of D/H or $^{18}O/^{16}O$. Unfortunately, to obtain this much fluid, uncontaminated and of known single lineage, may require manweeks of work per sample, and in some samples may even be impossible. Practically all the data reported above have been on composite samples of numerous inclusions and hence are averages that may or may not be representative. These problems should diminish as instrumental improvements reduce the needed sample size. Thus R. Kreulen (pers. comm.) reported determination of $^{13}C/^{12}C$ on 0.01 mg CO_2. Small samples, however, need increased vigilance to avoid serious experimental and instrumental errors. Thus, traces of certain solvents left in samples from cleaning prior to inclusion release can cause large errors in $^{18}O/^{16}O$ measurements (L.P. Knauth, pers. comm., 1983).

Several very different types of isotopic data are used in discussions of fluid-inclusion data. H_2O or CO_2 from fluid inclusions is sometimes extracted (by crushing or heating) and the ratios D/H, $^{18}O/^{16}O$, or $^{13}C/^{12}C$ of the fluids determined directly. These same ratios (and $^{34}S/^{32}S$) are also sometimes determined on minerals containing them (e.g., O in quartz, S in sulfides or sulfates, H from OH in mica, or C from CO_3 in carbonate). From these determined mineral values and laboratory determinations of the fractionation factors in various solid-liquid (or solid-solid) equilibria, the isotopic composition of the fluid from which a given mineral precipitated may be calculated. The validity of such calculations is, of course, limited by the applicability (and validity) of the temperatures and fractionation factors used. The temperatures are usually obtained from inclusions.

An additional problem is the possibility of exchange effects, between solids, between liquid and solid, or between molecular species in the fluid (e.g., Brenninkmeijer et al., 1983). Matsuhisa et al. (1979) found that plutonic igneous rocks typically have quartz-feldspar fractionations substantially larger than their laboratory equilibrium values at solidus temperatures, indicating substantial retrograde exchange effects. In some nonequilibrium solid/liquid pairs, some isotopic change toward the equilibrium isotopic fractionation can occur by actual diffusion and exchange, whereas other pairs seem to approach equilibrium only through a process of solution and reprecipitation. Wilkins and Sabine (1973) could exchange D for the H in some nominally anhydrous silicates within a few days, whereas $^{18}O/^{16}O$ equilibrium for the quartz/water pair is very sluggish unless recrystallization occurs (Matthews and Beckinsale, 1979; Matthews et. al., 1983). Ambiguity from retrograde exchange between host and fluid can be avoided completely only by the use of inclusions in a host that does not contain the element in question, e.g., D/H or $^{18}O/^{16}O$ in sphalerite or $^{13}C/^{12}C$ in quartz. The isotopic signatures of individual molecular species can provide useful information, but requires knowledge of the conditions under which isotopic exchange can occur. For example, the $^{13}C/^{12}C$ values for CO_2 and CH_4 in the same sample will probably differ, and may reflect different C sources, but exchange of C between these two can occur under geothermal conditions (Giggenbach, 1982).

Numerous gas analyses of fluid inclusions have reported Ar, but relatively few report its isotopic ratio. Atmospheric Ar contains some ^{36}Ar and some ^{38}Ar, but radiogenic Ar formed by the decay of ^{40}K is pure ^{40}Ar. Bakhanova et al. (1976; also Naydenov et al., 1978) have examined the Ar isotopic ratios from an Au deposit in Kazakhstan, USSR, and have found that inclusions in the richest ores were highest in ^{36}Ar, suggesting the presence of more surface waters containing dissolved light atmospheric Ar. In the future, we may also expect to see the use of ^{3}He/^{4}He in a somewhat similar manner. Here, the heavier isotope is also from radioactive decay, but the light isotope is derived mainly from deeper in the earth and comes, at least in part, from original nucleosynthesis. The large mass difference introduces new problems, however, as a result of significant differences in isotopic diffusion rates.

Beryl has been found to contain excess ^{4}He and ^{40}Ar, i.e., ^{4}He and ^{40}Ar in excess of that from radioactive decay since crystallization (Damon and Kulp, 1958, p. 449). This excess is probably present in two forms, as fluid inclusions, and in the relatively large channels in the structure, along with H_2O (<3%; Feklichev, 1963). Rama et al. (1965) reported large amounts of excess ^{40}Ar in the fluid inclusions of a metamorphic quartz vein, about 1000 times more than the known age (~250 my) and total K content (14 ppm) would have yielded. Similar excess ^{40}Ar from fluid inclusions has been reported for ultramafic xenoliths from Hawaii (Funkhouser et al., 1965; and Funkhouser and Naughton, 1968) and for hydrothermal fluorite (Lippolt and Gentner, 1963). Dalrymple and Lanphere (1969) reported >50 apparently anomalous ages attributed to excess ^{40}Ar, only a few of which were specifically assigned to ^{40}Ar from fluid inclusions.

In spite of the problems from possibly large initial ^{40}Ar contents in the fluid inclusions, Zentilli and Reynolds (1977) obtained a reasonable age on sylvite-bearing inclusions in quartz from a porphyry Cu orebody, using the ^{40}Ar/^{39}Ar age spectrum method. They used the behavior of the inclusions on heating under the microscope to evaluate their ^{40}Ar/^{39}Ar step heating data. Recent developments have greatly reduced the sample-size requirements for the method (Currie, 1982), and York et al. (1980, 1982a,b) used a laser probe mass spectrometer for dating of opaque ore minerals. It seems reasonable to suppose that a large part of the K and Ar involved in this determination is present in solution in fluid (or solid) inclusions rather than in the host mineral structures. York et al. used the isochron approach, separating various parts of the sample with varying K/Na ratios; by this means, they circumvent the initial Ar problem to some degree.

Release, Extraction and Analysis of Liquids

Many quantitative analyses of liquid inclusions have been reported, using a wide range of combinations of methods of release, extraction, and analysis, making comparison difficult. Furthermore, many of the published analyses contain obvious internal inconsistencies, such as gross cation/anion imbalance, or do not agree with optical observations (e.g., low total concentrations yet NaCl daughter crystals reported to be present). Others present analytical results that seem impossible to obtain with the stated methods, or do not state the methods, making evaluation impossible. References giving the several thousand published analyses up to 1972 are cited by Roedder (1972) and hence are not reviewed here. Also, the various qualitative and semiquantitative analytical procedures based on microscopy (Chapters 4 and 8) are not discussed here.

Most reported analyses of fluid inclusions are ratio analyses, in which the ratios of two or more ions are determined, commonly on solutions obtained by leaching a crushed sample. Ratios such as F/Cl, K/Na, or even those for the bulk of the solutes present in aqueous solutions, such as K/Na/Ca/Mg/SO$_4$/Cl, are obtained, but since the amount of H$_2$O is not determined (or is only crudely estimated) the concentrations of these solutes in the inclusion fluids are unknown. In contrast, a relatively few inclusion analyses are <u>quantitative</u> analyses, in which both solutes and solvent are determined.

Ratio analyses. Most of the several thousand inclusion analyses reported in the literature have been made on composite or "bulk" samples, by the water leach procedure. In this procedure, some of the millions or billions of inclusions in a sample (of perhaps 100 g) are opened by fine grinding (or decrepitation), and the sample is then leached with "pure" water. The leachate (or "extract") is then analyzed for such solutes as Na, K, Ca, Mg, SO$_4$, and Cl, and (mainly in the Soviet reports), also for pH and even Eh. Although the ion analysis is perhaps a valid procedure on a few very carefully selected samples, the results are so easily invalidated either partly or completely by any of a series of problems that the results must generally be viewed with suspicion. The biggest problem lies in the sampling, as multiple generations of inclusions are present in many if not most samples. Even if the inclusions in a given sample are products of only one process, it is exceedingly difficult to extract the fluid without gross loss or contamination, or both (Roedder, 1958, 1972). Too frequently this leaching and analysis is treated as a simple routine operation, and as a result of numerous potential analytical pitfalls along the way, much of the extensive early work is of dubious validity, entirely apart from the ever-present problem of the assignment of the inclusions in the sample to a single epoch of origin, and the added possibility of fluid changes during the formation of a given sample.

Sample cleaning is a particularly important and often neglected step. Most ordinary samples contain only perhaps 0.1% inclusion fluid, containing perhaps 10% total ions in solution, and in many studies, an ion of particular interest may constitute only 1% of the total ions present. This ion thus constitutes only <u>one part per million</u> of the whole sample. Significant contamination (and/or loss) of such a small amount of material in the sequence of processes that must be used is the rule rather than the exception, and the validity of the analytical results obtained will be a direct function of the care used in minimizing (and evaluating) the several sources of error.

Two types of sample cleaning need to be considered: (1) The elimination of all minerals but the one of interest, and (2) The elimination of surface impurities. The smaller the sample to be used, the easier it is to handpick clean grains under the microscope, free of other contaminating minerals. Larger samples, particularly of translucent or opaque materials such as vein quartz, are frequently selected by simple visual inspection of the pieces to eliminate those containing other minerals (on their outside surfaces). Such selection, although helpful, is far from sufficient. Grains of mineral contamination can be buried within the fragments, and finely divided impurity phases will almost always be found if a thin section of such material is examined.

If a 100 g sample is used, containing 0.1 wt % of fluid inclusions, with 1 wt % of element X in solution in the inclusion fluid, only 0.001 g of X will be present in the leachate. If even just a few milligrams of a mineral containing X are present as a buried crystal or as finely disseminated solid inclusions in the sample and are exposed during grinding (e.g., sericite flakes in quartz, when X is K, or calcite, when X is Ca), contamination from such sources can be severe. Even without discrete contaminant minerals, trace constituents in the structures may be extracted and contaminate. The standard rebuttal to suggestions of this possibility of contamination from solution of other minerals is that repeated leaches provide little more of the element of interest; this is a necessary, but far from sufficient proof. Such a dropoff in sequential leaches is characteristic in solubility determinations of most finely ground pure minerals.

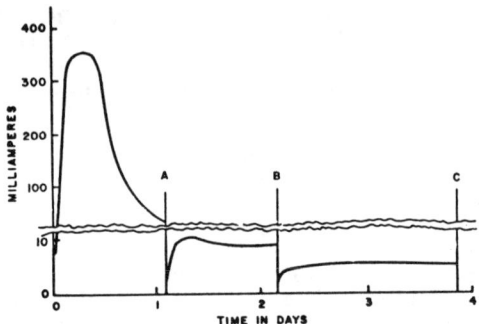

Figure 5-7. Typical plot of electrodialysis cell current flow vs time (diagrammatic) during the cleaning of a sample. The vertical lines at A, B, and C represent the times at which the solutions in the electrode chambers were removed and replaced with fresh deionized water. From Roedder (1958).

Surface impurities present a problem that is generally less obvious but that can be of similar magnitude. Included here are ions from soil or groundwater (or fingerprints) that have either adsorbed onto the mineral surfaces or have precipitated as minute grains of gypsum or other compounds in the surface cracks. I found that electrolytic cleaning was very simple yet effective in eliminating these contaminants (Roedder, 1958). The mineral fragments were placed in the bottom of a large Pyrex glass "U" tube with deionized water and Pt electrodes in each arm, and a constant DC voltage (e.g., 90 V) applied. An ammeter in the circuit provided control of the operation, as it monitored the conductivity of the fluid. The current flow increased rapidly as ions from the sample spread into the originally very low-conductivity water in the arms of the tube, and then declined with time (~1 day) as these ions clustered around the electrodes, leaving the main part of the water cleaner (Fig. 5-7). The contaminated water around the electrodes was then sucked off first, followed by the remainder of the water. New water was added to repeat the process, until there was essentially no change in conductivity with time, and the conductivity was approximately that of the deionized water used, modified by the solubility of the minerals. This took several days and three or four changes of water. (If slightly soluble minerals such as calcite are exposed somewhere on the sample, the conductivity will stay relatively high.) Once cleaned in this manner, it is safe to assume that the sample is free from any ionizable impurities except those in sealed solid inclusions or within completely closed and sealed fluid inclusions. The process has been subsequently been made more convenient by the addition of a water inlet at the bottom, permitting the contaminated water to be flushed over both tops simultaneously by the new water (D. Pinckney, pers. comm.).

Many investigators have used acid treatments to eliminate both extraneous mineral grains and surface contaminants. This treatment for acid-insoluble samples such as quartz will eliminate solid contaminants, but has some serious drawbacks. First, the acid does not dissolve extraneous mineral grains that are completely embedded. Second, on contact with the sample, the acid moves into the inevitable surface cracks in the grains, dissolving the impurities. This contaminated acid must subsequently be removed from these cracks, and even the clean acid must be removed quantitatively if it contains an anion that is to be determined. A sequence of water "washes," as generally described in the literature, will remove most of the acid on the outside surfaces of the grains, but when such a "clean" washed sample was subsequently put in the electrolytic cleaning cell, I found that considerable acid was still present in the cracks, and required days of cleaning. P. Eadington (pers. comm.) avoids some of these problems by using HBr, followed by HNO_3 and then washing.

Once the sample exterior is clean, the inclusion fluid in it must be released. Two methods have commonly been used, crushing (or ball milling), and thermal decrepitation. Some have ground the samples under water in an agate mortar.[5/] Others have used ball milling, but rarely report blank runs. I have found that standard high temperature porcelain mills and grinding media yield gross contamination, and even the use of a sintered alumina ball mill and alumina grinding media yielded significant contamination in grinding 1000 g samples of -10+48 mesh quartz (Roedder, 1958). Goguel (1963) used a special vacuum ball milling procedure on very small samples, particularly to obtain the gases for analysis. The ball milled samples were given a quick leach with water, filtered with a membrane filter and analyzed. Goguel (1964) revised some of his techniques to reduce contamination and obtained considerably lower values for B and Cl. These Cl values are actually the sum of Cl, Br, and I; this is probably true for many of the "Cl" analyses given in the literature, but it is generally not stated.

Some workers (e.g., Savel'yeva and Naumov, 1979) have used thermal decrepitation to open the inclusions, but I had no success with the method, at least for the ions in solution (Roedder, 1958, p. 263-266). I heated an 815 g sample of gold quartz from Grass Valley, California, for 23 hours at 488°C in a fused silica flask in a stream of pure N_2, and obtained 0.19 wt % H_2O, and 0.006 wt % CO_2 from the decrepitation (after flushing at 110°C). After decrepitation, I used the electrolytic cleaning cell (see above) to collect the ions remaining from evaporation of the fluids on decrepitation. I obtained only 7 mg total K+Na, corresponding to only 0.4 wt %, whereas another extraction procedure on a part of the same sample yielded nearly 25 times more. Several explanations are possible. First, when inclusions decrepitate, they frequently do not fly apart, but simply lose their contents through visible or invisible fractures. This could leave part of the nonvolatile salts inside the inclusions and hence relatively inaccessible to later leaching. Second, diffusion of alkalies in quartz is relatively rapid at the temperature used; any that diffused into the quartz would also be effectively lost.

A much more rapid heating could possibly minimize the second of these two problems. Some workers have heated samples to the decrepitation temperatures, and then crushed them before leaching; this might minimize both problems. Thompson et al. (1980) have used the decrepitation itself to inject an uncontaminated, dispersed sample into the detection device (see below).

Extraction of the inclusion ions from the ground host mineral might seem to be a simple filtering operation, but if the crushing has been fine, as is necessary to open small inclusions, it is not simple.[6/] Ball milling can yield fines that clog filters. Thompson (1981) used centrifuging to provide a clean filtrate. Contamination from all handling steps is also a constant problem. Thus even high-grade analytical filter papers were found to add far too much Cl (from the "acid washing" process), and asbestos filters had to be used (Roedder et al., 1963). To keep the blanks low, very high quality water and reagents are needed. More important than such contamination, however, is loss by adsorption on the large amount of new mineral surface in a finely ground sample ($\sim 10^4$ cm^2/g). An example of the possible magnitude of both contamination and loss from surface phenomenon is seen in the advertisements for a commercial laboratory glassware cleaner that make a sales point of the fact that after a

----

[5/] Agate is slightly porous, and one investigator found that his inexplicably high published results on Cl in his inclusions came about because a coworker in the laboratory was using concentrated HCl to clean the mortar.

[6/] Inclusions in cinnabar are the easiest, as they permit combining the release and extraction steps. The cinnabar can be simply sublimed away at low temperatures, thus leaving the inclusion solutes behind.

glass surface has been cleaned with it, and rinsed four times with distilled water, only 0.6 mg/m^2 of the cleaner remained on the surface.

Electrodialysis with organic membranes can eliminate surface adsorption problems completely (Roedder, 1958), but introduces a few new problems. Others have used an acid leach, which probably removes adsorbed ions satisfactorily, but can result in gross contamination from exposed impurity minerals. Obviously, all these various problems will be minimized if the amount of new surface is small relative to the amount of ions released. Thus, coarse crushing of samples selected to have a large amount of fluid, as large inclusions, would be optimum.

In a number of earlier papers, the leach solutions were evaporated to dryness and analyzed by qualitative or semiquantitative spectroscopy, but analysis of the extracted ions from samples in the 100 g range represents a relatively minor problem with the methods now available. Flame photometry is perhaps the most commonly used, but many others, such as atomic absorption spectroscopy (AA), ion chromatography (particularly for F, Cl, Br, and SO$_4$; see Thompson et al., 1983), ion-sensitive electrodes, direct-current plasma spectrometry (DCP), etc., have been tried and may be particularly suitable for a given application. Cation analyses are generally much simpler and more accurate than anion analyses. Cl presents relatively few problems, in large part because it is generally the major anion. Liquid chromatography has been used to detect amino acids (Kuznetsova et al, 1983).

The use of inductively coupled plasma (ICP) emission spectroscopy of elements released into a carrier gas during decrepitation has been reported in a series of papers from Imperial College, London (Alderton and Rankin, 1981; Thompson, 1981; Thompson et al., 1980; Alderton et al., 1982; and Rankin et al., 1982; see also Walsh and Howie, 1980). The method has very low detection limits for most metals (<10^{-10} g), as well as for B, C, S, and P, but F, Cl, and Br cannot be determined at all. As is common, the major uncertainties in the procedure may well lie in the sample extraction step. Chryssoulis (1983) has shown that the presence of feldspar and mica in the sample as decrepitated is of minor consequence, but I believe a more basic fallacy remains unresolved. When an inclusion decrepitates, part of the fluid is dispersed into the Ar carrier gas, and part evaporates, leaving a solid residue around the pit in the surface. These deposits have also been analyzed, by electron microprobe (see "Electron microprobe" section below). As the various solutes in inclusion fluids vary widely in their volatilities, these two portions of the original fluid, that in the carrier gas and that in the residue, will almost certainly differ significantly in composition, but in the analytical procedures used, each one is tacitly assumed to approach a representative sample of the solutes originally present.

The various S species that might be present in the original inclusion fluid, particularly SO_4^{2-}, HS^-, and S^{2-}, should be determined separately, but as even total S is frequently rather low, and much is normally in the form of sulfate, the analyses are generally of "total S as SO$_4$." Unfortunately, most of the analyses for S that have been reported, particularly in the Soviet literature, list "SO_4^{2-}" without either giving the method used or indicating whether the analysis is that of total S as SO$_4$ or of actual sulfate. As crushing and leaching have generally been done in an air environment, most such analyses may amount to total S as sulfate. Also, as small amounts of sulfide minerals are generally present and can oxidize quickly, most S determinations actually represent maximum values for the inclusion fluids. Only a very few workers have crushed their samples in an inert atmosphere and leached them with carefully deoxygenated solutions.

The interpretation of the analysis can be far more frustrating and difficult than the analysis itself. Most of the older analyses list Na, K, Ca, Mg, Cl, and SO$_4$. Generally, these six ions do not add up to electrical neutrality. If we assume that the analyses themselves have been made correctly, this unbalance can come about from any (or all) of a variety of reasons. Other (unanalyzed) ions are important constituents in some inclusions, and may cause unbalance. The most common of these are Al, Fe, B, P, F, Si, and OH. "Excess" cations can come from several sources. Thus excess alkalies can easily be leached from crushed feldspar, mica, and clay in the sample, and excess alkaline earths can come from exposed carbonates. Excess Cl is unlikely, but excess SO_4 can be released by oxidation of sulfides, which can proceed at an amazingly fast rate. If so, there may well be significant amounts of other cations in solution that are not normally determined, such as Fe or Cu. Silicate anions of one sort or another are frequently reported; whether these are artifacts from the leaching process or real constituents of the fluid is seldom obvious. Large amounts of silica have been moved around by fluids in the earth's crust -- probably more than any other material -- but the extrapolations from the amounts of silica detected by a given analytical procedure in a leach solution, to the fluids in the unopened inclusions, and on to these same fluids at the temperature of trapping, may be rather long steps.

The largest single source of ambiguity in inclusion analyses by far is a result of CO_2 in its various forms. Most analyses, if CO_2 is even mentioned, simply list HCO_3^-, to be taken at face value. Free CO_2 is a very common major constituent of inclusion fluids in a wide range of geological environments. On crushing under water it will form H_2CO_3. This will add to the cation/anion imbalance if HCO_3^- is determined. A simple crushing test to verify the absence of free CO_2 under pressure is an absolutely essential preliminary to any analyses of fluid inclusions. If solid carbonates are present, free CO_2 will dissolve them to add cations and twice as much carbon as HCO_3^- to the solution as the original CO_2 that was present. In addition, CO_3^{2-} and/or HCO_3 may be present in the original inclusion fluids. If HCO_3^- is high, it can dissociate during dry crushing to CO_3^- and free CO_2. In addition to these sources of ambiguity, some of the analytical procedures used for HCO_3^- or CO_3^{2-} are not always capable of distinguishing these two, and, in addition, the specific methods used are seldom stated.

An added problem is introduced by daughter minerals when present, and by the solubility of the solids precipitated on dry crushing. The leaching medium should take into solution all daughter phases, and all solids precipitated on opening (and nothing else), but this may not always be feasible. Thus if the inclusions in a sample from an ore deposit have visible chalcopyrite daughter crystals, any leach solution that will dissolve them would also dissolve the "minute amounts" of sulfide contamination commonly present in the sample to yield grossly erroneous values for Cu, Fe, and S. Loss of gases on opening may also precipitate solids that are relatively insoluble in pure leach water, and even without such loss, as the leach water is vastly more dilute in other salts than the inclusion fluids, it may not redissolve them.

A variety of special techniques have been used for specific problems, and many more may have to be devised, as the quantities of ions available from well documented samples commonly are toward the lower limits of currently available techniques. Thus thin-film anodic stripping voltammetry was used for Pb in inclusions (Miller and Shepherd, 1984), and a highly sensitive method recently developed for one rather difficult element, Al, may be applicable to fluid inclusions as well (Brady and Frantz, 1980).

Even a simple ratio of only two elements can be useful. Thus Holser (1963) and Kramer (1965) drew conclusions from the ratios Br/Cl and F/Cl in inclusions from saline deposits. Kozlowski and Karwowski (1974) used the ratio Cl/Br to aid in understanding the conditions of formation of some granite-gneiss massifs, and Kozlowski (1978) suggested that the ratio F/Cl might be useful as a prospecting tool. Many studies have reported Na/K ratios for inclusions from ore deposits, as these are relatively easy to determine (by flame photometer), and can be of value in understanding the sequence of wallrock alteration stages in ore deposits, as well as the temperature of rock-water equilibration (Montoya and Hemley, 1975).

Neutron activation analysis (NAA; also called INNA, for instrumental neutron activation analysis) is a particularly effective method for certain constituents. Analyses of Na, K, Rb, Cs, Cl, Br, Cu, Mn, and Zn have been made on inclusions by neutron activation; Touray (1976) added As, I, Dy, and W to this list (see also reviews by Laul, 1979; Muecke, 1980).

Czamanske et al. (1963) determined the Cu, Mn, and Zn contents of large inclusions in a transparent fluorite from southern Illinois, and in a translucent quartz sample from Creede, Colorado. They determined the volume of the inclusions with a vacuum crushing procedure (Fig. 5-2), leached the crushed fragments first with H_2O and then with cold dilute HNO_3, and the partly evaporated leachates were irradiated for 30 minutes at $10^{12}n/cm^2/sec$. Induced activities were counted after chemical separation procedures. The results for Cu and Zn in the fluorite were surprisingly high, particularly in the acid leach (they calculated to 0.9 and 1.1 wt % in the inclusion fluids, respectively) but the various possible sources of contamination seem to have been effectively excluded. Subsequent analyses of both similar and dissimilar samples by other methods and investigators have shown that the Cu and Zn values obtained by Czamanske et al. (1963) were high but not really exceptional.

Puchner and Holland (1966) also used neutron activation, but before opening, to determine Na, Cu, Mn, and Zn. The inclusions they used were in quartz from a Pb-Zn ore deposit at Providencia, Mexico, and were much smaller (6 - 770 μg). They used optical measurements to obtain the volumes. During the irradiation (5-25 h at ~1.3 x $10^{13}n/cm^2/sec$) synthetic liquid standards, sealed in fused silica capillaries, were run along with the unknowns. Na and Mn were found in all four samples but Cu and Zn were below their detection limits, which they estimated to be <12-24 ppm for Cu and <260-580 ppm for Zn. They suggested that their low values for Cu and Zn (compared with those of Czamanske et al., 1963, from different localities) were a result of their samples having been late-stage quartz, formed after ore deposition.

Krupka et al. (1977) devised a neutron activation method, involving separation of Na and K between irradiation and counting, to determine Na/K for inclusions from gold-quartz veins. Luckscheiter and Parekh (1979) used two sequential neutron activations to determine Na/Cl and Br/Cl (as well as K, Mn, and As) in the unopened inclusions in cylindrical samples drilled from three Alpine quartz crystals. As these two ratios were reasonably constant, they believed that the analyzed elements were almost totally associated with the fluid in the inclusions.

One notable recent development has been the analysis (by isotope dilution procedures) of Rb and Sr isotopes in fluid inclusions for age dating, in addition to that based on K/Ar mentioned earlier. Norman (1978) showed that valid Rb/Sr ages could be obtained from an isochron based on the amounts of Rb and Sr in the inclusions in quartz from a miarolytic granite (~300 ppm), and that multiple sources of Sr could be detected in the fluid inclusions from some ore deposits (see also Norman and Landis, 1980). Shepherd and Darbyshire (1981) made similar determinations on quartz from English W deposits. They obtained a geochron from 10 samples in which total Rb ranged from 166 to 1477 ppm, and total Sr from 164 to 301 ppm.

Complete quantitative analyses. The term "complete" should perhaps be in quotes, as no analysis is ever complete. By complete, I refer to an analysis that includes all major volatile and nonvolatile constituents. Usually this includes Na, K, Ca, Mg, Cl, SO_4, H_2O, and CO_2. HCO_3 is much more rarely analyzed. The major problem in such quantitative analyses is that of obtaining both the volatiles and the nonvolatiles on the same inclusions.

The vacuum crushing technique reported by Roedder et al. (1963), permits analyses of both the H_2O and the solutes in those samples with adequate inclusions. Although it has subsequently been used "blind" (i.e., on opaque samples), minimum ambiguity exists if the sample is in the form of transparent pieces, each containing one or more large inclusions, cut to have a minimum volume of pure mineral with a maximum volume of inclusion liquid (generally >1 mg total). The electrolytically cleaned samples were crushed in vacuum (Fig. 5-2). After the volatiles were measured, the metal sample tube was cut open, the crushed fragments (and tube walls) leached with a very few milliliters of high purity water, filtered through an asbestos filter, and analyzed microchemically for Na, K, Ca, Mg, Cl, B, and SO_4. Since the amount of water from those inclusions that were opened had been determined (and in some cases, also analyzed for D/H ratio), both the composition and concentration of solutes in the inclusions could be calculated.

Comparisons of the concentration of solutes determined by this procedure with that found by the freezing stage on the same or "similar" samples did not always agree very well, and Thompson et al. (1983) reported similar difficulties. Several possible reasons are evident. In addition to the normal experimental errors, possible formation of hydrated salts or insoluble precipitates on evaporation in the vacuum, etc., there was a special problem with some of these samples, since they contained some large, apparently empty inclusions, in addition to those with normal liquid/vapor filling. Any inclusion that is cut by a cleavage crack extending to the surface will eventually lose all of its water by evaporation, leaving the salts behind either in the inclusion or in the cleavage crack. These salts would be very difficult to remove during the electrolytic cleaning operation as diffusion through such small cracks would not only be slow, but the air in the inclusion and crack might prevent the water from contacting the salts, precluding any electrolytic cleaning. On vacuum crushing, however, these salts would be exposed for leaching along with the rest of the sample. Where possible, such dried-out inclusions were opened up (e.g., by use of a micro sandblast unit) prior to electrolytic cleaning.

Obviously, the easiest procedure would be to find inclusions large enough to simply pipette out aliquots for analysis. Such large inclusions have been analyzed (e.g., Maslova, 1961; Prikazchikov et al., 1964), but are far too rare to permit systematic comparisons of fluid composition. Also, large inclusions are likely to have leaked. Thus the analyses of the giant inclusions (<400 cm^3) from a huge quartz crystal in a Volynian pegmatite (USSR) reported by Prikazchikov et al. (1964) are very similar to many ground waters and hence probably have leaked.

Recrystallized salt beds represent a special environment, where larger inclusions are sufficiently abundant and well sealed that pipetting out inclusion fluid is frequently an effective procedure (Holser, 1963; Petrichenko and Shaydetskaya, 1968; and particularly Petrichenko, 1973). Although fluid inclusions are generated (and destroyed) rather easily by recrystallization of the salt, and may change their volume with changes in the ambient pressure and temperature, it seems that they are fairly reliable chemically in that they seldom leak.

Petrichenko (1973) used a tiny stream of water (under the microscope) to dissolve away the host to near the inclusion. The sample was then dried, and a very sharp tungsten carbide point used to break the remaining wall. Tiny tapering (cone-shaped) glass capillaries were used to suck up the fluid. From the capillary dimensions and length of the liquid plug, its volume was calculated. Appropriate precipitating reagents are then sucked up into the same capillary, mixed, and the capillary fused shut on both ends. The precipitate is centrifuged to the small end of the capillary and its volume measured optically, and compared with volumes for known standards. With this procedure, Petrichenko and his colleagues have made numerous analyses for K, Mg, Ca, Cl, and SO_4. Petrichenko also has used a modification to analyze (ratios only) the constituents in the precipitated residues from decrepitation of inclusions in nonsaline minerals. The decrepitated sample is placed in a damp box until droplets form from each decrepitated inclusion. These are sucked up and analyzed using the same procedures. This worked well even in nontransparent minerals.

Holser (1963) made microchemical analyses from large inclusions in Kansas salt by sucking up the liquid through a small drilled hole, using a micropipette that had a fritted glass end, to prevent halite fragments from being sucked up. Analyses were made for Mg, Ca, Br, and SO_4, all stated in terms of ratios to Cl.

Kramer (1965) made two novel but unpublished analyses of inclusions in halite (see Roedder, 1972, Table 6), using weight loss on evaporation to determine the amount of H_2O (see "Release, extraction and analysis of gases" above). The entire sample was then dissolved and analyzed, and corrected for Na and Cl from the host crystal. The necessary assumptions for this correction make the values for Na and Cl questionable, but the method certainly should provide good data on ions other than Na and Cl. The values for the concentration of solutes would be affected also by the fact that some saline brines, on evaporation, can form various hydrous phases, some with as much as 53 wt % H_2O (e.g., bischofite, $MgCl_2 \cdot 6H_2O$), thus reducing the weight loss (Roedder and Bassett, 1981).

pH and Eh determinations. It would be very desirable to know the pH and Eh of the inclusion fluids at the time of trapping. If these values could be obtained on the inclusion fluids at room temperature, together with inclusion composition, extrapolations to the temperature and pressure of formation might be possible, and these extrapolations certainly would be valuable in solving problems of ore transport and deposition (Barnes, 1969). The pH may be obtained either by calculation or measurement. One of the first attempts at calculation was by Uchameyshvili and Khitarov (1965), who used the analytical results on CO_2 and HCO_3^-. Sushchevskaya and Ryzhenko (1977), and Sushchevskaya et al. (1977) have attempted to calculate the pH of Sn-forming solutions from fluid inclusion data by several methods using an approximation for the necessary extrapolation to high temperature.

Although many measurements have been made of the pH of fluid inclusions, few of them are at all accurate. Most of the older measurements were merely qualitative -- large inclusions were opened and litmus paper was applied. Thus Newhouse (1932) found that liquid inclusions in galena from Leadville, Colorado, and Joplin, Missouri, were neutral to litmus paper. Zakharchenko (1950) opened a 1.5 cm^3 inclusion in quartz and found the liquid to be "alkaline to litmus" (the color of which changes between pH 4.5 and 8.3), but he noted that the inclusion fluid boiled violently on opening. The effects of such gas evolution on the pH of the fluids will vary with the gas and could be large. Kalyuzhnyi (1957, 1960, 1961) gave some results of microcolorimetric pH measurements of large inclusions in which the pH increased as much as one pH unit in the first few seconds after opening. He found pH values as low as 4.3 in multiphase inclusions in pegmatitic topaz. Maslova (1958) reported a pH of 4 for the liquid of a large (0.03 ml) inclusion in fluorite from the vuggy pegmatites of the Kermet-Tas deposit, using an unspecified method. Skropyshev (1957) tested the fluid from a large inclusion with three different indicators, and Prikazchikov et al. (1964) used a pH meter on the fluid from very large (~400 cm^3) inclusions. Yushkin and Srebrodol'skii (1965) also used three indicators, bromthymol blue, bromcresol purple, and cresol red, which they introduced into large inclusions in sulfur with a needle; the pH was estimated to be 7 to 7.5. Petrichenko and Shaydetskaya (1968) used three different methods on large inclusions in recrystallized halite; their results ranged from pH 4.95 to 6.2.

Erickson (1965, p. 527, and pers. comm.) determined the pH of large inclusions in calcite from the Upper Mississippi Valley Pb-Zn deposits to be about 7.5, by letting the inclusion fluid wet sensitive pH-indicating paper under a binocular microscope. Although no effervescence was noted, the inclusion fluid escaped almost instantly from the inclusion upon opening, possibly signifying merely the expansion of compressed gas in the bubble. Machairas (1963a,b,c) obtained a measure of the pH and the free CO_2 content of inclusions by a titration procedure; he crushed the samples in an alcoholic solution of phenophthalein at pH of 9.3-10, and measured the decolorization.

Other than this early work on large individual inclusions, in practically all the hundreds of inclusion pH determinations being published every year in the Soviet literature, the pH of dilute water leaches is reported as though it were the pH of the <u>actual inclusion fluid</u> (e.g., Kostyleva and Sukhushina, 1957; Trufanov, 1967). Some have claimed that the measured pH of individual inclusions was the same as that obtained from leachates. However, Khetchikov et al. (1966, 1968) found that the pH of the fluid in individual inclusions in synthetic quartz corresponded to that of the original fluid charge in the autoclave, but the pH of aqueous extractions from the quartz did not.

There are several major fallacies in the assumption that the pH of a leach solution is the same as the pH of the inclusions themselves, in addition to the loss of gas. Even if the diluting fluid were pure water at pH of 7.0 (the pH of the water used for leaching is seldom stated), it is very unlikely that the inclusion fluids are sufficiently buffered in composition that they can be drastically diluted without serious changes in pH, yet dilution factors (volume of leach solution/volume of inclusion fluid) as high as 5000 are implicit in some of the pH data reported, and these factors are seldom as low as 1000. In addition, there may be large effects on the pH of a leach solution just from contact with the air, or the host mineral or other solid phases. It has long been known that mineral surfaces react with water, and this reaction affects the pH to the extent of providing a diagnostic field test for certain minerals (Stevens and Carron, 1948), yet the pH values for extracts from a wide variety of minerals such as calcite, fluorite, spodumene, microcline, and quartz are frequently assumed to be the pH values for the inclusion fluids themselves, and to be sufficiently accurate that genetic conclusions may be drawn from them -- even from small differences between individual samples.

Numerous measurements of the oxidation potential (Eh) of inclusion leachates have also been published (e.g., Vasilenko and Kurshev, 1977; Moiseenko, 1976; Boyko and Markova, 1976; Lkhamsuren, 1976; Dzhumailo et al., 1973) in general with the implicit or explicit assumption that this measurement was also the Eh of the inclusion fluid. Unfortunately, the reasons for doubting the validity of such an assumption mentioned above for pH determinations on leachates are even more appropriate for Eh measurements.

Very few reports exist in which actual measurements of the Eh of inclusion fluids themselves have been made. Most of these are on large inclusions in water-clear recrystallized halite from the Donbass, USSR, and were made by O.I. Petrichenko and coworkers (e.g., Petrichenko et al., 1974; Petrichenko and Shaydetskaya, 1968, 1973; Shaydetskaya, 1975; Petrichenko and Shaidetskaya, 1976; Petrichenko and Slivko, 1973a,b). The methods are summarized by Petrichenko (1973). In part of this work, a pair of electrodes (platinum and calomel, using Zobell's procedure; no reference given) were inserted in 0.8 to 1 mm holes drilled into the large fluid inclusions. In other studies, Petrichenko reports Eh values obtained on inclusions as small as 0.1-0.2 mm. The measurements obtained varied with the nature of the sample, over a wide range (-210 to +515 mV). Two-phase inclusions gave higher Eh readings, "indicating that their bubbles represented infiltrated air." The acid reducing conditions during halite recrystallization are corroborated by the common occurrence of pyrite in such salt, and contrast with the weakly oxidizing environment of salt basins. Kovalevich (1976) also measured Eh on inclusions, in Miocene salt from the Forecarpathians, USSR. The interpretation of all such measurements is difficult at best, even when made using much larger sample volumes (pers. comm. D.C. Thorstenson).

Any large inclusions could be measured with existing equipment, as miniature and subminiature Eh electrodes are available, but the problems of extraction of the fluid precludes most such measurements; in addition, the other problems of pH measurement generally are applicable here.

It should be possible to calculate the Eh from detailed studies of the composition of the inclusions. Any multivalence state component or group theoretically could be used -- Fe^{2+}/Fe^{3+}, CO/CO_2, or HS^-/SO_4^- -- together with pH measurements. Mironova et al. (1973a) used pH determinations (presumably from water leaches) plus gas chromatographic data on the ratio of CH_4 and CO_2 (presumably released by decrepitation) to calculate Eh for some H_2S-bearing inclusions from the Klichkinskoe ore field, E. Zabaikalia, USSR, and Sushchevskaya et al. (1978) made similar calculations for inclusions from some Sn deposits. Interpretation of such data is subject to ambiguity since it is probable that neither the original high temperature equilibrium will be quenched in, nor that true room temperature equilibrium will be obtained. Much more commonly, the mineral assemblage in an ore deposit is used to calculate pH and Eh (e.g., Nedachi, 1974; Barton et al., 1977). In a few samples, daughter minerals may be useful in estimating the state of oxidation.

It would be valuable to know the S species present, particularly S^{2-} versus SO_4, but no known analyses of inclusions list both. Usually only SO_4 is determined. Except in rather rare instances, the concentration of S^{2-} is very low, at least as determined by the lack of evolution of H_2S on opening or by calculation from the high contents of heavy metals (Czamanske et al., 1963). Daughter minerals found in inclusions in various Brazilian quartz crystals include sulfides such as pyrite and sulfates such as gypsum. Simultaneous occurrence of sulfates and sulfides in the same inclusions (or even in the same mineral samples) would help to place constraints on the state of oxidation of the fluids.

Unfortunately, the Eh of an inclusion fluid would be strongly affected by any loss of H or H_2S from the system. Although the evidence against gross leakage of major constituents from inclusions is commanding (See Chapter 3), under some circumstances H may leak out and increase the oxidation state of the remaining fluids. One of the major difficulties in calling on this mechanism to explain the presence of hematite and anhydrite "daughter minerals" that do not dissolve on heating the inclusions is that of maintaining an adequate "H sink" outside the inclusion to drive the diffusion process. As many ore-forming processes apparently take place at low oxidation states, that is, high partial pressures of H, the most expectable change in the state of the fluid bathing the exterior of the crystal would be toward more oxidizing conditions, particularly as the erosion surface approaches the deposit and oxygenated surface waters are involved. Even this should have only a small effect, however, because diffusion rates are greatly reduced at surface temperatures, and because ground waters penetrating ore deposits are rather effectively buffered with respect to oxygen by reaction with sulfides. Probably most such hydrogen leakage occurs earlier, at higher temperatures.

In view of the numerous difficulties still inherent in determinations of the pH and Eh from inclusions, and in spite of all the efforts that have been made, these data may best be obtained from the mineral assemblage in the sample whenever it permits, rather than from the inclusions.

Analysis Using Various Microbeams

Laser microprobe. Several groups have used a laser, focused by a microscope, to decrepitate a given inclusion or group of inclusions and hence release the gases for analysis. Apparently Mikhaylova et al. (1973) were the first to use a laser for gas inclusions. Tsui (1976) and Tsui and Holland (1979) were the first to use the method for nonvolatile constituents, analyzing for Na, Ca,

Mg, Mn, and Cu. A large (>100 μm diameter) inclusion <500 μm below a polished surface was located with the microscope, and then an IR (CO_2) laser pulse was directed downward through the same microscope onto the inclusion. The heating from the pulse "vaporized" (decrepitated) the host mineral and the inclusion contents, if the latter was within the crater (a few hundred μm in diameter) formed by the pulse. The slightly ionized vapor from the inclusion contents passed between two electrodes placed close to the specimen surface that were maintained at a voltage close to air breakdown, and hence triggered a discharge that further ionized the vapors. The emitted characteristic radiation was resolved by a spectrograph and recorded photographically or photoelectrically. If the impulse did not explode the inclusion, additional impulses were used. Calibration was by use of synthetic "inclusions" made by filling holes in silica glass with known solutions.

The individual cation concentrations were determined with an uncertainty of a factor of 3 to 7 by combining the calculated weight ratios with the measured fluid inclusion salinities. The range of values calculated thus were: Ca XOO-XOOO ppm; Mg XO-XOO ppm; Mn X-~100 ppm. Very little Cu was detected, and this was probably from contamination, thus placing a rather low upper limit on the Cu concentration in the inclusions (<~50 ppm). The rather large uncertainties probably stem mainly from the unpredictable excitation conditions (e.g., depending on how the fractures open up as the inclusion contents explode, the bulk of the gases might escape effective ionization by the electrical discharge).

Ishkov and Reyf (1980) reported similar studies. They used an Nd glass laser in a "free-running" mode. As transparency of quartz to this wavelength is high, they believe the inclusions absorb enough energy to build up pressure and explode. This energy level resulted in very little evaporation of (and hence contamination by) the host phase. They used natural inclusions of >5x10^{-9} g mass that were <25 μm below the surface, and calibrated with synthetic inclusions made by mixing a drop of the standard (water solution) with premixed epoxy resin and curing. The result was spherical inclusions of known composition and measurable volume and mass.[7]/ Using such standards they found detection limits as follows (g): Mn 4 x 10^{-10}; Be 3 x 10^{-12}; B 3 x 10^{-11}; Cu 7 x 10^{-11}; Zn 5 x 10^{-10}; and Fe and Al 3 x 10^{-10}. The limit for Cu was somewhat lower than that found by Tsui and Holland, 1979). Note, however, that Ishkov and Reyf (1980) reported that the standard and unknown should have similar physical properties and be analyzed under identical conditions. Inclusions in quartz from the W and Mo stages of deposition at the Dzhida, USSR, deposit showed the following in approximate analysis (in g/l): Fe 30; Mn 40; B 3; Zn 15.

Bennett and Grant (1980) also investigated the method, using quartz samples from several Sn deposits and a porphyry Cu deposit, and a pulsed (or Q-switched) ruby laser at 694.3 nm. They reported qualitative data on Fe, Mn, B, Al, Ca, Ti, Cu, W, K, and Sn in terms of three categories, "strong," "weak," and "tentative" response. Only Fe showed a "strong" response, and only K and Sn showed "tentative" responses; all others were "weak." The main problem they encountered was too great a transparency for the wavelength used; inclusions with daughter minerals gave best results.

Deloule and Eloy (1982) and Eloy et al. (1983) used many shots from a pulsed Nd/YAG laser at very high irradiance levels (~10^9 w/cm^2) and very short duration (~3 ns) to erode into and release and ionize the inclusion contents for detection in a mass spectrometer (hence the abbreviation LPMS for laser probe mass spectro-

----

7/ Some of their synthetic inclusions contained a vapor bubble. Whether this is a "primary" air bubble, trapped in the mixing, a shrinkage bubble from cooling after the exothermic epoxy reaction (or from shrinkage during the polymerization itself), or a result of diffusion of H_2O out into the epoxy during or after the hardening is unknown. If the last, the concentration would be changed. When this procedure was tried using a saturated solution of NaCl, daughter crystals of NaCl were found in the inclusions, so some H_2O must be lost (H.E. Belkin, pers. comm.).

meter). They reported semiquantitative results for several nongaseous elemental ratios. Later studies of the laser ionization and other experimental parameters (Eloy, 1984) have resulted in quantitative analyses of solids, liquids, and gases. Others have used ionization by electron impact to enhance sensitivity. A widely-advertised commercial instrument (termed "LAMMA" for laser microprobe mass analyzer) uses a time-of-flight ("TOF") spectrometer. It might be appropriate for qualitative analysis of daughter crystals, but may present problems in calibration and particularly in the preparation of appropriate samples, since it is used in a transmission mode. Gibson et al. (1982) and Kotra and Gibson (1982, 1983) used a laser pulse to release gases from samples in a stream of He leading to a gas chromatograph (see earlier section on Gas Chromatography).

Electron microprobe. Knowledge of the true composition of the melt that was trapped in an inclusion would be exceedingly valuable to petrology. Barring local disequilibrium effects, for example, from fast crystallization, a melt inclusion represents a real point on the liquid line of descent for that magma, rather than those inferred points obtained from analyses of apparently genetically related rocks believed to represent a series. Such melt inclusion analyses might also reveal the existence of magma mixing. Melt inclusions also contain a partial and possibly a full complement of volatile materials, just waiting for suitable analytical techniques. As in so much geological research, however, things are never quite as simple as they may seem at first.

The electron microprobe, invented by R. Castaing, has provided petrology with a powerful tool matched perhaps only by that of Sorby's introduction of the petrographic microscope 120 years ago. It is sometimes abbreviated EPMA, for electron probe microanalysis. The electron microprobe and its relatives, the ion microprobe and the scanning electron microscope, permit analyses that were simply not even dreamed of earlier (K.F.J. Heinrich, 1980).

The first application to any inclusion problem was by Carron (1961) who used it to study the composition of glass (i.e., melt) inclusions in quartz phenocrysts that were truncated by the polished surface and hence accessible to the beam. As in the application of several other new analytical tools to geochemical problems, however, major difficulties generally lie not in the tools themselves, but in the selection of truly appropriate samples for analysis. One of the most serious problems in the analysis of melt inclusions is the gross sample heterogeneity caused by daughter phases. Daughter phases cut away in sample preparation or buried beneath the surface are lost to the analysis. Even when such phases are cut by the polished surface, they would rarely be present on that surface in the correct ratios to give a valid total analysis. Superimposed on this is the difficulty of obtaining a valid broad-beam analysis from a coarse-grained mixture (Albee et al., 1977).

After an inclusion is trapped, the host mineral generally continues crystallizing onto the walls, as is sometimes indicated by the apparent embedding of daughter crystals in the walls.[8/] If the host is a solid-solution series such as olivine, there should be either continuous reaction of the included melt and the crystal, if diffusion permits attainment of equilibrium, or continuous fractionation, if it does not. One way of investigating this problem is to examine the inclusion/host crystal interface, by making electron microprobe traces across it. If two phases are each homogeneous up to the contact, the true composition trace across the contact should have the shape of the dotted line on Figure 5-8. The electron microprobe trace across this contact will have the shape shown by the solid curve, however, because the beam excites a finite

----

8/ It is important to remember also that considerable crystallization is required to cause visible embedding. Thus the "minor" embedding sketched in Figure 3-1 would require crystallization on the walls of about one third of the original trapped melt.

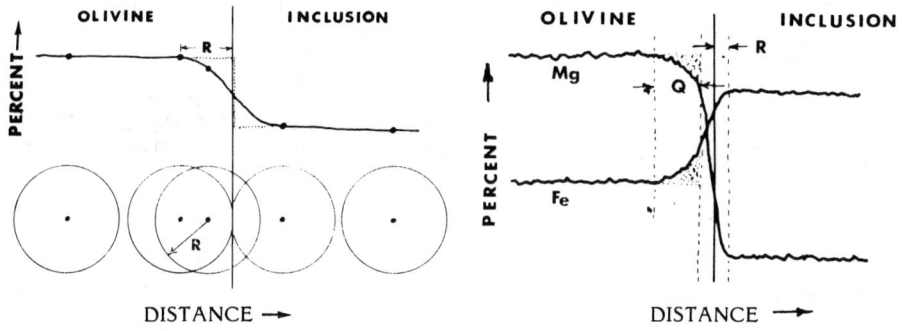

Figure 5-8. (Left) Electron-microprobe trace (diagrammatic) across inclusion/ host contact, in which each phase is uniform in composition and the effective circle of excitation of the microprobe has radius R. From Roedder (1979a).

Figure 5-9. (Right) Electron-microprobe traces (diagrammatic) across contact between a zoned olivine and a uniform melt inclusion in which there is a zone of olivine at the contact of width Q, with a more Fe-rich composition. The effective circle of excitation of the microprobe beam has radius R. From Roedder (1979a).

volume of sample, here illustrated as a circle of radius R. The dimension R will change with the element studied, and with the operating and sample conditions, and of course represents only an arbitrary cutoff in a continuously declining function. In practice, R is seldom under a few μm at best.

If we scan over a contact across which there has been some exchange of material after trapping, or some precipitation of a border of more Fe-rich olivine, or both, the traces will look as shown in Figure 5-9. The actual contact is shown by the solid vertical line. Only that part of the curvature indicated within zone "Q" is solely from a real compositional gradient. The apparent gradient within zone "R" is mainly instrumental in origin, as shown in Figure 5-8. Anderson (1974a) has shown that calculating the composition of the melt that was originally trapped in such an inclusion is not a trivial problem. One method that avoids many (but not all) of these problems involves homogenizing the inclusions in the laboratory before quenching and sectioning for analysis (Roedder and Weiblen, 1970b, p. 810; 1971, p. 515). An interesting version of this procedure involves examining the crystal/melt interface after the inclusion has been heated to various temperatures (Potter, 1975). Figure 5-10 shows traces for Fe across inclusions that have been heated, before sectioning, to the temperatures indicated. At low annealing temperatures, Fe is enriched in the wall of the inclusions; at high temperatures, it is depleted, suggesting that these inclusions were trapped at some intermediate temperature (see also Roedder and Weiblen, 1971, p. 516).

Analyses of glass inclusions by electron microprobe, by their very nature, are apt to be among the microprobe analyses that have the poorest accuracy and precision for several reasons: (1) The samples are generally small, so that contamination by host mineral (Fig. 5-8) is common. (2) The samples are frequently nonhomogeneous. Thus they not only show zonation relative to the contact with the host (Fig. 5-10), but also throughout, for reasons that are not always evident (e.g., Roedder and Weiblen, 1970b, p. 832; 1972a; 1975, p. 159; 1977b; 1978). (3) Alkali elements (especially Na but also K to a lesser extent) migrate in glasses under the electron bombardment, thus yielding count rates that decrease with time, sometimes to a value one half that at the start (see also Graham et al., 1983). The two most common procedures to minimize this, a larger, defocussed beam, and/or translating the specimen during counting, can normally not be used with small inclusions. Nielsen and Sigurdsson (1979) have described two other procedures. (4) Truly valid standards are difficult

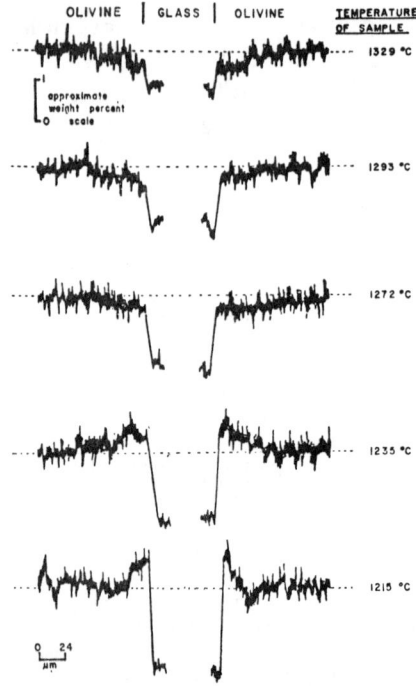

OLIVINE | GLASS | OLIVINE

TEMPERATURE OF SAMPLE

1329 °C

approximate weight percent scale

1293 °C

1272 °C

1235 °C

1215 °C

0 24
μm

Figure 5-10. Electron-microprobe traces for Fe showing olivine compositional variation adjacent to melt inclusions, after laboratory heating at the temperatures indicated. From Potter (1975).

to obtain. (5) The crystal-chemical constraints that can be used with mineral analyses to recognize erroneous data are not available with glasses. Inaccuracies in electron microprobe analyses of synthetic glasses have also caused difficulties in interpreting laboratory silicate phase equilibrium data (Roedder, 1983a).

The problem of host mineral contamination of the analysis could, in theory, be avoided by the use of samples reduced to extremely thin foils (Champness et al., 1981), but because the ion bombardment normally used for this thinning may show greatly different rates for different materials, (G.L. Nord, Jr., pers. comm.), sample preparation might be difficult.

One mode of operation of the electron microprobe that is particularly useful in silicate melt inclusion studies is the areal element scanning mode, in which the beam is made to scan over a small rectangular area of the sample while the count signal from a given detector, set for the desired element, is fed into an image-holding CRT (cathode ray tube or "scope") in a parallel scanning mode. A "picture" of the distribution of the element is thus gradually "painted" with count spots on the scope. Scanning is continued until the contrast between areas of interest is optimum for photography, or for placing the beam, manually, on a given spot on the display on the scope, for a point analysis. This scannning mode of operation, though only qualitative, is particularly effective for small inclusions that are below the normally rather poor resolution of the microprobe optics, and for those whose recognition in reflected light in the probe is made difficult by the reflectivity of the C coating that must be applied. On many occasions this scanning mode has revealed details such as inclusion inhomogeneities that were of major concern in attempts at analysis, yet were otherwise invisible. A more quantitative evaluation of such inhomogeneities can then be obtained by a line scan, in which the signals from several detectors are plotted against position as the sample is slowly translated under a stationary beam (see series of references to melt-inclusion inhomogeneities above). Differences in cathodoluminescence, viewed with dark-adapted eyes, can also be used to resolve some otherwise perplexing samples.

Still another mode of operation involves the use of a display on the scope of the magnitude of the current of back-scattered electrons as the beam scans over the surface. This current is mainly a function of average atomic (Z) number of the material being hit by the beam, and hence is a useful mapping procedure, particularly as it is quickly obtained and may recognize the difference between materials that might have the same high (or nil) concentration of the element plotted in the element scanning mode. (The distinction between positive and negative photographic images of such displays is essential in the interpretation and is sometimes left ambiguous in the published captions.)

The most frustrating aspect of melt-inclusion studies under the electron microprobe is the frequency with which what appears to be a straightforward problem under normal transmitted and reflected light turns out to be much more difficult "under the probe." Thus some microprobe optical systems invert the image and others do not. Combine this inversion with the poorer optics, and the more difficult mechanical movement of the sample under the probe, and even the simple problem of locating an inclusion of interest in a fine-grained field can be extremely frustrating. Locating photographs at low, medium, and high magnification, sometimes both in transmitted and reflected light, are essential. Low-magnification prints can easily be made by putting the section directly in a photo enlarger and projecting it onto a Polaroid film holder. Care must be used in deciding whether the section should be put in right side up or upside down. Even when the inclusion can be found, the appearance on the holding scope of the probe can be very different than the optical images, particularly from thin layers of overlapping phases. For this last problem, and for selection of suitable spots for analysis while scanning a slide on the microscope, I find that the alternation (and sometimes combination) of reflected and transmitted light, operated by two foot switches, is a great help. A variety of other helpful mechanical and analytical techniques for microprobe analyses of melt inclusions in normal polished thin sections (i.e., "probe mounts") are given in the series of papers by Roedder and Weiblen noted above. Specific applications of such probe analyses will be found in Chapters 16-18.

Although the presence of daughter minerals causes ambiguity and error in analyses of silicate melt inclusions by electron microprobe, the probe can also be used to identify these daughter minerals in melt inclusions. When the section cuts through a daughter mineral, and it is large enough, it can be analyzed, and even if it is too small, the major elements can still be determined qualitatively (see Chapter 18).

One of the first applications of the electron microprobe to aqueous fluid inclusions was to identify daughter minerals on the walls of unopened inclusions in quartz. Dolomanova et al. (1966; 1968) determined the presence of Fe, Cu, Ti, Ca, Zn, and Cr among the 11 daughter minerals on the walls of a fluid inclusion in an early smoky quartz from an ore deposit in Transbaikal. They believed that the Fe and Cr were present as chromite. E.I. Dolomanova stated (pers. comm., 1967) that she analyzed only those solid grains on the upper surface of unopened inclusions that were sufficiently close to the polished surface to permit electron beam penetration.

The electron microprobe has extremely low detection limits (10^{-13} to 10^{-14} g), and hence, it would be highly desirable if the instrument could be applied to the solutes in aqueous fluid inclusions. Eadington (1974) used the electron microprobe to obtain partial qualitative analyses of the solids formed on a broken mineral surface by the evaporation of inclusion fluids during decrepitation. He recognized the elements Na, K, Cl, Fe, Mn, S, Sn, and W in the evaporated residues in fluorite samples from a pegmatitic W deposit. Chryssoulis and Wilkinson (1983) reported very high Ag values in the "decrepitation craters" from inclusions from a granite in Mexico near a large Ag mine. They found 0.25 ± 0.07 mole/liter (corresponding to nearly 3 wt %). Such analyses can be made in a scanning mode, so that the distribution of a given element can be displayed, or in a spot mode, with multiple-element detection on that spot. As the evaporated deposits are highly irregular, and as the more volatile constituents may be lost during decrepitation, the method is only qualitative. Statistical handling of many data may minimize random errors, but cannot correct for systematic errors from differential volatility (see earlier section on the use of ICP). The same is true for the procedure described by Petrichenko (1973) for such deposits from evaporation, in the section entitled "Complete quantitative analysis."

The irregularities in the size and spatial distribution of the various crystals in the dried residue are such that it would be impossible to obtain a quantitative analysis with the above technique. For that reason, I have attempted to develop a technique for taking advantage of the electron microprobe's great sensitivity to make quantitative analyses of inclusion fluids, by preparing the samples as a uniform glass (Roedder, 1978a). The method involves taking a very tiny mineral fragment containing an inclusion of interest, cracking it open under clean conditions, adding (under the microscope) a micro-drop ($\sim 10^{-7}$ g) of pure water to the crushed debris, transferring this liquid by micropipette to a hydrophobic-coated planchet of Be, Cu, or C, adding an appropriate amount of a water-soluble organic glassformer, evaporating to form a tiny disk of organic glass containing the water-soluble ions from the inclusion fluid, uniformly distributed, and then probing the whole disk with the electron beam for Na, K, Ca, Mg, S, and Cl. As the most effective glass-former found to date is corn syrup (a mixture of sugars that effectively interfere with each other's crystallization, thus forming an organic glass), the resulting tiny sample disk is essentially a sugar glass. As such, it is in effect a miniature lollipop, hence the procedure has been dubbed the "minimillimicrolollipop method." Several other organic glass formers have been or are still to be tried, but the requirements of the method are surprisingly restrictive in this selection. The most important function of the glass former is to prevent the crystallization of the inclusion salts on drying, thus yielding a sample that contains a maximum concentration of salts homogeneously distributed throughout a minimum volume of uniform glass. The glass former should be of low average atomic number, and contain none of the elements to be analyzed. The volume minimum is set by the depth of penetration of the electron beam into the glass. Preliminary studies indicate that valid ratios of Na:K:Ca:Mg:S:Cl can be obtained with the method, but considerably more developmental and calibration work will be required before it can be made routine. The method seems to work well on $\sim 10^{-9}$ g solutes (i.e., inclusions in the 30 μm range, and if the sugar glass can be "ashed" without loss of volatile elements such as Na, it might become even better.

Eadington (1978, p. 11) reports that "microlitre quantities of solution can be freeze-dried onto an aluminum substrate to produce tiny disks of dried salts of reproducible geometry. Calibration is by freeze-drying standard solutions." Although the microliter volumes mentioned are considerably larger than those involved in the minimillimicrolollipop method, this freeze-drying procedure would seem to be an even better method of getting the inclusion solutes into the microprobe, and should be explored fully.

Scanning electron microscope. This instrument, abbreviated SEM, has several useful applications in inclusion studies. It differs from the electron microprobe in several respects. One is that the electron beam may be focussed much more sharply, permitting studies of much finer details, but the analytical results, however, are more often qualitative or semiquantitative. Like the electron microprobe, it can be operated in several modes. In the backscattered electron scanning mode, an image can be obtained from a polished surface that provides a map of the spatial variation of average atomic number, as reviewed by Caruso and Simmons (1984). Electrically-conductive coatings are generally needed (C or Au are most common), and these coatings must be grounded to the sample holder. Small grains can be grounded by cementing to the holder with electrically conductive paint. Uncoated samples can be used in some cases (Robinson, 1983).

The secondary electron mode is most frequently used in fluid-inclusion studies, on inclusions that are exposed on a broken surface. (Controlled breaking is best achieved by use of a modified miniature machinist's vise, e.g., as illustrated by Hallbauer (1983).) In this mode, the average atomic number in various portions of the sample has relatively little effect on the signal, but the topography of the sample has a great effect. As a result, the shapes of exposed daughter crystals, the walls of the inclusion, and incrustations from the evaporation of inclusion fluids are revealed, highly magnified and in delightful detail, permitting recognition of some phases on the basis of crystal habit and symmetry alone. Tilting of the sample in the SEM permits viewing hidden corners, and stereopair photography. After such an image is obtained, the instrument can be shifted to the point mode and the beam placed on a given daughter crystal. By use of an energy dispersive X-ray spectrometer (variously termed "Edax," "EDA," "EDX" or "EDS"), the characteristic X-rays emitted from that point on the daughter crystal can be converted to a qualitative or semiquantitative analysis of the elements present.

As inclusions are generally small, electron-microscope techniques would seem to be very appropriate for their study, but the very shallow penetration of the beam precludes its use on most unopened inclusions. The secondary electron topographic images of broken surfaces, or of various replicas from them (e.g., Akizuki, 1965a,b,c, 1966, 1967a,b, Kurshev and Trufanov, 1965) have been used to study the structure of the walls of inclusions and the changes in shape with recrystallization. The shapes and distribution of minute fluid inclusions on quartz grain boundaries of quartzites were studied by Sella and Deicha (1962a,b, 1963). During the following 12 years, the only use of SEM for inclusions was in the secondary electron "topographic" mode. Then, however, with the introduction of the nondispersive detector, semiquantitative analyses of minute areas picked out from the secondary electron images could be made. Hallbauer (1982, 1983, and pers. comm.) used this procedure to examine the composition of the residues from the evaporation of opened fluid inclusions, from quartz pebbles in the Witwatersrand Au deposit, Republic of South Africa. He showed that some inclusions contained significant Sn and Cs; the combination of these data with the occurrence of cassiterite daughter crystals in other inclusions suggested derivation of this quartz from a pegmatitic Sn deposit perhaps similar to those which are known to occur in the Archean Sn belt of Swaziland. Quite different trace element assemblages were found in the evaporated residues from inclusions in quartz from the nearby Rooiberg Sn deposit, indicating that such analyses may provide valuable "fingerprints" for provenance studies. Similarly, Bray (1980) found 13 elements in inclusions in quartz from a Cornish china clay pit, even though many of these elements were not present in minerals in the pit. Haynes et al. (1983) used SEM/EDA to determine the ratios of Na:Ca:K:Mg:Cl in the residues from evaporation of inclusion fluids on decrepitation; freeze-dried residues yielded similar ratios.

Although the analyses are only qualitative or semiquantitative at best, they permit much more unambiguous identifications of daughter minerals than mere crystal shape alone. Many additional applications have been made. Metzger et al. (1975, 1977) reported on the application of the technique to several geologic environments, Nesbitt and Kelly (1975, 1977) reported on liquid and melt inclusions in a carbonatite, and Le Bel (1976) described a series of daughter crystals from a porphyry Cu deposit. Anthony et al. (1983) showed that the procedure could be made sufficiently quantitative that data reduction programs could be applied to yield the stoichiometric ratios adequate for mineral identification. Figures 5-11 through 18 present some examples showing the amazing potentialities of the SEM procedure.

Although the identification by SEM of daughter crystals (as well as identification of the incrustations from evaporation of inclusion fluids) is a real breakthrough, it has certain limitations that must be remembered in using the data: (1) Although the samples can be tilted, they are seldom a flat plate perpendicular to the beam, so truly quantitative analyses are almost impossible, regardless of the sophistication of the data-reduction computer programs used. (2) Also as a result of irregular geometry, the direction of the secondary electron detector from the beam point must be adjusted to minimize absorption.   (This direction, and/or the takeoff angle, may preclude analysis of minerals in a cavity.)   (3) even though the beam may be only a small fraction of a micrometer in diameter, it penetrates at least several micrometers (depending upon the accelerating voltage used), so the "analysis" may well include material behind the crystal of interest.   (4) Even if low-energy electrons are used, to yield minimum penetration, secondary X-ray excitation of neighboring areas will con-

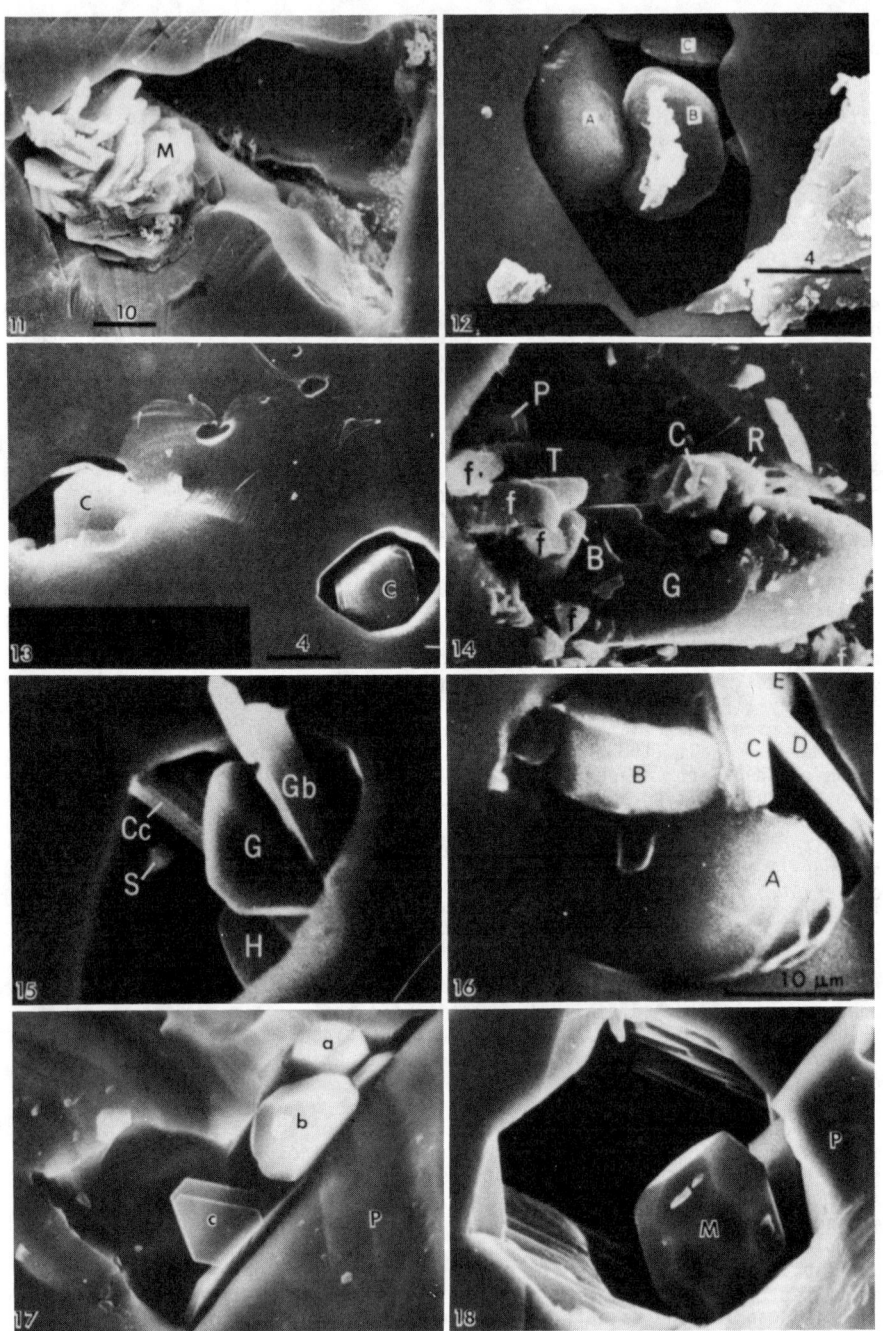

Figure 5-11 to 5-18 (on opposite page). SEM photographs of opened inclusions containing possible daughter crystals. Scale bars in μm.

tribute to the signal to the detector. (5) Many daughter crystals are covered with a film of salts from evaporation of the inclusion fluid (or the crystal itself may have formed on evaporation). (6) Unless handled quickly, hygroscopic daughter crystals may disappear before evacuation (Fuzikawa, 1982). (7) One cannot easily preselect the inclusion to be opened, and many lose their daughter crystals on opening. (8) Elements lighter than Na cannot be analyzed, and even Na and Mg give rather weak signals with present detectors.

These eight aspects are not meant to discredit the method; as will be seen in later chapters, the data derived even in these first few years have been invaluable. The last two items above deserve some additional comment. Loss of daughter crystals on fracturing the host might be minimized by cooling to LN_2 (liquid nitrogen) temperatures before breaking, followed by slow warming to let any solid CO_2 evaporate quietly (Fuzikawa, 1982; Kelly and Burgio, 1983). In connection with item 8, inclusions in Li pegmatite minerals present a special problem. Thus London et al. (1982) reported some daughter minerals in spodumene that gave no X-ray signal (later shown to be Li borate by means of the Gandolfi X-ray camera), and others that show only Si and Al. The Si/Al ratio in the latter permits inferences as to the presumed Li-Al-silicate phase involved, but not identification.

An interesting possibility for the use of the SEM to obtain the composition of the fluid part of inclusions would involve freezing the inclusion solid, cracking to expose the inclusion, and then using the SEM on the frozen inclusion contents. Pesheck et al. (1981) have applied this procedure to studies of inter-granular fluids in rocks and it should be equally applicable to intragranular fluids (i.e., fluid inclusions). They found that sublimation under the beam was negligible for oils below -140°C, and for salt solutions below -179°C. Kelly and Burgio (1983) and Kelly et al. (1983b), have adapted this procedure to obtain qualitative and semiquantitative analyses of frozen inclusions. Stage design and sample preparation steps are both critical and complex.

Even with qualitative SEM compositional data, the identification of the daughter crystals in inclusions is generally subject to considerable ambiguity. Some of the "identifications" or "suggested identifications" that have been pub-lished are based on unwarranted assumptions, and others are obviously wrong. Thus in one report, a phase that showed Ca and Cl by SEM was "identified" as

Figure 5-11. Quartz from Bingham porphyry Cu deposit, Utah, USA, showing platy assemblage of crystals tentatively identified as muscovite (M; i.e., sericite) on basis of morphology and compo-sition (major K, Al, and Si). From the sporadic occurrence of such crystals in the inclusions, and the relatively large volume percentage here, I presume this group constitutes an accidental solid inclusion, rather than a daughter phase.

Figure 5-12. Quartz from porphyry Cu deposit at Cerro Verde/Santa Rosa, S. Peru, showing probable halite (A), sylvite (B), and an Fe (and K) chloride (C). From Le Bel (1976, his plate III, Fig. 3).

Figure 5-13. Same locality as Figure 5-12, showing a chalcopyrite (C) crystal in each cavity. From Le Bel (1976, his plate V, Fig. 2).

Figure 5-14. Fluorite from the Emmett mine, Colorado, USA, containing six daughter minerals--gyp-sum (G), ferroan rhodochrosite (R), celestite (C), thenardite (T), phlogopite (P), and barite (B). Numerous cleavage fragments of fluorite (several are labelled f) are strewn over the surface. From Metzger et al. (1977).

Figure 5-15. Hydrothermal type inclusion in calcite from Magnet Cove, Arkansas. Several daughter minerals were definitely identified as sylvite (S), halite (H), and gypsum (G) and others tenta-tively as glauberite(?) (Gb) and chlorocalcite(?) (Cc). From Metzger et al. (1977).

Figure 5-16. Olivine from the famous Red Sea locality (Zabargad) for gem peridot, showing NaCl (a), gypsum (b), cancrinite (c), talc (d), and goethite (e). From Clocchiatti et al. (1981).

Figure 5-17. Pyrite (P) from Ba-Cu-Zn-Pb-Ag-mineralized volcanic rock, from Bien Venue farm, Bar-berton District, South Africa, showing arsenopyrite (a), quartz (b), and muscovite (c). From Hallbauer and Kable (1982).

Figure 5-18. Pyrite (P) from nonmineralized volcanic rock from an area adjacent to that of Fig. 5-17, showing magnetite(?) daughter mineral(M). The magnetite contains traces of Mn, Ti, and Mg. From Hallbauer and Kable (1982).

CaCl$_2$·6H$_2$O. However, CaCl$_2$·6H$_2$O melts at 29.92°C, yet these particular daughter crystals melted at 365-538°C! Solubility and particularly its change with temperature can also be instructive, but must be used with care in multicomponent systems. In this example, note that all of the calcium chloride hydrates are very soluble, and so even if CaCl$_2$·6H$_2$O did not melt at 29.92°C, since its solubility is 74 wt % at 0°C and 84 wt % at 20°C, it is hardly likely that it would persist in contact with solution as high as 538°C. Similarly, FeCl$_2$·4H$_2$O has a solubility of >80 wt % at 100°C, so would not be expected to persist to 570°C, as reported. (This also raises a question concerning the validity of the identification of the dihydrate, FeCl$_2$·2H$_2$O.) Others have thought they had FeCl$_3$ as a daughter phase, yet it melts at 282°C, and even at 100°C, a saturated solution of it is an orange syrup containing 84 wt % FeCl$_3$. Both calcium and iron chloride compounds in the inclusions may be hydroxy chlorides, chlorocarbonates, oxychlorides, or other similar phases that might be expected to have higher melting points and lower solubilities, but still show only the one cation and anion by SEM.

Transmission electron microscope. In this instrument, called TEM, the sample must be thin enough for the electron beam to penetrate through, and hence it must be either a very thin edge or a specially prepared "foil." Foils are made by eroding away the surface of a thin section with an ion beam ("ion thinning" or "ion milling"). The thickness used can vary with the voltage accelerating the beam of electrons. The foil can be observed in simple transmission (i.e., a shadow picture, like medical X-ray photographs); in this mode minute fluid inclusions (or voids?) can be seen as low density spots, e.g., in heat-treated citrine and amethyst (McLaren and Phakey, 1966). The TEM can also be used to reveal host-crystal dislocations associated with inclusions, and in the diffraction mode, has been used to identify the crystals in evaporated inclusion residues (e.g., Kalyuzhnyi and Mikolaichuk, 1968; Trufanov and Kurshev, 1968). Dimov et al. (1980) used it to report 15 different minerals, some quite rare, in evaporated residues from aqueous extracts from galena, sphalerite, quartz, and pyromorphite from the Madjarovo ore deposit in Bulgaria.

Ion microprobe. This instrument, which uses a focussed beam of heavy, charged ions to sputter off single atoms (or in some cases groups of atoms) from the surface of the sample, which are then analyzed by mass spectometry, has been used in only a few inclusion studies, so its full potential is still not adequately explored. Havette and Weiss (1976; see also Castaing et al., 1978) showed that it can be very effective in the study of silicate melt inclusions, in which daugher crystals >20 µm can be identified. The first use of the ion probe on aqueous inclusions was by Nambu et al. (1977). As the ion beam can remove significant amounts of material from the sample area impacted, Nambu et al. used the beam to bore into and then sputter away each inclusion (<30 µm diameter). To immobilize the inclusion contents, they kept the sample at -90°C. They obtained significant ion intensities for Li$^+$, B$^+$, Na$^+$, Mg$^+$, Al$^+$, K$^+$, Ca$^+$, Mn$^+$, Fe$^+$, Cu$^+$, and Zn$^+$; quantification could only be in terms of ion intensities relative to ^{23}Na$^+$. Yurimoto and Sueno (1984) recorded the spectra as the beam (of O$^-$ ions) sputtered through the contact between a silicate-melt inclusion in olivine and the host. One of the advantages of the ion microprobe over the electron microprobe is that it can be used on elements down to mass 1 (H). Thus water in silicate melts and melt inclusions can be determined (Delaney and Karsten, 1981), but calibration is difficult (Steele, 1983).

Proton microprobe. In this technique, called PIXE for proton induced X-ray emission, a highly collimated beam of protons is used to induce emission of characteristic X-rays from the sample. It is thus similar to the electron microprobe, but at present it requires larger sample areas (4-6 µm). Although it has not yet been used on fluid inclusions, its use on sulfides and oxides (Blank et al., 1982; Cabri et al., 1983) shows that it is particularly effective for many trace elements, and will certainly be appropriate for a variety of melt inclusion studies.

Chapter 6

INCLUSION SAMPLE SELECTION, PREPARATION, PETROGRAPHY and PHOTOGRAPHY

"There are no quick and easy ways to obtain fluid inclusion data"
(Hollister et al., 1981, p. 301)

CONTENTS

INTRODUCTION

It is all too common in most scientific literature that the important practical aspects of fluid inclusion research are omitted and have to be learned by each individual worker by experience (i.e., at a considerable waste of time). In this chapter, I discuss the methods and mechanics of obtaining an appropriate fluid inclusion sample for study -- what should be done, and the methods and sequence to be used -- and the problems most likely to be encountered. Thus the chapter includes criteria for sample selection and preparation from the outcrop to the microscope, and the various petrographic procedures for examining and photographing fluid inclusions. Only the most commonly used procedures are described here; many other special techniques, applicable only under limited conditions or using equipment not normally present in a fluid inclusion laboratory, will be found in other chapters.

The first review in English of laboratory techniques for fluid inclusion studies was a brief section in Roedder (1976a), aimed mainly at studies of ore deposits. Some of the laboratory problems involved in the study of melt inclusions were described in a review of magmatic inclusions (Roedder, 1979a); the latter review was followed by a more extensive discussion (Hollister et al., 1981) in a book (Hollister and Crawford, 1981) that dealt with petrologic applications of fluid inclusions, most particularly the study of the very small inclusions commonly found in igneous and metamorphic rocks. Parts of the present chapter are based on these reviews.

SAMPLE SELECTION IN THE FIELD

The selection of material to be studied is the most important aspect of fluid inclusion research. Optimum material will vary, of course, with the specific study, but, for normal optical determinations, the optimum would be large, clear, euhedral zoned crystals. Material of this type is not generally available, and even when it is, it may represent only the very latest stage of mineralization, as in many ore deposits. Sometimes, therefore, ordinary rocks must be studied. How should such material be sampled? Because the study of fluid inclusions is only a tool to help solve geological problems, development of an appropriate sampling procedure for a given fluid inclusion study should be based primarily on a specific understanding of the problem to be solved.

Before any extensive field sampling is attempted, one should verify (if possible) the presence of usable inclusions in a few samples from the field area. In many situations, this can be done without making special efforts for sample preparation, and one does not need access to a heating/cooling stage. Because most inclusions are <10 μm, they can generally be seen in a thin section

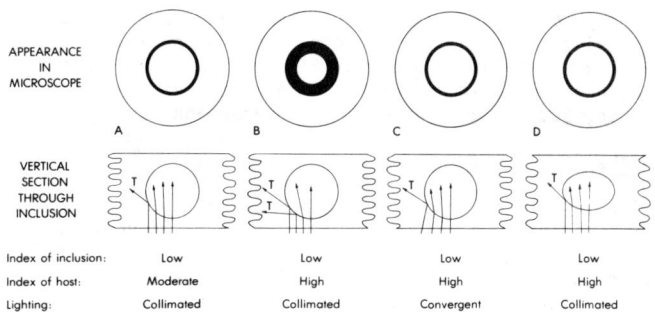

Figure 6-1. Diagrams illustrating the qualitative interpretation of the black bands from total reflection at the edge of an inclusion. Total reflection (light rays marked T) increases with difference in index ((A) vs (B)), with collimation of the light ((C) vs (B)), and with inclusion sphericity ((D) vs (B)). Additional reflection and refraction occurs at upper surface of inclusion. The diagrams are just as applicable to bubbles in a fluid inclusion as to an inclusion in a host mineral.

of normal thickness (30 μm). Therefore, many studies of fluid inclusions begin with observations of thin sections that have already been prepared for routine petrography. Such sections can provide guidance for subsequent sampling but are not adequate for most inclusion studies.

Another good way to make a preliminary check for fluid inclusions is to view small mineral fragments immersed in an oil with an index of refraction close to that of the mineral. The crushing stage is also exceedingly useful for reconnaissance if inclusions with high internal pressure are expected (see Chapter 7).

A few fluid inclusions can be found in quartz in virtually any rock, but, as they are much lower in index of refraction than the quartz, total reflection occurs at the inclined interfaces. As a result, when viewed with ordinary microscope illumination (i.e., rather strongly collimated), at low magnification, they generally appear as opaque specks. At higher magnification, and particularly when the substage condenser is inserted to provide convergent illumination, the broad dark borders resulting from total reflection are greatly minimized (Fig. 6-1), and the phase boundaries within the inclusion become visible.

Sampling for inclusion studies may differ significantly from other sampling. Hollister et al. (1981) provided some general guidelines for sampling in sedimentary, metamorphic, or igneous terranes; these are summarized below. Additional suggestions along such lines are implicit in many of the following chapters, in which fluid inclusion studies on a variety of more specific geologic environments are detailed. Thus sampling for studies of ore deposits may involve many procedures that are inapplicable in other types of materials.

Although some fluid inclusion studies can be reduced to the routine accumulation of data by the application of standard techniques, with results that are essentially as expected, most studies do not follow such a dull path. The available material may simply not be suitable to provide the answers sought, for any of a number of reasons; such an outcome is not uncommon in fluid inclusion work and one must learn to be stoical. The most common reasons for such disappointment are the absence of: (1) primary inclusions; (2) inclusions of any sort in the minerals of greatest interest; or (3) inclusions large enough to study with the available techniques. Sometimes perseverance in observation may eliminate these problems, but it is obvious that certain rocks are simply not suitable. More commonly, suitable inclusions are present, but the processes that have formed them and the host minerals turn out to be more complex than originally thought. Probably most samples contain some primary inclusions, but proving such an origin may be impossible. On the other hand, examination of the samples may reveal the existence of new, unexpected problems or phenomena, requiring redirection of the research, so one must also be flexible. More than one of my own studies has been a result of just such serendipity; the original problem remained unsolved, but new avenues were discovered and explored.

Sedimentary Rocks

In sampling sedimentary rocks, the major and often insurmountable hurdle is finding material coarse enough to study. It is relatively unusual to find a fluid inclusion as large as a tenth the size of the host grain, so fine sediments may have no usable inclusions. Most fine sediments do have some more coarsely crystalline material, such as fossil fillings, veins, concretions, or vugs, containing inclusions that may provide information on the post-depositional history. Grains of coarser detritus often contain inherited inclusions, resulting from processes that took place in the provenance, and overgrowths on detrital grains, although small, can yield usable inclusions that record conditions at the time of the overgrowth. Any recrystallization of the original sedimentary material, whether it be termed "authigenesis" or "low-grade metamorphism," may trap samples of the interstitial fluid at that time, and hence sampling should be concentrated on such material. The sequence of such events is frequently the most difficult aspect to determine, and all pertinent field data should be recorded.

Metamorphic Rocks

The sampling problems for metamorphic rocks are in part similar to those for sediments, except that the minerals in metamorphic rocks are usually coarser, and generally have had a much more complex history. An extensive structural and petrologic study of a metamorphic terrane is essential before a meaningful selection of samples can begin. As in sediments, inclusions from earlier stages can sometimes be found in appropriate samples, such as pebbles from a conglomerate, or boudins from a preexisting vein or layer, or a residual porphyroblast in a retrograded rock. Recrystallization in metamorphic rocks, whether in pressure shadows, gash veins, pod-like segregations, or other types of environments, is frequently a multistage process, and fluid inclusion studies will help in unravelling the sequence of events only if the sampling is done with careful attention to the relations of the samples to other geological features at the outcrop. Fluid compositions will also vary with the nature of the host rock.

Igneous Rocks

Igneous rocks involve rather different processes than sedimentary and metamorphic rocks, and hence sampling may differ considerably. Thus the igneous rock with the finest "grain size," i.e., glass, may contain fluid inclusions (vesicles) that can yield valuable information. Numerous studies have concentrated on the inclusions in phenocrysts or xenocrysts. Many other studies have involved xenoliths (sometimes ambiguously termed simply "inclusions") brought up with basaltic magmas. Still others have used fluid inclusions to help in following segregation processes such as the separation of an immiscible fluid phase (sulfide melt, hydrosaline melt, dense CO_2, another silicate melt, or an ore fluid) or in recognizing magma mixing or other changes in the liquid line-of-descent. Inclusions in pegmatites or pegmatitic segregations may also provide important information on such processes.

Hydrothermal Ore Deposits

Before sampling a hydrothermal ore deposit for fluid inclusion studies, it is most important to clarify, as far as possible, the sequence of all igneous, metamorphic, tectonic, and hydrothermal phenomena. This is required to relate results from fluid inclusion studies to specific events and thus to understand the temporal evolution of the process of mineralization. (Not uncommonly, these are the very questions we are trying to answer by studying fluid inclusions.) Careful observation and interpretation of systematic cross-cutting relationships between vein types, igneous intrusions, contact metamorphic effects, hydrothermal alteration zones, and structural features should be followed by detailed hand specimen, thin section, and polished section descriptions of the paragenetic sequences of crystallization and observations of superimposed metasomatic mineral assemblages (e.g., in skarns). If possible, material should be collected to ensure representative horizontal and vertical sampling so that spatial variations in fluid inclusion properties may also be determined. In such respects, sampling of hydrothermal deposits is similar to that of any other geological environment.

SAMPLE SELECTION FOR CUTTING AND POLISHING

Preparation of the doubly polished plates normally used in fluid inclusion work is more time consuming than preparing standard thin sections and requires more care than is generally given to these steps. In order to save time, a careful hand specimen study with a low power binocular microscope can find those areas or individual crystals that are the most likely candidates for further study. Cleavage plates are particularly useful. A coverglass stuck on with an index liquid that matches the mineral helps in looking into imperfect crystal faces or cleavage surfaces. The larger inclusions in clear crystals can be readily seen with a 10X hand lens, but lack of such inclusions should not exclude the material from further study. Less transparent minerals can be crushed and examined in a matching index liquid. Small loose crystals are best examined in a bath of matching index liquid; for this purpose, I cement thin rings of glass (2- to 5-mm sections cut from glass tubing) onto ordinary glass slides with epoxy. The crystals to be examined are placed in the ring, and index liquid added until the reflection from the top surface shows a change from concave to very slightly convex. A coverglass is then placed on top and is held in place by capillarity. Too little liquid results in a bubble, that invariably lies over the grain; too much can also cause trouble but can easily be removed with a thin strip of blotting paper. These ring slides can be used also for hand-picking grains under the binocular microscope for further preparation by filling with an index liquid to the point where the meniscus is flat, eliminating the need for a coverglass.

As only an approximate refractive index match is needed, several inexpensive liquids can suffice: kerosene of n = 1.45 (for fluorite); "HB 40 oil" made by the Monsanto Chemical Co., Wilmington, Delaware, adjusted with kerosene to n = 1.55 (for quartz); and α-monobromonaphthalene of n = 1.64 (for olivine). Methylene iodide saturated with sulfur (n = 1.788) is useful for many minerals of higher index of refraction, but there is no inexpensive matching liquid for sphalerite. For groups of tiny grains, ordinary liquid mounts with a coverglass suffice. When a single tiny grain is to be examined, however, it is easiest to put it in a tiny U-shaped "corral" of appropriately thin wire under the coverglass (a piece of paper clip will do for large grains). The coverglass should be only slightly larger than the corral, ~3-5 mm on an edge. The grain can then be manipulated for observation from various directions by using a probe tipped with a single tapering camel's hair, inserted under the coverglass in the open side of the corral.

Rankin and Aldous (1979) use tritolyl phosphate[1] as a mounting medium.

----

[1] Other names by which this material is sold include tri-cresyl phosphate, phosphoric acid tri-(methyl-phenyl) ester, and the trade names Lindol, Celluflex, and Kronitex. As the ortho form is hazardous if ingested, the "ortho-free" grade should be specified. The pure isomers are expensive; a suitable oil, e.g., was a technical grade mixture of 80% para and 20% meta isomer, available from Fisher Scientific Co.

It has a refractive index of 1.55 and is inexpensive. The main advantage is that it stays an optically clear metastable glass on cooling, even to -100°C, hence permits small grains or unpolished plates of quartz to be run on the freezing stage without polishing.

Large crystal size increases the probability of finding adequate criteria for establishing fluid inclusion origin, as well as the probability of finding the larger inclusions that are easier to study. On numerous occasions, however, crystals <1 mm have been found to contain adequate inclusions. These are difficult to prepare for study if they are free or projecting into vugs but present no problem if they are embedded in a hard matrix. Sections made of crystal-lined vugs are usually disappointing in that the cross-sectional area of such crystals in any given section is much smaller than expected.

Inclusions in opaque minerals are commonly analyzed chemically. Although the analytical procedures must be applied essentially blindly, some indication of the abundance of inclusions can be obtained by examination of the broken surfaces for cavities. Although single inclusions in the millimeter range could be located by X-ray procedures, even in a slab of galena 1 cm thick, the density contrast would not be adequate to differentiate between a full and an empty inclusion, particularly when the third dimension is unknown. The large neutron cross section for hydrogen permits the location of even small water inclusions as shadows on neutron radiographs; their position can even be determined in the third dimension through the use of stereo pairs (J.R. Dooley, Jr. and J.R. Shoptaugh, oral comm. 1969).

CUTTING AND POLISHING

The next step is the preparation of thin (or "thick") plates, mounted or unmounted, that are polished on both sides (for improved optics) and hence suitable for use on a heating/cooling stage. Most such plates are 0.1-1.0 mm thick. I cannot overemphasize the importance of adequate preparation of these plates. They are an absolute necessity for any serious fluid inclusion study, for several reasons. First, as demonstrated by Heald-Wetlaufer et al. (1982), study of such plates permits recognition of many features that would ordinarily escape detection by conventional petrographic methods. Furthermore, optical examination of inclusions, particularly under the less-than-optimum optical characteristics of heating and cooling stages, is sufficiently difficult without being penalized by additional problems from poor sample polish, which causes serious image degradation. For ordinary microscopy (but not for heating or cooling), a coverglass with a closely matching index liquid is a poor substitute for a good polish.

Since sample preparation is so costly and time consuming, much effort can be saved if plates are made from only a small group of the best samples first. These will point out any needs for changes in plate thickness, cutting direction, or other preparation features. If these first plates provide good inclusions, a systematic preparation of additional samples can proceed. If, however, these plates of the "best" material yield nothing of use, the "next-best" samples can be prepared, and so on. This will minimize wasted effort. Before abandoning any project for this reason, it is important to remember that the occurrence of "good" inclusions in any given sample is extremely difficult to predict. Different minerals in a given rock may vary widely in the size and frequency of occurrence of inclusions, and different grains of a given mineral in a given hand sample may vary similarly. Perseverance may pay off, even in some seemingly hopeless cases.

Plates are usually prepared by cutting a slice in the "optimum" direction for the specific sample. What is "optimum" must be based on experience, and the nature of the material. Individual crystals are generally cut through the center, but the specific direction will vary. Thus apatite and beryl frequently bear long tubular inclusions parallel to the c axis, and hence the cut should be parallel to c, but quartz may have inclusions elongated parallel to a or c, or perpendicular to {1011}. This cutting should be done with a thin continuous-rim sintered diamond cutoff wheel and adequate coolant. These wheels can be obtained in a variety of grits and grades specifically suited for cutting various types of materials. They cause much less fracturing of the surface adjacent to the saw cut than occurs with the standard notched-edge diamond blade used in slicing for thin-section preparation. The thinner the blade, the less pressure is needed in the cutting and the less sample is lost but the more chance there is of blade breakage. For small crystals, and for cutting wedges out of prepared slices, I use a blade 0.015 inch thick and 5 inches in diameter. Modern slow-speed diamond saws, using thin blades and gravity feed, cut slowly but automatically and yield very smooth, unfractured surfaces.

The quality of the polish, especially on the top surface, is critical. Many sample preparers do not polish the bottom surface of plates cemented onto glass, since the cementing material appears to eliminate the need for a polish. However, this is actually true only for those minerals having the same index of refraction -- and dispersion -- as the mounting medium, and the greater the difference in index, the greater the degradation of the image, particularly for small inclusions near the ground surface.

Sample preparation should be planned to avoid both excessive heating and mechanical shattering of the sample. In general, temperature during sample preparation should be kept below 100°C to avoid decrepitation of inclusions with high filling densities (i.e., those with low Th). Heating to 100°C may be too much for inclusions in some minerals formed at <100°C, and may even be too much for some dense CO_2 inclusions formed at 1200°C. The maximum actual sample temperature during sample preparation should always be recorded for interpretation of low Th results. If Th measurements turn out to be in the vicinity of the maximum temperatures used for sample preparation, some of the inclusions may have been damaged by stretching or lost by decrepitation (see Chapter 7), and new samples should be prepared, using lower temperatures. Several cements that require no heating are available. Thus cellulose nitrate base cements such as Duco can be used, but they are only effective if a porous backing such as wood is used, rather than glass, to permit evaporation of the solvent. Room-temperature-setting epoxy resins are frequently used, and the cyanonitrile-type contact cements such as "Super-glue" are particularly convenient because the set cement is soluble in acetone. UV-polymerized cements can also be used, but only with a slide material (or sample) that is transparent to UV.

Impregnation, which is required for porous or friable specimens, should be done at very low or room temperature. Several impregnation steps may be required to prevent the sample from disaggregating during the polishing procedure. Grinding steps used to thin the chip should be performed with care. Impregnated samples may disaggregate when heated on the stage. A record should be kept of the impregnating and cementing materials used and their indices of refraction. Some epoxy resins, particularly the "5-minute" types, turn very dark and blister at about ~150°C as they start to decompose; others, such as DER 332 with TETA hardener (Dow Chemical Co.) and Araldite 105 (CIBA Corp.), turn brown in thick section at 300°-350°C, but are still transparent and usable. Also, the hot-plate temperatures needed (and the even higher temperatures generally used) for the thermoplastic cements such as Crystalbond, Lakeside 70, and Canada balsam are well above the Th of the low temperature inclusions characteristic of some samples.

Some (improper?) mixing and curing procedures on some epoxies form immiscible liquid inclusions in the cured resin, many of which contain a moving vapor bubble, and excess epoxy can appear to be indigenous to the sample (see beyond, under "Artifacts"). Bubbles in the cementing materials should be carefully avoided, as they generally spoil the plate for photography, and if numerous, they confuse the eye in searching for suitable inclusions.

A perfect polish is first produced on one side of the sample. The size of the area to be polished should be selected with care, to avoid waste of time. An area that will fit on a normal polished circular 25 mm "probe mount" or a regular petrographic thin section is convenient to use but requires considerable time for polishing (if the mineral is hard) and may be excessive (if the mineral is inclusion-rich). Careful cleaning in an ultrasonic bath between steps is essential. The polished side is then attached to a glass slide with a cement that may be easily dissolved or melted. The specimen is then sawed and ground down to the desired thickness, and this second surface is polished.

The thickness of the doubly polished plate must be appropriate for the microscope freezing/heating stage to be used and for the particular sample. Plates that are too thick prevent adequate illumination and usually result also in degraded images from birefringence or in the superposition of other inclusions, cracks, etc., upon the image being examined. They also are slower to equilibrate in the heating/cooling stage. On the other hand, plates that are too thin present less actual volume of mineral for examination, for a given amount of sample preparation. Average material is usually suitable at ~1 mm thickness, but, if a dark and a clear mineral are present together (e.g., sphalerite and quartz), the quartz should be scanned when the slide is still thick, or the slide can be polished and cut in half at this stage and only half ground down for the darker material. Very dark minerals may require doubly polished sections of less than standard thin-section thickness (<30 μm); warpage in the cement can be a serious problem here. Standard thin sections, even if prepared with care, are satisfactory for examination for very tiny inclusions only, as all larger inclusions are necessarily lost in preparation. Further, as inclusions even in the center of the section are at maximum only 15 μm from a normally rather roughly prepared surface, they are apt either to have leaked or to leak during heating (Roedder and Skinner, 1968).

Thin crusts of crystal druses and fragile or porous samples are best prepared by coating or embedding in epoxy and curing, grinding down to the point where a maximum amount of the desired crystals are intersected, and then polishing. This polished surface is cemented to a slide with a soluble cement and a second parallel saw cut is made. Embedding is best done under vacuum. In most preparation laboratories, the sample is immersed in the embedding resin and then placed in a vacuum chamber until it stops bubbling, and then air is let in. Although frequently effective, this procedure is basically wrong, since the vacuum must work to pull air out against capillarity drawing the fluid into the sample. Much more effective embedding is obtained if the air in the pores of the sample is evacuated, and the resin is added while still evacuated. One of several procedures for achieving this is shown by Emons et al. (1982, their Fig. 4.7). After the sample is completely immersed, air is let in, and the vacuum in the pores then works in concert with capillarity to pull in the resin. A tippable container in the vacuum chamber for the resin mix and a rotatable carrousel for a series of samples permits multiple embedding with a single evacuation.

Before any section is thinned by grinding, it should be examined with a "temporary polish" in the form of a coverglass and matching index liquid. In this way, one can frequently recognize a large inclusion that is near the surface and that can be saved only by grinding and polishing from the opposite side.

All impregnating and cementing materials must be chosen with the nature of the sample and its final use in mind. Thus, many epoxies can stand both the heating and cooling operations necessary for samples from low-temperature environments, and hence, if the stage permits, the sample plate can be left on the glass slide during these operations. This is particularly important with very dark minerals like some sphalerites that must be cut into very thin (and, hence, fragile) plates. For several reasons, I prefer to complete all normal microscopy with the plates still cemented. If at all possible, however, the sample plates should be removed from the glass slides before heating or cooling runs. If thermoplastic resin is present during heating runs at ~150°C, even just as films in cracks, it will creep out over the surface, spoiling the optics, and will fog the stage windows. Ultrasonic cleaning in a suitable solvent is almost mandatory.

Heald-Wetlaufer et al. (1982) presented a comprehensive review of the preparation procedures used in the U.S. Geological Survey, including a detailed comparison of available impregnating and mounting media (Table 6-1). Additional details on polishing techniques are given by Cameron (1961), Saager (1967), Allman and Lawrence (1972), Simmons and Richter (1976), Holland et al. (1978), and numerous others (see particularly the references in Heald-Wetlaufer et al., 1982). Most fluid inclusion laboratories have developed their own procedures, but it is important to note that the procedures will (and should) vary not only with the laboratory but also with the mineral and the kind of study. Most important, the procedures used in ordinary commercial thin section preparation, including "polished thin sections," are generally not sufficiently gentle in the sawing and grinding stages to yield an unaffected inclusion sample.

INCLUSION PETROGRAPHY AND SELECTION

General Inclusion Petrography

Fluid inclusion petrography is very unlike standard petrography in many respects. The most important difference is that one must view inclusions in a three-dimensional context and not in the normal two dimensions as in ordinary petrographic studies of thin sections. Also, to quote T.J. Reynolds, one does not simply flip in and out the substage condenser and the upper polar. An experienced fluid inclusionist must constantly use each available adjustment, particularly those affecting illumination, to maximize the quality of the optical image.

It is particularly important to spend enough time on microscopy before any runs are made. Many samples require more time at the microscope before making thermometric runs than is needed for the runs themselves. Not only will additional time frequently yield inclusion data not recognized at first, but a more careful search will almost always result in finding bigger, clearer, or more obviously primary inclusions than were found at first and hence will save much otherwise wasted laboratory time. This is particularly important because the optical images with the heating stage are so much poorer than with normal microscopy. It is even more important with freezing work, as ice is much more difficult to see than a bubble.

Although the temptation is great to start immediately with high magnification, this is very wasteful of time. The entire plate should be scanned at low power (or even better, with a 10x hand lens first), then at intermediate power, before the observer "homes in" on individual inclusions at high power. It is easy to focus on the details of the many small inclusions of a given type and not even see the much larger inclusions of another type present in the same field. The location of particularly good inclusions seen during the scanning should be noted, of course, but the assumption should always be: "There are probably other, better, inclusions". It is also imperative to view at low power before assigning a primary or secondary origin to a given inclusion, as many of the critical features (faint traces of fractures, alignments with other

Table 6-1. Characteristics of several impregnating and mounting media used in the preparation of thin sections. Modified from Heald-Wetlaufer et al. (1982).

MOUNTING MEDIUM	SOURCE	SETTING CHARACTERISTICS		MIXING RATIOS OF COMPONENTS (by volume)
		At elevated temp.	At room temp.	
Crystalbond 509	Aremco Products, Inc. P.O. Box 145 Briarcliff Manor, NY 10510 (914) 762-0685	Melts at ~70-80°C. Heat slide to melting temp. before applying. Hardens rapidly on cooling.	Not possible	N.A.
Lakeside 70C	Wards Natural Science Establishment, Inc. P.O. Box 1712 Rochester, NY 14603 (716) 467-8400	Melts above 80°C. Heat slide before applying mounting medium. Hardens rapidly on cooling.	Not possible	N.A.
Dow Epoxy Resin 332	Dow Chemical Co. 1603 Santa Rosa Rd. Richmond, VA 23288 (804) 288-1601	Mix and set at >70°C. Sets in 3-10 mins.	Resin needs to be heated because it crystallizes at room temp. Sets overnight.	5:1 at 70°C Hardener is TETA (triethylenetetramine) (Dow Epoxy Hardener 24)
Epofix	Bunton Instrument Co. 615 S. Stonestreet Ave. Rockville, MD 20895 (301) 762-5115 Also Struers Sci. Instr., Denmark.	Mix and set at 50°C. Sets in 2 hrs.	Sets in 8 hrs when mixed and set at room temp.	8:1 (9:1 by wt). Less hardener slows down the reaction and delays setting. Stir ~2 min.
Epoxide	Buehler, Ltd. 2120 Greenwood St. Evanston, IL 60204	Mix and set at ~50-65°C. Sets in 4 hrs at 50-65°C.	Sets in 4 hours at room temp. if mixed at 50-65°C. Sets in 6-8 hrs if mixed at room temp.	6:1 (by wt) at 50-65°C. 5:1 (by wt) at room temp. Mix by wt. only.
Petropoxy 154	Palouse Petro Products Rt. 1, Box 92 Palouse, WA 99161 (509) 878-1848	Mix and set at ~125°C. Sets in 3 min.	Will not set at room temp.	10:1 at 125°C. Mix for 2 min.
Spurr Low-Viscosity Embedding Media	Ernest F. Fullam, Inc. P.O. Box 444 Schenectady, NY 12301 (518) 785-5533	Sets in 8 or more hrs at 80-85°C. Temp. can be as low as 60°C. Longer setting times are recommended.	Will not set at room temp.	Mix at room temp. (g): VCD-10.0; DER 736-4.0; NSA-26.0; DMAE 0.6. Mix VCD, DER, and NSA thoroughly, then add DMAE and stir again.
Devcon "2-Ton" Clear Epoxy	Local hardware or variety store.	Sets in several minutes at 100°C.	Sets in 30 min.	1:1 (dispensed automatically).
Araldite-AY-105 resin	Chemical Coatings & Engineering Co., Inc. 221 Brooke St. Media, PA 19063	Sets in 3 to 8 minutes at 780°C.	Sets in 3 to 5 hrs but unsuitable because it will be cloudy.	1:1 (by wt) with hardener 935F. Mix thoroughly.

MOUNTING MEDIUM	ADVANTAGES	DISADVANTAGES	PARTICULARLY GOOD FOR....	COMMENTS
Crystalbond 509	Sets rapidly. Transparent (but pale yellow) when hardened. Soluble in acetone.	Friction from high-speed laps may melt Crystalbond, removing section from slide. Picks up lap contaminants readily. High viscosity.	Samples that need to be removed from slide later.	Overheating causes bubbles to form; they reduce adherence and optical quality.
Lakeside 70C	Sets up rapidly. Strong bond to slide bond. Soluble in methanol, or in acetone followed by a dilute solution of sodium borate. n = 1.54.	Same as for Crystalbond. Amber-colored.	Samples that need to be removed from slide later.	Overheating causes bubbles to form; they reduce adherence and optical quality.
Dow Epoxy Resin 332	Impregnates samples readily. Can be made up in large batches (100-200 ml). Sets up quickly if hot. Forms a thin bond. Nearly colorless.	Resin crystallizes at room temp; heat to ~60°C to melt. Short pot life.	Thin sections, probe mounts, impregnation.	Excess epoxy may peel off the slide with time. Viscosity 900 cps at 25°C.
Epofix	Impregnates sections quickly and thoroughly when mixed and set at 50°C.	30 min. pot life at 25°C. 15 min. pot life at 50°C.	Impregnation, probe mounts, cold setting.	Resin soluble in alcohol and acetone. Liquid hardener soluble in alcohol, acetone, & water. Exothermic reaction. Viscosity 550 cps at 25°C & 150 cps at 50°C.
Epoxide	Impregnates samples readily. Nearly colorless.	Will turn brown to black at T >250°C. If not mixed completely, bubbles form where hardener concentrates. Can mix only small batches (<50 ml.). Short pot life.	Impregnation, thin sections, probe mounts.	Highly exothermic reaction makes mixing of small (50 ml or less) batches of epoxy preferable.
Petropoxy 154	Good for mounting sections. Sets rapidly. Strong bond. Long pot life (~5) days, extendable indefinitely if refrigerated. Nearly colorless. n = 1.54. Slides need not be frosted.	Epoxy shrinks up to 1% and excess epoxy around specimen can crack slide. Must be mixed and set at elevated temp. Rock must be completely dry or bubbles form.	Thin sections, repairing broken slides.	High temperature set may destroy low-temperature fluid inclusions.
Spurr Low-Viscosity Embedding Media	Very low viscosity (60 cps). Impregnates samples readily. Good for fine-grained samples. Colorless to slightly tinted. 3-4 days pot life (must be refrigerated).	Four components to be mixed. Must be mixed and set at elevated temp.	Impregnation of fine-grained samples.	More D.E.R. will make softer mounts. Less DMAE (to as little as 0.1 g) will lighten color but will increase cure time.
Devcon "2-Ton" Clear Epoxy	Easy to mix due to double piston dispenser. Inexpensive. Nearly colorless.	Too viscous for effective impregnation. Turns medium brown above 250°C.	Thin sections.	Uncured mixture dissolves in acetone.
Araldite- AY-105 resin	If hot, sets up quickly, forming a thin strong bond.	Must be heated; too viscous at RT.	Thin sections.	Uncured mixture dissolves in acetone.

inclusions, vague growth color banding, etc.) are only visible at low power. On the other hand, high magnification is absolutely essential for all small inclusions and for small daughter phases. In this connection, it is important to note that optical data from inclusions too small to run on heating or cooling stages may still provide much valuable corroboration of the meager quantitative data from a few larger inclusions, and in some samples, such optical data may be all that are available from certain zones. I use a 100x oil-immersion objective (plus 12.5x or 16x oculars) regularly in my own microscopy and would recommend it as the most important single special tool for inclusion work. If its depth of focus is inadequate to reach a desired inclusion, try turning the sample plate over and focusing through the bottom.

Lighting is an important consideration in searching for good inclusions. If the plate is very clear and transparent and contains rare small inclusions, the best technique is to close the substage diaphragm down to a very small diameter; this shows up small inclusions over a considerable depth of focus. When searching for good inclusions in samples that are crowded with inclusions, cracks, or solid debris (and for close examination of any given inclusion), the diaphragm should generally be wide open. This eliminates much of the troublesome superposition of images. An efficient IR filter should be used in the illumination path routinely. Thus, I have found that the heating effects on CO_2 inclusions from the IR in the microscope light can be significantly reduced by inserting a second IR filter.

Total reflection of light, at the intersection of a sloping interface and a medium of lesser index of refraction, is sometimes considered to be an unavoidable burden in inclusion work, and it can cause errors (see below, under "Artifacts"). Without it, however, most phase boundaries in fluid inclusions would be invisible. Also, it provides important qualitative data on the nature of the inclusion (Fig. 6-1) and permits quantitative determinations of the index of refraction of the fluids (Chapter 4). Total reflection can be demonstrated, qualitatively, with any flattened two- or three-fluid-phase inclusion and a hand lens or a low power binocular microscope. Viewed in transmitted light from a distant, small source, all fluids will transmit light when the inclusion planar surface is perpendicular to the light. As the sample is gradually tipped, the gas phase will become black (totally reflect) first, then the liquids, in sequence. In liquid plus gaseous CO_2 inclusions, it is even possible to watch the total reflection of the gas phase disappear, at constant tilt, as the warmth of the fingers vaporizes liquid CO_2 and increases the density and hence index of refraction of the dense CO_2 gas.

Phase microscopy has been used in some instances to enhance the visibility of materials of nearly identical index, but generally in inclusion studies the differences in index are so large that this technique is not helpful. In some cases, however, interferometry has been used to advantage (Tolansky and Morris, 1947; Lemmlein and Kliya, 1952a; Loskutov, 1959). Ingerson (1947) showed that dark-field illumination improved the visibility of small bubbles during homogenization experiments.

I find that reflected light, preferably from flexible fiber optics, sometimes helps to distinguish between vague fluid-fluid interfaces and miscellaneous irregularities in the cavity walls (e.g., Roedder 1971b, his Fig. 24). Placing a slit in the illumination system, rotated parallel to a "difficult" interface, can also help. Meyer (1950) devised an ingenious method of combining reflected and transmitted light for high-temperature microscopy, by using vertical illumination and a heated metallic reflecting surface below the polished mineral plate.

Birefringent minerals should always be examined with one polarizer in place, set parallel to a vibration direction of the sample. This eliminates the annoying double images seen when looking deep into a mineral of low birefringence and visible at almost any depth in a highly birefringent mineral. The polarizer should be set parallel to the ordinary ray on highly birefringent, uniaxial negative minerals such as the rhombohedral carbonates, unless the inclusions are very close to the surface. This is because the extraordinary ray image is always severely distorted and fuzzy. (As the ordinary ray in these minerals has a much higher index of refraction, this also permits one to focus significantly deeper into the section and hence brings more of the sample within range of the oil immersion objective.) If coverglasses have to be used, it is best to use the thinnest possible grade ("00"), to obtain the maximum depth of focus into the sample.

Faint growth banding in colored minerals can sometimes be enhanced by the use of a wedge interference filter, adjusted to give maximum visual contrast between adjacent bands. Some growth banding in colorless minerals such as quartz is evident only as very minute differences in index of refraction for the individual bands. These differences may be seen most readily by using highly collimated lighting. Growth-banding and other planar features with which inclusions may be associated are best viewed first at very low power on a binocular microscope, where the working distance is large enough to permit tipping the plate up at high angles. When using the petrographic microscope on samples containing solid opaque inclusions outlining growth-bands, one should view the sample alternately in transmitted and reflected light; for this purpose, two foot switches, one for each light source, are particularly convenient, as they leave the hands free (P.M. Bethke, pers. comm. 1971). Growth-banding that is otherwise invisible (Ebers and Kopp, 1979) and zones of recrystallization (Sprunt and Nur, 1979) may sometimes be vividly revealed when the plate is viewed in cathodoluminescence. Poty (1969) made effective use of radiation coloration from intense X-ray dosage to reveal very fine growth-banding, otherwise invisible, in quartz plates. Similarly, Wilkins and Bird (1980) have revealed beautiful growth features in fluorite by proton and α-particle bombardment. Some opaque minerals can be examined in IR illumination (Plumlee et al., 1983; Fig. 4-3).

Many workers record and place some emphasis on the percentages of each of several types of secondary inclusions in each sample. In a very few examples, these percentages may have some significance. Thus a low percentage of a given type in a sample shows that that particular rock was not subjected to as much fracturing and rehealing in the presence of a given fluid as some other rock. Unfortunately, the presence of such secondary inclusions merely indicates that such fracturing and healing took place and gives absolutely no indication of the volume of fluid that took that particular route. And unless the assumption that fracturing was a time-dependent process can be justified, such inclusions also give no indication of the relative lengths of time such fluids were in that vicinity. Note, however, that coexistence of several different types of inclusions, whether recognized by visual observation or by careful examination of the numerical thermometric results, is of great importance.

One should always be aware of the gross differences in the abundance of inclusions in different phases. Thus in some samples of the Oka, Canada, carbonatite, I found that the very large calcite crystals making up the bulk of the rock contained almost no usable inclusions, but the tiny interstitial apatite crystals, only 10-20 μm in diameter, had many large primary inclusions. Similar observations are reported by others on other carbonatites. In such cases, mineral separates may provide the best material to study (see, e.g., Rankin, 1977).

Relocating Inclusions

Relocating an inclusion, particularly a small one in a large plate crowded with inclusions, can be very frustrating. More than once I have had samples brought to my laboratory by visitors who were then unable to relocate their "best" inclusion because their documentation sketches or photographs were inadequate. The general location of an area in which good inclusions have been found can be circled with a felt-tip pen, but exact relocation is a different matter. I usually combine rough sketches (starting at actual plate size), to show the approximate location, and photographs or sketches of the area as seen through the microscope at one or more stages of magnification. The area enlarged in any stage should be outlined on the previous one. It is important to record at which point in any such series of sketches one switches from the erect image as seen by the naked eye to the inverted image as seen through the microscope, as this difference can cause much confusion when one returns to the sample at a later time. Quick, rough pencil sketches of each selected inclusion showing its shape, bubble position and size, and the actual inclusion size are all useful in relocating. These sketches should always be drawn as seen with the stage rotated to a standard reference position; this can be of great help in relocating. The sketches should also indicate unambiguously which polished surface is up. Neglecting this simple matter can (and does) cause considerable waste of time. Mechanical-stage vernier coordinates can be used as long as the section is still mounted on a glass slide. Poor reproducibility in the position of the slide when seated against the stops in the stage, and backlash in the stage mechanism, make such coordinates useless at high magnification, and, if the plate is to be removed from the slide, other relocating procedures will eventually be needed anyway.

Not infrequently a sketch can be more useful than a photograph, because it can readily show features such as inclusions at slightly different levels of focus that cannot be simultaneously photographed and can omit the clutter of unwanted and confusing detail that the camera insists on reproducing. Record photographs of special inclusions should always be made before running in case they decrepitate. Enlarged photographs of the whole slide (see section below on Inclusion Photography) can provide a useful primary "finding map-photograph," on which the approximate areas of more detailed sketches, or higher magnification photographs, can be indicated.

Selection by Size

Most mineral plates will have at least hundreds of inclusions, and some will have millions or billions, yet it may require many hours of study to obtain full thermometric data on a tiny group of a dozen inclusions. How does one decide which inclusions to study? Application of several "filters" will help. One of the easiest filters to apply, and one that will usually eliminate most inclusions, is a minimum size for further studies. The larger an inclusion is (up to ~1 mm), the easier it is to study and the more information it can usually supply. In part this stems from nucleation problems, since the smaller inclusions are less apt to have nucleated a full complement of those phases that should be present. It is also a result of simple problems of optical resolution. Practically all aspects of the petrography of daughter phases become much simpler as the inclusion size increases from 5 to 50 to 500 μm. Some phase identification is possible in inclusions <5 μm, as shown beyond, but dispersion causes spurious colors and diffraction interferes with resolution and makes determinations of Becke line movement and birefringence increasingly difficult.

Generally there is a very rapid decrease in the number of inclusions with increase in size. Many samples show more than an order of magnitude decrease in number of inclusions with each order of magnitude increase in size. As a result, one simple rule of thumb will often reduce the number of inclusions from staggering to manageable: if there are larger fish in the pond, throw the minnows back. Decide what minimum size inclusion one wishes to deal with on a given problem, after a quick scanning to estimate the apparent abundance vs size relationship, and then force yourself to ignore the smaller ones. It may seem difficult on first viewing of a sample with 10^9 inclusions per cubic centimeter, but with a little training the eyes can do an amazing job of subconscious data filtering. (It is important to check, however, that there are no apparent differences in the nature of the small versus the larger inclusions; such differences have been found and may be very significant in terms of origin and interpretation. As a result, it is wise to select a few small ones also.)

For the determination of inclusion size, and for many other purposes, it is best to have a graduated ocular reticle that has previously been calibrated with a stage micrometer for each of the lens combinations used. (Estimates of image size and of magnification based on the magnification of the individual lenses in the system are commonly in error.) For low-power binocular microscope use, a similar graduated ocular can be conveniently simulated by a simple piece of transparent millimeter scale on the end of a hand-held probe.

It is also desirable to know the actual amount of vertical movement of the microscope tube with each graduation or revolution of the fine focus knob on the microscope. Many microscopes have a vernier scale on the fine focus knob and some convenient unit amount of vertical movement per revolution, such as 100 μm. In order to keep my eyes at the microscope, I fasten a metal protrusion on the knob so I can count turns without looking. (All such measurements of true vertical movement must, of course, be multiplied by the appropriate index of refraction for the host mineral to obtain the equivalent movement of the plane of focus within the plate, i.e., the actual thickness of mineral traversed.)

Selection by Origin

This is the single most critical step in all inclusion studies and hence must be handled with utmost care. The difficult problem here generally is verifying the primary origin of the inclusions. Pseudosecondary and secondary inclusions are frequently run along with the primaries, but all should be specifically selected and recorded in advance as examples of these types (see Table 2-1 (p. 43) for criteria for distinguishing). Advance recording is important because it is far too easy to bias the data subconsciously by assigning inclusions of questionable origin to the "proper" category on the basis of subsequent measurements. It is also important to record the basis or bases on which the origin is assigned, since these criteria will vary from one inclusion to another and, with them, the degree of confidence in the assignments. Even if only primary inclusions are of interest, one should also obtain some data on secondary inclusions, as they may provide some surprises.

Many planes of secondary inclusions delineating fractures are obvious at a glance. These may be almost flat planes, particularly in a cleavable mineral. The plane may happen to be parallel with the plate section; if so, the inclusions outlining it may appear to be randomly arranged through the host, falsely implying a primary origin. Focusing up or down a distance equivalent to the average spacing between such inclusions should bring new inclusions into focus if they are randomly arrayed in three dimensions but will not if they are in a plane. Planes at an angle to that of the section are much more obvious. At successive levels of focus, particularly at higher magnifications and with the substage diaphragm wide open to yield a minimum depth of field, one will see in focus a row of inclusions that appears to move across the field with focusing. With practice these planes can be visually recognized even when there are many other unrelated inclusions in the field or when the number of inclusions delineating the plane is small. Curving planes of secondary inclusions are found in both cleavable and noncleavable minerals, but flat planes generally signify healed cleavage fractures. Note, however, as shown in Table 2-1 (p. 43), that healed cleavage fractures are normally parallel to common crystal forms of that phase and hence must be carefully distinguished from planes of primary inclusions that are also parallel to such forms.

Any given plane of secondaries tends to consist of inclusions that are generally similar in size, shape, and phase composition. If individual inclusions in such a plane are at all flattened, the flattening will be in the plane of the whole group. The major exceptions are in host minerals that are strongly acicular or bladed; in these, the individual inclusions may become elongated parallel to the crystal elongation. (A related exception is found in celestite; see Fig. 11-16.) Because secondary inclusions tend to change with time, the inclusions in different secondary planes in a given sample may vary widely in size, shape, and phase composition, depending upon their relative ages and possibly on their fluid compositions (see Chapter 3). Within a given plane, there may be major phase differences from necking down, and, although most inclusions in a plane will be similar in size (e.g., Figs. 2-16 and 3-13), a few much smaller ones may be found between larger ones. An occasional extra large inclusion in a plane may result from the fracturing and refilling of a preexisting primary inclusion. The most obvious size variation within a plane is the decrease to very small sizes that takes place toward the bottom of a crack that pinches out in a single crystal.

Selection by Original Phase Assemblage

Although the original phase assemblage in any given inclusion can be established most reliably by measurement of the phase changes that occur in the microscope heating/cooling stage, it is important to learn as much as possible about the inclusion composition and hence the nature of the original phases before any actual measurements are made. Such information is useful in inclusion selection and in guiding the later measurements and their sequence. Many procedures are available (see Chapter 4), but because five types of inclusions comprise perhaps 99% of all fluid inclusions,[2] the distinctions among these five are most important for sample selection. The five types are:

Water solution plus vapor bubble. Except in volcanic rocks, this is by far the most common inclusion type. Any two-phase fluid inclusion consisting of a spherical bubble in a clear fluid, particularly if the bubble is moving, can usually be safely assumed to be of this type until proven otherwise.

Silicate melt plus vapor bubble. Any two-phase inclusion consisting of a motionless spherical bubble in a clear "fluid," and occurring in an extrusive or shallow intrusive rock, is probably a silicate melt inclusion. Distinguishing between such inclusions and aqueous inclusions is not always easy and may cause serious problems (Fig. 6-2). Sorby (1858), Zirkel (1866), and Ermakov (1950, p. 80) listed several criteria as follows (somewhat modified):

(1) Since the index of refraction of the host mineral is closer to that of glass than that of water, the borders of glass inclusions are not as dark from total reflection as those of water inclusions; conversely the relief of the bubble is greater in a glass inclusion than in a water inclusion (Fig. 6-1).
(2) Glass inclusions frequently have more than one gas bubble, whereas water inclusions very rarely do and then only because of extremely irregular shape.
(3) A bubble in a glass inclusion does not move with gravity and often is not round. Bubbles in large aqueous inclusions, if free of the walls, will move like the bubble in a level, and if they are very small (<2 μm), they will generally be in constant pseudo-Brownian movement.
(4) Glass frequently is somewhat colored but water solutions in inclusions are almost always colorless.

----

[2] The remaining ~1% consists mainly of one-phase liquid water (or steam) inclusions, oil inclusions, light hydrocarbon gas inclusions (±CO, H_2, N_2, etc.) and, in extrusive rocks, immiscible sulfide inclusions.

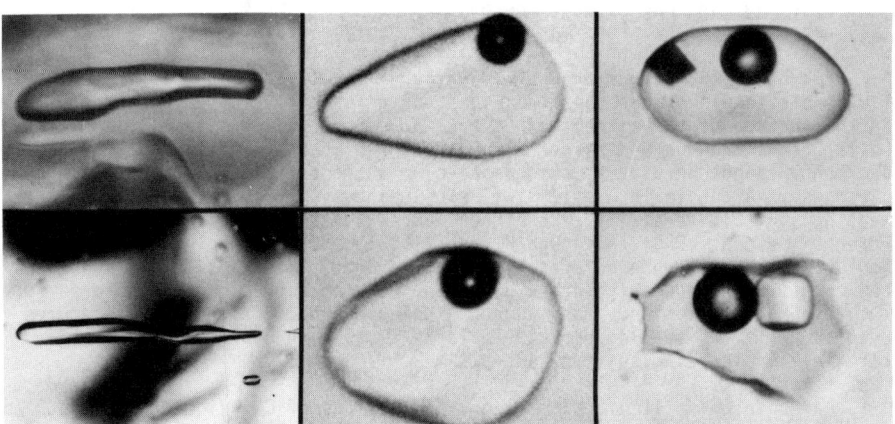

Figure 6-2. Photomicrographs of single-phase, gas-"liquid," and gas-"liquid"-daughter crystal inclusions from trapped aqueous fluids (top row) and silicate melt (bottom row). Criteria other than "general appearance" are needed to distinguish aqueous solutions from silicate melts (see text).

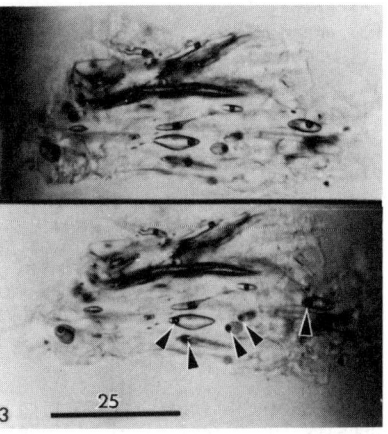

Figure 6-3. Grain of rhyolitic pumice containing elongated vesicles almost completely filled by water that has diffused in through the glass host. The small bubbles (vacuum) may be moved about by the minute thermal gradients set up by the shadow of a finger in the substage (dark areas). Only five bubbles (arrows) are free to move. The bubble volumes in these inclusions indicate that this pumice is probably $>10^6$ years old (see Chapter 16). Scale bar in μm.

(5) Glass inclusions maintain their bubbles when truncated in slide preparation, or even when intersected by a crack.

I do not agree with an additional criterion based on inclusion shape proposed by Ermakov (1950), but I can add the following criteria:

(6) For a given size, glass inclusions are much less apt to have a bubble than are aqueous inclusions.
(7) Glass frequently shows some degree of natural devitrification, recognizable as crystallinity or as a "granular" appearance under well-collimated illumination. The small grain size of such devitrification generally results in the scattering of blue light. Hence, such glass is yellow-brown in transmitted light and milky or bluish in reflected light.
(8) The bubble in aqueous inclusions frequently will move in the inclusion when a sufficient thermal gradient is impressed on it (as by touching a warm or cool probe to the section).
(9) Some bubbles in aqueous inclusions may even move from the minute changes in transverse thermal gradients that occur on tilting or moving the substage condenser slightly in and out of its normal coaxial position, or from the shadow of a finger held in the substage (Fig. 6-3).
(10) Even when the bubble in aqueous inclusions is held immobile by the walls, small solid particles in the liquid may move with gravity (Brewster, 1823b), in a thermal gradient, or, if magnetic, with a moving magnetic field from

a strong hand magnet near the objective.
(11) Large glass inclusions may show cracks in the glass.

Very tiny inclusions are particularly difficult to identify. Ikorskii
(1962) distinguished between minute gas and glass inclusions by putting oils
on the surface of the slide; opened gas inclusions filled with oil, and the
walls became transparent rather than dark. Two other criteria involve heating
and hence cannot be used at this particular stage but are listed here for com-
pleteness. A bubble in a glass inclusion changes only imperceptibly on moderate
heating (600°C); generally much higher temperatures are needed for appreciable
effects (Deicha, 1955), although Dekate (1963) mentions bubble movement at
590°C. Similarly, when held at high temperatures, glass inclusions may grow
new crystalline phases (though slowly) and require very high temperatures for
complete homogenization. Many aqueous inclusions homogenize at temperatures
over 500°C (e.g., see Roedder, 1972, Table 7), but Th over 1,000°C almost
certainly indicates glass inclusions.

Carbon dioxide liquid plus vapor. Inclusions of CO_2 liquid plus vapor
resemble aqueous inclusions but can be distinguished by several easy tests.
The critical temperature of pure CO_2 is ~+31°C, so all such inclusions will
homogenize with slight warming.[3]/ If the inclusion is easy to relocate, the
slide can be warmed briefly (e.g., on the microscope lamp housing, or even with
the fingers) and reexamined. A small hand-held hairdryer can also be used to
blow warm air over the slide. If the microscope is at all near to the inclusion
Th, the efficient absorption of infrared light from the microscope illumination
by the liquid CO_2 can be sufficient to cause homogenization. Thus many CO_2 inclusions will homogenize almost instantly when an IR filter is removed from the
light path. If the microscope is a little cooler, and particularly when examining with an oil immersion objective, homogenization may require more heating of
the inclusions. For maximum optical heating, I insert the analyzer (to cut
light intensity to the eyes) and the high power condenser, remove the polarizer
and IR filter, and turn up the voltage to the microscope light. Usually this
will homogenize CO_2 inclusions even on a relatively cool microscope stage. It
is important to remember that no liquid CO_2 will be found if the microscope
stage is above ~31°C.

CO_2 inclusions with filling densities very different from the critical
density (on either side) will not show two phases at room temperature. They
will be monophase, and they need to be cooled to form the second phase (Fig.
8-9). Inclusions with filling densities greater than critical will nucleate a
vapor bubble; those with densities less than critical will condense some liquid
phase. In either case, if the deviation from critical is not too great, simple
cooling with a drop of acetone or an ice cube may suffice. Greater cooling can
be obtained easily with a spray of liquid ethyl chloride from a small hand-
operated, 100 g cylinder (available as a medical supply for freezing skin for
minor surgery; Gebauer Chemical Co., Cleveland, Ohio, 44104). This is particularly useful when cooling is needed on a small inclusion that has been found
with the oil immersion objective; by lowering the substage condenser and swing-
ing out the high power condenser, ethyl chloride can be sprayed up against the
underside of the slide itself. Too much spray must be avoided, as it results
in condensation that blurs the image. A small flow of cooled N_2 from the intake
or exit tubes of the freezing stage can also be used.

Water solution plus liquid CO_2 plus vapor bubble. This type, sometimes
ambiguously termed simply a "three-phase inclusion," consists of three fluids:
a liquid water solution against the walls (i.e., wetting them); a large "bubble"
of another liquid (CO_2) within the water and generally not touching the walls;

----
[3]/ The presence of other gases may affect this critical temperature. N_2, CO, H_2, and CH_4 will drop it, and
H_2S and SO_2 will raise it.

and another smaller "bubble" within the liquid CO_2 (gaseous CO_2). The two CO_2 phases homogenize as described above for pure CO_2 inclusions.

Water solution plus solid crystal plus vapor bubble. This type is also sometimes ambiguously termed simply a "three-phase inclusion." If the crystal is an isotropic cube, it is normally NaCl, but its composition can be verified by several procedures (see Chapters 4 and 8). The presence of a crystal of NaCl indicates that the fluid is saturated with NaCl, but this saturation does not mean the solution must contain 26% NaCl, as in the pure system $NaCl-H_2O$; generally it will contain less than 26%, and sometimes much less (see Chapter 8).

Selection by Elimination of "Divergent" Inclusions

A problem is presented by occasional inclusions that differ visually in their phase ratios from the bulk of the inclusions, either in gas/liquid ratio or in the presence (or nature) of "daughter crystals" precipitated from the liquid after trapping. (It is important to keep in mind that the apparent gas/liquid ratio is strongly affected by inclusion shape, so inclusions that appear to be different may have identical contents and homogenization temperatures; e.g., see Bodnar and Beane, 1980, p. 881.) Generally, such visually divergent inclusions (sometimes called "anomalous" or "abnormal") will be found to have grossly different Th. Avoiding them, as recommended by Ermakov and Kalyuzhnyi (1957), and others, represents an attempt to eliminate those inclusions whose Th data would not be valid. The problem lies in deciding whether they are valid samples of preexisting fluids (e.g., formed from rapidly changing, inhomogeneous, or immiscible fluids) or samples that have been made divergent by some later (secondary) process and hence should be eliminated from the group to be run. If they are valid but are eliminated under the assumption that they have been altered, valuable information is lost; conversely, grossly misleading numbers will confuse and dilute good data from valid inclusions. The line is sometimes difficult to draw, and the subjectivity involved has caused some doubts about the validity of the method ever since the 1850s, when Sorby was accused of introducing bias into his samples by this selection procedure. The main secondary processes that must be considered when one finds such divergent inclusions are: (1) leakage and stretching (naturally or laboratory induced); (2) later refilling; and (3) necking down (see also Chapter 3).

Leakage. Leakage occurs in nature for a variety of reasons but is usually from cracking related to some deformation of the host crystal. These cracks are not always visible, as their width must be an appreciable fraction of a wavelength of the light to affect it. Leakage from such fracturing is fairly evident when a series of primary-appearing, but divergent, inclusions (often empty of all liquid) occurs as a planar array through a group of otherwise uniform liquid-rich primary inclusions. In thin sections, inclusions within a few micrometers of the top or bottom surfaces are frequently empty, particularly if their contents had been under high pressure. Overheating of inclusions beyond their Th, whether from later dikes (Lokerman, 1962, 1965) or in the laboratory (Roedder and Skinner, 1968; Larson et al., 1973; Bodnar and Bethke, 1980, 1984), can build up high pressures and cause decrepitation or stretching, particularly of the larger inclusions. Freezing can also cause similar damage if the vapor bubble is small.

Later refilling. Later refilling is a special case of leakage in which fracturing and rehealing of the inclusions occurred in the presence of a later fluid, under a different set of conditions. The opened inclusions may retain their original shape and spatial arrangement but will yield a different Th.

Necking down. Necking down is the process whereby a long thin tubular or flat inclusion spontaneously reduces its total interfacial energy by selective solution and redeposition, to yield two or more separate, smaller, more equant inclusions of the same total volume (see Chapter 3 and Fig. 3-9). Nothing is lost or gained from the system as a whole, but if a phase separation has occurred previous to the necking down, the separated inclusions will differ in phase ratios and Th. Necking down of homogeneous, one-phase inclusions will have no such consequences. Necking down occurs commonly in nature but it generally takes place near the original trapping temperature. It is probably responsible for much of the scatter of results that is apparent in any careful thermometric study. In the laboratory one must be alert to recognize the existence of necking down. When it has been interrupted while in progress, the long tails between inclusions are easy to see (Fig. 3-18). After sufficient recrystallization, however, it may be recognized only by the adjacent occurrence of two or more inclusions that are divergent in opposite directions (e.g., too much and too little vapor relative to other nearby inclusions), and sometimes by abandoned daughter crystals, left surrounded by host crystal.

Selection Based on Differences in Otherwise Similar Inclusions.

Several other criteria for selection or rejection of inclusions exist in addition to the above. Thus, all the inclusions of a given apparent origin may not be the same. The primary inclusions in a given crystal may, and frequently do, show significant differences between the several zones of a zoned crystal. If zoning is visible within a crystal, inclusions from each zone should be run, if possible. Zoning that is not visible in a crystal might still be revealed by examining inclusions from core and rim. Similarly, if the host occurs in two different crystal habits or size ranges, these two may have formed at different times. Inclusion shape, although a transient feature of little diagnostic value for the determination of origin (Roedder, 1968b), can reflect differences in composition or crystal habit at the time of trapping. Thus, the shapes of the primary inclusions in some zoned fluorite changes from negative tetrahedra in the core (that grew as an octahedron) to negative cubes in the rims (that grew as a cube; Roedder, 1977a). Secondary inclusions in different planes in a given crystal may differ in time of origin and hence in composition and shape.

The behavior of the vapor bubble in an inclusion when subjected to a slight thermal gradient can provide useful information on the compositional equivalence of two inclusions. The bubble can respond in four ways: (1) it can move up the gradient, toward the source of heat (most common behavior); (2) it can fail to move at all (even though free of the walls); (3) it can move down the gradient; or (4) it can oscillate with a periodicity complexly related to the ambient temperature.

The basis for all these phenomena appears to be variation in the surface tension over the bubble surface, causing flow of the surface that drags fluid with it. Although the phenomena are still not fully explained, the motions of the bubbles can be detected very quickly and simply by hot or cold probes (Roedder, 1965c; 1966; 1967c; 1972, p. 19). These motions are useful in inclusion studies in a variety of ways, only one of which is pertinent in the present context. The specific behavior is obviously controlled by the composition of the fluid. Even though the nature of the effects of these compositional parameters is complex and not understood, differences in the behavior of the bubbles in adjacent inclusions that otherwise seem to be identical indicate differences in composition and presumably in origin, and these differences frequently correlate with differences in freezing and Th. (Swanenberg, 1976, believes that certain bubble motions can be assigned to the presence of CO_2; although this may be true in some inclusions, I have found exceptions.)

I use a simple, homemade hot wire probe (Fig. 6-4) to check these movements. As the reaction is almost instantaneous, and may have to be viewed with objectives of short working distance, the probe heater must be very thin and should be at room temperature when first put in place above the plate but in the field

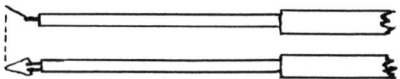

Figure 6-4. Diagram (approximately full size) of electrically heated probe for determination of the thermal response of inclusion bubbles. Power is supplied to the 0.019-mm Pt wire heating element from a variable voltage transformer via a foot switch. The probe body is a piece of 2-hole porcelain "spaghetti." The leads in the spaghetti are a larger diameter wire, so heating is concentrated in the heating element.

of view through the microscope. Once the probe is in place and its position relative to the inclusion is seen, the foot switch is tapped briefly, sending a pulse of current through the probe heating element. Any "attraction" or "repulsion" of the bubble will generally be instantaneous.

Number of Inclusions to be Selected

No fixed number of inclusions can be set as the appropriate number for any given investigation. If variations are found to exist in composition and/or density within apparently cogenetic groups of inclusions, this fact may be important in the interpretation and its firm establishment may require many more data. Many studies of metamorphic terranes have shown that about 200 inclusions are adequate to establish a pattern and provide a basis for a preliminary interpretation of the results. Smaller numbers of measurements may be adequate in ore deposit studies, where a single generation of inclusions, all primary, may be available, but even here, significant variations may be found in primary inclusions from different zones in a single crystal.

RECORD KEEPING

The problem of "bookkeeping" in fluid inclusion studies -- simple recording of the data -- may seem trivial at first, but it involves a sufficiently complex matrix of variables that it can interfere seriously with progress of the work and even with the recognition of important major trends or features if not done properly. The design of appropriate procedures will obviously vary with the nature of the investigation, but, as changes in the procedures made in the middle of a project, forcing transcription or improvement of old records, can be time consuming, it is important to design well. Two aspects should be considered separately: sample identification and coding, and data recording.

Sample Identification and Coding

Sample identification and coding includes all pertinent observations concerning a given inclusion other than the actual thermometric data derived from it. The obvious parameters include the host mineral, the original sample number, the inclusion size, shape, and vapor bubble diameter, the apparent phase assemblage and estimated phase volume percentages, and the assignment of origin (and the basis for assignment). But it must also permit recovery of such important facts as the inclusion's position in and relationship to its host crystal and the host crystal's position and relationship to the host rock. As an example that is not extreme, primary inclusions from various zones in each of a series of zoned crystals from each of several different samples may be selected for study, as well as a series of secondary (or possibly pseudosecondary?) inclusions from each of several planes crosscutting each of these crystals. Superimpose on this matrix the need to cut up the original crystals into fragments that will fit in the apparatus used, and the common requirement of a field of view of only a 3 mm circle during any run, and it is obvious that sample coding can become cumbersome.

Recording of Thermometric Data

Data recording is relatively simple compared with sample identification but it has some problems of its own. The various thermometric measurements, as detailed in the next chapter, generally result in a specific temperature (or range) for each phase change observed in a given inclusion. A genetically related group of many inclusions (e.g., a cluster of a given or estimated number of primaries, or similarly, a plane of secondaries) may behave so nearly the same that temperatures or ranges can be given for the group. Repeat measurements are frequently made on part or all of the inclusions, for several reasons. Each such measurement, or group of measurements, should have an assigned observational uncertainty (quite apart from the instrumental uncertainty).

The above involves straightforward, simple data recording. However, some less obvious items of data may prove to be either essential or desirable and are seldom superfluous. The possible need for each of these is discussed in Chapter 7; they are simply tabulated here because they are an integral part of the problem of record keeping:

(1) Previous heating and cooling history of the plate, in both sample preparation and runs; maximum and minimum temperatures achieved, and duration at each extreme.
(2) Heating (or cooling) history during the specific run yielding a given temperature of phase change; overall rates of temperature change from original ambient.
(3) Heating (or cooling) rate at the moment of the determined phase change.
(4) Date of determination and observer.
(5) Other unknown or calibrant samples in the stage during the same run.
(6) Equipment variables such as stage, window thickness, thermocouple, microscope objective, instrument mode, etc.

Some of these items may be identical for many individual inclusions or groups of runs, and hence need to be recorded only once, but they should be recorded. Thus, on numerous occasions, I have learned that some seemingly inconsequential change in observational technique or equipment variable in the past is pertinent to the interpretation of the results obtained before or after the change. Such records are particularly important in trying to unravel the mass of apparently irreproducible and conflicting data that may arise when some of the data represent stable equilibria, and other data represent metastable equilibria, and the observer does not know which are which.

PHOTOGRAPHY OF INCLUSIONS

Photography is a nearly indispensable part of most inclusion studies. Photographs provide an excellent record of the interrelationships of the inclusions and the host crystals; they provide documentation of the steps in disaggregation of a polished plate into the chips that are actually run; they provide a reference against which changes in the inclusions during runs, such as dissolution of the walls, recrystallization of daughter crystals, or changes in bubble diameter, can be measured; and some are used as illustrations for publications. Most of the procedures described above for inclusion petrography are useful for inclusion photography. Commonly the eye "sees" more than the film, however. No single best procedure for photography exists; the following section describes some of the procedures I have found useful in my own work.

Depth of Focus

As most inclusion features are three dimensional, the procedures used in normal photography of petrographic sections must generally be modified. This problem of the third dimension is the major hurdle in inclusion photography and commonly the results are disappointing, especially at high magnification. The eye is capable of instantly integrating several levels of focus through an inclusion into a three-dimensional mental image, but the camera stubbornly insists on recording only what is actually in focus at a single setting. Similarly, a planar group of inclusions that is almost parallel with the plane of focus may seem very photogenic, but only one small zone through the plane is actually in focus at any given setting. Such problems are most apparent at high magnification, where the depth of focus is at a minimum.

The obvious solution of increasing the depth of focus by "stopping down" the apertures in the microscope is seldom effective. The more collimated lighting from stopping down results in widening of the black shadows from total reflection around each inclusion (and each internal phase boundary), thus eliminating many of the features to be illustrated. Stopping down at very high magnification greatly increases diffraction effects, seriously reducing resolution. The most common undesired result of stopping down, however, comes from the fact that it emphasizes the many inclusions, cracks, or other imperfections in the plate above and below the inclusion of interest. Once again, the eye conveniently ignores these dark, out-of-focus smudges, but the camera is honest.

Except for low-magnification photography (see below), few options for resolving this dilemma are open. One is simply to compromise, by adjusting the lighting to minimize the undesirable features. Another is to look for a more photogenic field to photograph, one in which one or more inclusions are flattened perfectly parallel to the plane of focus, so that all phase boundaries are visible and not superimposed. Such a field should also be in a clear area, with no disturbing imperfections above or below, thus permitting more stopping down. This interference from superimposed images is particularly disturbing in inclusion-rich areas. It can be greatly reduced by making a thinner plate. Still another option is to photograph at lower magnification on a high-resolution film (such as Polaroid type 655) and then enlarge. Perhaps someday microholography will provide the ultimate solution.

Photography at Low Magnification

Documentation and finding photographs. Photographs of each entire polished plate are useful for several purposes. They document the size, zoning, and interrelations of the host crystals, and they are the first, indispensable step in relocating previously selected inclusions. Although a sketch will sometimes suffice for the latter purpose, particularly if the plate has few inclusions or is so shaped that relocation is simple, a photograph becomes almost essential if the rock is fine grained or the inclusions abundant. Such a photograph will represent an enlargement of only perhaps 3x to 10x, a low range that has not been common in petrographic laboratories,[4] but is becoming more so. I have used several procedures, depending on available equipment:

(1) Normal photography. Here a negative would be made and subsequently enlarged to an appropriate size. The resultant print can provide excellent resolution of detail, and the printing can be controlled to provide optimum contrast. The two-step process is slow, however, and the results are not immediately available. Color transparency film provides an alternative. Ektachrome 64 film can normally be developed in one day, and only the best frames need be enlarged. Relocation diagrams can be made by projecting the film and tracing the desired features and inclusions.

(2) Instant photography. Although most petrographic microscopes do not

----
4/ Even the 1X objective available for petrographic microscopes does not usually suffice to go this low.

permit viewing of large areas at low magnification, if a Polaroid camera attachment is available, an overlapping mosaic of photographs of adjoining areas can be composited. A mechanical stage will help insure complete coverage. A useful version of this compositing procedure that permits almost any desired degree of magnification involves putting the sample plate in an ordinary photographic enlarger and projecting onto Polaroid film in a holder that is indexed against the normal enlargement easel. Care must be used in orienting the plate. The plate must be inserted in the enlarger with the correct side up, depending upon whether the resultant finding photograph is to be used in connection with later optical systems (on microscope or microprobe) that do or do not invert the image. Most significant is that the results are immediately available for examination, but the cost per exposure is higher.

(3) Enlarging xerography. Some xerographic equipment designed to enlarge standard 2" x 2" (5 cm) photo transparencies will permit direct enlargements of a mounted sample plate in the desired range of magnification. The resulting prints do not have the resolution or contrast control of photography, but are obtained quickly and inexpensively.

Photographs for illustrations. Photography for scientific illustrations has much more stringent requirements than simple record photography. When the inevitable degradation of photographic quality upon reproduction in a journal is superimposed on the problems of depth of field described above, selection of the most appropriate field becomes increasingly difficult, and various procedures need be used to make the best photograph of the available inclusions.

If the color zoning in the host is pale, it is normally difficult to record. It can frequently be intensified by placing the appropriate color filter in the light path. The easiest procedure is to use a simple, inexpensive, hand-held "wedge interference filter," which provides the entire range of the visual spectrum. Merely adjust the wavelength by translating it in the light beam until a maximum contrast is seen.

Conversely, if the contrast between colored zones is so large that detail in one or another zone is lost (e.g., a fluorite with deep blue and colorless zones), simply adjust the wedge until the field is bathed in blue light to minimize the contrast.

Quite commonly, the desired field of view varies greatly in transparency from side to side. The eye adjusts easily to this on moving from one point to another but the film has fixed characteristics. This problem can be minimized by shifting the substage condenser off-axis, or by the standard darkroom "dodging" procedure, moving a card in the microscope illumination system to cut down the exposure on the brighter part of the field.

A poor polish can be partly corrected by adding a matching index liquid and a glass plate or coverglass with care to avoid trapping bubbles. Dark pits from plucking of grains during polishing will also be minimized by this procedure. Both top and bottom surfaces should be treated. Large samples can be immersed in a bath of the fluid (e.g., see Rankin and Greenaway, 1978). Even if the sample is not immersed, distracting black bars from total reflection on cleavage cracks can be minimized by filling the cracks with the appropriate index liquid and then cleaning off the excess.

Inadequate lighting on a specific feature, such as the contents of a large inclusion at low magnification, can sometimes be corrected by the use of supplementary light from one or more fiber optic illuminators, from above or below the sample. Sometimes several exposures on the same film, with the auxiliary lighting in different places, yield good results.

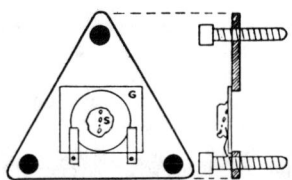

Figure 6-5. Tripod for tilting samples in a fluid bath for optimum photography of noncoplanar inclusions. The sample (S) on glass slide (G) is held over hole in tripod plate.

Polycrystalline plates sometimes yield much more illustrative photographs if they are taken with slightly crossed polars. This permits both the crystal boundaries and the internal details to be seen in all grains in a single picture. The position of the two polars relative to each other and to the sample can be adjusted to provide the optimum degree of contrast in the most important grains.

If a group of otherwise photogenic inclusions lie in a plane that is slightly inclined to the polished plate, the focus can be improved by tipping the slide appropriately. The simplest method is to mount the plate or slide on a larger glass plate, with lumps of modelling clay to provide the needed tilt. More control is possible with a simple homemade tilting device such as that shown in Figure 6-5. Optical problems from the tilt, particularly at larger angles, can be eliminated by immersing the tilting device plus sample in a bath of liquid with an index similar to that of the host. Waves in the rather large upper surface of this liquid from the inevitable vibrations can blur the images. The simplest remedy is to minimize the effective surface area through which the photograph is made by encircling it with a metal ring that sits on the top surface of the sample and extends above the liquid surface. Better results can be obtained by photographing through a flat-bottomed glass immersed in the liquid over the sample.

This sample tilting procedure turns out to be much less effective in practice than in theory. The problem is in that the magnitude of the tilt needed is frequently much larger than the slight difference in levels of focus would seem to indicate. The amount of tilt needed can be estimated quickly as long as the microscope has been calibrated both horizontally and vertically (see above, under "Selection by size").

Photography at Intermediate to High Magnification

Most of the procedures detailed in the previous section are just as appropriate here and need not be repeated. Some additional procedures that I have found useful are given below.

When the bubble in an inclusion is in an inconvenient position, it can sometimes be moved into a more photogenic position by application of a thermal gradient with a hot wire probe (Fig. 6-4) or even by putting a shadow across part of the field but out of the camera's field of view (Fig. 6-3). Similar results can be gotten by adjusting the substage condenser. If the bubble is partly obscured in a dark corner, fiber-optic light from above or below will generally help. Daughter crystals sometimes can be moved into a better position by a heating and cooling procedure. (However, some return only as a confused tangle of crystallites.)

For most of my own photomicrography, I use an instant film (Polaroid P/N 665) that provides both a positive (print) and a negative. The contrast of such prints is normally rather low, but the high-resolution negative can then be printed on higher contrast paper. This procedure permits immediate verification that a satisfactory photograph has been obtained and yet permits later enlargement, reduction, and cropping of prints made with appropriate density and contrast. These can then be trimmed and mounted together into composites that provide the maximum of photographic illustration in a minimum of space.

The scale of photographs is a subject that is sometimes treated rather too lightly. I object strongly to the (still too prevalent) use of a simple magnification value (e.g., "x250" or "250x") on illustrations. Such an expression is usually ambiguous. Does it refer to the magnification on the original negative, to the print that was sent to the journal, or to the photograph as finally printed? On checking carefully, I have found examples (but not so stated) of each of these three usages. In addition to this ambiguity, such magnification values become difficult to use when dealing with microfiche, microfilm, or other photocopies of journals. A simple bar scale on the photograph, its length labelled on it or in the caption, is the obvious solution. This bar scale can be established several ways. Some workers calibrate their equipment so that a known combination of lenses represents a given area. The number of possible lens combinations in microscope and camera, and the usual enlargement or reduction and cropping of the prints in my own procedures, makes this process unwieldy. Instead, I note, on the original photograph, the length of some easily measured feature, as determined using a calibrated ocular reticle scale. When the final print is made, the length can be used to establish the length in millimeters of an appropriate bar scale on the photograph. (This is easily cut from a thin strip of black press-on tape stuck on a metric ruler.)

Most optical microscopes are corrected to yield maximum resolution and minimum aberration in green light. All photographs should be taken with such green light, using an appropriate narrow-band green interference filter made for the purpose. This increases exposure times but helps improve resolution, particularly at high magnification. Longer exposures may require photography late at night or on weekends, when most buildings are much less aquiver.

If birefringent daughter phases are present, partly crossed polars usually provide far better results than completely crossed polars. If the birefringence of the host permits, the pair of polars (or the sample) can be rotated for optimum contrast. (See Chapter 4 for other recommendations.)

Optical microscope lenses are normally corrected to provide optimum resolution when there is a standard coverglass between the object and the objective. As a result, most inclusion photographs must be made under less-than-optimum optical conditions. As usual, the problem is most acute at high magnification. Frequently, the inclusion is too deep in the mineral; this depth effect is exacerbated by a high index of refraction for the mineral, since the "optimum" coverglass is assumed to be normal glass of index ~1.51. Little can be done about this except to try inverting the plate (or grinding it thinner). At the other extreme, if one looks at an inclusion very near the surface of a polished plate, there will be insufficient apparent coverglass thickness. In such cases, adding a drop of oil and a coverglass can improve the image.

The interrelationships of various photographs and the relocating of photographed areas can be very simple if an adequate series of numbered photographs are taken at increasing magnifications. On each photograph is drawn the approximate outline of the area covered by subsequent photographs at higher magnification, with their photograph numbers indicated (see also "Relocating Inclusions," above).

Fast moving bubbles, as in very small inclusions and particularly in CO_2 inclusions, may be blurred in ordinary photography. Using a good IR filter, and reducing the temperatures, may help, particularly for CO_2 inclusions. Increased light intensity cuts exposure time but speeds up the movement. If flash microscope illumination is not available, use of high speed film greatly cuts exposure time. As the pseudo-Brownian movement is very erratic, sometimes one of a series of exposures will happen to have occurred during a time of little movement. Such bubble movement can sometimes be made use of, to permit photography of the bubble in the optimum position.

CUTTING UP PLATES FOR MICROTHERMOMETRY

After the inclusions to be studied have been selected and all necessary sketches and photographs made for documentation, final preparation of the plates for the heating/cooling stage can be started. Although two heating/cooling stages (the U.S.G.S. and the Linkham stages) will accept large samples, only one (the U.S.G.S.) will permit viewing of the whole sample; the others have a very limited field of view (only a few millimeters). Thus, unless the U.S.G.S. stage is used, the doubly polished plates, whether on a glass slide or not, must generally be broken or cut into smaller pieces for microthermometry. To avoid loss of valuable data, each stage in disaggregation of a plate (sawing, breaking, etc.) should be well documented, so that interrelationships of inclusions to each other, to growth zones, and to the whole crystal can be reconstructed. If the plate is thick enough to be handled with care, it is best to remove it from the glass slide. Thermoplastic cements can be softened and the plate slid sideways, followed by cleaning with an appropriate solvent and a soft swab (e.g., Q-tips). The gentlest procedure is simply immersion in solvent. After removal, ultrasonic cleaning is best if the plate will not break up, as it helps to get the mounting resins, which might spoil the run, out of cracks and pores. Methyl alcohol is the best solvent for Lakeside cement; acetone tends to convert it into a soft gum and eventually dissolves most of it but leaves a residue. Complete removal of all resin is essential and may be surprisingly tedious, as any small portions left tend to coat the polished surface during the final cleaning. After cleaning and sawing (or breaking) the plate, the surfaces of the chip above and below the selected inclusions should be checked with the microscope, as nominally "invisible" fingerprints or residues from dirty solvent can wreak havoc with the optical image at high magnification.

Thin plates, particularly of minerals that are friable and those that contain only relatively low-temperature inclusions, can be left on the glass slide. Mark the parts on which runs are to be made (e.g., with India ink), and then remove them by cutting out wedges of sample (and glass slide) with a diamond circular saw, holding the slide in the hand rather than in a clamp. A diamond wire saw is even more gentle than a circular saw for this operation but is slow. The cuts are made so as to place the desired inclusions near the point of the wedge, to permit several chips to be placed in the heating stage simultaneously, still mounted on the glass wedges, all with their selected inclusions within a radius of several millimeters. To keep the thermal mass to a minimum, a coverglass can be substituted for the original glass slide (R.E. Bennet, Jr., pers. comm.).

ARTIFACTS AND THE PROBLEMS THEY CAUSE

Henry Baker, in an early book "The Microscope Made Easy" (1743), indicated that caution was needed in microscopy:

> "Beware of determining and declaring your Opinion suddenly on any Object; for Imagination often gets the the Start of Judgment, and makes People believe they see Things, which better Observations will convince them could not possibly be seen: therefore assert nothing till after repeated Experiments and Examinations in all Lights and in all Positions.
>
> When you employ the Microscope, shake off all Prejudice, nor harbour any favourite Opinions; for, if you do, 'tis not unlikely Fancy will betray you into Error, and make you think you see what you would wish to see.
>
> Remember that Truth alone is the Matter you are in search after; and if you have been mistaken, let not Vanity seduce you to persist in your Mistake."

During optical examination, a variety of spurious images and apparent or real optical artifacts are commonly seen; the microscopist soon learns to ignore them, but they can cause considerable confusion and wasted time when first encountered. These effects become most notable at higher magnifications, particularly at 1000x and above. The following are a few of the more common examples that I have encountered either in my own work or in working with students.

Inclusions that are intersected by a polished surface appear very different from those fully within the plate, and hence such "strange" inclusions are commonly spotted by beginners and professionals alike. Tiny droplets of oil from fingerprints or former mounting media on the surface are also common sources of confusion, as are polishing pits and abrasive grains or other debris within them. Cork dust and even dandruff flakes, particularly in an oil mount, provide some odd apparent "inclusions." Mounting media, both impregnating resins and immersion fluids, yield a separate set of artifacts. Thus if an index liquid or an embedding medium or cement flows into the cavity of an opened fluid inclusion or surface pit, particularly if incomplete filling leaves a bubble in the fluid, very realistic but spurious inclusions can be formed. Leaching with solvents before observation, particularly in an ultrasonic bath, can avoid some but not all such artifacts. Opened inclusions that have become completely filled with such fluids may look like normal single-phase inclusions. Comparison of the index of refraction of the "inclusion" fluid with the host will sometimes help clarify matters, but only if one has the data on the specific mounting medium used. If the recognition of such artifacts presents a serious problem, a coloring dye may be added to the impregnating or mounting resin.

All such problems are easily eliminated once the observer realizes his plane of focus is at the upper or lower polished surface, or in the mounting medium above or below. At low magnification, and when the sample is poorly polished, this is simple. But in a well polished sample of a clear mineral, at high magnification in a medium of matching index, this upper (or lower) surface is sometimes difficult to recognize. Usually there are a few imperfections in the polish, dust specks on the surface, etc., that mark it, and any crack or grain boundary can be followed by focussing up (or down) to the surface.

A spherical bubble in liquid (or even a roughly spherical fluid inclusion) acts as a negative lens in the optical system and may bring another lower inclusion (or even a substage diaphragm) into sharp focus, simulating another bubble within the first, and this frequently leads to an erroneous conclusion that liquid and gaseous CO_2 phases are present (Fig. 6-6). (Simple heating of the slide to above 31°C, where a liquid-gas meniscus is no longer possible in a CO_2 system, can remove this ambiguity.) Gemologists frequently use dark-field microscopy when examining gems for inclusions. Although darkfield illumination does show the inclusions brilliantly, it has some shortcomings. Thus, these illumination devices generally result in an optical artifact of an additional apparent bubble inside each real bubble, and one type results in two such apparent bubbles (Fig. 6-7).

Total reflection from a somewhat planar, inclined phase boundary can simulate the appearance of an opaque phase, until the sample plate is sufficiently tilted to place the phase boundary more nearly perpendicular to the line of sight (Fig. 6-8). Very small colorless objects (<1 μm) generally appear colored in pastel hues, particularly when there are differences in dispersion between object and host (Wilcox, 1983).

If the birefringence of the host is high (or the plate is thick), confusing double images can result. At worst, they can simulate a second "meniscus" at the side of a bubble, such as would form from a small amount of liquid CO_2. These problems are eliminated if one of the polarizers is inserted and rotated

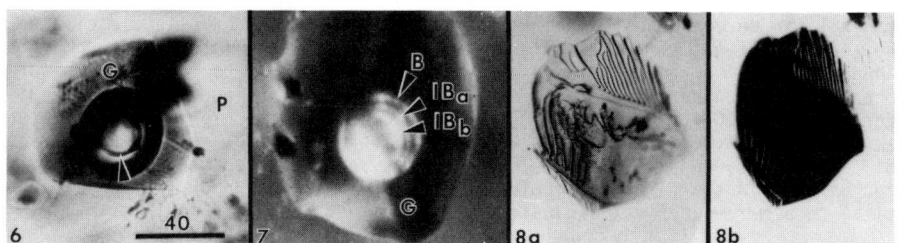

Figure 6-6. Photomicrograph of inclusion of glass (G) in pyroxene (P) from mafic nodule N42 from Vesuvius, Italy, showing what appears to be a two-phase bubble, hence presumably liquid plus gaseous CO_2 (as many other inclusions in this sample do contain). Actually the inner bubble (arrow) is an optical artifact due to the illumination procedures used. Scale bar in μm. Photo courtesy of H.E. Belkin.

Figure 6-7. Photomicrograph of a glass (G) inclusion in peridot (olivine), as viewed with a gemologist's darkfield microscope. This particular illumination procedure forms <u>two</u> apparent menisci within the actual bubble, both artifacts. The true wall of the bubble is at B; apparent inner bubbles IB_a and IB_b are both artifacts. Photo courtesy of E. Gübelin.

Figure 6-8. Internal 2 mm long crack in amber in transmitted light, showing the effects of total reflection. In (a), the line of sight is essentially perpendicular to the crack; in (b), the sample has been tilted somewhat, so that most of the crack reflects totally. With a few more degrees of tilt, even the bright lines (from undulations in the crack) become black and appear opaque. Flat aqueous inclusions behave similarly, but need a higher angle for total reflection.

to yield a single clear image. With highly birefringent uniaxial negative minerals, it is best to have this polarizer parallel to the <u>ordinary</u> vibration direction of the mineral, for both maximum depth of field and image clarity.

During impregnation of fragile samples before polishing, some very realistic artifacts may be formed. Thus during the setting of some epoxy resins (particularly if incorrectly mixed or formulated), minute globules of a second immiscible phase may form under some conditions. The presence of such visible globules in a layer of cement beneath the sample is a serious source of distraction or error in searching for fluid inclusions (as are the air bubbles caused by improper mixing or mounting procedures).

Such artifacts are most difficult to ignore when they occur in resins used for impregnation, because then the spurious globules appear <u>within</u> the sample. The lunar samples provide a memorable reminder. I had spent weeks at the microscope in a fruitless search for normal two-phase liquid-vapor inclusions of H_2O and/or CO_2 in the Apollo 11 lunar samples, as even a single such inclusion would have considerable petrological significance, but all I could find were glass inclusions. Many of these had a vapor bubble, but such bubbles are immobile. Then I discovered a crystal of anorthite with a plane of secondary two-phase inclusions, each with a rapidly moving bubble and a constant liquid/vapor ratio! Unfortunately, these were artifacts from the embedding epoxy filling a sloping crack in the anorthite, which had set up to yield <u>three</u> phases. The main body of the epoxy formed a clear, rigid resin with an index of refraction almost exactly matching the anorthite, and hence essentially invisible. Embedded in this resin were tiny globules of an immiscible liquid of much lower index of refraction, which, in turn, had formed vapor bubbles, either on cooling or on extraction of some of the components by reaction with the epoxy walls. This explanation became obvious only when I found a small mass of the same embedding epoxy adhering to the outside of the sample at the edge of the slide, complete with hundreds of excellent fluid inclusions, each with its vapor bubble in rapid pseudo-Brownian movement.

Depending upon inclusion geometry, the determination of the apparent index of refraction (i.e., index) of a fluid inclusion relative to that of the host mineral can be erroneous. Thus it might seem easy to make the simple distinction between an all-gas and an all-liquid inclusion, on the basis of the much higher apparent relief of the gas inclusion (i.e., dark borders from a wider zone of total reflection at the contacts). Most liquids have an index over 1.3 (1.33 for H_2O and 1.38 for NaCl brines), and most gases, unless highly compressed, have an index near 1.0. However, the apparent relief of an inclusion is a function of two factors, index difference between host and inclusion and the shape of the inclusion. Thus a flattened inclusion full of low-index vapor can appear identical to a more spherical inclusion full of higher index liquid (Fig. 6-1).

When one attempts to estimate the index of a given phase in an inclusion relative to that of other inclusion phases or to the host mineral, the inclusion geometry is also very important. When the match in index is close, the effect is unambiguous, striking, and useful (e.g., Figs. 2-26 and 4-8). However, when the index difference is larger, problems can arise. Total reflection at the near-vertical interfaces between various phases can yield line-like reflections that are so offset that they may appear as Becke lines from an entirely different phase boundary. Estimates of the index of phases in multiphase inclusions, with many phase boundaries, may be impossible, unless one can find a flattened inclusion, with each phase visible and isolated. Such lucid inclusions are useful, and admirably photogenic, but are unfortunately all too rare in the real world of inclusion study.

The subject of artifacts in fluid inclusions would not be complete without at least a mention of the several uses of fluid inclusions in the study of early life on Earth. Several studies have been made of the amino acids in fluid inclusions (e.g., Drozdova et al., 1964), and the recent developments in analytical techniques may make this procedure an important source of organic geochemical data. Two other such studies, however, fall more nearly into the category of artifacts. Thus Dombrowski (1960) has reported the occurrence of bacteria and bacterial spores, in part still viable, in inclusions in halite from several deposits, including some of Paleozoic age. The report was widely quoted in the popular press. The nature and occurrence of the fluid inclusions in these samples, however, would not preclude a relatively modern origin for the fluids, and for the bacteria, in these inclusions (Roedder, 1972, p. 61).

Some objects found in Archean rocks, which (after extensive studies) were originally reported to be the oldest known fossils, were later claimed to be fluid inclusions (Bridgwater et al., 1981). I have since found (Roedder, 1981c) that the objects are neither microfossils nor fluid inclusions. The lesson (just as pointed out by Baker in 1743) is that optical microscopy, although a powerful tool, provides data that require considerable care in interpretation.

Last but not least, the student should not be surprised to see rapid (but nonbiogenic) movement in some inclusions. The bubbles in most small inclusions are in constant, rapid, but irregular movement. Such pseudo-Brownian[5] movement is actually a result of extremely rapid flow on the surface of the bubble in response to surface-tension differences caused by the minute, statistically random thermal gradients from variations in local temperature of the liquid at this very small scale. Not all bubbles respond this way, even though small, and even the direction of the movement in a thermal gradient may be reversed (Roedder, 1965c, 1966, 1967c). The common disappearance of a small bubble into the black border at the side of an inclusion before homogenization is a similar manifestation but reflects the presence of thermal gradients in the heating stage.

----

[5] True Brownian movement is commonly visible in all sub-micrometer-sized daughter crystals. It is particularly striking between crossed polars when the daughter crystals are highly birefringent.

Occasionally, however, very _regular_ movements, of three types, are seen. All are based on minute local, _static_ rather than random thermal gradients in the host sample, established by absorption of infrared from the microscope light. First, the rapid surface flow on a fixed vapor bubble in an inclusion in a thermal gradient may drive internal "whirlpools" throughout the liquid, and any tiny solid speck (e.g., ~1 μm) in the liquid may be carried along in a very regular circular path. Second, such solid specks commonly are found adhering to the surface of the bubble itself (apparently the lowest energy configuration). Surface flow on the bubble can carry the speck in a regular circular path so fast as to be almost a blur to the eye. Third, a few bubbles _oscillate_ across the entire length of the inclusion with extreme regularity, but with a periodicity that is a complex function of the ambient temperature. These rather amazing phenomena provided the basis for an amusing motion picture taken through the microscope (Roedder, 1966).

Chapter 7

INCLUSION MEASUREMENTS --
HEATING, COOLING, DECREPITATION and CRUSHING

"One is tempted to conclude that the price of success in fluid inclusion studies is eternal vigilance." (Wilkins, 1979, p. 8)

"It is surprisingly easy to get beautifully consistent, reproducible, but incorrect numbers." (Roedder, 1976a, p. 84)

CONTENTS

INTRODUCTION

This chapter describes the equipment and procedures most frequently used to obtain numerical data on fluid inclusions. Most inclusion data reported in the current literature result from heating and freezing studies. The next most common data are those obtained by decrepitation (reported particularly in the Russian-language literature), and a still smaller amount of data results from use of the crushing stage. Currently, several types of apparatus are used to obtain each of these types of data. Each of the designs of such apparatus must represent a compromise based on various constraints such as flexibility, speed of operation, available accessories, accuracy, range, and cost, so that no particular design can necessarily be considered "best."

I summarize here these design and operation variables, the resultant limitations, and the more general practical problems in obtaining inclusion measurements and evaluating their validity. Particular attention is paid to the inevitable pitfalls that beset any such laboratory study, as well as some of the methods of avoiding or at least recognizing them. The theoretical aspects of interpretation of inclusion data will mainly be found in Chapters 8, 9, and 10.

MICROTHERMOMETRY USING HEATING/COOLING STAGES

General Characteristics and Requirements

The development of suitable equipment for heating or cooling inclusions, particularly while samples are under the microscope, has been a major hurdle to the progress of inclusion research, and a generally optimum design for such work has not been found. It is very difficult to achieve an adequately high (or low), known, and controllable temperature in the sample under conditions that permit sufficiently high magnification and adequate illumination and yet also permit flexibility and speed in operation. Most earlier models of stages described in this chapter were designed for a single mode of operation, either heating or cooling, whereas most recent designs (see later section) permit both modes.

To set up a homemade heating and/or cooling stage on the microscope is comparatively simple, and many different models have been described, but none is completely adequate and universal in application. The essential requirement for any heating/cooling stage design is that it permits one to maintain a sample at a constant, measurable and adjustable temperature while it is under observation in transmitted light with a microscope. This requirement imposes numerous constraints on the design of an effective stage, but there are additional factors, because the stage, its "peripherals" (such items as controllers, sensors, readouts, etc.), and the microscope must be compatible with one another. Compatibility needs to be considered only if substitutions or additions are made to a complete commercial unit. The following listing of aspects that should be considered and evaluated before any equipment is designed or bought is generally applicable but is mainly directed toward combination heating/cooling equipment. Other details to be considered will be found in the section describing such equipment.

A - The Stage

Temperature range(s) (-185° to +600°C covers most inclusion studies, excluding melt inclusions).
Ease of operation.
Response time of temperature control.
Rates of temperature change possible (manual or programmed, in both direction

Possibility of use of added fiber-optic illumination.
Provisions to avoid frosting of optical system during cooling.
Sensitivity of calibration to experimental changes, such as the heating
  rate or the objective used.
Sensitivity of calibration to sample size and position on stage, i.e.,
  thermal gradients.
Adaptability to change in design or use, e.g., in temperature sensors.
Amount of heating of objective, substage condenser, and microscope stage.
Ease of access for sample changing and calibration.
Ease of changing objectives during a run (a convenience, not a necessity).
Ease of changing between heating and cooling modes.
Maximum acceptable diameter (and thickness) of sample.
Maximum diameter of sample field observable at any given time.
Maximum diameter of field observable on translating sample during run
  (where translation possible).
Ease of scanning over this field, if sample translation is possible.
Required maintenance and replacement schedule.
Down time and costs of repairs.
Consumption of expendibles, such as liquid nitrogen (LN_2).

B - The peripherals

Quality of temperature control at any given temperature.
Readout capability -- visual vs recorded, and visibility.
Possibility of remotely actuated (i.e., foot switch) recording of spot
  temperatures or of placing a "hold" on the readout.
Short-term precision of results (reproducibility).
Long-term precision of results (drift).
Accuracy -- stability of instrumental calibration.
Need for pressure regulators for compressed air and/or N_2 gas.
Volume of LN_2 dewar and hence running time permitted with one filling.
Compressed air filter and convenience of replacement.
Power-supply stability.

In addition to the stage and its peripherals, the installation is usually
such that the microscope must be dedicated to this work; it must be compatible
with the stage and adequate optically. As a result, the microscope represents
the major part of the total cost of the installation. The following aspects
must be considered.

C - The microscope

At least 25x objective; must have the required minimum working distance
  and be compatible with stage.
Minimum working clearance required between objective and microscope stage
  to accommodate heating/cooling stage.
Maximum working clearance required between objective and top of heating/
  cooling stage, to permit focusing low-power objectives.
Adaptability of each objective to cooling coils and antifrost measures.
Substage condenser numerical aperture and working distance compatible with
  stage design.
Light source (50 watt halogen bulb preferred) and quality of infrared
  filtering.
Rotation of polars.

Optical requirements. As mentioned earlier, the microscope is the major
tool in all fluid-inclusion research; hence, "saving money" in its selection
and purchase may be poor economy. As very high magnifications cannot be used on
samples within heating/cooling stages anyway, a microscope that is dedicated for
use with the stage need not be of highest optical quality. However, that which
is used for inclusion petrography should be. Adequately convergent illumination

is a particular problem, as the distances involved in such equipment preclude the use of most standard microscope high-numerical-aperture condensers that are ordinarily so helpful in microscopy of inclusions. Illumination via flexible (and hence adjustable) fiber-optics illuminators is appropriate for many types of operations but is not adequate for very small inclusions. These require the highly convergent light of an appropriately designed substage condenser of high numerical aperture, which should be matched to that of the objective used. Most condensers are designed for a sample position 1 mm above the microscope stage, but heating/cooling stages raise the sample many millimeters above that position.

Another major problem with most heating stages is the working distance needed for the objective lens on the microscope. In order to keep the objective cool in the heating mode and to minimize thermal gradients in the sample in either mode from heat flow into or out of the objective, long-working-distance lenses must be used. All stage designs require at least some millimeters of clearance between sample and objective. Most commonly, the objectives used are those designed for use with the universal stage, and as they must be used here without the hemispheres, the effective magnification is considerably below their normal rating. The optical resolution under such conditions is also not optimum. For high magnification, the most satisfactory are the UTK 50/0.63 Leitz, the L32/0.40 Leitz, or the Zeiss 30/0.60 objectives. (Note that the actual magnification of the 50x UTK, without the hemispheres, is only 32x; see also footnote 5). Resolution can be improved by using monochromatic light; a green filter effectively eliminates the residual chromatic aberrations present in all optical systems, but the optic images from long-working-distance objectives are seldom as good as those obtained during ordinary microscopy. Optically, an ordinary objective of lower magnification, together with higher magnification oculars, is frequently better than the universal-stage objectives. I seldom use oculars less than 12.5x, and on occasion prefer to go to 25x oculars, in order to get magnification and still use long-working-distance objectives. The increased magnification may be "empty" and provide no increase in resolution, but it reduces eye strain in dealing with minute moving black bubbles, as in homogenization studies. Resolution is a serious problem in the recognition of ice in small inclusions. Wide-field oculars are always preferable. As long-working-distance objectives are expensive (>$1000) and can be damaged by heat, effective cooling coils are needed around them for high-temperature operation. These coils must usually be made in the individual laboratory, as the commercial ones frequently provide inadequate contact between objective and coil.

An effective infrared filter should always be kept in the light path because infrared radiation is absorbed by CO_2, causing internal heating of CO_2-rich inclusions, which is not detected by the temperature sensor. The normal "heat-absorbing" filter in many optical systems is frequently not adequate; doubling this filter is commonly found to have visible effects on the phase assemblages in CO_2 inclusions near Th. If infrared radiation is not removed, measured Th of CO_2-rich inclusions will be low by several degrees; even H_2O-rich inclusions can be affected (T.L. Woods, pers. comm., 1981).

Study of the behavior of birefringent daughter minerals, either original or formed during cooling, normally involves rotation of the sample between crossed polars. As most heating/cooling stages cannot be rotated, a microscope that permits rotation of both polars about the stationary sample should be used. A bar that connects the two polars for simultaneous rotation is useful but not required, as they can be moved individually by hand.

Thermal gradients. The inevitable thermal gradients within the stage and sample are particularly troublesome. Some stage designers object to the use of the term "inevitable," but I believe it is apt. Normally, a sample is heated electrically from the sides, and the total electrical energy dissipated there must flow away; i.e., it must flow down gradients. The distances involved are such that these thermal gradients may be as large as 100°/mm, at least in certain directions in the stage. Thermal gradients, both within the sample and between the sample and the temperature sensor, or even in the temperature sensor itself, cannot be avoided; they can only be minimized. Thus, thick thermocouple wires can conduct heat rapidly through an otherwise well-insulated cell wall (Larsen et al., 1973). In theory, the effect of such gradients on the accuracy of the results can be eliminated by an appropriate calibration procedure, but in practice this is difficult to achieve (see "Calibration"). This omnipresent problem of thermal gradients and hence differences between measured and actual sample temperature was recognized early in the history of fluid-inclusion studies, and some ingenious designs were developed to minimize it. Thus more than 100 years ago, Vogelsang and Geissler (1869) designed a heating stage in which the sample was placed in the center of a torus-shaped glass tube that was actually the bulb of the measuring thermometer.

The magnitude of the thermal gradients, and their inevitable variation with change in sample size, nature, placement, etc., make the use of any standardized "dynamic" procedure, such as the uniform rate of increase in temperature that is so frequently used in homogenization studies, hazardous at best. A useful motto for inclusionists, as in most phase-equilibrium studies, is "dynamically-derived data are doubtful." As in all microthermometry, a stepwise operation, involving at least an approach to static conditions after each change of temperature, is necessary to avoid serious errors.

Temperature sensors. One of the major uncertainties in all heating stages is the measurement of the sample temperature. This measurement must be done in such a way that either the value is correct (that is, it is the actual temperature of that part of the sample under observation), or provision must be made for calibration runs, permitting a suitable correction to be applied to cover gradients in and between sample and thermocouple.

Many options exist in the choice of temperature sensors for control and/or readout. The obvious requirements include electrical compatibility with existing controllers and readouts, accuracy, precision, stability, size and response time. Several commercial stages have been offered that are based on thermistors as sensors. These devices measure temperature differences by change in electrical resistance, can be made moderately small, and are sensitive, as they have a large change in signal (electromotive force, emf) per °C, but have generally shown poor stability in terms of emf vs temperature. Many stages use thermocouples. Many combinations have been tried, but the most common (with their trade designations) are chromel/alumel (type K); chromel/constantan (type E); Fe/constantan (type J); Cu/constantan (type T); Pt/Pt-10% Rh (type S); and Pt/Pt 13% Rh (type R). These various couples differ widely in their signal strength (emf/°C), linearity of signal with temperature, stability, and operating temperature range. The linearity of the signal is important only if the electronics package does not have adequate internal correction procedures built in for that specific thermocouple.

Thermocouples can be used in several ways. In older stages, the total current flow was sensed by means of an ammeter reading in temperature units. This procedure requires a given size (and hence resistance) of thermocouple wire and hence normally requires heavy wires, resulting in large junction leads and the possibility of major heat flow along the leads. Most modern units operate by measuring the voltage (emf) by means of a null procedure that eliminates current flow. As long as the unit has sufficient sensitivity, very fine wires can be used, and thermocouples made of wire 0.0125 mm in diameter are available. At elevated temperatures e.g., >600°C), in air, these wires may burn out quickly. Shielded thermocouples would not give this problem, but are much thicker and hence conduct much more heat. The Pt thermocouple has good stability over a large temperature range but relatively small signal strength. Problems of matching sensors and readouts can be eliminated by the procedure of simply measuring

the emf output of the thermocouple by nulling with a potentiometer. Standard
tables of emf vs °C for the thermocouple can be used, along with a calibration
correction. This procedure can yield high accuracy and precision but is much
slower; for a given degree of precision and accuracy, however, it is much cheaper
than the electronic readouts.

A thermocouple responds only to the _difference_ in temperature of two junctions, the unknown and the reference. A length of Cu wire (e.g., the measuring
circuit) can be put into the circuit with no effect if both junctions of the
Cu with the thermocouple wires are at the same known reference temperature.
This reference temperature is usually an ice-water mixture in a thermos bottle;
unless it has become stratified, this mixture is correctly assumed to be at 0°C.
Several commercial alternatives to the thermos bottle ("ice-point devices") are
much more convenient but can malfunction, and the malfunction may not be recognizable.

Pt resistance thermometers ("RTDs") -- essentially just a coil of very fine
Pt wire -- work on the principle of measuring the rather large change in resistance of Pt with temperature. They provide high accuracy, precision, and stability and are preferred by many workers, but as they cannot be made very small,
they are only effective in applications in which thermal gradients are small.

Equipment for Heating or Cooling Inclusions

Stages for heating only--moderate temperatures. Most early microscope
heating-stage designs have been for heating only, without provision for cooling
to below room temperature. Many designs have been described in the literature,
starting with a simple paraffin bath, Phillips (1875). (Sorby, 1858, also used
a "paraffin bath," but gave no particulars.) Most of the designs have good to
excellent _precision_, but to obtain good _accuracy_ is extremely difficult. Anyone
interested in designing his own apparatus would do well first to look over the
design features (and flaws) in the many versions that have been described, particularly in the last few decades.[1/] The unit described by Meyer (1950) is
novel in that it makes use of vertical illumination, the sample plate sitting
on a heated metal mirror. Similarly, the designs of Hayakawa et al. (1969,
1973; see also Nambu et al., 1978) and Trufanov (1972) are also novel, in that
they provide for hydrostatic pressure on the sample during heating, to prevent
decrepitation.

Large static thermal gradients present the most obvious source of trouble
in heating stages, particularly those in which air convection provides the heat
transfer. These gradients are to be expected in any sample that is heated only
from the sides. Thus, Ermakov (1944; 1950, p. 86) described an electrically
heated microscope stage (subsequently widely applied in the USSR) that operates
to 650°C, using convecting air as the heat-exchange medium. The considerable
amount of data that Ermakov (1950; e.g., his table 36, p. 249) presented on the
continuous formation and unidirectional "streaming" of gas bubbles in some inclusions at a constant high temperature apparently was not recognized as being
evidence of rather severe static thermal gradients within the samples on this
stage. Kalyuzhnyi (1958b) found that all temperatures determined using the "air-
heated thermochambers of the old design" (presumably Ermakov's) were "30° to

----

[1/] These include Ermakov (1944; 1950, p. 86); Bailey (1949); Meyer (1950); Bailey and Cameron (1951);
Skinner (1953); Lemmlein (1953); Richter and Abell (1953); Little (1955); Loskutov (1955); Pomarleanu
(1959); Kormushin (1960); Nadeau (1967, 1968); Miller (1968); Pashkov et al. (1968); Bazarov (1968);
Groshenko (1968a,b); Ohmoto (1968); Hayakawa et al. (1969, 1973); Kelly and Goddard (1969); Ohmoto and
Rye (1970); Trufanov (1972); Hughes and Lynch (1973); Trufanov and Rodzyanko (1973); Kalyuzhnyi (1973);
Balasubramaniam et al. (1975); Chepurov (1975); Sobolev and Kostyuk (1975); Durney (1976); Poty et al.
(1976); Zhovtula (1976); Nambu et al. (1978); and Yu and Lin (1978).

Table 7-1. Summary of commercially available heating/cooling stages

	Temp. Range (°C)	Price (dollars)[1]	Available from
		Heating only	
Leitz 350[2] Digital	R.T. to 350	4000[3]	E. Leitz, Inc., Rockleigh, NJ 07647, USA
Leitz 1350	R.T. to 1350	8500	E. Leitz, Inc., Rockleigh, NJ 07647, USA
Mettler FP 5/52	R.T. to 300	7055	Mettler Inst. Corp., Box 71, Hightown, NJ 08520, USA
Thomas-Kofler Model 40	R.T. to 350	1100	A.H. Thomas Co., P.O. Box 779, Philadelphia, PA 19105, USA
		Heating/Cooling	
Linkam TH 600	-180 to +600	4400	Linkam Sci. Instr., 37 Pine Ridge, Carshalton Beeches, Surrey, SM5 4QQ UK
Chaixmeca	-180 to +600	6450	Chaixmeca Ltd., BP3312-54014, Nancy Desilles, France
McLimans Model 3[4]	<-130 to >+500	2900	Dr. Roger McLimans, Conoco Explor. Research Div. 308, Ponca City, OK 74601, USA
U.S. Geological Survey[5]	-196 to +700	5000	Fluid Inc., Box 6873, Denver, CO 80206, USA
U.S. Geological Survey[5]	-165 to +700	4500	SGE, Inc., Dept. of Geosciences, Univ. of Arizona, Tucson, AZ 85721, USA

[1] Prices will vary depending upon specific peripherals and are constantly increasing; those quoted represent a normal installation as of September 1982.
[2] The description of this stage in the text is for the old model 350 stage; photographs and drawings of a stage called "350" in recent E. Leitz literature differ somewhat, and the digital version listed here is as advertised by E. Leitz, Inc. in March 1, 1982. Apparently the basic elements of the mechanical design are similar.
[3] For digital readout compatible with this stage, add $1,300.
[4] No drawing available.
[5] Modified from design of Werre et al. (1979), and including different peripherals.

60°C too low," and he described an improved chamber, good to 700°C, that used heat-conducting plates to move the heat from the windings to the sample. Kerrich (1974) used a film of silicone oil between the sample and the supporting plate to improve heat flow.

To avoid at least some of these troubles, Richter and Abell (1953) designed a high-temperature stage, operating up to 700°C, by surrounding the sample, above and below, with flat heating elements of wire, each having a central gap sufficient to permit viewing in transmitted light. Theoretically, flat heaters above and below a flat sample should give good results, but the heater dimensions were such that very large gradients still existed, as described in a later section. The concept was subsequently used and improved in the Mettler stage, discussed below. Obviously, thermal gradients could best be minimized by immersing the sample in a thermostated heat exchange fluid. As a result, I designed a heating stage (Roedder, 1962a) in which the sample was immersed in rapidly circulating thermostated silicone oil as the heat-exchange medium. Although the results were precise and probably also as accurate as any, the procedure was relatively slow and was limited to <~250°C by the nature of the silicone fluids available at the time.

Although several instrument companies (particularly in Japan) have manufactured heating stages in the past, only three moderate-temperature heating stages are still on the market, the Leitz Model 350, the Mettler FP 5/52, and the Thomas-Kofler Model 40 (Table 7-1). The Leitz was originally designed as one component of an assembly for chemical melting-point determinations up to 350°C, but most fluid-inclusion laboratories using the Leitz 350 stage have only the stage itself. It consists of a frame, fastened to the microscope, holding a cylindrical electric heater that has a central aperture. Heat is conducted to the sample through a removable metal disk on top, which has a 4 mm viewing hole over which the sample is placed and a horizontal hole that accepts a small, special, very short-bulb mercury thermometer. A metal slide with a coverglass window covers the sample. The entire unit can be translated over the width of the viewing hole relative to the microscope axis by two screws. Three different thermometers cover the range from room temperature to +350°C. Current for the furnace comes from a separately supplied variable voltage transformer. (The stage has provision for achieving low temperatures using a liquid CO_2 tank, but the single hand-operated valve on the CO_2 line provides inadequate control for inclusion studies.) At any appreciable heating rate, the difference between the thermometer reading and the actual sample temperature (i.e., the calibration correction) can be large.

The Mettler stage uses a somewhat novel heating procedure to minimize thermal gradients. The slide carrying the sample is effectively enclosed in a heated metal block, so it is heated from above and below. A small hole through top and bottom of the block permits viewing in transmitted light.

The Thomas stage is a modified Kofler stage (Kofler, 1934), also operating to 350°C, which consists of an electrically heated plate having a central 1.5 mm hole to permit viewing in transmitted light, and a heat-shield cover. A built-in condenser lens improves the illumination. Temperature control is by an auto-transformer. The sample, mounted on a microscope slide, can be moved around over the central hole by means of an external control.

Stages for heating only -- high temperatures. The design and operation of stages for high temperatures (e.g. >~700°C) present many more problems than do those for lower temperatures because of limitations on available materials for construction, heat-transferral to the microscope optics, radiation from the sample obscuring the transmitted light image, and a large increase in the inevitable thermal gradients. The sample is at a high temperature, perhaps 1000°C or higher, and various essentially cold surfaces must be within ~1 cm of this hot sample. Laterally, insulation outside the heating element can be used to minimize the heat flow in this direction. If this insulation is at all effective, most of the heat generated in the heating element must flow vertically, up or down, through a gradient of perhaps 1000°C/cm. Part of this flow is by conduction to the heating-stage assembly and the microscope stage, and part is by radiation through a rather large angle to the relatively cold and even water-cooled optical elements of the objective above and the condenser below. On this basis, there must be large thermal gradients in parts of the apparatus immediately surrounding the sample; hence, gradients in the sample itself are inevitable. Radiation increases exponentially with temperature, so at high temperatures, this heat loss can (and does) make for large thermal gradients, particularly in the exposed radiating surface of a relatively nonconductive sample. Many designs have been described, particularly in the Russian literature, but I know of no complete solution to this problem.

Some of the earliest high-temperature studies were by Barrabé and Deicha (1956, 1957) and Barrabé, et al. (1957, 1959). They heated minute polished spheres, cut from samples of quartz, to study the behavior of glass inclusions. Other hot stages have been described for use up to 1100°C (Brock, 1962) and 1200°C (Kalyuzhnyi, 1960). Dolgov and Bazarov (1965) described a 1400°C stage with which many Soviet high-temperature inclusion studies were made. A special microscope furnace using a Mo sheet as the heating element, in vacuum, has been described by Kalyuzhnyi (1965) for use on glass inclusions, in the 600 to 1600°C range.

Of all the models described in the Russian literature, one of the most widely used is the 1400°C stage of Dolgov and Bazarov, first described in 1965.

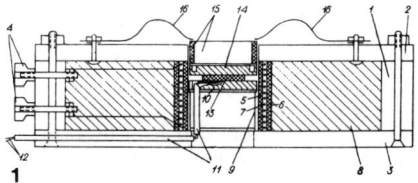
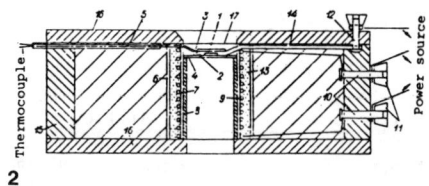

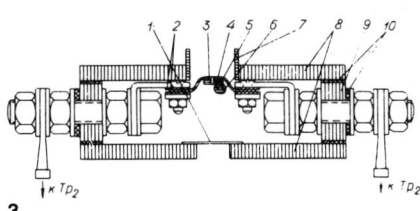
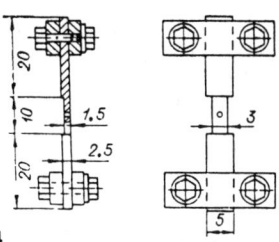

Figure 7-1. Microscope heating stage for temperatures <1400°C, first described by Dolgov and Bazarov (1965). From Sobolev and Kostyuk (1975; their Fig. 6). See text.

Figure 7-2. Microscope heating stage for temperatures <1650°C, first described by Bazarov (1968). From Sobolev and Kostyuk (1975; their Fig. 7). See text.

Figure 7-3. Microscope heating stage for temperatures <1500°C, first described by Chepurov and Pokhilenko (1972). From Sobolev and Kostyuk (1975; their Fig. 9). See text.

Figure 7-4. Silicon carbide heating strip used in microscope heating stage for temperatures <1600°C, first described by Mikhailov and Shatskii (1975). From Sobolev and Kostyuk (1975; their Fig. 10). The sample is placed over the hole. This construction has been used in almost all the recent investigations at Novosibirsk, USSR. Dimensions in mm.

As shown in Figure 7-1, heat is supplied by a vertical cylindrical coil (7). The sample (13) is placed betwen fused silica plates (10 and 14); a thermocouple (12) is in a depression in the lower plate. All heat lost by radiation to the microscope objective above, which is generally water cooled, and to the condenser below, must be supplied by conduction or radiation through these silica plates.

Figure 7-2 shows a later stage designed by Bazarov (1968). It features two-stage heating, an external cylindrical coil (9), held at intermediate temperatures, and a tiny Pt loop resistor (2, 14) surrounding the sample (1), which can be heated to 1650°C, with a stated measurement precision of ±10°C at 1450°C.

A stage designed by Chepurov and Pokhilenko (1972) (Fig. 7-3) uses a Pt heater strip (3) with a central hole, over which the sample is placed for observation in transmitted light, and under which a thermocouple is attached (4). A similar stage, but one using a silicon carbide (i.e., "silite") resistance heating rod (Figure 7-4) instead of the Pt strip, was described by Mikhailov and Shatskii (1975). This latter design has been used in much of the extensive recent very high temperature work done at Novosibirsk, USSR. Sobolev and Kostyuk (1975, p. 27 in original) stated that the "overall error in measurement [for the stage of Mikhailov and Shatskii, 1975?] is usually estimated at ~±10-15°C." When any such designs are used, a carefully standardized procedure can lead to a reasonably high precision of measurement, and the effect of thermal gradients on a measurement can, in theory, be eliminated by calibration procedures. This ideal is realized only if the calibration runs have the identical technique and thermal regime (including thermal transmissivities and emissivities) used in the running of the unknown sample. To set up a truly valid calibration of this type is difficult[2/]; hence, the <u>accuracy</u> of such measurements is generally unknown.

All the foregoing may give the false impression that I believe it impossible to obtain valid high-temperature homogenization data. Such data can be obtained, but the true accuracy is difficult to assess. Thus, one seldom discussed problem is the loss of contrast in the image from the radiation emitted by the sample itself. Several procedures can minimize but not eliminate this problem. It is best to use high intensity illumination, plus a neutral filter above the stage to cut the total light down to a comfortable level. As the radiation from the sample is enriched in yellow, a blue filter above the stage also helps to improve contrast. A dark-blue filter fills both requirements.

Only one high-temperature microscope stage suitable for inclusion studies is commercially available at this time, the Leitz Model 1350, designed for use up to 1350°C. The field of view is 2.5 mm in diameter. Temperature control of the small Pt-wound furnace is manual, using a variable voltage transformer. The calibration correction at the melting point of Au (1064°C) was 92°C. A replacement furnace provided by the manufacturer, of <u>slightly</u> different design, yielded a correction of only 51°C.

<u>Stages for cooling only</u>. Cooling inclusions under the microscope involves a surprising number of seemingly minor but experimentally very troublesome details. The essential requirements are very similar to those involved in heating stages, but as the temperatures used are seldom even as low as -150°C, the thermal-gradient problems are numerically smaller. Unfortunately, however, the nature of freezing data on inclusions is such that usually higher accuracy is needed than in heating-stage studies. Some few studies only require cooling to -20 or -30°C, but most require much lower temperatures. Many metastable supercooled aqueous inclusions refuse to freeze until -50°C or even -100°C, even though their stable equilibrium phase changes are all at -30°C or warmer. Gas-rich inclusions frequently require temperatures of ~-150°C. Optical problems are generally more limiting in freezing than in heating work because of the faintness of some phase boundaries. A seemingly trivial detail, condensation on cold surfaces in the optical system, can cause considerable trouble, and provisions to avoid it can cause considerable inconvenience in operation.

Although many earlier reports mentioned the use of subambient temperatures in the study of inclusions, the experimental difficulties generally prohibited acquiring quantitative data. To avoid some of these troubles, I designed a stage (Roedder, 1962a) that was apparently the first to provide reasonably accurate data. (A review of the development of the cooling method is given in that same paper.) In that stage, the sample is immersed in rapidly circulating thermostated acetone as the heat-exchange medium.[3/] Many other cooling procedures have subsequently been described in the literature, some involving novel designs.[4/] Thus, Sawkins (1966) described a cooling stage based on the use of several

----

[2/] As a possible example, Bakumenko et al. (1967, p. 143) reported a Th of 1300 ±10°C for melt inclusions in synthetic diopside crystals that were grown from a melt containing 5% CaF$_2$ flux that "was kept ... at 1350°C for 50 min and then allowed to crystallize at 1300°C for a further 20 min." Note, however, that because a pure diopside melt crystallizes at 1391.5°C, there is a question as to the actual temperature at which crystal growth and inclusion trapping occurred.

[3/] This stage was actually the same unit as the circulating-fluid heating stage mentioned above, but with with acetone in place of silicone oil as the circulating heat-exchange fluid.

[4/] For further details, see: Kern and Mattern, (1963); Mel'nikov, (1965); Velchev and Mel'nikov, (1965); Myaz' and Simkiv, (1965); Kormushin, (1965); Bazarov, (1966) (see also description by Sobolev and Kostyuk, 1975); Sawkins, (1966); Poty, (1968); Mel'nikov (1968); Takenouchi and Imai (1968); Dolgov et al. (1968); Kelly and Goddard (1969); Barnes et al. (1969); Takenouchi (1970); Kalyuzhnyi (1973); Kharlamov (1973); Smith (1973); Anonymous (1974); Poty et al. (1976); Kalyuzhnyi and Gigashvili (1976); Voznyak and Galaburda (1977); Khitarov (1978); Freckman (1978); and Nambu et al. (1978).

thermoelectric (Peltier) cooling units; it cooled to only about -18°C and could be controlled to ±1°C. I have also tried using thermoelectric cooling but found it to be difficult to control and surprisingly small in cooling capacity. Microscope cooling stages based on Joule-Thompson cooling are commercially available, but are not yet adequate for most fluid inclusion studies.

One small problem inherent in all freezing studies using circulating refrigerated acetone is the possibility that cracking of the host mineral will admit a small amount of acetone, which will lower the apparent freezing temperature of inclusions drastically. Such cracking is most likely in those samples containing inclusions having very small vapor bubbles, which are hence subjected to high internal pressures because of expansion on freezing. I found that valid freezing data on such samples can be obtained by the use of a miniature double-windowed cell, filled with oil, inside the regular cell of the cooling stage (Roedder, 1968c, p. 441). As the oil is immiscible with the inclusion fluid, this technique permits reasonably accurate determinations even on inclusions that do fracture on freezing. The high viscosity of most oils at low temperature precludes using them directly as the circulating fluid.

These cooling devices are generally rather complex in design and operation. I described a much simpler, very inexpensive, manually operated and controlled cooling procedure that gives accurate and useful but relatively low-precision results over a wide temperature range (down to -78°C), at the expense of a little more time (Roedder, 1962a, p. 1051). The operator can readily maintain two or three such units at separate and easily varied temperatures. With a little practice, temperature can be held constant to ±0.5°C by the periodic addition of small lumps of solid CO_2 to the acetone baths. Simple insulated containers, made of nested pairs of glass beakers with glass wool between, are adequate for the cold baths.

For temperatures in the range 0 to +40°C, an even simpler procedure can be used (Roedder, 1965a). The thin section (or even liquid immersion oil mount) is merely hung on a wire sling just under the surface of water in a flat-bottomed glass dish on the stage. Temperatures, read on a thermometer, are controlled by small additions of hot water or ice. Equilibration is obtained rapidly by pumping the water with a rubber ear syringe. Surface waves from vibration are minimized by a shallow ring placed over the slide.

The thermal mass of the 7 liters of acetone in my circulating fluid stage was so large that temperature changes could not be made rapidly. To permit faster operation over a greater range, I designed and built some crude models of a new stage, cooled by rapidly moving, single-pass N_2 gas, which was first cooled by LN_2, using some features of the cold stages described by Bailey (1949), Rhodes (1950) and Monier and Hocart (1950). One feature was the use of dichroic coatings on the windows, made to reflect infrared but transmit visible light, to reduce thermal gradients from radiation to or from the sample surface. Evolving out of these attempts, after several false steps over a number of years at the U.S. Geological Survey, came the "USGS" heating/cooling stage of Werre et al. (1979), as described below.

Commercially Available Stages for Heating and Cooling

Some of the stages described in the previous sections permitted operation in both heating and cooling modes, so that a given sample could be examined both above and below room temperature without a change of equipment. Although this capability presents a considerable advantage, the compromises necessary to make a piece of equipment more versatile usually result in sacrifices in other aspects. In 1976, Bernard Poty and colleagues at Centre de Recherches Pétrographiques et Géochimiques, Nancy, France, described a new stage (Poty et al., 1976; briefly described earlier -- Anonymous, 1974) that combined both versa-

Table 7-2. Comparison of design and operating features of three commercial heating/cooling stages.

[Data from advertising brochures, various purchasers, and personal experience. Some of these features vary depending upon the date of purchase.]

Feature	Chaixmeca	Linkam	USGS
Sample heating mechanism for above ambient	Conduction from electrical resistance heater in metal block; from below and ~15mm from sample, via fused silica top element of condenser.	Conduction from electrical resistance heater in silver block, from below and ~7 mm from sample, via sapphire window.	Forced convection of moving heat exchange gas surrounding sample above and below, heated by electrical resistance heater.
Sample cooling mechanism for below ambient	Conduction, from annular circulation of precooled N_2 in metal block below and ~17 mm from sample, via fused silica top element of condenser.	Conduction, from annular circulation of precooled N_2 in metal ring below metal block, ~7 mm from sample, via sapphire window.	Forced convection of moving heat-surrounding sample above and below. Some models permit LN_2 circulation.
Sample heating mechanism for below ambient	Old models - control rate of flow of N_2. New models - have two ranges: Below ~-52.5°C. Stop N_2 flow and let warm up. Above ~-52.5°C. Oppose N_2 cooling with electrical heating as above.	Oppose N_2 cooling with electrical heater as above.	Major changes: change ratio of cooled and noncooled gas, and/or rate of flow. Minor changes: warm cooled heat-exchange gas mixture to desired temperature by resistance heater.
Temperature sensor and location	Platinum resistance thermometer embedded in metal block ~6 mm from sample.	Platinum resistance thermometer embedded in silver block ~ 3 mm from sample.	Thermocouple touching sample.
Temp. range	-180 to +600°C	-180 to +600°C	-196 to +700°C
Control method	Manual or automatic on-off N_2 flow, with large oscillations (see text)	Automatic only; temperature set point stable to better than 0.5°C.	Manual only, but very precise.
Max. sample size	18 mm diameter circle.	Standard 25x45 mm section	22 mm diameter circle.
Max. sample area visible	3 mm diameter	2.2 mm diameter	22 mm diameter
Max. sample thickness	1.8 mm	1.5 mm	2 mm
Time for heating to 250°C	125°C/min	~10 min.	<1 min (Fluid Inc.) 2 min (USGS)
Time for heating to 450°C to 600°C	8 min 17 min	4 min 6 min	<2 min (Fluid Inc.) 10-15 min (USGS) <3 min (Fluid Inc.)
Time to change sample at room temperature	<1 min	<1 min	< 1 min
Time to reach -50°C	~1 min	~1.5 min	<1 min (Fluid Inc.) 3.5 min
Time to reach -150°C	~2 min	~2.25 min, but can be achieved in seconds if pressurized LN_2 is used.	<2 min (Fluid Inc.) ~10 min (USGS)

tility and ease of operation, and that included relatively few unfortunate compromises among the desired features tabulated at the start of this chapter. The subsequent commercial production and wide use of a stage based on this design (the "Chaixmeca stage") has been a major factor in the current renaissance in fluid-inclusion study. Several other stage designs, each based on somewhat different principles and goals, have been described subsequently and are now also widely used.

In this section, the major design characteristics, operating procedures, and limitations of the three units commercially available in the West are described (see also Table 7-2; another new type heating/cooling stage design, not listed in Table 7-1 or 7-2, was announced by Durney (1976) but apparently has not been described). Part of this material has been adapted from Hollister et al. (1981). In addition to the obvious need to compare the various units carefully with each other and with the nature of the planned investigations, another caveat is important to remember before purchase. Most such equipment is not truly of fixed design. Many design changes, some of them of major import, have been made in some units; these changes may seriously affect the usefulness of the equipment for certain applications.

Although the design of the stage itself is a central concern, a very important aspect of setting up an effective microthermometry laboratory lies in the selection of the "peripherals" -- the various devices that must be used with the stage (or several stages in a larger laboratory). These devices must be compatible with each other and with the stage, and much of the convenience and speed of operation, and the accuracy of the results, hinges on these selections. They also comprise a major share of the cost of any installation. The most obvious such peripheral is a temperature readout. A wide variety is available; most are, in effect, digital volt meters ("DVMs") that convert thermocouple emf to a digital temperature readout, either by an assumption of a simple linear relationship of thermocouple emf vs temperature, or by a more accurate nonlinear conversion designed for that sensor. (It is a great temptation to let oneself be fooled by the high apparent precision of the numbers on such readouts. Various sources of error make the accuracy of these numbers much poorer than the apparent precision, and, at best, these numbers are only the temperature of the sensor in the stage, not that of the particular inclusion in the sample. Thus instabilities in the electronics of the digital readout devices provided with some commercial stages may introduce sudden and random 0.5°C shifts in calibration.

In addition to a temperature readout, many other peripherals are either required or desirable: gas-pressure reducers, flow meters, valves, dewars, ice-point devices, gas filters, variable voltage transformers, low-temperature tubing and insulation, fiber-optic illuminators, voltage regulators, and thermocouples. Some workers like to use a minicomputer to provide temperature programming, which permits following a preset cooling or heating rate, but most use manual adjustments to control the rate, on the basis of the changes seen in the microscope. A printer that prints the temperature visible on the DVM readout when a footswitch is activated is useful to record temperatures of phase changes; the worker does not have to look away from the microscope. A less costly equivalent involves a switch-operated "hold" on the readout on the DVM; the number can then be recorded manually. Either of these devices presents a temptation, however, to obtain data from dynamic rather than from essentially static situations; dynamically derived data can be obtained very fast and can be very precise, but can be very inaccurate.

A precision temperature controller, to permit one to hold a given temperature for the time necessary to level out transient thermal gradients from a preceding temperature change, and to achieve phase equilibrium within the inclusion, might seem to be essential; however, most installations rely on the operator for such control, which keeps wasted time to a minimum. Most temperature goals cannot be known in advance because they must be determined by the operator watching the phase changes take place. In effect, any reasonably

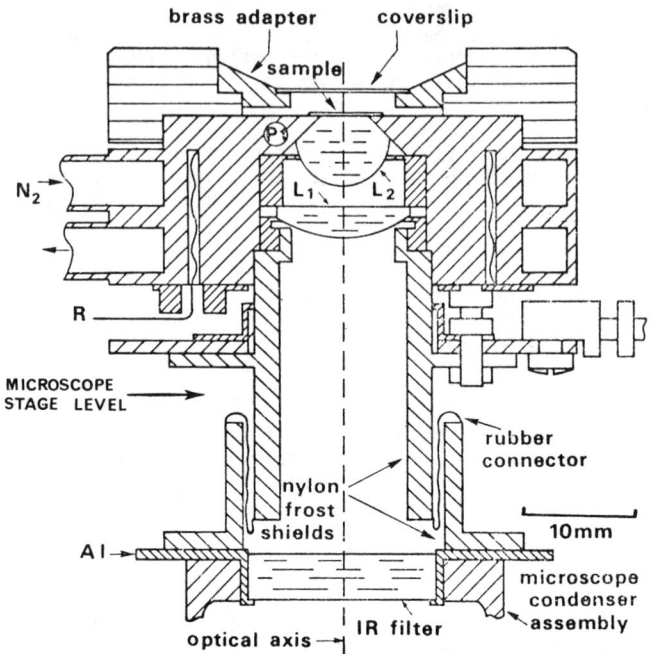

Figure 7-5. Cross section of Chaixmeca stage as modified by Burruss (1977). Except where noted, components are made of brass. Al, aluminum adapter for IR filter; N_2, nitrogen cooling gas inlet and outlet; Pt, platinum resistance temperature sensor; L_1 and L_2, condenser lenses; R, resistance heater element.

skilled operator, once experienced on the equipment, serves the purpose of what otherwise would be a rather sophisticated and expensive controller, quickly bringing the stage to what is found <u>during the run</u> to be the desired temperature, and levelling it out there. Obviously, this last requires stability of gas flow and line voltage, as well as an operator who can act like a "feedback loop" that dampens rather than amplifies oscillations from the desired temperature.

The Chaixmeca stage. The Chaixmeca stage was built on the basis of an original design by Poty et al. (1976) and is sold in France. It is designed for use with a Leitz H32/0.40 (6.6 mm-working distance) objective[5], and hence has a 1.8 mm quartz plate as the upper stage window. Many users have found that modifications are necessary in both the top and bottom of the stage. Thus, when using the Leitz UTK 50/0.63 objective, Burruss (1977) replaced the 1.8 mm quartz plate upper window of the stage with a coverslip; Cunningham and Carollo (1980) used a pair of coverslips separated by a dead air space. Figure 7-5 shows a schematic cross section of the Chaixmeca stage, as modified by Burruss (1977). These modifications result in improved contrast and resolution, and in appreciably smaller thermal gradients. Cunningham and Carollo (1980) have also added considerable insulation and have made other changes. For small inclusions, 16x

----

[5] This objective was designed for use with a 1.8 mm thick cover glass. The Leitz universal stage objective UTK 50/0.63, which has an optical rating of 32/0.40 when used without the hemispheres, or the Leitz L32/0.40 can be substituted. The last two are identical except that the UTK 50/0.63 is strain free, permitting the use of crossed polars. If only simple gas/liquid or gas/liquid/NaCl inclusions are to be run, the L32/0.60 will suffice. The prices (May, 1983) were: H32 $1395; UTK 50 $1134; L32 $873.

or even 25x oculars may be desirable, even though the magnification may in part be "empty." The stage is now supplied with Au-plated interior parts permitting at least temporary operation at temperatures up to 600°C. At high temperatures, a cooling coil of Cu tubing around the objective is used to prevent damage from heat radiated from the sample chamber.

An important design feature of the Chaixmeca, which is not present in the other commercial heating/cooling stages, is a built-in condenser of fused silica specially designed to stand the temperature range of the stage. This results in greatly improved illumination and resolution, particularly for small inclusions (<~15 μm). In spite of this special condenser, however, most users have found it necessary to modify their microscopes to get more light into the rather small aperture designed into the lower side of the stage. Thus, H.E. Belkin (pers. comm.) has found that a simple lens of ~3 cm focal length, mounted in place of the substage condenser, increases both the convergence and the intensity of illumination in the stage.

Moisture must be kept out of the optical path at low temperatures. Most users accomplish this by placing a Teflon, plastic, or rubber sleeve between the objective and the sample chamber and between the stage and the condenser (see Fig. 7-5 for the substage modifications of Burruss, 1977). Stopcock grease can be used between the base of the sleeve and the stage to help keep moisture out of the chamber. Silica gel or other absorbents may be sprinkled liberally in and around the sample chamber if moisture problems persist; some individuals use a flow of N_2 or carefully dried air directed at the sample-chamber windows (above and below). As in all cooling-stage studies, it is best to avoid condensation to begin with. Once ice forms, its sublimation at low temperatures is slow because of its low vapor pressure at low temperatures. After the the stage is used for cooling runs, it should be carefully dried (one way is to heat the stage to >100°C) to remove moisture, which, if allowed to remain, will result in corrosion and deterioration of the stage.

For the Chaixmeca stage, the temperature correction (the difference between temperature sensor and sample) increases steadily from 4°C at a stage temperature of about 150°C to about 15°C at 400°C. A modification of the Chaixmeca stage, described by Cunningham and Carollo (1980), reduces the correction to about 8°C at 400°C (C.G. Cunningham, pers. comm.). It features, among other items, an insulation ring that reduces heat loss by radiation, thus allowing higher temperatures (>200°C) to be achieved more rapidly and more accurately. Use of a double coverglass window in place of the thick quartz upper plate improves optics and also reduces the temperature corrections.

The controlling apparatus for the older models of the Chaixmeca stage can be set for automatic heating or cooling to a preset temperature, but the heating or cooling rate must be controlled by hand. Such control is relatively easy if the readout is connected to a strip chart recorder. These older models are cooled by passing N_2 gas, or air that has had CO_2 and H_2O removed by absorbents, through liquid N_2 and then through the stage. Heating is by a circumferential resistance heater in the main body of the stage, below the level of the sample.

Many model changes have been made in the Chaixmeca stage since it was first marketed. Changes in the design of the stage are minor, but two changes involving the mode of operation are major. First, the new models now use direct pumping of liquid nitrogen (LN_2). Cooling of the transfer tube eventually results in injection of a mixture of liquid and gaseous N_2 into the stage. This allows more rapid cooling of the stage to low temperatures (e.g., to -150°C in 2 minutes), which sometimes is very useful. However, because the rate of flow of LN_2 cannot be controlled (it is either on or off), the rate of warming when operating at low temperatures is difficult to control. This is exacerbated by the second design change that does not permit operation of the resistance heater

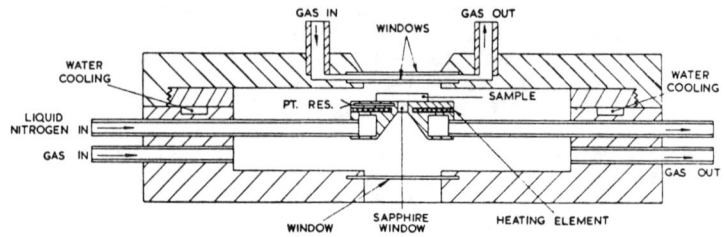

Figure 7-6.   Cross-section of Linkam stage (Shepherd, 1981). Pt, platinum resistance temperature sensor.

at temperatures less than ~-50°C. Users of the newer model have not been able to achieve a heating rate of less than 5°C/min at the melting temperature of CO_2, whereas it is necessary to have a heating rate of ~0.1°C/min to measure, for example, the effect of lowering of the CO_2-melting temperature by additional components such as CH_4 and N_2 (Henry, 1978).

The controller on this stage (e.g., a 1979 model bought in 1979) can only be used for very rough work, as it is based on on/off operation. As a result, the temperature cycling, in °C at the stated temperatures in °C, was found to be: 18° at -100°; 22° at -40°; 27° at -20°; 6° at +100°; 9° at +200°; and 10° at +300°.

The Linkam stage. The Linkam TH600 stage (Fig. 7-6) made in Surrey, England, was described by Shepherd (1981). It differs from the Chaixmeca stage in several important ways. First, it uses a flow of N_2 gas between double windows and under the chamber containing the Ag thermal block. This feature prevents the condensation of moisture during cooling runs and serves to prevent heating of the objective during heating runs, eliminating the need of a cooling coil. Secondly, it does not have a built-in condenser near the sample; hence, the contrast and resolution for small inclusions (<15 μm) are not quite as good as those provided by the Chaixmeca stage. The design features result in minimum correction over the entire temperature range. The overall uncertainty is mainly due to problems pertaining to the nature of the standards (see "Calibration"); it is ±2°C between -95 and +200°C, increasing to ±8°C at 592°C (MacDonald and Spooner, 1981). When a change is desired, the rate of change of temperature is programmed by the operator and implemented by an automatic circuit, which amplifies and linearizes the signal from a Pt resistance thermometer attached to the heater. Thus, heating rates can be automatically programmed at any of 27 different rates between 0.1°C/min and 90°C/min. Shepherd (1981, p. 1245) indicated, however, that "On approaching a phase transition it is good practice to reduce the heating-freezing rate to ≈0.5°C/min. to insure thermal equilibration." Low temperatures are attained and controlled by a flow of cold N_2 through an annulus in a Ag block holding the sample and a resistance heating element; heat from the element (via the controller) opposes the cooling of the gas flow to yield the temperature desired. In the heating mode, only the heater is used. In both modes, the sample temperature is controlled by thermal conduction between the sample and the Ag block. The sample sits on a 2 mm diameter sapphire window mounted in the Ag block. One inconvenient aspect is that the heating circuitry requires 5-10 min of warmup time before it will take effect on each first use. Another problem is introduced by the use of a sapphire plate on which the sample is placed. As this plate is not oriented with its c axis vertical, birefringence of daughter crystals can only be observed at the extinction positions.

The U.S. Geological Survey stage. The U.S. Geological Survey (USGS) stage (Fig. 7-7) was designed at the USGS and described briefly by Werre et al. (1979; it has sometimes been called the "W3B" stage from the names of the four authors). Two different suppliers offer commercial versions of the USGS stage (see Table

196

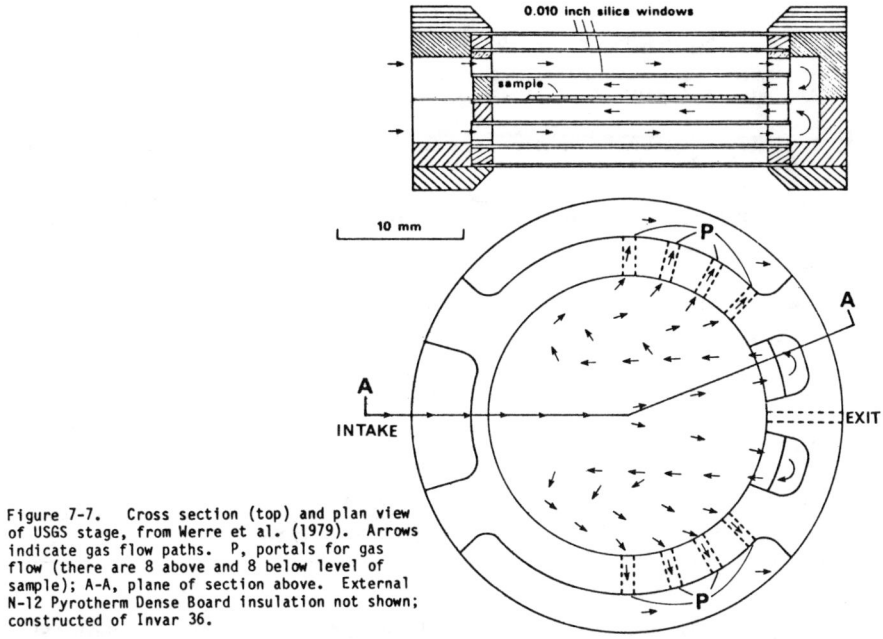

Figure 7-7. Cross section (top) and plan view of USGS stage, from Werre et al. (1979). Arrows indicate gas flow paths. P, portals for gas flow (there are 8 above and 8 below level of sample); A-A, plane of section above. External N-12 Pyrotherm Dense Board insulation not shown; constructed of Invar 36.

7-1). Additional details of construction and operation have been given by Woods et al. (1981). This stage is substantially different from the others, because the two major design concepts on which it is based are different. The first of these is that thermal gradients within a sample and between sample and temperature sensor can be most effectively reduced (they can never be eliminated) by totally immersing both in a rapidly moving heat-exchange fluid. The second is that to obtain rapid, precise, and accurate control of temperature changes in the sample, the total thermal mass (sample plus stage interior) should be kept to an absolute minimum. The heating or cooling of the sample is done by passing preheated or cooled N_2 gas directly over the sample. From above room temperature to about 500°C, ordinary compressed (but carefully filtered) air is generally used. Very rapid heating is achieved because the mass of material to be heated (essentially only the thin metal core and the sample plate) is very small, in contrast to that in the other stages. At temperatures above room temperature, the rate of heating or cooling can quickly be changed, making it easy to determine Th precisely by temperature cycling near Th. That is, by quickly cooling the sample from an apparent Th, one can see whether a gas bubble immediately expands into sight (Th has not been achieved) or if undercooling is required to cause sudden nucleation of the gas bubble (Th has been achieved). This process can be repeated several times to determine the precise Th of inclusions that would otherwise be impossible to measure because of the presence of large dark borders. Quick reversal for melting temperatures can also be achieved. Such temperature cycling may be necessary to establish Tm of solids, especially ice, which has an index of refraction close to that of water. Early versions required more time to reach very low temperatures (e.g., ~10 min to -150°C), but this has been improved by better insulation. Frost is prevented from accumulating on the stage windows by directing a flow of cold N_2 from the exhaust back onto the windows. Pasteris (1983) described a modification that permits entry of LN_2 directly into the stage, to obtain temperatures as low as -196°C; the stages sold by Fluid Inc. have similar modifications.

The continuous flow of N_2 around the sample in the stage minimizes thermal gradients within the sample, allowing measurements on inclusions throughout a

Table 7-3. Recommended terms and abbreviations for fluid inclusion studies

[To simplify typing and printing, subscripting has been avoided; context usually precludes any chance of ambiguity with the symbols for the elements.]

Th - Temperature of total homogenization. The phase into which homogenization occurs should also be stated (e.g., Th (L) or Th (V)).

Th L-V, Th CO_2 L-V, etc. - Temperature of homogenization of the stated pair of phases only. The phase into which homogenization occurs should also be stated [e.g., Th CO_2 L-V (V), or Th CO_2-H_2O (CO_2)].

Tt - Temperature of trapping. (Sometimes also called "temperature of formation" or "Tf", but as this results in ambiguity with "temperature of freezing", both are avoided here.)

Td - Temperature of decrepitation. The specific definition varies with the individual report.

Tm - Temperature of melting (or dissolving).

Tm NaCl, Tm dms, Tm ice, etc. - Temperature of melting (pure phase) or of apparent melting, i.e., solution (mixed system) of specific phase indicated.

Te - Temperature of eutectic, i.e., first recognized formation of liquid on warming a completely crystalline inclusion; only an approximate or "practical" value at best, as traces of other components will always result in traces of melting at lower temperatures.

Tn - Temperature of nucleation in fluid, generally on cooling (i.e., in a normally supercooled metastable fluid). Thus, Tn for a supercooled aqueous inclusion would be Tn ice; the heterogenization of a previously homogenized inclusion would be Tn V; similarly, Tn NaCl, Tn CO_2 V, etc.

---

relatively large sample (22-mm diameter) without major temperature corrections. Such large samples cannot be used in the other stages. The Chaixmeca stage, which has a 3 mm observation window, permits only 2% of the USGS stage area to be covered in a given run, and the Linkam, which has a 2 mm window, permits only 1% of the area to be seen.[6/] The USGS stage, like the Linkam stage, has no built-in condenser, which makes study of small inclusions more difficult than it is when using the Chaixmeca stage, but even though the sample is near to the microscope stage, the normal microscope substage condenser cannot be used. A lens of longer focal length must be substituted.

The thermocouple for the USGS stage rests directly on the sample, thereby minimizing temperature-correction errors; as for all thermocouples, calibration is required. The thermocouple also holds the sample down so that it is not dislodged by the N_2 gas flow. Very small samples are placed under an additional coverglass, which, in turn, is held down by the thermocouple. This and other design features of the stage makes the process of changing samples slightly slower than it is on the Chaixmeca or Linkam stages.

Measurements - Standard Routines

Standardization of terminology. Semantic arguments can never be won, but in microthermometry, so many different terms and abbreviations have been used in the literature to refer to the various phase changes that some standardization was needed. A consensus of active inclusion workers was reached, and the recommended terms, along with the rationale behind the choices (Table 7-3), was first published in Fluid Inclusion Research -- Proceedings of COFFI, vol. 10 (1977). No terminology can cover all situations, but the recommended terms adequately cover normal usage.

Sequence of measurements. For unknown inclusions, the sequence of measure-

----

[6/] Since the first heating run may stretch some inclusions, and will generally destroy most daughter crystals, inclusions in the remainder of the sample plate, outside that which was visible in the stage, generally should not be used again.

ments must be based on assumptions about the probable composition and phase assemblage at room temperature (detailed in Chapter 8), yet these assumptions must in part be based on inclusion behavior in preliminary microthermometry runs. Once the probable composition is established, the operation can become somewhat more routine. The following refers mainly to "ordinary" liquid inclusions (aqueous and CO_2); various procedures for special situations or special compositions will be found in Chapters 11 to 18.

With one exception, low-temperature phase changes should be measured first on all inclusions available in a chip because, when heated, inclusions may decrepitate or stretch the host and thus change the volume of the inclusion (see Chapter 3). The exceptions to this rule are high-density H_2O inclusions in easily deformed minerals such as fluorite, calcite, and barite; expansion of the fluid on freezing can cause such inclusions to stretch their walls. As many measurements as possible should be made in as limited a range of stage temperature as practicable on each chip before any major temperature change is made. Thermal gradients in the stage are thus kept to a minimum and accuracy is improved.

Inclusions of water solutions should be cooled until at least some are frozen, as normally evidenced by a sudden change to a granular gray or brownish color caused by the very fine grained ice dendrites, and a diminution of the diameter of the vapor bubble. Very low salinity and very small inclusions frequently freeze to clear ice. Ice can be identified by the following criteria: (1) an index of refraction appreciably less than that of the solution into which it melts; (2) very low birefringence, in the correct range for ice; (3) characteristic growth of two parallel plates from opposite sides of rounded grains, as temperature falls; (4) volume decrease during melting (i.e., increase in bubble size); (5) the general crystallization temperature range; and (6) the general prevalence of water, as determined by analysis of numerous other samples. Only rarely do the optical properties of the host mineral permit determination of optical sign and orientation relative to crystal habit, and then only after the initial very fine grained ice has recrystallized to coarser crystals. The most common phase that might be mistaken for ice is hydrohalite, $NaCl \cdot 2H_2O$, which is strongly birefringent, length slow, and exhibits an extinction angle of 35° from the prism face (Craig et al., 1975). It has an index of refraction well above that of the solution. It also is very slow to melt (in contrast to ice) and hence may persist well above 0°C.

T.J. Reynolds (pers. comm., 1982) recommends watching the vapor bubble intensively; on freezing, the slight change in volume results in a sudden movement (a "jerk"). He also finds that some inclusions that had Tn ice ~-40°C on the first freezing showed Tn ice at -10 to -20°C on subsequent freezings. A rapid warming, to establish the approximate temperatures (or ranges) for Te and Tm ice will suffice to guide a second run. During this run, the inclusions are frozen again and then warmed through the range first found for Te. As Te can only be approximated at best, it can be determined "dynamically" by slow heating. The sample can then be heated rapidly to a few degrees below the approximate value found for Tm ice. During this heating, the amount of melting of ice with temperature increase provides a monitoring criterion; when the amount of ice in any visible inclusion becomes small, the temperature rise should be stopped to insure equilibration, after which further increase should be at a rate controlled by the accuracy needed, the nature of the inclusion, and the nature of the sample. If high accuracy is not needed, "dynamically derived data" may be adequate and can be determined much more easily. If higher accuracy is needed, warming should proceed slowly and should stop when only a tiny grain of ice remains in that inclusion in the group that is apparently going to have the lowest value for Tm ice. After equilibration is assured, the temperature should be allowed to creep up until that last ice melts (i.e., Tm ice; see Fig. 8-11b). After a scan of the field to pick out the next probable inclusion to melt, further warming is permitted.

The rate at which the temperature can be taken up before Tm ice is actually reached requires some judgment, or the value for Tm ice will be too high. If the sample is large, there will be more thermal inertia. If the inclusion is large, there will also be more thermal inertia, for several reasons. First, the large heat of melting of ice must flow into the inclusion, down the miniscule thermal gradient between the stage and the host sample, and between the host and the inclusion. This problem becomes most obvious for inclusions that have sharp melting points (see "Calibration"). Second, the melting of ice results in the remaining ice crystal being bathed in a fluid that is more dilute than the bulk of the fluid, until diffusion levels out the concentration gradients. This can cause major problems in elongated or very irregularly shaped inclusions, where the distances for diffusive mixing become appreciable. It is even more serious for large three-dimensional inclusions, in which the last ice floats to the top of the inclusion and melts there, yielding a stratified fluid.

Mixed CO_2-H_2O inclusions should be cooled to ~-110°C and then warmed to about -60°C. The heating rate should then be progressively slowed in order to measure Tm CO_2 at a heating rate of about 0.1°C/minute. The stage can then be cooled again, and Tm CO_2 of a second inclusion (or group of simultaneously visible inclusions) can be measured. Hollister et al. (1981) reported that CO_2 crystals show the rather odd behavior of rapid recrystallization to a single crystal after freezing and that this recrystallization is noticeably faster than that for ice crystals. Next, Tm CO_2 clathrate, if it can be recognized, is measured (generally between +8° and +10°C; Tm CH_4 clathrate may be >+10°C), after refreezing down to about -50°C. Finally, Th CO_2 L-V can be measured (if above Tm clathrate). In some samples, Th CO_2 L-V will be found to be at a lower temperature than Tm CO_2 clathrate, and occasionally both Tm CO_2 clathrate and Tm ice can be measured.

After the freezing determinations are complete, homogenization determinations at above room temperature can proceed. The mode and sequence of homogenization must always be recorded; far too often in the literature these observations have been left ambiguous. Most ordinary G-L inclusions containing small bubbles homogenize in the liquid phase (i.e., by expansion of the liquid to eliminate the bubble). Low-density aqueous inclusions and many CO_2 inclusions homogenize in the vapor phase (i.e., by evaporation of the liquid). It is important to record also the behavior on heating of all solid phases originally present -- whether they seem unaffected, get smaller (or larger), or dissolve completely, and, if so, at what temperature(s). If homogenization occurs at a critical point (i.e., by fading of the meniscus), the behavior of this meniscus just before homogenization should be noted; if the meniscus moves toward the "gas" phase, the density of the inclusion was slightly greater than the critical density, and vice versa (see CO_2 in Chapter 8).

Because daughter crystals may not renucleate on cooling after homogenization, Tm dms should be determined carefully on the first try; note also that Tm dms may be above or below Th L-V. Similarly, to avoid the effects of stretching from overheating beyond Th L-V, particularly in soft minerals, those inclusions having the lowest values for Th should be determined first. Any inclusion first seen after the sample has been heated may have been stretched. Even if stretching is not a problem, decrepitation of overheated low-temperature secondary inclusions may ruin the sample before the highest temperature data are complete.

In order to avoid overheating any inclusion to >Th before Th is determined, visual estimates of the gas/liquid ratios are normally used to guide the sequence in which the inclusions are run. If the inclusion having the smallest ratio looks as though it would homogenize at 150°C, one may be tempted to set the heating stage to level out at ~125°C and to creep up from there. However, because the coefficient of expansion of water solutions varies inversely with the salinity, an inclusion filled with pure water at a given temperature will have a larger bubble (at room temperature) than one filled at the same temperature

with saline brine. Hence, all the selected inclusions should be checked peri-
odically, even when the concentration is on one.

Inclusions having wide dark borders present special problems, because
small bubbles can disappear in these areas of total reflection. The use of a
flexible fiber-optics illuminator from above or below can highlight the bubble
in some but not all such inclusions. Because some hysteresis (i.e., difference)
normally exists between Th L-V on heating and Tn V on cooling, a quick cycling
of temperature near Th with an increasing maximum temperature on each cycle can
still permit relatively accurate determinations of Th (see "The USGS stage").
Because the heating is very slow, the magnitude of the hysteresis is mainly a
function of the rate of cooling. This procedure permits an accurate determination of Th on inclusions for which only an estimate could be made otherwise.
The small thermal mass of the USGS stage makes such cycling relatively fast
(Woods et al., 1981).

The presentation of inclusion data must be adapted to the nature of the
problem and the data. Histograms of Th or Tm ice on which specific inclusion
types or samples are separately indicated, sequences of histograms arranged in
the order of time of formation or sample depth, plots of Tm ice vs Th for the
same inclusions, etc., may be appropriate. Far too often, only ranges are
given for several measurements, such as Tm and Th, without the actual data on
individual inclusions, which may show a positive or negative correlation between
the two variables for individual inclusions. One or more groups of inclusions
with similar characteristics will sometimes be evident in histograms. If this
grouping is not a result of sample selection bias, it can provide important
evidence of specific pulses or stages of fluid movement. Unfortunately, it is
very difficult to avoid sample selection bias. Great care is needed to avoid
the introduction of bias into the data by the selection of the intervals to be
plotted, particularly with small numbers of random(?) points. As Wise (1982,
p. 887) has so aptly noted (in a different but very closely parallel context):
"Unwanted peaks on bar-graph histograms are easily removed by shifting the boundary between two bars. This splits the peak into two smaller bars and relegates
the data to 'noise'. Conversely, new peaks can be generated." A continuum of
data points in a histogram can mean that a continuum of changing fluids was
involved; it can also result from poor precision in the measurements or other
procedural problems.

Television options. Commercially available television cameras, monitors,
and videotape recorders can be used with a trinocular head to record and to
demonstrate phase behavior in fluid inclusions to students and colleagues, and
also to preserve tapes of the phase transitions for later use (e.g., Halsor et
al., 1983). Commercially available "title generators" (or a second camera and
screen-splitting device) may be used to allow the temperature readout to be
displayed along with the optical image. Audio tracks on the tape allow verbal
comment. For fluid-inclusion work, cameras, monitors, and video tapes having
more than 700 lines/frame resolution (monitors having 1000 or even 2500 lines/
frame are available) should be used, because many details cannot be seen in
fluid inclusions at 500 lines/ frame. Color cameras and monitors are limited
to about 400 lines/frame. Black and white cameras are much less expensive and
generally adequate for teaching and laboratory record purposes. The resolution
of all the recorders is limited (in practical terms) to about 500 lines/frame,
so better resolution is obtained in live time than when played back from the
recorder.

Calibration

The sources of error to be corrected. The calibration of heating/freezing

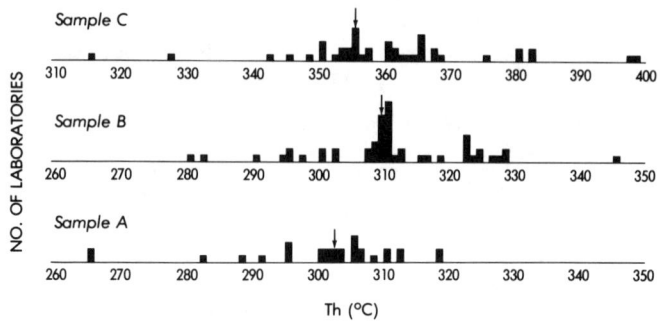

Figure 7-8. Plot of Th data reported by Naumov (1976) on a interlaboratory test of fluid-inclusion laboratories in the USSR. Three standard samples, A, B, and C, were run by 53 different workers in 24 organizations. The "accepted values" for each sample are indicated by arrows. See text for details.

stages is an onerous, time-consuming, and frequently frustrating task[7] that yields no usable data -- how necessary is it? At the extreme, some have indicated that calibration is probably needed only if one is interested in the highest accuracy; if the data are reproducible, they are probably good enough. Others believe that calibration is something that must be done only once, when first setting up the stage. If only numbers and a statement of reproducibility, are needed, this approach is fine, but it would be hazardous indeed to even try to guess the accuracy of such numbers. Usually, when careful calibration is attempted, the results are poorer than expected, and errors of 50°C or more in Th can sometimes be found.

One indication of the importance of calibrations (and presumably also a good indication of their general inadequacy) is found in a report of an interlaboratory study of inclusions in the USSR (Naumov, 1976). In this study, samples of synthetic quartz containing inclusions homogenizing in the range 302-355°C were run in sequence by 53 different workers in 24 organizations. Even though each of these workers must have known that this was a'test, the results showed a remarkably wide spread (Fig. 7-8). Naumov stated that he checked the value of Th before and after each test and hence knew there was no evidence of leakage or stretching, so the data are simply an indication of the sum of the effects of equipment problems, operator errors, and calibration procedures in the USSR. No similar test has been run in the West.

The degree of care given to the calibration of the stage should vary depending on the temperature range and phase changes under consideration. For example, errors of a few degrees in Th L-V should not affect conclusions based on extrapolation of isochores, because errors from uncertainties in the P-V-T properties of the fluid, from uncertainties in the composition of the fluid itself, and from the estimate of pressure of entrapment are greater.

Accuracy of calibration is generally more important for measurement of Tm than Th, and for other studies such as the effect of gas composition on Tm CO_2 (e.g., Hollister and Burruss, 1976), accurate calibration of the stage at Th

----
[7] J. Goss, after a summer spent calibrating, made this observation:
                             "There once was a guy on the stage
                             Its heating he tried to gauge.
                             But unless he sees the light
                             Before it's done right,
                             He'll be dead of extreme old age."

is extremely important. Thus, just 0.2°C lowering of Tm CO_2 implies a significant amount of another component in the system that fractionates into the fluid
phase relative to solid CO_2 at low temperature. Errors in measuring Tm of
clathrates, hydrates, and ice can also lead to significant errors in determining
both the presence and the amounts of components dissolved in aqueous phases.
Accuracy is particularly important in studies of the composition of aqueous
inclusions of low salinity. Thus, Browne et al. (1976) experienced considerable
difficulty in getting valid salinities from some low-salinity geothermal fluids.
D.M. Smith et al. (1982) reported Tm ice values for samples from Tayoltita,
Mexico, that ranged from -1.7 to +2.9°C (sic) but that had a "±2°C" error
assessment. From these data it is not clear whether the problem is stage calibration, measurement technique, clathrate formation, metastable superheated
ice, or a combination of these four factors.

Three quite independent temperature differences must be considered under
calibration. The first difference is between the true temperature of the inclusion, which I will call Ti, and the true temperature of the temperature sensor
(Ts). This difference (Ts-Ti) is a function of the stage design and the nature
of the sample. The second difference is between Ts and the temperature indicated
on the temperature readout (Tr). This difference (Ts - Tr) is a function of the
nature of the thermocouple itself (or other sensor), the reference junction, the
readout device, and any spurious emfs from various sources. (The readout device
must be calibrated internally before any measurements are made.) The third difference is between the true temperature of the inclusion (Ti) and some other
inclusion (Ti') in the same sample, at the same instant, but at a different
vertical and/or horizontal location in the sample (and stage). This difference
(Ti-Ti') is usually termed the "gradient" in the stage. For any given setup,
these three sources of error can be combined into a single correction appropriate
for that inclusion and setup, but it is instructive first to examine them
separately.

Temperature of sensor minus temperature of readout (Ts-Tr). One component
of the (Ts-Tr) difference is the true "thermocouple correction." No thermocouple
or other temperature sensor is perfect. Thermocouples made from a given pair of
uniform batches of high-quality thermocouple wire will generally be very similar
in emf vs temperature. However, they will change as a result of recrystallization, strain, and, particularly at elevated temperatures, with contamination, but
it is best to use separate thermocouples for low- and high-temperature measurements. These deviations from standard tables of emf vs temperature are generally
not great. Much greater potentialities for error stem from inaccuracies (or
worse, malfunctions) of the electronics in the readout, as well as in the reference junction (not used with Pt resistance thermometers). Electronic instabilities can introduce intermittent and unpredictable errors. Such electronic devices are rarely better than their company specifications and can be much worse.
Some workers calibrate thermocouples outside the stage, but I prefer, if at all
possible, to calibrate the thermocouple in its normal position in the stage.

Temperature of sensor minus temperature of inclusion (Ts-Ti). The (Ts-Ti)
difference, combined with (Ts-Tr), is sometimes called the "stage correction,"
but only (Ts-Ti) is related to stage factors. Stages such as the Chaixmeca and
Linkam, which have their temperature sensors embedded in the stage structure or
offset from the sample, may have a larger value for (Ts-Ti) than stages such as
the USGS, which have the sensor very near to the sample and "embedded" in the
same flowing heat-exchange medium as the sample. Thus, Macdonald and Spooner
(1981) reported a value of 4°C for the sum of (Ts-Ti) plus (Ts-Tr) at 200°C for
the Linkam stage, and smaller values at both higher and lower temperatures.
Poty et al. (1976) also reported about 4°C at 200°C for their original stage,
with increasing differences to 600°C. In the Leitz 1350°C stage, corrections
of as much as 90°C were needed at the melting point of Au (1064.4°C). The value
of (Ts-Ti) will vary not only with the specific stage design but also with the
size and nature of the sample, the microscope objective used, the substage condenser used (and its position), the intensity of the illumination, etc., all

of which affect the thermal regime. Thus, Fuzikawa (1982), in a careful calibration of the Chaixmeca stage, found a difference of 8°C at -60° between the
apparent melting point of a given standard, depending upon whether a 10x or a
25x objective was used. The 25x objective, being closer to the sample, warmed
it more. In contrast, Konnerup-Madsen (1977), also using the Chaixmeca stage,
found only a 1°C difference between the 10x and the 25x objectives at -60°C.
An even more extreme example was reported by my colleague H.E. Belkin, who tried
determining the triple point of CO_2 in very small inclusions by removing the
upper glass window of the Chaixmeca stage and using an oil-immersion objective
directly on the sample plate on the stage. The triple point was was found to
be "-87±0.5°C" in four runs -- reasonable precision but 30°C too low! Even
at moderate temperatures, one can actually watch the migration of inclusions
in soluble minerals such as halite in the thermal gradients in some commercial
stages.

    Temperature of inclusion "i" minus temperature of inclusion "i'"(Ti-Ti').
Thermal gradients in the stage are basically responsible for the difference
(Ts-Ti), but usually gradients are thought of only in terms of (Ti-Ti'). (Sometimes the change when the sensor is moved from one place to another in the stage
is used as a measure of "the gradient," but the only unambiguous measure is (Ti-
Ti').) Both lateral and vertical gradients are important. Thus, in the Leitz
350 stage at the melting of KNO_3 (396°C), a 10°C difference was found between
top and bottom of a quartz plate 0.65 mm thick, and similar differences are found
when a plate that has an inclusion near the bottom is inverted and rerun. Moving
a given inclusion 1 mm away from the center of the field of view yielded an apparent increase in Th of 4°C (i.e., the stage was 4°C cooler off center). The
difference from center to edge was 8-10°C. Poty et al. (1976) indicated a range
of 1.3°C across the Chaixmeca stage at 380°C, Macdonald and Spooner (1981) found
a 1.5°C maximum difference at -95°C in the Linkam stage, and Werre et al. (1979)
reported 0.2° gradients across the USGS stage at low temperatures. (The use of
a temperature-sensitive paint helps considerably in mapping gradients, even
though these paints are not accurate as calibration substances.)

    Such gradient measures, unfortunately, are very subject to the specific
individual setup, including such variables as the size, thickness, and thermal
conductivity of the sample, etc.

    Calibration-run procedure. How can all these sources of error be eliminated
or at least minimized? The only possible answer lies in calibration runs, using
materials of known melting point, under conditions as similar as possible in all
respects to those during the run itself, and repeated as often as necessary.
The similarity must also include the heating or cooling rates and schedule used.
Whenever possible, the standards should be run simultaneously, along with the
unknown inclusions (with due consideration of lateral thermal gradients). Such
calibration runs in effect provide the summation of all sources of temperature
differences. Calibration involves measuring the apparent melting temperatures
of substances having known melting temperature, which are placed in the same
position on the stage as the samples. The calibrations themselves, however, must
be done with considerable care. The first comprehensive study on calibration of
heating/cooling stages (for the Chaixmeca stage) was by Jehl (1975).

    The major causes for trouble in calibration are: (1) differences between
the geometry and thermal characteristics of the position of the standards for
the calibration run and that of the inclusions in a host mineral; (2) impurities
in the standards; (3) inaccuracies in the reported melting temperatures of commercially available substances; (4) thermal decomposition of the standards, (5)
recognition of what really does constitute melting as one watches substances
being heated; (6) changes in composition of the standards between delivery and
use; and (7) differences in the effective thermal mass between calibrant and
inclusion. All these problems are nontrivial; many of the problems and solutions
have been discussed in the literature (Jehl, 1975; Poty et al., 1976; Macdonald

and Spooner, 1981; Roedder, 1976a; Burruss, 1977).

Several procedures have been used to minimize the differences in geometry and thermal characteristics. Burruss (1977, p. 121) sealed the standards in

> "...rectangular cross-section capillary tubes (50 µm x 150 µm internal dimensions, Microslides #5005, Vitro Dynamics, Inc., 114 Beach Street, Rockaway, N.J. 07866). The length of the sealed tubes varied from 6 to 12 mm. The flat sides of the capillaries insured good thermal contact with the stage. The small internal dimensions insured the mass of material observed was small and thereby limited problems of dissipation of the heat of fusion of the solids. In most cases, melting of small, polycrystalline masses of the solids with dimensions approaching individual fluid inclusions (~30 µm largest dimension) could be observed in sections of the capillaries away from the main mass of solid. These small masses gave the sharpest final melting temperatures at slow heating rates (~0.5°C/min.)."

Liquid standards for low temperatures can be sealed in such flat capillaries, or can be run by using a quartz plate, about the thickness and diameter of the samples being studied, that has cavities (opened fluid inclusions) ~30 µm deep on the surface. The liquid standard compound is spread on the surface and the whole covered with a glass coverslip (Jehl, 1975; Poty et al., 1976). For solid compounds, small grains of the standard material can be placed between two cover slips, or better, in the cavities in a quartz plate. Care should be taken to use grains about the size of the fluid inclusions being studied. To avoid contamination, each quartz plate should be reserved for only a single calibrant substance.

Some standard materials work well using the capillary method, and some work better using a quartz plate. Compounds that are volatile, sublimate, or oxidize can only be used in sealed capillaries; most of those used by Burruss (1977) do sublimate and hence were not used by Macdonald and Spooner (1981).

The use of evacuated glass capillaries permits determination of the triple point, rather than the 1 atm melting point. With a little practice, preparation of such calibration standards is not difficult. A piece of thin-walled, several-millimeter-diameter soft glass tubing is carefully cleaned, then heated and pulled to form a capillary of appropriately small diameter (preferably ~0.5 mm). Fused silica tubing must be used for high-temperature standards. The tapering parts should be made as short as possible, for convenience later. Break at the center of the capillary and seal the capillary end, using an oxy-hydrogen micro-flame, but avoid forming a ball of glass at the end. An appropriately small crystal or sliver of standard substance is dropped in the open end and worked into the sealed end of the capillary by shaking or vibration. (Avoid unnecessary abrasion of the crystal beforehand to minimize sticking from static charges.) Evacuate the tube and seal off a short section of capillary containing the crystal.[8/] Heating of the crystal can be avoided by wrapping the end with wet asbestos or by the use of wet asbestos-wrapped tweezers for pulling. Highly volatile substances can be kept frozen in the end of the capillary by a tiny LN$_2$ bath. If the capillary is thin, and the flame small, calibration tubes can be made as short as 5 mm without melting the crystal. After such sealing, another fragment of the same material may be similarly sealed into the next part of the same capillary. Only one material is loaded into the capillary pulled from a given tube, to avoid contamination.

[8/] R. Lewis (pers. comm., 1979) has suggested pulling on the ends of a thin wire loop to seal off the heated part quickly.

Sublimation in these tubes may take place during the runs, particularly the longer more careful runs, the new crystals growing on the cooler parts (an additional proof of heating-stage gradients). If the tubes are too long, these cooler parts can be out of the field of view, and if the entire crystal transfers, that calibration point is lost. Hence, short tubes are desirable. This sublimation may move part of the substance to the top of the capillary, immediately above, indicating a vertical temperature gradient in the stage. When using the Richter-Abell (1953) heating stage, I found that these sublimed crystals (of Te or Sb) at top melted at a thermocouple reading 25°C or more higher than the main crystal at the bottom of the tube, just 0.3 mm below! (As the two elements were 99.99 and 99.999+% percent pure, the sublimation itself should have had no measurable effect.)

Transfer by sublimation in the tubes can be useful, however, as it permits one to avoid the ever-present problem of the normal large difference in effective thermal mass of standard vs inclusion. Rather than try to make a synthetic H_2O standard "inclusion" small, I freeze some water in one end of a capillary, seal it closed, and then, before use, I place the tube in a thermal gradient to evaporate some water and recondense it as minute beads on the cool end. These beads are then watched, rather than the main mass of water. Another procedure involves sealing an open capillary shut with an oxy-hydrogen flame; tiny droplets of pure water will condense in the cold end during the sealing.

P. Radomsky (pers. comm., 1983) has found that capillaries can be charged with liquid standards by holding the closed end of a 6 to 12 mm capillary in self-locking tweezers cooled by LN_2 and slowly feeding the standard from a microliter syringe to the open end; the liquid congeals in the closed end. While the standard is still frozen, the open end is sealed with a microtorch. If carbon spots form (representing destruction of some of the standard), the tube is discarded.

Some reports indicate that Th determinations (and presumably calibrations, if any) were made dynamically at rates as much as 3-5°C/min. Macdonald and Spooner (1981) have recommended using a dynamic procedure on calibrations (heating at 0.4°C/min). Although this method is certainly to be preferred if the unknowns are also run by the (same) dynamic procedure, I believe that calibrations, like the samples, should be run by the slower temperature-equilibration procedure, edging up slowly to the final melting temperature. The thermal gradients under static heat-flow conditions will generally be much less than those under dynamic equilibrium. In addition, a major difference exists between the heating rates that can be used and still obtain accurate results on ordinary, moderately saline inclusions, and those needed for inclusions that have very narrow melting ranges, such as nearly pure water or CO_2. When the melting range is narrow, as is characteristic of all good calibration standards, the entire heat of melting must be added while the stage is in this narrow range, or there will be thermal lag and erroneously high apparent melting temperatures. Thus, Henry (1978) found that a 0.1°C/min heating rate was necessary for accurate measurement of CO_2 melting. My colleague, H.E. Belkin, and I have found that inclusions 50 μm in diameter of nearly pure water from a geothermal area (Tm ice = -0.3°C) yielded an apparent Tm ice = +0.2°C at heating rates >0.14°C/min, even though the thin sample plate was immersed in rapidly circulating acetone of controlled temperature. Increase in inclusion size also exacerbates the problem. Thus, we found that Tm-ice for a large synthetic inclusion (~3 mm diameter) of pure water was +2.7°C at 0.77°C/min and was still in error (+0.4°C) even at 0.05°C/min. Stages having less efficient thermal-transfer mechanisms (e.g., flowing gas or static conduction) would yield even larger errors.

Standards for calibration. The problem of choice of standard substances for calibration is not trivial. The optimum substance should have a sharp, easily observed melting point (preferably in air) that is both reproducible and known and that is in the desired temperature range. The substance must be available in adequate purity, at reasonable cost. It should not decompose, or alter easily. Some workers add that it should not melt incongruently, must be reusable, must not sublime at or near its melting point, must be transparent, and it should be nontoxic. Few substances fit all these criteria, so compromises are needed. Many papers each list a few substances that were found to be suitable or unsuitable and the procedures used in running them, but several recent studies have compared the merits of numerous substances (Jehl, 1975; Burruss, 1977, and Macdonald and Spooner, 1981; see also Hollister et al., 1981 and Kuhnert-Brandstätter, 1982). When these various reports are compared, only a very few materials are generally agreed to be suitable; considerable disagreement exists concerning the suitability of many others. (Those rated as "++++" in Table 7-4 come closest to general acceptance.) In addition, one should be aware that the "accepted" melting point for a given compound differs significantly.[9] Some of these differences are explained in the following discussion of the criteria for choosing mentioned above.

(1) Sharp melting point. A pure compound should have a sharp melting point, but impurities are always present, so the melting point becomes a melting range. Although the melting of the last crystal (the highest temperature in the range) should generally be nearest to the true melting point, some compilations of melting points use the temperature of first evidence of melting, or that of melting of "all the smaller crystals," rather than that of the last crystal. I suggest that this usage comes from dynamically derived data, where the heat of fusion yields a thermal lag that can be empirically corrected, in part, by such arbitrary definitions of the "melting point."

(2) Easily observed. If the liquid has an index of refraction near that of the crystal, the melting of the last crystal can be hard to recognize. Crossed polars help on many substances.

(3) Reproducible. Irreproducibility is probably a result of irregularly distributed impurities. Because the amount of pure substance used may be only $\sim 10^{-9}$ g, a very tiny speck of contaminant, either in the substance as provided, or introduced in the preparation procedure, can have major effects. Only the highest grade should be obtained; "technical grade" substances should never be used. Considerable care is needed to avoid contamination of both the individual calibration-run material and the main supply bottle.

(4) Known temperature. Most standard melting points, as reported in both the chemical and the inclusion literature, are not actually standardized, and some differ by several degrees. These differences probably result from differences in product purity and the melting-temperature procedure used. For example, a chemist generally melts 3 to 10 orders of magnitude more material in his determination than is customary on a heating stage.

(5) Desired temperature range. Some ranges simply have no known adequate standard. P. Radomsky (pers. comm.) uses a series of highly purified alkanes to cover the range -56 to +70°C. Certain commercial preparations -- sticks, pellets, or paints -- are available throughout all ranges. These preparations are useful for thermal-gradient studies, but many have a melting point that is grain-size dependent and hence unsuitable for calibration (Macdonald and Spooner, 1981).

(6) Ease of decomposition. Some compounds may give a good sharp melting point if the duration of the run is short but gradually decompose, and hence yield spurious data, if run slowly or repeatedly.

----

[9] Emons et al. (1982) have recommended the eutectics of various soluble salts with water as standards, from -1.1 °C (Na_2SO_4) to -65.0 °C (KOH). Whether these will work well in the very small masses for inclusion stage calibration is not known.

Table 7-4. Calibration (melting point) standards.

[The "ratings" are based on a subjective combination of the evidence presented by various other workers (see text for references and discussion) and personal preference. Unsatisfactory substances have been omitted.]

M.P. (°C)	Substance1/	Formula	Rating 2/	Source 3/	Notes
-112.0	Ethyl alcohol	C2H5OH	+++		Also reported as -117.3°C.
-95.35	2-Propanone (acetone)	CH3COCH3	+++		Also reported as -94.6°C.
-95	Toluene	C6H5CH3	+++	c, d	May be difficult to freeze.
-95.0	n-Hexane	CH3(CH2)4CH3	+++	d	Also reported as -94.3°C.
-90.61	n-Heptane	CH3(CH2)5CH3	+++	d	....
-83.578	Acetic acid, ethyl ester (Ethyl acetate)	CH3CO2CH2CH3	+++		Also reported as -85.9°C & -82.4°C.
-77.9	Acetic acid, butyl ester (Butyl acetate)	CH3CO2(CH2)3CH3	+++		Also reported as -76.3°C.
-63.4 -63.5	Methane, trichloro (Chloroform)	CHCl3	++++		
-56.79	n-Octane	CH3(CH2)6CH3	+++		Also reported as -56.5°C.
-56.6	Carbon dioxide T.P.	CO2	+++	d	
-47.86	Benzene, 1,3-dimethyl (m-Xylene)	C8H10	++		Also reported as -47.4°C.
-45.6	Chlorobenzene	C6H5Cl	+++		
-45.0	Acetic acid ester		+++		
-39.9	Diethyl ketone		+++		
-38.86	Mercury	Hg	++++		Reflectivity changes on melting.
-34.7	Ethyl benzoate		+++		
-29.68	n-Decane	CH3(CH2)8CH3	++++	a, d	
-28.5	Nitromethane		+++		
-26.0	Benzaldehyde		+		
-25.18	Benzene, 1,2-dimethyl (o-Xylene)	C8H10	++	a, b, d	Also reported as -55.6 & -56.9°C 5/
-23.12	1-Tridecene	CH3(CH2)10CH:CH2	++	a	Also reported as -13°C 5/
-22.99	Methane, tetrachloro-(Carbon tetrachloride)	CCl4	+++	a	Also reported as -22.65°C & -22.96°C
-16.30	1,3-Dimethyl-2-ethylbenzene				
-15.9	Quinoline		+		Some find unsatisfactory.
-15.3	Benzyl alcohol	C6H5CH2OH	+		
-13	Benzoic acid, nitrile (Benzonitrile)	C6H5CN	+	a	
-12.91	1-Tetradecene	CH3(CH2)11CH:CH2	+	a	
-12.61	2,2,5-tetramethylhexane		++++		
-12.3 -12.4	Methyl benzoate		+++	a, b, d	
-9.60	n-Dodecane	CH3(CH2)10CH3	+++		Also reported as -8°C.
-9.0	2,5-Hexanedione (Acetonyl acetone)		+		
-7.2	Bromine		+++		
-6.5	Butanoic acid (Butyric acid)	CH3CH2CH2CO2H	+++		Also reported as -6.2°C.
-6.1	Aniline		+++		
-5.5	Tridecane	CH3(CH2)11CH3	+++		
+0.01	Water T.P.	H2O	++++		
+2.5	N,N-Dimethyl aniline	?	+++		
+5.2-5.7	Methane, diiodo (Methylene iodide)	CH2I2	++		
+5.5	Benzene	C6H6	+++		Toxic; also reported at +5.2°C.
+5.86	Tetradecane	CH3(CH2)12CH3	+++		
+17	n-Decylamine	?	++		
+40	Merck 9640 4/	?	+++	e	
+41.0-41.5	A.H.T. A		+	f	Sharp melting point.
+54.5	4-Nitrotoluene	CH3C6H4NO2	++		Also reported as 51.7°C.

M.P. (°C)	Substance [1]	Formula	Rating [2]	Source [3]	Notes
+68	Azobenzene	C_6H_5NO	+++		
+68	Nitrosobenzene	?	+		Minor sublimation
+70	Merck 9670			e	Minor sublimation. Also reported as 80.22, 80.25 & 80.72
+80.55	Naphthalene T.P.	$C_{10}H_8$	+++		
+95	Benzil	$C_6H_5COOC_6H_5$	++		Minor sublimation
+100.	Merck 9700	?	++	e	
+112.8	Sulfur	S	+++		Also reported as 113-114°
+114.5	Acetic acid, amide, N-phenyl (Acetanilide)	$CH_3CONHC_6H_5$	++++		Considerable sublimation
+122.36	Benzoic acid T.P.	C_6H_5COOH	+++		
+132.5	Urea	H_2NCONH_2	+		
+134.5-134.7	Acetic acid, amide, N(4-ethoxyphenyl) (Phenacetin)	?	+++		
+135	Merck 9735	?	+++	e	Minor sublimation
+135	P-Acetophenetide	$CH_3CONHC_6H_4OC_2H_5$	++		
+147.5-148	A.H.T. D	?	+	f	
+148	Phenylurea (mono)	?	+		
+150	Acetic acid, diphenyl-(hydroxy) (Benzilic acid)	$(C_6H_5)_2COHCO_2H$	+		Also reported as +151°C.
+151.46	Hexanedioic acid (Adipic acid) T.P.	$HO_2C(CH_2)_4CO_2H$	++++		Also reported as +153°C.
+163	Benzoic acid, amide, N-phenyl (Benzanilide)	$C_6H_5CONHC_6H_5$	+++		
+170	Benzene, 1,4-dihydroxy (Hydroquinone)	$C_6H_6O_2$	+		
+172.5-173	A.H.T. E	?	++	f	Minor sublimation; toxic.
+180	Merck 9780	?	+++	e	Sublimes
+185	Benzoic acid, 4-methoxy (Anisic acid)	$CH_3OC_6H_4COOH$	+++		
+190	Racephedrine hydrochloride		++		
+199.8	O-Toluidine	$CH_3C_6H_4NH_2$	++		
+200	Merck 9800	?	+++	e	Minor sublimation
+208.45	2-Chloroanthraquinone		+++		Sublimes, toxic
+210	Dicyandiamide		+++		
+216-216.5	A.H.T. F	?	++	f	Sublimes
+228	Saccharin		+++		
+231.97	Tin	Sn	+++		
+243-243.5	A.H.T. G	?	+++	f	Sharp melting point.
+247	Merck 9847	?	+++	e	Toxic
+247-248	Carbazole	$C_{12}H_9N$	+++		Major sublimation
+263	Phenophthalein		+++		Also reported as +261-262°C
+271.44	Bismuth	Bi	+		
+284.23	9,10-Anthraquinone	$C_{14}H_8O_2$	+++		Sublimes
+306.8	Sodium nitrate	$NaNO_3$	+++		Also reported as +308°C.
+321.11	Cadmium	Cd	+		
+327.43	Lead	Pb	++		
+327-328	A.H.T. H	?	+	f	Sodium acetate?
+333	Potassium nitrate	KNO_3	+		
+385	Cadmium iodide	CdI	+		
+392	Sodium dichromate	$Na_2Cr_2O_7$	+		
+398	Potassium dichromate	$K_2Cr_2O_7$	++++		
+402	Lead iodide	PbI_2	+	d,e	Toxic
+419.58	Zinc	Zn	+		

209

M.P. (°C)	Substance1/	Formula	Rating 2/	Source 3/	Notes
+434	Silver bromide	AgBr	+		99.99%
+449.5	Tellurium	Te	+++	g	
+455	Silver chloride	AgCl	+		
+498	Cupric chloride	CuCl2	+		
+501	Plumbous chloride	PbCl2	+		
+520	Cadmium fluoride	CdF2	+		
+529.4	Potassium fluoborate	KBF4	+		Toxic
+563.7	Sodium cyanide	NaCN	+		
+568	Cadmium chloride	CdCl2	+		Decomposes?
+592	Barium nitrate	Ba(NO3)2	+		
+614	Lithium chloride	LiCl	+		
+618	Lithium carbonate	Li2CO3	+		
+630.5	Antimony	Sb	+++	g	99.999+%
+634.5	Potassium cyanide	KCN	+		Toxic
+651	Sodium iodide	NaI	+		
+690	Vanadic oxide	V2O5	+		
+723	Potassium iodide	KI	+		Also reported as +733°C.
+730	Potassium bromide	KBr	+		
+755	Sodium bromide	NaBr	+		Also reported as +776°C.
+790	Potassium chloride	KCl	+		
+800.4	Sodium chloride	NaCl	+++		
+844	Lead chromate	PbCrO4	++		Formerly "+870°C," then "+843°C".
+845	Lithium fluoride	LiF	+		
+851	Sodium carbonate	Na2CO3	+		
+880	Potassium fluoride	KF	+		Toxic
+884	Sodium sulfate	Na2SO4	+		
+891	Potassium carbonate	K2CO3	++		
+961.93	Silver	Ag	+		
+962	Barium chloride	BaCl2	+		Also reported as +975°C.
+968.3	Potassium chromate	K2CrO4	+		
+980.8	Arsenic	As	+		Toxic
+992	Sodium fluoride	NaF	+++	h	
+1064.43	Gold	Au	+		Also reported as +1069°C.
+1076	Potassium sulfate	K2SO4	+		Also said to decompose at +1124°C.
+1185	Magnesium sulfate	MgSO4	+		Also reported as +1360°C.
+1330	Calcium fluoride	CaF2	+		Synthetic
+1391.5	Diopside	CaMgSi2O6	+++		
+1396	Magnesium fluoride	MgF2	+++		Synthetic
+1544	Pseudowollastonite	CaSiO3	+++		Synthetic

1/ Both the spelling and the naming may differ, depending upon the country of origin; where the meaning is clear, names approved by the International Union of Pure and Applied Chemistry have been used, but considerable ambiguity remains in some. T.P.=Triple point.

2/ +++ = highest recommendations; + = possibly useful or shows potentialities.

3/ Only listed for special sources; others generally available from chemical supply houses.
    a - American Petroleum Institute   d - Aldrich Chemical Co.   g - American Smelting & Refining Co.
    b - U.S. Bureau of Standards   e - Merck and Co. "Schemlzkorper"   h - U.S. Mint ("proof gold")
    c - Matheson, Coleman, & Bell   f - Arthur H. Thomas Co.

4/ Reproducibility decreases with age (Jehl, 1975).

5/ Two different compounds?

210

(7) Ease of alteration. Any substance that absorbs water from the air will yield spurious data if it is damp when sealed in a capillary. Similarly, surface oxidation of metal chips or splinters can keep the material from coalescing into a ball at the melting point. Some compounds decompose on the shelf and hence must be purchased fresh (Jehl, 1975).

(8) Incongruent melting. An incongruent melting point can be just as sharp as a congruent one and need not be grounds for rejection, so long as the incongruency is noted.

(9) Reusability. The amount of standard used is trivial, but if considerable efforts are needed to make the container (e.g., sealed evacuated capillaries), it is helpful if the unit can be reused. Thus, most metals that form a ball on melting are not reusable, as the melting of a crystalline ball may not be obvious. Mercury is an exception; the reflection from the ball changes appreciably on melting.

(10) Sublimation. If the substance sublimes rapidly, it may leave the stage before the melting point is reached. The problem is eliminated by the use of the sealed evacuated capillary (see above).

(11) Transparency. Although a convenience, transparency is not a necessity. The coalescence of splinters of pure tellurium or antimony into balls at melting is as obvious as the melting of any transparent substance.

(12) Toxicity. The amounts of substances actually used are so small that toxicity is probably a trivial concern in most standards. The main supply bottle should, of course, be kept secure, as should any chemical supply, to maintain its integrity, as well as for safety.

Fluid inclusions used as standards. In view of the difficulties of making a calibration run sufficiently similar to the runs of the unknowns, it might seem an obvious solution to use actual fluid inclusions as "standards." Some natural inclusions of liquid CO_2 are pure enough to be used as "primary" standards for calibration at the triple point of $-56.6°C.$[10]/ However, all other natural inclusions can only be used as secondary standards, having accepted temperatures based on determination in a series of laboratories, but they presumably could still be used for interlaboratory comparisons and for routine periodic checks in any given laboratory, to monitor possible instrumental drift or the effect of changes in technique. Although an interlaboratory comparison would be time-consuming, its value is undoubted; the establishment of such a test has been discussed at several international meetings, but it has yet to be made. As shown earlier, fluid-inclusion laboratories throughout the USSR cooperated in such a test in 1976, and the results were rather shocking. Particular care would be needed to avoid error from permanent damage to the standard inclusion by stretching. Stretching is not as probable in a quartz standard as in soft minerals, but it can occur (see Chapter 9).

Fluid inclusions in synthetic crystals, grown at known P, T, and X, might seem to be an obvious answer to the calibration problem. A study of primary inclusions in such crystals, started as an evaluation of fluid inclusion thermometry itself, showed that the largest uncertainty was in the temperature of growth in the autoclave, at the crystal face (Roedder and Kopp, 1975). Although thermocouple measurements on the outsides of the autoclaves are normally used for crystal growth control, the rapid convection currents inside the autoclave, driven by the thermal gradients (established to speed crystal growth), make the

----

[10]/ E.g., inclusions of CO_2 (+H_2O) from Calanda, Switzerland (Touray, 1968), are supplied with the Chaixmeca stage for calibration.

extrapolation from the outer surface temperature to the actual growth surface hazardous. R.J. Bodnar (pers. comm., 1983) has reported a procedure for manufacturing secondary inclusions in quartz over a wide range of P, T, and X that eliminates most of these difficulties and yields inclusion standards for almost any of the types and temperatures of phase changes normally observed in natural inclusions. These provide excellent teaching material as well.

DECREPITATION METHOD OF MICROTHERMOMETRY

If the difference between the high pressure inside an inclusion and the lower external pressure on the host mineral exceeds the strength of that mineral, the host will fracture or deform around the inclusion in order to let the inclusion expand and relieve the pressure difference. When the host fractures, the process is called "decrepitation," and the temperature is referred to as Td.

Far more geothermometric determinations have been made by the decrepitation method than by any other procedure. The method was first proposed by Scott (1948) and was developed by Smith, Peach, and coworkers (see Smith, 1953, and a series of their papers from 1948 on). Chapter 9 contains a discussion of its limitations. It is used rather extensively in the USSR (see Roedder, 1968a, 1972) but has been nearly abandoned in the West. It is based on the rate of explosion or decrepitation of inclusions in a coarsely crushed sample during a continuous rapid increase in temperature (~10°/min). In many samples, this rate of explosion increases rapidly above Th of the inclusions; in some, pressures may be high enough to cause decrepitation below Th.

Instrumentation for heating samples and particularly for detecting decrepitation runs the gamut from a simple stethoscope (Ermakov, 1950) to special microphones and various electronic amplifying, integrating, counting, and recording circuits (see Roedder, 1972, p. 28, and indices in Roedder, 1968a). The resultant data are usually plotted as a curve of time-integrated decrepitation intensity vs temperature. In some applications, the sample is under vacuum (Maiskiy, 1973). Pressure increase from the released gases permits distinction between decrepitation of fluid inclusions and "anomalous decrepitation" from thermal expansion of solids or other phenomena, and the gases can be analyzed as well. A special design for use at low temperatures, e.g., with saline samples, is given by Montoriol-Pous (1967). Karwowski and Kozlowski (1972) used a thermogravimetric balance to detect the decrepitation (weight losses <1.5 wt %). Pulou and Baudracco-Gritti (1978) designed a special electronic circuit to eliminate instrumental noise. The Russian literature includes many reports on the use of decrepitation; a number of these are aimed at possible uses in mineral exploration (e.g., Mel'nikov and Yudin, 1981, and references in Roedder, 1977b).

During homogenization runs, decrepitation is easy to observe under the microscope. If the sample does not blow apart, the inclusion darkens abruptly when low-index vapor replaces the fluid, which escapes through a crack leading to the sample surface. Decrepitation may be heard as a popping noise and may destroy the sample. Vapor-bubble expansion on decrepitation can be mistaken for Th L-V (V), so Th must be checked by repeating the measurement when decrepitation is suspected; if the inclusion leaked, Th will not be reproducible.

CRUSHING STAGE

Ever since Davy (1822) first drilled into inclusions, the behavior of fluid inclusions when they are opened to the atmosphere has provided much geologically useful data on the gas pressures in them. A "bubble" in an inclusion can be nearly a vacuum or can contain dense, high-pressure gas, and these two types of bubbles can appear identical. Deicha (1950) showed that by the simple but ingenious procedure of crushing a mineral grain in oil between glass plates on the stage of the microscope, the presence of gas under pressure in even minute inclusions could be detected by the sudden expansion of the gas to form bubbles in the viscous oil surrounding the grain. Since then, several new versions of Deicha's crushing stage have been developed (Roedder, 1970a) that are inexpensive, easy to use, and adaptable to a variety of situations.

By means of the crushing stage, one can determine two important aspects of the inclusions: a semiquantitative but exceedingly sensitive measure of the vapor pressure of noncondensable gases[11], and a rough qualitative estimate of the composition of these gases. If an inclusion contained only a water solution of nonvolatile salts, the bubble in it at room temperature will consist of water vapor at ~20 mm pressure. When the host mineral is crushed in liquid, such a bubble will collapse and disappear instantly. If there are noncondensable gases present in that vapor, however, the bubble will collapse (or expand) until the pressure in it is 1-atm; because a bubble of air 2 μm in diameter contains only about 10^9 molecules (~10^{-14} g), this test is extremely sensitive. If inclusions contain gas at high pressure, they will practically explode out into the mounting fluid when the first fracture touches them. A rough estimation of the composition can sometimes be obtained by watching the diameter of the evolved gas bubble change with time, when using mounting fluids that react with a given gas, or that are either very good or very poor solvents for a given gas.

Design

The crushing stage is simply a device to press two glass plates together while observing a mineral grain in oil between them. As the mineral is commonly harder than glass (quartz, olivine, etc.), deep pits and cracks form in the glass which interfere with later viewing. For this reason, I designed my stage to make use of ordinary expendable microscope slides.

The essential details of design are shown in Figure 7-9a. The device consists of a pair of metal plates, hinged together at one end and forced together at the other by a knurled nut on a screw. The screw is pivoted on the base plate and fits into a notch in the end of the upper plate for ease in cleaning and assembly. A nylon washer (N) under the nut makes the action smooth. Before heat treating, the upper plate (steel) is machined into a dish shape (D) about the vertical viewing hole to permit quick rotation of the microscope turret head from one objective to another, as necessary. The sample (S) is mounted in oil between two glass slides and placed between the metal plates and centered over the viewing hole. The crushing stage is then placed on the stage of the microscope and the screw tightened slowly, increasing the pressure on the mineral grains.

To avoid the necessity of clamping, the entire device is made heavy enough to stay in place by weight alone. If simple radial motion about the hinge pin were used to push the glass plates together, the plates would tend to tip, particularly if several grains of different size were mounted. For this reason, the support for the lower plate on the stage is equipped with a segmented horizontal pivot pin. The axis of this pin parallels the hinge axis and crosses the optical axis of the microscope near the level of the sample. Movement about this axis automatically aligns the plates, correcting for various grain sizes in the sample and, in effect, converts the simple rotary motion of the metal

----

[11] The term "noncondensable" as used here refers to gases such as N_2, O_2 and CO_2, that cannot be condensed by 1 atm at room temperature. The same term is used in vacuum studies to refer to gases that do not condense on a cold finger held at a given temperature, usually that of solid CO_2 (-78.5°C).

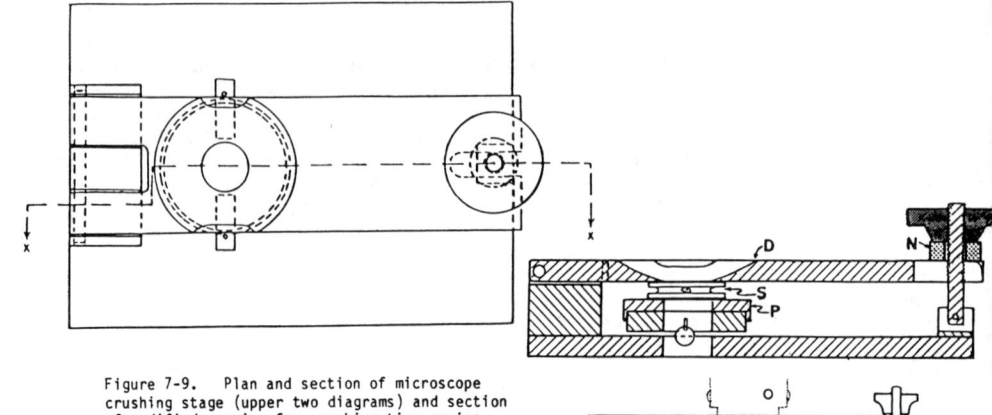

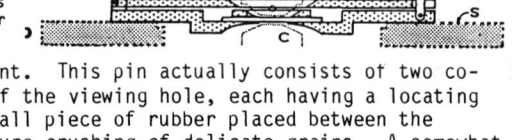

Figure 7-9. Plan and section of microscope crushing stage (upper two diagrams) and section of modified version for crushing tiny grains while observing at high magnification (lower diagram). From Roedder (1970a).

plates to partly translational movement. This pin actually consists of two co-linear segments, one on either side of the viewing hole, each having a locating pin for horizontal positioning. A small piece of rubber placed between the plates near the hinge prevents premature crushing of delicate grains. A somewhat modified version of this stage has been marketed by Chaixmeca, Ltd. (address in Table 7-1). An inexpensive homemade version can easily be made by boring a hole (for observation) through a large strap-hinge.

For crushing very tiny grains, a modified version of the crushing stage is needed to permit the use of condensers and objectives of very short working distance and high magnification. In principle, this stage is very similar to the preceding version but can be much lighter in construction (Fig. 7-9b). Adequate illumination is very important, and as the high-power condenser is generally designed to illuminate objects at a level just 1 mm (the thickness of a normal glass microscope slide) above the microscope stage surface (S), the lower plate of this "microcrusher" is stepped down in the center to place the slide at this level. (The central plate on the microscope stage is removed before placing the instrument.) The lower side of the viewing hole must be beveled strongly to permit the approach of the condenser (C). The pivot-pin assembly for automatic aligning is inverted (i.e., placed above the sample) and made very thin and beveled strongly to permit the use of a short-working-distance objective (O). Ordinary coverglasses are adequate.

Another modification of the stage, designed for crushing crystals as large as 1 cm, perhaps should be called a "macrocrusher." It consists of an open steel cup that has a small central viewing hole in the flat bottom, which sits on the stage of the microscope. Surrounding this hole is an annular groove in the inside bottom surface, into which is placed a flexible plastic O-ring for sealing. A heavy glass disk (~7 mm thick) is placed over the O-ring and viewing hole, and the cup is filled with an oil such as silicone or even motor oil. The sample is placed on this disk and another thick disk, mounted in a horizontal metal plate, is placed on top. The oil level is made barely high enough to contact the lower surface of the upper disk. The metal plate slips down over two vertical threaded rods welded to the bottom of the cup, one on either side of the viewing hole. Pressure is applied by thumbscrews on these rods. Light compression springs on the rods keep the plate from falling. The space between the nearest approach of the wings on the thumbscrews must be adequate for the microscope objective. This version of the stage has been particularly useful where only a very few inclusions were available in small crystals that had relatively clear faces (e.g., Roedder, 1963, p. 190). It is limited, of course, to inclusions large enough to be seen in spite of the rather long optical path through the thick plates. As the plates need only be roughly circular and are cut from scrap pieces of broken plate glass, they are inexpensive and may be discarded as soon as scratched or cracked.

A much simpler procedure can be used for water-soluble minerals such as salt. Simply place the chip in water on the stage of the microscope and watch as the solution front approaches the inclusion (e.g., Fig. 7-10; Roedder and Belkin, 1979b, 1981). Petrichenko (1973) reported using a tiny stream of water under the microscope to dissolve into a selected inclusion.

Operation

To use the crushing microscope stage, the sample grain or grains are placed on an ordinary glass microscope slide (about 1 mm thick) with an appropriate mounting fluid (discussed below) and are generally covered with a half of another microscope slide as a substitute for a coverglass. Care must be used to avoid trapping air bubbles in the mount. The slide is then placed over the viewing hole, the grain centered, and pressure applied gradually by means of the screw. Some grains break in a series of subparallel fractures that form in sequence from the outside inward. This behavior is particularly suitable for inclusion studies, as it permits accurate observations of the phenomena at the instant that the first crack reaches the inclusion. If the pressure is only a few atmospheres, the bubble in the inclusion will expand (Fig. 7-11). If the pressure is very high, it may result in a rapid train of bubbles, or an explosive expansion.

Other grains may build up elastic stress and yield with what appears (through the microscope) to be considerable violence, [12] leaving a group of gas bubbles (Figs. 7-12,13) whose volumes can be measured to estimate the original pressure. Even though a part of the grain may fly out of the field of view, the main part of the grain usually stays in place. Care must be used in the interpretation of bubbles left in the liquid whenever such violent movement has occurred, as the velocities may be high enough to cause cavitation of the liquid, and as this liquid is generally air saturated, the resulting exsolved air bubbles have a finite lifetime and may seem to come from the inclusions. Also, if the grain is near the edge of the mount, so that an air-oil meniscus is close, air bubbles may be sucked in, causing ambiguity.

As a result of the relatively large unsupported glass area of the viewing hole, the glass slides do occasionally break. Inconvenience from this breaking can be avoided if the spacer plate (P) is removed and an extra plate of glass is used beneath the slide. This prevents the slide breaking downward and contaminating the substage with broken glass and oil. If the upper glass breaks, oil may be splattered on the objective lens. To avoid this, a thin circular glass plate (coverglass) is placed over the viewing hole, on top of the upper plate. Breakage can be minimized if the viewing hole is made as small as convenient for the particular purpose, and if the glass plates are as thick as permitted by the optics and grain size. Scraps of various thicknesses of window and plate glass give considerable leeway in choice. As the breaking of grains may cause sudden movements of the glass plates toward each other, droplets of the mounting oil may be splattered sideways. These droplets are caught by a loop of soft foam plastic or tissue placed around the glass plates after the first contact with a grain is made.

----

[12] When using the crushing stage, the operator may notice a peculiar soreness of the jaw muscles after an hour or so. This is a perfectly normal symptom resulting from the almost involuntary clenching of the teeth in anticipation of the visually violent "explosion" while the screw is slowly being tightened. Also, when crushing, the operator must be prepared for a fairly high percentage of disappointments, because many grains form fractures that obscure the inclusion before it is opened; even more frequently, the grains merely withstand the pressure buildup until they suddenly collapse into a useless pile of rubble and bubbles of unknown source.

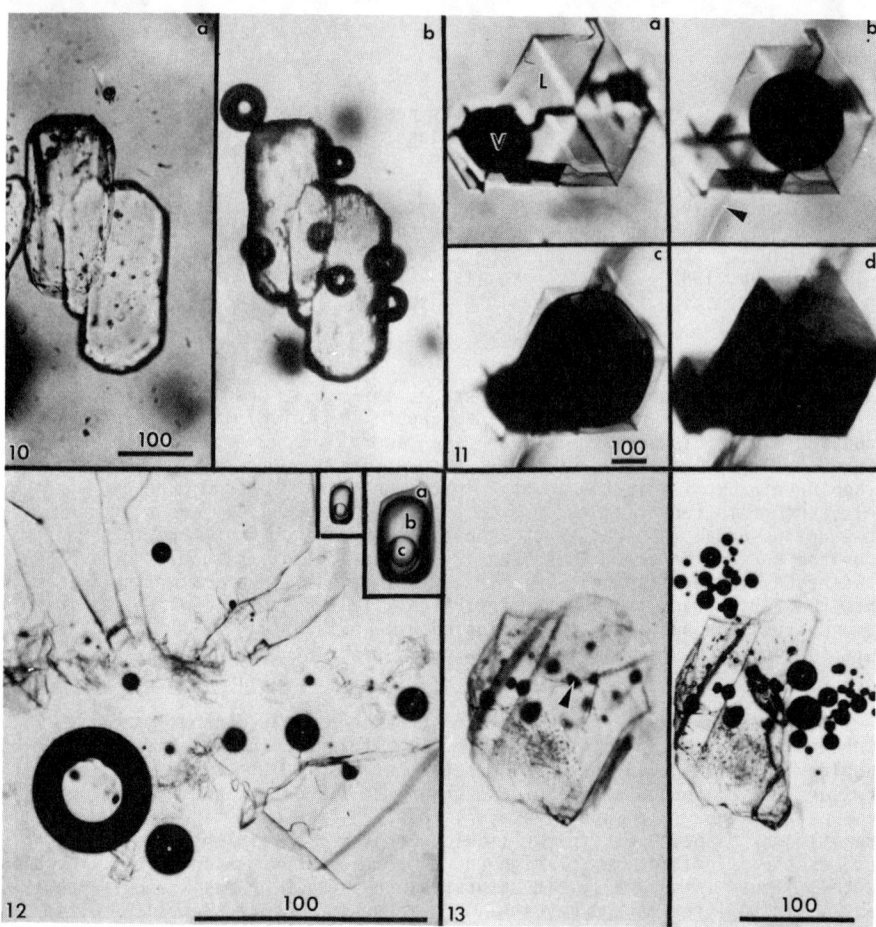

Figure 7-10. Inclusions of high-pressure gas (presumably CH₄) at interface between anhydrite crystal and embedding single salt crystal, during release of pressure by an advancing solution front (coming down from above). Note that large gas bubbles are evolved (right) even though no obvious gas inclusions were present before solution (left). Scale bar in μm. Sample from Asse, W. Germany (Roedder and Belkin, 1981).

Figure 7-11. Crushing of primary negative crystal inclusion in fluorite, containing brine saturated with CH₄ under pressure. (a) Inclusion before crushing. The brine (L) contains approximately 28 wt % salts. The dark vapor bubble (V) is slightly below the plane of focus. In (b), a crack (arrow) has intersected the inclusion, and the bubble is enlarging rapidly. (c) and (d), taken just seconds after (b), show the expulsion of the last of the brine. Shortly after (d) was taken, the gas in the inclusion contacted the mounting medium (kerosene), and kerosene was sucked in to fill the inclusion cavity quickly and completely as the gas dissolved in it. The whole sequence of events usually occurs in less than a second; in this example, the crack was apparently very tight. Scale bar in μm. Sample ER 59-6, Hill mine, southern Illinois fluorite-Pb-Zn district, Illinois, USA. From Roedder (1970a).

Figure 7-12. Crushing of pseudosecondary inclusion, from pegmatitic smoky quartz, containing liquid CO₂. The inclusion before crushing is shown in the insets at the same scale as the main photograph (left), and enlarged (right). It consists of aqueous solution of moderate salinity (a), liquid CO₂ (b), and a bubble of CO₂ gas (c). The main photograph shows the broken fragments and gas bubbles after crushing in a mounting medium of chlorinated hydrocarbon oil somewhat higher in index of refraction than quartz. Both photographs are at same magnification. The volume expansion is 150-200-fold. The difference between this and the predicted 300-fold expansion for liquid CO₂ may well be due to solution in the mounting oil in the time before the photograph was taken (15 seconds). Scale bar in μm. Sample ER 61-28a, Volta Bala, Teófilo Otoni, Minas Gerais, Brazil. From Roedder (1970a).

Figure 7-13. Photomicrographs of an olivine grain, embedded in oil of n = 1.64, before (left) and after (right) cracking on the crushing stage to prove the presence of highly compressed gas (CO₂) in the inclusions. The explosively evolved bubbles come from the single inclusion indicated by the arrow. Scale bar in μm. Sample ER-63-33a, olivine nodule from 1801 Kaupulehu flow of Hualalai, Hawaii, USA. From Roedder (1965a).

If very tiny inclusions are to be studied, high magnification is needed, precluding the use of a 1 mm thick "coverglass." For such inclusions, only a single very tiny grain is used, and a coverglass of 0.3 mm thickness (usually sold as poor quality "grade 3") suffices for crushing. A very small upper viewing hole is necessary; it can be achieved by an auxiliary disk, clipped onto the upper plate. Better still, the modified "microcrusher" version (Fig. 7-9, bottom) may be used.

When there are a number of grains in the same mount, they usually break in sequence and rarely simultaneously, as the largest one supports practically all the pressure until it breaks. Obviously, the observer's eyes should be focused on the grain that is going to break next, and it is frustrating to guess wrong, time after time. Two procedures help in eliminating this guesswork. First, as the plates gradually move together, the viscous oil, as it is squeezed out, carries with it any grains not actually being squeezed, so these moving grains may be ignored. Second, if hard minerals such as olivine or quartz are being crushed, and the optics permit it, the spacing plate may be removed and the grains mounted on a stack of two slides rather than one. Faint transmission-interference color fringes will then appear around the next grain to crack, as the distortion of the glass at the point of pressure locally reduces the air gap between the two plates.

When very soft or small grains are to be crushed, a much simpler and quicker procedure is adequate. A needlepoint gently placed at the correct point on the coverglass will crush the grain and yet not interfere with the viewing. Low-quality thick coverglasses are adequate. Even hard minerals such as olivine may be crushed by using a piece of microscope slide as a cover and a dentist's steel probe for pressure.

When crushing tiny grains of pumice (detailed in Chapter 16), it was necessary to watch the behavior of many grains. Even the needle technique was tiresome and time-consuming. To avoid finger-ineptness at this scale of operation, as well as the trouble and expense of a micromanipulator, I found that the needle procedure could be streamlined further. A uniform sample mount was made that had a fairly dense distribution of grains, and a large but very thin flexible ("grade 0," or better, "grade 00," at 0.080 mm) coverglass was used. This thin coverglass bends readily under the needle and hence crushes only the grains in the immediate vicinity of the needle point. The needle is braced against the objective and tipped down until it just starts to crush grains in the center of the field; it is held in that position while the other hand is used to move the mechanical stage slowly. The viscous drag of the mount carries the coverglass along with the slide under the needle. By this simple expedient, one can watch hundreds of grains marching dutifully by the point near the needle, where they are crushed in sequence, right at the point of focus of the observer's eyes.

The use of the "macrocrusher" is similar in general to that of the other microscope crushing stage. The oil must be poured slowly to avoid bubbles. An oil nearly matching the index of refraction of the crystal is best, in an amount that will cover the crystal. The crystal is then placed on the lower glass disk. Care should also be used here to avoid bubbles, but some may form anyway around irregularities in the crystal. These cause no trouble if the crystal is first dipped or inserted into the liquid off to the side of the cup; adhering bubbles may be brushed away there, before the crystal is moved laterally to the glass plate.

In both modifications, when a single relatively large grain is to be crushed, there is an annoying tendency for the upper glass (either thin coverglass or plate glass) to tip and pull all the mounting fluid away from the grain, forming bubbles. This tipping can be avoided by placing three tiny lumps of plasticine (modelling clay) in a triangle to support the plate temporarily.

Bubbles can also be avoided by mounting the grain in a very small drop of oil on the lower plate, and then wetting the center of the upper plate with another small drop before lowering it. After contact and the application of slight pressure, a plastic squeeze bottle that has a very thin outlet tip is used to add more oil beside the grain. Care must be used to squeeze and fill the tip before inserting it, or bubbles will be injected beside the crystal, where they are very difficult to remove. Similarly, the whole device can be placed vertically and the oil with its bubbles, unclamped fragments, and loose debris from partial crushing flushed out and replaced with clean oil for further crushing of a given grain, e.g., for photographic illustrations.

A fragment of a doubly polished plate gives excellent results, but it must be very small in diameter, preferably not much more than the thickness, or the glass pressure plates will break first.

Applications

The stage has proved useful in a variety of applications such as the rough dating of pumice (Chapter 16) and determining the degree of degassing (by boiling) of hydrothermal fluids before trapping (Chapter 15). However, two other applications are of more general interest and hence are given here.

Providing evidence against leakage into fluid inclusions. The pressures in most fluid inclusions at surface temperatures are very much less than atmospheric, and if partial leakage occurs, it should be inward. The rocks containing such fluid inclusions have been immersed in ground water for geologic periods of time, at pressures of approximately 100 atm/km depth (and in air at 1 atm since they were brought to the surface), yet when the inclusions are examined on the crushing stage, each now contains a vapor bubble that is nearly a vacuum and contains no noncondensable gases, proving that inward leakage under these conditions has been minor or nonexistent.

Composition of the gases in fluid inclusions. Although the crushing procedure can only give crude information on the quantity and nature of the gases evolved from any inclusion, this procedure can provide qualitative information that would otherwise be very difficult to obtain. Evidence on the composition of the evolved gases comes mainly through the use of specific solvents, or nonsolvents, for expected gases. The choice of a suitable mounting medium is difficult at best and must always be a compromise (e.g., Fig. 7-11). In preliminary work, optical visibility is most important; hence, the use of a matching index liquid is the best choice. The ideal fluid for simple detection of compressed gases should match the mineral in index of refraction, be able to wet it, be easy to free of bubbles during mounting, be transparent, and not absorb even very minute quantities of the expected gases. The mass of the emitted gas bubble is usually so small (10^{-11} to 10^{-14} g) relative to the mass of the surrounding liquid that even a very low solubility will permit the bubble to dissolve in a relatively few seconds. Thus, the liquid within a zone equal in width to the bubble radius weighs 4000 times as much as the gas, so a solubility of only 250 ppm would cause it to dissolve completely, yet the average diffusion distance into this fluid would be only 5 μm. Similarly, a solubility of only 1 ppm will permit a gas bubble 4 μm in diameter to dissolve in the fluid within a radius of only 20 μm.

If CO_2 is expected, a water solution of barium chloride can be used. When acidified, this is a rather poor solvent for CO_2; when alkaline, it dissolves CO_2 almost instantly. If the bubbles are large enough, a visible precipitate of $BaCO_3$ will form. Kerosene is an excellent solvent for methane and other organic gases. Other solvents have been discussed by Shugurova (1968). Sphalerite is particularly difficult to work with, because of its very high index of refraction. High-index liquids are almost required here, even though the solubilities of various gases in them are completely unknown.[13]/

The process of solution of the bubble in the various fluids should be watched carefully. If the diameter of the bubble is measured accurately at various times, and plotted against time since release, breaks in the slope of the plot may indicate the presence of several different gases. Bubbles from brine inclusions in Illinois fluorite similar to those shown in Figure 7-10 show several such breaks, although the bulk of the absorption of the bubble follows a smooth curve, presumably indicating that one constituent dominates (e.g., Fuzikawa, 1982). The actual mechanism of solution (and hence shape of the plotted curve) of such a composite bubble is complicated, for as the concentration gradients are changing from diffusion, the radius of the bubble is also changing, and with it, the width of the zone of liquid reached by diffusion, as well as the ratio of bubble surface to volume.

Most of the bubbles of CO_2 from crushing olivine crystals from the olivine nodules that are found in alkali basalts throughout the world (Roedder, 1965a) are moderately soluble in 1.64 index oil (alpha monobromonaphthalene) down to a radius corresponding to about 1% of their original volume, after which the rate is much slower. K/Ar determinations on some of these nodules (Funkhouser and Naughton, 1968) yield high ages, so possibly this second gas contains "excess" Ar.

All the qualitative data obtained by crushing on the microscope stage can, in theory, be obtained in quantitative form by the more elegant method of opening the inclusions in a mass spectrometer. Several factors favor crushing under the microscope, however. First, the quantity of gas released is generally orders of magnitude less than the amounts required by any but a very special gas mass spectrometer. The alternative of crushing larger samples in the mass spectrometer to get larger quantities of gas requires a homogeneity of sample that is rarely found. Second, the quantitative values obtained by a mass spectrometer are frequently subject to large uncertainties because of problems of calibration, loss by absorption, and contamination from many sources, all problems that become larger as the sample size decreases. Third, the crushing procedure requires little or no expenditure for special equipment (microscopic study would be required in either procedure) and can be performed in a few minutes.

R. Bodnar (pers. comm., 1983) has reported that crushing studies are starting to be made routinely on samples from epithermal precious-metal deposits. A correlation seems to exist between mineralization and the presence of gases in the inclusions. Hence, crushing studies may prove useful in exploration for such deposits, but much is still to be learned. The Geologic Research Group at Exxon Minerals Co., Houston, TX, has developed a hand-held crushing stage to be used with a handlens in the field or on a microscope (J.P. Lawler, pers. comm., 1983).

13/ Deicha (1952a) published on an ingenious technique for detecting the presence of water in inclusions by decrepitation in a hot viscous silicone liquid in a transparent vessel, permitting microscopy. This technique could be adapted to crushing as well, by first crushing in room-temperature silicone oil to observe the noncondensable gases and then crushing further at a temperature above the boiling point for the expected fluid salinity to observe the steam bubbles.

Chapter 8

INTERPRETATION and UTILIZATION of INCLUSION MEASUREMENTS: COMPOSTIONAL DATA on LIQUID and GAS INCLUSIONS

CONTENTS

INTRODUCTION

The various measurements commonly made on a given fluid inclusion or suite of inclusions, such as composition, Th, Tm, Te, or density, are not an end unto themselves. The goal is to use these and other inclusion data to understand the nature and sequence of the geologic processes that have been in operation. A brief review of this nature was given by Weisbrod et al. (1976). In Chapters 8-10, the most common procedures and problems in the interpretation and utilization of inclusion measurements are reviewed, including both the validity of the measurements themselves and the more general aspects of their use in evaluating the nature of the geologic processes. Later chapters cover the use of inclusion data in studying specific geologic environments. Because the problems of interpretation of the composition of silicate-melt inclusions are so different from those of liquid inclusions, such problems are omitted here and discussed in Chapters 14, 16, and 18. The interpretation of the less commonly obtained compositional data, such as chemical or isotopic analyses, has already been discussed in Chapters 4 and 5.

PHASE ASSEMBLAGE AT ROOM TEMPERATURE

The interpretation of phase changes on heating or cooling an inclusion might be assumed to require an understanding of the phase assemblage at room temperature. However, sometimes the procedure must be reversed, so that the phase behavior on heating or cooling provides data necessary to understand the room-temperature assemblage. In either case, guesses as to the inclusion composition must be consistent with both the room-temperature phase assemblage and the behavior on heating and cooling. Unless an inclusion fluid is actually analyzed, evidence as to its composition, based on physical behavior of the phases at various temperatures, as detailed below, is always ambiguous. Thus, even if the phase behavior of the inclusion fluid matches that of some fluid of known composition, this match can only be taken as a necessary but not sufficient proof that the inclusion fluid has that composition. The better the match and the greater the number of phase changes that agree with those of the known system, the more confidence one can have in the identification of the unknown. However, one must always remember that natural systems can be complex, and small amounts of residual fluids in particular can concentrate major amounts of rather exotic constituents. Thus the phase behavior of some inclusions will simply not fit any of the simple systems described below. However, as many natural fluids consist of a relatively few major constituents, "educated guesses" can be made for most inclusions.

FIRST ASSUMPTIONS AS TO COMPOSITION, BASED ON ROOM-TEMPERATURE ASSEMBLAGE AND GENERAL BEHAVIOR ON HEATING/COOLING

Because a large fraction of all fluid inclusions (other than melt) consist almost entirely of H_2O, CO_2, and NaCl (or other solutes) in various ratios, a few preliminary working assumptions should be made, based on the phases present at room temperature and the phase behavior during a few preliminary runs. Such assumptions may have to be revised as a result of later studies, but until then, they provide a relatively useful basis for collecting preliminary data. The major categories, based on the phase assemblage at room temperature and the behavior on heating/cooling, are as follows (in part from Hollister et al., 1981).

One Phase at Room Temperature

Three different behaviors are possible on cooling, depending on the gross composition. If a one-phase inclusion contains a liquid-water solution, a small bubble may nucleate on cooling, but will remain relatively small and will shrink abruptly or even disappear when the inclusion freezes (normally this freezing requires temperatures of -30°C or lower). The bubble may reappear on warming and persist up to room temperature. If so, the original inclusion was presumably a metastable stretched-water solution (see Chapter 10), or the inclusion was trapped near room temperature, and the expansion on freezing to form ice has stretched the walls.

If the inclusion consists of CO_2 fluid near to but less than the critical density for CO_2, liquid will precipitate on the inclusion walls on slight cooling; if more than the critical density, a gas bubble will appear and grow rapidly in size on cooling. If the density is farther from the critical in either direction, greater cooling will be needed to obtain the same effect. If the temperature of nucleation of the vapor bubble is only a degree or so below the temperature of homogenization on warming, CO_2 is also indicated, as water inclusions do not nucleate a vapor bubble easily at low temperatures.

The presence of major amounts of CO_2 is further confirmed if removal of the IR filter in the microscope lighting causes rehomogenization of an inclusion held at a temperature slightly below that at which it normally homogenizes. CO_2 is also suggested if the liquid solidifies suddenly on further cooling to ~-90 to -120°C (the inclusion goes dark momentarily from internal reflection of light caused by many CO_2 crystals, followed by rapid recrystallization to a single CO_2 crystal plus vapor). As only solid plus vapor are now present, there is no meniscus. Further cooling may result in nucleation of a second gas bubble or condensation of a new liquid from the vapor phase. If so, much CH_4 (and/or N_2) is probably present as well. Warm to ~-60°C, and then continue to warm slowly. If the CO_2 crystal melts below -56°C, an additional component (e.g., CH_4, N_2, SO_2, H_2S) is probably present that is more soluble in the fluid phase than it is in solid CO_2.

Some inclusions may show no change on cooling, even to ~-180°C, which normally indicates that the inclusion now consists of one of the following: (1) a gas of very low density, (2) a solid (crystal or glass), (3) a metastable supercooled fluid, or (4) normal ice from low-salinity aqueous fluids. Distinction between these alternatives sometimes is difficult. The crushing stage (see below) will permit the recognition of (1). Careful petrography, and in particular, observation of inclusions intersected by the polished surface, will generally suffice to recognize (2), but care is needed to differentiate from inclusions at the surface that are filled with mounting resins. Supercooled fluids (3) are most commonly formed from very strong brines, in very small inclusions, during fast cooling (see Chapter 10). Vitreous or amorphous water (Yannas, 1968) is essentially the pure-water equivalent. It has been reported to form from condensation of water vapor at very low temperatures, but its formation has never been verified in fluid inclusions. Slower cooling and larger inclusions may yield crystals. One of the most common explanations of "no change" behavior on cooling is simply that the inclusion contained low-salinity, near-surface aqueous fluids (4) trapped at near-surface temperatures. When cooled, such inclusions, particularly if small, generally freeze to form a single clear ice crystal. Unless the freezing is actually observed (once freezing starts, it is completed in a small fraction of a second), the inclusion appears unchanged, and the low birefringence of ice generally precludes its recognition on that basis. On reheating, such inclusions show melting over a very small temperature range near 0°C (or even above, if metastable) and hence must be watched carefully in this range or the melting will also be missed. Often this melting is only evidenced by a rapid increase in bubble size and/or movement of the bubble to a different location in the inclusion.

Two Fluid Phases at Room Temperature

If the inclusion contains two fluids at room temperature, the outer phase

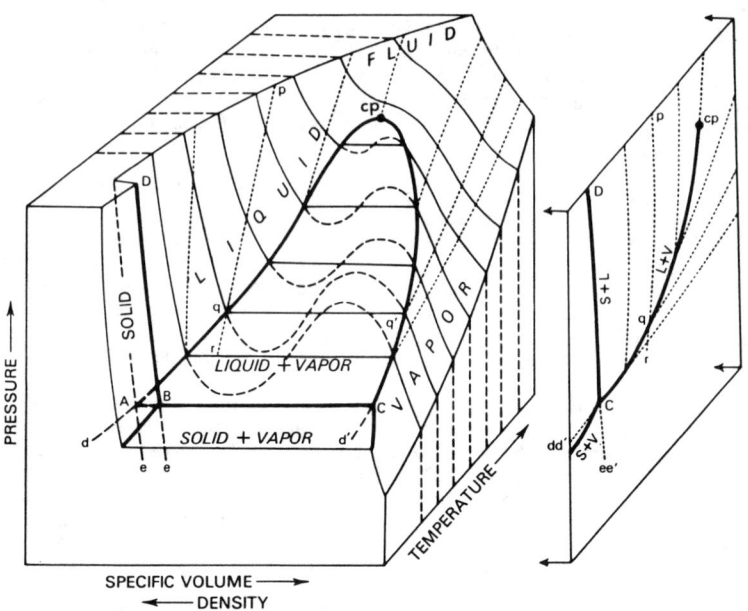

Figure 8-1. Schematic P-V-T diagram for the system H_2O, showing the surface representing stable one-phase equilibria (curved light solid lines), the hypothetical projected metastable one-phase surface (S-shaped surface of light dashed lines; only partly realizable; discussed in Chapter 10), and the surface representing stable two-phase liquid + vapor equilibria (ruled surface A-B-C-CP). One-phase isochores (lines of equal volume) are dotted (e.g., pq). P-V and P-T projections, plotted to scale, are seen in Figures 8-2 to 8-5. Point C is the triple point.

will be liquid and usually the inner phase (a centrally located bubble) will be vapor. Such inclusions normally will consist of H_2O or CO_2, and can be distinguished on the basis of the behavior of the vapor bubble on heating or freezing. Simple warming (on the microscope lamp housing or with a hand-held hair dryer) will cause CO_2 liquid and vapor to homogenize at temperatures <31°C, whereas most water inclusions will require much higher temperature for homogenization. If the bubble contracts when the liquid freezes between about -30°C and -60°C (or lower), the inclusion is H_2O-rich. Generally, the lower the temperature of freezing (Tn ice; always metastable - see Chapter 10), the more dissolved salt the inclusion contains. All H_2O-rich inclusions except those of very low salinity freeze to fine-grained aggregates that may be clear, yellowish, brownish, or dark, depending on thickness. The ice phase recrystallizes and coarsens slowly before it melts on heating. The bubble will increase in size during the warming, because the volume change on melting ice is much larger than the thermal expansion of the contents, and the water solutions are not far from maximum density.

If the inclusion freezes below ~-70°C, with expansion of the bubble, the inclusion cannot be H_2O rich and is probably CO_2 rich. If, on warming, the crystals melt <-56°C and homogenize to form a single fluid phase <+31°C, the inclusion is probably CO_2 rich. Also, frozen CO_2 recrystallizes to single grains rapidly after freezing.

Mixed H_2O-CO_2 inclusions that have two fluid phases at room temperature show two additional types of behavior on cooling. The central vapor bubble may precipitate a rim of a new liquid phase (Fig. 4-9). This suggests that the central "vapor bubble" was low-density CO_2 ("gas"), which precipitated liquid CO_2 at temperatures <31°C. On the other hand, a new smaller vapor bubble may form inside the original "vapor bubble", suggesting that the original "vapor bubble" was high-density CO_2 ("liquid") at room temperature. In both, the inner CO_2-fluid may contain other constituents and will behave accordingly when cooled to still lower temperatures (see above under "One Phase at Room Temperature"). The outer fluid is probably H_2O rich and may freeze and melt as described above, independent of the presence of the CO_2. Most such behavior is actually metastable, however; under stable equilibrium, which may require considerable time to achieve, clathrate crystals may form (see below).

Three Fluid Phases at Room Temperature

An inclusion consisting of three fluids at room temperature very likely contains either CO_2-rich fluids or organic-rich fluids (inner two phases) and an H_2O-rich fluid. Organic-rich fluids may behave in various ways (e.g., see Chapter 11), but the CO_2-rich and H_2O-rich fluids will behave as described above.

Fluid(s) Plus Solid(s) Present at Room Temperature

If a single isotropic cube is present, it is almost always a crystal of NaCl, and the fluid is an NaCl-saturated aqueous solution. If other constituents are not present, the existence of this daughter crystal at room temperature requires the liquid phase to contain 26.5 wt % NaCl, but the actual amount of NaCl present in solution may be much less than this if divalent cations such as Ca^{2+} or Mg^{2+} are also present. Solids other than NaCl are much rarer, and intelligent guesses as to their composition require much more work. Very commonly, the composition of inclusions containing a daughter crystal of NaCl are estimated not from measurements of the physical size of the crystal and inclusion, and hence the relative phase volumes at room temperature, but from the temperature of dissolution of the salt crystal (Tm NaCl) and the frequently invalid assumption that the inclusion composition lies in the system NaCl-H_2O. Measurement of the phase volumes at room temperature combined with dissolution temperature (Tm) can also be used to provide information on the fluid density. However, the presence of other unseen constituents in solution can greatly alter these equilibria.

EXPERIMENTAL DATA ON PERTINENT UNARY SYSTEMS

General Nature of Unary Systems in P-V-T Space

The interpretation of fluid-inclusion data is based in large measure on a comparison of the behavior of the studied inclusion with that expected of a similar inclusion containing a simplified composition, for which experimental pressure-volume-temperature-composition (P-V-T-X) data are already available. Such experimental data for the appropriate ranges of P and T are surprisingly limited, even in the simplest, most pertinent systems. In this section, these data are summarized first for the simple unary systems for H_2O, CO_2, and CH_4. As these systems have no compositional variable, they can be plotted in three-dimensional P-V-T space. After this section, some of the available data on binary, ternary, and higher systems are given, plus some of the more common applications to fluid inclusion study.

H_2O

The system H_2O is illustrated diagrammatically in P-V-T space in Figure 8-1, and the two most useful projections, P-T and T-V, are plotted to true scale

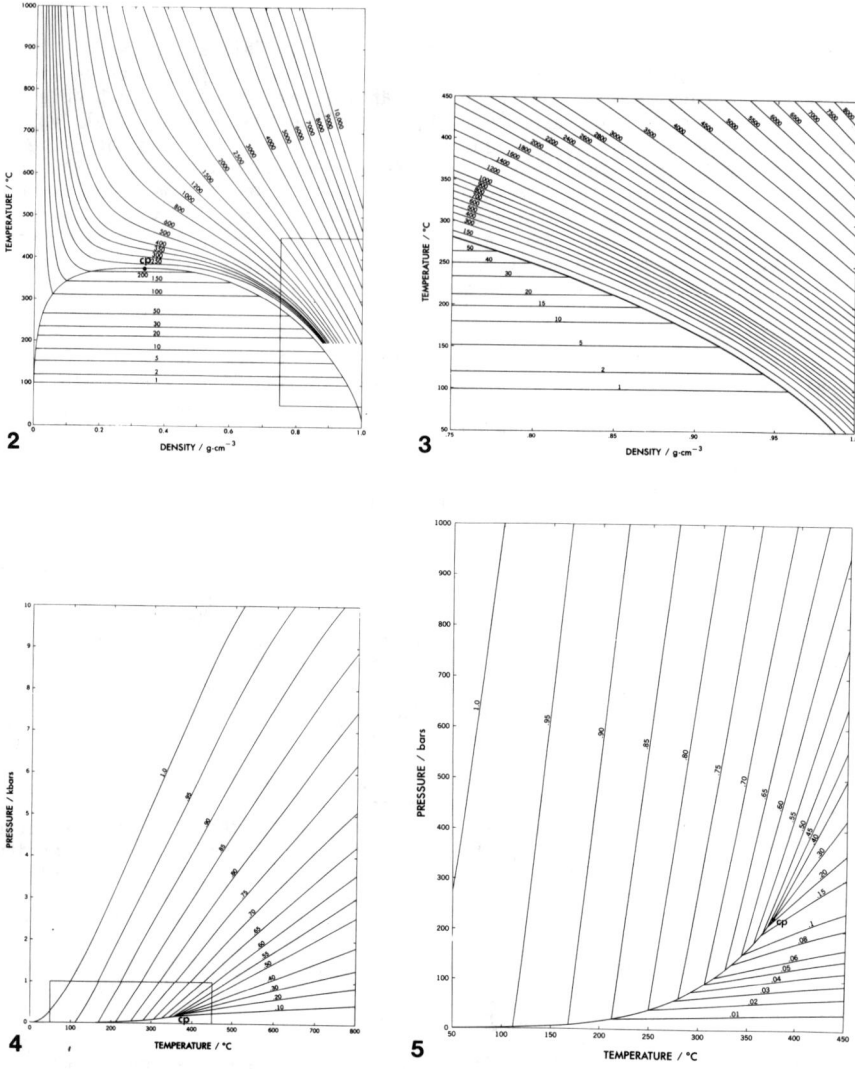

Figure 8-2.   T-V (i.e., density) plot for the system H₂O, from Fisher (1976).   Contours represent pressure, in bars.   CP = critical point, at 374°C, 220 bars, and density = 0.32 g/cm³.   The dome-shaped "two-phase curve" encloses the two-phase region where vapor (along the left of the curve) and liquid (along the right of the curve) coexist.   The entire one-phase area above this curve is best designated simply "fluid."   One may start with low-density steam at 100°C and 1 bar on the left, raise the temperature to >374°C first, then compress isothermally to form high-pressure, supercritical fluid at density = 0.96, then cool to yield hot water at 100°C, all without any break in properties or phase change.   For this reason, the "liquid" and "vapor" can be defined, and have real meaning, only in conjunction with each other, i.e., along the two-phase curve; all other regions are "fluid."   The area outlined by the rectangle is expanded in Figure 8-3.

Figure 8-3.   T-V (i.e., density) plot for the lower temperature, higher density part of the system H₂O from Figure 8-2, from Fisher (1976).   Contours represent pressure, in bars.   The heavy curve terminating the horizontal isobars separates the two-phase liquid-vapor coexistence region (on the left) from that of a single-phase fluid.

(See next page for legends for Figures 8-4 and 8-5.)

226

in Figures 8-2 to 8-5.[1/]   The high and low ranges of P-V-T space are normally
discussed separately, 0°C and 1 bar constituting the dividing line. Data on
the high range from several sources have been collated by Fisher (1976), who
published P-T and T-V plots covering a wide range of conditions (Figs. 8-2 and
8-4), as well as showing the geologically most important part of each of these
plots at expanded scales (Fig. 8-3 and 8-5). Although the T-V plots are useful,
the isochores (lines of constant volume and hence constant density) on the P-T
plots are most commonly used in the interpretation of fluid inclusion data. The
isochores have very much lower slopes in the lower right part of the diagrams
than in the upper left. These slopes reflect the fact that low-density fluids
are more compressible than high-density ones, but both are parts of a continuum
that has no phase boundary beyond the critical point (compare Figs. 8-2 and 8-4).
For actual graphical application, these four plots were presented in larger
format (16 x 20 cm) in the original publication.

Figure 8-6 is a P-T plot of the low-temperature, low-pressure range of the
system H_2O. This plot is like the familiar "triple point" diagram for most substances, with the exception of the negative slope (not visible at the scale of
Fig. 8-6) for the two-phase boundary for ice plus liquid. This negative slope
results from the volume expansion of H_2O on freezing and is involved in several
aspects of the interpretation of fluid-inclusion behavior. The triple point
(B), where ice, liquid, and vapor coexist in the pure system, is at ~4 mm Hg
and +0.01°C.

Inclusionists more and more commonly are using empirical or theoretical
equations of state to calculate densities of fluid inclusions, rather than
graphical or tabular data. Although the densities obtained are frequently
stated to three or four significant numbers, they are actually no more accurate
than those that can be obtained from graphs and tables, but are usually more
convenient to use. Variations in the assumptions made in the equations result
in significant differences in calculated density, but even larger uncertainties
are inherent in the assumptions that generally must be made about inclusion
composition. Table 8-1 gives the more commonly used equations of state for H_2O
and other geologically important systems.

CO_2

Except for the negative slope for the ice plus liquid boundary in the system H_2O, the P-T diagram for the system CO_2 (Figs. 8-7 and 8-8) is topologically
similar to that for H_2O. However, the slopes of the isochores are much lower
for CO_2 than for H_2O, a difference that has been widely used in fluid-inclusion
studies (e.g., see Chapter 10). The isochores in Figure 8-8, from Bergman
(1982), are noticeably concave toward the temperature axis. Other calculations
of the equation of state for CO_2 (e.g., Shmonov and Shmulovich, 1974; Touret
and Bottinga, 1979; Bottinga and Richet, 1981) show them to be essentially

----

[1/] Some plots in the inclusion literature use specific volume units, and others use density. Density
is generally more convenient for the study of inclusions, but much of the chemical and thermodynamic
literature uses specific volume.

Figure 8-4.   P-T plot for the system H_2O, from Fisher (1976). Contours represent density, in
g/cm^3 (i.e., isochores). The two-phase curve for coexistence of liquid plus vapor (near abcissa)
terminates at the critical point (CP) at 374°C, 220 bars, and density = 0.32 g/cm^3. The area outlined by the rectangle is expanded in Figure 8-5. The irregularities of the contours are probably
the result of inadequate constraints on the equations of state (polynomials of high degree) along
their boundaries.

Figure 8-5.   P-T plot for the lower pressure part of the system H_2O above 50°C. From Fisher
(1976), with correction of a minor drafting error. Contours represent density in g/cm^3 (i.e.,
isochores). The two-phase curve for coexistence of liquid plus vapor (near abcissa) terminates
at the critical point (CP) at 374°C, 220 bars, and density = 0.32 g/cm^3.

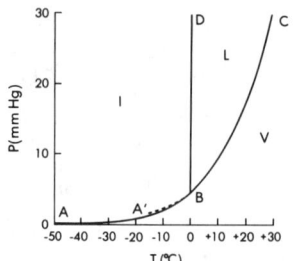

Figure 8-6. P-T plot for the system H_2O, in the low-temperature range, from data of Weast (1980). B = triple point; A-B = vapor pressure of ice; A'B = vapor pressure of metastable super-cooled water; B-C = vapor pressure of liquid water; B-D = ice-liquid equilibrium (the negative slope is not discernable at this scale, because it amounts to only 0.008°C/bar (Bridgeman, 1912); see Fig. 9-1).

straight, up to 20 kbar and 1200°C. The differences can be significant (e.g., ~1 kbar at 700°, 8.5 kbar, and density = 1.2). Still other calculations show the isochores to be concave toward the pressure axis, particularly in the range 0-100°C and 0-1 kbar. Several aspects of the use of CO_2 fluid inclusions are better served by T-V (or temperature-density) plots such as Figure 8-9. This figure is useful in determining the density of CO_2 inclusions from their low-temperature behavior. An inclusion having less than the critical density of 0.4 will precipitate liquid (i.e., heterogenize) on contacting the two-phase field; one having greater than the critical density will nucleate a bubble on contacting the two-phase field. The diagram is valid only for pure CO_2 incl-usions. Because water is essentially insoluble in CO_2 at low temperatures, the behavior of the CO_2 phase in a mixed H_2O-CO_2 inclusion will be essentially in-dependent of the presence of the H_2O (barring formation of clathrates as de-scribed below), but other gases will raise or lower the critical point and distort the boundary of the two-phase field.

The behavior of inclusions of CO_2 having near-critical density is illustra-tive of the general behavior expected from inclusions showing homogenization at a critical point. If the density is exactly the critical density, the meniscus between gas and liquid will simply fade to invisibility. If the density is slightly greater than critical, in theory the inclusion should show a vapor

Table 8-1. Equations of state for some geologically important systems
[Other equations are available, but these are among the most commonly used.]

System	T (°C)	P (bars)	X	Reference
H_2O	0 to 1,000	0- 1,000	---	Kennan et al. (1978)
	0 to 900	0-10,000	---	Haar et al. (1979)
	0 to 1,000	0-10,000	---	Helgeson & Kirkham (1974)
	20 to 1,000	100-10,000	---	Burnham et al. (1969)
CO_2	0 to 1,200	0-20,000	---	Touret & Bottinga (1979)
	-50 to 1,000	1,000-10,000	---	Bottinga & Richet (1981)
H_2O-CO_2	300 to 1,000	0-10,000	$XCO_2=0-1$	Kerrick & Jacobs (1981)
	400 to 1,000	0-10,000	$XCO_2=0-1$	Holloway (1981)
	100 to 1,000	0-10,000	$XCO_2=0-1$	Bodnar & Connolly (1984)
$H_2O-NaCl$	0 to 500	0- 2,000	0-30 wt% NaCl	Potter & Brown (1977)
	80 to 325	(vapor sat.)	0-40 wt% NaCl	Haas (1976)
	20 to 700	(vapor sat.)	0-70 wt% NaCl	Bodnar (1983)
$H_2O-CO_2-CH_4$	~300 to 1,000	0-10,000	$XCO_2=XCH_4=0-1$	Jacobs & Kerrick (1981a)
	400 to 1,000	0-10,000	$XCO_2=XCH_4=0.1$	Holloway (1981)
H_2O-CO_2-NaCl	~200 to 600	0- 3,000	0-30 wt% NaCl $XCO_2=0-1$	Bowers & Helgeson (1983)

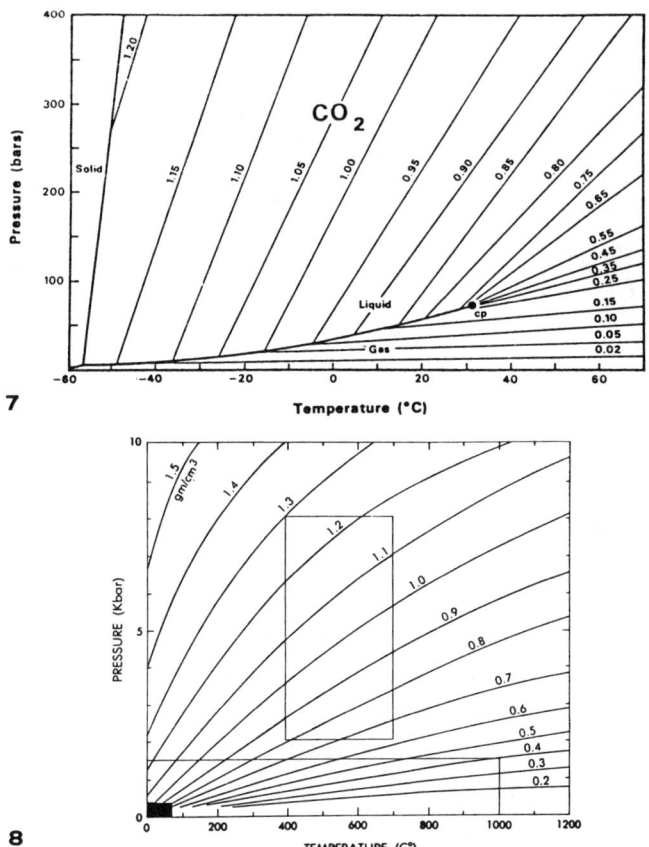

7

8

Figure 8-7.   P-T plot for the system CO_2, in the low-temperature range, from data of Angus et al. (1976).

Figure 8-8. P-T plot for the system CO_2, in the high-temperature range. Data calculated from a Redlich-Kwong equation of state (Bergman, 1982). The ranges of experimental data of Shmonov and Shmulovich (1974; upper rectangle) and of Kennedy (1954; lower rectangle) are indicated. Swanenberg (1980) presented a similar plot, but other equations of state yield straighter isochores. The black rectangle in the lower left represents the range of Figure 8-7.

bubble getting smaller and smaller and finally disappearing. In practice, however, because the two-phase field is so flat on top (Fig. 8-9), a change of a small fraction of a degree in temperature may cause a change from a large bubble to none. If the temperature rise is not extremely slow, this shrinkage will not be seen, and the inclusion will be thought to homogenize at a critical point. Similar phenomena occur when densities are slightly less than critical. Thus, the better the temperature control, the smaller the fraction of inclusions that will appear to have critical density and hence show critical phenomena.

Those few inclusions that have densities falling in a narrow range near the critical density may show a sudden reversal of the direction of motion of the meniscus on heating near to Th. This reversal has been described by Ermakov (1950, p. 246) and considered by him to be a special kind of "homogenization with inversion." Actually it is simply a result of asymmetry of the two sides of the top of the two-fluid-phase field for the specific system involved (compare Figs. 8-2 and 8-9). Because the H_2O solvus becomes more symmetrical upon

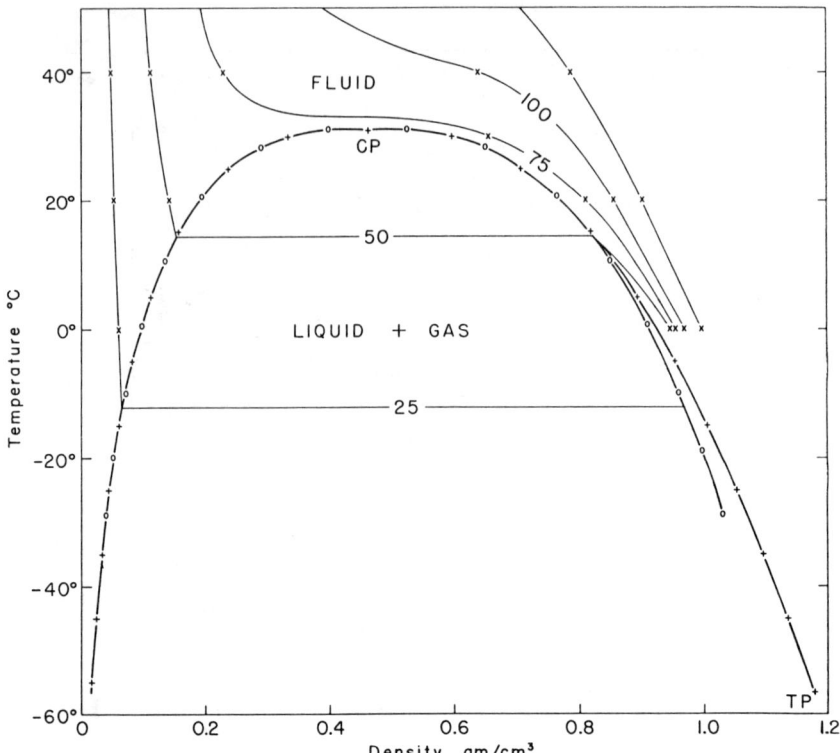

Figure 8-9. T-V (i.e., density) plot for the system CO_2 in the low-temperature range, from Roedder (1965a), from data of Plank and Kuprianoff (1929; crosses) and Hodgman (1953; circles) plus isobars from Kennedy (1954). The exact location of the critical point on such a flat-topped curve is difficult to determine; the more recent data place it at about 0.40 rather than 0.46 as shown here.

addition of NaCl, homogenization with inversion should be much less common in saline inclusions.

The data for the high P-T range (Fig. 8-10) are useful in estimating the pressure of trapping of CO_2 inclusions in metamorphic and igneous rocks from determinations of the density based on the low-temperature behavior (Chapters 13, 17).

CH_4

Swanenberg (1980) described the phase diagram for CH_4 and tabulated the properties of CH_4 fluids in the ranges of T and P of pertinence to inclusion studies (i.e., both high T and P for environment of trapping, and low T and P for microthermometric studies). He also discussed the systems of CH_4 with N_2 and with CO_2 (see below).

EXPERIMENTAL DATA ON PERTINENT BINARY SYSTEMS

General Nature of Binary Systems in P-V-T-X Space

Although most geologists have learned to "read" and use the binary and ternary phase diagrams normally presented in petrology courses, the phase diagrams most useful to fluid-inclusion studies, particularly some of the binary or higher systems involving one or more volatile components, may seem

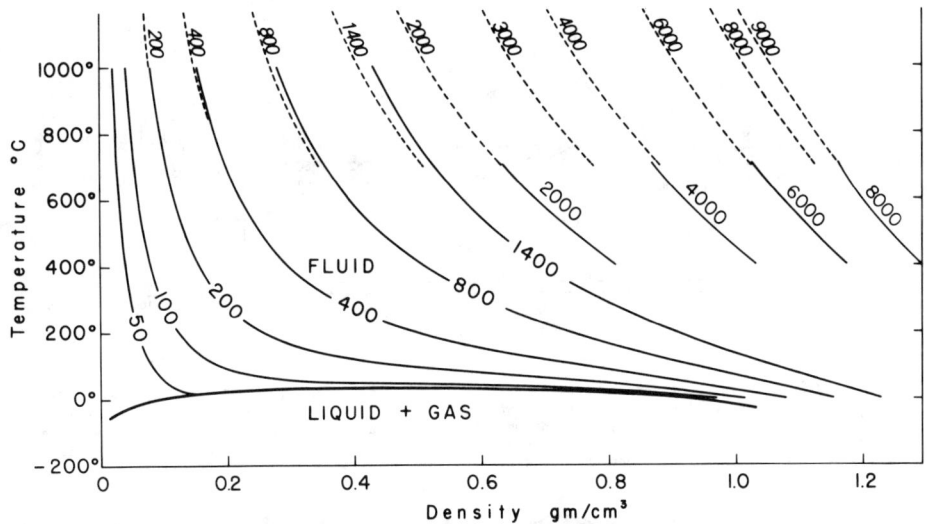

Figure 8-10. T-V (i.e., density) plot for the system CO_2, based on experimental data (solid lines) of Kennedy (1954; up to 1400 bar) and Shmonov and Shmulovich (1974; 2-8 kbar); dashed lines represent calculated data of Kerrick and Jacobs (1981).

rather strange. In addition, fluid inclusions represent constant volume (iso-choric) systems, so their behavior is best considered in terms of a constant volume line or plane through the more general system, quite unlike many other geologically important systems.

Most binary systems involving volatile constituents can be categorized in terms of a few important general characteristics, even though they differ in detail. Because most applications to fluid-inclusion studies involve constant volume, the description here will deal solely with P-T-X space. The simplest example would involve two gases, each having a unary P-T diagram such as that for CO_2 (Fig. 8-8). The critical point in each unary system, which is the limit for two-fluid-phase behavior, is affected by the addition of another component, and a critical-point curve will connect these two critical points. Most commonly, the critical pressure along this curve will increase from both ends, but at each end, the critical temperature will initially change toward that of the added component, as in the example of CO_2-CH_4 (see beyond), and possibly in the system NaCl-H_2O. In such a system, only unsaturated solutions show critical phenomena. In some silicate-volatile systems, this critical-point curve is broken by an intersection with solid-liquid equilibria, yielding two critical end points, upper and lower. Still other types of critical end points exist, for example, in the system CO_2-H_2O, and Seward and Franck (1981) have shown that "gas-gas" immiscibility exists in the system H_2-H_2O. These various diagram topologies can result in some rather bizarre inclusion behavior. Thus, in systems having liquid-vapor loops such as H_2O-CO_2, "retrograde conden-sation" (Ricci, 1951) can occur, which can cause a liquid phase to "evaporate" into the vapor phase on cooling.

Many of the older discussions of volatile-bearing systems (e.g., Morey and Ingerson, 1937) involve use of three-dimensional diagrams plotting T, P, and X, and T-P, T-X, and P-X projections. However, the diagrams for actual binary systems are apt to be much less lucid and obvious, for several reasons. Most important, fluid inclusions are isochoric and isoplethal systems (i.e., their volume and composition are fixed); these two restrictions result in simpler but even less familiar phase diagrams. The solubilities of a given component in any given phase may be either extremely small or large, distorting large parts of

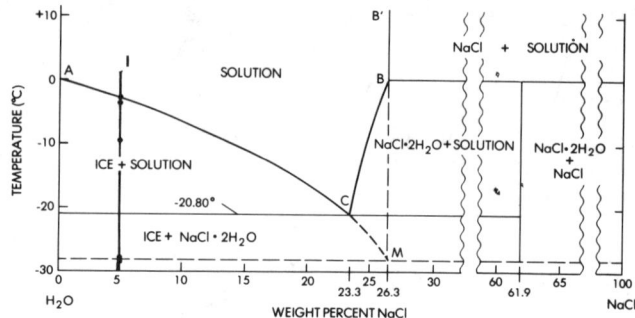

Figure 8-11a. T-X plot for the low-temperature part of the system NaCl-H$_2$O, in equilibrium with vapor, at 1 bar total pressure (Roedder 1962a). More recent data on the ice plus liquid curve (Potter et al., 1978; Linke, 1965) plot essentially the same. Dashed lines represent metastable extensions and meet at the metastable eutectic of ice plus NaCl plus liquid M at ~-28°C. Dots on vertical line "I" represent the freezing data for inclusion shown in Figure 8-11b.

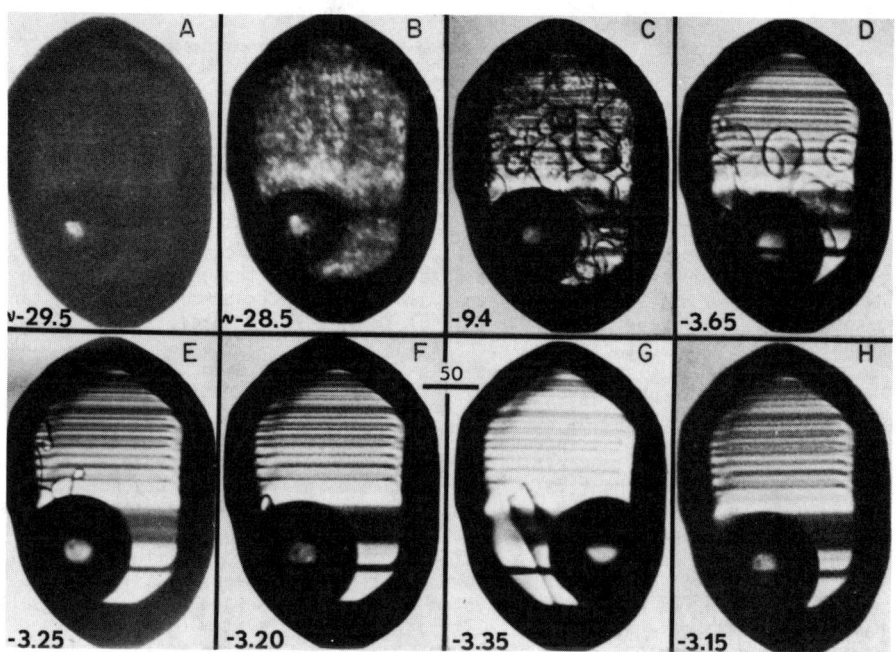

Figure 8-11b. Serial photomicrographs of an inclusion in a transparent plate of sphalerite from Creede, Colorado, USA, (sample ER 57-34), taken after equilibration at the temperatures indicated (°C). Assuming the fluid composition to be in the system NaCl-H$_2$O (an oversimplification), the behavior corresponds to composition I in Figure 8-11a. A and B show that Te is at ~-29°C; C, D, and E show a decreasing group of rounded ice crystals in a solution; in F only one tiny crystal remains; after F was taken, the temperature was dropped 0.15°C, causing typical growth of the ice crystal as two parallel plates, and the movement of the gas bubble, as shown in G; this crystal melted completely on raising the temperature (H) to 0.05°C above that of F; Tm ice is thus taken to be -3.15°C and corresponds to ~5.2 wt % NaCl equivalent (Fig. 8-11a). Te probably corresponds to the metastable eutectic point M (Fig. 8-11a), modified by the presence of other components. The expansion of the water on freezing results in a reduced bubble diameter in A and B; another photograph taken at +20°C also shows a slightly smaller bubble than that in H, caused by the thermal expansion of the liquid solution; this expansion makes the gas bubble disappear at slightly over 200°C. The horizontal bars are from oscillatory striae on the cavity walls. Scale bar in μm. From Roedder (1962a).

Table 8-2. Relation between depression of the freezing point θ and wt% NaCl (W_s) [from equation (4) of Potter et al.(1978)].

θ (°C)	W_s (Wt %)	θ (°C)	W_s (Wt %)	θ (°C)	W_s (Wt %)
0.000	0.000	7.000	10.508	14.000	17.893
1.000	1.698	8.000	11.728	15.000	18.767
2.000	3.343	9.000	12.886	16.000	19.606
3.000	4.922	10.000	13.985	17.000	20.412
4.000	6.430	11.000	15.032	18.000	21.189
5.000	7.862	12.000	16.029	19.000	21.939
6.000	9.221	13.000	16.982	20.000	22.663
				20.500	23.016

the diagram against one or the other side plane. As an example, the solubility of solid NaCl (or of most other solids) in low-density water vapor is very low. Similarly, if a large difference exists in the critical points of the two components, coexisting liquid and vapor can lie at almost opposite ends of the diagram. In addition to the inevitable liquid-vapor equilibria, other complications may be involved, such as liquid immiscibility at lower temperatures, intermediate compounds, and expansion on freezing. Some of these complications have been discussed in books by Sage and Lacey (1949), Ricci (1951), King (1969), and Rowlinson (1969). Excellent reviews by Burruss (1981a,b) of equilibria in the systems CO_2-CH_4 and CO_2-H_2O are based in part on the pioneering work by Ypma (1963) on constant volume phase transitions in CO_2-rich inclusions.

NaCl-H_2O

Phase equilibria in the system. The system NaCl-H_2O is usually thought of as the T-X projection of the condensed system at 1 atm (Fig. 8-11a). The single most commonly used part of this diagram is the line A-C, representing the liquidus for ice. This line is normally referred to as representing the depression of the freezing point of ice with addition of NaCl. Fluid inclusions always show metastable supercooling, so the determination is always made on warming from below, with ice already present, and hence is termed the temperature of melting of ice, or Tm ice (Fig. 8-11b). Potter et al. (1978) have determined this curve experimentally by a new method and have collated their new data with data from some earlier publications, using a linear least squares regression method. Of the four equations that they derived, the two most useful are:

$$W_s = 0.00 + 1.76958 \, θ - 4.2384 \times 10^{-2} \, θ^2 + 5.2778 \times 10^{-4} \, θ^3 [\pm 0.028].$$

$$θ = 0.00 + 0.581855 \, W_s + 3.48896 \times 10^{-3} W_s^2 + 4.314 \times 10^{-4} W_s^3 [\pm 0.03°],$$

where θ = the freezing point depression in °C, and W_s = the weight percent NaCl in solution. Table 8-2 gives a selected series of values derived from the second equation.

The system has one intermediate (i.e., binary) compound, the mineral hydrohalite, NaCl·$2H_2O$. It melts incongruently at +0.1°C to form solid NaCl (48.3 wt %) and liquid "B" (51.7 wt %; Fig. 8-11a), and has a eutectic with ice and liquid "C" at -20.8°C. The compound is unusual in two respects. First, it is very sluggish to crystallize, and hence the metastable assemblage ice plus NaCl may form on cooling; on warming, this metastable assemblage will show metastable eutectic melting to form liquid of composition "M" at ~-28°C. Second, hydrohalite is also very slow to melt and thus may persist for an hour at temperatures as much as 4° above its true melting point (Adams and Gibson, 1930).

Another novel aspect of the system is the very steep liquidus for NaCl, i.e., a very small change in solubility with temperature, particularly in the

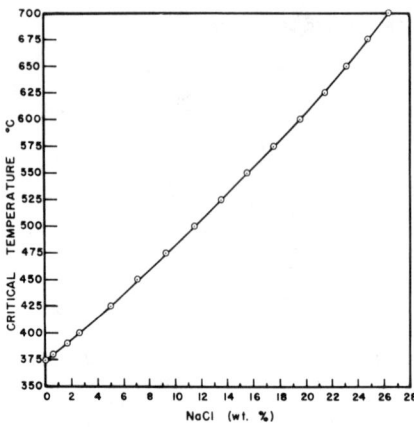

Figure 8-12. Plot showing critical temperatures of unsaturated aqueous sodium chloride solutions. From Sourirajan and Kennedy (1962).

Figure 8-13. P-X diagram for system NaCl-H₂0, showing isotherms for 600-700°C and composition of coexisting gases and liquids. From Sourirajan and Kennedy (1962). ⟶

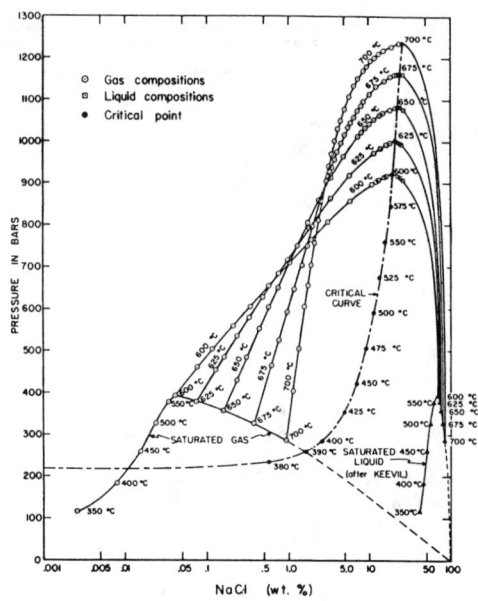

low-temperature range (below 100°C). This feature is useful in distinguishing cubic daughter crystals of NaCl from similar cubes of KCl, which has a much higher temperature coefficient of solubility (Fig. 8-14, but see also footnote 4). The liquidus curve for NaCl bends over with rising temperature and finally intersects the NaCl sideline at the melting point of NaCl, 800.4°C. The higher temperature part of the system has been the subject of intensive study (e.g., Lemmlein and Klevtsov, 1961; Sourirajan and Kennedy, 1962; Haas, 1970; Urusova and Ravich, 1971; Potter and Brown, 1975; Urusova, 1974, 1975; Hilbert, 1979; Rogers and Pitzer, 1982). Although these and other studies have provided many data, the system still has not been adequately explored throughout the entire volume of P-V-T-X space of prime geologic interest. An extensive tabulation of the volumetric properties of aqueous sodium chloride solutions up to 300°C and 1000 bars has been given by Rogers and Pitzer (1982).

NaCl in solution, like most other salts, raises the critical point of the mixture. Sourirajan and Kennedy (1962) measured this increase up to 700°C and 26.4 wt % NaCl (Figure 8-12). They also determined the NaCl concentration in the vapor phase in equilibrium with liquid or liquid and solid NaCl (Figure 8-13). As this plot has a logarithmic abcissa, it is evident that the partitioning of NaCl between vapor and liquid is strongly in favor of the liquid phase. This partitioning is pertinent to the interpretation of various high-temperature fluid inclusions, particularly those in porphyry Cu deposits (see Chapter 15, particularly Figs. 15-14 through 18). Also, because NaCl is the major "salt" present in most saline brine inclusions, it is the major contributor to the density of such solutions (see Fig. 9-11).

The change in solubility of NaCl in H₂0 with temperature is very small in the low-temperature range (line B-B' in Figure 8-11a) but becomes much greater at high temperatures. Figure 8-13 shows that the vapor pressures of the saturated liquids go through a maximum (~600°C and 400 bars), but as the composition axis is logarithmic on Figure 8-13, the data of Benrath et al. (1937) and Keevil (1942) have been replotted in Figures 8-14 and 8-15. In Figure 8-15, isopleths are shown for three compositions (Gunter et al., 1983); to the left

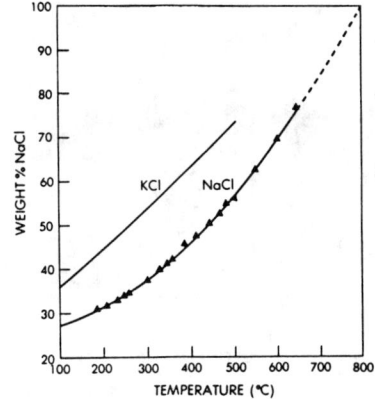

Figure 8-14. Compositions of the liquid phase of saturated aqueous sodium and potassium chloride solutions in equilibrium with vapor, i.e., the three-phase curves. Data for NaCl from Benrath et al. (1937) and Keevil (1942), plotted by Sourirajan and Kennedy (1962); data for KCl as summarized by Potter et al. (1977).

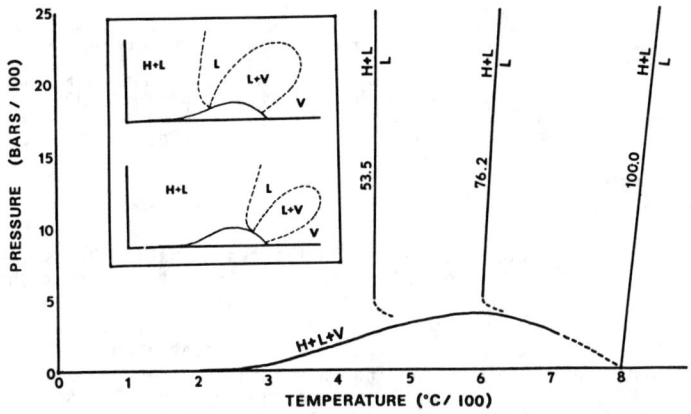

Figure 8-15. P-T plot of part of the system NaCl-H₂O. H = halite; L = liquid; V = vapor. Pressures on the three-phase curve H+L+V to 700°C are from Keevil (1942), with extrapolation to the melting point of NaCl at 800°C. The three liquidus curves (wt % NaCl as given) are from experimental work of Gunter et al. (1983), based in part on interpolation from adjacent curves. The curvature of these liquidi as they approach the three-phase curve is shown diagrammatically (Gunter et al., 1983); the reason for this curvature has not been determined. The inset (diagrammatic) shows the topology of the diagrams for high (upper) and intermediate (lower) salinity isopleths.

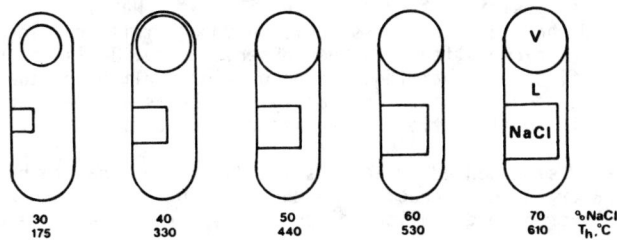

Figure 8-16. Diagram showing the appearance at 20°C of cylindrical fluid inclusions with hemispherical ends, each of a given total wt % NaCl, and each consisting of vapor (V), saturated NaCl solution (L, 26.5 wt % NaCl, density 1.20 g/cm³), and an NaCl cube (shown in cross section; density 2.2 g/cm³). The approximate homogenization temperature (Th) is also given. The diagram was constructed from the data of Khaibullin et al. (1980) by R.J. Bodnar (pers. comm.).

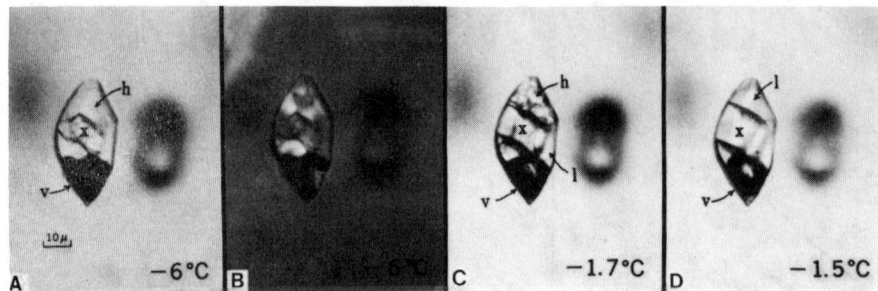

Figure 8-17. Serial photomicrographs of an inclusion, taken at the temperatures indicated, to illustrate the identification of halite by the formation of crystals of hydrohalite (NaCl·2H₂O) on cooling. The inclusion at room temperature appears almost identical with that in the photograph taken at -1.5°C (D). It contains saturated brine (1), vapor (v), and a large daughter crystal (cube) of halite (x). After freezing at -78°C, and then equilibrating at -6°C (A), all available solution has reacted with the halite to form a solid mass of crystals of hydrohalite (h), causing the "meniscus" between the former liquid and vapor (v) phases to be irregular and jagged. A small corroded but unreacted mass of halite remains. The new phase is birefringent, as seen in B, also taken at -6°C but using partly crossed polars. On warming, the hydrohalite decomposes (melts incongruently) to form halite plus solution (C); the last hydrohalite crystal disappears at -1.5°C (D). ER 63-133, quartz, from granite block in volcanic breccia, Ascension Island, South Atlantic Ocean. Scale bar in μm. From Roedder and Coombs (1967).

of each of these isopleths the assemblage is NaCl plus fluid, and to the right, fluid only.

Inclusion Behavior

The actual behavior of most inclusions in the system NaCl-H₂O, as discussed below, can be illustrated by Figures 8-11a through 8-15, but note that these descriptions and data do not apply to inclusions containing CO₂ or other solutes in addition to NaCl. A solution with less than 23.3 wt % NaCl will have Tm ice along the line AC on Figure 8-11a.[2] Slightly more concentrated solutions will, in theory, precipitate NaCl·2H₂O on cooling, although this rarely happens. Such inclusions generally show a metastable Tm ice, along the line CM. An NaCl-H₂O inclusion containing a daughter crystal of NaCl at 20°C must have more NaCl than that of a solution saturated at 20°C (26.47 wt %); the amount of NaCl, for pure NaCl-H₂O solutions, can be roughly estimated from appearance of the inclusions (Fig. 8-16). If such inclusions are frozen and heated quickly, the metastable eutectic NaCl plus ice at ~-28°C (point M, Fig. 8-11a) will be obtained. If they are cooled slowly, or frozen and then warmed to temperatures above M, the liquid will react with the NaCl crystal to form ~62 wt % more hydrohalite than the original NaCl crystal. This hydrohalite may form an insulating armor over the corroded remnant of NaCl and stop the reaction. If the inclusion contains more NaCl than the composition of hydrohalite (61.9 wt % NaCl), it should form a solid mass of hydrohalite plus excess NaCl. Although this reaction, using up all the water to form hydrohalite, has been observed (Fig. 8-17), it requires long equilibration times. On warming, the incongruent melting of the hydrohalite to form solution B and solid halite is plainly visible as a disappearance of the birefringent crystals and formation of new isotropic grains of NaCl.

A special case is presented by aqueous inclusions in a halite host crystal. On cooling, the water in these reacts with the salt in solution and with the walls, forming a layer of hydrohalite. Eventually, all the water reacts with

----

[2] Figure 8-11a is a diagram of the "condensed system," i.e., at a pressure greater than the vapor pressure, so no vapor phase is present. It is actually an isobaric section at 1 atm, but fluid inclusions are isochoric, fixed-volume systems, and hence are polybaric. In this range, the effects of the pressure differences are insignificant and can be ignored.

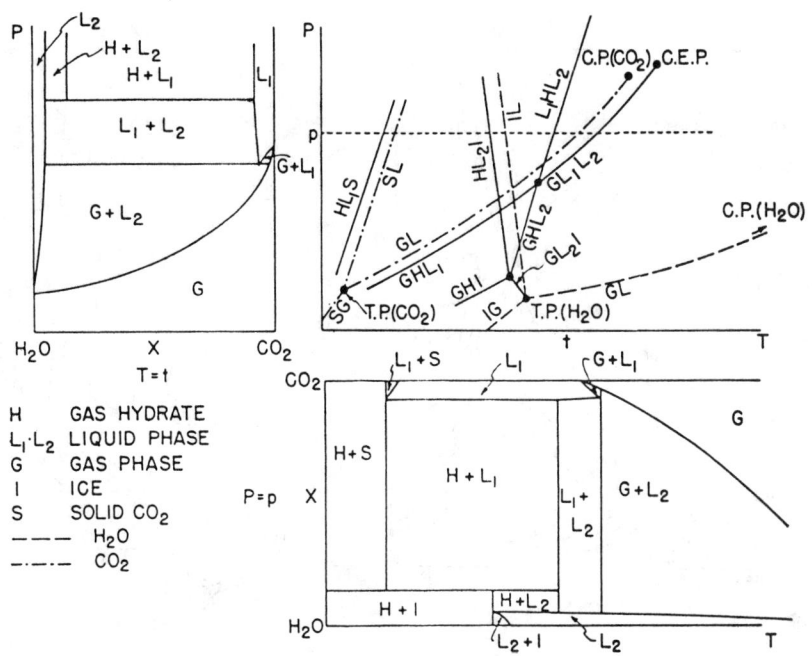

H   GAS HYDRATE
$L_1 \cdot L_2$ LIQUID PHASE
G   GAS PHASE
I   ICE
S   SOLID CO_2
– – – H_2O
–·–·– CO_2

Figure 8-18. Schematic projections of the system H_2O-CO_2, from Takenouchi and Kennedy (1964). C.P. = critical point, C.E.P. = critical end point; T.P. = triple point; L_1 = H_2O-rich; L_2 = CO_2-rich; H = CO_2 hydrate. Note that the T-X plot (lower right) is at P=p in the P-T plot.

the walls to form a solid mass of hydrohalite considerably larger than the original inclusion. Once hydrohalite forms, however, the persistence of the phase, even some degrees above its stable melting point, can cause much confusion. In addition, such inclusions will now have deeply corroded walls, and hence the optics are poor.

On heating a fluid inclusion containing a vapor bubble and a daughter crystal of NaCl, it will follow the "three-phase curve" (Fig. 8-15) as long as a vapor bubble remains, until the daughter crystal dissolves. The behavior thereafter, and the behavior of inclusions in which the bubble disappears first, are discussed in Chapter 9.

H_2O-CO_2

Phase equilibria at elevated temperature. Since the system H_2O-CO_2 involves two very different critical points, two very different molecules (one with no dipole moment (CO_2) and another (H_2O) with a strong dipole moment), and an intermolecular compound, carbon dioxide hydrate, the phase relations are rather complex (Figure 8-18). However, as many fluid-inclusion compositions come close to this system and it has been studied extensively, it will be discussed in some detail. At room temperature and below, liquid CO_2 and liquid water are relatively immiscible. Thus, water saturated with CO_2 at its critical temperature and pressure (31.1°C and 75 bars) contains 5.45 wt % CO_2, and liquid CO_2 will dissolve only 0.10 wt % H_2O at 22.6°C (Wiebe and Gaddy, 1940; Stone, 1943). At higher temperatures, the solubilities increase dramatically, and above a minimum

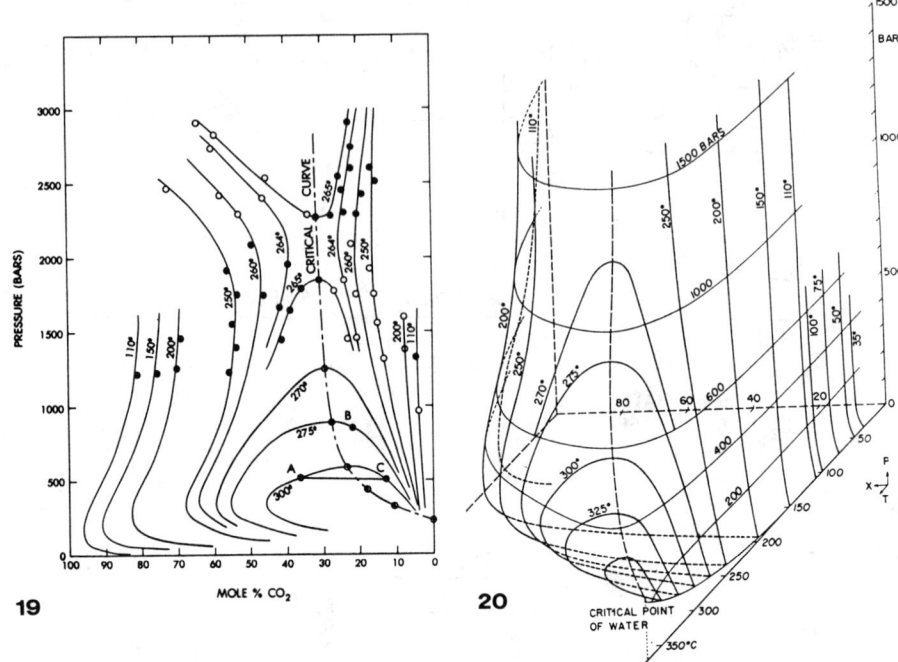

19

20

Figure 8-19. P-X plot of preliminary isothermal data for the system H_2O-CO_2, adapted from Figure 8 of Takenouchi and Kennedy (1964). Filled circles show composition of samples taken from upper part of autoclave; open circles show samples from lower part. The figure has also been reversed from the original, in order to be comparable with Figure 8-20. More recent data on this system, but without the sampling locations for high- and low-density phases, will be found in Fig. 9-9.

Figure 8-20. P-T-X diagram, in perspective, but true scale, for the system H_2O-CO_2 below 350°C and 1500 bars, from Takenouchi and Kennedy (1964). Compare with Figure 8-19.

of 265°C at ~2000 bars (Fig. 8-19), miscibility is complete. At very high pressure, immiscibility may occur again at ~400°C (Kerrick and Jacobs, 1981). The volume of P-T-X space over which immiscibility occurs below 1500 bars is shown in perspective in Figure 8-20. Figure 8-19 suggests that as pressures increase above the point of the "saddle" on the critical-point curve, the temperatures along the critical curve will start to increase again. Presumably, this critical-point curve ends at a critical end point at very high pressures (Todehide, 1963). The other segment of the critical curve presumably ends at a critical end point very close to the critical point of CO_2 ("CEP" on Fig. 8-18).

One interesting aspect of the immiscibility volume involves the density of the two fluids, as indicated by the data points on Figure 8-19, and described by Takenouchi and Kennedy (1964, p. 1062):

"Samples represented by filled circles were taken from the upper part of the bomb, and samples represented by open circles are from the bottom part of the bomb. It is interesting to note that inversion takes place in the system. At high pressures the water rich phase floats on top of the carbon-dioxide rich phase, whereas at lower pressures the carbon dioxide rich phase floats on top of

238

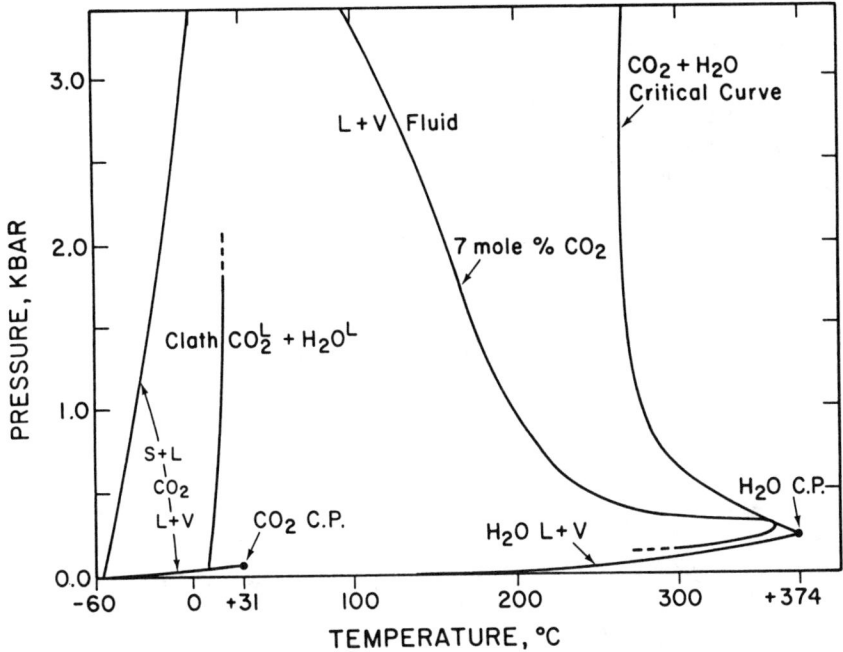

Figure 8-21. Phase equilibria in part of the H_2O-CO_2 system. The high-pressure clathrate-stability curve (clath/CO_2 liquid plus H_2O liquid) is dashed above 1.8 kbar because of lack of experimental data. From Burruss (1981c).

the water rich phase. Along each isotherm there is a pressure at which the density of the two coexisting phases is the same even though the bulk composition of the coexisting phases may be grossly different. The pointlessness of continuing the normal 1-atm distinction between the gas and liquid on to higher pressures is strikingly evident here."

This density inversion explains the strange observation that perplexed early investigators: on heating, the "bubble" sometimes sank to the bottom of the "liquid" within the inclusion (e.g., Hartley, 1876). This density inversion is an additional reason to consider any combination of two or more fluid phases that do not mix under the specified conditions as immiscibility. The scientific and semantic rationale behind this seemingly odd usage has been detailed by Roedder and Coombs (1967, p. 419).

A P-T projection of some of the equilibria in this system is shown in Figure 8-21. This diagram does not show the critical-point curve and critical end point near the critical point of CO_2, because data on these two are lacking, but it does show the critical-point curve originating from the critical point for H_2O. Only part of one of a family of bubble-point and dew-point curves, that for 7 mole % CO_2, is shown (compare with Fig. 8-23b to see how the isochores must radiate from each point along each of these curves).

Kerrick and Jacobs (1981) developed a Redlich-Kwong equation for these homogeneous mixtures of H_2O and CO_2, from which they suggest that a field of immiscibility comes in again at ~400°C and pressures >10 kbar. The properties of the homogeneous fluids at high temperatures and pressures have been studied experimentally by Greenwood (1973), Chou and Williams (1977, 1979), and Gehrig

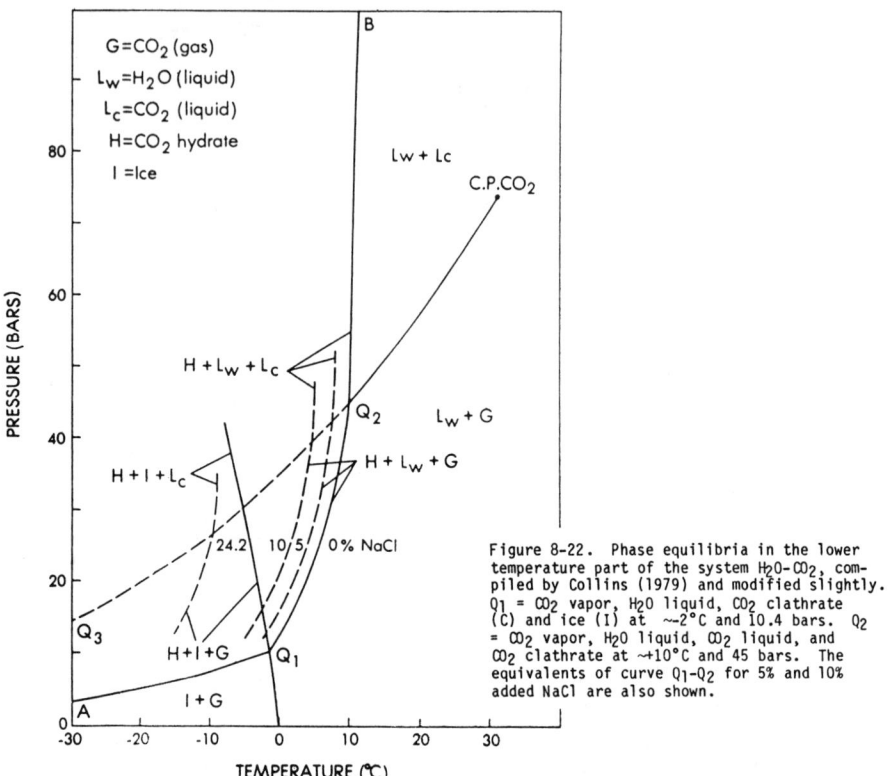

Figure 8-22. Phase equilibria in the lower temperature part of the system H_2O-CO_2, compiled by Collins (1979) and modified slightly. Q_1 = CO_2 vapor, H_2O liquid, CO_2 clathrate (C) and ice (I) at ~-2°C and 10.4 bars. Q_2 = CO_2 vapor, H_2O liquid, CO_2 liquid, and CO_2 clathrate at ~+10°C and 45 bars. The equivalents of curve Q_1-Q_2 for 5% and 10% added NaCl are also shown.

et al. (1979), and theoretically by Shmulovich et al. (1980), Kerrick and Jacobs (1981), and Ziegenbein and Johannes (1982).

As a result of the very limited miscibility of CO_2 and H_2O at low temperatures, and the complete miscibility at high temperatures, some inclusions have two critical homogenization points. Thus, Roedder (1962, p. 191) described three-fluid-phase CO_2-H_2O inclusions that showed a critical point for CO_2 liquid plus vapor at +30.15°C (the deviation from the 31.1° critical point of pure CO_2 presumably is due to the presence of other gases), and a critical point for CO_2 fluid plus the water solution, yielding a homogeneous fluid, at 314°-317°C. The two most common behaviors on heating mixed CO_2-H_2O inclusions involve the homogenization of the CO_2 phases ≤31.1°C, followed by the water phase expanding to "dissolve" and eliminate the CO_2 phase (if the CO_2 phase is small), or the converse, if the water phase is small.

Phase equilibria at depressed temperature. On cooling immiscible mixtures of H_2O and CO_2, the two substances act independently, as the two liquid phases are essentially immiscible. The two CO_2 phases, liquid and gas, follow down the L-V curve for CO_2, and the liquid water essentially follows the L-V curve for H_2O; the small vapor pressure of H_2O at these temperatures corresponds to a small amount of water in the CO_2 vapor phase. When point Q_2 (Fig. 8-22) is

240

reached at +10°C, CO_2 should react with H_2O to form the CO_2 hydrate phase (i.e., clathrate) $CO_2 \cdot 5.75H_2O$.[3/] The crystals may be almost invisible, depending on the index of the liquid phase (Roedder, 1963). This reaction is generally sluggish; if so, the L-V equilibria may follow the metastable extension Q2-Q3. The structure of the clathrate is such that its nominal composition is $CO_2 \cdot 6H_2O$, but not all the "cages" are normally filled with CO_2, hence the odd formula. At $5.75H_2O$, the crystals contain 30 wt % CO_2. On the basis of the density of liquid CO_2 at +10°C (Quinn and Jones, 1936), this corresponds to 33 vol % liquid CO_2 and 67 vol % H_2O. From the data of Makogon (1974) and Quinn and Jones (1936), the formation of the hydrate at +10°C results in a 14.5% volume decrease, so the vapor bubble should increase in size when the hydrate forms, even if the crystals cannot be seen (see also Takenouchi and Kennedy, 1965a; Bozzo et al., 1973). CH_4 forms a similar clathrate, but this compound has dissociation temperatures considerably higher than those for the CO_2 hydrate. It has been reported in inclusions at temperatures as high as +18°C.

If excess liquid H_2O is present, it (theoretically) should solidify at ~0°C (the 20 bars pressure will drop the melting point only ~0.16°C; see Figure 10-1), but the presence of CO_2 in solution will result in a larger drop. If excess liquid CO_2 is present, it should solidify at -56.6°C. Takenouchi and Kennedy (1965a) showed how the phase relationships observed in such inclusions could be interpreted in terms of their studies of the dissociation pressure of carbon dioxide hydrate. In most actual inclusions, these phase relations are normally complicated by metastability and by the presence of ionic solutes in the water and other gases in the CO_2 (see below). Thus, Mullis (1979) has explored the system H_2O-CH_4 in order to explain the behavior of some Alpine inclusions. Methane also forms a hydrate, and similarly, Beny et al. (1981) recognized the formation of H_2S hydrate. The physical properties and solid solution relationships of many of these hydrates were reviewed by Makogon (1974); they exist in nature in some natural gas fields and ocean sediments (Kvenvolden and McMenamin, 1980). The literature on clathrates has been summarized in an extensive annotated bibliography (Hollister and Burruss, 1976, p. 165; see also Cady, 1983).

Other Two-Gas Systems

The existence of occasional gas inclusions under relatively high pressure has been known since the work of Davy (1822). However, the relatively easy access to low temperatures during microscopy, starting two decades ago (Roedder, 1962a, 1963), revealed that some apparently "empty" inclusions contained dense supercritical gas mixtures. Since then, the much lower temperatures attainable by means of the stage designed by Poty et al. (1976) and increased use of the crushing stage have resulted in the discovery that many single-phase, empty-appearing inclusions actually contained high-pressure gas. Actual analyses have shown that the major constituents are generally CO_2 and CH_4, but N_2, H_2, C_2H_6, CO, and H_2S can also be important or even major constituents. Swanenberg (1980) discussed the available data on isochoric sections of nonionic systems, including those involving CO_2, CH_4, and N_2. Although the specific composition of a given gas inclusion cannot be obtained unambiguously from freezing data alone, such data can permit some educated guesses. Some of the procedures for making such

----

[3/] The intermolecular compound, $CO_2 \cdot 5.75$ H_2O (structural formula: $8CO_2 \cdot 46H_2O$) is one of several such "clathrate" compounds. The term comes from the Greek for "bar" and refers to the latticelike cages of water molecules in the structures, into which the various gas molecules fit. Most of these compounds are stable only below room temperature and have variable compositions, both in terms of the nature of the caged molecules, such as CH_4, CO_2, and H_2S, and the degree to which the cages are filled. Note, however, that if the CO_2 pressure is <10.4 bar (point Q_1 on Fig. 8-22), no CO_2 hydrate will form and the CO_2 in solution in the liquid water will merely act as another solute, depressing the freezing point slightly.

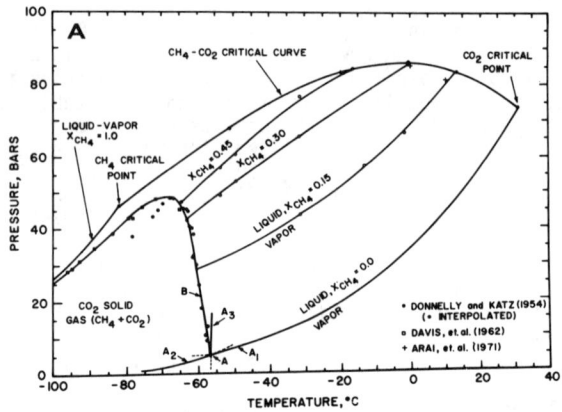

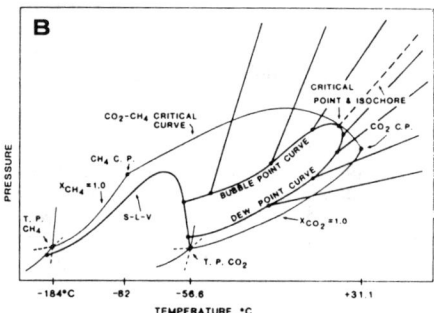

Figure 8-23. (A) Experimentally observed P-T phase equilibria for the system CO_2-CH_4, slightly modified from Hollister and Burruss (1976). Points B_1 and B_2 of Figure 8-24 fall on the univariant, solid-liquid-vapor coexistence curve B. The three curves labeled $XCH_4 = 0.15$, 0.30, and 0.45 are the bubble-point curves for these bulk compositions in the system. The equivalent dew-point curves are not shown. (B) Schematic representation of (A) with addition of the dew-point curve and single-phase isochores for a constant composition system with $XCH_4 = 0.15$, from Burruss, 1981c. (C) Composite phase diagram for the system CO_2-CH_4, from Swanenberg (1979). Light lines within the two-phase field L-Cm-V and heavy lines are isochores for the densities listed, in g/cm^3, bubble-point, and dew-point curves at $XCH_4 = 0.1$. Curves (a) and (b) are the L + V curves for the pure CO_2 and CH_4 systems, respectively; (c) is the critical curve for mixtures of CO_2 and CH_4; (d) is the solid-liquid-vapor curve, from the triple point of CO_2 (TCO$_2$) to that of CH_4 (not shown--at lower temperatures); (e) is the bubble-point curve, and (f) the dew-point curve for mixtures in which $XCH_4 = 0.1$.

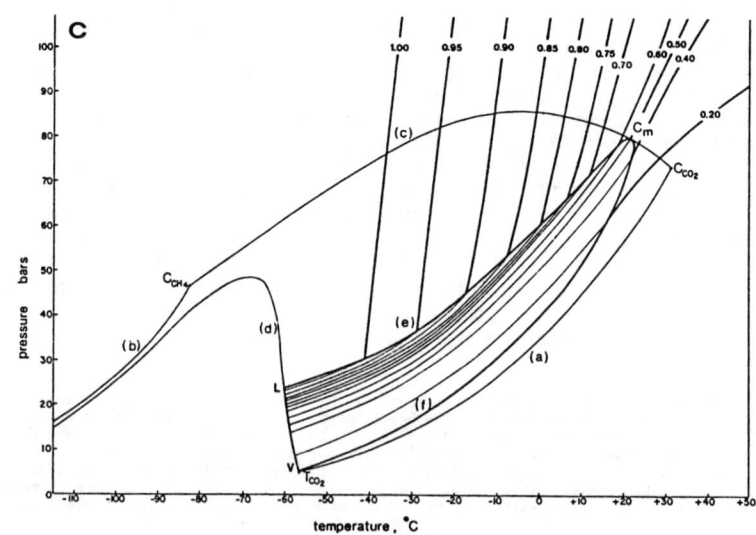

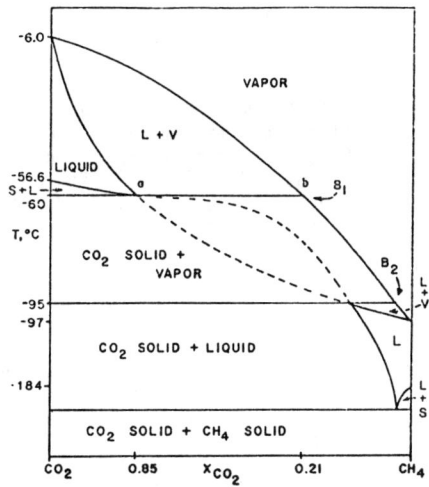

Figure 8-24. Isobaric T-X section through Figure 8-23A at about 30 bars. The T axis is not to scale. Dashed curves are the metastable extensions of the contiguous phase boundaries. From Burruss (1981c), constructed from the data of Donnelly and Katz (1954).

guesses, particularly for inclusions consisting mainly of CO_2 and CH_4, have been given by Burruss (1981a,b) and can only be briefly summarized here. Thus, the presence of small amounts of other gases in CO_2 inclusions will affect both the critical temperature and the triple point, generally shifting them in the direction of the equivalent point for the impurity gas itself. (The critical pressures for these intermediate mixtures may be higher than that for either pure end member.)

CO_2-CH_4. The general behavior of binary gas systems of interest in inclusion study is best illustrated by the one binary gas system most commonly found in nature, the system CO_2-CH_4, for which appreciable data are available. Berdnikov and Tomilenko (1983) illustrated the freezing behavior of CO_2-CH_4 inclusions (see also Fig. 11-18). Just as for the binary NaCl-H_2O, a full representation of all pertinent aspects would require a multidimensional figure in P-V-T-X space, but the major aspects can be illustrated with a P-T plot (Fig. 8-23) and a T-X section through it (Fig. 8-24), taken, along with the following description (with only minor changes), from Hollister and Burruss (1976) and Burruss (1981c).

The available equilibria in the CO_2-CH_4 system were first applied quantitatively to fluid-inclusion observations by Hollister and Burruss (1976). Figure 8-23, from that work, contains two P-T projections of the major features of the CO_2-CH_4 equilibria. Figure 8-23A is a quantitative P-T diagram, modified from Donnelly and Katz (1954), with addition of more recent solid-liquid-vapor coexistence measurements from Davis et al. (1962) and liquid-vapor data from Arai et al. (1971). The most important features are the shift of univariant, solid-liquid-vapor equilibria (curve B) and the critical curve to lower temperature with increasing XCH_4. Point A is the solid-liquid-vapor invariant point (i.e., triple point) for pure CO_2, and curves A_1, A_2, and A_3 are the liquid-vapor, solid-vapor, and solid-liquid CO_2 coexistence curves, respectively.

Figure 8-23B schematically illustrates the relationships of one-phase isochores to both the bubble-point and dew-point curves of a constant composition system in which $XCH_4 \approx 0.15$. This figure is similar to Figure 2 of Swanenberg (1979), which illustrates the same relations for $XCH_4 \approx 0.10$ (Fig. 8-23C).

Several significant features of this system are more obvious in an isobaric T-XCO_2 section (Fig. 8-24) at the pressure of solid-liquid-vapor (S-L-V) equilibria for a vapor-saturated liquid with $XCH_4 \approx 0.15$ (point B_1 in Fig. 8-23a). The boundaries of the liquid and vapor field shown in Figure 8-24 illustrate the strong partitioning of CH_4 into the vapor phase. Furthermore, the compositions of coexisting liquid and vapor at S-L-V equilibrium at B_2 in Figure 8-24 (see also Fig. 8-23a) clearly show that the partition coefficient is a function of temperature. Note, however, that very significant amounts of CH_4 can dissolve in liquid CO_2, dropping the bubble-point curve. Thus, if a CO_2 inclusion containing some CH_4 is assumed to be pure CO_2, the apparent density obtained from Th CO_2 L-V will be too high, as will be the apparent pressure of trapping.

The strong partitioning of CH_4 into the vapor phase makes it difficult to estimate the bulk XCH_4 in an inclusion when only the temperature of disappearance of the solid CO_2 phase (the final melting temperature) is known. In the example of Figure 8-24, the solid CO_2 phase has a final melting temperature of -60°C (B_1 in Figs. 8-23a and 8-24), thereby fixing the compositions of the coexisting liquid and vapor phases at points a and b. However, the composition of the <u>bulk</u> system could fall <u>anywhere</u> between a and b. At the temperature of disappearance of solid CO_2, the bulk composition of the inclusion is therefore a function of the relative volumes of the two phases a and b. Swanenberg (1979) approached this problem by using an estimate of the relative abundance of liquid and vapor phases at the final melting temperature to calculate bulk XCH_4. Swanenberg's (1979) technique is apparently limited to bulk $XCH_4 \leqslant 0.3$, and is subject to error from the difficulty of estimating the relative volume percent of a phase within an irregular volume.

By explicit use of the constant volume property of inclusions, Burruss (1981a,c) has developed an ingenious procedure based on a display of the phase equilibria, using the Helmholtz free-energy variables V, T, X. When these diagrams are compared with the low-temperature phase behavior of pure CO_2-CH_4 inclusions, relatively unambiguous values for density (molar volume) and composition of the inclusion can be obtained. So long as the original assumption as to the nature of the constituents of the gas mixture is valid, and the experimental data on such systems are available, such methods will work.

Other gases. The methods described above will not provide unambiguous quantitative analyses of multicomponent gas mixtures, but when combined with other data that may provide compositional constraints, they can provide semi-quantitative or quantitative estimates of probable composition. Furthermore, discrepancies between the actual behavior and that predicted by a given simple model system can suggest the presence of at least some components other than those of the model, and can place constraints on the nature of these added components. These constraints will become more explicit with every improvement in the experimental and theoretical data on the pertinent systems. Thus, Jacobs and Kerrick (1981a) have shown, from a new equation of state for CH_4, that small amounts of CH_4 in the ternary system H_2O-CO_2-CH_4 slightly increase the activity of H_2O and significantly decrease the activity of CO_2; equations of state have been proposed for the systems CO_2-CH_4 (Heyen et al., 1982b) and CO_2-CH_4-C_2H_6 (Heyen et al., 1982a) and applied by these authors to understanding the behavior of fluid inclusions containing such mixtures below 50°C and 100 bar. Drummond (1981) has measured the solubilities of CO_2 and H_2S in 0-6 m NaCl solutions between 25 and 400°C, and Touray and Guilhaumou (1984) have discussed the low-temperature phase behavior of H_2S-bearing inclusions.

In many low-temperature environments, both salts and gases are present in significant amounts, and the phase behavior can only be interpreted in terms of the quaternary system CO_2-CH_4-NaCl-H_2O (Ramboz, 1980); unfortunately, practically nothing is known about this system.

Similarly, major amounts of N_2 are present together with CO_2 in some gaseous

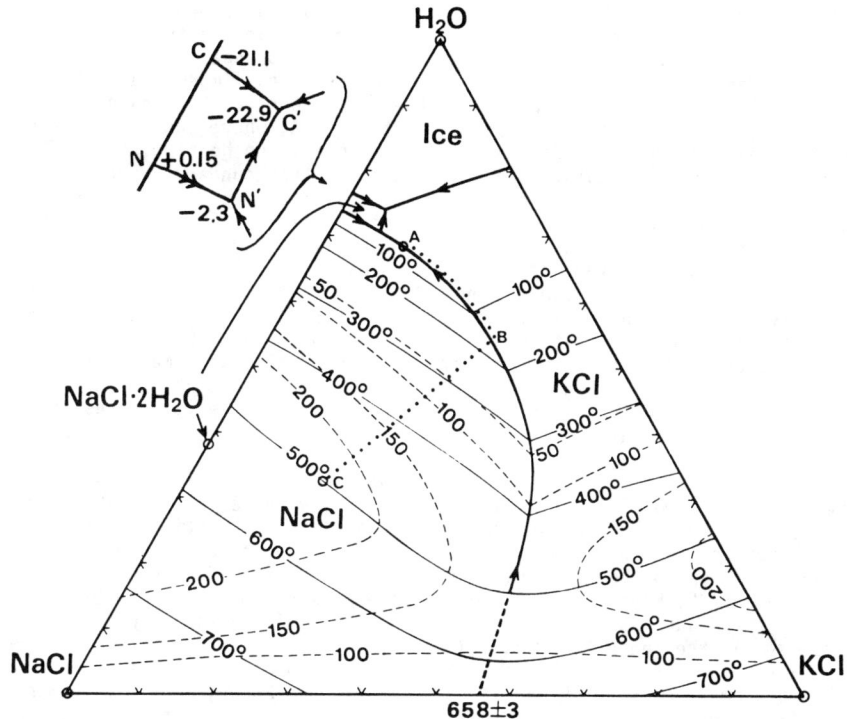

Figure 8-25. Phase diagram for vapor-saturated system NaCl-KCl-H₂O, in wt %, showing isotherms (°C; thin solid lines) and isobars (kg/cm²; thin dashed lines). Data compiled and smoothed where necessary by R.J. Bodnar (manuscript in preparation) from Cornec and Krombach (1932; isotherms <200°C), Ravich and Borovaya (1949, 1950; isotherms >200°C), Sourirajan and Kennedy (1962; NaCl-H₂O binary), and Roedder (1971b).

NaCl-KCl eutectic from Chou (1982). Details in the low-temperature part are shown schematically at the left. The path taken by the fluid phase on heating an inclusion containing daughter crystals of KCl and NaCl is shown by the dotted line. It starts at (A); the last KCl dissolves at 150°C (B), and the last NaCl at 500°C (C) (see text).

inclusions (Guilhaumou, 1982), requiring a knowlege of the system CO_2-N_2. This system is very similar to that for CO_2-CH_4 (Fig. 8-23), causing considerable ambiguity in the interpretation of inclusion phase behavior if independent analytical data are not available (Swanenberg, 1980; Touret, 1982). In particular, N_2 can lower Th greatly but leave Tm CO_2 near -56°C. In addition, numerous other topologies are possible in such systems (Schneider, 1978). Such large amounts of higher hydrocarbons are miscible with CO_2 at elevated temperatures (Orr et al., 1981) that it can be used effectively in oil well flooding (Stalkup, 1978; Orr, 1983). A number of essentially nonvolatile organic compounds are also soluble in supercritical CO_2, so it is used to decaffeinate coffee and in a variety of other commercial processes (Schneider et al., 1980).

EXPERIMENTAL DATA ON PERTINENT TERNARY AND HIGHER SYSTEMS

NaCl-KCl-H₂O

Of all the possible ternary salt-water systems, NaCl-KCl-H₂O represents by far the closest approach to many actual inclusion fluids. Figure 8-25 shows the phase diagram for the vapor-saturated surface in this system. This diagram is applicable, in theory, to any inclusion containing H₂O, Na⁺, K⁺, and Cl⁻, but as this composition is generally not known to be present unless daughter

245

crystals of NaCl and KCl are visible, the diagram is generally used only for such two-daughter-mineral inclusions. The dotted line on this diagram represents the path of the vapor-saturated fluid composition on heating an inclusion in which we assume that the KCl crystal dissolves (i.e., Tm KCl) at 150°C and Tm NaCl is 500°C. At room temperature, the presence of the two daughter crystals and liquid requires that the liquid is at point A, where the 25°C isotherm intersects the boundary curve for NaCl + KCl + liquid. On heating, the inclusion fluid dissolves KCl rapidly and hence moves up the boundary curve toward B. Because Tm KCl is 150°C and the 150° isotherm intersects the boundary curve at B, the last of the KCl melts at the moment the fluid reaches this point.[4]/ Because the inclusion now consists of only fluid B and an NaCl daughter crystal, further heating, causing dissolution of NaCl, must move the fluid directly toward the NaCl corner, i.e., along BC. When Tm NaCl is reached, the temperature is 500°C and the fluid is at point C, where the 500° isotherm intersects the line BC. The composition of point C, the original vapor-saturated fluid that was trapped, can be read off the diagram as 49.0 wt % NaCl, 18.6 % KCl, and only 32.4 % H_2O. Such fluids are not uncommon in porphyry copper deposits. Note also that even though the temperature is 500°C, the vapor pressure is only a little over 200 bars in this fluid, which is best called a "hydrosaline melt."

Some workers (e.g., Kalyuzhnyi, 1976; Cloke et al., 1978; Eastoe, 1979) believe that some of the fluids were saturated with respect to NaCl before trapping; i.e., solid NaCl phase was actually present at the time of trapping. The best proof of this would be NaCl crystals enclosed in other minerals, but they generally have not been reported, except for ones that formed by necking down (Roedder, 1972, Plate 10). Obviously, the presence of other constituents will modify the diagram shown in Figure 8-25, but P-V-T-X data on such multicomponent systems, at the concentrations and temperatures of interest here, are simply unavailable. Unfortunately, the densities and compressibilities of fluids in this system at elevated temperatures are also not known, so the five-dimensional P-V-T-X diagram for the system cannot be drawn; hence, the behavior of the vapor bubble in such inclusions cannot be anticipated. As a result, Figure 8-25, although valuable for establishing composition, is useful in thermometry only insofar as the solution of daughter crystals provides some temperature limits.

NaCl-H_2O-CO_2

Phase equilibria in the system. Although the system NaCl-KCl-H_2O just described represents a close approach to many salt-water inclusions, the prevalence of CO_2 in natural fluids makes the system NaCl-H_2O-CO_2 even more commonly representative. Unfortunately, much less is known of the phase relationships in this ternary, and because two volatiles are involved, a complete plot of the system would require five dimensions. In contrast, much of the pertinent information in the system NaCl-KCl-H_2O can be given on a two-dimensional projection of a three-dimensional T-X plot (Fig. 8-25).

The first major study of the system of direct pertinence to fluid inclusions was that of Ellis and Golding (1963), who determined the solubility of CO_2 in NaCl solutions up to 2 molal (10.47 wt %) at temperatures as high as 330°C. The most obvious effect is the large decrease in CO_2 solubility with increase in NaCl (the "salting-out" effect). Ellis and Golding showed that as temperature increases, the salting-out effect falls to a minimum at 150°C, but at higher

----

[4]/ It is common knowledge that the thermal coefficient of solubility for KCl is approximately 8 times greater than that for NaCl; this difference can be used to distinguish the two isotropic daughter crystals in inclusions. These data refer to the pure binary systems only. In the ternary, the difference is more extreme. Note that because the motion of the fluid composition along A-B on heating (Fig. 8-25) is almost exactly toward the KCl corner, essentially only KCl is being added to the fluid along this path, i.e., the increase in solubility of NaCl with temperature from 25° to 150° is effectively zero in these solutions. In fact, if the boundary curve is exactly as shown, a little NaCl actually precipitates at first on heating, i.e., NaCl shows retrograde solubility up to ~110°C, according to data of Akhumov and Vasil'ev (1931), or up to ~160°C, according to data of Cornec and Krombach (1932).

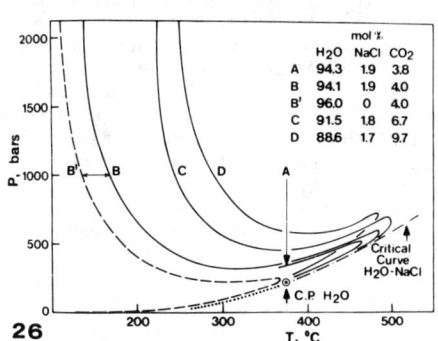

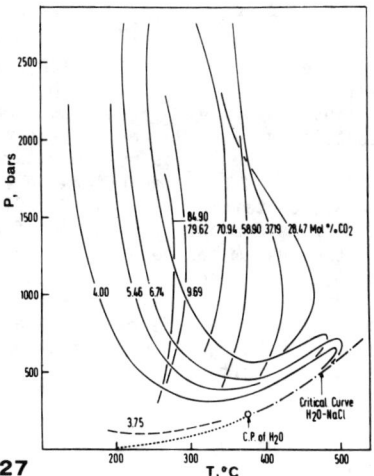

Figure 8-26. Isoplethic phase-boundary curves of the H₂O-rich side of the ternary system NaCl-H₂O-CO₂ in a P-T plane, from Gehrig et al. (1979). To the left and below each curve, two fluid phases exist; to the right and above, only one. Compare with Figure 8-21. The ratio of H₂O to NaCl was that of a 6 wt % NaCl solution for all curves but B'.

Figure 8-27. Same as Figure 8-26, but plotted for a wider range of CO₂ contents, from Gehrig et al. (1979).

temperatures, it increases again as the critical temperature of the system is raised by the dissolved salt. Takenouchi and Kennedy (1965b) measured the solubility of CO_2 in 6 and 20 wt % NaCl solutions up to 1400 bars and 450°C. Their data show that even 6 wt % NaCl reduces the solubility to about half the value for pure water at the same pressure and temperature (Fig. 9-9). Drummond (1981) determined the solubility of CO_2 in 0-6 m NaCl solutions between 25 and 400°C.

Gehrig et al. (1979) have extended these experimental data for 6 wt % NaCl to 500°C and 3000 bars. Some of their results are shown on Figure 8-26. This figure illustrates several important features that are probably common to many such systems: (1) Note that isoplethic (composition = constant) phase-boundary curves in the pressure-temperature plane show a temperature maximum and a pressure minimum. The four bubble-point/dew-point curves A, B, C, and D, for 6 wt % salt solutions, are the ternary equivalents of those in the binary system (curve B' and Fig. 8-21). (2) A comparison of curves B and B' shows that the addition of 6 wt % NaCl raises the bubble-point curve only about 30°C in the high-pressure range but increases the maximum for the two-fluid-phase field nearly 100°C. (3) Although the critical point for a simple 6 wt % NaCl solution is about 440°C (Figure 8-12), the addition of 9.7 mole % CO_2 to such a solution results in two-phase conditions extending to almost 480°C (Figure 8-26, curve D). As CO_2 contents increase, the shape and position of these curves changes considerably (Figure 8-27).

As will be seen in Chapter 9, the limits of the field(s) of two fluid phases in CO_2-bearing aqueous inclusions are of great concern in the application of inclusions to geobarometry. For such considerations, most commonly a solvus is plotted in terms of temperature vs XCO_2, which shows the limits of solubility (i.e., the boundary between one-phase and two-phase equilibria), <u>for a given salinity and pressure</u>. Konnerup-Madsen (1977) showed that a series of mixed CO_2-H_2O inclusions from a granite homogenized <u>above</u> the solvus at 1.75 kbar in

247

the pure system H_2O-CO_2, as a result of the presence of other constituents (such as NaCl). On the basis of observations of the phase into which homogenization occurred (H_2O-rich or CO_2-rich), the crest of the solvus for these compositions must be at ~28 mole % CO_2. Hendel and Hollister (1981) showed an empirical solvus for 2.6 wt % NaCl, based on the behavior of natural inclusions (probably containing some CH_4 and salts other than NaCl), and other workers have tried to establish the limits on the water-rich side. Sisson (1979) and Sisson et al. (1981) found that inclusions in calcareous metasedimentary rocks suggest that brines containing 23-24 wt % NaCl equivalent are immiscible with a CO_2-rich phase at the metamorphic conditions of approximately 600°C and 6.5 kbar. Immiscibility in the system has been calculated by Bowers and Helgeson (1983) to high temperatures and pressures, based on a modified Redlich-Kwong equation of state fit to Gehrig's (1980) data. Deviations from the pure ternary system due to other constituents, a wide range of salinities and pressures, recognition of homogeneous vs heterogeneous fluid trapping, and the complications introduced by the pressure minima in the pure system (Fig. 8-26), can result in considerable ambiguity.

Inclusion behavior. Depending upon the bulk density and composition, as well as on the presence of other constituents, the behavior of inclusions in this system can be complex (e.g., Pichavant et al., 1982). On heating inclusions whose composition falls in this system, if two CO_2 fluids are present, normally their homogenization will be the first phase change. Th CO_2 L-V can be used to obtain an estimate of the density of the homogeneous CO_2 phase; this Th, combined with optical measurements of the phase volumes, can yield an estimate of the overall CO_2 concentration in the fluid. (The solubility of CO_2 is low in water and particularly low in saline solutions.) Further heating will result in solution of the NaCl daughter crystal (if one was present) and eventually will yield complete homogenization (if the inclusion does not decrepitate first).

The behavior on cooling is commonly complicated by the several compositional variables and by metastable equilibria. It is also complicated by observational problems; Collins (1979) presented an excellent series of nine photographs of an inclusion at various low temperatures (his Fig. 4). If the CO_2 at room temperature is present as a single phase at greater than critical density, cooling will cause nucleation of a vapor bubble. If, on the other hand, the CO_2 is less than critical density, a liquid rim will condense. Normally this phase assemblage will be metastable, as CO_2 hydrate should form, generally in the range 0 to +10°C (Fig. 8-22 -- note that curve Q1-Q2 is displaced to lower temperatures by NaCl in solution; Collins, 1979). The actual stable temperature is a function of both CO_2 pressure and salinity. If cooling is fast, the freezing of liquid CO_2 and its (metastable) triple point at -56.6°C can be recognized on warming. (If not -56.6°C, either the calibration of the stage is in error or the CO_2 is not pure, or both.) After complete freezing, Te (also probably metastable) may be determined on the frozen-water phase.

Eventually, though, the CO_2 will react with the water isothermally to form CO_2 hydrate, with consequent volume reduction and bubble increase (particularly large if the water had been present as ice, because of a 14.5% volume reduction to form the hydrate even from liquid water). If CO_2 is in excess, all H_2O will be used to form hydrate; if H_2O is in excess, ice will still be present and Tm ice may be determined. Note, however, that as some water has been withdrawn from the liquid to form the hydrate, Tm ice will be low and will yield too great a value for NaCl wt% equivalent, as described by Collins (1979). Furthermore, the CO_2 is just another solute in solution, which will also act to depress the freezing point, even if no hydrate forms. In low-salinity fluids like those found in some geothermal areas, CO_2 is the major contributor (Hedenquist, 1982). On further warming, Tm CO_2 hydrate can be determined, if the crystals can be seen. In some inclusions, even though the crystals are invisible, several features signify their presence and mark their melting. An irregular "jagged" meniscus between solution and vapor usually indicates the presence of the crystals. An apparently clear area in the liquid, free of floating crystals of ice, may be considered suspicious. Melting may occur over a very short temperature range (just as in pure-water inclusions) and be indicated by a sudden smoothing of the meniscus and formation of a liquid CO_2 film. The exact interpretation of the value of Tm CO_2 hydrate is somewhat ambiguous, as it will be a complex function of the salinity, the CO_2 pressure, and the CO_2 purity.

NaCl-H_2O-CH_4

Methane is surprisingly soluble in sodium chloride brines (Price, 1979), and many of the highly saline pore waters in deep sedimentary piles, like those along the Gulf of Mexico coast, contain appreciable methane. Hanor (1980) has examined this system, and particularly the effect of methane on the P-V-T properties of inclusions containing such fluids. The solubility of methane in 150,000 ppm NaCl solution at 300°C and 1 kbar is ~20,000 ppm (Haas, 1978), but it drops off rapidly as pressure decreases. If methane is present but unknown, large errors in the pressure correction can occur (see Chapter 9). The unrecognized presence of methane in aqueous inclusions will cause a minor error (increase) in the apparent salinity by the effect of the pressure on the ice-water equilibria and will cause an additional error in the same direction as a result of methane combining with water to form methane hydrate (CH_4·$5.75H_2O$). This process removes water from the liquid solution, leaving a more saline brine having a lower Tm ice (Collins, 1979). An added problem is the difficulty in distinguishing ice melting from hydrate melting (Hollister and Burruss, 1976). Methane is common, particularly in low-temperature environments, so crushing stage tests should always be tried first (Chapter 7).

Four- and Five-Component Salt Systems

Most aqueous fluid inclusions contain significant amounts of ions other than Na, K, and Cl; hence, only multicomponent systems can hope to act as good models. Problems in handling multicomponent systems preclude exact modelling, but Crawford (1981) has shown how data on parts of the system NaCl-$CaCl_2$-$MgCl_2$-H_2O can be used to interpret certain inclusions in metamorphic rocks. Probably the single most important general feature of these multicomponent systems is that the addition of other components will lower the liquidus and eutectic temperatures below those for pure NaCl-H_2O inclusions. (The only exception comes about when solid solutions are involved.) Thus, the addition of a small amount of KCl to the system NaCl-H_2O lowers the incongruent melting of NaCl·$2H_2O$ from +0.1 to -2.3°C, and the NaCl·$2H_2O$-ice eutectic from -21.1 to -22.9°C (see Fig. 8-25). It must also change the metastable NaCl-ice eutectic from ~-28°C to some lower value, but this has not been experimentally determined.

The amount of lowering of the NaCl-H_2O liquidus is much greater when divalent chlorides are added than when KCl is added, and these salts have much lower eutectics with ice (e.g., $MgCl_2$-ice at -35°C, and $CaCl_2$-ice at -52°C). Although some exotic salts such as LiCl can yield eutectic liquids at temperatures of -75° to -78°C (Borisenko, 1977), the only geologically common constituent in aqueous inclusions that will result in eutectic melting at temperatures as low as -50°C is $CaCl_2$ (see, e.g., Brass, 1980, his Table 1). In addition, however, it is important to note that the systems $AlCl_3$-H_2O and $FeCl_3$-H_2O both have eutectics at -55°C (Linke, 1958). Neither Fe nor Al is routinely analyzed in most inclusion studies, but the several reports of daughter minerals containing them suggests that these possibilities should be considered whenever low eutectic melting is encountered.

The specific behavior of multicomponent inclusion fluids on freezing, such as the temperature and amount of eutectic melting and the nature of the various hydrate phases, cannot be fully understood without adequate experimental data on closely matching compositions. However, some of the properties of such solutions can be predicted without such direct compositional data. Thus, Clynne and Potter (1977) showed that the density and thermal coefficient of expansion of the multicomponent fluids as commonly found in inclusions could be estimated, within ±1%, from that of a simple NaCl solution having the same depression of the freezing point (i.e., Tm ice) as that of the unknown fluids.

Another feature of many systems involving NaCl and a divalent cation such as Ca or Mg (e.g., Crawford, 1981, her Fig. 4.8; Brass, 1980) is the very strong "salting out" effect; addition of the divalent cation greatly decreases the solubility of NaCl. Thus, a solution containing ~33 wt % ($CaCl_2$ + $MgCl_2$) will dissolve only ~2 wt % NaCl to be saturated at the eutectic. Harris et al. (1979) showed that in the system $NaCl-CaCl_2-H_2O$, the lowest eutectic (between hydrohalite ($NaCl \cdot 2H_2O$), antarcticite ($CaCl_2 \cdot 6H_2O$), and ice) contains only ~1.7 wt % NaCl, and the solubility of NaCl at higher temperatures and higher calcium content is even less. Thus, data from Linke (1958) shows a solubility (at saturation) of only 1.02 wt % NaCl at 25°C in the system $CaCl_2-NaCl-H_2O$. Clynne and Potter (1977), Clynne et al. (1981) and Potter et al. (1977) determined the solubility of NaCl in solutions of KCl, $MgCl_2$, $CaCl_2$, and mixed $CaCl_2-KCl$ in the range 10-100°C, and found that it varied widely. Although most of their determinations were for <15 wt % of salts other than NaCl, they found that 18.6 wt % $MgCl_2$ reduced the solubility of NaCl to 7.2 wt % at 19.3°C. For this reason, it is never safe to assume that the presence of a daughter crystal of NaCl in an inclusion at room temperature requires that the fluid in contact with it contains the same amount of NaCl as a saturated fluid in the pure system $NaCl-H_2O$ (~26 wt %). Although this erroneous assumption is commonly made, all that can be safely assumed is that the fluid is _saturated_ with respect to NaCl.

Burruss (1981c) pointed out that one should be able to apply the phase rule to inclusion systems to determine the number of components from the number of phases present under univariant conditions, but that recognition of phases present in small amounts might cause practical problems. As an example, Weeks and Ackley (1982) reported that seawater starts to freeze (Tm ice) at -2.0°C (i.e., ~3.3 wt % NaCl equiv.), but at -70°C still has brine present, plus at least 5 solid salts, and ice. Large inclusions of seawater might thus yield Te values below -70°C. In addition to this problem, however, three others add to the ambiguity in such interpretations: solid solution phases, the all-too-common metastable phases, and the almost universal presence of numerous minor components in natural inclusion fluids. Which of these various problems may have significant effects on the _visible_ phase assemblage is difficult to assess.

Chapter 9

INTERPRETATION and UTILIZATION of INCLUSION MEASUREMENTS: TEMPERATURE, PRESSURE and DENSITY at TRAPPING

CONTENTS

INTRODUCTION

Most published fluid inclusion studies deal with determinations of the temperature of trapping (i.e., Tt; also called "temperature of formation"). These determinations are normally made using the <u>homogenization</u> method, by watching the inclusion phases homogenize while heating under the microscope. Other papers involve the <u>decrepitation</u> method, in which an attempt is made to relate the minute explosions recorded during heating to the homogenization of the inclusions. Still other studies try to apply these and other inclusion data to the determination of the pressure of trapping (Pt), and a relatively few measurements have been made of the density of the fluids at the time of trapping.

All these studies involve a series of somewhat interconnected procedures and concepts. In this chapter, I detail the steps that must be taken in obtaining the original raw data, in arriving at conclusions that may be drawn from these data as to the temperature, pressure, and density of the fluids at the time of trapping, and in determining the validity of these conclusions.

TEMPERATURES OF HOMOGENIZATION (Th) AND TRAPPING (Tt)

Principles and Assumptions of Inclusion Thermometry

Sorby (1858) showed that the gas bubbles in the fluid of most inclusions were the result of differential shrinkage from Tt to the temperature of observation, and that Tt can be estimated by heating the sample until the bubble disappears. <u>Any</u> reversible phase change that occurs on cooling an originally homogeneous fluid phase -- formation of a gas phase, crystallization on the walls or as daughter minerals, splitting into immiscible fluids, etc., similarly provides a potential thermometer. The following discussion will center on the homogenization method as commonly used on inclusions of liquid plus vapor, but the remarks are generally applicable to any phase change.

The limitations on the use of shrinkage bubbles were detailed first by Sorby. Since his time, many restatements of the assumptions and limitations of the homogenization method have been made. The method and the data obtained have been hotly debated at times, and in the heat of this "debate," ignorance and partisanship have resulted in an unfortunately large number of erroneous statements. The limitations and assumptions are relatively simple; the problems arise in attempts to estimate the validity of given samples and the magnitude of the possible errors. The major assumptions are as follows:

1. <u>The fluid trapped when the inclusion was sealed was a single homogeneous phase.</u> As shown in Chapter 2, a single inclusion is inadequate to fulfill this requirement, but when many inclusions in a sample all show apparently the same phase ratio, we can safely presume that a homogeneous fluid was trapped. The distinction between trapping of a homogeneous vs a heterogeneous fluid <u>must</u> be made <u>before</u> any thermometric data are interpreted. If the system was heterogeneous, ordinary thermometry is difficult but barometry becomes possible (see section entitled "Heterogeneous Systems -- Boiling vs Effervescence").

2. <u>The cavity in which the fluid is trapped does not change in volume after sealing.</u> Change can and does occur. The several mechanisms include: crystallization on the walls or in the fluid itself, thermal contraction of the host mineral (and precipitated minerals) on cooling, and dilational change from internal or external pressure. These mechanisms are detailed in Chapter 3. Ermakov (1950) and others placed much emphasis on reduction of cavity volume by crystallization on the walls or by the formation of daughter minerals and presumed that this reduction can cause rather gross inaccuracies in Th. Two factors tend to minimize error from this cause. First, precipitated daughter minerals (and precipitated matter on the walls) will have no effect on the apparent volume if they redissolve on reheating. Second, the volume change upon crystallization or dissolution of a solute is not simply equal to the volume of the solid solute. For example, the behavior of NaCl in supercritical water is remarkably anomalous in that it has a very large and negative molal volume, on the order of magnitude of 2 liters per mole at low concentrations, but the molal volume approaches zero at high concentrations (Benson et al., 1953; Copeland et al., 1953). Thus, the volume change when NaCl crystallizes is highly variable. Ermakov also laid great stress on the thermal expansion of the mineral host. This expansion would have a very small effect on the size of the bubble at room temperature (minerals with larger thermal coefficients have somewhat smaller bubbles), but as the contraction on cooling is exactly reversed on heating, there should be no effect on Th. The last mechanism, dilational change, theoretically has an effect on Th but can generally be neglected in practice. It is discussed in more detail under "Changes in Volume" in Chapter 3.

If the laboratory technique is not correct, or the sample has been overheated by later thermal events in nature, the walls of low-temperature inclusions in some soft minerals may be permanently stretched, yielding <u>reproducible</u> but erroneously high Th values (see "Experimental Problems" below).

3. <u>Nothing is added or lost from the inclusion after sealing.</u> This assumption has also been discussed in Chapter 3. I believe that leakage occurs but that it is relatively rare except in those rocks that have been crushed or otherwise deformed, or in which very high pressure gradients have been set up (as in inclusions in the vicinity of a near-surface dike).

A special case, equivalent to leakage, is the necking down of large inclusions into several smaller ones (see Chapter 3). When this necking down and resealing has occurred after the formation of a gas bubble (or of any other new phase), the two resulting inclusions will have different Th values (Fig. 3-9). The inclusion that traps the bubble will have Th greater than the original Th, and perhaps have Th>Tt -- just as though it had trapped a primary gas bubble (which, in effect, it has); the other one will have Th<Th of the original inclusion (actually, it will show the temperature at which the necking down sealed it). Necking down (and other changes in inclusion shape) is very common in nature and accounts for a good part of the scatter in results of practically all careful thermometric studies. Necking down should be much faster at elevated temperatures. Thus, the amount of necking down during the first 50°C of cooling after trapping could be far greater than that during the next 50°C, even though the time involved for the second interval might be much greater. If such a process occurs, it might not result in large enough variations in the gas/liquid

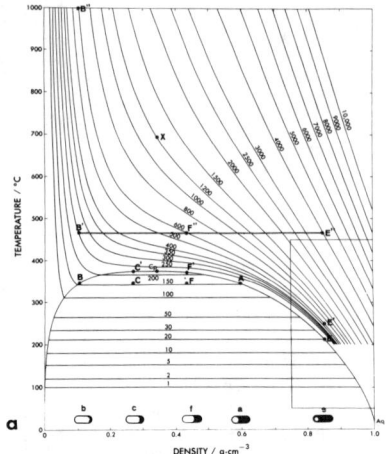
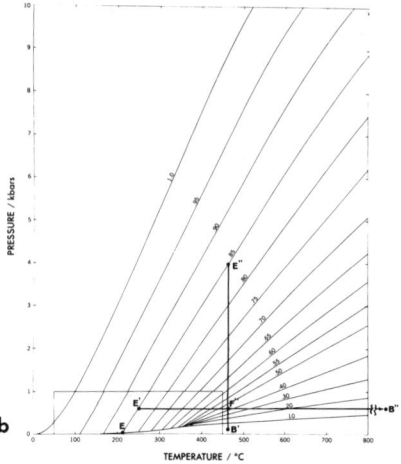

Figure 9-1. Temperature-density plot for pure water, (9-1a), and equivalent P-T plot (9-1b), illustrating thermometry by use of inclusions trapped from fluids boiling at 340°C (inclusions a,b,c,f), and the nature of the pressure correction (inclusions b and e). Inclusion (e) could have been trapped from boiling solutions at E, at 20 bars, or at higher pressures (e.g., E' at 300 bars or E" at 4000 bars and 470°C). Inclusion (b) similarly could have been trapped from boiling solution at B, or any higher pressure (e.g., B" at 1000°C and 600 bars, and hence have a pressure correction of 660°C). Cp = critical point; Aq = room-temperature water. The line Aq-Cp on Figure 9-1a is sometimes called the "boiling curve." See text for details.

ratios to cause these particular inclusions to be disregarded in advance.[1] The many groups of inclusions that give very consistent temperatures would merely be inclusions that, by virtue of their original shape, did not neck down (at least not after formation of the vapor phase). The balance of the scatter found in the Th values of apparently contemporary inclusions is believed to be a result of occasional leakage, experimental errors, lack of true contemporaneity, etc.

4. The effects of pressure are insignificant, or are known. Much confusion exists about the effects of pressure. In evaluating thermometry, pressure is of concern mainly in that it controls the density of the fluids and hence the magnitude of the pressure corrections that must be added to Th to get Tt. If a fluid inclusion is trapped from a homogeneous fluid along the boiling curve[2] (i.e., under P-T conditions such that it was in equilibrium with either a vapor or gas phase, or almost so), Th = Tt, and no pressure correction is needed. Such an inclusion would start to form a gas bubble as soon as it cooled below Tt. It would make no difference whether the gas bubble that formed had the same composition as the liquid (e.g., a steam bubble in pure water) or had a completely different composition (e.g., a CO_2 bubble in an aqueous inclusion). Even more important, one would not need to know anything about the composition of the trapped fluid to obtain a correct value for Tt, because now Tt = Th, regardless of composition.

Most inclusions have trapped fluids at a P-T combination above the liquid/vapor curve (see inclusion E' on Fig. 9-1). In these, a bubble does not form on cooling until the pressure and temperature have dropped to the liquid/vapor curve. As the density stays essentially constant under these conditions, the change in P and T inside the inclusion must be represented by a vertical line

----
[1] See footnote 6, p. 65.

[2] That part of the boundary of the two-phase field shown on Figure 9-1a between Aq and Cp is frequently termed the "boiling curve." To avoid excluding inclusions homogenizing in the vapor phase, the terms "two-phase curve" or "liquid-vapor curve" are preferable.

on Figure 9-1a and by a sloping isochore on Figure 9-1b. This difference in temperature (Tt-Th) is called the "pressure correction"; it must be added to Th to obtain the true formation or trapping temperature, Tt. The correction is actually a temperature correction as a result of pressure; however, the term "pressure correction" is in such general use today that it is accepted here even though it is semantically wrong. Some reports refer to "corrected Th values" without further clarification. This terminology introduces ambiguity, as it might refer to pressure, instrumental calibration, or other corrections. In view of the possible complexities, the specific pressure assumed in making the pressure correction must always be specified.

The actual value of the pressure correction (in degrees Celsius) is obtained by dividing the difference between the pressure of trapping (i.e., of formation) (Pt) and the internal pressure at homogenization (Ph), by the slope of the isochore originating at Th, Ph. Thus, pressure correction = $(Pt - Ph)/(\Delta P/\Delta T)$.

If Pt can be calculated from geologic field data on the depth of cover, or from some other independent geobarometer, the measured Th can be corrected for pressure to yield Tt. Too frequently, this pressure correction is made under the assumption that the fluid in the inclusion is either pure water, or has the same vapor pressure, thermal expansion, and compressibility as pure water. This assumption is generally not valid, as most natural fluids in inclusions are NaCl solutions. For these fluids, the liquid/vapor curve is raised above that of water roughly in proportion to the concentration of salts. As the system NaCl-H_2O is reasonably close to natural systems, and as this is the only one for which extensive P-V-T-X data are available, it is usually used. The specific procedures for making the pressure correction are given below. In addition to this normal pressure correction, a small correction should be made for the sum of two dilational changes (as detailed in Chapter 3, under "Reversible changes in volume"); it may amount to ~4 to 6°C/kbar.

5. The origin of the inclusion is known. Inclusions can only give information on the environment under which they were formed. If they are primary, they indicate the conditions under which the enclosing mineral formed; if secondary, they indicate some later conditions. Conclusive evidence of secondary origin is easy to find for many inclusions; as shown in Chapter 2, equally conclusive evidence for primary origin is relatively rare. In many studies, however, data on known secondary inclusions have proven useful.

6. The determinations of Th are not only precise, but accurate. This assumption is usually given rather short shrift in the literature, frequently only a brief statement of precision. As detailed in Chapter 7, the uncertainty in accuracy, if it can be estimated at all from the details given, is much larger than that in duplicate measurements (i.e., the precision); therefore, high precision does not necessarily indicate high accuracy.

Although the limitations and assumptions detailed above may seem prohibitive, many samples, from many geologic environments, are apparently adequate to fulfill the requirements, and I believe that of all the geologic thermometers now known, fluid inclusions provide perhaps the most accurate, and certainly the most generally applicable method. Of the six items discussed above, number 4, pressure corrections, is probably the greatest single source of numerical error; number 5, origin of the inclusions, is probably the greatest single source of ambiguity in interpretation; and number 6, determination of Th, can also be a serious source of error under some circumstances.

Heterogeneous Systems -- Boiling vs Effervescence

Unless the heterogeneity was very uniform and of small dimensions relative to the inclusion dimensions (e.g., a colloidal dispersion), individual inclusions formed from heterogeneous systems can be expected to trap different ratios of the phases present, and all further phase changes on cooling will be superimposed on these original differences. As discussed in Chapter 2, such evidence of original two-phase systems gives us valuable information on the environment, but if an inclusion has trapped both phases, it will give erroneous temperature, pressure, and density data. If the two phases of an originally heterogeneous fluid system are trapped in separate inclusions (Fig. 9-1a), both temperature and pressure may be obtained from them, but only under special circumstances (see section entitled PRESSURE OF TRAPPING).

Two situations occur in nature in which two fluid phases are present and may be trapped as inclusions. In both situations, one fluid has high density and is liquid-like, and the other has low density and is gas-like. The two situations are true boiling, in which the low-density fluid is the vapor of the liquid (e.g., steam bubbles in water), or true effervescence, in which the low-density fluid is compositionally different (e.g., CO_2 bubbles in water). In both situations, the term "primary gas" is used to refer to the low-density phase. Unfortunately, in much inclusion literature, the term "boiling" is applied loosely to any example in which evidence of trapping of two fluids of different density is found, under the assumption that true boiling took place. This false assumption has caused many erroneous estimates of pressure. The trapping of primary gas, either from boiling or effervescence, can also cause particularly large positive errors in Th (Th > Tt) in the more common homogenization in the liquid phase, illustrated by inclusion (F) in Figure 9-1a. Inclusion (a), which trapped only liquid, homogenizes in the liquid phase at (A) and yields the correct Th (340°C, and Th = Tt). Inclusion (b), which trapped only vapor, also yields the correct Th, but by homogenization in the vapor phase at (B). Inclusion (c), which trapped a steam bubble plus some liquid, homogenizes in the vapor phase at (C') (too high, Th > Tt). Similarly, inclusion (f), which trapped liquid plus a small vapor bubble, homogenizes in the liquid phase at (F') (also too high). In my experience, trapping of primary gas, although common in geothermal samples, certain types of ore deposits, and some other environments, is relatively rare. Where it does occur, large numbers of inclusions may show evidence of it.

In studies of those ore deposits in which inclusions with widely variable gas/liquid ratios are found, it is important to verify whether or not "boiling" has actually occurred, since both true boiling and particularly effervescence can cause ore deposition. In addition to "boiling," variable gas/liquid ratios can be caused by trapping at different times from fluids under different P-T conditions, by leakage of part of the inclusions, or by necking down (discussed in Chapter 3), but such processes can usually be recognized by microscopy. In the simplest case, "boiling" will result in two types of inclusions, representing the trapping of either the liquid phase or the vapor phase. Sometimes the latter are called "steam" inclusions, solely on the (inadequate) basis of low density. These inclusions contain a very little liquid and a very large bubble at room temperature. If they trapped only the vapor phase, on heating they will homogenize in the vapor phase by evaporation of the liquid (i.e., Th L + V(V)), at the same temperature as the liquid inclusions homogenize by expansion of the liquid to eliminate the vapor bubble (i.e., Th L + V(L)). (The determination of Th L + V(V) will be accurate only when the steam inclusion has a narrow reentrant into which the last bit of fluid phase has been concentrated by capillarity, e.g., Fig. 9-2.) It is apparent from studies of fluid inclusions that small amounts of liquid are more commonly trapped with the steam than are small amounts of vapor (i.e., a small bubble) trapped with the liquid. It is possible, although hazardous, to use the minimum temperatures determined on a large number of presumably coeval inclusions (e.g., a,b,c, and f on Fig. 9-1a), as a maximum value for the true Tt, as such inclusions are likely to have trapped pure liquid or pure vapor.

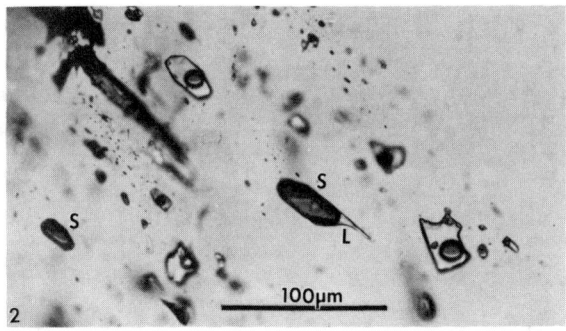

Figure 9-2. Two pseudosecondary inclusions of "steam" (S) and 4 of liquid, trapped in quartz from a boiling geothermal system. If the steam inclusion in the center did not have the narrow reentrant in which the liquid phase (L) is concentrated by capillarity, the liquid would form a thin film on the walls and be invisible. Sample Ka-19 from Kawerau, New Zealand, courtesy of P.R.L. Browne.

Experimental Problems

In addition to the normal experimental problems of sample selection, temperature control and measurement, calibration of stages, etc., discussed in earlier chapters, several problems may make the experimentally best of measurements grossly inaccurate. One such problem, caused by surface tension, can yield Th values that are far too low (see Chapter 10); it becomes serious only with very small inclusions (i.e., 2-3 μm). A series of special problems in melt-inclusion studies are covered in appropriate later chapters. Two other problems, much more general in occurrence, are <u>stretching</u> and <u>leakage</u>.

The pressure within a fluid inclusion can vary greatly, depending upon the composition and density of the contents and the temperature. As an example, a pure-water inclusion containing liquid and a vapor bubble has an internal pressure of only 24 mm (~0.03 bars) at 25°C -- in other words, the bubble is a moderately good vacuum. Figure 8-5 shows that if the inclusion homogenizes at 215°C, the vapor pressure at that temperature is ~20 bars. The pressure in the inclusion on heating above Th will depend upon the bulk density and hence on the way the inclusion homogenizes. If it homogenizes at 215°C in the vapor phase (by evaporation of the liquid), the density is 0.01 g/cm^3, and the pressure in the inclusion will increase very slightly with temperature along the isochore in Figure 8-5 marked ".01." If, however, the inclusion homogenizes at 215°C in the liquid phase (by expansion of the liquid, eliminating the vapor bubble), the density is 0.85 g/cm^3, and the pressure will increase rapidly, at about 14 bars/°C, along the ".85" isochore. Hence even a small amount of "overheating" -- heating beyond Th -- can yield high internal pressures. Substances other than pure water have different rates of pressure increase (isochore slopes), and some isochores are strongly curved, but most behave as shown in Figure 8-5. The initial pressure of liquid CO_2 inclusions is higher, e.g., 70 bars at 31°C, but the slopes of the isochores are much lower than those for pure water. The slopes of the various isochores are discussed in more detail later in this chapter, under "PRESSURE OF TRAPPING."

As the internal pressure builds up during continued heating, four different events may take place. The inclusion may stretch, it may undergo partial decrepitation, it may leak, or it may decrepitate completely. The first two events are discussed in this section; the second two are covered in the next.

<u>Stretching and partial decrepitation.</u> Stretching refers to a permanent deformation of the host crystal around an inclusion, generally <u>without</u> visible

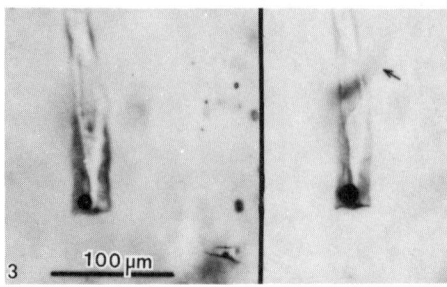

Figure 9-3. Inclusion A/4 from fluorite sample ER 65-115, from Larson et al. (1973), photographed in transmitted light at room temperature before (left) and after (right) overheating. A cleavage crack (arrow) is visible as a result of the expansion on overheating, but such cracks frequently are not visible. Note also the larger bubble after heating. Although the pressure in this inclusion at the moment of cracking may have been as high as 1300 bars, it has not cracked to the surface, as it (and many other similar inclusions) continued to give consistent almost reproducible Th determinations, 40-70°C too high. A small inclusion in lower right (not run) also shows a faint crack after the heating.

cracking, that may occur in soft minerals such as fluorite if the internal pressure exceeds a certain finite limit, e.g., by overheating beyond Th (in the laboratory or in nature), or by expansion of ice on freezing an inclusion that had a bubble so small that it was eliminated on freezing (Lawler and Crawford, 1983; see also Chapter 10). This expansion or stretching may be evidenced only by a rise in Th in subsequent runs[3]/, and once the specific overheating limit is exceeded (this limit will vary with inclusion size as well), the increase in Th is a surprisingly regular function of the amount of overheating and hence internal pressure (Bodnar and Bethke, 1980, 1984). As a result of such stretching, either in nature or the laboratory, such inclusions can yield reproducible but erroneously high values for Th that may well go undetected.

Partial decrepitation is a closely related phenomenon. A very few of the overheated inclusions may show a small fracture out into the surrounding crystal but not to the surface. The volume increase represented by this crack partly relieves the pressure. The crack may subsequently heal, trapping numerous inclusions of this expanded fluid in the form of a halo of tiny secondaries around the larger central original inclusion, which now has a lower density. Recognition of such partial decrepitation, first described by Lemmlein (1956a), is relatively simple and seldom causes errors (Fig. 3-22). Touret (1977) described "exploded" inclusions from granulites, and Swanenberg (1980) found numerous examples, which he called "decrepitation clusters," in the high-grade metamorphic rocks (granulites) of southwest Norway.

Other than the isolation of some of the inclusion fluid in new secondary inclusions, I believe that no real difference exists between stretched and partially decrepitated inclusions. However, the major mechanisms involved in the deformation may be different and may range from dislocation creep to open fractures. Without an obvious halo of small secondaries, or a visible fracture, the stretched inclusions present a much more serious hazard to thermometry. Stretching obviously is related to the amount of overheating and to inclusion size, but many other factors may well be involved, and hence its effects may be variable and erratic.

The concept of stretching was introduced by Larson et al. (1973) when it was discovered as a result of an interlaboratory standardization. In their work, they found that inclusions in fluorite from east Tennessee had Th raised by as much as 71°C (from 118 to 189°C) as a result of unspecified amounts of overheating during early reconnaissance Th runs. Only later were a few such inclusions found to have tiny cracks (Fig. 9-3). Cracks in general are invisible if their

----
[3]/ Theoretically, the stretching should be evident in an increase in the diameter of the vapor bubble at room temperature after the run. Although such bubble measurements should always be made before and after a run, particularly to detect leakage, note that the stretching that results in a 10°C rise in Th of a water inclusion from 215 to 225°C will cause an increase in the diameter of the bubble at room temperature of only 0.5%, an amount far too small to detect by most bubble-measurement procedures.

width is a small fraction of the wavelength of the light used, and this lower limit is largest if the difference in index of refraction of host and the fluid filling the crack is small (e.g., fluorite with n = 1.43 and aqueous brines of n = 1.36). The few cracks seen were far too small to form the halo of secondary inclusions characteristic of partial decrepitation, and many inclusions showed no visible cracks. However, the walls of the inclusions had expanded enough to increase the volume of the (normally small) bubble measurably (Fig. 9-3).

Stretching represents a permanent deformation of the walls, permitting expansion of the fluid contents, until the pressure is reduced to the point that elastic deformation of the surrounding undisturbed host prevents further propagation. Probably several mechanisms are involved; small cracks are obvious in some examples (Fig. 9-3), but plastic deformation may be involved in others. Presumably some of this expansion is reversed on cooling, which closes the cracks appreciably but not completely (as indicated by the increase in bubble size at room temperature). Repeat determinations of the new higher Th values on the fluorite from east Tennessee yielded results that were either identical or increased one or a few degrees. The smallest inclusions showed the smallest amount of stretching.

Inclusions in sphalerite from Laisvall, Sweden, were also believed to have stretched. The high index of refraction of sphalerite causes heavy black borders on the inclusions. Roedder (1968d, p. 394) noted during the original Th runs that "...Some inclusions in sphalerite ... in which the very small bubble had seemingly disappeared (into the dark borders) at about 150°C suddenly showed a very tiny bubble in rapid Brownian [sic[4/]] movement at 180-190°C; these homogenized at temperatures up to 223°C, and the process [i.e., homogenization] could be repeated." This behavior was misinterpreted at the time. I now believe that the reappearance of the bubble at ~180°C was due to stretching and that the true Th for these inclusions should have been about 150°C (Larson et al., 1973).

One of the most extreme examples of stretching of inclusions is seen in halite (Roedder and Belkin, 1979a, 1980a). Such inclusions decrepitate if they are heated rapidly to temperatures above Th. However, salt is so plastic, particularly in the presence of water, that decrepitation will not occur if the heating rate is slow; instead, once Th is reached and the internal pressure starts to increase rapidly (~14 bars/°C), the host salt simply expands by permanent plastic deformation. Thus, inclusions having tiny bubbles at room temperature and an original Th = 40°C, after heating to 250°C, will have large bubbles at room temperature and will have a new Th near 250°C (Fig. 3-23). This expansion can happen even at low temperatures: a large inclusion containing a tiny bubble, homogenizing at 20°C, was overheated to 40°C and cooled; on redetermination, Th had risen to 39°C. Obviously, salt samples must not be heated above Th before Th is determined. However, I do not believe that measurements of Th on inclusions in salt have much meaning anyway. If inclusions in salt can expand that easily under a few hours of internal pressure in the laboratory, I must assume that they have also expanded (or contracted) in response to natural differences in internal and external pressure, due to changes in P and T. Therefore, the inclusion volume we now see represents some rather unknown and arbitrary "quench" conditions after a long period of sequential "anneals."

Stretching can occur either in nature or in the laboratory. Sabouraud et al. (1980) suggested that some fluorite samples from Pb-Zn ore deposits in France have been stretched by natural overheating, and Bodnar and Bethke (1980, 1984) believe that natural stretching is a potentially serious problem. The latter investigators made a systematic laboratory study designed to determine how much overheating is needed before stretching starts, the amount of stretching (in

----

4/ Since shown to be pseudo-Brownian (see Chapter 7).

°C change in Th) for any given overheating, and the criteria that might be used to recognize stretched inclusions. This study showed the value of the USGS gas-flow stage (see Chapter 7). The combination of very high precision and very rapid thermal response of this stage permitted the necessarily large number (1300) of precision measurements for statistical validation of the complex relationships studied. This task would have been enormously time consuming if the older conduction-heated stages had been used.

Bodnar and Bethke (1980, 1984) found that fluid inclusions in fluorite and sphalerite stretched systematically and somewhat predictably. The amount of overheating needed to initiate stretching depends upon the properties of the inclusion fluid, the inclusion size and shape, and the physical properties of the host mineral. Essentially no stretching occurred in fluorite on overheating <30°C. They reported that sphalerite was more resistant to stretching than fluorite, although the Laisvall sphalerite reported above apparently stretched on overheating ~30-40°C. They made reconnaissance studies on barite which indicated that stretching in barite took place on little or no overheating. They also suggested that the systematic relationship between the internal pressure necessary to initiate stretching and inclusion volume provides a means of recognizing previously stretched inclusions and estimating the magnitude of postentrapment thermal events.5/

Thus the frequently used criterion for establishing the validity of Th values, consistency of results on repeated runs, is a necessary but not sufficient criterion. Observation of the temperature of the first disappearance of the bubble on gradual heating is also necessary but not sufficient. This temperature should be reproducible. Any appreciable overheating of the sample beyond this temperature may cause stretching and give new higher Th values that are also reproducible but erroneous. The amount of overheating that will cause such permanent effects will vary with the sample but will be smallest for large inclusions of fluids having steep isochores (i.e., those with the lowest Th) in soft cleavable minerals.

If the mineral is very soft, or the pressures are high, or both, even the first disappearance of the bubble on gradual heating is not necessarily a valid Th. Therefore, I have considerable doubts concerning the validity of most measurements of Th in halite, unless they are very low (see also Chapter 11).

In addition to the above, the constancy of the volume of the bubble before and after the first run should be verified, either photographically or by measurements of its diameter. Once again, lack of change is a necessary but not sufficient criterion, as the precision of such measurements is poor at best, and they would show no change if the crack closed completely on cooling. The best way to avoid error from stretching is to avoid accidental overheating. This can be accomplished by running the lowest temperature inclusions first in any given sample plate (i.e., those with the smallest volume-percent vapor bubble) and then proceeding to higher temperature inclusions.

Leakage and complete decrepitation. If a crack forms in the walls of an inclusion and eventually connects with the sample surface (or with other invisible natural fractures or those induced by careless sample preparation), inclusion fluid can leak out, ruining the inclusion for thermometry. The probability of this type of leakage, like that of stretching, will naturally vary with the pressures involved (and hence with the density and nature of the fluid filling the inclusion and the temperature) as well as with the size of the inclusion and the nature of the mineral. Leakage can be a major source of frustration, because the largest and hence most easily and accurately studied inclusions are

----

5/ Unfortunately, in the small size ranges, surface tension results in similar effects (see Chapter 10).

most likely to leak. In some samples, the pressure at Th is greater than the strength of the inclusion walls, and all inclusions leak before Th is reached. Several workers have used externally applied pressure to prevent such leakage. Lemmlein et al. (1962) heated samples of topaz containing inclusions with numerous solid phases to 700 or 740°C for hours under external pressure to prevent decrepitation and then examined the quenched(?) samples (but see also Voznyak, 1968, for a rebuttal). London et al. (1982) have used the same proce- dure on inclusions in spodumene from the Tanco pegmatite in Manitoba, Canada.

Leakage of an inclusion during a Th run usually shows up as an increase (or no change) in the size of the vapor bubble at constant (or rising) tempera- ture. As most gas inclusions are rather dark, complete leakage usually results in the inclusion turning dark. Not all samples leak all their fluid. Sometimes the crack is so small that only part is lost. When such inclusions are cooled back to room temperature the vapor bubble will be larger than before. Measure- ment of the bubble diameter before and after each run is a necessary but insuf- ficient proof of no leakage (insufficient because it is insensitive to small but significant leakage). Flattened masses of gas and liquid commonly stream out of cracks when inclusions are leaking. These are strikingly visible when the crack angle to the·line of sight results in total reflection (black) for the steam bubbles and light transmission (clear) for liquid masses. Often this motion is seen at some distance from the inclusion, but with no visible connection.

Decrepitation, by definition, involves a cracking or breaking up of the sample, and many decrepitations are indeed explosive, throwing the fragments around and emitting sharp noises. (Detection of these explosions is the basis for the decrepitation method of geothermometry, discussed in a later section in this chapter.) The violent decrepitation of many minerals when heated[6]/ is due, in large part, to the explosion of the fluid inclusions, and the internal pressures are occasionally adequate to cause explosion without heating. A quartz crystal decrepitated spontaneously and broke the glass in a museum display case (pers. comm., R. Parker, Zürich, 1965); similar spontaneous decrepitation on a much larger scale has occurred in salt mines (see Chapter 11). Some samples present evidence of decrepitation by heating in nature (Deicha, 1961). Such information may provide useful information on the age relations of dikes and ore mineralization (Lokerman, 1962, 1965; Ermakov and Kholmskii, 1965). Similar reasoning holds for the emptying of inclusions as a result of cataclasis (Ypma, 1963). In the heating stage, decrepitation of the all-too-common planes of secondary inclusions, having low Th, may disrupt the plate on the stage and ruin adjacent primary inclusions before their (normally higher) Th is reached. How- ever, many decrepitated inclusions lose their fluid contents quickly but not explosively, through tiny and frequently invisible fractures, leaving the sample intact but perhaps more friable. The distinction between such leakage and decrepitation is purely semantic and of no real consequence.

Most fluid-inclusion studies of ore deposits show Th declining with para- genetic stage, but reversals, indicated by a temporary increase, are not rare. Too frequently, such reversals are described without consideration of the prob- able effect on the earlier, lower temperature inclusions. Either the evidence for the sequence of inclusion formation, or the experimental data, may be wrong in those examples where the early inclusions should have decrepitated during the reported temperature rise, but apparently did not.

----

[6]/ Decrepitation can also occur on cooling. Killingly and Muenow (1975b) noted considerable decrepitation of CO_2 inclusions in olivine occurred during cooling from high temperatures, presumably because the rapid temperature change resulted in the addition of differential thermal stress to grains already strained by high internal (inclusion) pressure.

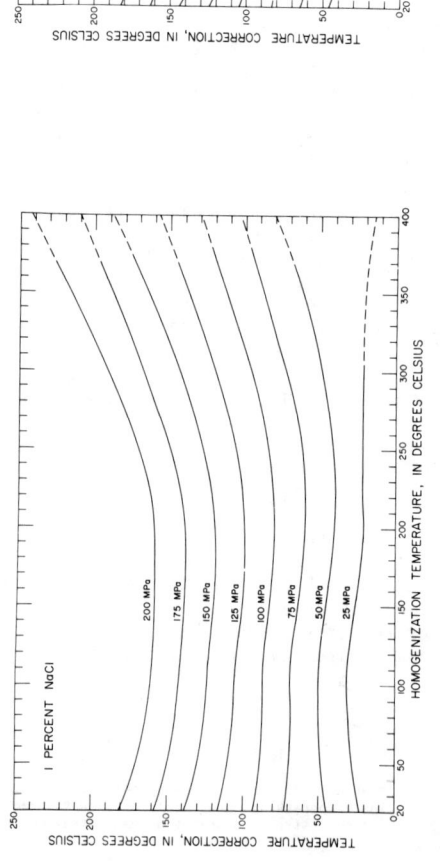

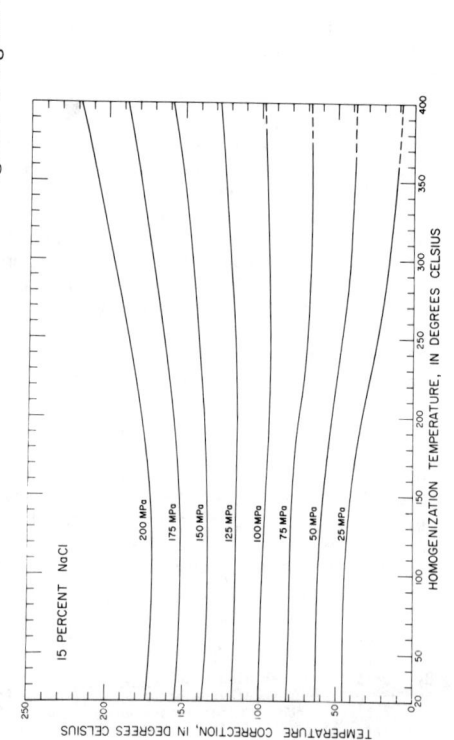

Figure 9-4 legend on opposite page.

Pressure Corrections for Aqueous Inclusions

Sorby (1858) attempted to determine the thermal expansion of brines up to 200°C by making synthetic inclusions in heavy-walled glass capillaries, but the first thorough study of the properties of NaCl brines that would permit determination of the pressure correction was by Lemmlein and Klevtsov (1961). They presented a series of graphs of the pressure correction (up to 30 wt % NaCl at 150-500°C and pressures up to 1750 atm) that have been used in practically all fluid-inclusion studies published from 1961 to 1977.

Potter (1977) found that the data of Lemmlein and Klevtsov (1961) were inconsistent with later much more precise data in several respects and that some were in error by as much as 33°C, particularly near the critical point curve. (Only one of these inconsistencies is related to an apparently previously unreported error in the high-pressure portion of the tabular data of Lemmlein and Klevtsov, an error that was pointed out by Roedder and Bodnar (1980, p. 277); graphical intercomparisons reveal that all data entries for pressures ≥1500 atm and degrees of fill <85% should have a degree of fill 2.5% larger than stated in the original Russian text and the English translation). Potter (1977) presented a new set of pressure-correction graphs covering the range 20 to 400°C, for 1, 5, 10, 15, 20, and 25 wt % NaCl solutions to 2000 bars (i.e., 200 megapascals). Four of these graphs are given here in Figure 9-4. The volumetric data for aqueous sodium chloride solutions were compiled by Potter et al. (1977) and then evaluated by a weighted least-squares regression procedure (Potter and Brown, 1975) to generate the graphs. Not only do these new graphs correct the discrepancies in the previous ones and extend coverage to the important range 20-150°C, but in the original publication (Potter, 1977) they are printed in a large format, permitting more precise reading and interpolation.

These corrections are precisely appropriate only if: (1) the inclusion contains a pure NaCl solution; (2) the salinity of this fluid has been correctly determined; (3) the estimate of the pressure of formation is correct; and (4) the inclusion homogenizes in the liquid phase. The third of these requirements is discussed later in this chapter, but a few comments on the others are in order. Although Na and Cl are the major ions in many if not most inclusions, other ions are normally present. However, Potter and Clynne (1978b) have shown, by experimental studies of brines in the system Na-K-Ca-Mg-Cl-Br-SO_4-H_2O, that the P-V-T-X properties of these brines are predicted to within ±1.0% by the properties of an NaCl solution having the same value for Tm ice. Although this is not true for any composition, of course, it is true for compositions most frequently found in inclusions. Potter and Clynne (1978b) showed that the P-V-T-X properties of the brines need to be corrected slightly if the atomic ratios of Ca/Na, K/Na, and Mg/Na are greater than 0.5, 0.3, and 0.2, respectively, but the magnitudes and directions of the corrections were not given. The observed ratios in most inclusions are within these limits. Therefore, if the salinity is determined in terms of "weight percent NaCl equivalent" (i.e., the weight percent NaCl that would be needed to cause the same value for Tm ice), as has long been the practice simply because Na and Cl are the most common constituents, Potter's curves will apply.

Several special situations are not covered by the above procedure. If the inclusion shows evidence of exotic or very highly concentrated solutions, the pressure correction can only be assumed to be similar to that for 25 wt % NaCl (note, for example, that the differences in pressure correction between 15 and 25 wt % NaCl are not large, even at 400°C). If appreciable amounts of CO_2 or other gases are present, the isochores and hence the pressure corrections will

Figure 9-4. "Pressure correction" for inclusions containing NaCl solutions of the weight percent NaCl indicated, as a function of Th and pressure in megapascals (i.e., bars x 0.1), from Potter (1977). Note - in the original publication, these diagrams are given in a larger format (each is 10 x 15 cm) that make for easier interpolation.

be very different, as shown in a following section "PRESSURE OF TRAPPING." If a daughter crystal is present and dissolves in part or completely at Th L-V, the isochores may be flatter, i.e., the pressure correction much larger, particularly if the daughter crystal is NaCl, since under some conditions NaCl shows a negative molal volume change on solution (see under "PRESSURE OF TRAPPING").

During reconnaissance work, one may have some values for Th and an estimate of the probable pressure of trapping from geologic considerations, and perhaps only a crude guess as to salinity. Can an estimate of the pressure correction be made under such conditions, and how serious is the effect of salinity on the pressure correction? Using data on the system NaCl-H_2O, Roedder (1971b, Fig. 31) plotted the magnitude of the error in the pressure correction if the wrong salinity is assumed. From this it is apparent that so long as Th is 300°C or less and the actual salinity is 5% or more, the errors involved in using the data for the wrong salinity are minor so long as any number in the range 5-20% is assumed. The specific choice makes little difference, but the errors are probably minimal if 10% is assumed. If the actual inclusion salinity is zero, or if the data for pure water are used for saline inclusions, the errors can be large and either positive or negative. The errors resulting from assuming the wrong salinity are largest if Th is 350°C or higher, as expected from the much greater compressibility of such hot fluids.

Another problem is presented by values of Th >400°C. On Figure 9-4, the spreading of the isobars toward higher temperatures show that the pressure corrections for Th >400° are going to be still larger than at 400°C. Figures 8-2 and 9-1 show that the pressure corrections for pure-water inclusions with Th >350°C become very large. Thus an inclusion of essentially pure water that formed at 700°C and 1200 bars pressure (point X on Fig. 9-1a) would have Th = 375°C (i.e., at Cp. on Fig. 9-1a) and a pressure correction of 325°C. Such pressure corrections are very strongly affected by composition, so it is very hazardous to try to estimate their magnitude unless the composition is known.

Finally, the pressure corrections of Potter (1977) as well as those of Lemmlein and Klevtsov (1961), although not so stated, are only valid for inclusions homogenizing in the liquid phase. Figure 9-1 shows that the pressure corrections for inclusions homogenizing in the vapor phase, particularly those trapped at relatively low pressure, will be vastly greater than those with the same Th and Ph, but homogenizing in the liquid phase. Thus, an inclusion homogenizing in the liquid at 320°C and trapped at 250 bars would have a pressure correction of ~20°C (i.e., it was trapped at ~340°C), whereas a similar inclusion also trapped at 250 bars but homogenizing in the vapor phase at 320°C would have been trapped at 700°C, yielding a pressure correction of 380°C.

Decrepitometry

Considerable discussion in the literature concerns the theoretical significance and practical usefulness of the decrepitation method. Decrepitation can result from a variety of factors, such as the buildup of internal stress in a single grain between two minerals that have differing rates of thermal expansion; most commonly, however, it is a result of the internal pressure within fluid inclusions exceeding the tensile strength of the brittle host mineral[1] at that temperature, causing it to break. Inclusions having a high degree of filling (and hence low Th) have a very abrupt increase in $\Delta P/\Delta T$ at Th (as much as 14 bars/°C). This abrupt increase is the theoretical basis for the method, since it is assumed that Td ≈ Th, or at least that Td is a known function of Th. However, Td is also a function of so many variables other than Th that any agreement of Td with Th is considered by some to be essentially only coincidental. In

----

[1] Decrepitation data have even been published for grains of ductile metallic Au.

addition, the change in ΔP/ΔT is much less abrupt or even·negligible at Th for lower density, higher temperature inclusions, and ΔP/ΔT actually <u>decreases</u> (but stays positive) above Th in those inclusions containing less than the critical density of filling (i.e., Th L-V(V); see Fig. 8-5).

Most of the objections to the decrepitation method are founded on this lack of a theoretical basis for at least some inclusions, the commonly observed lack of agreement with actual Th data (e.g., Little, 1955, 1960), and the decrepitation in samples apparently free of visible fluid inclusions (e.g., Kennedy, 1950a; Stephenson, 1952). Also, some degree of subjectivity is involved in eliminating "anomalous" decrepitation (Stephenson, 1952; Smith and Little, 1953) and in selecting the appropriate spot on the decrepigram for the "start of decrepitation" of a given generation of inclusions (Kennedy, 1950a). In addition to instrumental problems, scatter in the results can stem from the irregularity in the size of the inclusions (large ones have lower Td); their shape (jagged irregular ones decrepitate more easily than smooth equant ones); variations in composition (e.g., $NaCl:H_2O:CO_2$) that cause gross differences in internal pressures; abundance of inclusions (if abundant they can act in concert); arrangement of inclusions in planes, which makes decrepitation easier; variations in heating rate; grain size of the host mineral; and variations in the toughness or brittleness of the mineral (i.e., the "snap" of Smith and Little, 1953) and their changes with temperature. These factors all result in large variations in the amount of overshoot (heating above Th) before decrepitation. The literature contains many records of overshooting of 50 or 100°C without decrepitation or leakage, and Roedder (1970b) reported some in quartz that could be overheated 600°C. Certain liquid-CO_2 inclusions in olivine may be heated 1200°C above their filling temperature without decrepitation (Roedder, 1965a). Inclusion size is a particularly important variable, as shown by studies of inclusions in quartz, which found that inclusions ~35 μm in diameter will decrepitate at ~850 bars internal pressure, but inclusions ~1 μm in diameter can withstand ~6000 bars (Fig. 3-21). Even at a given size of inclusion, the wall thickness (i.e., the position of the inclusion within the grain) is an important parameter.

Mechanical stress from thermal gradients in the grains can also cause decrepitation and hence must affect the temperature at which inclusions decrepitate. In a study of the evolution of gases from decrepitating CO_2 inclusions in olivine, Killingley and Muenow (1975b) found an increased rate of decrepitation during <u>cooling</u>. Apparently the sample thermal gradients in their apparatus during cooling were larger than those during heating, and hence the stress was greater.

Attempts to quantize the many variables and phenomena involved have been only partly successful. Khetchikov and Samoilovich (1970) have examined the difference between Tt and Td of synthetic quartz grown at known temperatures and pressures and have found that Td can range from 100°C below to 160°C above Tt, depending on Pt. Wilkins and Ewald (1982) have reexamined the decrepitation process in an attempt to develop a theoretical basis for it and suspect that at least four distinct mechanisms of decrepitation appear to be involved.

The above discussion shows that the decrepitation method has serious problems in both theory and practice, that interpretation of its results is complex and difficult, and that the method probably will never be truly calibratable to yield a valid measure of Th (or Tt), free of error and subjectivity. The only possible exception may be runs made on rather uniform sample suites that have been previously standardized by the homogenization method (e.g., Khoteev, 1980). However, the method is rapid, inexpensive, requires little training, and integrates the results of many hundreds or thousands of inclusions. It probably is most useful for purely empirical screening, to recognize <u>differences</u> between samples (e.g., Burlinson et al., 1983; Wilkins et al., 1983), or the presence of <u>several different generations</u> of inclusions, particularly in the opaque minerals and in the low-temperature range. Since the method can readily be adapted to field use (Ermakov, 1966b, reported handling 90 samples per working day for a field unit), it has been used most effectively in exploration. Some examples from the literature have been given by Roedder (1977b), in which a "steam halo" of high-temperature inclusions surrounding a deposit provided a larger target than the deposit itself and revealed blind deposits (see also Demin, 1970; Ermakov and Kuznetsov, 1971; and numerous other references in Roedder, 1968a). Workers in the People's Republic of China have recently expanded the use of decrepitation in ore-deposit studies (e.g., see many of the 54 papers in a recent symposium - Academia Sinica, 1981), and they have prepared a standard sample for decrepitation work (same reference, p. 115).

As a warning to the reader, I must admit my personal bias against the use of decrepitation data in any quantitative connotation. More than 400 papers have been published, mainly Russian, in which decrepitation data are reported, but very few of these are referred to in this volume. However, not all Russian workers accept decrepitation temperature data at face value (e.g., see Khetchikov and Samoilovich, 1970; Dmitriyev, 1970; Butuzov et al., 1971; Pal'mova, 1972; Bobolovich, 1972; Barsukov and Sushchevskaya, 1973; Pashkov and Piloyan, 1973; Sharonov et al., 1973; and other entries in indices in Roedder, 1968a).

PRESSURE OF TRAPPING (Pt)[8]

Introduction

The use of fluid inclusions as geobarometers, to determine the pressure of past environments, is intimately related to their use as geothermometers. Since Sorby's 1858 classic paper on the use of fluid inclusions for geothermometry, inclusions have been the subject of numerous papers, mostly in the Russian literature, and have provided a significant part of the data in many more. In these many papers, two aspects are most commonly mistated or misunderstood: (1) the effect of the hydrostatic pressure at the time of trapping of a given inclusion on the thermometric results obtained, and (2) the use of inclusion data themselves to obtain an estimate of this pressure. In many, if not most, inclusion investigations, the pressure is not determined from the inclusions. Most inclusions have trapped fluids at pressures higher than their vapor pressures (i.e., in the one-phase "fluid" field above the two-phase curve shown in Figure 9-1a). Generally, the pressure is estimated from independent evidence of the depth of cover at the time of trapping (e.g., from geologic reconstructions of the thickness of material since removed by erosion or faulting); then, this pressure is used, along with P-V-T data on supposedly appropriate solutions, to calculate the pressure correction. The following section shows why some commonly used procedures for evaluating pressure from inclusion studies (i.e., inclusion geobarometers) are wrong and can yield very erroneous pressure values. It also reviews the variety of valid inclusion geobarometers, along with their precision, accuracy, limitations, and applications.

Evidence of boiling and data on the pressures existing during geologic processes are of far more than academic interest. Boiling can cause ore deposition, and because pressure data provide some evidence of the depth beneath the surface at which the process occurred, they can provide the exploration geologist with valuable information on the amount of cover that has been faulted or eroded away and the possible nature of the deposit. Pressure differences may eventually also yield information on the direction of flow of the ore-forming fluids.

----

[8] This section taken with some modification from Roedder and Bodnar (1980).

General Principles and Nature of Available Data

   Relationship to inclusion geothermometry. The six assumptions that form
the basis of geothermometry using inclusions (see beginning of this chapter) are
equally pertinent to geobarometry. The first assumption, the trapping of a
single fluid phase, is essential to geothermometry, but as will be shown later,
evidence of the trapping of a nonhomogeneous mixture of two fluid phases, particularly liquid and vapor, can provide an excellent geobarometer.

   The second assumption, no change in volume, is of relatively minor concern
in most geobarometry because the most common volume change, that due to precipitation on the walls during cooling, is generally not only small but also easily
reversed during heating in the laboratory, except in silicate melts. Very significant errors may occur, however, if care is not used in the determination of
Th, because of permanent deformation (stretching) of the host-mineral walls from
overheating. The errors from dilation (see "Changes in Volume," Chapter 3) are
well below the errors from other sources, and they can generally be ignored at
this time.

   Assumption number three, that nothing is added or lost, may be pertinent
to geobarometry in several situations. The loss of H from inclusions by diffusion through the host mineral has been suggested as a possible explanation for
some puzzling daughter minerals that do not homogenize (Roedder and Skinner,
1968). As the hydrogen presumably comes from the dissociation of inclusion
H_2O, the volume change would be effectively controlled by the chemical behavior
of the oxygen left behind. This oxygen may oxidize components of the fluid,
such as sulfide to sulfate, or, in hydrous silicate-melt inclusions, it may
diffuse into the host (Anderson and Sans, 1975, and pers. comm.). The volume
changes within the inclusion in these examples might be difficult to predict.

   Assumption four, that the effects of pressure are insignificant, or are
known, refers to the fact that Th of the vapor and liquid phase of an inclusion
establishes only a minimum value for the Tt. All inclusions of a given composition trapped along the isochore originating at the Th point on a P-T plot (Fig.
9-1b) will have the same homogenization behavior, so either P or T must be known
to determine the other. This interrelationship is examined in more detail below.

   The fifth assumption, concerning the origin of the inclusion, causes problems in many inclusion studies. This ambiguity, frustrating to the student of
inclusions, does not affect the validity of most of the geobarometers presented
here, but it is of major concern to some.

   Practically all estimates of either the temperature or the hydrostatic
pressure at the time of formation of the host mineral from inclusions require
two types of data: (1) the composition of the fluid phase (or phases) trapped,
and (2) the phase behavior and P-V-T-X properties of that composition in the
range involved. Unfortunately, these two main requirements place severe constraints on the accuracy of all pressure determinations based on inclusions.

   Obviously, the difficulty in estimating the depth of cover, plus the uncertainty of hydrostatic vs lithostatic pressure described above, and the errors
in estimations of the composition of the fluid, and hence its P-V-T characteristics, can cause relatively large uncertainties in the pressure correction.
Fortunately, absolute geothermometric accuracy is sometimes only of minor concern, e.g., in current studies in ore deposition, where the relative temperature
values for various samples may provide the most valuable data. When the pressure
corrections are large, the limitations are such that most commonly the geothermometric or geobarometric data from fluid inclusions are compared with independent evidence from other sources in an attempt to arrive at a consensus and to
recognize which data or assumptions are invalid, and why.

Composition of the fluid phase(s) trapped. Many inclusion studies involve
only a qualitative identification of the phases present (e.g., liquid-water
solution, halite or other daughter crystals, and vapor bubble) and a crudely
quantitative estimate of the relative volumes and hence masses of these various
phases. As seen in earlier chapters, these volume estimates can be precise but
seldom accurate, and quantitative data, from heating and cooling studies, are
absolutely necessary for pressure estimates.

The fluids present around a growing or healing crystal are not always homogeneous. In many places, two immiscible fluids, such as H_2O and CO_2, or water
and steam, were present. Fluid inclusions forming in such a heterogeneous fluid
environment may trap only one of the fluids, or both of them in random ratios.
Such compositionally divergent inclusions provide several of the geobarometers
described below, but compositional data are required on both fluids. The use
of data from more than one kind of inclusion to provide a geobarometer must be
based on the assumption (often unstated) that the several inclusions used are
truly contemporaneous (i.e., cogenetic). Validation of this assumption is a
far from trivial matter and has led to many dubious or erroneous data in the
literature.

Phase behavior and P-V-T-X properties of fluid phase(s) trapped. Any
determination of pressure based on study of a single inclusion requires compositional data on the inclusion contents and P-V-T-X data on such a composition.
The available data on the phase behavior and P-V-T-X properties of appropriate
compositions in the range involved in fluid-inclusion studies are meager, and
even at best require considerable extrapolation. P-V-T data are most complete
on the two pure compounds that are major components of many inclusion fluids,
H_2O and CO_2 (for details and references, see Chapter 8), and these data are
frequently used in interpreting fluid inclusions. Natural fluids are not pure,
however, because the water inclusions often are at least one molal in ionic
solutes, and CO_2 inclusions frequently contain appreciable CH_4 and N_2, so the
properties of the pure compounds are useful only as an approximation.

For water solutions, the most extensive experimental data available are for
the system $NaCl-H_2O$, but even this system has not been studied in the entire
range of interest. In addition, most natural fluids are not simple solutions
of $NaCl$ and H_2O but contain significant amounts of other solutes, such as K^+,
Ca^{2+}, SO_4^{2-}, and HCO_3^-. Although Potter and Clynne (1978b) showed that many
of the solutes present in natural inclusion fluids result in fluids that have
thermodynamic properties rather close to those measured for simple $NaCl-H_2O$
solutions that have the same value for the depression of the freezing point
(i.e., Tm ice), extrapolation to fluids containing significant quantities of
CO_2 is hardly warranted. The available experimental data on the system $NaCl-H_2O-CO_2$ are even more limited.

Depth vs pressure; hydrostatic vs lithostatic. Most pressures determined
from fluid inclusions are stated to represent either the "lithostatic" or the
"hydrostatic" pressure, or some intermediate value. The local pressure at the
site of the inclusion at the time of trapping is actually hydrostatic, in any
case, because it is a fluid pressure, but the two terms are used to indicate
the source of the pressure on the fluid. "Lithostatic" is generally used to
refer to the pressure from a column of country rock of density and height appropriate for the depth of the sample below the surface at the time of trapping;
"hydrostatic" usually refers to the pressure of a column of fresh water above
the sample, of the appropriate temperature and hence density, corrected for depth
to water table. Although the pressures are probably between these two limits in
most natural situations, the actual values in nature may span a considerably
wider range at both ends, and even within this range numerous factors may affect
the specific pressure.

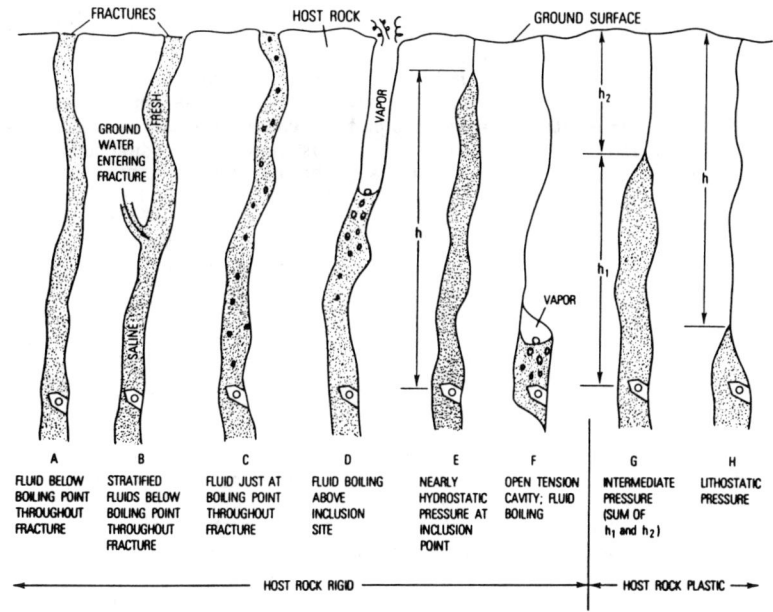

Figure 9-5. Diagram showing range of possible pressure conditions on a fluid inclusion trapped in a crystal growing freely into a fluid in a vein. See text for discussion. From Roedder and Bodnar (1980).

In Figure 9-5, diagram A represents the simplest possibility, in which the hot fluid moves up an open fracture. The fluid will expand as pressure decreases, but the cooling will generally be much greater than adiabatic, as the fluid is losing heat to the cooler adjacent rock before flowing out on the surface as a hot spring. The pressure at the inclusion would be that from a column of fluid of appropriate composition and temperature. The actual integrated density of the fluid column would be higher than that of the fluid trapped in the inclusion because the density increases toward the surface. A modification of this is shown in diagram B, where the upper part of the column is ground water. Because density increases with both salinity and cooling, the pressure in A and B might be identical if the integrated density of the cool dilute ground water is the same as that of the hotter, more saline fluids over the same interval of the upper portion of the vein length. If the fluid in the vein above the inclusion is at the point of boiling throughout its length (diagram C), and the salinity is known, the data of Haas (1971) permit an unambiguous pressure determination. Unfortunately, however, there is a fourth and probably common possibility, shown in diagram D, which may result in pressures at the inclusion that are considerably less than simple "hydrostatic" (as in diagram A), depending upon where in the column the switch from liquid to vapor occurs. A vapor-dominated system, in which liquid is above vapor, would be treated similarly.

The four possibilities described above all involve a vein system assumed to be open to the surface. In diagrams E through H (Fig. 9-5), a tight, almost closed vein or throttle (Toulmin and Clark, 1967) is assumed. Where the host rocks are rigid, as shown in E, the pressure of a flowing fluid at the inclusion site will be the hydrostatic head, h, plus an unknown overpressure limited only by the vein constriction. This overpressure could result from the presence of an igneous body below or from a hydrostatic load not directly above the vein, such as an artesian-type regional system. Diagram F illustrates an extreme condition, in which fault movements have increased the volume occupied by a relatively fixed volume of fluid; this increase causes boiling. The boiling would cease when the pressure in the cavity reached the vapor pressure of the fluid at that temperature, even though this chamber might be deep in the earth. This low-pressure condition could persist as long as the surrounding rock remains rigid. The same mechanism has probably been operative in many ore deposits, but only enough to cause migration of fluids and not enough to cause boiling. Thus, the well-known concentration of ore or vein matter in the apices of folds is frequently ascribed to the lower pressures there, due to the mechanics of folding; such low pressures may well be reflected in the inclusion densities.

In diagrams G and H, depths are assumed to be such that the country rocks are plastic. In G, the pressure would be the sum of the hydrostatic pressure of fluid column, h_1, plus the lithostatic pressure of rock column, h_2; in H, it would be simply lithostatic for column h.

In any lithostatic pressure environment such as that shown in diagram H, the true pressure at the inclusion can be considerably less than lithostatic if the country rocks are not completely plastic during the time involved. Also, the pressure in H can be more than lithostatic if the horizontal extent is small relative to the depth h, or if these country rocks are under compression from regional forces. In such cases, the pressure in the chamber could be greater than lithostatic without breaking out to the surface. The only situation in which the pressure would be truly "lithostatic" would be the rather rare horizontal vein having a large lateral extent relative to its depth, so that a flat slab of rock is effectively "floating" on fluid under pressure. "Overpressures" can also be caused by the increase in vapor pressure during crystallization of magmas; when roof rocks yield suddenly, explosive volcanism may result. Similar overpressures in sedimentary rocks (Barker, 1972) can result from the heating of formation waters trapped under impermeable beds. Owing to the large lateral extent, such fluids are not likely to be pressurized much above the lithostatic pressures.

The pressure environment within the pores of a rock, e.g., during metamorphism, is much more difficult to quantify than that within a vein or open fracture. Obviously, the pressure within disconnected pores between crystals, or even within single crystals, will be close to the regional lithostatic load if the host rocks or crystals are completely plastic during the time involved. Thus, in the interpretation of inclusions in deep-seated rocks that have been unloaded over a significant period of time, an important but unanswered question is the following: what amount of change, if any, has occurred in the inclusions during the slow decrease in pressure and temperature from the maximum values? This problem is equivalent to that of the apparent "quenching temperature" during the annealing of various compositional differences between minerals used as geothermometers. Such annealing of inclusions apparently does occur in rocksalt (Roedder and Belkin 1979a); does it occur in quartz? (See discussion in Chapter 3).

Although low pressures within tension fractures in veins (Fig. 9-5F) may be relatively rare, such low pressures may be the prevalent environment during the healing of the small disconnected tension fractures that have yielded the abundant secondary inclusions in many metamorphic rock samples. If so, this mechanism could explain some otherwise discrepant data, but apparently it has not previously been suggested. Distinction between low-density inclusions formed at relatively great depths by this mechanism and similar low-density inclusions formed by boiling in effectively open fractures at much shallower depths would be impossible from the inclusions alone and would require additional data.

Geobarometry based on vapor pressure of solution. If a group of two-phase, liquid-plus-vapor inclusions in a given sample all have the same composition and Th, the general presumption is that a homogeneous fluid was trapped. Thus, we can safely assume that the hydrostatic pressure at the time of trapping was greater than the vapor pressure of that particular solution at that temperature, or the fluid would have been boiling (i.e., two-phase). Such simple inclusions provide no evidence concerning how much the pressure might have exceeded the vapor pressure. Simple two-phase, liquid-plus-vapor inclusions are the most common by far; hence, this minimum (vapor pressure) value is generally the only constraint on the pressure that can be obtained from the inclusions themselves. For ore deposits formed at low temperatures, this vapor pressure may not result in much of a constraint. Thus, only 37 m of hydrostatic head of cold water (at density = 1.0 g/cm^3) is required to prevent boiling of the typically 150°C fluids that formed the Mississippi Valley-type ore deposits (Roedder, 1976a).

If there are gases in solution, the vapor pressure of the mixed solution would be higher than that of the pure solvent liquid, but appropriate P-V-T-X data are generally lacking. If such inclusions are studied on the crushing stage, however, an estimation of the internal pressure, at room temperature, can be obtained. Such an estimate must be less, of course, than the pressure at Tt.

Geobarometry based on comparison of Th with an independent geothermometer or geobarometer. If the composition and Th of a simple two-phase, liquid-plus-vapor inclusion are known, and another independent geothermometer is available to determine the (higher) temperature of formation (i.e., Tt) of the host or associated minerals, Pt may be determined from P-V-T data on the fluid of the inclusion. The uncertainty in the values derived from any such geobarometer is obviously at least equal to the sums of the uncertainties in the precision and accuracy of the determinations by the two geothermometers used, as well as other factors. The independent thermometer must be very accurate in order to provide useful results since a relatively small pressure correction can be equivalent to a rather large pressure. Thus, for low-temperature ore deposits, an error of 25°C in the temperature obtained from the independent geothermometer is equivalent to the hydrostatic pressure from >3 km depth of burial (Roedder, 1971d). Some published geobarometry data have been based on such poor "thermometric" data that the pressure values obtained (and the geologic speculation based in turn on them) are virtually meaningless. The limitations imposed by inadequate P-V-T data on the inclusion fluid can also be severe.

Essentially, the Th for the fluid inclusion, along with knowledge of its composition, limits the possible conditions of trapping to a single isochore on a P-T plot for the appropriate composition fluid. If another independent geothermometer-geobarometer can also be used on the same samples, and it yields a line having a different slope on a P-T plot, the point of intersection of these two lines will be the Pt and Tt; because of experimental uncertainties in the determination of both lines, the intersection becomes an area. Bethke and Barton (1971) showed that the distribution of Mn between sphalerite and galena could thus be combined with fluid inclusion Th to yield both Pt and Tt. Fortunately, the slopes of the data on Mn distribution and the inclusions on a P-T plot are strongly inclined to each other, thus reducing errors from this source. The accuracy of this determination would be limited not only by that of the experimental measurements involved but also by the validity of the necessary assumption that the sphalerite and galena and the fluid inclusions were all truly cogenetic. Validation of this assumption is not a trivial matter (Barton et al., 1963).

Coveney and Kelly (1970) described a similar technique for use on inclusions trapped in quartz that permits some limits to be placed on the P-T conditions at trapping. Quartz has two modifications, with a rapid inversion, at 573°C (at 1 atm). Quartz that crystallized as the high-temperature (β) form can be recognized as such (by crystal morphology, etching, etc.), even though it has inverted on cooling to the low-temperature (α) form. The inversion temperature is raised by pressure, but only $28°C/kbar$. Most isochores for fluid inclusions in quartz are much flatter on a P-T plot; hence they intersect the α/β line at a large angle. If a given inclusion were trapped by growth of an α-quartz crystal, the conditions of trapping would still have to lie along the isochore defined by its homogenization but below the intersection of this isochore with the α/β-quartz inversion line. If the host crystal can be shown to have grown as β-quartz, the growth conditions must lie along a different isochore at a higher T than that of the intersection. However, at the instant of inversion, the internal pressure in the inclusion will decrease owing to the ~1% volume increase on going from α to β. Because the thermal expansions of α- and β-quartz are quite different, the P-T path for an inclusion trapped in quartz will consist of two lines having a discontinuity at the α/β inversion, and hence is not isochoric.

Geobarometry based on simultaneous trapping of two immiscible fluids. Where two essentially immiscible fluids are present, each with known P-V-T properties, and separate inclusions of each fluid were trapped simultaneously, both P and T can be determined from the values for the Th of the two inclusions. In addition to the requirement that the P-V-T properties of both fluids be known, the slopes of the appropriate isochores for the two fluids on a P-T plot must be significantly different, or the accuracy of the determination will be poor. Most of the large number of reported geobarometric determinations (mainly in the Russian literature, see Roedder 1968a) are based on the use of the two immiscible fluids, CO_2 and H_2O. Unfortunately, these two fluids are not truly immiscible (see below), and severe problems exist in establishing the contemporaneity of trapping. Furthermore, many of the published reports do not make clear the all-important point, whether the CO_2 and H_2O phases studied were in separate inclusions or were two immiscible phases in a single inclusion.

The only pair of fluids that give promise of providing good geobarometric data by this method are the oil and brine inclusions found in some Mississippi Valley-type ore deposits (Roedder 1963, p. 176-177), but the P-V-T data on the oil phase are unknown and can only be guessed at the present. Oil that shows postentrapment degradation or maturation will not provide good data. Evidence of such changes is obvious in some oil inclusions from these Mississippi Valley-type ore deposits (Roedder, 1972, Plate 9), but no such visible evidence is found in other oil inclusions in the same deposits. When coeval inclusions of brine and oil are found, Th is always higher for the brine inclusions. Some hydrocarbon fluids have P-V-T properties similar to H_2O (R.C. Burruss, pers. comm., as quoted by Narr and Currie (1982)), but many are very different (see Fig. 11-25), thus explaining the different Th values (i.e., the pressure corrections differ).

Geobarometry based on simultaneous trapping of two partly immiscible fluids. If two partly immiscible fluids, each saturated with respect to the other, are present at the time of trapping, inclusions of these two fluids can provide some constraints on the pressure, so long as the appropriate P-V-T-X data are available and the mutual solubilities decrease from the conditions of trapping to those of observation. The method was originally proposed by Smith and Little (1959). Guilhaumou et al. (1981) suggested use of the immiscible pair brine-(CO_2,N_2) mixture. The nature of the fluid-inclusion evidence on the phase condition (homogeneous vs heterogeneous) at the time of trapping is crucial. If, for example, a series of inclusions all have the same ratio of the two fluids, presumably a homogeneous phase was trapped, and the P-T conditions at trapping are limited to the appropriate one-phase area of the pertinent diagram. If different ratios are found in different coeval inclusions, the fluids were presumably immiscible at the time of trapping, but a variety of problems make it rather difficult to determine the pressure from such data.

Problems exist even when the inclusions are proved to be coeval. If necking down occurs in inclusions that contain more than one phase (from either the trapping of immiscible fluids or subsequent phase separation), the resulting inclusions can be very misleading. Similarly, the changes in mutual solubilities between Tt (and Pt) and those at the time of observation can be rather complex. Such problems may have invalidated many of the published geobarometric data sets where this method was used.

Geothermometry based on trapping of boiling fluids. The boiling of a fluid is merely a special example of immiscibility under the prevailing conditions, in which the composition variable is eliminated. If a boiling liquid and its coexisting vapor phase are trapped separately in a pair of inclusions, these two inclusions will homogenize in the liquid and in the vapor phase, respectively. These two homogenizations must be at the same temperature, and if the boiling curve is known for that fluid, the pressure can be determined from this Th.

Experimental difficulties are involved (e.g., inaccurate Th values may be recorded because the small amount of liquid phase that condenses in the vapor inclusion after trapping usually coats the walls as film and hence is difficult to see), but, more importantly, individual inclusions may have trapped a mixture of two phases, rather than a single homogeneous phase (Fig. 9-1a). As the immiscibility "solvus" generally closes at higher temperatures, this trapping of a mixture will result in inclusions with higher Th than would be obtained on inclusions that trapped only liquid or only gas. Thus, where a group of such inclusions can be proven coeval, the minimum Th for inclusion homogenization in vapor and in liquid is provided by inclusions trapping pure end members and should be equal and represent Tt. All other inclusions, from the trapping of mixtures, would yield higher, and spurious, Th values. This method is, however, subject to many serious pitfalls.

The identity of these two Th values is not only a necessary requirement, but also can be taken as relatively unambiguous proof of boiling. Because the pressure environment under which boiling can occur in nature may be rather variable and transient, and such pairs of inclusions are seldom exactly coeval, small differences may be expected. Of course, two separate fluids, one liquid and one vapor, could be trapped at different times and at a fortuitous combination of P and T values that would yield such similar Th values.

Not uncommonly, however, individual inclusions grown from heterogeneous mixtures have trapped only the dispersed phase (which occurs as isolated droplets in the other, continuous phase). This is particularly expectable when most of the host crystal growth occurs from the continuous phase, e.g., from a water solution that contains dispersed steam bubbles. In theory, it makes no difference whether the dispersed phase is the same composition as the continuous phase (as in true boiling of a one-component system) or whether it consists merely of bubbles of a minor, more volatile constituent in a solution. The geobarometry method should work in either case, but solubility reversal and observational problems effectively preclude the latter.

Evidence of boiling provides us with some of the most accurate and unambiguous geobarometry data available and has been reported in numerous ore deposits. However, the most important part of the evidence, the proof of contemporaneity of trapping of the two types of inclusions, is frequently poor or lacking. The mere existence of the two inclusion types is inadequate, as this can also come about by sequential trapping of different fluids at different times, by necking down, and by leakage (Roedder, 1979b). Careful microscopy is necessary to minimize or eliminate these sources of ambiguity. The distinction is far more than merely academic, because several varieties of rich ore deposits ("bonanzas") are widely believed to have formed as a result of the gross change in the chemistry of the ore fluids upon boiling or effervescence.

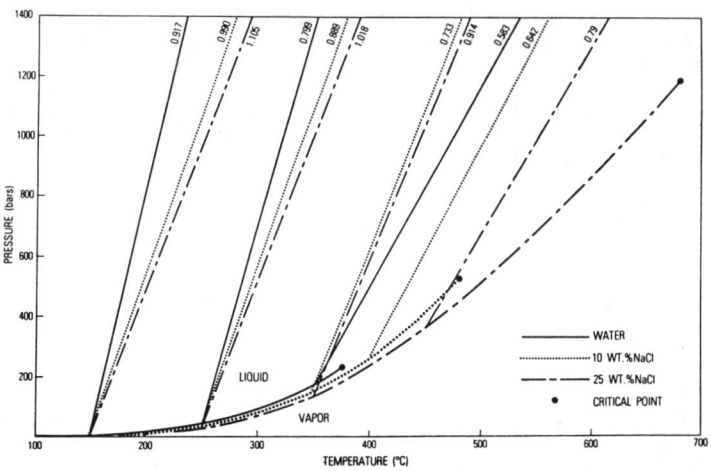

Figure 9-6. A portion of the phase diagrams for water and 10 and 25 wt % NaCl solutions, showing
the liquid-vapor curves and several isochores (g/cm³). Data for the 10 wt % isochore originating
at Th 400°C are from Urusova (1975); those for the 25 wt % isochore originating at Th 450°C are
extrapolated from Potter and Brown (1977). Other data from Sourirajan and Kennedy (1962), Burnham
et al. (1969), Keenan et al. (1969), and Haas (1976). From Roedder and Bodnar (1980).

Not uncommonly, one may find inclusion evidence of boiling at a given
pressure, along with other inclusions, in the same deposit, indicating possibly
higher pressures. In this situation, the low-pressure limit established by the
boiling inclusions is most informative, because it places an upper (hydrostatic)
depth limit on the crystal in a vein open to the surface (if we ignore the
interval of unsaturated ground above the water table). The higher pressures
are not as meaningful because of mechanisms for generating lithostatic or greater
pressures at shallow depths (Fig. 9-5). The only way in which such an interpretation of the evidence of boiling could yield misleading depths is shown in the
relatively rare example F in Figure 9-5.

Geobarometry based on inclusions containing daughter minerals. Many fluid
inclusions contain one or more solid phases in addition to liquid and vapor; if
the P-V-T-X properties of that particular salt-H_2O system are known, the phase-
disappearance temperatures may be used to determine a Pt for the fluid inclusion.

A fluid inclusion containing solid salt, saturated liquid, and saturated
vapor at ambient temperature might follow any one of three possible paths to
homogenization, indicating three different trapping pressures. If the vapor-
bubble-disappearance temperature (Th L-V) is higher than the temperature of
solution of the salt (Tm NaCl), the inclusion trapped an unsaturated solution.
The minimum Tt and Pt are Th L-V and Ph for a solution of the salinity determined
from Tm NaCl. If Tm NaCl and Th L-V are the same, the temperature and pressure
on the solid-liquid-vapor curve at that point are also minimum values for Tt
and Pt. Furthermore, where one of these inclusion types coexists with inclusions
that homogenize at the same temperature but in the vapor phase, then both were
probably trapped from a boiling solution, and Th = Tt and Ph = Pt. Finally,
if Th L-V occurs below Tm NaCl, the latter provides a minimum value for Tt.
The minimum pressure is the pressure along the solid-liquid-vapor curve at Tm
NaCl. Examples of pressure determinations on this basis are given in the next
section.

Application to Inclusions of Aqueous Salt Solutions

Fluid-inclusion-pressure determinations are based on the volumetric properties of the inclusion fluid. Therefore, the composition of the fluid must be
known and experimental P-V-T-X data for that particular fluid must be available
before the results of heating/freezing runs can be used to calculate Pt. This
section describes the procedures that can be used for given compositions and
some of the errors that have been made in geobarometric studies in the past.

Pure-water inclusions. Inclusions of essentially pure water are relatively
rare. They are most commonly found in samples from low-pressure, near-surface
hot-spring environments (Roedder 1977a) but are also found in samples from geothermal systems formed at considerably higher pressures. In the latter, they
may permit some reconstruction of the former P-T regime of the geothermal system
(Browne et al., 1976). If the salinity is low (less than a few percent), Figure
9-1 provides the basis for interpreting the data. (To a first approximation,
each 0.1°C depression of the freezing point (i.e., Tm ice) corresponds to 5850
ppm of salts.) CO_2 in geothermal fluids can significantly influence both the
apparent salinity and the pressure, as discussed below.

Low- to moderate-salinity inclusions.[9/] To extend the methods of pressure
determination described above to solutions of low or moderate salinity, we need
to determine only the composition of the inclusion fluid from Tm ice and then
use the volumetric properties of that fluid to determine the correct isochore.
Combining this with an independent mineralogical geothermometer will provide
Pt. The isochores for $NaCl-H_2O$ solutions differ from those for pure water (Fig.
9-6). Note that, contrary to some published statements of mine, the isochores
for pure water at low temperatures are steeper than those for NaCl solutions,
but this relationship is reversed at higher temperatures.

High-salinity multiphase inclusions. If a homogeneous fluid of sufficiently
high NaCl concentration is trapped as an inclusion, the saturation limit may be
exceeded as the fluid cools; the cooling causes precipitation of a daughter crystal of halite. When this multiphase inclusion is reheated, it once again becomes
homogeneous. Ph depends on the temperature, salinity, and order of disappearance
of the phases, as indicated by the three inclusions shown in Figure 9-7, all of
which are assumed to be in the system $NaCl-H_2O$ and to homogenize completely at
400°C.

Inclusion A follows the solid-liquid-vapor curve until the halite dissolves
at 158°C (Tm NaCl), corresponding to a 30 wt % NaCl solution (Benrath et al.,
1937). With continued heating the inclusion follows the liquid-vapor curve for
a 30 wt % NaCl solution (Haas, 1976; Urusova, 1975) until the vapor bubble disappears at 400°C (Th L-V). At this Th, Ph, and thus the minimum Pt, is 222 bars
(Urusova, 1975). This inclusion may have been trapped at any higher P and T
along the isochore originating at A (Fig. 9-7).

Inclusion B follows the solid-liquid-vapor curve until, at 400°C, both the
halite and the vapor bubble disappear simultaneously (Tm NaCl = Th L-V). At
this temperature, the inclusion contains a 46 wt % NaCl solution (Benrath et
al., 1937) under a pressure of 182 bars (Keevil, 1942; Sourirajan and Kennedy,
1962), corresponding to the minimum Pt. The inclusion could have been trapped
at any higher P and T along the isochore originating at B (Fig. 9-7).

Inclusion C follows the solid-liquid-vapor curve until the vapor phase
disappears at 310°C (Th L-V) at only 66.4 bars (Haas, 1976). From this Th L-V

[9/] For simplicity, low- to moderate-salinity inclusions are arbitrarily defined here as those that are
unsaturated at room temperature, i.e., less than ~26 wt % NaCl equivalent.

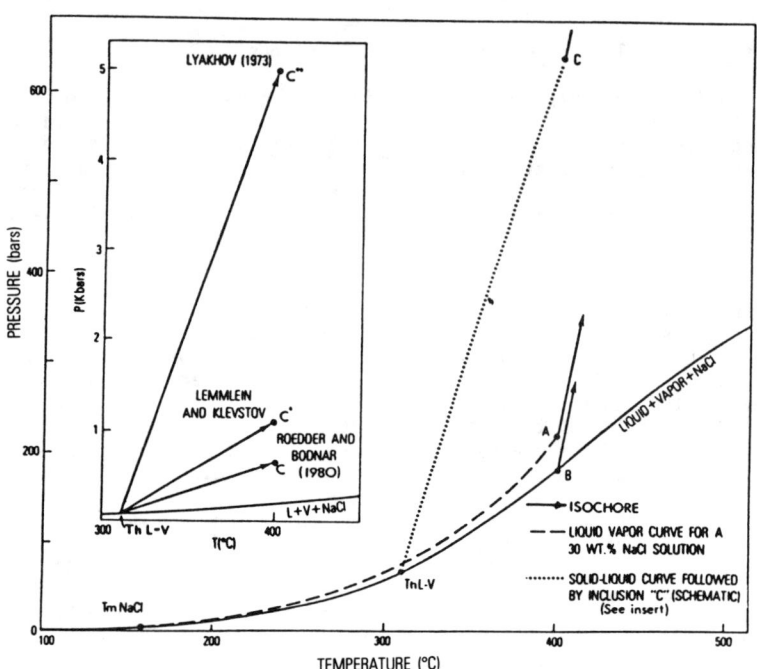

Figure 9-7. Pressures at homogenization for three halite-bearing fluid inclusions (A, B, C), from Roedder and Bodnar (1980). All are assumed to be in the system NaCl-H₂O and to homogenize at the same temperature (400°C) but to exhibit differing modes of homogenization, resulting in different pressures at Th. Data are from Sourirajan and Kennedy (1962), Urusova (1975), Haas (1976), and Potter and Brown (1977). The inset shows the pressures obtained for inclusion C using the various methods described in the text.

to Tm NaCl at 400°C, the inclusion follows a solid-liquid curve (liquidus), shown schematically in Figure 9-7. The pressure at this point, which is the <u>minimum</u> Pt, is calculated to be ~650 bars, as shown below. Other ions in solution, such as Ca^{2+}, could reduce this minimum even farther (Stewart and Potter, 1979). The inclusion could have been trapped at any higher P and T along the isochore originating at C (Fig. 9-7).

Although fluid inclusions homogenizing by halite disappearance are common, much confusion concerns their origin and interpretation. Assuming that this behavior is not a result of kinetic effects (Chivas and Wilkins, 1977; Eastoe, 1978) or that the inclusions have not trapped the halite as solid phase, as suggested by some data (e.g., Bodnar, 1978; Nagano et al., 1977; Wilson, 1978), the most common explanation is that the inclusions were trapped at "high pressure" (e.g., Piznyur, 1968; Dolgov et al., 1976; Kamilli, 1978; Milovskiy et al., 1978; and Andreyev and Shvadus, 1977), on the basis of what is generally called the "Lemmlein and Klevtsov method." As detailed below, this method is found to yield spurious results.

The "Lemmlein and Klevtsov method" was apparently first used by Klevtsov and Lemmlein (1959a) to determine minimum Pt for fluid inclusions homogenizing by halite disappearance. The pressure was obtained by conducting experimental P-V-T studies on aqueous solutions thought to correspond to the composition of the inclusion fluids and then plotting a pressure-correction diagram from the data. However, Roedder and Bodnar (1980) found several errors in the experimental and theoretical data of Klevtsov and Lemmlein (1959a) that invalidate their results. The net result was that Klevtsov and Lemmlein did not have the phase assemblage (i.e., with solid NaCl) in their pressure vessel that they thought they had, and the isochore they measured was for a single-phase (fluid) system without the volume change from a dissolving crystal of halite; the resulting isochore was therefore much steeper than it should have been. The actual path along the solid-liquid surface cannot be determined because of the lack of volumetric data in this region. However, if the apparent molar volumes of NaCl are <u>negative</u> in this region, as Urusova's (1975) data indicate, the actual pressure change between Th L-V and Tm NaCl might be very small, though the temperature difference may be very large.

Lyakhov (1973) presented a different method of calculating pressures from inclusions homogenizing by halite disappearance, which is also based on the P-V-T-X properties of NaCl-H$_2$O solutions. Unfortunately, this method assumes a constant volume but allows the mass of water in the inclusion to vary as the density of the water changes. This results in the erroneous conclusion that above 320°C, the solubility of NaCl decreases rapidly. In Lyakhov's method, the effect of this erroneous conclusion is that an inclusion following a path to Th described by inclusion C (Fig. 9-7) would have been trapped at a fictitious minimum pressure of 5000 bars.

Roedder and Bodnar (1980) concluded that the "Lemmlein and Klevtsov" and "Lyakhov" methods of determining pressures are <u>both</u> invalid, not only for the theoretical reasons given above, but also because the pressures obtained from these methods are not consistent with other data. Thus, using the "Lemmlein and Klevtsov method," Kamilli (1978), indicates that the inclusion data "require pressures much greater than any reasonable lithostatic load," and Bodnar and Beane (1980) reported pressures an order of magnitude higher than those obtained from other fluid-inclusion data and from geologic reconstruction. Also, pressures of 5000 bars (Lyakhov 1973) and 6000 bars (Dolgov et al., 1976) in quartz and 2700 bars in fluorite (Erwood et al., 1979) are in gross disagreement with measured pressures required to decrepitate inclusions in quartz (850 bars; Naumov et al., 1966) and to stretch fluid inclusions in fluorite (100-700 bars; Bodnar and Bethke, 1984).

Roedder and Bodnar (1980) showed that a good approximation of Ph in such multiphase inclusions at Th can be calculated . They illustrated this method by calculating the pressure of the same inclusion C (Fig. 9-7) in which Th L-V is 310°C and Tm NaCl is 400°C. To simplify calculations, they assumed that the inclusion contains 1000 g of H$_2$O and that the fluid properties are adequately represented by those of the NaCl-H$_2$O system. Assuming that NaCl solubility is independent of pressure (Adams, 1931), they obtained the bulk composition (46 wt % NaCl) from Tm NaCl (400°C) and the solubility data of Benrath et al. (1937). Hence they needed to calculate only the density of the homogeneous fluid and to refer this value to the P-V-T-X measurements of Urusova (1975) to obtain the pressure at final homogenization (Tm NaCl).

At 310°C, the inclusion volume is simply the volume of saturated aqueous solution plus the volume of solid halite remaining. The solution volume is the total mass of solution divided by its density. A saturated solution on the liquid-vapor curve at 310°C is 10.8 molal NaCl and has a density of 1.07 g/cm^3 (Haas 1976), which, when combined with the previous assumption of 1000 g of H$_2$O and a molecular weight of NaCl of 58.44 g/mole, gives a solution volume of 1524 cm^3. The volume of halite remaining can be calculated from the difference in solubility between 310° and 400°C (Benrath et al., 1937) and the molar volume at 310°C. When a value for the solubility of NaCl at 400°C of 46 wt % or 14.6 molal is used, calculated from Benrath et al. (1937), the difference in solubility is 3.8 moles/1000 g H$_2$O. The molar volume of halite at 25°C is 27.018 cm^3

(Robie et al., 1966), which increases 3.92% upon heating from 25°C (Skinner 1966) to give a molar volume at 310°C of 28.08 cm^3. Therefore, the volume of halite at 310°C is 106.7 cm^3, and the total inclusion volume at 310°C is 1631 cm^3.

The inclusion volume at homogenization is the volume at 310°C plus the increase due to thermal expansion upon heating and dissolution of the inclusion walls. Where the host mineral is quartz, the inclusion volume will increase 0.54% or 8.8 cm^3 from thermal expansion when heated from 310 to 400°C (Skinner, 1966). The increase in volume due to dissolution of quartz from the walls when the inclusion is heated from 300 to 400°C was found to be insignificant and was ignored in the calculations. The two dilation effects (see Chapter 3) were incorrectly assumed to cancel each other, but the error introduced by this misassumption is small.

From the above, the density of the homogeneous fluid at 400°C calculated from the mass of solution (1853 g) and the inclusion volume (1639.8 cm^3) is 1.130 g/cm^3. Extrapolating Urusova's (1975) data at 400°C along isochores from 45 wt % NaCl (her maximimum concentration at 400°C) to 46 wt % NaCl provides a pressure at final homogenization of 640 bars. This pressure is compared with pressures at homogenization obtained by the "Lemmlein and Klevtsov" method (~1025 bars) and the "Lyakhov" method (5000 bars) for this same inclusion (Fig. 9-7). These pressures are all minima; Pt could be at any higher value along the appropriate isochore.

The accuracy of pressure determination by this method is difficult to establish and, of course, depends on the accuracy of the P-V-T-X data used. In this respect, Roedder and Bodnar (1980) noted that where the data of Urusova (1975) and Sourirajan and Kennedy (1962) overlap, Urusova's data are always higher by ~25 bars, and Sourirajan and Kennedy's data are thought to be high because of the presence of H_2 gas derived from corrosion (Liu and Lindsay, 1972; J. Haas, Jr., pers. comm., 1979). A complete analysis of the accuracy and precision of the method must await experimental volumetric data along the liquid-solid curve of the NaCl-H_2O system at high P and T.

Although halite is by far the most common solid phase in fluid inclusions, sylvite is often found along with halite, especially in porphyry-type ore deposits. Combining data on the NaCl-KCl-H_2O ternary diagram with the data of Ravich and Borovaya (1949) (see Fig. 8-25), these inclusions bearing sylvite and halite may be used to estimate fluid compositions and minimum pressures attending entrapment. Other ions in solution, such as Ca^{2+}, would significantly lower the minimum. The vapor pressures of such "hydrosaline melts" are surprisingly low. Thus, Stewart and Potter (1979) showed that a fluid saturated with respect to NaCl, KCl, and $CaCl_2$ at 350°C has a maximum vapor pressure of only ~25 bars.

Most other solid phases are more rarely found, but their solubilities and volumetric properties so poorly known that inclusions containing them provide no useful data on Pt.

Application to Carbon Dioxide-Bearing Inclusions

Carbon dioxide-bearing fluid inclusions occur in rocks from a wide range of geologic environments, and many methods have been suggested for determining Tt and Pt from these inclusions. Considerable confusion exists, however, mainly from failure to consider the trapping conditions. The three most commonly used methods are described below. The first requires separate inclusions of CO_2 and H_2O; the other two are based on mixed H_2O-CO_2 inclusions.

Intersecting isochores in pure CO_2 and H_2O systems. Kalyuzhnyi and Koltun

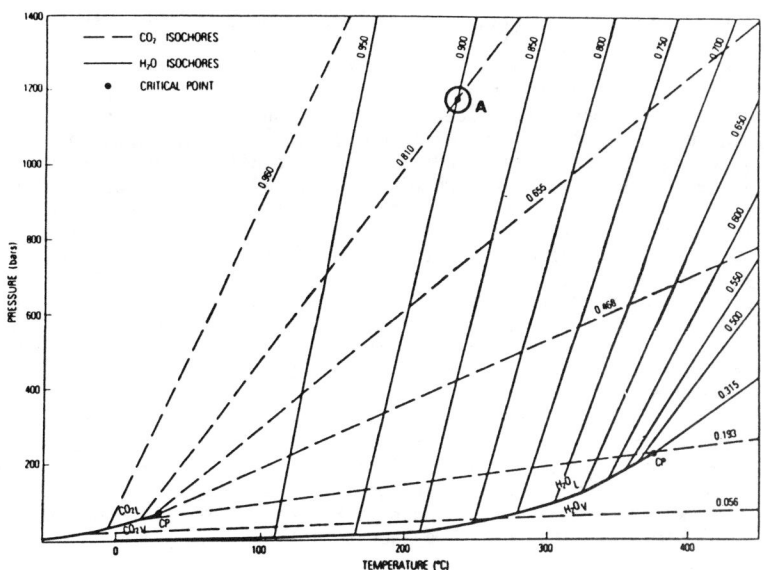

Figure 9-8. Combined P-T diagrams for CO_2 and H_2O, illustrating the Kalyuzhnyi and Koltun (1953) method of geothermobarometry using separate CO_2 and H_2O inclusions trapped at the same T and P, from Roedder and Bodnar (1980). Point A represents the common trapping temperature (237°C) and pressure (1180 bars) for a CO_2 inclusion and an H_2O inclusion homogenizing at 11°C and 167°C, respectively, both in the liquid phase. Data are from Kennedy (1954), Burnham et al. (1969), Keenan et al. (1969), and Weast (1980).

(1953) applied a method of geobarometry, first described by Nacken (1921), which is applicable to separate CO_2 inclusions and H_2O inclusions trapped at the same P-T conditions, either at the same location at different times or at the same time but at different locations. By plotting the P-V-T diagrams for H_2O and CO_2 in the same plane, P_t and T_t are defined by the intersection of the isochore corresponding to the CO_2 inclusion with that corresponding to the H_2O inclusion (Fig. 9-8). Thus, a CO_2 inclusion homogenizing in the liquid phase at 11°C and an H_2O inclusion homogenizing in the liquid phase at 167°C would both have been trapped at 237°C and 1180 bars (A, Fig. 9-8). This method of obtaining trapping conditions is valid if, and only if, the two inclusions were separately trapped as essentially pure components at the same temperature and pressure.

Pressure determinations from CO_2-bearing fluid inclusions are much more complex if the carbon dioxide and water were not physically separated at the time of trapping. The addition of CO_2 to water causes the critical locus to migrate from that of pure H_2O to lower temperatures and higher pressures, reaching a minimum temperature of 266°C at 2450 bars and 41.5 mole % CO_2 (Todheide and Franck, 1963). Upon further addition of CO_2, the critical temperature slowly rises, reaching 268°C at 43.5 mole % CO_2 and ~3600 bars; this rise results in a large two-phase field spanning a wide range of P-T-X conditions (Figs. 8-19, 20; 9-9).

Naumov and Malinin graphical method. Naumov and Malinin (1968) proposed a graphical technique for determining pressure that uses the T_h and T_d of inclusions containing both liquid-water solution and CO_2. The method is based on a straightline extrapolation, in P-T space, from the point representing partial homogenization (liquid and gaseous CO_2, i.e., T_h CO_2 L-V), through T_d (assumed to be at 850 atm for inclusions in quartz), to T_h, and thus P_h. The validity of this technique is in question because there is no basis for assuming that the pressure increases linearly with temperature between Th CO_2 L-V and total Th. In fact, because the slopes of the "isochores" and mutual solubilities (see below) of the two phases are constantly changing as the system within the inclusion travels through P-T space, it would be quite surprising if the pressure did vary in such a simple fashion with temperature. Gehrig (1980) has shown that isochores in the two-phase field (CO_2-rich vapor and (H_2O + NaCl)-rich liquid) for the system H_2O-CO_2-NaCl are strongly curved. Further, the assumption that decrepitation begins at 850 atm is based on experimental studies (Naumov et al., 1966) using synthetic quartz crystals that normally contain very large fluid inclusions relative to natural quartz and that do not contain lower temperature secondary inclusions. As Naumov and Malinin (1968) themselves pointed out, pressures necessary to decrepitate inclusions in synthetic quartz range from 850 to >3000 atm, depending on inclusion size, and Swanenberg (1980) reported pressures of 6000 bars without decrepitation. Therefore, if the smallest inclusions in synthetic quartz (decrepitating at >3000 atm) correspond to the largest inclusions studied in natural quartz, as is possible, the onset of mass decrepitation of natural quartz samples will be >3000 atm. As a result of the extrapolation procedure, the difference in the assumed pressure at Td is magnified in the estimate of Ph. Finally, and most important, at many of the P-T-X combinations obtained by this technique, that were reported in the literature, water and CO_2 form two immiscible phases (Chapter 8), which raises the possibility that the inclusions did not trap homogeneous fluids.

Mixed H_2O-CO_2 inclusions. At ambient temperatures, inclusions consisting purely of CO_2 and H_2O commonly contain three phases -- liquid and gaseous CO_2, and liquid H_2O. These phases are essentially pure (relative to each other), so by measuring the volumes of the three phases at a known temperature and using the density data for CO_2 and H_2O, the mole % CO_2 in the inclusion may be calculated. Such a procedure was used to obtain the mole % CO_2 (and mole % H_2O) in three hypothetical inclusions, shown in the inset of Figure 9-9. In practice, volume measurements are generally very imprecise, for reasons previously mentioned, and caution is needed when compositions are calculated by means of this technique.

The complexities of pressure determinations from inclusions containing both CO_2 and H_2O might best be illustrated by examining the phase changes upon heating of three hypothetical pure H_2O-CO_2 fluid inclusions and referring these observational data to the CO_2 solubility diagram shown in Figure 9-9. First, consider a fluid inclusion assumed to have trapped a homogeneous fluid containing 25 mole % CO_2 and having the volume percentages of phases at 25°C shown by inclusion B (Fig. 9-9). On heating to 27°C, the CO_2 liquid and vapor phases homogenize in the liquid phase, which then has a density of 0.672 g/cm^3 (Quinn and Jones, 1936; Kennedy, 1954; Newitt et al., 1956). Continued heating causes the mutual solubilities of CO_2 and H_2O to increase until, at 275°C, the inclusion contains a homogeneous fluid phase composed of 25 mole % CO_2 and 75 mole % H_2O (B in Fig. 9-9). These data require that the internal pressure,[10] and thus the minimum trapping pressure, Pt, is ~1000 bars. Note that homogenization at 275°C and 1000 bars would also occur in an inclusion containing 45 mole % CO_2.

If the pressure on this fluid dropped below 1000 bars, CO_2 and H_2O would no longer be completely miscible. Thus, at 275°C and 575 bars, a CO_2-rich fluid containing 40 mole % H_2O and an H_2O-rich fluid containing 11 mole % CO_2 coexist (Fig. 9-9), and two separate inclusions trapped at these conditions would have phase relations as shown by inclusions A and C in the inset on Figure 9-9. Upon heating from 25°C, the CO_2 phases in inclusions A and C would homogenize at

----

[10] The high internal pressure developed during homogenization of many CO_2-H_2O inclusions is the reason that decrepitation is common before Th is reached.

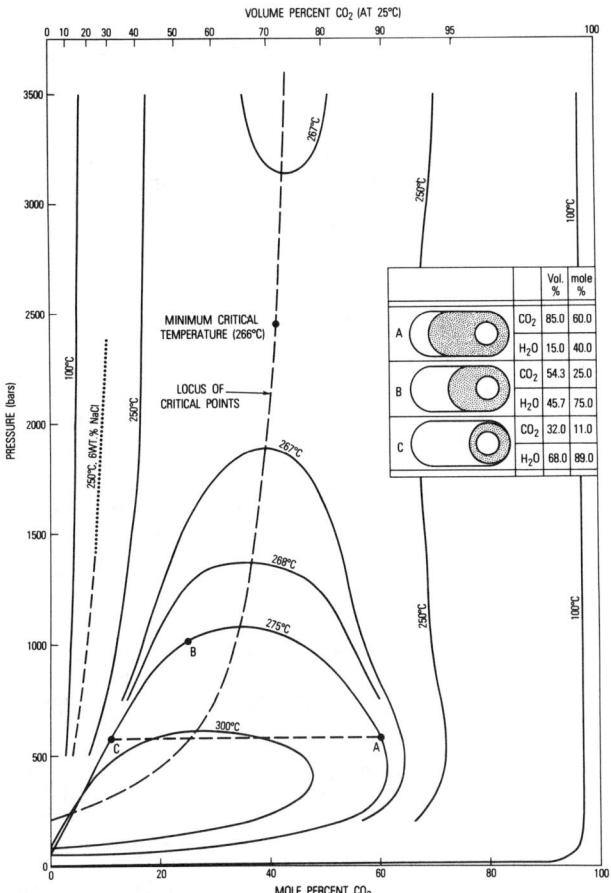

Figure 9-9. P-X plot of isotherms showing compositions of coexisting phases in the system H_2O-CO_2, using data of Todheide and Franck (1963) and Greenwood and Barnes (1966). The upper abscissa shows volume percent CO_2 at 25°C along the CO_2 liquid-vapor curve (64 bars), assuming densities of CO_2 liquid, CO_2 vapor, and H_2O liquid of 0.71, 0.24, and 1.0 g/cm^3, respectively (Newitt et al., 1956; Keenan et al., 1969). The inset shows the two-dimensional appearance at the stated conditions for three cylindrical inclusions having compositions as given (liquid CO_2 shaded), which are also shown on the diagram. The 250°C isotherm for a 6 wt % NaCl solution from Takenouchi and Kennedy (1965b) is shown for comparison. From Roedder and Bodnar (1980).

26 and 28°C, respectively, and complete homogenization would occur in both inclusions at 275°C and 575 bars. In most fluid-inclusion work, the composition is unknown or is imprecisely known from volume measurements. In this situation, the critical pressure along the 275°C isotherm determines a maximum Pt of 1080 bars. Ypma (1963) pointed out, however, that noncoeval CO_2-H_2O inclusions, trapped under different P-T conditions, that can homogenize (fortuitously) at the same temperature.

Most CO_2-bearing fluid inclusions contain an aqueous salt solution rather than pure H_2O, adding further complexity to pressure determination. When NaCl is added to the CO_2-H_2O system, the critical solubility of CO_2 at a given temperature and pressure decreases (see Chapter 8), and the miscibility gap is widened.

The 250°C isotherm for CO_2 solubility in a 6 wt % NaCl solution, from Takenouchi and Kennedy (1965b), is shown in Figure 9-9 for comparison with pure H_2O. Note that at 250°C, a 10 mole % CO_2 fluid at 750 bars would be just within the one-phase region if we assume a pure H_2O-CO_2 system, but well into the two-phase region if the H_2O phase actually is a 6 wt % NaCl solution.

If a 10 mole % CO_2 inclusion that had a Th of 250°C were trapped as a homogeneous fluid, Figure 9-9 shows that it would have had a minimum Pt of 750 bars, if we assume that the inclusion consists of pure CO_2 and H_2O. However, if the H_2O phase contained 6 wt % NaCl, the minimum trapping pressure would have been ~2500 bars, determined by extrapolating the 250°C, 6 wt % NaCl isotherm of Takenouchi and Kennedy (1965b) to higher pressure.

Methane is completely miscible with both liquid and gaseous CO_2 and is a common component of CO_2-bearing fluid inclusions in metamorphic rocks. Its presence is usually implied when the triple point for the CO_2 phases is found to be at a lower temperature than that for pure CO_2. Hollister and Burruss (1976) suggested that the addition of CH_4 to the H_2O-CO_2 system raises the top of the solvus to higher temperatures, and the miscibility gap is therefore widened. Swanenberg (1979) indicated that data from CO_2-CH_4 inclusions may be used in pressure determinations if the fluid density is expressed as an "equivalent CO_2 density." This method is based on the observation that, over the temperature range 200-800°C, the slopes of the isochores for the CO_2-CH_4 system are similar to those of the pure CO_2 system.

Application to Inclusions in Igneous Rocks

Silicate-melt inclusions. Although many determinations of Th have been made on silicate-melt inclusions (generally they fall in the range 800-1200°C), Ph determinations based on such inclusions normally are not made. In large part, this stems from the fact that silicate melts are relatively incompressible, so there is little need for a pressure correction to be applied to most such Th determinations. Murase and McBirney (1973) determined the adiabatic compressibility of a series of silicate melts from their densities and longitudinal-wave velocities and found them to fall in the range of 2 to 7 x 10^{12} cm^2/dyne at 1000-1200°C. Combining these data with thermal-expansion data on crystals and glasses (Skinner, 1966), one finds that the effect of 1 kbar at the time of trapping would correspond to ~20°C, an amount far smaller than the probable experimental error alone on most determinations of silicate melt Th (Roedder, 1979a). Most determinations of melt Th are made on glassy inclusions, from rocks formed under intermediate or shallow depths, so the pressures are seldom >1 kbar.

Silicate-melt/CO_2 inclusion pairs. If the silicate-melt inclusions give evidence of having been trapped from an immiscible mixture of silicate melt with another, more compressible fluid, pressure estimates become feasible. Here the silicate-melt inclusions, containing a relatively incompressible fluid, permit a fairly close estimate of the true Tt (i.e., the necessary "independent geothermometer"). This value can then be used, along with P-V-T data on the other, more compressible fluid (in a separate inclusion), to obtain Ph. The major limitation in the practical application of the procedure lies in the difficulty of proving contemporaneity of the two inclusion types. Many of the published pressure determinations from inclusions in igneous minerals are based on such an assumption of immiscibility but lack the necessary evidence to prove contemporaneity of trapping.

The abundant CO_2 inclusions found along with silicate-melt inclusions in the olivine of olivine nodules from basalt occurrences all over the world (Roedder, 1965a) provide an example of the application of the method. The silicate-melt inclusions in these samples, and actual observations on lavas, show that these inclusions were probably trapped at ~1200°C. If we assume this

Tt, and obtain the density of the CO_2 (0.85 g/cm^3) from Th CO_2 (10°C; see Fig. 8-9), the P-V-T data on CO_2 (Fig. 8-10) permit an estimate of 6 kbar for Pt, corresponding to the hydrostatic pressure from ~20 km of liquid basalt. This is not a minimum or maximum value, but the actual pressure, under the given assumptions. Unfortunately, this pressure estimate must be based on extrapolations from experimental data (Fig. 8-10) in which maximum temperatures and pressures are 1000°C and 1400 bars (Kennedy, 1954), or 707°C and 8000 bars (Shmonov and Shmulovich, 1974). Other inclusions in these samples were presumably trapped at similar temperatures but greater depths; these inclusions had such high internal pressures that they have generally decrepitated upon eruption at the surface at ~1200°C. Therefore, the range of application of the method is limited (Roedder, 1965a), but most important, the decrepitation of those inclusions formed at higher pressures will certainly result in a biased sample.

Bilal and Touret (1977) reported Th in the liquid phase as low as +10°C on presumably pure CO_2 inclusions in phenocrysts from a basalt, and estimated Pt = 5 kbar. However, using the molal volume data on CO_2 of Shmonov and Shmulovich (1974), this density CO_2 at 5 kbar would indicate a trapping temperature of only 900°C. Extrapolation to an assumed Tt of 1200°C indicates a trapping pressure >>6 kbar.

Lower pressure CO_2 inclusions in basaltic glass from submarine flows were studied by Moore et al. (1977), who showed a constant relationship between the pressure of CO_2 in gas vesicles (i.e., bubbles) and the known depth of water in which the eruption occurred. Using this method, one can estimate the pressure at the time of eruption of unknown basalt samples simply by piercing the bubbles under a liquid in which CO_2 is not soluble, and measuring the volume expansion (Chapter 16).

The above examples are based on the relatively immiscible pair of fluids, CO_2 and basalt. Inclusions from somewhat less immiscible fluids, such as silicate melt and water, can also be used. If the second fluid phase is a hydro-saline melt, containing >50% of NaCl and other salts, as in the Ascension Island granites (Roedder and Coombs, 1967) or in the various hypabyssal granites reported in the extensive Soviet literature (see many entries in Roedder, 1968a, particularly those by A.I. Zakharchenko), inclusion data could, in theory, provide an estimate of pressure. Such an estimate requires, however, that the composition of the aqueous fluid phase be known and that P-V-T data on such fluids be available; neither of these requirements can be met satisfactorily at present.

Where the concentration of salts in the aqueous phase is low, the P-V-T data for H_2O can be used, and the accuracy of the resulting geobarometric values will increase. The simplest case involves silicate-melt inclusions containing H_2O but having no evidence of a second, volatile-rich phase. These assumptions require that the pressure at the time of trapping was above that of the vapor pressure of water over that melt. Thus Anderson (1974a) and Anderson and Sans (1975, and pers. comm.) devised several methods for estimating the H_2O content of magmas from melt inclusions and found some high values, as much as 12 ± 2% H_2O (Roedder, 1979a). Such inclusion data can be combined with experimental data on the vapor pressures of hydrous melts, when they become available, to provide at least a lower limit for Pt.

Application to Siting of Nuclear Reactors

One interesting and important new facet of fluid-inclusion barometry, one that may well have more extensive application in the future, is its use in evaluating a site for a nuclear reactor. Cunningham (1974) provided such an evaluation, based on a study of fluid inclusions in euhedral crystals protruding into vuggy cavities along a fault that cut a proposed site. The important question was to date the last movement on the fault and hence to evaluate the possibility of renewed movement. The crystals must have grown since the last movement on the fault. An elevated pressure at trapping, shown by thermometric data on the inclusions in the crystals, established a minimum depth below the surface at the time of their formation. Subsequent erosion had removed this overburden, so when an estimate was made of the rates of denudation, a minimum time since the last movement was obtained. Numerous other similar applications have been published subsequently (Chapter 13).

Summary of the Present Status of Geobarometry

Most fluid-inclusion studies today involve the determination of Th and a search for geologic evidence concerning depth of burial in order to estimate the pressure correction to be added to Th to obtain Tt. Assuming that the composition of the inclusion can be determined, the four interrelated variables here are Th, Ph, Tt, and Pt; if any three are known, the fourth can be obtained. Th can be determined with relatively high accuracy at present, and, assuming experimental data are available, this determination also fixes Ph. Estimates of Pt, however, have very large error bars resulting from the necessarily large uncertainties of the geologic reconstruction and the problem of hydrostatic vs lithostatic pressure mentioned in a previous section. Independent geothermometers for obtaining Tt are not always available and are seldom very accurate. However, I believe that a good possibility exists of using them in the future as these other geothermometers are refined, and hence of determining Pt from (Tt-Th) and the P-V-T-X properties of the fluid with an accuracy at least comparable with that from geologic reconstruction.

As in all science, we can expect that more accurate thermometric determinations on inclusions will certainly become available in the future. Will they help in geobarometry? In theory, it is possible to obtain both P and T from careful studies of the small differences that should exist between otherwise identical coeval inclusions in two host minerals with differing thermal expansions. Such an effort seems doomed, however, by the requirement of coeval inclusions; whenever detailed studies are made, it is apparent that the fluids present in most geologic environments have varied in P, T, and X even during the formation of small parts of a single crystal (e.g., Roedder, 1977c). Such data emphasize the difficulty of proving two inclusions to be truly coeval, and show that even small errors in such an assignment of origin can result in major errors in pressure by any method requiring coeval inclusions. Although immiscible fluid pairs now provide us with some of our best geobarometers, this same problem of finding truly coeval inclusions in material formed in a changing environment makes future refinements in inclusion thermometry on such inclusions of relatively little value for geobarometric determinations.

More accurate determinations of the composition of inclusions are also forthcoming, as a result of the application of a variety of new techniques. Although such improved data on the composition of the former fluid phase are invaluable in other investigations, such as the causes of ore deposition or the nature of the various reactions that have yielded metamorphic rock assemblages, they are of rather limited immediate value to geobarometry, because experimental P-V-T-X data are generally lacking for such compositions. Even the simple systems, such as $NaCl-H_2O$, are not known to the P, T, and X limits necessary, and as the complexity of the composition increases, the available data decrease very rapidly. After such P-V-T-X data become available, however, we may expect that the accuracy of geobarometry based on inclusions that have trapped a single, homogeneous phase may exceed that from the two-fluid methods because the difficult requirement of coeval inclusions is not involved.

The effects on an inclusion of events after it is trapped, although generally a source of possibly major error in both geothermometry and geobarometry

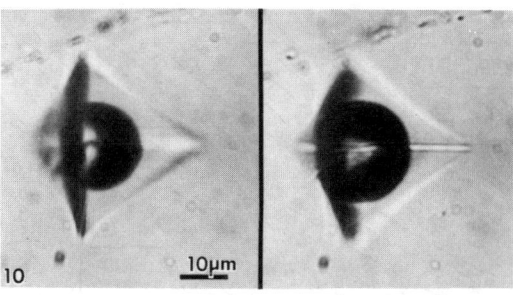

Figure 9-10. Photomicrographs of the same facet-ted tetrahedral inclusion in fluorite at two levels of focus. The bubble in the photograph on the right has a diameter corresponding to 63% greater volume than that on the left. Presumably, if viewed perpendicular to one of the tetrahedral faces, the correct diameter (whatever it is) would be seen. Sample from Serra S'Ilixi, Sardinia, courtesy of B. De Vivo.

of the original environment, may provide useful data on the pressures during this post-trapping history. Although I do not believe that decrepitation of inclusions in the laboratory provides much quantitative thermometric data, natural decrepitation can prove useful. The rocks containing fluid inclusions, formed deep in the earth at elevated P and T, have dropped to present-day surface P and T via a generally unknown route on a P-T diagram for the appropriate fluid. If the path taken is such that the pressure within an inclusion exceeds the external pressure, the inclusion may deform or rupture its host crystal. Evidence for such natural decrepitation, reported by numerous investigators (e.g., see Voznyak and Kalyuzhnyi, 1976, and various entries in Roedder, 1968a), may provide valuable information on depths of burial and rates of uplift and denudation; it should be looked for routinely. Perhaps the most serious problem lies in the development of petrographic criteria for assigning relative ages to the various generations of secondary inclusions that may reflect such decrepitation in many metamorphic samples.

One aspect of natural decrepitation that may hold considerable promise for clarifying the time sequence in certain geologically complex terranes has not been adequately applied in the past. This is the recognition of natural decrepitation from the pressure rise due to local heating, e.g., at the intersection of veins and dikes (Roedder, 1977b). Similarly, some constraints on the P-T path taken by magmas during ascent and eruption may be obtained from data on those high-pressure CO_2 inclusions that have decrepitated and on those that have not (Roedder, 1965a).

A study of the systematics of stretching of overheated fluid inclusions (Bodnar and Bethke, 1984) suggests the possibility of recognizing naturally stretched inclusions. If so, we can use such naturally stretched inclusions as geobarometers. Bodnar and Bethke (1984) showed that the internal pressure necessary to initiate stretching of fluid inclusions in fluorite varies in a systematic manner as a function of inclusion size but that, once begun, the amount of stretching is a linear function of the internal pressure. When a fluorite crystal has been overheated by some later thermal event, its large inclusions might have been stretched but other smaller inclusions might not have been, and heating tests would reveal an increase in filling temperature with inclusion size for inclusions exceeding a certain minimum size. Then, knowing the pressure necessary to initiate stretching of the smallest observed stretched inclusion, and assuming that this pressure represents the difference between the internal and external pressures existing when stretching occurred, the external pressure might be calculated. This pressure is the external pres-

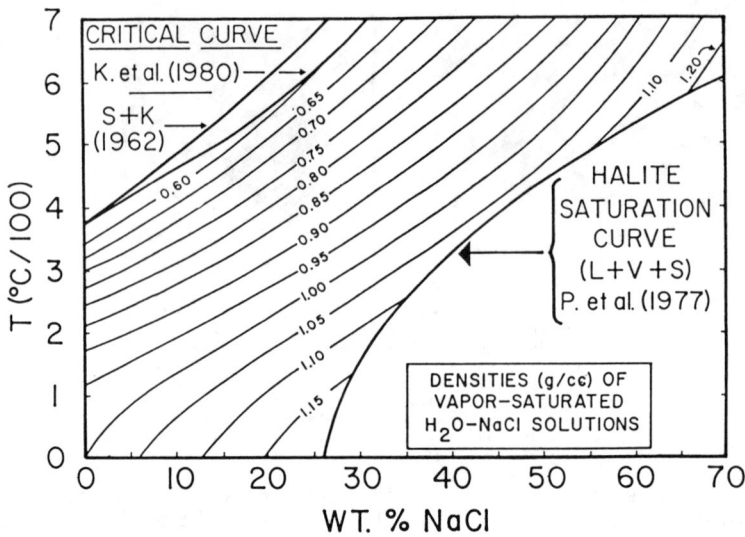

Figure 9-11. Density of vapor-saturated 10 and 25 wt. % NaCl fluids in the system NaCl-H₂O, calculated by Bodnar (1983) using a stepwise multiple regression from literature data and plotted on
his Figure 4. References used are Sourirajan and Kennedy (1962), Potter et al. (1977), and
Khaibullin et al. (1980).

sure at the time of stretching and not necessarily the pressure during trapping
of those particular inclusions. Furthermore, evidence suggests that for a
given pressure difference, the stretching is dependent on the confining pressure
(Poland, 1982).

    The above discussion shows that of the numerous inclusion geobarometers
used in the past, some are wrong in concept and can yield grossly inaccurate
pressures, and others need an independent geothermometer to yield an estimate
of the pressure. Single inclusions can yield only data that are functions of
both P and T, but pairs of inclusions trapped from immiscible fluids (e.g.,
liquid water and steam, or water and CO_2) can yield both P and T. Inclusions
of an immiscible water (or CO_2) phase along with a silicate-melt phase are a
special example: the silicate-melt inclusions provide, in effect, an independent
thermometer, and hence the pair can provide a geobarometer.

    Any determination of the pressure of formation from inclusions requires the
following:

    1. Good evidence of the time of trapping of the inclusion relative to the
geologic process being studied, from careful microscopy.
    2. Good evidence of freedom from various secondary effects on the inclusion
since trapping, both in nature and the laboratory (necking down, leakage, decrepitation, stretching, etc.).
    3. Good thermometric data (Tm ice, Tm daughter minerals, Th) on the phases
in the inclusion (and sometimes volumetric measures of the individual phases).
    4. Good compositional data on the inclusion from thermometry and other
procedures.
    5. Good experimental P-V-T-X data covering the necessary range of condi-
tions, on appropriate systems for the composition found.

    The most serious handicap to accurate geobarometry at the present time is
the lack of good experimental P-V-T-X data.

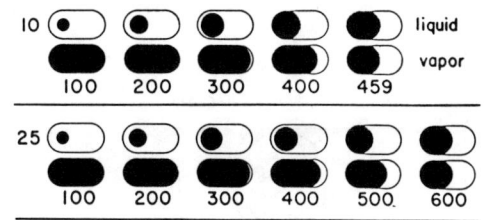

Figure 9-12. Room-temperature phase relations of fluid inclusions having salinities of 10 and 25 wt % NaCl and Th from 100°C to the critical temperature for each composition. Also shown are the phase relations of inclusions that trapped the vapor phase that would have been in equilibrium with the liquid phase at each temperature. From Bodnar (1982).

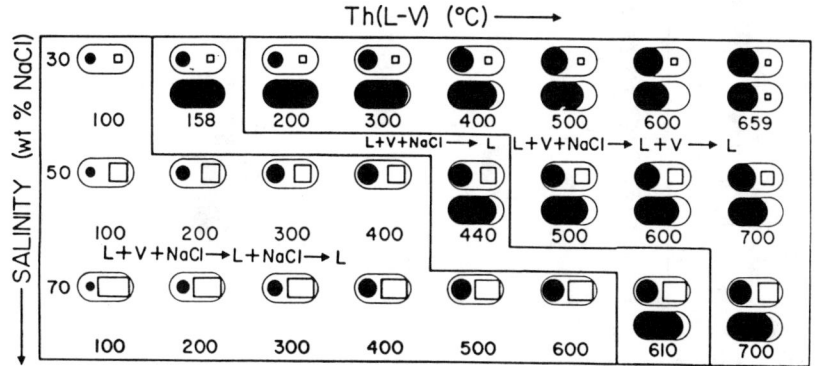

Figure 9-13. Room-temperature phase relations of fluid inclusions having salinities of 30, 50, and 70 wt % NaCl and Th of 100 to 700°C. Also shown are the room-temperature phase relations of inclusions that trapped the vapor phase that would have been in equilibrium with the liquid phase at each temperature. Vapor-rich inclusions are not shown with inclusions homogenizing by halite disappearance (left side of figure) because these could not have been trapped in equilibrium with a vapor phase. From Bodnar (1982).

DENSITY OF FLUIDS[11]/

Estimates of Density from Room-Temperature Phase Ratios

Unlike the estimates of pressure and temperature, estimates of the density of the ore-forming fluids based on fluid inclusions can be reasonably accurate and unambiguous and no other source for such information seems to exist. If the relative volumes of the liquid and gas phases are determined at room temperature (by using geometrically regular inclusions), and if the salinity is known (from freezing data), the density of the originally homogeneous fluid can be estimated (see also p. 103). Thus, simple linear intercepts of the phase boundaries on long tubular inclusions give an approximate phase ratio; thin flat inclusions can be photographed or sketched onto paper, using a camera lucida, and the relative volumes of liquid and vapor can be obtained by using a pantograph or (much more easily) by cutting out and weighing the two "phases" on a good balance. Although these methods can be highly precise, their accuracy can be very poor when the inclusion walls are not truly parallel, as is common. Inclusions that

----

[11]/ In the Russian literature (e.g., Ermakov, 1950), this subject is usually discussed in terms of the "state of aggregation," an arbitrary dividing line being drawn between "gas" and "liquid" at 50 vol %. Naumov and Naumov (1980) presented a summary plot of density vs Th for 1015 studies from the literature; unfortunately, the procedures used, and hence the validity of the data, as well as the geologic environments, probably vary widely.

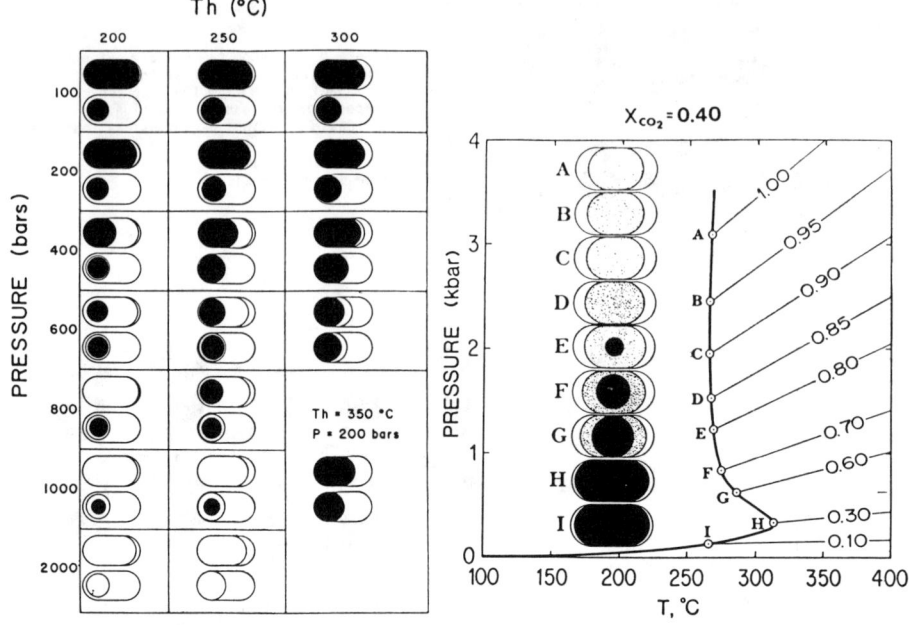

Figure 9-14. Phase relations at 25°C of coexisting H_2O-CO_2 inclusions trapped in the two-fluid-phase field. From Bodnar (1982).

Figure 9-15. Calculated phase ratios at 25°C of nine H_2O-CO_2 fluid inclusions, all with the same composition of 40 mole % CO_2, but with different temperatures and pressures of homogenization (points A-I along solvus). Inclusion densities (g/cm³) were calculated from the PRK equation of state and are shown on the isochores originating at the temperature and pressure of homogenization and extending into the one-phase field. Inclusions represent cylinders with hemispherical ends (capsules) as viewed at the level of maximum width of the cylinder. The unshaded area represents the liquid H_2O-rich phase, the stippled area represents liquid CO_2, and the shaded area represents CO_2 vapor. Solvus data is from Takenouchi and Kennedy (1964). From Bodnar (1982).

have an appreciable third dimension can also be measured, but the errors involved become much larger unless a universal stage is used.

At first glance, it might appear that the volume of the spherical gas bubble could be obtained accurately, from its diameter, but that the volume of the liquid phase (total inclusion volume minus gas-bubble volume) would be inexact, except in very geometrically regular inclusions, whose volumes can be assumed to consist of the sum of measurable cylinders, cones, pyramids, or prisms. Unfortunately, not only may the inclusion volumes have gross errors, but the volume of the spherical gas bubble may also be surprisingly inexact. The negative-lens effects of a curved inclusion wall may cause an error of as much as 50% in the estimate of the bubble volume in minerals of having a high index of refraction, such as sphalerite (Roedder, 1972, Plate 11, Figs. 7, 8). Measurements of vapor-bubble diameters can be very inaccurate even in flat-sided inclusions, and to decide what level of focus (if any) gives the true diameter of very small bubbles is sometimes difficult. Even a mineral like fluorite that has a low index of refraction can give inaccurate data if the inclusions are tetrahedral and are not viewed perpendicular to a flat face (Fig. 9-10). Even larger errors can occur if the bubble is deformed by contact with the walls. Volume changes from external or internal pressure, and from thermal expansion of the host, are negligible compared with the effects of fluid composition and inaccuracies in phase-volume measurement on density estimates.

Most solutes increase the density of water solutions over the nominal 1.0 g/cm^3, and the effect of the vapor bubble is in the opposite direction. Thus the saline brines found in Mississippi Valley-type deposits have room temperature densities as high as 1.2 g/cm^3; when such brines are heated to Th, the expansion to eliminate the vapor bubble (~10 vol %) still results in a density of ~1.08 g/cm^3 at Th. In environments in which the overlying rocks provide a confining pressure on the fluids, fluid movement will be controlled by pressure differentials independent of gravity, but in a system at hydrostatic pressure, the density of the fluids will control the flow. In the Mississippi Valley-type deposits, the fluid densities are almost always greater than 1.0 g/cm^3, even at Th. As a result, the hot ore-forming fluids can displace fresh cold meteoric and ground waters at a density near 1.0 g/cm^3.

The room-temperature density of the liquid phase must be estimated from the salinity (as determined on the freezing stage), or, if chemical analyses have been made, it can be measured experimentally on synthetic fluids of the same composition (see Potter et al., 1975). Brown (1942) has shown that even very small compositional differences may result in density differences that can be significant in surface-water circulation patterns. However, the density increase from dissolving 1 wt % salt would be cancelled out by a 30°C rise in temperature. Rubin and Roth (1979) have given a theoretical discussion of instabilities in ground water resulting from the opposing effects of temperature and salinity gradients.

Bodnar (1983) has shown that the volume and density of even irregular inclusions can be estimated from bubble diameter, Tm ice, Th L-V, and the volumetric properties of the fluids. The last involves a least-squares-fit polynomial equation describing the relationship between density, temperature, and salinity of vapor-saturated H$_2$O-NaCl solutions (Fig. 9-11). Bodnar's method makes use of an important relationship discovered by Clynne and Potter (1977), who showed from experimental studies with brines in the system Na-K-Ca-Mg-Cl-Br-SO$_4$-H$_2$O that the P-V-T-X properties of these brines are predicted within ±1% by the properties of an NaCl solution having the same depression of the freezing point (i.e., Tm ice). The method does not require the assumption of the same density for the unknown and the equivalent NaCl solution, which would be invalid, but only that the percent change in density from room temperature to Th will be closely equivalent. The major inaccuracy in the use of this method will probably result from the difficulty of measuring the true diameter and hence volume of the bubble, as mentioned above.

Phase Ratios at Room Temperature from Various P and T of Trapping

Bodnar (1982) has calculated, from the best data available, the 25°C phase volume ratios in inclusions of various compositions trapped at various P and T. These phase ratios have then been presented both in tabular form and in the form of the appearance under the microscope of a "standard" cylindrical inclusion with hemispherical ends. Figures 9-12 to 9-15 give a few of Bodnar's many diagrams. These diagrams make it easier to recognize possible pairs of inclusions from an immiscible separation (e.g., boiling) and to estimate the possible P and T of trapping. The angle of view of such an inclusion under the microscope will make a large difference in the apparent phase ratio.

Chapter 10

INTERPRETATION and UTILIZATION of INCLUSION MEASUREMENTS: METASTABILITY

CONTENTS

INTRODUCTION

The term _metastability_ is used here in a broad sense, to include almost any temporary configuration other than the most stable, lowest energy, or equilibrium state. I say "almost any" because inclusions are imperfections in otherwise more perfect crystals; hence, they themselves must generally be considered as examples of metastability in the broadest sense. Much of the following has been adapted from Roedder (1971a).

Inclusions are very small systems to consider, even from an atomic viewpoint. Thus, an inclusion $10\mu m$ in diameter containing a 30 wt % solution of a salt may have only 10^{11}-10^{12} "molecules" of that salt; at 10 ppm PbS it would contain only 25 million "molecules" of PbS, enough to form a crystal only 200 unit cells on an edge. The smaller the system, the more common metastability becomes. Hence, not surprisingly, metastability -- resulting from failure to nucleate new but stable phases -- becomes a problem that can be observed in (and frequently interferes with) many types of inclusion studies. It is common in samples as found, even though they have had geologic time periods to equilibrate, and it is even more commonly observed in the phase changes during the much shorter time spans of laboratory experiments on inclusions.

Metastability can cause serious errors in a variety of inclusion measurements and has not always been given the attention in inclusion studies that it deserves. A large fraction of the confusing and mutually contradictory data that students have brought to me arises from their (unknowing) combination of data from metastable and stable equilibria, or from their attempts to interpret data on metastable phase assemblages in terms of stable equilibrium phase relations. Metastability thus can cause problems, but it can also be useful, and it does add a little piquancy to otherwise plodding, dull, inclusion data gathering.

The most common type of metastability arises from the absence of one or more phases that should have nucleated in an inclusion after some change in conditions. Nucleation processes are generally classified as homogeneous or heterogeneous. The former are based on the probability of spontaneous formation in one phase of a cluster of atoms, ions, or molecules in the configuration of a new, more stable phase, of a size adequate to preclude immediate elimination or dispersal by forces such as surface tension or simple thermal motion. Hetero- geneous and homogeneous nucleation processes are similar in many respects, but heterogeneous processes occur at an interface between phases that may include either integral parts of the system or extraneous matter such as dust particles.

As a result of the random processes involved, the probability of homogen- eous nucleation increases with time, volume, and degree of supersaturation. How- ever, the inclusion "systems" under examination are so small that even geologic time is often inadequate to compensate fully for the small volume; hence, meta- stability is common in many inclusions as found, and even more common in those that have been altered by preparatory or laboratory conditions. The probability of heterogeneous nucleation also increases slightly with time, but as the phase boundary responsible may be caused by an impurity as small as a dust particle, such nucleation is also volume-dependent; systems of large volume are more likely to contain at least one such nucleus, whereas many small systems may contain none.

In addition to a metastable phase assemblage within the inclusion, inclu- sions may be metastable (i.e., have higher than minimum energy) as a result of their shape, their position in the crystal, or other features. Natural elimina- tion of such metastability is a continuous process, in contrast to the discon- tinuous nature of nucleation, but as long as the rates are slow enough to permit observation, such metastability can be considered as equivalent to that from failure to nucleate.

The aims of this chapter are to show that the validity and significance of most inclusion studies are based in part on the degree to which these various types of metastability are recognized and correctly evaluated, and that although these deviations from stable equilibrium conditions can cause severe experimental difficulties, many of them provide valuable data, sometimes otherwise unavail- able, on the composition of inclusions (Roedder, 1963, 1967b, 1971a; Coolen, 1980).

METASTABLE PHASE ASSEMBLAGES IN LIQUID INCLUSIONS

Failure to Nucleate in Nature

Nucleation of a vapor bubble. The most frequent and obvious example of metastability in inclusions is the failure of some to form a vapor phase (bubble) on cooling. As the bubble is a measure of the differential shrinkage of the liquid and the surrounding crystal host on cooling, inclusions trapped at surface temperature never have bubbles formed in this manner[1]. However, even in min- erals formed at elevated temperature, the smallest inclusions in any cogenetic group commonly lack a bubble. The size at which the division occurs between those inclusions with bubbles and those without, though never precise, is sur- prisingly consistent for a given sample and varies inversely with the volume percent vapor that should be present. Thus, all inclusions larger than 1 μm (and some smaller ones) in minerals formed at high temperatures (>350°C) may have bubbles, whereas inclusions as large as 20 μm in some minerals formed near 100°C seldom show bubbles. Aqueous inclusions formed at 70°C may be as large as 100 μm and still not nucleate a vapor bubble (Roedder, 1967a).

----

[1] The converse of this statement (i.e., inclusions without bubbles must have formed at surface tempera- ture) is not generally valid, as detailed below.

292

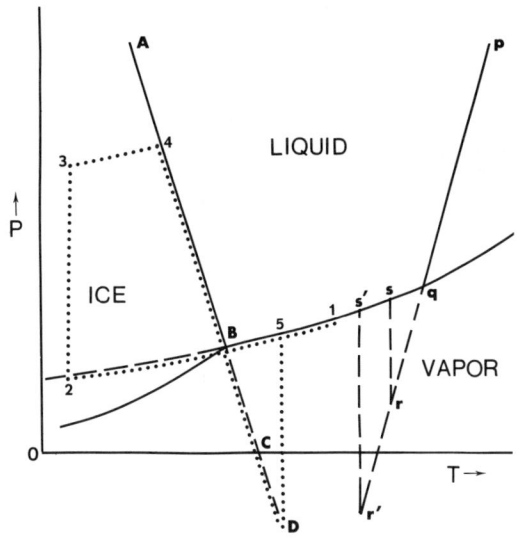

Figure 10-1. P-T plot of the phase diagram for water (not to scale) showing the stable phase boundaries (solid lines) and metastable extensions (dashed) in the immediate vicinity of the triple point B (0°C and 4 mm pressure). (See Figs. 8-6 and 10-4 for true-scale plots.) A homogeneous inclusion cooled from point (p) should nucleate vapor at (q); if it does not, it continues on the metastable extension of isochore (p-q) to some point (r) or (r'), where vapor nucleates in the metastable stretched liquid and the inclusion returns to the L + V line. The several types of metastability commonly encountered on cooling inclusions are illustrated by paths 1 through 5. See text for details.

The explanation for the lack of a vapor bubble is usually assumed to be simply metastability from lack of nucleation. Thus, an inclusion trapped at point "p" on Figure 8-1, on cooling, would follow isochore "p"-"q". At point "q," it should nucleate a vapor bubble of specific volume "q'," but if it does not, it follows "pq" extended along the (hypothetical) metastable curved surface until finally at some point "r" it nucleates a bubble and hence rises to the surface ABCcp. Figure 10-1 shows the behavior of the same inclusion on a P-T plot. The point "r" at which nucleation occurs can fall anywhere along the line pq extended and may even fall in the area of negative pressure ("r'" on Fig. 10-1; discussed below).

The full explanation probably involves several superimposed phenomena. First, surface-tension forces result in higher pressures inside bubbles, in inverse ratio to the radius of the bubble. As the solubility of gases in liquids increases with pressure, and surface tension causes high pressures in small bubbles, very small bubbles in a free liquid or foam collapse instantaneously by dissolution of the gas in the liquid. Somewhat larger bubbles dissolve rapidly, effectively transferring their gas to other, still larger bubbles[2]. Fluid inclusions are essentially isochoric (fixed-volume) systems, however, and hence yield very different results. Whether the bubble is undissolved gas or merely the vapor of the surrounding fluid, the only way surface tension can cause it to collapse under isothermal isochoric conditions is to stretch the liquid, perhaps even into the range of negative pressures (Roedder, 1967a). The size division between inclusions with and without bubbles thus reflects the balance between surface tension tending to collapse the bubble and shrinkage (internal tension) of the liquid trying to form it. As the surface tension varies inversely with the radius of the bubble (and hence inversely with the cube root of the bubble volume), but the internal tension is independent of volume, there is a bubble volume (and, hence, an inclusion volume for a given composition and total density) below which the fluid will remain stretched indefinitely (i.e., without nucleation of a bubble), as the stable configuration. Even if a bubble did

----
[2] Such phenomena are very easily observed under the microscope in soapsuds or in an effervescing microchemical reaction.

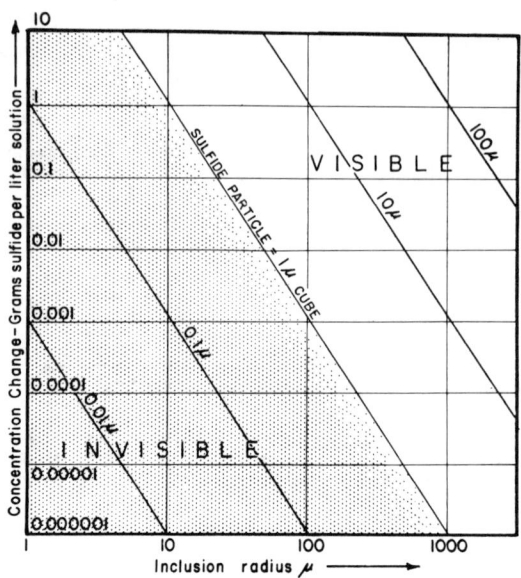

Figure 10-2. Diagram showing the size of the sulfide particle (cube edge) that should precipitate from a given fluid-filled inclusion (spherical) with a given concentration change. Based on density of solution = 1, sulfide = 5. The shaded area represents region below the limit of resolution of the light microscope. From Roedder (1960a).

nucleate, the surface tension would cause it to "blink out" instantly, thus putting the liquid under tensile stress.

Second, superimposed on this size effect is the phenomenon of difficulty of nucleation of a bubble in those inclusions that are slightly above the limiting volume. Several lines of evidence indicate that this nucleation is probably heterogeneous, thus accounting for some if not most of the irregularities in the size limit between those inclusions that have nucleated bubbles and those that have not.

Nucleation of other phases. The fact that inclusions are isochoric systems is thus very important in the stability of a bubble, once nucleated. However, other phases that may nucleate and grow, such as daughter crystals of NaCl, generally result in liquid volume changes that are small or even negative (Roedder, 1967b, p. 557); hence, their presence or absence is almost completely controlled by nucleation phenomena alone. Halite apparently nucleates with relative ease, as it is found even in very tiny inclusions, and only rarely are fluids found to be supersaturated in halite (Touray and Sabouraud, 1970). Many groups of strongly saline inclusions contain salt crystals in the smallest inclusions resolvable under the light microscope (<1 μm). However, a crude but nonetheless real size threshold can be observed in the cogenetic groups of multiphase inclusions. These thresholds are not simply a matter of difficulty of observation, as size estimates based on the phase volumes in larger inclusions indicate that many of the missing phases should be present as crystals plainly visible under the light microscope (Fig. 10-2). Because all the larger inclusions in such groups appear to have the same assemblage of daughter minerals in the same volume ratios, none show evidence of necking down, and all occur in the same healed fracture, the bulk composition of each individual inclusion is presumed to be the same.

Another possible interpretation is based on the suggestion of significant compositional differences between large and small inclusions, due to surface

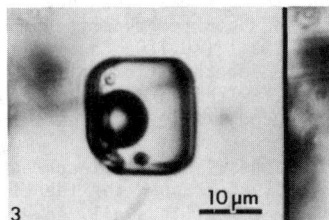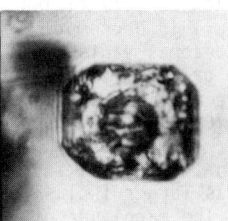

Figure 10-3. Primary fluid inclusion in anhydrite from DW-5 core, Hohi field (near Hachobaru), Kyushu, Japan, as found (a) and after a freezing run (b); both photographs taken at room temperature. New daughter crystals (gypsum?) have formed by reaction of the fluid with the walls, resulting in ~25 vol % enlargement of the inclusion (b). Scale bar in μm. Photographs courtesy of M. Sasada.

phenomena in the fluid at the time of the trapping. This explanation of these differences is probably not valid because of the lack of intermediate types between inclusions containing a full-sized crystal of a given phase and those having none, and the fact that in multiphase inclusion groups, different phases are missing in the various inclusions of the same size.

Reaction of the inclusion fluid with the walls during cooling in nature should always be considered as a possible cause for the formation of new phases, but such reaction has seldom been observed. Thus, one might expect that the CO_2 of inclusions in the olivine of basalts (Roedder, 1965a) might react to form magnesite on cooling, and that scapolite might form on the walls of saline inclusions in alkali feldspar (Roedder and Coombs, 1967), but neither mineral was found. The lack of such reactions may be the result of relatively rapid cooling in nature, yielding metastable assemblages. The formation scapolite may require more saline fluids than are commonly found (Vanko and Bishop, 1982).

One particularly interesting example of metability from failure of the inclusion fluid to react with the walls to form a new solid phase was found in the Hohi geothermal field in Japan (Fig. 10-3). Anhydrite crystals in core samples from this field have trapped primary inclusions of low-salinity fluids with Th ~265°C (Taguchi, 1981). On cooling to room temperature, these fluids should react with the walls to form gypsum, but this has not taken place. New crystals (presumably gypsum) formed in some of these on freezing; on return to room temperature these crystals persisted and the volume of the inclusion had increased greatly. Similar behavior for inclusions in anhydrite has not been reported before, probably because most anhydrite crystals with usable inclusions have been from porphyry copper deposits, where the solutions are strongly saline, thus precluding the reaction with the walls on cooling to form gypsum.

Failure to Nucleate in the Laboratory (Other than Superheated Ice)

Nucleation of a vapor bubble. Most workers on inclusions have reported a difference between the temperature at which an inclusion homogenizes on heating (Th L-V) and the (lower) temperature at which it renucleates a bubble on subsequent cooling (Tn V). The magnitude of this difference, or hysteresis, is generally decreased but never eliminated by the use of greater care and time in the experimental technique, to eliminate temperature lag. The magnitude of the hysteresis varies but is generally in the range of 1°C to several tens of degrees for ordinary saline fluid inclusions. Nucleation occurs rapidly in most inclusions on cooling, but after homogenization, some low-temperature inclusions may be cooled back to room temperature and held for hours or even weeks without a bubble nucleating. In close analogy with the natural nucleation of bubbles discussed above, the larger the inclusion in a given group, the smaller the hysteresis. Several workers have reported small differences in Th of inclusions of a given cogenetic group, the smaller inclusions always homogenizing at slightly lower temperatures. Tn V in the smaller inclusions is also lowered by an even larger amount.

Since hot, highly expanded, low-density fluids are more gas-like they should show a smaller hysteresis. This presumption is supported by most homogenization data, and the hysteresis becomes vanishingly small (<0.1°C) for inclusions homogenizing near or at the critical temperature for that fluid. Thus CO_2 inclusions homogenizing in the range 29-31°C (i.e., within a degree or two of the critical temperature of 31°C) show almost no recognizable hysteresis.

The above discussion shows that all small inclusions will yield incorrect (low) values for Th L-V(L), whereas larger inclusions that trapped the identical fluid will yield correct values. The magnitude of the effect is small, however, and can generally be ignored. Thus, if we assume that a bubble of 1 μm diameter is the approximate lower limit, below which surface tension will cause bubble collapse, even though in so doing the liquid is stretched to less than its correct density, a spherical inclusion of 2.3 μm diameter, containing a bubble and 10% NaCl solution, trapped on the boiling curve at 350°C, would yield a Th of only about 300°C, 50° too low. The effect is extremely dependent on the volume of the total inclusion; thus, a spherical inclusion of 13 μm diameter, filled under identical conditions, would have an error of only ~1°C in its Th. Accurate calculations of this effect are not possible now because the necessary data are not available (surface tension and extensibility of metastable inclusion fluids at Th, under "negative pressure")[3]. In fact, inclusion data themselves may eventually provide the best estimates of these missing data.

All these observations are compatible with the same two controls on the formation and growth of a gas bubble discussed in an earlier section -- the bubble can nucleate successfully, and be stable, only when there is an appropriate nucleus and the final radius is adequate to reduce the surface tension to less than the internal tension. Conversely, on heating any inclusion near to its Th, the bubble decreases uniformly down to a certain size and then suddenly vanishes, at the point at which the surface tension can stretch the fluid and permit the collapse. The smaller size of the bubbles in small inclusions of a given group simply permits this to occur at a temperature much further below Th (i.e., greater hysteresis). Although this explanation seems adequate to cover the known facts, several studies of the behavior of water in contact with various surfaces, as in capillaries, have shown poorly understood discontinuities or differences with temperature, capillary radius, or other variables (e.g., Muller and Schufle, 1968) that may also occur in inclusions.

Irregularities in the magnitude or time of duration of the hysteresis within a group of inclusions are attributable to the vagaries of heterogeneous nucleation (Brinson, 1966; Apfel, 1970), which is probably the most common type. Thus, inclusions that always renucleate the vapor bubble at the same spot on the wall (or on a daughter mineral) are not rare. Unfortunately, if the nucleation center or "mote" responsible for nucleation is invisible and free floating, the bubble can nucleate anywhere in the inclusion, as if from homogeneous nucleation.

The smallest stable bubbles found in any inclusions are those in very small, low-temperature, highly saline inclusions close to homogenization. These bubbles can be well under 1 μm in diameter. Although such bubbles are difficult to measure, as they are generally moving rapidly, they are obviously much smaller than any that can be maintained in a free fluid. The specification of high salinity is important because it decreases the surface tension of the fluid.

Nucleation of other phases. A variety of metastable phenomena are observable on cooling fluid inclusions below room temperature. Gross supercooling of aqueous solutions occurs universally (along lines Ad and Cd' in Fig. 8-1); for

----

[3] Although true negative pressures are rather commonly observed in inclusions as discussed below, the pressures in the above example are positive but less than these particular compositions would show under stable equilibrium, as pointed out by R.J. Bodnar (pers. comm.).

this reason, it is necessary to freeze an inclusion to form ice first and then determine the temperature of <u>disappearance</u> of the phase ice during heating (i.e., Tm ice), rather than trying to determine the temperature of <u>crystallization</u> of ice (i.e., Tn ice). This is also the reason for discontinuing the use of the terms "freezing temperature" or "depression of the freezing point," because these terms are misleading in their implications of the procedure used. Coolen (1980) described extreme metastability in CO_2 inclusions (see Fig. 13-8) that permitted obtaining otherwise unavailable density data.

As in the crystallization of most substances, there is some degree of supercooling at which nucleation is easiest; above and below this temperature, nucleation becomes less likely. Water solutions seem to freeze most readily roughly 40°C below Tm ice. On further cooling, however, viscosities increase greatly, and both nucleation and growth of crystals become very slow. Hence, rapid cooling to liquid-nitrogen temperature may take the inclusion through the optimum temperature range too fast to freeze. Even very slow cooling (<7 days from room temperature to -78°C; Roedder, 1963, p. 206) is inadequate to freeze some aqueous inclusions (e.g., see Fig. 13-1D). I have found that a solution saturated at room temperature with respect to the chlorides of Na, K, Ca, and Mg, which is syrupy at room temperature, will actually crack with a sudden blow at -196°C -- it is a clear brittle <u>glass</u>. Even at intermediate temperatures (-78°C) ice is extremely difficult to nucleate and grow in such bitters (Roedder, 1963, p. 205). Once a glass is formed, further cooling is counterproductive; crystallization will proceed, however, <u>on warming</u> to the range for optimum nucleation and crystal growth. Angell and Sare (1970) have determined the glass transition temperature for various binary salt-water systems. It is lowest for eutectic compositions, and for LiCl and $ZnCl_2$ solutions (~-130°C).

Supercooling is most troublesome experimentally in small inclusions (Roedder, 1962a; 1963, p. 206) and in strong brines. The supercooling provides some useful information, however, in that it suggests extremely slow rates of flow of the fluids prior to trapping (Roedder, 1963, p. 175). When ice forms, the increase in concentration of solutes in the liquid phase may cause the precipitation of crystals of new daughter minerals that persist on warming to room temperature, indicating that the fluids were originally supersaturated at room temperature. Such processes have yielded halite or unidentified solids in some aqueous inclusions, and a variety of unknown solids in some oil inclusions.

During the melting or solution of daughter minerals, compositional gradients are established that must be eliminated by diffusion before equilibrium is obtained. (Some phases, however, such as $NaCl \cdot 2H_2O$, are much more sluggish in their approach to equilibrium than others, so presumably other rate-limiting steps exist in addition to diffusion.) Large and particularly elongated or amoeboid inclusions that have nucleated a crystal of ice (or salt) in one end may require 10 times as long to reach equilibrium as smaller, more equant inclusions. Similarly, gravitational stratification of inclusions occurs commonly, due to the floating of ice crystals and their subsequent melting to form a lower concentration and hence lower density brine. These types of nonequilibrium are particularly misleading in that a reversal in the direction of temperature change will lead to an equivalent reversal in the direction of phase change (e.g., from ice melting on warming to ice crystallizing on cooling). Such behavior represents, of course, only local equilibrium within part of the inclusion, but it is difficult to avoid the assumption that such behavior represents equilibrium for the whole inclusion system, and gross errors in Tm ice will result if these conditions are not recognized. This is the main reason why I do not believe in the use of NaCl solutions as freezing standards in the calibration of stages; as normally made, such synthetic standards are large, as compared with the volumes of most natural inclusions, and stratification introduces major errors. If the procedure described by Ishkov and Reyf (1980), in which an emulsion of the standard solution in epoxy resin is made, could be shown to yield valid <u>small</u> synthetic inclusions, this objection would be removed, but it seems to have short-comings (see Calibration, Chapter 7).

The equilibrium nucleation and growth of some hydrate phases is difficult to achieve experimentally, but the sluggishness of the process permits some useful observations. Thus, the slowness of the reaction of halite + ice (or solution) to form $NaCl \cdot 2H_2O$ [4] sometimes permits the recognition of this mode of formation (and hence the identification of the original NaCl) in inclusions that are otherwise hopelessly complex. Similarly, the almost instant separation of metastable liquid CO_2, instead of the stable but sluggishly crystallizing, almost invisible clathrate compound $CO_2 \cdot 5.75H_2O$, may permit the identification of CO_2 in inclusions where it would otherwise be missed (Fig. 4-9).

Superheated Ice Under High Negative Pressure

Metastability that results from failure to freeze on cooling can be frustrating but normally does not lead to misunderstanding or error. However, one aspect of metastability -- the occurrence of metastable superheated ice in liquid-water inclusions at high negative pressures -- can lead to serious errors in Tm ice if not recognized. The term "negative pressure" is somewhat esoteric[5], but the effects are of considerable practical significance to inclusion studies. Although negative pressures are not encountered in many inclusion studies, they are very commonly observed in some, particularly studies of low-salinity inclusions formed at low to moderate temperatures.

Liquids can be metastably stretched, under certain conditions, to occupy a larger volume (at a lower density) than that called for at equilibrium. The pressure on the liquid is then less than the (hypothetical) vapor pressure and often less than zero (i.e., negative pressures). Such metastable pressure will persist as long as nuclei for a vapor phase are absent or do not form. Negative pressures are surprisingly common and are of considerable significance in various biological systems and in many processes involving fluids, such as superheating in steam boilers and cavitation around ship propellers. Certain fluid inclusions show more extreme static negative pressures than ever recorded in any of the rather extensive biological, medical, and engineering studies of the subject (Roedder, 1967a).

In inclusion studies, this type of metastability can occur only if the vapor bubble in an H_2O-rich inclusion is eliminated by expansion on freezing, and it will cause error in the determination of Tm ice only if a vapor bubble is absent when the last ice melts on warming. It is a result of the negative slope of the boundary curve for ice plus liquid (DB and DC on Fig. 8-1). This curve is shown in greater detail and true scale in Figure 10-4.

The phenomenon is best illustrated by following a hypothetical pure-water inclusion on Figure 10-1, starting with liquid plus vapor at room temperature, point 1. On cooling, the system follows the liquid-plus-vapor curve to the triple point B; the temperature is now 0°C, and the pressure in the inclusion is the vapor pressure of water at this temperature, 4 mm. The inclusion should freeze on further cooling but instead always supercools, following the metastable extension of the liquid-plus-vapor curve (dashed). At some point 2, ice nucleates, and in this example, its expansion on freezing eliminates the vapor, leaving only ice in the inclusion. The inclusion is now at some indeterminant point 3 in the ice field. The pressure is high (perhaps hundreds of bars) but is not

---

[4] Some other aspects of the metastability of $NaCl \cdot 2H_2O$ (hydrohalite) have been discussed in Chapter 8.

[5] The concept of negative pressure may seem contradictory, but it is not taken from Alice's Wonderland, as one might think. The synonym "overexpanded liquid" is sometimes used. The concept started with studies of the cooling of water-filled glass "Berthelot tubes" (Berthelot, 1850).

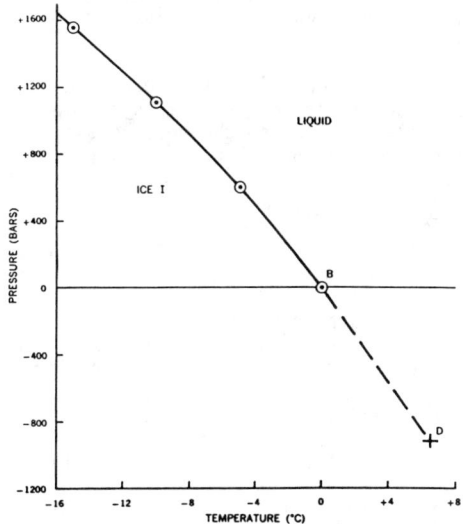

Figure 10-4. P-T plot of the melting curve for ice I. Circles are data points from Bridgeman (1912). B-D is a possible (straightline) extrapolation into the region of negative pressure to the maximum observed temperature of superheated ice (D). Figure 10-1 shows the detail in the immediate vicinity of triple point B.

readily determined, as it is controlled by the original G/L ratio and the compressibilities of ice and the host crystal. On warming, the inclusion moves along some course toward AB, and on intersecting the ice/liquid boundary, at 4, it develops some liquid water. On further warming, more ice melts, forming the denser liquid phase, and the pressure drops rapidly to 4 mm at point B, where a vapor phase should (but generally does not) nucleate. If vapor does not nucleate, the inclusion follows the metastable extension BC down to zero pressure, at a temperature very slightly above 0°C, and continues on CD into the range of negative pressures and to temperatures as high as +6.5°C. Along BCD, the inclusion still has only stretched water plus ice in it, and its behavior gives no hint of metastability; on slight cooling, the ice crystal grows, and on slight heating, more ice melts. This metastable equilibrium may be maintained for seconds, minutes, or even hours. Then, at point D, which is at some nearly arbitrary temperature and time, a vapor bubble suddenly nucleates and expands to relieve the metastable stretching of the liquid. The ice crystal is now unstable and melts almost instantaneously to yield the stable assemblage of liquid plus still more vapor, at point 5. Various other sequences of phase changes may be followed, as detailed below.

The magnitude of the negative pressure within inclusions having superheated ice is estimated to be at least 900 bars, and possibly more than 1000 bars, but there are numerous uncertainties. A straight-line extrapolation (Fig. 10-4), derived from the Clausius-Clapeyron equation and data on water and ice at 0°C (Bridgeman, 1912) yields a value of more than 900 bars at +6.5°C, and any curvature is likely to increase this value. The validity of this extrapolation is uncertain, because it requires the assumptions that the heat of melting and the thermal expansions and extensibilities of ice and water under negative pressure are comparable to the equivalent values under positive pressure. Although these assumptions are at least approximately true, some complications exist (Roedder, 1967a). Unpublished data of Murck and coworkers (Hollister et al., 1981) indicate that superheated ice can be obtained in very small inclusions at temperatures as high as +20°C.

The above example, for pure water, showed ice at temperatures above 0°C, and hence was obviously strange. However, saline fluids can behave similarly,

299

and in such fluids, the liquid-plus-ice curve ABCD in Figure 10-1 can all be offset to temperatures as low as -20°C; in such examples, the existence of metastability may be much less evident.

Five different sequences of phase changes are possible in frozen aqueous fluid inclusions, only one of which (sequence (1) below) represents stable equilibria throughout. The equilibrium sequence of phase changes, usually obtained when a frozen inclusion (consisting of saline solution and vapor bubble at room temperature) is warmed, is:

$$
\begin{array}{ccccccc}
 & & & & Tm & & \\
I+S+V & \to & I+S+L+V & \to & I+L+V & \to & L+V \\
\sim -50°C & & & & & <0°C &
\end{array}
\tag{1}
$$

where I represents ice crystals, S represents salt crystals (mainly $NaCl \cdot 2H_2O$), L is liquid, V is vapor, and Tm is the (stable) temperature of melting of the last ice crystal. When the vapor bubble is eliminated, any of four different in-part-metastable sequences may be followed, depending upon the quantities of the various phases and the temperatures attained before nucleation of vapor, and a given inclusion may follow different routes on successive runs. Each of the following four sequences has been found in natural inclusions (Roedder, 1967a):

$$
\begin{array}{ccccccccc}
 & M & R & & & Tm & & & \\
I+S & \to & I+S+L & \to & I+S+L+V & \to & I+L+V & \to & L+V
\end{array}
\tag{2}
$$

$$
\begin{array}{ccccccccc}
 & & M & R & & Tm & & & \\
I+S & \to & I+S+L & \to & I+L & \to & I+L+V & \to & L+V
\end{array}
\tag{3}
$$

$$
\begin{array}{ccccccc}
 & & M & R & & & \\
I+S & \to & I+S+L & \to & I+L & \to & L+V
\end{array}
\tag{4}
$$

$$
\begin{array}{ccccccccc}
 & & M & & M & R & & & \\
I+S & \to & I+S+L & \to & I+L & \to & L & \to & L+V
\end{array}
\tag{5}
$$

where M indicates a metastable assemblage under negative pressure, and R indicates a rapid phase change.[6] The latter half of the example discussed above and shown in Figure 10-1 follows sequence (4), but with added salt, and is probably the most common.

Sequence (2) is difficult to recognize with certainty, because of the mass of solids present, but the moment of nucleation in (3), (4), and (5) is readily recognized by an apparently instantaneous appearance of a vapor bubble in the liquid and an almost instantaneous melting of ice. Because the metastable assemblages may persist for as long as 4 hours, the actual melting of ice when the negative pressure is released is difficult to observe. On several occasions, however, other observers and I have witnessed it, and I am certain that there is a definite interval of time (estimated at 0.1 to 0.2 second) over which the melting occurs after nucleation. The melting front must advance through the ice crystal at rates on the order of 200 µm/sec.

Sequences (2) and (3) permit valid determinations of Tm ice after nucleation of vapor, but (4) and (5) provide only maximum values. Even though the ice in sequences (2) and (3) before nucleation is at subzero temperatures, it is still "superheated," in that it is in metastable equilibrium with a brine at a temperature too high for the salinity.[7] The presence of superheated ice is most

----

[6] On rare occasions, the assemblage I+S+V is seen to form I+S+L on heating from still lower temperatures, when the vapor phase was small enough to be eliminated by thermal expansion of the solid and liquid phases.

[7] The temperature above which vapor should be present at equilibrium can be determined by recooling, after vapor nucleation, to the point where expansion from growth of ice just eliminates the vapor bubble.

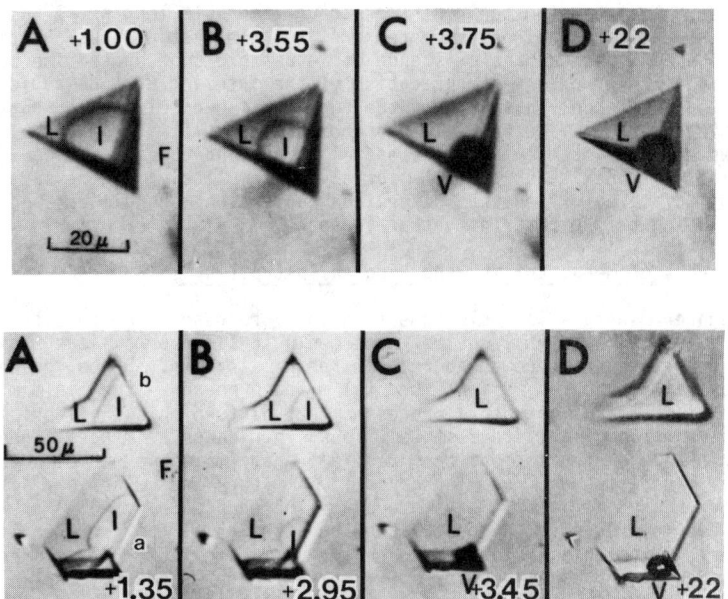

Figure 10-5. Sequential photographs showing metastable ice in inclusions in fluorite from Poncha Springs, Colorado, USA, equilibrated at the temperatures indicated (°C). The upper series shows sudden nucleation of a vapor bubble between B and C, via sequence (4) at +3.65±0.1°C. The time between individual photographs was 5 to 15 minutes. I, ice; L, liquid water solution; V, vapor bubble; F, fluorite host crystal. In the lower series, inclusion (a) also follows sequence (4), nucleation taking place between B and C, at +3.15±0.25°C; however, inclusion (b) follows sequence (5), last melting being between B and C at +3.15±0.25°C, with no nucleation of a vapor bubble. From Roedder (1967a).

obvious in inclusions of low salinity that follow sequence (4) or (5) and that have true Tm ice values near 0°C. In these, metastable ice crystals have been found to persist at temperatures as high as +6.2°C via sequence (4), and +6.5°C via sequence (5). Most such inclusions contain an estimated 90 vol % ice at 0°C, and many show as much as 70% at +2°C and 50% at +3°C (Fig. 10-5).

Nucleation of vapor appears to be somewhat random in time and temperature, as evidenced by considerable inconsistency on reruns of a given inclusion. However, as expected from the phase-volume relationships involved, those inclusions that yield metastable assemblages on one run are most likely to show them on another run. Also, the maximum temperature of existence of metastable ice via sequences (4) and (5) is seldom 10°C above the stable Tm ice, and usually is only a few degrees higher. The presence of nucleation sites with specific properties is suspected in some inclusions that show a tendency to nucleate in a relatively narrow temperature range. Although inherent variation among inclusions invalidates direct comparisons, for any given cogenetic group of inclusions, the probability of nucleation increases greatly with increase in temperature.

Repeat runs that follow the same sequence give essentially identical temperatures for phase changes involving reversible equilibria (stable or metastable); however, because the temperatures of the irreversible rapid phase changes by means of sequences (2), (3), (4), or (5) are controlled by nucleation, repeat runs that follow the same sequence generally give differing temperatures (for specific data, see Roedder, 1967a).

Other than extending the run times to unreasonable lengths, I know of no way of avoiding metastable superheated ice, but the most obvious question is the practical one -- how can the <u>existence</u> of metastable superheated ice be recognized, and hence error in interpretation of the freezing data be avoided? Careful observation is the only answer. If the bubble has been eliminated on freezing, metastable equilibria are probable (but not proved). If a vapor bubble nucleates and grows almost instantly under essentially constant-temperature conditions, and particularly if the ice phase melts (partly or completely) in the same instant, metastability certainly existed before the nucleation, and all data obtained on that inclusion before that instant are suspect.

Special Types of Metastability and "Frozen Instability"

Daughter minerals in inclusions frequently represent the results of nucleation and very slow growth (cooling over geologic time) or at least long periods of time at near-surface temperatures for reequilibration. Hence, daughter minerals generally occur as a single crystal of each phase -- the minimum energy configuration -- and laboratory processes that disturb this equilibrium are frequently not completely reversible. Thus, when daughter minerals renucleate and grow after homogenization, they generally have multiple nuclei and grow as feathery, dendritic, or irregular crystals. If they are relatively soluble in the fluid, recrystallization can take place in seconds (e.g., CO_2) or may require months, but less soluble phases may show no signs at all of recrystallization to the original single euhedral daughter crystal. In fact, the presence of a large number of very small crystals of a given daughter phase in an inclusion as found in nature normally may be used as valid evidence of a low solubility of that phase in the inclusion fluids at surface temperatures, and perhaps even at trapping temperatures. Similarly, the presence of multiple vapor bubbles in an aqueous inclusion (most commonly in elongated or amoeboid ones) usually signifies that they have formed recently, e.g., by leakage or laboratory manipulations, because the stable configuration in all but a few special geometric cases requires a single bubble, and the necessary reequilibration to cause coalescence is relatively rapid.

Several other less obvious types of disequilibrium should be mentioned. (1) The molecular species present, particularly in the gas phase, may represent equilibrium at the elevated temperature of trapping, or partial reequilibration on cooling, or no equilibrium at all (Barker, 1965a). Mississippi Valley-type deposits commonly have fluid inclusions containing appreciable quantities of both sulfate ion and methane, which Barton (1967) has shown to be a metastable assemblage. (2) In several examples the present phase assemblage may represent equilibrium under the present conditions but is not reversible to the original state. Inclusions that were originally a homogeneous organic fluid (oil) commonly split into two or more phases after trapping. The evidence for the original homogeneity lies in the uniform ratio of phases in many inclusions after splitting. I do not know whether the splitting signifies that the molecular configurations in the original oil were metastable, or that the inclusion has been subjected to some external influence (such as heat or natural radioactivity), or that it has changed composition (as through loss of hydrogen). The presence in the inclusions of some ore deposits of crystals of hematite or anhydrite that do not redissolve on heating similarly suggests loss of hydrogen at those sites (Roedder and Skinner, 1968). (3) Inclusion fluids may not maintain complete retrograde isotopic exchange, and hence isotopic equilibrium, with their host mineral during cooling (Rye and O'Neil, 1968). Similarly, isotopic equilibrium between molecular species in the fluid may not be maintained.

METASTABLE PHASE ASSEMBLAGES IN SILICATE MELT INCLUSIONS

In many respects, silicate-melt inclusions behave as do the aqueous inclusions just described, but at higher temperatures. The smaller inclusions frequently do not nucleate a bubble on cooling, but when nucleation does occur, the bubbles commonly form on the walls or on daughter minerals. As melt inclusions are usually held in a temperature range in which they can crystallize for a much shorter time than aqueous inclusions, they generally consist of metastable glass, without crystals, although the relative lack of completely crystallized melt inclusions in intrusive rocks may be merely a result of difficulty in recognizing them.

The viscosity of the hot melt and its great increase on cooling (Fig. 18-19) result in some differences between melt and aqueous inclusions. Multiple bubbles are common. They are quenched in place by rapid cooling. Crystallization sometimes is arrested after only very minute crystals have formed. This crystallization, from vast numbers of nuclei, yields a slightly brownish glass, sometimes with a very faint granularity in strongly collimated lighting. A few inclusions have feathery dendritic crystals in the glass, and only rarely are there large coarse single crystals of each daughter mineral, presumably signifying slow cooling (if regular in occurrence) or trapping of solid inclusions (if irregular). The olivine of the lunar samples and of some terrestrial basalts contains melt inclusions having perhaps the lowest energy state. These not only show single crystals of each daughter phase but these crystals are epitaxially oriented on the walls of the host (see Chapters 16 and 18). Heating experiments on these inclusions (Roedder, 1971c) show that melting of such daughter minerals, as well as renucleation and growth, requires hours, days, or weeks, rather than seconds as in most aqueous inclusions.

METASTABILITY RESULTING FROM SHAPE OR POSITION

The equilibrium shape of any inclusion is that representing a minimum total surface energy. This does not usually mean a minimum surface area (i.e., a spherical inclusion), but means instead negative crystal shape. However, the crystal faces frequently form a subspherical cavity. Such equilibrium shapes are seldom formed at the moment of trapping, and most inclusions at this stage are far from equilibrium shape. With time, the material of the walls dissolves and recrystallizes into a lower energy configuration, and those inclusions that have not changed shape are probably still in a metastable configuration. Most aqueous inclusions larger than 100 μm in relatively insoluble minerals have not changed shape much, evidently because the compositional gradients that are induced in the solution by differences in surface energy are simply too small to cause much material transfer over such distances, even in geologic time. Smaller inclusions in the same samples may recrystallize to globular shapes outlined by facets from many different crystal forms.

If the host mineral is essentially insoluble in the fluid of the inclusion, even small inclusions will retain their original shape from the time of trapping. Thus, some fluorite samples from southern Illinois have a few faceted negative crystal cavities containing brine, and thousands of almost perfectly spherical oil inclusions, each having a small flat side toward the crystal center marking the spot where the droplet first touched and wet the cube surface (Figs. 2-27, 28). Similar almost spherical outlines are shown by the CO_2 inclusions in the olivine of olivine nodules and phenocrysts in basalts (Fig. 17-2), presumably from the same cause. All such examples, however, provide no proof that the spherical form is not the lowest energy configuration.

The driving forces causing recrystallization and change in shape of an inclusion are miniscule and would generally be ineffective except that the distances involved are very short and the time is long. The process is fast enough to be actually observed in soluble salts, as described by Lemmlein (1951) and Lemmlein and Kliya (1952a; see Fig. 3-8). Simple elimination of surface area is probably the major driving force. Many inclusions initially have long tubular or amoeboid shapes. Their surface area can easily be cut at least in half by the necking-down process, yielding one or several smaller, more equant individual inclusions. When necking occurs in inclusions that have had a phase separation of either a bubble or a daughter mineral, the new inclusions have different individual densities and compositions. Not infrequently, the daughter mineral is left almost without liquid (Figs. 3-15 to 18). Such recrystallization, if not recognized, may result in a misunderstanding of the gross composition of the mineral-forming fluid, and is probably the major cause for the scatter observed in all careful studies of Th of supposedly coeval inclusions. Fortunately, the process of recrystallization is most rapid at elevated temperature; hence, most necking down occurs near to Tt, reducing the errors (Fig. 3-14). The effects of recrystallization are most important to the interpretation of the origin of inclusions. Many published criteria for inclusion origin are based on inclusion shape; I believe that recrystallization after trapping makes these criteria invalid, except under certain limited conditions.

In addition to a net reduction of surface area, recrystallization tends to change curving nonrational surfaces, such as those resulting from conchoidal fractures, quickly into surfaces consisting of several sets of crystal faces. These and other causes for change in shape or position have been discussed in Chapter 4.

Chapter 11
SEDIMENTARY ENVIRONMENTS

CONTENTS

INTRODUCTION

In this chapter, I discuss the fluid-inclusion studies made on sedimentary environments other than ore deposits. As only a few sedimentary environments provide material suitable for fluid inclusion studies, and most of these have not been studied extensively, the coverage of this chapter is necessarily spotty. Most of these sedimentary materials have been buried deep enough, long enough, and/or hot enough to significantly change the original sediment (e.g., chalk into crystalline limestone), but not to be metamorphosed in the usual usage of the word. A general indication of the temperatures and pressures in the sedimentary environment is given by Figure 11-1 on which a variety of different geothermal gradients are plotted. The pressure on a fluid about to be trapped in a fluid inclusion in a sediment is probably always intermediate between the two pressure scales given in Figure 11-1, that for lithostatic loading and that for hydrostatic loading (see also Fig. 9-5), but can be very high (e.g., Friedman et al., 1981; Friedman, 1983). Even in soft sediments, "aquathermal pressuring" (Barker, 1972), the large-scale equivalent of overheating an inclusion above Th, can result in abnormal pressure zones. The composition of such early fluids is controlled by a variety of processes (Hanor, 1981).

Studies of Th of low-temperature sedimentary environments should be made only on specially-prepared samples that have not been overheated, and hence stretched. Thus, Th values of 100°C, or even 50°C, in samples that have been cemented with Lakeside-70 (m.p. 70°C but usually heated far higher), as have been reported, are probably invalid.

Most sedimentary diagenesis involves at least some recrystallization, overgrowth, or growth of new phases. These new crystals may trap fluid as inclusions that provide data on the nature, composition, pressure, temperature, and density of the fluids present during diagenesis. Most optical microscope methods of study require inclusions >1-2 μm in diameter, and these are rare in fine-grained products of diagenesis. Thus, Clayton (1982) reported that as much as 1% H_2O was present in flint in the form of minute fluid inclusions. The possibilities of finding inclusions of useful size increase as the size of the host crystal increases. In spite of this limitation, reasonably valid quantitative or qualitative physical and chemical data have been obtained on inclusions from a series of specific diagenetic environments covered in this chapter. The first section below, however, deals with an entirely different aspect of inclusions in sediments, inherited inclusions.

PROVENANCE OF DETRITUS

Nature of the Problem

The individual grains in most sand-sized or coarser detritus carry an observable assemblage of inherited fluid inclusions, trapped during the processes leading up to the formation of the individual detrital grains before sedimentation. Although these inclusions do not provide information on the temperature and pressure of the processes forming the sediment, they can provide valuable

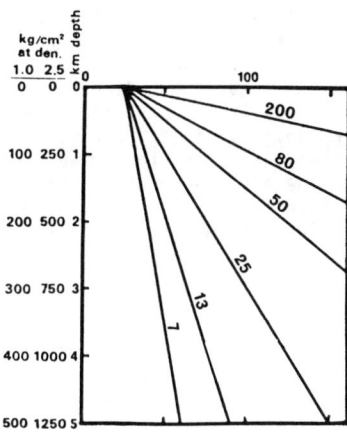

Figure 11-1. Temperature-depth relations with various geotherms. Some geologists would exclude all temperatures above 200°C from "sedimentary environments." From Roedder (1979c.)

data on the provenance of the detritus. Relatively few such studies have been made, as detailed below, but I believe that many questions on provenance in stratigraphy and sedimentation might be answered by careful studies of the inherited inclusions in the detritus. Once a given detrital sediment has gone through diagenesis or a metamorphic event, the problem of distinguishing the original (inherited) inclusions from the newly formed metamorphic ones may become difficult.

Gold-bearing Conglomerates

A large part of the world's total Au production has come from the rather enigmatic Au-bearing Precambrian quartz-pebble conglomerates. Fluid inclusions in these deposits are small and difficult to study, and only four reports exist. Krendelev et al. (1972, 1973) reported extensive studies of the inclusions in a large number of such deposits, including three samples from the "Rand" (Witwatersrand), South Africa. Unfortunately, these two reports contain too many ambiguities and inconsistencies for much interpretation of the results (Roedder, 1984b).

The third report on fluid inclusions in the Rand is by Shepherd (1977), who distinguished five principal types of inherited (i.e., "predepositional") inclusions in the pebbles. These types differ in the amounts of CO_2 and collectively indicate a moderate- to high-pressure, high-temperature environment of original presumed vein-quartz formation. Systematic variation in the relative abundance of these inclusion assemblages for different sections of the ore field demonstrated the presence of several different, well-defined provenance areas or multiple entry points into the basins. A marked sympathetic relationship between uraniferous banket ores and the presence of vein quartz containing inclusions rich in liquid CO_2, together with a corresponding antipathetic relationship for Au, strongly suggests separate sources for the two metals. Post-depositional inclusions are subordinate and offer no support for the alternative epigenetic model, showing only a later interaction of relatively cool circulating ground waters.

A series of papers by Hallbauer (1982, 1983) and Hallbauer and Kable (1979, 1982) indicates that data from the fluid inclusions in quartz pebbles have permitted the recognition of another distinct class of "blue opalescent quartz pebbles." These authors have identified numerous solid phases in the inclusions in both quartz and pyrite by SEM, including orthoclase, muscovite, calcite, apatite, iron-rich phyllosilicate, barium feldspar, chlorite, rutile, corundum, anhydrite, cassiterite, and various mixed chlorides, including $CaCl_2$. Although called "daughter minerals," some of the crystals in their photomicrographs actually look more like solid inclusions. Whichever they are, these crystals, provide valuable insight into the original environments of formation of the quartz pebbles.

Other Detrital Sediments

The environments from which detrital sand grains originate can range from veins in low-grade metamorphic rocks to granite gneiss and granite, and the inherited inclusions in these grains reflect these environments. Other quartz grains, from quartz phenocrysts in lava and pyroclastic rocks, have still another suite of characteristic (melt) inclusions. As a result, numerous workers have reported use of the types and nature of inclusions -- liquid, melt, and solid -- as "fingerprints" to identify the provenance of quartz grains in sandstone (Ermakov, 1950; Keller and Littlefield, 1950; Dolgov, 1954; Davidenko, 1968; Kornilov, 1968; Lofoli, 1972; Simanovich and Ivensen, 1972; LeRibault, 1974; Shchepetkin et al., 1976; and Prone, 1981). Shatagin and Dorogovin (1978) traced the erosion of a granite massif by means of the inclusions in conglomerate pebbles. The resistance of quartz to weathering permitted Clocchiatti and Mervoyer (1976; see also Clocchiatti, 1970) to distinguish between several possible origins of pyroclastic material, even after lateritization. Bauman et al. (1976) found such studies useful in understanding the genesis of commercial glass sand, and Kholief and Hamed (1976) used inclusions to help in correlating Miocene hydrocarbon-productive horizons (see also Kholief, 1975). As one additional example, Konev and Chalov (1972) used the fluid inclusions in quartz pebbles to propose the source of the Takatin diamonds in the USSR.

In many mature sandstones, erosional and sedimentary processes have eliminated most of the weaker inclusion-rich grains, so few inclusions remain. Considerable care is needed in the interpretation of statistical data on the numbers of grains with various recognizable types of inclusions. Thus a sandstone which was derived mainly from a volcanic terrane might have few or no glass inclusions visible, as such inclusions generally place the surrounding quartz under high stress from the $\beta \rightarrow \alpha$ inversion. A small percentage of detrital quartz from a metamorphic terrane in the same sandstone may have retained its recognizable liquid-water inclusions, and hence give the false impression that all the quartz is from a metamorphic terrane.

SALINE DEPOSITS

As will be shown later, the fluid inclusions in saline beds and domes are of more than mere academic interest (see also Roedder, 1984a). Their modes of formation and destruction differ from those of inclusions in most minerals and merit special study. Saline formations can be very sensitive indicators of diagenesis, as they recrystallize rather easily. Although this sensitivity provides valuable data, it also provides numerous possibilities for error. Disruption by coring procedures causes many natural inclusions to leak, and causes the formation of many new secondary inclusions, i.e., artifacts (Roedder and Belkin, 1981; Roedder and Bassett, 1981). Much bedded salt has recrystallized, but it is not uncommon to find within bedded halite deposits single crystals that have very clear multiple planes of primary inclusions (Fig. 11-2), corresponding to growth zoning parallel to the faces of original crystal cube corners. Unfortunately,

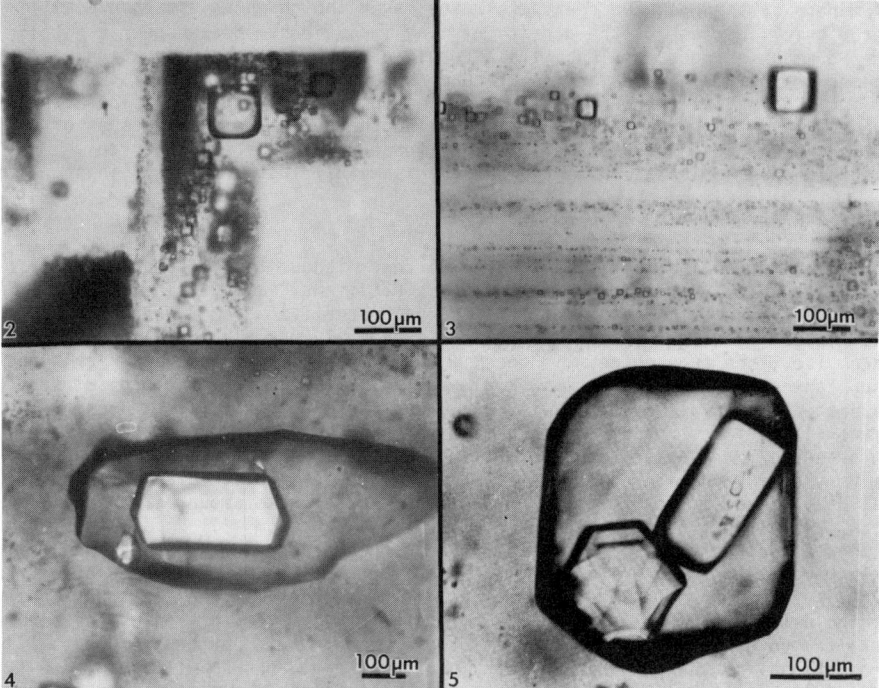

Figure 11-2. Crystal of NaCl from ERDA No. 9 borehole, Waste Isolation Pilot Plant site, Carlsbad, New Mexico, USA, showing sharp crystallographically controlled boundaries of dark primary inclusion-rich zones. Such boundaries probably represent primary crystallization features rather than recrystallization phenomena. From Roedder and Belkin (1979a).

Figure 11-3. Crystal of primary salt from Palo Duro basin, Texas, USA, showing banded primary inclusions outlining features of "chevron" salt (see text).

Figure 11-4. Primary inclusion in recrystallized salt from Gibson dome, Paradox basin, Utah, USA, showing large unidentified birefringent daughter crystal (carnallite?), photographed with partly crossed polars.

Figure 11-5. Three-phase inclusion in sylvite from the Stebnikskoye potash deposit, pre-Carpathian depression, USSR. From Petrichenko (1977, his Fig. 88). The cubic daughter mineral is halite; the hexagonal one is carnallite.

two quite different processes, <u>hopper</u> growth and <u>chevron</u> growth, can yield halite crystals that have such planes of inclusions, and these two processes should not be confused (as they have been in some of my own work, as well as in that of others; see also Holser, 1979a).

Hopper Growth

Hopper growth occurs when the surface fluid of a body of saline water becomes so concentrated by evaporation that thin flat salt crystal nuclei form there and hang suspended (Dellwig, 1955). Further growth on the four edges exposed to the supersaturated brine extends the flat crystal. As the crystal mass increases, it sinks slightly, but the edge is still at water level and continues to grow laterally. Continuous rapid growth at the edges, and continuous sinking, results in a very shallow square pyramid, floating point down. Many of these hoppers may aggregate into floating crusts. Each such hopper, because it has grown rapidly, has trapped large numbers of fluid inclusions, arrayed as planes parallel to {100}. Some hoppers that I have collected from the surface

of modern salt ponds have been pure white, even though much less than a millimeter thick, because of the density of tiny inclusions. Most hopper crystals
are ~5 mm on an edge, or less. As a result of the process of formation, a
hopper crystal may have a series of cube corners outlined by parallel groups
of inclusions, but the whole assemblage remains a nearly two-dimensional entity.
Eventually, all such hopper crystals sink to the bottom, either because of wave
action, or because they became too heavy for the surface-tension forces. The
special hopper orientation at the surface is mainly lost when they sink.

Chevron Growth

Chevron growth takes place on the bottom, without the influence of an air-
brine interface (Wardlaw and Schwerdtner, 1966). Halite crystals, like all
others, have certain crystal directions that normally[1] grow faster than others.
The fastest growing direction in halite is commonly normal to the octahedron
face, i.e., the cube corners. Hence, in a layer of randomly oriented salt nuclei
(spontaneously formed or from sunken hoppers) on the bottom of a body of supersaturated brine, all crystals will grow, but those on which {111} faces the
brine (i.e., faces up) will eventually overtake those neighboring crystals not
so oriented. Grigoriev and Zhabin (1975) have shown that this general process
is very common in nature; it yields, for example, the spectacular oriented crystal druses seen in most mineral collections. After some amount of crystallization, but without the addition of new nuclei, the bottom will be a carpet of
elongated salt crystals, each with a cube corner protruding upward. Such crystals may be of any length, but a width of 5-10 mm is common. If they grow very
slowly, from very slightly supersaturated fluids, they will be clear and free
of inclusions. However, as in all commercial crystal growth, inclusions become
more common as growth rate increases, their numbers being perhaps exponentially
proportional to the growth rate. Any period of rapid growth will result in a
triangular pyramidal cap of inclusion-rich salt on each crystal, covering the
three exposed {100} faces. This inclusion-rich zone (and its host crystal) is
three dimensional and hence differs from such zones in the hopper crystals.
Evidence of chevron growth is very much more common than that of hopper growth.

Primary Fluid Inclusions in Unrecrystallized Bedded Salt

Both hopper and chevron growth would be essentially unrecognizable without
the primary fluid inclusions, as these planes outline the growth stages. Most
such inclusions are liquid filled and roughly cubic (Fig. 11-3). The cubic
cavities are generally 3-30 µm on an edge, but many are <1 µm, and a very few
are as large as 300 µm. Although highly variable, the inclusions in the most
inclusion-rich parts of such crystals have been found in quantities of ~10^{10}/cm^3
for salt from Goderich, Ontario, Canada (Roedder, 1963), Carlsbad, New Mexico,
USA (Roedder and Belkin, 1979a), Wieliczka, Poland (Roedder, 1972, p. 43), and
the Palo Duro basin, Texas, USA (Roedder, unpublished data).

The fluid in these primary inclusions represents brine that formed by evaporation at the time of formation of these strata. Inclusions in hopper salt
presumably trapped brine that was right at the interface with the air; those
in chevron growth trapped brine from the bottom levels, but these brines probably
had been in contact with the atmosphere at some time. If we assume no leakage,
the atmospheric gases that were in solution in those brines should still be in
solution. Although the various gases have different solubilities, and are much
lower in brine than in fresh water (Smith and Kennedy, 1983), and subsequent
diffusion of He would be expectable, the isotopic signature of atmospheric xenon,

----

[1] "Normally" must be added, since this generalization holds only for growth from ordinary natural
fluids. When fluid composition or temperature change, crystal growth rates, and the resulting crystal
habit, can change considerably; if they did not, morphological crystallography would be a simple and
dull subject indeed.

for example, should still be recognizable. Such inclusion gases may be the best samples we have of early atmosphere noble gases.

Primary Fluid Inclusions in Recrystallized Bedded Salt

Most of the crystals we see now in many bedded salt deposits are not the original ones. Recrystallization at one or more stages in the history of the deposit has resulted in the growth of new (and usually larger) crystals at the expense of the older ones. Some of this recrystallization may take place directly on the bottom of the depositional site as a result of periods of temporary undersaturation, followed by slight supersaturation; this process permits the irregular solution cavities to be refilled with clear halite, which is in crystallographic continuity with the inclusion-rich remains of the older crystal (Wardlaw and Schwerdtner, 1966; Shearman, 1970, 1978). Recrystallization can also occur at any later time; some bedded salt contains evidence that fluids of different composition were present at various times in the history of the beds (e.g., Roedder and Belkin, 1979a). Such recrystallized salt, although generally much clearer than the original salt, since it contains few inclusions, may still have large amounts of fluid, but as a very few large inclusions, 1-10 mm or even larger. These inclusions are also primary, as they were trapped during the growth of the host crystals, i.e., during the recrystallization.

If the recrystallization took place on the sea floor at the time of the original deposition, the fluids in these large inclusions could be essentially the same as those in the planes of primary inclusions outlining the chevron growth. One might think, however, that if the recrystallization occurred millions of years later, the fluids might be very different. This is not necessarily so. Salt beds are relatively impervious, and hence the recrystallization could possibly occur in an essentially isochemical system; the new larger inclusions could contain the same fluid, merely remobilized. Such details may be important; careful study of both the inclusion fluids and the host salt will be required for clarification. The partitioning of Br between salt and fluid during evaporation has been used extensively (Holser, 1979b) and may help in deciphering the origin of some of these inclusions. The major difficulty in using Br values lies in the multiplicity of original sources, and the unknown amounts of exchange of the fluids present during the original crystallization and later recrystallizations. There may also be serious kinetic problems (Land and Prezbindowski, 1981; Stoessell and Moore, 1983; Kelly et al., 1983a).

The large inclusions in recrystallized salt may contain daughter crystals (e.g., Figs. 11-4, 5), indicating the presence of materials other than NaCl in the solution that was trapped. If the inclusions are merely the coalesced equivalent of many tiny inclusions from primary salt crystals, such daughter crystals could represent the accumulation of other ions from the evaporation of seawater, i.e., bitterns, and hence could indicate that evaporation at that point had proceeded almost to the stage of precipitation of K and Mg salts. The daughter crystals also could represent the diagenetic dissolution and migration of such K and Mg salts from another part of the saline sequence and hence could have no genetic significance relative to the local host salt. Whatever they represent, the daughter crystals suggest the possibility of K-Mg deposits somewhere in the system, and hence have potential value in exploration for potash deposits.

In addition to obvious primary inclusions, some planes of secondary inclusions may be found in bedded salt, but they are not common. Intergranular inclusions -- along grain boundaries -- are very common, and may constitute the only method of recognizing the existence of the grain boundary. In situ, intergranular inclusions are presumably filled with brines, but in samples that have been exposed to the air for a length of time, the fluids have normally migrated to the surface and evaporated, leaving the white incrustations so commonly seen along grain boundaries on the outside of salt cores. This evaporation leaves a series of gas inclusions along the grain boundaries. In recrystallized salt, these boundaries commonly show the 120° intersections (Fig. 11-6) characteristic of many recrystallized metamorphic rocks. Details on the nature of these grain boundaries and their inclusions in samples from a salt anticline are given by Roedder and Belkin (1981).

Inclusions in Domal Salt

As a result of flowage during the formation of a dome, the original salt crystals have been thoroughly recrystallized, perhaps repeatedly, and most of the original fluid inclusions they may have had have been kneaded out. Most domal salts contain <0.1 wt % H_2O, whereas bedded salts commonly contain >1 wt % (Roedder and Bassett, 1981). Although most of the inclusions in bedded salt are recognizably primary, in either original or recrystallized salt, the origin of the relatively few inclusions in domal salt is essentially unknown. In the Rayburn and Vacherie domes in Louisiana, Roedder and Belkin (1979b) found most fluid inclusions at the contact between solid anhydrite crystals and the host salt. This contact is apparently the lowest energy location for brine inclusions (Fig. 11-7). Similar inclusions were found in the salt from the Oakwood dome, eastern Texas, USA (Dix and Jackson, 1982), and between solid inclusions of halite in sylvite from the USSR (Fig. 11-10). Flowage of the salt can draw such inclusions out into long tubes (Fig. 11-8). High-pressure gas inclusions are also found at anhydrite/salt interfaces in Louisiana (Fig. 11-9), and from the Asse salt anticline, West Germany (Roedder and Belkin, 1981). Dix and Jackson (1982) found that the fluid inclusion data suggested that the unfoliated salt at the crest of the dome was once strongly foliated, but that this fabric was destroyed by solid-state recrystallization.

Composition of Liquid Inclusions in Salt

When inclusions in halite are studied on the freezing stage, their compositions are frequently found to range widely, from bitterns with high concentrations of calcium and magnesium, with Te = -50 to -60°C, and even as low as -69.8°C, from the Williston basin, North Dakota, USA (P.M. Radomsky, pers. comm.), to essentially pure solutions of NaCl, with Te = -21°C (Roedder, 1963, p. 182). Inclusions in salt from Asse, W. Germany, (Roedder and Belkin, 1981), after freezing at -115°C, showed visible recrystallization at -80°C, suggesting that a trace of liquid was probably present. Pure NaCl-H_2O inclusions are apparently missing in the Carlsbad area of New Mexico, USA (Roedder and Belkin, 1979a). A wide range is expectable as a result of normal events in the history of many salt beds. During the original crystallization, evaporation continuously concentrates the more highly soluble ions such as K, Ca, and Mg into the residual liquids, and unless refluxing is effective, inclusions will be trapped representing various stages in this process. Under diagenesis, formation waters penetrating the recrystallizing halite bed may dissolve the more soluble minerals and gradually flush out residual and grain-boundary fluids, until only NaCl (and $CaSO_4$) remain. Any fluids trapped during subsequent recrystallization or upon the healing of fractures will contain only NaCl (plus any solutes originally in the waters passing through).

Inclusions in salt are particularly difficult to freeze, and when frozen, frequently exhibit various metastable phase assemblages, causing considerable ambiguity in the interpretation of the freezing data. As concentrated brines are viscous when cold, most phase changes also tend to be sluggish, and the crystals formed can be submicroscopic. Thus, on warming apparently unfrozen inclusions from very low temperatures, sometimes visible dendritic ice crystals nucleate and grow, and cause a decrease in the bubble size (Fig. 11-11), but if these ice crystals were not visible, the decrease in bubble size on warming might be thought to be simply from thermal expansion of the liquid.

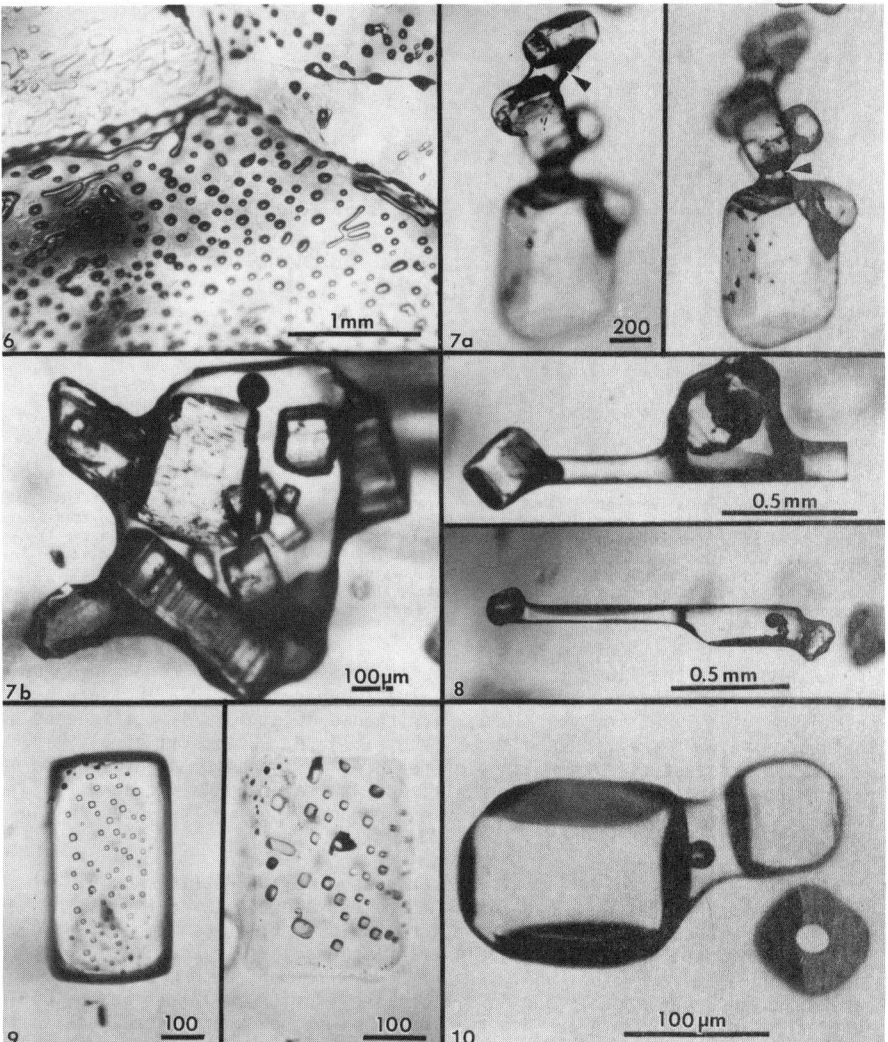

Figure 11-6. Fluid inclusions (gas) on interface outlining 120° junction between recrystallized salt crystals in core from ERDA No. 9 borehole, Waste Isolation Pilot Plant site, Carlsbad, New Mexico, USA. From Roedder and Belkin (1979a).

Figure 11-7. Anhydrite in salt. (a) Group of anhydrite crystals in salt from Rayburn dome, Louisiana, USA, showing brine inclusions adhering to anhydrite, making a fillet between the crystals (arrows). The two photos were taken at different levels of focus, in strongly convergent light; in normal collimated microscope lighting, these fillets are hidden in broad black shadows. From Roedder and Belkin (1979b.) (b) Brine inclusion, with vapor bubble (black), wetting several anhydrite crystals in salt from Oakwood dome, east Texas basin, USA. Scale bar in μm. From Dix and Jackson (1982).

Figure 11-8. Brine inclusions (seminegative crystal shape) between anhydrite crystals, from Rayburn salt dome, Louisiana, USA. Several such inclusions were all parallel, suggesting direction of stretching or shearing of host salt (now a single, clear crystal). From Roedder and Belkin (1979b).

Figure 11-9. Group of semioriented inclusions of compressed organic(?) gas at interface between rectangular anhydrite crystals and host salt, from Rayburn salt dome, Louisiana, USA. Scale bars in μm. From Roedder and Belkin (1979b.)

Figure 11-10. Solid inclusions of halite, bridged by an aqueous inclusion, in sylvite from the Permian El'tonskoye deposit, pre-Caspian basin, USSR. From Petrichenko (1977, his Fig. 99). An adjacent inclusion, out of focus, contains only gas.

313

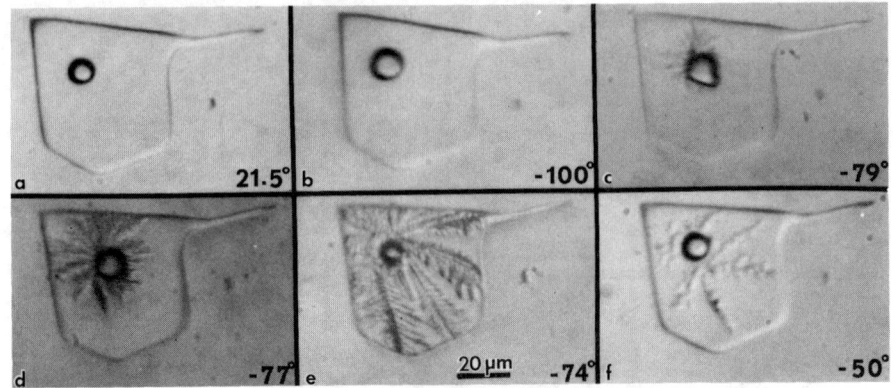

Figure 11-11. Serial photomicrographs of one of a plane of secondary inclusions in salt from 1690.8', No. 1 Mansfield bore, Oldham County, Palo Duro basin, Texas, USA, at the temperatures indicated (°C). On cooling to -135°C and warming to -100°C, the only visible change is the larger bubble. On further warming, to ~-79°C, very weakly birefringent dendritic crystals nucleate on the vapor bubble and as they grow, the vapor bubble diminishes in size, suggesting that the crystals are ice. On warming to -54°C, the crystals start melting, and are gone at -48°C. The approximate elapsed times between these photos, in minutes, are: (a)-(b),5; (b)-(c),4; (c)-(d),2; (d)-(e),2; (e)-(f),6. Photos courtesy of H.E. Belkin.

Analyses of the fluid in inclusions in halite have been reported by several workers (see Tables 4 and 6 in Roedder, 1972; Derevyagin, 1973; Holser, 1963; Kovalevich, 1975; Petrichenko, 1973; Petrichenko et al., 1974; Petrichenko and Shaydetskaya, 1973; Petrichenko and Slivko, 1973a,b; Sedletskii et al., 1973). Several of these studies show large differences in composition between inclusions in primary salt and those in recrystallized material, and Petrichenko (1973) and Vorob'ev (1978) suggested the use of these compositional differences in the exploration for potash deposits.

The pH of such fluids in salt beds ranges widely. Petrichenko (1973) reported pH determinations, made with microelectrodes on large inclusions in salt from the Donbass, USSR, to range from ~3.5 to 6.4 (mostly 4.5 to 6.4). I have found that large (>300 µm) fluid inclusions in recrystallized salt from the Palo Duro basin, Texas, USA, range from 3.1 to 5.5, with most values (40 determinations) between 3.4 and 4.3. (Samples from the Delaware basin, Carlsbad, New Mexico, USA, were higher (4.5 to 6.0) as might be expected from these since various magnesium silicates were found in that salt. Bodine (1976) has shown that even very small amounts of some magnesium phases can effectively buffer the pH of such brines.) Significant differences were noted between pH values for different types of inclusions at Palo Duro. The determinations were made under the binocular microscope by touching a sharp tapered point of an appropriate narrow-range pH test paper into the opened inclusions, and are subject to an uncertainty of visual match of ~±0.5 pH unit). The procedure was checked using standard buffers, but the validity of the test paper results must still be verified. The data reported by Petrichenko are probably some of the most valid pH determinations made on inclusions; because the inclusions available were huge, the otherwise very serious experimental problems in measurement were minimized.

Although several thousand analyses of inclusions have been reported in the literature, on many types of host minerals, the only actual determinations of Eh (-400 to +515 mV) of the fluids have been on inclusions in halite, by Petrichenko and coworkers (see Chapter 5).

In addition to the major ions Na, Cl, K, Mg, and Ca, several of the minor constituents are useful in understanding the environment of formation. The Cl/Br ratio in both the saline water and the precipitated salt decreases during evaporation and precipitation of NaCl. Holser (1963) used this ratio to indicate the degree of evaporation of seawater. Sabouraud-Rosset (1973, 1974, 1976) found that the Cl/Br ratio in fluid inclusions in gypsum crystals from various types of saline environments ranges from 150 to more than 1000, and she has been able to recognize leaching by later waters in some of these environments on the basis of neutron-activation analyses for Cl and Br. This ratio is particularly suitable, as it is not apt to be affected seriously by precipitation and reactions with clays, etc., which can seriously influence other minor elements (Petrichenko and Slivko, 1974). Such reactions particularly affect the ratio K/Mg and the content of SO_4. Organic matter is also present in the inclusions in many saline deposits, as bitumen, as separate gas, liquid, and/or liquefied hydrocarbons, under pressure and as solutes in the brine. Data are just becoming available on the δD and $\delta^{18}O$ ratios of the water in fluid inclusions in salt; these data are discussed below under "Geologic history of the deposit."

Composition of Gas Inclusions in Salt

Many salt domes and anticlines contain at least some gas inclusions under pressure, and some have inclusions at pressures up to several hundred atmospheres (e.g., Fig. 11-9). Some bedded salts have similar gas inclusions under pressure (Roedder and Belkin, 1979a). Sometimes the abundance of the inclusions and pressure of the gases in them is enough to make the salt decrepitate either spontaneously or under minor relief of stress (e.g., during mining). This material, called "popping salt," can become a major hazard in mining (Roedder, 1972, p. 43) because serious mine accidents can result when large volumes of salt explosively and spontaneously decrepitate into the mine openings. Belchic (1961), Hoy et al. (1962), Thoma and Eckart (1964), and Thoms and Martinez (1980) and Mahtab (1982) have described mine "blowouts" in which as much as 7500 tons of salt suddenly decrepitated and resulted in fatalities.

The gas responsible for popping salt varies greatly in composition from one locality to another. Bunsen (1851, p. 251) found the gas in salt from Wieliczka, Poland, to be nearly 85% CH_4, and Ackerman et al. (1964) reported that the gas in the Werra potash deposits, East Germany, contained 84% CO_2 and 14% N_2. I have found that popping salt from a salt dome in Mexico contained an easily liquefiable mixture of hydrocarbon gases and CO_2 (Roedder, 1972, p. 43) and that the salt from Asse, West Germany, contained gas inclusions, presumably hydrocarbon, under high pressure (Fig. 7-10), as did that from the Rayburn and Vacherie domes in Louisiana (Roedder and Belkin, 1979b). Vil'denberg et al. (1978) reported gas containing 85-95% N_2 in salt from the Caspian basin, USSR, and Norman and Bernhardt (1984) reported high pressure N_2 (30-500 bars) in samples from the potash deposits of Carlsbad, New Mexico, USA.

There is no consensus about the origin of these gases; they have been termed both syngenetic and epigenetic. As the composition varies so widely, probably different processes are involved. Some gas inclusions come from a source completely apart from the salt. Petrichenko (1978) reported dense CO_2 inclusions in halite, presumably derived from a nearby intrusion of basalt (Fig. 11-12a). Roedder and Belkin (1978) found high-pressure gas inclusions, presumably mainly CO_2, in a salt sample taken 1-2 cm from a 4 m igneous dike crosscutting potash ore beds. Although melting of presumed CO_2 crystals in these inclusions took took place at -56°C, liquid was still present at -68°C; this liquid may result from organic components. The CO_2 probably came from the dike, and the organic components from the salt.

Homogenization Temperatures of Inclusions in Salt

Temperature measurements on inclusions in salt by the homogenization method are generally suspect and frequently impossible, because of the rather common

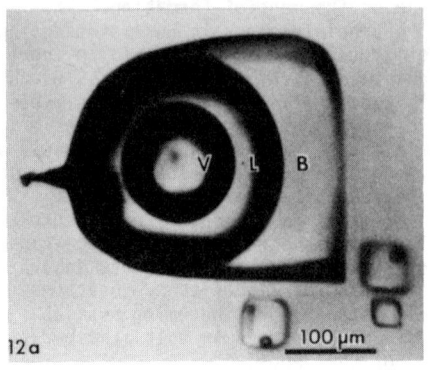

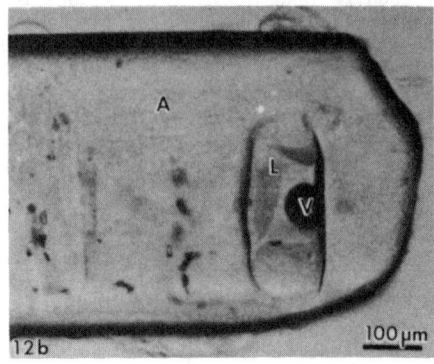

Figure 11-12a. Mixed CO_2-H_2O inclusion in vein halite, photographed at 0°C, containing liquid CO_2(L), gaseous CO_2(V), and brine (B). Note associated H_2O inclusions containing small bubbles, indicating that the trapped fluid was two-phase. In Devonian salt from the Novosenzharskaya structure, Dnepr-Donets basin, USSR. From Petrichenko (1977, his Fig. 208).

Figure 11-12b. Fluid inclusion containing liquid (L) and vapor (V) in anhydrite crystal (A) from Oakwood salt dome, Texas, USA, from Dix and Jackson (1982).

leakage of the larger inclusions, because of necking down, because of metastable stretched liquid under negative pressure, and because of volume changes, both in nature and during the laboratory measurements (Chapter 3). Such processes may explain some of the reports of exceedingly high Th values (e.g., 240-360°C; Panov, 1975). Petrichenko (1973) reported that Th increased with each determination until it finally "stabilized" at 23-25°C higher (than on the first run) after three or more runs. P.M. Radomsky (pers. comm.) reported that on homogenizing inclusions in salt from the Williston basin, North Dakota, the bubble reduced with temperature increase to a small size and then persisted for tens of degrees more before homogenizing. I would interpret this behavior as suggesting the presence of noncondensable gas in the bubble. As this is compressed by expansion of the liquid, the internal pressure rises to a point at which the walls of the inclusion yield somewhat.

Although extensive compositional studies have been made on inclusions in gypsum (Sabouraud-Rosset, 1973, 1974, 1976), it may give unreliable thermometric results (Kul'chitskaya, 1974). Gypsum from a few halite deposits has yielded seemingly reasonable values. Kovalevich (1975) found that homogenization of sylvite daughter crystals in some primary inclusions in halite from Stebnik in the USSR took place at 38-60°C, whereas secondary gas/liquid inclusions in recrystallized halite homogenized in the range 56-86°C, the average of 80 determinations was 71°C. Petrichenko and Slivko (1973a) established that diagenetic alteration of Permian salt in the Donbass took place at ~60°C, and Roedder and Belkin (1979a) showed that inclusions in both primary and recrystallized salt in a core from the Delaware basin (Carlsbad area, New Mexico, USA) homogenized in the range 25-45°C. In view of the data on volume change presented in Chapter 3, I have doubts as to the significance of _any_ of these values, my own included.

For determination of Th with a minimum of ambiguity, other minerals that have formed in the halite beds are generally more tractable (e.g., Fig. 11-12b).

Sedletskii et al. (1973) measured Th of 40-110°C[2] for inclusions in authigenic quartz crystals from Upper Jurassic saline deposits in the Hissar Range, USSR.

Inclusion Data in Nuclear-Waste Management

Salt beds or domes have been under consideration for some years as one of the potential geologic environments for the safe storage of nuclear waste. Fluid-inclusion data are needed in two aspects of any discussion of the engineering problems and possible hazards of such a storage site:

Geologic history of the deposit. Since the goal is to keep the wastes out of the biosphere for as long as they are hazardous, storage for geologically significant periods of time (10^3-10^5 years) is required. Prediction of the future of any environment on earth for such periods is a difficult geologic problem; both uniformitarianism and catastrophism must be considered. In uniformitarianism, the geologic evidence of the past is the best guide to the future. Fluid inclusions provide one source of information on this past. For example, in some salt beds, organic gas under pressure in some inclusions and not in others gives evidence of the presence of several different fluids at different times, suggesting that fluids can and have been moved through such beds (Roedder and Belkin, 1979a).

Only a few isotopic studies of the H and O of the H_2O in fluid inclusions in salt beds have been made, but they reveal a rather wide range of different fluids have been present in the salt, possibly at various times in the past. Values for δD ranged from -5 to -55, and $\delta^{18}O$ from -9 to +4 per mil. Although individual data subsets seem explicable on one or another ad hoc basis, numerous problems remain unsolved at this time. At the WIPP site (Waste Isolation Pilot Plant) in the Delaware basin at Carlsbad, New Mexico, USA, waters from various formations were examined. Powers et al. (1978) presented isotopic data of J.R. O'Neil (U.S. Geological Survey). Many of these have δD ~-50 and $\delta^{18}O$ ~-7 per mil (relative to SMOW), but one water (from ERDA bore No. 6) had δD 0.0 and $\delta^{18}O$ +10.4. O'Neil et al. (ms in preparation) also analyzed nine samples of water extracted from relatively large liquid inclusions in recrystallized salt from these rock salt beds, all of which fell along a line between the two isotopic signatures noted above (Fig. 11-13). In a similar study of 39 inclusions in salt from the Palo Duro basin in Texas, USA, Beeunas and Knauth (1983) found them to range, in δD and $\delta^{18}O$ per mil (SMOW), respectively, from -5 and +4.3 to -55 and -6.8, i.e., along a line with a very different slope than that shown in Figure 11-13 for WIPP inclusions.

The simplest explanation is that these data arrays represent mixing lines between fluid pairs. The interpretation of such data is still ambiguous, however, as many processes may affect the position of any given sample, in addition to possibly major laboratory problems of sample extraction and analysis. Ocean water[3] becomes enriched in both D and ^{18}O during evaporation, and hence moves toward the upper right on Figure 11-13. The specific slope for this movement will vary with the environment under which the evaporation occurs, but a slope of 3.2, as shown in Figure 11-13, is reasonable for arid climates. Meteoric waters will follow a similar slope upon evaporation, but generally start with a much lighter signature. The water evaporating in a given evaporite basin will be some unknown mixture of sea water, meteoric water, and ground water that may have precipitated or reacted with various phases during percolation

----

[2] These authors erroneously subtracted 15-20°C from these values as a "correction for salinity" to obtain trapping temperatures.

[3] The Standard Mean Ocean Water is used as the reference value, at δD = 0 and $\delta^{18}O$ = 0; it is generally assumed to have been the same as this (or nearly the same) in the past (see Fig. 11-13).

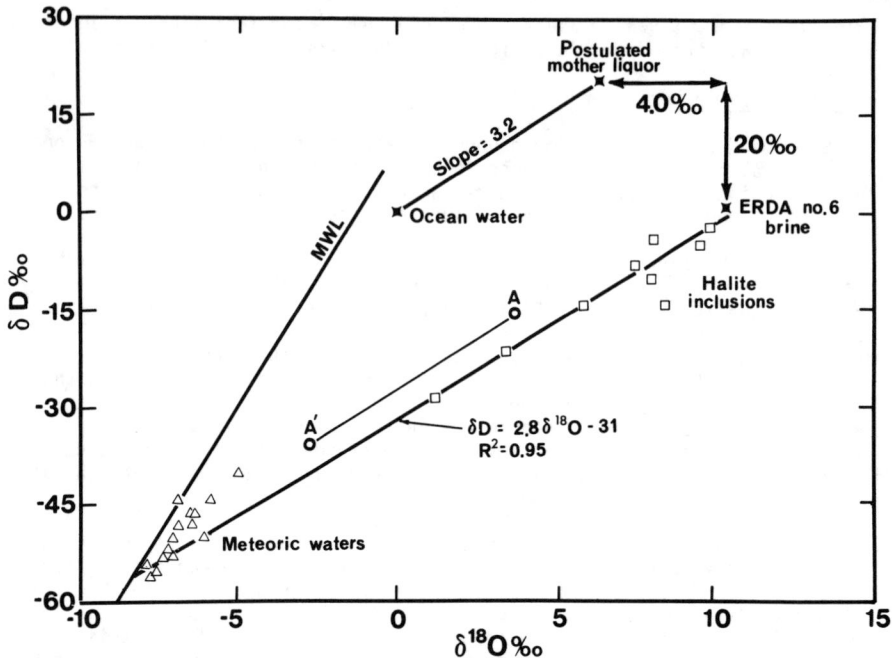

Figure 11-13. Plot of δD vs δ¹⁸O for waters from the Delaware basin, New Mexico, USA. Data from Powers et al. (1978) and O'Neil et al. (in preparation). "Meteoric water" samples were taken from wells in various formations in the area. Inclusion data are from large inclusions in recrystallized salt, released by crushing. Evaporation will cause ocean water to move to the upper right, e.g., along some slope such as that shown. A least-squares line fit to the ERDA No. 6 brine and the inclusion data extrapolates to the lower part of the field of local meteoric waters and suggested a mixing line to O'Neil (see text).

through adjacent sabkhas. Any formation water that is later introduced into the salt beds has been in contact with detrital sediments and hence may exhibit an "oxygen shift." Most sediments contain heavier O than the waters, so any isotopic O exchange will drive the water to the right. Sedimentary rocks also contain appreciable H, particularly in clays, that may exchange with waters moving through (or be driven off during diagenesis). Sofer (1978) showed that the water in gypsum precipitated from sea water during evaporation will have δD 20 per mil lighter and δ¹⁸O 4 per mil heavier than the original water. Subsequent dehydration of such gypsum to anhydrite during diagenesis would release this highly modified water. Two additional variables that cannot be ignored involve the composition of meteoric water in the past in general (i.e., the position of the meteoric water line (MWL) on Fig. 11-13), and the specific position of the local meteoric water (and ocean water) along that line, which will vary with local climatic and topographic conditions at that time.

In some dome salts, the presence of inclusions having Te at ~-21°C requires a source of freshwater, without significant ions other than Na and Cl (Roedder and Belkin, 1979b). Nearly pure NaCl-H₂O fluid in a salt bed can have three possible origins: (1) It can be seawater trapped in the first-formed salt crystals, before evaporation and crystallization caused much buildup of other ions. (2) It can be water evolved by the dehydration of clay minerals or the conversion of other phases such as gypsum to anhydrite during diagenesis. (3) It can be freshwater that has penetrated the salt at some unknown time in the past and been trapped.

The third origin has the greatest potential importance to the possible use of these domes for nuclear-waste storage. At the depth from which these origi- nally bedded salts have flowed to form these domes (~10-12 km), the fluids in the pores of the surrounding sediments were almost certainly highly saline for- mation waters, containing significant ions other than Na and Cl. However, during the rise to the surface, the salt dome must have penetrated aquifers containing essentially freshwater. If fracturing occurred during this rise, or subsequently (and we have no knowledge of the age of these inclusions), freshwaters could enter the salt and become trapped. The corollary is that if this happened in the past, it might also happen in the future, and Knauth and Kumar (1983) showed from isotopic studies that water in current brine leaks at the Avery Island salt dome mine in Louisiana, USA, were essentially meteoric water.

Engineering of a waste storage site. The containers (canisters) for the waste are to be designed to prevent any contact of the waste with ground waters, at least for the early part of the storage term, when the radioactivity and temperature are highest. Salt solutions are rather corrosive, but bitterns, containing divalent cations, are much more corrosive (Stewart et al., 1980). Furthermore, when a fluid inclusion in a soluble mineral such as salt is sub- jected to a thermal gradient, it will move, generally up the gradient (Fig. 11-14). The radioactivity of the nuclear waste will develop a thermal gradient in the surrounding salt, and hence the fluid inclusions in that salt will migrate toward the canisters. The amount of inclusion fluid present, its compo- sition, and its rate of migration should all be known in order to design adequate safeguards into the canister and its immediate surroundings. The amount of fluid in the salt deposit is a surprisingly difficult number to obtain with even order-of-magnitude accuracy (Roedder and Bassett, 1981). The amount may be >2 wt % in some salt (Bassett and Roedder, 1981). The composition has been esti- mated from freezing data, but actual analyses are needed. Finally, the rate of movement of the inclusions in a thermal gradient is a function of many variables. Measurements have been made in both the laboratory and field (Roedder and Belkin, 1980a,b; Jenks and Claiborne, 1981), but the conclusion reached was that we can- not at this time make a truly valid calculation of the amount of brine expected to arrive at the canister in any given situation (Roedder and Chou, 1982; Chou, 1983; Roedder, 1984a).

In addition to the above most immediate applications of inclusion data, studies of fluid inclusions help in understanding the environment of formation and diagenesis of salt deposits; hence, they may aid in the exploration of a given basin, bed, or dome for mineral resources and for potential storage sites. As one example, some geologic studies of the Palo Duro basin, Texas, USA, sug- gested that the salt might have formed in relatively deep water (e.g., Bein and Land, 1982). However, at least some of the salt in these beds was found to have a regular periodicity in the abundance of inclusions that could only be explained in terms of diurnal variations, which, in turn, required very shallow water (Fig. 11-15; Roedder 1982b).

CEMENTS AND OVERGROWTHS ON DETRITUS

Anyone who has looked at detrital carbonate rocks in thin section is aware of the carbonate overgrowths on the grains that commonly act as a cement. These overgrowths are so small that they would seem to be very unlikely places to find usable fluid inclusions, but a study by Nelson (1973) has shown that small but usable primary fluid inclusions can indeed be found in such cement. In a crinoi- dal biosparite, the Fernvale Limestone (Upper Ordovician), from samples across northern Arkansas, USA, such inclusions show Th ranging from 85-170°C, with a distinct mode in the range 110-150°C. The salinity ranged from 5-25 wt % NaCl equivalent. These data are very similar to those found for inclusions in the barite-Pb-Zn deposits of central and northern Missouri, USA (Leach, 1973), and it is presumed that the cementation fluids and the ore-forming fluids represent a

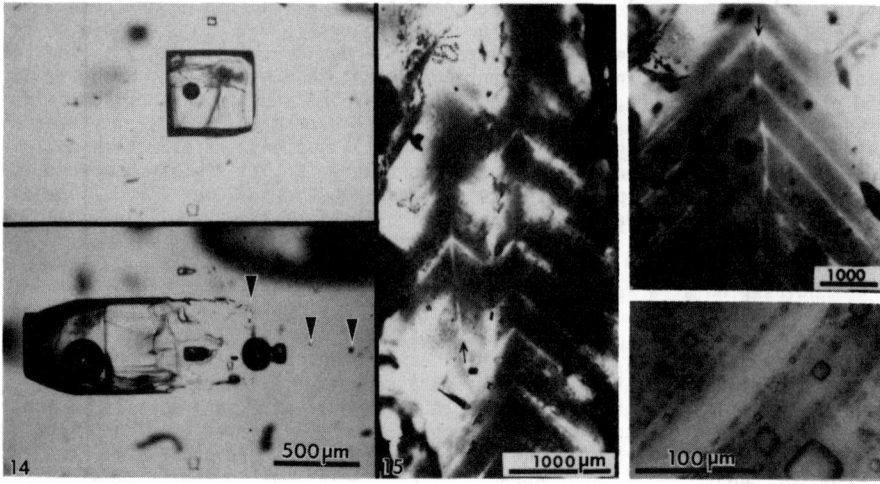

Figure 11-14. Inclusion in salt from ERDA No. 9 borehole, Waste Isolation Pilot Plant site, Carlsbad, New Mexico, USA, before (above) and after (below) a 156-hour run at 202°C ambient and 1.5°C/cm gradient. The large inclusion has split into a large liquid-rich part and a small dumb-bell-shaped gas-rich part that moved in opposite directions relative to the thermal gradient, which increased to the left. The fiducial mark (a vertical scratch) is visible left of the inclusion in the upper photo; it is almost invisible in the lower photo, because of the illumination needed to see the (much larger) bubble now present, but the original position of the inclusion can still be seen, outlined by a series of small specks to the right, which act as internal reference points (arrows). From Roedder and Belkin (1980a.)

Figure 11-15. Photomicrographs in transmitted light of doubly polished plate of salt from the Palo Duro basin, Texas, USA, showing regular, presumably diurnal banding. Presumed original growth direction upward in all. (a) Part of two elongated single crystals (~2x8 mm) showing 12 bands, averaging 0.39 mm/band (clear septum to clear septum, perpendicular to the bands). Note that the separate crystal on the left has an additional clear septum along its centerline (arrow). (b) Part of an elongated single crystal (~5x12 mm), showing 14 bands, averaging 0.54 mm/band. Note additional clear septum along centerline (arrow). (c) Detail of salt across one clear septum shown in (b) in same orientation. From Roedder (1982b).

single episode of fluid flow through these rocks. The possibility of multiple periods of circulation (and cementation), however, must always be considered, and Lohmann (1983) has shown that fibrous carbonate cements are replaced with low magnesium calcite cement in the presence of a water phase (recorded in fluid inclusions). Nahnybida et al. (1982) were able to show that calcite, dolomite, and anhydrite cements in upper Devonian dolostones formed at 140-155°C and >2.5 km depth, from 2.4-3.7 molar solutions (NaCl equivalent).

Klosterman (1981) made a similar study of carbonate cements in the Smackover Formation of Late Jurassic age in Arkansas. She found Th to range from 58 to 158°C, and very low Te (-51 to -68°C), corresponding to a complex calcium chloride brine. Similar results were reported by McLimans (1981). Moore et al. (1983) have shown that present-day brines in the Smackover Formation are 10 times more saline than seawater, and have a $\delta^{18}O$ composition near +5. As the cements have $\delta^{18}O$ of -6.5, the equilibrium temperature was estimated to be 90°C, in agreement with the inclusion data. The inclusion data should also be compared with the results of a study of the $^{87}Sr/^{86}Sr$ ratio in the brines and various diagenetic phases in this same formation (Stueber and Puskar, 1983).

Fluid inclusions may help clarify the "dolomite problem," i.e., when and in what sequence did various types of dolomite form? Many mining areas in carbonate rocks show several stages of dolomitization. Generally the fluid inclusions in such carbonates are difficult to work with, but Leach et al. (1983)

were very successful in such work on dolomite in the Viburnum Trend, Missouri, USA. Freeman (1973) has reported that the fluid inclusions in some epigenetic dolomite-calcite cement show that the fluids forming the dolomite were both hotter and saltier than those forming the associated calcite cement.

Quartz cement may provide similar opportunities for finding inclusions that can delineate the environment of sandstone diagenesis. R.C. Nelson (pers. comm., 1975) has studied quartz overgrowths in the Crystal Mountain and Blakely Sandstones (Lower and Middle Ordovician) from central Arkansas, USA. Most of the inclusions are trapped at the original grain boundaries of the detrital quartz grains, but some occur within the overgrowth, and considerable care must be used to avoid confusion with the inclusions of the original detrital grains. Th for inclusions in the overgrowths ranged from 97.5 to >150°C (possibly as high as 175°C). More recently, Pagel (1977) has made use of the inclusions in quartz cement in a study of U deposits.

Studies such as these show that fluid inclusions can provide useful quantitative data on problems that heretofore have been plagued by ambiguous qualitative data. The inclusions will generally require extensive and careful microscopy, but the possibilities of getting answers to otherwise rather intractable problems make the effort worthwhile.

VUGS, VEINS, GEODES, AND OTHER DIAGENETIC MINERALS IN SEDIMENTARY ROCKS

In many areas throughout the world, what were originally soft and porous Paleozoic and younger carbonate sedimentary rocks have been strongly modified postdepositionally to yield hard, dense crystalline limestone and dolomite. Although the individual sample porosity may be low, such rocks commonly contain open vugs of varying size, sometimes lined with crystals of calcite, dolomite, quartz, fluorite, barite, and celestite. Some also contain sulfides, particularly very large sphalerite crystals, and less often galena, pyrite, marcasite, chalcopyrite, wurtzite, millerite, etc. Many localities are known, but those at Clay Center, Ohio, and Herkimer, New York, USA, have provided spectacular mineral specimens for many collections.

In many of these geodes, vugs, and veins in carbonate rocks, one or more minerals may be relatively rare, yet these few crystals may be euhedral, clear, and sometimes several centimeters on an edge. Some sphalerite crystals in vugs in Indiana limestones are tens of centimeters in diameter. The nucleation and growth of such a sparse distribution of large, essentially perfect crystals of a given mineral, particularly relatively insoluble phases such as sphalerite, requires that both nucleation and growth take place exceedingly slowly, from solutions having a very low degree of supersaturation, under essentially static conditions.

In the laboratory, even very slow precipitation of ZnS from solution usually results in a milky solution that contains many millions of submicrometer crystals per milliliter. The distribution of nuclei formed in the fluid in rocks may have been about one per cubic meter, twelve orders of magnitude less. Furthermore, growth was so slow that there was time for diffusion to occur to transport material through the rock pores to these few nuclei, instead of forming new nuclei. Even where the nuclei were more abundant, the degree of perfection of the crystals is sometimes phenomenal, e.g., the Herkimer, New York, USA, quartz crystals (commonly known as "Herkimer diamonds"). Generally no solid evidence exists on the specific time at which this slow crystallization took place, except that it was postlithification, but the widespread occurrence of such crystal-lined vugs indicates that the process was relatively common.

Some indication of the elapsed time involved in the growth may be provided

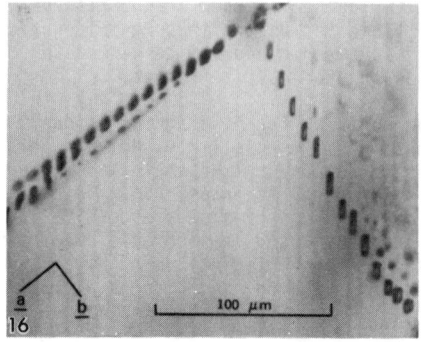
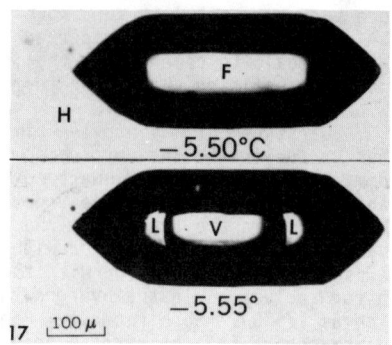

Figure 11-16. Varved celestite crystal, viewed looking along c axis, taken in transmitted, strongly collimated light and oriented so that varves parallel to (210) appear N-S (see indicated positions of crystal axes). Two planes of secondary inclusions, formed by healing of fractures crossing the varves, reveal differences in solubility of the different parts of the individual varve. From Roedder (1969).

Figure 11-17. Photomicrographs of a negative crystal-shaped inclusion in quartz. At room temperature and down to -5.50°C, inclusion appears to be filled with a single phase -- a colorless "gas," as seen in the upper photo (taken at -5.50°C). On slight further cooling (lower photo, taken at -5.55°C), the "gas" is seen to have been a supercritical fluid (F), as it separates abruptly into liquid (L) plus vapor (V). It is believed to be a mixture of organic compounds, probably mainly CH_4 and C_2H_6. Sample ER 61-26, "Herkimer County diamond," Herkimer County, New York, USA. From Roedder (1963, p. 202; see also Rosasco et al., 1975b, Fig. 2c).

by celestite (Fig. 11-16). Crystals from several localities show a microscopic, very regular compositional banding, which has been interpreted as annual "varves," representing variations in the Ba content due to annual variations in the salinity from mixing of saline brines with annually variable amounts of meteoric water (Roedder, 1969). If this interpretation is correct, these celestite crystals grew in $\sim 10^4$ years. The solutions from which they grew were hot ($\sim 100°C$) and moderately to strongly saline (~ 25 wt % NaCl equivalent). Leach (1980) has found similar banding in the central Missouri barite district, but reversed, i.e., regular bands of high Sr content in barite.

Some studies on inclusions from geodes, vugs, and veins have been summarized in tabular form by Roedder (1979c), but many of the studies show special features too complex to tabulate. For example, Touray and Barlier (1975) have been able to correlate the map distribution of the phase composition of multiphase inclusions, rich in organic matter, in quartz with the degree of metamorphism of clay, and with the metamorphism of the opaque organic matter in the enclosing rocks (as measured by its reflectivity; see also Barlier et al., 1974).

The quartz crystals from Herkimer and adjacent Montgomery County, New York, USA, are in vugs in a crystalline petroliferous dolomite. Inclusions in them contain, in addition to brine, several types of organic matter: high-pressure supercritical mixtures of organic gases (Fig. 11-17); colorless to yellow, highly fluorescent oils; and broken fragments of anthraxolite, a black organic material having a conchoidal fracture, which is also found in the vugs with the quartz (Dunn and Fisher, 1954). Individual inclusions may have one or more of these phases, in any ratio, suggesting that all were separate phases at the time of trapping. Small fragments of anthraxolite (also called pyrobitumen) sometimes are found adhering to the vapor bubble, on the surface of which they move rapidly when minute thermal gradients, caused by asymmetric illumination, result in surface-tension differences. The oily liquids show a variety of colors and Tm (Roedder, 1963, p. 201) and hence presumably indicate a variety of source materials or histories.

The physical properties of the gas inclusions show that they also have a

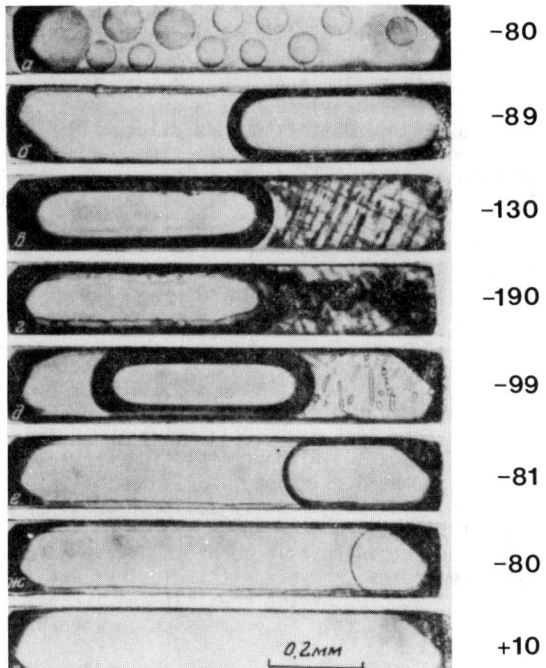

	-80
	-89
	-130
	-190
	-99
	-81
	-80
0.2мм	+10

Figure 11-18. Photomicrographs of a CH_4-rich inclusion in quartz from a a vein in Carboniferous sandstone in the USSR, taken at the temperatures indicated (°C). From Lazarenko et al. (1976).

variety of compositions, although the major constituents are probably CH_4 and C_2H_6 (Fig. 11-17 and Rosasco et al., 1975b). Good compositional data on these gas inclusions, together with Th of cogenetic aqueous inclusions, might permit a valid estimate of the depth at the time of trapping, as their presence shows that these were trapped as actual bubbles of dense supercritical fluid in the brines from which the crystals grew. Hence, these gas inclusions should record the ambient hydrostatic pressure. Whether that pressure was simply hydrostatic, from a fluid column to the surface, or had a lithostatic component (e.g., Barker, 1972), is unknown.

Such CH_4-rich inclusions are not uncommon in quartz from sedimentary rocks. Touray and Sagon (1967) and Touray and Jauzein (1967) described several occurrences in France, and similar quartz crystals ("Marmarosh diamonds") have been described in the USSR (Voznyak et al., 1974; Bratus' et al., 1975; Lazarenko et al., 1976). The last reference has a particularly striking series of photomicrographs of such CH_4-rich inclusions (Fig. 11-18), but these inclusions have a much lower critical temperature than the inclusion shown in Figure 11-17, hence presumably a much higher CH_4 content.

The water inclusions along with the organic fluids illustrated in Figure 11-17 showed Te at -23 to -27°C, and at the (variable) Tm, they showed an $NaCl \cdot 2H_2O$ liquidus. These data suggest that the solution was near saturation with respect to NaCl and contained very little other material. The host rock in the area, the Little Falls Dolomite (Upper Cambrian), is overlain by halite-bearing Silurian formations; hence saturation of the fluids with NaCl is expectable.

Sphalerite, quartz, and fluorite from geodes, vugs, and clay seams in some

Paleozoic limestone and dolomite of Indiana contain inclusions of brine, in which Tm ice is in the range -10 to -16°C (14-20 wt % NaCl equivalent) and Th is ~100°C. These salinities are even greater than those in modern oil-field brines in the area and may represent mixing of strong brines from the adjacent Illinois or Michigan basins, USA, with more dilute fluids or ground water (Shaffer, 1981). The data on inclusions in fluorite, sphalerite, and celestite from similar Paleozoic occurrences in Ohio, New York, Iowa, Missouri, and Tennessee, USA, are similar (Kinsland, 1977, 1979; Haynes and Mostaghel, 1979; Roedder, 1979c; Coveney and Goebel, 1983; Leach, 1979). Coveney and Goebel (1983) warn that postentrapment generation of CH_4 within such inclusions, as described by Hanor (1980), can lead to erroneously high values of Th. The high values for Th (even 260°C) obtained from such samples also may be erroneous as a result of the common use of oven-drying of drill cuttings.

Many suggestions have been made that these scattered occurrences of sphalerite, etc., have formed from the same kind of fluids as did the Mississippi Valley ores. The general geologic environments and mineralogy are very similar; only the $^{87}Sr/^{86}Sr$ values (Kessen et al., 1981; Grant and Bliss, 1983) and the volumes of ore minerals are different. Although the fluid inclusions in these minerals are similar to those in the Mississippi Valley deposits (see Chapter 15), in that they were formed from hot, strongly saline brines, containing much organic matter, the temperature range and salinity are both a little lower. Dilution of hot brines has been suggested as a mechanism for the formation of the Mississippi Valley-type ore bodies; the present inclusions may reflect more such dilution, but detailed studies of the chemical and isotopic composition of the fluids in inclusions are required to prove such an interpretation.

Often the inclusions in barite have the lowest Th and contain the most dilute fluid. Barite is known to dissolve and reprecipitate with surprising ease, so inclusions in it may represent the latest fluids (perhaps even a separate episode, Leach, 1980). Barite inclusions tend to leak and to neck down; both phenomena yield erroneous Th values. However, a few studies of carefully selected inclusions in barite from such occurrences have yielded consistent and reasonable data that are probably valid (Roedder, 1979c; Leach, 1980).

In addition to the above, numerous geodes in shale and limestone are lined with quartz or calcite and rarely contain sulfides other than pyrite or marcasite. I have examined many of these geode minerals, generally with disappointing results, as they are remarkably free of recognizable primary inclusions. Many of the inclusions from geodes exhibit metastable superheated ice, so Tm ice cannot be determined. Calcite in such geodes, e.g., from the famous Keokuk, Iowa, USA, locality, is commonly the last stage of deposition and is found to have inclusions filled only with liquid, indicating that it probably grew at <~40°C.

The quartz of some geodes from Kentucky, USA, has many gypsum crystals embedded in it. This fact may reflect a complex history of sequential replacements, the last occurring at ~60°C, from saline fluids (Roedder, 1979c).

One aspect of inclusions in geodes that is frequently asked about pertains to the very large "fluid inclusions" visible or audible in unopened geodes from various localities. The amplified sloshing sounds made as these natural containers are shaken has attracted the attention of many visitors to mineral shows (Sutton, 1964). Agate geodes containing visible liquid water constitute a collector's item called "enhydros," and are found particularly in the soil from the weathering of basalt in Brazil. Because such geodes have polycrystalline walls, they will probably leak in time, and most museum specimens are either kept in water or given impervious coatings to keep them from drying out. The rate of leakage may be very slow; one 10 cm enhydro in my possession has been steadily losing a few milligrams a year to the laboratory air. As a result of such leakage, the fluid in these geodes has probably been replaced by passing fluids many times since the geodes were first formed, and the present water in them is probably moderately recent ground water. This has been verified in one occurrence by studies of the isotopic composition of the waters (Matsui et al., 1974).

SPECIAL ENVIRONMENTS

Inclusions in Amber

During the solidification of natural resins to form amber, a variety of inclusions were trapped, in addition to the well-known fossil insects. These inclusions have attracted the attention of various scientists, starting with Brewster (1820), and add much to the fascination of amber as a "gem" material (e.g., Koivula, 1981). However, these inclusions are also of some interest scientifically. Geochemists could learn much about the evolution of the atmosphere, and about the many geochemical cycles that were strongly affected by that atmosphere, if one or more samples of early atmospheres were available for analysis of the major, minor, and trace gases, and the isotopic ratios of several of these gases, particularly xenon in very old samples (e.g., Thomsen, 1980). Some of the various inclusions in amber may provide samples of previous atmospheres, albeit very young (Tertiary) and possibly modified by various processes, but still they merit careful study. Older samples (calcite fossils and salt, discussed below) may have even greater potential.

Gas inclusions. The most notable fluid inclusions in amber are gas bubbles, which vary greatly in size, shape, and abundance. Very commonly, planes of gas bubbles are seen, parallel to banding in color and/or index of refraction. Other inclusions are purely random in distribution or are attached to fossil insects. Some are plainly visible to the naked eye, but "blue" amber derives its bluish cast from large numbers of very tiny inclusions (Trofimov, 1974). Whether these inclusions are gas bubbles or iron sulfide grains (Flamini et al., 1975) is not clear.

Perhaps the most striking aspect of these various gas inclusions is their variability in shape (Fig. 11-19). Many are essentially spherical, but adjacent inclusions in the same sample may have bizarre shapes, e.g., spherical with a long tapering tail, or highly irregular. Other inclusions are disk-shaped, commonly having semiconcentric internal ringlike structures.

Several studies have been made of the composition of the gases in inclusions in amber. Trofimov (1974) found that "ivory"-type amber from the Baltic area contains inclusions of modified air, consisting of N_2, CO_2, O_2, H_2, and CH_4, with traces of Ar, Kr, Xe, Ne, and He. My own studies (unpublished) have been mainly on the gas pressure in the larger gas bubbles (10 to 300 μm) in amber from Yantarny, USSR, on the Baltic coast, and from Colombia, as estimated on the crushing stage. A few of these bubbles showed no gas, a few showed 5-20 vol % gas, several had 50-75%, and one very large bubble, 100%. The large bubble may well have contained air, from leakage, but the range in pressure of the other bubbles, within a single sample, is difficult to explain by leakage.

Two-phase, gas/liquid inclusions. Dr. Lev Jacobson of the National Institutes of Health, Bethesda, Maryland, asked me to examine some amber samples he had obtained on loan from the Dominican Republic. These samples had some millimeter-size, two-phase inclusions, consisting of a larger mobile bubble and a liquid of apparently low viscosity. Subsequently, I have found smaller but similar partly liquid-filled inclusions in Baltic amber (Fig. 11-20). The G/L ratio varies greatly, even in adjacent inclusions. Freezing data indicate that the liquid phase is essentially water.

Possible interpretations of the inclusions. The origin of these various

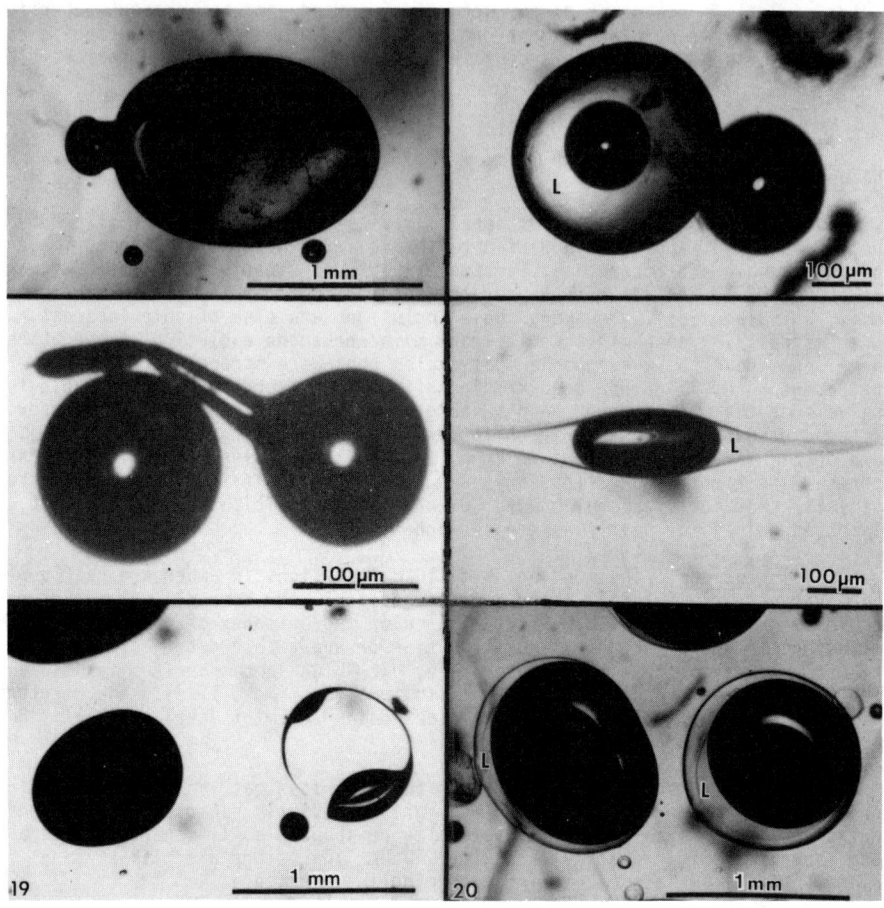

Figure 11-19. Gas inclusions in amber from Yantarny, Baltic Sea coast, USSR (U.S. Natl. Museum sample no. 115850).

Figure 11-20. Two-phase, liquid-plus-vapor inclusions in amber from Giron, near Bucaramanga, Santander, Columbia. The liquid (L) is essentially water. (U.S. Natl. Museum sample no. R7313).

types of inclusions is basic to any possible use to which they might be put, but at present is almost purely speculation. During the hardening of a mass of resin, through evaporation of volatile compounds, one might expect that the outer surface might lose its volatile compounds first and become rigid, while the interior is still soft. Further loss from the interior should yield abundant shrinkage cavities filled with the more volatile constituents (bubbles) throughout the core. The fact that such features are generally not seen in clear amber, suggests that diffusion through the solidified amber is not the major rate-limiting step in the drying process; i.e., amber dries throughout rather than from the surface inward. However, many of the features suggest some volume shrinkage and/or flowage of the host after a given bubble has formed, yielding collapsed bubbles like that shown in Figure 11-19. The planes of bubbles parallel to striae suggest possible trapping of air at the interface between a new flow of resin and an older solidified mass.

The two-phase inclusions may represent droplets of moisture enclosed by resin, possibly together with some air. All such inclusions could then be modified in volume and shape by shrinkage of the host as it hardened, and modified in composition by various reactions with the host material and diffusion through it to (or from) the atmosphere. Meteoric (or ocean) water could also leak in. The processes of amber formation are complex and differ from one type of resin to another, as evidenced by the extensive studies of Savkevich (1970) and the nuclear magnetic resonance studies of Lambert and Frye (1982). Certainly some of the gases (and perhaps liquids) in the inclusions could be indigenous, and admittedly, considerable effort would be needed to use the inclusions to obtain information on previous atmospheres; however, in view of the lack of better samples, they may have to be used.

Inclusions in Speleothems

Fluid inclusions in speleothems, the various carbonate formations precipitated in limestone caverns, provide an interesting but unexpectedly complex possible source of paleoclimatic data. During the precipitation of the calcite crystals making up these formations, the cave seepage waters are trapped as single-phase, pure-water fluid'inclusions, sometimes in surprisingly abundant amounts. Petrographic examination of these materials reveals a wealth of textural detail, the correct interpretation of which is vital to any use of the fluid inclusion data. However, several of the studies of such material differ significantly in their interpretations (e.g., Folk and Asserto, 1976; Kendall and Broughton, 1978), and a full understanding of the effects of the several processes involved in the formation and subsequent history of speleothems (and cements) lies in the future. From an examination of some of these materials, and some of the published data, evidence for leakage in at least some samples is obvious. Part of this leakage is probably from laboratory sample handling procedures, but the possibility of earlier leakage in nature is not excluded. Thompson et al. (1976) and Schwarcz et al. (1976) made a study of both D/H in the fluid inclusions and $^{18}O/^{16}O$ in calcite, together with $^{230}Th/^{234}U$ radioactive dating of the calcite, in speleothems from several North American localities. The fluid-inclusion samples (5-10 g) were crushed under vacuum to release the water. (Oxygen was measured on the calcite rather than on the inclusions to minimize errors due to probable reequilibration of the inclusions after trapping, but this procedure requires the added assumption that the modern meteoric water $\delta D/\delta^{18}O$ line applies in the past.) Since D/H in the inclusions did not change appreciably during at least part of the interval involved (200,000 B.P. to the present), Thompson et al. and Schwarcz et al. inferred that changes in $\delta^{18}O$ are related to deposition temperature, $\delta^{18}O$ increasing with temperature decrease, rather than with time. Deposition rates appear to be greatest during summer and may have fallen to zero during glacial advances. Maxima in curves of relative paleotemperature, on the basis of secular changes in $\delta^{18}O$ of calcite as dated by the $^{230}Th/^{234}U$ method, correspond to maxima in summer insolation in the Northern Hemisphere, as well as to high sea stands marked by raised coral reefs and to thermal maxima observed in speleothems from other regions of North America. Relative changes in temperature are clearly indicated by analyses of individual growth layers, but the calibration to yield actual temperatures hinges on precise evaluation of two competing effects -- change, with temperature, in the isotopic fractionation (1) between water and calcite, and (2) between water vapor and precipitation (Schwarcz et al., 1976).

Harmon et al. (1979a,b) claim that the three basic assumptions on which this work is based have been verified by studies of modern, actively forming speleothems. These assumptions are: (1) that deep cave temperatures approximate the mean annual regional surface temperature; (2) that the isotopic composition of ground-water seepage in a cave is representative of the average of meteoric precipitation falling on the ground above the cave (in spite of the likelihood of variable degrees of rock-water reequilibration and hence variation in $\delta^{18}O$ before precipitation of the speleothem); and (3) that speleothems were deposited under conditions of isotopic equilibrium. Harmon and Schwarcz (1980) and Harmon and Atkinson (1980) then reversed the procedure, using detailed δD data from the inclusions and $\delta^{18}O$ from calcite to show that some unreasonable calculated cave temperatures (i.e., below 0°C) may actually be the result of globally different δD - $\delta^{18}O$ relationships in meteoric waters during the Pleistocene ice ages. Yonge (1981; see also Schwarcz and Yonge, 1983) has also explained some of the discrepancies in the previous work as resulting from incomplete extraction of the water after crushing of the samples, and has revised the method to give more reproducible results. The procedures described by Yonge are also applicable to other inclusion studies.

Inclusions in Sulfur

One of the most interesting minerals that acts as a recorder of diagenetic conditions in saline deposits is sulfur. Sulfur is found in many saline deposits, where it generally is assumed to have formed from bacterial oxidation of organic matter and reduction of sulfate from anhydrite or from aqueous solution. The time at which this action took place is frequently a problem, and the fluid inclusions in the sulfur may help. Commonly, these inclusions are large, and as no evidence of leakage in inclusions in sulfur has been reported, these inclusions should provide good material for analysis. The inclusions in sulfur and associated gypsum in the famous Sicilian deposits were so large that they were some of the first to be analyzed quantitatively and reasonably completely (Silvestri, 1882; Sjogren, 1893). Sjogren found the gypsum to contain a sodium-chloride-rich solution containing 4 wt % salts, but the fluid inclusion in sulfur analyzed by Silvestri contained only 0.1 wt % salts. The latter may be an unfortunate example of an inclusion in sulfur that did actually leak, in view of the very low salinity and the very large size (6 cm^3).

Yushkin and Srebrodol'skii (1965) studied large inclusions in sulfur from the Rozdol and Shorsui deposits in carbonate-sulfate rocks in Uzbekistan, USSR. The inclusions are single phase (i.e., full of liquid) except for a film of bitumen in some. The presence of large single-phase inclusions indicates that the sulfur grew at low temperatures, as even cooling from 40°C will cause a bubble to form in most large inclusions. The fluid released some H_2S when the crystals were crushed and had a pH of 7-7.5, as determined by the use of various organic indicators. It contained about 6% salts, mainly (Na,K) and Ca chloride, bicarbonate, and sulfate. Merlich and Datsenko (1972) found that the fluid in inclusions in sulfur deposits from the Little Carpathians in southern USSR was an NaCl solution similar to seawater.

Beskrovnyi and Lebedev (1971) studied various minerals from the Gaurdak sulfur deposit, which is in a sequence of gypsum-anhydrite-halite rocks in limestone in southeastern Turkmenia, USSR. This is a different type of sulfur deposit, however, as it also contains celestite, fluorite, barite, danburite, hematite, gypsum, calcite, solid bitumens, quartz, sphalerite (high in Hg and Pb), oil, and gas. Inclusions in early fluorite homogenized at 130-160°C, but those in late gypsum and calcite homogenized at <50°C. (Presumably the sulfur must have crystallized after the fluorite, as sulfur melts at 113°C. Nothing precludes the sulfur forming first as a liquid.)

Inclusions in Ice

When snow coalesces to form ice, some air is trapped between the ice crystals as gas inclusions. These gas inclusions in ice have been studied extensively (see discussion in Chapters 3 and 5). Scholander et al. (1956 and subsequent papers; see Morse and Coachman, 1983) showed that the bubbles in glacial ice provided a sample of earlier atmospheres, particularly the CO_2

328

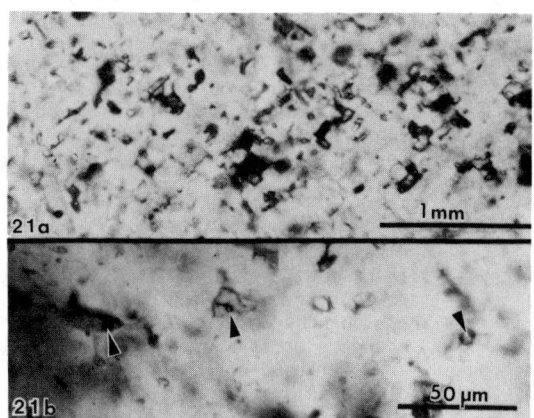

Figure 11-21. (a) Cross section of crinoid spine from sediments of Burlington age (Mississippian), Keokuk, Iowa, USA. The spine is a single crystal of calcite; the c axis of the calcite is parallel to the length of the spine. The crystal is crowded with gas/liquid inclusions ranging from all gas (dark) to all liquid (the bulk of the inclusions in this field). (b) A detail of part of same spine as in (a), showing 3 inclusions with gas bubbles (arrows). Sample courtesy of P. Kier.

content, and made numerous measurements. Although evidence exists that the air in such inclusions dissolves in the ice during deep burial (Gow and Williamson, 1975) and can form air hydrate inclusions (Shoji and Langway, 1982), Delmas et al. (1979) have shown that the CO_2 content of these gas inclusions can be used to estimate the variation in atmospheric CO_2 content during the last 100,000 years. Contamination with carbonate dust and during sample preparation was found to be the cause for unreasonably high CO_2 values reported by previous workers (see also Barnola et al., 1983).

Inclusions in Living Organisms

Some marine organisms precipitate single crystals of calcite within their structures. The most common examples are the crinoids, the stem segments (columnals) of which consist of single crystals of calcite as large as 1 cm, with the c axis parallel to the length of the stem. The calyx, individual plates, and spines may also be single calcite crystals. I have found that such crystals (from various crinoids and echinoids) contain such large numbers of fluid inclusions that they are nearly opaque except in very thin sections (Fig. 11-21). Some of these inclusions are empty (gas filled), others are liquid-filled, and some contain gas and liquid in various ratios. If the liquid is standard 35 per mil seawater, it should have Tm ice at -2.0°C (Riley, 1975; Weeks and Ackley, 1982). The gas phase may represent merely varying degrees of leakage, but the origin and nature of the fluid now present in these inclusions is possibly important. Certainly it is to be expected that the organism would have effected chemical and isotopic fractionations, (particularly for C, H, and O) and the inclusion fluid might be more appropriately termed "crinoid juice" than seawater. But a careful study of both living animals and fossils should be made to determine particularly the relationship between the liquid and the surrounding seawater, and when the actual sealing occurs, and hence whether some (or none) of these inclusions are valid samples of ancient seawater and perhaps of the noble gases dissolved in it.

HYDROCARBON-BEARING ENVIRONMENTS

Sphalerite in Bituminous Coal Beds

All coal contains at least some sulfides, and the environmental significance has aroused considerable interest in the nature and origin of these sulfides. Iron sulfides are by far the most abundant and are mainly of authigenic origin. In some coal, however, sphalerite is a significant constituent, and as it is a transparent mineral, fluid-inclusion studies on it may yield information on the conditions of origin. Leach (1973) studied inclusions in sphalerite from coal mines in central Missouri, USA, and found Th = 80-110°C and strongly saline brines (>22 wt % NaCl equivalent). As the sphalerite from the adjacent northern Arkansas, USA, zinc district had similar fluid inclusions (83-132°C and >22 wt % NaCl equivalent), Leach suggested that these two mineralizations formed from a single episode of fluid flow.

In the northwest part of the Illinois basin, some bituminous coal has very coarsely crystalline, banded yellow and purple sphalerite both as vertical veins ("cleats") and as crystals in clay dikes cutting the coal seams (Hatch et al., 1976). Fragments of coal are found embedded in the sphalerite. The sequence of color banding is regular throughout a large part of the basin. Samples of this sphalerite show many secondary or pseudosecondary inclusions and only a very few that might be primary. A group of 25 inclusions from the cleat sphalerite and 19 from the claydike sphalerite, had Th ranging from ~90-95°C, and a maximum of 102°C. Tm ice of all samples ranged from -15.6 to -18.9°C and averaged -16.9°C, and all showed Te <-26.5°C (Roedder 1979c). Cobb (1981) measured a much larger number of inclusions from four localities and found the maximum range of Th to be 75 to 113°C.

The interpretation of these results is not at all clear, but obviously the sphalerite postdates the coal. Cobb (1981) showed that some coal compacted after sphalerite deposition began. Bituminous coal itself is a sensitive indicator of "diagenetic" (low-grade metamorphic) conditions. Bostick (1973, and pers. comm., 1976) has shown from coal petrography that the high-sphalerite coals of the Illinois basin were never subjected to long-continued burial temperatures >40-65°C, and any anomalous temperature events on the order of 100°C could have lasted only 1-3 million years. Thus, the maximum temperatures from the inclusion data and the coal-petrography data may not be in conflict. Cobb (1981) suggested that the data fit a geothermal gradient of 23.5°C/km (see Fig. 11-1) and a minimum depth of 2.2 km.

A connection might exist between the fluids depositing this sphalerite and those forming the sphalerite in the Upper Mississippi Valley Zn deposits ~150 km to the north. The Th values found in the Upper Mississippi Valley Zn deposits by earlier workers (see Roedder, 1976a, Table IV) range from 75-121°C, and Tm ice was found to be ~-20°C (Roedder, 1967d). These fluids were therefore in the same temperature range but somewhat more saline than those in the coal beds. As pointed out by Hatch et al. (1976), from the color banding, the deposition of this sphalerite was a basinwide event, making it evident that large-scale fluid movements must be involved.

Hydrocarbons in Inclusions

In the preceding part of this chapter, organic materials have been mentioned frequently, as most, if not all, fluids in sedimentary processes have at least some hydrocarbons present. CH_4, at least in small amounts, is almost ubiquitous in inclusions in sedimentary rocks, as might be expected in view of its common occurrence as a gas and its solubility in brines (Duffy et al., 1961; Price, 1979), and it may have considerable effect on the behavior of inclusions that have trapped such fluids (Hanor, 1980; see also Chapter 8). In this section, I deal with inclusions filled with hydrocarbons.

After its exposure to surficial O_2 and other weathering agents, organic matter in contact with fluids moving through a sediment is subject to a continuous series of changes with time, depth, and temperature. Oxygenated surface

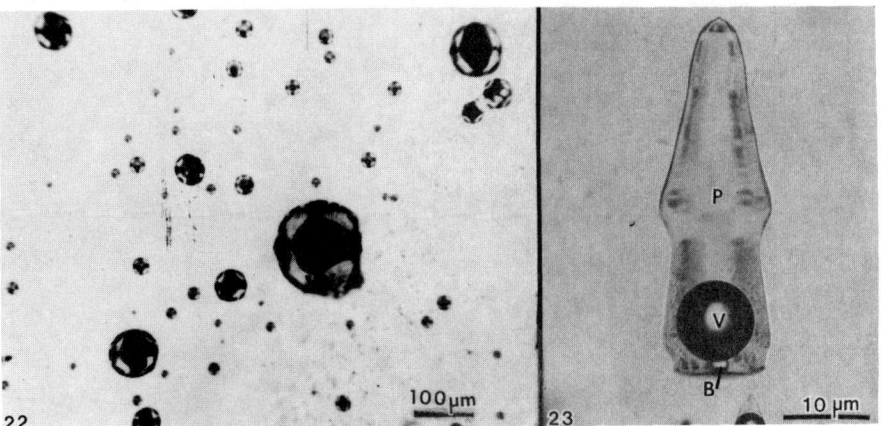

Figure 11-22. Photomicrograph of petroleum inclusions in fluorite from southern Illinois fluorite-Pb-Zn deposits, Illinois, USA, that have formed three phases since originally being trapped as a homogeneous phase (see text). A globular dark brown, slightly birefringent mass with four appendages is a phase (bitumen?) that preferentially wets certain crystallographic planes of the fluorite, thereby appearing to divide the lighter phase into four quadrants. Actually, the "attachment" of each arm to the wall is cone shaped, as the center of the cone also does not wet the wall. Additional blebs of dark material are stuck on the walls halfway between each pair of arms. The dark circle in each inclusion is a gas bubble under high pressure. On crushing, the light phase vaporizes instantly. From Roedder (1962b; see also Fig. 2-27,28).

Figure 11-23. A primary petroleum inclusion (P) in fluorite from the Cave-in-Rock fluorite-Pb-Zn district, southern Illinois, USA. Oil droplets floating in the brines from which these fluorite crystals grew adhered to the growing surface of the fluorite cube (here horizontal). More oil was added to the globule as the fluorite grew up around it, yielding the tapered shape. A small amount of brine (B) was also trapped in the inclusion. After trapping, a very small amount of dark material precipitated on the walls. Even though the shrinkage bubble (V) appears large, Th was between 98 to 101°C for a series of such inclusions; adjacent brine inclusions had Th as high as 148.3°C. The tip of another similar shaped inclusion is at the bottom. Photograph and data courtesy of C.G. Cunningham.

waters may penetrate to considerable depths (Winograd and Robertson, 1982), but many deep waters contain some organic matter. Organic matter in an inclusion is effectively isolated from reactions with subsequent external oxygenated environments.[4] Many oil inclusions now contain only a clear yellow oil and a shrinkage bubble, and sometimes a small dark globule of bitumen(?). Presumably the bubble and bitumen are daughter phases, formed since trapping, and the bitumen may have been precipitated as a result of isochemical changes after trapping. The most obvious evidence of such post trapping changes is found in oil inclusions in fluorite from some fluorite-Pb-Zn deposits in southern Illinois, USA. These inclusions now consist of major amounts of two (or three) immiscible hydrocarbon phases (Fig. 11-22), sometimes in a relatively uniform volume ratio. Why such internal chemical change occurred in the inclusions in only certain samples is not clear. Not all oil inclusions in samples of fluorite from these deposits exhibit these post trapping changes to such a degree (Fig. 11-23).

Individual primary inclusions trapped from heterogeneous, two-fluid-phase systems such as this can give no reliable information on the overall ratio of the two fluid phases, brine and oil, as they may consistently trap only one (or the other) of the two fluids (see Chapter 2). However, if a fracture forms in a crystal while it is surrounded by such a two-phase system, the fluid that enters the crack and becomes trapped may be a valid sample of the fluid present at the opening of the crack at that moment. An example of the trapping of an emulsion of oil and water is shown in Figure 11-24. This plane of pseudosecondary inclusions contains roughly equal amounts of oil and brine. Other planes of pseudosecondary inclusions from these same samples contain essentially only

----

[4] True isolation requires an unbroken single crystal. In some fluorite samples from southern Illinois, USA, large primary oil inclusions are clear yellow-brown, but an occasional inclusion in the same plates appears black and opaque. Although I presume that these black inclusions have been oxidized, only a few have visible cracks out to the surface.

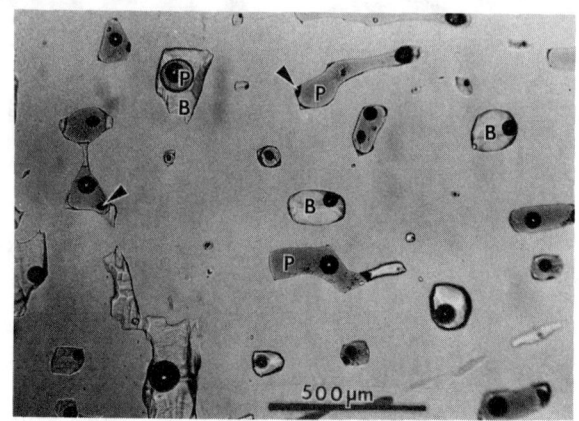

Figure 11-24. Photomicrograph in plain transmitted light of a plane of pseudosecondary inclusions on a (111) cleavage fracture in a pale-colored fluorite crystal from a vug, which trapped an immiscible mixture of yellow petroleum (P) and brine (B). Recrystallization of the fluorite walls of the fracture to isolate individual inclusions occurred almost entirely via the brine. As a result the inclusions of brine have become three dimensional and are lined with negative crystal facets, whereas the petroleum inclusions are still flat and are almost two dimensional. In addition, the index of refraction of the petroleum is much closer to that of fluorite, yielding low relief. Small irregular masses of dark-brown birefringent matter (arrows) have formed in the petroleum since trapping, and the round gas bubbles (probably methane, under pressure) characteristically occur in it rather than in the brine, where both phases are present. Sample ER 59-3, Hill mine, Cave-in Rock, southern Illinois fluorite Pb-Zn district, USA. From Roedder (1972).

brine, so the example shown in Figure 11-24 may represent only a limited pulse of oil-rich fluids, or perhaps a small amount of oil-rich emulsion concentrated by gravity in the top (or bottom?) of the vug in which the host fluorite grew.

Insofar as the environment of trapping can be estimated or at least constrained by data on the organic compounds present, these various hydrocarbon inclusions can be useful. Thus, in theory at least, the many changes that take place during the migration and maturation of petroleum should be recorded in oil inclusions trapped at various stages in time or distance along the way. Very little has been done with these inclusions, but as a result of the recent tremendous improvement in appropriate analytical techniques, such as micro-Raman spectrometry, micro-gas chromatography, mass spectrometry, and particularly the combination of the last two, the approach has considerable potential.

Nature of Organic Inclusions

Organic gases, liquids, and solids have been recognized in inclusions in many samples, and these materials have been characterized to various degrees. The literature is extensive, and only a few items can be reviewed here. A pioneering study was made by Murray (1957), who reported a detailed mass-spectrometric analysis of the molecular constituents in organic liquid-gas inclusions in quartz crystals from vugs in a dolomite core from a gas-productive depth interval in Mississippian rocks in Alberta, Canada. The inclusions contained mainly CH_4 and C_2H_6 under pressure and small amounts of many other constituents, but no water. On heating, the two phases homogenized in the gas phase at about 100°C. The trapping of such inclusions might seem to require growth of quartz from the hydrocarbon fluid; this problem has been discussed in Chapter 3.

The specific nature of organic inclusions is difficult to assess without such analyses, since a wide range of compounds can be present (e.g., Kvenvolden and Roedder, 1971; see also Fig. 12-4), and the phase behavior can vary widely. Burruss (1981b) has shown how the composition of mixtures of CO_2 and CH_4 can be determined by their low-temperature phase equilibria, but major amounts of other gases such as N_2, H_2, H_2S, or C_2H_6, as well as heavier hydrocarbons, may confuse matters. CH_4 alone can have a major effect on the phase behavior of inclusions (Hanor, 1980). Considerable interest has been shown recently in the complexities in the phase behavior and the very significant solubility of various oils in dense CO_2, as a result of the use of CO_2 flooding in oil fields (Orr et al., 1981). As an example, when the crushing stage is used to determine the pressure in gas inclusions that are single phase at room temperature, sometimes many volumes of compressed gas may evolve and leave behind a major amount of a liquid that is not volatile at 1 atm. (The solubility of crude oil in CH_4 was determined by Price et al. (1983)). Similarly, daughter crystals are sometimes found in organic inclusions, and organic solids such as bitumen, but neither these daughter crystals, nor the new crystalline solids that form on freezing, have been identified.

In the North Derbyshire Pb-Zn-fluorite district, England, organic materials of several types have been found associated with the ores in Carboniferous limestone. Although these various materials are found there as inclusions, particularly in fluorite, they are also found free in larger masses, and have been studied extensively in an attempt to determine their origins (e.g., Nooner et al., 1973; Pering, 1973). Evidently, these materials have undergone several stages of selective leaching, transport, and deposition, and possibly recent partial microbial oxidation. This district is only one of many such Pb-Zn-fluorite districts in carbonate rocks that contain organic matter (see Chapter 15). Some mines have active oil or tar seeps; others that seem to be free of it have been found to contain measurable of hydrocarbons (<100 ppm) in certain minerals, such as sphalerite (Rickard et al., 1975).

Relationship to Petroleum Formation and Migration

Even if no organic phase is visible in the aqueous fluid inclusions, a growing body of evidence shows that very significant amounts of a wide variety of hydrocarbons can dissolve in water or brine (Price, 1976, 1981a,b); these compounds should be looked for. The "hydrocarbon peaks" frequently seen in mass-spectrometric studies of inclusions, which are usually passed off as normal atmospheric contamination, may in part be real. Although two-phase migration of oil and brine does take place (Magara, 1981), the migration of petroleum compounds will certainly be easier if they are in true solution in brine. Other compounds, such as amino acids (K. Kvenvolden, pers. comm., 1975), should also be looked for, as they might define the maximum temperatures to which a sample has been subjected in the past. Kuznetsova et al. (1983) reported analyses of amino acids from inclusions in Ukrainian Shield rocks, but the chances of obtaining usable data are essentially restricted to the Pleistocene or Upper Miocene (T. Hoering. pers. comm.; see also Hare et al., 1980).

An important application of inclusions in connection with the history of oil-bearing rocks is in dating the formation of fracture porosity at depth. Currie and Nwachukwu (1974) studied inclusions in mineral fillings of fractures in oil-field formations in Canada. They found that Th ranged from 45-120°C, which they interpreted as indicating that fractures opened progressively as an accompaniment to tectonism, regional uplift, and erosional unloading. They also suggested on this basis that incipient fracture porosity at depth could gradually become a network of open fractures under conditions of continued uplift and erosional unloading. Organic coatings on some of the minerals studied suggest the possibility of establishing the time relationship between fracture-porosity formation and oil migration. McLimans (1981) dated the generation and migration of oil in the Fateh field, Dubai, on the basis of aqueous and oil-bearing inclusions in calcite cements. Similar studies were reported by Seyful'-Mulyukov et al. (1978) and in several other Russian papers. Tillman (1983) and Tillman and Barnes (1983) have attempted to decipher the fracturing and fluid migration histories in the Northern Appalachian basin on the basis of Th of aqueous inclusions in calcite and quartz. Unfortunately, no freezing data were obtained in this work.

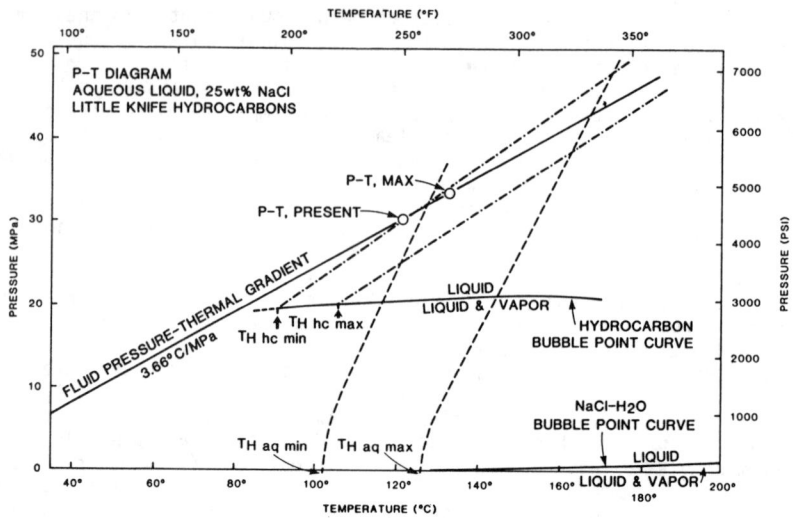

Figure 11-25. P-T diagram showing the isochores for fluid inclusions of hydrocarbons and of aqueous solutions in fracture fillings from the Little Knife field, North Dakota, USA, (see text). From Narr and Burruss (1982).

Narr and Currie (1982) examined the fluid inclusions (of both hydrocarbons and water) in quartz and calcite from fracture fillings in the Altamont field, Utah, USA. As they found a range of Th values at a given depth in each well, they suggested that fracture filling continued throughout an episode of decreasing host-rock burial. The inclusion data also suggested that fractures in the youngest strata were filled at higher temperatures (i.e., greater depths) than were fractures in older beds, and that the latest, lowest-temperature deposition of quartz and calcite occurred under remarkably uniform conditions of temperature throughout the field. They also used the inclusion data to establish paleo-geothermal gradients and the depths at which fracturing occurred, and hence the amount of overburden removed since fracture filling.

Narr and Burruss (1982) reported another interesting application of inclusions from carbonate fracture fillings in carbonate units in the Little Knife field, North Dakota, USA. Hydrocarbon inclusions had Th = 90-106°C, and aqueous inclusions had Th = 102-126°C. From these data and the P-V-T properties of Little Knife reservoir fluids (Fig. 11-25), they concluded: (1) the fractures formed after the strata were buried to at least their present depth of 9800 ft (3000 m), which indicates their age is post-Mesozoic; (2) the pore-fluid pressure gradient was normal hydrostatic immediately after, if not during, fracture system development; (3) formation-water salinity has remained fairly constant since fracture initiation; (4) migration of hydrocarbons into the reservoir probably preceded or accompanied fracture genesis; and (5) CH_4 concentration may have decreased since fracture initiation. The potential for using fluid inclusions to document changing CH_4 concentration within a reservoir could be significant to studies of hydrocarbon migration.

Ypma (1979b) showed that a sedimentary basin in Namibia, even though consisting of very old rocks (500-600 my), has petroleum potential, in part on the basis of fluid-inclusion studies. The inclusions showed that the highest temper-

334

ature achieved by these sediments was 150°C, i.e., below the temperature at which liquid hydrocarbons are destroyed. Furthermore, he showed by gas chromatography that inclusions in a Zn-Pb deposit in the area, which had Th values between 170° and 230°C, still had C_2H_6, CH_4, C_3H_8, and C_2H_4, in that order of abundance, suggesting that the ore deposit represented a localized (or temporary) temperature anomaly. Ypma also made use of data reported by Kvenvolden and Roedder (1971) on inclusions of organic compounds in quartz crystals from South-West Africa (e.g., see Figs. 4-2,4). After detailed studies of those inclusion fluids, Kvenvolden and Roedder concluded that the molecular composition and distribution of hydrocarbons (long-chain alkanes and isoprenoids) suggested biological precursors for these components. The host crystals are in still older rocks (>700 my), but the organic materials could well have been introduced by fluid circulation from overlying younger rocks, of ages from late Precambrian to at least Jurassic. In another somewhat similar application, Visser (1982) used Th determinations to establish the onset of petroleum migration and the maximum temperature (145-160°C) of a Venezuelan petroleum source rock. Borak and Friedman (1981), however, reported oil well bottom-hole temperatures of 218°C at 9.1 km depth (2.5 kbar) in the Anadarko Basin, Oklahoma, USA; this places these rocks at the edge of greenschist-facies metamorphism (see also Friedman et al., 1981; Friedman, 1983).

One of the most striking developments in hydrocarbon-inclusion study is the application of ultraviolet fluorescence microscopy (Burruss et al., 1980, 1981; Burruss, 1981b). This equipment, now commercially available, allows observation of thin sections under vertical UV illumination. As in any new aspect of microscopy, there are several problems with regard to phase recognition, artifacts, etc.; these have been discussed by Burruss (see also van Gijzel, 1979). Various hydrocarbons fluoresce differently, and when combined with petrographic recognition of primary vs secondary origin during a diagenetic sequence, such studies place limits on the timing of hydrocarbon migration relative to diagenetic mineralization during cementation and fracture filling. Burruss et al. (1983a,b) used such techniques to decipher the sequence of events in the Oman foredeep on the basis of hydrocarbon inclusions in carbonate fracture filling cements.

Chapter 12

LOW- to MEDIUM-GRADE METAMORPHIC ENVIRONMENTS

CONTENTS

INTRODUCTION

Metamorphic rocks represent an approach to equilibrium between various solid phases at some new pressure (P) and temperature (T). Reactions in the solid state are notoriously slow compared with those mediated by a fluid phase, and the presence of such a fluid phase during metamorphism is generally assumed. As pointed out by Crawford et al. (1979a), the nature of this fluid phase (in high-grade metamorphic rocks) is generally determined on the basis of theoretical or laboratory studies of metamorphic assemblages, under the assumption that such are the only lines of evidence available. Fortunately, small amounts of this fluid, or sequence of fluids, are sometimes trapped in the fluid inclusions in the minerals of these rocks. Although such inclusions are usually present, most are very small, making study difficult, and they are almost always secondary, raising the possibility of their formation during retrograde metamorphism.

In this chapter, I discuss the lower part of the P-T range of environments normally considered under the term "metamorphism," roughly 0-5 kbar and 150-450°C. I have excluded the following series of other environments, even though they may fall in this range of P and T, because they are best discussed in other chapters, as indicated in parentheses: contact metamorphism (14), basalt-water interaction (16), most ore deposits in metamorphic rocks or formed by metamorphic fluids (15), and geothermal fields, including deep sea basalts (16). In addition, many metamorphic terranes include rocks extending from low- to high-grade metamorphism, and most of the divisions are vague and arbitrary at best. Much more extensive treatments of inclusions in various metamorphic rocks will be found in Weisbrod et al. (1976), Touret (1977), and in a series of chapters in a book edited by Hollister and Crawford (1981). See also a series of 10 papers on the fluids in metamorphism in the Journal of Geological Society (v. 140, part 4, 1983).

NATURE OF VOLATILES IN LOW- TO MEDIUM-GRADE METAMORPHIC ENVIRONMENTS

Origin of the Volatiles

Simple physical devolatilization (i.e., dewatering). What is the source of the fluids now found in fluid inclusions in metamorphic rocks? Certainly the major and most obvious source of the volatiles in low-grade metamorphism is the interstitial pore water in the original materials, whether they be normal detrital sediments, volcanic flows, or tephra. Materials that may start with more than 50 vol % H_2O, e.g., some muds, lose most of this interstitial fluid by compaction while still in the category of sedimentary rocks. The "line" between sedimentary rocks and low-grade metamorphic rocks might fall at ~10 vol % pores. Aa flows and vesicular flow tops may maintain much greater porosity during low-grade metamorphism.

In the temperature range considered in this chapter, most of these interstitial fluids are expelled, and the porosity decreases from >10% to <~1%. This dewatering represents a major geochemical separation process involving huge quantities of material. Thus, a sedimentary basin 500 x 500 km, filled with an average of 10 km of sediments having 10% porosity, contains ~250,000 km^3 of interstitial water. Much of this water is expelled to the surface during the low-P-T part of the compaction process, and hence may come out with very little change from the original seawater (or other) composition present during sedimentation. As the temperatures rise during the last stages of compaction, the compositions of the much smaller amounts of fluid expelled at this time will change dramatically, as discussed below. Several authors have suggested using the Cl/Br ratio to recognize the source of such waters; it may help, but it is only a single parameter in a multiparameter equation. The path followed on a P-T diagram will vary with the burial rates, and the time and the degree to which dewatering takes place will have major effects on the resultant metamorphism (Norris and Henley, 1976).

A noteworthy exception to the above scenario is provided by retrograde metamorphism of massive igneous rocks or preexisting high-grade metamorphic rocks. These rocks will have only very minor amounts of free water, mainly filling joints and shears, and may even take in water during metamorphism, rather than expel it. Thus, to convert an anhydrous basalt to a chlorite schist requires the addition of ~13 wt % H_2O, or ~2 wt % to form a hornblende schist.

The dewatering scenario above deals with connate water only, e.g., original seawater, modified by various reactions, as detailed below. Even though this connate water is being expelled during compaction, the introduction of meteoric water is possible or even probable. This is particularly true within a few thousand feet of the surface if appropriate hydrologic gradients are established by elevation differences, or if the fluid-density decrease due to temperature overrides density increase due to salinity, permitting the establishment of convection cells with cold ground water.[1]

Chemical devolatilization. Somewhat less voluminous, but in many respects more important, are the volatiles present in the original solid materials. Thus, the various clay minerals contain ~8.5 wt % H_2O for air-dry material; on conversion of such clays to shale, this drops to 5, to slate ~4, and to schist, ~2. (The nature of the new phases formed is influenced by the composition of the fluid present at the time; e.g., see Güven and Carney, 1979.) Similarly, any carbonate that breaks down releases ~40 wt % CO_2. Degradation of organic materials yields a variety of volatiles, but H_2O is the major one. Cellulose by itself will break down to form 56% H_2O and 44% free C, but other organic materials form CH_4 and higher hydrocarbons (HHC), H_2, CO_2, and many more complex compounds. Some degree of interaction of these materials with inorganic compounds in sedimentary rocks can certainly be expected, because many of the components are not in equilibrium. Transition elements such as Fe and Mn may be present in oxidized form, and reduced C (plus much H and some S) will be present in the organic matter. The Eh of the resulting fluid cannot be established on the basis of simple stoichiometry, because expulsion of the breakdown products of the organic matter may take place at low temperatures, precluding adequate rates of reaction with the inorganic matter. During this physical and chemical devolatilization process, these fluids will be involved in major chemical reactions with the detritus (e.g., extensive dissolution and albitization of K-feldspar; Land and Prezbindowski, 1981; Land and Milliken, 1981), and in isotopic exchange with the various solid phases. The approach to isotopic equilibration in such reactions at low temperatures may be slow even in geologic terms but becomes more rapid as the temperature increases.

Trace volatiles. In addition to the above major sources of volatiles, a variety of minor sources are of geochemical importance, even though they are quantitatively miniscule. The interstitial fluids, if they have been in contact with the atmosphere, will contain gases in solution. O_2 from this source will probably be minor compared with that from oxidized Fe and Mn, and the N_2 from the air will merely add to that from the decomposition of proteins in the organic matter.

Argon in fluid inclusions presents a more complex problem but may provide valuable data. Some atmospheric Ar, containing its normal small ^{36}Ar component

----

[1] Note that the density increase from dissolving 1 wt % NaCl in water at room temperature would be cancelled by a 30°C rise in temperature.

(0.3%), will be present in any water exposed to the air (Potter and Clynne, 1978c). K is present in fluid inclusions, however, and after trapping occurs, radiogenic ^{40}Ar formed from the K will build up in the inclusion. Measurement of the $^{40}K/^{40}Ar$ ratio, with a correction for the atmospheric Ar contamination (based on the amount of ^{36}Ar), should thus yield the time that has passed since trapping, and some attempts have been made to use this method (Zentilli and Reynolds, 1977).

The major problem in such K/Ar age determinations lies in the "excess" radiogenic Ar, from the devolatilization of the sedimentary rock. This contribution becomes very significant once the bulk of the interstitial water is expelled. Thus, a schist with 1 vol % porosity and 3 wt % K_2O currently generates enough radiogenic Ar to add ~ 0.02 ppm ^{40}Ar to the pore fluid each 10^6 years; since the percentage of ^{40}K decreases with time, this rate was considerably greater in the Precambrian. Only a small part of this new ^{40}Ar will be free to dissolve in the pore fluid, as most will be trapped in the K-bearing minerals where it is formed, but if these minerals are recrystallized, e.g., when clays convert to mica, most of their contained Ar might be expected to be released to the pore fluids. Thus, if the volume of pore fluid is small during recrystallization of the bulk of the rock, the concentration in the fluid of this "excess" ^{40}Ar can be high. (Note, however, that any new phases that form must grow in the presence of a high-Ar fluid and hence may take in appreciable ^{40}Ar.) In a study of the excess ^{40}Ar in fluid inclusions, Rama et al. (1965) made a standard "K/Ar age determination" on some quartz from metamorphically derived white "bull" quartz veins in low-grade metamorphic rocks in Bethesda, Maryland, USA. These veins were probably formed during the Appalachian orogeny, but the K/Ar "age" was far too large, as the amount of ^{40}Ar was ~ 1000 times larger than that which could have formed in the intervening time by decay of the ^{40}K in the inclusion fluids plus that in the quartz structure itself (14 ppm total K). This huge discrepancy is the result of large amounts of "excess" ^{40}Ar sweated out of the schists and into the pore fluids during metamorphism. These pore fluids then formed the quartz veins, and some of the fluid was trapped in the large numbers of apparently secondary fluid inclusions. This fluid had to contain ~ 40 ppm Ar to yield the observed results.

The amounts of alpha-active elements (mainly U) in the inclusion fluids are probably too small for an age determination by this procedure, but He is special, in that degassing of the deep interior of the Earth apparently is introducing new He, the signature of which can be recognized from its very high ^3He content. Developments in mass spectrometry make the detection of the isotopic ratio of He possible even on extremely small samples, so the $^3He/^4He$ ratio in fluid inclusions is currently under study. The problem is somewhat similar to that involving Ar, in that there can be an atmospheric contribution, plus radiogenic ^4He from alpha-active elements in the rock (plus later decay in the inclusion itself).

Since some Xe is from the original planetary materials, and some is formed in rocks by spontaneous fission of U, and Xe should be sweated out of metamorphic rocks along with He and Ar, the Xe isotopic ratio in inclusions may help in understanding the origin of such fluids. This ratio should be examined as soon as instrumental developments permit.

Forms of Volatiles in Low- to Medium-Grade Metamorphic Rocks and Their Significance

The volatiles (mainly H_2O and CO_2) now in metamorphic rocks are in five major forms: (1) fluid inclusions (i.e., intercrystalline inclusions); (2) pore fluids (i.e., intracrystalline inclusions); (3) hydrous (and carbonate) minerals; (4) absorbed surface molecules; and (5) random molecules or atoms in certain mineral structures. The fluid inclusions record the composition of the pore fluids at the time of their trapping. Since these pore fluids have presumably changed considerably in composition with time, and the T and P of the rocks have changed during their history, successive generations of inclusions should record these changes. At present, the pore fluids are generally missing from the (air-dried) samples studied, but must be present in situ; they almost certainly differ, both chemically and isotopically, from the earlier pore fluids trapped in the fluid inclusions. The volatiles present as hydrous and carbonate minerals may constitute the great bulk of the total volatiles; hence, even a very small fraction of these volatiles can cause major chemical and isotopic errors during an analysis of the fluid inclusions.

The volatiles present as random molecules represent an important and fascinating source of information but can cause major problems in the interpretation of some inclusion analyses. Beryl and cordierite (and perhaps quartz) contain channels in their structures in which various volatile molecules such as H_2O and CO_2 can be trapped. This phenomenon would generally be of little concern quantitatively, except during attempts at analysis of fluid inclusions in such minerals. However, the compositional ratios of the volatiles in such channels (just as, e.g., the ratio of structural constituents like (OH)/F in phases such as topaz), may possibly reflect the ratios in the fluid phase present at some time in the past. To use these channel volatiles in this manner, however, requires knowledge of the partitioning expected due to the nature of the channel itself, as well as some evidence that no subsequent exchange took place between channel and fluid.

The possible presence of random molecules or atoms of volatiles in quartz is much more significant in several aspects of inclusion study than are the volatiles in cordierite or beryl. First, the total amount of H in quartz in forms other than fluid inclusions (mainly OH⁻) can be in the same range as that in the fluid inclusions in the same quartz (see references and discussions of "Exsolution inclusions" in Chapter 2, and "Irreversible changes" in Chapter 3); hence, this hydrogen must be carefully considered in any extraction of such fluid inclusions using heat. Second, the existence of such H and the evidence (Chapter 2) that, at least under some conditions, H (as H_2O) may exsolve from the quartz structure to form small inclusions (and the converse, inclusions may dissolve in the quartz), make the interpretation of the origin of small (i.e., several micrometers or less in size) inclusions in metamorphic quartz subject to serious ambiguity that I fear will take years of work to clear up. Third, since Li, Na, and K are each normally present in the quartz structure in amounts even larger than those of H (in part charge-balancing for Al substitution for Si; Roedder, 1958), and since these alkali ions are known to migrate easily through quartz (possibly along with OH⁻?), the nature and concentration of solute ions in such metamorphic inclusions may only reflect their relative diffusion rates. (The specific charge-balancing process necessarily accompanying alkali movement is not certain, but oxygen and silicon must move through the quartz structure if the inclusions change volume, either direction.) Fourth, since the mechanism for the migration of H through quartz, via OH⁻ ions formed en route, does not exist for molecules of CO_2 or CH_4, the ratio $H_2O/CO_2/CH_4$ in inclusions may also change with time. As a result of these last three aspects, I believe that some (or most?) small inclusions in metamorphic quartz, as found today, quite possibly do not represent an actual fluid existing in the host rocks at the time of healing. It is even possible that these "new" fluids might coalesce to form larger inclusions (i.e., larger than several micrometers). As very different processes are involved, the composition of fluids from exsolution will almost certainly be very different from any normal residual intergranular fluid. These reservations, however, may not hold for large inclusions, particularly those that are in definite spatial arrays outlining healed fractures. Although such a healed fracture is a logical site for large numbers of dislocations that could act as nucleation centers for the precipitation of new inclusions, the normal interpretation of such planes of secondary inclusions as representing fluid trapped during healing of a fracture seems more likely.

ORIGIN OF INCLUSIONS IN LOW- TO MEDIUM-GRADE METAMORPHIC ROCKS

Primary Inclusions in New Crystals

Ambiguity as to inclusion origin is one of the major impediments to inter-pretation of inclusion data from many studies of metamorphic rocks. Some inclu-sions, primary or secondary, are generally present in detrital grains, and such inherited inclusions may persist through at least the lower grades of metamor-phism (particularly in coarse minerals and conglomerates), but, by definition, the only possibility for primary inclusions in metamorphic rocks would be those trapped during the growth of new metamorphic crystals. Where such new minerals exist and can be recognized as such, a careful search may reveal primary inclu-sions. Any phase that grows as metamorphic porphyroblasts may trap primary fluid inclusions. Because these crystals grew in essentially a solid matrix, they often have such large numbers of solid inclusions that recognition of pri-mary fluid inclusions is difficult. Using partly crossed polars, rather than alternating the use of crossed polars with plain light, may help in the search.

In many metamorphic rocks, all minerals presently observed have formed during the metamorphism and hence may contain primary inclusions, but the like-lihood of identifying primary inclusions will be greater in cases where new metamorphic minerals can be unequivocally identified, as in the phases replacing original sedimentary features (fossils, etc.), overgrowths on preexisting grains, quartz segregations formed in the "pressure shadow" of a boudin or por-phyroblast, porphyroblasts themselves, and of course all crosscutting veins and segregations. Many metamorphic rocks contain no recognizable primary inclusions, even though the entire rock consists of new crystals. The explanation lies in the probable mode of growth of such crystals -- by very slow addition of material along grain contacts via either diffusion across the contact, as in grain coarsening, or via exchange of material along an intergranular film. In con-trast, the several examples of more likely sites noted above have at least had the possibility of a significant amount of free fluid phase.

The most obvious places for finding primary inclusions are the faceted crystals projecting into vugs. For many years, fluid-inclusion studies in the USSR were mostly aimed in support of the exploration effort for domestic supplies of radiograde quartz crystals for military use, and many of the major early papers dealt with primary inclusions in such crystals. The Swiss Alpine cleft minerals, particularly quartz, have similarly provided material for a series of important studies in the West by M. Frey, N. Guilhaumou, J. Hoefs, J. Mullis, B. Poty, H. Stalder, J. Touray and their colleagues.

One special vein type that has provided much valuable data on metamorphic environments and their changes with time is the cross-fiber quartz vein (Fig. 12-1) and the possibly related "white stripes" of inclusions seen in some Alpine quartz crystals (e.g., Figs. 12-7,8). I found similar "white stripes" (consist-ing of planes of inclusions) in some needle-like quartz crystals from the Jeffrey quarry, Pulaski County, Arkansas, USA (see Roedder and Skinner, 1968, their Fig. 14 and p. 729). Durney (1972) has shown, from Th studies of zones of fluid inclusions marking the growth stages in cross-fiber veins, that some opened (and grew) along a center line crack (Fig. 12-1), whereas others opened and grew along the edges of the vein. Several studies of such inclusions have been made and various theories of origin have been proposed (Lemmlein, 1937, 1946; Stalder and Touray, 1970; Durney, 1972; Durney and Ramsay, 1973; Kerrich, 1974; Mullis, 1975, 1976; Rykart, 1977). The details of the various occurrences are such that no one theory seem adequate for all, but it seems necessary to invoke growth of cross-fiber quartz to yield many, if not most, of the "white stripes."

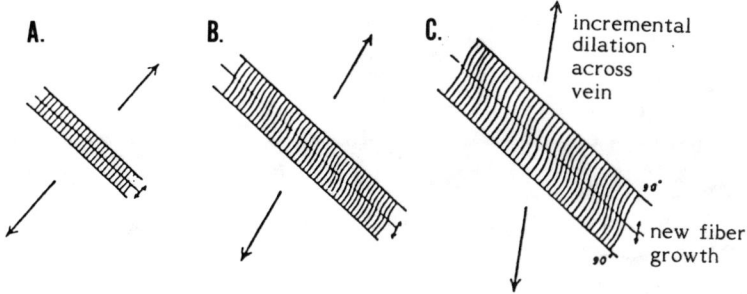

A.   B.   C.

incremental
dilation
across
vein

new fiber
growth

Figure 12-1   Stages in the formation of a fibrous quartz vein, opening along the centerline. (Others can open along the side of the vein.) From Durney and Ramsey (1973).

Secondary Inclusions

Many of the studies of inclusions in metamorphic rocks have had to deal solely with secondary inclusions, generally in quartz, as no primary inclusions were found. The interpretation of such secondaries is made difficult by the long sequence of events in the history of many metamorphic rocks, usually involving at least several possible periods of deformation during which secondary inclusions may have been trapped.

Origin of the fractures yielding secondary inclusions. Many of the curving planes of inclusions in metamorphic quartz clearly represent healed fractures. The physics of crack formation in quartz has been extensively studied experimentally, but the interpretation of the data obtained has been confused (Kekulawala et al., 1981). This confusion results from considerable changes in both the solubility and the diffusivity of water in quartz with T and P, precipitation-hardening during deformation (i.e., formation of fluid inclusions), and other such problems causing experimental difficulties due to nonequilibrium conditions. Superimposed on these problems is the very large extrapolation necessary ($\sim10^{10}$) from rate experiments in the laboratory to deformation in nature.

Once fractures are formed in quartz, laboratory studies have shown that they can start to heal and trap fluid inclusions within hours (Shelton and Orville, 1980). The origin and nature of the fractures yielding natural planes of secondaries has recently been investigated extensively, using special sample-preparation techniques and the scanning electron microscope, particularly by G. Simmons and colleagues (e.g., Simmons and Richter, 1976; Padovani et al., 1982). Most of the microfractures they studied were formed during the last stages of retrograde metamorphism. These studies have shown that after formation of such fractures, in addition to the recrystallization of the host phase (as discussed in Chapters 2 (Fig. 2-16) and 3 (Fig. 3-9)), new phases grow within many of the fractures, thus forming solid inclusions along with the fluid inclusions.

Origin of the inclusions outlining fractures. Although the exact nature of the process (or processes?) yielding them is in doubt (see Chapters 2 and 3, and discussion of Fig. 9-5), the inclusions on healed fractures originally trapped a fluid. Since the fluid present in metamorphic rocks has generally changed over the duration of the metamorphic event, one of the most important goals of inclusion study in such rocks is to place the various secondary planes, each containing fluids having different compositions and/or densities, in chronologic sequence. Many samples provide abundant inclusions but little evidence of sequence (Fig. 12-2).

Some of the criteria that might be found for the establishment of a chron-

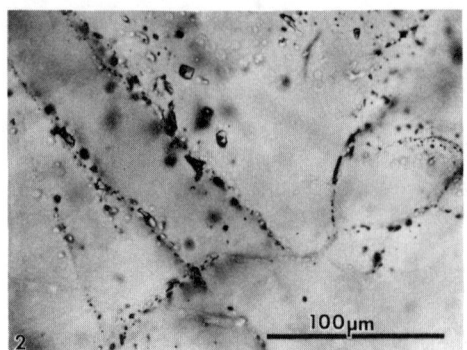

Figure 12-2   Typical group of planes of secondary inclusions, of unknown sequence, in metamorphic quartz. Largest inclusion is 7.5 μm wide.

100μm

ologic sequence for secondary inclusions have been discussed in Chapters 2 and 3 and will only be summarized here:

(1) Crosscutting relations, in which an earlier plane is offset by a later one. This is not common, as lateral movement on the fractures yielding secondary inclusion planes is frequently undetectable.

(2) Crosscutting relations in which the fluid from the earlier plane has been replaced by the later fluid at the intersection. Touret (1977), quoting a thesis by Pagel (1975), showed that the nature of the inclusion at the intersection provides the test. In order to apply this criterion, some distinction in size or shape must exist between the inclusions in the two intersecting planes, or ambiguity will remain. If the two fluids are not visibly different in composition or density, resolution of the sequence may require microthermometry on both planes and on the inclusions at the intersection (Touret, 1981).

(3) Crosscutting relations in which earlier cavities at the intersection have been obliterated (caveat similar to number 2 above).

(4) Degree of "maturity," representing stages in the evolution of the inclusions in any given plane with time (see Chapters 2 and 3 and Figs. 2-16, 3-13 and 3-14).

(5) Degree of divergence from a smooth plane. Roedder (1971a) has shown that inclusions in some planes of secondaries seem to move away from the original plane (see "Stress Gradients," under "Physical Changes" in Chapter 3). When this process has been arrested in midmovement, leaving a trail of inclusions (Fig. 3-19), it is obvious, but if adequate time (particularly at high temperature) has permitted the inclusions to move to something approaching "equilibrium" positions, they may appear to be truly random in array[2] and I know of no way to distinguish them. As a result, several workers (e.g., Hollister and Burruss, 1976; Konnerup-Madsen, 1977; Swanenberg, 1980) generally have interpreted the randomly arrayed "isolated" inclusions as earliest. Such inclusions are often called by the incorrect term "primary," or by the misleading and semantically contradictory phrase "most nearly primary."

(6) Orientation relative to deformation features of known chronology. Thus, if a rock has undergone two deformations, involving different principal stress directions, these deformations may each be reflected in healed fractures that have appropriate azimuths in oriented samples (e.g., Wise, 1964).

2/ Figure 15-26 may show a good example.

344

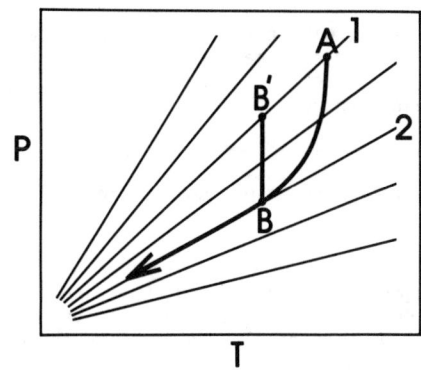

Figure 12-3 P-T diagram with isochores, showing a possible P-T path taken during uplift by an inclusion trapped at A, on isochore 1. If the pressure is decreased with little cooling (A → B), the internal pressure in the inclusion (on isochore 1 at B') may be sufficiently greater than the rock pressure (B) to cause decrepitation of the inclusion, releasing enough fluid to permit resealing with a fluid density corresponding to isochore 2.

(7) Nature of solid inclusions among the fluid inclusions compared with the mineralogy developed in the metamorphic sequence. This criterion follows from the SEM studies of minerals deposited in microcracks (e.g., Padovani et al., 1982). Once a mineral grain is completely embedded in quartz by crack healing, it is no longer involved in most mineralogical phase changes.

(8) Density of filling. Although the intergranular fluid at a given point within a metamorphic terrane can be expected to change with P and T as the point approaches the present erosion surface, Swanenberg (1980) and Hollister et al. (1981) have shown that the commonly made assumption of a simple relationship of density of fill with sequence is not necessarily valid. Depending upon the relative slopes of the isochore for the fluid involved and that of the P-T path followed, the density of the inclusions trapped may decrease or increase with the time of trapping.

(9) Correlation of inclusion parameters with those of nearby gash veins or other features of known origin. Although sometimes considerable differences may have existed between the intergranular fluids present in a rock mass and those within open gash veins in the same rock (Yardley, 1982), if the fluids in the gash veins do match those in one set of secondaries in the rock, this match provides a reasonably valid fixed chronologic point against which other secondary planes may be compared.

(10) Evidence of prior decrepitation of some inclusions. If the path taken by a metamorphic terrane after peak P and T drops too far into the low-pressure side of the isochore for earlier formed inclusions, the pressure difference (internal minus external) can result in partial or complete decrepitation (Fig. 12-3). In either case, the inclusion, and possibly new satellitic inclusions around it (e.g., Fig. 3-22), will heal with a lower density fluid content. Such groups were called "decrepitation clusters" by Swanenberg (1980); they should be distinguished from possibly similar-appearing "clusters" of primary inclusions. Such decrepitated inclusions are generally assumed to predate any inclusion (of the same size) that falls on a steeper isochore, but this assumption need not be true. Natural decrepitation may also be caused by a nearby intrusion, thus providing a local fixed point in the sequence.

(11) Relationship of the inclusions to recrystallization of strained quartz. Strain in quartz, as evidenced by wavy or blocky extinctions, is eliminated by recrystallization. Kerrich (1976) and Wilkins and Barkas (1978) have shown that new inclusions can form along the subgrain boundaries but that the recrystallization to form new quartz can eliminate both preexisting inclusions and these new subgrain-boundary inclusions.

(12) Nature of the crack. Simmons and Richter (1976) described "dPdT

cracks" formed where local strain is high (e.g., within a single grain); these cracks are contrasted with tectonic fractures crossing many grains. Swanenberg (1980) proposed that dPdT cracks, generally parallel to the basal plane {0001} form "earlier" than the tectonic cracks, but I suspect the distinction might not be generally applicable.

COMPOSITION OF INCLUSIONS IN LOW- TO MEDIUM-GRADE METAMORPHIC ROCKS

Volatile Constituents

H_2O and CO_2. Generally, the three major volatiles are H_2O, CO_2, and CH_4, in that order of abundance, but in a few localities, this sequence may be reversed, or other constituents such as N_2, or more rarely, H_2S, may be abundant. H_2O and CO_2 are by far the most commonly studied, only in part because their immiscibility at room temperature makes semiquantitative estimations of H_2O/CO_2 possible by microscopy. This ratio is of great importance as it is the main control on the chemical potentials of H_2O and CO_2, which in turn control the many hydration-dehydration and carbonation-decarbonation reactions that are the essence of most metamorphism (e.g., LeAnderson, 1981). When H_2O and CO_2 are present as a homogeneous fluid, the H_2O acts as a chemical diluent for the CO_2 and vice versa.

Since these two compounds are present as a fluid phase, they may move between different parts of a compositionally layered series. Thus, Ferry (1979) showed that over a single large outcrop of metasediments in Maine, the chemical potentials for FeO, MgO, K_2O, Al_2O_3, and CaO differed considerably from layer to layer, but those for CO_2 and H_2O differed very little. On the other hand, since these fluids can move considerable distances, the fluid present as inclusions in any given assemblage does not necessarily represent the equilibrium ratio for that P, T, and mineral assemblage. If little fluid movement has occurred, in a compositionally layered metamorphic terrane, the composition of the inclusions may differ from layer to layer, each being appropriate to the assemblage containing it.

The wide range of ratios observed in various environments, from <1 to >99 mole % CO_2, can arise in a variety of ways. Thus, CO_2-free fluids can be modified ground waters; they can arise because the original sediment (and its interstitial water) was essentially CO_2-free; the fluid may have arisen from a dehydration reaction (e.g., the dehydration of gypsum to anhydrite in an evaporite sequence); or any CO_2 present may have been used up in carbonation reactions during alteration of silicates containing Ca, Mg, or Fe. Similarly, pure CO_2 inclusions can be an introduced fluid (e.g., as associated with some basaltic magmatism); they might come from silication reactions in impure anhydrous carbonates; they might result from the dissolution of water into the host quartz structure (see "Exsolution inclusions," Chapter 2); or they might come about by the loss of all water to hydration reactions with silicate rock minerals. Immiscibility between CO_2 and H_2O can yield a CO_2-rich fluid (Fig. 8-19) but cannot by itself yield a pure CO_2 fluid.

Organic materials - HHC, CH_4, N_2, and other gases. By the time the sediments in a basin have been heated to such a degree that most geologists would call them metamorphic rocks, most higher hydrocarbon organic matter (HHC) originally present has decomposed to a relatively few simple constituents, mainly CO_2, H_2O, CH_4, and N_2, but including small amounts of higher hydrocarbons, graphite, and H_2S. The reaction $2C + 2H_2O \rightarrow CH_4 + CO_2$, involving the remains of organic matter, or graphite, may contribute to the CH_4 as well. As metamorphism increases, the average molecular weight of the hydrocarbons rapidly decreases toward that of methane, and the sum of all hydrocarbons other than CH_4 frequently totals much less than 1% of the CH_4 (Fig. 12-4). Recognizable liquid hydro-

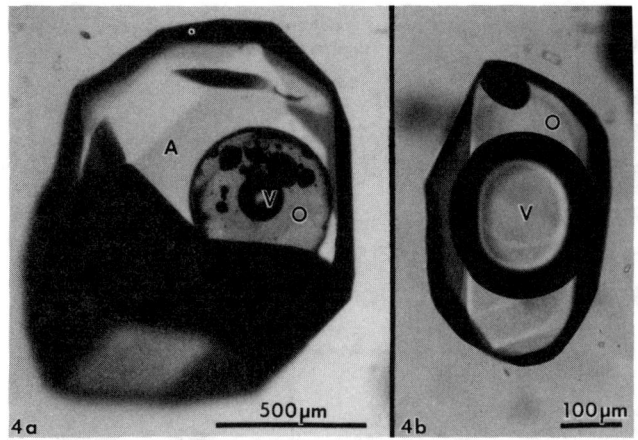

Figure 12-4   Large, primary, in-part-faceted inclusions of organic liquid and water solution in quartz from a vein in Precambrian rocks from South-West Africa, from Kvenvolden and Roedder (1971). Inclusion (a) consists of a globule of yellow organic liquid (O), which contains vapor (V) and many blebs of partly birefringent dark-red to opaque bitumen, formed after trapping(?). The aqueous solution (A) is colorless. Inclusion (b) contains no visible water phase, only colorless organic liquid (O), a large vapor bubble (V) and a bleb of bitumen(?). On crushing such inclusions, most of the organic liquid flashed into vapor; analysis of the evolved gases showed major CH_4, C_2H_6, and C_3H_8, plus various high-molecular-weight components, dominantly n-alkanes and isoprenoid hydrocarbons. The n-alkanes range from at least n-C_{10} to n-C_{33}.

carbons, as reported by Perthuisot et al. (1978), Frey et al. (1980b), Guilhaumou et al. (1981), Guilhaumou (1982), Visser (1982), and others usually imply temperatures <200°C, though Ypma (1979b) reported some C_3H_8 and C_2H_4 as high as 230°C, and Guilhaumou (1982) reported some liquid aliphatic hydrocarbon inclusions in quartz that she believes implies an upper limit of 250°C. One problem in setting an apparent upper limit for liquid hydrocarbons in metamorphism lies in the definition of liquid hydrocarbon; at room temperature, many fluid inclusions in metamorphic environments contain a hydrocarbon liquid that consists of major CH_4 and lesser HHC, under pressure. This "liquid hydrocarbon" may flash into vapor completely on opening, or a residue of nonvolatile HHC may remain.

It has generally been thought that under metamorphism at 250-300°C, over geologic time, most higher hydrocarbons would break down to form mainly CH_4 and graphite. Price et al. (1981) have shown that C_{15} hydrocarbons are still relatively abundant at 9590 m depth in Oklahoma, at an estimated bottom hole temperature of 252°C, and similarly, Price (1982) found high concentrations in cores of Lower Cretaceous rocks from depths of 6400-7500 m in South Texas, at present-day temperatures of 262-296°C. Apparently these hydrocarbons are thermally stable to high temperatures (at least 300°C) in abnormally-pressured semi-closed systems, over geologic time. Several independent indicators place these environments in the zone of greenschist metamorphism.

Using a series of gas-chromatographic procedures, Kvenvolden and Roedder (1971) reported the presence of a variety of both high- and low-molecular-weight compounds in large inclusions in quartz crystals from veins in a dike in Precambrian metasedimentary rocks in South-West Africa. The high-molecular-weight compounds were dominantly n-alkanes and isoprenoid hydrocarbons, the latter ranging from at least n-C_{10} to n-C_{33}. The molecular composition and distribution of hydrocarbons suggested biological precursors for these components. These samples consisted of clear quartz crystals containing large visible inclusions of yellow oil, and other inclusions filled with low-boiling organic fluids, so at least part of the gases obtained almost certainly came from these latter inclusions.

An interesting facet of the high CH_4 content in some of the fluids in quartz from Alpine clefts in Switzerland is the apparent influence it has on the crystal habit. Stalder and Touray (1970) and Mullis (1975) noted that bipyramidal skeletal crystals always contained inclusions very high in CH_4, whereas crystals with other habits contained much less CH_4 in their inclusions. The habit change

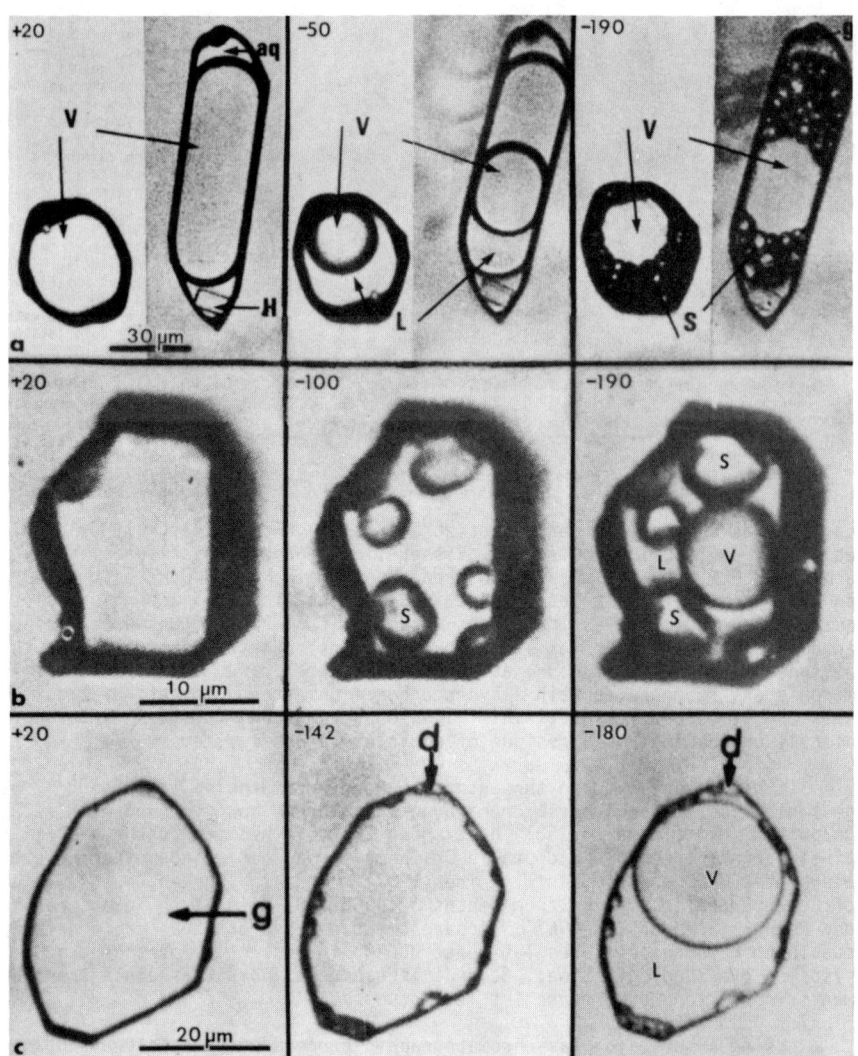

Figure 12-5   Fluid inclusions containing N_2-CO_2 mixtures in dolomite and quartz from Triassic outcrops in northern Tunisia, photographed at the temperatures indicated (°C), from Guilhaumou et al. (1981; Series c) and Guilhaumou (1982; Series a and b). The ratio $N_2/(N_2+CO_2)$ varies greatly: ~0.2 (series a), 0.56 (b), 0.76 (c). V = vapor (CO_2 and/or N_2); L = liquid (CO_2 and/or N_2); S (series a) and (b), and d (series c)) = solid CO_2; H = halite; aq = aqueous solution; g = ice (series a), and gas (series c).

may be related to an actual phase change (separation of an immiscible CH_4-rich phase) in the vugs where the crystals grew (see also "Inclusion studies in structural, thermal, and temporal problems").

The combination of freezing stages capable of achieving temperatures as low as -190°C and the laser Raman method permits the recognition of N_2 in the inclusions in more and more metamorphic terranes (e.g., see Guilhaumou et al., 1981; Guilhaumou, 1982; Kreulen and Schuiling, 1982; and Fig. 12-5). In some, it is the major constituent, for example, in the fluid inclusions in Colombian emeralds

(Dele-Dubois et al., 1980). Kreulen and Schuiling (1982) suggested three possible origins for N_2 in such metamorphic rocks: (1) breakdown of organic matter; (2) breakdown of biotite or other minerals where some K^+ is replaced by NH_4^+; and (3) degassing of deep-seated N_2. The abundance of N_2 in diamonds (Chapter 17) lends support to the third possibility, but the almost complete absence of N_2 from the high-pressure gases in inclusions in peridotite nodules from the mantle contradicts it.

Major Solutes in Aqueous Inclusions

General nature. The many reported analyses of the composition of aqueous inclusions in low- to medium-grade metamorphic rocks show them to be extremely varied in the nature and amount of their solutes. Some fluids are highly concentrated Ca-Mg chloride brines; others are equally rich in NaCl and now have a large daughter crystal of NaCl; still others are very low in salinity. Concentrations of salts are so high in the fluid inclusions of the ferruginous quartzites from the Krivoy Rog in the USSR that buildup of salts in recycled plant process water used during grinding causes serious problems (Pedan et al., 1978). Certainly, some of these various solutes represent the original components in solution in the interstitial fluids during deposition of the original sediments. However, these fluids must also reflect a series of chemical exchanges with the host-rock assemblage during the extensive mineralogical changes that must have occurred during metamorphism. Such fluids presumably had a long time to become equilibrated with respect to the surrounding metamorphic minerals before they were trapped.

The K/Na ratio varies greatly, and has been studied extensively, particularly in higher grade rocks (see Chapter 13). One of the important possible sources for cations in solution lies in the recrystallization of clay minerals; these reactions can absorb, exchange, or contribute large quantities of cations to the pore fluids and are certainly responsible for at least some of the changes in fluid composition as metamorphic grade increases. As in most metamorphic studies, superimposed on these various prograde reactions may be a series of retrograde reactions, depending mainly upon the chemistry and availability of pore fluids. The solutes in these various fluids do not originate simply as a result of interaction of a preexisting fluid with the present host metamorphic assemblage; their origin is complicated by two additional factors: fluid mobility and the obvious lack of evidence for phases that have been destroyed.

Evidence of evaporite dissolution. Obviously, a phase that has been destroyed will not be visible. It is easy to forget, however, that such phases may have existed in the past and may have made major contributions to the solutes in metamorphic fluids. The most obvious such phase is halite. The former presence of halite beds that have been dissolved out of a sedimentary sequence may be inferred from the residual mineralogy, the presence of collapse structures, and similar features. However, once the remaining residual beds have been metamorphosed beyond low grade, such evidence may become obscure or even obliterated, and the dissolved salt will be widely dispersed by fluid migration. Evidence from fluid inclusions of evaporite solution during metamorphism has been reported from northern Tunisia (Perthuisot et al., 1978), Saskatchewan (Pagel et al., 1980), the French Alps (Grappin et al., 1979), Namibia (Behr and Horn, 1982), the Swiss Alps (Fig. 3-4; Poty and Stalder, 1970), and elsewhere. If the evaporite sequence is gone completely, only in special cases will the presence of NaCl-rich fluid inclusions from solution of evaporites be so localized in a metamorphic series that the previous existence of an evaporite horizon can be suspected. Rich (1979) has found NaCl-rich inclusions restricted to a relatively narrow stratigraphic zone in Silurian and younger metasediments just above a major regional unconformity in New England (Fig. 12-6). The available evidence suggests that brine formed during pre-peak prograde metamorphism through the solution of evaporite beds. Although this sequence possibly correlates with a

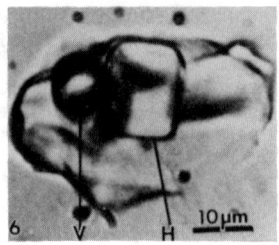

Figure 12-6   Fluid inclusion containing vapor (V) and large halite daughter crystal (H) from Paleozoic metamorphic rocks in New Hampshire, USA. These halite cubes, occurring in a stratigraphically limited range, are all that is left of a former evaporite zone. From Rich (1979).

known evaporite sequence elsewhere (evaporites of the Salina Formation in New York), these saline inclusions are the only remaining trace of an otherwise unknown evaporite sequence in New England. The solution of such preexisting evaporites may be responsible for much of the NaCl so common in the fluid inclusions of many metamorphic sequences, including many high-grade granite gneisses.

Apparently the flow of metamorphic fluids is not always adequate to disperse evaporite salt. Engel and Engel (1953) reported the presence of actual halite as a rock mineral in the Precambrian Grenville Series in the northwest Adirondack Mountains, New York. J.F. Whelan (pers. comm.) has reported the existence of several anhydrite layers in these rocks, almost certainly the residue from an evaporite sequence. One may wonder how common halite may be as a rock mineral, as it would normally only be found in cores taken from far below the water table; at the surface, such NaCl would normally be represented by pores in the rock. Halite in cores would normally not be preserved in ordinary thin sectioning, unless its presence were anticipated and special methods used to preserve it.

The fate of the thick beds of anhydrite that are so prevalent in evaporite sequences is not as obvious. Anhydrite is much less soluble than halite, but from the general absence of anhydrite layers in metamorphic sequences, it presumably also has been dissolved and dispersed. Some of the metamorphic quartz crystal veins in Brazil had anhydrite crystals, but most of these are now represented by hollow casts of quartz of anhydrite crystal shape. Evidence of former evaporites has also been been found in one high-grade metamorphic sequence (granulite facies; Leake et al., 1979).

Mobility of metamorphic fluids. The mobility of metamorphic fluids presents us with a paradox, in that some metamorphic terranes provide abundant evidence of widespread and relatively rapid movement of fluids, whereas others provide equally convincing evidence of an almost complete lack of fluid migration, over millions of years. In these latter terranes, the movement may have approached the minimum rate resulting from pure diffusion, without mass flow of fluid. Compositional differences should drive diffusion, but this process is slow, even in geologic time (e.g., Hikita et al., 1979). Metamorphic studies (e.g., Ferry, 1982; Etheridge and Wall, 1982; Schuiling and Kreulen, 1979) have shown that mass movement of fluid in metamorphic terranes from the several sources of pressure gradients is not only relatively fast at times but is probably responsible for a major part of the heat flow under some conditions.

In contrast, numerous other studies of inclusions in several parts of compositionally layered metamorphic rocks have shown clearly that the metamorphic fluids were spatially nonuniform at any given time. Although the scale may vary, abundant evidence has been obtained on compositionally variable metamorphic rock sequences showing that the mineral assemblage in any given layer or bed generally could not have been in equilibrium with that in adjacent layers (e.g., Ferry, 1979). As a consequence, the intergranular fluid through which and with which the minerals in these assemblages have come to local equilibrium must also have been different, e.g., in such major variables as the K/Na/Ca ratios. Many actual measurements have shown this to be true. Although this local control of the composition has been evaluated in all careful studies of metamorphic inclusions (e.g., Crawford, et al., 1979a,b), some studies seem to ignore it. The control can be in either direction, and is based on the effective rock/water ratio at the time of trapping. If this ratio is high, as in a metamorphic rock containing a trace of intergranular fluid, the fluid composition will be effectively controlled by the rock. If the ratio is low, as along the walls of an open fracture through which fluid is flowing, the reverse is true. This rock/fluid ratio can sometimes be calculated; Ferry (1982) has calculated the rock/fluid ratio needed to accomplish the observed mineral reactions in a regionally metamorphosed impure carbonate sequence.

Etheridge and Wall (1982) pointed out that during dewatering the fluid must move pervasively through the rock and hence equilibrate (at least locally) with the rock minerals. When a fracture is reached, however, fluid movement is faster, the fluid source may be distant, and rock-fluid interaction is restricted, causing the observed wide range of fluid-inclusion types seen in such vein samples. Extensive studies by Stalder (1974) have revealed that all constituents of the Alpine cleft assemblage in the Grimsel area of Switzerland, with the exception of H_2O and CO_2 and occasionally Na and K in the inclusions, have been leached from the adjacent wall rocks. The fact that the fluids moving through these veins have leached the wall rocks extensively is evidence that they can hardly result from local devolatilization reactions. The K/Na ratio in the vein fluids seems to have been buffered by the feldspars in feldspar-bearing rocks. Detailed studies of fluid-inclusion compositions, and calculations of equilibrium constants in the system adularia-high albite-KCl-NaCl-H_2O, indicate that these equilibria may be used tentatively as a geothermometric crosscheck on inclusion thermometry (Poty et al., 1974).

High concentrations of calcium and magnesium. Although Na is normally the most abundant cation among fluid-inclusion solutes, Ca and/or Mg can be more abundant than Na. Some Cl-rich metamorphic fluids were extremely high in Ca and Mg. Such inclusions normally have very small or no daughter minerals and may appear very ordinary but can yield rather unexpected results on the freezing stage. They may have extremely low Tm and Te (generally below -50°C and sometimes below -60°C; see Chapter 8) and may be very difficult to freeze, because such brines have high viscosities at low temperature. Such compositions are most common in metamorphosed carbonate-rich sediments (e.g., Crawford et al., 1979a,b) but can also form during low-grade metamorphism of basalt, as in the Lake Superior copper country (e.g., Roedder, 1963, p. 178), or of anorthosite (Guha et al., 1979). Crawford et al. (1979a) showed that although the composition of the inclusions in the calc-silicate samples were high-$CaCl_2$ brines (plus CO_2), the nearby carbonaceous semipelites contain dense CO_2 and CO_2-CH_4 inclusions in matrix quartz, suggesting that the individual compositions were locally controlled (and that fluid movement was negligible). Kreulen (1977, 1980) showed that on Naxos, Greece, the calc-silicate marble was a relatively closed system permitting local control of the fluids, whereas the surrounding peletic schists were completely open to fluid migration.

Isotopic Composition of Inclusions

Two procedures are used to obtain the isotopic composition of inclusion fluids (mainly on O, H, and C): (1) direct, via extraction and analysis of the fluid; and (2) indirect, via analyses of the same isotopes in the host minerals and use of previously determined isotopic-fractionation factors between crystal and fluid. Direct measurement requires volumes of inclusion fluid that are sometimes difficult to obtain (e.g., Hoefs and Stalder, 1977; Hoefs and Morteani, 1979). At lower metamorphic temperatures, the various solid phases are not in isotopic equilibrium with each other or with the fluid (e.g., Kerrich et al.,

1978), and even between fluid species such as CO_2 and CH_4, isotopic equilibrium may require millions of years at 300°C (Giggenbach, 1982). Rumble (1982) has shown that $\delta^{18}O$ in quartz from impervious interlayered rocks may differ by 1-2 per mil in samples separated by only 1 cm.

Since all three elements studied in inclusion fluids, O, H, and C, have multiple possible original sources and may have taken any of multiple routes between the source and the inclusion, and since each source and route has its own isotopic signature, unambiguous interpretation of such isotopic data is difficult at best. Several interpretations have been attempted, however, as described in this and later chapters. All isotopic studies of inclusions are subject to the limitation that they must be based on the average isotopic signature of a composite of many, usually small, inclusions. Considerably more isotopic work has been done on inclusions from higher grade metamorphic rocks (Chapter 13).

Isotopic studies have been made of the C from inclusions in fissure quartz from several Alpine localities (Hoefs and Stalder, 1977; Hoefs and Morteani, 1979). The $\delta^{13}C$ values for CO_2 from the inclusions range from -1.5 to -9.6 per mil (PDB), and are believed to represent possibly a mixture of juvenile CO_2 plus CO_2 from decarbonatization or decomposition of organic matter. The inclusions also contain considerable CH_4 (and possibly other hydrocarbons). The $\delta^{13}C$ values for total C (CO_2 + hydrocarbons) was found to range from -10 to -30 per mil. The conclusion was that the CH_4 was generated from carbonaceous matter in the surrounding rocks.

Kerrich et al. (1978) determined $\delta^{18}O$ of quartz and other minerals, and $\delta^{13}C$ of carbonates, from a series of low-temperature metamorphic veins in Britain and Switzerland. They found that the value of $\delta^{18}O$ of vein quartz was controlled by that of the detrital quartz in the country rock. From this they concluded that vein systems in rocks deformed under conditions of low-grade metamorphism form principally by local diffusional mass transport and do not require hydrothermal transport in solution. Isotopic disequilibrium was common, and several lines of evidence have suggested that the water/rock ratio was low during deformation and veining. In contrast, Rumble et al. (1982) presented evidence that the minimum water/rock ratio (by volume) was 1.5 to 4.0 (from mineralogic evidence) and 4.0 to 6.0 (from isotopic evidence) at a locality in New Hampshire USA. The fluids were water-rich ($X_{H_2O} > 0.8$), plus CO_2 and CH_4, at 3.5 kbar and 600°C. The species present in the fluid at P and T were not determined from inclusions, however, but were based on consideration of the mineral assemblage and the system C-H-O.

INCLUSION STUDIES APPLIED TO THERMAL, TEMPORAL, AND STRUCTURAL PROBLEMS

General Aspects

Since inclusions provide a sample of the fluids present at the time of trapping, and at least place some (possibly ambiguous) constraints on T and P, they can be used in a variety of ways to clarify the geology of complex metamorphic terranes, although sometimes they introduce new unexpected problems. Despite the numerous sources of ambiguity in interpretation, inclusion data combined with other laboratory and field data can provide valuable insights. The following examples are chosen from among many to illustrate some of the possible applications; better examples may well be found elsewhere in the literature. Although the items are categorized in the following sections, overlap is considerable. Thus, data that help clarify the sequence of events (i.e., a temporal problem) may also help clarify a structural problem. In these applications, one must always keep in mind the possibility that since the inclusions were originally trapped, some may have stretched, collapsed, or leaked.

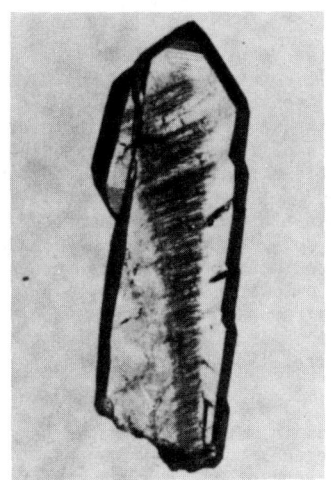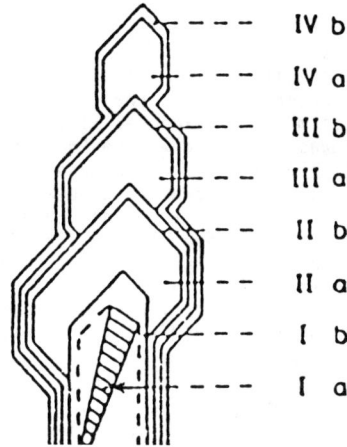

Figure 12-7 (left)   Quartz crystal (7 mm) containing a "white stripe" (generation Ia in Fig. 12-8) of inclusions (see text). From Mullis (1975).

Figure 12-8 (right)   Idealized sequence of quartz-growth generations proposed by Mullis (1975): Ia: Quartz containing a white stripe; IIa, IIIa, IVa: sceptres and skeletal quartz; Ib, IIb, IIIb, IVb: late stages of prismatic growth. Generations Ia, IIa, IIIa, and IVa contain CH_4-rich fluid inclusions (>95 vol % CH_4); generations Ib, IIb, IIIb, and IVb contain H_2O-rich fluid inclusions (<8 vol % CH_4).

Mapping of Isograds

Mullis (1979) studied inclusions in synkinematic quartz crystals from the external part of the central Alps, Switzerland. As many of these inclusions contained mainly CH_4 and/or H_2O, he made a detailed study of the possible use of the system CH_4-H_2O as a geologic thermometer and barometer. The quartz crystals in the clefts in these rocks show an alternation between several different crystal habits (Fig. 12-7, 8), and the composition of the inclusions found in them varies with the habit (Stalder and Touray, 1970; Mullis, 1975, 1976). Where the quartz grew as normal prismatic crystals, the inclusions are H_2O-rich (>92 vol %); where the quartz grew as scepter-shaped or skeletal crystals, the inclusions are CH_4-rich (>95 vol %). Mullis proposed that sudden pressure drops, under essentially isothermal conditions, caused separation of an immiscible CH_4-rich fluid from an H_2O-rich fluid and permitted trapping of CH_4-rich inclusions in the quartz forming under those conditions. By using the aqueous inclusions for temperature and P-V-T data on CH_4 for pressure, he established both P and T for a given crystal. (Why the quartz grows in skeletal or scepter habits in the presence of a CH_4-rich fluid, and whether the quartz actually grows from a CH_4-rich fluid are not known.)

Mullis (1979) concluded from these studies that the inclusion composition and density reflected the temperature zonation during the metamorphism. Using Grubenmann's (1904) categories for Alpine metamorphism (non-, anchi-, epi- and mesometamorphic), Mullis established the following minimum P-T conditions:

Zone	T	P
Nonmetamorphic zone to low-grade anchizone	>200°C	>1,200 bar
Medium-grade anchizone to high-grade anchizone	>270°C	>1,700 bar

Frey et al. (1980b) combined these inclusion data with measurements of the crystallinity of illite (IC) and coal rank (by means of reflectivity, R) from the same series of very low grade metamorphic rocks studied by Mullis. They found a general increase in IC and R from tectonically higher to lower units and from external to internal parts within the same tectonic unit (e.g., a given nappe), as might be expected as an indication of increase in metamorphic grade. In the lowest grade zone, the "deep diagenetic zone," the fluids contained >1 mole % of higher hydrocarbons and had Th <200°C. The low- and medium-grade anchizone was dominated by CH_4-bearing fluids and had Th minimum 200-270°C. In the higher grade anchizone and the epizone, H_2O-bearing fluids were encountered (see Fig. 13-2). Several examples of inverted metamorphism were recognized, in which higher grade units had been thrust onto lower grade units. In such cases, the metamorphism must have preceded the final transport of the nappes; hence, the inclusions helped clarify a structural problem. Unfortunately, the three measured parameters did not show similar correlations in two other areas (Frey et al., 1980a).

The salinity of aqueous fluid inclusions in metamorphic rocks generally seems to change with change in metamorphic grade, but these changes do not follow a consistent pattern. In view of the wide range of possible sources for both solutes and water, this inconsistency is not surprising. Thus, when a sedimentary rock that has been leached of all its interstitial NaCl by circulating fresh water is metamorphosed, dewatering will yield essentially freshwater until the temperature reached is adequate to have significant solutes contributed by various silicate-water reactions. Later, this same body of rock may be exposed to strong brines as a result of structural events that shift the hydrologic flow paths. The intergranular fluids of still other rocks may experience an increase in salinity as a result of extraction of water from the pore fluids to form hydrous phases. In a study in the Swiss Alps, Poty and Stalder (1970) showed that the salt concentration increased with T and P, from the zeolitic facies to the greenschist facies, but they showed that certain other field data were in conflict with this generalization.

Perhaps the most striking overall difference in inclusion composition with metamorphic grade is that between inclusions in low- to medium-grade rocks and those in high-grade rocks. As detailed in Chapter 13, inclusions in high-grade rocks much more commonly have CO_2 as the major component.

Source of Fluids

The ease with which fluids can move under pressure gradients makes it very difficult to identify the sources of the fluids in the inclusions in any given sample. When these fluids have compositions appropriate to the local mineral assemblage (e.g., Crawford et al., 1979a), they are either indigenous or have had time to come to equilibrium with the rocks. Hoefs and Morteani (1979) tried to determine the source of CO_2 in fluid inclusions from fissure quartz from the Tauern Window in Austria, on the basis of $\delta^{13}C$ values, but could not assign the origin of the CO_2 to an external source (juvenile) or an internal origin (organic matter in adjacent rocks).

O'Hara (1980) concluded that apparent intersections of isograds in central Arizona could be caused by exchange of CO_2-rich and H_2O-rich fluids. Similarly, Kreulen and Schuiling (1982) have proposed that the very high N_2 content they found in synmetamorphic quartz segregations from schist surrounding the Dôme de l'Agout, France, came from an external source and moved outward from the dome. The N_2 concentration increased with metamorphic grade. In a study of the Munchberg Gneiss Massif in Germany, Vollbrecht (1981) proposed that high-salinity fluids trapped in quartz from joints in the overthrusting gneiss-nappe were derived from very low grade sediments beneath the nappe. The movement of such fluids has very important effects on the deformation mechanisms.

Depth of Burial During Metamorphism; Uplift History

If the procedures for geobarometry using inclusions are sufficiently accurate, discrepancies between such pressure estimates and reconstructions of the apparent depth of burial may suggest that erosion (or faulting) has removed units for which there may be little or no other evidence. Thus, Poty (1969), in a study of the inclusions of Bourg d'Oisans, France, concluded that the densities were so high that a thick sedimentary cover, now gone, must have been present. Similarly, Touray and Tona (1974) have shown that the inclusions in fluorite deposits in Granada, Spain, were formed at a minimum depth that implies the existence of nappes now eroded.

Crawford et al. (1979a) have shown that the inclusions now observed in calc-silicate rocks in the Prince Rupert area, British Columbia, Canada, were not formed at peak metamorphic T and P; they have suggested that the discrepancy between predicted and observed densities may permit deductions as to the history of uplift for the samples. Most estimates of uplift history, however, have been based on inclusions in higher grade assemblages (Chapter 13).

Siting of Nuclear Reactors

Fluid inclusions provide one of the many examples of "pure" scientific research, pursued for what might have been considered purely "academic" reasons, that turns out to have social value. The siting of nuclear reactors involves careful evaluation of the hazards from various possible geologic processes that might occur after the reactor is in operation. One such process is movement on faults in or near the site. Movement on a fault actually cutting the site could cause major disruption, and seismic action from movement on a fault near the site could also cause damage. Not uncommonly, excavations for major construction projects uncover evidence of faults. When such is found, the geologist is immediately asked to predict the possibility of renewed movement on that fault. This question is not easy to answer. The probability of such renewed movement is assumed to be inversely related to the length of time since the last movement; if a fault has not moved for a very long time, it can be considered inactive. However, an old inactive fault may look very like a recently active one. Fortunately, fluid inclusions can sometimes help to place some constraints on the time since the last movement.

Cunningham (1974) first applied fluid-inclusion geobarometry to this question at the Ginna Project in New York State; his study was based on fluid inclusions in euhedral crystals protruding into vuggy cavities along a fault that cut the proposed site. The occurrence of the crystals suggested that they must have grown since the last movement on the fault. An elevated pressure at trapping, shown by thermometric data on the inclusions in the crystals, established a minimum depth below the surface at the time of formation. (The inclusions can be either primary or secondary for this procedure, and in fact, the latest inclusions, i.e., secondaries, are actually the most useful.) Subsequent erosion had removed this much material from the site, so by estimating the rates of denudation, a minimum time since the last movement was obtained. Since this first application, similar procedures have been used at several other reactor sites, as reviewed by Peck (1982). Kelly (1974) used Th of inclusions in quartz, along with an estimated geothermal gradient, to arrive at a depth of formation for quartz from a fault in a site near Limerick, Pennsylvania. Th plus several other criteria based on inclusions were used to preclude modern formation of the inclusions at near-surface conditions. Pottorf et al. (1977) made a similar study of the site at Nine Mile Island, New York. Several such other reports exist (e.g., Kelly, 1975), but although they are theoretically available for public inspection, they are very difficult to track down in the U.S. Nuclear Regulatory Commission's computerized document files.

INCLUSION STUDIES OF METAMORPHIC GOLD DEPOSITS

General Aspects

Although most fluid-inclusion studies of ore deposits are treated separately in Chapter 15, the metamorphic Au deposits are tied so intimately to the geochemical and tectonic environment of the associated metamorphic complexes that they are better discussed in this chapter. Although metamorphic rocks contain many types of Au deposits, I refer here mainly to the Au quartz veins that have been related to the metamorphic dewatering of a thick pile of sedimentary or volcanic rocks, with or without an added intrusive heat source, as in the Yellowknife district, Northwest Territories, Canada (Boyle, 1955; Kerrich, 1977); the Alleghany district, California (Radtke et al., 1980b); the Carolina slate belt, USA (Ford and Feiss, 1982); the Homestake gold mine, South Dakota, USA (Rye and Rye, 1974); and the New Zealand geosyncline (Henley et al., 1976).

Kerrich (1977), and Kerrich and Allison (1978) were able to show, from isotopic values on minerals, that the fluids which deposited the Au in the Yellowknife greenstone belt were metamorphic fluids derived from the dehydration of a metabasalt sequence, expelled along large shear zones. On the basis of isotopic analyses of both minerals and inclusions, Rye and Rye (1974) came to a similar conclusion on a metamorphic origin for the quartz and the Au at the Homestake mine in South Dakota, in which original syngenetic-exhalative Au was mobilized into dilatant zones during regional metamorphism.

Composition, Temperature, and Pressure

CO_2 is present in the fluid inclusions of many Au-quartz vein deposits, but its presence is not always recorded. Most of the reports specifically listing CO_2 indicated that a separate liquid-CO_2 phase was observed at room temperature (e.g., see Fig. 4-9). Considerable CO_2 can be present, however, without the appearance of such a separate liquid phase. No such liquid will be present if the vapor bubble contains CO_2 at less than ~70 atm pressure (and the aqueous phase will contain whatever amount of CO_2 will dissolve in it at such pressures).

Although many studies of inclusions in Au deposits have involved decrepitometry (e.g., references in Roedder, 1984b), those Au-quartz veins in which Th has been determined are generally in the range 200-350°C (Fig. 12-9). Most such inclusions show no NaCl daughter mineral, and salinities are generally low, but dawsonite [$NaAl(CO_3)(OH)_2$] has been rather commonly reported since it was first recognized as a daughter mineral in Au-quartz veins in the Alleghany district, California, USA (Coveney and Kelly, 1971), in amounts equivalent to 0.34 molar dawsonite in the original fluid (Coveney, 1981).

The pressure correction applicable to such Au deposits may be large. As an example, consider a Au-quartz vein formed during greenschist-facies metamorphism, containing inclusions that homogenize at 250°C. The pressure in such an environment would have been at least ~2 kbar. Assuming a pure 5 wt % NaCl solution and Th = 250°C, the pressure correction for these inclusions would be 170°C; i.e., Tt would have been 420°C (Potter, 1977). However, Coveney (1981) reported Th values as high as 300°C and suggested pressures as great as 2500 bars. Extrapolation from the data of Potter (1977) would put maximum Tt at 530°C for such inclusions, well above the estimates by Coveney (1981) from various geothermometers and those by Taylor (1981) from isotopic studies in the area. Metamorphic Au-quartz veins in greenstones commonly contain CO_2 (Kerrich and Fyfe, 1981). The pressure corrections for CO_2-rich fluids will be less than those for simple aqueous inclusions having the same Th. Even though their isochores have a lower slope than those for H_2O-rich fluids (i.e., they are more compressible fluids), they start at considerably higher pressure at homogenization (Roedder and Bodnar, 1980; R.J. Bodnar, ms. in preparation). Note, however, that the value of the

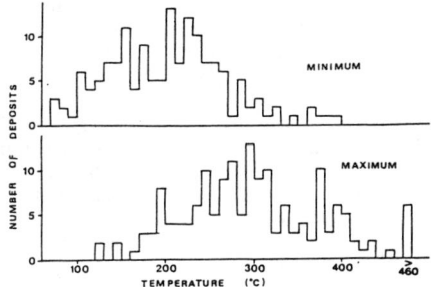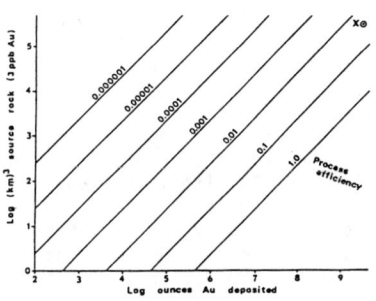

Figure 12-9 (left)   Histograms of the maximum and minimum Th for inclusions in the productive Au-deposition stage in 151 Au deposits. From Roedder (1984b).

Figure 12-10 (right)   Plot of amount of Au deposited from a given volume of source rock (density = 2.7 g/cm³), at various "process efficiencies," defined as product of: (weight fraction of total Au present that is extracted) x (weight fraction of extracted Au that is precipitated). Thus, "0.001" indicates that only one-thousandth of the total Au present has been extracted and subsequently precipitated. Plot concept from S. Romberger (pers. comm., 1982). See text for significance of point "x." From Roedder (1984b).

Henry's Law constant K for the solution of CO_2 in water solutions has a pronounced maximum at ~175°C (Ellis and Golding, 1963).

Relationship to Gold

The quartz of most Au-quartz veins contains large numbers of tiny, obviously secondary inclusions, but only rarely are primary inclusions recognized. This in itself is not as serious a problem as it might seem. Petrographic evidence indicates that Au has generally been introduced later than the bulk of the quartz; hence, at least some of these secondary inclusions may contain the actual Au-depositing fluid (Boyle, 1954). The major problem here is that careful examination of such quartz generally shows that different fluids have been present during several different stages of fracturing and healing of the sample. Features that are obvious under the microscope include the presence or absence of a liquid CO_2 phase or daughter minerals such as cubes of halite, as well as variation in the liquid/vapor ratio. Which (if any) of these fluids was the Au-depositing fluid? Very careful petrography may yield an answer; e.g., the presence of flecks of Au along healed planes of secondary inclusions of one type but not along planes that have other compositions. To my knowledge, except for a brief mention by Boyle (1954), such petrographic detail has not been obtained on most Au deposits. The problem is complicated by the many periods of crushing and recrystallization of quartz that are observed in almost all detailed studies of Au-quartz veins.

One seemingly obvious and straightforward approach to this dilemma is to look for Au in solution in the inclusion fluids. Such analyses have been reported (see below), but their interpretation is also ambiguous. Another seemingly obvious approach is to study inclusions actually in Au. Very little has been done in this area. Petrovskaya et al. (1971; 1973) reported gas inclusions containing 90% CO_2, together with some nitrogen and other gases, at pressures of "5-20 atm", in native Au. They suggested that immiscible CO_2 globules in the surrounding fluid adhered to and were enclosed by the growing crystal of Au. Several reports from the USSR list Td for alluvial Au grains, but the interpretation of such results would be exceedingly difficult.

Many analyses have been made of the gases evolved on heating samples of Au ore. These gases generally are assumed to come from the fluid inclusions, but proof of this is difficult. Perhaps the single most important value is the

H_2O/CO_2 ratio, as this has major effects on wall-rock alteration. The many uncertainties inherent in the analysis of a bulk sample for H_2O and CO_2 can be avoided by simple optical measurements of the phases present in the inclusions. Care must be used, however, if significant CH_4 is present, as is common in many metamorphic terranes.

Such optical studies sometimes reveal the presence of almost pure CO_2 inclusions, containing little water. However, whether such inclusions are: (1) a result of immiscibility of an earlier mixed CO_2-H_2O phase into H_2O-rich and CO_2-rich fluids before trapping or necking down (Roedder and Bodnar, 1980); (2) a result of loss of H_2O from earlier mixed CO_2-H_2O inclusions by various processes (Roedder, 1981d); or (3) samples of an actual Au-transporting fluid, is generally not known.

Gold Content of the Ore-Forming Fluids

The Au content of Au-depositing solutions in nature is important but relatively unknown. I am aware of only a few reports of analyses of fluid inclusions for Au; these show from 1×10^{-6} to 15×10^{-3} mole/liter (see Roedder, 1984b for details). The latter figure is so high that the Au <u>in the inclusions themselves</u> would contribute as much as 2 oz Au/tonne of quartz; hence, it seems unreasonable.

In view of the paucity of actual analytical data, another question may be asked: What amount of Au must be carried by the hydrothermal fluid in order to form a reasonably-sized Au deposit, assuming mass transport, rather than diffusion? Using geologically reasonable values for flow rates (0.1 to 0.001 mm/sec; Roedder, 1960a), ore containing 0.1 oz Au/tonne (i.e., 3 ppm) would require the <u>precipitation</u> of only 3 ppb of Au from the fluids. No necessary connection exists between the amount precipitated and the amount remaining in <u>solution</u>, as presumably determined by the analyses mentioned above, but the 3 ppb is two orders of magnitude smaller than the smallest value reported above, and six orders of magnitude smaller than the largest value reported. For comparison, this 3 ppb concentration is, in turn, one to three orders of magnitude larger than the "high" Au content measured in present-day waters of geothermal areas (Weissberg et al., 1979).

A comparison of the precipitation of Au and gangue is also instructive. If we assume that the quartz was precipitated from the same solution, the 3 ppb Au precipitation would require deposition of ~0.01 wt % SiO_2 from these fluids. This value is reasonable, in view of the solubility of quartz in various laboratory studies and the fact that we must deal here with only the <u>change</u> in solubility, not the <u>total</u> solubility.

Precipitation of only 3 ppb of Au might seem at first to require the passage of tremendous quantities of water, but such quantities are still geologically reasonable. The total amount of Au mined throughout the history of man is ~87,000 tonnes. To hold this amount of Au in solution at 3 ppb would require 3×10^4 km^3 of metamorphic fluids. At 10 vol % porosity, this amount of fluid would be present in the pores of a <u>single</u> basin 100 x 300 km in extent and 10 km thick (Fig. 12-10). Most surprising, perhaps, is the fact that this amount of Au would correspond to only ~2% of the total Au present in the rocks of the basin, i.e., only ~2% of the total Au present would have to be extracted from the rocks by the liquids as they leave, and precipitated in economically workable deposits (Roedder, 1982a; see point "x" on Fig. 12-10).

The value of 3 ppb, if correct, illustrates the difficulties inherent in attempting to determine the nature of the Au species transported in solution from determinations of the compositions of fluid inclusions, as is commonly done. <u>Each</u> of the various ions found in the inclusion analyses is generally present in the 10,000-ppm range, six or seven orders of magnitude more concentrated than the Au. If a highly effective ligand is present and responsible for Au solubility, it <u>could</u> be present in concentrations many orders of magnitude less than those of the normal solutes determined in inclusion analyses.

Regardless of the mechanism(s) involved in the transport and deposition of Au in metamorphic terranes, one aspect seems to have had very little attention. The fluids evolved during metamorphism are presumed to have transported and deposited many different elements, in a variety of different ore-deposit types. Yet most of those that formed the Au-quartz veins have seemingly moved (and deposited) essentially only quartz, Au, carbonate, K, and sometimes minor Na (Gallagher, 1940) and W (scheelite). The explanation of this basic distinction is critical to an understanding of the processes of ore formation but has not been addressed.

Chapter 13

MEDIUM- to HIGH-GRADE METAMORPHIC ENVIRONMENTS

CONTENTS

INTRODUCTION

Many of the processes and concepts described in Chapter 12 are pertinent here as well, but the high-grade metamorphic rocks discussed here have several notable features that distinguish them from the low- to medium-grade metamorphic rocks discussed in Chapter 12. In this chapter, I discuss the upper half of the P-T environments normally considered under the term "metamorphism" ($>\sim5$ kbar and $>\sim450°C$ to magmatic temperatures). High-grade metamorphic rocks are much lower in total volatiles than are the lower grade rocks, as the devolatilization process has gone much further. Their porosity is lower and is often close to zero, the volatile-bearing phases they have are lower in volatiles and/or in amount, and they generally have relatively few fluid inclusions. Ambiguity about inclusion origin in these rocks can be as great as or greater than that for the lower grade rocks. Some rocks may be essentially free of inclusions. Most of the inherited inclusions (e.g., in detrital grains in metasediments) will have been swept out by recrystallization, along with most of the original depositional textures. Furthermore, as a result of higher temperatures, and perhaps longer times at those temperatures, the minerals may be much closer to equilibrium with each other and with the small amount of fluid phase present as pore fluids and inclusions, in terms of both chemical and isotopic composition. The porosity and permeability are near zero, and even the microcracking appears to be a retrograde phenomenon (Padovani et al., 1982), yet differences in fluid composition as a result of differences in local host-rock mineral assemblages over an outcrop are smaller than those in the lower grade rocks, or even immeasurable. Over distances of kilometers, however, the differences can still be striking. Commonly superimposed on this picture are the effects of retrograde reactions, which may involve the introduction of new fluids from external sources. Thus, adjacent planes of secondary inclusions may have grossly different compositions or densities, just as they do in the lower grade rocks. In some terranes equilibrated under high T-P vapor-absent conditions, the inclusions may only be from late-stage events, but even such data can constrain the uplift path in P and T (Selverstone et al., 1983). Lastly, the uncertainties in estimates of P and T are generally larger than those in the lower grade rocks, mainly because the fluid-inclusion data must be extrapolated further to the higher P and T values.

As in Chapter 12, I have excluded contact-metamorphic rocks, most ore deposits in metamorphic rocks or those formed by metamorphic fluids, and kimberlites and ultramafic nodules, even though some of these may have formed in this same range of P and/or T; they are discussed in Chapters 14, 15, and 17, respectively. The dividing line between high-grade metamorphic environments and igneous processes is much more nebulous than that between medium- and high-grade metamorphism (Chapters 12 and 13). Even if an arbitrary division based on temperature is chosen, the data to which it needs to be applied are seldom accurate. The selection of a significant upper dividing line, e.g., in the series gneiss-granite or granulite-granite, is even more difficult and risks a surfeit of semantic arguments as well. Touret (1981) considers the first formation of "mobilisates" -- migmatitic to pegmatitic stringers -- as a convenient boundary.

Touret (1981) has pointed out that the study of inclusions in such high-grade rocks started a relatively few years ago, with the publication of three important papers: Dolgov et al. (1967) on dense CO_2-H_2O inclusions in kyanite from metamorphic rocks and pegmatites in the USSR; Touret (1971) on Norwegian granulites; and Hollister and Burruss (1976) on the Khtada Lake complex in Canada. Since then, still other major studies have been made, but much is still unknown.

FLUID COMPOSITIONS VS MINERAL ASSEMBLAGE

Nonmineralized Rocks

The common achievement of equilibrium between fluids and solids at higher temperatures, or at least a close approach to it, in contrast to the lack of phase equilibrium in lower temperature environments, permits detailed comparison among results from experimental metamorphic petrology, theoretical predictions of equilibria, and the natural mineral and inclusion assemblage. Eugster (1981, 1982) provided excellent reviews of the present status of this comparison (see also Loomis, 1983; Bowers and Helgeson, 1983b).

In a study of fluid inclusions in calc-silicate rocks from Prince Rupert, British Columbia, Canada, Crawford et al. (1979a) showed that not only did the CO_2/H_2O ratio of the inclusions generally agree with the calculated ratio for the mineral assemblage but that the aqueous inclusions in quartz grains in the calc-silicate rocks were brines containing a very high proportion of $CaCl_2$ (some had Te as low as -60°C), whereas quartz in nearby carbonaceous semipelites had inclusions containing dense CO_2 and CO_2-CH_4 mixtures. This difference suggests internal control of the fluid composition. Yardley (1979) and Yardley et al. (1983) reported that the dominant fluid in inclusions from the upper sillimanite zone (temperature maximum ~640°C) in some Dalradian metasediments is CH_4. These inclusions had Th near -110°C. Bodnar and Connolly (1983) showed that fluid pressure could be derived from integration of fluid inclusion and mineral phase equilibria, particularly for assemblages in equilibrium with CO_2/H_2O mixtures.

In addition to CO_2/H_2O, the ratio K/Na is particularly useful. Poty et al. (1974) calculated the K/Na ratio in the fluid phase that should be in equilibrium with two feldspars (adularia and high albite); they showed that analyses of K/Na in fluid inclusions in quartz from such assemblages could be used as a geothermometer. Luckscheiter and Morteani (1980a) applied this geothermometer to analyses they made of inclusions from quartz from Alpine veins in gneiss, amphibolite, and mica schist. Their results (435-490°C) are lower than estimates made by others on the basis of $^{18}O/^{16}O$ studies (500-600°C). They ascribe the differences to a combination of sampling problems (primary vs secondary inclusions) and the influence of CO_2 on the equilibria (because this geothermometer "is valid only for CO_2-free aqueous inclusions"). They also concluded that the very high Ca/Na and Mg/Ca ratios they obtained on extraction and analysis of fluid inclusions in quartz are probably spurious, as a result of contamination from solid inclusions. Poty et al. (1974) got much lower ratios in otherwise equivalent quartz samples from the Western and Central Alps.

Touret (1981, p. 190) reported that the aqueous inclusions in high-grade metamorphic rocks (in which most inclusions are CO_2) range widely in salinity from halite-bearing inclusions to almost pure water in a single specimen. Apparently this range cannot be explained by necking down. In large part it may be caused by multiple sources of introduced fluid, presumably at different times, or it might result from fluid immiscibility, but because no experimental data on the system H_2O-NaCl-CO_2 are available for these P-T conditions, quantitative evaluation is not possible.

For many years before the abundance of CO_2 in metamorphic fluids was recognized (as evidenced by the fluid inclusions), the common assumption in metamorphic dehydration reactions was that $P(H_2O) \approx P(total)$. We now know that H_2O and CO_2 each act as major diluents for the other in such fluids, and many current studies have concentrated on the calculated H_2O/CO_2 ratio and its many important implications (e.g., Jacobs and Kerrick, 1981b), but several other constituents can be examined by similar procedures. Any mineral whose structure can accept several different volatile species has the potential of providing a record of the ratio between those species in the fluid, to be compared with that obtained from the fluid inclusions, provided the necessary partitioning data are available. The substitution of F for OH in micas and amphiboles provides an example. Petersen et al. (1981) have been able to estimate the fugacities of H_2O, HF and F (as well as CO_2, CH_4, CO, H_2, S_2, and O_2) in the fluids during metamorphism of Grenville marbles near Balmat, New York, USA, on the basis of the mineral assemblage. They showed that failure to consider possible F substitution can lead to large errors in estimated P, T, and fluid compositions. Rozen et al. (1977) suggested that the composition of scapolite and apatite can be used as indicators of the composition of the volatiles during metamorphism of the Kola Peninsula granulites, USSR. Cordierite, with its well-known channels, can accept both CO_2 and H_2O (as well as organic compounds), and the CO_2/H_2O ratio in it has been investigated by several groups (Armbruster and Bloss, 1980; Johannes and Schreyer, 1981; Zimmermann, 1981). The interpretation is complicated by variation in the partition coefficient with variables other than merely the ratio of the surrounding fluid. Thus, Coolen (1980) showed that the S/O ratio in scapolite from Tanzanian granulites is a sensitive pressure indicator under S-saturated conditions.

The presence or absence of NaCl-rich scapolite has been used as an indicator of the nature of the fluids. The presence of highly saline fluid inclusions in or associated with sodic plagioclase suggests the possibility of reaction to form scapolite, perhaps during cooling. Such reaction products have been looked for but not found at Ascension Island (Roedder and Coombs, 1967). Vanko and Bishop (1982) showed that this reaction proceeds, but only at 700°C and above in very strong brines [X(NaCl) >0.64 or >85 wt % NaCl].

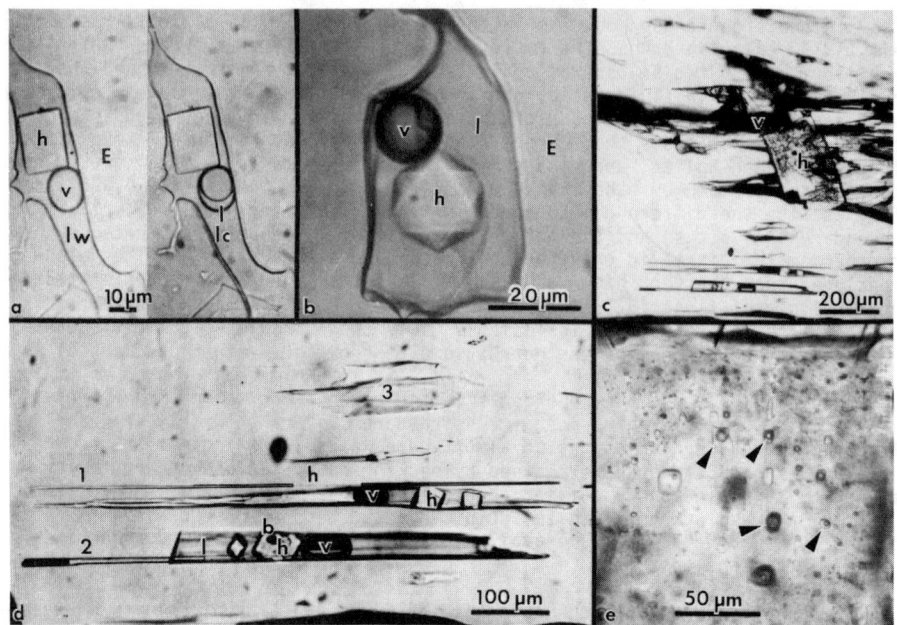

Figure 13-1. Photomicrographs of inclusions in emeralds from Chivor and Muzo emerald mines, Colombia, taken in plain transmitted light. Samples courtsey of Banco de la Republica, Bogota, Colombia. (a) Photomicrographs taken at +27.5°C (left) and +10.0°C (right) of a multiphase inclusion in emerald (E), showing a relatively small vapor bubble (v) and a large daughter crystal of halite (h) in saturated brine (lw). The bubble contains CO_2 gas at moderately high pressures, as shown by a crescent-shaped fillet of liquid CO_2 (lc, right photo) that evaporates into the bubble when warmed to room temperature. Sample ER 63-139b, Muzo. (b) Photomicrograph of inclusions in emerald (E), showing an isotropic daughter crystal (h) of apparently hexagonal outline (actually it is an octahedron) that is believed to be halite, plus vapor (v) and liquid (l). Why this particular halite daughter (and some others) should be an octahedron rather than a cube is unknown. Sample ER 63-139b. Muzo. (c) Photomicrograph of a cluster of multiphase inclusions in emerald. The amoeboid mass in the center right is all one large inclusion containing a black vapor bubble (v) and a large daughter crystal of halite (h; enlarged in photo (e)). Several tubular inclusions with the same phases are shown at bottom of photo; in photo (d) they are shown enlarged. Sample ER 61-6, from an unspecified locality in Colombia (probably Chivor). (d) Enlargement of the group of tubular inclusions shown at bottom of photo (c). Each contains one or more crystals of halite (h), a small anisotropic grain (b), a gas bubble (v) with considerable CO_2 under pressure (see also (a)), and a very strongly saline solution (l) containing other ions in addition to NaCl. In the two partly overlapping very thin, acicular inclusions (1) and (2), the darkest band is vapor, liquid is lighter, and the NaCl crystal (h, in contact with the emerald walls), is bright and almost invisible (this is shown particularly effectively in inclusion 1). Inclusion 3 is an all-liquid inclusion, presumably formed from necking down of the larger inclusion visible in photo (c). (e) Enlargement of part of the halite daughter crystal visible in photo (c), showing many fluid inclusions in the daughter crystal, some containing liquid and vapor in various ratios (arrows), as a result of growth at various stages during the cooling of this emerald or of necking down of a larger amoeboid daughter crystal, or both. Scale bars in μm.

Although reference was made in Chapter 12 to the obliteration of evidence of former evaporite sequences during metamorphism, and to the dispersal of the soluble chemical constituents, an occurrence of anhydrite in granulite-facies rocks has been reported and is suggested to represent the metamorphosed remains of an evaporite sequence (Leake et al., 1979). As these rocks are older than 2000 m.y., this anhydrite is one of the earliest evidences of the salinity of Precambrian oceans.

The amount of H_2O is the largest unknown in the composition of fluids available during high-grade metamorphism. In part this will be evident in the amounts of hydrous minerals present, such as hornblende or mica. However, Wilkins and Sabine (1973) showed that a series of nominally anhydrous silicates such as kyanite, andalusite, andradite, pyrope, diopside, rhodonite, andesine, adularia, and olivine, all have OH contents equivalent to at least 0.008 wt % H_2O (specifically excluding H_2O as liquid inclusions). A water content of 0.008% may seem insignificant but amounts to 8×10^5 tons of H_2O in every cubic mile of such rocks. Perhaps even more significant to fluid-inclusion studies was that these values, obtained by IR absorption spectra, were verified by observing the changes when H was replaced by D after treatment with 100 bars D_2O pressure at 750°C for <5 days. H diffusion and isotopic exchange through these minerals (except olivine) was surprisingly rapid.

Friedrichsen and Morteani (1979) studied the O and H isotopes from vein and host-rock minerals from high-grade gneisses of the Western Tauern Window in Austria. That only a small variation was found in the H isotope composition of the biotites from both the gneiss and the fissure fillings (δD = -54 to -59 per mil) suggested a common water source of probably deep-seated origin and no detectable contribution from isotopically light meteoric water. Oxygen isotope fractionations between coexisting quartz and biotite of 3.5 to 7.0 per mil were found, indicating equilibrium temperatures of 640 to 450°C, respectively, when the fractionation curve of Hoernes and Friedrichsen (1978) was used. These equilibration temperatures decrease for samples to the north across the Window. The metamorphism of the host rocks and the filling of the fissures occurred at the same temperature in a given sample locality.

Colombian Emerald Deposits

Some metamorphic assemblages provide relatively few constraints on the composition of the fluids. The Colombian emerald deposits, for example, occur as veins and pockets of quartz, albite, calcite, pyrite, and emerald in albitized carbonaceous shale, siltstone, and carbonate rocks (Johnson, 1961a,b; Feininger, 1970). Parisite [Ce,La]$_2$Ca(CO$_3$)$_3$F$_2$] occurs in some of the veins (as at Muzo but not at Chivor). Although the host sediments show little visible evidence of elevated temperatures, the fluid inclusions in the emeralds and the associated quartz contain exceedingly saline brines, some of which are very difficult to freeze (Roedder, 1963), plus a halite daughter crystal (>15 vol.%), a small amount of liquid CO_2, and a moderate-sized vapor bubble (~11%).[1] These fluid inclusions provide excellent illustrations of a variety of aspects of inclusion study (see also Figs. 3-16, 4-8). As Te for these inclusions ranged from -63 to -58°C, and Tm ice was ~-34°C (Roedder, 1963), major amounts of Ca (and probably other ions) must be present. These inclusions were exceedingly difficult to freeze, and some never froze (see Chapter 10, p. 297). Although liquid CO_2 is present (e.g., Fig. 13-1a), the water phase showed no dissolved CO_2, HCO_3^-, or CO_3^{2-} by Raman spectroscopy (Rosasco and Roedder, 1979). The absence of such spectra suggest that the concentrations of these constituents are probably each less than a few thousand ppm. The cubic daughter crystals (and more rarely, octahedra; Fig. 13-1b) of halite are so common in these emeralds that they have been used to establish that (1) the stone is natural and not synthetic, and (2) that it probably came from the Colombian emerald mines (Gübelin, 1953).[2] The larger daughter crystals themselves very commonly contain fluid inclusions (Fig. 13-1e).

----

[1] A small birefringent daughter crystal commonly found in these inclusions was originally thought to be parisite but was shown by Raman spectroscopy (Chapter 4) to be some phase (unidentified) other than parisite.

[2] I have found NaCl crystals in primary inclusions in Rhodesian (Sandawana) emeralds, however, and they have also been reported in emeralds from Western Australia and India (Roedder, 1982c). There have been several reports of "three-phase inclusions", presumably liquid plus gas plus NaCl crystal, in some synthetic emeralds (Webster, 1952; Wells, 1953). As some of the procedures used to grow synthetic emeralds are still secret, such inclusions could be important, but the true source of any given stone is difficult to verify, and some synthetic emerald is crystallized on natural beryl seed plates. See also Roedder (1982c, p. 501).

those that did not had Tm NaCl at ~330°C and still had a vapor·bubble when the run was stopped at 355°C (Roedder, 1982c). The pressure correction to be added to these values may be large. Escobar and Mariano (1976) suggested 6000 m of sedimentary overburden, requiring a pressure correction of 163°C if the fluids are pure NaCl-H_2O solutions (Potter, 1977), and yielding Tt >518°C; even greater depths of overburden have been proposed by others. The high salt concentrations in the fluid may be from the solution of sedimentary beds and domes of salt that Oppenheim (1948) reported nearby.

Metallic Mineral Deposits

Although numerous ore deposits have been either formed by, or subjected to, high-grade metamorphic events, a genetic link is difficult to prove (see also p. 461). Many of the massive sulfide deposits of the world have been subjected to moderate- to high-grade metamorphic conditions, but very few studies of them have involved any investigation of the fluid inclusions. Many of the ores themselves do not appear very appropriate for inclusion study, but the host rocks might yield useful data.

Wilkins (1977) concluded that at Broken Hill, New South Wales, Australia, inclusions from the period of ore formation have been eliminated by repeated deformation and recrystallization. He found that the inclusions provided only a record of the sequence of metamorphic fluids since the period of high-grade metamorphism, and suggested (p. 187) that there is no easy way to prove that "...early-formed inclusions have not developed their present compositions by diffusive loss of water or other components during the long period of retrograde metamorphism. This applies especially to the highly saline inclusions...." He noted, however, a consistency of composition within individual healed fractures and even over distances of some kilometers. Fluid flow affected the composition, however, as certain shear zones were characterized by a distinct assemblage of inclusion compositions. The compositions range from high-salinity aqueous to high-density CO_2 ±CH_4. Both (1978) has shown that some of the deposits at Broken Hill have been remobilized during retrograde metamorphism by fluids having Th = 187-204°C and salinities ~23 wt %.

Similarly, Konnerup-Madsen (1979b) described inclusions from molybdenite mineralization associated with a zone of amphibolitic high-grade banded gneiss in Norway. These fluids ranged from nearly pure CO_2 to entirely aqueous and moderately saline. The latter he believes represent fluids present during repetitive episodes of microfracturing and progressive introduction of meteoric water during uplift and cooling. The high-density, CO_2-rich fluids are considered to approximate most nearly the fluids present during peak conditions of metamorphism, at 3.5 to 4.5 kbar and ~650°C.

The high CH_4 fluids so commonly found in the lower grade metamorphic terranes may well be involved in the formation of graphite vein deposits. Frost (1979) has shown, from calculated equilibria in the system C-O-H, that graphite veins can form by oxidation of a CH_4-rich fluid phase, thus permitting the transport of C from one place to another (e.g., from a graphitic gneiss to a graphite vein).

FLUID INCLUSIONS, METAMORPHIC ZONATION, AND UPLIFT

The concept of metamorphic facies does not involve the composition of the rocks involved but refers only to those rocks that have equilibrated within a given P-T range, usually as identified by some specific (index) minerals. As a result, one should not expect the fluids in equilibrium with each of the various assemblages within a given facies to be the same. However, if a given type of starting material is followed through a progression of metamorphic facies, as long as equilibrium is obtained, the composition of the fluids should be relatively consistent for each mineral assemblage. This progression will be

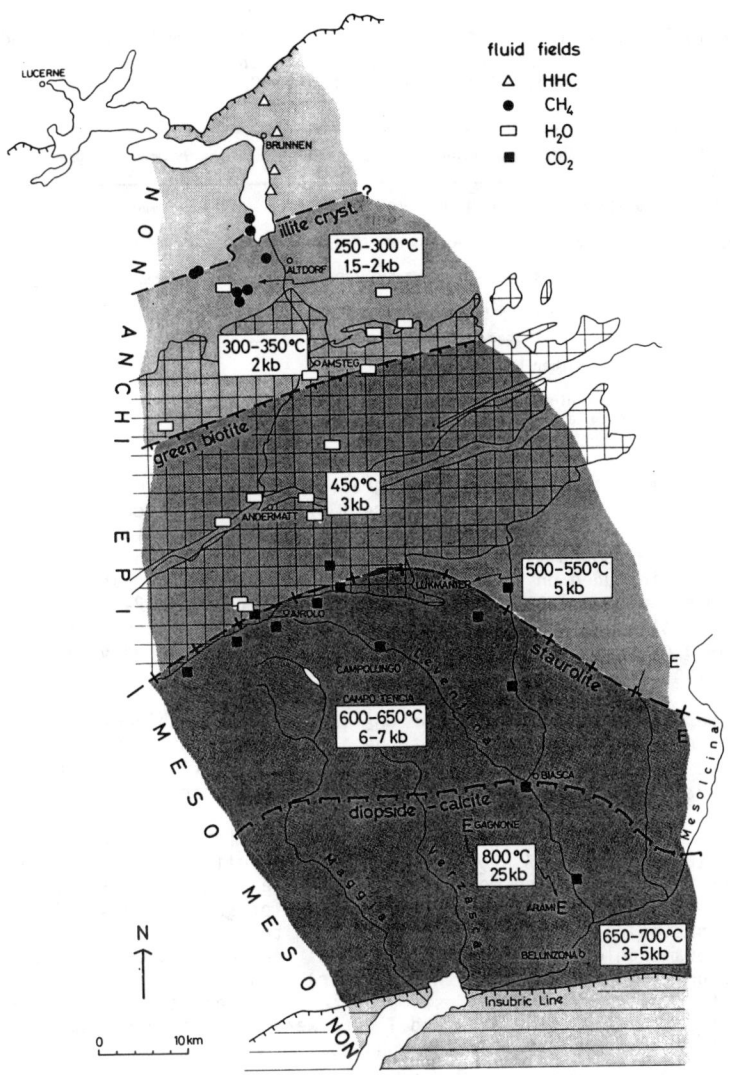

Figure 13-2 Map of metamorphic zones and fluid-inclusion data from fissure quartz crystals along the "Geotraverse" of Frey et al. (1980a, their Fig. 2) in the Swiss Alps. E, eclogites.

illustrated with two examples that may well be typical.

Central Alps of Switzerland

Several of the most detailed studies of the relationship between fluid inclusions and other criteria for metamorphic grade in medium- to high-grade rocks have been made in the Swiss Alps. A detailed study of a belt 50 x 100 km in size ("Geotraverse") across a series of metamorphic zones in the Central Alps has been made by a large number of individuals, over several decades, as summarized by Frey et al. (1980a). The study extended from very low grade rocks in the north to granulite, eclogite, and garnet peridotite in the south (Fig. 13-2), covering the entire range of Grubenmann (1904), from his non- to

Table 13-1. Fluid fields, fluid composition and metamorphic grade of earliest fluid inclusions in the "Geotraverse" (see Fig. 13-2). From Frey et al. (1980a).

Fluid field (see Fig. 13-2)	Fluid composition**	Metamorphic grade
HHC*	~1 to >80 mole % HHC (CH_4, H_2O, CO_2, NaCl)	Nonmetamorphic zone
CH_4	~1 to >90 mole % CH_4, <1 mole-% HHC (H_2O, CO_2, NaCl)	Low- and medium-grade anchizone
H_2O	~90 to >99 mole % H_2O, <1 mole-% CH_4 (CO_2, NaCl)	High-grade anchizone and epizone
CO_2	~10 to >60 mole % CO_2 (CH_4, H_2O, NaCl)	Mesozone

* HHC = higher hydrocarbons

** In parentheses: additional species found in inclusions. Small amounts (<3 mole %) of H_2S and N_2 may be present as well.

anchi- and epi- to mesometamorphic grades. Some of the data on the low-temperature end of the area shown in Figure 13-2 have been discussed in Chapter 12. The change in fluid-inclusion composition with metamorphic grade is quite evident. The compositions of the four types of inclusions are given in Table 13-1. Not included on Figure 13-2 or Table 13-1 are the various retrograde reactions, such as late development of laumontite. The P-T estimates are combinations of a variety of geothermometers and geobarometers, only part of which are based on inclusion data; Frey et al. (1980a) presented various caveats concerning these estimates.

The difference between major CH_4 in the lower anchizone and major H_2O in the upper anchizone is believed to be a result of equilibria involving graphite. Near the quartz-fayalite-magnetite buffer, a C-O-H fluid at 2 kbar contains major CH_4 at 300°C and major H_2O at 450°C (Eugster and Skippen, 1967).

The most striking change is that from H_2O-rich (>90 mole % H_2O) to CO_2-rich (~10 to >60 mole % CO_2) at about the staurolite isograd (the boundary between the epi- and the meso-metamorphic zones). Frey et al. (1980a) listed the three possible origins for this CO_2 put forward by Hoefs and Stalder (1977): decarbonation reactions, oxidation of graphite, or juvenile gases. All high-grade metamorphic rocks such as granulites characteristically have CO_2-rich inclusions (see last part of this chapter), and Frey et al. also reported some eclogites having the highest estimated temperatures and pressures (800°C and 25 kbar). These particular P and T data were obtained not from inclusions but from element partitioning between phases.

Frey et al. (1980a) calculated mean geothermal gradients for these areas and found that they decrease southward, from ~40 to ~25°C/km. Attempts to establish the dip of the isotherms at the peak of metamorphism on the basis of these data are subject to numerous problems. For example, as shown by England and Richardson (1977), the maximum in P is probably reached earlier than the maximum in temperature, so there may be errors inherent in these gradient estimates. Also, metamorphism, and the trapping of inclusions, was probably continuing during the thrusting and folding (and possibly even during the sedimentation).

Khtada Lake Complex, British Columbia, Canada

Hollister and Burruss (1976) presented an extensive study of fluid inclusions in matrix quartz from a high-grade metamorphic gneiss complex, the Khtada Lake complex, British Columbia, Canada. This complex is composed of various hornblende and/or biotite gneisses (some almost "plutonic") and migmatites, the most leucocratic of the gneisses being tonalitic in composition. Field evidence suggests a quasibatholithic environment. Abundant sillimanite and absence of primary muscovite, coupled with the occurrence of K-spar, suggests temperatures >650°C at 6 kbar. Some graphite is present. Although the inclusions contained only three phases (H_2O-rich fluid, CO_2-rich fluid, and gas, each from 0-100 vol %), the studies permitted an internally consistent interpretation of the thermometric results in terms of the geologic work in the area. Only a few of the major conclusions are discussed here. Most important, the work showed that the inclusion contents consisted essentially of H_2O, CO_2, CH_4, and "NaCl." Although Hollister and Burruss have collated widely scattered data from the literature on parts of the compositional range of this four-component system, in particular over parts of the P-V-T range of interest here, considerable physico-chemical reasoning was needed to interpolate or extrapolate to fill the gaps. By appropriate simplifying assumptions, and by the existence of inclusions whose compositions permitted treatment as a system of less than four components, they were able to interpret the phase behavior of these inclusions both at low temper-atures (melting phenomena) and at higher temperatures (homogenization tempera-tures). The basic data they used were: estimated vol % liquid CO_2, Tm CO_2, Tm ice, Tm clathrate, Th CO_2 L-V, Th CO_2-H_2O, and the phase in which homogenization took place. These measurements provided not only compositional data on the in-clusions but also density data, thus defining the appropriate isochores.

The lowest temperatures found (-56.5 to -69°?C) were for equilibria among CO_2 solid, CO_2-rich liquid, and CH_4-rich gas, in inclusions essentially in the system CO_2-H_2O (see Fig. 8-23 and discussion of it). Since CH_4 is miscible with both the gas and liquid phases but not with the solid, the three-phase assemblage changes the invariant triple point for CO_2 to a line, which shifts to lower temperatures with increasing CH_4.[3] For such inclusions, the tempera-ture of this three-phase assemblage, in conjunction with the G/L ratio upon melting, provides a measure of the mole fraction of CH_4. The next higher temper-ature was Tm ice, observable in only eight inclusions. It ranged from -1 to -7°C in mixed CO_2-H_2O inclusions, suggesting moderately low salinities. (The actual salinity may be less than indicated, due to clathrate crystals removing H_2O from the system, as pointed out by Collins, 1979.) A very few aqueous inclu-sions without visible CO_2 showed higher salinity. At higher temperatures (+6.5 to +17°C), the incongruent melting temperature of CO_2-CH_4 hydrates (clathrates) provides some constraints on inclusion composition, but only·with considerable ambiguity. Not only was the measurement necessarily of low accuracy, but these equilibria are affected by the nature of the gas mixture, the pressure, and the salinity (Fig. 8-22).

The next higher temperature observation was Th CO_2 L+V (L). This homogen-ization occurred at -35 to +30°C, corresponding to CO_2 densities from 1.09 to ~0.35 g/cm^3 (Fig. 8-7); two inclusions homogenized in the vapor phase and hence contained lower density fluid. Th H_2O + CO_2 was next. This occurred, in either the CO_2-rich or the H_2O-rich phase, at temperatures from 159 to 339°C (for those inclusions that did not decrepitate first). Interpretation of the pressure of trapping of such CO_2-H_2O inclusions was based on an estimation from the solvus at 1 kbar for the pure system CO_2-H_2O (Todheide and Franck, 1963) and the water-rich leg of the solvus for 1 kbar and 6 wt % NaCl in H_2O (Takenouchi and Kennedy, 1965b), as shown on Figure 13-3 (see also next section). The magnitude of the effects of CH_4 on raising this solvus could only be surmised.

----

[3] The alliterative memory aid: "If the low-temperature form has less, the 'inversion' temperature is low-ered" is useful here. Solid CO_2 is the low-temperature form, and hence the "inversion" (solid ≠ liquid + vapor) is lowered by the addition of CH_4.

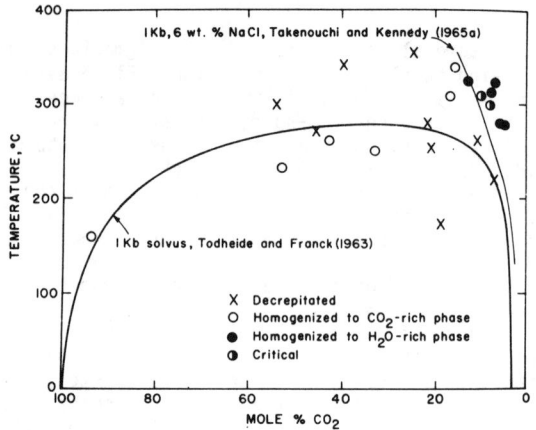

Figure 13-3 1-kbar solvus in the H_2O-CO_2 system. Data points are observed high-temperature inclusion-homogenization phenomena. Note cluster of data points near the experimental curve for CO_2 solubility in 6 wt % NaCl solution at 1 kbar. Since this cluster includes both types of homogenization, it implies that the crest of the solvus for the inclusion fluids is very near the H_2O side, at ~10 mole % CO_2. From Hollister and Burruss (1976).

Although quartz grains from numerous rock types were examined, the intra-sample composition range was sometimes as large as the intersample range. Some distinctions could be made in the nature of the inclusions vs their composition: The highest density CO_2 inclusions were small (<10 µm) and isolated, particularly near garnets that may have provided a "pressure shadow," protecting them from subsequent "tectonic traumas;" inclusions of intermediate density occurred in nonplanar groups; and those of low density occurred mainly along healed fractures and were probably secondary; inclusions of negative crystal shape rarely had a visible H_2O phase. Most inclusions were <~10 µm in diameter. Many of the parameters measured showed two or more groupings of values, suggesting several inclusion-forming events at specific times and hence differing fluid characteristics.

The pressure of trapping of pure CO_2 inclusions (purity verified from Tm CO_2) could be established from their densities. Hollister and Burruss plotted isochores from literature data for these dense inclusions and labelled the two isochores with their respective Th $CO_2(L)$ values rather than the normal density labels (Fig. 13-4). These two isochores limit the range of metamorphism in P-T space. The temperature limit is only a <u>minimum</u> value from independent evidence (Fig. 13-4). Hollister and Burruss also pointed out the problem of volume change on the α-β inversion of quartz, and the fact that they may not have had inclusions trapped at maximum metamorphic conditions.

The data obtained imply the existence of at least two compositionally different fluids, essentially pure CO_2, and CO_2 + H_2O ± CH_4. These two fluids could represent "primary" fluids that equilibrated at different P, T, and fO_2 conditions. Another possibility is that all inclusions were originally a CO_2-CH_4-H_2O mixture and that some leaked H, leaving pure CO_2. A third possibility is that the fluids became immiscible, perhaps during cooling, after the temperature of the crest of the solvus was reached.

Uplift Rates

Weisbrod and Poty (1975), dealing with inclusions in pegmatites, pointed out that if a series of inclusions of different composition and density, and hence having different isochores, occur in the same sample, and if some independent evidence is available to place constraints on the parts of each of these isochores along which those inclusions were trapped, a path in P-T space can be delineated, representing the rock environment from the metamorphic maximum through retrograde metamorphism during the uplift toward the surface. Hollister et al. (1979) applied these procedures to five metamorphic terranes in Canada and the USA, in part from the literature. The constraints placed on the iso-

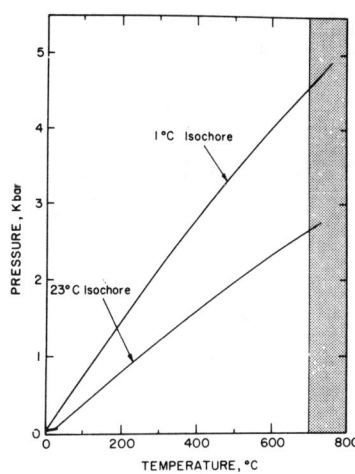

Figure 13-4   Isochores for pure CO_2 inclusions homogenizing at the temperatures indicated. The shaded area brackets the estimate for the minimum thermal conditions at the peak of metamorphism, as deduced from the mineral assemblages. From Hollister and Burruss (1976).

chores were of several sorts, all necessarily rather inexact. Thus, the estimated P and T of the top of the CO_2-H_2O solvus, for low-salinity fluids (see next section), plus evidence of heterogeneous or homogeneous trapping, were used as a constraint. The original (maximum) metamorphic conditions were estimated from the mineral association, which provides various constraints. The rupture strength of quartz presented another constraint. Although the rupture strength is obviously dependent on inclusion size and various other parameters (see Fig. 3-21), the presence of an inclusion having a given isochore necessarily precludes any excursion toward the temperature axis throughout the subsequent history of the sample that would have caused the pressure differential (P_{inc}-P_{rock}) to exceed the strength of the host quartz. Decrepitated inclusions indicate that the limit was exceeded. The use of essentially instantaneous laboratory decrepitation strength values for samples held for geological periods of time ($\sim 10^{15}$ times longer) makes this constraint very crude at best.

The slope of the resultant path in P-T space represents the uplift of the sample toward the surface. In four of the five examples given by Hollister et al. (1979), this path is convex toward the temperature axis, implying that initially the rate of uplift exceeded the rate of heat loss. When these data are combined with mineral and rock ages, rates of uplift can be obtained. Using such data, Hollister and Sherwood (1980) estimated that the initial uplift rate for the Coast Ranges, British Columbia, Canada, beginning from about 30 km depth, was very high, 10 mm/yr, averaging 5-8 mm/yr for the total uplift. Hollister (1982) later decreased the estimate of the average rate to 2 mm/yr, which is still unusually high.

FLUID IMMISCIBILITY -- EVIDENCE AND NATURE

One of the serious sources of ambiguity in the interpretation of a complex set of compositionally varied secondary inclusions in many metamorphic rocks is the possibility that at some time in the past, the fluids present in the pores have reached a T-P-X condition wherein they unmix to form two compositionally distinct fluids, which can be trapped in separate inclusions[4]. In a relatively

----
[4] Pichavant et al. (1982) presented a theoretical and geometrical approach to the analysis of phase equilibria in systems involving immiscibility, and Ramboz et al. (1982) discussed the interpretation of inclusion data in terms of immiscibility. Note, however, that these two papers deal, in large part, with the special constraints on equilibria under isoplethal-isochoric conditions (i.e., _after_ sealing of an inclusion); the fluids in the pore spaces of a rock are not necessarily under these same constraints.

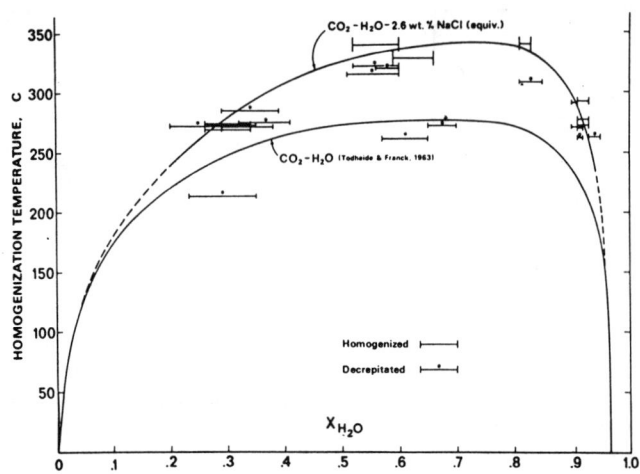

Figure 13-5   Empirical solvus for CO_2-H_2O plus 2.6 wt % NaCl equivalent in the aqueous phase at room temperature. The solvus for pure CO_2-H_2O of Todheide and Franck (1963) at 1 kbar is shown for comparison. The length of the horizontal and vertical bars gives the estimated errors of measurement. Th for inclusions that decrepitated would have been at a temperature greater than that indicated. From Hendel and Hollister (1981).

few metamorphic examples, individual healed fractures can be found containing inclusions filled with various ratios of the two fluids (however, these can also be from later necking down). Most such rocks present us simply with different fluids in different healed fractures. Are these from immiscible fluids, or are they from the trapping of different fluids at different times? The question is important, but the answers are frequently nebulous. Some evidence can be obtained from the nature of the fluids and the probable conditions of trapping. Assuming that data are available on the appropriate system, does immiscibility actually occur in the P-T range involved, and do the conjugate liquids in the synthetic system have compositions approaching those found in the inclusions?

Most examples of possible inclusion-fluid immiscibility in high-grade metamorphic rocks involve CO_2-rich and H_2O-rich fluids, but a few involve CH_4 or N_2. Tomilenko and Dolgov (1978) described inclusions in some jadeite-albitite bodies in a hyperbasite massif that had Tm -182.5°C and Th -165°C, indicating major CH_4, plus other gases. They proposed that a CH_4(+CO_2) fluid split from an H_2O-rich fluid at ~400°C and 4-8 kbar. High-N_2 fluids can also be expected to be immiscible with H_2O-rich fluids over a wide range of P and T.

The conditions under which immiscibility exists between a CO_2-rich phase and an H_2O-rich phase cover a very wide range, from that of a glass of beer to temperatures probably >>400°C and pressures >>3.5 kbar (Todheide and Franck, 1963). The limits on the field of immiscibility, or solvus, are well known for the pure system CO_2-H_2O (see Figs. 8-19, 20), but the changes in solubility of CO_2 in H_2O with various solutes (including other gases) are known for very few combinations of P, T, and X (see also Bowers and Helgeson, 1983a,b).

Actually, in view of the paucity of experimental data, the evidence for the extent of the field of CO_2-H_2O immiscibility in the "system" H_2O-CO_2-CH_4-NaCl at various pressures and temperatures lies mainly in the petrographic evidence from the inclusions themselves. Ypma and Fuzikawa (1980) recognized evidence of immiscibility between a CO_2-rich phase (some containing <45 vol % CH_4) and very saline brines in some Australian uranium deposits. They proposed that mixing of a dilute CO_2-CH_4-rich fluid with very saline brines caused the exsolution of a relatively pure CO_2-CH_4 fluid. The brines vary in salinity, containing as much as 20-30 wt % $CaCl_2$, 5-10 wt % $MgCl_2$, 10-20 wt % NaCl, and

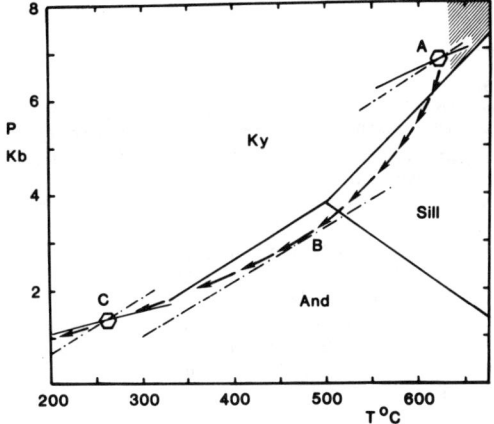

Figure 13-6 Diagrammatic representation of possible uplift path for the Lessard Formation near Flinton, Ontario, Canada. A, Intersection of densest CO_2 and H_2O-NaCl isochores. B, Intersection of uplift path with the highest T and lowest P H_2O-NaCl isochore. C, Intersection of CO_2 and H_2O-NaCl isochores associated with the low-pressure and low-temperature "event." Shaded area: estimated peak metamorphic conditions. From Sisson et al. (1981).

minor amounts of KCl and $FeCl_2$. Inclusions of this H_2O-rich solution have a small vapor bubble (Th = 110-160°C) and some CH_4 in the vapor. Pressures were probably in the range of 250-350 bars.

Hendel and Hollister (1981) were able to propose an empirically determined solvus for the "system" CO_2-(H_2O + 2.6 wt % NaCl) (Fig. 13-5). This solvus was based on 25 inclusions from three healed fractures. These inclusions showed a wide range of ratios of liquid CO_2 to liquid H_2O solution at the reference temperature chosen (40°C to insure homogenization of CO_2 phases). The purity of the CO_2 was verified by Tm CO_2, which ranged from -56.7 to -56.9°C. The salinity (2.6 wt % NaCl equivalent) of the H_2O-rich phase was determined (with several assumptions) from Tm clathrate (+8.6 to 8.8°C) on the basis of data of Bozzo et al. (1973). The density of the CO_2 phase was determined from Th CO_2, and from this and the phase volumes at 40°C, the mole fractions of CO_2 and H_2O were estimated, and Th CO_2 + H_2O plotted (Fig. 13-5) to outline the solvus indicated. Note that the 2.6 wt % NaCl equiv. salts in the water phase raised the solvus ~70°C from that in the pure system CO_2-H_2O (Todheide and Franck, 1963).

The solvus in the system CO_2-H_2O varies with pressure (see Figs. 8-19, 20), but as it varies little in the probable region of interest (1-3 kbar), Hendel and Hollister (1981) used the solvus at 1 kbar for comparison. The minimum temperature at ~2 kbar (Fig. 8-19) would be only ~10° lower. Extrapolation to the higher pressures and much higher salinities and temperatures of many natural systems would be hazardous without additional evidence. Some such evidence was obtained by Crawford and Sisson (1981), and Sisson et al. (1981), who studied the matrix quartz of a Precambrian calcareous psammite and a carbonate schist in Canada. This work suggested that brines containing 23-24 wt % NaCl equiv. are immiscible with CO_2 at the metamorphic conditions of ~600°C and 6.5 kbar. Some of the aqueous inclusions show Te values as low as -63°C, suggesting that Ca and probably other ions are also present. The inclusion evidence is that the two fluids occur as separate inclusions, and where planes of the two types cross, there is no apparent fluid mixing. Using the density information from the inclusions, they inferred that the path in P-T space taken by these rocks during cooling and uplift was strongly convex downward (Fig. 13-6). Since the path shown is as much as 2 to 3 kbar below the pressure along the isochore for the earliest inclusions, one might ask why these inclusions have not decrepitated. The answer is probably that they were all small (<12 µm), and the smaller an inclusion is, the larger internal pressure differential it can stand without decrepitating (see Fig. 3-21).

GRANULITES AND OTHER HIGH GRADE ROCKS

General Nature of Inclusions in Very High Grade Rocks

Most inclusions in most high-grade rocks, whether they be sillimanite schists, eclogites or granulites, contain high concentrations of CO_2 or are almost pure CO_2; frequently, they are only $< \sim10$ μm in size. Under ordinary petrographic examination, such inclusions could easily be ignored, since they appear as small black specks and even at high magnification seem empty. On the island of Naxos, Greece, Schuiling and Kreulen (1979) reported that synmetamorphic quartz segregations from the zoned metamorphic complex of schist and marble, wrapped around a central core of migmatite, contain CO_2-rich (50-80 mole % CO_2) fluid inclusions. The temperature gradient from the granulite-grade core outward was 300°C in 15 km (700 to 400°C) and took place at 5-7 kbar. The authors suggested that the quartz segregations represent the main passageways of escaping fluids, fluids that also were responsible for an important transfer of heat and may have been the prime cause of metamorphism and updoming of the complex. On the basis of $\delta^{13}C$ studies, Kreulen (1977) estimated that approximately two-thirds of the CO_2 must be of deep-seated origin, the remainder resulting from decarbonation reactions and oxidation of graphite in the local rocks.

The fluid inclusions in eclogites have been studied by several groups. Gorokhov et al. (1975) reported that melt inclusions with daughter minerals in omphacite and garnet had Th of 1180-1360°C (using an Ar atmosphere and long heating runs). CO_2 inclusions in garnet, possibly trapped at a later time, have high densities (Th +5 to +12°C) and Tm CO_2 -57°C. Fluid inclusions in matrix quartz and from veins and fissures in eclogites and glaucophane-bearing rocks from the Central Tauern Window in the Swiss Alps were studied by Luckscheiter and Morteani (1980b). Inclusions in the eclogites contained only CO_2 plus 8-15 mole % CH_4, at a very high density (<1.15 g/cm³; Th CO_2 L-V(L) at -54°C.)[5] Assuming eclogite formation at 500-550°C, a trapping pressure of ~8 kbar was estimated. Some of the aqueous inclusions in these rocks contained 50 wt % NaCl equiv., and showed Tm NaCl as high as 450°C, whereas others were as low as 3.5 wt %. The composition of the fluids changed from early high-CO_2 to later more H_2O-rich, as also reported by Luckscheiter and Morteani (1980a). This H_2O is assumed to have been introduced into these essentially anhydrous rocks from dehydration reactions in surrounding schists. In places, this has resulted in amphibolitization of the eclogites. In this respect, Luckscheiter and Morteani (1980b) disagreed completely with the interpretation of Holland (1979, 1983), who proposed, from mineral-assemblage considerations, that the kyanite-bearing eclogites of the Tauern Window formed under high H_2O activities and low CO_2 and NaCl ($XCO_2 < 0.1$; $XNaCl < 0.02$). Luckscheiter and Morteani assumed that the CO_2 is of "juvenile" origin, because CO_2 is known to be commonly associated with basaltic volcanism (Roedder, 1965a), and the eclogites presumably represent former basic and ultrabasic rocks.

The first paper on high-density inclusions in high-grade metamorphic rocks was that of Dolgov et al. (1967). They studied inclusions in kyanite from kyanite-garnet-mica schists and from plagioclase-quartz-kyanite pegmatites and quartz lenses in these rocks, from the Mamsk region in the northern Baikal highlands, USSR. To determine the density of the CO_2, they averaged the Th CO_2 L+V(L) and Tn CO_2 V. The former ranged from -7 to -22°C, and the latter, from -18 to -32°C. From this average they calculated the density to range from 1.00 to 1.06 g/cm³. H_2O-bearing inclusions in the same kyanite showed Th $\sim240-280$°C. Since the isochores for the dense CO_2 inclusions show a pressure of 2400-3200 atm at 240-280°C, they assumed that this was the pressure of formation of the

----

[5] Note that this corresponds to a density near to the limit of detection of CO_2 density from stable Th CO_2, since the isochore originating at the triple point of -56.6°C has a density of ~1.16 g/cm³ (Fig. 8-7). Coolen (1980) has shown that metastable phenomena can be used to estimate even higher densities (next section). Touret et al. (1982) reported similar high CO_2 densities (<1.15 g/cm³) in eclogites from western Norway.

Table 13-2. Studies of inclusions in some granulite terranes

Locality	Reference	Inclusion composition	Estimated Conditions T (°C)	P (kbar)
Bamble granulite, southern Norway	Touret (1969, 1971)	CO_2	800	8
Garnet gneiss granulite, Malagasy	Berglund & Touret (1976)	CO_2+40 wt % H_2O	700-800	6-8
Chogar complex, USSR	Tomilenko et al. (1977)	CO_2	1000-1100	10-11
Charnockite massif, southern India	Janardhan et al. (1979)	(Probably high CO_2)	600-700	<9
Central highlands of the Adirondack Mountains, New York, USA	Hollister et al. (1979)	$X_{CO_2} < 0.5$	~800	~8-9
Granulites of northern Lapland, Finland	Klatt (1979)	CO_2	700-750	5-7
Rogaland complex anorthosite-granulite, southwestern Norway	Swanenberg (1980)	CO_2-, H_2O-, and N_2-rich	750-850	~5-9
Furua granulite complex, southern Tanzania	Coolen (1980, 1982)	$CO_2 \pm CH_4$	750-850	~7-9.5
Vredefort Dome, South Africa	Schreyer & Medenbach (1981)	CO_2 and CO_2+ <50 vol % H_2O	~850	3
Granulite xenoliths in alkali olivine basalt, southern Chile	Selverstone (1982)	CO_2	~850	5-7

kyanite. Although several of the assumptions in this procedure are wrong,[6] and the true pressure was probably much more than 2.4-3.2 kbar, this paper was seminal, in that it was the first to report that small, seemingly empty inclusions in high-grade metamorphic rocks actually contained high-density CO_2.

Studies in Granulite Terranes

The fluid inclusions in many granulite terranes have been studied since the pioneering work of Touret (1969, 1971, 1974) in the Bamble region of southern Norway, in which he discovered the common presence of relatively pure CO_2 inclusions. Several of these studies are listed in Table 13-2; all reveal CO_2 to be the most abundant volatile, and sometimes practically the only volatile, in such terranes. Fluid inclusions have been used to one degree or another in geologic studies ever since Sorby (1858); why should it have taken 111 years after Sorby for this important discovery to be made? Several aspects probably contributed: (1) Deep-seated metamorphic rocks, formed at high P and T, and particularly H_2O-deficient types like granulites, might seem to be a rather unlikely place to look for "fluid inclusions." (2) Inclusions in most metamorphic rocks are small and easily missed. (3) Even if found, under ordinary petrographic examination, most high-density CO_2 inclusions are one phase and hence look empty. (4) Until the advent of the Poty stage (Anonymous, 1974), and until P-V-T data on CO_2 became available, little could be done with such inclusions anyway. In the brilliant light of retrospect, however, it still seems strange that the significance of this association was not realized earlier. Thus, many (or most) of the world's gem sapphires and rubies come from high-grade metamorphic terranes, in large part from the granulite facies (e.g., the gem-bearing gneisses and granulites of Sri Lanka: Dahanayake and Ranasinghe, 1981; Munasinghe and Dissanayake, 1981). Ever since the work of Brewster more than 150 years ago (e.g., Brewster, 1827; see also various entries in Table 1 of Roedder, 1972; and Gübelin, 1953, 1974), we have known that many gem sapphires and rubies contain liquid CO_2

----

[6] Note also that if the H_2O inclusions were trapped at 240-280°C, and 2.4-3.2 kbar, they should have Th ranging from 80 to 150°C; if they were trapped at 2.4-3.2 kbar and have Th = 240-280°C, Tt was 400-530°C.

inclusions. As they are gem material, they are relatively free of flaws and hence can have relatively large inclusions without decrepitation, yet the significance of this in terms of metamorphic petrology was not recognized for more than 100 years. It makes one wonder how many other important "new" geologic facts lie buried among the published descriptive data of the past.

Touret (1977) has pointed out that a distinct relation exists between maximum CO_2 densities and the specific metamorphic type. Thus, the low-pressure, cordierite-bearing granulites contain CO_2 inclusions that have densities corresponding to Th (i.e., Th CO_2 L-V(L)) values in the range 0 to -10°C (e.g., in the Bamble region of Norway (Touret, 1971)). The medium- to high-pressure granulites, however, show Th values, i.e., Th CO_2 L-V(L), of -30°C and lower, corresponding to considerably higher densities. Furthermore, in many granulite terranes, the inclusions are on isochores beneath the probable P and T conditions of formation, as determined by other petrologic geothermometers and geobarometers. In such examples, the inclusions presumably were trapped after the peak in P and T. In some, the probable pressure at trapping of the inclusions was ~1/2 that present during the formation of the host rock.

The P-T conditions for granulite formation estimated by various methods involving inclusion data (Table 13-2) are generally in the same range (~800°C and 8 kbar) as the P-T conditions estimated from various mineralogical, geochemical, and isotopic procedures (e.g., Touret, 1974, Table 1). Generally, these estimates are based on those very few inclusions yielding the highest P-T values; most inclusions yield much lower values. As pointed out by R. Kreulen (pers. comm., 1983), the possibility should be considered that these inclusions may have adjusted their density by partial decrepitation followed by recrystallization, eliminating the evidence of decrepitation. The listing in Table 13-2 is not at all complete, and only a few features of these occurrences can be discussed in any more detail here.

Furua granulite complex, Tanzania. The reports by Coolen (1980, 1982) deal with a 1000 km long belt of Precambrian banded granulite (hornblende-pyroxene, ± garnet and scapolite) and various associated high-grade metavolcanic and metapelitic rocks. Microthermometric studies showed that CO_2 was the major constituent of the fluid in inclusions in quartz, plagioclase, and garnet from calc-silicate granulite. These inclusions were generally very well faceted (Fig. 13-7), and show some remarkable doubly metastable phase behavior, suggesting extremely high CO_2 densities (Fig. 13-8). Ordinarily, on cooling high-density CO_2 inclusions, a gas bubble forms at some low temperature, and on further cooling to below ~-56°C, the triple point is reached and the liquid CO_2 freezes (Fig. 8-7). As in H_2O inclusions, freezing of the liquid CO_2 may require 40 or 50 degrees of supercooling (i.e., to -90 or -100°C). Inclusions with densities greater than ~1.16 g/cm^3 intersect the univariant solid/liquid boundary first, and hence freeze without a vapor phase.[7] Coolen found that on cooling some of his densest CO_2 inclusions to temperatures <-56°C, the liquid would first fail to nucleate crystals and hence would continue as metastable liquid CO_2 into the field of solid CO_2 (Fig. 13-8). On still further cooling, the metastable liquid would fail to nucleate a vapor bubble when it crossed the metastable extension of the L-V curve. Eventually, a vapor bubble would nucleate at temperatures as low as -81.8°C, and at still lower temperatures, the stable solid CO_2 would nucleate. The density of the CO_2 can be estimated from such data, as shown in Figure 13-8. Considerable complications arise, however, as some of these inclusions probably contain appreciable CH_4 and N_2, as evidenced by the separation of the "gas phase" coexisting with solid CO_2 (along line 6 in Fig. 13-8) into liquid and vapor at ~-155°C.

----

[7] The stated density at the triple point varies depending upon the source of the data. Figure 13-8 shows it at ~1.165 g/cm^3, but Angus et al. (1976) listed it as 1.178 g/cm^3.

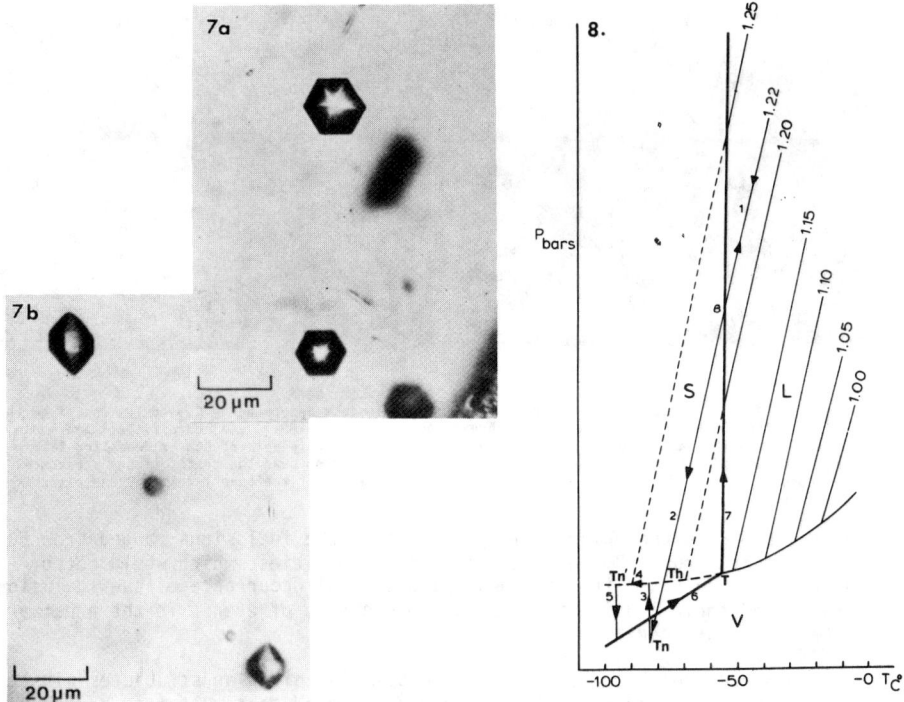

Figure 13-7   Two views of negative-crystal-shaped CO_2 inclusions in quartz, looking along the c axis (a) and perpendicular to the c axis (b), from biotite-garnet granulite, Tanzania (Coolen, 1980). These inclusions are filled with high-density CO_2, as they show Th CO_2 L-V(L) at -35.8°C (a) and -51.5°C (b).

Figure 13-8   Schematic representation of the multiply-metastable freezing-heating path of inclusions with CO_2 densities higher than 1.16 g/cm³, found by Coolen (1980) in granulites from Tanzania. Indicated with arrows: path for inclusion with density 1.22 g/cm³, showing metastable homogenization at Th = -78°C; P scale is relative. Generally, upon cooling, the P-T path will follow an isochore (1). Failure to nucleate solid at (8) results in a supercooled liquid inclusion following the metastable extension of the isochore into field of solid (2). Failure to nucleate a vapor phase in the metastable supercooled liquid on intersecting the metastable extension of L-V equilibria (dashed line to left of T) at Th (-78°C) results in inclusion (now a supercooled and stretched liquid) following isochore into field of (metastable) vapor below metastable Th. At some temperature Tn, a vapor bubble nucleates, taking the inclusion up along (3) to the metastable L+V curve. Further cooling along (4) finally results in nucleation of stable solid at Tn'; pressure immediately drops along (5) to that of stable assemblage solid + vapor. Subsequent heating moves the inclusion along stable equilibrium path (6) to triple point (T), where expansion during partial melting eliminates the vapor phase and inclusion then follows S+L curve (7) to (8), where last solid melts. Further heating results in inclusion retracing movement along 1.22 isochore (1). As a result of such metastable extensions, Th (metastable) for CO_2 inclusions with densities greater than 1.16 g/cm³ can be measured, but only if cooling beneath the triple-point temperature (T) is stopped after the appearance of a gas bubble but before solid has formed (4). Upon reheating at this stage, the gas bubble disappears from the metastable liquid at Th; subsequently, the isochore path (2)-(1) will be followed. Contamination will shift all these isochores and phase boundaries. Modified from Coolen (1980).

From these studies, Coolen suggested that the granulite-facies metamorphism in the area took place at 750-850°C and 7 to 9.5 kbar, in agreement with estimates made from element partitioning between garnet and orthopyroxene.

Vredefort Dome, South Africa. The Vredefort Dome, an enigmatic large circular structure in South Africa, has been the subject of many studies. Schreyer and Medenbach (1981; Schreyer, 1983) have studied the inclusions in these rocks

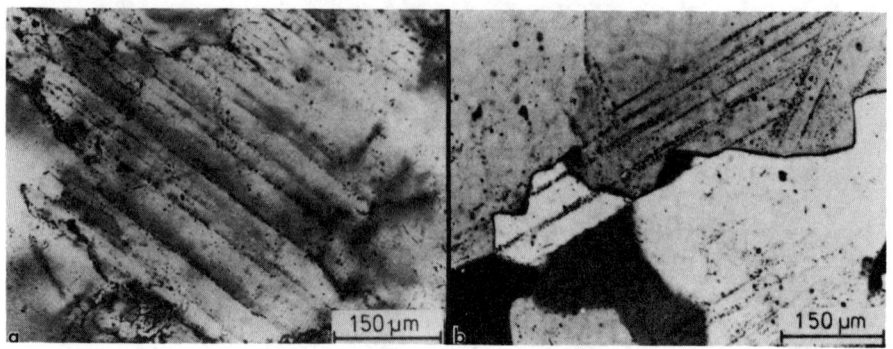

Figure 13-9a,b   Quartz grains showing planar {0001} elements decorated by CO_2 inclusions, from the
the Vredefort Dome, South Africa, from Schreyer and Medenbach (1981). (a) Unrecrystallized grain
from near the margin of the structure, in plain light. (b) "Polygonized" quartz annealing fabric
having relict planar elements that cross grain boundaries, from near the center of the structure,
in crossed polars. Some of the newly formed grain boundaries of the quartz follow these planar
elements.

and have added to the enigma. They found that the inclusions ranged from 50 vol
% CO_2 to nearly pure CO_2, at moderately high densities, which would not be
unusual for such granulite terranes. The spatial occurrence of the inclusions,
however, and their implications as to the sequence of events in the area, are
unique and rather strange.

The structure is one of the most conspicuous circular structures within the
crust of the earth and commonly has been considered to result from a meteorite
impact. Various shock features, such as pseudotachylite veins, shatter cones,
stishovite, and coesite give abundant evidence of an extensive area subjected to
a very high pressure event. As in many shocked rocks, the quartz shows shock-
induced deformation lamellae along {0001}. These planes are now decorated
with rows of fluid inclusions (Fig. 13-9a). Considerable recrystallization of
formerly shocked quartz has taken place, but the original orientation of the
planes of inclusions has been preserved (Fig. 13-9b).

These inclusions consist mainly of pure CO_2, have densities ~0.6 to 0.7
g/cm^3, and occur throughout the rocks of the dome, though they are most abundant
in the rocks of the core. In addition, the rocks in the center of the dome have
been most extensively recrystallized, and on the basis of independent petrologic
studies, have been subjected to the highest metamorphic temperatures. But the
evidence presented by Schreyer and Medenbach (1981) and their colleagues is that
the high-temperature metamorphic center predated the shock event. They have
suggested that the postshock CO_2 fluids might have been derived from the release
of older fluid inclusions from the lower crustal granulites affected by the cat-
astrophic shattering event, or from a direct mantle source that might have been
genetically connected with the Vredefort event itself. Furthermore, they have
suggested that the inclusion evidence supports an endogenous mechanism for the
shock event, possibly even internal CO_2 pressure. Admittedly, no precedent
exists for any but a meteorite-impact origin for the shock features seen at the
Vredefort Dome, but the spatial and chronological evidence seen in the fluid
inclusions is difficult to explain by meteorite impact.

Anorthosite-granulite complex, southwest Norway. Swanenberg (1980), in a
study of a complex of high-grade metamorphic to plutonic rocks in Norway, found
CO_2-rich inclusions (±N_2) more abundant in the granulite-facies rocks and
H_2O-rich inclusions more abundant in the amphibolite-facies rocks, just as has
been found elsewhere. Rocks where H_2O-rich inclusions are abundant do not yield

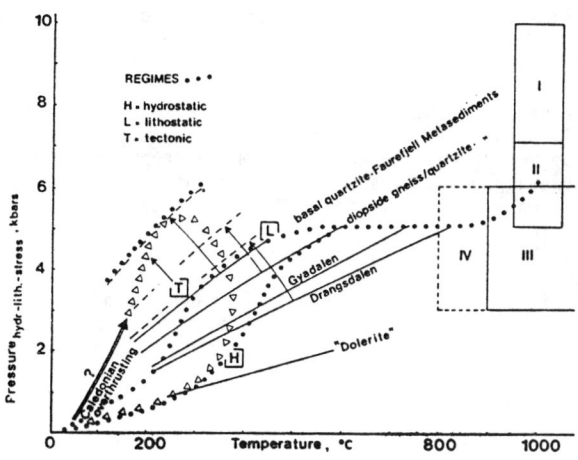

Figure 13-10 Tentative P-T trajectories for elements of the Rogaland complex, southwest Norway, as deduced from fluid inclusions (Swanenberg, 1980). The solid lines are isochores corresponding to the most abundant type of CO_2 inclusions in the stated environment. The possible P-T effect of Caledonian overthrusting is indicated by triangles.

reliable whole-rock Rb-Sr isochrons, suggesting partial movement of parent Rb or daughter Sr. Te values as low as -80.2°C and Tm ice as low as -59°C suggest high concentrations of divalent cations, perhaps 30 wt % $CaCl_2$. Some N_2-rich inclusions (0.45 to 0.75 g/cm³) were found in the vicinity of mafic rocks and in pegmatites, and some N_2 is present in many of the CO_2 inclusions. From a consideration of the available data on the various gas systems, and calculations of the equation of state for various systems, Swanenberg concluded that if the depression of the melting point of CO_2 in inclusions is assumed to be from CH_4, but actually is from N_2, significant errors will be result. Both N_2 and CH_4 lower Th CO_2 L-V(L), so if their effects are neglected, pressure estimates will be too high. In reference to this specific locality, Swanenberg challenged the currently accepted assumption that high-density CO_2 inclusions closely approach a synmetamorphic granulite-facies fluid. In part, his challenge is based on a detailed study of the criteria for relative ages of various planes ("trails") and clusters of inclusions.

The P-T trajectories for these rocks during cooling and uplift for different areas apparently varied widely (Fig. 13-10). Most interesting, Swanenberg has suggested that possible Caledonian overthrusting has resulted in an excursion into high-pressure, low-temperature conditions, during which CO_2-rich and N_2-rich inclusions were reequilibrated to considerably higher densities and hence new isochores (Fig. 13-10). Some inclusions have densities as high as 1.23 g/cm³; these are mostly in rocks that have been deformed while under this tectonic load.

Granulite xenoliths in basalt. Selverstone (1982) studied a series of xenoliths from alkali basalt flows in the Pali-Aike volcanic field in southern Chile. Essentially pure CO_2 inclusions are ubiquitous in the gabbroic granulite xenoliths and in part occur in modes that permit chronologic assignment relative to various petrographic features of the granulites, such as symplectic intergrowths and exsolution lamellae in the pyroxenes. Selverstone found some CO_2 inclusions (<2 µm in maximum size) decorating pyroxene exsolution lamellae. These inclusions are presumably associated with the high elastic strain from dislocation concentrations along such planes; they could have formed at any time during or after the exsolution. Some clusters of CO_2 inclusions appear to obliterate inclusions on lamellae, hence postdate them. Some of these inclusions apparently were formed from fluids released by the consumption of their former host minerals during symplectite formation.

Granulites of other composition from the region and the granitic xenoliths contain mixed CO_2-H_2O inclusions. Selverstone has suggested several mechanisms by which the CO_2 inclusions could have become entrapped: (1) an intergranular fluid phase between growing crystals; (2) healing of later fractures; (3) precipitation of fluid as an impurity phase (Green and Radcliffe, 1975); and (4) oxidation of solid inclusions, such as C atoms reacting with lattice or free O_2 (Freund et al., 1980). Actually, the last two mechanisms are essentially the same. Although some of the inclusions Selverstone described are fairly obviously secondary inclusions, formed by mechanism 2, many are of unknown origin, and recrystallization of inclusions may have eliminated most of the evidence of origin. Some of these inclusions have decrepitated as a result of heating to ~1200°C by the basalt during transit to the surface. This heating results in a very unusual path in P-T space, essentially perpendicular to the isochores.

Origin of the CO_2. Several suggestions have been made in this chapter about the source of the CO_2 in these various high-grade metamorphic terranes. As Touret (1981) has pointed out, two different aspects must be considered simultaneously -- the high CO_2, and the low H_2O. If the water can be removed from typical mixed CO_2-H_2O fluids, CO_2 percentages will increase. Hydration of silicates is a very obvious mechanism for increasing the CO_2 content of fluids in lower grade metamorphic terranes, but it is not appropriate for the rather anhydrous assemblages typical of granulite terranes. Touret suggested that a mechanism exists in the formation of the migmatites that so often accompany granulites. These migmatites represent anatectic melts ("mobilisates"), and although they have formed essentially by the melting of anhydrous minerals, H_2O has been shown to be much more soluble in such melts than is CO_2. Thus, the formation of such melts would extract H_2O from any mixed fluids, but this water would be released again during crystallization of these melts.

Other possible sources include decarbonation reactions, though several C isotopic studies seem to preclude that as a major source, and the oxidation of sedimentary C, which is similarly inadequate, except locally. The prevalence of CO_2 in these terranes, throughout a wide range of rock compositions, and all over the world, suggests a nonlocal source. Newton et al. (1980; see also Green, 1972; Glassley, 1983a,b) have postulated a major role for CO_2 in the evolution of the crust of the Earth and in the formation of charnockites and granulites in the continental roots. This mechanism involves the dehydration of metamorphic assemblages by a rising stream of CO_2 that flushes out the H_2O to produce dry granulite residues. Certainly the almost universal presence of essentially pure CO_2 inclusions in the ultramafic nodules brought up with alkali basalt intrusions (Roedder, 1965a; see Chapter 17) suggests that a world-encircling deepseated source is available to supply CO_2 to all deep metamorphic terranes.

Chapter 14

INTRUSIVE ROCK and PEGMATITIC ENVIRONMENTS

"To many petrologists a volatile component is exactly like a Maxwell demon;
it does just what one may wish it to do."   (Bowen, 1928)

CONTENTS

INTRODUCTION

The environments discussed in this chapter cannot be nicely representative
of those that might be found in a textbook on igneous petrology, because they
are necessarily limited to those on which fluid-inclusion studies have been
made. Many intrusive-rock types have not been studied adequately, and others
seem to have no fluid inclusions to study. However, intrusive rocks do show

the widest range of inclusion types of any of the environments discussed in this book. Thus, the following fluids have all been reported as having been present in one or another intrusive rock: immiscible silicate melts; low- to high-density CO_2, H_2O, and mixtures of the two; hydrosaline melts (salts plus <20% H_2O); carbonate melts; dense CH_4 gas; and possibly even sulfide-, metal-, or Fe-oxide-rich melts. Because igneous rocks, by definition, start as magma, most such inclusions represent one or another type of immiscibility, as shown in Figure 2-25. This wide range results only in part from the very wide range in composition of the host rocks or melts. Another source of diversity is the wide range of pressures and temperatures involved. Thus, some inclusions form as a result of trapping of a gas phase or of silicate or sulfide melts at magmatic temperatures, yet many other inclusions in intrusive rocks form during late-stage deuteric alteration or even later low-temperature hydrothermal activity.

Unfortunately, not enough work has been done on these rocks to show the interrelations of these various types of inclusion fluids and their host rocks, and the relations suggested in Figure 2-25 must remain as labelled -- "hypothetical." As a result of this situation, several parts of this chapter lack coherence. We can hope that some additional years of study of these inclusions will reveal the significance of some of the differences that today can only be surmised. In view of the great importance of volatiles in igneous petrology, and the great potential of inclusion studies toward an understanding of melt/volatile interactions, it is rather surprising that so little has been done in the study of the fluid inclusions in igneous rocks.

Intrusive-rock environments that have yielded metallic ore deposits (e.g., many skarns and the Sn-W greisens), and those that yielded the deep-seated peridotite and dunite xenoliths found in alkali basalts are reserved for discussion in Chapters 15 and 17, respectively. Although carbonatites may have some genetic links with deep-seated rocks such as kimberlite, they and their associated alkalic rocks seem to have crystallized where we now find them; hence, they are discussed in this chapter. As in previous chapters, the dividing lines between environments are arbitrary and diffuse, e.g., between intrusive and high-grade metamorphic (Chapter 13) and between instrusive and extrusive (Chapter 16). In this chapter, I deal mainly with inclusions trapped during igneous processes (i.e., "magmatic inclusions"; see review by Roedder, 1979a), even though some of the constituents may not have an igneous origin; however, one must always remember that in many igneous rocks, most inclusions are secondary and may have no connection with magmatic processes.

HIGH-GRADE METAMORPHIC VS IGNEOUS ENVIRONMENTS

The distinction between high-grade metamorphic and igneous environments normally rests on the assumed presence of some arbitrary percentage of a melt phase in the latter, usually with the added proviso that intrusive igneous rocks have probably moved by fluid flow, as contrasted with subsolidus plastic flow in metamorphic rocks. In systems containing no volatiles, the dividing line would be the solidus, but in natural multicomponent rocks containing volatiles, the concept of a specific solidus temperature must be based on some arbitrary ratio of nonvolatile to volatile constituents. Appreciable melt phase obviously makes flowage easier but is not a requirement, as evidenced by salt domes and diapirs. Also, unless inclusions of glass (± crystals) are found, what actual textural evidence exists for the former presence of a melt phase in a well-annealed igneous rock to distinguish it from an equivalent metamorphic one? Good textural evidence of an igneous origin is best seen in those igneous rocks formed under shallow conditions, where cooling rates are faster.

The problems of extending these concepts to the deeper rocks led to two schools of thought, the "magmatists" and the "granitizers." The former have

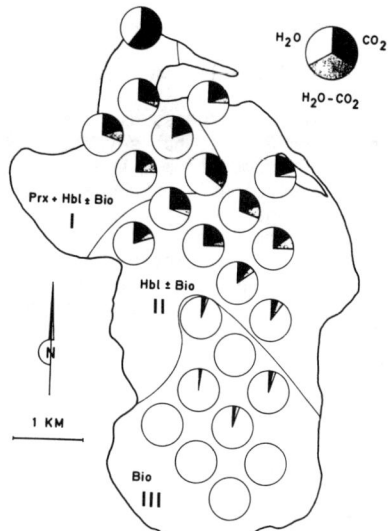

Figure 14-1. Fluid-inclusion distribution in the Kleivan Granite, southern Norway, as mapped by Konnerup-Madsen (1977). The centers of the circles mark the sample locations on which relative inclusion abundance was determined, in terms of percentages of three categories: essentially CO_2 (dark segment), mixed CO_2-H_2O (gray), and essentially H_2O (white). The granite is a biotite granite in the south (area III), and an orthopyroxene (opx)-hornblende (hb) granite (i.e., a typical granulite assemblage) in the north (area I). The hornblende and orthopyroxene isograds are drawn.

held that one of the common lines of textural evidence for a former melt is the presence of migmatitic stringers of granitic composition; these stringers are considered to be essentially the result of crystallization of a silicate-melt phase -- a "mobilisate" -- that may have moved from its place of origin (i.e., to form "injection gneiss;" such "mobilisates" should also have compositions in the minimum-melting trough in the system leucite-nepheline-silica). The granitizers have suggested, however, that such features form by processes of recrystallization in an essentially solid medium. What evidence of the nature or even of the existence of any fluid phase can be found in the form of fluid inclusions in such migmatitic "mobilisates?"

Touret (1981) reviewed the literature on the inclusions in "mobilisates" from several regions of high-grade granulites and migmatites. In general, the granulites and other metamorphic rocks contain the dense CO_2-rich inclusions ($\pm CH_4$ and N_2) typical of granulites, whereas the associated "mobilisates" contain H_2O-rich inclusions. Touret (1981, 1982) explained this difference as stemming from the expected differences in the solubility of CO_2 and H_2O in a silicate melt, once it forms. (Melt will form preferentially where H_2O is available.) H_2O will dissolve much more readily in the melt, leaving the volatile fluid residually enriched in CO_2. Powell (1982) has shown that the formation of such a melt decreases the activity of H_2O in the system and hence drives dehydration reactions further. Thompson (1982) suggested that a free fluid phase may not even be present during some partial melting in the lower crust and upper mantle (i.e., H_2O-undersaturated melting). Wendlandt (1981) constructed a model for the formation of charnockites, in which an influx of CO_2 from an external source through an anatectic melt-crystal system flushes out the vapor-phase buffer and extracts H_2O from the melt, causing crystallization of charnockite. Such a mechanism might have been recorded by the inclusions in the Kleivan granite in southern Norway (Konnerup-Madsen, 1977; Fig. 14-1). As discussed by Touret (1981), the variables are such that the degree to which the

383

several inclusion types overlap the isograds on this map is perhaps expectable.

H_2O will partition preferentially into the silicate liquid from a preexisting mixed CO_2-H_2O fluid during anatectic partial melting, as in the formation of migmatites. Touret (1981, 1982) suggested that this process will leave a CO_2-rich residue that may explain some CO_2-rich inclusions, but several questions remain to be answered. First, if these migmatites formed from a hydrous silicate melt, why are there no reports of silicate-melt inclusions in the quartz and feldspar that formed from the melt? Second, as a separate CO_2-rich fluid phase was present, why should the "mobilisate," which is frequently only centimeters from its former host, exsolve and trap essentially only H_2O inclusions as it crystallizes, rather than newly mixed CO_2-H_2O inclusions? Part of the problem probably lies in the inherent ambiguity about the origin of most inclusions in such rocks and in the difficulty in recognizing small melt inclusions, even if they were trapped and preserved from subsequent recrystallization.

INCLUSIONS OF HOMOGENEOUS SILICATE MELT

Th Measurements on Melt Inclusions in Intrusive Rocks

One of the major problems in the study of melt inclusions in intrusive rocks is the difficulty in recognizing the inclusions. Glass inclusions are found only in those rocks in which trapping occurred relatively near the surface of the earth. In many intrusive rocks, melt inclusions are not only small (5-10 μm), but, as a result of slow cooling in nature, they are now usually completely crystalline, thus changing their appearance. Unless specifically expected and looked for, these little clusters of crystals can easily be missed. The vapor phase is no longer a spherical bubble but merely a reflective film of vapor between the crystals, which also adds to the difficulty in optical resolution of the phases present. Distinction between such crystallized-melt inclusions and accidentally trapped solid inclusions is also difficult and can be a common source of problems.

Most studies of melt inclusions have been made on extrusive rather than intrusive rocks, and the procedures used for the two types of rocks are identical. Many problems exist in such studies, in both the laboratory procedures themselves and in the interpretation of the data; as these problems are detailed in Chapter 16, they need not be given here. Most of this work has been done in the USSR; the book by Sobolev and Kostyuk (1975) is an excellent summary of the data obtained to that date. The most striking single feature of these studies is the high homogenization temperatures reported, sometimes even >1400°C. As discussed in Chapters 7 and 16, I have serious doubts about these high values.

The faster the cooling rate of the host rock, the more easily melt inclusions may be detected. Thus, recognizable inclusions are not only more abundant in the shallow intrusives, but they are frequently larger. The faster cooling rates tend to cause the skeletal growth forms responsible for most of the larger inclusions in magmatic rocks. Apatite, in particular, frequently grows as hollow tubes that trap relatively large tubular inclusions (e.g., from gabbros; Gardner, 1972). Shukaylo (1979) determined Th for melt inclusions in apatite from diorite porphyries to be in the range 1250-1300°C. In some igneous rocks, I have found that the only recognizable fluid inclusions were those in the small prisms of apatite, whereas the major rock-forming phases were inclusion-free.

Composition of Melt Inclusions in Intrusive Rocks

In theory, melt inclusions in magmatic minerals, trapped from a homogeneous melt, constitute valid samples of the magma, complete with its volatile constituents; hence, analysis of the melt for such volatiles should provide answers to the important questions about the nature and amounts of volatiles in magmas. Although the laboratory problems are exceedingly difficult, attempts to obtain such data have been most extensive and successful in studies of extrusive rocks and hence are reviewed in Chapter 16 rather than here.

Studies of the bulk composition of glass inclusions in granitic rocks by electron microprobe are relatively few. Reyf (1973) described some muscovite-rich crystalline inclusions in potash granites from Buryatiya, USSR, and Skryabin (1976) found similar ones in granite of the Voronezh massif, USSR. Such inclusions dissolve considerable quartz from the walls on heating. Skryabin reported 61-69 vol % muscovite and 29-37 vol % quartz (presumably <u>after</u> solution of the walls). Melting started at 590°C, and the vapor phase homogenized at 825°C. The last of the muscovite melted at 1025°C, after 6- to 11-hour runs. Since the muscovite should contain ~5 wt % H_2O, the trapped melt should have at least 3.2-3.6 wt % H_2O, suggesting $P(H_2O)$ of ~500 bars at 1025°C. In a later paper (1978), Skryabin made an electron microprobe analysis of one of these homogenized inclusions; it showed essentially the same composition as the bulk rock, but with more K_2O and less Na_2O and CaO. He suggested that these differences are caused by crystallization of plagioclase before the melt inclusions were trapped. (Note, however, that Huang et al. (1973) reported that nearly 30 kbar pressure is needed to keep muscovite from melting to sanidine, corundum, and liquid at 1000°C; at 10 kbar the temperature is 825°C. In the presence of excess SiO_2, this temperature is reduced still more (Kerrick, 1972).)

INCLUSIONS OF COEXISTING IMMISCIBLE FLUIDS

Other than the simple silicate-melt inclusions that were trapped as the homogeneous fluids mentioned above, all primary fluid inclusions in intrusive rocks arise from heterogeneous systems of two (or more) fluid phases. Each of the following fluid phases has been found as inclusions, apparently signifying immiscibility with silicate melt: sulfide, metal, high- to low-density CO_2 (± other gases), high- to low-density H_2O (± salts), carbonate, hydrocarbon, and possibly even Fe-rich oxide ± Ti or P (Philpotts, 1967; Weidner, 1982). Still other pairs of fluid phases represent a second stage of immiscibility, resulting from one of the above fluids evolving into a P-T-X realm where it again had to split. Thus, the silicate melt can split to form two different silicate melts, and H_2O-bearing phases can split (i.e., boil) to form a low-density phase. The mechanisms of trapping inclusions of coexisting fluids, and criteria for their recognition, are discussed in Chapter 2. The evidence for some of the immiscible fluids mentioned above is reviewed in the following sections; that for sulfide and metal will be found in Chapters 16 and 18; that for high-density CO_2 in Chapter 17; and that for low-density CO_2 and silicate/silicate immiscibility in Chapters 16 and 18.

INCLUSIONS IN GRANITIC ENVIRONMENTS

General Nature of the Processes Involved

Although the composition of granite falls in the low-temperature trough of the system $NaAlSiO_4$-$KAlSiO_4$-SiO_2, which in turn was termed "petrogeny's residua system" by Bowen (1937), granite is only an intermediate stage in a long evolutionary line of melt to "fluid" compositions. Crystallization of volatile-bearing granitic magmas can result in a wide range of fluid compositions, at various P, T, and degrees of crystallization. These residual fluids may react with earlier formed crystals (deuteric alteration); they may react with intruded sediments, forming skarns; they may transport metals, forming various types of ore deposits such as the porphyry coppers; they may carry rare trace elements, to form pegmatites bearing Li, Be, Ta, Cs, etc.; or they may remain in the rock to form such features as miarolitic cavities. (In this connection, it is important to remember that if 1 wt % H_2O remains as a fluid phase of density 1.0 g/cm³, the solidified rock will have 2.7 vol % porosity; if this H_2O is present as a fluid of density 0.2 g/cm³ at the time of crystallization, the rock will have 12 vol % porosity.)

In this section, these various processes are discussed in the light of the evidence from fluid inclusions. A reasonably complete explanation of the fluid P-T-X conditions under which these processes have occurred, and their spatial and temporal relationships, must await the results of future studies, but such explanations will certainly have to draw heavily on the inclusion evidence.

H_2O and CO_2 in Magmas

The amounts and nature of the volatile materials are perhaps the most important compositional parameters obtainable from magmatic inclusions. The several analytical procedures used to determine the H_2O and CO_2 content of such inclusions in presumably early-crystallizing phenocrysts (detailed in Chapter 16) yield some surprisingly high values, as much as 12 ± 2 wt % volatiles, probably mainly H_2O! In contrast, Burnham (1979) believes that many magmatic liquids initially may have <1 wt % H_2O; Day and Fenn (1982) suggested, however, that unresolved problems exist in the Burnham model in the undersaturated region, and IR studies by Stolper (1982a,b) have shown that molecular water exists in various silicate melts. Although H_2O can be considered just another component in a magma, it does have some special properties that set it apart from the other oxides: it has a high vapor pressure and is highly compressible, even small amounts can cause major changes in the mineral assemblage, and it causes greater changes in such physical properties as viscosity and liquidus temperature per unit weight percent (and even mole percent) than do most other oxides. CO_2 was thought to be a relatively inert constituent of magmas, and its concentration to be mainly important in changing the activity of water in mixtures (Holloway, 1976; Swanson, 1979), but recent experimental work, summarized by Burnham (1979), showed that it has considerable effect on the properties of magmas, although in different ways than does H_2O. (Whitney and Stormer (1983) showed that sulfur fugacities may be high in calc-alkaline magmas and may lead to separation of SO_2-rich gases.)

The concentration of volatiles in the residual liquid must increase on crystallization of volatile-free phases from a volatile-bearing melt (or crystallization of phases with a lower volatile content than that of the melt). By itself, this increase in concentration will increase the vapor pressure of the residuum,[1] and if this vapor pressure rises to more than the pressure on the system, a new, relatively low density fluid phase will form (i.e., vesiculation will occur). In the simplest hypothetical melt consisting of quartz, feldspar, and water, this fluid phase will be essentially dense steam, and the magma will be said to have undergone "second boiling" or "resurgent boiling" (Burnham, 1979; Burnham and Ohmoto, 1980). Whether such boiling actually occurs is controlled by a series of factors: original H_2O content; bulk composition of the melt; pressure on the system (and its changes with rise toward the surface, and from tectonism); how much water becomes tied up in solids or is lost by other processes; and the magma temperature and the several factors that control it.

In terms of possible inclusions, the "steam" phase (with some dissolved silica and alkalies) could form low- to high-density aqueous inclusions, depending

----

[1] During this crystallization, the temperature generally is also decreasing. This decrease opposes the vapor-pressure increase from increase in concentration of the volatiles in the residual liquid but is of minor significance in the early stages; eventually, however, temperature decrease becomes the dominant effect and vapor pressure decreases with further crystallization.

on the pressure and temperature at the time of trapping. The steam could also
form miarolitic cavities in the mush of crystals (in effect, large fluid inclusions with polycrystalline walls), and could cause such common deuteric-type,
late-stage effects as the recrystallization and replacement of earlier aplite
(Kozlowski, 1978) and the uralitization of early pyroxene or the chloritization
of biotite. Hibbard (1979) suggested that myrmekite could be used as a marker
between preaqueous and postaqueous phase saturation in granitic systems. Late-
stage H_2O-rich fluids can produce major mineralogical changes by altering and
replacing earlier phases with albite, tourmaline, muscovite, topaz, etc.
Weisbrod and Poty (1975) presented a very extensive study of the thermodynamics
and mineralogy of these hydrothermal effects at the relatively simple Mayres
pegmatite in the French Massif Central, based in large part on detailed microthermometric and chemical analytical studies of the inclusions present. Unfortunately, natural processes and "systems" are generally not simple, and no
single scenario can suffice to describe the possible sequences of events.
Natural magmas contain a variety of constituents in addition to H_2O and CO_2 that
will enter into any such silicate-volatile immiscibility, and may have dramatic
effects on the nature of the volatile-rich phase, the stage at which it may
form, and even its very existence as a separate phase.

Other Constituents in a Volatile-Rich Residua -- Common Pegmatites vs Rare-Element Pegmatites

During the crystallization of various magmas rich in silica (e.g., monzon-
itic or granitic magmas), the most abundant solid phases that form initially
are anhydrous quartz and alkali feldspars. Separation of these from the melt
enriches the residual homogeneous melt in all constituents that do not enter
these particular crystal structures. The most obvious such constituent to
become concentrated in the melt by crystal fractionation is H_2O, but many other
less abundant species can also be concentrated, and this enrichment process
sometimes produces sufficient amounts of these trace elements to make ore
deposits. Most ores of Be, Li, Sn, W, Rb, Cs, Nb, Ta, and Mo have formed by
such processes, as have some ores of Cu, Pb, Zn, Ag, Sb, U, Th, and other
metals. The major chemical changes in the residual melt during crystal fractionation include a depletion in Ca, Mg, and Fe, and an increase in alkalies, H_2O,
CO_2, Cl, S, F, and Si, as well as increases in the trace metals listed above.
Although the temperature declines during fractionation, the concentration and
hence pressure of H_2O and other volatile components in the residual melt
increases. If the vapor pressure of the homogeneous melt (between the crystals)
is sufficient, an immiscible, low-density volatile phase may separate at any
time (i.e., vesiculation may occur), as indicated along the top of Figure 2-25,
bleeding off some of the volatile constituents. Whether such vesiculation occurs
or not, the evolution presumably (and generally) leads to a silicic, alkali-rich,
somewhat hydrous melt, which, when it crystallizes, forms a rock similar to
granite but coarser grained (i.e., "simple pegmatite" or "common pegmatite"),
owing to the increased concentrations of volatiles (Jahns and Burnham, 1969).
Kosukhin (1977) found that hydrous silicate-melt inclusions in various common
pegmatite minerals have homogenization temperatures of 540-690°C. Such common
pegmatites have little in them other than the normal elements making up the
granite.

Most geologists would probably agree up to this point. Most might also
agree that the ultimate fluid phase is probably an aqueous solution at moderate
to low temperatures (200-100°C); such fluids are saturated with silica and
deposit quartz veins. Perhaps the most generally accepted concept of the several
stages of evolution between granite and quartz veins is still that summarized
by Turner and Verhoogen (1960), as follows:

Stage	Phases present during formation	Temperature range (°C)	Most characteristic mineral phases
Magmatic	Crystals, silicate melt	----	-------
Pegmatitic	Crystals, silicate melt, gas	800-600	Allanite, monazite, tantalite, columbite, uraninite
Pneumatolytic	Crystals, gas	600-400	Tourmaline, muscovite, beryl, topaz, albite, lepidolite, phosphates
Hydrothermal	Crystals, aqueous solution, gas	400-100	Cryolite, fluocarbonates, sulfides and zeolites

Turner and Verhoogen (1960, p. 429) ask a perplexing question: "Why are pegmatites uniformly simple in composition throughout one granite province and predominantly complex in another?" The extensive studies of the mineralogy and structure of the rare-element pegmatites by Cameron et al. (1949), Jahns (1955), Jahns and Burnham (1969), and coworkers have shown that although many types exist, the zoning and mineralogy among the representatives of a given group, wherever they occur, are remarkably similar (see also Brown, 1982). The evolution of the fluids forming these several stages is an open question, with many possible answers. In the material below I add some of my personal prejudices (in large part from Roedder, 1981b,e), based in part on the evidence from inclusions.

Obviously, the nature of the fluids must change drastically in composition because geologic evidence clearly indicates that after the common pegmatites formed, subsequent fluids (in some localities only) carried and deposited relatively large quantities of the rather rare elements listed above (Li, Be, etc.). In many areas, there is good evidence that these deposits (sometimes called "zoned pegmatites" or "complex pegmatites" but here called "rare-element pegmatites") formed by alteration and replacement of the preexisting minerals. Some contain huge crystals (more than 1 m long) of otherwise rare Li minerals such as spodumene, and even a Cs mineral (pollucite), along with relatively large crystals of many other rare minerals. The main volume of such pegmatites usually consists of more ordinary minerals such as quartz and feldspar.

Some of the crystals are not only large but are very perfect and are even of gem quality. All the world's gem tourmaline, kunzite, and aquamarine, and most of its topaz, come from such pegmatites (see, e.g., Taylor et al., 1979). Extensive exploration has gone on in the Soviet Union for optical-grade fluorite and quartz crystals from pegmatites. I believe that both the crystal size and perfection argue for extremely slow crystallization, which avoids the necessity of invoking the highly concentrated "rare-element melts" that some have suggested. Apparently, adequate supplies of solvent (water) are available, and the concentration of any given rare element in the fluid need not be great if time is adequate for the passage of large quantities of fluids through the deposit (Fig. 15-1). The problem of the lack of evidence in some pegmatite bodies of a large enough "feeder channel" through which such fluids could have entered and left becomes more apparent than real under these conditions. For example, a flow rate of only 1 cm/hr through a crack 1 cm wide and 10 m long would pass ~10 tons of fluid per year. (Other logistical problems remain, however.)

The physical nature and the chemical composition of the fluids responsible for the formation of the rare-element pegmatites have been much debated. Suggestions have included essentially hydrous silicate melts, hydrosaline brines with as much as 84 wt % salts (Kozlowski and Karwowski, 1973), and even a low-density vapor phase (i.e., "pneumatolytic processes"). What evidence on the nature of these fluids can be found in the fluid inclusions in these minerals? This general subject has been pursued actively in the Soviet Union for many years, but little of this literature has been translated. The problem of the nature of these fluids is difficult to approach, however, because the fluids obviously have changed drastically with time at a given place, as many rare-element pegmatites form as crosscutting bodies or replacements of preexisting common pegmatites. It is equally obvious that the evolution has been different in different places. Some of these differences could well be based on differences in the original composition of the magma. Thus, if during the original crystal fractionation, two elements are both rejected completely by the growing crystals, their ratio in the residual liquid will be the same as in the original magma. Because many of the elements involved are originally present as minor and trace elements, and the degree of concentration involved is very large, even slight deviations in the partitioning between crystals and liquid can cause major differences in the final results. The actual enrichment factors are unknown, but whatever the process, they must be high in order to permit the precipitation of such minerals as uraninite and bismuthinite by fluids from magmas that originally contained only 4.0 ppm U and 0.2 ppm Bi.

Unfortunately, the preceding argument for the concentration of trace elements into the residual liquid by crystal fractionation is simply inadequate to explain all the data, even though it has been standard fare for generations of geology students and it explains beautifully the observations on major-element differentiation in many igneous bodies. If vesiculation occurs, anything that partitions into the low-density phase will be carried up (notably alkalies; Orville, 1972; Lagache and Weisbrod, 1977; Sakuyama and Kushiro, 1979), but such vesiculation is also not adequate. Careful studies of the sequences of material erupted from some major volcanic ash-flow centers, including accurate and precise analyses of the trace elements present, can not be explained by simple fractional crystallization. The enrichment of certain parts of the magma in various trace elements (including many of those listed in the start of this section) would require impossibly large amounts of crystallization, and the assemblage of crystals that would have to be extracted to yield the measured degree of concentration of one element is incompatible with the data on another element (R.L. Smith, 1979, and pers. comm.). Obviously, some other mechanism must be at work, in addition to crystal fractionation, and it must be highly efficient. A mechanism that appears to explain the data is a modified Soret effect, termed thermogravitational convection-diffusion (Shaw et al., 1976; Hildreth, 1977, 1979). In this process, the progressive enrichment of the upper parts of the magma chamber in water, and other changes in melt structure, result in a redistribution of trace elements on a grand scale. This process is a combination of convection and Soret diffusion in a thermal gradient and a gravitational field, and does not require the presence of a separate vapor phase (Shaw, 1974). An understanding of the process will require extensive experimental studies, as the effects of most of the pertinent parameters are unknown at present. For the present discussion, however, the above process represents a "preconcentration" stage that eases the problems and increases the possible range of compositions in the later stages of particular interest here.

The difference in the behavior of F and Cl provides a good example of the problems in following such a process. F and Cl are present in ordinary rocks (and hence presumably were present in the hypothetical granitic magma) at levels of ~800 and 300 ppm, respectively.[2/] Much of the Cl, at least, is present as fluid inclusions (e.g., Terashima and Ishihara, 1980). Recent experimental studies have shown that F, in addition to B and H_2O, may have a very great effect on the phase relationships in the system quartz-albite-orthoclase (Manning and

----
[2/] A chloride-rich solution has even been suggested as a late-stage fluid in a platiniferous dunite pipe (Schiffries, 1982).

389

Pichavant, 1983). Although the distribution coefficients for these two elements between a silicate melt and silicate crystals are unknown for most P-V-T-X combinations, and the mass ratios of the two phases are also unknown, the bulk of both elements probably precipitated with the crystals during crystal fractionation, mainly as minor constituents of several common rock-forming minerals, as well as in larger amounts in mica, amphibole, and apatite. (Thus the F and OH contents of apatite have several potential uses; Stormer and Carmichael, 1971; Candela, 1983.) Some unknown but probably small fraction of each element, however, is concentrated into the residual liquid. Even though this fluid is probably greatly enriched in Cl relative to F, when this fluid forms a rare-element pegmatite, several F-bearing minerals may be formed, such as topaz or fluorite, but rarely do any of the minerals that form contain significant amounts of Cl.

Rare-element pegmatites represent environments in which the residual fluid from a larger mass of rock has been channeled into the structure and has precipitated some part of its complement of rare elements, as at the fluoride pegmatite at Ivigtut, Greenland. If such a movement does not take place, the residual fluid will simply remain between the crystals, yielding miarolitic cavities lined with crystals like those found in the pegmatite. An excellent example was described by Raade and Haug (1980), who found a Na-rich granite that had miarolitic cavities lined with a large number of rare fluorides, including such minerals as sellaite [MgF_2], gagarinite [$NaCaY(F,Cl)_6$], thomsenolite [$NaCaAlF_6 \cdot H_2O$], neighborite [$NaMgF_3$], bastnaesite [$(Ce,La)(CO_3)F$], and fluorite. Such phases obviously should also be expected as daughter crystals in inclusions.

Several workers (e.g., Nedachi, 1980; Munoz and Swenson, 1981; Tsusue et al., 1981) have used the distribution of Cl^- and OH^- between biotite and apatite, or between biotite and a hydrothermal fluid, to estimate the conditions of genesis and the nature of the fluid phase. Later alteration of phases, and the possible effects of the numerous other compositional variables in both apatite and biotite, make such data difficult to interpret, but there is hope that eventually analyses of fluids in inclusions can be obtained to corroborate the results.

The precipitated solids in a rare-element pegmatite represent materials that must have been present in the fluid, but they give us almost no measure of the gross composition of the fluid, because these pegmatites were, in my view, almost certainly "open systems," through which fluid moved. The fluid inclusions in such minerals, however, provide us with a sample of the fluids. The "Zanorsh" or "chamber" pegmatites of Volynia and Kazakhstan, USSR, provide good examples of pegmatitic fluorides. In these pegmatites, crystals of fluorite and topaz as much as 1 m long, together with other phases, line central cavities as large as 200 m^3 in volume (N.P. Ermakov, pers. comm., 1960). I found that the fluid inclusions in such fluorites from the optical-fluorite-bearing chamber pegmatites of the Kayib pluton in northern Kazakhstan (Roedder, 1963) contain daughter crystals, indicating that the originally homogeneous fluids that were trapped contained at least 42 wt % dissolved solids (Table 14-1), mainly chlorides (~220,000 ppm Cl). The fluorine content could not be determined but must be much lower. Many other investigations of inclusions in minerals from rare-element pegmatites report aqueous fluids ranging in concentration of salts from "hydrosaline melts," as described above, to much more dilute fluids. Some contain high concentrations of CO_2 as well. Taylor et al. (1979) have shown that isotopic data on the inclusion fluids from the gem-bearing pockets in pegmatite-aplite dikes in San Diego County, California, suggest that this water, at least, was magmatic in origin, even though the hydrogen in the wall-rock hornblende was meteoric in isotopic composition.

One very simple but exceedingly important line of evidence about the general nature of the fluids that formed any given euhedral crystal found lining a cavity is consistently ignored in discussions of silicate melts vs aqueous fluids. This

Table 14-1 Calculated composition of known materials present in inclusions in fluorite from Kazakhstan, USSR. [Based on optical measurements; average of four similar inclusions; 27 vol % gas ignored; from Roedder, 1963.]

		Weight percent	
KCl	in daughter crystal[a]	11.3	
	in solution[b]	8.8	
	total		20.1
NaCl	in daughter crystal[c]	2.5	
	in solution[b]	17.3	
	total		19.8
Opaque minerals[d]			1.2
Quartz			0.9
	Minimum total solids[e]		42.0
H₂O			58.0

[a] Yellowish isotropic cube having high thermal coefficient of solubility; recrystallizes to sharp cubo-dodecahedron on cooling after partial solution.

[b] Solubilities in the pure system NaCl-KCl-H_2O are assumed; significant divalent cations could reduce these.

[c] Colorless isotropic cube having low thermal coefficient of solubility.

[d] Several other phases of doubtful identification, present in small amounts, have been ignored.

[e] Does not include unknown amounts of salt in solution containing other ions such as Ca, Mg, F, CO_3, SO_4, and BO_4. At least some Ca or Li is presumed to be present, as Te = ~-49°C.

evidence is the nature of the surface of the crystal as found. If those surfaces are clear, bright, and free from any layer or coating of various silicate minerals, it seems impossible to avoid the assumption that the last fluid (other than ground water) filling the cavity, and hence in contact with those faces, could not have been a silicate melt. Whatever fluid was left at the end of crystallization must have been completely miscible with or soluble in later ground water, as I know of no way in which the residual silicate melt in such a cavity could be drained away completely, leaving clean mineral faces. The crystals lining shrinkage vugs in various metallurgical slags are usually slag coated; only where a later vapor-phase transport has occurred does one find bright crystals. Many fine museum specimens of crystals of a large variety of minerals come from the linings of cavities or "vugs" in pegmatites. The only way that such crystals could have formed from a silicate melt would be for some still later fluids to dissolve all traces of the coating or to convert it to the clay minerals that coat some of these crystals as found. I think it highly unlikely, however, that later fluids could completely alter whatever silicate minerals crystallized from this surface film to form clay and not even dull the shiny faces of the large silicate minerals lining the cavity. I suspect that the clay simply precipitated from solution or washed into most cavities. Even if the crystal faces are rough, the same concept holds true; so long as the exposed faces consist only of the mineral, and have no coating of glass or other crystalline phases, the crystal could not have grown from a silicate melt, hydrous or not. The fluid from which the crystal grew must have been a water-soluble fluid. This concept seems just as applicable to a small 1 cm "miarolitic vug" in a granite as to a 200 m³ cavity in a chamber pegmatite lined with 1 m crystals.

In addition to such considerations, the nature of the fluid inclusions themselves in such cavity-lining crystals provides compelling evidence for aqueous fluids. Very extensive studies of the fluid inclusions in such pegmatite minerals have been made in the Soviet Union. In addition to numerous unidentified

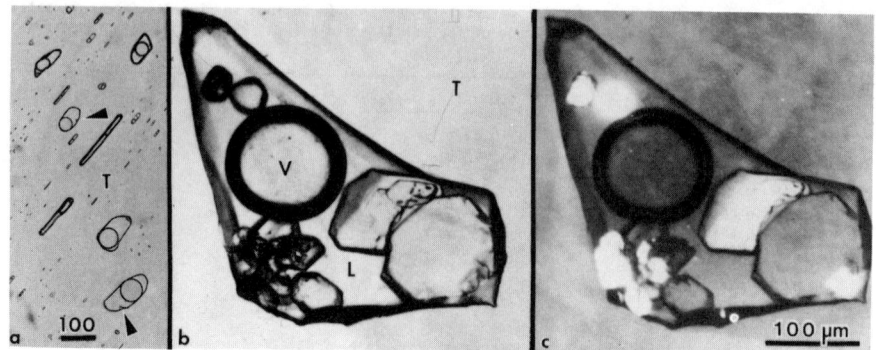

Figure 14-2 (a) Photomicrograph of plane of probably pseudosecondary inclusions in topaz (T), parallel to {001}, each containing a large vapor bubble in the liquid, and three tiny daughter crystals. (For examples, see arrows; in other inclusions they are hidden.) These crystals are unidentified, but are recognizably different phases. Scale bar in μm. From Roedder (1963, p. 173). USNM specimen 96595, Rukuba Sn mine, Nigeria.

Figure 14-2 (b) Photomicrograph of a large multiphase inclusion, in plain light, and (c) with almost crossed polarizers, set to place the enclosing topaz (T) almost at extinction. This and other inclusions in the sample contain at least 16 daughter minerals, presumably all different phases, plus liquid (L) and vapor (V). Ten daughter crystals are seen to be birefringent in (c); presumably some of those crystals that appear isotropic are not but merely have their extinction positions parallel with those of the enclosing topaz. There may be as many as five opaque phases. From a cleavage flake of topaz from a Volynian pegmatite, USSR, courtesy of Bernard Poty. All the inclusions are strongly flattened parallel to the (001) cleavage and may be primary or pseudosecondary in origin. Lemmlein et al. (1962) reported on the homogenization of similar inclusions in topaz from Volynia, USSR, and recorded the presence of large daughter crystals of quartz and muscovite, lesser cryolite, and still smaller amounts of various fluorides and chlorides of Na, K, and Ca. Lyakhov (1966) gave X-ray powder-diffraction data on 6 of the 14 different solid phases, including hydrous ferrous chloride [$FeCl_2 \cdot 2H_2O$], which he extracted from inclusions in morion (dark smoky quartz) from these same pegmatites. Other inclusions have an even higher percentage of solids.

and possibly new species, a remarkable assortment of exceedingly rare and frequently water-soluble daughter crystals has been found, such as villiaumite [NaF], hydrous ferrous chloride [unnamed; $FeCl_2 \cdot 2H_2O$], borax [$Na_2B_4O_5(OH)_4 \cdot 8H_2O$], elpasolite [$K_2NaAlF_6$], cryolite [$Na_3AlF_6$], caracolite [$Na_3Pb_2(SO_4)_3Cl$], teepleite [$Na_2B(OH_4)Cl$], avogadrite [$(K,Cs)BF_4$], and a new, unnamed chloride of Al and Zn (see particularly: Kalyuzhnyi, 1956, 1958a; Lemmlein, Kliya, and Ostrovskii, 1962; Ermakov, 1965; Lyakhov, 1966; and Kalyuzhnyi and Voznyak, 1967; and other references in Roedder, 1972). A fascinating consideration is that huge daughter crystals of such minerals probably formed in the central chamber of the pegmatites and have since been leached away by ground water. These chambers, the largest-of-all "inclusions," were enclosed by polycrystalline rocks, not by a single crystal "bottle," and hence have leaked. A 200 m³ "fluid inclusion" of this sort could well have originally had such phases as water-soluble rare-mineral daughter crystals many meters long![3]

Although some such pegmatitic fluids are dilute (e.g., Fig. 14-2a), many are exceedingly rich in a wide variety of solutes. The number of separate daughter phases forming in a single such inclusion is remarkable. Figures 14-2b,c show an assemblage of at least 16 daughter crystals, presumably all different phases. So many different phases are present in these pegmatitic topaz crystals that Dr. N.P. Ermakov would offer prizes to any student in his laboratory who identified a new one (pers. comm., 1970). Because the residual fluids forming such pegmatites represent a concentration of Na, K, C, H, O, Cl, and S, plus probably significant amounts of at least some of the following: F, Al, B, Li, Rb, Cs, Zr, Si, Ti, Ca, Mg, Fe, Zn, Cu, Sn, Be, P, Sr, and REE, no problem

----

[3] Mark Barton (pers. comm.) has pointed out that since some of the daughter-phases (such as cryolite) are not highly soluble, yet are generally not found as crystals in the vugs, they may have been lost by reaction with later fluids.

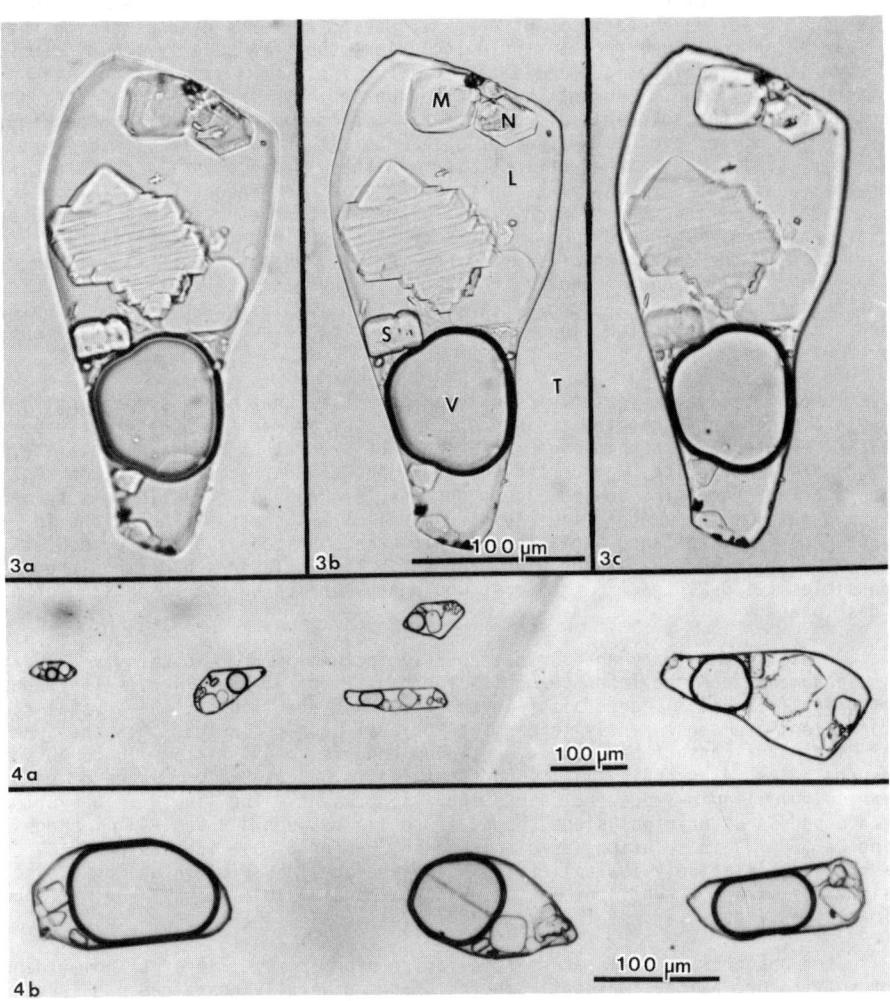

Figure 14-3.   Detailed photomicrographs of one inclusion from the group shown in Figure 14-4a, taken with the plane of focus at three different levels. In (b), the plane of focus cuts through most of the phases in this very flat inclusion. In (a), the plane of focus is raised, and in (c), it is lowered relative to (b). Note that the movement of the Becke lines in (a) shows that many of the daughter minerals are higher in index than is the enclosing liquid (L), and some are very much higher (S). In (c), however, note that the two larger crystals at the top (M and N) have an index of refraction less than that of the liquid, which, in turn, has an index of refraction less than that of the enclosing topaz (T). These specific daughter minerals have not been positively identified, but various low-index fluoride minerals such as cryolite (n=1.34) and avogadrite [K,Cs)BF₄, n=1.32] have been reported in the extensive Soviet work on such daughter minerals from this and similar localities. Same sample as Figure 14-2b and c.

Figure 14-4.   Photomicrographs of two planes of inclusions, (a) and (b), only a few millimeters apart, that have trapped two entirely different densities of fluids. In each plane, all inclusions apparently have a uniform ratio of phases, and all appear to have the same phases, but the volume percentage of gas bubble differs grossly between the two planes. The small but plainly visible differences in gas-bubble ratio between adjacent inclusions (particularly in (b)) may be only apparent and come from irregularities in the third dimension, or they may be real and stem from bubble nucleation before necking down occurred. The differences between the inclusions in (a) and (b) may be a result of differences in the confining pressure at the time of trapping.

exists in terms of the phase rule in justifying such a complex assemblage. Although sometimes these lists of daughter phases are presented as though they all were (or could be) from a single inclusion, they are usually from inclusions from a single locality, and could thus be from inclusions from various stages of topaz deposition. Many of the fluoride phases have indices of refraction below that of the interstitial brines (which may be as high as 1.36; see Figs. 4-7 and 14-3) and even below that of pure water (thus avogadrite, (K,Cs)BF$_4$, has n = 1.32). Many planes of inclusions in these topaz crystals contain a very uniform phase ratio, signifying the trapping of a homogeneous fluid (Fig. 14-4). The major variations between individual planes that were presumably trapped at different times is in the apparent density of this fluid (Figs. 14-4a,b). Simple variation in the confining pressure, from tectonic events, could yield such differences. Still other secondary inclusions in these crystals -- actually the most common type by far -- consist simply of liquid plus a large vapor bubble, and no daughter crystals. The age relations of these two types of inclusions are not known.

One of the varieties of rare-element pegmatite that has a surprisingly uniform mineralogy wherever it is found is the Li pegmatite. These zoned pegmatites contain major tonnages of one or more of several Li minerals (petalite, spodumene, eucryptite, and/or lepidolite), plus concentrations of Be, Rb, Cs, Sn, and Ta. According to the older concepts (see above), such minerals formed from a gaseous "pneumatolytic" fluid. A preliminary study of inclusions in spodumene from the Tanco pegmatite, Bernic Lake, Manitoba, Canada (London et al., 1982), showed that the fluids trapped in the inclusions had far higher densities ($\sim$1.8-2.0 g/cm^3) than most would consider to be characteristic of a "gas" phase.

Presumably primary and pseudosecondary inclusions are exceedingly abundant in spodumene from the Tanco pegmatite (perhaps 1 vol %; Fig. 14-5). They consist of vapor ($\sim$10 vol %), crystals ($\sim$10 to 80 vol %), and liquid. The crystal-rich inclusions are consistently larger than crystal-poor inclusions, but the types are intimately associated, and a continuum appears to exist from $\sim$10 to 80 vol % crystals. A few contain small amounts of liquid CO$_2$ at 25°C. Planes of secondary simple liquid-vapor inclusions are also present. The liquid in the primary and pseudosecondary inclusions showed Te to lie between -49 and -45°C, suggesting Ca ($\pm$ Li?) in solution, as discussed in Chapter 8. Tm ice was $\sim$-4.9°C, suggesting relatively low salinities. The average Th L-V(L) value for 42 inclusions was $\sim$+260°C. Only one inclusion containing visible CO$_2$ was run; it showed Tm clathrate at -1.7° and Th L-V at +287°C.

The daughter crystals are particularly informative. There was no evidence of sylvite or halite. Crystal-rich inclusions generally contained one large highly birefringent grain (hexagonal or trigonal?), one large anhedral isotropic grain, and sprays of tabular anisotropic crystals of low birefringence, plus other less easily characterized phases (Fig. 14-6). SEM studies of opened inclusions permitted recognition of some of these daughter phases as silica, albite, Cs-analcime, and pollucite (Fig. 14-7). Another daughter phase showed only Al and Si and hence may be a Li or Be aluminosilicate. Films containing K, Ca, Mg, S, and Cl, from evaporation of the liquid phase, precipitated on all inclusion surfaces on opening.

One major phase consists entirely of elements too light to determine by SEM with the available energy-dispersive unit (i.e., only elements lighter than Na). On this basis, the unknown could be a Li- or Be-carbonate, borate, or fluoride, or a combination such as fluocarbonate or fluoborate. H also cannot be detected by this procedure, so (OH) or H$_2$O could also be present. M.E. Zolensky (pers. comm., 1982) extracted a large daughter crystal from one of these inclusions that appeared to be this same phase and obtained an X-ray diffraction pattern using the Gandolfi camera (Zolensky and Bodnar, 1982).

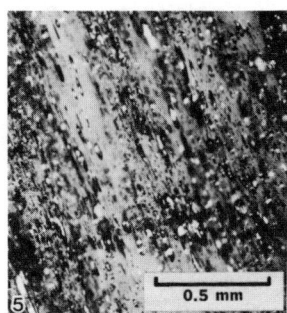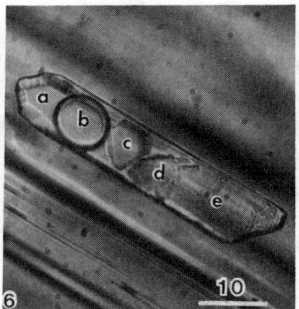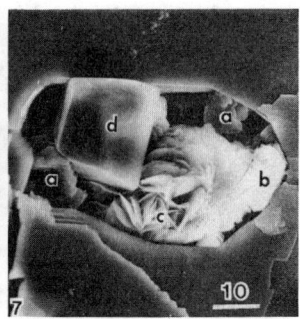

Figure 14-5. Photomicrograph in crossed polars of a doubly polished chip (0.5 mm thick) of coarse-grained spodumene from the Tanco pegmatite, Manitoba, Canada. This photograph illustrates the abundance, distribution, and variations in size of inclusions that were examined. The highly birefringent (white) grains are daughter minerals that may be $Li_2B_4O_7$ (see text). From London et al. (1982).

Figure 14-6. Photomicrograph (in plane-polarized light at 25°C) of a crystal-rich inclusion in spodumene from the Tanco pegmatite. The inclusion contains (a) H_2O-rich liquid; (b) H_2O vapor ($\pm CO_2$); (c) a highly birefringent, possibly trigonal daughter crystal [$Li_2B_4O_7$?]; (d) a spray of tabular crystals having low birefringence; and (e) an aggregate of apparently isotropic and anisotropic grains. Scale bar in μm. From London et al. (1982).

Figure 14-7. SEM photograph of a crystal-rich inclusion in spodumene from the Tanco pegmatite. This inclusion appears to contain (a) quartz; (b) albite; (c) a Li- or Be-aluminosilicate with Al/Si = 1 (possibly cookeite, eucryptite, or euclase); and (d) a phase that consists entirely of elements lighter than Na (possibly $Li_2B_4O_7$; see text). Scale bar in μm. From London et al. (1982).

The pattern was that of $Li_2B_4O_7$, a new mineral.

On heating to <350°C, the daughter crystals showed no signs of melting or dissolution, thus eliminating many of the normal soluble-salt daughter phases as candidates.[4] Most of these inclusions will decrepitate between 290 and 350°C, so the samples (15 runs) were heated in conventional cold-seal hydrothermal bombs at $P(H_2O)$ = 2 kbar (to prevent decrepitation), at temperatures of 325 to 500°C for 5 to 24 hours. Quenching in water dropped the temperature to <300°C in 5 to 10 seconds. All solid phases in 90-95% of the inclusions melted to a quenchable glass plus an aqueous phase between 375°C ("solidus" or start of solution) and 440°C (liquidus or complete homogenization). Figure 14-8 shows the behavior of a typical solid-rich inclusion in a sequence of three runs.

These preliminary data cannot distinguish whether the original fluid at the time of trapping was (1) a single homogeneous hydrous borosilicate, or (2) a heterogeneous system of immiscible borosilicate melt and water-rich fluids. The crystalline phases within the inclusions are unlikely to represent accidentally trapped solids. If a single fluid phase was trapped, then the gross differences between crystal-rich and crystal-poor inclusions could be explained by extensive necking down after most daughter phases had crystallized but before separation of a vapor phase (i.e., between 260 and 375°C). If two immiscible fluids were trapped, then the daughter crystals probably would represent the crystallization of the borosilicate melt phase. The variable amounts of daughter crystals would reflect entrapment of variable amounts of a silicate phase from an emulsion of silicate melt dispersed in the aqueous fluid. Differences in the wetting characteristics of aqueous and silicate fluids might also have produced the observed differences in size of crystal-rich and crystal-poor inclusions.

----
[4] This behavior also indicates that these inclusions differ from the inclusions formed in quartz from a synthetic Li-pegmatite melt by Bazarov (1975).

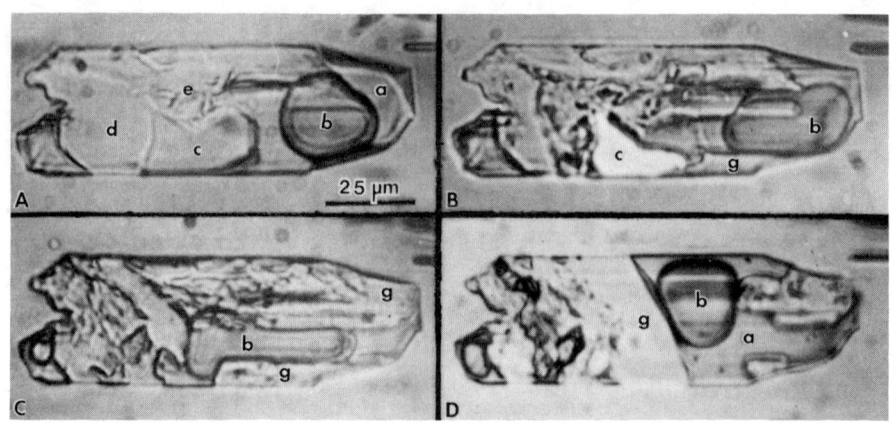

Figure 14-8.   Series of photomicrographs of a crystal-rich inclusion in spodumene from the Tanco pegmatite before and after hydrothermal runs.   (A) Before the runs (taken in plane-polarized light), the inclusion contained (at 25°C):   (a) H_2O-rich liquid; (b) H_2O vapor; (c) a highly bire- fringent grain; (d) a large, apparently isotropic grain; and (e) a spray of tabular anisotropic crystals having low birefringence. (B) After quench from 375°C (crossed polars); the daughter minerals, especially the highly birefringent phase (c), show evidence of melting or solution.   The vapor bubble (b) is constricted by a small amount of glass (g) that lines the walls of the inclu- sion.   (C) After quench from 420°C (crossed polars); the highly birefringent phase has disappeared completely, most other phases show evidence of appreciable melting or solution, and the vapor bub- ble (b) is highly distorted by large amounts of glass (g) along the walls of the inclusion.   (D) After quench from 450°C (crossed polars); the solid contents of the inclusion appear to have melted or dissolved completely, and the glass (g) appears to have begun to coalesce into a bead.   From London et al. (1982).

    Full interpretation of such preliminary data is not now possible, but London et al. (1982) showed that extrapolation of an estimated isochore for the water solution up to Tm daughter crystals (375-440°C) indicates (with several caveats) a pressure of ~2100-2600 bar.   When the inclusion data are superimposed on the P-T diagram for the lithium aluminosilicates, London et al. found that the apparent environment of inclusion trapping was in the petalite field, near the triple point for petalite-spodumene-eucryptite, and hence agrees with earlier interpretations of the genesis of lithium pegmatite mineralogy.

    An additional feature of the inclusions in these lithium pegmatites, pointed out by David London (pers. comm.), indicates a need for caution in the interpre- tation of the environment of formation of the lithium pegmatites, and in the interpretation of inclusions in general.   This feature is the nature of the in- clusions in "sqi", a miner's term for a relatively fine-grained aggregate of spodumene and quartz that is found in many lithium pegmatites and is generally accepted as resulting from the almost isochemical breakdown of earlier petalite to spodumene plus quartz (Černý and Ferguson, 1972; Černý, 1975; London et al., 1982).   The spodumene of such aggregates contains abundant multiphase inclusions similar to those shown in Figures 14-5 through 8.   The closely associated, inter- grown quartz, however, has almost solely simple inclusions of liquid plus vapor. In addition, inclusions in quartz contain CO_2-rich fluids, whereas the fluid in the inclusions in spodumene contains little or no CO_2.   Several explanations may be offered, but the most plausible at present is that the decomposition of peta- lite to spodumene and quartz took place in the presence of very solute-rich fluids represented by the multiphase inclusions now present in the spodumene, but that the quartz recrystallized later in the presence of CO_2-rich low density fluids.   The salinity of the aqueous component of inclusions in quartz actually is higher (as determined from Tm ice, corrected for CO_2 clathration) than that for the aqueous fluid in P inclusions in spodumene (prior to homogenization of the daughter minerals). Thus examination of inclusions in the quartz only would provide an erroneous and incomplete interpretation of the conditions of genesis of "sqi", yet the textural evidence would generally be accepted as indicating the quartz and spodumene formed simultaneously.

Fluid Evolution -- Continuous or Discontinuous?

As discussed above, even though granites formed from a silicate melt, the fluid inclusions tell us that at least parts of the rare-element pegmatites formed from solute-rich aqueous solutions. An important but unanswered question is whether the composition of the fluid changed continuously between these extremes or changed suddenly because of immiscibility. Here I refer to immiscibility at depth between a somewhat hydrous aluminosilicate melt and a relatively dense, H_2O-rich and probably salt-rich fluid, and not to the separation of an immiscible, low-density vapor phase (i.e., vesiculation) at near-surface conditions. This question of immiscibility vs a continuum in the evolution is important to several fields of petrology and economic geology and has been widely debated (e.g., see references in Roedder and Coombs, 1967; Moore and Lockwood, 1973) but remains open. Laboratory experiments show that either route is possible, depending on the values chosen for the variables P, T, and X, but which route was followed by nature in any given occurrence? Because at least some of the "minor" elements such as F, Cl, and Rb, which are greatly enriched at this stage, are known to increase the solubility of water in silicate melts, or of silicates in water solutions, by orders of magnitude, the problem is not very amenable to calculation at this time.

Fluid inclusions provide answers to these questions, but like those from the Delphic oracle, the answers are sometimes obscure and generally equivocal. If the composition of the fluid changed continuously, and if appropriate samples were examined, one should find inclusions ranging in composition from essentially silicate melt in the minerals of granites and perhaps common pegmatites, through inclusions having various intermediate ratios of silicates to water, to essentially simple water solutions. If, however, immiscibility has occurred at some point in this evolution, there should be a break in the range of compositions of the inclusions. In spite of the increase in mutual solubility with high F and Cl, Eadington (1983) reported primary inclusions of silicate melt and others of highly saline brines (Th 550 to >620°C and 54-65 wt % salts) in a magmatic quartz-topaz rock, which he ascribed to immiscibility.

One of the major problems in using fluid inclusions for this purpose is that such intermediate-composition inclusions, except under special circumstances, are not only scarce but are difficult to study. Rocks of granitic composition that crystallized near the surface and hence crystallized rather quickly may contain glassy or subsequently devitrified glass inclusions and, on analysis, may show some water (e.g., Sommer, 1977). Granites that have crystallized more slowly, at depth, normally reveal no glass inclusions. This lack is expectable, because a hydrous-silicate-melt inclusion should form a mass of crystals (i.e., "stony inclusions") during such slow cooling. However, even such crystalline inclusions are not normally reported. In small part, this lack of reports might result from the difficulty in recognizing such inclusions, but I suspect that it primarily results from the fact that the textures of the minerals in deep-seated granites are mainly a product of their recrystallization, as suggested by Tuttle (1952). A complete recrystallization would eliminate early-formed melt inclusions. Where such melt inclusions are reported from common pegmatites, or from the early formed outer zones of chamber pegmatites, which are essentially identical with common pegmatites, they are found to contain much water. Thus, Kosukhin (1977) reported silicate-melt inclusions having homogenization temperatures of only 540-690°C in the common pegmatite border zones of several chambered pegmatites.

Granites formed under relatively shallow conditions apparently have not recrystallized, and numerous studies, particularly by Soviet workers (see various entries in indices in Roedder, 1968a), have been made of the inclusions to be found in them. Such inclusions generally have trapped either silicate melts having homogenization temperatures of ~900-1100°C, or hydrosaline melts yielding large cubes of NaCl, liquid, and vapor at room temperature. The hydrosaline melt inclusions are sometimes accompanied by low-density, aqueous inclusions, presumably representing boiling or condensation (e.g., Watanabe, 1981).

The presence of high-salinity fluids in granitic rocks has been known, qualitatively, for more than 100 years (e.g., Zirkel (1876) provided detailed descriptions of such inclusions in several granitic rocks in the western United States). These inclusions contain important clues concerning the evolving fluids, but as they were trapped from fluids that remained in the granite and did not move to form pegmatite deposits elsewhere, there is the added problem of determining the origin of the inclusions, i.e., the stage at which they were trapped. In particular, the question of a continuum vs immiscibility is generally not addressed. In any such discussion, however, it is important to note that even if immiscibility had occurred, studies of a series of random inclusions, selected without adequate paragenetic control and microscopy, might falsely appear to reveal a continuum. Thus, if various combinations of immiscible silicate melt and aqueous fluid were trapped in different inclusions at the same temperature, the differences in compressibilities of these two fluids could cause the homogenization temperatures to have a wide spread (Roedder and Bodnar, 1980), and the inclusion bulk compositions would suggest a continuum. Once the melt has crystallized, the mineralogy would also suggest a continuum.

Most inclusion papers that present homogenization temperatures high enough to suspect possible intermediate compositions between silicate and aqueous (>500°C), provide insufficient information about the actual bulk composition of the fluid trapped, and inadequate evidence that this fluid was actually homogeneous. For example, Lemmlein et al. (1962), in a widely quoted paper, showed that some primary inclusions in topaz (from a pegmatite in Volynia, USSR) that were crowded with crystals of quartz, silicate minerals, and various fluorides and chlorides of K, Na, and Ca at room temperature, could be homogenized to a melt by heating to 740°C (also given as 700°C) for 3.5 hours (under external pressure to prevent decrepitation). From this study, they concluded that these Volynia "chamber" pegmatites formed from a melt. However, Voznyak (1968, and pers. comm., 1970), after an intensive study of inclusions in Volynian topaz, showed that the inclusions studied by Lemmlein et al. were composites, in which accidentally trapped solid crystals were enclosed along with some liquid. He reported that primary inclusions in these topaz crystals contain only liquid and gas at room temperature and homogenize ~400°C. Gigashvili (1969) came to similar conclusions (see also Kalyuzhnyi et al., 1966). The fact that such topaz crystals have clean faces, uncoated by silicate melt, also contradicts the conclusion of Lemmlein et al., and corroborates those of Voznyak. However, at least some of the above-listed solid phases that Voznyak thought were accidentally trapped occur in uniform ratios in groups of inclusions in some Volynian topaz (Fig. 14-5), hence solutions of high concentration were present at some time during the history of these pegmatites. These solids probably consist mainly of water-soluble phases.

To my knowledge, only one individual, A.I. Zakharchenko of Leningrad, has specifically indicated that on the basis of detailed studies of fluid inclusions (in granites from Kazakhstan and elsewhere in the USSR), a continuum of compositions was found. A series of 29 papers by Zakharchenko (and colleagues) from 1964 through 1976, all deal with the general problem of fluid inclusions in late-stage processes related to granitic massifs (see references in Roedder, 1968a and 1972). Although many of these papers describe inclusions having a wide range of compositions (from silicate melt through hydrosaline melts to aqueous solutions) and homogenization temperatures (900-250°C), and include transitional types, only two papers (Zakharchenko, 1973, 1976) have in the available English abstracts a

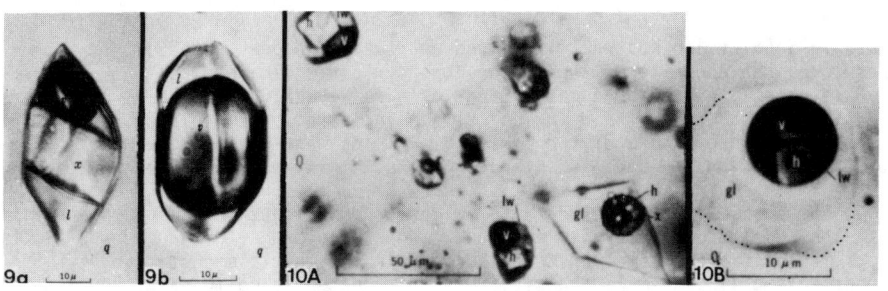

Figure 14-9. Adjacent inclusions of hydrosaline melt (a) and of CO_2-bearing, low-density aqueous fluid (b), from a plane of secondary inclusions in quartz from a granitic block in Ascension Island volcanic rocks. Inclusion type (b) is believed to represent the trapping of a vapor phase from the boiling of hydrosaline melts such as inclusion type (a), during the eruption of these blocks. From Roedder and Coombs (1967).

Figure 14-10. Photomicrographs of inclusions in granite from Ascension Island in transmitted plain light, showing evidence of immiscibility between silicate melt and hydrous saline fluids. From Roedder and Coombs (1967). During the growth of the host quartz crystal (Q) from a brine-saturated hydrous silicate melt of rhyolitic composition, some parts of this melt were trapped as primary inclusions. Immiscible globules of highly saline aqueous fluid in the melt were also trapped in these inclusions. On cooling, the melt formed a glass (gl). The saline fluid formed at least three phases: as shown in the lower right of (A), a wet mass of crystals (x) lines the cavity (one larger cube is presumably NaCl(h)). Several other droplets of the saline fluid were trapped without silicate melt, forming inclusions now containing a large crystal of halite (h), vapor (v), saturated liquid-water solution (lw), and several unidentified crystals. As shown in (B), the aqueous fluid formed an isotropic cube, presumably NaCl (h), a thin, almost invisible layer of saturated liquid-water solution (lw), and vapor (v). The strongly convergent lighting needed to see within the bubbles makes the glass/quartz contact almost invisible in (B), so it has been partly outlined.

specific mention of "continuous" or "gradual" changes between the silicate and saline types. Zakharchenko indicated that the final hydrothermal fluids were the fluids responsible for forming various metallic ore deposits. He also suggested (Zakharchenko, 1971) two evolutionary paths: the adamellites and plagiogranites had late fluids enriched in Na and Cl and yielded ores of Pb, Zn and Cu, and the potassium granites had late fluids enriched in F, B, etc., and yielded ores of W, Sn, Mo, etc.

The presence of both silicate-melt inclusions and aqueous inclusions in the same crystal signifies immiscibility between these two fluids only if the two types of inclusions are shown to be coeval (i.e., trapped simultaneously). Evidence for simultaneous trapping is difficult to find and is commonly ambiguous. If inclusions can be found that have trapped both silicate melt (now glass) and aqueous fluids, immiscibility is almost surely indicated.

Such inclusions were found in the vuggy granitic blocks ejected from Ascension Island together with trachytic breccias (Roedder and Coombs, 1967). The feldspar of these blocks contains vast numbers of hydrosaline melt inclusions, containing <70 wt % NaCl, and having fluid bulk densities estimated as high as 1.4 g/cm^3 (Figure 14-9a). These inclusions are so abundant (~10^{10}/cm^3) that powder-diffraction lines for NaCl were also obtained when single-crystal X-ray-diffraction photographs are made of these feldspars. Although this feldspar probably grew from a silicate melt, not from the hydrosaline melt, it contains only rare melt inclusions. Presumably the feldspar was preferentially wetted by the hydrosaline melt. Many more melt inclusions are found in the quartz of these granitic blocks, along with inclusions of hydrosaline melt and some closely associated low-density inclusions believed to result from boiling of the hydrosaline melt (Fig. 14-9b). This quartz also contains relatively few inclusions that have trapped both silicate melt (now glass) and hydrosaline melt (now saturated solution plus salt crystals; see Fig. 14-10). Similar inclusions have been reported elsewhere but usually without detailed documentation (see

data in Sobolev and Kostyuk, 1975; and entries in Roedder, 1968a). At room temperature, the Ascension Island inclusions contain silicate glass and a large gas
bubble, partly filled by a mass of one or more wet halite crystals. The ratio
of silicate to saline fluid varies between inclusions, as would be expected if
a heterogeneous mixture of two immiscible fluids were trapped. Various alternative explanations to the assumption of immiscibility were examined and discarded.

The possibility that the salts, water, and silicate melt might homogenize
at high temperatures was also examined. When quenched after 42 hours at 1005°C,
such inclusions consisted of three phases: silicate glass containing two
spherical bodies, a small spherical mass of apparently solid salt, and a small
vapor bubble. Presumably the water dissolved into the silicate at melt temperature, leaving a droplet of immiscible salt melt; the vapor bubble probably formed
during quench (Roedder, 1970b). Koster van Groos and Wyllie (1969), in studies
of the system $NaAlSi_3O_8-NaCl-H_2O$, verified the immiscibility of silicate melt
and an $NaCl-H_2O$ fluid at 1 kbar and obtained fluid inclusions similar to those
of Ascension Island in some of their experiments.

INCLUSIONS OF MAGMATIC VS METEORIC FLUIDS

Early Stages of Cooling of Intrusion

The evidence cited above indicates that the residual fluids within a granitic intrusion may range from a hydrosaline melt, at a density of ~2 g/cm^3, to
low-density steam at a density $<<1$ g/cm^3. These fluids may have similar or very
different isotopic values (e.g., Kuroda et al., 1982) and K/Na ratios (e.g.,
Orville, 1972) than those present in the magma. If these fluids are given off
by the intrusive body at a pressure greater than the local hydrostatic pressure,
they will mix with and flush out preexisting fluids. As such fluids leave their
source, the pressure drop may result in a new stage of immiscibility (boiling,
and evolution of gases, or condensation of a liquid phase from lower density
fluids), but as such low-density fluids contact cooler country rock, they may
eventually recondense. As a result of such processes, inclusions of early magmatic fluid in both the intrusive body and in surrounding skarns may have an
extremely wide range in salinity and density, even without any mixing with
meteoric waters. Thus, Konnerup-Madsen (1979a) found salinities ranging from 2
to 60 wt % NaCl in inclusions from deep-seated granitic intrusions; he suggested
these inclusions were trapped during final solidification at 5-6 kbar and 700-
800°C. Erwood et al. (1979) reported fluid inclusions consisting of extremely
saline fluids in skarn at the Naica mine in Mexico, having Th = 500-800°C.
Highly saline fluids are sometimes ascribed to simple concentration by boiling,
but unless rather special conditions are envisaged, whereby heat for continued
boiling can flow into a given packet of fluid, relatively little increase in
concentration can be caused by the heat in the fluid itself (see also discussion
of Porphyry Copper deposits, in Chapter 15). Condensation of a liquid phase
from a low-density aqueous fluid is, however, an effective mechanism for generating high-salinity fluids.

Late Stages of Cooling of Intrusion

An igneous intrusion introduces a major thermal anomaly extending well
beyond the limits of the intrusion itself. Heat flow by conduction, and the
much more rapid heat flow by the movement of hot fluids, results in a large
mass of heated rock. The solidified intrusive body generally will have fractures, and the intruded country rock will have its original porosity, plus the
additional fractures formed during the intrusion. This porosity will be filled
with fluids. Because most fractures are interconnected (i.e., the rocks are
permeable), one or more thermal-convection cells will start, which will continue
to run until the source of energy disappears as the temperature of the intrusive

body approaches that of the intruded country rocks. A series of model studies of such circulation, involving extensive computer simulations, have been made by Cathles (1977), Norton (1979), and others; these studies provide quantitative solutions to the direction and rates of fluid flow and the resultant long-term thermal history. However, most such simulations use the properties of pure water and do not involve fluids that have densities greater than that of ground water; hence, they are not applicable to the early stages of cooling, when dense fluids must have been present, as evidenced by the inclusions. Such densities would drive convection in the opposite direction to that described by Cathles, Norton, and others.

Isotopic studies of O, and of H from hydrous minerals and fluid inclusions in minerals from both the intrusive body and the country rock have shown that these convection cells have in fact been very extensive (see, e.g., Taylor, 1977). Thus, the inclusions in minerals of the intrusion, as well as in skarns or other altered country rocks, may contain either magmatic water from the intrusion, or ground water, or a mixture. Inclusions (usually secondary) that were trapped late in the cooling process are most likely to contain a high percentage of meteoric water, or to be essentially pure meteoric water, ± some solutes leached from the rocks traversed. Isotopic evidence for such processes has been reported from many ore deposits, as detailed in Chapter 15.

Postcooling-Stage Inclusions

Even after an intrusion has cooled completely, additional secondary inclusions can be trapped at any time as a result of tectonic fracturing. Thus, many granite bodies have one or more sets of healed fractures, now marked by fluid inclusions, that represent trapping of whatever fluids were present in the pores of the rock at the time of fracturing and healing (see also Chapter 12). Inclusions are so abundant in such sets of healed fractures that they make planes of weakness and are responsible for some of the differences in ease of fracturing of granites in various directions (the "rift" and "grain" of quarry workers; Dale, 1923).

The fluid in inclusions in a granite, particularly planes of secondary inclusions, will be released in part during fracturing. Such release of inclusion fluids presents a possible explanation of the increase in salinity of ground water with depth observed in boreholes in granite at the Stripa mine site in central Sweden (Nordstrom, 1983). The concentrations of Cl (and various cations) increase from near zero at the surface to more than 600 ppm Cl below 800 m and the Br/Cl ratio is ~3 times that of seawater. The available data do not permit a choice between several possible scenarios: (1) the granite has these ions as a residual intergranular fluid, which is presently being leached out by the very slowly moving ground water; (2) the ions were present in fluid inclusions, which were opened by tectonism in the past and are now being leached; or (3) present-day fracturing of inclusions, as from glacial rebound, is releasing a continuous new supply of ions to the ground water. As this granite is under consideration as the site for a nuclear-waste repository, the details of the ground-water origin and movement must be clearly understood.

A similar mechanism might explain the "high" total dissolved solids (avg. 1300 ppm) reported by Mack and Ferrell (1979) in wells along an 85 km long line in the Sierra Nevada foothills, California. Because deep ground waters move very slowly, even a small amount of shearing movement might well open enough inclusions to provide an adequate supply of ions for such waters. Couture et al. (1983) showed, however, that the brines in deep cores in Precambrian granite in northern Illinois came mainly from overlying sediments.

A series of studies have been published in which various transient geochemical anomalies (positive or negative) have been detected in seismic areas. Some

of these anomalies may presage fault movement and hence have been investigated
as possibly useful earthquake precursors (Carapezza et al., 1980). Such anomalies are usually assumed to arise from the release of material by increasing
stress on the rocks; fluid inclusions provide one possible source for the
materials. The $^3He/^4He$ ratio has been shown to be "one of the best tools"
(Carapezza et al., 1980, p. 96). Measurements have also been made of the He/Ar
ratio (Sugisaki, 1981); He concentration (Borodzich et al., 1979); Rn concentration (Sultanxodjaev et al., 1976); $^{13}C/^{12}C$ (Kravtsov et al., 1979b); CO_2 concentration (Irwin and Barnes, 1980); Ar concentration, $^{40}Ar/^{36}Ar$, and reducing
capacity of volcanic gases (Carapezza et al., 1980); and H_2 concentrations
(presumably released from reactions with freshly broken mineral surfaces; Wakita
et al., 1980 and Sugisaki et al., 1983; or from serpentine masses near the fault;
Sato et al., 1982). Kita et al. (1982) proposed that hydrogen forms along faults
by reaction between H_2O and freshly broken Si-O bonds. Giardini et al. (1976)
reported that all the rocks they studied emitted various gases when they were
stressed to failure under high vacuum (H_2, CH_4, H_2O, N_2, CO, O_2, and CO_2 were
most abundant). They reported volumes of most gases "equivalent to several hundred liters each per ton of rock" (p. 355). Such amounts of gases seem unexpectedly high. Thus for CO_2 alone, assuming 200 liters CO_2, this corresponds to the
release of ~400 ppm CO_2, a very large amount indeed.

INCLUSIONS IN THE ALKALIC ROCK-CARBONATITE ASSOCIATION

Alkalic Rocks

Early silicate-melt inclusions. Many studies have been made of the fluid
inclusions in the numerous alkalic massifs in the USSR (see entries in Roedder,
1968a). The most notable massifs are Botogol and Nyurgan (E. Sayan); Dakhunur
(S.E. Tuva); Middle-Tatarien; Synnyr (N. Pribaikal'e); Borgoi (S. Pribaikal'e);
Kiya (Kuznetsk Alatau); Il'men (Urals); Oktjabr' (E. Priazov'e); and, in the
Kola Peninsula, the Kovdor, Lovozero and Khibiny massifs. Very similar rocks
(and inclusions) have been found in the Ilímaussaq massif in Greenland
(Sørensen, 1974). These various alkalic rocks (including nepheline syenites,
ijolites, urtites, naujaites, foyaites, theralites, etc., but excluding alkali
basalts) differ from granite in having a lower silica content and in the molar
ratio of alkalies to alumina (>1 vs ~1 for granites), and they contain a very
different but characteristic assemblage of minor and trace constituents, which
is found in these rocks worldwide (Sørensen, 1974). As indicated by their different composition, alkaline rocks have a different liquid line of descent from
that of granite; these differences are particularly obvious in the late-stage
fluids and "alkalic pegmatites." Some of the studies of Th of inclusions in
both the intrusive rocks themselves and their late-stage pegmatitic segregates
have indicated rather high temperatures.

Sobolev et al. (1974) and Sobolev and Kostyuk (1975) presented extensive
reviews of the Soviet studies of melt inclusions in alkaline rocks and reported
a very wide range of Th values. Many Th measurements were >1100°C. Some of the
higher values they reported (as much as 1260°C for inclusions in plagioclase
from theralite at Kiya, USSR; Chepurov et al., 1975) have been challenged by
Volokhov (1975) on two major points: (1) the possibility of leakage of H^+ ions
(from the dissociation of water), and (2) the possibility that a volatile-rich
phase had already separated before trapping, and hence that the silicate-melt
phase was not representative of the whole magma. Volokhov believed that both
of these sources of error raise the observed Th above the true value. I question
the logic of the second point,[5] but the first may well be valid (see discussion

----

[5] Even if a magma loses its volatiles before crystallizing, inclusions of such devolatilized magma are
still valid samples of that magma.

of "Leakage," Chapter 3). Sobolev and Kostyuk (1975, p. 54) minimized the possibility of leakage by reference to some experiments on inclusions in natural quartz and published a more detailed rebuttal (Sobolev·et al., 1976), but it is not at all obvious that the experiments on quartz are pertinent to the problem at hand.

Chepurov et al. (1975) reported numerous high Th values for melt inclusions in plagioclase, apatite, pyroxene, and nepheline for four Kuznetsk Alatau thera-lites and other alkalic rocks. They ranged from 1240-1260°C for the earliest mineral (plagioclase), through 1160-1200°C for apatite, then 1100-1180°C for pyroxene and <1100°C for nepheline, except one that showed 1100-1140°C. Glasses in the more rapidly cooled inclusions and those in laboratory-homogenized inclusions were analyzed by electron microprobe. The latter glasses were similar to the bulk rock in composition, but the former corresponded to feldspars plus nepheline in composition.

Others report lower values for Th on melt inclusions in alkalic rocks. Andreyev and Shvadus (1977) measured Th values for the Synnyr and nearby massifs; some pyroxene inclusions yielded 1150°C, but all other results were <1100°C, including potassium feldspar at 840-890°C. They also noted inclusions "with a salt phase" in some nepheline syenites. Some of these are reported as "silicate-salt melt solution," having a solid-gas-liquid ratio (presumably volume percents) of 80:15:5. Melting of the three to seven daughter crystals present (both aniso-tropic and isotropic) into the liquid started at ~200°C, and Th was at 840 to 890°C. On cooling, mainly isotropic daughter phases reappeared, and a transparent isotropic rim was seen, presumably of glass from the melting of the aniso-tropic (silicate?) daughter phases. Other inclusions apparently consisted solely of gas, liquid, and NaCl; these had Th L+S(L) at 210°C. The origin of these last inclusions is not stated, but the others are all described as primary. Hence, I would suggest that the "inclusions with a salt phase" may be cogenetic with the more ordinary silicate-melt inclusions and represent an immiscible hydrosaline melt phase (see next section).

Kogarko and Romanchev (1973, 1977) studied crystallized melt inclusions in nepheline and apatite from urtites, foyaites, juvites, lujavrites, ijolites, and apatite ores of the Lovozero and Khibiny massifs. They reported Th values in the range 800-990°C. The electron microprobe was used (along with optical pro-perties) to identify the individual phases in the intergrown masses of daughter minerals that made up the crystallized inclusions. They found nepheline, soda-lite, potassium feldspar, various pyroxenes including aegirine, biotite, iron sulfide, villiaumite (NaF), and apatite. On heating, inclusion melting started (i.e., Te) at ~700°C and homogenization occurred in the silicate-melt phase. The compositions of the original melts (presumably calculated from mineral modes) were judged to be similar in the two host minerals; that from which the apatite crystallized at Khibiny resembled the ijolite-urtites there, and that at the more differentiated Lovozero massif resembled the melanocratic foyaite there. Kogarko and Romanchev suggested a deep origin for the parent magmas, on the basis of the inclusion evidence of very low water content.

Late aqueous brine inclusions. The aqueous brine inclusions found in the alkalic rocks differ in several respects from those in the ordinary silica-satu-rated igneous rocks. As might be expected, those from the alkalic rocks may have alkali-rich daughter minerals; they also commonly have very high concentra-tions of solids, and high Th. Thus, Sobolev et al. (1970) reported that inclu-sions in chkalovite [$Na_2BeSi_2O_6$] in the late-stage alkalic pegmatites from the Ilímaussaq alkalic complex in Greenland have daughter minerals that started to dissolve at 300-360°C and finally homogenized at 860-980°C.[6] Konnerup-Madsen

----

[6] Some of these same inclusions had a second immiscible fluid that separated out as a high-index globule during cooling and then redissolved on further cooling (see Chapter 3).

(1980), and Konnerup-Madsen and Rose-Hansen (1982), also working on chkalovite from Ilímaussaq, reported Th values of <980°C and salt concentrations of <46 wt %. Many papers on such high-temperature aqueous inclusions in silicate minerals have reported significant solution of the walls near Th, indicating that the fluids as trapped contained significant silicate materials in solution, i.e., they are water-chloride-silicate melts. Aspden (1980), working on primary inclusions in pegmatitic apatite from the Alno ijolite in Sweden, found that most solid inclusions consisted of calcite but that aqueous brine inclusions had as daughter minerals nahcolite [NaHCO$_3$], kalicinite [KHCO$_3$], and alkali halides. These saline fluids were responsible for the late-stage alteration of the host ijolite at Alno. In various papers on the Soviet alkalic massifs in the USSR, similar aqueous brine or hydrosaline-melt inclusions are mentioned, but usually the origin of the inclusions is unstated or ambiguous, as is their age relationship to the silicate-melt fluids. Not infrequently, the saline inclusions are just tacitly assumed to be later, although if they are generated by immiscibility, they would be cogenetic.

Hydrocarbon-rich gas inclusions. The largest single fraction of the several thousand published analyses of the gases obtained from crushing or heating of rocks has been of hydrocarbon-rich gases obtained from alkalic rocks in the USSR. Many of these were obtained by ball milling of rocks from the Kola Peninsula, but similar methane-rich gas has been reported from several other alkalic massifs in the USSR. Most of these results are given in a series of 23 papers by I.A. Petersil'e (see also "Petersilje" and "Petersilie"), S.V. Ikorskii, and coworkers published in 1958-1968 (for references, see Roedder, 1972). These gases are distinctive in that most combustible gases found in igneous rocks are high in H but these are high (<92 vol %) in methane (or CO$_2$ and methane) under high pressure, have very appreciable concentrations of the higher hydrocarbons C$_2$H$_6$, C$_3$H$_8$, C$_4$H$_{10}$, and higher, and have only small and variable contents of H, N, and He. Thus, they are very similar to the gases from many ordinary natural-gas fields. In addition to those obtained by vacuum ball milling, similar gases, presumably air contaminated, have been obtained from boreholes in the complex, where they present a hazard to mining for apatite and are discharged continuously to the atmosphere. The oil fraction in cold chloroform extracts from these rocks consists of pure paraffinic hydrocarbons; a variety of bituminous compounds containing nitrogen, oxygen, and sulfur has also been found in small but significant amounts (0.002 wt %). The associated mafic and ultramafic intrusive rocks in the area show in general very low gas contents, without heavy hydrocarbons.

The amount of gas has little real significance; it merely reflects the abundance and size of the inclusions. Zakrzhevskaya (1964) found that many of the inclusions in nepheline are of several generations, possibly because of repeated crushing of the rocks, and are present in adequate volume and frequency to yield the gases found on vacuum ball milling. However, gross and inexplicable differences have been found in the compositions of the gases from different samples from the same complex (representative analyses are presented by Roedder, 1972, Table 2). I doubt that analytical problems are the sole cause of these differences, as the ball milling of large samples releases tens of milliliters of gas per 100 g of rock, but the later analyses do show less variation than the earlier ones. In contrast, the gases extracted from three single inclusions in pyroxene from a theralite from Kuznetsk Alatau were reported by Chepurov et al. (1975) to consist of N$_2$ plus inert gases (5-18%) and CO$_2$ (70%); two contained HCl, and one contained (H$_2$S or SO$_2$), but no hydrocarbons, H$_2$, or CO were found. These three analyses (all reported to three significant numbers) were made on gas bubbles of 6, 9, and 10 μm in diameter, by N.A. Shugurova, using the selective absorption technique (see Chapter 5 for caveats).

Microscopy of the gas inclusions in alkalic rocks is apparently difficult. Both Ikorskii (1962) and Zakrzhevskaya (1964) reported both glass and gas inclusions in these rocks. Ikorskii and Romanikhin (1964) and Ikorskii (1965, 1968)

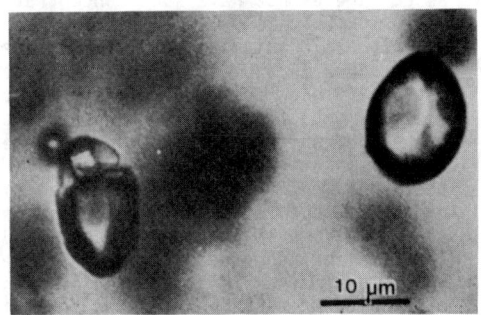

Figure 14-11. Photograph of hydrocarbon-rich inclusions in chkalovite from hydrothermal veins in the Ilímaussaq intrusion, Greenland, taken at +20°C. A healed fracture contains hydrocarbon-rich inclusions and one exceptional mixed brine-hydrocarbon inclusion. From Konnerup-Madsen et al. (1979).

10 μm

presented evidence that most of the gas inclusions are primary, having formed early, during the crystallization of the nepheline. Dudkin (1964, p. 84) reported that the highest concentrations of liquid-gas inclusions occur in the borders of the nepheline grains in the Khibiny pluton; this distribution implies a late-stage primary origin, but the evidence of primary origin is seldom convincing. As will be seen below, the origin of these inclusions is of critical importance in understanding their strange composition.

Petersil'e (1961, 1962) indicated that the hydrocarbon gases of these alkalic intrusions do not occur in the surrounding effusive, sedimentary, and older metamorphic rocks and that the geologic setting of the massif excludes migration from any sedimentary formation. Thus, the crystalline schists surrounding the Khibiny intrusion are practically free of hydrocarbon gases (Petersil'e, 1963). Petersil'e believes that the gases are of inorganic origin and that the heavier hydrocarbons and the aliphatic ester type of bitumens are a result of a catalytic polymerization. According to Petersilje and Pripachkin (1979), isotopic studies of the carbon confirm an abiogenic origin.

After these extensive studies were made, the Ilímaussaq intrusion in Greenland was found to have similar high-pressure, methane-rich, gas inclusions, as well as "disseminated bituminous substances composed mainly of oil, alcohol-benzol resins and asphaltenes" (Petersilie and Sørensen, 1970, p. 59). Detailed studies of these hydrocarbon-gas inclusions were then made by Konnerup-Madsen et al. (1979, 1981) and by Konnerup-Madsen and Rose-Hansen (1982). This work showed that these inclusions were formed as a result of preferential trapping of droplets of immiscible inorganic hydrocarbon-rich fluid in a highly saline aqueous solution (and not in a silicate melt), at a late stage in the crystallization of rather dry nepheline syenites under very low oxygen fugacities, at temperatures of 800 to 500°C and pressures of 1.4 to 0.8 kbar (Fig. 14-11).

This interpretation avoids a seeming contradiction presented by the earlier work, discussed by Roedder (1981e). French (1966) had shown that a C-H-O gas in equilibrium with graphite at 1000 bar and the magnetite-wüstite buffer will have 95 mole % CH$_4$ only at temperatures of <570°C. Although graphite does occur in some alkalic intrusive complexes, there is little if any evidence for graphite in these particular rocks. More important, numerous Soviet studies seemed to show that at least some of the hydrocarbon-rich inclusions in the nepheline of normal magmatic alkalic rocks (as opposed to late-stage pegmatitic nepheline) were primary[7] and that melt inclusions also found in the minerals of these rocks indicated crystallization temperatures for nepheline of 800-1180°C, and only rarely as low as 700°C. At these high temperatures, the equilibrium gas

----

[7] In later studies, Kogarko and Romanchev (1977) indicated that the hydrocarbon-rich inclusions they studied in Khibiny apatite and nepheline, except from late pegmatites, are clearly secondary.

assemblage cannot have such concentrations of CH_4.

To avoid this paradox, it might be suggested that a less methane-rich gas was trapped in the nepheline at high temperatures and shifted toward a more methane-rich composition on cooling, according to the equilibria calculated by French (1966). However, once a gas inclusion is trapped in a crystal bottle of nepheline, which has almost no variable-valence elements in its structure, the elemental ratio of the gas is fixed. Hydrogen might possibly diffuse through the nepheline,[8] but without an oxygen sink, the concentration of total oxygen present in the inclusion must remain the same as that of the gas originally trapped. However, the gas in the inclusions was essentially CH_4 and had a very low total oxygen content. The interpretation of Konnerup-Madsen et al. above, involving immiscible separation and hence concentration of CH_4, avoids this difficulty by assuming that the inclusions are either secondary in main-stage minerals or primary in late-stage minerals, thus permitting the trapping of the CH_4-rich inclusions at the lower temperatures required. Those inclusions trapped at higher temperatures have considerably less CH_4 and more CO_2; some primary inclusions in very early phases of alkalic magmatism are relatively pure CO_2. Gerlach (1980) and Konnerup-Madsen et al. (1981) have attempted to relate the gas compositions found to the available thermodynamic evidence on equilibrium gas assemblages. The problem is not simple, as only estimates are available for such important parameters as temperature, total pressure [and $P(H_2O)$], fO_2 buffering by the mineral assemblage (e.g., the presence of graphite is sometimes assumed), and particularly the C-H-O ratios. Since bulk inclusion analyses must be used, several generations of inclusions may well be sampled, together with gases released from the mineral structures themselves. Konnerup-Madsen et al. (1981) reported 12 analyses of the gases obtained on vacuum ball milling of sodalite, nepheline, arfvedsonite, eudialyte, and chkalovite from Ilímaussaq. Presented on a water-free basis, these analyses show an average of 64 vol % CH_4 and only moderate variation, except in H_2 (3.5-44%). N_2, Ar, CO_2, He, H_2, CH_4, and seven other hydrocarbon gases were analyzed. The overall composition of these gases was similar to that of the gases from the Khibiny and Lovozero massifs. Konnerup-Madsen and Rose-Hansen (1982) also analyzed gases from nearby granites and quartz syenites and found them to be characterized by CO_2 and only minor CH_4. They suggested that the difference is a result of several features of the alkalic intrusions: low oxygen fugacities (between magnetite-wüstite and quartz-fayalite-magnetite), a wide temperature range of crystallization and a relatively low solidus temperature, and the retention of the volatiles in the melt during crystallization of the rocks.

In contrast with the origin proposed by Konnerup-Madsen, Gerlach (1980) has suggested that the CH_4 and bituminous matter have formed in these alkaline intrusives by subsolidus reactions between CO_2, H_2O and the rock during fracturing and recrystallization events on cooling, at low fO_2 values buffered by the rock assemblage.

Carbonatites

The origin of the carbonatites is controversial. Origins that have been proposed include ultrametamorphism of sedimentary carbonate rocks, carbonate metasomatism of preexisting rocks, and carbonate magmas. An examination of the fluid inclusions in carbonatites may be useful in understanding these rocks, and may well have more than academic interest. The Palabora carbonatite of South Africa is one of the top 10 sources of copper in the world (Aldous and Rankin, 1979). A book on the geology of carbonatites (E.W. Heinrich, 1980) contains a

----

[8] Note, however, that the gases released on ball milling individual minerals from the Ilímaussaq rocks showed as much as 3.68 vol % He (Konnerup-Madsen et al., 1981). U is high in this intrusive body, but not in these minerals, so presumably this He represents primary gas, formed before trapping.

short review of inclusion studies on them.[9/] Part of the controversy over the origin of carbonatites may arise from the possibility that not all carbonatites formed in the same way. Certainly some connection must exist between kimberlites, carbonatites, and alkalic rocks, in view of the close spatial and temporal associations of these three otherwise relatively rare rock types, but the mechanism involved is in question (e.g., Le Bas, 1977; Wyllie, 1979a,b; Bailey, 1980; Bachinski and Scott, 1979). Studies of the fluid inclusions in carbonatites have revealed a wide range of temperatures, and compositions that include carbonate melts, various carbonate-silicate-salt-H_2O fluids, dense CO_2, and simple water solutions. The multiplicity of fluids found in each of several different occurrences lends credence to the concept that several processes were involved. Thus, in the Wet Mountains of Colorado, some carbonatite rock originates from replacement of silicate dike rocks, and some is primary magmatic (Armbrustmacher, 1979).

Melt inclusions. Inclusions of essentially carbonate melt have been reported in minerals from various carbonatites, but at every locality in which melt inclusions in both silicate rocks (various alkaline picrites, nephelinites, pyroxenites, melilites, etc.) and associated carbonatites have been studied, those in the carbonatites have lower Th. The differences are as much as 700°C in some localities (e.g., 1250 vs 550°C; Romanchev, 1972; Romanchev and Sokolov, 1979). Aldous and Rankin (1979) reported that primary inclusions in olivine from the earliest carbonatite facies (phoscorite) at Palabora, South Africa, are assemblages of calcite, dolomite, phlogopite, magnetite, and sulfides of Cu and Fe plus traces of Ni. These olivine crystals precipitated from a true carbonatitic magma.

Nesbitt and Kelly (1977; see also Metzger et al., 1977) studied primary inclusions in several minerals from the carbonatite at Magnet Cove, Arkansas. Inclusions in apatite were most abundant (<17 vol % inclusions!), and although some contained large amounts of solids, the phase ratios between solids, liquid, and gas (CO_2) were highly variable. Melt inclusions in monticellite were much more systematic. The major solid phase was calcite, together with lesser amounts of larnite(?) or merwinite(?) and still smaller amounts of apatite, diopside, and magnetite (or magnesioferrite). Water solution plus a small bubble (CO_2-rich?) made up <50 % of the inclusion volume. The bulk composition of the fluid trapped was estimated to be (wt %): CaO 49.7, MgO 1.0, SiO_2 15.7, P_2O_5 1.1, H_2O 11.4, CO_2 16.7, Fe oxides 4.4. Heating for 1 week at 800°C had little effect on these inclusions. Similar inclusions, but obviously in secondary planes, cut through some monticellite crystals but did not extend into the surrounding calcite. This feature and several others suggest a complex history.

Puzanov and Partsevskiy (1978) and Puzanov et al. (1978) reported on melt inclusions in apatite and anhydrite from the Seligdar apatite deposit in the Central Aldan, USSR. This deposit is associated with a carbonatite(?). Melt inclusions in the apatite, containing calcite, hematite, and probably sodium and potassium carbonates, homogenize at high temperatures (>1000°C). Inclusions in anhydrite homogenize between 795 and 830°C. Vozniak et al. (1981) also reported high Th values for melt inclusions in baddeleyite (ZrO_2) from carbonatites in the Sea-of-Azov area.

Water-rich and CO_2 inclusions. A wide range of compositions has been reported for multiphase inclusions from a series of carbonatites. They may contain high concentrations of various silicate, carbonate, and chloride daughter phases in a water solution ± CO_2. Others are simple gas/liquid aqueous inclusions, and (apparently) pure CO_2 inclusions. Calcite rarely contains good inclusions; most of the inclusions studied from carbonatites have been in apatite crystals.

[9/] Note, however, that this material is mainly in Appendices 1 and 2 of Heinrich's book, and hence is not covered in the index to the book.

In the Sokli carbonatite, Finland, Haapala (1980) reported primary poly-phase aqueous inclusions in apatite with 60-80 vol % daughter crystals. The daughter phases include carbonates (effervescence when dissolving after crushing in HCl-glycerol), and opaque minerals. The gas phase homogenizes first, at 250-350°C. The daughter crystals start to dissolve well below 500°C, and most dissolve between 500 and 550°C, but near 600°C the inclusions leak, before total homogenization. Simple aqueous liquid plus gas primary inclusions have Tm ice between -15.4 and -21.4°C, and Th 168-238°C. Liquid CO_2 was rare, but some gas pressure was found in the inclusions.

Rankin and Le Bas (1974a,b) and Rankin (1975) described the Wasaki carbona-tite in western Kenya, and Rankin (1977) described the Tororo carbonatite in eastern Uganda. Inclusions in these rocks showed <80 vol % daughter crystals, of which nahcolite ($NaHCO_3$) was the major phase, plus kalicinite ($KHCO_3$), pyrrhotite, halite, and unknowns. The nahcolite dissolved at 99-183°C, and the inclusions from Wasaki homogenized in the range 360-490°; some homogenized in the liquid phase, some in the vapor phase, and some with critical behavior. On further heating, as much as 5 vol % of a new liquid phase formed and then redis-solved at still higher temperatures (500-595°C). Rankin (1975) believes this phase is an Na_2CO_3-rich liquid. Solid inclusions of calcite were also present, as were primary silicate-melt inclusions. At the Tororo complex, the inclusions also contained nahcolite daughter crystals (~30 vol %), plus pyrrhotite and some unknowns. The daughters dissolved before homogenization, which was at 208-466°C. No silicate-melt inclusions or second liquid blebs were found.

At the Amba Dongar carbonatite in India, Roedder (1973) reported some daughter-crystal-rich inclusions in apatite crystals (~50 vol % solids, many birefringent), but most of the inclusions were simple H_2O liquid/vapor inclusions of either low or high but rarely intermediate density. Some contained a rather dense CO_2 phase as well (>0.5 g/cm^3). Fluorite from the large (~10^7 tonnes) associated fluorite deposit contained only very low salinity aqueous fluid inclu-sions containing a vapor bubble of <25 atm CO_2 pressure which homogenized at 115-150°C. Fluorite from the equally large fluorite deposit associated with the Okorusu carbonatite in South West Africa had similar aqueous inclusions with small highly birefringent daughter phases.

Interpretation. The fluid inclusions in carbonatites are evidence for a complex history involving fluids of various compositions. Some pairs of these fluids obviously represent fluids that were mutually immiscible at the time of trapping. The temporal sequence of the various fluids is difficult to establish at any given deposit. Although some carbonatites may have formed by replacement processes, the intrusion of a carbonate magma is the commonly accepted mode of origin for most carbonatites, and the report of volcanic carbonatitic lapilli by Keller (1981) leaves little doubt that calcitic carbonatite magmas exist. Hamilton et al. (1979) reviewed the three ways in which such a magma (that had correct major- and trace-element content) could form: partial melting of the upper mantle; fractional crystallization of a CO_2-rich alkaline silicate magma; and separation of an immiscible carbonate melt from an initially homogeneous CO_2-rich alkaline silicate magma. The possible range of composition of the silicate part of the pair has not always been explicit. Haggerty and McMahon (1979) presented evidence from natural occurrences that kimberlite was a possi-bility for the silicate part of the pair. Le Bas and Handley (1979) showed that apatite crystals in the early stages of alkaline-carbonatitic evolution have similar compositions but that these later diverge, suggesting that ijolite and sövite are the crystalline products of conjugate immiscible liquids. Treiman and Essene (1981) included the okaite, a melilite-nepheline-calcite rock, and the ijolites at Oka, Quebec, and other workers would include various types of nepheline syenite as possible products of immiscibility.

Studies of immiscibility in synthetic laboratory systems have been success-

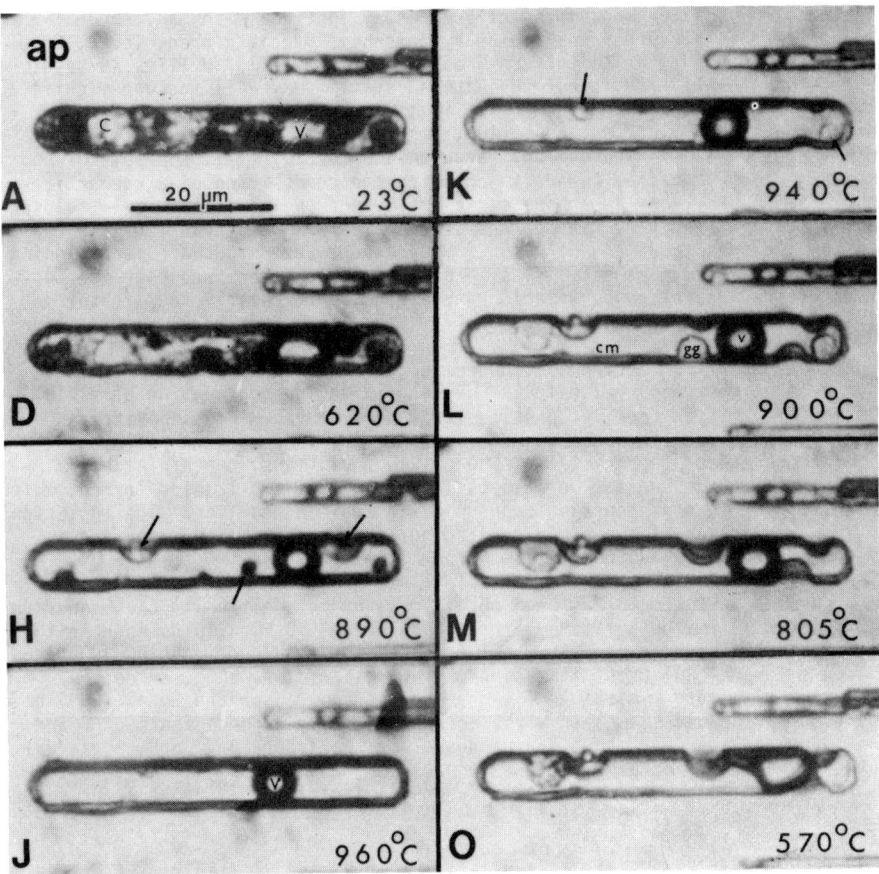

Figure 14-12. The behavior of a complex multisolid carbonate-rich silicate-melt inclusion in apatite from the Usaki carbonatite complex in West Kenya (Rankin, 1975; Rankin and Le Bas, 1974b), at the temperatures indicated (°C), during heating (A,D,H,J) and subsequent cooling (K,L,M,O). (A) At 23°C, the inclusion contains a large vapor bubble (V), colorless carbonate-rich solids (c), and numerous, small, crystalline specks of green, brown and black mineral phases. Heating to 422°C causes very little change, but at 565°C melting of the colorless carbonate-rich material started, and at 620°C (D) much was melted. Progressive heating to 850°C caused only slight solution of the remaining mineral phases (black, brown, and green specks), but above 850°C these specks dissolved noticeably, and at 890°C (H) pale green colored globules formed (some of which are arrowed). Thus, two immiscible liquids are present in the inclusion at this temperature. Continued heating caused these globules to dissolve in the carbonate melt. Complete solution took place at 960°C (J) at which temperature only one single homogeneous melt existed, together with vapor bubble (v), now smaller. This bubble failed to dissolve even at 1100°C.

On cooling from 960 to 940°C (K), green colored globules (some of which are arrowed) separated from the melt and grew in size, until at 900°C (L) two separate co-existing melts were present: green globules (gg) and colorless carbonate-rich melt (cm), plus a vapor bubble (v). On cooling to 805°C (M) no phase changes occurred except that some of the globules coalesced to form larger ones. Some small, black crystals formed in the green globules by 705°C, and by 570°C (O) some of the globules crystallized, but others remained liquid. The colorless carbonate-rich melt started crystallizing at 553°C. Additional crystallization was seen as low as 451°C, but below that temperature observation became difficult. Scale bar in μm.

ful but the results are not necessarily directly applicable to many natural systems.[10] Several carbonate-bearing systems have been explored, (e.g., Freestone and Hamilton, 1980; Koster van Groos and Wyllie, 1973; Wyllie, 1979a,b; and Boettcher et al., 1980), but the natural systems may well include significant amounts of Na_2O, K_2O, CaO, MgO, Al_2O_3, SiO_2, P_2O_5, CO_2, and H_2O, as well as lesser but possibly important FeO, TiO_2, Nb_2O_5, F, Cl, and S. In fact, three liquid phases have been proposed by Ferguson and Currie (1971) for some dikes at Callander Bay, Ontario, Canada -- one with the composition of a kaersutite-olivine-lamprophyre, one a carbonate-rich melt, and one equivalent to a feldspar-zeolite-rich syenite. The rare earths are particularly important in some carbonatites, and Wendlandt and Harrison (1979) have determined their partitioning between silicate and carbonate melts. They also have determined that the CO_2-rich vapor phase carries appreciable rare earths, particularly the lighter end of the series.

Perhaps the most pertinent evidence for silicate-carbonate-melt immiscibility is that observed by Rankin and Le Bas (1974b) during heating of fluid inclusions in apatite from carbonatites from West Kenya. These apatites showed silicate-melt (glass) inclusions and carbonate-rich inclusions containing 60 vol % carbonate daughter crystals. When mixed inclusions, containing both silicate melt and carbonate, were heated (Fig. 14-12), two liquids formed, which homogenized at higher temperatures and unmixed again on cooling. The inclusions also showed evidence of two other coexisting fluids -- CO_2-rich vapor and a carbonate-bearing aqueous phase (i.e., four fluid phases).

Once a carbonate melt has been obtained, a major question is the manner in which the various mixed silicate-carbonate-chloride-H_2O-CO_2 fluids seen in the inclusions could have formed from it. Are they the result of continuous processes such as fractional crystallization of the bulk of the $CaCO_3$, or do they arise via one or more immiscible fluid separations? Dawson and Fuge (1980) have shown that carbonatites contain variable but generally high contents of F and Cl. If the correct constituents are chosen for the laboratory work, a gradual transition is possible (e.g., Dernov-Pegarov and Malinin, 1976), but I believe that the inclusion evidence generally favors immiscibility -- probably at several stages.

As a working hypothesis, I suggest (Fig. 2-25) that immiscibility may well be involved between each of the following pairs, perhaps even in sequence, in at least some carbonatites:

Silicate melt (alkalic, high in CO_2)
    Immiscibility stage 1.
Carbonatite melt (essentially calcite composition, plus minor silicate, etc.)
    Immiscibility stage 2.
Alkali-carbonate-rich, H_2O-bearing melt (~50-80% solids)
    Immiscibility stage 3.
Water-rich fluid (low solute concentrations, minor CO_2)
    Immiscibility stage 4.
CO_2 fluid (moderate to high density).

This hypothesis does not require simultaneous immiscibility between all five of these fluids, and I presume some compositional evolution in each fluid (by crystal fractionation, rock interaction, etc.) before the next variety of immiscibility occurs. In some carbonatites, this compositional evolution might span the composition gap represented by immiscibility in other carbonatites.

----

[10] Some of these have been most directly applicable to a unique form of carbonatitic lava, the alkali carbonate lavas of Oldoinyo Lengai volcano, Tanzania, in which a natrocarbonatite melt containing Na_2O:CaO: K_2O ~ 4:2:1 separated from phonolitic or nephelinitic magmas (Hamilton et al., 1979).

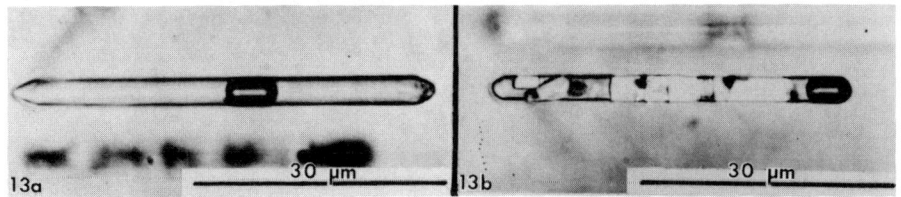

Figure 14-13. Photomicrographs of a pair of adjacent primary inclusions in apatite from the Sokli carbonatite in Finland (Haapala, 1980). In each photograph, the other inclusion is visible but out of focus. (a) An aqueous inclusion of low salinity; (b) an adjacent daughter-crystal-rich inclusion. Scale bar in µm.

The hypothesis also does not require that all steps in the sequence be taken in the sequence given; e.g., a CO_2 fluid might evolve at several stages. Immiscibility stages 1 and 2 were discussed above. The evidence for immiscibility stage 3 is seen in several carbonatites. The inclusion data that I reported (1973) from the Amba Dongar (India) carbonatite is suggestive of such immiscibility, but perhaps the best evidence is in the Sokli carbonatite in Finland, where Haapala (1980) showed the existence of two very different populations of primary inclusions in the same apatite crystals (Fig. 14-13). Haapala discussed several alternate explanations, but they seem unlikely. In view of the wide range of daughter crystals reported in the multiphase water-bearing fluid inclusions (e.g., see Fig. 5-15), including all the constituents listed above, plus SrO, BaO, SO_4, and CuO, experimental verification of the extent of immiscibility (and the P-V-T-X properties of the fluids) seems impossibly complex. The evidence for immiscibility stage 4 is found at Sokli, Wasaki, Amba Dongar, and a series of other carbonatites (see Indices in Roedder, 1968a).

Chapter 15

ORE DEPOSITION ENVIRONMENTS

"Just as with any other exploration tool, even an occasional success [based on inclusion study] can justify the effort. But one should always keep in mind that lots of ore has been found by wrong ideas and even wrong data, so Caveat emptor!" (Roedder, 1977b)

CONTENTS

INTRODUCTION

The study of the environment of ore deposition as evidenced by fluid inclusions started as a purely academic pursuit, but in a relatively few decades it has evolved into an aid to mineral exploration, in part because the better we know how ore deposits form, the better we know where to look for them. The validity of this statement is evidenced by the numbers of commercial enterprises that have funded and staffed their own private fluid-inclusion laboratories. In attempting to understand the nature of the ore fluids from a study of the inclusions present in ore bodies, one should never forget that these fluids can change drastically prior to trapping, as a result of processes within the ore deposit. Hence, the specific environment of origin of the inclusion is important, as discussed in Chapter 2.

In this chapter, I present a few examples[1] from each of the major types of

----

[1] An indication of how selective these examples must be: each yearly issue of Fluid Inclusion Research--Proceedings of COFFI (Roedder, 1968a) lists 200-400 studies of fluid inclusions from ore deposits.

ore-deposition environments that have been studied, and then summarize the ways in which such fluid-inclusion data can be used as a tool in mineral exploration. Several types of ore deposit are more conveniently discussed in other chapters (as indicated): Au-bearing conglomerates, saline, and S deposits (11); gold quartz veins in metamorphic rocks (5 and 12); Nb, Ta, and Cu in carbonatites and Li, Be, etc., in pegmatites (14); metalliferous sea-floor vents (16); and diamonds (17). Not all geologists will agree with the assignments of given ore deposits to the various categories used below, but such assignments are not critical to the significance of the data reported. As much more extensive fluid-inclusion studies have been made on Mississippi Valley, epithermal vein, porphyry Cu, and Sn-W deposits than on other types of ore deposits, these four types will be emphasized.

MISSISSIPPI VALLEY-TYPE DEPOSITS[2]/

The Differences

"The origin" of the Mississippi Valley type of ore deposit, using the type term in its most general meaning, has been the subject of an extensive and occasionally heated discussion, only a small part of which stems from problems of exactly what types of deposits are to be included. The range of deposits covered is well illustrated in the volume stemming from a symposium on the genesis of stratiform Pb-Zn-barite-fluorite deposits, held at the United Nations in New York in 1966 (Brown, 1967). In an excellent review and sequel to this volume, Brown (1970, p. 104) summarized 11 distinctive characteristics of these deposits, as they are found in the three main districts in the Mississippi Valley proper.

Other apparently similar deposits in the United States and throughout the world differ most drastically from these characteristics in terms of the isotopic composition of the Pb. Brown therefore suggested (1970, p. 117) that the broad class of Mississippi Valley-type deposits might be divided into three categories, based on the presence of normal type, B-type, and J-type Pb. These three leads yield model ages that are, respectively, approximately correct, "too old," and "too young."

These differences are particularly significant in discussions of "the origin" of these deposits, in that they are one of the most obvious indications that there can hardly be only one source of Pb and one mechanism of origin. Indeed, one of the most important causes for differences of opinion as to genesis is that a multiplicity of mechanisms must have operated in the various specific deposits. One does not need to go to isotopic studies to see this, however; the tremendous range in gross composition of the individual deposits -- from essentially pure Pb with only traces of Zn, F, and Ba, to essentially pure Zn, to mixtures containing major amounts of all four (and other) elements -- suggests significant differences in the chemistry and perhaps in the mechanism(s) operating.

Most ore deposits result from one or more processes in which an ore element is dissolved from a (presumably) dispersed or dilute source area, transported to the future site of the ore body, and deposited there in a relatively concentrated form. Thus, Pb ores have several thousand times as much Pb as ordinary rocks. The many theories of origin of the Mississippi Valley-type ores differ in part on the source or sources chosen for the ore elements and on the source and original valence state of the S. The term "ore element" used here includes not only the obvious metals Pb and Zn (and to a minor extent, Cu) but also the F and Ba that occur in even greater amounts. Thus, the production of the Illinois-

----

2/ Taken, in part, from Roedder (1976a).

Kentucky district, USA, reported by Grogan and Bradbury (1968), is Pb, 60,000; Zn, 200,000; and CaF_2, 10,000,000 tonnes (ratio 1:3:170). Other districts are equally biased but toward high barite or high Pb.

The Similarities

Inclusions have yielded at least some data bearing on each of the following features of Mississippi Valley ore fluids: density, rate and direction of movement, pressure, temperature, gross salinity, pH, noncondensable gases, isotopic composition, and solute composition. The fluid-inclusion data from a large number of Mississippi Valley-type ore deposits all over the world are remarkably uniform, in view of the differences mentioned in the previous section. The data are consistent even though the deposits cover a wide range of elemental composition (e.g., from almost pure Pb at Laisvall, Sweden, and southeastern Missouri, USA, to almost pure Zn at Friedensville, Pennsylvania, USA, to almost pure fluorite in some deposits in southern Illinois, USA); isotopic makeup (normal Pb, B-type Pb, or J-type Pb); structural setting; geologic age; host-rock type; etc. All deposits studied, from many places in the world and with almost no exceptions, show that the ore-forming fluids had the following characteristics (summarized from 52 entries in Table IV in Roedder, 1976a; see also Roedder, 1979c, footnote 2, for 10 more recent references).

Density -- Always >1.0 and frequently >1.1 g/cm^3 at the time of trapping, so they are always more dense than surface waters.

Rate of fluid movement -- Very slow, perhaps in the range of a few m/yr.

Total pressure -- Presumably low but always greater than the vapor pressure of the brines (i.e., no boiling has occurred), although gases may have been in solution at pressures as much as 20 bars.

Temperature -- Generally 100-150°C, seldom as high as 200°C. (Some of the highest reported numbers, 280°C at East Tennessee, USA (Miller, 1968, 1969) and 223°C at Laisvall, Sweden (Roedder, 1968d), were later shown to be invalid (Larsen et al., 1973).) Late calcite commonly has Th <100°C.

Gross salinity -- Usually >15 wt % NaCl equiv. and frequently >20%, yet daughter crystals of NaCl are almost never found, implying appreciable amounts of cations other than Na (see Chapter 8).

Organic matter -- Frequently but not always observed, as gases such as CH_4 in the vapor bubble and in solution, as immiscible oil-like droplets, and as various other compounds in solution in the brines.

Solutes (i.e., "salts") in solution -- Highly concentrated solution of mainly Na and Ca chlorides, with very minor K, Mg, B, and low or extremely low in all S species (Rosasco et al., 1975b). Heavy metals (Zn and Cu) may be high (Czamanske et al., 1963). The relative abundance by weight percent of the major fluid inclusion ions is Cl>Na>Ca>>K>Mg>B in almost all Mississippi Valley-type deposits.

Although generally considered to be variants on the normal Mississippi Valley deposits, the barite deposits sometimes associated with them are frequently quite different in terms of fluid-inclusion data. Necking down and leakage are common problems with barite, but in most deposits, those few inclusions that do look valid generally have much lower salinities and/or Th than the sphalerite or fluorite in nearby possibly related mineral deposits (Roedder, 1979b,c).

Table 15-1. Proposed sources for the ore elements and for the fluids responsible for their dissolution, transport, and deposition in Mississippi Valley-type deposits

Ore elements

1. Seawater (hence originally from weathering of rocks)
2. Leached from sedimentary rocks or mud in place
3. Expelled from sedimentary minerals, particularly from shale, during diagenesis
4. Expelled from recrystallizing metamorphic minerals
5. Leached from metamorphic rocks in place
6. Leached from igneous rocks in place
7. Expelled from crystallizing magma
8. Volcanic exhalations (via fumaroles, hot springs, etc.)

Fluids

1. Meteoric (surface) waters (± deep circulation and heating to form "hydatogenic" fluids)
2. Seawater (± evaporation to cause enrichment in salts)
3. Connate and compaction fluids in sediments (± dissolved evaporites and possibly modified by osmotic processes)
4. Metamorphic waters, expelled during dehydration reactions
5. Magmatic waters, evolved from crystallizing magma ("magmatic-hydrothermal" or "telethermal")

Discussion of the Inclusion Evidence

The quality and quantity of inclusion data available on the various deposits vary greatly as a result of the gross differences in the availability of suitable inclusions to study. The material from some deposits, such as Friedensville, Pennsylvania, USA, and the Southeast Missouri Lead Belt, USA, provided almost no usable inclusions, whereas that from other deposits such as southern Illinois provided large (~1 mm) primary inclusions in abundance, and Rankin and Greenaway (1978) presented a series of photomicrographs of huge primary inclusions in British fluorites, some containing >1 ml fluid.

Nature and source of the ore fluids. Central to any understanding of the processes of dissolution at the source, transport, and redeposition of the ore elements is the nature and origin of the responsible fluid -- the ore-forming fluid. A very important fact is that so long as these fluids can move through distances measured in kilometers during geologic time (e.g., 1 km equals only 300 years flow at 1 μm/sec), the source need not be enriched. As a good approximation, each part per million of an element in a rock corresponds to about 10,000 metric tonnes of that element per cubic mile of rock (or 2400 metric tonnes/km^3). Thus, the extraction process can remove enough metal to make a big ore deposit from geologically reasonable amounts of very dilute source materials; geology kindly provides such vast amounts of materials and time that the extraction (and concentration) processes can be (and probably are) exceedingly slow and inefficient. Some of the factors in the hydrodynamics and geochemistry of these fluids are discussed by Kelly et al. (1983a).

A variety of sources has been suggested for the ore elements and for the ore fluids forming the Mississippi Valley-type deposits (Table 15-1). Also, some theories of origin invoke combinations of several sources for the fluids, i.e., mixing. This mixing mechanism, involving separate sources for metal- and S-bearing fluids, was first detailed, for the Pine Point, Canada, deposits, by Beales and Jackson (1966; see also Jackson and Beales, 1967). Hoagland (1973) has since applied it to the Central Tennessee deposit, USA, and Zimmerman and Kesler (1981) have done the same for the Sweetwater district in Tennessee, USA. Certain features of some deposits are much more compatible with a single ore fluid than with mixing (McLimans et al., 1980). Note also that the divisions in Table 15-1 are rather indistinct; there may well be a continuum between sea, connate, compaction, and metamorphic water, and several varieties may be combined in the water involved in any given deposit.

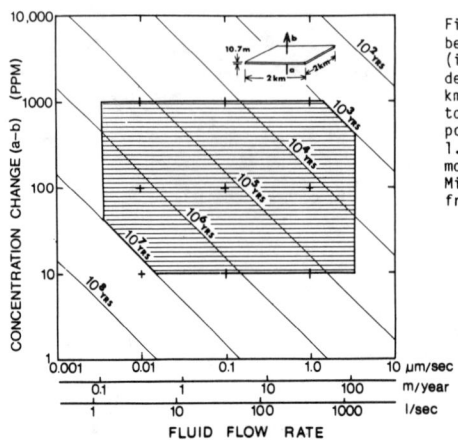

Figure 15-1. Diagram showing the interrelations between time, flow rate, and concentration change (incoming minus outgoing fluid) needed for the deposition of 10^8 tons of 10% ore in a stratum 4 km^2 in area and 10.7 m thick. The rock is assumed to have a bulk density of 2.34 g/cm^3 and a bulk porosity of 10 vol %, and the liquid, a density of 1.0 g/cm^3. The shaded area is the geologically most probable range of conditions under which most Mississippi Valley-type deposits formed. Adapted from Roedder (1960a).

Obviously the ore itself must be the product of a much larger volume and weight of ore fluid; the ore fluid, therefore, must not only be transported from the site of dissolution to the ore body, but many successive volumes must pass through the site of the ore body. For a given amount of ore, the volume of this ore fluid will bear a simple inverse relationship to the concentration change in the ore element (input minus output). Thus, if this concentration change is only 1 ppm, 1 million times more fluid than ore is required, and at 100 ppm, 10,000 times more fluid. These volumes may seem large, but when examined in geologic terms they are not unreasonable. Even when geologically very small segments of time are assumed for ore deposition (Roedder, 1960a), the flow rates required are low (<1 µm/sec) and the quantity of flow is also reasonable (e.g., 38 liters/min or 10 gallons/min would be adequate to form the main ore body at Pine Point, Canada in 10^5 years; Roedder, 1968c, p. 447).

In Figure 15-1, the interrelations of the three variables, concentration change, flow rate, and deposition time, are illustrated graphically on a log-log plot for a very large hypothetical tabular deposit of 10^8 tonnes of 10% ore (Pb, Zn, or any other element). For simplicity, the model has been set up for flow perpendicular to the beds; in nature, most of the flow is probably parallel with the beds, but the numerical results will be identical.

The central shaded area on this diagram represents what I believe to be the geologically most reasonable range of these variables for the formation of most Mississippi Valley-type ore deposits. The right- and left-hand limits of this area are based on arbitrary guesses as to the maximum and minimum flow rates to be expected in such sedimentary terranes. The lower and upper limits of 10- and 1000-ppm concentration change (i.e., the amount of precipitation) are based on data from fluid inclusions (Roedder, 1960, 1972) and brine analyses (White, 1967; Carpenter et al., 1974). The diagonal time limits are chosen somewhat arbitrarily at 1000 and 10 m.y. on the basis of several lines of evidence (Roedder, 1976a). Other data may result in these boundaries being moved one way or the other, but it seems unlikely that they can be moved very far. Thus, the single neutron-activation analysis for Zn in inclusions in fluorite from southern Illinois giving the highest heavy-metal content ever reported from inclusions (Czamanske et al., 1963) would move the upper boundary to 10,000 ppm at a maximum, assuming complete precipitation of all heavy metals. (Note, however, that this analysis was for metal that had not precipitated.) Similarly, White (1971) has suggested that the ore fluids at White Pine, Michigan, USA, flowed for 10^6-10^8 years. Regardless of the position chosen for the boundaries of the shaded area, the diagram gives the quantitative relationships between the variables.

The final unknown in the process is the actual cause for deposition. Here also, many possibilities have been proposed, including mixing of waters, changes in T or P, loss of gases, reaction with country rock, internal changes within the liquid itself (e.g., sulfate being reduced slowly by organic compounds in solution), and biological activity.

The possible combinations of these variables in the construction of a theory of origin for a given ore body are almost infinite, and the evidence to select or discard any given concept or model is frequently nebulous, almost always ambiguous, and sometimes completely lacking. The various data from fluid inclusions alone cannot determine the origin or origins of stratiform ores, but regardless of this origin, the data place severe limitations on the mechanisms of ore deposition that could have been operative. Any theory of origin must be compatible with the inclusion data, or the data must be refuted, if the theory is to stand.

Relation of inclusions to the ore fluids. Brown (1970) pointed out that although fluid-inclusion evidence on Mississippi Valley deposits is widely quoted and is generally believed in North America, many students of this type of deposit in Western Europe, particularly in France and Germany, tend to ignore the inclusion data, under the assumption that it gives no information on the original ore fluids. The common assumption there has been that the inclusions were formed during later reworking of the deposits. On the other hand, if the fluid inclusions provide true data on the ore fluids, they preclude several of the low-temperature and syngenetic concepts of origin most generally held in Europe. Part of this problem may arise from different usage of terminology by various researchers, but a major difference remains between the paleokarst, sabkha, lagoon, and similar environments at essentially surface temperatures (although perhaps sunheated) commonly proposed in Europe and the 150°C or higher reported for inclusions from many Mississippi Valley-type deposits. Although rarely discussed in any detail in the literature, the crux of the "reworking" problem lies in exactly which reincarnation of the ore deposit one is referring to as the origin. If an ore body is formed from fluids at one set of P-T-X conditions (and presumably contains inclusions representative of such conditions), and then is completely reworked (i.e., all crystals are dissolved and reprecipitated via a different fluid medium), then the new inclusion data will give only the P-T-X parameters of the reworking process (Bernard, 1973, p. 53-55). The necessary assumption here is that primary inclusions reveal the conditions of formation of the host crystal where we see it now, and obviously can tell us nothing of any previous deposit. Recognition of the possibility of such former deposits would then become a field problem rather than an inclusion problem.

In this connection, Bernard (1973, p. 54) has made the point that fluid-inclusion specialists sample only the late "recrystallized" idiomorphic ore minerals. As a result, they get P-T-X data corresponding only to the fluids present during the last "recrystallization" of the ores, under conditions of deep burial, whereas the original fine-grained ores formed from fresh waters under near-surface conditions. Good primary inclusions are larger and easier to find, and more photogenic, in the coarser crystals from any deposit; hence, these have received the most attention. The new crystal containing the inclusions may have formed by growth at the expense of immediately adjacent crystals via an intergranular pore-fluid (i.e., typical recrystallization) or by direct free crystallization from a moving mass of liquid that had previously dissolved other crystals at some distance. The fuzzy line between "recrystallization" and "reworking" (or "remobilization") depends upon the interpretation of this distance. Regardless of the distance moved, the inclusions in the new crystals represent samples of the fluid from which the present host crystal actually formed. If any fluid inclusions containing earlier fluids were present before, they would be effectively dispersed and lost by either process.

I believe that the recrystallization argument is refuted by the inclusion evidence itself, as well as by the field data. Many examples in the literature on inclusions verify that the tiny inclusions in very fine grained ores have salinities and Th as high as those of the coarse crystals (e.g., Roedder, 1967d, 1968c,d, 1971d). Also, the very tiny inclusions found in fine-grained ores, although frequently too small to use for homogenization studies, still show gas/liquid ratios visually similar to inclusions in coarse-grained ores; only the very tiniest inclusions may remain single phase, and these can generally be shown to be metastable (Roedder, 1971a). If the hot saline inclusions are a result of later reworking, every sample, from every Mississippi Valley-type deposit studied, would have to have been recrystallized in the presence of hot saline fluids; this would seem to require a very unlikely coincidence. As shown in Chapter 11, minerals formed near the surface do trap inclusions, but I know of no data to show that any primary inclusions in any Mississippi Valley-type deposit were either filled with fresh water or were formed at surface temperatures. The only possible exceptions are some barite deposits that may well have been reworked by surface waters.

Rate of flow of ore fluids. Except for possible surges from large-scale solution collapse of the wall rock (as in eastern Tennessee, USA, Roedder, 1967d, p. 353-353), I believe that the ore fluids have moved very slowly at the site of deposition. Several lines of evidence support this concept, the first two of which are based on metastable equilibria in inclusions:

(1) Freezing studies of fluid inclusions in Mississippi Valley ore and gangue minerals generally report gross supercooling on the microscope freezing stage. This is taken to signify a complete freedom from such solid nuclei in suspension as are normally present in surface waters; these nuclei in surface waters usually preclude more than ~10°C supercooling, whereas inclusions in Mississippi Valley-type samples (like most inclusions in ore deposits) require supercooling of 20-40°C below their equilibrium freezing temperature before they will freeze.

(2) Metastable superheated ice is commonly encountered in freezing studies on inclusions in Mississippi Valley ore samples. As in item 1, this metastability requires that the inclusion liquid be exceptionally clean and free of suspended solid nuclei.

(3) Crystals of ore and gangue minerals growing in open fractures or vugs may contain crystals of other phases that nucleated and grew simultaneously, but they are very clean and free of inclusions of clay or other debris that might have been carried in suspension, in contrast to the more clearly dynamic systems responsible for some magmatic-hydrothermal veins (Barton et al., 1971).

(4) The inclusions and other microscopic features reveal exceedingly minute regular oscillatory growth bands in several Mississippi Valley ores and related types of occurrences. If these bands are truly annual varves, as suggested (Roedder, 1968c, e; 1969; Leach, 1973), they indicate crystal-growth rates in the range of only 10 μm/year. This, in turn, suggests very quiet conditions, fluid flows being in the range of perhaps 1 μm/sec over tens of thousands of years, unless the concentration change (amount precipitated) is very low (<1 ppm, see Fig. 15-1).

Although they do not involve inclusions, four other features, described by Roedder (1976a, p. 89), also suggest slow flow rates.

Temperature of formation. Many serious problems bedevil the use of Th for geological thermometry, particularly for deposits formed at high temperature and pressure, but most of these problems are minor for samples from low-temperature stratiform deposits. Within a given deposit, the temperatures for various paragenetic stages show little variation (Cunningham and Heyl, 1980), and apparently large volumes of rock were at thermal equilibrium (Leach et al., 1983). In addition, the temperature corrections needed for the effects of hydrostatic pressure are small and fairly well known (Chapter 8), so I believe that the values obtained on these deposits are probably the most precise and accurate in the field of geologic thermometry. This does not mean that erroneous data cannot be obtained. For example, Sabouraud et al. (1980) believed that extensive decrepitation of low-temperature inclusions (plus refilling at higher temperatures) took place in some Pb-Zn deposits. However, Touray and Ziserman (1981) have disagreed with this interpretation. Other caveats stem from post-trapping generation of CH_4 within some inclusions, increasing Th (Hanor, 1980), and from overheating of inclusions (Coveny and Goebel, 1983).

A necessary corollary of the slow flow and deposition rates and the large volumes of liquid passing any given point in the ore is that a considerable volume of rock in the vicinity of the ore body must have been heated to the temperatures recorded in the inclusions. The heat capacity of water is so high that the fluid from five complete changes of the water in the pores of a rock with only 10% porosity has a total heat capacity greater than that of the rock. As such, any temperature differences will soon be levelled and the temperature of the fluids will effectively control the temperature of the rock; hence, the inclusion temperatures cannot be considered to represent merely some local late-stage recrystallization environment, as some have argued.

pH of the ore fluids. An extensive literature in Russian concerns procedures to determine inclusion pH (Chapter 8; Roedder, 1972, p. JJ39), but only one modern attempt has been made in the Western world (Erickson, 1965; pH about 7.5, using pH-sensitive papers, on samples from Upper Mississippi Valley Pb-Zn deposits, USA). Another procedure for determination of inclusion pH is that of crushing the host mineral in water and measuring the pH of the slurry. Hundreds of such measurements will be found in the Russian inclusion literature (Roedder, 1968a), but for several reasons, in particular the effects of various mineral surfaces, loss of gases, and the several-thousandfold dilution, I believe that they are practically worthless (Roedder, 1972, p. JJ38-JJ41; see also p. 136).

Noncondensable gases and organic matter. Although very little information is available on noncondensable gases (i.e., not condensed by 1 atm at room temperature) and organic matter, these materials could be of considerable import to problems of the chemistry of ore deposition. Thus, unless the mixing of two fluids is assumed to take place (and there are problems with that hypothesis too), one is faced with the difficult task of simultaneous transport of reduced S and metals in a given fluid. Organic matter should, in general, reduce sulfate to sulfide, although these reactions are very slow. (Ohmoto and Lasaga (1982) showed that inorganic reductions of sulfate appear to be fast enough to become geochemically important at temperatures above about 200°C). Barton (1967) has proposed that the Mississippi Valley-type ore metals have been transported by a fluid bearing organic matter and containing S as sulfate; slow reduction of the sulfate in this nonequilibrium solution avoids the problems of simultaneous transport of sulfide and metal ions in the same fluid. The solubility of CH_4 alone in 50,000 ppm brines is adequate to provide, at saturation under pressures equivalent to a depth of only 300 m (Hanor, 1980), reducing capacity to form about 2 g S^{2-}/liter of brine. In addition, many subsurface waters are known to be saturated with respect to CH_4, and studies using the crushing stage reveal appreciable gas in solution, presumably mainly CH_4, in the inclusion fluids (perhaps ~800 ppm; Roedder, 1967d). In contrast, magmatic hydrothermal fluids also contain CH_4, but usually in much smaller amounts, which could well be from simple inorganic equilibrium reactions in the system C-H-O (Roedder, 1972). In addition to the obvious CH_4, a surprisingly high percentage of Mississippi Valley-type ores contain small amounts of yellow-brown, fluorescent oil, which presumably comes from the sediments traversed, as well as a wide variety of other organic compounds (Gize and Hoering, 1980).

Organic materials in solution may play several roles in the origin of Mississippi Valley ores. In addition to the reducing capacity of the CH_4 mentioned above, the liquid-hydrocarbon phase, present as immiscible globules in the brine and as rather surprising amounts in solution (Price, 1976), would have been a continuing internal source for reductants of sulfate to form sulfide. (It might also serve as the source for the S, as some such organic material is high in S.) In some deposits, some semisolid organic debris may also have been present in the solutions (e.g., Hansonburg, New Mexico, USA; Roedder et al., 1968, their Figs. 9-11), but one must remember that the most effective reductants may have constituted a very minor fraction of the total organic matter, and as a result of being effective reductants, they may well be used up and hence not found on analysis. Many different organic compounds may have been present. Thus, the oxalate ion, which is a commonly used reductant in aqueous oxidation-reduction titrations, was present in at least some connate brines, as the oxalate mineral whewellite $[CaC_2O_4 \cdot H_2O]$ has been recognized in several Mississippi Valley-type occurrences. Miller et al. (1972) found fatty acids in bedded barite that they believe are from sulfate-reducing bacteria. (Such bacterial action would also explain the common occurrence of free H_2S in the inclusions in this "fetid" barite.)

Organic matter may have an additional role to play in providing complexing agents for heavy metals. Giordano and Barnes (1981) reported that the salicylate anion may be an effective ligand for Pb, and Veitch and McLeroy (1972) found that amino acids in solution greatly increased the solubility of heavy metals in carbonate environments.

Hanor (1979) has pointed out that the presence of a separate CH_4 phase may be instrumental in precipitating the ore. Dissolved CO_2 in the fluid would partition into such a CH_4 phase and hence raise the solution pH.

Isotopic ratios. Roedder et al. (1963) and Hall and Friedman (1963) measured $\overline{D/H}$ on inclusion samples from the southern Illinois fluorite-Pb-Zn district, and the Upper Mississippi Valley Pb-Zn district, USA. The D/H ratios (and the concentration and gross chemical composition of the salts present) for the main ore stage are all similar to present connate waters in those formations, but some late minerals were deposited from fluids of different composition.

The original isotopic signature of the water (e.g., sea or meteoric) can be changed by isotopic exchange with the rocks through which the water percolates only if the exchange is sufficiently rapid and if there is a significant reservoir present. Oxygen will exchange in geologic environments at temperatures as low as 150°C (Clayton et al., 1968; Pinckney and Rye, 1972), and rocks contain a large reservoir of oxygen. Hydrogen is probably exchanged more readily, but the reservoir of H in the rocks is so small that it will be controlled by the water composition, rather than the reverse. Thus, even a normal shale contains less H than would be present in the water filling its pores, but it contains about seven times as much O as the water. Several changes of water could thus exchange the H of the rock completely, but far more changes would be needed for the O. No O-isotopic determinations have yet been reported for Mississippi Valley-type inclusion liquids, but as the improvements in this technique have been rapid, more work of this kind may be expected. Although it would be of considerable interest, the isotopic composition of the S in the inclusion fluids is not within reach of present experimental methods, because of the low concentrations and available sample size.

Kessen et al. (1981) showed that the $^{87}Sr/^{86}Sr$ ratio of gangue minerals is higher than that of the host carbonate beds, but that the Sr probably came from silicate minerals in the sedimentary succession. No evidence was found for a magmatic Sr component.

Table 15-2 Typical fluid-inclusion analysis: Sphalerite, Tri-State
District, Oklahoma (Roedder, 1967b)

	Parts per million	Moles per liter
Na^+	57,100	2.47
K^+	2,700	0.07
Ca^{2+}	18,000	0.45
Mg^{2+}	2,400	0.10
Cl^-	124,600	3.51
SO_4^{2-}	<3,300	<0.03
$B_4O_7^{2-}$	107	0.0007
HCO_3^-	?	?

Total salts	208,000 (does not include heavy metals)
H_2O	792,000 (probably includes approximately 800 ppm CH_4)

Overall chemical composition. The solute composition is surprisingly uniform among the various districts examined and consists mainly of Na and Ca
chlorides. Table 15-2 shows a typical analysis. Almost without exception, such
analyses show the descending weight-percent sequence Cl-Na-Ca-K-Mg-B. Bicarbonate is probably low. Total S, stated as sulfate, seldom exceeds a few thousand
parts per million. The value given for total S as sulfate in Table 15-2 is a
maximum, because an appreciable but unknown amount of S was contributed by oxidation of the sphalerite during the laboratory leaching. Unfortunately, the
determination of microgram quantities of S is one of the most difficult analytical problems and at present does not permit distinction between the various
valence states. Some evidence, however, indicates that both sulfide and sulfate
S must be very low. Laser Raman spectroscopy of a large inclusion in fluorite
from the southern Illinois deposits (Rosasco et al., 1975b) revealed no detectable
HS^- or SO_4^{-2}. Most significant, however, is the concentration of salts. The
freezing data indicating very strong brines have been corroborated by those few
analyses in which actual concentrations were determined (rather than just ratios). Analyses of this type have been reported for southern Illinois, USA; Tri-
State district (Kansas-Missouri-Oklahoma, USA); Santander, Spain; and the Upper
Mississippi Valley, USA. Some studies (e.g., Hall and Friedman, 1963) showed
that systematic changes occurred during ore formation. Additional studies of
this sort are needed on other deposits, but most deposits simply do not provide
suitable sample material.

A single analysis, by neutron activation, of an inclusion in fluorite from
southern Illinois, USA (Czamanske et al., 1963) showed unexpectedly high heavy-
metal contents. The fluid contained approximately 1% each of Cu and Zn, and
0.4% Mn. Pinckney and Haffty (1970) analyzed some other samples from the same
locality by atomic-absorption spectroscopy and found much lower concentrations
(maxima in ppm: Zn, 1040; Cu, 350; Cl, 152,000); the cause for the difference
is not known, and it should be resolved.

As Na and K are among the most abundant constituents present in inclusions
and may be determined by flame photometry with relative ease, precision, and
accuracy, the Na/K ratio is one of the most useful parameters. The fluids that
formed the Mississippi Valley-type ore deposits seem to be characterized by much
higher Na/K ratios than are those having magmatic-hydrothermal affiliations.
(Na/Ca and Na/Cl are also higher.) Even though the Mississippi Valley fluids
have very high Na/K ratios and are very similar in many respects to "normal"
connate and oil-field waters, the most striking difference is that the inclusion
fluids have lower Na/K ratios than the lowest ratio reported in oil-field waters
(Roedder et al., 1963; Roedder, 1979b). Oil-field brines normally show Mg >> K,
but in these inclusions, K > Mg. Hall and Friedman (1963) suggested that the
extra K may represent a magmatic contribution. Land and Milliken (1981) have
shown that dissolution or albitization of K-feldspar in sediments is widespread

so perhaps the differences in fluid compositions merely represent different degrees of fluid/rock equilibration.

With few exceptions, Ca exceeds Mg, frequently by a large factor. Although little attention has been given it, the Ca/Mg ratio in the ore-forming fluid controls (or is controlled by) dolomitization. Care is needed to obtain valid determinations of this ratio in inclusions, however, because of contamination from embedded carbonate crystals exposed during crushing. Mississippi Valley-type ores formed from fluids with a rather uniform Ca/Mg weight ratio between 4 and 8, but other types of deposits deviate widely at both ends of this range. Geronsin (1980) analyzed water leachates from galena from a series of Mississippi Valley-type deposits and compared the results in terms of various cation ratios. Haynes et al. (1983) reported large differences in cation ratios determined by SEM/EDA on decrepitation residues from samples from East Tennessee, USA.

Density of the ore fluids. Most Mississippi Valley-type fluids have densities very close to or slightly >1.0 g/cm^3, whereas the magmatic fluids, with a few notable exceptions, are significantly <1.0. In this connection, Hanor (1973) has shown that subsurface brines in many areas have in situ densities for which the increasing salinity and temperature with depth just compensate, to yield a gravitationally stable column.

Conclusions concerning Mississippi Valley-type ore deposits. When all the surprisingly uniform inclusion data are considered -- the high total concentrations of salts, the composition of these salts, the presence of oil and CH_4, and the low temperatures of formation -- together with the numerous other distinctive structural and mineralogical features of this type of deposit (Ohle, 1959), the immense areas over which the deposits are found, and the isotopic data, it seems necessary to invoke processes grossly different from those responsible for most normal hydrothermal deposits. The above inclusion data place many constraints on the source, transportation, and deposition of the Mississippi Valley-type ores, but they do not provide specific answers to many questions. Taken together, these data indicate a slow-moving, moderately hot (~150°C), saline ore fluid that differs somewhat in chemical composition from normal connate and oil-field waters. None of the data preclude a purely deep diagenetic sedimentary environmental origin, nor do they require a magmatic origin, but they do preclude the cave, karst, sabkha, and lagoonal environments proposed by some. The low to undetectable S and the high metal contents reported by various workers suggest that, at least for some deposits, S was quantitatively precipitated out of the solution, leaving an excess of metal still in solution.

Scattered Pb-Zn mineralization of Mississippi Valley-type is found in many cratonic carbonate sequences (see Chapter 11). The inclusions in these minerals are similar to those found in the ore deposits and suggest a genetic similarity (Roedder, 1979c). Although many deep basins now contain moderately hot saline brines, particularly below 3 km, some geologists are reluctant to accept such an environment for the thinner piles of cratonic sediments. Admittedly, little evidence of such brines may be visible in the usual outcrop, once the brines have been replaced with surface water, as most of the minerals present are relatively inert, but the inclusions present in the minerals can and do preserve evidence of these hot brines for geologic time. The presence of such hot fluids under pressure in present-day sedimentary piles, and their migration, whether it is due to compaction or other gradients (Magara, 1973; Price, 1975; Brecke, 1979), is exceedingly important to an understanding of both ore and oil deposits. Furthermore, hot saline brine will leach heavy metals from shale (Long and Angino, 1982).

However, as stated by Ohle (1980, p. 161) "... there is still great disagreement among geologists on most facets of the mechanism [of genesis of these deposits], such as the source of the metals, the timing of their release, the

Table 15-3 Summary of primary inclusion data on some Kuroko deposits, Japan [cpy = chalcopyrite; gn = galena; sp = sphalerite; py = pyrite; qtz = quartz.]

Locality	Mineral & ore type examined	Th (°C)	Tm ice (°C)	Salinity (wt % NaCl equiv.)	Reference
Various	Quartz of cpy-bearing veins in silicified zone	---	-2.7 to -3.4	4.5 to 5.5	Takenouchi & Imai (1968)
Furutobe mine	Barite (Keiko)	99-211	---	---	Homma & Miyazawa (1969)
	Quartz (Keiko)	117-190	---	---	
	Sphalerite (Ohko)	245	---	---	
	Fluorite (Ohko)	159	---	---	
	Barite (Kuroko)	116-212	---	---	
Kosaka mine	Quartz	225-310*	-1.5 to -5.3	2.5 to 8.3	Lu (1969)
	Sphalerite	190-245			
	Barite	120-300	---	---	
Kosaka & other mines	Siliceous ore	180-290	---	---	Tokunaga et al. (1970)**
	Yellow ore	220-290	---	---	
	Black ore	80-200	---	---	
Kosaka mine	Quartz, lower siliceous ore	264-305	---	5.7-8.4	Urabe & Sato (1978)***
	Quartz, upper siliceous ore	225-300	---	2.1-3.9	
	Barite, black ore	130-300	---	2.7-5.4	
Kosaka mine	Upper Kuroko (gn-sp-barite)	100-150	---	---	Sato (1970)
	Lower Kuroko (sp-py-cpy-barite)	150-200	---	---	
	Oko (py-cpy)	~200	---	---	
	Keiko (py-cpy-qtz)	200-300	---	---	
Shakanai mine	Quartz	132-274	---	2.4 to 5.2****	Enjoji (1972)
Iwami mine, Shimane Pref.	Sphalerite of stockwork	200-310	---	1 to 5	Yoshida (1979)
	Quartz of stockwork	230-295	---	---	
Fukazawa	Unspecified from stockwork	250-310	---	---	Takenouchi (1980a)*****
Matsumine	Quartz of stockwork	230-335	---	---	Takenouchi (1980a)
Uchinotai West, Kosaka	Quartz of stockwork	248-340	---	---	Takenouchi (1980a)
Uwamuki mine	Quartz of bedded or	260-320	---	2.5 to 5.5	Marutani & Takenouchi (1978)
	Quartz of stockwork	280-320	---	---	

*The range of temperature for quartz was interpreted as due to trapping of primary gas bubbles and that of barite to leakage.
**These Th ranges were subsequently extended somewhat by Tokunaga and Honma (1974).
***Quoting Lu (1969), and his unpublished data.
****Enjoji, 1972, p. 117.
*****Scaled from his diagrams.

origin of the saline solutions, how the metals are carried, the causes of deposition, or why it occurs where it does."

KUROKO DEPOSITS

The Kuroko-type deposits are spectacularly abundant in Japan and have been the subject of rather intensive geologic study (e.g., Ishihara, 1974; Pisutha-Arnond and Ohmoto, 1983; Ohmoto and Skinner, 1984). Many of these studies involve inclusion work, the data from which show that the fluids responsible for the Kuroko deposits were quite different from those forming the Mississippi Valley-type deposits in that they were roughly similar to seawater in concentration of salts and were hot (some were >300°C; Table 15-3). For some time, the general model of Kuroko ore deposition has involved volcanic exhalations in a submarine volcanic environment, in part on the basis of fluid-inclusion data (such as those shown in Table 15-3). Furthermore, a series of Upper Tertiary vein-type polymetallic deposits found in Japanese volcanic rocks are frequently indicated as the probable deeper equivalents of the Kuroko ores. As shown by Hattori and Sakai (1979, 1980), however, when isotopic data on inclusions in these vein-type ores are put on a plot of δD vs $\delta^{18}O$ (e.g., Fig. 11-13) they are found to be very much like those of the local meteoric waters, whereas the inclusions in Kuroko samples have apparently evolved from mixtures of seawater plus meteoric water modified by rock-water interaction. Some magmatic water may have been present in both.

The distribution of inclusion data within individual Kuroko ore deposits generally shows an increase in temperature and salinity with depth in the deposit. Various lines of evidence, including lack of evidence of boiling, have been used to estimate the depth of water above the deposits during ore formation. Yoshida (1979) estimated ⩾50 m for the Iwami deposit; Marutani and Takenouchi (1978) estimated >1000 m for the Uwamuki deposit; and Pisutha-Arnond and Ohmoto (1980) estimated >~1500 m for Kuroko deposits in general.

One interesting aspect of the inclusion data that does not seem to have been addressed is the difference between the observed salinities and that of seawater. At 3 wt % NaCl, inclusions of undiluted seawater should have Tm ice of ~-1.8°C, yet most of the available data (Table 15-3) indicate that many inclusions are more concentrated than seawater. Because no evidence for boiling (which could increase salinities) has been found, and because meteoric water (of very low salinity) is the major diluent, these inclusion salinities should all be less than that of seawater. This discrepancy might be explained by assuming that the proposed (possible) small magmatic contribution was such a highly saline fluid that it could increase the salinity of the mixture adequately without major effects on the isotopic ratios. Also, heated seawater could have increased its solute content by reaction with rock and by loss of water to hydration reactions. J.W. Hedenquist (pers. comm., 1983) suggested that up to ~1.5°C of depression of the freezing point could be accounted for by dissolved CO_2 alone.

EPITHERMAL DEPOSITS

General Features

Epithermal ore deposits have been a fruitful area for inclusion studies, in large part because the vuggy nature of the ores favors the formation of large crystals and recognizable primary inclusions. These deposits are characterized by temperatures generally <300°C, low salinity (frequently <5 wt % NaCl equiv.), evidence of boiling,[3] and evidence of meteoric water as the major source of the fluids. Circulation of the fluids was usually driven by an intrusive or volcanic heat source, which may also have contributed water and/or metals and may well have been very similar to present-day geothermal systems (Wetlaufer et al., 1979; Henley and Ellis, 1983). Table 15-4 lists some of the characteristics of a representative series of 11 deposits from the extensive inclusion literature. Hayba (1983) presented a compilation of fluid inclusion and stable isotope data on a series of epithermal deposits.

Most epithermal deposits are assumed to have formed near the surface, but the evidence for estimates of depth may not always be on a sound basis. Thus, Radtke et al. (1980a) estimated the depth of ore deposition at the Carlin, Nevada, Au deposit to be 300-500 m, but Bodnar and Kuehn (1984) have shown that significant CO_2 in these same fluid inclusions results in a depth estimate of 500 to 3000 m.

The salinity of the fluids is generally low but varies widely. The measurement of low salinities can cause problems in the laboratory. D.M. Smith et al. (1982) reported "freezing point" (presumably Tm ice) values from -4.7 to +2.9°C [sic] at Tayoltita, Mexico. If we assume that some (unstated) experimental error

----

[3] The nature of the evidence is sometimes quite ambiguous, e.g., Buchanan (1979) and Fahley (1979). The actual frequency of occurrence of boiling in these deposits may have been overstated by some. Heald-Wetlaufer et al. (1983) compiled detailed data on 15 epithermal precious- and base-metal districts and found boiling clearly associated with precious-metal deposition in only 2 of the 15.

Table 15-4 Recent studies of inclusions in some epithermal ore deposits

Deposit	Type and assemblage	Minerals studied	Th (°C)	Salinity (wt % NaCl equiv.)	Meteoric water involved	Notes	Reference
Emperor deposit, Fiji	Epithermal Au-Ag-Te in Tertiary volcanics	Quartz	317–170	~5.5	Major	Some boiling at top of mine. Th on barren stages.	Ahmad (1979)
Carlin deposit, Nevada	Disseminated Au replacement in sediments	Quartz, barite, & calcite	350–152	2–5	Major	Isotopic temperatures higher; boiling in later stages.	Radtke et al. (1980a)
Tonopah, Nevada	Epithermal Au-Ag	?	300–250	<1	Major	Are temperatures Th values? Some boiling.	Taylor (1979a)
Tonopah, Nevada	Epithermal Au-Ag	Quartz	290–140	1–3	Probably	Ore stage Th 280–220°C; boiling seen.**	Fahley (1979)
Tayoltita, Mexico	Epithermal Ag-Au veins in volcanics	Quartz	310–250	1.9–13.8*	Probably	Boiling seen.	D.M. Smith et al. (1982)
Las Cuevas, Mexico	Fluorite at rhyolite/limestone contact	Fluorite	130–60	0–3	Major	Some inclusions of petroleum.	Ruiz et al. (1980)
Lake City district, Colorado	Polymetallic veins in Tertiary volcanics	Quartz, sphalerite, & fluorite	272–183	0.1–6.6	?	Values are sample averages; boiling seen.	Slack (1980)
Finlandia vein, Colqui district, Peru	Polymetallic vein in Tertiary volcanics	Quartz, sphalerite	270–140	0–13	Major	Boiling & mixing with connate waters yield high salinity.	Kamilli & Ohmoto (1977)
Sunnyside mine, Colorado	Polymetallic vein in Tertiary volcanics	Quartz, fluorite, & rhodochrosite	320–170	0–3.6	Major	Boiling seen.	Casadevall & Ohmoto (1977)
Various deposits, Nevada	Au-quartz-andularia veins & disseminated	Quartz	330–245	0.9–7.3	Major	Late calcite has low Th. Most <2.1 wt % salinity.	Nash (1972)
Creede, Colorado	Epithermal Ag-polymet. veins in volcanics.	Quartz, sphalerite, fluorite	270–190	5–12	Major	Boiling in top of vein seen.	Barton et al. (1977)
Pachuca, Mexico	Epithermal Ag-Au veins in volcanics	Quartz, calcite	260–180	0–7.5	?	Boiling seen.	Drier (1976)

*Data contradictory -- see text. **Evidence for boiling is quite ambiguous, however.

made all these values too high by 2.9°C, Tm ice would range from -7.6 to 0°C, corresponding to salinities of 11.3 to 0 wt % NaCl equiv., but Smith et al. reported a salinity range of 13.8 to 1.9 wt %.

Several of the epithermal fluorite deposits seem to have formed from essentially heated surface waters, as they show Tm ice to be essentially 0.0°C. These inclusions commonly show metastable superheated ice, however, so many yield invalid determinations of Tm ice. Veins of coarsely crystalline fluorite plus minor opal, calcite, barite, etc., at Browns Canyon, Chaffee County, Colorado, USA, provide a good example of the problems presented by the inclusions in such deposits (Roedder, 1977a). Th, measured on 179 inclusions that had a bubble at room temperature, ranged from 119 to 161°C, and averaged 141°C, but many liquid, one-phase inclusions were found (i.e., no bubble). These one-phase, liquid-filled inclusions are believed to represent metastable equilibrium (at room temperature); i.e., they were trapped at ~140°C but now contain stretched liquid under negative pressure, from failure to nucleate a vapor bubble since the original cooling ~7 m.y. ago. Three facts support this contention: First, no two-phase inclusions were found that had intermediate Th (i.e., <119°C). Second, the liquid-filled inclusions did not decrepitate even at 160°C, nor become stretched (Larson et al., 1973). If these liquid-filled inclusions had actually formed at room temperature, they would have developed internal pressures of 1500-2000 bars at 160°C, and Bodnar and Bethke (1984) have shown that inclusions in fluorite stretch when internal pressure exceeds 200-700 bars. Third, some previously liquid-filled inclusions developed a vapor bubble at room temperature after a freezing cycle, a bubble similar in size to those that homogenized at 119-161°C. Similar reasoning precludes formation of the liquid-filled inclusions by necking down at near-surface temperatures.

On warming the frozen inclusions to near 0°C, only 20 of the 179 were found to hold the stable assemblage ice + water + vapor bubble. Tm ice, the melting of the last ice crystal under stable equilibrium, occurred in 5 of these inclusions at 0.00±0.05°C, in 1 at -0.07±0.03°C, in 11 at -0.10±0.05°C, in 2 at -0.13±0.03°C, and in 1 at -0.15±0.10°C. The average stable freezing temperature was -0.08°C, corresponding to ~1400 ppm NaCl equivalent salts in solution.

Expansion on freezing eliminated the small vapor bubbles in the rest of these inclusions. Failure to renucleate a vapor bubble on reheating resulted in the persistence of metastable "superheated" ice, at temperatures as high as +6.6°C and for as long as 4 hours. The apparent value of Tm ice under such conditions cannot be used to determine salinity (see Chapter 10).

Since most hot waters from deep in the earth contain appreciable solutes, the very low solute concentrations found in this type of deposit (Roedder, 1977a, reported similar data from five other similar fluorspar deposits) suggest that circulation was shallow or that the time for rock-water interaction was short. But in either case, how was fluorite picked up, without major quantities of other solutes?

The cations in inclusions in epithermal deposits vary widely, in large part as a function of the nature of the wall rocks and the alteration taking place, but Na is normally dominant, and Na/K (atomic) is generally in the range 2-10, as is Ca/Mg. Although alteration processes may result in large shifts in cation ratios and may explain the source of the cations found, the origin of the anions is somewhat of an enigma. Numerous isotopic studies generally show meteoric water to be the major or even sole source of the water, so where can all this Cl come from? Several potential sources should be considered. Small amounts of Cl (generally <100 ppm) are present in all meteoric water. Even a small contribution of connate waters (which may contain 20-30 wt % chlorides in solution) will add much Cl, and magmatic fluids can contain even higher concentrations. In light of the evidence, however, much of the Cl must have been leached out of rocks during deep circulation of the water. CO_2 may be more common than previously assumed and may be as high as 4 mole %, but still not enough to be readily recognized without use of the crushing stage (Bodnar and Kuehn, 1984), and cannot be present if Tm ice is near 0°C.

As seen in Table 15-4, the temperatures of formation for these deposits (which are near to Th, as the pressure corrections will be low) are usually between 200 and 300°C, but waning stages of late calcite, etc., can be <200°C. In those deposits for which data have been obtained on each of a series of paragenetic stages, the general trend is usually toward lower temperatures and salinities during successive stages. Minor temperature "reversals" (i.e., an increase during a paragenetic stage) do occur, but because major reversals can cause natural decrepitation of the earlier inclusions (see Chapter 9), the evidence for such processes can be self-destructing.

The gases in solution in epithermal ore-forming fluids are of great importance in the chemistry of ore deposition, but are poorly known. Casadevall and Ohmoto (1977) reported semiquantitative mass-spectrometric analyses of the gases in inclusions from the Sunnyside mine, Colorado, USA, that condensed at liquid N_2 temperature (CO_2, H_2S, SO_2), as distinct from the noncondensable gases (Ar, N_2, CH_4, H_2). H_2O constituted 0.993 to 0.999 mole fraction of the sum of H_2O plus condensable gases in 11 of 12 samples. CO_2 constituted the bulk of the condensable gases, but minor amounts of H_2S, SO_2, and hydrocarbons were also found.

The density of the ore fluids found in the liquid-rich fluid inclusions in these deposits rarely exceeds 0.95 g/cm^3 and may extend to <0.8 g/cm^3 in the higher temperature deposits.

Creede, Colorado, USA[4]/

The Creede epithermal Ag-Pb-Zn-Cu vein deposit will be used as an example of an epithermal deposit on which a considerable body of fluid-inclusion data has been obtained, along with many other sorts of data that should all be evaluated together, as a unit. The veins at Creede are open-space fillings in fractures in devitrified welded tuffs. Five stages of ore deposition, designated A to E, have been recognized, and inclusion data have been obtained on the first four; the last ("gel" pyrite) is of minor significance. Much of the work has concentrated on various localities along one vein, the OH vein, but similar data have been obtained on the P vein, the Amethyst vein, and other veins in the district. The ores are relatively simple in mineralogy and structure and generally consist of crustified vuggy veins of sphalerite, galena, and chalcopyrite, together with pyrite, hematite, chlorite, quartz, adularia, and sericite. Minor amounts of fluorite, rhodochrosite, siderite, native Ag, barite, and tetrahedrite-tennantite are present. The sphalerite is low but highly variable in Fe content and occurs as relatively coarse, multiply-zoned crystals containing numerous large primary inclusions (Fig. 15-2) and relatively few small secondary inclusions, and hence is excellent for detailed studies of the changes in fluid inclusions from zone to zone.

The inclusion data from Creede provide good examples of several aspects of epithermal ore-deposition conditions. One such aspect concerns the presence and significance of gases in solution in the ore fluids. Both field and inclusion data indicate that boiling occurred at the top of the ore body. The inclusion evidence is the existence of apparently empty "gas" inclusions from the trapping of steam bubbles, adjacent to other inclusions having the normal liquid-rich

----

[4]/ This description is taken, in large part, from Woods et al. (1982), which also includes reviews of the geologic and hydrologic settings, and summaries of all published (and many unpublished) data from fluid-inclusion studies of Creede samples, plus a full bibliography.

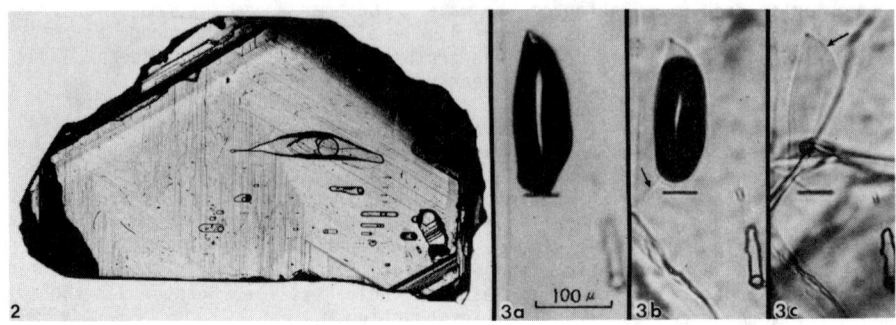

Figure 15-2. Primary inclusions in yellow, color-zoned sphalerite crystal from a vug in the Ag-Pb-Zn-Cu deposit at Creede, Colorado, USA. The large inclusions (containing mildly saline brine and a small vapor bubble) lie in a plane parallel to and only about 0.3 mm under the striated crystal surface, so that both the inclusions and the crystal face are in focus. Such inclusions must be primary. Presumably they form as a result of the covering over of irregularities (stepwise growth) on the crystal face. Sample PMB-AA-107-59, from the OH vein. From Roedder (1972).

Figure 15-3. Crushing-stage test of quartz sample ER-57-25 from the Amethyst 5 level, OH vein, Creede, Colorado, USA, that apparently grew in a boiling solution, as it contains many large apparently empty inclusions (trapped steam bubbles), as well as normal two-phase liquid-rich inclusions homogenizing at about 200°C (one of these is visible at lower right). (a) Before crushing. (b) Just after a very tiny fracture (arrow) let in a little of the mounting medium (1.55 index liquid). (c) After a slight additional crushing, which opened several larger cracks and filled the inclusion almost instantly, leaving a small bubble (much less than 1 vol %), which then dissolved in a few seconds, before this photo was taken. Only a faint trace of the inclusion outline is visible (arrow). The small dark spot at the intersection of the two cracks is crushed mineral and not a bubble. From Roedder (1970a).

filling, in samples from the uppermost levels. On cooling, the small amount of liquid water that condenses may not be visible, and the inclusion will appear empty and indeed will be essentially a vacuum. The crushing stage provides a good test to distinguish between such an inclusion and one that has merely leaked its liquid and is now filled with air (Fig. 15-3). The crushing test illustrated in Figure 15-3 reveals that the concentration of "noncondensable gases" (i.e., at room temperature and one atm) in these steam inclusions is extremely low. Because noncondensable gases such as CO_2, CH_4, N_2, H_2, Ar, and H_2S should be present in many terrestrial waters and will partition strongly into the vapor phase on boiling, and yet no such gas was found, these results suggest that before these steam inclusions were trapped, boiling had flushed all noncondensable gases out of the fluids. Rama et al. (1965) reported no detectable Ar in the fluid inclusions in a similar sample from Creede.

Most of the studies of inclusions at Creede have been microthermometric. Data have been obtained on 2575 fluid inclusions in ore and gangue minerals. Many of these individual inclusions were run by both heating and freezing procedures. In summary, they show Th between 200 and 270°C, and Tm ice between -3 and -8°C, corresponding to 4.9 to 11.7 wt % NaCl equivalent salinity. Although the salinities are higher than those of many other epithermal deposits (Table 15-4), in most respects, Creede is a rather typical epithermal deposit. The inclusion data on the deposit are most interesting in terms of the variations of Th and Tm with time and with respect to each other.

Ore stage A, which is recognized district-wide, consists primarily of quartz together with minor chlorite, rhodochrosite, and sulfide. Relatively few data on Th and Tm for the quartz and rhodochrosite show them to be in the lower parts of the total range of Th and salinity.

Ore stage B, also district-wide, consists of relatively fine grained

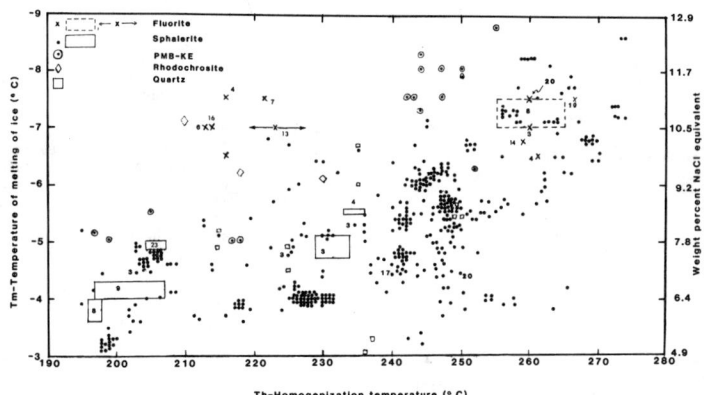

Figure 15-4. Graph of Th versus Tm ice for all primary and selected pseudosecondary fluid inclusions in sphalerite, quartz, fluorite, and rhodochrosite from all studied paragenetic stages at Creede, Colorado, USA. Modified from Woods et al. (1982). Rectangular fields represent groups of fluid inclusion data that were reported only in terms of ranges for Th and Tm. The numbers in the fields indicate the number of data points included. Numbers beside individual points indicate the number of fluid inclusions having identical values for both Th and Tm. One point for sphalerite at Th = 225°C and Tm = -2.8 to -2.95°C falls just out of the figure and has not been plotted.

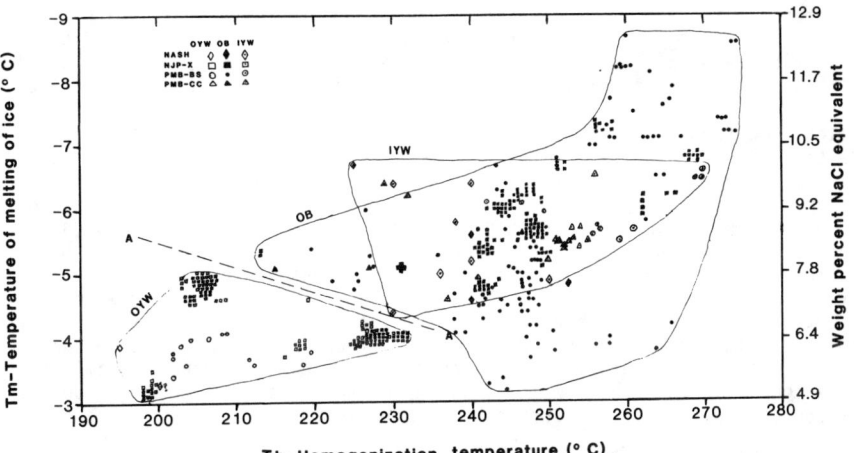

Figure 15-5. Graph of Th versus Tm for all data points for primary and selected pseudosecondary fluid inclusions in sphalerite from Figure 15-4 that can be assigned to one of the three major stratigraphic zones of D-stage ore deposition: Inner yellow-white (IYW), Orange-brown (OB), or Outer yellow-white (OYW). Numbers beside individual symbols indicate the number of fluid inclusions having identical values for both Th and Tm. The symbol for each inclusion indicates which of the four sets of data and three zones it represents. Line A-A' separates data on inclusions from the Outer yellow-white zone from data on inclusions in the other two zones. From Woods et al. (1982).

sphalerite, galena, chalcopyrite, and gangue minerals; it makes up much of the total mass of the ore and most of the Ag mineralization. The following stage C is volumetrically minor and restricted in extent; it consists of fluorite overgrowing siderite-manganosiderite and quartz. Much of this fluorite was subsequently deeply etched and commonly was completely removed, leaving octahedral casts.

The D-stage ore deposition consists of relatively coarse sphalerite (zoned euhedral single crystals, some 10 cm or more in diameter), galena, chalcopyrite,

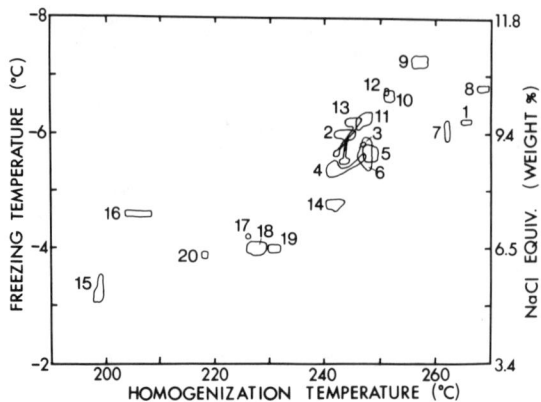

Figure 15-6. Plot of Th vs Tm ice for 221 primary inclusions in a 5-cm band of zoned sphalerite (sample NJP-X-1-59) from Creede, Colorado, USA. The 20 numbered areas include all data points from each of the 20 zones sampled, numbered in sequence from zone 1 (deposited earliest) to 20 (latest). The individual points could not be plotted at this scale because of overlap (i.e., duplicate results from different inclusions in a given group, and, in part, in different groups). The numbers of inclusions in each of the areas outlined are as follows: 1(2), 2(18), 3(1), 4(21), 5(27), 6(9), 7(4), 8(9), 9(8), 10(4), 11(15), 12(2), 13(7), 14(12), 15(14), 16(11), 17(4), 18(32), 19(13), 20(8). See text for significance. From Roedder, (1977c).

quartz and hematite. The generally minor illite alteration probably belongs here. The D-stage has been divided into three substages on the basis of color banding in sphalerite: inner yellow-white (IYW), orange-brown (OB), and outer yellow-white (OYW). The following E stage resulted in deposition of fibrous ("gel") pyrite and some marcasite and stibnite.

When all the available matched Tm-Th data on inclusions in sphalerite (the bulk of the data), plus a smaller number of inclusions in quartz, and still fewer in fluorite (and only three in rhodochrosite) are placed on a single plot of Tm vs Th, they cover a rather large area (Fig. 15-4) and are not very instructive. The great bulk of the points representing individual inclusions fall along a curving band from upper right to lower left. Most of the points scattered outside this band are from fluorite or rhodochrosite, or are from veins other than the OH. However, when only the data from specific stages are plotted, all from the same vein, several interesting features become apparent. Most of the data are on primary inclusions in D-stage sphalerite, and hence can be categorized into the three relatively easily recognized substages, IYW (earliest), OB, and OYW (latest). Figure 15-5 shows the three envelopes enclosing all data points from each of these three stages. These data are from samples from wherever in the entire length and depth of the OH vein that D-stage sphalerite was selected for study (~1500 m length and ~230 m vertical depth). This breakdown shows that the D-stage sphalerite-forming ore fluids were hotter and generally more saline (but in almost the same ranges) during the first two substages and became much cooler and less saline during the later OYW substage.

The multiple banding of the D-stage sphalerite crystals permits an even more detailed breakdown than this simple threefold one, particularly if only one sample locality is studied. Thus, 221 points on Figure 15-6 are all from a single sample ("NJP-X-1-59") from ~350 m below the present surface and probably no more than 600 m below the surface at the time of deposition. It was a group of zoned subparallel sphalerite crystals, in which the total depositional thickness was ~5 cm. The sample was cut into a series of doubly polished plates, mostly ~3 mm thick: special care was taken to avoid damage to the inclusions. The color of individual zones in transmitted light varies from almost colorless through yellow and orange-red to brown, in plate thicknesses of 2-5 mm. Analyses of similar zones in other crystals from the deposit show that variation in Fe content is correlated with the color variation but is not the sole cause; the range for these particular zones is from about 0.1 to 3 mole % FeS.

The complex "stratigraphic succession" in this crystal (and in other cogenetic crystals from the sampling area) had been established previously by P.M. Bethke and P.B. Barton, Jr., of the U.S. Geological Survey. This was done by using many doubly polished plates from a series of crystals, as no individual vug was necessarily a site of continuous deposition throughout the history of the deposit. The criteria used to recognize and correlate zones have been detailed elsewhere (Bethke and Barton, unpublished data). As in any complex stratigraphic correlation, no single criterion is necessarily definitive; hence, Bethke and Barton used a combination of criteria, including transmitted light color and its variability, zones of accidental solid inclusions, birefringence, codeposited minerals, minor-element concentrations, and leach zones (i.e., disconformities), and, particularly, the sequence of occurrence of such features. Each leach zone represents a time period when the solutions passing this particular point were no longer saturated with respect to sphalerite; these solutions may well have been or become saturated elsewhere, and may have deposited sphalerite before and/or after this point. As a result of the vagaries of ore-fluid movement through the tortuous and possibly changing channels of rubble in the vein, the relative and absolute thickness of given zones in individual crystals may show large differences. By means of the procedures listed above, the 5 cm band was divided into 20 "stratigraphic horizons" or zones; material equivalent to all 20 zones together makes up only the latest quarter of the ore at Creede.

The inclusions selected were almost certainly primary, mostly irregular and smoothly rounded to subhedral, and in the size range 15 to 450 μm. As these samples have very few planes of pseudosecondary or even secondary inclusions, the possibility of refilling of earlier inclusions with new fluid at a different set of values for P, T and X is very unlikely. Approximately 10 inclusions (to a maximum of 32) were selected for runs from each of the 20 sequential zones (230 inclusions in all, of which nine leaked during homogenization and had to be discarded). This selection is an important part of the process, as only a very small percentage of the available inclusions could be used. To be selected, an inclusion had to fulfill all three of the following criteria: (1) good evidence of primary origin; (2) optimum optical quality; and (3) clean-cut evidence of occurrence in a specifically identified zone. The evidence of primary origin generally consisted of the occurrence of large isolated inclusions, particularly at irregularities in the visible zoning, indicating imperfect growth. The optical quality of an inclusion is mainly a function of its shape and the absence of confusing detail in the plate above it. Although the small bubble in an inclusion near Th can usually be revealed with a fiber optics illuminator, the same is not true with respect to small crystals of ice near Tm ice, particularly in sphalerite. As both Th and Tm ice were to be obtained on the same inclusions, only inclusions with smooth and flat or only gently curving upper and lower surfaces could be used. The problem of identifying with certainty the specific zone in which an inclusion occurs is minimal in sample plates that have vertically oriented broad zones of contrasting features, but this problem becomes very difficult in plates that have strongly inclined thin zones and little contrast. The problem is greatest where growth has been irregular and the zone boundaries are convoluted in three dimensions, so that simple tilting of the section relative to the line of sight is inadequate. An additional problem arises from the occasional presence of reentrants, either from solution or growth phenomena, which may result in a given inclusion, formed in such a reentrant, appearing to be in an earlier zone.

After inclusion selection, small parts of the plates, each containing one or more selected inclusions, were cut out for the runs. Some of the zones were represented by samples from more than one plate, or from several places on a given plate.

Th for these inclusions were run using a Leitz early model 350 stage and very slow heating schedule, essentially letting the temperature level off after each increment, particularly near Th; a fiber optics illuminator helped prevent loss of the bubble in dark corners. Tm ice was determined using a circulating refrigerated acetone stage (Roedder, 1962a). No evidence of carbon dioxide hydrate was seen. The possibility of unintentional operator bias was avoided by making the runs essentially as "blind" determinations by several operators over a period of months, without the operators knowing the interrelationships of the individual chips or how the data fitted together.

The results of Th and Tm determinations on these 221 inclusions are plotted in Figure 15-6. These results are surprisingly consistent. All inclusions for each given zone, even though they may have been selected from parts of the zone at some centimeters distance from each other in the sections, form very tight clusters on the diagram. The total range for all inclusions is 198 to 269°C (Th) and -3.1 to -7.3°C (Tm ice, corresponding to a salinity of 5.1 to 10.9 wt % NaCl equiv.).

The loci of the individual groups on Figure 15-6, numbered there in the order of their depositional sequence, show that during the deposition of this 5 cm layer of sphalerite, the salinity and temperature first increased somewhat and then decreased significantly. Superimposed on these general trends, however, are several smaller reversals. Even these minor reversals are statistically significant, however, and are not just due to experimental error, because each zone's location on Figure 15-6 (except zone 3, in which only one suitable inclusion was found) is based on multiple points within the areas outlined. With three exceptions, the data points from one group do not overlap with those from adjacent groups in the sequence. One exception is in zone 4; four of the 21 inclusions selected as being in zone 4 have values for Th and Tm within the tight cluster of 27 outlining zone 5 on the figure. Whether these four represent real variation in the fluids during deposition of zone 4, or whether they were actually zone 5 inclusions that only appeared to be in zone 4, is not known. The second and major exception is the almost total overlap of the area of zone 5 with that of zone 6. Although the distinction between the host sphalerite for these two zones was fairly obvious, the fluids depositing this sphalerite appear to have changed very little in temperature and salinity. The third exception is a single inclusion, which is in zone 11 along with the 14 others, but which was originally identified as being in zone 10. The other four inclusions identified as zone 10 form a tight cluster, so theoretically the small circle for zone 10, as plotted, should have a long extension down to overlap zone 11 to include this one inclusion, presumably misidentified as to zone.

The maximum range of Th for inclusions from any given zone is only 4°C, except for zone 4, in which the range is 6°, because of the four anomalous inclusions mentioned above. The average range is only 2°C, including the data for zone 4. The equivalent maximum range of Tm ice is 0.6°C, although the average range is only 0.2°C.

A series of conclusions can be drawn from these results:

(1) The variations used to establish the sphalerite "stratigraphy" are paralleled by changes in the ore fluids trapped in inclusions. This parallelism may indicate but does not require a cause-and-effect relationship.
(2) The fluids forming this deposit did not simply decrease in salinity and temperature with time, as is frequently considered to be implicit under the assumption of eventual quenching and dilution with cold ground water.
(3) Even the less distinct boundaries can be significant in that the ore fluids changed appreciably from one zone to the next. These changes are well above experimental errors and generally are larger than the variation within a given zone.
(4) The ore fluid changed abruptly at each zone boundary but was constant during the deposition of each zone. This places some constraints on the hydrology of the system and implies that the cause or causes of the sphalerite zonation were surprisingly static throughout the deposition of each given zone, since the variations in Th and Tm within each zone are small.

(5) If such a fine structure is ignored by blind sampling, serious errors can occur in the assignment of thermal gradients during deposition, and in the correlation and interpretation of fluid-inclusion data with the results of chemical analyses of the host.

(6) Primary inclusions can be recognized adequately in this material. Any pseudosecondary or secondary inclusions would have degraded the consistency seen in Figure 15-6.

(7) The inclusions have not necked down since they were trapped. If they had, there would be a gross horizontal dispersion of the individual groups on Figure 15-6.

(8) The experimental techniques used have excellent precision. No real proof of the accuracy of these numbers exists, but in the present study precision is much more important than accuracy. As these inclusions were, in all probability, trapped under very shallow conditions, Th is essentially equal to Tt. The maximum estimated pressure correction to be added to Th to obtain Tt for these samples is only a few degrees Celsius.

(9) No recognizable difference was noted between large and small inclusions from the same group. This observation seems to contradict some experimental data presented by Barnes et al. (1969) that they interpret as evidence that fluid inclusions trap a compositionally nonrepresentative sample of the fluid present.

(10) These fluid inclusions obviously did not leak. If they had leaked, the leakage would have to have been essentially identical for all inclusions in each given group, an unlikely event.

Much smaller sphalerite crystals in this same mine have formed from the same range of fluids. The smaller size makes detailed stratigraphic assignment of any given inclusion difficult. Detailed correlation of growth zones from one crystal to another in the same sample and eventually with other samples and other parts of the mine is essential to understanding many aspects of ore deposition, such as the nature of the hydrologic flow in the system, the establishment of the contemporaneity of deposition of several different minerals, the chemistry of the ore-forming fluid, and, eventually, the cause of the ore deposition. Sphalerite crystals from other localities along the same vein show similar "stratigraphic sections" and inclusion data, but the degree of correlation possible with those shown in Figure 15-6 decreases with distance, and finally even the IYW-OB-OYW division is not recognizable. However, as pointed out by D. Hayba (pers. comm.), the steady states separated by abrupt changes will, in the long range, permit good spatial mapping of T and P.

The fluid-inclusion data can best be explained by a mixing model involving at least two fluids -- one hot and saline, the other cool and fresher. Sudden changes in the mixing ratio, presumably from changes in the plumbing, punctuated long periods of remarkably uniform conditions of ore fluid flow and deposition. The effects of other processes such as convection and heat exchange with wall rocks also must have been superimposed on this simple mixing model. Several aspects of the inclusion data may be interpreted as suggesting exceedingly slow ore deposition (Woods et al., 1982).

Bethke and Rye (1979) reported studies of the isotopes of C, H, and O in fluid inclusions and their host minerals from Creede. Establishing the source of the fluids was not simple, and more study is needed. Meteoric waters were a major source, but the data suggest the probable involvement of two different meteoric sources, plus a magmatic contribution, and considerable isotopic exchange with the wall rocks. Foley et al. (1982) then showed that some of the complications in the δD data were explicable by a novel mechanism. The vein minerals (particularly quartz) grew essentially from saline fluids (4-12 wt % NaCl equiv.) having δD of ~-70 per mil, which were trapped in large primary inclusions. Occasional incursions of slightly cooler almost fresh groundwater resulted in cracking of the quartz and the trapping of large numbers of pseudosecondary fluid inclusions having δD ~-100 per mil. Some of the large preexisting primary inclusions also were apparently opened and flushed at this same time. Previous inclusion and isotopic studies of Creede had not recognized the nature of these pseudosecondaries, and hence the samples run for isotopic measurements were presumably mixed samples.[5]

DEEPER, HOTTER VEIN AND REPLACEMENT DEPOSITS

The boundary between the epithermal deposits above and those considered here is arbitrary and diffuse and is based mainly on temperature of formation (>300°C) and depth of formation (generally >1000 m). Also the abundance of vugs and crustification in veins declines as depth increases. Only a few examples will be discussed to reveal the range and nature of the inclusion data obtained.

Casapalca, Peru

Inclusions from the Casapalca, Peru, Ag-Pb-Zn-Cu deposit were first studied by Rye and Sawkins (1974). The first and main stage of deposition accounts for >80% of the total vein filling of sulfides and quartz. This stage was followed by a late sulfide-sulfosalt stage (15%) and a postore calcite stage (5%). Vugs in the vein provided crystals from the several stages, permitting recognition of both primary and pseudosecondary inclusion origin; sampling was possible over ~1000 m vertically. Secondary inclusions were rare and hence could contribute only minor contamination to the samples analyzed isotopically.

Four types of inclusions were recognized. Type I consisted of liquid plus <30 vol % vapor; Type II was similar but also had a cube of NaCl (plus possibly KCl); Type III consisted of liquid plus >70 vol % vapor; and Type IV, like Type I but also containing radiating birefringent needles, assumed to be dawsonite. Types II and III were in close association and only in main-stage material from the 1700 level in the mine and above; they were thus ascribed to boiling. Several other daughter minerals, hematite(?), a sulfate(?), and a carbonate(?) were also observed. A few relatively large tetrahedral opaque daughter crystals (chalcopyrite?) were found in some inclusions, but Tsui and Holland (1979), on analyzing Casapalca fluid inclusions for Cu, found only 1-50 ppm. The method used (laser release plus optical emission spectroscopy) was such that the Cu determinations probably represent the sum of Cu in daughter minerals and in solution.

Although some color zonation was seen in the sphalerite (1.5 - 8.1 mole % FeS), a color-zone stratigraphy could not be established, and the lack of sufficiently precise paragenetic control from one sampling locality to another prevented a meaningful time-space analysis of the temperature patterns, but inclusions in sphalerite and quartz of the main stage had Th ranging from ~370 to 320°C, and the later stages were lower. The cores of some crystals yielded Th values ~30°C higher than the rims. As the available geologic data provided no good indication of depth, no pressure correction was applied (note, however, that none would be required for inclusions trapped during boiling). The vertical interval over which mineralization occurred (>1000 m) corresponds to ~100 bars hydrostatic pressure difference. The pressure at the point of trapping of some inclusions believed to represent boiling conditions was calculated (on the basis

----

[5] This work points up the value of careful observations in fluid inclusion work and the trouble that can result from misinterpretation of the data obtained. I had studied the large primary inclusions in the sphalerite from Creede (Roedder, 1960b) and had reported (1963, p. 207) that "Except for two that apparently had leaked as they showed about 0° freezing temperatures, all yielded temperatures [Tm ice] of -3.3 to -9.4° [°C], in a systematic zonal pattern related to the color zoning." If I had also taken time to run some of the small but abundant "secondary" (actually pseudosecondary) inclusions in the quartz at that time, and found that they too were fresh water, at least some of the problems of interpretation of the later isotopic data on quartz could have been cleared up more easily.

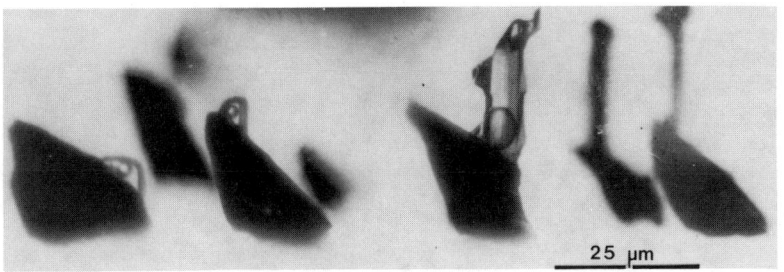

Figure 15-7. Primary fluid inclusions attached to a row of chalcopyrite crystals on a growth zone in a color-banded sphalerite crystal from the Providencia mine, Zacatecas, Mexico. The three fluid inclusions in the plane of focus each show liquid and a vapor bubble. Tt (i.e., pressure corrected) was 308±1°C. From Sawkins (1964).

of data of Haas, 1971) to be only 80 bars, corresponding to ~800 m hydrostatic head. Similar data on the depth of various samples are inconsistent. No liquid CO_2 was observed, but part of the discrepancy might be explained by the variable amounts of CO_2 found to be present in these inclusion fluids (Rye and Sawkins, 1974, p. 190, 199), which would drastically lower the apparent "boiling" temperature.

Salinity estimates, based on Tm ice and Tm NaCl, ranged from 4 to 40 wt % NaCl equiv. during the main stage and 4 to 12 wt % during the late stage. Sharp fluctuations in salinity occurred during the periods of boiling, and Rye and Sawkins suggested that these periods corresponded to discrete pulses of extremely saline intermittently "boiling" solution passing upward through the system.

Isotopic studies were made on fluids from inclusions in various minerals. The δD values are compatible with a deep-seated postmagmatic origin for the fluids forming the main- and late-stage ores, followed by an incursion of meteoric waters (from high elevations) during the postore calcite formation. Values for $\delta^{18}O$, $\delta^{13}C$, and $\delta^{34}S$ corroborate this model.

The gases in inclusions from Casapalca were analyzed by Graber et al. (1979). The suite of gases found during main-stage mineralization also suggested a magmatic source, buffered by the assemblage pyrite-pyrrhotite-magnetite. Episodes of boiling had higher SO_2 levels, in agreement with the Fe content of the sphalerite.

A series of studies of the inclusions in the pipe-like Pb-Zn ore bodies at the Providencia mine, Zacatecas, Mexico (Sawkins, 1964; Rye 1966; Rye and O'Neil, 1968; Rye and Haffty, 1969), showed very similar results in all respects, except for a lack of any evidence of meteoric-water incursion. The evidence for a primary origin for at least some of these inclusions was irrefutable (Fig. 15-7). A study of the Bluebell mine in Canada (see below) showed that it contrasts strongly with both Casapalca and Providencia.

Jamestown, Colorado, USA

The Jamestown district has been a major producer of Au and fluorspar from a series of deposits apparently associated with an Na-granite stock. Au-Te mineralization occurs as veins, which have formed in two stages: an early quartz-fluorite stage (Th < 375°C; salinity very high), and a later Au stage (Th 205-270°C; salinity ~4 wt %), according to Kelly and Goddard (1969) and Nash and Cunningham (1973). Boiling occurred, presumably due to variable CO_2 contents (Fig. 15-8), at many locations that were at relatively high altitudes or near the stock.

The associated large fluorspar deposits form stockworks, breccia zones, and pipe-shaped bodies (Nash and Cunningham, 1973). Fluid-inclusion studies

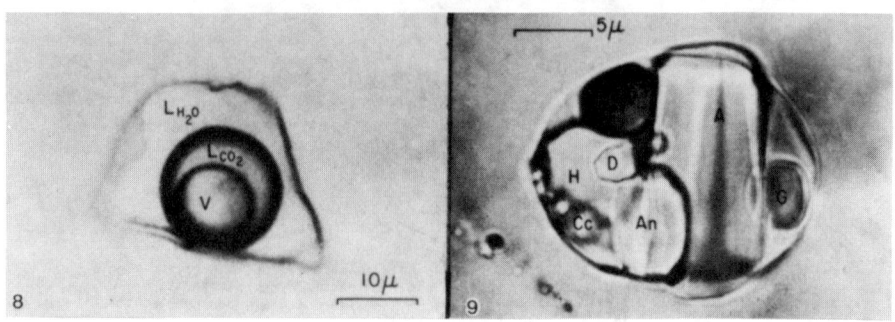

Figure 15-8. H_2O-CO_2 inclusion from a fluorspar deposit in the Jamestown district, Colorado, USA. On warming, the inner bubble of CO_2 shrinks and disappears (i.e., Th CO_2 L-V(L)) at 29.5°C; Th L-V (L) occurs at 280°C. From Nash and Cunningham (1973).

Figure 15-9. Multiphase inclusion in fluorite from the Emmett mine, Jamestown district, Colorado, USA. From Nash and Cunningham (1973). A = prismatic, orthorhombic morphology, moderate relief, parallel extinction, length fast, low apparent birefringence, biaxial, 2V ~60-80°, positive, does not dissolve on heating; An = anhydrite; Cc = calcite; D = unknown hexagonal colorless plate, moderate relief, moderate birefringence, dissolves by 125°C; G = unknown, equant, low birefringence, high relief (on heating, it dissolves incongruently between 150 and 180°C); H = halite; V = vapor.

indicate an early quartz stage formed by boiling fluids having Th of 250-375°C and salinities of 20-30 wt %. Inclusions in the main fluorite stage show Th of 250-350°C, some CO_2, and very high salinities, even >50 wt %, reflected in as many as 10 daughter crystals. Nash and Cunningham (1973) reported (by optical study) halite, sylvite, fluorite, hematite, anhydrite(?), calcite(?), and 9 unknowns (Fig. 15-9). Detailed SEM and microchemical studies were made by Metzger et al. (1977) on these daughter crystals, which may constitute >50 vol % of the inclusions (see Fig. 5-14). The major phases identified were gypsum, a calcium aluminum silicate (probably anorthite), barite, celestite (normally celestite shows retrograde solubility, but apparently it is prograde in these strong solutions), halite, sylvite, thenardite [Na_2SO_4], and possibly arcanite [K_2SO_4] and glaserite [$(K,Na)_2SO_4$]. Other phases found include ferroan rhodochrosite and phlogopite. Metzger et al. (1977) could not confirm the presence of hematite, anhydrite, or calcite. Nash and Cunningham also noted that some daughter crystals grew when the inclusions were heated to 150-180°C, and others did not dissolve in 3 or more hours of heating. Whether the crystals that grew were a phase that switched from retrograde to prograde solubility with temperature increase or a new phase from some incongruent solubility is not clear, but they did report a different new phase that formed at ~120°C and redissolved by ~185°C.

Bluebell mine, British Columbia, Canada

Ohmoto (1968) and Ohmoto and Rye (1970) made very extensive and detailed mineralogical, geochemical, inclusion and isotopic studies of this Pb-Zn-Ag limestone replacement ore deposit. Although most (~90 %) of the ore is massive, the latest part is represented by euhedral crystals, including quartz, in vugs, suitable for inclusion study. Only one specimen of quartz and light sphalerite from the massive ore was suitable.

In >5000 inclusions studied, no evidence of boiling or NaCl daughter crystals was found in any sample. Small carbonate and possibly chlorite and/or dickite crystals are present in some inclusions, and CO_2 hydrate crystals were observed on cooling in most of the larger inclusions. The few inclusion data on the massive ore showed Th ≤ 410°C (Tt estimated to be ≤ 445°C and the pressure correction was estimated at <80°C. The later vuggy ore (Period III) could be subdivided into numerous stages and substages, and the data plot as a relatively smooth curve on a Tm ice vs Th diagram (similar to Fig. 15-6). They show a general drop in salinity (from ~10 % to ~3 % NaCl equiv.) and concomitant drop in Th (400 to 220°C), but with three temporary reversals. $P(H_2O)$ in the hydrothermal fluids was estimated to be ~300-800 atm, and the depth of ore deposition ~6 km.

Isotopic analyses of quartz and carbonate samples showed that $\delta^{18}O$ in the hydrothermal fluids changed gradually from +5 per mil to -13 per mil (SMOW) during a period when Tt dropped from 450 to 330°C, but δD and $\delta^{13}C$ were nearly constant during this period (at -152±5 per mil (SMOW) and 5.5 per mil (PDB) respectively). Both $\delta^{18}O$ and $\delta^{13}C$ became more positive in still later fluids. Ohmoto and Rye (1970) pointed out the similiarity of the Bluebell hydrothermal fluids and the Salton Sea (California, USA) brines, suggesting that during at least the later stages of mineralization at Bluebell, the waters were largely meteoric in origin, though possibly of two sources. The range of $\delta^{18}O$ values could be merely a result of differing degrees of isotopic equilibration with the country rocks. Several different sources for the C are suggested by the isotopic data.

PORPHYRY COPPER AND RELATED DEPOSITS

Types of Ore Deposit Covered

The term porphyry copper was originally used to describe a series of large Cu deposits in the western USA, in which the Cu minerals form disseminated grains or thin veinlets in stockwork or breccia in or associated with shallow siliceous porphyritic igneous intrusions (Titley, 1982). Many of these deposits contain minor but economically significant amounts of Mo (the porphyry Cu-Mo deposits), and many also contain appreciable Au, as exemplified by Bingham, Utah, USA, which to 1972, produced ~8 million oz Au (Gilmour, 1982). The Cu/Mo/Au ratio varies widely in different deposits (Kesler, 1973; Hollister, 1975; Titley, 1978). Genetically related Cu skarns are present in many deposits. Some deposits contain major Mo and very little Cu, and Sillitoe et al. (1975) have shown that the major Sn deposits of Bolivia are so closely related as to warrant the term "porphyry Sn deposits." Although future work presumably will reveal significant differences between these types, the available fluid-inclusion data are generally similar throughout all, so they are treated here as a group.

General Nature of the Fluid Inclusions Found

The last decade has seen considerable work on fluid inclusions from various types of porphyry deposit, and the similarities among the inclusion data are rather surprising in view of the many other significant geologic parameters that vary widely from one porphyry deposit to another. Most of these studies have been on secondary inclusions, either knowingly or unknowingly, because most porphyry deposits have been repeatedly brecciated and subsequently undergone extensive annealing, so that the very abundant secondary inclusions may appear primary (Fig. 15-26; Roedder, 1971b, Figs. 25, 26). Although a few inclusion data were reported earlier (e.g., Roedder, 1963, and various Russian papers summarized by Roedder, 1968a), the first study of any extent on inclusions in a Cu porphyry was at Bingham, Utah, USA (Roedder, 1971b). In essence, I found that the fluids present at some time during the formation of this porphyry deposit (and possibly during the precipitation of the Cu, Mo, and Au) were (1) very hot (as much as 725°C), (2) very saline (>60 wt % salts), and (3) "boiling" (or, more precisely, consisting of two phases, liquid and vapor; Bodnar, 1981; Fig. 15-10). In addition, a few inclusions from the Climax, Colorado, USA, porphyry Mo deposit and the Butte, Montana, USA, porphyry Cu deposit were also studied and were found to show similar evidence of "boiling" but lower salinity (35 wt %), and temperature (460°C for Climax). Similar inclusions were also found in samples from five porphyry Cu deposits in Arizona, USA: Sierrita, Esperanza, Mission, Twin Buttes, and Silver Bell. Theodore and Menzie (1984) have shown that the fluid inclusions

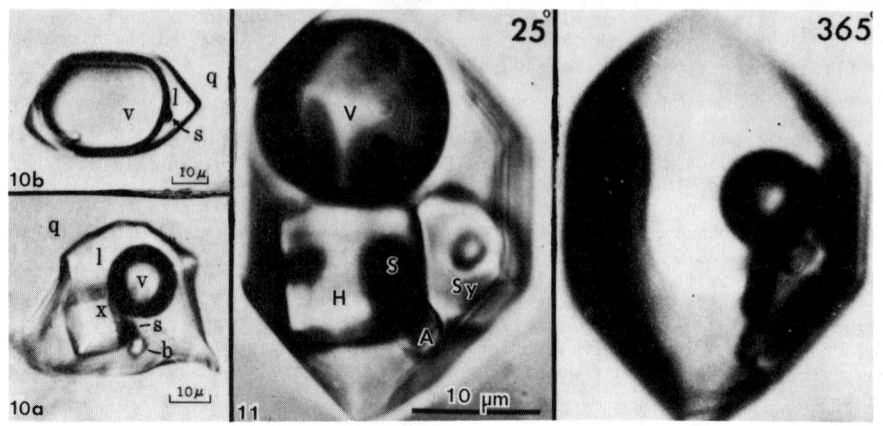

Figure 15-10a,b. Photomicrograph of the two most common types of secondary inclusions at Bingham, Utah, USA, from a 5-10-cm quartz-molybdenite vein from near the center of the deposit. Type (a) contains saturated liquid (l); a small vapor bubble (v); a red or opaque "daughter mineral," presumably hematite (specularite, s); and a large crystal of halite (x). Many of the larger inclusions also contain a small birefringent crystal, possibly anhydrite (b), and other phases. Type (b) usually contains a small opaque grain, presumably hematite (specularite, s), low-salinity liquid (l) and a large bubble of compressed gas (v). The composition of type (b) is as expected if it represents the dense steam phase from the boiling (or condensation) of fluid of the gross composition of type (a). From Roedder (1971b).

Figure 15-11. Photomicrographs of multiphase type (a) inclusion from Bingham, Utah, USA, taken at the temperatures indicated. At 25°C, the phases visible are vapor (V), halite (H), sylvite (Sy), hematite? (S), anhydrite? (A), and several unidentified crystals. The sylvite is dissolved by 80° C (i.e., Tm KCl), and the halite at 365°C. The hematite(?) and anhydrite(?) remain. From Roedder (1971b).

in the F-deficient type porphyry Mo systems contrast with the others, in that they are lower in salinity (4-16 wt %), higher in CO_2, and homogenize in the liquid phase at 250-400°C.

The salinity of the fluids at Bingham was based on the temperatures of dissolution of the large daughter crystals in the polyphase, crystal-rich inclusions (Fig. 15-11) that were found only in the central part of the deposit, in or near the igneous stock, and not in the peripheral Pb-Zn deposits. Measurements of Tm KCl and Tm NaCl on these inclusions, interpreted in terms of the system NaCl-KCl-H_2O (Fig. 8-25), yielded the exceptionally high values for salinity. The exact composition of the original fluid can only be approximated, but the results from two independent methods, optical and thermometric, agree fairly well, at least for the major constituents NaCl and KCl. The optical measurement of the volumes of phases present permits calculation of the total composition, but <u>only</u> in terms of the recognized constituents. Thus, ions in solution other than Na^+, K^+, and Cl^- can only be surmized. The Ca^{2+} content of the fluids now present must be small, as the present SO_4^{2-} content may be large (12,000 ppm in one sample; Rosasco and Roedder, 1979); the commonly found daughter crystal of anhydrite(?) amounts to only ~0.1 vol %, corresponding to only ~600 ppm Ca in the original fluids (Roedder, 1971b). Significant amounts of other ions must be present, since inclusions like those shown in Figure 15-11 have Te <-32°C (Roedder, 1963).

The K/Na ratio of the fluids reflects the temperature and the mineralogy of the source region and the alteration reactions that are so widespread around porphyry deposits. Thus, the highest temperature and highest salinity inclusions are associated with the pervasive potassic alteration so characteristic of the cores of porphyry Cu systems. Hence, chemical analyses for K and Na would be desirable. Unfortunately, most samples from porphyry deposits contain a mixture of different generations of secondary inclusions. In addition, disseminated white mica is so common that it would be difficult to avoid in any bulk leaching study. However, if the system NaCl-KCl-H₂O is appropriate, the K/Na ratio can be obtained from $\overline{Tm}$ KCl and Tm NaCl, as described (Fig. 8-25). Thus, the inclusion illustrated by point C on Figure 8-25 has 18.6 wt % KCl and 49.0 wt % NaCl (and only 32.4 wt % H₂O), yielding K/Na (atomic) of 0.30. ·Similar values are obtained from measurements of the volume percentages of KCl and NaCl, and estimates of the volume of liquid phase, assuming it has the composition of point A (Fig. 8-25).

Not all apparent daughter minerals in the polyphase inclusions dissolved on heating. The most common assemblage of solids in these polyphase inclusions consisted of a large halite crystal, a smaller sylvite crystal, a very small birefringent prism having moderate birefringence and parallel extinction (anhydrite), and a red to black hexagonal flake (hematite). The anhydrite has been verified, for Bingham samples, by laser Raman spectroscopy (Rosasco and Roedder, 1979). Neither the anhydrite nor the hematite redissolve on heating, even if the sample is held at temperature for extended periods. The regularity of occurrence of these two phases at Bingham precludes an origin as accidental solid inclusions. The only apparent resolution of this dilemma seems to require the leakage of H₂ out of the inclusions (Roedder, 1967b; Roedder and Skinner, 1968). The assumption here is that the original inclusion fluid was trapped in quartz under low redox conditions (iron as Fe^{2+} and S as S^{2-}), and, subsequently, the fluid surrounding the quartz host became more oxidizing, as might be expected from late incursion of oxygenated surface waters. Dissociation of the water within the inclusion formed some H₂, which diffused through the quartz down the concentration gradient to the more oxidizing external environment, leaving behind the O₂ to oxidize Fe^{2+} to Fe^{3+}, precipitating hematite, and S^{2-} to SO_4^{2-}, precipitating anhydrite. The present state of oxidation of these inclusion fluids, as determined by laser Raman analysis (Rosasco and Roedder, 1979) that detected no S^{2-} but very high SO_4^{2-} (12,000±4000 ppm), is more highly oxidized than is appropriate for the mineral assemblage in the ores, and hence supports the concept of post-trapping oxidation. However, as noted by T.G. Theodore (pers. comm.), unoxidized sulfide daughter crystals are recognized in many deposits (Table 15-5). The failure of the precipitated daughter crystals to redissolve on heating might also be explained by sluggish kinetics.

The very high Th values found for both polyphase and simple liquid-vapor inclusions from Bingham (many in the range 645-725°C) were several hundred degrees above what most would have expected; hence on several occasions suggestions were made that these data were spurious (e.g., resulting from the trapping of mixtures of vapor and liquid, leakage, or necking down), but the nature and occurrence of the inclusions from Bingham that were measured effectively excluded these other interpretations. Also, the high temperatures were not as exceptional as they were first thought to be. Nash and Theodore (1971) studied the Copper Canyon, Nevada, USA, porphyry Cu deposit and also found evidence of boiling, but at lower salinities (~40 wt %) and considerably lower temperatures (<380°C). (This work focused on the East ore body, which is well outside the outer limit of the intrusive rock at Copper Canyon.) Since then, a large number of other apparently related but not identical porphyry deposits from around the world (including those containing Mo or Sn) have been studied (Table 15-5; see also reviews by Titley and Beane, 1981 and five papers in Titley, 1982). Inclusions from most of these deposits show maximum Th values in the range 400-800°C. Eastoe and Eadington (1982) reported Th ~1000°C for multiphase inclusions (halite, sylvite, opaque minerals, etc.), and Wilson et al. (1980) reported many Th values >800°C, but which they suggested are a result of heterogeneous trapping. In addition, the earlier extensive study by Kelly and Turneaure (1970) of the fluid inclusions in veins of the Bolivian Sn deposits (subsequently considered to be part of the porphyry Sn deposits of Sillitoe et al., 1975; see also Grant et al., 1980) showed that these deposits also formed from very hot, very saline, "boiling" fluids. The uniformity of the combination of the three inclusion

Table 15-5   Reports giving fluid-inclusion data on porphyry Cu-Mo-Sn deposits or prospects

Deposits	Type	Th[1/] (°C)	Salinity (wt %)[2/]	Boiling[3/]	Daughter phases[4/]	Reference
USSR and Asia						
Nine deposits in USSR	Cu-Mo	500-600	high	---	---	Berzina et al. (1976)
Various deposits in USSR	Cu	>600	60	yes	---	Sotnikov et al. (1973)
Kal'makyr, Central Asia	Cu-Mo	860	high	yes	---	Sotnikov & Berzina (1975)
Zhireken, USSR	Cu	500	high	yes	S, A, hem, cpy, CO_2, & biref. dm.	Ermakov & Piznyur (1974)
Sar Cheshmeh and Darreh Zar, Iran	Cu	>650	55	yes	S, and hem.	Etminan (1977)
Yulong, China	Cu-Mo	600	56	yes		Y. Li et al. (1981)
Pacific and Indian Ocean						
Koloula, Guadalcanal	Cu	500-700	~60	yes	S, hem, C?, CO_2 & pyrite?.	Chivas (1976); Chivas & Wilkins (1977)
Panguna, Papua New Guinea	Cu	800	76	yes	---	Eastoe (1978; 1983)
Frieda, Papua New Guinea	Cu	>600	>60	yes	S, A, hem, cpy, moly(?): pyrite or bornite?; 4 unknowns. CO_2 not found.	Eastoe (1976; 1983)
Yandera, Papua New Guinea	Cu	470	60	yes	Hem.	Watmuff (1978)
Mamut, Malaysia	Cu	460	50	yes	---	Imai et al. (1976)
Santo Tomas II (Philex) and Tapian (Marcopper) Philippines	Cu	>600	70	yes	S, hem, 1 opaque, & another dm.	Takenouchi (1980b)
Baguio district, Philippines[9/]	Cu	>700	85	yes	S, A?, hem?, C, & sulfides.	Balce (1979)
South America						
El Salvador, Chile	Cu	>600	>50	yes	S, hem?, cpy?, & CO_2.	Gustafson & Hunt (1975)
Cerro Verde/Santa Rosa, Peru	Cu	450	high	yes	S, A, cpy, & Fe & Mn chlorides; also Cu, Ba.	Le Bel (1976, 1980)
Porphyry Sn deposits, Bolivia	Sn	530	47	yes	S, hem, CO_2, nonmag opq, magnetite, Na-amphibole, 18 unknowns.	Kelly & Turneaure (1970)
Porphyry Sn deposits, Bolivia	Sn	>450	40	yes	S, hem, & hydrous Fe chloride, dawsonite, <8 dms; also Cu, Sn & B.	Grant et al. (1980)
Rio Pisco, Peru	Cu-Mo	---	60	yes	S, A, hem, opq, & 4-11 unknowns.	Agar (1981)
British Columbia						
Endako	Mo	>500	high	---	---	Dawson (1972)
Endako	Mo	>600	60	no	S, hem, moly & 1 unknown.	Bloom (1981)
Hudson Bay Mountain	Mo	>600	80	yes	S, A?, hem, cpy, & erythrosiderite(?): 1 unknown. CO_2 not found.	Bloom (1981)
Granisle and Bell	Cu	<800	<70	yes		Wilson et al. (1980)
Western United States						
Climax, Colorado	Mo	600	~35[8/]	yes	A?, C, CO_2, & opq (nonmag); 4 biref dms. No sylvite.	Hall et al. (1974)
Climax, Colorado	Mo	460	35	yes	Several transparent & opq dms.	Roedder (1971b)
Henderson, Colorado	Mo	500-650	65	yes	Hem, C, & moly.	White et al. (1981)

Table 15-5   Reports giving fluid-inclusion data on porphyry Cu-Mo-Sn deposits or prospects (concluded)

Deposits	Type	Th[1]/ (°C)	Salinity (wt %)[2]/	Boiling[3]/	Daughter phases[4]/	Reference
Questa, New Mexico	Mo	>600	70	yes	S, hem, moly & 1· unknown.	Bloom (1981)
Butte, Montana	Cu-Mo	---	35	yes	Hem? C? & both biref & opq dms.	Roedder (1971b)
Bingham, Utah	Cu-Mo	725	>60	yes	S, A, hem, C?, CO₂, & pyrite?	Roedder (1971b)
Bingham, Utah	Cu-Mo	600	high	yes	S, A, hem, C, CO₂, & nonmag opq.	Moore and Nash (1974)
Southwest Tintic, Utah	Cu	700	~40	yes	S, A, cpy, CO₂, muscovite or phengite, & 3 unknowns.	Ramboz (1979)
Santa Rita, New Mexico	Cu	>800	~64	?--	S, hem, cpy, pyrite, magnetite.	Reynolds & Beane (1979)
Santa Rita, New Mexico	Cu	>550	>60	yes	S, A?, hem, C?, & 2 opq.	Ahmad & Rose (1980)
Red Mountain, Arizona	Cu	425	50	yes	S, A, hem, cpy & nonmag opq; no CO₂.	Bodnar & Beane (1980)
Bagdad, Arizona	Cu	310	35	yes	Hem, cpy?, CO₂, & a biref dm.	Nash & Cunningham (1974)
Ely, Nevada	Cu	600	42	yes	---	Huang (1976)
Ely, Nevada	Cu-Mo	>550	50	yes	Cpy.	Goss & Cathles (1983)
Sierrita-Esperanza, Arizona	Cu	>500	45	moderate	S, hem, cpy, & CO₂.	Denis et al. (1980)
Sierrita, Arizona	Cu	430	41	yes	S, A, hem, low- to zero-biref unknown, & opq.	Preece and Beane (1982)
Copper Canyon, Nevada	Cu	500	high	---	S, A?, hem, C?, CO₂, & chlorite or epidote.	Nash & Theodore (1971); Theodore & Blake (1975)
Other Areas						
37 deposits in US & Canada	Cu	250-700	high[5]/	yes	S, A, hem, C, CO₂, & dawsonite?	Nash (1976)
Various United States deposits	Cu	500	60[6]/	yes	---	Cunningham (1975)
Ballachulish, Scotland	Cu-Mo	300	---	yes[7]/	S. CO₂ not found.	A.M. Evans et al. (1979)
Sapo Alegre, Puerto Rico	Cu	400	---	---	---	Cox et al. (1975)
Tanamá & Helecho, Puerto Rico	Cu	>500	>50	yes	Fe chloride, S, hem, Na-K-Fe-Ca-chlorides, cpy.	Cox (1984)
Mines Gaspé, Québec, Canada	Cu	506	47	yes	S, A, CO₂, & 3 opaques	Shelton (1983)

1/ Maximum temperature of homogenization (some may be temperature of trapping).

2/ Maximum salinity in NaCl equivalent weight percent.

3/ "Boiling" refers to the coexistence of two phases of grossly different density, regardless of how this condition was achieved (see text).

4/ NaCl present in all. Not all phases coexist within individual inclusions. S-sylvite; A-anhydrite; Hem-hematite; C-carbonate; Cpy-chalcopyrite; CO₂-recognized CO₂ gas or liquid; dms-daughter minerals; moly-molybdenite; biref-birefringent; nonmag-nonmagnetic.

5/ High salinity in all but 3 of 37 deposits.

6/ Mainly 0-13 wt % NaCl equiv. Inclusions containing daughter halite very rare.

7/ But based on the (invalid?) criterion of coexistence of liquid-rich and gas-rich inclusions "with similar homogenization and salinity values" (A.M. Evans et al., 1979, p. 47; emphasis added).

8/ Mostly 0.7 to 12 wt % NaCl equivalent, but some have a halite daughter phase.

9/ Includes the Santo Tomas II, Santo Nino, Kennon, and 4 other deposits.

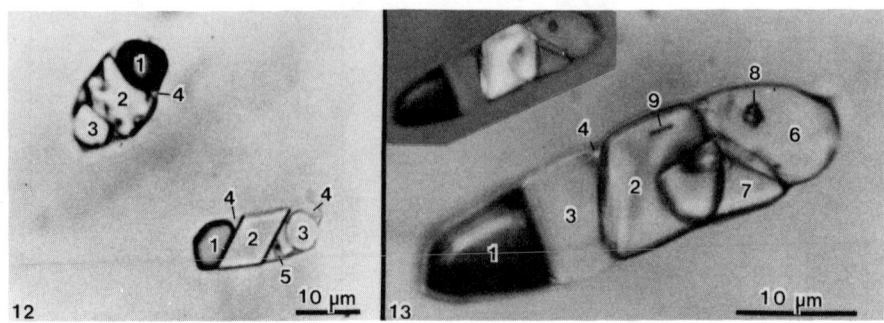

Figure 15-12. Polyphase inclusions in vein quartz from Chorolque porphyry Sn deposit, Bolivia. From Grant et al. (1977, 1980). The identified phases are: 1-vapor bubble; 2-hydrous Fe chloride; 3-halite or (NaK)Cl; 4-fluid; 5-unidentified.

Figure 15-13. Polyphase inclusion in quartz of quartz-tourmaline altered vent rock from Chorolque porphyry Sn deposit, Bolivia. From Grant et al. (1977, 1980). Phase identification as in Figure 15-12, 6-probably sylvite, and 7, 8, and 9, all unidentified. Inset shows birefringence of hydrous Fe chloride in crossed polars.

characteristics of most porphyry deposits, very high temperatures (~400-800°C), very high salinities (~40->60 wt % salts), and "boiling " sets them apart from most other ore deposits as completely as do the geological characteristics. Noticeable differences in the fluid-inclusion characteristics among the Cu-Au, Cu, Cu-Mo, Mo, Mo-Sn, and Sn-W porphyries certainly have some genetic signifi- cance but are not completely understood at this time. Regardless of the physical and chemical significance of these three attributes of the fluids, the presence of inclusions indicating such conditions can apparently be used as an empirical "fingerprint" for porphyry Cu deposits, and hence can be of value as an explora- tion tool for this one large and important class of deposit (Roedder, 1971b, 1977b; Cunningham, 1975; Nash, 1976).

In contrast to the above data, strongly saline, but apparently somewhat lower-temperature inclusions were found in a series of barren intrusions having similar chemistry and level of emplacement in Colorado (Bradbury, 1975); Steve Ludington and T.G. Theodore (pers. comm.) also have found such inclusions in other barren intrusions. The relationship between such high-salinity inclusions in barren intrusions and apparently similar inclusions in productive porphyry Cu intrusions is an open question. R.J. Bodnar (pers. comm.) has suggested that there may be no significant difference in the major characteristics such as salinity, Th, boiling, composition, etc.; only the metal content of the fluids may differ, for reasons still to be established.

Heavy-Metal Content of Inclusions

The heavy metal content of the inclusion fluids in the porphyry deposits has been estimated by various methods, including chemical analysis of leachates (Hall et al., 1974), and by calculations based on the volume percentage of var- ious heavy-metal-bearing daughter phases, particularly those identified by SEM studies of opened inclusions (see Fig. 10-2). Fe is most commonly recognized in the form of hematite[6], generally corresponding to <1 wt % Fe in the fluids,

----

[6] Ramboz (1979) has suggested that the common identification of hematite may be in error, on the basis of evidence presented by Le Bel (1976) that the pseudohexagonal platelets he found at Cerro Verde, Peru, were muscovite or ferruginous phengite. However, the optical properties reported for platelets from Bingham (Roedder, 1971b) and the SEM analysis showing only Fe, are almost irrefutable. Also, the identification of hematite from inclusions from the porphyry Cu prospect at Copper Creek, Arizona, USA, using the Gandolfi X-ray technique, is irrefutable (Zolensky and Bodnar, 1982). Both identifications are probably correct. Eastoe (1978) suggested that such hematite flakes at Panguna, Papua New Guinea, may be solid inclusions.

but magnetite, pyrite, and chalcopyrite are not rare. Grant et al. (1980) reported hydrous Fe chloride as a major daughter crystal in some Bolivian porphyry Sn deposits; the amounts of this phase are so large (~50 vol %; Figure 15-12, 13) that the solutions must have held far more Fe than Na. Le Bel (1976, 1980) reported Fe, K, and Mn chlorides as daughter phases present in many of the inclusions from Cerro Verde, Peru (see Fig. 5-12). Cox (1984) reported an Fe-Cl phase forming hexagonal prisms to make up a major portion of the large amounts of daughter crystals in a Puerto Rican porphyry copper. Wilson et al. (1980) suggested that the K-Fe-Cl phase that they found in inclusions from the Granisle and Bell porphyry Cu deposits (British Columbia, Canada) may be erythrosiderite $[K_2Fe^{3+}Cl_5 \cdot H_2O]$. With these high concentrations of Fe^{3+}, hydromolysite $[FeCl_3 \cdot 6H_2O]$ has also been suggested, but note, however, the caveats in the section on SEM in Chapter 5. The Fe concentrations suggested by these inclusion studies are orders of magnitude higher than the "extremely high" values reported by Whitney et al. (1979) from experimental studies of chloride solutions equilibrated with synthetic quartz monzonite assemblages.

Even though porphyry Cu deposits themselves are characterized by remarkably low Zn contents (Johan, 1980), Zn has been reported in the inclusions from several porphyry deposits (e.g., Eastoe, 1978), and Pb-Zn deposits commonly occur peripheral to the Cu deposit. Hall et al. (1974) reported <4600 ppm Zn in leachates from Climax, Colorado, and Le Bel (1976) reported a Zn-bearing phase at Cerro Verde but believed it was a solid inclusion.

Although Fe is the most abundant metal, the most important heavy metal in these inclusions is Cu. Chalcopyrite is commonly reported as a daughter phase (Table 15-5) and has been positively identified from one locality by the Gandolfi technique (Zolensky and Bodnar, 1982), but some of the many reports are based only on tetrahedral shape as seen by SEM (Fig. 5-13), and hence the mineral could also be tetrahedrite or other phases. (Some reports of "pyrite," based on finding opaque "cubes," might actually refer to chalcopyrite sphenoids, because such sphenoids will appear square if viewed along a twofold axis.) The volume percentage of chalcopyrite daughter crystals shown in the illustrations of these various reports indicates that very appreciable amounts of Cu have precipitated in the inclusions and that additional amounts of Cu (generally unknown but probably low) must still be in solution. Measurements of the daughter-crystal size have resulted in estimates of the Cu content of the inclusions that range from 250 ppm for Sierrita-Esperanza, Arizona, USA, (Denis et al., 1980), to 500-1000 ppm for Butte, Montana, USA, (Roberts, 1973), about 1900 for Panguna, Papua New Guinea (Eastoe, 1978), 1000-2000 for southwest Tintic, Utah, USA, (Ramboz, 1979), 2000-5000 for Sar Cheshmeh, Iran (Etminan, 1974, 1977), and 4000-16,000 for Cu- and Mo-bearing breccia pipes in Mexico (Sawkins and Scherkenbach, 1981). Hall et al. (1974) reported 40-4600 ppm Cu in leaches from Climax, Colorado, USA.

Origin of Porphyry Deposits

A complete consensus has not been reached on the origin of the fluids or the Cu in these deposits or on the role played by the igneous rocks. The mineralogy of the extensive alteration of large volumes of host rock indicates that very significant amounts of hot fluid must have moved through, beyond and above the volume now occupied by Cu ore. Some workers have suggested that the heat, the Cu, and the fluids came from the igneous rocks; others have suggested that the igneous rocks merely provided the necessary heat energy to drive circulation of ground water, which removed Cu from surrounding country rocks or the igneous rocks and redeposited it in more concentrated form near the apex of the intrusive body. Although these questions are of considerable importance to the economic geologist, only the nature of the fluids that have been present in such bodies, as evidenced by the fluid inclusions found, can be discussed here. The discussion is biased in favor of a magmatic origin, as I believe the fluid-inclusion evidence requires. This conclusion does not conflict with the isotopic evidence (e.g., Sheppard et al., 1971) that significant amounts of meteoric water were involved in the later extensive sericitization and argillization.

Although certain aspects of the fluid inclusions found in the porphyry deposits are remarkably uniform among the many occurrences, the geologic parameters are numerous and wide-ranging, as discussed by Burnham and Ohmoto (1980) and Burnham (1981). The most important parameter is the original H_2O content of the intrusive rock. Additional important parameters include the bulk composition of the magma (and, hence, its tectonic setting) and particularly the nature of the solid phases that form during ascent and cooling. If enough hydrous phases form, such as hornblende or mica, the remaining silicate fluid may be so depleted in H_2O that little if any volatile phase may form and leave the intrusive, whether at fixed pressure (i.e., "second boiling"), or by pressure release during ascent, or by a combination of these processes. Any fluid that does leave under these conditions, however, may be high in Cl. The temperature of the magma, its metal, Cl, and S contents, and its oxidation state are also critical, as is the overlying pressure at each stage of intrusion and crystallization. The interrelationships of some of these variables were discussed by Chivas (1981).

In at least some silicic magmas, these variables have been such that an immiscible hydrosaline melt separated during the magmatic stage. Evidence for this separation has been found in experimental studies (e.g., Koster van Groos and Wyllie, 1969) and in primary inclusions of such melts in magmatic minerals (e.g., at Ascension Island, Roedder and Coombs, 1967), but the general assumption is that a much lower salinity fluid is normally evolved. The pressure, temperature, and salinity of the fluids as they are evolved, and hence their density and the appropriate isochores, have been widely discussed.

Superimposed on these unknowns, however, is the possibility or, perhaps better, the probability, of phase or composition changes in the fluid after its first separation from the silicate magma as an immiscible fluid. Possible processes include (1) reaction with preexisting minerals, which withdraws H_2O from the fluid and changes the cation values (including particularly the acidity); (2) precipitation of NaCl on cooling; and (3) splitting into two immiscible phases. The last can come about by the formation of bubbles of a new, lower density, lower salinity phase (boiling or effervescence), or by the formation of a new, higher density, higher salinity phase (condensation), as detailed by Henley and McNabb (1978). Since continuous brecciation is common, any or all of these modified fluids may be trapped. Note, however, that any fluids that were undersaturated with respect to the inclusion-bearing phases would not be trapped (M. Barton, pers. comm.).

Although the fluid inclusions in most porphyry Cu deposits are remarkably uniform in the three characteristics of high maximum temperatures, high salinities, and evidence of "boiling," perhaps the single most striking aspect of the fluid inclusions in most porphyry deposits is the extreme variability in each deposit of the apparent sequence of formation, volume percent vapor, salinity of the liquid phase, presence or absence of daughter minerals, and their number, type, and volume percentages. Figure 15-14 provides a good indication of this variation. Ahmad and Rose (1980) have added several boundary lines from the pure system NaCl-H_2O to this diagram. Two features bear special note: (1) some of their inclusions contain visible daughter crystals of KCl and hence all could have as much as 10 wt % KCl (Fig. 8-25); as a result, data on the pure system NaCl-H_2O are not necessarily applicable. (2) The temperatures recorded on the abcissa of Figure 15-14 are Th L-V only, and do not give Tm NaCl, whether Tm NaCl was above or below Th L-V.

Ahmad and Rose (1980, p. 239) listed the following seven possible causes for the scattering of the data for Santa Rita, Arizona, USA, shown on this

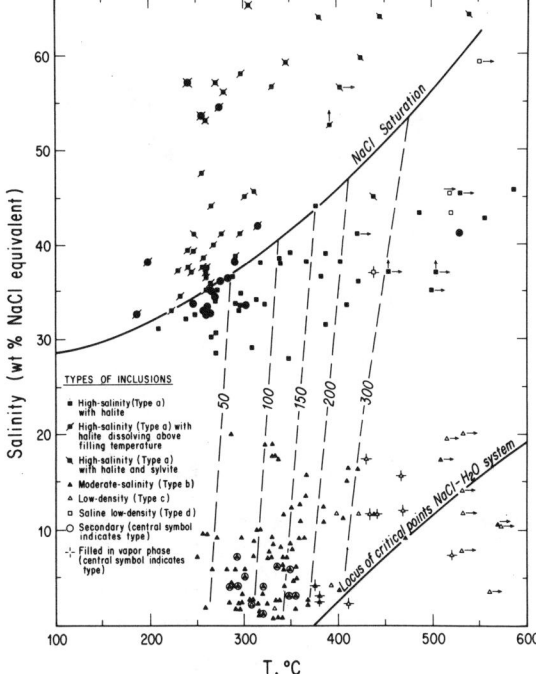

Figure 15-14. Plot of salinity vs Th L-V ("filling temperature") for inclusions on which both were measured, from the Santa Rita, New Mexico, USA, skarn-porphyry Cu deposit. Modified from Ahmad and Rose (1980). All points shown above 25 wt % NaCl have halite or halite plus sylvite daughter crystals at room temperature. Note that abcissa temperatures are for Th L-V only, and are not for complete homogenization. Dashed lines refer to vapor pressures of NaCl solutions at the indicated temperatures and salinities.

figure; presumably these causes are applicable to other porphyries as well:

(1) The inclusions have leaked either naturally or during heating in the microscopic stage.
(2) The pressure varied during mineralization, so that different pressure corrections are needed for different inclusions.
(3) The fluid was boiling, and many inclusions trapped mixtures of liquid and vapor.
(4) The inclusions have necked down into two or more inclusions since their original formation and do not represent the original condition.
(5) The [presumed primary] inclusions are [actually] secondary and of a variety of ages.
(6) The inclusions formed in the liquid + solid salt ± vapor field, and trapped mixtures of three phases.
(7) The quartz in individual small segments of these veins was deposited from fluids with a large range of temperatures and salinities.

They suggested that numbers 1, 2, 4, and 5 are either unlikely, too rare, or too small in effect to explain the major variations, leaving numbers 3, 6, and 7 as appropriate explanations for some or much of the variation. Number 6 is discussed below. At Bingham, Utah, USA, I would consider only numbers 5 and 7 to be the major sources of variation, superimposed on the effects of liquid-vapor equilibria under conditions of changing P and T. Number 7 represents, in effect, a lack of good paragenetic control; such control may be difficult or impossible to obtain in many porphyry systems.

Ahmad and Rose (1980) believe that the data gap in salinities shown in Figure 15-14 between 20 and 27.5 wt % NaCl is real and that it indicates the involvement of two different fluids. I would suggest that the apparent gap may

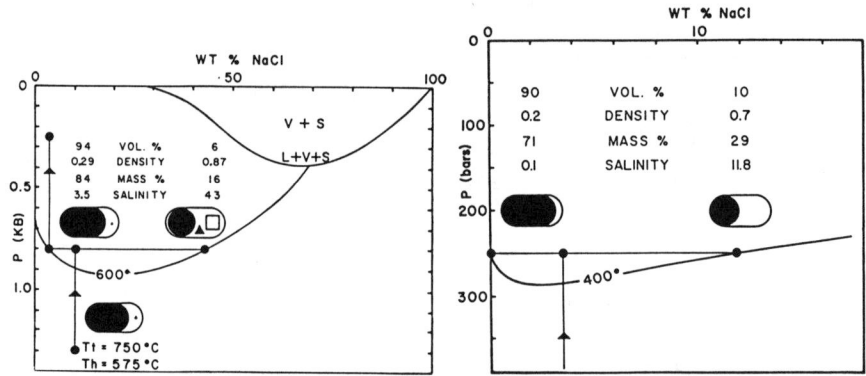

Figure 15-15. P-X projection of the 600°C isotherm in the H₂0-NaCl system, showing the composition, density, and mass and volume percents of the coexisting phases at 800 bars. Also shown are the room-temperature phase relations of fluid inclusions that trapped either the liquid phase or the vapor phase at 600°C and 800 bars and the room-temperature phase relations of an inclusion that trapped a 10 wt % NaCl fluid at 750°C and 1300 bars. From Bodnar (1982).

Figure 15-16. P-X projection of a part of the H₂0-NaCl phase diagram, showing the compositions and densities of coexisting liquid and vapor phases at 400°C and 250 bars. Also shown are the mass and volume percents of each of the phases and the 25°C phase relations of fluid inclusions that trapped the liquid or the vapor phase at 400°C, 250 bars. From Bodnar (1982).

be an artifact introduced by one or more varieties of metastability. Failure on cooling to nucleate what should be very small daughter crystals of halite, and particularly failure to nucleate NaCl·2H₂O, could yield this gap. Additional problems may stem from lack of stage calibration at low temperatures, and particularly from viewing multicomponent natural systems in terms of the pure system NaCl-H₂O. Ahmad and Rose indicated that on heating some of these inclusions above Th a second liquid droplet having a high index of refraction formed; this phenomenon suggests a complex system, as no geologically likely simple system is known to show such immiscibility (see Chapter 4).

Some idea of the very major changes that fluids can undergo in P-T-X space for the pure system NaCl-H₂O, under conditions appropriate for porphyry systems, can be seen in the calculations by Bodnar (1982). Although this work ignores the effects of KCl, CO₂, and all constituents other than NaCl and H₂O, it shows what the phase assemblage at room temperature will be for inclusions trapped under a variety of conditions, and particularly what pairs of inclusions, representing the trapping of liquid and vapor phases (from boiling or condensation), will look like.

Figure 15-15 is a replotting by Bodnar (1982) of the 600°C isotherm data of Sourirajan and Kennedy (1962), using a linear rather than a logarithmic abcissa (compare with Fig. 8-13). To illustrate the possible behavior of porphyry-type fluids, Bodnar has shown the room-temperature appearance of "standard" capsule-shaped inclusions trapped under various conditions. In this example, he assumed an original magmatic fluid containing 10% NaCl, trapped at 750°C and 1300 bars. This fluid would have Th = 575°C, and inclusions of this fluid would contain a large vapor bubble (bottom inclusion). If such a fluid were cooled along an isochore to 600°C and 800 bars, it would split into two immiscible fluids (in the example, it would condense some liquid phase), and inclusions that trapped one or the other of these two fluids would look as shown. To obtain these figures, Bodnar (1983) used calculations of the densities of NaCl-H₂O fluids under various P-T-X combinations, based on a variety of data, to evaluate the fluid-inclusion parameters shown on Figure 15-15. The original low-density, low-salinity fluid split to yield 6 vol % of a much higher density, higher salinity fluid, which, on cooling to room temperature, would contain a daughter crystal of halite and a smaller bubble. The partitioning of Cu between these two phases, shown in the form of a black triangle representing a chalcopyrite daughter crystal, is purely diagrammatic.

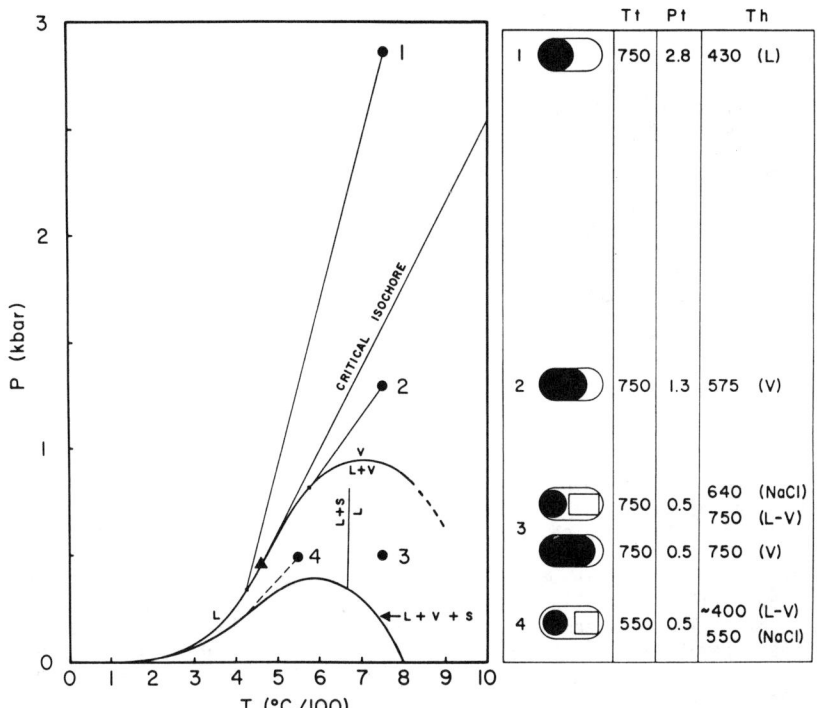

Figure 15-17. P-T projection showing phase boundaries for a 10 wt % NaCl bulk composition. The right side of this figure shows the room-temperature phase relations of fluid inclusions trapped at points 1-4 on the P-T projection. From Bodnar (1982).

The behavior of the "vapor" phase shown in Figure 15-15 involves the additional assumption that it is then cooled to 400°C and 250 bars (black dot in upper left of Figs. 15-15,16). Under these assumptions, the fluid again splits into "liquid" and "vapor" phases, having the widely different salinities and densities shown.

Perhaps the most important single point of this work is the concept that Th in porphyry-type inclusions is not a good evidence of Tt and hence of paragenetic stage. This concept is seen more clearly in some similar examples in a P-T plot (Fig. 15-17). This plot shows the room-temperature phase relations of four inclusions. Inclusions 1, 2, and 3 were all trapped at 750°C from an original 10 wt % NaCl fluid, but at three different pressures. On cooling, inclusion 1, trapped at 2800 bars, follows down the isochore shown and forms a vapor bubble at 430°C (i.e., it would show Th L-V(L) at 430°C). Inclusion 2 (taken from Figure 15-15) will contact the two-phase curve at 575°C and hence shows Th L-V(V) at 575°C. A composition containing 10 wt % of NaCl at 750°C and only 500 bars pressure will consist of two fluids, a dense highly saline fluid and a low-density, low-salinity fluid, as shown. Inclusions of the low-density phase will show Th L-V(V) at 750°C; those of the dense phase will show Tm NaCl at 665°C and Th L-V(L) at 750°C.

Inclusion 4 on Figure 15-17 provides an example of fluids having Tm NaCl>

449

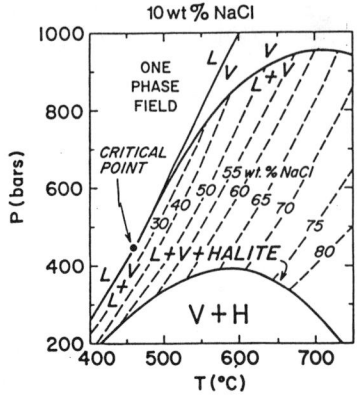

Figure 15-18. Enlargement of the central part of Figure 15-17 (400-750°C, 200-1000 bars), showing the composition of the liquid phase in the liquid+ vapor field for a bulk composition of 10 wt % NaCl. From Bodnar (1982).

Th L-V. If the fluid from point 3 is cooled to 550°C, still at 500 bars, and trapped in an inclusion, it will have Th L-V ~400°C and Tm NaCl 550°C. The details of the field in which this inclusion falls are shown in Figure 15-18.

White et al. (1971) showed that during formation of a porphyry deposit, a low-density fluid phase filled much of the available porosity, and this vapor-dominated part of the system was overlain by cooler liquid-filled rocks. The vapor phase from "boiling" at the bottom of the vapor-dominated part would travel upward, transferring heat and volatiles, and would finally condense. If the condensed fluid flowed down again, the process would become essentially a giant heat pipe and hence could result in extremely rapid heat flow.

Although concentration by true boiling is frequently used as an explanation of the high salinities, the maximum possible increase in concentration is small, unless an external source of heat is available to flow into a given parcel of brine. However, high concentrations can be achieved easily if a small amount of liquid condenses from a vapor phase. Henley and McNabb (1978) have discussed the quantitative aspects of the process, assuming that a magmatic plume of low-density, low-salinity fluid is evolved and subsequently condenses in part to yield the high-salinity fluids found in the inclusions. These essentially magmatic fluids are responsible for the potassic alteration characteristic of the cores of these systems. Subsequent mixing with larger volumes of ground water yields the lower temperature, lower salinity fluids that result in the extensive phyllic, argillic, and propylitic alteration that characterizes the outer fringe of many of these deposits and is also superimposed on the core materials as the thermal pulse dies down. Inclusions of such fluids yield the lower values for Th and salinity that are found throughout most porphyry deposits, and have the isotopic signature of local ground waters, possibly modified by some oxygen exchange (e.g., Hall et al., 1974; Preece and Beane, 1982).

The density of the highly saline fluids, on the basis of measurements of multiphase inclusions, was well above 1.0 g/cm^3, and even as high as 1.3 g/cm^3 at the time and temperature of trapping. The problem is how such a fluid could have been held in the obviously permeable, fractured rocks of the deposit when the surrounding rocks at some distance must have been saturated with cold ground water at a density of 1.0 g/cm^3. Perhaps these hot fluids precipitated enough quartz on cooling (or anhydrite on heating) to be self-sealing and hence made their own impervious container. In any case, the high densities add a new and severely limiting parameter to the extensive computer calculations of flow lines in hydrologic models of flow patterns in the vicinity of cooling intrusive bodies, made to simulate the conditions during formation of porphyry deposits (e.g., Cathles, 1977, 1981; Norton, 1979; Norton and Cathles, 1979). These models generally deal with cold ground water, heated water, and steam, all at densities <1.0, and are not extended to environments in which the ore-forming fluids are more dense than the surrounding ground water. I pointed out (Roedder, 1971b, 1977b) that the inclusion evidence of densities >1.0 at temperature would reverse the flow patterns from those produced by normal thermal-convection cells, and hence would severely limit the applicability of simulations based on the assumption of ore-fluid densities <1.0. Such simulations are much more realistic once the high temperatures have been quenched and the high salinities have been flushed out or diluted; as such, they are particularly applicable to the later stages of alteration and to any remobilization and redeposition of ore that may occur at that stage.

The chalcopyrite daughter crystals in such dense fluids might suggest that these fluids were responsible for the transport of the Cu. Although they obviously contained Cu, this does not require that the Cu was deposited by these fluids at the high temperatures indicated by their Th values. R.J. Bodnar (pers. comm.) has pointed out that the precipitation of Cu probably occurred at <400°C, as the solubility of Cu in these brines increases greatly above 400°C. If the actual deposition of Cu from these fluids occurred at these lower temperatures, the daughter crystals of chalcopyrite should dissolve at such temperatures. Unfortunately, most reports do not indicate the temperature of dissolution of chalcopyrite daughter crystals, and many indicate only that some (or all) unknown opaque daughter crystals did not dissolve.

Several studies of fluid inclusions in porphyry Cu deposits have considered the compositions in terms of the pure system $NaCl-H_2O$, particularly the P-X plot of Sourirajan and Kennedy (1962; see Fig. 8-13). In some reports, this is done even when the inclusions were stated to contain more KCl than NaCl! In many studies the compositions of the multiphase inclusions have been plotted on the diagram for the condensed system $NaCl-KCl-H_2O$ (Figure 15-19), on the basis of Tm KCl and Tm NaCl, as described above. The resulting compositions are in different fields for individual deposits; for some deposits, the compositions are spread over a rather wide range. Some of the individual fields are elongated, and Cloke and Kesler (1979; see also Erwood et al., 1979) have extended lines drawn through these "halite trends" back to the NaCl-KCl sideline. On the basis of the exact position of this extrapolation, and the inclusion behavior, these workers concluded (in contrast with the condensation model of Henley and McNabb, 1978) that there are four different trends, all formed by the crystallization of KCl-bearing halite from the hydrothermal solutions before the remaining liquid was trapped in the inclusions, i.e., saturated fluids were trapped. The differences are based on the specific effects of pressure before the trapping, resulting in various sequences of liquid, vapor, and crystals. Kalyuzhnyi (1976), Eastoe (1978, 1979), Wilson et al. (1980) and others also proposed that some salt-rich magmatic fluids had precipitated NaCl before trapping. R.J. Bodnar (pers. comm.) has noted that if both halite and vapor disappear at the same Th, the inclusion could have trapped saturated brines on the 3-phase, L+V+S curve (Fig. 15-17).

The precise point on the NaCl-KCl sideline at which such extrapolations of "halite trends" intersect is strongly dependent on the accuracy of Tm KCl, and on which inclusion data are plotted. For example, inclusions in which the KCl is small and hidden[7] or that failed to nucleate KCl on cooling, or that should nucleate it, but only at less than room temperature, are automatically excluded, but could well fill in the area between the bulk of the data points for a given trend and the NaCl-H_2O sideline. Kwak and Tan (1981) showed that Ca can have major effects. The direction of the "halite trend" must also be based on the

----

[7] For example, even KCl-rich samples in the Granisle-Bell envelope on Figure 15-19 would have only ~15 vol % KCl at room temperature.

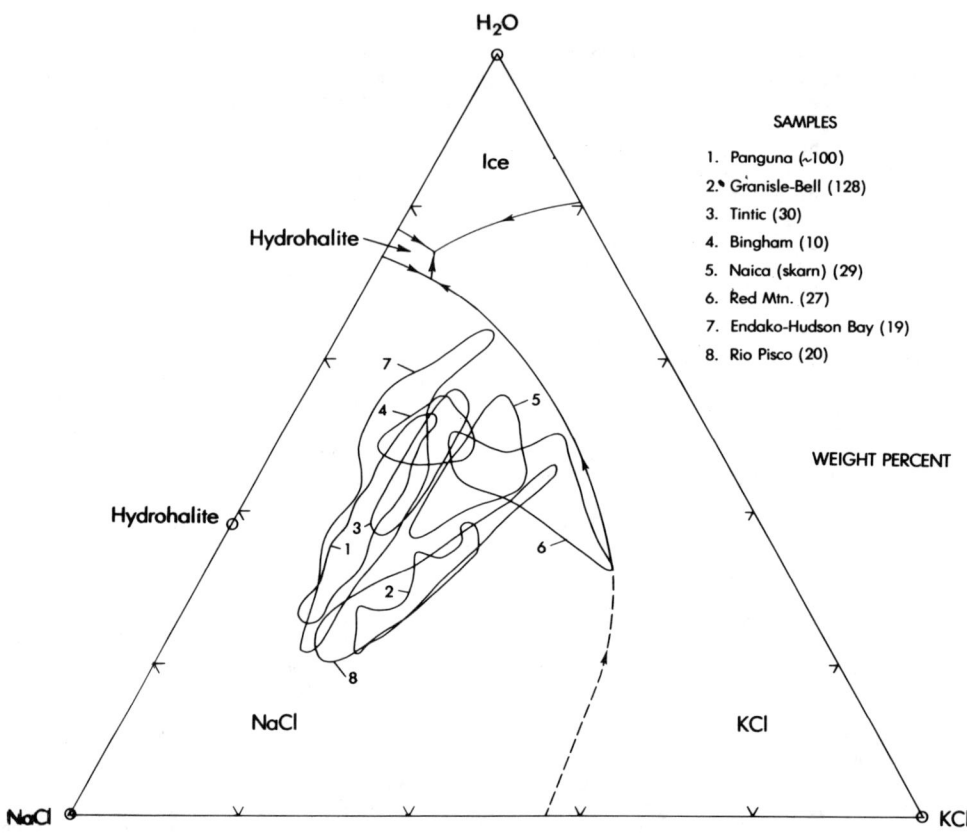

H₂O

Ice

Hydrohalite

Hydrohalite

NaCl   KCl

NaCl   KCl

Figure 15-19. Plot of the system NaCl-KCl-H₂O, showing the composition ranges for fluid inclusions from several porphyry deposits, from the following sources: 1. Eastoe (1979); 2. Wilson et al. (1980); 3. Ramboz (1979); 4. Moore and Nash (1974); 5. Erwood et al. (1979); 6. Bodnar and Beane (1980); 7. Bloom (1981); 8. Agar (1981). The number of inclusions measured is given in parentheses. The low-H₂O part of the NaCl-KCl cotectic from I-Ming Chou (pers. comm.). See Fig. 8-25 for isotherms and isobars, and text for explanation.

"best fit" to a somewhat diffuse group of data points, and unfortunately these plotted points sometimes constitute a selected data set. The differences between the positions of the points for various deposits obviously are "trying to tell us something." However, a valid interpretation of these trends must await adequate data on the system NaCl-KCl-H₂O and the effects on it of the many other solutes that are almost certainly present, in highly variable amounts, as well as on adequate procedures to place the inclusions in a given sample in correct time sequence. Obviously, accurate analyses of individual inclusions would be preferred to the indirect thermometric method used above, which requires several rather dubious assumptions.

The problem of whether most, some, or none of the fluids in inclusions in the porphyry deposits were saturated with NaCl at the time of trapping is particularly difficult. Any inclusion that has Tm NaCl<Th L+V presumably was not saturated in NaCl at trapping, unless both saturated liquid and vapor were trapped simultaneously (as has been suggested by some). An inclusion having Tm NaCl>Th L+V (i.e., "halite homogenization" of Wilson et al., 1980) may have been formed in various ways:

(1) It could have been trapped as a homogeneous fluid, undersaturated with

452

respect to NaCl; on cooling, it moved along the appropriate isochore in P-T space for a fluid of that density and composition until it reached the halite liquidus, at which halite formed. Tm NaCl will thus be some unspecified amount less than Tt.

(2) It could have been trapped as an NaCl-saturated fluid, i.e., halite crystals were present at Tt but were not trapped in the inclusion. On cooling after trapping, halite will form immediately (barring metastability), and on reheating, Tm NaCl = Tt.

(3) It could have been trapped as a saturated fluid, along with some solid halite. If so, Tm NaCl>Tt.

(4) It could have been trapped during necking down of an inclusion containing a daughter crystal. If so, Tm NaCl is essentially meaningless, and Tt cannot be determined.

If NaCl was present as a solid phase in the fluids present during the extensive brecciation and rehealing that occurred in the formation of most porphyry deposits, it should be found as solid inclusions. Also, any given fluid that forms a daughter crystal at Tm NaCl after being trapped in an inclusion should precipitate NaCl in the bulk of the fluid outside the inclusion host crystal, once it cools to Tm NaCl (unless it has been diluted or flushed away by this time). However, such solid inclusions of NaCl are rarely reported. In part, this lack might be simply a result of the similarity of index of refraction (e.g., see Fig. 4-8), but if so, cubic casts from the solution of such solid inclusions should be common; to my knowledge, these casts have never been reported from porphyry deposits. Also, if solid NaCl is trapped in fluid inclusions (case 3 above), all ratios of NaCl to fluid should be found. Table 15-4 shows, however, that the maximum wt % NaCl in a large number of careful studies is 85, but not higher. The same reasoning effectively excludes necking down (case 4 above).

Whether the high-salinity inclusion fluids form by direct immiscibility from a silicate magma or by partial condensation from a larger volume of lower salinity fluid evolved from the magma, the variables are such that a wide range in the NaCl-H$_2$O ratio is not only possible, but probable, and hence could yield the "halite trends" shown in Figure 15-19. Thus, the liquid-plus-vapor field on the P-X diagram for the pure system NaCl-H$_2$O (Fig. 8-13), shows that both pressure and temperature changes will result in major changes in the salinity of the "liquid" phase. In a shallow intrusive environment, in which abundant brecciation is taking place, major fluctuations in pressure should be expected. In part, such changes could well be brought about by the precipitation of quartz in the existing fractures reducing the permeability until the internal pressure rises enough to cause new brecciation. From the above considerations, I suggest that there is no need to assume that the high-salinity fluids were saturated at the time of trapping, and I do not believe that they generally were, but much is still to be learned about porphyry deposits.

TUNGSTEN AND OTHER SKARN DEPOSITS

A continuum exists between the porphyry-type and the skarn-type of ore deposit, and at least some skarn ores appear to be simply porphyry deposits in which some limestone or dolomite was present in the country rock. However, the mineralogy of many of the skarn deposits differs significantly from that of the porphyry type. The many types of skarn deposits have been reviewed systematically and thoroughly by Einaudi et al. (1981); they include ores of Fe, W, Cu, Zn-Pb, Mo, and Sn. Relatively few of these various types of skarn deposits have been the subject of extensive fluid-inclusion studies, and of those discussed by Einaudi et al. (1981), four are listed on Table 15-5 as "Porphyry" deposits. As in the porphyry deposits, the central early high-temperature mineralization is commonly overprinted with later, lower temperature phases of mineralization

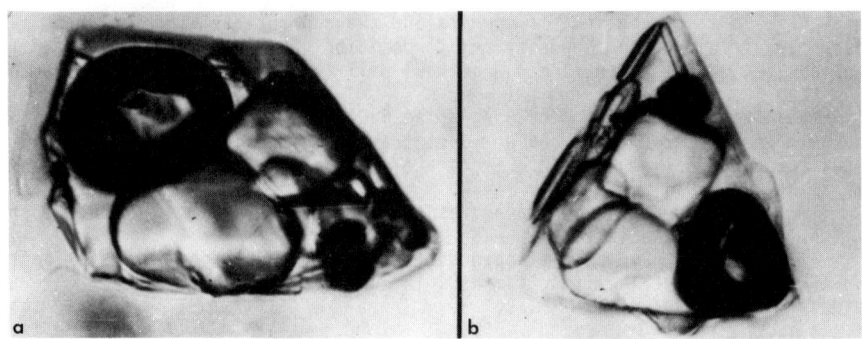

Figure 15-20. Multiphase high-salinity inclusions in fluorite from lower levels of the Tyrnyauz skarn ore deposit, USSR. In both photographs, the large opaque crystal is a rhombic dodecahedron of magnetite. (a) from Kulikov (1982b); (b) courtesy of Igor Kulikov.

(sometimes involving F and B minerals), and the distal parts may grade into vein-type Pb-Zn-Ag deposits.

Many of the fluid-inclusion studies of skarn deposits indicated relatively high temperatures (400-650°C), intermediate salinity (10-45 wt % NaCl equiv.), relatively low CO_2 content ($X\ CO_2 < 0.1$), and boiling only in the upper parts of the system (Einaudi et al., 1981). Most such studies have involved skarn (gangue) minerals rather than ore minerals, but the ore minerals have generally been introduced later than the formation of the skarn, and probably at lower temperatures. Possible genetic relations between the Sn-skarn deposits and the Sn-greisen and Sn-replacement deposits have been discussed briefly by Einaudi et al. (1981).

Not all the heavy-metal-bearing fluids involved in the formation of skarn deposits were at "lower" temperatures. Kulikov (1982a,b, and pers. comm.) has described high-Fe polyphase inclusions in fluorite from the lower levels of the Tyrnyauz deposit, USSR, that homogenize at 400-700°C (some may have Th >800°C, but leakage became a problem). Almost all these inclusions contain large daughter crystals of halite and sylvite,[8] and most also contain a crystal of calcite and of magnetite (Figs. 15-20a,b). As many as 20 different solid phases have been recognized (many of them opaque phases), as well as the phase CaFCl (verified by X-ray diffraction), from various samples. As the specific daughter-mineral assemblage varies, presumably the ore fluids have varied as well. The amount of Fe precipitated as a daughter crystal of magnetite in the two inclusions shown in Figures 15-20a,b is estimated to be ~1.5 and 2.7 wt %. The total Fe content of the original fluid should also include the unknown amounts of Fe in other unidentified daughter phases, and that still in solution. These two values can only be guessed at this stage, but Dr. Kulikov reported (pers. comm.) that the fluid turns brown on exposure to the air. Naumov and Shapenko (1980) reported an estimate of 7% Fe in these inclusions, in part on the basis of estimates of 2.5 to 4.6 vol % magnetite as a daughter phase.

Very high Fe contents are also apparent in the daughter phases reported elsewhere. Tan (1979) reported hematite, magnetite, Fe sulfides, and possibly molysite [$FeCl_3$] and amarantite [$Fe(SO_4)(OH) \cdot 3H_2O$] as daughter phases at the

----

[8] Kulikov (1982b) noted that the concentration of soluble daughter crystals in fluorite from Tyrnyauz was so high that it seriously affected the percentage recovery of fluorite on flotation unless the pulp was first washed with water.

Table 15-6 Data from some recent studies of fluid inclusions in ore-bearing skarns
[See also entries for Santa Rita, New Mexico, and Ely and Copper Canyon, Nevada, USA,
and Mines Gaspé, Quebec, Canada, in Table 15-5.]

Deposit	Type	Mineral	Th (°C)	Salinity (wt % NaCl equiv.)	Reference
Tyrnyauz, Caucasus, USSR	W-Mo	Fluorite	615-645	>45	Naumov & Shapenko (1978)
Tyrnyauz, Caucasus, USSR	W-Mo	Fluorite	400-700	60 < 80	Kulikov (1982a,b)[1]
Yaogangshan, China	W	Garnet	400-425	--	Lu et al. (1982)
Yaguki mine, Fukushima Pref., Japan	W-Cu-Fe	Quartz & scheelite	200-360	1.0-10.5	Muramatsu & Nambu (1982a)
Fifteen Japanese deposits	W-Fe-Cu	Various	160-430	(low)	Takenouchi (1980a)
Kamaishi district, Iwate Pref., Japan	Fe-Cu[2]	Quartz	247->380	30-50	Muramatsu & Nambu (1982b)
Shinyama deposit, Iwate Pref., Japan	Fe-Cu	Quartz & calcite	290-334	4.2-15	Muramatsu & Nambu (1982c)
Salau, Pyrenees, France	W	Quartz	190-230	13	Fonteilles et al. (1980)
King Island, Tasmania, Australia	W	Garnet	300-800	<64[3]	Tan & Kwak (1979)
King Island, Tasmania, Australia	W	Scheelite	100-430[4]	(high)	Tan (1979)
Patagonia, Arizona, USA	?	Garnet	380-480	>10	Surles (1978)
Sierra Nevada, California, USA	W	Garnet	600-650[5]	4-8	Kerrick (1977)
Naica, Chihuahua, Mexico	Polymetallic	Fluorite	119-684	20-63	Erwood et al. (1979)
Dongpo scheelite, China	W	Various	200-510	<40	Lu & Crawford (1982)

[1] Also pers. comm.
[2] From igneous rocks of the district containing the Shinyama deposit.
[3] Includes major $CaCl_2$ (Kwak & Tan, 1981).
[4] Tm NaCl only.
[5] Corrected for pressure.

King Island, Tasmania, Australia, scheelite-skarn deposit. Studies of the chemistry of Fe in the hydrothermal solutions responsible for depositing the contact-metamorphic magnetite deposits (Cornwall, Pennsylvania-type) verify the
high concentrations of Fe in such fluids (e.g., Eugster and Chou, 1979; Whitney
et al., 1979; Boctor et al., 1979).

Although most inclusion studies of skarns have dealt with ore-related occurrences, some studies have dealt with skarns from non-ore-related contact metamorphism, since the composition of the fluid inclusions can aid in understanding the
complex mineralogical changes in such zones. Kerrick (1977) reported very low
salinities (3-8 wt % NaCl equiv.) and low XCO_2 (<0.1) in the Sierra Nevada, California, USA, trapped at 600-650°C (pressure corrected), and Feldman and Papike
(1981) also found that the fluids in skarns at a dolomitic limestone-quartz monzonite contact had very high H_2O/CO_2 ratios but that the salinity was high and
variable (~20 to ~40 wt % salts). A study of a similar contact by Mogk (1979)
showed higher CO_2 contents (<30 vol %), but the mineral assemblages present still
required H_2O-rich fluids. These differences may merely be reflecting the time of
trapping of the inclusions in relation to changes in the nature of the magmatic
fluids with time and ground-water incursion, and to the evolution of CO_2 from
decarbonation reactions.

Table 15-6 lists some of the more recent studies of inclusions in skarn ore
deposits. Many more studies have been reported (Roedder, 1968a). In Table 15-6,

the values for Th (except those listed for garnet) refer to the ore stages. Later gangue minerals, formed at much lower temperatures and salinities, are present in most of these deposits.

The extreme variation in salinity seen in Table 15-6 is most notable and may result from various mixtures of highly saline original magmatic fluids (either exsolved directly from the magma or condensed, as discussed above, in the section "PORPHYRY COPPER AND RELATED DEPOSITS") and later circulating ground waters. A correlation frequently seen between the higher salinities and the higher values for Th would support this concept. Most of the studies have not included the isotopic analyses that might help resolve this question, but those few that have included such measurements (e.g., Theodore and Blake, 1978) generally suggest a mixture of magmatic and meteoric waters.

TIN-TUNGSTEN GREISENS AND VEINS

General Features

The group of Sn-W greisen and vein deposits covers a wide range of environments that appear to have affiliations with and grade into the Be-bearing pegmatites, the polymetallic sulfide veins, and the porphyry Sn deposits. Thus, a sharp line cannot be drawn between the sericite-quartz stage in the porphyry-type ores and greisenization. In some districts, many different types of W deposits are found in rather close association. Lu et al. (1982) listed 10 types in the W district of southern China that have such a wide range of mineralogy that the specific nature of the fluids responsible for forming these various deposits must also have varied widely. Burt (1981) has attempted to explain the features of greisen and Mo-porphyry deposits in terms of equilibria in the system K_2O-Al_2O_3-SiO_2-H_2O-F_2O_{-1}, since F, unlike Cl, is incorporated in greisen minerals. The effects of Ca, Fe, Mn, Be, W, O_2, S_2, and CO_2 are then considered.

Both salinity and Th vary rather widely in these deposits. A representative selection of the major recent studies is given in Table 15-7, and two of these studies are discussed in more detail below. As in Table 15-6, where available, the Th values are those for the ore stage; Th values for inclusions in the associated pegmatite and granite are generally higher. Paragenetic sequences are not always easy to recognize, but Sn and W deposition is frequently assumed to overlap or to be synchronous with the greisenization stage. Evidence of "boiling" is common, though how much of the evidence of "boiling" is from evolution of CO_2 is not certain, and a CO_2 phase is commonly reported in the inclusions.

The series of Sn-W deposits of the Erzgebirge, near and on the border between Czechoslovakia and the German Democratic Republic, has been the subject of several major studies (Table 15-7). All these deposits show a very wide range of salinity. Some of the highest salinity fluids, as evidenced by the daughter crystals, are found in presumably pre-ore topaz crystals (Fig. 15-21). R. Thomas (1979, and pers. comm.) and Thomas and Bauman (1980) reported major amounts of halite, sylvite, ulexite [$NaCaB_5O_9 \cdot 8H_2O$], and borax [$Na_2B_4O_5(OH)_4 \cdot 8H_2O$], as well as anhydrite, hematite, a carbonate, and numerous unidentified minerals. The borate phases were verified by X-ray diffraction and microchemical tests for borate ion. Collins (1981) found no evidence of saturation with NaCl or KCl but did report as many as seven different daughter(?) phases, including phases that were isotropic, anisotropic, and opaque, in individual inclusions in the Cleveland replacement Sn deposit in Tasmania, Australia. Most of the inclusions were low in salinity.

Brown (1983) studied the F-Sn-W skarn at Mt. Garnet, Queensland, Australia. It occurs at a granite/marble contact. Some of the inclusions in quartz and fluorite of the greisenized granite, and in skarn minerals contain large volumes

Table 15-7 Data from some recent studies of fluid inclusions in Sn-W-bearing greisens (G) and veins (V).

Deposit	Type	Th (°C)[1]	Salinity (wt % NaCl equiv.)	Reference
Dzhida, USSR	Mo-W(G)	350-400[2]	---	Kolonin & Kosals (1977)
Akchatau, Kazakhstan, USSR	W(G)	260-440	27-65	Doroshenko & Pavlun' (1977,1978)
Mongolia (various)	Sn-W(G)	294-368	high[3]	Ivanova (1976)
Malaysia (western)	Sn(G)	300-330	---	Foo (1977)
Dongpo, China	W(G)	208-269	---	Lu et al. (1982)
China, 19 W-sulfide veins	W(V)	240-328	---	Lu et al. (1982)
Ani mine, Akita, Japan	Cu-Sn(V)	150-300	18-22	Nambu et al. (1978)
Suzuyama Sn veins, Japan	Sn(V)	200-350	[4]	Takenouchi (1980a)
Takatori, Ohtani, & Kaneuchi mines, Japan	W-Sn(V)	215-360	1-10	Takenouchi & Imai (1971)
Ohtani mine, Japan	W-Sn-Cu(V)	164-324	1.5-9	Kim (1981)
Eurajoki, Finland	Sn-Be(G)	260-390	3-17	Haapala & Kinnunen (1979)
Panasqueira, Portugal	W-Sn(V)	230-360	5-10	Kelly & Rye (1979)
Izerskie Mtns., Poland	Sn-W(V)	300-450	[5]	Karwowski (1977)
Karkonosze massif, Poland	Sn-W(V)	150-405	(low)	Kozlowski (1978)
Preisselberg, Krupka, CSSR	Sn(V,G)	300-427	3-40	Durisova (1978)
Krupka & Cinovec, CSSR	Sn-W(G)	300-500	2-35	Durisova et al. (1979)
Erzgebirge, DDR	Sn(G)	390-450	3-34	R. Thomas. (1979)
Erzgebirge, DDR	Sn(G)	270->550	0-45	Thomas & Bauman (1980)
Brioude-Massiac district, France	Sn-W(V)	~400	(low)	Bril & Ramboz (1982)
Cligga Head, Cornwall, UK	Sn-W(V)	200-400	2-12	Jackson et al. (1977)
St. Michaels Mount, Cornwall, UK	Sn-W(V)	260-430	4-20	Moore & Moore (1979)
Moina, Tasmania, Australia	Sn-W(V)	330-480	>26	Kwak & Askins (1981)
Renison Bell, Tasmania, Australia	Sn	300-350	~10.5	Patterson et al. (1981)
Cleveland, Tasmania, Australia	Sn-W-Cu[7]	320-360	8-14	Collins (1981)
Mt. Garnet, Queensland, Australia	Sn-W(G)	300->600	70	Brown (1983)
New England, NSW, Australia	Sn-W(V)	250-420	20-40	Eadington (1983)
Bolivia (various)	Sn(V)	<530	<47	Kelly & Turneaure (1970)
Pasto Bueno, Peru	W-Pb-Zn(V)	175-500	>40	Landis (1972); Landis & Rye (1974); Norman & Landis (1983)
Soktuy, E. Transbaikalia, USSR	W(V)	150-680	(high)	Bazheyev (1980)

[1] Excluding samples from associated pegmatite and granite.
[2] Greisenization at 400-650°C.
[3] In associated topaz.
[4] Polyphase, high-salinity inclusions in which Th was as high as 500°C, were present in the associated granodiorite porphyry but absent in the Sn-quartz veins.
[5] NaCl daughter crystals noted.
[6] Replacement body in dolomite.
[7] Replacement lenses in sedimentary rocks.

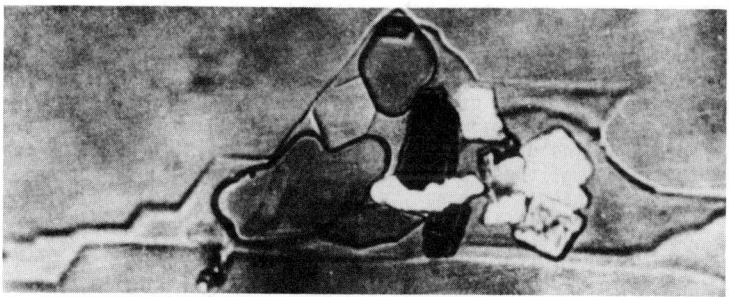

Figure 15-21. Multiphase inclusion in topaz (Th = 554°C) from Schneckenstein Sn deposit, Erzgebirge, DDR. From Thomas and Bauman (1980).

of daughter minerals(<70 wt %), consisting of many phases. The following were recognized: halite, sylvite, pyrite, sphalerite, $CaCO_3$, and phases identified by SEM as bearing Ca-Cl; Mg-Cl, Fe-Cl; Mn-Cl, Ca-Si, and numerous others.

The easiest explanation for the wide range of salinities found in these various deposits would involve mixing of early high-temperature magmatic brines and later fresh meteoric waters. However, Collins (1981) interpreted the uniform, relatively low salinity that he found at the Cleveland deposit and the uniformity of $\delta^{18}O$ and $\delta^{34}S$ values as evidence of magmatic fluids from a granitic source, involving no mixing of meteoric water. Both magmatic and meteoric, plus possibly a third water were involved in formation of the Pasto Bueno deposit in Peru (Landis, 1972; Landis and Rye, 1974). Further studies on the multiple sources for this deposit included Sr isotopic data on inclusion fluids (Norman and Landis, 1983). Higgins (1980) found fluid inclusion evidence that W was transported by fluids rich in CO_2 at the Grey River W prospect, Newfoundland, Canada, and suggested carbonate or bicarbonate complexes may be important in W transport at very high pressure.

Tin-Tungsten Deposits of the Eastern Andes, Bolivia

Kelly and Turneaure (1970), in what might well be termed a tour de force in both inclusion study methods and their application, examined the fluid inclusions in the Sn-W deposits of the Eastern Andes of Bolivia by a wide range of procedures. Originally, geothermometry was the main goal. Many potential mineralogical geothermometers were applied to these ores, with disappointing or conflicting results. Study of the fluid inclusions provided not only the best geothermometric data on the deposits but also a wealth of other information on the nature of the ore-forming environment. The individual deposits vary, but certain highly systematic trends in both salinity and temperature were observed in the formation of the Sn and W ores. The quartz and cassiterite of the Sn ores, in particular, formed from hot, highly saline (<47 wt %), low-CO_2 fluids (Fig. 15-22); later stages of mineralization involved much more dilute fluids (down to nearly fresh water) and temperatures even <70°C (depositing the late hydrous phosphates such as vivianite). The thermal maximum, during cassiterite deposition, was about 530°C. Liquid densities ranged from 0.70 to 1.15 g/cm^3 during the early stages but gradually approached 1.00 in later stages.

Active boiling of these early vein fluids was recorded in the inclusions (Fig. 15-23) and resulted in partitioning of most of the CO_2 into a vapor phase, which may have carried a small amount of salts in solution. Most important, however, the evolution of this vapor phase apparently changed the ore-fluid chemistry enough to cause the precipitation of quartz and cassiterite and may explain the restricted vertical distribution of very high grade Sn ore in several deposits of shallow origin. Kelly and Turneaure preferred a magmatic source for the brines and metals but suggested that progressive cooling and dilution of these brines may reflect gradual influx of meteoric water into the magmatic hydrothermal system.

A particularly detailed study was made of the daughter crystals in the highly saline inclusions. As many as eight apparently different phases were found in individual inclusions (Fig. 15-24) and a total of 26 in all the inclusions studied. Most of these phases were not identified, but halite, sylvite, a rhombohedral carbonate(?), magnetite, and hematite were recognized. Samples of the dry salts formed by leaching crushed quartz and cassiterite gave X-ray patterns for halite only, yet the sylvite daughter crystals in some inclusions were ~1/3 the volume of the halite (Kelly and Turneaure, 1970, p. 656). The magnetite crystals lost their magnetism promptly at the known Curie point (578°C) for magnetite, verifying their identity, and have volumes corresponding to at least 3000 ppm Fe. The magnetite does not redissolve on heating and was considered to be a result of hydrogen leakage. Most of the daughter crystals dissolve before Th, and many of the remainder dissolve very close to Th, suggesting the trapping of fluids that were boiling and saturated with respect to those particular phases. This evidence of boiling corroborates that obtained from paired gas-rich and liquid-rich inclusions that had very similar values for Th.

One daughter phase, a light-green fibrous silicate(?) mineral, seemed not to dissolve during normal runs. After 23 hours at Th, however, the crystals became rounded; they dissolved completely after 48 hours at temperature. The

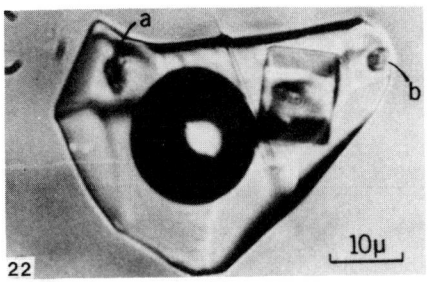

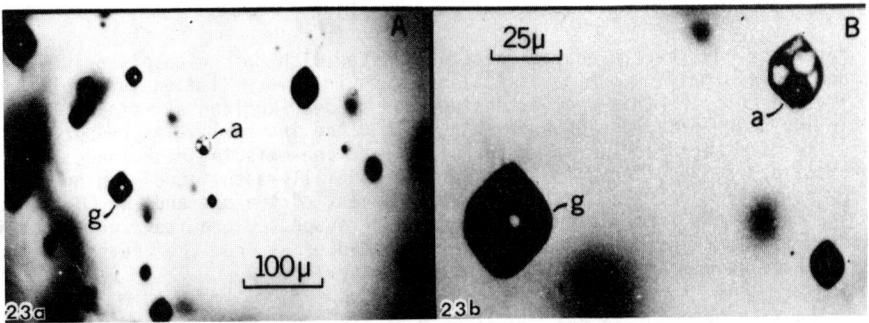

Figure 15-22. Primary inclusion in quartz containing large halite cube and unidentified daughter salts ("a" and "b") from the Laramcota mine, Bolivian Sn district, from Kelly and Turneaure (1970). Total salinity is ~46.7 wt %; Th L-V(L) is 430°C.

Figure 15-23. Gas-rich, totally reflecting inclusions (one labelled "g") interspersed with salt-bearing inclusions (one labelled "a") throughout a quartz crystal, without any relation to fractures, from Araca Sn deposit, Bolivia. Both types of inclusions in this field homogenize at ~426° C, but precise Th values cannot be determined, even in the largest gas-rich inclusions. The photograph on the right is an enlargement of two inclusions at locations "g" and "a" in the left photograph. From Kelly and Turneaure (1970).

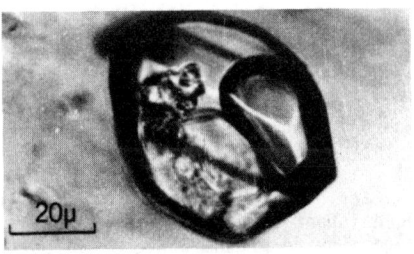

Figure 15-24. Primary multiphase inclusion in cassiterite from the Laramcota Sn mine, Bolivia, that is jammed with several unidentified daughter salts. The salts are completely dissolved at 353°C; Th L-V(L) is at 485°C. From Kelly and Turneaure (1970).

mineral crystallized as fibrous radial spherulites on cooling and hence was considered to be a true daughter phase, which equilibrated slowly. The magnetite grains failed to dissolve at all and slowly converted to hematite (presumably from further loss of hydrogen?). The conversion was complete after 7 days at temperature. Even in these long heating runs, no dissolution of the quartz or cassiterite host-crystal walls was seen.

The inclusion data, along with the field and mineralogical data, support the concept of a single prolonged mineralizing "event," in which the major paragenetic stages evident in the vein mineralogy are explicable in terms of decreasing temperature with time.

Tin-Tungsten Deposits of Panasqueira, Portugal

Kelly and Rye (1979) studied the geologic, fluid-inclusion, and stable-isotope features of the large Sn-W deposits at Panasqueira, Portugal, which consist

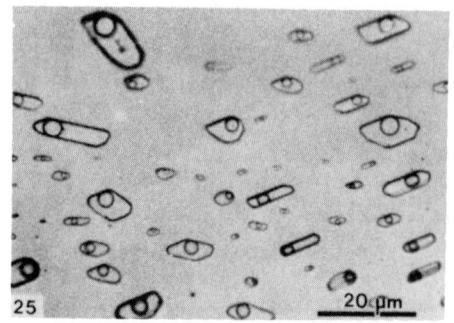

Figure 15-25. Plane of pseudosecondary inclusions in apatite from Panasqueira Sn-W mine, Portugal, showing typical G/L ratio, and traces of daughter phases (black specks). From Kelly and Rye (1979).

of ferberite-cassiterite-quartz veins associated with Hercynian plutonism. Economically significant mineralization occurs in a very limited vertical zone, 100 to 300 m thick, in a very extensive zone of near-horizontal veins. The fluid-inclusion data were used to suggest that the internal vein pressure at the time of mineralization was adequate to lift the existing rock load, i.e., vertical hydraulic dilation. The veins are spatially associated with one or more greisenized granite cupolas. The Sn content of the ore and the CO_2 content of the inclusions increase toward the known or suspected granite cupolas. Vein apatite is fluorapatite, and all white mica studied -- from the greisen, veins, and altered wall rocks -- is F-bearing muscovite.

The vein paragenesis was complicated by repeated depositional cycles, but four main stages can be recognized. The fluids forming the vein minerals were relatively low salinity, NaCl-dominated brines that were well below their critical temperatures throughout the mineralization period. During the bulk of the ore deposition, the fluids were at 230-360°C, had salinities of 5-10 wt %, and densities of 0.74 to 0.93 g/cm^3. Daughter minerals are small and relatively rare, or nonexistent (Fig. 15-25), but inclusions in the associated Panasqueira granite are crowded with daughter phases. The last (carbonate) stage took place at ~120°C, from fluids having <5 wt % salinity and densities ~1.00 g/cm^3. Some CO_2 effervesced early in the mineralization, when as much as 9 mole % CO_2 was present, but effervescence then ceased throughout the remainder of the mineralization, when the liquids contained <2 mole % CO_2. Pressures during most of the mineralization were estimated to have remained <~100 bars, corresponding to depths of 600-1300 m below the ground water table at the time.

Various O and S isotope thermometers were also tested but generally gave inconsistent or unreasonable results. The ^{18}O contents of the late carbonate stage require predominance of meteoric water at that time, but similar studies of the earlier stages are ambiguous, because of extensive exchange with the host rocks. Hence, the fluids could be predominantly magmatic, highly exchanged meteoric water, or a mixture. Studies of the H in these samples failed to resolve this ambiguity and introduced two new complications. Apparently, two waters were present during the first (oxide-silicate) stage, one having δD values of -67 to -124 per mil, and the other, -41 to -63 per mil. Each of the two waters apparently deposited different minerals, but texturally these different minerals appear penecontemporaneous. Various explanations of these data were attempted, but none fully solved the enigma. The $\delta^{13}C$ of CO_2 extracted from inclusions in quartz and wolframite (Bussink, 1981) suggests a strong contribution of C from organic matter in the host schist.

Snee at al (1983), using high-precision dating procedures, showed that gangue mineral deposition (and that of the ore) took place in several discrete events over a very long period of time (minimum of 3 million years), and suggested that these mineralizations were associated with different intrusive centers through time.

A unique combination of fission-track dating and fluid-inclusion data was also used (Kelly and Wagner, 1977) to show that one and possibly two reheating events occurred long after ore deposition. This heating was to temperatures >150°C.

MISCELLANEOUS ORE-DEPOSIT TYPES

Metamorphosed Ore Deposits

Although numerous ore deposits have been either formed by, or subjected to, high-grade metamorphic events, a genetic link is difficult to prove (see also a discussion of Broken Hill on p. 366). Many of the massive sulfide deposits of the world have been subjected to moderate- to high-grade metamorphic conditions, but very few studies of them have involved any investigation of the fluid inclusions. Many of the ores themselves do not appear very appropriate for inclusion study, but the host rocks might yield useful data. Carignan et al. (1979) reported Th values of 100-480°C in quartz from the Millenbach massive sulfide deposit in Canada.

During later metamorphism of an ore deposit, the original inclusions will be lost if the host is recrystallized. If the host mineral is not recrystallized, and if the inclusions are trapped at a sufficiently high temperature and low density that internal pressures during metamorphism do not cause decrepitation, they may survive the metamorphism (e.g., the inherited inclusions in some of the quartz pebbles in the Witwatersrand Au conglomerates of South Africa (Chapter 11). Ohmoto and Rye (1970) and Ripley and Ohmoto (1977a,b) reported that some fluid inclusions in quartz from the Bluebell mine (Canada) and the Raul mine (Peru) withstood metamorphic temperatures as much as 150°C above Tt with no significant change. Most commonly, complete recrystallization and even transport and redeposition of the ores takes place, and new inclusions are trapped that register the metamorphic environment. Relatively few studies have been made of metamorphosed ore deposits, in large part because they frequently are not very amenable to study. Rutherford (1963) examined dense CO_2 inclusions in a Canadian massive sulfide deposit, and Walshe and Solomon (1981) reported some rather low Th values (130-150°C; Tt estimated to be 200-225°C for 1 kbar) for fluorite from veins in the Cu deposits at Mt. Lyell, Tasmania, Australia, that had been metamorphosed to the lower greenschist facies.

A few very small inclusions were found in quartz of the metamorphosed stratabound Cu deposit at the Raul mine, Peru (Rye and O'Neil, 1977; Ripley and Ohmoto, 1977a,b). These had Th = 350±10°C and were interpreted as being representative of the original depositional conditions, because the host rocks of the deposit have only been metamorphosed to the upper greenschist-lower amphibolite facies at temperatures of 400-500°C (Ripley and Ohmoto, 1977b). Since the inclusions were all <5 μm, they were presumed to be capable of withstanding a metamorphic cycle including ~100°C of overheating without decrepitating or stretching.

In addition to the Au-bearing ores in a number of the deposits described in this chapter, many of the Au deposits of the world occur in metamorphic rocks and are generally believed to be a result of the metamorphism, and hence are discussed in Chapter 12.

Uranium Deposits

Extensive discussions of the origin and chemistry of the fluids responsible for the sandstone-type U-V deposits of the Colorado Plateau, USA, might have been greatly simplified if some usable fluid inclusions were available in these ores, but all the primary ore and gangue minerals are very fine grained, opaque, or both. However, the vein-type U deposits are much more amenable, and a variety

Deposit or occurrence	Reference	Results
Cluff Lake & Rabbit Lake, Saskatchewan, Canada	Pagel & Jaffrezic (1977); Pagel et al. (1980)	Inclusions consist of diagenetic highly saline Na-Ca-Cl brines derived from the host sandstones, modified by reactions with basement rocks. Cl/Br (wt) = 55-65. K/Na (atomic) = 0.09. Th L-V = 90-185°C; Tm NaCl 100-250°C.
Margnac & Fanay, France	Leroy (1978)	CO_2-rich fluids, from the unmixing of complex CO_2-H_2O mixtures following a release of pressure, circulated through faults in granite after emplacement of lamprophyre and micaceous episyenitization of the granite and deposited pitchblende and pyrite. Tt ~345°C. Later, cooler, more water-rich fluids converted pitchblende to coffinite and pyrite to hematite, and finally deposited fluorite at <135°C, indicating a decay of the geothermal system. CO_2 contents of inclusions were estimated in the field by use of the crushing stage.
Bois Noirs-Limouzat, France	Cuney (1978)	Late-stage fluids from crystallization of granite released U by altering sphene, zircon, monazite, and xenotime, and precipitated easily leachable uraninite in a quartz-muscovite assemblage. Meteoric water, plus deep CO_2 and significant H_2S, remobilized the U and redeposited it along shears on CO_2 loss, at 77-100°C; subsequently, part was converted to coffinite or mobilized again.
Rössing uraniferous alaskite, SW. Africa	Cuney (1980)	Boiling of the magma by increase in CO_2 partial pressures resulted in formation of dense CO_2-rich and dense, highly-saline fluids at 625°C and 6 kbar, which redistributed U, forming uraninite-fluorite veins.
Nabarlek, Northwest Territories, Australia	Fuzikawa (1982)	Mixing of several fluids, resulting in a wide range of salinities, from 0.4 to 37 wt % $CaCl_2$ equivalent (Ca>>Na) may be responsible for U deposition. The high Ca results in unique freezing behavior. Th = 70-150°C. Solutions were two-phase, the gas phase consisting of hydrocarbons and CO_2.
Marysvale, Utah, USA	Cunningham et al. (1979 and pers. comm.); Steven et al. (1981).	Fluid inclusion data indicate that the primary U minerals were deposited from very dilute (<1 %) hydrothermal fluids at ~150 to 225°C, with a vertical thermal gradient. Other data suggest that the fluids were reducing, acid, and F-rich. U^{4+} - F complexes became unstable on pH increase as the fluids rose, and hence fluorite, uraninite, and coffinite were precipitated.
Bokan Mountain U-Th, southeastern Alaska, USA	Thompson et al. (1979)	Vein and pipe-shaped zones of U-Th mineralization (uranothorite), with calcite, fluorite, quartz, pyrite, and hematite in a peralkaline granite-syenite complex. Preliminary studies of inclusions in quartz and calcite show Th = 320°C. S and C isotope analyses on contemporaneous pyrite and calcite indicate the ore fluids had a pH of 4.5 and were moderately reducing.
Vendée and South Brittany, France	Cathelineau (1982)	The deposits are located in joints, spatially associated with syntectonic leucogranites and surrounding metamorphic rocks. The primary U deposits were formed by low-salinity, low-CO_2 fluids at P = few hundred bars, T = 150-400°C, but were extensively reworked by later fluids.
Many deposits in the USSR	Khitarov et al. (1980)	Many chemical analyses (of water leachates) are presented for examples of various types of U deposits. Constituents report-

have been studied (Table 15-8). Also, U, in the range <1 to >1000 ppm, has been determined by inductively coupled plasma-emission spectrometry in fluids evolved from decrepitating inclusions from quartz of some granites (Rankin et al., 1982).

Special Environments

A limited number of fluid-inclusion studies have been made in each of such a wide range of deposit types that they cannot be summarized or even listed here (see indices in Roedder, 1968a). Some of these studies have revealed rather unusual environments and suggest that future inclusion studies may still provide some surprises. Three examples should suffice to illustrate.

Sulfur-rich environments. Bény et al. (1982) found a solid and three fluids in a study of fluid inclusions in fluorite and quartz from two Spanish localities. The daughter crystal was S, and the fluids were liquid H_2O, liquid H_2S-CO_2, and an H_2S-CO_2 gas. In H_2S-rich inclusions, H_2S hydrate (probably $H_2S \cdot 5.75H_2O$) formed at low temperature (Touray and Guilhaumou, 1984). The liquid H_2S-CO_2 and gaseous H_2S-CO_2 homogenize, generally in the liquid phase, at +24 to +48°C. Total homogenization, including the dissolution of the molten sulfur droplet, occurs at >300°C.

Organic matter in mercury deposits. One of the characteristics of many of the Hg deposits of the world is the combination of rather low inclusion temperatures and various kinds of organic matter. Zatsikha et al. (1973) described an odd Hg deposit (the Slaviansk deposit, in the Ukraine, USSR) that is confined to a circular breccia in a Devonian salt "stock" (i.e., dome?). Lyubinetskaya et al. (1980) reported that in this deposit, early fluorite had Th 410-415°C, but most cinnabar formed with calcite having Th 130-120°C. Some cinnabar crystals were trapped in fluid inclusions in quartz having Th 245-205°C. Organic matter in the form of inclusions of solid ("kerite" and "asphalt"), liquid (light petroleum), and gas (CH_4 and others) was also paragenetically associated with the cinnabar. The petroleum was found in small variable amounts (i.e., trapped as an immiscible fluid?). IR spectra revealed a wide range of n alkyl compounds, paraffinic acids, and "bicyclic aromatic structures." The "kerite" presumably formed as a result of hydrothermal solutions reacting with bituminous matter in sedimentary rocks, but the preservation of the various compounds at temperatures as high as 300-400°C seems unrealistic, so presumably these were introduced later. (Cinnabar has a high vapor pressure, but requires 584°C for sublimation.) Zatsikha et al. (1973) reported much lower temperatures for the high-temperature Hg mineralization, with metacinnabar (250-210°C), and for the main ore stage, with cinnabar (150-70°C).

Aqueous fluid inclusions in chromite. Johan and Le Bel (1978, 1980), and Johan et al. (1982, 1983) have reported that stratiform and podiform chromite from ophiolite complexes contain aqueous fluid inclusions (primary?), also containing CO_2, CH_4, C_2H_4, and C_2H_6. The density of filling is low (~40% vapor), as is the salinity (5 wt % NaCl equivalent). $CH_4/CO_2 = 0.1$, and the $\delta^{13}C$ values for C in these two compounds were said to correspond to the isotopic fractionation expected at high temperatures. K/Na is ~0.7.

For microthermometry on such samples, the plates must be very thin, so that the chromite appears yellow-brown or orange. An interesting aspect is that these aqueous inclusions occur systematically in the massive chromite but not in the disseminated chromite. Watkinson and Mainwaring (1981) have also reported saline fluid inclusions in a series of stratiform and podiform chromite deposits in Canada. It would be interesting to know whether there might be a connection of some sort between the origin of these aqueous chloride-bearing fluids in ophiolites and the origin of the chloride solutions proposed by Schiffries (1982) as causing major metasomatic reactions -- and transport of platinum -- in the Bushveld Complex of South Africa. Inclusions of various solid silicate minerals in chromite from the Bushveld were interpreted by MacDonald (1979) as being the result of replacement.

USE OF FLUID INCLUSIONS AS TOOLS IN MINERAL EXPLORATION [9]

The study of fluid inclusions in ore deposits has generally resulted from "academic" interest, at least in the Western world. However, this study has many applications that can be either directly or indirectly helpful in the search for,

----
[9] Modified and abbreviated from Roedder (1977b).

and the development of, new deposits or extensions of known deposits. Thus, the study of fluid inclusions, although no panacea for exploration, should be considered a _potentially_ useful tool in exploration for many types of ore deposits.

Like every exploration and resource-evaluation tool, the study of inclusions should not be used blindly or by itself, but the resulting data should be interpreted in conjunction with careful geologic and paragenetic studies. I do not claim that inclusions provide us with a magic black box that _finds_ ore or even mineralization, but merely that they may be _helpful_. The potential usefulness of inclusions should not, however, be considered as limited to the types of deposits in which they have already proven useful.

The interpretation of most inclusion data is necessarily subject to various degrees of ambiguity with respect to its validity. Misunderstanding of the degree of this ambiguity, in either direction, is unfortunate. If significant ambiguity exists but is ignored or minimized, the resulting over-interpretation may be seriously in error; conversely, if the degree of ambiguity is exaggerated, valuable data may be ignored or not even obtained. Both these situations have occurred all too frequently in the past. The use of fluid inclusions in exploration requires (1) knowledge of the general features of inclusions and the types of data available from them and (2) careful attention to their significance and limitations, as discussed in these chapters.

Applications of Inclusion Data to Clarify the Regional or Local Geology

An understanding of the geologic or structural setting of an ore body can be important to exploration. Fluid-inclusion data can contribute to such an understanding in several ways. For example, an ore-bearing vein near several different possibly related plutons has been genetically related to one of the plutons, on the basis of horizontal gradients in inclusion temperatures rising toward that pluton (N.P. Ermakov, pers. comm., 1970).

A second example is found in a study by Piznyur (1957), of the quartz crystal deposit at Barsukchi, USSR; he showed that the inclusion temperatures, pressures, and CO_2 contents were higher and densities lower in those veins that were closest to the assumed granitic intrusive source. The temperature range was large, from 515°C in a vein near the intrusion to 80°C in one far away. The density difference was also reflected in the mode of homogenization -- in the vapor phase for inclusions near the intrusion and in the liquid phase at some distance. The density of CO_2 can also be used as an indication of the depth of formation (Theodore et al., 1982).

Laz'ko (1957) showed, by mapping Th of fluid inclusions in quartz from the Kurumkan, USSR, quartz crystal deposit, that large Archean pegmatite bodies north of the deposit could be excluded as the source of the quartz vein fluids and that, instead, the deposits seem to be related to Proterozoic granodiorite intrusions to the south.

Any parameter that could indicate which of several intrusions is most likely to have ore deposits associated with it would be useful. For example, Stollery et al. (1971) suggested, on the basis of studies at the Providencia stock, in Mexico, that high Cl contents in intrusive rocks may correlate with ore formation and hence provide a possible prospecting tool. They showed that at Providencia, this Cl is present in biotite and as highly saline fluid inclusions in various minerals.

However, neither leachable nor total Cl (or F) is necessarily a good measure of the halide content of the fluid in the inclusions, because both will be influenced by exchange with later hydrothermal fluids (this is particularly true for biotite), by ground-water leaching, by contamination from airborne salt, and most importantly, by the volume of saline fluid inclusions opened in the analysis. Optical estimates of the actual salinity of individual inclusions present, regardless of their abundance (even though only semiquantitative), should provide a much more accurate and valid proof of the former presence of a saline-rich fluid and hence of the possibility of ore deposits formed by such fluids.

In a structurally complex area, cut by many generations of barren quartz veins, differences in the composition or temperature of the inclusions from the several generations of quartz-precipitating solutions may permit reconstruction of the sequence of vein formation, faulting, and intrusion.

The search in the USSR during World War II for quartz for piezoelectric use spurred many studies of the fluid inclusions in quartz crystal veins and their country rocks, several of which indicate significant differences in the inclusions between those veins that have productive "nests" of high-grade crystals and those that produce only poor crystals. The differences found were based on either Td or composition. Although the empirical data derived in these various studies may be directly applicable only in those specific deposits, the principle should be applicable in any deposit. Differences must have existed between the fluids that formed a barren (or ore-poor) vein and those that formed a rich vein in any given metalliferous district. Inclusions of these two fluids should differ in one or more parameters, and intuitively one might expect that these differences would even be larger than those between the two types of quartz veins described above, but these differences might have been only in minor constituents, e.g., the ore metal content. Similar studies of various districts show differences between the fluids forming the Au-bearing and the all-too-frequent barren quartz veins (Roedder, 1977b, p. 507-508).

Even within the ore stage, however, at every deposit where the inclusion host minerals were characterized by their positions in a zoned and/or paragenetic sequence (i.e., spatially and/or chronologically), readily recognized correlations are found between the inclusion data and the positions in the sequence. In most examples, temperature and salinity drop as both time and distance increase, but there are exceptions (see, e.g., above discussion of Creede, Colorado, USA). All these variations may tell us important things about the formation of the deposit, and even though we are far from understanding the wealth of data which nature has provided, these variations can be put to empirical but practical use in many ways in exploration by helping to clarify the chronology of events in a complex district, including structural events, mineralization, alteration, etc.

One seemingly rather obvious but apparently seldom-used application of fluid inclusions lies in the recognition of age relationships of intersecting dikes and veins. Although these relative ages may be important in unravelling a complex structural picture, they may not be obvious in the field or mine because of later shearing and alteration. Even without such later effects, the macroscopic and mineralogical effects of a dike on a quartz vein would be minimal; hence, the age relations would be unknown unless the actual intersection is exposed. However, such dikes may cause decrepitation of inclusions in the earlier quartz vein. Such natural decrepitation of the fluid inclusions in quartz veins resulting from the heat of later dikes, as evidenced by empty inclusions or greatly diminished decrepitation of samples near the dike, has been recognized and could provide a relatively unambiguous assignment of relative ages (Lokerman, 1962, 1965); similar findings have been reported by Prokhorov et al. (1968) and by Korobeynikov and Matsyushevskiy (1973).

Guha et al. (1983) reported an entirely different application. They showed that the evolution of the fluid compositions (particularly the Ca/Na and Ca/Mg ratios) correlated with changes in the mechanical development of a shear zone at the Henderson Cu-Au deposit, Chibougamau mining district, Quebec, Canada.

Isotopic age determinations ($^{40}Ar/^{36}Ar$) on fluid inclusions are becoming feasible (see Chapter 5), and some have been made on ore minerals (York et al., 1982b); although some of the Ar found in the latter may come from fluid inclusions, part must be either in the ore mineral itself or in solid inclusions.

Applications of Inclusion Data in the Search for "Blind" Ore Bodies

Because an ore body is a rather small geologic feature, many exploration techniques are based on attempts to recognize the existence of some halo or anomaly around it that is larger than the ore body itself, thus improving the chances of finding it. This technique is particularly important in finding "blind" ore bodies. As was first detailed by Ermakov (1966b), fluid inclusions provide such halos.

The hot fluids that have formed many ore deposits heat the surrounding wall rocks and penetrate them, either by mass flow or through diffusion, causing various mineralogical changes. The resultant wall-rock-alteration halo obviously presents a larger exploration target than does the ore body itself. During this alteration process, however, the fluid in the pores of the wall rocks is also changed chemically and eventually may approach or actually reach equilibrium with the original wall-rock minerals. Any additional movement of this fluid (or diffusion of its constituent ions or heat) farther into the wall rock may not cause visible alteration, but such changed fluids can be found there as inclusions trapped during recrystallization of the wall-rock minerals, or, more commonly, as secondary inclusions from healing of fractures. Recognition of such new inclusions by the measurement of any appropriate parameter may thus provide an even larger target for exploration than does the alteration. Thirty papers, almost all Russian, that deal with such data have been summarized in a table in Roedder (1977b). Of course, the halos defined by fluid-inclusion data are thermal (or chemical) anomalies that do not necessarily reflect the presence of an ore fluid.

The most commonly used parameter for recognizing halos is Td (or Th) of the inclusions, as the new inclusions generally have been formed at higher temperatures toward ore. These inclusion zones are sometimes termed "steam bath," "steaming-through," or "streaming-through" halos, or "aureoles of evaporation" in the Russian literature, and are reported to be recognizable tens or hundreds of meters from ore. Fortunately, there is little need for the accuracy afforded by the normal optical determination of Th, and hence relative parameters, obtained quickly by the decrepitation method, may be adequate, so long as all the limitations of such data are kept in mind.

Several examples of such use of Td data have been illustrated by Roedder (1977b), together with some of the possible hazards in their interpretation. Thus, it would be good to have some independent source of evidence that the several reversals noted there (i.e., decreases in Td or activity toward ore) were actually due to later leaching, shearing, recrystallization, or dikes, since many other ad hoc explanations could be suggested. Many variables, including concentrations of salt and CO_2, and degree of fill, will have a major influence on Td. Similarly, verification that decrepitation activity anomalies do not occur in unmineralized rocks would be useful. Only rarely is such a clear statement made to this effect, although such verification may be commonly obtained. Also, in view of the charges that have been made in the past of subjectivity in the interpretation of decrepigrams, it is unfortunate to find that the supporting data and even basic details on the procedures used in data acquisition and reduction are frequently missing, so that a real evaluation of the work is simply not possible. The relationships between decrepitation activity, Td, and Th is complex. It may well be that decrepitation can only be used as a guide to local exploration in such deposits after the general geologic (and inclusion) features are known, permitting in effect a locally valid "calibration" of the significance

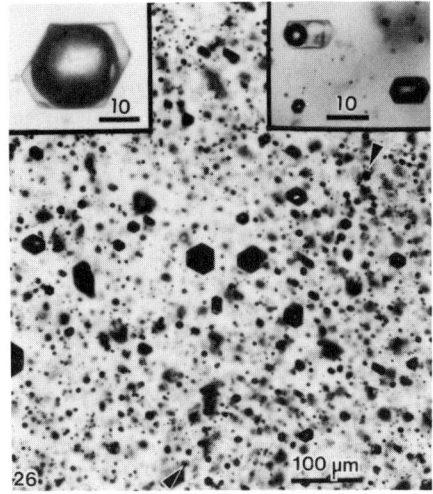

Figure 15-26., Quartz plate from Bingham, Utah, USA, sample ER 69-16, cut normal to c axis, showing complete recrystallization of all inclusions to hexagonal negative crystal shape. Inclusions out of focus are also sharply euhedral. Although these appear randomly distributed they are probably secondary, as vague curving planar arrays can be found (one rather obvious plane is indicated by the arrows). Most of these inclusions are hexagonal bipyramids with only a large bubble, just tangent to the walls (upper left inset), but a few contain a salt crystal as well (upper right inset, which is a magnified view of a small group in the center). Scale bars in μm.

of each type of decrepigram for that deposit.

In addition to Th, Td, and decrepitation activity, other inclusion parameters can be expected to change on approaching ore, although these may differ from one ore deposit to another. Examples include the total concentration of salts in the inclusions; the abundance of halite daughter crystals and gas-rich inclusions, particularly toward porphyry Cu deposits; the presence of liquid CO_2 (e.g., Rankin and Alderton, 1983); the ratio of Na to other ions in analyses of inclusions; the sample weight loss on heating; the sample H_2O content; and visual estimates of the number of inclusions. Bodnar (1981) showed that the nature and composition of fluid inclusions can be used as an exploration guide for epithermal Au-Ag deposits. Because inclusions tend to coalesce with time and temperature (Chapter 3), inclusion size might be expected to increase toward ore; a very rough trend of this sort has been noted by T.G. Theodore (pers. comm.). Similarly, the core of a porphyry Cu system represents a large mass of rock that would cool slowly. Such long cooling may permit complete recrystallization of the inclusions to sharply euhedral negative crystals (Fig. 15-26).

Thompson et al. (1980) have shown that decrepitating carefully prepared samples in the Ar supply for an inductively-coupled plasma spectrometer can provide analytical data on the fluids in the decrepitating inclusions. Alderton et al. (1982) have shown that the sensitivity of the method for a variety of heavy metals is so high that its greatest potential is in its use for routine mineral exploration. Similarly, the use of gas chromatography for analysis of the gases evolved on heating samples (in part from inclusions) has been suggested as an empirical exploration tool (Norman, 1981; Palin and Norman, 1982; Clifton, 1983a,b). Correct interpretation of such data may be very difficult, and the possibilities for obtaining misleading data are numerous.

As the main anion in fluid inclusions is almost always Cl, the amount of water-leachable Cl should also increase toward ore, particularly in typical porphyry Cu deposits. To my knowledge this has not been tested, but, as in the evaluation of various plutons as ore bearers on the basis of water-leachable Cl described earlier, optical detection of the presence of even a few highly saline inclusions would be much less ambiguous.

Applications of Inclusion Data to the Evaluation of Altered or Weathered Material in Outcrop or Sediments

When ore deposits weather, resistant gangue minerals such as quartz will persist into the gossan or soil, their original fluid inclusions unchanged; in fact, the inclusions in quartz grains in sediments have been used in provenance studies (Chapter 11). The first use of soils over blind ore bodies was reported by Ermakov (1957). Several later Russian papers reported success in the use of the decrepitation method on gangue minerals, such as quartz panned from steam sediments, to recognize the presence of particular kinds of inclusions indicative of ore in the watershed. The technique is identical with that used to pinpoint geochemical anomalies by working upstream.

Optical identification of the presence of the more unusual "boiling" or polyphase inclusions typical of some ore deposits in quartz grains in stream sediments should be possible, even after all ore minerals are long gone. Such optical investigations are also helpful in evaluating soils, leached cap rocks, and gossans to determine the probable nature of the preexisting (and buried) ore. Acid-insoluble residues are particularly helpful here, as they greatly concentrate the useful part of the sample. Thus, I have found (unpublished data) that the inclusions in a few percent of insoluble residues from acid leaching of some replacement Mn ores permitted identification of the nature of the material that had been replaced. Similarly, silicate-melt (glass) inclusions may persist unchanged in quartz from igneous rocks long after complete hydrothermal sericitic or argillic alteration has eliminated all other evidence on the nature of the original rocks. Such procedures have been used to recognize the source of quartz grains in laterite (Clocchiatti and Mervoyer, 1974) and shale (Kozlowski, 1981). Studies of the inclusions present in the ore pebbles and boulders found in glaciated terrains has been useful in delineating the nature of the target ore-body outcrop upstream (Kinnunen, 1981).

Applications of Inclusion Data Toward an Understanding of the Environment of Ore Deposition

W.C. Kelly summed up this aspect of the use of fluid inclusions in mineral exploration in a letter to me as follows:

> "The greatest contribution of fluid inclusions lies in the subtle but pervasive impact they have had on the basic philosophy of explorationists. The most effective exploration geologists I know operate with 'models' (whether they call them that or not), entering any new terrane with established models in mind, and sniffing out those bits of ground that show the critical ingredients symptomatic of model ground that has paid off elsewhere. To the extent that fluid inclusions have served to refine those mental models, they have honed the basic skills and mental outlook of the explorationist. It is here that they have had their greatest impact."

The inclusion data on practically any of the groups of ore deposits discussed in this chapter should help to refine the model of how such deposits formed.

Mississippi Valley-type deposits. The combination of Th and salinity data and other data from inclusions in the Mississippi Valley-type deposits effectively preclude both the freshwater karst and the seawater lagoonal environments of origin and hence invalidate any exploration concepts based on them.

The inclusions also tell us that the ore-forming fluids had densities greater than 1.0 g/cm^3 at Tt. These densities place important constraints on the paleohydrology at the time of ore formation. Because the times involved are almost certainly long, even small density differences can drive fluids long distances through rocks of average permeability. Note, however, that any ore fluid in a hydraulic gradient will obviously follow the most permeable paths, so ore deposition is automatically related to various sedimentary or structural features such as reefs, breccia zones, or paleokarst. These paths also provide the opportunity for the mixing of different fluids, mixing that has been proposed as one possible cause for the actual precipitation. An exploration program that concentrates on searching for permeable sedimentary features, on the basis of a syngenetic model for the origin of the ore, may thus be eminently successful even though basically wrong.

One of the most striking convergences of scientific lines of thought has been the rather recent recognition of the similarity of the problems of the petroleum geologist and the minerals-exploration geologist in looking for deposits in sedimentary rocks around the margins of basins. This aspect has been discussed by many (e.g., Dunham, 1970; Macqueen, 1976) and places additional importance on the full characterization of the organic compounds found in inclusions from each of these environments.

Kuroko deposits. Several models have been proposed for the formation of this important group of deposits. Some have suggested the deposition of these ores from volcanic exhalations on the floor of shallow ocean lagoons, on the basis of field and textural data. However, fluid-inclusion evidence indicates an ore fluid that was relatively low in salinity (generally less than 5 wt %) and hot (<300°C). These inclusion data have two important consequences. First, such ore fluids would boil instantly at or below the ocean floor unless the water was rather deep. This minimum depth is 122 m for 200°C fluids, and 730 m for 300°C fluids -- hardly shallow lagoons. More important, perhaps, is that these fluids have a much lower density than seawater, even without boiling; hence, if they came out at such depths, they would rise rapidly through the sea and mix with it. Although the details of the mixing are complex functions of salinity and temperature (Sato, 1972), such mixing could well result in precipitation of various metals on the sea floor and yield the field and textural evidence of sedimentary sulfides but could not yield the crystals of various gangue and ore minerals from the deeper ores that contain the inclusions studied. These inclusions almost certainly could not have formed in a shallow lagoon; hence, it may be necessary to assume two quite different conditions of ore formation for different parts of these ores.

Porphyry deposits. The inclusions in the porphyry Cu deposits are so characteristic that they might well be used for exploration, as has been suggested (Roedder, 1971b, 1977b; Cunningham, 1975; Denis, 1974; Nash, 1976). Bodnar (1981) claimed that a high Th is not important for exploration but that halite and chalcopyrite daughter crystals, and homogenization to vapor, are favorable criteria. Fortunately, the inclusions are in quartz and hence will persist in gossans and stream sediments and are optically recognizable at a glance. As discussed earlier, however, these guides are not infallible.

Thermal gradients. Regardless of the type of deposit, fluid-inclusion data can be useful in the recognition of lateral or vertical thermal gradients within the deposit at a given time during the ore deposition. These in turn may delineate the feeder channels and the directions of ore fluid movement. Telescoped ore deposition and changes in the "plumbing" of the deposit with time make it imperative that all such temperature studies be combined with very detailed paragenetic studies on the ore. Lateral and vertical temperature gradients have been detected in those deposits where the density of sampling was adequate. Such gradients can be expected generally, and their significance as a prime cause for ore deposition has been discussed extensively.

If the salinity is uniform, vertical gradients in Th indicate density gradients and hence the direction of convective flow. The opposite situation, in which little or no change is seen in Th of the inclusions with depth, also places significant constraints on the thermal and hydrologic regime at the time of trapping that may be important. Vertical changes in inclusions may also signify changes in ore type, or bottoming, that can be of great importance in development work.

Laterally uniform Th values over a large area imply a broad source for the fluids. Thus, the absence of a strong thermal or salinity zonation in the inclusions at the Jamestown, Colorado, USA, fluorite and Au deposits relative to an exposed stock caused Nash and Cunningham (1973) to conclude that the fluids emanated from a larger intrusive body at depth.

Boiling of the ore fluids. Another significant application of fluid-inclusion data to exploration lies in the recognition of boiling (i.e., immiscibility) of the ore fluids, through the presence of both vapor-rich and vapor-poor fluid inclusions. Because such boiling gives a measure of the maximum hydrostatic pressure at the site of ore deposition, it places at least some limits on the depth of cover at the time, which can be also useful in exploration.

A difficult but not insoluble problem in such studies is that of differentiating between normal boiling (i.e., steam) and the effervescence of dissolved gases; either may be pertinent to ore deposition, but the pressures involved may be quite different. Data from the crushing stage can permit differentiation between these two cases (see Chapter 7). Evidence of boiling (or effervescence) has been found in many ore deposits (see indices in Roedder, 1968a) and should be looked for in all, because it has many important consequences.

Evidence of boiling is found in the occurrence of low density inclusions in some granites toward the apices and cupolas, and hence such occurrences might provide useful structural guides in places where the shape of the intrusion is unknown.

An extensive literature exists, particularly in Russian, on the use of fluid inclusions to characterize ore deposits as formed from pneumatolytic vs hydrothermal solutions. As explained in Chapter 2, the division is made on the basis of homogenization behavior (into the gas vs the liquid phase), and the tacit but invalid assumption that this is a real boundary. In differentiating true boiling from effervescence, however, the distinction between gas and liquid is real and important.

Practical Problems

Every exploration tool considered for any given application must be evaluated in terms of (1) the probability of obtaining useful data; (2) when, where, and at what scale the tool should be applied; and (3) the cost of these data in time and money. If the study of fluid inclusions is being considered as a tool, how should the individual exploration geologist or company management decide these questions? The first question must be decided on the basis of the available information on the rocks involved, the nature of the possible targets, and the evidence in the literature or company files of previous success. The second aspect is a little more nebulous. Obviously there can be no set time in an exploration program when such an application would be fruitful, but it should be early. The most effective use of fluid-inclusion data should always involve first a reconnaissance examination of a few samples to determine the general nature of the inclusions present. This stage may represent one of the most useful applications of inclusion study, if the exploration geologist is able to classify a fluid-inclusion population as belonging to a particular ore-deposit mode. Thus, if core drilling has intersected a barren quartz vein of unknown affiliation, a few minutes examination of quartz grains in oil under the microscope will suffice to distinguish, by bubble and daughter crystal size alone, among inclusions having Th of ~150, 350, or 550°C.

If reconnaissance reveals the presence of inclusions that are physically suitable and potentially useful, that information has value in planning further work. Even if the inclusions in these reconnaissance samples are very poor or nonexistent, do not give up hope, as gross differences may exist in the suitability of even adjacent samples for inclusion studies. This inherent variability of sample material is an unfortunate fact of life and must always be kept in mind. When reconnaissance samples are selected for inclusion work, they should be the best material available for the problem at hand. As an example, in "porphyry country," the so-called "greasy" gray varieties of quartz should be sampled (T.G. Theodore, pers. comm.).

The cost of inclusion studies will vary widely depending upon the nature of the specific work involved. Sample preparation for decrepitation studies in the field or laboratory can be set up on an assembly-line basis, similar to geochemical tests. Decrepitation requires equipment which, outside the USSR, is usually homemade; Ermakov (1966b) reported handling 90 samples per working day for a field unit. Optical microscopy, except reconnaissance work, normally requires the preparation of doubly polished plates or thick sections, although grain mounts in oil and ordinary petrographic thin sections are of some use. The single most important equipment cost, however, is that of the microscope. To try to economize on quality here would be shortsighted indeed because this instrument is the major tool for all inclusion work.

Most chemical analyses of inclusion fluids require fairly extensive laboratory work, but a simple cleaning, crushing, and leaching (Roedder et al., 1963) will provide a solution adequate for analysis of K/Na ratio by the conventional flame photometer present in many company laboratories. However, a series of caveats apply (see Chapter 5).

Commercially available heating/cooling stages are expensive (~$5000-7000), but before any decision is made on such equipment, I suggest that consultants first be considered, mainly because the equipment is not the only expense. Far more important in the long run is the time required to learn the techniques, to find suitable inclusions, and then to use them. Although the newly available commercial heating and freezing stages have greatly simplified the mechanical problems in making the runs, the total time involved can still be large, and much of it should not be delegated to technicians. If preliminary work by an academic consultant or his graduate students shows that more extensive inclusion studies are desirable, then the decision can be made to dedicate manpower and money for equipment to establish in-house expertise. I personally believe that any exploration geologist who has access to a microscope would do well at least to learn to recognize the presence of fluid inclusions in his slides. Far too often I have found geologists who have used a microscope extensively and have "never seen a fluid inclusion," yet their slides were full of them.

Chapter 16

EXTRUSIVE ROCK and VOLCANIC ENVIRONMENTS

CONTENTS

INTRODUCTION

Although most extrusive magmas have cooled and crystallized rather rapidly and hence consist of rather small crystals, the rapid crystal growth results in the trapping of numerous magmatic inclusions. In addition, most such magmas are erupted at subliquidus temperatures, so they contain crystals of one or more early-formed phases as phenocrysts. These also contain magmatic inclusions, trapped during the earlier growth of such crystals. The inclusions in shallow intrusive rocks differ but little from those in extrusive rocks, and hence are included here. As the pressures during crystallization are relatively low, water-rich phases, when present at all, must be either high in salinity or low in density, and are of considerable value in understanding the processes of volcanism. Evidence of high pressures is found only in inherited inclusions in strong crystals that have withstood transport from a deep source (e.g., dense CO_2 inclusions in xenocrystic olivine from the mantle; see Chapter 17). Following magma emplacement at or near the surface of the earth, fluid transport by thermal convection and consequent cooling and mixing can result in the formation of new minerals, e.g., in altered deep-sea basalts, in deep-sea thermal vents, and in other geothermal features. The inclusions trapped during these various processes under surface or near-surface pressures are also considered in this chapter. Many of the features reported in this chapter are also found in the lunar basalts (Chapter 18); in fact, some such features were found in the lunar rocks before they were recognized in terrestrial samples.

SILICATE-MELT INCLUSIONS

The study of silicate-melt inclusions began early in the history of petrography with the keen observations of Sorby (1858), Vogelsang (1867), and especially Zirkel (e.g., 1866, 1873). Sorby (1858) first recognized the similarity between glassy inclusions in lava and those found in the minerals of metallurgical slag. Fouqué and Michel-Lévy (1879, 1881), made important observations on the resemblance of melt inclusions in synthetic minerals formed by fusion processes to those found in fresh igneous rocks. After these papers, little was published anywhere on melt inclusions for many years, until Tuttle (1952) made an important observation on the origin of granite, based on the lack of glass inclusions in granite. Then Barrabé and Deicha (1956, 1957) and Barrabé et al. (1957) homogenized melt inclusions by heating spheres of volcanic quartz from Guadeloupe containing melt inclusions. Subsequently, Kalyuzhnyi (1961, 1965) studied inclusions in phenocrysts in dacites, and since then, interest in such studies has rapidly increased. Much of this later work is in the Russian literature and has been reviewed by Sobolev and Kostyuk (1975). The extensive French work on magmatic inclusions has been thoroughly reviewed by Clocchiatti (1975), and Roedder (1979a) has presented a general review of magmatic inclusions.

Data from Petrography

Silicate-sulfide immiscibility. If the silicate melt (i.e., fluid) from which a crystal is growing forms a new, immiscible fluid phase, globules of this new phase may be trapped as fluid inclusions (see Chapter 2 and Fig. 2-29) and thus provide a valuable record of this former immiscibility. Much can be learned by simple petrographic examination of such inclusions. The most common example is the formation of an immiscible sulfide melt, visible in the inclusions as an opaque sphere. Where this is intersected by the polished surface, it is generally found to be pyrrhotite or troilite, plus smaller amounts of other phases, on the basis of ore microscopy. Because the partitioning of Ni and Cu, and the Pt-group elements (e.g., Buchanan and Nolan, 1979), strongly favors the sulfide melt over the silicate under at least some conditions, the formation and gravitative concentration of such immiscible sulfide melts has yielded important magmatic sulfide ore deposits. The existence and nature of sulfide-melt immiscibility, as evidenced in sulfide-bearing silicate-melt inclusions, can thus be of possible importance in exploration. One of the most important aspects is to determine at what stage the immiscibility occurred. If, for example, it occurred early, when the magma had few crystals, gravitative settling and concentration of the globules of relatively dense sulfide melt would be possible; it would not be possible if immiscibility occurred later, in a crystal-rich mush.

The occurrence of immiscible sulfide blebs in silicate-melt inclusions in various minerals can be used to determine when sulfide-melt immiscibility first

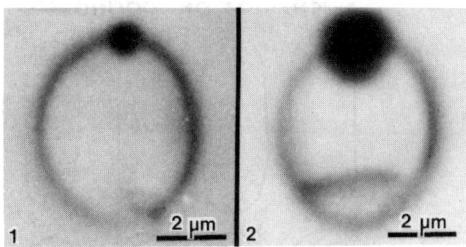

Figures 16-1, 2.   Adjacent tiny inclusions in olivine from the prehistoric Makaopuhi lava lake in Hawaii.   Figure 16-1 shows a typical immiscible sulfide globule, believed to be a true daughter phase, whereas the globule in Figure 16-2 probably represents trapping of a primary sulfide globule. From Roedder (1976b).

took place (i.e., when sulfide saturation was achieved).[1] The occurrence of sulfide globules in glassy basalts is valid evidence of early attainment of sulfide saturation. Pedersen (1979) studied immiscible sulfide together with native Fe in basalt from Diskø Island, western Greenland. Czamanske and Moore (1977) found small amounts (0.0022 vol %) of such globules in glassy basalt from the Mid-Atlantic Ridge rift valley. These globules contain <20-26 wt % (Cu + Ni); this content varies in part with the degree of differentiation of the host magma. Crystals of silicates growing from such a heterogeneous two-phase fluid may trap inclusions either of sulfide or silicate melt, or mixtures in any ratio. In contrast, if immiscibility is late, silicate-melt inclusions trapped earlier will develop an immiscible sulfide globule after trapping, as a daughter phase. As in all daughter phases, the primary criterion for recognition as a daughter phase is the uniformity of phase ratios in various inclusions (see Figs. 16-1, 2). On further cooling, these various sulfide melts crystallize to a mixture of sulfide (and oxide) phases, depending on composition, but usually preserve their external rounded or spherical shapes. A few, formed at low fO_2, as in the Fe-bearing basalts from Diskø Island, western Greenland, may form internal globules of immiscible Fe metal within the immiscible sulfide melt (Pedersen, 1979).

The amount of sulfide dissolved in the silicate melt at the point of satur-ation is controlled by a series of variables, particularly oxygen fugacity, FeO content, P, and T (e.g., Naldrett et al., 1979; Wendlandt, 1982). In glassy rocks, a minimum value for S solubility can be obtained from analysis of globule-free glass. (Since gravitative segregation may have depleted or enriched any given sample in sulfide phase, estimates of the sulfide/silicate phase ratio in a glassy rock may be misleading.) Once a sample of a homogeneous melt is trapped in an inclusion, however, the sulfide/silicate phase ratio becomes meaningful, though difficult to measure precisely. The total S in such an originally homo-geneous inclusion can be determined by heating until it again becomes homogene-ous, quenching, grinding down until the inclusion is intersected, polishing, and determining S by electron microprobe (Roedder and Weiblen, 1970b, p. 810). The time needed to attain homogeneity will be an unknown amount longer than that necessary to have the sulfide globule disappear, because the chemical gradients must also level out by diffusion. This may require several hours.

Numerous analyses have been made of the silicate-glass phase in melt inclu-sions, some of which also contained immiscible sulfide globules. Comparison of the S content of melt inclusions with that of the whole rock should permit an estimate of the S lost during eruption. Clocchiatti et al. (1979) found 1800 ppm S in glass inclusions in basalts from Réunion (Indian Ocean). This value is greater than that for the whole rock, suggesting loss of S during or after eruption (but after trapping of the inclusions). However, interlaboratory

----

[1] Whitney and Stormer (1983) showed that the fugacity of S in a quartz latite tuff was very near to the S condensation curve, implying that liquid S may have been present in the parent magma.

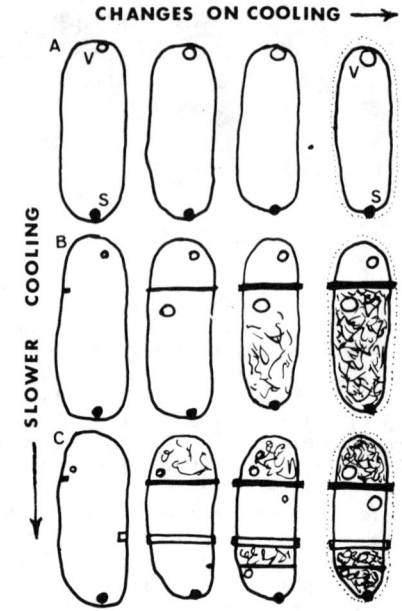

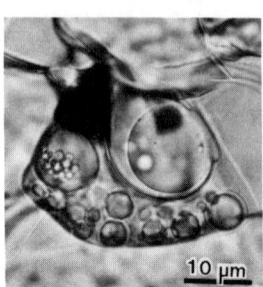

Figure 16-3 (above). Melt inclusion showing silicate liquid immiscibility in olivine crystal from 1965 Makaopuhi lava lake, Hawaii. From Roedder and Weiblen (1971).

Figure 16-4 (to the right). Changes within inclusions after trapping -- crystallization of daughter minerals. V = vapor; S = sulfide globule. See text. From Roedder (1979a).

comparisons have raised some questions concerning S determinations by electron microprobe (pers. comm., D.M. Harris).

Silicate-melt inclusions in olivine in samples of basalt from the 1965 lava lake in Makaopuhi, Hawaii, USA, that were quenched as collected, show sulfide blebs only in those quenched from 1135°C or lower (Roedder and Weiblen, 1971), suggesting that saturation occurred at ~1135°C in that basalt. The relatively large sulfide globule shown in Figure 16-2, also from olivine from that same volcano but from a prehistoric eruption, represents the trapping of a preexisting sulfide globule. Either this host olivine crystal grew at relatively low temperature or the host basalt reached saturation at a higher temperature.

Silicate-silicate immiscibility. For many years in the early history of petrology, the concept of silicate-silicate immiscibility, the splitting of one magma to form two contrasting magmas, was considered by some to be a petrologically important process. Most petrologists considered the evidence for its occurrence in nature to be ambiguous at best (Roedder, 1979d). The discovery of silicate-silicate immiscibility in the late stages of crystallization of lunar basalts (Roedder and Weiblen, 1970a; see Chapter 18 for discussion of the petrologic significance) reopened the search for evidence in terrestrial rocks. Roedder and Weiblen (1971) reported similar evidence in basaltic rocks from the 1965 lava lake in Makaopuhi, Hawaii, USA; the Modoc lavas, California, USA; high-iron basalts at Diskø Island, western Greenland; and upper Precambrian basalts, Minnesota, USA. In a few samples, immiscibility was found as globules of one glass in another, of different index of refraction, within a melt inclusion in a crystal (e.g., Figure 16-3). Similar silicate immiscibility within single silicate-melt inclusions has also been reported by Krasov and Clocchiatti (1979), Clocchiatti et al. (1980), and Philpotts (1981). In most such rocks, however, immiscibility was evident only in the late-stage 2/, normally high-

----

2/ However, one relatively fresh, fine-grained basalt flow, from the upper Precambrian North Shore Volcanic Group, Minnesota, USA, showed immiscibility even though the rock contained 44% glass (Roedder and Weiblen, 1971).

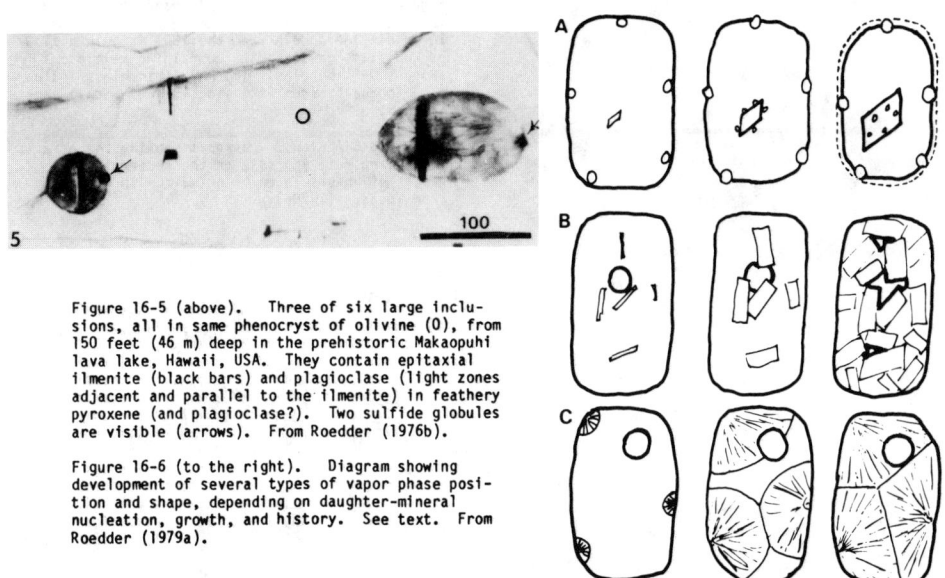

Figure 16-5 (above). Three of six large inclusions, all in same phenocryst of olivine (O), from 150 feet (46 m) deep in the prehistoric Makaopuhi lava lake, Hawaii, USA. They contain epitaxial ilmenite (black bars) and plagioclase (light zones adjacent and parallel to the ilmenite) in feathery pyroxene (and plagioclase?). Two sulfide globules are visible (arrows). From Roedder (1976b).

Figure 16-6 (to the right). Diagram showing development of several types of vapor phase position and shape, depending on daughter-mineral nucleation, growth, and history. See text. From Roedder (1979a).

silica mesostasis or, as I believe such areas of mesostasis should be considered, _interstitial melt inclusions_. Electron microprobe analyses showed that the two glasses have compositions resembling potassic granite and ferropyroxenite, but the composition of the Fe-rich phase was highly variable from one locality to another. Similar immiscibility has since been reported in a wide range of terrestrial rocks (Philpotts, 1982; Naslund, 1983), though Biggar (1979) claimed that in some terrestrial samples the separation occurred metastably, at subliquidus temperatures (see also Freestone, 1979).

Nucleation and growth of daughter phases. Several typical features of daughter-phase formation are well illustrated by inclusions in olivine phenocrysts from both the terrestrial and the lunar mare basalts (Fig. 16-4). Some terrestrial and lunar basaltic melts precipitate an immiscible sulfide globule even in 5 μm inclusions that are too small to form a vapor bubble (Fig. 16-1). Sulfide globules generally nucleate on the wall and may become partly enclosed in the host material (e.g., Fig. 16-1; 16-4A).

The zone of precipitated olivine, the so-called "border of cognate substance" in the Russian literature, is crystallographically continuous with the host crystal and may go unnoticed unless such features as the partly enclosed sulfide globule are recognized. Minor differences in refractive index can sometimes be detected, e.g., in topaz (Voznyak and Kalyuzhnyi, 1974b), and faint ghosts of the former outlines, in olivine (Anderson, 1974a). Anderson also suggested procedures for estimating the magnitude of such postentrapment crystallization (see also Harris, 1981a).

Figure 16-4B shows opaque daughter crystals of ilmenite that nucleated epitaxially on the olivine walls of the inclusion. The resulting flat plates of ilmenite are always parallel to (100) of the enclosing olivine, forming bulkheads that effectively divide the inclusions into isolated parts that subsequently each have their own history of nucleation and growth of daughter phases. Continued precipitation of olivine causes the ends of the plate to become embedded. Pyroxene may nucleate and form a mass of feathery crystals in one or more of the isolated parts, crystallizing dendritically (and presumably rapidly) from a

477

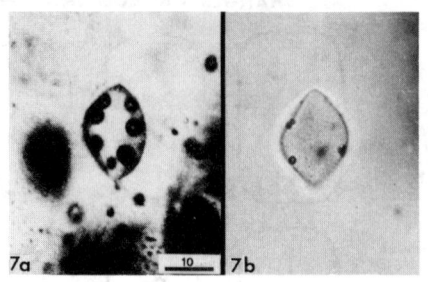

Figure 16-7 (left). Multiple vapor bubbles in melt inclusions. (a) Basaltic glass inclusion in an olivine nodule from the 1801 Kaupulehu flow of Hualalai, Hawaii, USA, containing multiple vapor (CO_2?) bubbles. (b) Rhyolitic glass inclusion in quartz from Ascension Island. Scale same in both.

Figure 16-8 (below). Diagram showing the effects of cooling rate and size on the nucleation and growth of phases in melt inclusions in olivine. From Roedder (1979a).

COOLING RATE

FAST   **SLOW**

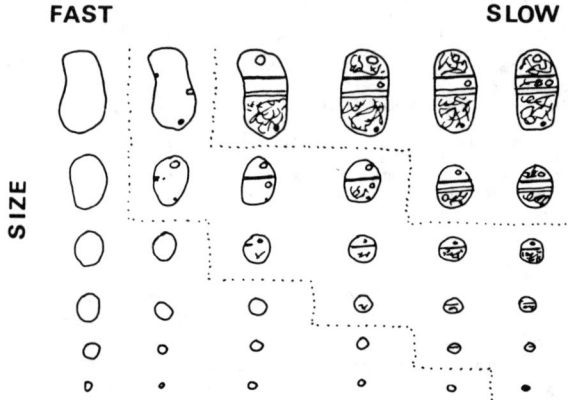

SIZE

strongly supersaturated melt. In larger inclusions, several plates of ilmenite and/or epitaxial plates of anorthitic plagioclase may form (Fig. 16-4C), which may result in multiple isolated chambers (Figs. 16-5 and 18-4b). On detailed examination of the edges of inclusions such as those in Figure 16-5, the ilmenite daughter crystal is always found to be embedded in the walls to some degree (e.g., Fig. 3-1), as are the sulfide globules mentioned above.

Several other nucleation phenomena are illustrated in Figure 16-6. The nucleation of multiple bubbles on the inclusion walls and on daughter minerals is common in the more viscous silicic melts but is seen more rarely in basaltic inclusions (Figs. 16-6A, 7). Most small silicate melt inclusions have no bubble, but apparently nucleation of a bubble is easier in the more hydrous melts. Growth of daughter crystals can deform the vapor bubbles into irregular shapes that are difficult to recognize in an almost opaque crystalline-melt inclusion (Fig. 16-6B). If early cooling is fast and later cooling is slow, fine feathery radial or spherulitic daughter crystals may form (Fig. 16-6C). The included glass is essentially a solid at this stage, so the bubble remains round, even though it is now enclosed by crystals.

Relative cooling rates. If one makes the (possibly hazardous) assumption that the bulk compositions are similar, some estimates of relative cooling rates can be obtained from petrography of melt inclusions. Inclusions in olivine (and other minerals) show that the daughter phases present are mainly a function of inclusion size, cooling rate, and final quenching temperature (Fig. 16-8). Because nucleation is a random process, averages of many inclusions must be compared, and highly precise results cannot be expected. If cooling is sufficiently fast through the temperature range in which daughter phases should form (at equilibrium), only glass will be present. Slightly slower cooling may permit nucleation of a sulfide globule, a bubble, or epitaxial ilmenite and plagioclase. Still slower cooling, maintained through a lower temperature range, permits nucleation and growth of feathery pyroxene, and presumably of more plagioclase and ilmenite, from the remaining liquid (Roedder and Weiblen, 1971; Roedder, 1976b).

Nucleation in natural silicate melts may be either heterogeneous (i.e., at interfaces with other phases), or homogeneous (i.e., within the single pure fluid phase); the evidence for each of these modes has been much discussed (see Chapter 10). Berkebile and Dowty (1982) suggested that homogeneous nucleation probably never occurs in the laboratory or in nature in basaltic compositions. All nucleation that they observed was heterogeneous, as it occurred at interfaces with other phases. Other experimenters have come to different conclusions, and the specific mechanisms involved may be controlled by the melt speciation (Kirkpatrick, 1981). However, whether the nucleation in an inclusion is homogeneous or heterogeneous, it will be volume-dependent as well, so the smaller the inclusion, the less likely a given phase will nucleate. When cooling is very fast, all sizes of inclusions will consist solely of glass. When cooling rates are intermediate, the larger inclusions will, on the average, contain more phases than the smaller ones. When cooling is very slow, only the very smallest inclusions will still contain glass. These differences must be based on averages. For example, one of two adjacent inclusions (of presumably similar bulk composition and in the same olivine grain) may have nucleated epitaxial ilmenite and plagioclase, plus pyroxene, whereas the other, of the same size, contains only ilmenite and vapor in glass (Fig. 18-12). By establishing a sequence of more-or-less objective categories based on observable differences such as the presence or absence of epitaxial ilmenite or plagioclase, or of random crystals, one can plot the number of inclusions having a given assemblage versus a measure of the individual inclusion volume, the maximum inclusion length. Comparisons between such plots may reveal cooling-rate differences (see Roedder, 1976b, and Chapter 18), but the unknown effects of compositional variation (particularly the volatile content) must be kept in mind.

Provenance for volcanic ash and soil. The silicate melt and aqueous inclusions in phenocrysts are unaffected by weathering, provided the host mineral remains unaffected, and they may well provide the last unambiguous line of evidence concerning the processes yielding the original host mineral. For example, the melt inclusions in quartz crystals in tropical soils may be used as indicators of a volcanic origin (Clocchiatti, 1975). In addition, the melt inclusions in crystals derived from an airfall pumice (quickly cooled) will be glassy, whereas those from a thick ashflow may be completely devitrified. Glass inclusions in volcanic detritus in weathered rocks and soils can also be dated, using the fission-track method (Vincent et al., 1981).

Microthermometry of Melt Inclusions

Temperature of trapping. The use of melt inclusions to estimate the temperature of trapping of the melt is obvious, in view of the extensive use of aqueous inclusions for this purpose. This temperature is usually assumed to be given by Th, the temperature of homogenization of the inclusion contents -- gas, glass and crystals -- to a uniform single melt phase, when the sample is heated in the laboratory. This Th is a minimum temperature of trapping, as a pressure correction should be added to Th to obtain the true trapping temperature (Tt), as in aqueous inclusions. The relative incompressibility of silicate melts compared with aqueous fluids, however, and the low pressures of most igneous environments represented in melt-inclusion studies are such that these corrections can be ignored at this time (Roedder and Bodnar, 1980), particularly since several other problems (discussed below) result in much larger errors.

If the gas and melt phases homogenize at a lower temperature than that of

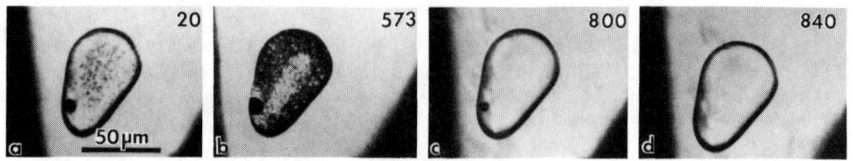

Figure 16-9. Photomicrographs of melt inclusion in quartz phenocryst from rhyolite, taken at the temperatures indicated (°C). From Clocchiatti (1975).

the disappearance of daughter crystals, the difference between these two temperatures represents a minimum pressure correction. Although this is a generally valid concept and readily applicable to aqueous inclusions, there are practical problems in making use of it in silicate-melt inclusion studies, particularly for inclusions that contain coarse daughter phases (Clocchiatti, 1975). Daughter phases may actually become coarser during the heating run if the original inclusion had only slightly crystallized (Fig. 16-9). If the homogenization run is of too short duration, the run time may be insufficient for dissolution of daughter crystals and of material crystallized on the inclusion walls after trapping. Perhaps more important but much less obvious (and frequently ignored) is the problem of allowing adequate run time for diffusion to level the local concentration gradients caused by the dissolution of such phases. The worst case would be that of a long tubular inclusion having several daughter phases along its length. The resulting compositionally zoned melts will show temporary (local) equilibrium, and apparently "reversible" phase behavior that may easily be mistaken for equilibrium for the whole inclusion system. Additional problems may occur if inclusions contain much water (discussed below) or if they yield particularly viscous, high-silica melts. In these examples, the final "Th" determined will be too high, perhaps by 100°C or more, if these diffusion-limited rate problems are not eliminated.

The obvious solultion to problems from slow diffusion within the inclusions is the use of longer runs; such runs, however, promote sample oxidation, which makes many samples opaque. Longer runs also exacerbate the possible loss of hydrogen by diffusion (see below) and hence may cause a large increase in Th. Obviously, equilibration problems can be reduced by the use of smaller inclusions, but this advantage may be offset by increased observational problems and the possibilities of increased error from the boundary-layer problem (Fig. 2-36).

Another method for estimating Th involves heating a sample to each of a series of sequentially higher temperatures, with intervening quenching and examination. Roedder and Weiblen (1970b) used this method on inclusions in the lunar samples; although the method is very tedious, it consumes a minimum amount of material. If sufficient material is available, groups of grains could be heated, each group to a specific temperature, and then examined. Such a procedure is particularly appropriate for studies of rhyolitic melt inclusions that may require many hours to achieve equilibrium (see, e.g., Takenouchi, 1972, and Takenouchi and Katsura, 1972). Although even this procedure is relatively slow, it minimizes the possible loss of hydrogen (since each sample is only heated once) and poses few experimental problems. Such problems are more difficult to evaluate and control in the seemingly unambiguous and certainly more common procedure of using a heating microscope stage and watching phase changes on slow heating. In addition, as indicated in Chapter 7, the "simple" experimental problems involved in making measurements using a microscope heating stage at high temperatures can result in large temperature errors.

The heating could be done in a reducing atomsphere to prevent loss of hydrogen. If the host mineral is permeable to hydrogen, it could move either way, so the problem of choosing a "correct" atmosphere becomes critical.

The foregoing may suggest that a valid determination of melt-inclusion Th is impossible. On the contrary, careful homogenization studies, particularly on inclusions having low volatile contents (i.e., <~0.3%), can produce valid estimates of the crystallization temperatures of the host minerals. Minute inclusions, as long as they already have the required phases, should provide the least difficulty. However, I urge caution in evaluating such data for any hydrous melt inclusions until experiments having appropriate external atmospheres have been performed on that specific mineral. In particular, one should remember that the three major sources of possible error in such determinations -- inadequate time for equilibration, loss of H_2 or H_2O, and thermal gradients in the stage -- generally will <u>all</u> tend to yield <u>high results</u>.

<u>Temperature and sequence of phase changes during cooling</u>. Melt inclusions can provide data on the temperatures of the liquidus, the solidus, and the phase sequences on crystallization. Here, however, we once again encounter the nemesis of many inclusion studies, metastability. Once a melt inclusion is trapped within a mineral (without solid inclusions), it is isolated from all other solid phases that might have been present in the same magma. Obviously, because the inclusion is trapped <u>within</u> a crystal, one can never determine the true liquidus for that magma[3]; however, one can sometimes determine the temperature at which the melt within individual inclusions became saturated, on cooling, with respect to additional phases. This determination cannot be made by simply watching a homogenized inclusion form new daughter phases during a cooling cycle, because gross undercooling is to be expected. However, if the inclusions already contain the phase or phases of interest, nonequilibrium nucleation is no longer a problem. The temperature of <u>disappearance</u> of these phases on reheating will be the temperatures at which they <u>should</u> have crystallized, at equilibrium, on cooling (e.g., Figs. 18-8 to 11). Although such data, if correct, may be very helpful in understanding the liquid evolution, particularly if the thermal properties of all the phases are known (e.g., Romanchev, 1977), the possibilities for large errors in both Th and the sequence of values of Tm for the various daughter phases are numerous (see Chapters 7 and 10).

Sobolev et al. (1974) studied melt inclusions in a series of effusive alkaline rocks, mainly from the USSR, and placed the liquidus at 1250-1290°C, but possibly higher than 1600°C in leucite-bearing rocks. Some earlier studies (e.g., Sobolev et al., 1972) had shown very high Th values (1600-1700°C), which were assumed to be valid measures of the temperature of crystallization. Bazarova and Krasnov (1975) then stated that 95% of all crystallized inclusions in some leucite-bearing basaltoids have leaked, and 99% of the balance leaked during homogenization studies, "with no characteristic evidence of leakage." (Apparently the 99% leakage was based on rise in Th on a second run, but the criteria for the "95%" is not stated.) They also found that inclusions in leucite in leucitite from the East African rift system and from the Leucite Hills, Wyoming, USA, had Th of 1230-1250°C, Tm of ore minerals 1150-1170°C, Tm of silicate daughter minerals <1100°C, and a solidus ~800°C. Subsequently, Sobolev et al. (1975) also studied samples from the Leucite Hills and reported on the crystallization sequence for inclusions in phlogopite, diopside, and leucite from wyomingite, a leucite-bearing rock. They found the following: diopside liquidus, 1320°C; first phlogopite crystallization, 1270°C; diopside crystallization stopped, 1220°C; leucite formed 1250-1150°C; second-generation phlogopite, 1120-1040°C; solidus, ~1040°C.

Th has also been determined for phenocrysts in various silicic lavas, particularly by R. Clocchiatti and his colleagues in France. Anorthoclase from quartz-free lavas of Pantelleria Island yielded 970-1020°C (Benhamou and Clocchiatti, 1976), whereas inclusions in quartz and anorthoclase of the quartz-bearing lavas showed minimum Th of 750-800°C. In some inclusions, acmite crystals formed at 850°C during cooling but were resorbed by 700°C. Cordierite

----
[3] If the host crystal is one of very few phenocrysts in the rock, Tt of inclusions in that crystal should be close to the actual liquidus.

crystallized at 950°C in some rhyodacitic lavas (Clocchiatti and Metrich, 1977). Similar temperatures (740-830°C) were reported for quartz and anorthoclase of pantellerites from Ethiopia (Metrich and Clocchiatti, 1979).

When the phenocrysts of basalts were studied, much higher temperatures were recorded. Bytownite megacrysts from basalts of the Asal Rift formed at 1220±25°C, and were soon joined by olivine, clinopyroxene, and oxides. Basalts and oceanites from Reunion Island (Indian Ocean) became saturated with sulfide at >1200°C (Clocchiatti et al., 1979).

The solidus can be estimated, very roughly, from the behavior of all-crystalline inclusions (crystallized either naturally or by laboratory heat treatments). If an inclusion is completely crystalline and hence has no interstitial glass to melt, heating below the solidus will result in only extremely slow recrystallization, via solid-state processes. Once some melt is present between the crystals, recrystallization of any fine-grained or feathery daughter crystals will proceed much faster. Magmas are multicomponent systems and hence will not possess the sharply defined solidus temperatures seen in many one-, two-, and three-component systems, but if recrystallization is noticeable at one temperature and not at some lower temperature, the apparent solidus can be said to be between the two temperatures.

A different procedure entails experimental determinations of Th for each phase in the rock and then comparing the sequence of temperatures thus established with that evident from the normal petrographic criteria (e.g., Bazarova and Kazaryan, 1977, on a leucite phonolite).

Electron Microprobe Analysis of Melt Inclusions

Significance of melt-inclusion composition. One of the most important uses of melt inclusions in petrology is to provide data on the compositions along the liquid line of descent of the magma. Once again, one must consider the possible differences between melt within the host crystal and that outside the crystal in the main mass of magma. These differences between the composition of such melt inclusions and the surrounding glass (e.g., in a glassy rock) or the residual melt or mesostasis (e.g., the interstitial inclusions in an essentially holocrystalline rock) have been used in a number of studies to help understand the differentiation processes.

Knowledge of the true composition of the original melt trapped in an inclusion would be exceedingly valuable to petrology. Barring local disequilibrium effects, for example, from fast crystallization, a melt inclusion represents a real point on the liquid line of descent for that magma, rather than those inferred points obtained from analyses of apparently genetically related, partly crystalline rocks assumed to represent a series of natural liquids. Melson (1983) was able to trace the changes with time of the volcanic products from Mount St. Helens, Washington, USA, using the composition of melt inclusions. Such compositions can also reveal the existence of magma mixing (Rhodes et al., 1979; Basaltic Volcanism Study Project, 1981).

The electron microprobe is the obvious tool to apply to the analysis of melt inclusions. However, as in so much research, things are never quite so simple as they may seem at first. A variety of experimental problems (discussed in Chapter 5) may combine to yield potentially large inaccuracies in electron microprobe analyses of melt inclusions. Some of these problems can often be minimized or overcome by the use of appropriate samples and procedures, and under such conditions, the electron microprobe has provided a wealth of otherwise unavailable data on melt inclusions. Some of the most consistent electron-microprobe data on melt inclusions has been on inclusions in olivine from Kilauea volcano, Hawaii, USA (Harris and Anderson, 1983), and Fuego volcano, Guatemala (Harris and Anderson, 1984). In many inclusion studies, however, the inherent

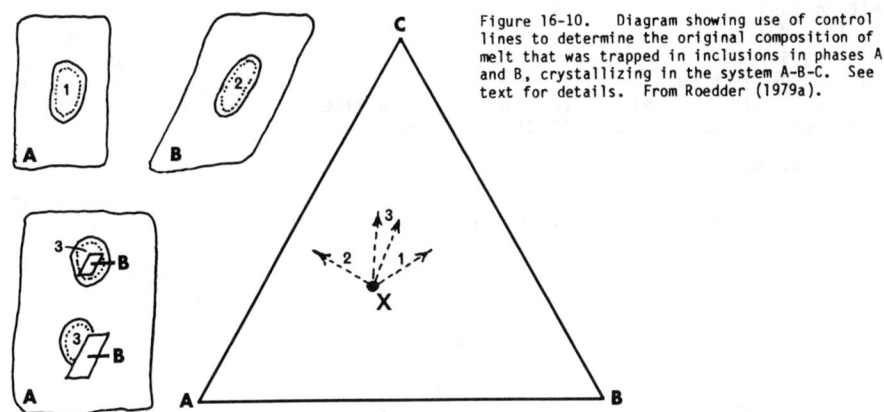

Figure 16-10. Diagram showing use of control lines to determine the original composition of melt that was trapped in inclusions in phases A and B, crystallizing in the system A-B-C. See text for details. From Roedder (1979a).

difficulties make the analyses considerably less reliable than comparable analyses of crystalline compounds (see also Graham et al., 1983). As discussed in Chapter 5, the ion microprobe has great potential for the analysis of melt inclusions, particularly for minor constituents.

Control lines. Any analysis of an inclusion in a given host mineral must lie somewhere along a "control" or "fractionation line" for that mineral on a compositional plot (Fig. 16-10). If the inclusion was too small for accurate analysis, some of the surrounding host mineral may be included in the analysis, and the apparent composition will be intermediate. If the inclusion has crystallized some host mineral after being trapped, the composition point for the remaining melt will have moved along the appropriate control line in a direction away from the host-mineral point. Thus in Figure 16-10, an inclusion of composition X, in phase A, would evolve along control line 1 by crystallizing A on the walls; one in phase B would evolve along control line 2 by crystallizing B on the walls. These two inclusions are illustrated in the upper left of the diagram. The intersection of the two control lines provides the actual composition (X) that was trapped. The method requires the melt to have been trapped at the same stage of differentiation in crystals of the two different phases, or, better still, three phases.

On the lower left of Figure 16-10, I show two melt inclusions in phase A, one trapped as a result of a solid inclusion of phase B (lower), the other containing a true daughter crystal of phase B (upper). In either inclusion, the residual melt will lie along some intermediate and generally curving control line 3, depending on the ratio of A and B crystallizing. For this purpose, where only the composition of the residual melt is considered, it makes no difference whether B is a daughter crystal or a solid inclusion. This difference would be very important to know, however, if the apparent gross compositions of these two inclusions were being obtained.

Watson (1976), Clocchiatti (1977a,b), Clochiatti et al. (1978), and Clocchiatti and Bizouard (1979) used the intersecting control-line procedure to estimate the composition of the original melt trapped in inclusions of crystals from basalts. When only one mineral phase contains usable inclusions, yet several minerals are crystallizing, it becomes increasingly difficult to interpret the inclusion analyses (Weiblen, 1977). In the interpretation of such solid/liquid equilibria, particularly where the host is a solid solution, the effects of the relative masses of phases available for reaction should be kept in mind, as discussed in Chapter 2 (p. 41).

Individual melt-inclusion analyses. Even single analyses of melt inclusions can be informative, however, particularly if the inclusion is homogeneous. If melt inclusions in phenocrysts are found to have compositions that do not fit in the liquid line of descent of the host-rock series, the host crystals may be xenocrysts rather than phenocrysts. Compositions of some melt inclusions are found to be bizarre and hence to suggest that unknown processes might have been involved, e.g., to yield K-rich inclusions in ilmenite from the Duluth complex (P.W. Weiblen, pers. comm., 1978).

Multiple analyses of many inclusions, from crystal hosts that have formed at several stages in the differentiation, can also yield important data. Most commonly, such analyses reveal the liquid line of descent of a fractionating magma. Examples of such studies include a pseudoleucitite (Chepurov et al., 1974); a pantellerite (Metrich and Clocchiatti, 1979); nuée ardente deposits at Soufrière, St. Vincent (Bardintzeff and Clocchiatti, 1980); plagioclase, pyroxene, and apatite from a dacite of the Stantiaquito dome in Guatemala (Bardinizeff et al., 1980); cordierite from a rhyodacite (Clocchiatti and Metrich, 1977); anorthoclase megacrysts from Erebus volcano, Antarctica (Clocchiatti et al., 1976); basalts and oceanites from Réunion Island (Clocchiatti et al., 1979); dacite from Guadeloupe (Clocchiatti and Mervoyer, 1976); dacite from Mount St. Helens, Washington, USA (Melson, 1983); and the lunar basalts (see Chapter 18).

Clocchiatti and Krasov (1979) were able to recognize a wide range of differentiation in the inclusions in plagioclase phenocrysts from the Karimski volcano in Kamchatka, USSR. Melts having the most evolved compositions were apparently rich in volatile elements and contained daughter crystals of amphibole and biotite, and of magnetite, potassic feldspar, and quartz. Some showed silicate immiscibility. Larhidi et al. (1980), in a study of melt inclusions in pantellerites and comendites of Sardinia, noted that ion probe and electron microprobe analyses revealed a significant loss of Na (and some K) wherever a crack intersected the glass, even though no alteration was visible by light microscopy.

Anderson and Wright (1972; see also Anderson, 1975) showed that magma mixing was a dominant process in the origin of the basaltic magmas of Kilauea volcano, in part through the analysis of melt inclusions in various silicate and oxide phases. Similar studies of inclusions in refractory megacrysts of Cr-Al spinel from oceanic tholeiites suggested magma mixing (Donaldson and Brown, 1977), and Rhodes et al. (1979) reported magma mixing as a fundamental process in the formation of midocean-ridge basalts, in part on the basis of analyses of melt inclusions in olivine and plagioclase.

VOLATILES IN INCLUSIONS

Amount and Nature -- Analytical Problems

Volatiles are important in almost every aspect of igneous petrology, and melt inclusions provide the best available samples. Hence, valid determinations of the volatile constituents in silicate-melt inclusions might be among the most significant and useful information obtainable from them. For more than a century, since the early experiments in France of Daubrée and Sainte-Claire Deville on the effects of mineralizers on rock and mineral behavior, debate has continued in the geologic literature on the subject of volatiles in magmas. Throughout this time, actual samples of magma have been available, in the form of melt inclusions, each having at least a partial and possibly a full complement of volatiles. All that was needed were suitable analytical techniques. As an example, Figure 16-11 shows a grain from an Icelandic tuff, containing many melt inclusions. The melt must have contained appreciable volatiles, because wherever a later fracture has intersected an inclusion, the opened inclusion has expelled the melt and has formed a large dark vapor bubble.

The first attempt at analysis of such volatile-rich and hence presumably

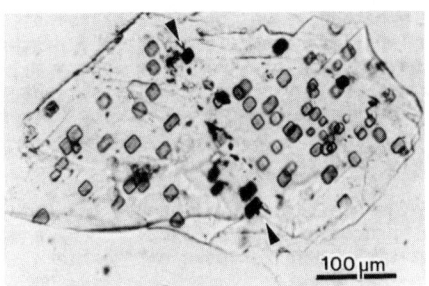

Figure 16-11. Crystal fragment from an Icelandic tuff, mounted in oil, containing many silicate-melt inclusions. Where these have been cut by a later fracture, now healed (arrows), they have expelled the melt and formed a large dark vapor bubble. From Roedder (1979a).

hydrous-melt inclusions was by Carron (1961), who used the newly developed Castaing electron microprobe to obtain analyses of glass inclusions in quartz phenocrysts. These analyses yielded totals <100%, presumably because of the presence of volatiles that could not be determined by electron microprobe. Clocchiatti (1971) used infrared absorption to prove the presence of (OH) in such glasses and hence to suggest a possible cause for the low totals. Since then, A.T. Anderson and coworkers (see below) have also used the microprobe analysis-by-difference method on melt inclusions in phenocrysts from explosively erupted tephra (hence presumably from water-rich magmas). Takenouchi and Imai (1975) presented 12 microprobe analyses of melt inclusions in silicic igneous rocks in which totals averaged 95%; part of this deviation from 100% might be ascribed to H_2O in the glass.

Anderson (1973, 1974a), and Anderson and Sans (1975, and pers. comm.) have devised several other indirect methods for estimating the H_2O contents of magmas from melt inclusions, including the amount of H_2O loss that is required by a distillation model to account for the decrease in Cl/K ratios with differentiation, and geothermometry of melt inclusion/host crystal pairs; they found some high values, as much as $12\pm2\%$ H_2O! Although high, these results appear consistent with the amounts estimated from the low totals found by microprobe analysis. Anderson (1974b) found the analysis-by-difference method to have an uncertainty of $\pm1.4\%$ absolute for a two-standard-deviation error. Sommer (1977) and Johnston (1978) have also used the anaysis-by-difference procedure, and Sommer has attempted to corroborate such results by using mass spectrometry. Many workers express little faith in any estimates based on low electron microprobe totals, particularly because the amounts of H_2O (or other volatiles) estimated from them are much higher than previously thought. However, it is difficult to avoid the conclusion that the low totals in these analyses are indeed produced, at least in large part, by volatiles, because when Anderson took aliquots of the same samples and drove off the H_2O by heating, reanalysis by the identical microprobe technique gave totals of nearly 100%.

Many attempts have been made to analyze the volatiles in bulk samples containing melt inclusions, but the many problems of extraction, contamination, loss, and analysis make many of the results of dubious value. In addition to these bulk-sample analyses, and their inherent ambiguities, various papers in the Russian literature over the past decade have presented analyses of the gases extracted from the gas bubbles in single melt inclusions in igneous minerals. Most of these analyses were performed at Novosibirsk, USSR, and are generally credited to N.A. Shugurova, Yu.A. Dolgov, or both. The problems inherent in this procedure have been discussed in Chapter 5.

In addition to studies of the composition of individual melt inclusions, numerous studies have been made of the H_2O and CO_2 contents of bulk samples of volcanic glasses that may be compared with the inclusion analyses. Although small sample size is not a problem here, the low concentrations make accurate determinations difficult. Some of the same methods used for melt inclusions have been applied to bulk glass samples. Thus, measurements of the CO_2 content of melt inclusions in submarine basalts (Harris, 1981b) yielded an approximate value for the minimum apparent solubility of CO_2 in tholeiitic basalts (containing 0.10 to 0.34 wt % H_2O) at temperatures of quenching on the sea floor given by the expression: CO_2 (wt %) = 0.0005 + 0.059P (kbar); this CO_2 concentration can be used as a geobarometer in a variety of contexts.

The elemental and isotopic ratios of the noble gases in igneous rocks may provide valuable insight into deep igneous processes (James, 1983). Many analyses have been made of these gases in basaltic glasses erupted on the ocean floor and in submarine hydrothermal fluids (see Chapter 17 for details), but such gases are not necessarily representative of the original magmatic gases. For this reason, analyses of these gases in melt inclusions in phenocrysts would be highly desirable.

Mass spectrometry has been used to determine the volatile contents of volcanic samples, both bulk glass and glass inclusions, but, as discussed below, the results are not always in agreement with other methods. Sommer (1977) measured H_2O, CO_2, and CO evolved on heating crushed samples of quartz phenocrysts (from rhyolite) containing silicate melt inclusions. His results (average of four determinations originally given as percentage to four significant numbers) are: H_2O 91.6, CO_2 3.5, and CO 4.8 %. Unfortunately, the volume of inclusions analyzed was only estimated (1%). In a later paper (Sommer and Schramm, 1984), this shortcoming was partly corrected, permitting quantitative (manometric) determinations of the volatiles in a rhyolitic magma (<3.99 wt % H_2O). Muenow (1973) studied basaltic glass samples by direct volatilization from a high-temperature effusion cell assembly into the spectrometer. The technique was subsequently modified (Killingley and Muenow, 1975a; Delaney et al., 1978; Muenow et al., 1979; Byers et al., 1983). The H_2O, CO_2, CH_4, CO, and trace amounts of C_2H_4 and/or C_2H_6 obtained from basalts were assumed to be actually present in the samples, and not modified during the heating (to 1200°C) since the experimental conditions "provide little opportunity for molecular collisions once the volatiles are released from the sample" (Byers et al., 1983, p. 1555).[4]/

Other techniques are being applied to obtain direct analyses. Stolper (1982a,b) has used near-IR transmission spectra to characterize the molecular species of H_2O in silicate melts and to determine the amounts present. Measurements can be made on samples 100 μm in diameter. Although the ion probe seemed at first to provide the ultimate method for determination of H in melt inclusions (e.g., Clocchiatti, 1975), it has some serious shortcomings, in addition to high cost. Delaney and Karsten (1979) found that they could analyze 10-20 μm inclusions for H_2O and its D/H ratio, but that calibration was rather involved (see also Steele, 1983). Both sulfide and sulfate S can be determined simultaneously on bulk samples of basalt by a special solution procedure (Sakai et al., 1983). Other studies have involved gas chromatography or mass spectrometry, and particularly the combination of these two techniques, but it is fair to say that numerous problems remain in the analysis of the volatiles in melt inclusions. Even in synthetic melts in the laboratory, the experimental problems are severe (Kuroda et al., 1982).

D.M. Harris (1979a,b, 1981a,b) described a new technique for directly determining H_2O, CO_2, SO_2, and total noncondensable gas released from samples heated to 1280°C in vacuum. The technique is based on Boyles Law and pressure measurements made using a diaphragm-type capacitance manometer, previously calibrated against a McLeod mercury pressure gauge standard. The gas evolved on heating the samples to 1280°C in a fused silica tube under the microscope is continuously collected

----

4/ Note, however, the several caveats on p. 120.

in a liquid-N_2 cold trap, together with background (blank) contributions from the apparatus. Gas components are identified by the characteristic vapor pressure and temperature ranges over which solid and vapor are in equilibrium during sublimation of individual components. The masses of CO_2, SO_2, and H_2O derived from samples and blanks are calculated by using the ideal gas law, the molecular weights of the components, and the gauge constant (i.e., the ratio of the number of moles of a gas to its partial pressure in the constant volume). The net amounts (after correction for the blanks) of H_2O, CO_2, and SO_2 released from the sample were linearly proportional to the inclusion sample mass, and the extrapolation to zero inclusion mass went through the origin. Comparisons of H_2O determinations by this technique with those obtained in four samples of basaltic glass and one nearly stoichiometric hydrous mineral by the Penfield, gas-chromatographic, microcoulometric, and vacuum-fusion techniques used elsewhere generally show excellent agreement. For two samples, Harris' results (1981c) differed from previously reported values by as much as a factor of three, but most of his own other results are in agreement with data obtained by others, who used very different methods. Thus, his CO_2 data on submarine basalts (1981b) are consistent with those of Moore et al. (1977). I suspect that the major cause for the reported differences is, as pointed out by Harris (1981b,c), in the selection, preparation, and handling of samples that are appropriate for the method used.

Determinations of SO_2 by this technique agree reasonably well with determinations of total S by X-ray fluorescence and wet-chemical techniques but only poorly with electron microprobe results (Harris, 1981a and pers. comm., 1983). Determinations of CO_2 by the present technique are reproducible but cannot be compared directly with measurements made in other labs because of differences in samples analyzed. The principal advantages of this analytical technique are the very small samples that can be analyzed, the simultaneous determination of H_2O, CO_2, SO_2, and noncondensable gas, the avoidance of calibration procedures dependent on chemical standards, and the visual observations that can be made during sample outgassing.

The crystals containing the carefully preselected individual inclusions are cleaned of externally adhering glass by treatment under sonication with concentrated fluoboric acid at room temperature for 30 min, the treatment being repeated as needed (A.T. Anderson, pers. comm.). A polished surface that cuts into the inclusion is then prepared, which permits electron microprobe analysis and provides a ready exit for the volatiles on heating (for details, see Harris and Anderson, 1983). Because the detection limit for each gas is 10 ng (i.e., 10^{-9} g) and the blanks are typically on the order of 50 ng H_2O, 10 ng S, and 10 ng CO_2, and are reproducible to <±5 ng, the minimum amount of any of the three gases that can be detected in a single cubic inclusion 150 μm on an edge is ~0.1 wt %; for larger samples, this amount will be proportionately smaller. The inclusions studied were much larger than 100 μm on an edge.

Although incomplete evolution of the volatiles under the conditions used might seem a valid possibility, the results for H_2O reported by Harris agree with those obtained by four other laboratories using four different methods on "duplicate" samples, and he obtained good results when analyzing appropriately small fragments of minerals of known water content such as diaspore. Two of the samples analyzed by Harris were also run by Moore (1965) for H_2O, using the Penfield method, and by Muenow et al. (1979) for both H_2O and CO_2, by mass spectrometry. None of the results agreed, and both other methods showed variable but two to five times more H_2O, and CO_2, than found by Harris. Some of these differences might be explained by sample differences, but I suspect that major analytical errors may be involved in both other methods, as applied to these materials.

Perhaps the greatest value of such analyses of H_2O and CO_2 in individual melt inclusions is that the escape or retention of these components from the melt during differentiation can be documented. H_2O and CO_2 contents can also be related to other geochemical criteria for differentiation, such as the composition of the inclusions and their host crystals, as well as to the stages of volcanism as evidenced by the nature of the formations from which the inclusions were obtained. Furthermore, phenocrysts in ash provide samples of the magma from violent eruptions that could not possibly be sampled directly. Under favorable circumstances, the geochemical evolution of the magma body inferred (ex post facto) from melt inclusions might be correlated with both eruptive mechanisms and related precursory phenomena.

If electron microprobe analyses are made of individual glass inclusions and their host crystals before heating and release of H_2O and CO_2 in a separate determination, the results can be compared with the volatiles obtained. In a study of basaltic-melt inclusions in olivine phenocrysts from the October 14, 1974, vulcanian eruption of Fuego volcano in Guatemala (Harris, 1979a), twofold increases in H_2O (1.6 to 3.5%), Cl, and K_2O in the glass were noted as the host olivine became more Fe-rich (Fo 77 to Fo 71) and as the SiO_2 in the glass inclusions increased from 51 to 54%. Similarly, the compositions of the melt inclusions are systematically related to their temperatures of entrapment, inferred from the distribution of MgO between olivine and melt (Roeder, 1974), made with corrections for crystallization of olivine on the walls after trapping (Harris, 1981a). Pressures of entrapment were estimated from the temperatures and measured volatile contents and were related to a physical mechanism and energy budget for explosive eruption (Harris, 1981d), as well as to budgets for S and Cl in the preeruption magma body and its eruption products (Rose et al., 1982). A similar approach was followed by Rose et al. (1984) to estimate the contribution of S, Cl, and H_2O to the atmosphere from the 1980 eruption of Mount St. Helens.

Harris and Anderson (1984) also determined the CO_2 content of the glass inclusions from Fuego, using the vapor-pressure method. The amounts released were small and mainly provided only upper bounds for the CO_2 content of the melt. This amount was estimated to be ~0.15±0.05%. The inclusion evidence indicates, however, that a separate gas phase was also present, making difficult any calculation of the quantity of gases available to drive explosive volcanism.

Conflicting data have been published on the water content of melt inclusions in some Hawaiian olivine phenocrysts. Using the method described above, involving thermal release from opened inclusions, Harris (1981a) and Harris and Anderson (1983) found some of these melt inclusions to have 0.03-0.10 wt % CO_2 and 0.23±0.04 wt % H_2O in replicate determinations for individual host crystals (Fo 87) from the 1959 eruption and higher values (0.33-0.56 wt % H_2O, 0.3±0.1 wt % CO_2) for a single inclusion in a Fo 88 host from a submarine eruption. These concentrations are the same or greater than the concentration of H_2O in the glass outside the crystals (Harris, 1981a). Delaney et al. (1978) and Muenow et al. (1979) investigated essentially similar samples by mass spectrometry. Since the volume of melt in their sample was not known, their results are qualitative, but although they found CO_2, the H_2O was below their detection limit (i.e., <0.004 wt %). They found this result "unexpected," but accepted it and suggested that the source rocks must have contained virtually no water at the time of partial melting.

An equally likely explanation for these discrepant results is evident from studies of CO_2 in inclusions in various basaltic samples (Roedder, 1965a). I suggest that two types of inclusions were present in the samples run by Delaney and Muenow: silicate-melt inclusions and a few high-pressure CO_2 inclusions. In the mass spectrometric technique, samples containing melt inclusions were heated slowly to 1250-1300°C to decrepitate the inclusions, and by rapid mass scanning, the authors detected "spiking" -- sudden evolution of CO_2 gas from the sample. Since the melt inclusions were probably trapped at ~1250°C, they probably did

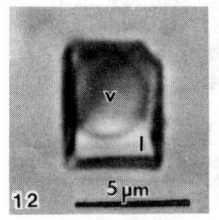

Figure 16-12. CO_2 inclusion showing liquid (l) and vapor (v), in sanidine from ejected granitic block from Ascension Island, as described by Roedder and Coombs (1967).

not develop enough internal pressures to decrepitate at 1250-1300°C, as olivine can withstand ~7000 bars internal pressure at 1200°C (Roedder, 1965a). Some of the CO_2 inclusions, however, would probably decrepitate at 1250-1300°C, or even on cooling (Killingley and Muenow, 1975b). In so doing, they would also release a small amount of the melt that normally lines the CO_2 inclusions. However, even if the instrumental detection limit is taken as 10^{-11} g H_2O, and the H_2O content as that found by Harris, an appreciable amount of glass would have to degas completely during the time of the mass sweep through m/e = 18(H_2O) to be detected. Quantitative measurement of the amount of inclusion melt actually degassed would help in interpreting the results of Delaney et al. (1978) and Muenow et al. (1979).

The gases present in the magma, as determined from analysis of melt inclusions, are largely lost during eruption. The gases present in volcanic plumes like those from Soufrière, St. Vincent (Cronn and Nutmagul, 1982) and Mount St. Helens (Casadevall et al., 1983) consist of such volatiles from the magma, but they may be modified by various processes during eruption (Gerlach, 1980, 1982). Cronn and Nutmagul reported significant amounts of two S species, COS and CS_2, in the plume at Soufrière; these are seldom even looked for in inclusion analyses. The compositions of the gases obtained from melt inclusions should be compared with those collected from volcanic vents and with the calculated equilibrium ratios of gaseous species.

Immiscible Phase Separation vs Vesiculation

Even if no quantitative measurements are made, qualitative evidence from inclusions of fluid immiscibility of any type in magmas has petrologic significance. The presence of even a single primary inclusion of a "gas" phase in an igneous mineral requires that saturation was achieved, and all the consequences of immiscible separation, including volcanic eruptions, formation of ore deposits, etc., become feasible. A.T. Anderson, Jr. has also suggested in several of his studies (e.g., Rose et al., 1978) that "bubble rafting" has occurred, in which a vapor phase preferentially nucleates on the surface of certain phenocrysts, and hence, just as in mineral-flotation technology, the phenocrysts are carried toward the top of the chamber even though they are heavier than the magma in which they are immersed.

Hydrosaline melt in granitic melts. Because the highly saline fluids (as are common in the shallow granitic porphyry intrusives associated with the porphyry copper deposits) contain little H_2O, they have relatively low vapor pressures and hence may be found in fluid inclusions formed under near-surface conditions. The typical example shown by point C on Figure 8-25, containing only 32.4% H_2O, would have a vapor pressure only slightly >200 bars at 500°C. Similarly, the hydrosaline melts that separated from the Ascension Island granite magmas as immiscible fluids (Fig. 14-10) could exist at very low pressure. However, the CO_2 inclusions in some of these same rocks (Fig. 16-12) suggest that the pressures were not low.

CO_2 fluid in basaltic melts. When a magma containing volatiles approaches the surface, it eventually reaches a level at which the hydrostatic pressure is

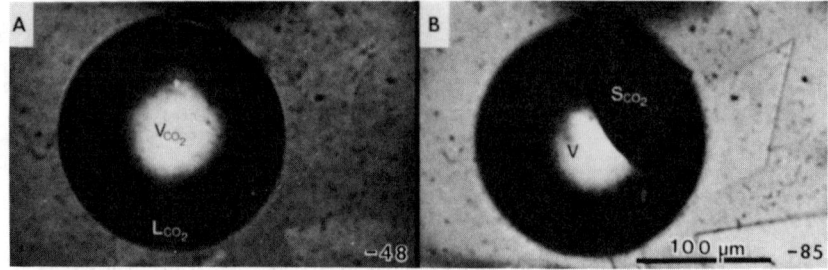

Figure 16-13. CO_2-filled vesicle in basaltic glass from submarine flow (Moore et al., 1977); photographs taken at the temperatures indicated (°C). At -48°C (A), an invisible film of liquid CO_2 (L CO_2) coats the walls and surrounds the vapor (V CO_2), and at -85°C (B), a mass of solid CO_2 (S CO_2) in vapor (V) is visible.

reduced to less than the vapor pressure. At some point slightly above this level, the degree of supersaturation becomes adequate to cause nucleation and expansion of a low-density gas phase (i.e., the lava vesiculates). Sparks (1978) suggested that the levels of volatile supersaturation pressure required for nucleation are low (<10 bars) due to surface active components in common magmas (i.e., heterogeneous nucleation), but this view has been challenged.

If the solubility of the exsolving fluid in the magma increases with pressure (as it does for ordinary gases), surface tension should result in a certain critical radius, below which bubbles will not form. The critical radius for gases in near-surface magmas is still an open question and may be important in applications of gas nucleation (e.g., in midocean-ridge basalts; Moore, 1979). Extrapolation of this concept to determine the critical radius at higher pressures (e.g., to the environment of trapping of the CO_2 inclusions in the ultrabasic nodules at >5 kbar) would be more difficult, due to a lack of data on the surface tension at such fluid-liquid interfaces.

Moore et al. (1977) found that vesicles in basaltic glass from midocean submarine tholeiitic basalts contained CO_2 at pressures >1 atm. (Hekinian et al. (1973) reported CO_2 vesicles in basalts dredged from the Mid-Atlantic Rift that had high enough pressure to break the rock when brought to the surface.) Although such vesicles are enclosed by a glassy rather than a crystalline host, they are essentially identical with normal fluid inclusions. The statement that the gas now in the vesicles is essentially pure CO_2 was based on several lines of evidence: (1) Observed values for Tm CO_2 (solid to gas) were -59±3°C, some of this range being a result of instrumental rather than compositional variables. (2) Gases were released by crushing samples in vacuum and were analyzed by gas chromatography, in part after cryogenic separation. The analyses showed ~95% CO_2[5], plus H_2O, N_2, SO_2, O_2, and Ar. H_2S, CH_4, C_2H_6, He, and H_2 were not detected on analysis. (No H_2S or SO_2 could be detected by odor.) (3) The rate of dissolution with time of the bubble released into glycerine showed an inflection point corresponding to a mixture of CO_2 with ~4 vol % of a less soluble gas. (Note, however, that in these larger gas inclusions, the cooling from adiabatic expansion of the gas changed the solubilities of CO_2 in glycerine and other liquids and resulted in an initial expansion of the bubble with time as thermal equilibrium was attained.) (4) The released gas bubbles reacted quickly in alkaline $BaCl_2$ solution to form a white $BaCO_3$ precipitate. At least some S gas was assumed to have been present originally (Moore and Calk, 1971) but has since been consumed by reactions to form the sulfide spherules that line the walls of some inclusions. The original C/S weight ratio was estimated to be ~6-8. (The occurrence of the spherules does not seem to preclude diffusion of sulfide from the melt to the interface.)

----

[5] Since H_2O is much more soluble in basaltic melt than CO_2, very little H_2O would be expected in these vesicles.

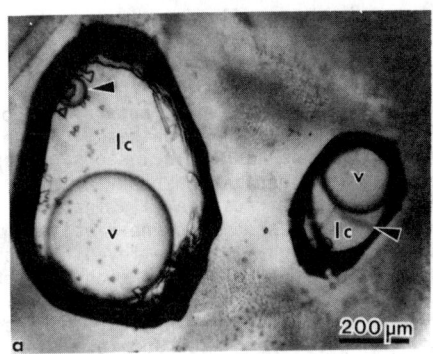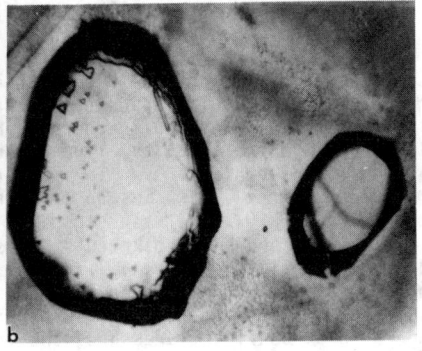

Figure 16-14. Typical behavior of large inclusions of essentially pure CO_2, at nearly the critical density of filling, from a cut gem sapphire from Yogo Gulch, Montana, USA, U.S. Nat. Mus. sample. The small triangles in the large inclusion and the line in the small inclusion (arrow) are markings on the walls of the inclusions. (a) Because of absorption of infrared light from the microscope light, the meniscus between CO_2 liquid (lc) and CO_2 gas (v) becomes fainter as their densitites approach each other at about 31°C. Such light absorption obviously cannot be perfectly uniform, and so even under static conditions, new bubbles of vapor form continuously at hot spots in the liquid and move to the cooler main vapor bubble (arrow in left inclusion in (a)). The large inclusion homogenizes in the liquid phase by a fading and shrinkage of the vapor bubble, indicating the fluid density to be slightly greater than the critical density; the small inclusion homogenizes by fading of the meniscus, indicating the fluid in it to be almost exactly at the critical density. The difference in filling density responsible for such a difference in behavior may be very small. After complete homogenization (b), slight undercooling of the stretched fluid (now. a supercooled, subcritical fluid) results in a sudden splitting into a liquid and many faint vapor bubbles. From Roedder (1972).

Various procedures were tried to determine the density of the vesicle CO_2 gas at room temperature. (1) Formation of CO_2 liquid was observed in only one sample. It evaporated at higher temperatures (Th CO_2 L-V(V) = 18.3±0.1°C). Both this temperature and the measured ratio of L/V provide an estimate of the density of this inclusion. Other inclusions that formed no visible liquid CO_2 could either have had too low a density to form liquid at any temperature, or the CO_2 liquid may have formed an invisible film on the walls. As most of the "inclusions" (i.e., vesicles) were spherical (Fig. 16-13), such films would be very difficult to see. (2) The volume of solid CO_2 formed at low temperatures provides a rough measure. (3) The expansion of the gas was measured upon release to 1 atm. This was done by measuring, then piercing, bubbles that were just under a polished surface, under a layer of glycerine, then measuring the volume of the bubble that evolved. Cooling to increase the viscosity of the glycerine helped to control the operation. More than half the vesicles had leaked part of or all their gas, which, along with several laboratory problems, complicated the measurements. However, 90 good measurements were obtained; these showed a correlation between the known eruption pressure (based on depth) and the volumetric expansion of the bubble. The expansion was 0.2±0.05 times the eruption pressure in bars, or 20±5 times the known eruption depth, in kilometers, below the surface of the ocean. As a result, these vesicles provide a possible means of estimating the depth of water at eruption for samples of unknown provenance. They also provide information on the "rigid temperature" of the host glass. During cooling of the lava, flowage of the host glass will permit collapse of the vesicles as their internal pressure drops, until further collapse is stopped by the rise in glass viscosity. Different samples gave evidence of rigid temperatures ranging from 800 to 1000°C. The amount of CO_2 in the vesicles corresponded to 400-900 ppm of the rock. Moore (1979) suggested that this CO_2 is probably of deep mantle origin (see also Chapter 18).

CO_2 is frequently found to be associated with near-surface basaltic volcanism, in many different geologic environments. In addition to the common evidence of carbonatization of the wall rocks of basaltic dikes, fluid inclusions in the associated minerals commonly show high CO_2 contents. Two examples will suffice: (1) in the Delaware basin, southeastern New Mexico, USA, Permian salt beds are cut by a basalt dike several meters thick, which may have melted the salt adjacent to it. Some fluid inclusions in salt 1-2 cm from the dike apparently contain dense but impure CO_2 (Roedder and Belkin, 1978). (2) Sapphires from the well-known sapphire locality, Yogo Gulch, Montana, USA, show inclusions of essentially critical-density CO_2. The near-critical homogenization behavior of a particularly large CO_2 inclusion (750 µm), from the microscope illumination, is shown in Figure 16-14. These sapphires are in an altered pyroxene-biotite-analcite dike. Although the dike cuts limestone, Clabaugh (1952) believed that the sapphire crystallized directly from the magma at depth. The inclusions indicate that the sapphire formed under high pressure, presumably in the presence of immiscible globules of dense CO_2 fluid carrying unknown solids in solution; later, under a lower confining pressure, the internal pressure caused some of the inclusions to crack and let some of the fluid escape into the fracture, where it caused healing of the fracture and the trapping of many tiny secondary CO_2 inclusions.

Steam vesicles in pumice and the diffusion of gases in rhyolitic pumice; 5 minute "age determinations." Like the CO_2 vesicles in basaltic glass just described, the vesicles in rhyolitic pumice provide an interesting and useful kind of "fluid inclusion." At the time of eruption, each of these vesicles contained gas at some pressure greater than 1 atm.[6] The bulk of this gas was probably steam (H_2O). On cooling to surface temperature, this steam condensed to form a very small amount of liquid water, which then diffused into the glass walls leaving essentially a vacuum (Roedder, 1970a). When grains of relatively recent pumice (10,000 years old or less) are crushed in oil, two categories of apparently empty vesicles are found: (1) some vesicles release a gas bubble of about the same volume as the original vesicle when the embedding oil enters; (2) others fill up completely with oil almost instantaneously (Fig. 16-15). Those vesicles containing gas at atmospheric pressure have simply leaked, and hence are full of air, but the others have maintained their original vacuum. This maintenance of a vacuum for geologic time is rather surprising because the average wall thickness between the vesicle and atmospheric pressure is only a few to a few tens of micrometers. This small wall thickness is caused by the many long tubular vesicles and cracks in pumice that act as channels to bring atmospheric pressure throughout the grains. (These channels are responsible for the fact that floating pumice eventually becomes waterlogged and sinks). Only the isolated vesicle, completely surrounded by unfractured glass walls, maintains a vacuum. Although no actual diffusion rates for atmospheric gases through volcanic glass have been determined, obviously these rates must be very low.

In order to place some limits on these rates, we must know the sensitivity of the crushing procedure. If gases, such as air, that are not condensed by 1 atm at room temperature are present in an inclusion, the smallest amount that can be detected by the crushing procedure is that which will form a visible bubble in the mounting medium that enters on crushing. As shown in Chapter 7, the limit on the sensitivity of the crushing procedure is the amount of gas in a 2 µm bubble at 1 atm. This amount of gas (4 $µm^3$), if it consists of N_2 at 1 atm, weighs $5 \cdot 10^{-15}$ g and contains only about 100 million gas molecules. As the surface area of the vesicles being considered is on the order of 10^3 $µm^2$, these data would indicate that the rate of diffusion of atmospheric N_2 through the glass walls was <300,000 molecules/cm^2/sec.

----

[6] Sparks (1978) has proposed that nonequilibrium effects produce very significant excess internal pressures; viscous resistance of the liquid can yield tens of bars at viscosities of 10^8 poise and above.

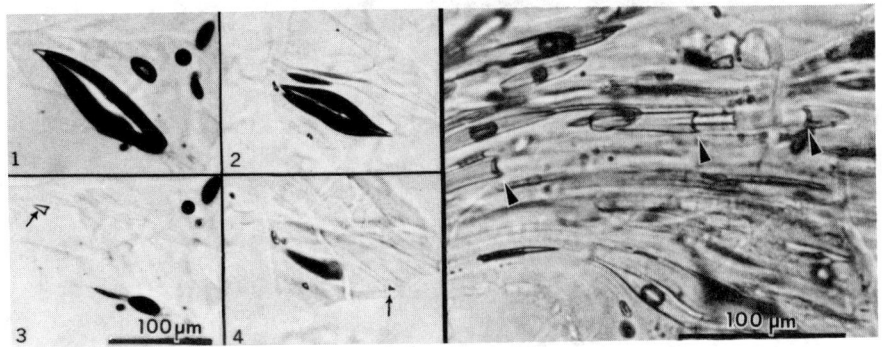

Figure 16-15 (left). Crushing of relatively large vacuum vesicles in a pumice known to be about 3 million years old. Photos 1 and 2 show fragments of pumice mounted in a nearly matching index liquid in transmitted light, on the crushing stage. The dark elongated areas are vacuum vesicles containing only water vapor and a trace of liquid water (in the pointed ends). Photos 3 and 4 show these same grains after crushing. The mounting oil has filled the cavities completely and instantly, leaving just the tiny fillets of water visible (arrows). The dark areas in 3 and 4 are other unopened vesicles. Other smaller vesicles in this sample contained a higher ratio of liquid water to vapor. Sample ER 64-49, Jemez Mountains, New Mexico, USA. From Roedder (1970a).

Figure 16-16 (right). Crushing of small vacuum vesicles in a pumice known to be about 10 million years old. A series of tubular vesicles, each containing a centrally located water-vapor bubble (dark) in liquid water are seen in transmitted light. Several cracks have transected a part of these inclusions and have permitted the mounting index liquid, which matches the glass, to replace the water-vapor bubbles, yielding a series of odd crescent-shaped menisci of water against oil (arrows). Sample ER 62-64, Nevada Test Site, Nevada, USA. From Roedder (1970a).

Several unknowns seriously limit the precision of this estimate. As the mounting medium flows in, some part of the gas present in the inclusion may dissolve in it. However, the mounting medium is generally saturated with air at 1 atm, and might even be expected to exsolve some gas as it flows into the vacuum environment of the vesicle. In addition, if the mounting medium flows in suddenly (as it generally does), the sudden compression of the gas in the vesicle causes transient but high temperatures. These various effects have been studied extensively in connection with the problem of negative pressures and cavitation in liquids (Hickling, 1965), but the unknowns involved in fluid-inclusion studies are too large to permit quantification. In any case, the thin volcanic glass walls of pumice vesicles evidently do hold a vacuum well.

An additional useful facet of this study of pumice vesicles is the discovery that they may be used to determine the age of the pumice. The method is crude and only semiquantitative at best, but it yields a "dating" in only a few minutes (Roedder and Smith, 1964).

On eruption, most rhyolitic glass is almost completely anhydrous. At surface temperatures, however, water from the atmosphere or ground diffuses into the glass to form a hydrated glass containing about 3% H_2O. The network of interconnecting channels and fractures through pumice results in essentially complete hydration of the glass after about 10,000 years. The unopened vesicles, however, still contain only water vapor; i.e., they are still close to a vacuum. Hence liquid water gradually accumulates in them, diffusing through from 1 atm pressure outside toward 20 mm pressure inside. Stages in this slow process are visible under the microscope. First, only the very tiny pointed ends of large vesicles show a fillet of liquid water (Fig. 16-15). This generally corresponds ~10^4 years. Within the interval 10^4-10^6 years, the smallest vesicles accumulate about 10% liquid, and after 10^6 years, the smallest vesicles become more nearly filled with water (Fig. 16-16; see also Fig. 6-3), and frequently fill completely, leaving no vapor bubble.

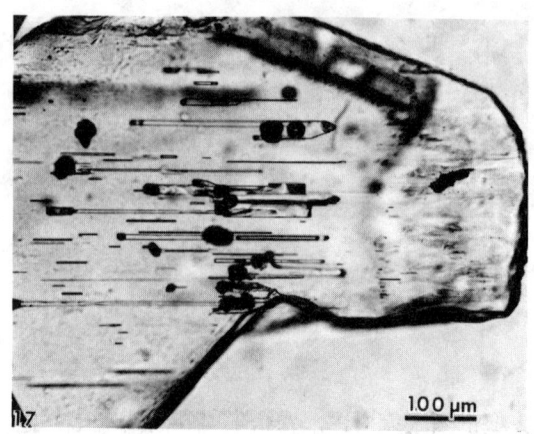

Figure 16-17. Anhydrite crystal from veinlet in Red Sea metalliferous sediments, containing numerous primary inclusions trapped as the anhydrite crystal, growing from left to right, enclosed opaque grains and liquid on the leeward side. Sample from core 268, collected during the R/V Valdivia cruise - MESEDA III - Red Sea Commission. Sample depth 1083-1087 cm, in the Atlantis II Deep (Red Sea). Sample courtesy of E. Oudin and the Saudi-Sudanese Red Sea Commission.

The crushing procedure is necessary to prove that any given partly filled vesicle is actually sealed and that it is not merely filled with ground water and air at 1 atm, from leakage. Crushing also permits recognition of the difference between sealed vacuum vesicles and air-filled vesicles that have become partly filled with mounting oil by capillarity.

It is interesting to note that vesicles in pumice known to be 10^6-10^7 years old still have held their vacuum, because the vapor bubbles in the liquid water within the vesicles collapse instantly and completely on crushing in oil (Fig. 16-16). If the solubility of air in the water in the vesicle is neglected, this evidence of vacuum reduces the maximum diffusion rate for atmospheric gases calculated above by another two or three orders of magnitude.

INCLUSIONS FROM GEOTHERMAL SYSTEMS

Submarine Geothermal Systems

The discovery of hot brine pools in the Red Sea, associated with metalliferous sediments (Swallow and Crease, 1965), and the subsequent discoveries of the TAG (Trans-Atlantic Geotraverse) hydrothermal field on the Mid-Atlantic Ridge in 1972 (Rona et al., 1976; Rona, 1980), of deep-sea massive sulfide deposits at 21°N on the East Pacific Rise (Francheteau et al., 1979; Hekinian et al., 1980), and of vents from which extremely hot (~350°C) fluids were issuing ("black smokers;" MacDonald et al., 1980), have resulted in many models of deep-sea-floor hydrothermal circulation patterns, basalt/water interaction, and the relationship between geothermal activity and ore deposition. Some of these models have made use of fluid-inclusion data.

The metalliferous sediments beneath the hot brine pools in the Red Sea are mainly too fine grained for inclusion studies, but Thisse et al. (1983) found large (50-100 µm) and abundant inclusions in anhydrite from veinlets cutting these sediments. The inclusions showed that boiling had taken place at ~340°C and that the fluid contained ~22% NaCl equivalent (Fig. 16-17). These temperatures are much higher than those of the "hot brines," which are well below 100°C.

As multiple hydrates form on cooling the inclusions, but Raman spectroscopy revealed no dissolved gases, the salinity must correspond to solutes other than simply NaCl.

Some of the deep sea hydrothermal fluids contain measurable amounts of CH_4, CO, N_2O, and H_2 (Lilley et al., 1982). These gases have generally been attributed to abiogenic water-rock reactions, and the accompanying ^3He (and possibly part of the CH_4) to mantle degassing (see also Rison and Craig, 1981). Baross et al. (1982) and Baross and Deming (1983) have reported, however, that they found living bacteria present in some of these waters at 306°C, and that they grew at 250°C and 265 atm and produced CH_4, CO, and H_2. This discovery raises numerous questions concerning the compositions of the gases in the fluid inclusions from many ore deposits and other environments whose chemistry has always been explained in terms of purely abiogenic reactions. The discovery should also be considered in connection with problems of the isotopic compositions of gases from such fluids.

Jehl (1975; summarized in part in Jehl et al., 1977) studied the hydrothermal metamorphism of the oceanic crust evidenced in dredge samples of basalt from the North Atlantic. From secondary inclusions in primary plagioclase and from primary inclusions in newly formed hydrothermal minerals, Jehl showed that the alteration was by aqueous fluids that ranged in salinity from 2 to 16 wt % NaCl equiv. Th(L) ranged between 124 and 335°C, and no evidence of boiling or CO_2 was found (crushing stage used), but an unknown hydrate was seen in one inclusion, and minor Ca and Mg were probably present, as estimated from Te. The fluids were presumed to be seawater that had flowed through the fractured basalt as a result of thermal convection. The mineral assemblages found ranged from the zeolite and prehnite-pumpellyite to the greenschist metamorphic facies. The estimated thermal gradient was 150±50°C/km, corresponding to a maximum temperature of 450°C and a maximum total pressure of 1200 bars. Bischoff (1980) argued that the maximum subsurface temperature at 21°N was ~420°C, using the P-V-T properties of pure water. Chen (1981) suggested 460°C as a more appropriate upper limit, "in view of the salinity of seawater". A series of studies has shown, however, that the fluid inclusions contain even higher salinities, as much as five times that of seawater. Delaney (1982), Delaney et al. (1982) and Delaney and Kelley (1983) adopted the process invoked by Henley and McNabb (1978) for the generation of highly concentrated brines from low-salinity fluids in the formation of the porphyry copper deposits -- partial condensation of a very hot fluid (originally seawater) of moderate salinity, to yield a higher salinity liquid and a lower salinity vapor. Delaney et al., however, did not report vapor-rich inclusions in their samples. Ito and Anderson (1983) reported Cl-bearing amphiboles in altered gabbros from the Mid-Cayman Rise, which they suggest indicates formation of concentrated brines in the oceanic crust.

Relatively little fluid-inclusion work has been published on the minerals of the very hot vents at 21°N since their discovery. Oudin et al. (1981), in an extensive study of the mineralogy and geochemistry of the materials, reported Th mainly in the range 250-330°C, and Tm ice ranging from -1.8 to -2.7°C.[7/] The pressure correction was estimated to be 17-23°C. Le Bel and Oudin (1982) reported a very good correlation between the fluid temperatures actually measured in the flowing vents and those obtained from primary fluid inclusions in minerals formed in the walls. A wide range of Th values was obtained, in part because a gradient existed through the walls where the crystals grew (<350°C inside and +2°C seawater outside). Some inclusion evidence also shows that individual vents change in temperature during their lifetime. The possible relationship of such vent sulfide deposits and the sulfide ores in ophiolite complexes, such as those

----

7/ An interesting anecdote concerning the variability of inclusion occurrence: I was privileged to receive a small sample of the beautifully euhedral hexagonal wurtzite plates from one of the "black smokers" to check for fluid inclusions. Although H.E. Belkin and I spent considerable time on sample preparation and search of these wurtzites, we did not find a single usable inclusion.

at Cyprus, has been discussed at some length (e.g., Francheteau et al., 1979; Henley and Ellis, 1983; Spooner, 1980; Reed, 1983). Henley and Ellis (1983) presented a comprehensive review of the geochemistry of geothermal systems in general.

Rona et al. (1980) reported a vug lined with quartz crystals in basalt dredged from 250 km south of the TAG hydrothermal field along the Mid-Atlantic Ridge. The host basalt had undergone hydrothermal alteration, forming chlorites, analcime, calcite, and quartz-rich rocks. Oxygen-isotope thermometry indicated that the quartz was deposited from hydrothermal solutions composed either of sea water at 200°C or "primary" water at 330°C. Primary fluid inclusions in these quartz crystals (Roedder, unpublished data) had Th values ranging from 148 to 181°C, and Tm ice ranging from -0.4 to -3.1°C, corresponding to 0.69 to 5.1 wt % NaCl equiv. The individual pressure corrections needed, for the appropriate salinities and assuming quartz crystallization at the depth of dredging (3200-3400 m), ranged from 34 to 38°C, yielding Tt of 183 to 214°C. These temperatures are well below those found in some active vent areas. Whether the lower temperatures reflect more admixture with cold seawater at vents some distance from the main circulation or the later stages of a dying hot vent is not known.

Terrestrial Geothermal Systems

Although extensive studies have been made of the geochemistry of geothermal fluids (e.g., Ellis and Mahon, 1977) and of the mineralogy of cores from geothermal wells, relatively few studies have been made of fluid inclusions in minerals from geothermal areas. This is unfortunate, because the fluid inclusions represent samples of the fluids present at some time in the past, when the host minerals grew, and hence can provide valuable evidence on changes in the system with time. Perhaps the ultimate such change is the complete quenching of a geothermal system. Thus, Robinson (1974) compared fluid-inclusion data from the Tui mine, New Zealand, with data on present-day geothermal fluids in the area, and Wetlaufer et al. (1979) showed that in many respects, the Creede, Colorado, USA, polymetallic vein deposit (see Chapter 15) is a fossil geothermal system. White (1981), and Henley and Ellis (1983) have compared various other ore-deposit types, including the porphyry Cu, vein-type Au-Ag, Carlin-type Au, and Cyprus-type massive sulfide deposits, with present-day geothermal systems. Many of the shallow fluorite deposits, in particular, have been considered in this light (e.g., Dominique et al., 1973, from the Voltenne deposit in France, and Roedder, 1977a, from the Browns Canyon deposit in Colorado, USA). The study of fluid inclusions in such ore deposits can document the detailed evolution in time and space of a small part of a fossil geothermal system, whereas studies of active geothermal systems reveal the "instantaneous" present-day patterns of systems analogous to some that have formed ores. Other localities show fluid-inclusion evidence of fossil geothermal systems but no ore deposits (e.g., Touray, 1973; Wodzicki and Bowen, 1979). Jefferis and Voight (1981) have analyzed the development of the fracture pattern near the midocean plate boundary in Iceland in part on the basis of Th of fluid inclusions in calcite and quartz from vugs in basalt, formed by geothermal activity. Fluid inclusions have also proved valuable in understanding the paleothermal gradients in a "dry geothermal" test well at Los Alamos, New Mexico, USA (Burruss and Hollister, 1979).

Studies of the fluid inclusions from drill core samples from active geothermal systems have considerable potential as a geothermal exploration tool. Leach (1982) showed that fluid inclusions can be used: (1) during drilling to give an indication of reservoir conditions; (2) in interpretations of the thermal recovery of wells after their completion; (3) in understanding the petrology of the reservoir rocks encountered by the wells; (4) in reservoir modeling; and (5) in aiding the determination of the stability ranges of certain hydrothermal minerals.

Fluid-inclusion study as an active exploration tool is now being widely used the world over, particularly in the Philippines and New Zealand. For example, core samples are flown from Philippine drill sites to an inclusion laboratory in New Zealand to provide rapid feedback on thermal conditions soon after drilling, potentially saving millions of dollars in funds for exploration and further drilling (J.W. Hedenquist, pers. comm., 1983).

Japan. Taguchi et al. (1979) reported Th values for hydrothermal quartz and anhydrite from the young, very active Hatchobaru field, on Kyushu, to be 130-170°C higher than the temperatures measured during drilling, and even 50-90°C higher than the temperatures obtained 120 hours after stopping the mud circulation down the well. As a result of these data, they suggested that even 120 hours is inadequate for a return to the original (predrilling) conditions. Later papers (Taguchi, 1981, 1982; Taguchi and Hayashi, 1982, 1983; Taguchi et al., 1981; and Hayashi et al., 1981) showed that the Th of the fluid inclusions at Hatchobaru was fairly close to the H_2O boiling curve down to 1100 m depth and in good agreement with present reservoir temperatures. In contrast, they reported that inclusions in quartz, amorphous silica, anhydrite, and calcite from the less active Kirishima field, Japan (see also Taguchi, 1983), have a wide range of Th (150-300°C) and that the minimum values are quite close to the present underground temperatures. These temperature data may indicate that the field has cooled down to some extent and that the inclusions have been trapped at various stages during the cooling. Taguchi et al. (1980) also reported that past underground temperatures were higher, from a comparison of the annealing of spontaneous fission tracks in zircon with inclusion Th values. Comparisons between Th for primary and secondary inclusions are useful in such places.

Kubota (1979) and Shimazu and Yajima (1973) studied inclusions from the Hachimantai field, Japan. Their reported values for Th were 30-50°C higher than well temperature.

Philippines. Leach (1982) reported many "homogenization temperatures" (actually temperatures of trapping, not homogenization, as they have been corrected for pressure) for anhydrite crystals from several fields. The values obtained for the Tongonan field (Leyte) were 10-20°C higher than the measured temperatures in the wells, suggesting that the field has cooled slightly since the inclusions were trapped, whereas those from the Okoy field (southern Negros) were only ~5°C higher.

Unpublished data by J.W. Hedenquist (pers. comm., 1983) and 3 students (M. Zaide, E. Napoles, and H. Aniceta) showed apparent salinities that are often 3 to 5 times the present salinities of ~1 wt % NaCl. These results, combined with Leach's (1982) evidence of an earlier period of alteration that is now being overprinted, suggests that an earlier, higher salinity fluid (more closely related to a "porphyry fluid"?) may have been present in both the Tongonan and Okoy systems.

Yellowstone National Park, Wyoming, USA. A series of wells in Yellowstone National Park was drilled for geochemical and mineralogical research purposes (Muffler and Hofeling, 1979). I have examined parts of cores from drillholes Y-1, Y-3, Y-5, Y-8, and Y-11, from depths of 167 to 508 feet, but found the material difficult to work with as most of the new hydrothermal crystals were rather small. Only a few values (unpublished) were obtained on quartz for Th (157-195°C), and Tm ice (-0.1 to -0.2°C), plus evidence of boiling conditions.

Salton Sea and Cerro Prieto, California, USA and Mexico. Huang (1977) studied inclusions in calcite, anhydrite, and quartz from depths of 590-1220 m in the Salton Sea field. Th was found to be within 30°C (generally higher) of the drillhole temperatures. Inclusions in one quartz crystal showed a systematic change (with growth stage?) in Th from ~261°C down to ~230°C, suggesting at least 31°C cooling of the reservoir. Barker (1979, 1983) and Barker and Elders (1981) were able to estimate the maximum reservoir temperatures accurately at Cerro Prieto, using the reflectance of vitrinite (one type of kerogen; see also

Frey et al., 1980b, and Chapter 12). As this reflectance changes as a function of both time and temperature, direct comparison with inclusion Th is not possible, but the approximate equivalence of measured temperatures, fluid-inclusion and oxygen-isotope geothermometry, and the absence of retrograde mineral assemblages, all indicate that the sediments penetrated are now at the highest temperature they may have yet experienced. Freckman and Olson (1978) measured Th of minerals in the Cerro Prieto and Salton Sea fields. These, and the mean temperatures calculated from oxygen-isotopic data for coexisting quartz and calcite (from data of Friedman and O'Neil, 1977), showed close agreement with well temperatures. Miller (1980) found a similar agreement between present temperatures and Th of fluid inclusions in calcite from the East Mesa geothermal field, Imperial Valley, California. However, minerals from some fracture fillings showed a wide range of Th, suggesting a polythermal history for vein-mineral deposition.

Even very thin veinlets can be used in this manner. Some years ago, I found excellent large primary inclusions in sphalerite crystals only 1 mm thick filling 1 mm fractures in core from a borehole in the Salton Sea area. These inclusions (reported in Skinner et al., 1967, p. 325) have Tt = 310°C, considerably below present well temperatures, but a salinity (~27%) in good agreement with that of present fluids from the well. The high salinity may be from solution of surface halite (Rex, 1983).

Although veinlet material may provide the best inclusions, it may yield some misleading data. Fluid inclusion studies were performed on calcite matrix and detrital quartz in samples from three wells at Cerro Prieto by M. Sterner and D.M. Kerrick (pers. comm., ms. in preparation). Three independent geothermometers were compared: fluid-inclusion geothermometry (Sterner and Kerrick), isotope geothermometry (Williams and Elders, 1981) and well-log thermometry (Mercado, 1976). Excellent agreement was found between these three independent geothermometers for well M84 located in the central (hottest) portion of the field where maximum temperatures reach 350°C. In cooler wells to the southwest (M6 and M9) fluid inclusion data yield temperatures higher than those obtained from the isotopic data. In well M9, a consistent difference of approximately 80°C was determined for depths below 800 m. It is postulated that temperatures recorded from fluid inclusions in calcite cement are indicative of the maximum temperatures achieved in the corresponding sample region. Elders et al. (1978, p. 41) reported that wide ranges in Th are sometimes recorded from fluid inclusions in calcite vein material from Cerro Prieto. Sterner and Kerrick suggested that Th values of inclusions in the calcite matrix more accurately represent the overall thermal profile because they are less susceptible to the local incursions of hot water responsible for vein formation.

The Geysers, California, USA. Salinities at the Geysers geothermal field in Sonoma County, California, USA, are much below those from the Salton Sea and are similar to those at Cerro Prieto. This is a vapordominated system, i.e., it discharges dry steam at essentially zero salinity. This condition might have been induced by production from the field, or might be of natural origin (pers. comm., R.W. Henley), but at least some liquid has been present since. Sternfeld (1981) reported salinities of 0.5 to 2.2 wt % NaCl equiv. and Th values of ~200 to 350°C (depending upon depth) for inclusions trapped as a liquid phase. Many of these samples had Th appreciably above the boiling curve for the depth sampled.

New Zealand. Hedenquist (1982) showed that Th values for inclusions in hydrothermal calcite and quartz in the Waiotapu system follow the hydrostatic boiling curve closely, even though some well temperatures do not. Inversions in the well temperatures (i.e., temperatures decreasing with depth) presumably occurred after the inclusions were trapped, by lateral movement of cooler water. CO_2 in the inclusion fluids (as determined by freezing studies, Hedenquist, 1983) was ten times greater than at present. Isotopic studies have suggested a dominantly meteoric fluid, plus fluid mixing and boiling at depth. At present,

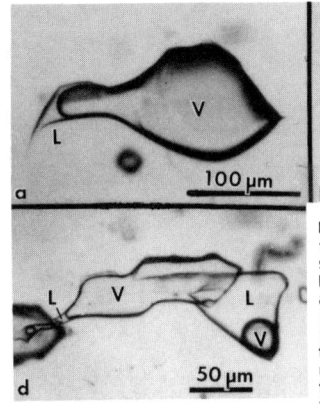
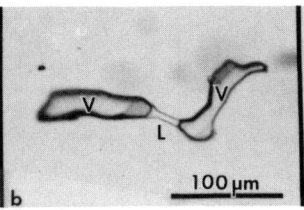
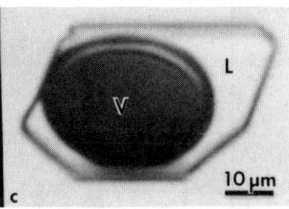
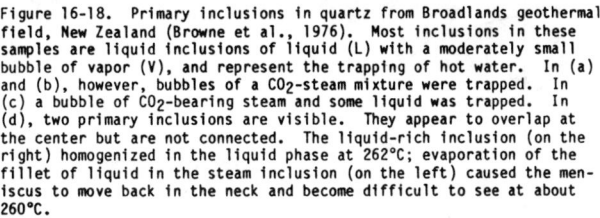

Figure 16-18. Primary inclusions in quartz from Broadlands geothermal field, New Zealand (Browne et al., 1976). Most inclusions in these samples are liquid inclusions of liquid (L) with a moderately small bubble of vapor (V), and represent the trapping of hot water. In (a) and (b), however, bubbles of a CO_2-steam mixture were trapped. In (c) a bubble of CO_2-bearing steam and some liquid was trapped. In (d), two primary inclusions are visible. They appear to overlap at the center but are not connected. The liquid-rich inclusion (on the right) homogenized in the liquid phase at 262°C; evaporation of the fillet of liquid in the steam inclusion (on the left) caused the meniscus to move back in the neck and become difficult to see at about 260°C.

some of these fluids are depositing ore-grade concentrations of precious metals near the surface.

Browne et al. (1976) studied inclusions in quartz from a series of wells in the Broadlands field. Small birefringent crystals were present in some inclusions, but were not affected by heating to Th, and were presumed to be accidental solid inclusions. Th was found to range from 185 to 295°C, and most measurements were within ±20°C of the well temperature. Tm ice, measured by means of the circulating acetone stage (see Chapter 7), ranged from -0.1 to -0.8°C, corresponding to 5100-13,600 ppm NaCl equiv. for the primaries and 1700-12,000 for the secondaries. Several inclusions showed Tm clathrate rather than Tm ice. Present-day salinities in the areas where the samples were collected are ~2000 ppm NaCl equiv. (Ellis and Mahon, 1977, their Table 9-9)

A gas phase ("boiling") was present during the formation of many of these inclusions, as evidenced by many seemingly empty inclusions. Studies of these "empty" inclusions showed pressures of ~0.25 bar (CO_2?) on the crushing stage, and those that had small reentrants showed a fillet of liquid (Fig. 16-18a,b,d). Even where such reentrants were present, the point of homogenization of such vapor-rich inclusions (in the vapor phase) was difficult to determine with any accuracy, but was always close to that of associated liquid-rich inclusions (Fig. 16-18d). Inclusions such as those shown in Figure 16-15c, that trapped liquid plus primary gas, would yield very high and meaningless values for Th.

Interpretation of these data illustrates several uncertainties common to many geothermal studies:
(1) Determinations of Tm ice (also termed "FPD," for freezing-point depression) need to be reasonably precise, because the entire range was only 0.7°C. Accurate determinations of such low-salinity fluids require very slow heating schedules and calibration samples of similar thermal inertia (i.e., heat of melting).
(2) One of the important questions concerning any geothermal field is that of its past history and hence speculations as to its future life. The large clear quartz crystals studied by Browne et al. (1976) contain fluids initially trapped at about the same temperature as fluids in the field at present, and almost certainly must have formed slowly, but there is no real evidence as to the age of these crystals. Present-day waters are precipitating amorphous silica and cristobalite at the surface, not multi-centimeter-long quartz crystals, but they might well be forming such crystals at depth. Isotopic studies of such quartz crystals from the Broadlands field (Blattner, 1975) suggest that at least some may have grown over a long period of time. All that can be said for certain

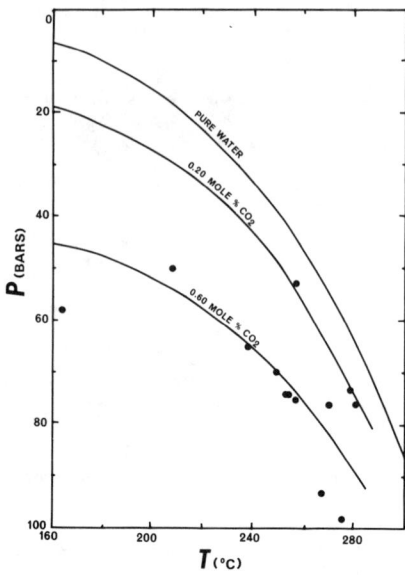

Figure 16-19. Vapor-pressure curve for pure water (Bain, 1964), and solutions containing 0.20 and 0.60 mole % CO_2 (0.5 and 1.5 wt %), modified from Ellis and Golding (1963). The measured present-day temperatures and pressures at depths where samples were collected in the Broadlands field, New Zealand, are shown by circles. From Browne et al. (1976).

is that the field has been essentially constant in temperature for at least the time since these crystals formed. (Elders (1980) has used fission-track anneal-ing studies to estimate the age and duration of heating in the Cerro Prieto field.)

(3) The pressure corrections to be applied to such values of Th, neglecting the effects of CO_2, are small (in this locality) but not simple to estimate (Browne et al., 1976, p. 148).

(4) If a hydrate forms on cooling (and it may be essentially invisible), it slightly increases the salinity of the remaining fluid (Collins, 1979) and hence the apparent total salinity. (Formation of CO_2 hydrate, however, also removes CO_2 from the fluid.)

(5) By convention, the "salinities" of fluid inclusions are expressed in terms of wt % NaCl equivalent, but CO_2, even if present in amounts less than that necessary to form clathrate, will also contribute to the FPD (i.e., the difference between Tm ice and 0°C). In low-salinity fluids, FPD from CO_2 can be greater than FPD from the total nonvolatile solutes (Hedenquist and Henley, 1984). For example, calculations show that deep water (salinity ~1400-2000 ppm, at 270°C) containing 0.6 wt % dissolved CO_2 will freeze at -0.43°C, giving an apparent salinity (NaCl equiv.) of ~7400 ppm. Similarly, present-day Broadlands water containing 2.1 wt % dissolved CO_2 (much more than found in the inclusions) will freeze at about -1°C, giving an apparent NaCl salinity equiv. of 18,300 ppm. Present-day Broadlands fluids are estimated to contain ~2-4 wt % CO_2 before exploitation (Sutton and McNabb, 1977), but the amount present at any point in space and time may vary widely, due to boiling.

(6) The presence of both liquid-rich and vapor-rich (steam) inclusions in some of these Broadlands samples suggests that at the time of growth of the host quartz crystals, the hydrothermal fluids in these parts of the field were close to "boiling" and did actually "boil" at times (i.e., a CO_2-rich fluid phase separated).[8]

(7) All mean Th values of liquid-rich inclusions are below the boiling point for pure water at the present measured well pressure, and the average discrepancy is 26°C, as they should be, considering the CO_2 contents. Similarly, the present

----

[8] Ellis (1970) has shown that the mineral assemblage found in vugs and veins in this core, adularia, cal-cite, and sulfides such as pyrite, is, in itself, adequate evidence of the occurrence of boiling, due to the concomitant increase in pH from loss of acidic gases.

temperatures in the wells average 31°C below the boiling temperature for pure water at the pressure of the sample point.

The differences between the inclusion data and present-day fluids are presumably a result mainly of the variability of CO_2 contents. Browne et al. (1976) estimated from the inclusion data that the CO_2 lowered the vapor pressure curve by about 30°C, and Sutton and McNabb (1977) and Mahon et al. (1980) have explored the effects of CO_2 on the boiling curve rigorously. The effects of salinity are in the opposite direction, but insignificant at these concentrations (e.g., 20,000 ppm raises the boiling point 0.3°C, Haas, 1971). Figure 16-19 shows the vapor-pressure curve for solutions containing 0.20 mole % CO_2 (i.e., the present deep-water concentration) and 0.60 mole % CO_2, from Henry's Law. Some samples plot on, or close to, the vapor-pressure curve for 1.5 wt % CO_2, indicating that this CO_2 concentration would have been sufficient to cause past boiling at present well temperatures, which are the same as Th (L-V) L. Present total discharge from these wells contains far more than 0.20 mole % CO_2, indicating that the wells must be drawing unequally from two phases; this is corroborated by data on the enthalpy of discharge. Browne et al. have no direct evidence of the CO_2 concentrations in the total fluids at the time of inclusion trapping, but indirect evidence suggests that at the time of quartz deposition, the CO_2 concentrations may have been high. It is worth noting that the present CO_2 concentrations of fluids in New Zealand geothermal fields vary tenfold or more between systems, and at least by a factor of 10 over time in some systems (e.g., Waiotapu; Hedenquist, 1983).

Italy. Belkin et al. (1983) studied fluid inclusions from five wells in the Larderello field, and one from the Piancastagnaio field, in southern Tuscany. These samples showed Th values ranging from 187 to 386°C, and salinities from 0.5 to 42 wt % NaCl equiv. The high salinities were believed to be the result of partial condensation. Some inclusions contained much CO_2 as well. Some very low values for Te (≥-52°C) suggest significant Ca in solution. Some homogenizations were in the liquid, and others were in the vapor phase, as might be expected for vapor-dominated systems.

Chapter 17

UPPER MANTLE ENVIRONMENTS

CONTENTS

INTRODUCTION

In this chapter, I discuss the data available from fluid inclusions in materials believed to originate deep in the earth. A sharp line cannot be drawn, but most of the samples discussed here have probably come from >20 km depth, and some are from much greater depths. These deep samples, particularly xenoliths erupted with alkali basalts and kimberlites -- and the host magmas themselves -- represent valuable windows into the·interior of the Earth. Without them, our knowledge of the nature of the lower crust and upper mantle would be limited to inferences from models designed to fit only the available indirect evidence from geophysical properties and measurements. As will be seen, diamonds present us with a truly unique and even more valuable type of window.

Much has been learned of these deep zones from mineralogical, geochemical, and isotopic studies of the erupted materials. Here I can discuss only one small facet of this study, the fluid inclusions of both melts and volatiles. Many recent studies point to gross mantle heterogeneity, and many changes may occur during the ascent of materials to the surface, particularly in the volatiles, so we must consider here also the various lines of evidence on the composition and nature of all volatiles, whether trapped in inclusions or not.

In addition to changes that may take place in the volatiles on the way up, the material erupted at the surface may be a composite from various levels. Magmas may lose some of their original constituents through crystallization or phase separation, or gain others by magma mixing or assimilation. Similarly, the xenoliths and xenocrysts found, as well as the phenocrysts, are unlikely to be from a single depth. This caveat must always be kept in mind. Fortunately, the best samples, diamonds, can only form at great depths.

VOLATILES IN MANTLE MATERIALS OTHER THAN
INCLUSIONS IN ULTRAMAFIC NODULES

Hydrogen, Carbon, Sulfur and Other Volatiles in Mantle Materials

All samples of mantle materials, whether brought to the surface as solids or as fluids, contain at least small amounts of H, C, S, and other volatiles. Some ultramafic nodules contain phlogopite (and rarely amphibole) and even the nominally volatile-free minerals such as pyroxene may contain gases in their structures (Serdiuchenko, 1981). Some kimberlites contain carbonate minerals as well and may be extensively serpentinized. Graphite is present in some (Pasteris, 1981a), and most nodules and kimberlites contain sulfide minerals (e.g., Pasteris, 1981b,c). Although the serpentine and at least part of the carbonate, phlogopite, sulfide, and graphite represent alteration of preexisting materials, alteration that took place at an unspecified depth, some or perhaps most of the volatiles (and the K) involved in the alteration presumably also came from depths comparable with those yielding the original solids.

Similarly, all magmas contain some volatiles, particularly H_2O and CO_2, which are of major importance in controlling the nature and amount of the partial melting that led to their formation. In discussing the origin of kimberlites, Wyllie (1980; see also Bailey, 1980) suggested, from a consideration of melting relationships in the system peridotite-CO_2-H_2O, that a minor thermal perturbation at depth might trigger release of reduced vapors with major components C-H-O. These volatile components in turn could result in partial melting at higher levels and eventually permit eruption of kimberlitic magma. The isotopic signatures of C, H, O, and N in the various volatile species have been reviewed by Matsuo (1984).

The ratio of H_2O to CO_2 and the specific effects of H_2O vs CO_2 on the amount of mantle melting and the composition of the resulting liquids have been discussed extensively. It is generally agreed that the two compounds yield grossly different results and that CO_2-rich volatiles are involved in the origin of kimberlites and carbonatites (e.g., see reviews by Mysen, 1977; Wyllie, 1979a,b). Water is generally more soluble in silicate melts than is CO_2, so the formation of melt may leave CO_2 as a solid (e.g., carbonate) or, at pressures

$<\sim20$ kbar,[1] CO_2 may form a vapor phase, thus permitting a partial physical separation of H_2O and CO_2. Crystallization of a water-enriched melt at higher levels could release aqueous solutions that are likely candidates for some of the metasomatic changes so commonly observed in mantle materials.

The separation of a CO_2-rich phase may have major petrologic consequences. Many petrologists have noted the association of kimberlites, carbonatites, and alkaline magmas (e.g., Bailey, 1980; Carswell, 1980), but the specific combinations of tectonic activity, volatile flux, temperature, and source materials that must be involved are still in debate. In addition to the presumed connection of such a CO_2 phase with carbonatites and kimberlites (and their diamonds?), CO_2 is the main volatile constituent in the formation of deep crustal charnockites and granulites (Newton et al., 1980), high-temperature thermal domes in metamorphic terranes (Schuiling and Kreulen, 1979), midocean-ridge basalts (Moore, 1979), and "phreatic" basaltic volcanism (Barnes and McCoy, 1979); CO_2 is almost the sole constituent in the inclusions in ultramafic nodules that have been erupted with alkali basalts throughout the world (Roedder, 1965a).

The oxidation state of the volatiles in these deep environments is of particular importance but difficult to evaluate. Measurements of the intrinsic O_2 fugacity of mantle materials (e.g., Arculus and Delano, 1981) showed that most mantle-derived magmas are initially highly reduced (near to the pressure-corrected Fe-wüstite buffer) and that the relatively much more oxidized values observed at the surface reflect late-stage alteration, perhaps by H_2 loss (Sato, 1978). One important aspect of the state of oxidation in these environments is the CO/CO_2 ratio in any separated volatile phase. Eggler et al. (1979) have shown that CO will behave like CO_2 in terms of melting relations of peridotite, but the CO content found in CO_2 inclusions in nodules (see below) is small or negligible (see also Bergman and Dubessy, 1984).

The exact nature of the occurrence of C in deep-seated materials in situ is also enigmatic. Although some C must be present in solid carbonates and diamond, and as a CO_2 fluid, Freund et al. (1980) have shown that C atoms can exist in significant amounts in solid solution in forsterite. If such C is present in mantle materials, its solution and exsolution could affect many petrologic processes (Freund, 1982; Freund et al., 1983). Of special interest is the unexpectedly high diffusivity of such C (Oberheuser et al., 1983), as found in natural olivine (containing 60-180 ppm C). This diffusion may well be involved in the formation of minute CO_2 inclusions decorating dislocations in olivine (Green and Radcliffe, 1975) and the exsolution lamellae in clinopyroxene (Selverstone, 1982), or even the much larger CO_2 inclusions attached to solid spinel inclusions in orthopyroxene (Roedder, 1983; see below). Pasteris (1981a) has suggested that the large amounts (6-9 wt %) of graphite that she found in serpentinized regions of olivine grains in several kimberlites might have formed from such free C atoms. Alternatively, this graphite could form by reduction of a CO_2 fluid.

The nature (and amount) of S in mantle materials has not had as much attention as has been given CO_2 and H_2O, in part because S does not separate as readily from the silicate phases. Some movement of S has occurred, however, because metasomatic alteration to form or remobilize sulfides is common (Pasteris, 1981b). Experimental studies of sulfide saturation in basalt and andesite melts at <30 kbar (Wendlandt, 1982) showed a strong negative pressure dependence of S solubility in silicate melts, so a magma may be sulfide-saturated in the source region but not on eruption, until the late stages of crystallization. These results would suggest that glassy basalt flows should presumably be undersaturated with respect to sulfide melt, but many of them contain immiscible sulfide blebs in the glass, indicating saturation. Some of these blebs are also trapped

----

[1] Ellis and Wyllie (1979) believe that a CO_2 vapor phase can exist at pressures <90 kbar.

as sulfide inclusions in early phenocrysts.

F and Cl are also present in small amounts in mantle materials and may become concentrated in late-stage metasomatizing fluids. Smith et al. (1981) found 0.43 wt % F and 0.08 wt % Cl in primary-textured phlogopites in coarse depleted garnet-lherzolite xenoliths from kimberlites. Apatite in such materials is fluorapatite and seldom contains as much as 1% Cl.

Although the noble gases are present in very minute quantities, isotopic studies of He, Ne, Ar, Kr, and Xe from samples from deep in the Earth may be particularly useful in understanding the processes involved (e.g., Lupton, 1983a,b). These gases may become concentrated in a gas phase if such a phase forms, as is recorded in the CO_2 inclusions in ultramafic xenoliths, but studies of these gases have also been made on solid samples without significant gas inclusions. Smith (1977) reported measurements of Ne, Ar, Kr, and Xe from a variety of samples, including peridotite, carbonatite, kimberlite, and MOR (midocean-ridge) basalts, and Stettler and Bochsler (1979) reported on the He, Ne, and Ar isotopes in basalts. The five sources of noble gases in such samples include: (1) primordial gas from undegassed parts of the earth; (2) undegassed radiogenic ^{129}Xe and $^{131-136}$Xe from extinct ^{129}I and ^{244}Pu; (3) radiogenic gases from continuing decay of U, Th, and K; (4) products of nuclear reactions such as (n, α) or (α, n); and (5) gases from the present or past atmosphere or hydrosphere. Superimposed on the ambiguity from these multiple sources is the problem of possible isotopic fractionation processes, particularly in the lighter isotopes. Kurz and Jenkins (1981) showed that for He in fresh basaltic glass at ocean temperatures, the diffusion rates are such that He loss (or gain, i.e., source 5 above) is insignificant. They also found the ^{3}He/^{4}He ratio remarkably constant in samples from five spreading centers. Other basalts, however, show wide variations in this ratio (James, 1983). Xe has a series of stable isotopes, formed by several mechanisms, and provides such a unique opportunity to reduce the ambiguity that its study has been dubbed "xenology" (see review by Staudacher and Allègre, 1982).

Inclusions in Diamond

Diamond is a fascinating and surprisingly complex mineral, as is abundantly verified by the perplexing results of numerous recent studies (e.g., see Field, 1979). If a volatile-rich phase was ever present at the great depths in the mantle at which diamonds have formed ($\geqslant\sim$150 km; $\geqslant\sim$45 kbar) and was trapped as fluid inclusions in the minerals growing there, probably the only fluid inclusions that could possibly survive eruption to the surface would be those in diamond; all other minerals almost certainly lack sufficient strength to contain these high pressures without decrepitation. Diamond thus represents our only hope for a "sampling device" for such deep-seated fluids. Since the stability field of diamond is limited, if inclusions decrepitated during eruption, they could not reheal at the lower pressures. As is common in science, however, the available actual data on the environment of origin of diamond are still somewhat inconsistent and enigmatic, in spite of numerous studies (e.g., see Sellschop, 1975, J.W. Harris, 1979, extensive reviews by Harris and Gurney, 1979; Meyer, 1982; and Bibby, 1978, 1982). Diamond is far from being simply C; it contains trace impurities of many types and sizes, in a wide range of concentrations, as summarized below.[2] Many of the enigmas presented by the data on diamonds may be related, in part, to inhomogeneities in the mantle. Thus, the normal value assumed for δ^{13}C of primordial Earth C is $\sim$-5 to -7 per mil, but δ^{13}C for diamonds ranges from $\sim$0 to -30 per mil (Koval'skii and Cherskii, 1973; Deines, 1980, 1982). Deines showed that an unambiguous selection from the many genetic models

----

[2] Only inclusions in single-crystal diamond are discussed here; numerous studies of polycrystalline diamond are ignored because of the ambiguity as to origin or modification of the included material.

(some involving a vapor phase) is not possible and that most of the very large range in $\delta^{13}C$ observed was inherited from the source C. This, in turn, may be a result of part of the C having come from ocean-sediment carbonate in deeply subducted oceanic plates (Ringwood, 1977; Boyd and Gurney, 1982). Milledge et al. (1983) showed that the large spread in $\delta^{13}C$ values is characteristic of the type II diamonds only. Swart et al. (1983b) showed that different growth zones in single diamonds differed by as much as 4 per mil. In addition to the fractionation processes discussed by Deines (1980), the original materials that coalesced to form the Earth may have been isotopically heterogeneous, as suggested by the very large differences in $\delta^{13}C$ from various meteoritic sources (Swart et al., 1983a).

Macroscopic solid inclusions. Numerous studies have been made of the solid mineral inclusions in diamond (see Harris, 1968; Meyer and Tsai, 1979; Harris and Gurney, 1979; and Meyer, 1982, for comprehensive reviews). Although olivine, garnet, and pyroxene are the most abundant minerals found, a large number of other phases have been reported. Even though these inclusions are solid and not fluid, they are pertinent in this discussion for several reasons. First, the common occurrence of such macroscopic solid inclusions causes difficulties in interpreting analyses of the submicroscopic "melt inclusions in diamond" that have been proposed by some. Second, many studies of the volatiles in diamonds have revealed large differences between stones from different localities, between different stones from the same locality, and even between different growth zones of a given stone. Because the solid inclusions in diamonds provide information on the several distinctly different environments in which the diamond actually grew (as contrasted with the minerals in the rock in which the diamonds are now found; Hervig et al., 1980), these solid inclusions may help unravel some of the intricacies of the associated volatiles. Deines (1982) was able to correlate the ^{13}C content of diamonds with the composition of their inclusions. The solid inclusions also can be used to place constraints on the P-T conditions of formation of the host diamond, by means of studies of strain birefringence in the host diamond. Cohen and Rosenfeld (1979) used such procedures to suggest that some garnet-bearing diamonds may have formed at 145 to 300 km. Gurney (1979, p. 106) suggested that the peridotitic suite of mineral inclusions in a diamond crystallized from a melt that was saturated with respect to H_2O and CO_2.

Macroscopic fluid inclusions. Bauer and Spencer (1904, p. 119) and Schlossmacher (1932, p. 348-351) referred to fluid inclusions in diamond that contained liquid water, liquid CO_2, and/or gas. These and similar statements have been widely quoted and requoted ever since (with and without attribution), but unfortunately the criteria used for phase identification are never stated. Sutton (1928, p. 42) calls such statements "a favorite myth." I have examined many inclusions in diamond, including the sample described and illustrated by Eppler (1961, Figs. 19 and 20), and have not seen any recognizable liquid, CO_2 or otherwise. Cavities are present in some crystals, but of these, most (or all?) seem to be connected with the surface. Such cavities could be the result of removal of solid silicate embayments in the diamond by weathering or by the strong acid (HF and/or H_2SO_4) cleaning treatment frequently given commercial diamonds. I suspect that the reports of liquid inclusions in the older literature (e.g., Brewster, 1862) are mainly a result of misinterpreting various colorless solid inclusions.

However, Giardini and Melton (1975a) reported that "cloud-like" regions in some Arkansas diamonds were due to large numbers of "empty cavities," 1 to 30 μm in size (Fig. 17-1). SEM studies showed that some of these cavities were multisided (6-, 8-, or 12-sided) and that others were amoeboid flattened cavities. The single most important question concerning these inclusions is whether any of them were actual isolated cavities, rather than being interconnected. In view of their large size, such inclusions should be amenable to ordinary optical petrographic techniques (including freezing studies), but, unfortunately, such

Figure 17-1. SEM photograph of broken surface of a 1.54-ct. Arkansas diamond containing a cloud of inclusions that consisted of relatively large flattened cavities. The scale bar is the average of the two somewhat different scales given in the original. From Giardini and Melton (1975a).

observations were not made[3]. Harris and Gurney (1979, p. 559) made IR spectra of cloudy white areas in diamonds from "West and South Africa" (sic., probably South West Africa, i.e., Namibia), and of adjacent clear parts. They obtained evidence of carbonate and hydroxyl groups from the cloudy areas that was not seen in the clear parts.

Submicroscopic "melt inclusions." When apparently inclusion-free diamonds are analyzed for nonvolatile impurity elements, the results frequently show high (positive) interelement correlations suggesting the presence of specific (but submicroscopic) mineral inclusions. Using this approach, Sellschop et al. (1974) and Fesq et al. (1975) were able to infer the presence, in specific analyzed samples, of olivine (or enstatite), garnet, diopside, rutile (or ilmenite), chrome spinel, and sulfides. Other analyses, however, even of the purest gem-quality diamonds, did not provide such identifiable correlations, and in these two papers the impurities in such "inclusion-free" stones were assigned to submicroscopic[4] inclusions "of the parental magma." Fesq et al. (1975) gave an estimated composition of this "trapped melt" as follows (wt %): MgO, 20-25; FeO, 11; Al_2O_3, 7.7; TiO_2, 1.1; MnO, 0.15; CaO, 5.0; Na_2O, 1.2; K_2O, 0.65; SiO_2, not given. Sellschop (1979) suggested that such impurities are from submicroscopic mineral inclusions of garnet-diopside composition, and several workers assume that such submicroscopic inclusions are present in all diamonds. If these impurities are indeed present as melt inclusions, I estimate, on the basis of 10 ppm Mg impurity concentration in "inclusion-free" diamonds (Fesq et al., 1975) that such diamonds should contain $\sim 10^{11}$ melt inclusions per cubic centimeter, each ~ 0.1 μm in size. The SEM technique should be adequate to verify the presence of such inclusions on broken surfaces, but to the best of my knowledge, such inclusions have not been reported.

Volatiles in diamond. An extensive literature exists on the volatiles in diamond -- probably more data than on the volatiles in any other mineral -- and many analyses have been made using state-of-the-art techniques (e.g., see Sellschop et al., 1977, 1979; Field, 1979; Bibby, 1978, 1982). Ozima and Zashu (1983b) and Ozima et al. (1983) reported extremely high $^3He/^4He$ ratios in diamonds, some

----

[3] I have recently examined the crushed fragments remaining from the experiments of Giardini and Melton (1975a) on these two diamonds (courtesy of U.S. National Museum). The crushed residue from both samples (USNM 125973, 2.06 carat, and 125974, 1.53 carat) contained numerous fragments <4 mm. Distinction between the frosted curving original exterior crystal surfaces and the bright new cleavage surfaces was relatively easy. Various fragments showed evidence that both original stones were in part an open porous mesh of generally single-crystal diamond. Irregular partly faceted cavities as large as 100x300 μm were found in numerous fragments; some of these cavities made holes all the way through the individual fragments, and some intersected the external crystal surfaces. Although individual multifaceted cavities as seen in Figure 17-1 were found in some cleavage surfaces, no such cavities, large or small, were found to be completely enclosed in any of the fragments, even though most of the weight of the samples was present as fragments in the 2-4-mm range. The origin of such cavities is unknown, although several ad hoc explanations might be considered. A vermicular (symplectitic) intergrowth of another mineral with the diamond might have been subsequently weathered out, or the diamond might have been partly dissolved along dislocation bundles before eruption.

[4] I.e., not visible using a 50X polarizing microscope.

approaching that for solar-type He, suggesting that it is primitive He that evolved very little since the formation of the Earth. H has been reported, by means of four different techniques, at values ranging from 4.5 to 2500 ppm; whether these are real differences between the samples used or artifacts of the techniques used is not clear. N is apparently present in all diamonds, and the characteristics of the four types of diamonds (Types IA, IB, IIA, and IIB, established on the basis of optical and electrical properties) have now been shown to result from the presence and form of N or B in the structure (Bibby et al., 1978; Clark et al., 1979). N ranges from 5 to >5000 ppm, and this sort of range can be found even in stones from the same kimberlite pipe. The bulk of the N is present either as presumed atom pairs or as single substitutional paramagnetic N atoms (Bibby et al., 1978; Bursill, 1983), but the distribution within a given stone may be inhomogeneous. The value for $\delta^{15}N$ of N from Namibian diamonds (Wand et al., 1980) differs considerably from previous values for mantle N.

As discussed by Bibby (1982), volatile constituents (including elements such as H, N, and O, as well as compounds such as H_2O, CO, and CO_2), may be present in diamonds in various forms: (1) substitutionally in the diamond structure; (2) interstitially in the structure; (3) in mineral or melt inclusions within the diamond (or, I would add, in the fractures sometimes observed in or around such inclusions); or (4) as actual fluid inclusions. H, N, and O are normally present in the first two forms,[5] but in the present context, the fourth of these is of main interest. Unfortunately, however, the problems in relating the results of analysis of the gases obtained from a given diamond sample to these various possible sources of such gases have not been fully resolved.

The bulk of the available quantitative analyses of gases actually released from diamonds are by Melton and Giardini (1974, 1975, 1976, 1980, 1981, 1982), Giardini and Melton (1975a,b; 1976), Giardini et al. (1982) and Melton et al. (1972), who have reported mass spectrometric analyses of the gases from diamond samples from various localities. They used either high-temperature graphitization (3000°C) or impact crushing of the stones between tungsten carbide plates to release the gases, both methods under high vacuum. These analyses are difficult and unique; hence, it is unfortunate that in these papers some important details are frequently missing. These missing details include calibrations (particularly in the low-mass range), magnitude of the background correction, blank correction (was it made in the data as reported?), some sample weights, some of the actual volumes of gas obtained, expected changes in gas species during release and analysis, and, most particularly, sample characterization both before and after the run. Minute amounts of H_2O are notoriously difficult to handle in vacuum apparatus without major losses by absorption on practically any and all surfaces, yet some of these papers report the volume percent of water to three significant numbers.

The species reported include major or significant amounts of N_2, H_2, H_2O, CO_2, CO, and CH_4, and minor amounts of Ar, O_2, ethylene, other hydrocarbons, methyl and ethyl alcohol, and butene. Alcohols in 36 natural samples averaged 0.6 vol %. Although these analyses are represented to be of "occluded fluids" (Melton and Giardini, 1981), determination of the source of the gases found is difficult. Gases within inclusions should be released completely if the crushing opens the cavity. Graphitization ought to release essentially all the gases present, both those present as inclusions and those in the lattice, whereas crushing of an inclusion-free stone should release only the very minute amount of gas atoms that were in the lattice adjacent to the fracture surfaces. Hence, one might expect that the volumes of gas obtained from an inclusion-free stone

----

[5] In one of the few studies in which multiple analytical techniques were applied to a single stone, Sellschop et al. (1979) reported the contents of H, N, and O in one sample to be 200, 180, and 77 µg/g, respectively.

would, on graphitization, be many orders of magnitude greater than that obtained on crushing, but volume data on gases released on graphitization are not given -- only the composition in terms of percentages.

The results of crushing the two Arkansas diamonds containing cloudlike inclusions mentioned above (Giardini and Melton, 1975a) were perhaps the most important, as these stones appear to have contained actual gas inclusions (Fig. 17-1). Giardini and Melton reported 5.4×10^{-4} and 3.4×10^{-5} cm^3 gas (mainly H$_2$O and CO$_2$) was obtained when they crushed these two cloudy diamonds. In comparison, when a series of clear, apparently inclusion-free stones were crushed, the volumes of gas obtained were in the same range or higher. If we assume that these cloudlike inclusions actually are high-pressure gas inclusions, the volume of gas that might be released on crushing can be estimated. Several of the variables involved can only be estimated crudely, and they may have been individually in error by an order of magnitude or more, but the result obtained, ≥0.3 cm^3 STP, was four orders of magnitude larger than that reported by Giardini and Melton. This result suggests that few or perhaps none of the inclusions in the cloud were filled with high-pressure gas. In another paper, Melton and Giardini (1980) reported on Ar isotopes released when they crushed a 6.3-carat Arkansas diamond crystal containing "a number of relatively small, totally-enclosed inclusions" (p. 462). They obtained 3.42×10^{-6} cm^3 (STP) of Ar when they crushed the stone into <2 mm fragments. The Ar had a ^{40}Ar/^{36}Ar ratio of 189 (as opposed to 296 for today's atmosphere), which they used to calculate a model age of crystallization for the diamond of 3.1 billion years, on the basis of several assumptions. Ozima and Zashu (1983a) attempted to obtain a K-Ar isochron age on diamonds but found only 0.7 ppm K and no radiogenic ^{40}Ar. One other diamond yielded an apparent K-Ar age of ~4.5 b.y.

Melton and Giardini (1974, 1975, 1980, 1981, 1982), Giardini and Melton (1981, 1983) and Giardini et al. (1982) used these various analyses to interpret the growth environment and the mechanism of formation of diamond, mantle inhomogeneities, large crustal accumulations of nonbiogenic petroleum, and the evolution of the Earth's atmosphere and oceans. I suggest, however, that several additional aspects should be explored before such speculations are in order, entirely apart from that of determining whether or not an actual fluid phase was present during diamond formation.

One aspect is the problem of reproducibility. Apparently the only check given in this series of papers on the reproducibility of gas evolution and the analysis on crushing of a given diamond is a pair of crushings reported by Melton and Giardini (1974). That paper reported two sequential crushings of a single sample by means of two different procedures. Although H$_2$O constituted >75% of the gas obtained in each crushing, the second crushing released five times more H$_2$ (in percentage of total) than the first; unfortunately, however, the stated amounts of C$_2$H$_4$ (and also O$_2$) obtained on the first crushing differ by a factor of 2 in the two tables (4 and 5) in which this one analysis is listed.

The other aspect to be explored is the important problem of the redox state of the environment of diamond formation (e.g., see Eggler et al., 1980). In theory, the analysis of the gaseous species present on release into the vacuum of the mass spectrometer at room temperature can be used to calculate the composition and hence the redox state of the gases present in a hypothetical fluid phase at the very high pressures and temperatures of diamond growth. The obvious question is this: Are the individual species so calculated (particularly H$_2$, CO, CO$_2$, H$_2$O, and CH$_4$) in equilibrium with each other and with diamond at mantle temperatures and pressures? This question arises because the redox state of the released gases, as analyzed, seems to vary greatly. Thus, for individual analyses, the volume ratio H$_2$/H$_2$O ranges from 6.0 to 0.01, and the ratio CO/CO$_2$ ranges from 23 to ~0. More importantly, however, no correlation exists between these two ratios, and the ratio (H$_2$/H$_2$O)/(CO/CO$_2$) ranges from 52 to 0.008. The implications of the occurrence of these individual species (H_2, CO, CO_2, CH_4, alcohols, etc.) are sufficiently important with respect to diamond genesis and the formation of the Earth's mantle that in the near future, independent studies should be completed to corroborate or refute the data of Melton and Giardini.

Inclusions in Kimberlite

Kimberlites are complex and hence difficult rocks to study, in part because they have generally been altered extensively (Pasteris, 1981c). Spherical sulfide inclusions in kimberlite minerals, suggesting an immiscible sulfide melt, have been reported by several workers (e.g., Merkel et al., 1976; Laz'ko, 1977), but in other examples, the mineralogy and textures are suggestive of later replacement processes (Boctor and Meyer, 1977) and even of segregation of volatiles (Clement and Skinner, 1982). Haggerty and McMahðn (1979) reported textures indicating that an immiscible carbonate phase was also present in some kimberlites.

Silicate-melt inclusions in kimberlites have been studied, particularly in the Mir and Udachnaya pipes in Yakutia, USSR. Popivnyak and Laz'ko (1979) found both glassy and partly crystallized primary melt inclusions in the two generations of olivine phenocrysts and in garnet from the kimberlite cement. Th usually ranged from 1150-1200°C. Melt inclusions in associated garnet peridotite xenoliths had Th = 1220-1270°C. Some inclusions have decrepitated on eruption, suggesting high internal pressures. Pokhilenko and Usova (1978) reported much lower Th values for secondary melt inclusions in olivine, also from the Udachnaya pipe (760-810°C, plus some >1000°C). They also reported electron microprobe analyses of a late melt, wetting surfaces in the xenolith (in wt %): SiO_2 34-37, MgO 31-40, FeO 6.0-7.8, Al_2O_3 0.5-1.4, CaO 9.6-13.6, Na_2O 0.7, K_2O 0.4.

The origin of the abundant carbonate phase that may act as the matrix for silicate crystals and xenoliths in some kimberlites has been debated extensively -- is it a primary magmatic phase, or is it a late metasomatic product? The textural evidence for existence of an immiscible carbonate melt mentioned above is corroborated by studies of melt inclusions in the carbonates themselves. Mal'kov and Bobolovich (1977) reported brownish (silicate?) melt inclusions in calcite from several other Yakutian kimberlite pipes. These inclusions had Th ~700-750°C. Late, metasomatic calcite contained liquid inclusions having Th 225-252°C. Mal'kov and Bobolovich also found glassy or partly crystallized melt inclusions in early apatite (large phenocrysts) from kimberlite; this glass turned almost opaque at 950-980°C, so Th could not be determined. Other types of melt inclusions in apatite had widely variable gas/glass ratios, suggesting the presence of a gas phase during trapping. Primary melt inclusions in associated silicate minerals had partly decrepitated during eruption.

Because both Mal'kov and Bobolovich (1977) and Popivnyak and Laz'ko (1979) reported visible gas inclusions in kimberlite minerals, the composition of the gases present in such kimberlites is of interest. Lutts et al. (1976) reported gas analyses (by gas chromatography of gases evolved on grinding) of samples of garnet peridotite xenoliths from kimberlites, spinel peridotite xenoliths from alkali basalts, eclogites, and three kimberlites. Fresh kimberlites produced the largest volumes of gas. These three kimberlite samples yielded (in cm^3/g): He 0.00012-0.00035, H_2 1.89-3.64, O_2 0.0077-0.16, N_2 0.30-3.20, CH_4 0.068-0.14, CO_2 0.031-0.91, H_2/CH_4 15-53. Kravtsov et al. (1979a) reported gas chromatographic (and isotopic) analyses of the gases evolved on crushing samples from the Mir pipe in Yakutia, USSR, as follows (vol %):

	H_2	N_2	CH_4	C_2H_6	C_3H_8	CO_2	C_2H_4	C_3H_6
Violet pyrope	1.92	97.95	0.09	0.02	0.01	NF	NF	NF
Garnetiferous porphyritic peridotite xenolith	41.93	48.91	7.46	0.99	0.38	NF	0.33	NF
Apomeymechite-type breccia	1.95	NF	0.33	0.08	0.02	97.53	0.05	0.04

(NF = Not found)

Several other Russian papers have reported hydrocarbons in kimberlite pipes. Kravtsov et al. (1976) reported that bitumens and hydrocarbons in the Udachnaya pipe have different $\delta^{13}C$ values than those in sedimentary wall rocks. Various samples of eclogite and kimberlite from 362.9 m depth had $\delta^{13}C$ (PDB) ranging from -5.0 per mil to -28.5 per mil, whereas the organic matter in the sedimentary rocks had ~-40 per mil. However, Vdovykin et al. (1979), who studied the bitumens in the kimberlite of the Mir pipe, claimed that the dispersed organic substances came from the sedimentary wall rocks.

Schulze (1981, 1984) studied the polyphase solid inclusions in garnet and diopside of discrete nodules in a kimberlite from Kentucky, USA, and concluded that these represented melt inclusions of the same composition as the liquid from which the nodules crystallized. The inclusions now consist of Ti-rich phlogopite, serpentine (formerly olivine), and calcite. In addition to these phases, the inclusions in the garnets also contain spinel and aluminous clinopyroxene. Schulze inferred that the inclusions in garnet were once the same as those in diopside, but through subsolidus reaction, spinel, aluminous clinopyroxene, and aluminous orthopyroxene (now serpentine) formed at the expense of host garnet and included olivine.

The studied inclusions are large (<2 mm), and apparently all have fractures that connect to the edge of the host grain. These fractures might have formed from pressure release during eruption, but in any case have subsequently admitted the fluids that altered olivine to serpentine (D.J. Schulze, pers. comm.)

By the use of least-squares mixing calculations (Wright and Doherty, 1970) and the known compositions of the various phases (plus an assumed composition for the olivine), Schulze established that the original trapped liquid had a composition equivalent to a mixture of phlogopite, calcite, and olivine. The exact ratio of these three components cannot be established, because of the nature of the inclusions, but the probable weight ratio was ~20 phlogopite, 15 calcite, and 65 olivine. These values compare favorably with some independent estimates of the bulk composition of kimberlites.

Mantle Methane

Gold (1979), Gold and Soter (1980), and MacDonald (1983) have proposed a widely publicized theory that large amounts of abiogenic methane were originally present in the interior of the Earth, and that this methane has been and still is gradually outgassing. Three important geologic consequences of this theory are that: (1) some (or many) present-day petroleum resources may be a result of polymerization of such methane during its transit; (2) some (or many) present-day natural gas resources may contain such methane; and (3) some undiscovered but possibly large resources of mantle methane may have been trapped during the outgassing in deep geologic environments that would not normally be drilled for natural gas, as they would be considered unlikely reservoirs for normal biogenic methane.

Gold and colleagues have marshalled considerable evidence to support this theory, but most is ambiguous, in that it does not preclude, and may even favor, a biogenic origin (see, e.g., Planetary Sciences Unit, 1982). In the present context, I discuss only the evidence presented by analyses of fluid inclusions and related fluids from various possibly pertinent environments where deep-seated CH_4 might be leaking out.

Geothermal field gases. The noncondensable gases from geothermal fields may represent a mixture of magmatic gases with gas from other sources. He of apparent mantle origin has been reported in many active systems (e.g., Torgerson et al., 1982; Poliak et al., 1982). Analyses of small primary fluid inclusions in quartz crystals from the Wairakei field in New Zealand by gas chromatography (Roedder and F.F. Andrawes, unpublished ms.) showed major CO_2 but no detectable CH_4. From the detection limits and the inclusion size, CH_4 was probably <70 ppm. Much larger samples of noncondensable gases are available after condensation of geothermal steam. Data on such gases from The Geysers, California, USA, and Showa Shinzan, Japan (A. Truesdell, pers. comm.), Wairakei, New Zealand (Ellis and Mahon, 1977), and various other areas generally show small amounts of CH_4, (~1 mole % for the sum of all noncondensable gases), corresponding to chemical equilibrium between H_2, H_2O, and CO_2 at elevated P and T. The maximum found (at Clear Lake, California, USA) was 6% CH_4, from fluids believed to have contacted organic matter in sedimentary rocks.

Midocean basalt vesicles. Moore et al. (1977), using both mass spectography and gas chromatography, were unable to detect CH_4 in the gases in vesicles from basalt glass dredged from the sea floor; from this, they estimate CH_4 to be <0.2 mole %.

Fluid inclusions in minerals from magmatic environments. Many of the thousands of analyses reported in the literature (Roedder, 1972) are of samples from igneous rocks or veins and mineral deposits associated with them. Many of these showed no CH_4; others showed small amounts (usually <0.1%), appropriate for the known equilibria between gas species in such mixtures at the conditions of trapping or later reequilibration on cooling, and do not require an external source for CH_4. (Inclusions from sediments or metasediments may have much higher CH_4 contents.)

Gases extracted from vesicle-free basaltic glass. Analyses of the gases released by vacuum fusion of basaltic glasses from Reykjanes Ridge, East Pacific Rise, Mid-Atlantic Ridge, and Kilauea volcano, Hawaii, USA (Harris, 1981b), show 0.9 to 3.2 mole % total noncondensable gases. This total includes major H_2 and CO, plus O_2, Ar, N_2, He (and CH_4 if present), but the molecular species present may be modified during release for analysis.

Gases present in East Pacific Rise hydrothermal fluids. Welhan and Craig (1979) found CH_4 as well as He in solution in the hot fluids issuing from vents on the East Pacific Rise. Exact calculation is impossible because of sample contamination with sea water, but the concentration found (2×10^{-5} cc (STP) of CH_4 per gram of water) corresponds to <2 ppm. The host fluids probably represent sea water, recharged through at least some organic matter in deep sea sediments, and subsequently altered by reaction with hot basalt. The CH_4 found might thus be biogenic in origin (e.g., Baross et al., 1982; Baross and Deming, 1983) or from straight C-H-O equilibria, but Welhan and Craig (1983) have suggested, on isotopic and chemical evidence, that the CH_4 might be of mantle origin. Lilley et al. (1982) reported CH_4 analyses from several vents and suggested that at least part was microbially produced.

Gases in inclusions in alkalic intrusives. The inclusions in the alkalic massifs (nepheline syenites and related rocks) from several localities differ from those in most other igneous rocks in that CH_4 is a frequent and sometimes prominent component among the inclusion gases (see Chapter 14). Lesser amounts of higher hydrocarbons are also found, including even small amounts of solid bitumens (see Roedder, 1972, for analyses). Although these hydrocarbons have been explained in terms of chemical equilibria in the system C-H-O, problems exist with such interpretations (Roedder, 1981e, p. 39). High CH_4 concentrations are only possible by reaction of the gases with the host rocks (including graphite) at relatively low temperatures (200-300°C); hence, such fluids must be from very late stage processes. An alternative, proposed by Konnerup-Madsen et al. (1979), involves concentration of CH_4 via liquid immiscibility (see Chapter 14).

Gases in CO_2 inclusions in ultramafic xenoliths. The gas inclusions in olivine and other minerals from ultramafic xenoliths represent a sample of a gas phase present and trapped at considerable depths (commonly 10-15 km and sometimes more). Roedder (1965a) presented evidence from crushing-stage studies (p. 1752) that some of these CO_2 inclusions contained a maximum of perhaps 1 vol % of another gas. This 1% represents the sum of CO, SO_2, COS, Ar, and other gases, including possibly CH_4. Additional details on the composition of these gases are given in the next section.

The above discussion shows that although CH_4 is present in many natural fluids, it is normally a very minor constituent. The low concentrations do not preclude such fluids as possible sources for CH_4-rich gases, since processes for concentration of the CH_4 can be envisaged. However, each such ad hoc process that must be invoked in sequence to achieve the desired result must also detract from the credibility of the theory of major emanation of primordial methane from the mantle.

CO_2 AND MELT INCLUSIONS IN ULTRAMAFIC NODULES

The Host Nodules

Geologic occurrence. Ultramafic nodules or xenoliths are such a common feature of alkali basalt volcanism throughout the world that their presence can be used as a field-mapping criterion to indicate that a given group of flows, plugs, dikes, or pyroclastics is indeed alkali basalt in composition. I reported (1965a) that almost all[6] samples of such nodules that I examined from localities scattered over the globe contained liquid CO_2 inclusions. Such inclusions, which have an internal pressure of 70 bars at room temperature, and must have had much higher pressures at the high temperatures existing during eruption, were totally unexpected. Since that paper was published, I have looked at nodules from many other localities and have found CO_2 inclusions in all. In addition, other workers have reported CO_2 inclusions from a number of other nodule localities, including some in nodules from kimberlites (Table 17-1). Inclusions in nodules from kimberlites have been studied much less than have those from alkali basalts. Not all such nodules have come from the mantle, but the difficult problem of distinguishing between nodules from the mantle and similar-appearing nodules formed at much shallower depths will be held in abeyance until the inclusions have been described.

General petrography. The nodules that have been studied for fluid inclusions generally consist of major olivine, clinopyroxene, and orthopyroxene, and lesser spinel, biotite, garnet, and apatite, etc. Wide variations in mineralogy result in numerous rock names, but dunites, peridotites, and lherzolites are generally the most abundant (see reviews by Jackson and Wright, 1970; Carswell,

----

[6] Liquid CO_2 was found in 64 of 72 localities; at 8 others, only high-pressure gas inclusions (CO_2?) were observed. Similar liquid CO_2 inclusions were found in phenocrysts of olivine, titanaugite, and oxyhornblende from some of the host basalts. I have since found (unpublished data) abundant high-density primary CO_2 inclusions in nodules of various composition -- olivine-spinel dunite, wehrlite, pyroxenite, and allivalite--from transitional olivine tholeiites from Reunion Island (provided courtesy of B.G.J. Upton). The nodules do not occur in the alkalic basalt sequences on Reunion. During formation of planes of secondary inclusions in fractures in olivine from the nodules, three immiscible fluid phases were present and trapped in various ratios: Basalt melt, sulfide melt, and CO_2 fluid. A few secondary fluid inclusions seem to require the presence of four immiscible fluids, the three above plus an immiscible high-silica melt.

Table 17-1 Studies of ultramafic nodules in which CO_2 inclusions were examined.

Type*	Locality	Reference	Maximum CO_2 density (g/cm^3)
A,K,P	Worldwide (72 localities)	Roedder (1965a)	0.89
A	E. Kamchatka, etc., USSR	Bakumenko (1975)	-
K	Bournac, Massif Central, France	Bilal & Touret (1976)	1.075
P	Puy Beaunit, Massif Central, France	Bilal & Touret (1977)	0.85**
A	Arizona, Hawaii, Germany	Murck et al. (1978)	1.14
A	Dreiser Weiher, W. Germany	Solovova et al. (1982)	1.18
K	Various, South Africa	Forestier & Touret (1980)	(~1.1)***
A	Various, Massif Central, France	Forestier & Touret (1980)	(~1.0)***
A	Loihi Seamount, Hawaii, USA	Roedder (1983b)	0.8
A	NW. Sardinia	De Negri & Touret (1981)	-
K	Lesotho	Bilal, 1978, as quoted by Touret (1981)	1.173
A	Eastern China	Qi (1983)	1.14
A	Assab, Erythrea (Ethiopia)	Clocchiatti et al. (1981)	1.02
A	Pali-Aike volcanic field, Chile	Selverstone (1982), and Selverstone & Stern (1983)	1.02
A****	Lunar Crater volcanic field, Nevada, USA	Bergman (1982), and Bergman & Dubessy (1984)	1.17
A	Hoggar, central Sahara	Miller & Richter (1982)	1.16
A	Vesuvius, Italy	Belkin et al. (1984)	0.75

*Types:
  A,- Xenoliths in alkali olivine basalt or similar rocks; K,- Xenoliths in kimberlite; P - Phenocryst in alkali olivine basalt.
**Less dense inclusions contain appreciable hydrocarbons.
***Estimated from Th CO_2. The plotted Th values are all for homogenization in the liquid (J. Touret, pers. comm., 1983).
****Also found in anorthoclase megacrysts - see Figure 2-34.

1980; and DePaolo, 1983). The mineralogy of nodules from kimberlites differs from that of nodules from basalts (e.g., Irving et al., 1980; Ahrens et al., 1975), but the compositions of the minerals in the specific nodules that were studied for inclusions were seldom determined.

These rocks vary widely in grain size. Except for occasional megacrysts, most of these rocks are <1 cm in grain size, and they average ~1-2 mm; sheared samples can be much finer grained. The textures also vary greatly (e.g., see review by Mercier and Nicolas, 1975), but most grains show smooth boundaries, and 120° junctions are common. Some of the nodules are porphyritic, and small grains of spinel are commonly found enclosed within olivine. Apatite and biotite, when found, are usually interstitial.

As might be expected for xenolithic nodules that were transported to the surface and erupted with basaltic magma, the nodules are coated with and frequently penetrated by stringers of basaltic glass. Individual nodules range from tens of centimeters to disaggregated clusters of just a few grains; when completely disaggregated, the individual xenocrysts may be difficult to distinguish from new phenocrysts, except by means of their inclusions.

Most grains are unfractured, but in some nodules, most of the olivine is cut by closely spaced fractures without visible offset. Apparently some nodules were sheared extensively in the presence of silicate melt and CO_2, either at their origin or on the way up, but others were simply broken loose and carried up gently.

The Inclusions

General petrography. Three types of fluids were trapped to form fluid inclusions in nodule minerals: silicate melt, sulfide melt, and CO_2. The three fluids occur together in various ratios, as primary, pseudosecondary, and secondary inclusions in most nodule minerals. The major exception is mica; no recognizable CO_2 inclusion has been found in either biotite or phlogopite in the nodules. (Empty inclusions, presumably representing decrepitated inclusions, occur in some of the mica.) In addition to the three normal types of inclusions, exsolution inclusions of CO_2 occur in some samples.

The CO_2 inclusions are rarely >30 μm in size, and usually are <5 μm; many are <1 μm. Silicate-melt inclusions cover the same range as those of CO_2 but also are occasionally as large as 100 μm. Sulfide melt usually occurs only as small spherical blebs in inclusions of silicate melt. The abundance of inclusions varies over many orders of magnitude. The presence of even a single liquid CO_2 inclusion a few micrometers in diameter is just as significant, in terms of its chemical and physical environment of formation, as a large number of such inclusions. On the other hand, the inclusions are so irregularly distributed and rare in some of the nodules that completely negative results on the search of a single slide or plate[7] cannot be considered as valid proof of their absence at that locality. The frequency of occurrence of recognizable CO_2 inclusions in the various minerals gave no useful guidelines in this search. In some nodules the pyroxenes had abundant inclusions and the olivine had few; in others, this was reversed. CO_2 inclusions were seen in plagioclase from only a few localities, but comparatively little plagioclase was found in the samples examined. (Presumably most of the nodules were formed in high-pressure environments where plagioclase was not stable.) Although a few dark inclusions, possibly gaseous, were seen in some of the dark-red garnets of the eclogitic nodules that were examined and in some spinel grains, the still higher index of refraction and the much smaller volume of garnet and spinel studied (as these are minor phases and much thinner plates are needed for transparency) would make the discovery of recognizable CO_2 inclusions unlikely. An exception was found in the rather transparent hercynitic spinels in nodules from Vesuvius (Belkin et al., 1984).

The volume percent of CO_2 inclusions varies widely, even from grain to grain in the same sample. Only a few tiny inclusions have been found in some large (25 cm^2) plates (after hours of search), corresponding to a few parts CO_2 per trillion of rock, yet individual small grains of olivine have been found with an estimated 3 vol % CO_2, corresponding to about 0.5 wt % CO_2 (see later section dealing with composition). Most of the nodules studied contain only about 1 part per billion of CO_2. In addition to nodule localities, I examined (1965a) olivine-rich and ultramafic rocks from other types of occurrences. Samples from 31 localities showed high-pressure gas inclusions. Ten other localities showed no evidence of CO_2 inclusions, but such negative evidence is not conclusive. This work indicates merely that gas inclusions are present in many mafic igneous rocks but does not prove that the gas is CO_2 or that the rocks are related genetically. The volumes of gas released on crushing were generally small.

As nodules had been studied intensively for many years prior to 1965 without a single report of CO_2 inclusions, some explanation is in order. Most of the inclusions studied appear, at first glance, to be merely opaque specks

----

[7] Ordinary petrographic thin sections of ultramafic nodules are particularly inappropriate for searching for CO_2 inclusions, except in material that is rich in inclusions. Doubly polished plates, prepared with care and generally with several impregnation steps, are much preferred for most work and are absolutely essential on difficult samples.

(solid inclusions) in the enclosing mineral, and hence are easily overlooked. During ordinary petrographic examinations of either thin sections or grain mounts, these inclusions would normally be seen as bothersome dust and ignored. They differ from the common opaque solid inclusions, however, in that even in ordinary lighting at moderate magnification, the larger ones usually show a pinpoint of light coming through the center of the otherwise dark mass. The low index of refraction of CO_2 (well under 1.2; Quinn and Jones, 1936), causes a rounded inclusion of CO_2 in a high-index mineral to act as a strong negative lens. Strongly convergent light, as from the normal high-power substage condenser, will counteract this (Fig. 6-1), but unless the light is used with a good infrared (IR) filter, it will cause the two CO_2 phases to homogenize, making the inclusion appear empty. Even at the optimum conditions of observation detailed below, and knowledge of what to look for, most of the CO_2 inclusions in the samples studied are unsuited for positive identification; hence, the earlier failure to recognize the CO_2 is understandable. Most CO_2 inclusions are rounded to almost spherical. Only rarely are they flattened permitting good delineation of the inclusion contents; many (or most) of the photomicrographs that have been published are of such flattened, atypical inclusions.

Primary inclusions. As will be obvious later, valid proof of a primary origin for even a single CO_2 inclusion is of considerable importance in the interpretation of the origin. I reported (1965a) fairly unambiguously primary inclusions of silicate melt and a very few "presumably primary" inclusions of CO_2 in a study of many nodules from Hawaii, USA (Figs. 17-2,3,4,5). In contrast, samples from the young Loihi Seamount, Hawaii, USA (Roedder, 1983b) provided abundant and good evidence for primary CO_2 inclusions in olivine, even though a total of only a few square centimeters of xenolith was present in the available slides.

Perhaps the best evidence for a primary origin for CO_2 inclusions found in the Loihi samples is the occurrence of relatively large isolated inclusions, well away from any other inclusion or healed fracture (Figs. 17-6,7,8). The inclusions shown in Figures 17-6,7 trapped only a CO_2 globule. That shown in Figure 17-8 trapped silicate melt, sulfide melt, and CO_2; it is a fairly large brownish glass inclusion containing an opaque spherule of sulfide melt and a "vapor bubble" that is too large to have formed by normal shrinkage of the silicate melt on cooling. Inside this "vapor bubble" is another bubble, which can be seen by using strongly convergent light and a highly efficient IR filter. The "bubble in a bubble" represents gaseous CO_2 in liquid CO_2. Circular optical artifacts of various types can be very misleading in such inclusions, but lateral movement of the bubble, and particularly its sudden disappearance when the IR filter is removed, provide adequate verification that this gas bubble is real and not an artifact. Other primary CO_2 inclusions were found adhering to spinel crystals that were enclosed by the growing olivine, with or without silicate melt (Fig. 17-9). Such preferential wetting by the dispersed globules of dense CO_2 is not uncommon.

Ultramafic nodules erupted from Vesuvius volcano, Italy (Belkin et al., 1984), also provided some examples of primary CO_2 inclusions. Some hercynitic spinel crystals in those nodules show one or a group of large isolated inclusions in the core of the crystal, and no evidence of any healed fracture (Fig. 17-10). Growth of olivine and pyroxene crystals from these nodules has occasionally trapped melt and a CO_2 globule when the crystal grew around and enclosed solid inclusions of biotite (Fig. 17-11).

Perhaps the most striking evidence for primary inclusions of CO_2 is found in the tubular inclusions present in some megacrysts (e.g., Fig. 2-34). These tubular inclusions show that CO_2 exsolved from the melt to form bubbles on the crystal surface. Continued exsolution of CO_2 from the melt to the bubble took place as the crystal grew, yielding a tubular inclusion (Chapter 2).

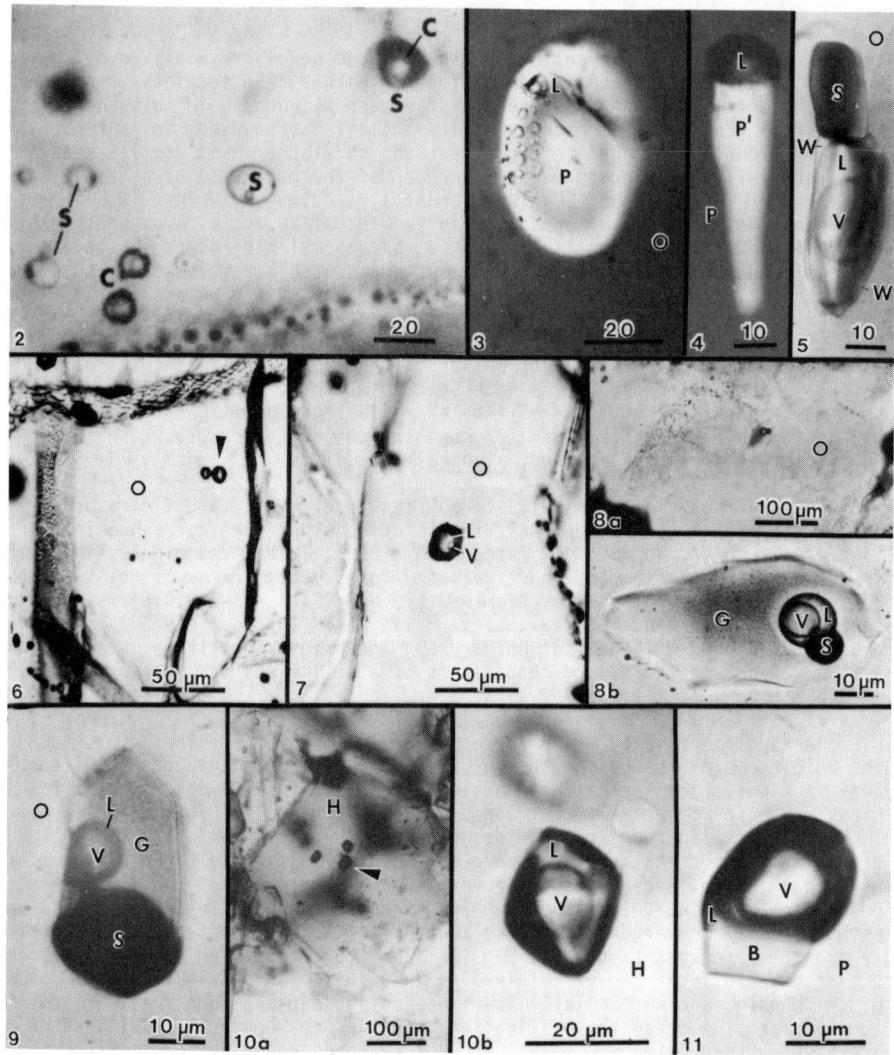

Figure 17-2. Isolated, presumably primary, inclusions of silicate glass (S) with small shrinkage bubble and opaque daughter mineral, and two, presumably primary, liquid CO_2 inclusions, showing characteristic dark borders and large, faint, rapidly-moving vapor bubbles (C). Inclusion in upper right shows all three phases; CO_2 in it homogenizes in the gas phase at $30.91\pm0.02°C$. This group of inclusions of melt and CO_2 could be a result of major recrystallization of a plane of secondary melt plus CO_2 inclusions originally parallel to the plane of the photo, but several features suggest that the inclusions are more likely to be primary. A plane of secondary CO_2 inclusions is visible at bottom. Sample ER 63-33a, from the 1801 Kaupulehu flow of Hualalai, Hawaii, USA. Scale bar in μm. From Roedder (1965a).

Figure 17-3. Presumably primary CO_2 inclusions trapped at interface between embedded crystal of pyroxene (P) and surrounding olivine (O). The largest inclusion (L) shows a circular bubble of vapor in liquid CO_2; it homogenizes in the liquid phase. These CO_2 inclusions could also be exsolution inclusions. Partially crossed polars. Ngatutura Point, Ohuka, New Zealand (sample ER 64-174). Scale bar in μm. From Roedder (1965a).

Pseudosecondary inclusions. Proof of a pseudosecondary origin for CO_2 inclusions is as petrologically useful as proof of a primary origin. The best evidence of pseudosecondary CO_2 inclusions was also found in samples from the Loihi Seamount (Roedder, 1983b). Several good examples were found in which the cores of the olivine crystals were badly fractured, as outlined now by many planes of CO_2 inclusions, but these fractures did not penetrate into the rim of the crystal (Figs. 17-12,13). Figure 17-12 shows a cored olivine crystal containing a combination of primary and pseudosecondary inclusions in the core and essentially no inclusions in the rim. The crystal shown in Figure 17-13 is similar but has only pseudosecondary inclusions in the core. Later fractures, outlined by planes of true secondary inclusions, cut both core and rim. In detail, the CO_2 inclusions in these planes of pseudosecondary and secondary inclusions generally appear to be oval cavities filled with just liquid and gaseous CO_2. A film of silicate melt is sometimes visible; it is probably present in all such inclusions but generally is invisible.

Secondary inclusions. Literally millions of small secondary CO_2 inclusions, outlining healed fractures, are present for every primary inclusion. Many olivine grains have one or more such planes (Fig. 17-14), commonly outlined by inclusions that are graduated in size, from the healing of a fracture that was wedge shaped (Fig. 2-15). Only rarely can such planes be traced across grain boundaries. Individual inclusions in a given plane range from smoothly ovoid to sharply faceted negative crystals, commonly have fillets of silicate melt in the corners (Fig. 17-15), and occasionally have corner spines elongated perpendicular to {100} of the enclosing olivine (Fig. 17-16). I have no explanation for these differences in shape.

In some xenoliths, the fracturing is intense, and although it has a general

Figure 17-4. Presumably primary CO_2 inclusion (L) attached to end of unknown crystal (P'; probably another pyroxene) embedded in pyroxene (P). A number of these embedded crystals, many with such CO_2 inclusions covering their larger ends, are parallel throughout the pyroxene, and although unconnected, show uniform inclined extinction. Their index of refraction is only slightly greater than that of the host, hence partially crossed polars were used for this photograph. Presumably the embedded crystals grew simultaneously with the host, but their growth was stopped by preferential adherence of CO_2 blebs to their exposed terminations. Alternatively, both embedded crystal and CO_2 could be from exsolution. Isla de Guadalupe, NE Pacific (sample ER 63-128). From Roedder (1965a).

Figure 17-5. Presumably primary CO_2 inclusion attached to spinel grain (S) in orthopyroxene (O) from olivine nodule from Ichinome-Gata, Japan (sample ER 65-8). Many elongated spinel crystals lie parallel to an extinction direction of the pyroxene, and such CO_2 inclusions are attached to many. V, gaseous CO_2; L, liquid CO_2; W, liquid water. Alternatively, spinel, CO_2, and H_2O could result from exsolution rather than from primary trapping. This sample contains the only recognized water in all the nodules that I examined. The CO_2/H_2O ratio is apparently constant throughout the sample. From Roedder (1965a).

Figure 17-6. Pair of isolated primary CO_2 inclusions (arrow) in olivine (O). Sample KK-31-1, from Loihi Seamount, Hawaii, USA. From Roedder (1983b).

Figure 17-7. Single large isolated primary inclusion containing liquid CO_2 (L) and vapor CO_2 (V) in olivine (O). Sample KK-31-1 from Loihi Seamount, Hawaii, USA. From Roedder (1983b).

Figure 17-8a,b. Single large isolated primary four-phase inclusion in olivine, at low (a) and high (b) magnification. Consists of silicate melt, now slightly devitrified glass (G), which was trapped along with a globule of an immiscible sulfide melt (S), and an immiscible globule of dense supercritical CO_2, now consisting of liquid CO_2 (L) and vapor CO_2 (V). The vapor bubble moves about within the liquid CO_2. Sample KK-27-9A from Loihi Seamount, Hawaii, USA. From Roedder (1983b).

Figure 17-9. Primary silicate-melt inclusion attached to solid inclusion of opaque spinel(?) in olivine (O). The silicate melt, now slightly devitrified glass (G), was trapped along with a globule of supercritical CO_2, now consisting of CO_2 vapor (V) plus a faint rim of CO_2 liquid (L). Sample KK-27-9A from Loihi Seamount, Hawaii, USA. From Roedder (1983b).

Figure 17-10. Primary CO_2 inclusions in hercynitic spinel crystal (H) in xenolith from Vesuvius, Italy. Three large CO_2 inclusions are visible in the core of the crystal in (a); in (b), a detail of one of these inclusions (arrow in (a)), taken at room temperature, shows CO_2 liquid (L) and vapor (V). Th CO_2 L+V(L) for this particular inclusion is at +29.4°C. Sample N-30, from Belkin et al. (1984).

Figure 17-11. A solid inclusion of biotite (B) in pyroxene (P) in xenolith from Vesuvius, Italy. Attached to the biotite is a primary CO_2 inclusion consisting of CO_2 liquid (L) and vapor (V). Th CO_2 L+V(V) is at +30.2°C. Sample N-34, from Belkin et al. (1984).

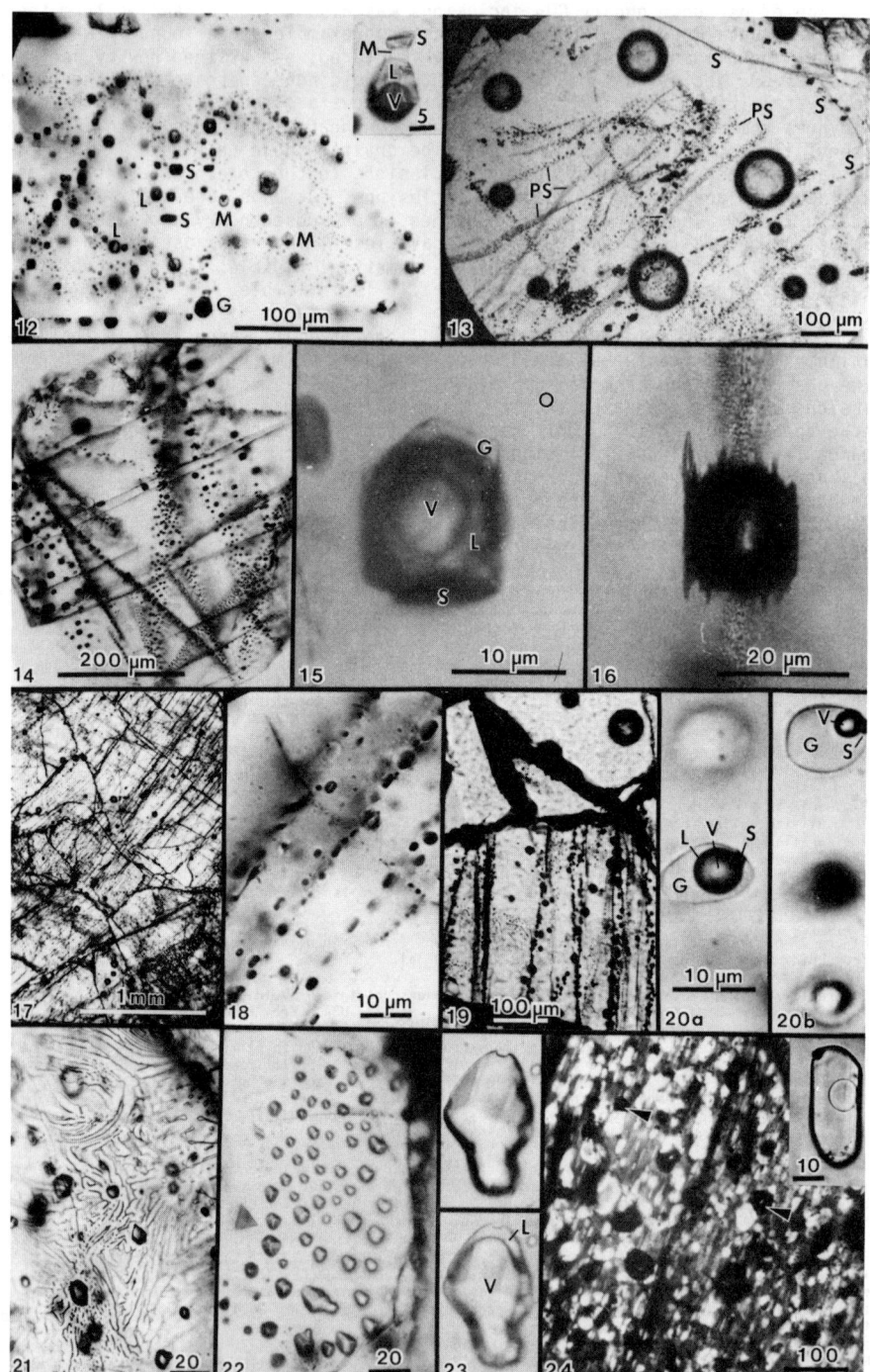

Figure 17-12. Single crystal of "cored" olivine. The core of the crystal (lower left part) grew first, trapping many primary CO_2 inclusions (L) and some solid spinel crystals (S). A few spinel crystals also resulted in the trapping of some silicate melt (M) and some CO_2, now liquid (L) and vapor (V); see inclusion (G), magnified in inset. The core also contains numerous planes of pseudosecondary CO_2 inclusions that represent the healing of fractures that formed before the growth of the inclusion-free rim (optically continuous with core). Sample KK-17-5B from Loihi Seamount, Hawaii, USA. Photographed with crossed polars. Inset scale bar in μm. From Roedder (1983b).

Figure 17-13. Single crystal of "cored" olivine, photographed with crossed polars. Large dark bubbles are artifacts in the mounting medium. At least two stages of growth and fracturing of olivine are evident here. First, the core of the crystal (lower part) grew; at some later time, it was fractured in the presence of a CO_2 fluid, and healing of the fractures formed planes of pseudosecondary (PS) inclusions; more olivine then grew on the core, in optical continuity, yielding the clear upper part; another stage of fracturing took place, yielding secondary CO_2 planes (S). Sample KK-27-9A from Loihi Seamount, Hawaii, USA. From Roedder (1983b).

Figure 17-14. Olivine grain, mounted in resin of n = 1.67, showing numerous planes of sharply faceted secondary CO_2 inclusions. See also Fig. 17-15. Sample ER 63-33a, from 1801 Kaupulehu flow of Hualalai, Hawaii, USA. From Roedder (1965a).

Figure 17-15. Faceted secondary inclusion from healed fracture in same olivine (O) sample as shown in Figure 17-14. Opaque plates, possibly of spinel (S) or graphite(?), line cavity, and recrystallization has presumably occurred through the glass (G). Most of the inclusion is filled with liquid CO_2 (L) and gaseous CO_2 (V). From Roedder (1965a).

Figure 17-16. CO_2 inclusion that has unexplained "spines." Although this inclusion is in a pyroxene crystal from a nodule, many similar inclusions are seen in olivine. Sample ER 63-55, Camargo, Mexico. From Roedder (1965a).

Figure 17-17. Sheared dunite xenolith. The field of view includes four optically distinct grains of olivine. The small dark spheres are bubbles in the mounting medium. A part of one grain is enlarged in Figure 17-18. Sample KK-27-4a, Loihi Seamount, Hawaii, USA. From Roedder (1983b).

Figure 17-18. Enlargement of small area from Figure 17-17, showing small CO_2 inclusions (mostly <2 μm) delineating closely spaced shears. From Roedder (1983b).

Figure 17-19. Two adjacent olivine grains (upper and lower parts of photomicrograph) in dunite nodule, showing large differences in amount of fracturing and hence CO_2 inclusions in adjoining grains. Note that some fractures are outlined by a relatively few large inclusions, whereas other fractures now contain large numbers of minute CO_2 inclusions. Large dark sphere at top is a bubble in mounting medium. Sample KK-27-9B, Loihi Seamount, Hawaii, USA. Scale bar in μm. From Roedder (1983b).

Figure 17-20a,b. Part of a plane (vertical in these photographs) of small secondary inclusions in olivine, from sample KK-31-11A, Loihi Seamount, Hawaii, USA, at two levels of focus, showing that adjacent secondary inclusions outlining this plane contain varying ratios of CO_2 and glass. This variation indicates the presence of immiscible globules of dense CO_2 in the melt at the time of fracturing. The uppermost inclusion in (b) trapped only silicate melt (now glass, G), which shrank to form a vapor bubble (V) and precipitated a small globule of immiscible sulfide melt (S). The middle inclusion in (a) and the lower one in (b) (out of focus in (a)) represent trapping of silicate melt (G) plus a globule of supercritical CO_2, which now consists of liquid CO_2 (L) and a faint vapor bubble (V). A small immiscible globule of sulfide (S) has also separated after trapping. From Roedder (1983b).

Figure 17-21. A fracture, marked by a plane of secondary CO_2 inclusions in olivine that has been opened a second time (presumably by decrepitation of the CO_2 inclusions during eruption). The original secondary inclusions now contain much lower density CO_2, and the healing of the new fracture has started; necking down has formed very thin flat vermicular inclusions. Sample KK-17-5BC, Loihi Seamount, Hawaii, USA. Scale bar in μm. From Roedder (1983b).

Figure 17-22. Plane of relatively large secondary or pseudosecondary inclusions in olivine grain that presumably have reopened at relatively shallow depths during eruption and hence now contain low-density CO_2. Largest inclusion (at bottom) is enlarged in Figure 17-23. Sample KK-31-11A, Loihi Seamount, Hawaii, USA. Scale bar in μm. From Roedder (1983b).

Figure 17-23a,b. Largest inclusion in Figure 17-22 at -45°C (lower) and at room temperature (upper). The last of the liquid CO_2 (L) evaporates into the vapor phase (V) at +2.7°C. All the inclusions in the plane in Figure 17-22 behave similarly, but the liquid CO_2 in them is more difficult to see than in this inclusion. From Roedder (1983b).

Figure 17-24. Possible exsolution CO_2 inclusions (dark spherical masses) in orthopyroxene. Some are pure CO_2 (inset); others are attached to exsolved clinopyroxene(?) blebs (e.g., see arrows). Photographed with crossed polars. Inset is a high-density pure-CO_2 inclusion from this field of view. Scale bars in μm. Sample KK-17-5Be, Loihi Seamount, Hawaii, USA. From Roedder (1983b).

orientation throughout the xenolith, the olivine crystals in the mosaic show
some individual control of the direction of fracturing (Figs. 17-17,18).
Certain grains have many such fractures, whereas adjoining grains have undergone
little or no deformation (Fig. 17-19). The fracturing of xenoliths presumably
suspended in a melt presents some problems, and although some fracturing can
occur from decrepitation of high-pressure inclusions during eruption, most
planes of secondaries could hardly form that way (e.g., Figs. 17-13,14,17,19).

A mixture of silicate melt and CO_2 was present at the time of fracturing,
as some individual former fractures now show adjacent inclusions, some containing
just silicate glass and a small bubble (from shrinkage) and others containing CO_2
liquid plus gas or a mixture of the two materials (i.e., glass and a bubble that
is too large to be from shrinkage alone and that now consists of liquid CO_2 plus
vapor (Fig. 17-20a,b). (Care is necessary here to recognize where necking down
has occurred.) Occasional sulfide-melt inclusions are also found in these planes
of secondary inclusions.

Some secondary planes of CO_2 inclusions have reopened (presumably from external pressure decrease during rise to the surface), and occasionally the process of healing of the new fracture surface has been halted by quenching upon
eruption (Fig. 17-21). Presumably, the plane of secondary CO_2 inclusions shown
in Figures 17-22 and 17-23 has reopened and healed at relatively low pressure,
but the minute vermicular inclusions on the fracture plane, as shown in Figure
17-21, have not had time to coalesce.

Some of the planes of secondary or pseudosecondary inclusions are delineated
by rather closely spaced, relatively large CO_2 inclusions (Fig. 17-22), whereas
others have only very small inclusions (Fig. 17-13). In the latter type, however, an abrupt cutoff is generally evident at ~0.3 μm. A large part of the size
range is simply a result of the width of the wedge-shaped fracture before healing, but if the interpretation derived from secondary aqueous inclusions in
metamorphic quartz can be applied here, some of the differences may be a result
of the time factor. Planes of secondary inclusions in quartz at metamorphic
temperatures apparently change with time, from a flat sheet of liquid at first,
to isolated flat inclusions, to small more equant inclusions, to a smaller number
of larger inclusions (Chapter 3). These changes are in response to a decrease
in surface energy, and the final result has been termed "mature" (Swanenberg,
1980). From this analogy, I suspect that planes of large CO_2 inclusions, such
as those shown in Figure 17-22, have had a long period of time since original
trapping to develop to their present state, whereas planes such as those seen in
Figure 17-13 have had a much shorter time and hence are far less mature. The
actual volume of inclusion per unit area will be approximately a direct function
of the width of the original crack. The magnitude of this time factor for olivine is unknown, and the concept is purely qualitative at this time; R.J. Bodnar
(pers. comm., 1983) has determined that in quartz, secondary inclusions mature
in weeks at 600°C

Exsolution inclusions. Exsolution of CO_2 from natural olivine to form inclusions was studied in the electron microscope by Green and Radcliffe (1975).
Most of the inclusions that they attributed to this process were below optical
resolution and ~5 orders of magnitude smaller in volume than those in olivine
studied by optical microscopy (e.g., Table 17-1), and hence may have formed by
another process. However, some of the orthopyroxene grains from several localities have abundant, coarse, physically and optically oriented blebs of another
pyroxene (presumably clinopyroxene) scattered throughout (Figs. 17-4, 24). These
blebs may be from simultaneous growth of the two phases, or from exsolution.
Some of these included pyroxene grains have relatively large attached inclusions
of CO_2 and/or spinel. Additional inclusions of spinel and CO_2, or just CO_2
(Fig. 17-24 inset), are randomly scattered through the orthopyroxene crystals;
these may also be exsolution inclusions.

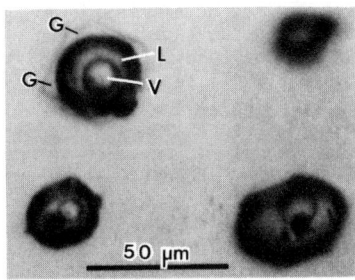

Figure 17-25. Photomicrograph of fluid inclusions in olivine from ultramafic nodule in basalt, showing glass (G), liquid CO_2 (L), and gaseous CO_2 (V). The two CO_2 phases homogenize in the liquid phase by the slight warming caused by absorption of infrared (IR) light on removal of the IR filter on the microscope light. During the growth or later fracturing and healing of the host olivine, at an estimated depth of 8-16 km and 1200°C, the CO_2 was present as homogeneous supercritical gas bubbles in the basaltic, CO_2-saturated, melt. Sample ER 63-33, 1801 Kaupulehu flow of Hualalai, Hawaii, USA. From Roedder (1965a).

Identification of CO_2 and microthermometry of CO_2 and melt inclusions. I applied a variety of physical procedures (1965a) to verify the identification of liquid CO_2 in the inclusions in the nodules. The crushing stage (see Chapter 7) revealed that the expansion of the presumed liquid CO_2 to form gas was ~350-fold (i.e., a sevenfold increase in diameter), as is appropriate for liquid CO_2 changing to vapor at 1 atm. As the pressure in an inclusion containing liquid CO_2 at room temperature is ~70 bars, the inclusions practically explode into the mounting fluid on crushing. When both melt and CO_2 are present in the same inclusion, the distinction is obvious (Fig. 17-25). In a few examples, slight strain birefringence is visible around the inclusions when the host olivine is put at extinction. Inclusions containing no visible liquid/gas meniscus were frequently found to be full of gas at 1 atm (i.e., no expansion or contraction on crushing) and hence are presumed to have decrepitated and are now filled with air. Similar appearing one-phase inclusions in other samples evolved large quantities of gas on crushing; these inclusions were filled with dense fluid CO_2 and would need to be cooled to below room temperature to develop a vapor bubble. (On Figure 8-9 they would fall in the "fluid" field at greater densities than the critical density.) The crushing tests showed that the evolved gas was relatively pure CO_2 (see next section).

The extremely high thermal expansion of the liquid phase in these inclusions matched that of liquid CO_2, but this property is not definitive of CO_2; almost any fluid near its critical temperature will show similar large expansivity. Much more definitive is the critical temperature (see next section).

The infrared absorption of CO_2 is so strong that it is readily used as an identifying criterion. Although CO_2 liquid and gas are both transparent and colorless in the visible spectrum, they show strong absorption in the IR (see Plass and Stull, 1963). Two-phase CO_2 inclusions can be made to homogenize in two ways, either by simple warming of the slide, or by increasing the intensity of the microscope lighting (first reported for CO_2 inclusions in pegmatitic beryl by Cameron et al., 1953). In fact, the size of the bubble can be adjusted rather closely by adjusting either a substage diaphragm or the voltage to the light source. When adjusted so that the inclusion just stays homogeneous, an interruption of the light beam for as little as 0.1 sec will cause momentary heterogenization. When an IR-absorbing ("heat-absorbing") filter is placed in the light path, a much higher intensity light is needed to cause homogenization. Olivine is rather transparent in the IR (Duke and Stevens, 1964); hence, heating of the whole mineral plate by the light is probably not significant. This is corroborated by the fact that inclusions in 3 mm thick polished plates seemed to react to light just as fast as those in 0.030 mm thin sections.

IR absorption in the inclusions has some significance in the practical problem of searching for them. To see detail inside most CO_2 inclusions in olivine, rather bright light, a high-power substage condenser, and a magnification of at least 300 are needed. Under these conditions, unless the stage of the microscope is rather cool and an efficient IR filter is used, the inclusions will be homogenized and hence appear empty. The IR absorption also has significance in any accurate Th determinations. By absorption, the inclusion is warmed slightly relative to the surrounding mineral and relative to the circulating heat-exchange medium. Thus, normal illumination, without filters, may give an apparent Th that is several tenths of a degree low, and intense illumination gave Th as much as 1.28°C lower than was obtained with low illumination plus an effective IR filter. The temperature differences vary with the size of the inclusion. If a reasonably good filter is used, the light intensity is significant only in precise determinations in the vicinity of the critical point, where very minor differences in the thermal regime in the neighborhood of the inclusion may have considerable effects. Thus, individual inclusions have been found, which have near to critical density of filling, that homogenize in the gas phase or by fading of the meniscus, depending on how the heating with the light is applied. In addition, absorption of all light by an opaque grain in a CO_2 inclusion can cause localized boiling of the CO_2 liquid.

The homogenization behavior and temperature range are useful for verification of CO_2 and permit a reliable estimate to be made of the density of the CO_2 fluid. As seen in Figure 8-9, inclusions of pure CO_2 with a density of ~0.40 g/cm³ will show critical behavior. Those with <0.40 density will homogenize in the vapor phase, and those with >0.40 density will homogenize in the liquid phase; in both types, Th will be <+31.1°C. Near the critical density, the measurement of phase volume percentages at a lower temperature (e.g., 20°C) will provide a fairly accurate density value, but away from the critical density (<~0.3 or >~0.7 g/cm³), measurement of Th will generally provide a more accurate density value. Very dense CO_2 fluids may require low temperatures to heterogenize, down to the triple point of -56.6°C for fluids having density 1.16 (see Fig. 13-8).

The triple point for pure CO_2 at -56.6°C is also an excellent diagnostic criterion (so long as the inclusions are large enough to measure it). In those relatively few studies where the triple point was determined, on inclusions from nodules, it was found to be near -56°C. Only one locality that I studied (1965a) showed visible liquid water as a phase in some CO_2 inclusions (e.g., Fig. 17-5). Another such inclusion from this sample formed CO_2 hydrate, which melted at +10.1±0.2°C.

Only rarely have determinations been made of Th of silicate-melt inclusions trapped presumably simultaneously with the CO_2 inclusions. Belkin et al. (1984) reported Th of ~1200°C for primary melt inclusions believed to be cogenetic with primary CO_2 inclusions, in olivine, pyroxene, and spinel from nodules from Vesuvius.

Composition of CO_2 and melt inclusions. The precise composition of the gas in the CO_2 inclusions (and its variation) is of considerable petrologic significance but is surprisingly elusive. The earlier crushing-stage studies showed that the gas was ~99 vol % CO_2, but that ~1% of some other constituent(s) was present (see Chapter 7, and Roedder, 1965a). Chemical analysis of the gases evolved on crushing or heating might seem to be the obvious answer but provides ambiguous results. The amount of CO_2 present in most nodules, estimated optically to be commonly in the parts per billion range, from the volume of the inclusions, is so small that large samples must be used, greatly increasing the problem of adequate sample purity. Most of the attempts at extraction and chemical analysis have been made on several dunitic nodules from one Hawaiian sample (Sample ER 63-33; Roedder, 1965a) that showed orders of magnitude more CO_2 as inclusions than most other samples examined. Most igneous rocks contain at least some (presumably free) C, and Hoefs (1965, p. 413) found 0.022 wt % C (and 0.050

wt % CO_2) in a split of nodule sample ER 63-33. Furthermore, a truly satisfactory extraction procedure, providing good yield without major loss and/or contamination, has not been found. A few of the difficulties have been discussed by Roedder (1965a) and are summarized on p. 528.

The low-temperature behavior of the CO_2 inclusions can provide an estimate of the purity of the CO_2. Murck et al. (1978) found, in a study of more than 200 fluid inclusions from nodules from Arizona, Hawaii, and Germany, that most contained nearly pure CO_2 but that some contained a small amount (0.05-0.10 mole fraction) of SO_2, H_2S, or COS.

If the inclusion fluid happens to have the critical density, the critical temperature provides a valid but slightly ambiguous constraint on purity. The critical temperature of pure CO_2, +31.1°C, is very specific for CO_2. Among common substances, only acetylene (36°C), ethane (32.1°C), and nitrous oxide (36.5°C) are near it. Other gases that might be found in a volcanic environment are either far below or far above it (e.g., in °C, H_2, -239.9; N_2 -147; CO -139; CH_4 -82.5; and HCl +51.4; H_2S +100.4; SO_2 +157.2; and H_2O +374). Mixtures of any of these with CO_2 would generally show critical phenomena at temperatures intermediate between those for the constituents. Although liquid water may be present in these inclusions, it is not very soluble in liquid CO_2 at room temperature (Stone, 1943, reported about 0.1 wt % at 22°C) and hence should have little effect. H_2S, HCl, and SO_2 all have strong odors, but no odors were detected on crushing large samples of the nodules in air. With one exception, all inclusions that I examined (1965a) that had critical density homogenized within 0.1° of +31.0°C.[8/] The only exception noted was a secondary inclusion in a sample from Hoher Hagen, Germany, (Fig. 17-26). Numerous measurements by others (Table 17-1) have shown critical behavior a degree or two below the critical temperature for pure CO_2 and presumbably indicate small but significant contents of impurities such as N_2, CO, or CH_4.

Raman spectroscopy has also been applied to analyze these CO_2 inclusions. Bergman (1982) analyzed gas inclusions from a series of samples from the Lunar Crater volcanic field in Nevada, USA. Special search scans were made over the characteristic bands for CO_2, CO, CH_4, H_2S, N_2, H_2, and SO_2. The only component detected in addition to CO_2 was CO, and that only in inclusions from a crosscutting vein of amphibolite; inclusions in minerals of this vein showed <12 mole % CO (and had Tm CO_2 as low as -61.4±0.3°C). Bergman and Dubessy (1984) discussed the thermodynamic significance of this CO, and of the composition of the CO_2 inclusions in general. The maximum concentrations of the other species not detected was estimated at 0.05-0.5 mole %. Perhaps most noteworthy among these results is the general absence of N_2, which might well be expected in such deep-seated gases. Although numerous studies of the amount and isotopic signature of N_2 in mantle materials have been made (Matsuo, 1984), to my knowledge, N_2 has only been reported once in mantle-inclusion gases, by R. Kreulen (pers. comm.), who found it in gases from an olivine nodule from West Germany.

Although most substances other than gases probably have negligible solubilities in CO_2 at room temperature, some lines of evidence suggest that at magmatic temperature, at least some relatively nonvolatile compounds are soluble. If this is so, CO_2-rich fluids might be involved in material transport in the mantle and in the extensive mantle metasomatism that apparently has taken place. Green (1979) showed, by ion-probe analysis, that in addition to ^{40}Ar and ^{40}MgO, the following masses were detected in CO_2 inclusions: without ambiguity, 39 (K), 47 (Ti); "suggestive evidence," 140 (Ce); and inconclusive, 1 (H), 85 (Rb), 86 (Sr), 87 (Sr), 138 (Ba), 142 (Ce), and 208 (Pb). Green suggested that these materials are present in a surface film lining the CO_2 inclusions and were formerly in solution in the CO_2. K in solution in CO_2 could explain the wide-

----

8/ These runs were made using the circulating thermostated acetone stage (Roedder, 1962a) and hence are believed to be as accurate as any other published results.

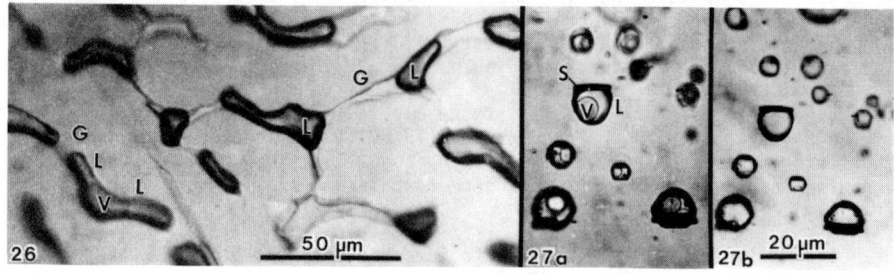

Figure 17-26. Typical partially healed fracture containing glass (G) plus CO_2 (L), in process of necking down to form individual secondary inclusions. One faint CO_2 vapor bubble (V) is visible. Although this healed fracture is typical, the inclusions are not typical, as they homogenize with near-critical behavior at temperatures above the critical point for pure CO_2. This inclusion homogenizes in the gas phase at $31.65\pm0.05°C$, indicating that constituents other than just CO_2 are present. Hoher Hagen, Göttingen, W. Germany (sample ER 63-138a). From Roedder (1965a).

Figure 17-27. Typical hemispherical secondary inclusion of CO_2 liquid (L) and vapor (V), one of many outlining a healed fracture. In several, the CO_2 is attached to a flat opaque plate, presumably of spinel (S) (or graphite?), crystallographically oriented in the enclosing olivine. Photographed at ~25°C (a) and just above Th in the liquid phase at $30.40\pm0.02°C$ (b). Some planes have dozens of such inclusions, in identical orientation with respect to the host olivine. Sample ER 63-32Z, 1801 Kaupulehu flow of Hualalai, Hawaii, USA. From Roedder (1965a).

spread evidence of K metasomatism, even without resorting to the aqueous phase proposed by Ryabchikov and Boettcher (1980) and Ryabchikov et al. (1982). Roedder (1972, plate 6) reported apparent daughter crystals in CO_2 inclusions in sapphire from Yogo Gulch, Montana, USA, and Wendlandt and Harrison (1979) found evidence of transport of rare-earth elements by dense CO_2 (see also Stosch, 1982). Similarly, Mysen (1981) reported vapor/crystal partition coefficients much greater than unity for Sm between CO_2 fluids and diopside at 20-30 kbar and 900-1100°C and Zindler and Jagoutz (1984) showed in a study of Nd and Sr isotopes that a major fraction of the nodule Ba, K, Rb, and Cs resides in minor mantle phases such as liquid inclusions and grain boundary materials. Halogens may also be present (Dreibus et al., 1980), although those reported by Shcheka et al. (1979) are presumably from later alteration. Rovetta and Mathez (1982) found possible daughter crystals(?) of magnesite and other phases in CO_2 inclusions from a lherzolite xenolith from the Canary Islands.

The equilibria between CO_2, H_2O, and ferrous silicates at high temperatures should result in the formation of some CO, H_2, and CH_4 in the inclusions, but the behavior of most of the inclusions at the critical point indicates that only very small quantities, if any, are now present. The nodules have generally been cooled rather fast, and some even come from volcanic bombs, but at the high pressures involved, internal gas reactions upon cooling might still be significant. At least part of any CO present at high temperatures probably has dissociated on cooling, according to the well-known reaction $2CO = CO_2 + C$. Mathez and Delaney (1981) have proposed that such breakdown of CO-bearing gases has precipitated C on many rock surfaces; such precipitates may be of importance in controlling the electrical conductivity of mantle materials (Duba and Shankland, 1982). The gases inside inclusions in olivine could thus precipitate C; the dark borders seen in some CO_2 inclusions (Fig. 17-2) or some of the odd opaque plates found in some planes of secondary inclusions (Figs. 17-25,27) could be graphite. These plates were tentatively identified as spinel rather than graphite in part on the basis of the crystal faces visible on some of the plate edges (Roedder, 1965a, p. 1749) under asymmetric dark-field illumination; graphite flakes do show such prism and pyramid forms, but not commonly. Another line of evidence favoring spinel was the occurrence of dark-greenish-brown isotropic cubic or octahedral crystals, epitaxially arrayed relative to the host olivine, in many of the associated silicate-melt inclusions. Even if the opaque grains are graphite, however, the bulk of the CO_2 could not be from the dissociation of CO, as many optically clear inclusions have no opaque material visible.

As H_2O is more soluble than CO_2 in basaltic melt, a separate immiscible CO_2-rich phase could form and be trapped in the growing crystals, while the bulk of the water stayed in the liquid melt. Only one locality yielded CO_2 inclusions containing recognizable H_2O (Fig. 17-5; Roedder, 1965a, Figs. 13, 14). Suzuoki et al. (1975) reported analyses for H in olivine from several peridotites. The amounts they found correspond to 4.7 and 0.6 ppm H_2O, but these values are for total H, and hence include H_2O in melt inclusions, etc. Obviously, the CO_2 inclusions must not be considered a "sample" of the volatiles present in the magma, but only of those evolved as a separate phase under those conditions, perhaps modified by internal reactions on cooling. (Walther and Althaus (1983) believe that they have evidence that the inclusions may have had more water originally but have lost it during the ascent.)

The noble gases might be expected to partition into a CO_2 phase, once it forms. Kurz and Jenkins (1981) showed that He does partition strongly into the vesicles in glassy basalt. The concentrations of He, Ne, Ar, Kr, and Xe could be of importance in a variety of geochemical studies (e.g., Thomsen, 1980; Staudacher and Allègre, 1982; and review by DePaolo, 1983). Inclusion-rich sample ER 63-33a was found to contain 5.5×10^{-7} cc/g STP of radiogenic Ar (Roedder, 1965a), and Xe has been determined in a series of ultramafic rocks; how much of this gas comes from the CO_2 inclusions is unknown. Studies of the isotopic ratios of He, Ne, and Ar in xenoliths and phenocrysts (Kaneoka and Takaoka, 1980; Kyser and Rison, 1982) and from basalts (e.g., Kurz et al., 1981; Hart et al., 1983) have shown that mixing of sources is very likely and that partial outgassing and mass fractionation processes have occurred, but Fisher (1983) has reported incomplete retention and significant atmospheric contamination of noble gases in such samples. Suzanne Wass (pers. comm., 1983, unpublished data from K. O'Nions' laboratory) has found a high $^3He/^4He$ ratio, indicating derivation of ^3He from a primordial reservoir. The He/Ne ratio varies widely in different nodule types from an Australian locality, suggesting different sources for the various types.

The composition of the silicate-melt inclusions associated with the CO_2 inclusions, and even in the same inclusions with the CO_2, has been studied very little. Roedder (1965a) found that the index of refraction of the glass in some was 1.61, suggesting a basaltic composition. Most melt inclusions had no daughter crystals. Shcheka et al. (1978) reported that the glass inclusions in the rims of nodules were of different composition from those in the cores. Murck et al. (1978) reported seven electron microprobe determinations of the composition of the glass lining CO_2 inclusions. When corrected for crystallization of host olivine or pyroxene on the walls during cooling, the composition as trapped was found to be that of a high-alumina basalt. Irving and Mathez (1982) suggest that the andesitic glass inclusions in such nodules form by "flash melting" on decompression during ascent.

Olivine crystals from ultramafic nodules in various parts of the world are sometimes of gem quality (the cut stone is called "peridot"). Relatively small cut or tumble-polished peridots in the jewelry trade normally come from such nodules and hence have inclusions of spinel, glass, and/or CO_2. However, most large gem peridots come from a single locality, on the island of Zabargad (or St. Johns) in the Red Sea, and show an entirely different type of inclusion. As reported by Clocchiatti et al. (1981), these peridots formed hydrothermally, from hypersaline fluids at 750-900°C. The partial pressure of CO_2 in these fluids was estimated to be only 0.4-0.8 kbar.

The $^{13}C/^{12}C$ ratio in the CO_2 of the inclusions from ultramafic nodules would be of obvious value in tracing the relationship of the CO_2 fluid to carbonatites, alkalic magmatism, and even diamonds, and might shed some light on the possibility of crustal carbonates as the source for the CO_2 (Glassley, 1983a). The significance of the $^{13}C/^{12}C$ ratio and its variations has been reviewed by Bergman and Dubessy (1983). Unfortunately, the analysis of this ratio is far more difficult than it might seem, as a result of a combination of problems:

(1) small quantity of CO_2 present; (2) difficulty in obtaining essentially quantitative release; (3) significant quantities of magmatic(?), free(?) C in the sample, as graphite or as carbon atoms; (4) significant quantities of contaminant (surface?) C in samples; and (5) significant adsorption (and isotopic fractionation) on any new olivine surfaces formed. Several workers have found by acid leaching techniques that most of the total carbon in these nodules is on the surface rather than within the silicate grains (e.g., Mathez et al., 1982). O'Neil and Jackson (1969) found that carbonate present in Hawaiian garnet peridotites and nephelinites, originally thought to be primary, had C isotope ratios typical of freshwater carbonate. R. Kreulen (pers. comm.) reported that after evaluation of laboratory data on the various factors, he believed that a value for $\delta^{13}C$ of -2.4 per mil vs PDB is probably reasonable for the CO_2 from Hawaiian dunite nodule sample ER 63-33, and others have found similar values (e.g., Watanabe et al., 1983, obtained $\delta^{13}C$ values of -3.2 for total C (released by oxidative pyrolysis) and -26.9 for graphitic carbon (residue from HCl solution), and Matsuo, 1984, reported -3.4 and -4.7 for C as CO_2, and -26.9 per mil for graphitic carbon in olivine from this same locality). In contrast, D.M. Thomas (1979) found -12 to -14 per mil, O'Neil (1971) found +0.6 per mil, and M. Sommer (pers. comm.) estimated ~-28 per mil for C extracted from presumably similar inclusions from this same locality. C from other presumably deep sources is generally in the same range:

Source	$\delta^{13}C$ per mil (PDB)	Reference
"Phreatic" explosions, Alaska	- 6.4	Barnes & McCoy (1979)
Submarine seeps, Alaska	- 2.7	Kvenvolden et al. (1979)
Granulites, Bamble, Norway (in silicate network)	-13.9	Pineau et al. (1981)
Alkalic intrusives, CH_4, etc.		
Ilimaussaq, Greenland	- 7.7	Petersile & Sörenson (1970)
Khibiny, USSR	-3.2 to -14.6	Petersile & Sörenson (1970)
Lovozero, USSR	- 5.3	Petersile & Sörenson (1970)
Presumed deep CO_2, Naxos, Greece	-16 to ~-5	Kreulen (1980)
Mid-Atlantic Ridge basalts	- 7.6	Pineau et al. (1976)

Depth of Formation of Inclusions vs Origin of Host Mineral

Fluid inclusions may be used to obtain either the temperature or the pressure of trapping, but rarely both. Normally, the laboratory-measured Th is used to determine the isochore on an appropriate P-T diagram, and the true Th is obtained from the intersection of this isochore with the pressure obtained by some independent method. The CO_2 inclusions in nodules permit this procedure to be reversed. The density of supercritical CO_2 at basaltic eruption temperatures and high pressure is strongly pressure dependent but relatively insensitive to temperature (Figs. 8-8,10). Because the magma temperatures could not have changed greatly during the last 20 km of upward movement (the only part that is pertinent to these inclusions, as discussed below), I assume that the trapping temperature was about 1200°C, as in most basaltic volcanism.

This assumption, available P-V-T data on CO_2 (Figs. 8-8,10), and an estimate of the density of CO_2 at the time of trapping, permit an estimate of the pressure at the time of trapping. Except for some relatively minor reversible dilational changes (see Chapter 3), the volume of the inclusion, and hence the bulk density of the CO_2, is essentially the same now as at the time of trapping. (Kirby and Green (1980) have suggested that olivine can easily deform plastically during eruption, leading to incorrect densities and hence pressures, and transmission electron micrographs of some minute inclusions in olivine show dislocation loops indicating expansion. However, the petrographic evidence on the much larger inclusions reported here suggests no recognizable expansion. Much more extensive TEM studies are in order.) The approximate density of the CO_2 originally filling

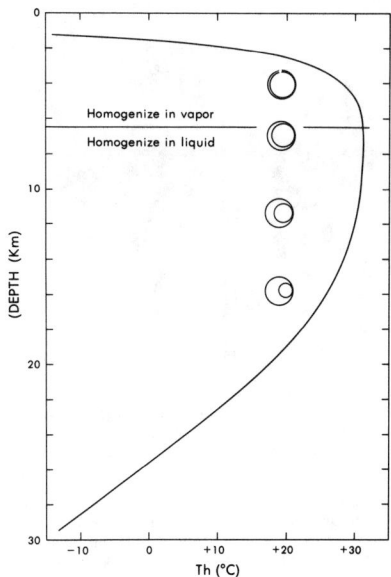

Figure 17-28. Plot of Th of CO_2 inclusions vs depth of trapping. Based on three assumptions: (1) inclusions contain pure CO_2; (2) inclusions were trapped at 1200°C; and (3) pressure was that of a hydrostatic head of basaltic magma with density = 2.7 g/cm³, plus 150 bar seawater pressure (appropriate for Loihi Seamount, Hawaii, USA). The appearance of four spherical inclusions at 20°C, and trapped at the depths indicated, is shown. From Belkin et al. (1984).

the inclusion can be determined from optical measurements to yield estimates of the volumes of the liquid and vapor. As the CO_2 is apparently fairly pure, a more accurate measure of the density of the CO_2 can be obtained from the nature and temperature of homogenization, except when the inclusion has near-critical density. If the density was found to be 1.0, Figure 8-8 indicates that the pressure at the time of trapping of that specific inclusion was 8 kbar. This is not a maximum or minimum pressure -- if the temperature estimate is correct, the inclusion must have formed at that pressure. If this pressure is assumed to be hydrostatic[9], from a column of basaltic lava of 2.7 g/cm³, the depth must have been 30 km. Belkin et al. (1984) presented a plot (Fig. 17-28) showing the relationship between Th CO_2 L-V and depth; as this plot was actually made for use at the Loihi Seamount, which is below the sea, it has been corrected for 150-bar water pressure and hence would not be accurate for use on land.

When numerous inclusions in a given suite of nodules are examined, however, or even those from a single nodule, the density is found to vary widely. If different primary inclusions yield different pressures, and if these differences are related to the mineralogy in some systematic way, they must be accepted as indicating mineral growth at different depths. Belkin et al. (1984) found this situation for spinel crystals in nodules from Vesuvius, formed relatively near the surface (i.e., 3-16 km depth). At Loihi, Roedder (1983b) found that primary inclusions in new olivine crystals forming during the eruption on some nodules at the interface with basalt contained lower densities of CO_2, corresponding to shallower depths, than even some secondary inclusions (Fig. 17-29). If primary inclusions reveal a random variation, partial decrepitation and healing at the new pressure may have taken place during eruption. Evidence of such decrepitation is most apt to be seen in the larger inclusions (Figs. 17-30, 31), though

----

[9] If one assumes that the magma at the point of inclusion trapping was an isolated pocket, under lithostatic pressure from rock with density = 3.3 g/cm³, the depth equivalent to a given CO_2 density will be 18% lower. Thus an inclusion calculated to have been trapped at 16 km under hydrostatic pressure could have been formed at ~13 km under lithostatic pressure.

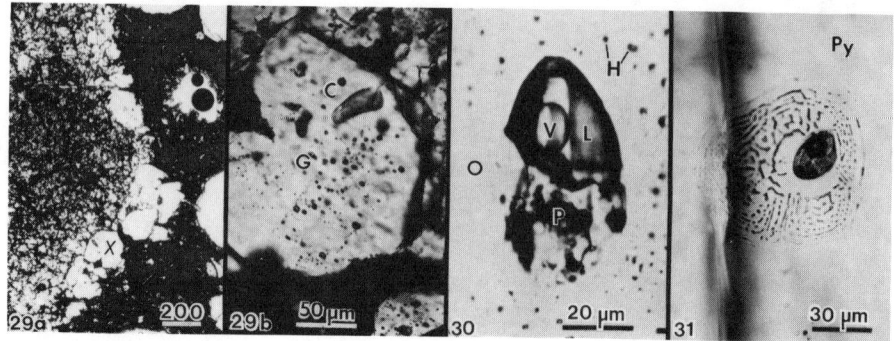

Figure 17-29a,b. Figure 17-29a shows a fine-grained xenolith in vesicular basalt (right side). New coarser olivine crystals have grown around the edge of the xenolith. Sample KK-27-4A, from Loihi Seamount, Hawaii, USA. Figure 17-29b is an enlargement of new olivine crystal X from Fig. 17-29a, showing primary silicate melt (G) and CO_2 (C) inclusions. From Roedder (1983b).

Figure 17-30. Presumed primary melt inclusion in olivine (O), now containing numerous birefringent crystals (P), possibly with glass, and liquid (L) and gaseous (V) CO_2. The two CO_2 phases homogenize in the liquid phase. A halo of tiny inclusions (H) surrounds the central inclusion, presumably indicating fracturing and partial leakage during pressure drop on ascent. From an olivine phenocryst (xenocryst?) in olivine leucite nephelinite, Mosenberg, Westeifel, W. Germany, sample ER 64-123. From Roedder (1965a).

Figure 17-31. Photomicrograph of melt inclusion in pyroxene (Py) from cumulate nodule N81 from Lagno Di Pollena, Vesuvius, Italy (<79 AD). Presumably this inclusion contained high pressure CO_2, which caused it to decrepitate during eruption. The fracture that formed permitted loss of some pressure to the adjacent major crack (vertical black zone), and was subsequently healed, trapping the the halo of small secondary inclusions. From Belkin et al. (1984).

it is most common for all larger inclusions to be completely empty.[10] Small inclusions may still persist, however, and provide accurate data, but the densest inclusions, representing the greatest pressures, may appear empty unless cooled. Partial decrepitation requires that the healing of the fracture must take place during eruption; attempts are being made to quantify this crack-healing process (D.L. Smith et al., 1982; Wanamaker et al., 1982), but as surface diffusion may be involved, setting up experiments under truly appropriate conditions (olivine at 1200°C, internal CO_2 pressure much greater than the ~8 kbar confining pressure, and possibly some silicate melt with the CO_2) will be difficult. The various degrees of maturity shown by the healed fractures provides qualitative evidence of time since healing first occurred, but this maturation may be even more difficult to model with validity.

When the inclusions in any given grain are considered, the apparently primary inclusions generally have a higher density (and hence greater depth of origin) than the secondaries. Apparent exceptions may be a result of partial decrepitation. As many planes of secondary inclusions have CO_2 densities equivalent to <~8 kbar internal pressure, but not higher, this 8 kbar seems to be approximately the upper limit that a healed fracture in olivine can stand on eruption.

Since CO_2 inclusions with densities of 1.2 have not even been reported, and since most larger inclusions with densities >0.9 seem to have decrepitated during eruption, it is obvious that in a study of the products of basaltic volcanism, the CO_2 inclusions we see now may be a biased sample. Any inclusion

----

[10] There is considerable evidence that some inclusions have leaked, as the nodules are usually very friable, and commonly the lava surrounding the nodule is partially separated from it by a layer of gas vesicles. As olivine phenocrysts in the same lavas generally do not have attached gas bubbles, one can assume that these vesicles represent gas evolved from the nodules, rather than merely collected from the lava.

with a density corresponding to >~8 kbar either must be very small or have been erupted at less than 1200°C to avoid decrepitation. Minerals stronger than olivine could also increase the pressure limit. Thus, spinel may be stronger than olivine, and diamond would be even better. As a result of this bias, nodule samples that may have formed at >100 km depth may have no inclusions, primary or secondary, representing trapping at depths >25 km. But they may also have primary inclusions in new crystals, grown during ascent, at even lower pressures. Care is needed in the interpretation.

Interpretation of the depth of origin of the possible exsolution inclusions is particularly difficult. As a result of its good cleavage, one might expect orthopyroxene to decrepitate more readily than olivine. Yet the presumed exsolution inclusions found in it are rather large (e.g., Figs. 17-4, 24), and the silicate exsolution features, if they are from exsolution and not from simultaneous growth, are coarse, representing long annealing times. Green and Radcliffe (1975) believed that the minute inclusions they found on grain boundaries, deformation defects, and exsolution features consisted of CO_2, and formed as solid-state precipitates. Freund et al. (1980) reported very appreciable amounts of C in xenolithic olivine, as atomic C in solid solution. Such C might exsolve at lower pressures (taking O from ferric iron or other variable-valence elements) and form the xenolithic CO_2 inclusions, but the apparent amounts shown in Figure 17-24 seem to be rather large for this mechanism. Selverstone (1982) has proposed this origin for some CO_2 inclusions in Chilean xenoliths. The very coarse pyroxene exsolution blebs (and the possibly exsolved spinel) suggest very long reequilibration times, so perhaps there was time for coalescence of CO_2 into these larger inclusions. The density data on these CO_2 inclusions (see Fig. 17-24 inset) shows that if such annealing took place, at Loihi, at least, it was at a depth of only ~10-15 km.

Other Petrologic Aspects

Vesiculation and volcanism. The CO_2 in the inclusions constitutes only an extremely small part of the total volatiles erupted from any given vent. Only a small part of the separate immiscible CO_2 phase at depth happened to become trapped in sufficiently strong olivine "pressure vessels" to hold it upon eruption. A much larger part of these minute globules presumably stayed suspended in the saturated liquid basalt that carried the nodules up to the surface. The expansion of any such gas bubbles along grain boundaries within a nodule should help in nodule disaggregation.

These globules represent incipient vesiculation, even at >10 km depth, and should prevent any supersaturation. Some have stated that it is unlikely that magmas at depth are often saturated with gas and cite as evidence the lack of visible vesicles in deeply eroded basalt dikes. Even if the magmas forming these dikes had been saturated with respect to a CO_2 fluid, several factors tend to minimize the significance of this evidence: (1) unless formed by second boiling, upon crystallization, the vesicles we see in a dike are only those bubbles that were accidentally trapped during their ascent, and the entire dike filling may have been rather effectively degassed earlier; (2) in glassy dikes, at least some of the gases evolved on pressure decrease at essentially constant temperature during ascent may have been resorbed by temperature decrease at constant pressure during solidification in place; and (3) "vesiculation" on a scale as indicated by the primary CO_2 inclusions in olivine would be difficult to recognize, as a CO_2 globule 10 μm in diameter at a depth of 16 km would expand to only 25 μm radius at 0.6 km depth.

During eruption, 20 μm diameter globules will have a negligible rate of ascent relative to the enclosing magma, but as the pressure decreases during ascent, the globule radius will increase, from simple expansion and from further outgassing of the saturated magma with pressure drop. The rate of rise of such

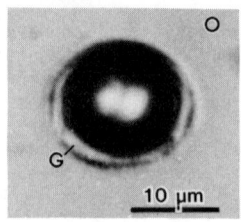

Figure 17-32. Gas inclusion lined with a thin film of glass (G), in olivine (O) phenocryst from basalt surface crust of Kilauea Iki lava lake, Hawaii, USA (sample ER 63-86). Most (but not all) of these inclusions show a vacuum on crushing.

bubbles, according to Stokes law, is a direct function of the <u>square</u> of the radius. Thus, the processes of expansion (resulting in a lower bulk density, which is controlled by the ambient pressure, and in an expulsion upward of an equal volume of magma), and of bubble rise through the magma, become self reinforcing and presumably result in vigorous lava fountaining and other volcanic phenomena.

Origin of phenocrysts. CO_2 inclusions in phenocrysts, either primary or secondary, can be used to establish a minimum depth for phenocryst crystallization and may permit distinctions between phenocryst and xenocryst. Although xenocrysts from the disaggregation of nodules are common, the large (10 mm) sharply euhedral titanaugite phenocrysts in basalt from Antarctica that I studied (1965a, sample ER 63-169) showed definite planes of secondary liquid CO_2 inclusions[11], and hence must have formed at considerable depth.

Some of the glass-plus-gas inclusions in the olivine phenocrysts in the lavas of the Kilauea Iki (Hawaii, USA) eruption of 1959 (Fig. 17-32) contain gas under pressure, and others contain essentially a vacuum bubble. The latter are common in volcanic samples and presumably have formed by the condensation of steam-rich gas bubbles trapped very near the surface, but as the hydrostatic pressure increases at ~1 bar/3 m, noncondensable gas inclusions that have moderate pressures at room temperature can form at a depth of a few meters.

High-temperature tensile strength of silicates. Those inclusions that do not decrepitate on eruption at the surface (at ~1200°C) have been exposed to 1 atm external pressure while holding 2.5-5 kbar internal pressure. This fact places a crude minimum on the ultimate high-temperature tensile strength of the enclosing minerals. With the assumption that these small "pressure vessels" behave like infinitely thick-walled containers, the tensile stress in the walls should be approximately equal to the internal pressure, i.e., ~4.7 kbar at 1200°C. Although room-temperature tensile strengths over 68 kbar have been determined for silica glass and single crystals of various materials, the tensile strength shown by olivine is far in excess of that of most materials at 1200°C. Only a very small volume of the olivine crystal surrounding an inclusion is under maximum stress, and it probably has few if any major dislocations, as those inclusions that had weaker walls probably decrepitated on eruption. However, many of the inclusions that do withstand eruption -- and subsequent reheating to 1200°C or more -- occur in planes of secondaries along healed fractures. Thus, not only is their stress somewhat additive, but the olivine holding the two parts together has grown as a fracture healing. Although this olivine presumably contains many dislocations and even discontinuities, it withstood the tension. Also, since these inclusions (in olivine, pyroxene, and plagioclase) have withstood heating to about 1200°C over their filling temperature without decrepitation, a "decrepigram" of such material would have a minimum correction of 1200°C for "overshoot."

----

[11] This sample also showed one plane of several dozen secondary inclusions in a large xenocrystic(?) olivine crystal, each inclusion containing a birefringent crystal, glass, and liquid and gaseous CO_2, all in apparently <u>uniform</u> ratios, implying that a homogeneous fluid was trapped. This observation has not been reconciled with field and laboratory data indicating low solubilities for CO_2 in basic melts.

Chapter 18

EXTRATERRESTRIAL ENVIRONMENTS

"The carbonaceous chondrites have been around for more than 4 billion years;
I suspect they will remain enigmatic, at least in part, for a few more years"
(Roedder, 1981f, p. 350).

CONTENTS

INTRODUCTION

Except for some earlier studies of the gas bubbles in tektites, and an occasional mention of glass inclusions in the minerals of some meteorites, the study of fluid inclusions in extraterrestrial samples did not really start until the return of the first Apollo samples in 1969. These various lunar studies have provided some surprises and have revealed much about the many lunar processes represented by the available samples. However, the extremely inadequate sampling density must be constantly kept in mind. Various authors have tried to describe how little would now be known of terrestrial geology if the same numbers of samples, taken from the same number of localities, plus the same amount of "time in the field" were all that we had from which to learn about the Earth. In spite of this very serious limitation, however, the study of the lunar samples has provided a wealth of information on the Moon, and the study of the inclusions has added its share of information on certain of the processes involved. These lunar inclusions also provide some important lessons that should not be ignored in the study of terrestrial-melt inclusions, particularly the misleading effects of preferential wetting, and gross differences in inclusion composition from problems of nucleation and from differences in the host mineral.

A rather extensive planning exercise was carried out in 1973 by the U.S. National Aeronautics and Space Administration (NASA) to consider what sample-handling protocol might be appropriate if a soil sample were to be returned to earth from Mars by an unmanned probe. The major question was what sort of sterilization procedure would provide the maximum assurance of biological safety on bringing the sample to Earth (in case viable microorganisms were originally present), and the minimum degradation of the sample for various geochemical and other scientific studies. A series of five "simulated Martian soils" consisting of terrestrial and meteoritic materials was subjected to one of the suggested sterilization protocols (275°C for 1 day in He at 1 atm), and then the treated and untreated samples were examined and compared, to see "how much science would be lost" by such a sterilization. In this test I found (Roedder, 1974) that this sterilization had almost no noticeable effect on the significance of the wide range of available data on the geologic history of these samples that could be obtained from a study of their fluid inclusions. These results, however, cannot necessarily be extrapolated to higher temperatures or to other types of inclusions or sample materials. An even more important result of this trial run was the evidence that a surprisingly large amount of geologic information concerning the source area can be derived by the nondestructive study of the 1-5 μm fluid inclusions in <1 mg of soil, sterilized or not. Much of this information can be obtained even if the soil is very fine grained (the minimum would be ~10 μm grains). In addition, the quantity of material from which some usable inclusion data can be derived, nondestructively, may even be in the microgram range.

After the investigations of the lunar samples, inclusions in meteorites came under study, in part as a result of an hypothesis that some meteorites may have originated on the Moon. In addition to the well-known and expectable glass inclusions, inclusions of aqueous fluids have been found in a series of meteorites (Fieni et al., 1978). This discovery was completely unexpected and still remains thoroughly enigmatic after several years of study.

The current status of research on inclusions in extraterrestrial materials is not very satisfactory. As will be evident in the tentative interpretations in this chapter, too many inexplicable aspects remain, but the field is young. Perhaps a second decade of research will provide a more cohesive and satisfying body of knowledge.

INCLUSIONS IN THE LUNAR SAMPLES

Inclusions of Homogeneous Melts in the Mare Basalts

At first glance, a section of a lunar mare basalt resembles many terrestrial basalts, but some of the differences are significant. I joined the group studying the samples brought back by Apollo 11 in the hopes of finding H_2O or CO_2 inclusions. If present, such inclusions could be of considerable importance for two reasons. First, they might tell us much about the processes that have occurred on the Moon. Second, if a permanent manned lunar base is ever to be established, a lunar source of H_2O and CO_2 would greatly simplify the logistics, since it would permit local production of food and oxygen. Fluid inclusions, released by simple decrepitation, might be adequate. Unfortunately, I was unable to find even a single minute H_2O (or CO_2) inclusion in these samples. They show none of the late-stage alteration features so common in terrestrial rocks (particularly in association with ore deposits); serpentine, actinolite, epidote, chlorite, sericite, secondary aqueous fluid inclusions (and the ore deposits) are missing completely. We know now that the lunar samples are the driest rocks known to man.[1] However, they do contain several interesting and seemingly novel types of silicate-melt inclusions. The discussion of inclusions in the mare basalts is best split into two parts -- inclusions of homogeneous melts (in the phases crystallizing early), and inclusions of heterogeneous (i.e., immiscible) silicate melts (in the late phases).

Inclusions in early olivine. Melt inclusions in the lunar basalts are most obvious in olivine (Roedder, 1979a; Roedder and Weiblen, 1970a,b,c, 1971, 1972a, b, 1973a,b, 1978; Weiblen and Roedder, 1973). The absence of the serpentinization so common in terrestrial olivine makes lunar olivine resemble the much more abundant pyroxene in these mare basalts, but the melt inclusions help in distinguishing the two. Except for very late stage crystals (see below), lunar pyroxene is essentially free of melt inclusions, but olivine grains commonly have one or more inclusions, sometimes relatively large. Even small melt inclusions in olivine have formed a vapor bubble on cooling, and many also have formed a globule of an immiscible sulfide melt (Fig. 18-1). The relatively large size of the vapor bubble in these inclusions, too large to be from differential shrinkage alone, presumably results from the crystallization of considerable olivine on the inclusion walls after trapping. A few show a globule of sulfide melt that is also too large to have exsolved from the amount of melt trapped in the inclusion, suggesting that in these particular samples immiscible globules of sulfide were present at the time of growth of the enclosing olivine (Roedder and Weiblen, 1978, their Fig. 2).

The large size of many of these melt inclusions relative to that of the host olivine grain is unusual. Some olivine grains contain inclusions that are one fifth the size of the host crystal (Fig. 18-3), and a few are one half the size (Fig. 18-2), presumably from growth on an original skeletal crystal, formed by rapid growth after nucleation in a strongly supersaturated magma (Fig. 2-3a,c).

The daughter crystals are the most striking feature of the melt inclusions in olivine. The most common daughter crystal is ilmenite (identified by electron microprobe analysis), which almost always forms a single thin plate, flattened parallel (0001), and arranged parallel to (100) of the enclosing olivine (Fig. 18-2). Only one or two such ilmenite plates have been otherwise oriented among the many thousands that I have examined. The ilmenite daughter crystal has nucleated epitaxially on the host olivine wall. Large inclusions may have numerous parallel ilmenite plates, yet adjacent inclusions, of differing bulk composition, may have none (Fig. 18-4). The host olivine grain for the particular inclusions shown in Figure 18-4 also contains several of the rather rare inclusions of low-Ni metallic Fe that have been found in lunar olivine phenocrysts (Fig. 18-4a), presumably signifying the presence of an immiscible metallic Fe liquid at the time of crystallization of the olivine.

In addition to the epitaxial ilmenite, sulfide-melt globule, and vapor bubble, the remaining melt in many of the inclusions in olivine has formed a mat of small pyroxene crystals (Fig. 18-5), and, in some, another parallel epitaxial plate of a transparent mineral. Electron-microprobe analyses of this transparent plate show it to be anorthite. Although common, it occurs in a smaller percentage of the inclusions than does ilmenite, and is almost never

----

[1] Although no aqueous or CO_2 inclusions have been found, the needed H_2O and CO_2 might still be obtained from lunar soil, which contains some C, and and also contains significant amounts of H, implanted by the solar wind (Roedder, 1981a).

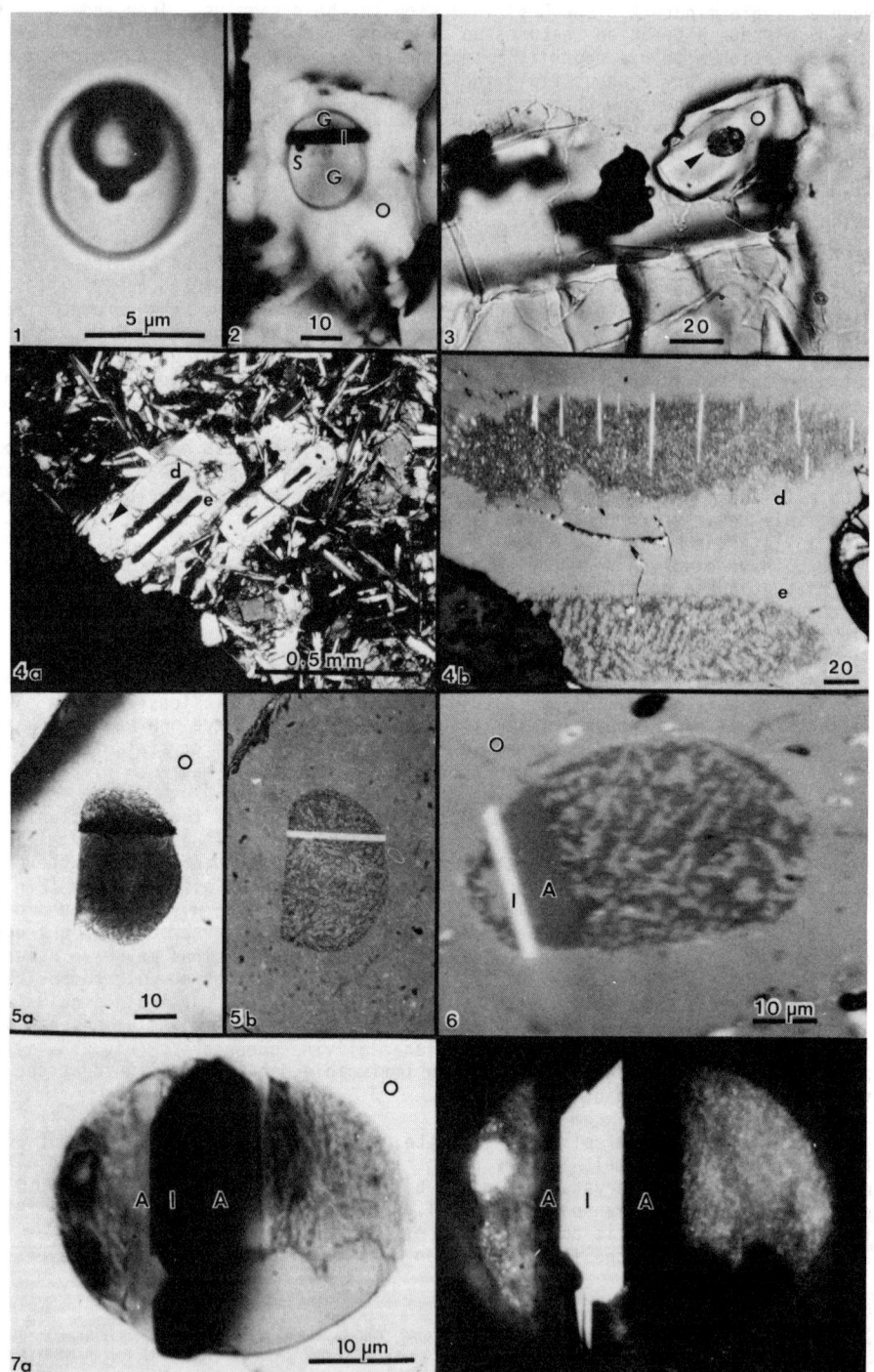

present without ilmenite. Most inclusions containing both phases show the two plates parallel and in contact, so I cannot tell whether the anorthite nucleated and grew epitaxially on olivine or on ilmenite (Figs. 18-6,7).

The ilmenite plate commonly appears to project into the wall of the host olivine (Figs. 18-3,5,6,7), but the anorthite does not. This apparent projection is a result of the nucleation of the ilmenite crystal on the wall at an early stage, when the inclusion was larger, before significant precipitation of olivine from the trapped melt (see also Fig. 3-1). The anorthite apparently nucleated later, after most crystallization of olivine from the inclusion melt was complete. The ilmenite plate apparently grew completely across the inclusion, thus isolating the two parts of the melt, as these two parts frequently show different daughter phases. Thus, one part may nucleate a shrinkage bubble, and the other, a mat of feathery pyroxene crystals (see also Fig. 16-4). Some inclusions show a plate of anorthite on each side of the ilmenite plate (Fig. 18-7b). Some of the larger plates of ilmenite show diagonal lamellae of an unidentified phase that is bluish in reflected light (e.g., that shown in Fig. 18-7b).

Tm of these various daughter phases can used as evidence on the sequence and temperatures of various phase changes that presumably occurred in these inclusions -- and perhaps in the host magma -- during cooling. The following procedure (from Roedder and Weiblen, 1970b) was developed and used on the Apollo 11 samples; similar studies were made of olivine from samples from Apollo 12 and terrestrial localities.

The inclusions in olivine required such high magnification (500-1250x) and such a high-numerical-aperture condenser that direct observation by high-temperature microscopy was not feasible. Instead, a series of heating experiments was performed, using the quenching technique to permit observation at room temperature. Also, because normal doubly-polished plates were unavailable and the probe mounts could not be destroyed, individual tiny grains of olivine (~0.2 mm) were used. Such grains, containing suitable inclusions, were selected from an uncovered oil mount of olivine fragments. Each selected grain was then mounted

Figure 18-1.   Small melt inclusion in olivine from lunar basalt 12018,83, containing only glass, an exsolved immiscible sulfide-melt globule, and an oversize vapor bubble, presumably resulting from the crystallization of considerable olivine on the walls after trapping. From Roedder and Weiblen (1971).

Figure 18-2.   Relatively large melt inclusion in the center of a small olivine grain (lighter colored; O) from lunar basalt 10071,32. Inclusion contains glass (G), opaque epitaxial ilmenite plate (I), and a small exsolved immiscible sulfide globule (S). Scale bar in μm. From Roedder (1971c).

Figure 18-3.   Small olivine grain (O) in ferrobasalt fragment 24109,57-6 (from Luna 24 mission), containing a relatively large crystalline melt inclusion (arrow), in transmitted light. Scale bar in μm. From Roedder and Weiblen (1978).

Figure 18-4.   (a) Two olivine phenocrysts in lunar olivine basalt 14321,25, each with several large silicate-melt inclusions (transmitted light; crossed polars). The arrow points to two apparently primary inclusions of metallic Fe and high-Si melt. (b) An enlargement of the central part of (a) in reflected light. Scale bar in μm. Note multiple epitaxially oriented ilmenite plates and fine-grained matrix of inclusion (d) and the coarse random dendrites of inclusion (e). These two inclusions have different bulk compositions (Roedder and Weiblen, 1972b, Table 1, analyses 2 and 3).

Figure 18-5.   Typical large crystallized melt inclusion in olivine (O) from low-Ti lunar basalt 12035,23, with epitaxial ilmenite plate in transmitted light (a) and reflected light (b). Scale bar in μm. From Weiblen and Roedder (1976).

Figure 18-6.   Crystallized melt inclusion in olivine (O) from lunar basalt 12036,8 in reflected light. Epitaxial daughter crystal plates of ilmenite (I) and anorthite (A) are visible. The bulk of the inclusion consists of pyroxene dendrites (light grey) in glass. From Roedder and Weiblen (1972b).

Figure 18-7.   Primary silicate-melt inclusion in olivine (O) from lunar basalt 10020, in transmitted light (a) and reflected (b). This relatively large inclusion (32 μm long) of an originally homogeneous melt has crystallized to yield a plate of ilmenite (I) and two parallel plates of anorthite (A) (on each side of the ilmenite). The glass in each end has crystallized to a feathery mass of pyroxene. From Roedder and Weiblen (1970b).

in a matching index liquid for detailed microscopy and photography. Several annoying problems in this operation were avoided by making these mountings in a 3 x 3 mm, three-sided "corral" of the appropriate thickness of wire covered with a similar size of an extra-thin (grade "00") coverglass. The open side permitted the insertion of the thin end of a single hair from a camel's hair brush to turn the grain over or move it for best observation and photography. The grain was then washed with acetone, loosely folded into a tiny Fe foil envelope, and placed in a fused silica tube which was then evacuated and sealed. The tube was hung at the center of the hot zone in a Pt-wound vertical tube furnace, controlled through a bridge circuit using the furnace-element resistance as a temperature sensor for control. Sample temperatures were measured potentiometrically with a Pt/Pt 10% Rh thermocouple, previously calibrated in the same furnace at the melting points of pure NaCl (800.4°C), pure Au (at that time considered to be 1062.6°C), and pure synthetic $CaMgSi_2O_6$ (1391.5°C). The reference junctions were held at 0.0°C. At the end of the run, the silica tube was quenched in water (visible glow from the Fe foil inside disappeared in 1-2 sec), and the grain removed, remounted, and reexamined. The inside of the iron foil envelopes remained bright and clean. Many runs were made simultaneously on new unheated grains and previously run grains, providing checks on the attainment of equilibrium. Although 1 hr at temperatures near the liquidus seemed to be adequate, most runs were in the range 4 hr to 2 days.

The results obtained on 137 runs on 80 different inclusions were remarkably consistent, indicating a uniform environment of trapping, as is also shown by the uniformity of the original phase assemblage in most of the crystallized inclusions. The only apparent inconsistencies in the laboratory experiments stem from failure to nucleate, at furnace temperature, a given phase in a previously homogenized inclusion, and from the nucleation, during quenching, of a vapor bubble in some of those inclusions that probably had none at furnace temperature.

Although all these runs were made on olivine from sample 10020-41, similar-appearing inclusions in the olivine of rocks 10003, 10045, 10071, and 10072 presumably should yield similar results. The phase changes that occurred in these inclusions are detailed below, starting with the lowest temperatures. The temperatures stated were reproducible, using new inclusions, to about ±5°C.

Beginning with a virtually holocrystalline inclusion, at some temperature below 1065°C, a sufficient amount of intergranular liquid formed to permit visible recrystallization of the pyroxene(?) mat in a matter of hours. After 19 hr at 1065°C, this recrystallization resulted in a coarsening and lightening of the originally almost opaque mass. This intergranular liquid has not been observed but is postulated on the basis of the recrystallization, which seems far too fast to have resulted only from the thermal acceleration of solid-state reactions, and on the basis of some experiments on synthetic melts. Thus, the solidus lies below 1065°C.

As temperature increased, the amount of intergranular liquid increased rapidly and became plainly visible as glass in the quenched product at 1085°C, and at 1100°C, the size of the plagioclase daughter crystal had visibly declined (Fig. 18-8B). Further heating resulted in the complete melting of plagioclase by 1110°C (Fig. 18-9B). The single crystals of ilmenite seemed to be only slightly dissolved at 1110°C. The mat of fine pyroxene crystals recrystallized rapidly to yield individual equant crystals >10 μm in diameter within hours at 1100°. Only a few percent of these pyroxene crystals remained at 1125°C, and none were present at 1145°C, so the last is assumed to melt at about 1130°C. The small amount of ilmenite remaining dissolved as the temperature increased, the last dissolving at 1210°C (Fig. 18-10). This phase may still be ilmenite at this stage, but could now be armalcolite, particularly at this temperature. This and the other temperatures given are all for olivine-saturated melts, as they were determined in olivine "containers"; the natural melts were also presumably olivine-saturated, however. A significant amount of olivine dissolved in these melts, particularly near 1200°C, because the inclusion cavity became visibly larger. The amount of olivine dissolved cannot be estimated with precision, as solution and redeposition cause considerable changes in the shapes of the inclusion (Fig. 18-10).

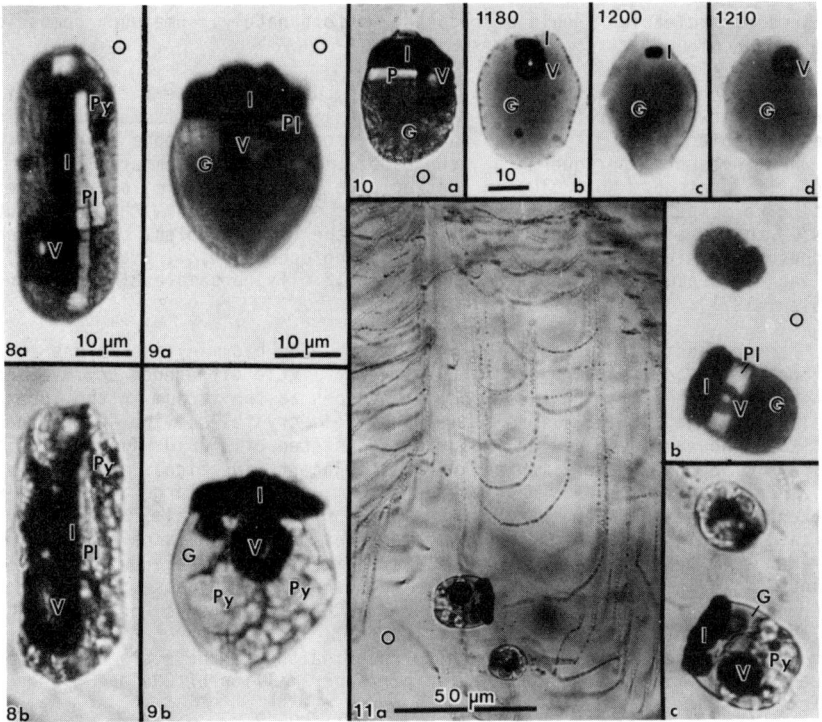

Figure 18-8. Primary inclusion in olivine (O) of basalt 10020,41. (a) As found, and (b) after heating to 1100°C for 2 days. Note coarsening of pyroxene (Py) and partial dissolution of the plagioclase daughter crystal (Pl). Ilmenite (I) and vapor (V) appear unaffected. From Roedder and Weiblen (1970b).

Figure 18-9. Primary inclusion in olivine (O) of lunar basalt 10020,41. (a) As found and (b) after heating at 1110°C for 2 days. Note that plagioclase (Pl) is gone in (b). From Roedder and Weiblen (1970b).

Figure 18-10. Primary inclusion in olivine (O) of lunar basalt 10020,41, as found (a), containing vapor (V), and epitaxial ilmenite (I) and plagioclase (P) plates in devitrified glass (G), and after sequential heating for 4 hours each (b-d) to the temperatures (°C) shown (see text). Note that the inclusion has changed shape by dissolving the walls. The bubble in (d) formed on quench. Scale bar in μm. From Roedder and Weiblen (1970b).

Figure 18-11. Primary inclusions in olivine (O) of lunar basalt 10020,41, as found (b), and after heating at 1110°C for 2 days (a, and enlarged in c). Note that the plagioclase is completely gone, the ilmenite reduced, and only a small amount of pyroxene remains. An opaque grain (sulfide?) has become visible in the larger inclusion, and a number of dislocation loops in the host olivine have become decorated with opaque particles. From Roedder and Weiblen (1970a).

The original inclusions show abundant textural evidence that plagioclase nucleated and grew before pyroxene, but we see that at equilibrium, pyroxene crystallizes at higher temperatures than plagioclase. Thus, in runs between 1105 and 1130°C, the plagioclase melted completely, but the original feathery pyroxene crystals recrystallized into stubby prisms.

I have compared such data with laboratory phase-equilibrium data for melts of similar compositions made under known and controlled conditions (e.g., Usselman et al., 1975; Nabelek et al., 1978) and with petrographic evidence on the crystallization sequence in the magma as a whole, because in this larger system, problems of nucleation should be reduced. Unfortunately, some such comparisons have been ambiguous.

As in some aqueous inclusions, the vapor bubble in these inclusions disappears at a lower temperature than that of the liquidus. Most small inclusions have no bubble in the glass after quenching from 1200°C or above (Fig. 18-10c), but where present after quenching, the bubble is of a size comparable with that originally present. No inclusion that originally had a bubble lost it after heating to temperatures below 1200°C, so it is safe to presume that at about 1200°C, the expansion of the melt eliminated the vapor bubbles, but that some of these inclusions renucleated a bubble during quench (e.g., Fig. 18-10d). Similar temperatures were obtained on Apollo 12 olivine samples (Roedder and Weiblen, 1971).

The crushed grains of olivine that were heated had many inclusions cut by the broken surface. At low temperatures, very little difference could be detected between the behavior of those inclusions open to the vacuum in the tube and envelope and that of inclusions sealed in single-crystal olivine "containers." At 1125°C, however, the sealed inclusions consisted of liquid plus a few small crystals of pyroxene and ilmenite, whereas adjacent, identical, and presumably contemporaneous inclusions, open to the vacuum, contained large amounts of pyroxene. (At higher temperatures a comparison was impossible, because the silicate melt in the open inclusions creeps out over the olivine grain.) This difference in behavior is most probably explained by the presence of small amounts of volatile substances in these inclusions, which cause a significant lowering of the liquidus in the sealed inclusions (as a guess, perhaps 10°C), but which are lost from the opened inclusions. This interpretation is in accord with chemical analyses of some of these rocks; although the rocks are indeed exceedingly dry, they do contain a few parts per million of CO_2 and H_2O or H (Friedman et al., 1970).

Many hours of heat treatments in appropriate temperature ranges generally failed to cause the nucleation of pyroxene or plagioclase crystals in those inclusions in which they had previously been present but from which they had been eliminated by melting. Only ilmenite and the vapor phase renucleated readily. The ilmenite formed as a mass of small crystals rather than as a single oriented plate after heating a formerly homogenized inclusion for 4 hr at 1130°C.

No determination was made of the fate of the sulfide globule on homogenization, but if this globule represents a separation of an immiscible sulfide phase after trapping in the olivine, as generally seems to be true, it should redissolve on heating. Crude estimates of the volumes of the various phases indicate that these inclusions contain approximately 0.6 wt % Fe sulfide, or 0.2% S. The presence of S in one homogenized inclusion was confirmed by electron microprobe (Table 18-1). The analysis indicates 0.15 ± 0.05 wt % sulfur, based on three sulfide standards and background intensities on the olivine host and olivine and pyroxene standards containing similar amounts of Fe. This amount of S is much greater than the 0.038% S reported by Skinner and Peck (1969) for a presumably sulfide-saturated Alae lava lake basalt at Kilauea, Hawaii, USA, but these two lavas have different bulk compositions.

A decoration of dislocations was found in the olivine after these vacuum heat treatments. Many of the olivine crystals, as returned by Apollo 11, show faint lines of very minute specks running through them in various directions. These lines are usually visible only at 1000x. After some heat treatments, for example, 2 days at 1110°C, in vacuum, these faint lines become plainly decorated, as they became rows of apparently opaque grains <1 μm or less in diameter (Fig. 18-11). Many of the rows go through 180° loops, and some appear to start at the melt inclusions. The particular significance of these apparent dislocation loops is not known at present, but they could yield significant data on lunar-crystal

Table 18-1. Electron-microprobe analysis of melt inclusion in olivine from lunar basalt 10020.

Analysis	A Inclusion	B Bulk rock	C Bulk rock	D Olivine host
SiO_2	40.8	39.92	39.95	37.5
Al_2O_3	9.8	10.04	10.19	<0.1
FeO	20.7	19.35	19.14	26.5
Fe_2O_3	---	0·00	0·03	---
CaO	11.1	11.24	11.31	0.6
MgO	7.0	7.81	7.87	34.5
K_2O	0.05	0.05*	0.05	0.0
Na_2O	0.2	0.37*	0.39	0.0
TiO_2	9.8	10.72	10.52	0.0
Cr_2O_3	0.4	0.40*	0.38	0.3
MnO	0.24	0.24	0.27	0.28
NiO	<0.01	<0.0020†	---	<0.01
S	0.15	0.15	0.18	---
P_2O_5	---	0.08	0.07	---
BaO	---	0.0067†	---	---
Subtotal	100.25	100.37	100.35	99.79
Less 0	0.07	0.08	0.09	---
Arithmetic Total	100.18	100.29	100.26	99.79

A. Homogenized melt, 20 μm in diameter, in olivine phenocryst from rock 10020-41, grain A-6, Paul W. Weiblen, analyst.
B. Analysis (conventional methods) of rock 10020-30 (Maxwell et al., 1970). Also found: H_2O^- 0.01; C 0.01.
C. Analysis of rock 10020-23 (Peck and Smith, 1970).
D. Host olivine crystal, Paul W. Weiblen, analyst.
* Atomic absorption spectroscopy.
† As element, not oxide.

growth phenomena, or possibly later deformation, and should be investigated further. These decorations disappear at higher temperatures. Similar decorated dislocation loops are common in some terrestrial olivines (e.g., Roedder 1976b, his Figs. 15-19; see also following section on symplectite inclusions).

The composition of the melt inclusions is of considerable interest. The melt inclusions are a unique sample of presumably the complete lunar magma present at the time and place of trapping, without the effects of any later changes due to loss of volatiles, crystal fractionation, or other process (except for crystallization onto the walls). The bulk composition of these inclusions, on the basis of measurements of the sizes of the phases present in a group of 12, in weight percent, was sulfide 0.6, ilmenite 23.8, plagioclase 10.5, and glass (plus feathery pyroxene) 65.0. The bubble amounts to 1.6 vol %. All these numbers may be only good to ~50% of the amount present.

An attempt was made to analyze two inclusions in sample 10020, 28-7667 by electron microprobe (see Roedder and Weiblen, 1970b, for analytical details). Daughter crystals of plagioclase and ilmenite were identified optically, and confirmed by electron-microprobe analyses for Si, Ca, Fe, and Ti. The volume of remaining glass proved to be too small for reliable analysis, but traverses across the inclusions indicate that the glass is about 10% higher in SiO_2 and 5% higher in CaO than plagioclase in the sample.

Phase relations suggest that increasingly Fe-rich olivine would precipitate on the walls of such inclusions during cooling. A total of six traverses for Fe and Mg across the host/inclusion boundaries showed a consistent gradual increase toward the inclusion in the host over a distance of 10 μm, reaching a maximum increase of 1-2 mole % Fa at the inclusion boundary.

A much more accurate procedure for determining melt composition involves homogenizing the inclusion in the laboratory at 1210°C, quenching to yield a uniform glass, and then analyzing that glass. This corrects for crystallization onto the walls and eliminates the uncertainties implicit in any volume estimates of small objects. A polished surface must be cut through the grain and its inclusion after heating. One homogenized glass inclusion 20 μm long in olivine of sample 10020-41, grain A-6 was studied (Table 18-1, analysis A). A check was made for gradients in Fe and Mg, similar to those described above, and none were found within the error of the analyses (~ 0.5 mole % Fe). The adjacent enclosing olivine has the composition $Fo_{69}Fa_{30}La_1$ (analysis D). Within the error of the electron-microprobe analyses ($\sim 5\%$ of the amount present), there is good agreement between the composition found for the inclusion and analyses of the bulk rock (analyses B and C). Similar procedures were used in analyzing homogenized inclusions in Apollo 12 basalt 12018 (Roedder and Weiblen, 1971). This latter rock has a very different bulk composition, and the inclusion analyses reflect it. When more accurate probe analyses are available, the recognition of real differences between inclusion and bulk rock analyses may be possible, thus permitting calculation of the composition and amount of crystals present at the time of trapping of the inclusion.

The only other published study of the thermal behavior of melt inclusions in lunar olivine is that by Sobolev et al. (1980). Rather than using the quenching technique as I used, they used high-temperature microscopy involving "a special micro-heater assembly" (p. 112) consisting of a Pt tube 3 mm in diameter and 10 mm long, in an atmosphere of Ar or He. The temperatures they obtained were higher than those I described for Apollo 11 olivines. I found complete homogenization (and ilmenite liquidus) at ~ 1210°C, and melting of the last pyroxene at 1130°C, whereas Sobolev et al. reported Th = 1320-1365°C, and the last pyroxene at 1285-1320°C. Since the samples they used are from a different locality than mine, however, there is no need for the results to be the same. They also reported electron microprobe analyses of low-Ca pyroxene crystals that formed in originally glassy inclusions during the microthermometric runs.

Inclusions in other early minerals. Melt inclusions trapped as a single homogeneous melt phase have been found in the other minerals of the mare basalts, particularly in chromian ulvöspinel (see especially Weiblen and Roedder, 1976; Roedder and Weiblen, 1978), and a few in chromite, plagioclase, and pyroxene. Analyses of inclusions in plagioclase and pyroxene would be particularly desirable, as these are the main minerals forming during the bulk of the crystallization of these basalts. Unfortunately, usable silicate-melt inclusions in these two minerals are rather rare, except for those parts that crystallized late, after immiscibility had taken place in the residual melt (see later section). In some lunar rocks, some of the feldspar crystals are distinctly cored (Fig. 2-4). This core now consists of coarse crystals of pyroxene and other minerals but probably was originally a large melt inclusion, trapped as a result of rapid skeletal growth after nucleation in a supercooled magma, followed by solid growth. The presence of the melt permitted the skeletal feldspar in the core to recrystallize onto the walls, producing a negative crystal shape. The walls of these inclusions sometimes show "reverse" zoning, Na increasing inward, as they were also in contact with a differentiating magma, i.e., the melt inclusion.

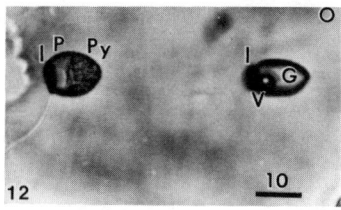

Figure 18-12. Two small melt inclusions in olivine (O) of lunar basalt 15555,34, in transmitted plain light. The left inclusion has nucleated ilmenite (I), plagioclase (P), and pyroxene (Py); the right inclusion contains only ilmenite (I), glass (G), and a shrinkage bubble. Scale bar in μm. From Roedder and Weiblen (1972b).

Constraints on maximum and minimum cooling rates. Melt inclusions in the lunar basalts provide some data on relative cooling rates as well as some constraints on the absolute cooling rates. Inclusions in lunar olivine show that the phases present are mainly a function of cooling rate and inclusion size (Fig. 16-8). When cooling is fast, or the inclusion is small, only glass will be present. Slightly slower cooling or larger inclusions may permit nucleation of a sulfide globule, a bubble, or epitaxial ilmenite and plagioclase. Still slower cooling, maintained through a lower temperature range, and/or a larger inclusion, permits nucleation and growth of feathery pyroxene, and presumably of more plagioclase and ilmenite, from the remaining liquid.

This sequence is the most common, but that for any individual inclusion may not agree with it. For example, Figure 18-12 illustrates two small adjacent inclusions in the same olivine grain, one of which nucleated epitaxial ilmenite and plagioclase, plus pyroxene, whereas the other, of the same size, contains only ilmenite and vapor in glass. By establishing a sequence of more-or-less objective categories based on observable differences such as the presence or absence of epitaxial ilmenite or plagioclase, or of random crystals, one can plot the number of inclusions having a given assemblage versus a measure of the individual inclusion volume, the maximum inclusion length (Fig. 18-13). Inclusions in these three lunar samples showed significant differences, suggesting a range of cooling rates, assuming that differences in bulk composition are not significant.

The epitaxial daughter-mineral plates of ilmenite and plagioclase in these lunar olivines provide an interesting scientific anecdote. Although many people had studied basalts over the years, to my knowledge, no one had ever reported such inclusions in terrestrial samples. After finding them in both Apollo 11 and 12, I examined at some old slides of Hawaiian rocks that had been around the U.S. Geological Survey for years and found that such inclusions were plainly visible. In fact, the first such slide I examined (Fig. 16-5) contained six inclusions in a single 1 mm olivine phenocryst, each with obvious plates of epitaxial ilmenite and plagioclase (plus feathery pyroxene and immiscible sulfide blebs). We simply had to go to the Moon to open our eyes to what was plainly visible on Earth.

As a result of this discovery of epitaxial daughter crystals in terrestrial as well as lunar samples, a comparison was desirable. I compared lunar olivines with olivines from the modern Kilauea Iki and the prehistoric Makaopuhi lava lakes in Hawaii, USA, where something is known about cooling rates (Fig. 18-14). The data have been normalized to percentages to eliminate the problem of comparing different numbers of inclusions. These data show that all-glass inclusions are essentially the smaller inclusions and that epitaxial ilmenite and plagioclase are much more abundant in the larger ones, but there are differences among the samples. How much of this difference can be attributed to the compositional range of the enclosed melts is unknown. The Kilauea Iki sample shown in the lower right of Figure 18-14 was surprising, however, as it has a reversed

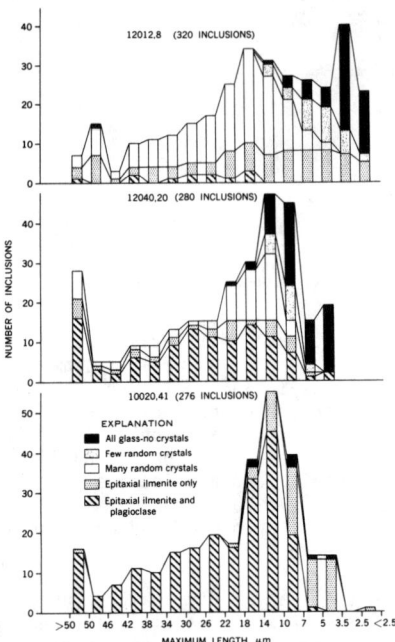

Figure 18-13. Frequency of occurrence of various phase assemblages in primary-melt inclusions in three different lunar olivines, plotted against maximum inclusion length. From Roedder and Weiblen (1971).

Figure 18-14. Histograms showing abundance of various inclusion types vs maximum dimensions for inclusions in olivines from lunar (top row) and Hawaiian (bottom row) lavas. The size brackets for the five sample subsets in each histogram, from left to right, are as follows: >40, 20-40, 10-20, 5-10, and <5 μm. The number of inclusions measured in each sample is given at lower right of each graph. The first three histograms in upper row are in large part adapted from Roedder and Weiblen (1971). Kilauea Iki sample (lower right) is from 1967 borehole 1, at 5.3 ft (1.6 m) depth. Pyroxene may be present in either of the two epitaxial categories.

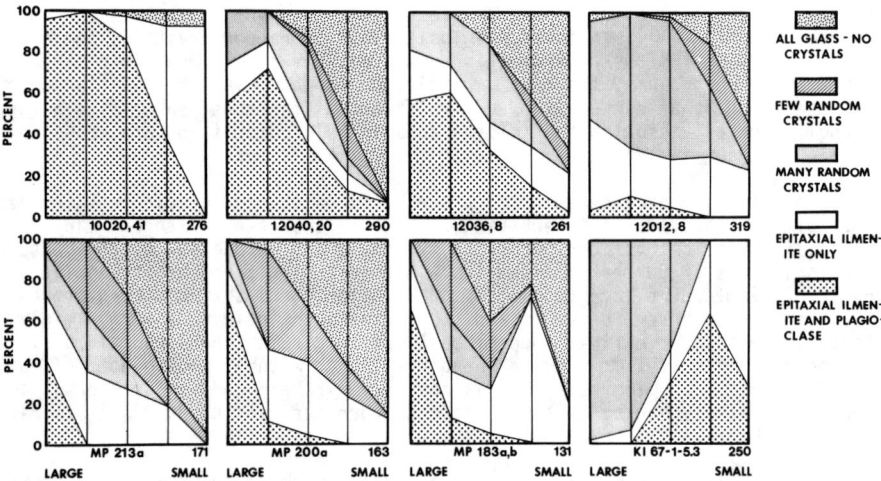

distribution. I offered an <u>ad</u> <u>hoc</u> and not very satisfactory explanation for these peculiar data based on a two-stage cooling model (Roedder, 1976b).

Laboratory heating of natural glassy inclusions, or previously homogenized ones, also permits us to place some constraints on the absolute cooling rates in the natural cooling cycle. Thus, if inclusions that were originally glass nucleate crystals during a given laboratory cooling cycle, the natural cooling cycle must have been faster than the laboratory one. Similarly, ilmenite nucleates as random crystals at one laboratory cooling rate and as epitaxially oriented plates

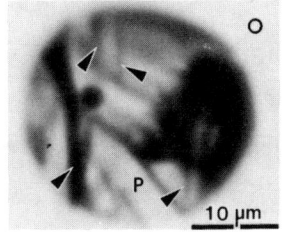

Figure 18-15. Inclusion in olivine (O) in sample from prehistoric Makaopuhi lava lake in Hawaii, USA. It was homogenized in the laboratory and then cooled slowly from 1120 to 1020°C over an 11-day period, causing crystallization of pyroxene (P) and renucleation of several black epitaxially arranged ilmenite plates (see arrows). The black sphere is presumably a globule of sulfide. From Roedder and Weiblen (1971).

at a slower rate. Figure 18-15 shows a formerly homogeneous glass inclusion in which several epitaxial plates of ilmenite formed during laboratory cooling from 1120 to 1020°C over an 11 day period.

Inclusions of Heterogeneous Melts in the Mare Basalts

History of the concept of silicate immiscibility in igneous petrology. The history of science is littered with the shells of cast-off concepts, each of which seemed eminently suitable to hold to perfection all aspects of the then-known body of facts. One such concept is that silicate liquid immiscibility in magmas is the mechanism by which certain rock types have formed. It was proposed by Rosenbusch (1887; also by F. Zirkel) early in the history of petrology to explain the juxtaposition of rocks having quite disparate compositions, usually without intermediate types. Rocks such as basalt and rhyolite, and various pairs of dike rocks, were generally and conveniently assumed to have formed simply by the "splitting" of a formerly homogeneous magma into two immiscible magmas of contrasting composition. These magmas were thought to have separated, like oil and water, because of density differences.

Early studies at the Geophysical Laboratory of the Carnegie Institution of Washington, D.C., verified that the systems $CaO-SiO_2$, $MgO-SiO_2$, and $FeO-SiO_2$ did indeed have extensive fields of immiscibility. Greig (1927) showed, however, that these fields were eliminated by the addition of just a few percent of alkalies or alumina as a third component. Inasmuch as the resulting ternary immiscibility fields were all similar, Greig used a pseudoternary plot, which I called a "Greig diagram" (Roedder, 1978b) of $(CaO + MgO + FeO)-SiO_2-(Na_2O + K_2O + Al_2O_3)$ to show the relationship of the field of immiscibility to the composition of igneous rocks. No igneous rock composition even approached the limits of the immiscibililty field on this diagram; even those rocks having the lowest total $(Na_2O + K_2O + Al_2O_3)$ contained twice as much of these components as could be found in any synthetic mixture showing immiscibility (~10 vs 5 wt %). Furthermore, immiscibility took place in these synthetic melts only at excessively high temperatures (~1700°C). Thus, experimental verification of immiscibility was found only in geologically unlikely compositions, at geologically unreasonable temperatures. Bowen (1928) then applied a temporary coup de grâce by showing that if immiscibility had indeed occurred in natural magmas, the expected lines of evidence for it had not been found in the rocks. In addition, Bowen demonstrated that crystallization differentiation could produce both continuous and discontinuous compositional variations in igneous rocks; he described ample field and textural evidence to support this view. Silicate immiscibility as a potential petrological process was thus put to rest.

Since then, however, new evidence has come to light that negates all three of the objections of Bowen and Greig: immiscibility has been found in the laboratory in geologically reasonable compositions and temperature ranges, and evidence for it has been found in natural rocks, most particularly those on the Moon. A more extensive discussion of the behavior of systems involving immiscibility and the various lines of evidence for it is given elsewhere (Roedder, 1978b, 1979d).

Immiscible silicate-melt inclusions in mare basalts. I wish I could say that I had hoped to find silicate immiscibility in the Apollo samples, but the

Table 18-2. Electron microprobe analyses of immiscible silicate-melt inclusions, now glass, from lunar basalts (Paul W. Weiblen, analyst). From Roedder and Weiblen (1971). Number of inclusions averaged shown in parenthesis; some minor constituents were not analyzed on all samples; s.d., standard deviation.

	High-Si melt				High-Fe melt			
	Apollo 11 (35)		Apollo 12 (15)		Apollo 11 (7)		Apollo 12 (5)	
	wt %	s.d.	wt %	s.d.	wt %	s.d.	wt %	s.d.
SiO_2	75.8	3.2	76.3	1.3	44.1	5.4	42.2	2.3
Al_2O_3	11.4	1.6	11.5	1.3	3.0	0.8	5.4	1.2
FeO	2.6	1.1	3.0	0.6	31.2	3.6	32.9	2.7
MgO	0.25	0.9	0.07	0.08	1.9	1.2	0.79	0.27
CaO	1.8	0.18	1.6	0.97	10.9	1.8	10.7	0.5
K_2O	6.4	2.4	6.7	0.99	0.25	0.08	0.41	0.27
Na_2O	0.35	0.31	0.14	0.26	0.11	0.08	0.16	0.35
P_2O_5	---	---	0.15	0.17	1.2	---	0.57	0.09
TiO_2	0.53	0.13	0.68	0.23	3.8	---	4.2	1.0
MnO	---	---	0.05	0.01	---	---	0.32	0.04
BaO	---	---	0.62	0.07	---	---	0.0	---
Arithmetic Total	99.03	2.3	100.81	1.7	96.46	4.3	97.65	1.0

true story is far from it, and, in fact, finding it was an example of serendipity. As mentioned above, I went to Houston to look for liquid H_2O and CO_2 inclusions in the first Apollo 11 samples. In the fruitless search for such inclusions, however, I spent several days staring at various odd textures in these slides. On careful examination, some of these odd textures turned out to be artifacts of various kinds, particularly from the several resins used in slide preparation (see Chapter 6). However, some single crystals of anorthite were packed with rounded masses of an unknown material and these, in turn, like the famous fleas, with still smaller globules, ad infinitum. Sometimes the masses appeared in sharply defined rows, but all were embedded in a single crystal of anorthite. Some late-forming crystals, later shown to be the new mineral pyroxferroite, contained what looked like a row of glass "molars" (Fig. 2-32). Then it suddenly dawned on me that all these and many other odd textures I had seen could be explained by late-stage silicate-liquid immiscibility, just as I had reported from the laboratory 18 years earlier and completely forgotten. Since then, it has become apparent that most silicate melt inclusions in most lunar basalts are a result of silicate immiscibility. Such serendipity makes science fun!

Crystallization of the major part of the lunar basalt magmas was normal, but eventually the residual melt between the crystals reached a composition that resulted in splitting to form globules of a second immiscible liquid; these globules were subsequently trapped in various crystals. This is stable silicate immiscibility, not the metastable subsolidus immiscibility observed by electron microscopy in many glasses. Microprobe analyses (Roedder, 1979e; Roedder and Weiblen, 1970a,b, 1971, 1972a,b, 1973a,b, 1977a,b, 1978; Weiblen and Roedder, 1973, 1976) showed that this liquid had the composition of a quartz-rich K-granite. The immiscible liquid globules nucleated on and stuck to the surface of the growing late-stage minerals, forming the bottoms of such rows of "molars" and marking the onset of immiscibility. Further exsolution of this immiscible granitic melt onto these nuclei, as the host crystallized from the Fe-rich melt,

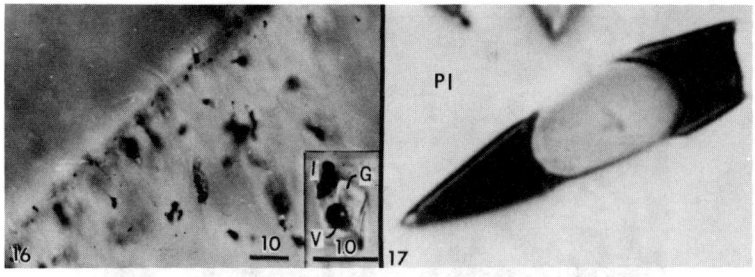

Figure 18-16. Pyroxene single crystal from lunar basalt 10072,36, which grew from the core (upper left) towards the rim (lower right). The core is inclusion-free, but many small melt inclusions are present in the rim. As shown in the inset, these inclusions now consist of glass (G), vapor bubble (V), and, in some, an opaque daughter mineral (I, ilmenite?). Scale bars in μm. See text for explanation. From Roedder and Weiblen (1970a).

Figure 18-17. Primary inclusion of immiscible high-Si (light) and high-Fe (dark) melts, now glasses, in plagioclase crystal (Pl) of lunar basalt 12057. Inclusion is 18 μm long. From Roedder and Weiblen (1971).

caused some globules to grow and coalesce, yielding the tops of the "teeth" (see Figs. 2-29b,32). Many lunar basalt pyroxene crystals contain a core that is almost inclusion-free and an outer rim densely packed with minute melt inclusions; the sudden change marks the onset of immiscibility (Fig. 18-16).

A very puzzling aspect was the lack of any evidence of the presumed other melt from which these granitic-melt globules must have separated. Not until considerably later did I find samples showing much more obvious textural evidence of immiscibility, such as two coexisting glasses separated by a sharp meniscus: a colorless, low-index glass ("high-Si melt") and a dark-brown, high-index glass ("high-Fe melt"). Part of the problem was that the high-index glass was never seen in large masses or in contact with pyroxene. The reason for this is now obvious, but only in retrospect: because the high-index glass has the composition of a titaniferous ferropyroxenite and contains about 80% normative pyroxene (Table 18-2), it crystallized readily, particularly wherever a pyroxene nucleus was available, and hence nicely eliminated the evidence for its own former existence. One might call it "self-destructing evidence," and, as such, this concept might be very pertinent to the search for immiscibility in terrestrial rocks.

Thus, as shown in Figure 18-16, some of the silicate-melt inclusions in a pyroxene host trapped only the high-Si melt and now consist of glass plus a vapor bubble. In those that trapped both high-Si and high-Fe melt, the high-Fe melt crystallized its pyroxene component onto the walls of the host pyroxene, leaving the ilmenite component as a daughter crystal. This process also explains the occurrence of isolated small ilmenite crystals in the pyroxene host, from the trapping of just the high-Fe melt. In contrast, when both melts were trapped in plagioclase, they are now found as two glasses (Fig. 18-17). One of the largest readily recognized inclusions of immiscible melts from the lunar samples is shown in Figure 18-18. Here, in contrast to the inclusion shown in Figure 18-17, the ilmenite component of the high-Fe melt crystallized out on the ilmenite walls, leaving only pyroxene daughter crystals.

Some lunar samples have cooled slowly enough to permit complete crystalli-zation (to K-feldspar and silica) of even the extremely viscous high-Si melt. These viscosities must be extremely high, comparable with that of obsidian (Figure 18-19), so the cooling rate must have been extremely low. (In spite of these viscosities, the evidence seems conclusive that the immiscibility was from stable and not metastable equilibrium.) The actual viscosity is difficult to establish. Experimental determinations of the viscosity of terrestrial obsidian presumably all were made on samples containing the ~0.2% H_2O typically present

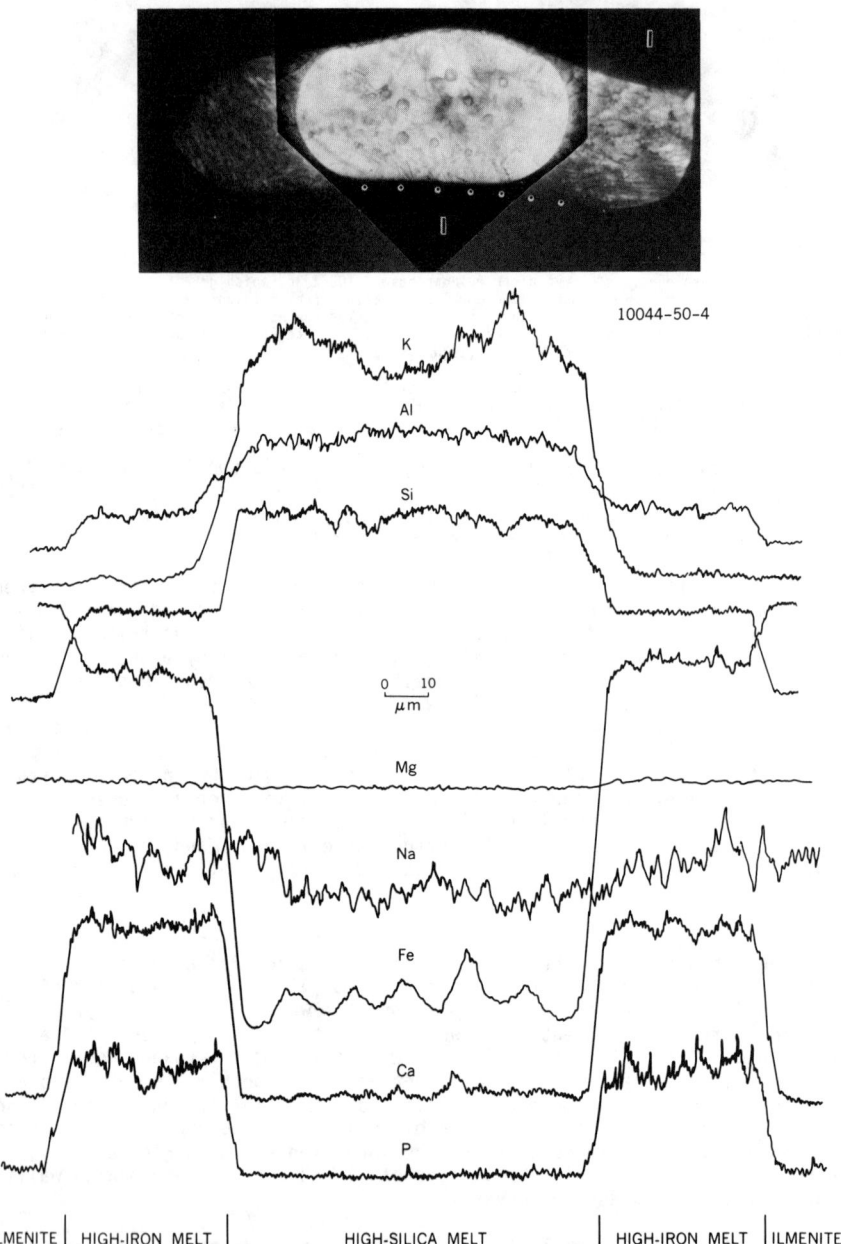

10044-50-4

ILMENITE | HIGH-IRON MELT | HIGH-SILICA MELT | HIGH-IRON MELT | ILMENITE

Figure 18-18. Large primary inclusion of immiscible silicate melts in ilmenite crystal (I) from lunar basalt 10044. The high contrast in transparency between the colorless central globule of high-Si melt (now containing a few small globules of high-Fe melt and many unidentified acicular crystals) and the surrounding dark feathery mass of pyroxene crystals (from the crystallization of the high-Fe melt) required the two different photographic exposures in this composite print. The lower contact of the inclusion with the ilmenite host crystal has been dotted in. Electron-probe traces from along the length of this inclusion are shown at the bottom of the figure. From Roedder and Weiblen (1971).

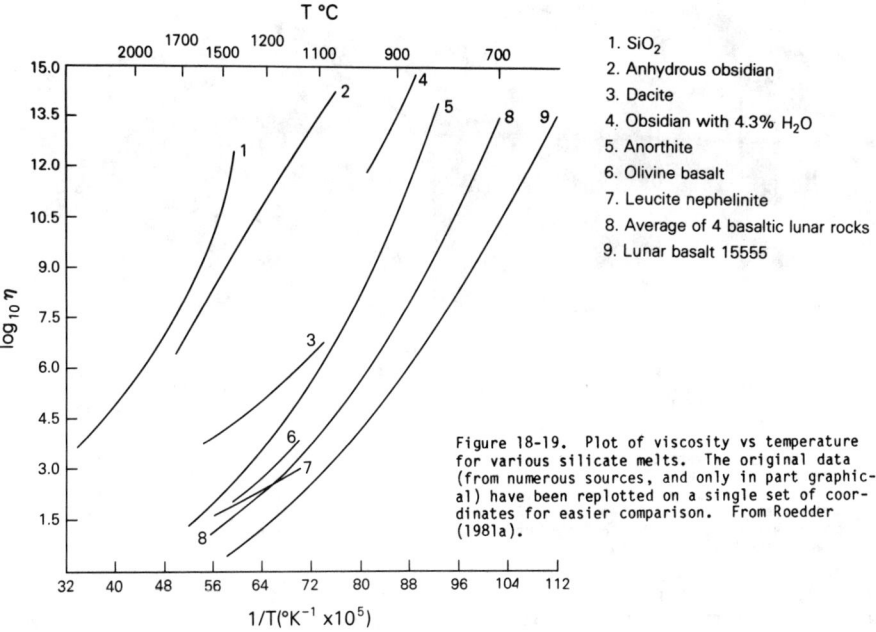

Figure 18-19. Plot of viscosity vs temperature for various silicate melts. The original data (from numerous sources, and only in part graphical) have been replotted on a single set of coordinates for easier comparison. From Roedder (1981a).

1. SiO₂
2. Anhydrous obsidian
3. Dacite
4. Obsidian with 4.3% H₂O
5. Anorthite
6. Olivine basalt
7. Leucite nephelinite
8. Average of 4 basaltic lunar rocks
9. Lunar basalt 15555

in obsidian, but the lunar melts presumably were even drier. However, the viscosity of alkali-aluminosilicate melts shows a sharp maximum at a 1:1 molar ratio of alkalies to alumina (Schairer and Bowen, 1947), but the lunar glasses are deficient in alkali and hence should have lower viscosities. (In contrast to these 4-billion-year-old glasses, terrestrial glasses more than a few million years old are seldom found.) In many samples, the high-Si melt contains only occasional acicular unidentified daughter crystals, but the high-Fe melt is a crystalline mass of pyroxene and minor ilmenite (Fig. 18-20). Only rarely are both melts found as glasses. In some of these rocks, <u>four</u> simultaneous immiscible melts were present: the two silicate melts, Fe sulfide melt, and Fe metal melt (Roedder and Weiblen, 1972b).

The compositions of inclusions of high-Si and high-Fe melts from the Apollo 11 and 12 samples show some differences between samples (Table 18-2). Considering the difficulties of obtaining accurate analyses on such small volumes of glasses, particularly for the high-Fe glasses, little significance should be attached to the differences shown, particularly those in the minor constituents. Ba is strongly concentrated in the high-Si glass, and P and Mn, in the high-Fe glass. Although the standard deviations are large and cannot now be accurately assigned to analytical difficulties or real sample variation, some of the differences are real. A comparison of the K_2O/Na_2O weight ratios for high-Si melt compositions obtained from various groups of inclusion samples from the Apollo 11, 12, 14, and the Soviet Luna 16 missions (Roedder and Weiblen, 1972a, their Table 3) shows this parameter to range widely, from 0.14 to 48, for unknown reasons.

The analysis of these small glass inclusions by electron microprobe encountered numerous difficulties not present in the analysis of crystalline phases. In addition to small size (frequently smaller than the volume excited by the electron beam), and the lack of the crystal-chemical constraints that are so useful in recognizing invalid analytical data on crystals, many of the glass

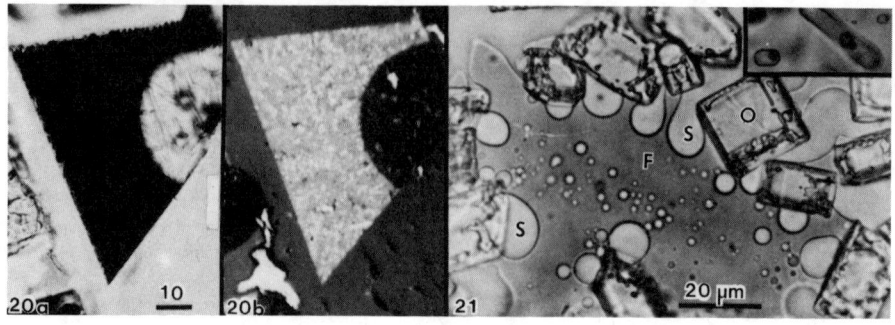

Figure 18-20. Interstitial inclusion of immiscible melts between plagioclase laths (P) in lunar basalt 14310,169, in plain transmitted (a) and reflected light (b). The globule of transparent high-Si melt contains a few acicular crystals. The high-Fe melt has crystallized to a fine aggregate of several phases, probably pyroxene and ilmenite. Bar scale in μm. From Roedder and Weiblen (1972b).

Figure 18-21. Synthetic melt, previously homogeneous, after 7-1/2 hr cooling cycle ending with quench from 1045°C. Globules of colorless high-Si glass (S) wet the surfaces of olivine (fayalite; O) crystals and extend out into yellowish high-Fe glass (F). As the fayalite crystals grew, they enclosed globules of high-Si glass, resulting in a skeletal appearance. Some of these inclusions in fayalite formed shrinkage bubbles on cooling (inset; same scale). From Roedder and Weiblen (1970b).

inclusions are inhomogeneous, as they either have visible daughter crystals, or show major chemical inhomogeneities in what appears optically to be uniform glass. (See a series of electron-microprobe traces across high-Si glass inclusions in Roedder and Weiblen, 1970b, and in other papers in that series referenced above, p. 535.) A variety of experimental procedures that were used to find suitable inclusions and to minimize the errors in analysis are also given in those papers and are generally applicable to melt-inclusion analysis. In addition, controversy concerning the compositions of immiscible silicate liquids in the system $K_2O-FeO-Al_2O_3-SiO_2$ (Biggar, 1983) has resulted in the recognition of the existence of serious problems in the analysis of such glasses by electron microprobe (Roedder, 1983a).

The best verification of actual silicate immiscibility in any natural system comes from the laboratory. If a synthetic composition of similar composition shows immiscible separation of two melts of different composition from an originally homogeneous melt, immiscibility is established. A synthetic charge that had a composition corresponding to a 50-50 mixture of Apollo 11 average high-Si and high-Fe melts (Table 18-2) was heated at 1350°C in a pure Fe container in vacuum. On quenching, it proved to be an optically transparent, homogeneous brown glass, n = 1.569±0.005. Parts of this glass, held at lower temperatures and then quenched, showed stable immiscibility at the liquidus temperature (Fig. 18-21).

Like the epitaxial ilmenite daughter crystals discovered in melt inclusions in lunar olivine and subsequently recognized in terrestrial basalts, as mentioned earlier, silicate immiscibility was also thought to be unique to lunar rocks until similar evidence of immiscibility was found in samples of various terrestrial basalts (Roedder and Weiblen, 1971; see also Chapter 16). In these rapidly cooled rocks, the immiscible liquids are retained as isolated globules. Under conditions of slower cooling, as in the early history of the Earth or Moon, significant separation of liquids might have occurred; such separation, and crystallization, effectively eliminate the best evidence of immiscibility -- menisci between compositionally different glasses.

<u>Petrologic significance of immiscibility.</u> The discovery of immiscible

silicate-melt inclusions in the lunar samples (and subsequently in terrestrial samples) raises a series of petrologic questions. Although in most of the lunar basalts, the immiscibility took place so late in the cooling history (~95% crystalline) that physical separation of the two melts would be difficult, immiscible separation of the residual melt was found in one terrestrial basalt containing only 56% crystals (Roedder and Weiblen, 1971). Furthermore, the possible extent of immiscibility in the multicomponent systems representing terrestrial igneous rocks is unknown and is amazingly sensitive to compositional variation (Roedder, 1979d).

The distribution of elements between two immiscible silicate melts depends on the intensive variables T, P and the chemical potentials of the components. When immiscibility occurs, the new melts produced may be expected to differ in composition from melts produced by crystal fractionation. This difference has implications in developing and evaluating hypotheses for source magmas, any proposed stratification of the interior of the Moon, and explanations of similarities and differences between meteoritic, lunar, and earth materials. Although one of the products of the immiscibility, the granitic high-Si melt, is similar in gross chemical composition to the end product of classical magmatic differentiation by crystal fractionation, there are important differences in the partitioning of some elements. In particular, P is strongly enriched in the granitic residue from crystal fractionation (Anderson and Greenland, 1969) but very depleted in that from immiscibility. This difference raises the intriguing question of whether the partitioning of P and other minor and trace elements might be used to determine if immiscibility has been involved in the petrogenesis of both plutonic and volcanic terrestrial rocks, as well as complex lunar rocks such as partly granitic samples like lunar sample 12013. Obviously, if the original basalt magma contained only 0.05% K_2O (as is true for a number of low-alkali Apollo 11 and Apollo 12 rocks), a maximum of only 1% of "silicic" magma (i.e., high-Si melt containing 5% K_2O) can be obtained from it. This 1% is reduced by the amount of K_2O that enters the earlier rock minerals, mainly plagioclase, and is further reduced by the predictably incomplete separation of such a viscous liquid from the crystal mush. In spite of these objections, several lines of evidence suggest that K, at least, has been concentrated toward the surface of the Moon and that rocks of essentially granitic composition (perhaps such as sample 12013), though scarce, do occur somewhere on the surface of the Moon.

Even if the high-Si melt is not segregated into a granitic rock, the unique distribution of elements produced by immiscibility is important in evaluating radioactive isotope data on the lunar rocks. The compositions of the high-Si glasses show that much of the K in the lunar rocks is in the glass inclusions. Obviously consideration of problems of Ar loss must be directed to the glass of these samples. High concentrations of U, Rb, and Sr would also be expected in the high-Si glass. Preferential leaching of parent or daughter elements from such glass during sample preparation may be an important factor to consider in the interpretation of data on these elements.

The differentiation trends for the lunar mare magmas have been discussed at length. What can the study of melt inclusions, from those in olivine through to high-Si melt, contribute to an understanding of these trends? For several reasons, the contribution is limited, even though many analyses have been made. Roedder and Weiblen (1972b) plotted the composition of all silicate-melt inclusions that they had analyzed from Apollo 11, 12, 14, 15, and Luna 16 on silica-variation diagrams. Although most of the data points could be enclosed in broad envelopes showing large decreases in FeO, MgO, and CaO as differentiation progressed (as evidenced by silica changing from 41 to >80%), the Al_2O_3 data were widely scattered, and K_2O showed almost no enrichment until immiscibility occurred. (The high-Fe melts would represent a drastic "reversal.") The lack of clear trends on such plots is probably a result of: (1) combination of data on different host minerals (plus the fact that the data on inclusions in ilmenite

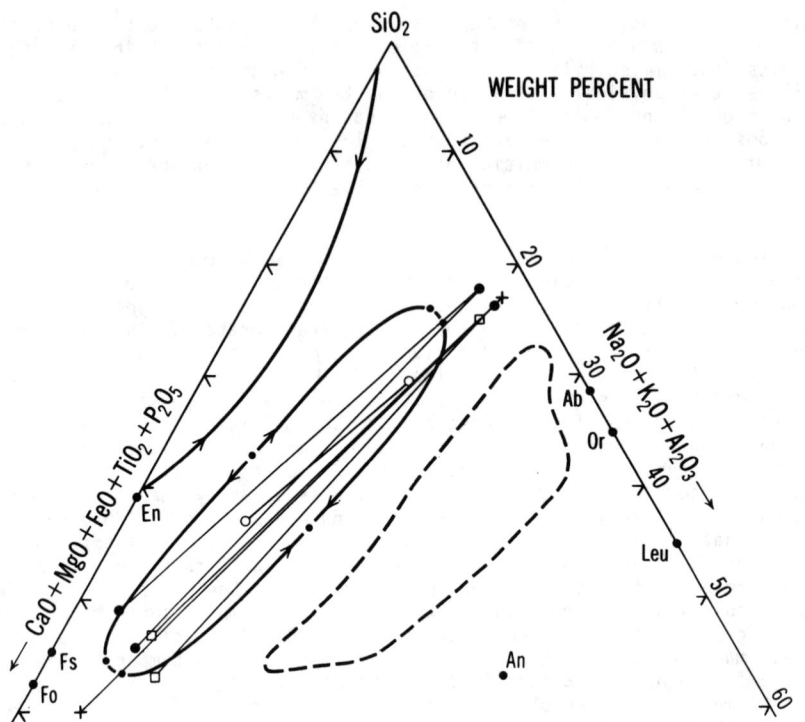

Figure 18-22. Pseudoternary Greig diagram showing field of low-temperature immiscibility in the system leucite-fayalite-SiO2, and tielines for various conjugate melt pairs, adapted from Weiblen and Roedder (1973). All compositions recalculated on the basis of plotted oxides only. Solid circles, coexisting glasses in lunar basalts from Apollo 11, 12, and 15; crosses, coexisting glasses in lunar basalt 14310; open circles, coexisting glasses in synthetic Apollo 11 sample after equilibration at 1045°C; open squares, coexisting glasses in Apollo 15 grain. The last item is from Switzer (1975); all other data from Roedder and Weiblen (1970b, 1971, 1972a,b) and Weiblen and Roedder (1973). Other lunar samples are similar but have been omitted for clarity. Most analyzed terrestrial volcanic rock suites fall within the dashed field (Brooks and Gelinas, 1975).

are particularly anomalous, as described below); (2) combination of data on various magma series (e.g., high- and low-K, high- and low-Ti basalts; (3) very few data points in the intermediate silica range; (4) daughter phases in some inclusions excluded in the analysis; and (5) analytical problems. Examination of a still larger data set (Roedder and Weiblen, 1977b) failed to resolve most of the problems. It verified increasing K_2O and $FeO/FeO+MgO$ with SiO_2 increase, and suggested that immiscibility occurred at ~2% K_2O, but many other questions remain unanswered.

One of the most interesting petrologic questions concerns the compositions of the immiscible melts. As if the discovery of the immiscibility itself were not sufficiently serendipitous, when these analyses were plotted on a Greig diagram (Fig. 18-22), the two melts plot very near to the plotted positions of the conjugate low-temperature immiscible melts in the system leucite-fayalite-silica that I had studied many years earlier! Now that the compositions of the two melts are known, these plotted positions should not be surprising because K_2O, FeO, Al_2O_3, and SiO_2 make up the bulk of these compositions (97 and 80% for the high-Si and high-Fe melts respectively). Since Apollo 11, similar analyses have been obtained from samples from all the other lunar missions, including the three Soviet ones (Roedder and Weiblen, 1977b). Only a few of these data are plotted on Figure 18-22, as the tight clustering would result in confusing overlap.

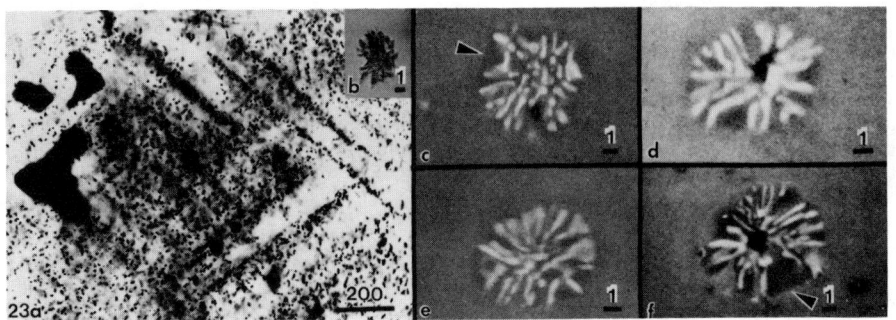

Figure 18-23a-e. Symplectite inclusions in olivine phenocryst from lunar mare basalt 12018,83, from Roedder and Weiblen (1971). (a) and (b) taken in transmitted light; (c) through (f) SEM images. Scale bars in μm. The inclusions consist of rosettelike vermicular groups of chrome spinel(?) rods (bright in SEM) in an unidentified matrix, arranged along presumed crystal-growth patterns in the host olivine (a). Figures (a) through (d) show unheated samples; (e) and (f) show samples that have been heated for 24 hours at 1123°C and then 3 hours at 1219°C and quenched, but show no visible change. Some of the inclusions (both unheated and heated) appear to have a central cavity (dark in (d) and (f)). Note that the matrix of the rods has a darker color (i.e., a lower average atomic weight) than the surrounding olivine, yielding an oval contact (arrows).

Figure 18-22 is <u>greatly oversimplified</u> and hence not necessarily valid for complex multicomponent melts. With that firmly in mind, note that the compositions of most rocks plot to the lower right of the field of immiscibility. The composition of the liquid will move as crystallization proceeds, however, and the plotted positions of pyroxene and anorthite show that crystallization of pyroxene will drive the bulk composition away from the field of immiscibility, and crystallization of anorthite will drive it toward the field of immiscibility. The compositions of some terrestrial examples of immiscibility have been plotted on this same diagram by Roedder (1979d).

When the K_2O/Na_2O ratio of a bulk rock is compared with that of its high-Si melt inclusions, a strong inverse relation is noted (Roedder and Weiblen, 1972a, b). This may be related to the amount and the time (i.e., temperature) at which plagioclase crystallized from the melt, as plagioclase crystallization increases the K_2O/Na_2O ratio in the residual melt.

Inclusions in Highlands Rocks

Samples from the lunar highlands (anorthosites, norites, troctolites, etc.) do not contain glass inclusions of the types described above. Many samples from the highlands are igneous or metamorphic rocks that have apparently had a long cooling cycle, so that all glass has been eliminated. (New glass has been generated or introduced by various impact processes, but not as melt inclusions.) Wormy intergrowths (symplectite) and possibly other related intergrowths of spinel and two pyroxenes, of several textural varieties and compositions, have been observed in olivine in many Apollo rocks, and four theories have been proposed for their origin (Bell et al., 1975). One of these theories, which may be applicable to certain of these symplectites, is that of crystallization of former melt inclusions, and hence should be discussed here.

Symplectite inclusions in olivine. Some symplectite inclusions of spinel plus two pyroxenes occur on olivine grain boundaries in lunar troctolite 76535. These inclusions were believed to be the result of crystallization of trapped late-stage melts at >1000°C (Bell et al., 1975). Possibly similar intergrowths of vermiform spinel plus pyroxene(?) are abundant <u>in</u> the olivine of some mare basalts (Fig. 18-23), as well as in olivine from terrestrial basalts (Roedder, 1976b, Fig. 5). These inclusions, which sometimes appear to decorate growth

layers in the host crystal, are almost opaque rounded masses 5-15 μm in size
that consist of curving radial rods ~0.2 μm thick of chrome spinel(?) in an unidentified host (pyroxene?) that is not the same as the surrounding olivine.
As possibly similar features, decorating dislocations in olivine, precipitated
on heat treatment (Fig. 18-11), I suggest that both the natural symplectites in
the olivine and those formed by heat treatment originated from exsolution of
trace elements in the host olivine, but others disagree with this hypothesis
(Bell et al., 1975).

Symplectite inclusions in plagioclase. Possibly similar inclusions of
spinel were found in the plagioclase of lunar highlands troctolite 76535 (Bell
et al., 1975) and in numerous plagioclase grains from various lunar soils and
breccias (e.g., Roedder and Weiblen, 1973, their Fig. 8; James and McGee, 1979,
their Fig. 3). In some of these grains, the core and rim of the plagioclase
have very different inclusion population densities. Some of these inclusions
in plagioclase appear to consist not of spinel but of pyroxene plus a vapor
phase; this result is exactly that to be expected if a melt whose composition
was essentially that of a mixture of pyroxene and plagioclase were trapped in
plagioclase and cooled slowly. However, some inclusions of this type in Luna
24 soil fragments that were thought to consist of pyroxene seem instead to consist of olivine, thus suggesting a troctolitic melt, if they are indeed melt
inclusions. Sobolev et al. (1980) performed heating experiments on possibly
similar inclusions in olivine from Luna 24, and reported homogenization at
1300-1360°C.

Significance. The distinction between the two modes of origin proposed for
these particular types of symplectite inclusions -- crystallization of melt inclusions or exsolution of minor elements out of the host structure -- is of considerable importance but is still an open question. In many samples, these
inclusions, both in olivine (e.g., Fig. 18-23) and in plagioclase, appear to be
present in high concentration, but when such materials are examined in reflected
light, the concentration is found to be <<1% (400 ppm by volume in one Luna 24
feldspar). Hence, if they are a result of diffusion of trace constituents, the
starting concentrations in the host need not be high. Also, if these inclusions
have formed by diffusion and exsolution, we should ask what similar diffusion
processes might have affected the composition of small true melt inclusions?

Inexplicable Inclusions

In the pressure of examination of the lunar samples as fast as they were
returned, a surprising number of inexplicable inclusion types and some inexplicable data were reported. Only a few of the more striking types of inclusion
problems can be mentioned below, in the hope of encouraging additional study to
resolve the problems. Additional puzzles are found throughout the literature on
lunar melt inclusions (e.g., Roedder and Weiblen, 1971, p. 519; 1972b, p. 278).

Inclusions in olivine with high K_D values for Fe/Mg. The equilibrium distribution of Fe and Mg between melt and olivine crystals has been the subject
of many studies. Equilibrium values of the distribution coefficient K_D [Fe
(crystal)·Mg(liquid)/Mg(crystal)·Fe(liquid)], determined from laboratory data
(Roeder and Emslie, 1970; Longhi et al., 1978; Walker et al., 1976; Lipin,
1976; Grove and Bence, 1977; Takahashi, 1978), are about 0.3, whereas most of
the Luna 24 inclusions (Roedder and Weiblen, 1978) and many of those from the
Apollo samples (e.g., Roedder and Weiblen, 1971, p. 515), have analyses yielding
K_D values between 0.5 and 1.5. Why? Several ad hoc (and hence not very satisfactory) disequilibrium mechanisms can be visualized to explain these high values,
invoking different diffusion rates for Fe and Mg, either in the main melt adjacent to the growing crystal before the inclusion was trapped, or in the crystal
adjacent to the inclusion. The measured K_D values also do not necessarily
reflect the local equilibrium values at the actual inclusion/host interface.

Low-K inclusions in ilmenite. Many of the large ilmenite crystals in the lunar high-Ti mare basalts have 50-100 μm melt inclusions (Fig. 18-24). Some of these are merely the normal immiscible K-granite melt (i.e., high-Si melt, now a glass) containing ~76 wt % SiO_2 and 6-7% K_2O. Other glass inclusions, even in the same ilmenite grains, also have ~76% SiO_2, and may look identical (Figs. 18-25 to 29), but instead of 6-7% K_2O, they contain an average of 0.04% K_2O, and many have no detectable K_2O by electron microprobe. Most of the difference is made up by CaO. About 70% of the 440 analyzed inclusions are low-K, and 30% are high-K, and only very few have intermediate compositions (Roedder, 1979e). When these glasses were first described (Roedder and Weiblen, 1975), 13 suggested mechanisms of formation were presented, and each in turn was shown to be invalid. To my knowledge, no one has come forward since then with a fourteenth and workable hypothesis. Still other glassy inclusions in ilmenite are high in K (~2% K_2O) but inexplicably low in SiO_2 (~35%). A very few inclusions in ilmenite have apparently been cooled slowly enough to crystallize large amounts of tridymite or cristobalite (Fig. 18-30).

Gas inclusions in single crystals. Many of the igneous rock fragments in the lunar breccias and soils contain intergranular gas inclusions (e.g., from Luna 20, Roedder and Weiblen, 1973a, their Figs. 9 and 10). Similar inclusions are found in many terrestrial samples. In hypabyssal rocks they are called miarolitic cavities and represent the volume originally occupied by a residual H_2O-rich phase that filled the interstices between early crystals. As the lunar magmas were extremely low in volatiles, however, I believe that most of the gas inclusions in them are essentially shrinkage cavities. Once a magma has crystallized to the point at which there is very little interstitial melt, the crystal mush can support itself, and further volume reduction within the mass (mainly during crystallization of the remaining melt) will result in low internal pressure on the liquid phase and the formation of shrinkage cavities between the existing crystals.

In these same soil samples, however, a few primary gas inclusions were found entirely within single crystals of ~An90 plagioclase (Fig. 18-31), and three small (2-4 μm) primary gas inclusions were found in Apollo 11 olivine (Roedder and Weiblen, 1970b, p. 833). The "sanctity" of most of these samples prevented any crushing-stage studies, but no visible liquid condensed when some were cooled to ~-5°C.

A primary gas inclusion within a single crystal in an igneous rock may form by several mechanisms. The most commonly considered mechanism requires the presence of a distinct gas phase at the time of solidification. Such a phase may form under a wide range of conditions (Roedder, 1972), so the interpretation of the physical significance of the resulting inclusion is difficult. A gas inclusion in a single crystal (Fig. 18-31), could form by exsolution of gas from the original melt, as a result of the crystallization of an essentially gas-free phase increasing the concentration of gas in the remaining melt, either at the growing crystal surface or throughout the melt.

Primary gas inclusions of identical appearance can also form in single crystals by an entirely different process. If the crystal (plagioclase in Fig. 18-31) grows from an essentially monomineralic melt (i.e., anorthositic), and some growth irregularity causes the trapping of some of this melt, subsequent crystallization of plagioclase from this melt on the walls of the cavity will form a gas inclusion. The "gas" will then be a shrinkage bubble resulting from the volume decrease on crystallization, since this takes place within a rigid container (the host crystal). As such, the bubble can be essentially a vacuum. Any components in the melt that have relatively high vapor pressures, such as alkalies and S compounds, will, of course, partition strongly into the vapor phase and may condense to solids at room temperature, once again leaving essentially a vacuum in the inclusion.

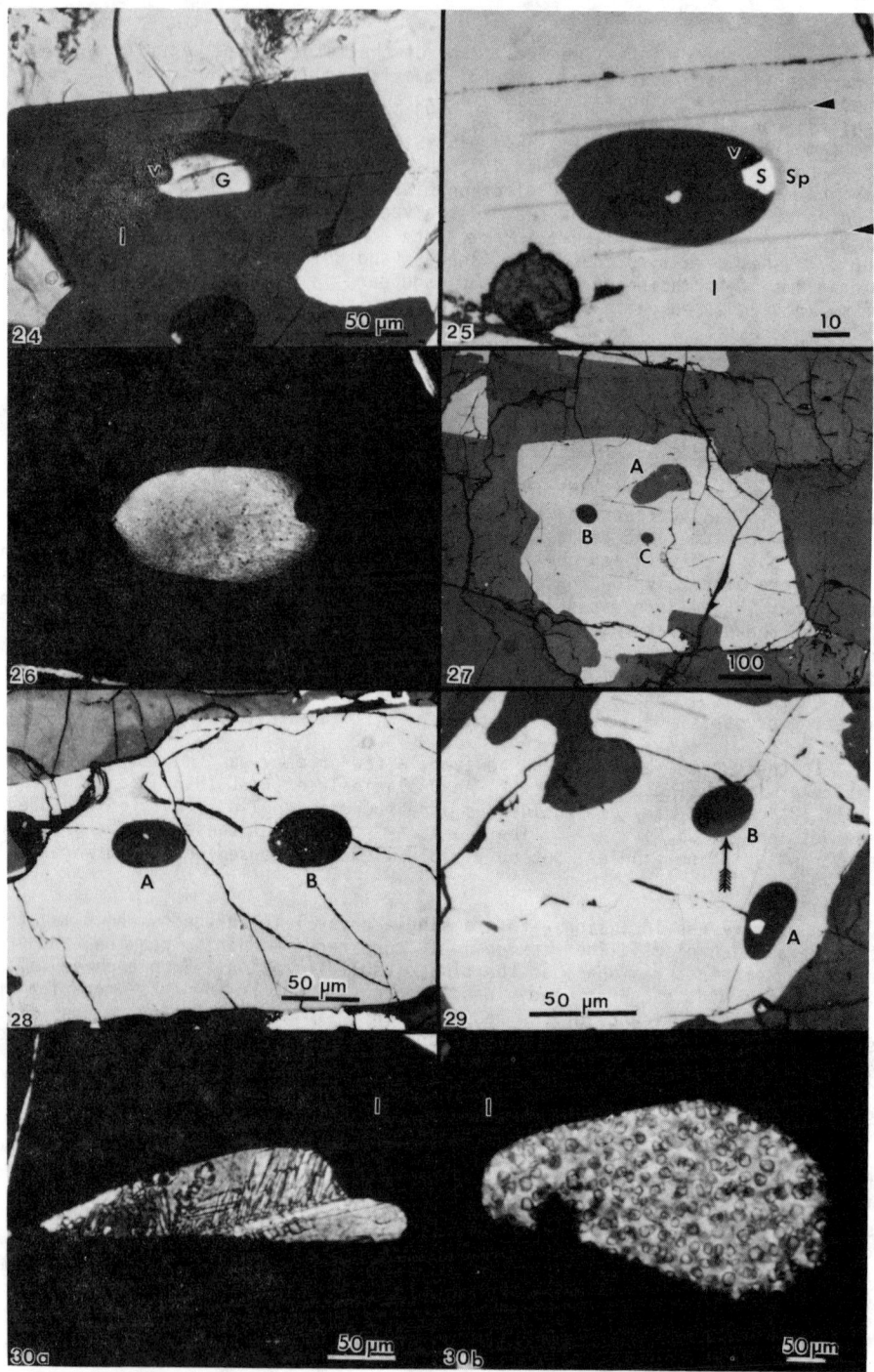

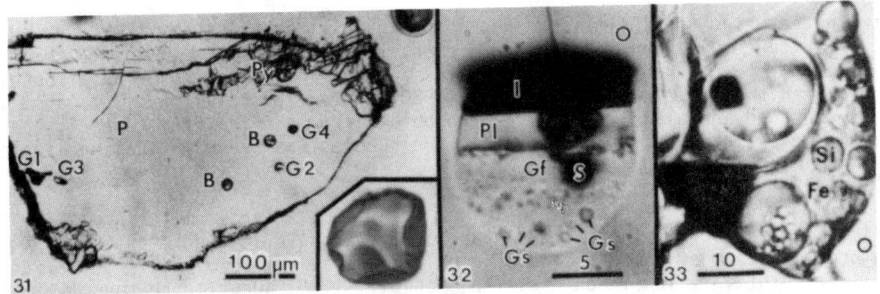

Figure 18-31. Single crystal of lunar plagioclase (P) from Luna 20 fragment 506.3 (~An90) enclosing a small area of pigeonitic pyroxene (Py) and several gas inclusions (G). Inclusions Gl and G2 are at the surface and hence could be merely cavities from plucking of pyroxene grains during slide preparation, but G3 and G4 (14 μm in size; enlarged in inset) are entirely surrounded by plagioclase crystal. Two circular areas (B) are artifacts (bubbles in the mounting media). Transmitted plain light. From Roedder and Weiblen (1973).

Figure 18-32. Early melt inclusion in lunar olivine (O) from mare basalt 10045,32, now containing epitaxial daughter crystals of ilmenite (I) and plagioclase (Pl), and a glass (Gf) from which numerous little blebs of very low-index, high-Si glass (Gs) have separated. An opaque sphere (presumably sulfide, S) is out of focus. The shrinkage bubble (over the plagioclase crystal) is apparently packed with abrasive compound from sample polishing. Scale bar in μm. From Roedder and Weiblen (1970b).

Figure 18-33. Melt inclusion in olivine crystal (O) from 1965 Makaopuhi lava lake, Hawaii, USA (sample M-13-12), showing splitting into immiscible liquids, one high-Fe (Fe) and one high-Si (Si). Scale bar in μm. From Roedder and Weiblen (1971).

Distinction between these two processes would be desirable but is difficult. In the second process, crystallization of plagioclase from the trapped, essentially monomineralic melt would concentrate the nonplagioclase constituents into a small amount of residual melt that might crystallize to form daughter crystals of olivine or pyroxene. Unfortunately, no such residue could be resolved on the walls of the small inclusion found (Fig. 18-31 inset), nor on the walls of several other similar inclusions. Even if found, however, this residue would not prove origin by the second process. Identical results could form by the first mechanism if, as is not uncommon, some of the surrounding melt was trapped along with the bubble. At present, I have no evidence enabling me to choose

Figure 18-24. Completely isotropic glassy high-K (4.37% K_2O) melt inclusion (G) in lunar ilmenite (I) grain 75075,88,P8. The inclusion is viewed in transmitted plain light plus a small amount of reflected plain light to show upper surface of ilmenite. Note shrinkage bubble (v) in glass. Another similar inclusion is at center bottom. From Roedder and Weiblen (1975).

Figure 18-25,26. Low-K melt inclusion in ilmenite (I) of lunar sample 71175,34,P7 (0.1% K_2O) viewed in reflected (Fig. 25) and transmitted (Fig. 26) plain light. It includes a faceted sulfide mass (S) and vapor(?) bubble (v). Note exsolution lamellae (arrows) and chrome spinel (Sp) at contact of sulfide with wall of inclusion. Figure 26 shows tiny crystals from devitrification. Scale bar in μm. From Roedder and Weiblen (1975).

Figure 18-27. Adjacent low- and high-K melt inclusions in lunar ilmenite 71175,34,P2. Inclusion A is pyroxene of the same composition as that outside and is probably a reentrant. Inclusions B and C are high-K (5.68% K_2O) and low-K (0.009% K_2O), respectively. Reflected plain light. Scale bar in μm. From Roedder and Weiblen (1975).

Figure 18-28. Adjacent high-K and low-K melt inclusions in a segment of a single tabular ilmenite crystal in lunar sample 70135,61,P3. Inclusion A contains 5.90% K_2O; inclusion B contains only 0.033% K_2O. The 55 μm of ilmenite between them shows no compositional gradients. Reflected plain light. From Roedder and Weiblen (1975).

Figure 18-29. Adjacent high-K and low-K inclusions in single ilmenite grain from lunar sample 75075,82,P26, in reflected light. The low-K inclusion (A) contains an immiscible sulfide bleb, and the high-K inclusion (B) has a vapor bubble (dark) and a partial rim (arrow) of pyroxene(?). From Roedder (1979e).

Figure 18-30a,b. Melt inclusions, in lunar ilmenite (I), now consisting of brown glass and silica crystals. The glass of the inclusion on the left contains only 49.3% SiO_2 but contains a large dendrite of tridymite (sample 10047,31); that on the right contains many slightly euhedral crystals of silica that may be tridymite or cristobalite (sample 10044,7652). From Roedder and Weiblen (1970a).

between these two possibilities, and, unfortunately, problems exist with both. Little evidence has been found of sufficient volatiles in lunar rocks to permit the first process, and the second requires the unlikely trapping of an almost pure anorthite liquid. Sobolev et al. (1980) cooled 38 inclusions (1 to 6 μm in size) that had anomalously large vapor bubbles, from olivine from Luna 24, to -198°C; no evidence for phase change in the bubbles was seen, indicating little if any H_2O or CO_2 could be present.

High-Si glass inclusions in olivine. Many of the larger melt inclusions in early phenocrystic olivine in the lunar mare lavas have, in addition to epitaxial daughter crystals of ilmenite and plagioclase and an immiscible sulfide bleb, small beads of very low-index glass (n probably <1.50) adhering to the olivine walls and protruding into the essentially basaltic glass of the inclusion (Fig. 18-32). Similar low-index beads were found in olivine from three Apollo 11 samples (Roedder and Weiblen, 1970b, p. 816), and presumably similar immiscibility was found in a terrestrial basalt (Fig. 18-33). Why such melts did not react with the host olivine is unknown. These beads were put aside as an unsolved enigma and forgotten, until samples of lunar soil from the Soviet Luna 24 mission were examined. In these, Roedder and Weiblen (1977a, 1978) reported a series of inclusions in olivine that consisted only of clear high-Si glass plus a vapor bubble and, in some, a sulfide globule (Figs. 18-34 to 37). These are isolated, apparently primary inclusions in unshocked, unzoned, single-crystal fragments of intermediate olivine (Fo 51 to Fo 73). The glass is optically uniform (Figs. 18-34,35) but compositionally zoned (Figs. 18-36,37). The glass is all high in silica, the maximum being 94%! The glasses are also very low in Al and Ca. How such extreme melts could form, how they got trapped in the olivine, and why they did not at least react with the walls to form pyroxene are still unanswered questions. The only feasible mechanism of formation suggested so far (Roedder and Weiblen, 1978) invokes metastable silicate immiscibility of grossly super-cooled liquids. Although feasible, this mechanism seems so unlikely that it is far from satisfactory.

INCLUSIONS IN TEKTITES AND METEORITES

Tektites

Tektites are strange glassy blebs and masses found in certain areas of the Earth (Czechoslovakia, Philippines, southeast Asia, Australia, etc.). Considerable controversy exists about their place of origin. Various theories have been proposed, the most widely accepted being that they were formed by large meteorite impacts on the surface of the Earth that fused surface materials and blasted the melt into high Earth trajectories. A cometary origin has also been considered (Suess, 1951), and O'Keefe (1976) argued that tektites are ejecta from lunar volcanoes. The various arguments involve extensive studies of the geochemistry of the various groups of tektites (and their possible trajectories) that are not directly pertinent here, but many tektites and possibly related materials contain plainly visible gas inclusions (bubbles) in the glass. These inclusions have been studied rather extensively.

Suess (1951) crushed gas-rich tektites in an evacuated tube and measured the pressure of the gas released by means of a McLeod gauge. He found no measurable amount of gas; i.e., the content of the bubbles must represent a fairly good vacuum (<10^{-3} atm). Suess indicated that an internal pressure of ~1 mm Hg would be required to form bubbles against the surface tension of the hot glass at zero external pressure. Water vapor might have formed the bubble and subsequently dissolved in the glass, but Suess also reported evidence suggesting that tektite glass may be very low in H_2O. The temperatures may have been high enough to form SiO vapor bubbles.

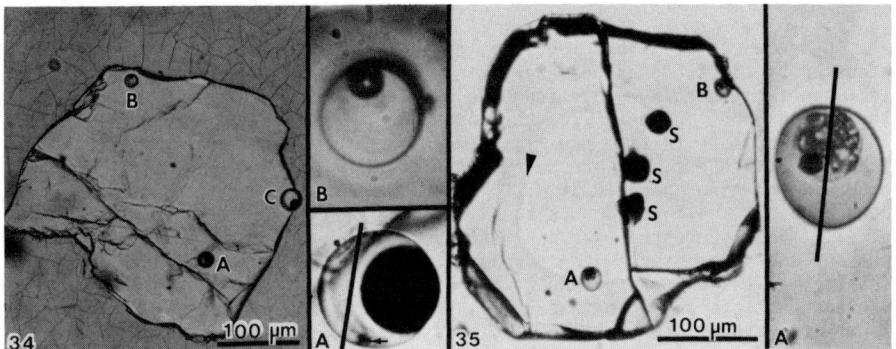

Figure 18-34. Single crystal of apparently unzoned Fo-73 olivine grain in Luna-24 sample 24109, 50-2, containing three silicate-glass inclusions, in transmitted plain light. At right are enlarged views of inclusions A and B. The curved shadows in A are optical artifacts; the glass is optically featureless except for a small speck (sulfide?; arrow). Inclusion B is completely surrounded by the olivine crystal. An electron-microprobe traverse across the glass of inclusion A (along the line) is shown in Figure 18-36. From Roedder and Weiblen (1978).

Figure 18-35. Single-crystal, apparently unzoned Fo-51 olivine grain in Luna-24 sample 24174,50-93, containing two silicate glass inclusions, A and B, and three spinel crystals (S), in transmitted plain light. At the right is an enlarged view of inclusion A. Both glass inclusions contain sulfide globules. Only A is at the surface. An electron-microprobe traverse across A (along the line shown) is given in Figure 18-37. The curving line paralleling the grain edge (arrow) is an artifact from the bottom edge of this grain. The vapor bubble is packed with abrasive grains. From Roedder and Weiblen (1978).

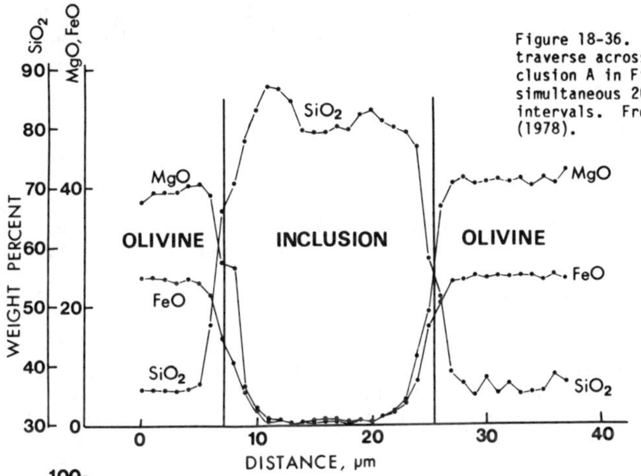

Figure 18-36. Electron-microprobe traverse across the glass part of inclusion A in Figure 18-34 made with simultaneous 20-second counts at 1 μm intervals. From Roedder and Weiblen (1978).

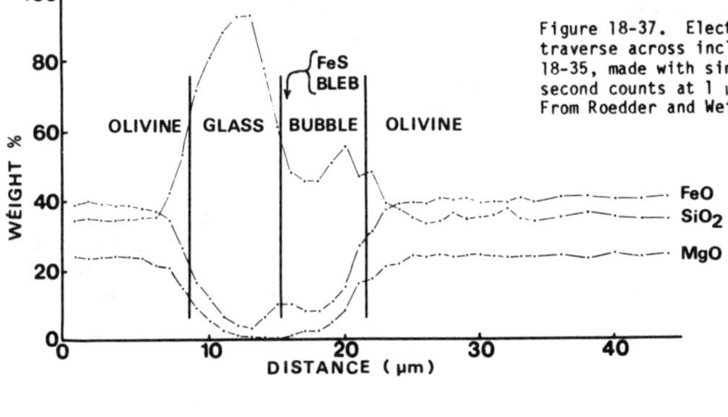

Figure 18-37. Electron-microprobe traverse across inclusion A in Figure 18-35, made with simultaneous 20-second counts at 1 μm intervals. From Roedder and Weiblen (1978).

559

O'Keefe et al. (1962, 1964), using spectrographic analysis of the light produced by electrodeless discharge in a large bubble (0.98 cm^3) in a bediasite tektite (from Georgia, USA), found Ne, He, and O_2. O'Keefe et al. assumed that the Ne and He diffused in from the atmosphere. Another large bubble (0.89 cm^3) in a philippinite tektite gave lines for H_2 only. If this H_2 had diffused in from the atmosphere, its pressure in the bubble could not exceed the pressure of H_2 in the atmosphere (0.5 x 10^{-6} atm), but the electrodeless discharge method used probably could not detect such low concentrations. As a result, O'Keefe et al. (1964) suggested that a H-bearing compound, such as water, was originally present and decomposed in the discharge. Müller and Gentner (1968) used high-sensitivity gas chromatography to analyze the gases from individual vesicles in various tektites and other natural glasses. They found empty vesicles adjacent to others with N/O ratios like air, and suggested that the air was trapped "...during the fusion of the tektite forming material or during the flight of the tektite through the lower atmosphere..." (p. 410).

Yu.A. Dolgov and coworkers have reported many analyses of the gases from bubbles in tektites, using sequential ultramicrochemical absorption procedures (for description and caveats, see Chapter 5). Dolgov et al. (1969a) studied bubbles in a series of moldavites (from Czechoslovakia) and reported major CO_2 and H_2, minor N_2 and acid gases, and no O_2 or CO. Dolgov et al. (1969b) listed 20 analyses (reported to four significant numbers) of vesicles in various types of tektites, for acid gases, CO_2, H_2, and N_2 (plus inert gases). No CO or O_2 was found in any vesicles, and N_2 was found in only four. Very large differences were reported between individual bubbles of a given type in a given tektite, and between different types. The total ranges found were (vol %): acid gases, 0-21.10; CO_2, 53.02-100.0; H_2, 0.0-40.10; N_2, 0.0-6.10. The reported decreases in bubble size on opening correspond to original bubble pressures (at room temperature) of 0.01 to 0.0002 atm.

Dolgov et al. (1971) reported additional analyses of bubbles in bediasite and Ivory Coast tektites, and of "Darwin glass" and "Libyan desert glass" (two enigmatic glasses, from Australia and Libya, respectively, possibly related to tektites). The bubbles in the Darwin glass contained CO_2 and air at pressures as large as several percent of atmospheric, which the authors believed confirmed a terrestrial origin. A larger number of additional analyses were given by Dolgov and Shugurova (1976a,b), Dolgov and Simonov (1976), Dolgov and Vishnevskiy (1976), and Shugurova et al. (1976).

Jessberger and Gentner (1972) crushed samples of Muong-Nong (Indo-China) tektites and Libyan desert glass at room temperature and analyzed the evolved gases by high-sensitivity mass spectrometry. The N_2:Ar:Kr:Xe ratios as well as the rare-gas-isotope ratios were found to be atmospheric, indicating a terrestrial origin for these glasses (or, I would suggest, possible later contamination by leakage of air into some of the vesicles). They report that the concentration of the active gases O_2, $\overline{CO_2}$, CO, and SO_2 varied widely between adjacent bubbles. Total gas pressure in the bubbles of these glasses was in the range of 100 mm of Hg, much higher than that found for other types of tektites. However, when using the crushing stage, I have found that the pressure even in adjacent bubbles in Muong-Nong vesicles may vary widely, suggesting air leakage into some through microcracks.

In addition to the studies of gases in vesicles noted above, a series of studies has been made of the gases in tektites by neutron activation (e.g., Shukla et al., 1979) and of the gases released on heating tektite glass (with or without bubbles?), in part to determine the ages by the K/Ar method (e.g., see O'Keefe, 1976). The composition of the evolved gases varies widely, suggesting that much is still to be learned about tektites.

Melt and Gas Inclusions in Stony Meteorites[2]

Glass, as interstitial inclusions or mesostasis and as that trapped within individual crystals from meteorites, has been noted in many meteorites since the early days of petrography, but relatively few detailed studies have been made of these glasses. Highly metamorphosed meteorites may have no recognizable glass inclusions or even the crystallized equivalents; less highly metamorphosed meteorites may have recognizable devitrified glass inclusions, and undevitrified glass can be a major component of nonmetamorphosed meteorites. Just as in terrestrial igneous rocks, the larger masses of glass (both interstitial and normal melt inclusions) commonly may be completely devitrified, whereas small melt inclusions within single crystals in the same sample will be glassy, as a result of nucleation probabilities (Chapter 3). Electron-microprobe analysis of such glasses usually shows them to be rather siliceous, and because such melts were presumably essentially anhydrous, viscosity may have been high and nucleation rates low. In other samples, the glass is high in normative anorthite.

The most common occurrence of silicate-melt inclusions in meteorites is in the chondrules, in chondritic meteorites. Chondrules, as defined by Fredriksson et al. (1973, p. 476), are "once wholly or partly liquid, individual bodies which cooled, solidified or crystallized rapidly." In spite of extensive studies over a long period of time, the enigma of chondrules in meteorites is still with us. Many theories have been proposed to explain the origin of these millimeter sized spheroids and the manner in which they have been incorporated in the chondritic meteorites, yet many questions remain unanswered. Study of the melt inclusions in them may eventually provide some additional insights into the origin of the host chondrules.

Most chondrules consist of both crystals and glass, in various ratios. The crystals may be randomly oriented and euhedral but commonly are in one or more groups of fine feathery or bladed crystals that may be subparallel (i.e., "barred" chondrules) or radial (usually radiating from a position on the edge of the chondrule). Many textural and mineralogical varieties have been described, but the crystals are usually forsteritic olivine or pyroxene. As a result of the apparently rapid crystallization, melt inclusions may be found as rows along the center lines of individual laths and as thin to thick septa between the laths, which are commonly subparallel crystallographically as well as physically.

As if chondrules themselves were not sufficiently enigmatic, several objects have been found in sections of lunar igneous spinel troctolite 62295 (an igneous rock) that resemble certain meteoritic barred olivine chondrules (Roedder and Weiblen, 1977c). Each object consists of an apparently spherical single crystal of Fo90 olivine, ~0.6-0.8 mm in diameter, containing a set of ~30-40 subparallel septa or "interstitial inclusions" that appear to be glass but that actually are parts of a single crystal of An95 plagioclase, whereas the septa in ordinary meteoritic chondrules actually consist of glass or devitrified glass. Presumably, the original melt was essentially an olivine-anorthite melt; after rapid crystallization of most of the olivine as laths, the septa of residual melt crystallized slowly enough to form single crystals of plagioclase. The olivine of the 62295 chondrules is radially zoned, having a relatively Fe-rich core and rim and an Fe-poor intermediate zone. Several possible origins for these objects have been proposed. They may be impact-generated melt globules that solidified in flight, spherical phenocrysts, or meteoritic chondrules, but none of these proposals seem adequate to explain the detailed observations.

----

[2] The general term "magmatic inclusions" has been used for all types of inclusions of former fluids within crystals formed during magmatic processes (Roedder, 1979a). Liquid inclusions are covered in a separate section (following) at end of footnote. "Silicate melt inclusions" is an inadequate term in view of the range of phase compositions involved in individual inclusions.

Many of the inclusion features common to most meteorites can be illustrated by work done on the Murchison carbonaceous chondrite. The Murchison meteorite was studied in detail by Fuchs et al. (1973) and has been the subject of considerable study and debate ever since (e.g., see Beckett et al., 1979; Grossman and Olsen, 1974; Jarosevich, 1971; MacDougall, 1979; McSween, 1977; Olsen and Grossman, 1974, 1978; Richardson and McSween, 1978; Roedder, 1981f). Murchison consists of two quite different materials: dispersed light-colored particles (diameters <4.5 mm) of a relatively coarse grained crystalline fraction formed over a range of high temperatures, and a very fine grained matrix formed at low temperatures. The matrix is a black hydrous carbonaceous material consisting mostly of hydrous layer-lattice silicates (phyllosilicates) high in Mg and Fe, plus calcite, whewellite [$CaC_2O_4 \cdot H_2O$], gypsum, and a poorly defined sulfide phase. The low-temperature hydrous part of the sequence of environments or events that yielded this obviously bimodal distribution is itself a question of considerable consequence in understanding the origin of the solar system, but here I deal only with the high-temperature part.

Most of the high-temperature part consists of isolated grains of individual euhedral or broken olivine crystals plus a few fragments of pyroxene or glass; ~23 vol % of the meteorite consists of clusters of loosely packed grains, mostly olivine and pyroxene; and <2 vol % of the meteorite consists of true chondrules (Grossman and Olsen, 1974).

The olivine crystals (mostly Fo99-100) from all three types of material contain silicate-melt inclusions. Fuchs et al. (1973) and Grossman and Olsen (1974) interpreted both the olivines and their melt inclusions as "primitive" high-temperature condensates from low-pressure gas (mainly H_2) in the original solar nebula. They proposed that the Ca-Al-rich melt-inclusion material condensed from the vapor phase first, as liquid droplets; then, at a much lower temperature and pressure, the host olivine crystals grew, also from the vapor phase (mainly H_2), at about 1170°C and <10^{-3} atm, and enclosed the round droplets of (now metastable, supercooled) melt to yield the melt inclusions.

Lord (1965), following several earlier workers, laid the foundations for the concept of a condensation sequence during cooling of a solar nebula; these calculations were extended by Grossman (1972) and others and have been a powerful tool for explaining many otherwise puzzling features of the origin and evolution of the solar system. However, my studies of the silicate-melt inclusions in Murchison (Roedder, 1981f) indicated that this theory was incorrect and that the inclusions and the olivine formed from a gas-bearing melt. The various lines of evidence need not be detailed here, but, in essence, the inclusion data and the lines of evidence presented by other workers,[3] are most compatible with a relatively high-temperature, two-stage formation of the olivine crystals from a gas-bearing, olivine-rich silicate melt containing some Ca and Al, rather than from a vapor phase. This evidence pertains only to these specific parts of the Murchison meteorite; other components in it (e.g., Grossman et al., 1979) may well be primary condensates from the solar nebula, in part subsequently altered. Much of the disagreement in the literature may well be based on these differences.

The silicate-melt inclusions in the Murchison olivine crystals consist of clear colorless glass, <36 μm in size, have no sign of devitrification, and generally have no daughter crystals. One type (in cored olivines) contains spinel daughter crystals (~5 vol %). Some melt inclusions contain a larger octahedral spinel crystal, but as the composition of the glass in such inclusions is the same as in those without spinel (Fuchs et al., 1973), and solid octahedra of spinel also occur embedded in the olivine, I assume that these spinel crystals

----

[3] McSween (1977) and Richardson and McSween (1978) reached the same conclusion but on other textural and chemical evidence.

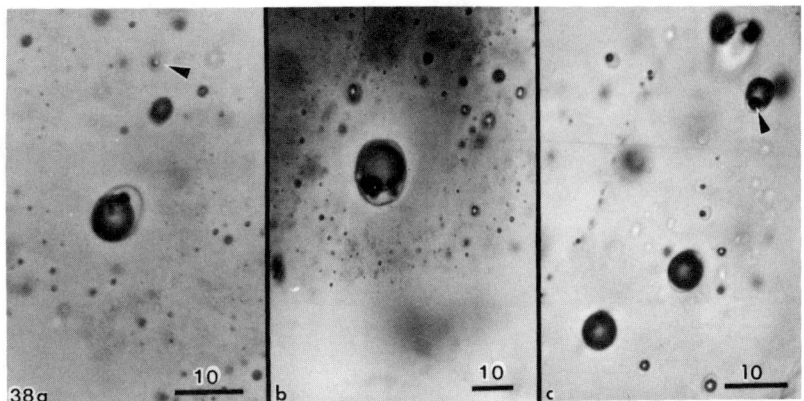

Figure 18-38. Inclusions in cores of cored olivine grains from Murchison carbonaceous chrondrite meteorite, from Roedder (1981f). (a) One 9 μm inclusion showing ~65 vol % vapor (dark, lower left), 2% opaque spherule (probably sulfide), glass (clear, upper right), and surrounding inclusion-free "halo." (Note - Although this particular inclusion is not at the surface, the polished surface intersects three other similar opaque spherules; all three were Fe sulfide with significant Ni). Some other smaller inclusions (arrow) have only 5-10 vol % vapor. Inclusion-free rim of crystal at lower left. (b) A 15 μm inclusion at edge of core, showing ~45 vol % vapor (dark, upper part), ~52% glass (clear), 3% opaque spherule, and a pronounced inclusion-free halo in the surrounding cloud of tiny primary inclusions. Inclusion-free rim of crystal at bottom. (c) Group of inclusions of three types in core of olivine grain. Many of the smaller (primary) inclusions are silicate glass with a small (5-10 vol %) vapor bubble (dark), but the larger (primary) inclusions are essentially all vapor, with a coating of glass that actually may amount to 25-30 vol %. Some also have an opaque spherule (arrow). A plane of pseudosecondary inclusions crosses the upper left corner; this plane stops abruptly at the edge of the core (beyond the area of this photo). Scale bars in μm.

are trapped solid inclusions. Some melt inclusions contain a spherule of Fe metal or Fe sulfide (see below). A vapor bubble is present in many but not all; it may occupy <75 vol %. Such "vapor"-rich inclusions are small (mainly <1 μm) but exceedingly abundant (~5 x 10^{10}/cm^3) in the cores of some cored olivine grains (Fig. 18-38). (The odd halo of inclusion-free olivine surrounding the larger inclusions seen in Figure 18-38 is paradoxical. Perhaps the best explanation is that when these cored olivines formed, they had a relatively uniform distribution of small inclusions, but by diffusion of ions (and holes) through the crystal during a very long period of heating, driven by the resulting small reduction in surface energy, slightly larger inclusions "ate up" the surrounding small inclusions, leaving an inclusion-free zone. Boland and Duba (1983) have shown that such diffusion in olivine is particularly fast along dislocations, at temperatures >1200°C.)

A bubble formed by shrinkage on cooling of a silicate melt in an olivine "bottle" would be essentially a vacuum, but such a bubble would appear visually identical with a bubble formed as a result of trapping of gas under pressure, so a series of inclusion-bearing grains were crushed in oil on the crushing stage. All were found to contain a vacuum, i.e., <10^9 molecules of noncondensable gas (Figs. 18-39,40).

The average composition of the 14 glass inclusions analyzed by Fuchs et al. (1973) is (wt %): SiO_2, 52.1; CaO, 18.1; Al_2O_3, 21.6; FeO, 1.3; MgO, 4.2; MnO, 0.03; K_2O, 0.05; Na_2O, 0.3; TiO_2, 0.8; Cr_2O_3, 0.2; sum, 98.7. The CIPW norm for such a composition is ~60 wt % anorthite, 27% diopside, and 13% quartz. With few exceptions, the compositions deviate little from this average. Silicate inclusions having similar composition have been reported from the Niger I C2 meteorite (Desnoyers, 1980).

Many olivine crystals contain one or more nearly spherical blebs of metal, and some contain many, mostly in the range of 5-10 μm diameter (Fig. 18-41).

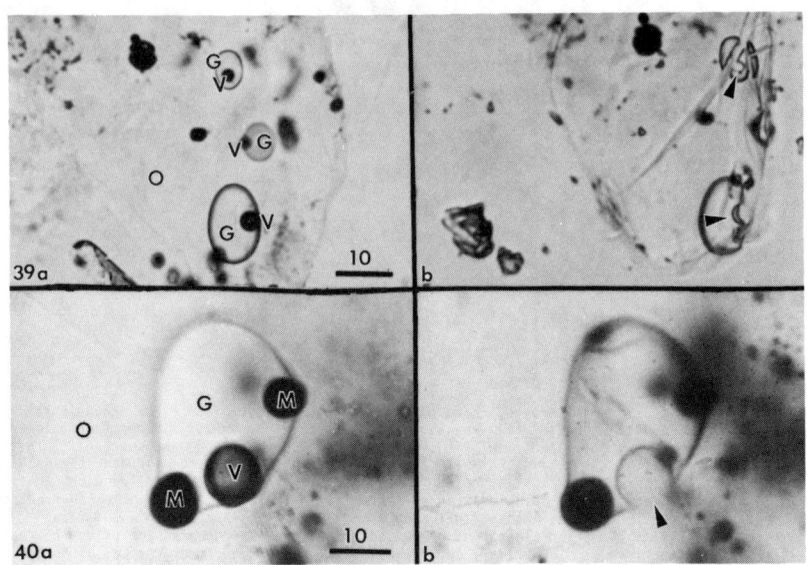

Figures 18-39,40. Primary silicate-melt inclusions in isolated olivine grains from Murchison meteorite (sample USNM 5347), mounted in matching index oil (n ~1.64) on the crushing stage. Left pictures, before crushing; right pictures, after crushing. The dark "vapor" bubbles (V) in three glass + "vapor" inclusions shown in Figure 39 and in the one inclusion consisting of glass (G, clear) + "vapor" (dark gray) + two opaque spherules (M, metal?) shown in Figure 40 all filled instantly completely with oil (arrows in right pictures), indicating that they contained no detectable noncondensable gas (i.e., <10^9 molecules). Scale bars in μm. From Roedder (1981f).

All gradations exist between pure silicate melt and pure metal inclusions (Fig. 18-42), and some silicate melt has apparently been trapped as the host olivine crystal grew around one or more metal blebs (Figs. 18-41,43). (I have found abundant similar metal/silicate inclusions in the olivine of other chondrite meteorites, including Iota, Tribune, Selma, Ochansk, Saratov, and Murray (unpublished data).) Fuchs et al. (1973) reported that these metal globules in Murchison contain (wt %): Cr, 0.20-0.96; P, 0.28-0.37; Ni, 4.0-7.4; and Co, 0.32-0.74. Si, Mn, and Cu were below detection limits (~0.05). Although most of the opaque globules in Murchison olivine are Fe-rich metal, the very few small opaque spherules within silicate-melt inclusions (Fig. 18-38) available at the surface for analysis were found to consist of Fe sulfide, plus minor Ni (Roedder, 1981f).

Most of the inclusions studied in Murchison are primary, as are most inclusions in meteoritic minerals, but planes of secondary and pseudosecondary inclusions are present in many grains, and such planes provided an important line of evidence against the condensation hypothesis. A few grains of olivine were found, each of which contained a number of inclusions that had a regular ratio of glass:"vapor":sulfide. In one group of 12 inclusions, this ratio was ~75:23:2 (vol %); in another group (of five), it was ~66:32:2 (Fig. 18-44). Such inclusions presumably trapped a homogeneous melt, which then crystallized considerable olivine on the walls (to yield the large "vapor" bubble) and became saturated and exsolved an immiscible globule of sulfide. Melt inclusions in olivine crystals in most chondritic meteorites that I have examined have smaller vapor bubbles (frequently ~15 vol %), and many have a small sulfide(?) bleb. However, because the vapor/glass ratio varied widely in inclusions in Murchison olivine,[4/] I pro-

----

[4/]Melt inclusions with large vapor bubbles, and apparently all-gas inclusions, were also found in Clovis, Murray, Al Rais, Tribune, Selma, Ochansk, and Saratov (unpublished data).

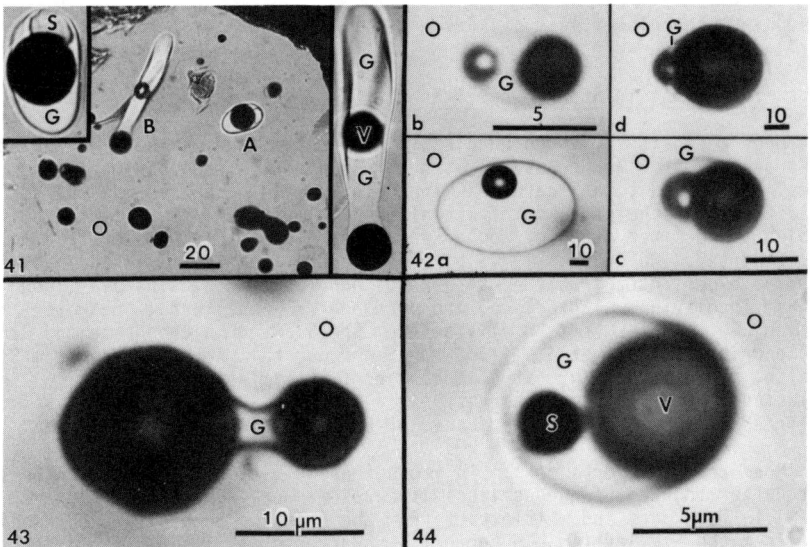

Figure 18-41. Isolated olivine grain (O) from Murchison meteorite (sample USNM 5377) with many opaque spherules of Fe metal, some of which have attached glass (G) ± vapor (V). Inclusion A contains a spinel crystal as well (see inset at left). Silicate melt in inclusion B seems to have been trapped as the crystal grew from bottom upward and surrounded the Fe metal spherule (see inset at right, taken at a different plane of focus). Scale bar in μm. From Roedder (1981f).

Figure 18-42. Four inclusions from isolated olivine (O) crystals in Murchison meteorite (sample USNM 5347), showing a wide range of ratios of glass (clear, G) to opaque spherule (presumably mostly metal). Each inclusion has a dark shrinkage bubble as well. Inclusion (a) has no opaque spherule, only glass and a bubble estimated to be ~4 vol %; (b) has ~40 vol % opaque spherule and ~8 vol % vapor; (c) has ~70 vol % opaque spherule; and (d) has perhaps 90 vol % opaque spherule. Scale bars in μm. From Roedder (1981f).

Figure 18-43. Pair of opaque spherules in olivine (O) of Murchison meteorite (sample USNM 5347) connected by a fillet of glass (G). From Roedder (1981f).

Figure 18-44. Primary inclusion in isolated olivine grain (O) from Murchison meteorite (sample USNM 5347) containing ~32 vol % "vapor" (V) and ~2.5% opaque spherule, presumably Fe sulfide (S), in glass (G). This inclusion is one of five in the grain, each having the same phase ratio (within the precision of the measurement), suggesting trapping of a homogeneous melt. The volume of "vapor" cannot be from shrinkage alone; it indicates considerable crystallization of olivine on the walls after trapping.

posed that these olivines grew from a two-phase, gas-bearing melt. Vapor bubbles (of S or possibly alkalies) formed on some of the growing olivine crystal surfaces and were enclosed as "vapor" bubbles, in part along with silicate melt. Subsequent condensation within these bubbles yielded the vacuum found. Gravitational pressure was presumably negligible or zero, but the vapor pressure at this time had to be high enough to permit nucleation in spite of the surrounding nebula gas pressure and the small-bubble surface-tension barrier.

The silicate melt from which the olivine grew was very olivine rich, and the melt inclusions crystallized still more olivine on the walls, driving the composition toward the plagioclase primary-phase field on the appropriate plagioclase-olivine-silica phase diagram. Failure to nucleate plagioclase when the composition reached the olivine-plagioclase boundary permitted the composition to continue to evolve along a metastable olivine-extraction line ("olivine control line") into the plagioclase primary-phase field, as described by McSween (1977).[5]

Liquid Inclusions in Stony Meteorites

Most meteorites, and chondrules in particular, seem to have formed at high temperatures, from anhydrous melts, out in space at presumably near-zero pressures. As a result, silicate-melt inclusions are expectable and apparently ubiquitous, but the presence of actual liquid inclusions (i.e., with moving bubbles at room temperature) would seem almost impossible. However, a few have been reported. Walenczak (1977a,b) reported CO_2-rich liquid inclusions in olivine and hypersthene from the Pultusk H5 chondrite, and Yasinskaya (1969) reported them in olivine, plagioclase, and pyroxene from the Yurtuk howardite and Chervony Kut eucrite, and in olivine from the Saratov L4 chondrite and Ochansk H4 chondrite. Warner et al. (1983) examined one thin section each from Saratov, Ochansk, and Chervony Kut but could not confirm Yasinskaya's report of liquid inclusions. In fact, the observations of Yasinskaya and Walenczak have not been confirmed in the literature and have not been generally accepted.

More recent reports of liquid inclusions in meteorites appear to be somewhat better founded. Aqueous liquid inclusions were reported by Fiéni et al. (1978) in feldspars and whitlockite from the Peetz L6 chondrite and in whitlockite from the St. Severin LL6 chondrite. Warner et al. (1983) confirmed those in Peetz.

The Kweiyang Institute Research Group (1978), Lu et al. (1978), and Z. Li et al. (1981) all reported the presence of "fluid inclusions" in the Jilin (formerly "Kirin") H5 chondrite, and Z. Li et al. (1981) also reported them in the Dongtai chondrite. In part, these reports may refer to silicate-melt inclusions. Thus, the first two papers refer to homogenization at 1050-1200°C and 502°C for glass and "gas-rich" fluid inclusions, respectively, from Jilin, but Z. Li et al. (1981) specifically referred to melt, gaseous, and fluid inclusions in Jilin and Dongtai.

Liquid inclusions in a diogenite. Warner et al. (1983)[6] reported unambiguous two-phase, liquid-vapor inclusions in En77 orthopyroxene in achondritic meteorite ALHA 77256 from Antarctica (a diogenite). The inclusions (<100 μm) were evidently secondary in origin. The possibility that the liquid inclusions were terrestrial artifacts was considered and rejected by Warner et al., on the basis of their occurrence. The locations of liquid inclusions were mapped to insure that they were not artifacts in the mounting medium or were otherwise introduced during sample preparation. Inclusions occurred at all levels in the thin sections and were not confined to the bottom of the rock chip, where epoxy was used to bond the sample to the glass slide. Liquid inclusions were present in samples that were vacuum-impregnated and in those that were not impregnated. Finally, if the observed liquid inclusions were artifacts introduced during thin-section preparation or during impact on Antarctic ice, inclusions should be present in other Antarctic meteorites prepared in a similar manner. Warner et al. (1983) searched about 25 additional Antarctic achondrites for liquid inclusions, without success. (Later they were found in "lunar meteorite" ALHA 81005, both in vesicles in the glass fusion crust and in pyroxene and feldspars.)

----

[5] Note, however, that in these olivine-rich compositions, only ~2.5 wt % additional olivine must crystallize, under metastable conditions, after the olivine-plagioclase boundary is reached, to achieve the glass compositions found. Much experimental silicate work is bedeviled by (or sometimes is only possible as a result of) such metastable extensions of liquidus surfaces into regions where other stable phases failed to nucleate.

[6] Ashwal et al. (1981) reported a few preliminary data, and more details were given in three abstracts published in 1982, but the 1983 paper greatly expanded the data base and considerably modified the previous interpretations concerning the pressure of formation.

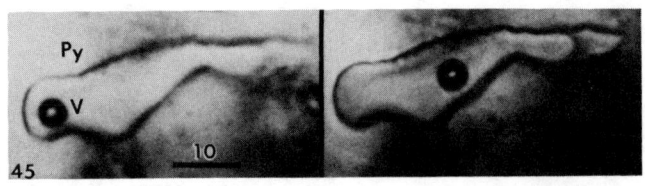

Figure 18-45. Two views of large flat liquid inclusion with moving vapor bubble (V) in orthopyroxene (Py) from diogenite meteorite ALHA 77256, taken at room temperature before and after heating to <225°C. Scale bar in μm. From Warner et al. (1983).

The vapor bubbles did not generally move spontaneously at room temperature, or rapidly upon heating, suggesting that the included fluid may be viscous, but the existence of movement was unambiguous (Fig. 18-45). The vapor bubble ranged from 1 to >90 vol % of these inclusions, and many presumed inclusions consisted of one phase only; whether these contain glass, liquid, or gas is not easily determined. In all two-phase inclusions, the vapor bubble continuously decreased in size as they were heated from -180°C to a maximum of ~225°C. Th was determined for a series of 49 inclusions containing relatively small bubbles; all homogenized to liquid in the range ~25-225°C with no significant preference for a given range. (Fifteen other inclusions having larger bubbles did not homogenize at ~225°C, the maximum temperature that could be used without damaging these samples.) As Th was approached, the vapor bubble decreased in size and moved sluggishly. Some bubbles showed rapid motion within 20°C of Th.

Freezing in the conventional sense of precipitation of a crystalline phase was not observed on cooling to temperatures as low as -180°C. However, in some runs, the vapor bubble became visibly deformed at temperatures below -50°C, and in some, the deformation became extreme at temperatures below about -100°C, suggesting formation of an invisible solid phase. Final "melting" of this phase appeared to be about -20 to -25°C. This phase could not be pure water ice because when ice melts there is a reduction in the combined volume of the liquid plus solid phases and hence an increase in the volume of the vapor bubble. The nature of the solid phase is unknown.

During some cooling experiments, a small amount of one or two equant discrete solid phases (~1 μm) appeared below temperatures of about -80 to -90°C; these may be salt.

In other cooling runs, a rather faint second vapor bubble appeared around the existing vapor bubble. This second vapor bubble was not always symmetric with the first vapor bubble, suggesting that it was real and not merely an optical artifact. The second vapor bubble appeared on cooling at about -40°C, and within the rather large error placed on these measurements by the observational difficulties, this second bubble disappeared on warming at about the same temperature.

Preliminary laser-activated Raman microprobe spectroscopy verified that the inclusions are aqueous; symmetric and antisymmetric stretch bands characteristic of H_2O (3200-3600 cm^{-1} $\Delta\nu$) were found (Fig. 18-46). This spectrum is characteristic of H_2O, but its overall quality was reduced due to fluorescence interferences. A broad band at ~2900 cm^{-1}, characteristic of C-H stretching, was interpreted as indicating the presence of higher aliphatic hydrocarbons than CH_4 (Ashwal et al., 1982). Other Raman data suggested possible graphite (Adar et al., 1982). Raman vibration bands characteristic of CO_2, CH_4, H_2, O_2, N_2, and S-bearing species were not identified. However, the extreme fluorescence of these inclusions may have masked vibration bands characteristic of the above species.

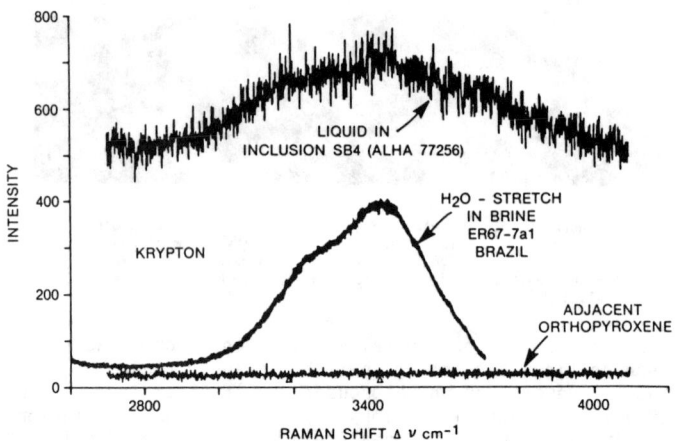

Figure 18-46. Raman spectrum obtained from liquid inclusion SB4 from ALHA 77256,52SB4. The observed bands between 3200 and 3600 cm-1 are characteristic of symmetric and antisymmetric stretch of H2O, as can be seen in the reference spectrum of H2O from a brine inclusion in pegmatitic quartz (Rosasco and Roedder, 1979). The low background level of the adjacent orthopyroxene host for inclusion SB4 is also shown. Spectra were obtained by using the 5682A line of a Kr ion laser. From Warner et al. (1983).

The most reasonable interpretation of these data is that the liquid inclusions contain a highly concentrated brine. This interpretation would explain the apparent viscosity and the behavior on cooling. The mixture of phases formed on cooling must have a molar volume such that during the melting, the net ΔV of liquid + crystals must be positive. Thus, if a solid hydrate is present that has $\Delta V > 0$ on melting (or transformation), this expansion, plus normal ΔV on warming for all solid and liquid phases, could permit the presence of some normal water ice in the mixture, even though it has $\Delta V < 0$ on melting.

Warner et al. (1983) also reported and illustrated some three-phase inclusions in the orthopyroxene of ALHA 77256. Presumably the three phases are glass, liquid water, and vapor. The liquid-water(?) phase is much less viscous than that in the inclusions described above, as the vapor bubble (consistently ~1 vol % of the water(?) phase) moves spontaneously at room temperature. Th (in liquid) occurred at ~80-100°C, but as the ~20 such inclusions were all very small, the results are only rough approximations. Freezing was not observed.

Liquid inclusions in chondrites. As a diogenite is, in effect, an igneous rock, the presence of aqueous liquid inclusions, trapped during late stages of crystallization of a "magma chamber" of some unknown (but presumably small) size, is not too difficult to accept. However, chondrules almost certainly must have formed at very high temperatures, and suspended in space rather than in a magma body, making the possibililty of liquid inclusions that much more remote. Nevertheless, Warner et al. (1983) reported liquid inclusions, with moving bubbles, in chondrites Bjurbole (L4), Faith (H5), Holbrook (L6), and Jilin (H5). Ashwal et al. (1982) reported additional discoveries in chondrites ALHA 77230 (L4) and ALHA 77299 (H3). These fluid inclusions occur in olivine both in and out of chondrules and also in pyroxene in Jilin. They have morphologic and microthermo-metric characteristics similar to those of the two-phase fluid inclusions in the diogenite. That is: (1) the vapor bubble generally does not move spontaneously at room temperature and barely moves at higher temperatures, except in Jilin, where fluid inclusions were discovered as a result of vapor-bubble movement (sometimes rapid) at room temperature; (2) Th values range from room temperature to a few hundred degrees; (3) freezing, i.e., deformation of the bubble on

cooling, was only observed in some experiments; and (4) the vapor bubble increases in size with decreasing temperature, to -180°C. From these data, the
authors inferred that the composition and origin of the included fluid in the
chondritic meteorites were similar to those of the included fluid in the diogenite.

Interpretation of liquid inclusions in meteorites. These liquid inclusions
present us with an enigma. On the basis of an experimental phase-equilibrium
study of the system $Fe-MgO-SiO_2-O_2$, Larimer (1968) noted that if a vapor phase
was present during crystallization of olivine, it would be H_2-rich. Such a
phase must be H_2O-bearing and not CO_2-bearing, on the basis of redox conditions
calculated from observed Fe/Mg ratios in olivine in chondrites.

Unfortunately, to get from an H_2-bearing vapor, presumably at high temperature and very low pressure, to liquid aqueous inclusions, either primary or
secondary, in olivine and pyroxene of chondrules or achondrites, requires some
flights of imagination. All meteorites in which Warner et al. (1983) found
liquid inclusions also contain gas inclusions; in Jilin, gas inclusions (i.e.,
vacuum) outnumber liquid inclusions about 20-fold.[7] Gas inclusions with or
without CO_2 have been reported in unspecified meteorites by Yasinskaya in a
series of three papers on the classification of such inclusions (see Yasinskaya,
1978). Chupina (1976), Chupina and Dolgov (1977), Kolomenskii et al. (1978),
and Dolgov and Vishnevsky (1978) reported gas inclusions under pressure and gave
gas analyses from the olivine of the Bragin pallasite and from olivine, plagioclase, and pyroxene of the Nikol'skoe and Elenovka chondrites and the Norton
County enstatite achondrite. As detailed earlier in this chapter, I found primary gas (i.e., vacuum) inclusions in olivine from the Murchison CM2 chondrite
and have found similar gas inclusions in the Saratov (L4) and Ochansk (H4)
chondrites (unpublished data). Occurrences of hydrous minerals in meteorites
(other than in carbonaceous chondrites) include kaersutitic amphibole in melt
inclusions in the Chassigny achondrite (Floran et al., 1978), possible preter-
restrial iddingsite alteration of olivine in the Nakhla and Governador Valadares
achondrites (Reid and Bunch, 1975; Berkley et al., 1980), and biotite, tremolite,
and serpentine (also siderite and calcite) from Jilin (Kweiyang Institute
Research Group, 1978). Robert et al. (1981) and Mattey et al. (1983) reported
isotopic analyses of the hydrogen evolved from various meteorites at various temperatures. Values of -300 (and possibly -800) per mil for H from feldspar from
the St. Severin meteorite were attributed by Robert et al. to the decrepitation
of fluid inclusions; -800 is close to the present estimates of primordial D/H
in the solar system. However, how aqueous fluids could be condensed, and how
they could be trapped as inclusions in the olivine of apparently unmetamorphosed
chondrules, without reaction, is paradoxical.

The origin of these fluid inclusions is unknown, the physical and chemical
characteristics of the included fluid are currently poorly constrained, the
apparent unmixing of the vapor phase at low temperature is unexplained, and
Warner et al. (1983, p. 734) stated "...a number of additional, unconfirmed, and
somewhat puzzling observations each of which may have compositional significance.
We are continuing our characterization of the fluid inclusions in the hope that
we can understand the genesis of their extraterrestrial fluids." R. Rudnick
(pers. comm., 1983) found that liquid inclusions with moving bubbles filling
some of the gas bubbles in vesicular glass in lunar sample 61538 (that almost
certainly are artifacts from laboratory fluids being sucked into vacuum bubbles
through minute cracks during sample preparation) had low-temperature behavior
similar to that reported above for meteorite ALHA 77256. However, not all the

----
[7] I have found that the cores of some grains of olivine in Jilin are densely packed with gas inclusions,
but that the rims of the grains are essentially inclusion free, and D. Henry and R. Rudnick (pers. comm.,
1983) reported identical textures in olivine crystals in Bjürbole chondrules. Unlike the cored olivines
from Murchison, described above, the inclusions in the Jilin olivine seem to be in secondary planes
(unpublished data).

data on liquid inclusions in meteorites can be easily explained away by the "artifact hypothesis," and the question of whether all, some, or none of these liquid inclusions in meteorites are real is still open.

Chapter 19

FUTURE of INCLUSION STUDIES

"There is no necessary connexion between the size of an object and the value of a fact, and ... though the objects I have described are minute, the conclusions to be derived from the facts are great."   Sorby (1858, p. 497).

CONTENTS

The above quotation from the father of fluid-inclusion study, H.C. Sorby, is just as appropriate now as it was 125 years ago, when Sorby was having trouble convincing a skeptical geological fraternity that although inclusions are small, they can provide much evidence about geologic processes, if only one tries to understand what they are telling us.

In this chapter, I briefly outline a few of the anticipated significant trends in inclusion study and some of the unresolved problems in interpretation of inclusion data that either are currently under active investigation or remain as challenges for future workers[1]. Foretelling the future is always risky, unless one is adept at the art of using the fortune teller's cunningly ambiguous generalities. Throughout this book, I have suggested many possible avenues for future research. Some have been stated explicitly, but others are only implicit in the discussions of the earlier work. The following pages outline some of the

----

[1] The numerous references documenting these various aspects are given in the earlier chapters (see Subject Index) and need not be repeated here.

routes that might be taken. Undoubtedly, further exploration will reveal some completely new and unexpected avenues and prove other routes to be dead ends, but so be it.

POSSIBLE DEVELOPMENTS IN INCLUSION METHODS AND EQUIPMENT

Chemical Analytical Procedures

Recent dramatic progress in the development of analytical instruments and procedures for small samples is particularly applicable to inclusion studies. This progress can be measured in terms of decreases in both the required sample size and the detectability limits, and of increases in the number of elements detectable at a given limit. McCrone (1970, 1971) summarized these trends in three diagrams (Figs. 19-1,2,3); since the time of McCrone's reports, most of the methods shown in those diagrams have been improved. In addition, various other procedures have been developed, and then applied to inclusions. Among these procedures are the proton probe (PIXE), inductively coupled plasma spectroscopy (ICP), ion chromatography, synchrotron radiation X-ray spectroscopy, isotope dilution, time-resolved Raman spectrometry, and laser microprobe mass analysis. Each of these methods may have specific applications in inclusion studies, but no single technique can be adequate in itself. These new developments make possible the analysis of individual inclusions selected to be of a known origin, and previously tested by nondestructive microscopy. As a result, interest will decline in the analysis of bulk samples containing large numbers of inclusions of possibly multiple origins. However, decreases in the amount of fluid analyzed normally result in increases in the cost of the equipment, in the difficulty and uncertainty of the analysis, and in the consequences of possible sample contamination or loss in extraction and handling.

Unfortunately, no real breakthroughs seem to be on the horizon for the problems of determination of inclusion pH and Eh at the time of formation. Improvements in the sensitivity and particularly in the calibration of laser-activated Raman spectroscopy may eventually provide sufficiently accurate determinations of the various S and C species present[2] to permit valid calculations of both pH and Eh, but such calculations require experimental data on speciation at elevated T and P, and the assumptions of equilibrium and no loss of H_2. Such determinations must also involve evaluation of the significance of the host and any daughter minerals containing variable-valence elements (particularly those that do not dissolve on heating). Identification of such phases by scanning electron microscopy with an energy-dispersive detector, by Raman spectroscopy, and by Gandolfi-camera X-ray diffraction, will certainly become more nearly routine, but such methods should always supplement, not replace, the determination of physical and optical properties by simple microscopy.

Because fluid inclusions may represent the last residual fluid from geologic processes such as crystallization of a magma, they may contain high concentrations of various rare elements. They also preserve highly soluble daughter crystals that normally could not persist under surface conditions. This combination provides a "window to a new mineralogy"; I am sure that many new minerals will be reported from fluid inclusions in the future.

In addition to the determination of the isotope ratios of individual elements such as C, H, O, N, S, Sr, Ar, and He, the inherent sensitivity of the mass spectrometer makes it particularly appropriate for analyses of the volatile molecular species present. The wide range of inclusion sample sizes and possible

----

[2] The laser-activated Raman method may eventually provide analytical data on metal complexes in solution. Also, since some ions and gaseous species have characteristic Raman spectra only when in solid compounds, there is considerable potential for low-temperature Raman spectroscopy.

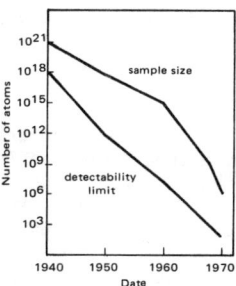

Figure 19-1. Progress in ultra-microanalysis, 1940-1970. From McCrone (1970).

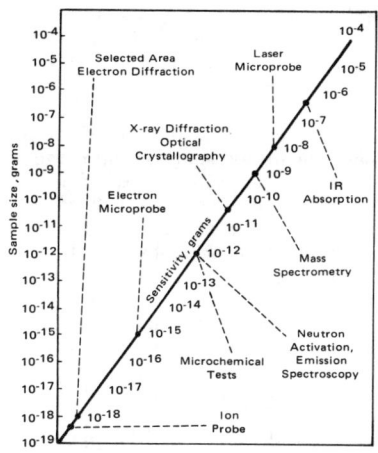

Figure 19-2. Sensitivity of microanalytical methods as of 1970. From McCrone (1970).

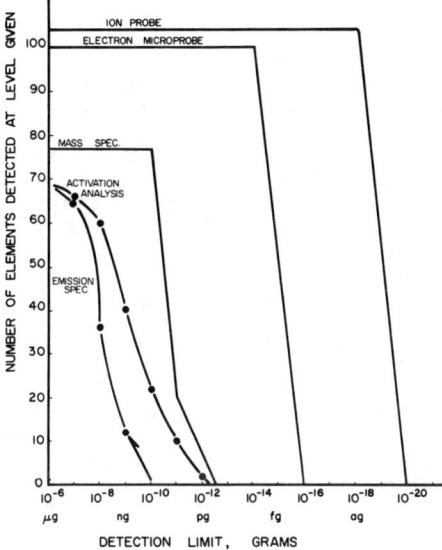

Figure 19-3. Versatility and sensitivity of physical analytical methods as of 1971. From McCrone (1971).

molecular compositions may cause changes in the yields of various fragments formed during ionization that can seriously affect the accuracy of analysis, regardless of the sophistication of the deconvolution procedures used in the data reduction. Such procedures can resolve some of the problems from the over-lapping of elements or molecules that have almost identical mass/charge ratio (i.e., isobars; e.g., $^{14}N^{14}N^+$ and $^{12}C^{16}O^+$, or $^{12}C^{16}O^{17}O^+$ and $^{13}C^{16}O^{16}O^+$), but in the future, ambiguity from overlaps of isobars will most likely be avoided by more extensive use of high-resolution mass spectrometry, which is capable of resolving composite peaks from a series of isobars.

The major new development in stable-isotope studies of fluid inclusions is the reduction in sample size requirements. Several milligrams of H_2O are common-ly required for determinations of D/H and $^{18}O/^{16}O$, or of CO_2 for $^{13}C/^{12}C$ and $^{18}O/^{16}O$. However, M. Sommer (pers. comm., 1983) reported determinations on 0.02 mg H_2O. R. Kreulen (pers. comm., 1983) reported determinations of $^{13}C/^{12}C$ on 0.01 mg CO_2, and Kazahaya (1983) reported on 0.04 mg CO_2. These values correspond to single inclusions in the 200-300 μm range. However, since a milligram of H_2O contains $\sim 3 \times 10^{19}$ molecules, and some techniques for counting individual molecules are known (e.g., Smalley, 1982), we can presumably expect still further improvements in the sample requirements; the major limitation will remain in the extraction procedures, where problems from absorption (and contamination) increase as sample size decreases.

The noble gases are particularly suitable for mass spectrometric techniques. They can be extracted, purified, and analyzed even though present in minute quantities (e.g., $<10^{12}$ atoms for He), so I expect that numerous studies of both the isotopic and elemental ratios of He, Ne, Ar, Kr, and Xe in fluid inclusions will be undertaken; these studies may help in many petrologic problems. Ar isotopic determinations will be particularly useful in various applications of K/Ar dating techniques. Similarly, Sr isotopes and Rb/Sr age determinations on fluid inclusions are now feasible and should become common. When the nature of the inclusions or their host phase permits, fission-track age determinations can also be obtained and may help in clarifying the chronology of multiple thermal events.

Microthermometric Equipment and Procedures

The major future thrust in microthermometry will probably be in techniques for the study of smaller inclusions (i.e., <5 µm), and over a wider temperature range. Small inclusions are more likely to be preserved even when internal pressures are high and hence may provide a record of unusual trapping conditions that would otherwise be lost. One of the major obstacles to microthermometry of small inclusions, at high or low temperatures, is the current lack of better long-working-distance, high-magnification, high-numerical-aperture objectives. The needs for good substage illumination are similar and may be filled by a similar objective, inverted. Such objectives would greatly enhance the effectiveness of currently available heating/freezing stages. Perhaps future observations may be made by using wavelengths other than visible (and IR) light, by ultrasound, or possibly even by nonoptical procedures. Improvements in computer image-analysis systems may even permit automation of microthermometric data acquisition. Currently available equipment could handle simple liquid-vapor homogenization; whether the computer can supplant the eye in resolving the complex three-dimensional images of phase changes in an irregular multiphase inclusion is far less certain.

The currently available microthermometric stages are vastly better than those formerly available, but they do not go to high enough temperatures. There is need for a new stage, particularly for the range 600-1400°C, a range in which many important geologic processes take place. Although the precision of the stages for >600°C may be high (as described in the literature), the accuracy is probably low, but above 1000°C, the need for accuracy is less.

The problem of decrepitation of inclusions during laboratory measurements is always bothersome and actually precludes microthermometric studies on some material. A high-pressure heating stage that would permit visual monitoring of the behavior of an inclusion during heating, while under an external pressure of perhaps 1 kb to prevent decrepitation, is obviously needed. Small windows, made of sapphire, could be used, and Ar or N_2 has been suggested as the pressure (and heat-conducting) medium, but such high-pressure gases would constitute a small but real explosion hazard; the pressure medium might better be a nonvolatile fluid such as a melted salt. Some of the technology of the present-day diamond anvil cell or the high P-T spectrographic cell might be adapted.

Many experimental problems contributing to the inaccuracy of thermometric stages may be resolved if truly valid calibration procedures can be devised. R.J. Bodnar (pers. comm., 1983) has reported a procedure for manufacturing primary and secondary inclusions that consist of pure H_2O or H_2O plus any desired concentration of CO_2, CH_4, NaCl, KCl, $CaCl_2$, etc., all in quartz and formed at known P and T, and of known density. This procedure yields inclusion standards for almost any of the types and temperatures of phase changes normally observed in natural inclusions, as well as composition standards for ICP, Raman, mass spectrometric, and other types of analytical studies.

Possible Developments in Inclusion Interpretation

Origin of inclusions. Obviously, much is still to be learned about the mechanisms of formation of inclusions, and this problem is basic to the use of fluid inclusion data to understand the geologic process being investigated. Even in the relatively "simple" trapping of primary inclusions during crystal growth from and into a free fluid phase, many problems and enigmas remain:

(1) Why do some crystals contain many inclusions, yet adjacent, presumably coeval, crystals of the same mineral have none?

(2) Why do some minerals trap inclusions much more commonly than do others?

(3) How are large primary inclusions of dense CH_4 (or CO_2) finally sealed shut in quartz without evidence of even minor amounts of a presumably coexisting fluid phase in which the host mineral can be assumed to be at least slightly soluble?

(4) Are there any unambiguous criteria for primary origin?

(5) How thick is the boundary layer (at the wall of an inclusion) from which the crystal grew during trapping under a given set of steady-state growth conditions?

(6) During continuous growth, what are the actual (rather than inferred) mechanisms and the critical growth parameters that cause an inclusion cavity to form originally and later cause it to be sealed closed?

(7) How can inclusions formed by such a process of continuous growth be distinguished from those formed by the covering of earlier dissolution pits?

(8) How can one distinguish between primary inclusions in original phases and primary inclusions in recrystallized phases (e.g., in salt and in lithium pegmatites)?

The possible processes of formation of primary inclusions in metamorphic environments, in which only a trivial amount of intergranular fluid phase may be present, are much less well understood than are those in growth from a free fluid. Furthermore, possibilities exist for dissolution of fluid inclusions in the host mineral -- and for the formation of new "exsolution inclusions" when conditions change. New grain growth during recrystallization generally sweeps out preexisting fluid inclusions, but occasionally does not do so; why?

One can hope that current developments in the study of imperfections in crystals and particularly in dislocation theory, plus experimental procedures to permit study of crystal imperfections (such as X-ray topography, decoration of dislocations, and transmission electron microscopy and diffraction) may eventually answer many of the above questions; until such answers are clear, inclusion study must remain, to some degree, subjective.

The problems of the mechanism(s) of formation of secondary inclusions are even less well understood. Although the general features of the healing of a fracture, leading to the trapping of a plane of secondary inclusions, were described 60 years ago, little has been learned since of the actual physical mechanisms that lead to trapping. Many of these questions concerning the origin of inclusions might be resolved through careful experimental studies of the formation of synthetic inclusions under known physical and chemical conditions.

Changes in inclusions after trapping. The present status of knowledge about changes in inclusions after trapping is perhaps even poorer than that on the origin of inclusions and generally involves problems of short-term (e.g., laboratory) or long-term (metamorphic) rate studies that have barely been touched. The following are only some of the questions that should be addressed:

(1) Under what circumstances (e.g., heating by a nearby dike or drop in pressure during metamorphic uplift) will a given inclusion stretch (yield plastically), rather than decrepitate?

(2) Can inclusions stretch (or collapse) in strong minerals such as olivine or quartz as they do in salt, either in nature or in the laboratory?

(3) What tools can be developed for both recognition and quantification of volume change?

(4) If such tools become available, what is the magnitude of the change under given conditions of time, temperature, and pressure difference?

(5) Under what conditions can H_2 or other constituents leak into or out of inclusions?

(6) How important are dislocations and other crystal imperfections in both leakage and deformation of inclusions?

(7) What is the lowest surface free energy configuration for a given host/fluid-inclusion interface (i.e., what are the presumed adsorbed species that cause some inclusions to be smoothly globular, whereas other inclusions in the same crystal, of different composition, are sharply faceted negative crystals)?

(8) What are the rates for such changes under given P-T-X conditions?

(9) Solution-reprecipitation is presumably the dominant mode of material transfer during necking down of many inclusions, possibly via an extremely thin surface film of a separate fluid phase, but is true surface diffusion involved in the recrystallization of gas inclusions in igneous minerals at high temperatures, where solubilities may be very low?

(10) What are the mechanisms responsible for the apparent "maturing" of planes of secondary inclusions, from many small to a few large inclusions?

(11) How common is the reopening of healed fractures, and what is the evidence for reopening?

Resolution of at least some of these problems may be forthcoming from the laboratory production of secondary inclusions mentioned earlier.

Metastability. By their small size and general freedom from spurious nuclei, fluid inclusions provide us with "visual autoclaves" that permit extensions of assemblages much farther into metastable fields than is possible in most laboratory studies. Such extreme cases of metastability may provide otherwise unobtainable P-V-T data and may help in quantifying the surface-tension correction that should be added to Th determinations on small inclusions. I also believe that statistical studies of daughter-phase nucleation in different metastable supercooled inclusion populations may have numerous applications for both melt and aqueous inclusions. Partial loss of volatiles from a magma during trapping of melt inclusions might be documented by a greater frequency of daughter phases in earlier inclusions that contain a higher concentration of volatiles (and hence nucleate more readily). Similarly, the different cooling rates for different parts of a single flow may be recognized or possibly even measured.

New P-V-T-X data. The most critical need for the interpretation of micro-thermometric data on fluid inclusions is a firm experimental data base[3] on the P-V-T-X properties of the various pertinent systems, particularly parts of the system $H_2O-CO_2-NaCl-KCl-CH_4$. Even the best known of the applicable systems, $NaCl-H_2O$, has not been determined over the full range of P, V, T, and X of interest to geologists. Fluid-inclusion data can (and should) be used to antic-ipate the phase behavior of the experimental systems and, hence, to design effec-tive experiments. Similarly, synthetic fluid inclusions of known composition, trapped at known P and T (see p. 212) can be used as miniature "visual auto-claves" to determine phase equilibria at P-T-X conditions that are difficult to investigate using available high P-T technology. An obvious example would be thermal-expansion (density) data on corrosive brines. Experimental and theore-tical thermodynamic data are needed to understand the phase changes in complex inclusions on heating or cooling and to derive estimates of the temperature and pressure of trapping from inclusions. As discussed in Chapter 9, new independent geothermometers (or geobarometers) should be developed, that can be used along with inclusion data to provide more accurate geobarometry (or geothermometry).

POSSIBLE DEVELOPMENTS IN INCLUSION STUDIES OF SPECIFIC ENVIRONMENTS

Sedimentary Environments

Although weathering and transport processes tend to eliminate the inclusion-rich detrital grains of any given mineral, careful studies of the fluid inclu-sions in detrital rocks can provide much geologically valuable information. Studies of the inclusions in the quartz pebbles of the Witwatersrand Au-U depos-its in South Africa have shown that important data on provenance can be obtained; many other conglomerates should be so studied, even though the immediate economic significance may be absent. Certainly this aspect has been almost totally ignor-ed by sedimentary geologists, yet it has potential in predicting the trend of sandstone bodies for use in petroleum exploration.

Future improvements in the study of very small inclusions would help in the presently difficult work of unravelling diagenetic history by the use of fluid inclusions in cements and overgrowths on detritus. In theory, a large number of other important questions in sedimentary petrology and ore research can be answered by studies of inclusions, such questions as primary vs diagenetic vs hydrothermal chert (or dolomite), but until new techniques are developed, most inclusions of <2 μm diameter are simply not usable. A problem that must be addressed in tracing such diagenetic history is that of possible volume changes in the inclusions (and/or leakage), brought about by increasing temperatures during deeper burial (see next section).

The study of the hydrocarbon chemistry of inclusions in overgrowths, in secondary fractures through detrital grains, and in new joint fillings in sedi-ments, combined with knowledge of thermal-maturation processes for hydrocarbons, has great potential as an aid to understanding the fluid migration, thermal pat-terns, and uplift history of petroleum-bearing terranes. Similarly, the newer highly senstive techniques of hydrocarbon analysis should be applied to the hydrocarbons known to be dissolved in many aqueous inclusions (an area that has hardly been touched.) A related question is whether biogenic molecules (or even fossil bacteria) might still be recognized in fluid inclusions. If such bacteria were living at the time of trapping, how might they have altered the chemistry of the included fluid before they died?

----

[3] The existence of significant differences between recent results of various theoretical (calculated) equations of state, even for simple one-component systems such as CO_2, and based on the same original experimental data set, makes it evident that many experimental data must still be determined.

The relatively rapid degradation of the various amino acids with time and temperature imposes severe limits on their use, but these compounds should certainly be searched for in inclusion fluids in relatively young samples such as speleothems. These amino acids may provide needed corroboration for the radioactive disequilibrium ages on the host carbonate that now must be used to interpret the isotopic signatures of the inclusion fluids.

Fluid inclusions provide poor but perhaps still the best available samples of early atmospheres (in amber for young ages and in salt for Permian or possibly even Devonian), particularly for studies of noble-gas isotopic ratios. The high-pressure gases in "popping salt," although probably unrelated to the foregoing, present some interesting geochemical problems: where do these gases come from, and why are their compositions so different in different salt deposits?

Metamorphic Environments

Study of the fluid inclusions in metamorphic rocks is difficult primarily because the inclusions frequently are small, secondary, and generally cannot easily be assigned to a specific stage in the rock history. These problems exist even in those simple rocks that have undergone only one prograde metamorphic sequence, followed by a retrograde stage during uplift. Additional studies of the variation in inclusion composition with the host-rock assemblage, and particularly the study of fracture-filling assemblages, will help to clarify such problems. The necessary extrapolations from Th to Tt are generally so large that independent geobarometers (or geothermometers) will almost certainly continue to be needed to obtain Tt and Pt.

Waters expelled early in metamorphism of marine sediments start as essentially seawater and become increasingly modified. The origin of the volatiles present later, at higher P and T, is sometimes much more obscure, and much work is needed on metamorphic inclusions before it will be possible to chart the probable change in composition of the pore fluids with time in any given terrane, because too many simultaneous and sequential processes are at work. With increasing grade, organic matter decomposes to form H_2O, CO_2, H_2S, hydrocarbons, and nitrogen (as N_2 and as complex compounds); hydrocarbons break down in turn; evaporites dissolve; osmotic-membrane filtration takes place; fluids in equilibrium with one assemblage may move and be trapped in another; H_2 (or H_2O) may dissolve in or exsolve from quartz; H_2O (± solutes) and CO_2 (± CH_4, N_2, etc.) may become immiscible; hydrous minerals, graphite, and carbonates form or are destroyed; and evolving CO_2 from decarbonation reactions may sweep H_2O out as it leaves. Although the last of these processes probably has played a large role, no combination of these processes seems adequate to explain the almost ubiquitous switch from hydrous fluids in moderately high grade (e.g., amphibolite) metamorphism to essentially pure CO_2 fluids for all granulite-grade rocks. In this connection, rocks that may have formed at relatively high pressure and low temperature should be searched carefully for extreme-density CO_2 inclusions (>1.2 g/cm^3); such inclusions will only become obvious with extensive and careful studies at low temperatures and with laser Raman spectrography, particularly of small inclusions.

Perhaps the single most pressing problem in the study of inclusions in metamorphic rocks is not the existence, but rather the magnitude of the possible long-term effects of placing inclusions in a pressure gradient, which occurs whenever the P-T path of the rock does not happen to follow the isochore for the fluid trapped in the inclusion. Under what conditions do the volumes of such inclusions expand or contract, thus changing Th and the isochore, and eliminating evidence of the former conditions? Also, does any given plane of secondary inclusions represent the original fluid and conditions of trapping, or have the inclusions decrepitated and refilled with a new fluid at a new P and T? How important is the internal fluid pressure in the mechanisms of deformation, particularly in shear-zone environments?

The composition of the inclusions presents other questions. If the mineral becomes zoned, or the intergranular (pore) fluid changes from that within the inclusion, a chemical gradient is established. Under what conditions do the inclusions gain or lose materials, and how are these materials transported? What isotopic fractionations occur during these processes?

These questions must be answered before unambiguous interpretations can be made of the significance of P, T, or X data from fluid inclusions in metamorphic rocks. Until such ambiguity is removed, fluid-inclusion studies on such rocks should continue to be compared with data from other independent methods, as recommended by Touret (1977). The lesson geologists learned from the Lord Kelvin affair (Gould, 1983) should never be forgotten. Many lines of geological evidence as to the great age of the Earth were temporarily overridden by Lord Kelvin's calculations of a much younger age, based on the then-known laws of physics. Although his calculations were mathematically correct, his assumptions were not.

Intrusive Rock and Pegmatitic Environments

Ordinary silicate-melt inclusions trapped during the crystallization of igneous rocks might seem to provide few problems, but numerous questions remain. Problems in microthermometry and chemical analysis of such melt inclusions are covered below, under "Extrusive-Rock and Volcanic Environments." The lack of evident silicate-melt inclusions in many igneous intrusive rocks, particularly those formed at greater depths, is puzzling. Three possible reasons are: (1) inclusions were simply not trapped, because of the very slow crystallization; (2) the inclusions have also crystallized and hence are not easily recognized as such; or (3) the original magmatic minerals (e.g., a single alkali feldspar solid solution) have lost all their melt inclusions on recrystallization to new minerals (e.g., separate Na- and K-feldspars). Little evidence exists to choose among these three, and still other reasons may exist.

Many of the inclusions in most intrusive rock and pegmatitic environments can best be viewed in terms of fluid-fluid immiscibility, as these various immiscible separations control the compositions of the fluids and, in many examples, are the cause of the trapping of the inclusions. Only a few of the many proposed compositional types of immiscibility have been tested using synthetic compositions.

A major question that could (in theory) be readily answered by inclusion study is whether (and at what stage or depth) silicate-aqueous fluid immiscibility occurs in the presumed evolutionary sequence granite → common pegmatite → rare-element pegmatite. In spite of many studies of the inclusions in granites and their associated pegmatites, mainly in the Soviet Union, the answer is still ambiguous. The immiscibility may hinge on the original concentration of volatiles and what fraction was lost during intrusion and crystallization; these questions also are far from settled. The pegmatite "system" is so complex, however, that much work will be needed to understand the effects of even a few of the many compositional variables (plus temperature and pressure) on the extent of the immiscibility. Inclusion evidence points toward the existence of at least some dense hydrous silicate-bearing Li-pegmatite melts at temperatures below 500°C.

Inclusions in the carbonatite intrusions seem to suggest a series of stages of immiscibility, possibly sequential, and still another type of immiscibility may have occurred in the origin of the hydrocarbon-rich fluids associated with some alkalic rock massifs. Each of these various "systems" should be explored experimentally, to aid in understanding the significance of the data from the natural rocks.

Ore-Deposition Environments

Many avenues are open for further inclusion work on ore deposition. Further analytical data, both chemical and thermometric, on inclusions from samples closely tied to the available paragenetic and structural data may reveal flow directions, and hence be useful in delineating the environment -- and possibly the cause -- of ore deposition. Lower temperature deposits such as the Mississippi Valley type, and the epithermal veins, appear to offer the best opportunity for determining flow directions. Many more inclusion analyses are needed on the Mississippi Valley-type ores to corroborate the scanty evidence now available for high-metal, low-S fluids. Detailed study of the organic compounds present in these inclusions might prove valuable in understanding the source environments for the fluids. The improvements in analytical techniques for stable-isotope analyses of inclusions, and in the understanding of the processes affecting the various fractionation factors, are such that we can expect many new isotopic studies of inclusions in Mississippi Valley-type ores.

The same features are equally appropriate for other types of ore deposits. Many of the epithermal vein-type deposits have formed from fluids that appear to include waters from two or even three sources, and there is also the possibility of thermal shock yielding pseudosecondary inclusions. Greater resolution of the isotopic signatures of such mixtures should ultimately do much to clarify these processes. It might also help solve the problem of the source(s) of the major solutes and the metals in the ore fluids.

As the studies of inclusions in the various paragenetic stages of ore deposits become more detailed, the processes of ore deposition are found to be more fascinating and frustratingly complex than originally envisaged. Deposition of even a single crystal has sometimes taken place in a large number of successive stages, each with its own fluid T, P, and X. The recognition of this fine structure indicates that much of the older work must now be redone in greater detail. Most important, this fine structure must be considered in trying to reconstruct the "plumbing system" for the deposit, both in time and space, and the environment in which the ore was transported and precipitated. Eventually, it may also provide some constraints on the time required for ore deposition. This time value, in turn, might be used in calculations of intergranular fluid diffusion in the development of wall-rock-alteration zones.

In view of the extensive literature on the application of the decrepitation method to ore deposits, particularly in the USSR, an important question concerns the usefulness of decrepitation anomalies in the search for ore. I believe that in some situations, decrepitation can be useful as a qualitative empirical field method. However, for the reasons given in Chapter 9, I doubt that decrepitation can ever provide calibratable data of any real value in understanding the environment of ore deposition.

Many important questions remain concerning the geochemistry of the ore-forming fluids. Some fluids evolved from the dewatering of volcanic-metamorphic terranes seem to have formed base-metal deposits; other fluids have formed simple Au-quartz veins (± some W); still others have formed no ore deposits. Why?

The data on inclusions in the porphyry Cu deposits have raised many problems, all of which may be solved, at least in part, by additional inclusion study. (An equal amount of effort spent on the study of inclusions in other types of ores would probably have raised just as many problems.) Among these problems are the following:

(1) Why are some stocks barren of Cu ores, even though they have highly saline

fluid inclusions?

(2) When in the evolution of any given ore deposit were the Cu and other metals introduced and when and why did these metals precipitate?

(3) What inclusion evidence exists for differences in the P-V-T-X conditions during deposition of the many types of porphyry ores?

(4) Similarly, what inclusion evidence exists to link porphyry deposits genetically, spatially, or chronologically with skarn and epithermal deposits?

(5) What sources of H_2O (magmatic, meteoric, or other) contributed to the alteration and the ore-forming fluids, and if multiple, where and how did mixing occur?

(6) How prevalent was an early, low-density, low-salinity "plume," from which high-salinity fluids could form by partial condensation, as compared with evolution of a dense NaCl-rich fluid directly from a magma?

(7) At what stage in the evolution of the fluids did they become saturated with respect to NaCl, and was this before or after the inclusions under examination were trapped?

(8) Can better criteria be developed to determine the relative ages of the many healed fractures?

(9) Are the cation ratios in the inclusion fluids (particularly K:Na:Ca) appropriate to the solid-mineral assemblage presumably present at the time of trapping?

(10) Are the nondissolving solids commonly found in such inclusions accidentally trapped solids, true daughter minerals, or a result of autooxidation from loss of H?

(11) If they are the result of autooxidation, why are the sulfides not oxidized?

(12) If these daughter minerals do not result from oxidation, are the dissolution rates too slow to be detected on a laboratory time scale?

(13) Which, if any, of the microthermometric data might be invalid because of heterogeneous trapping of a vapor bubble or a crystal with the fluid?

(14) Did boiling (or condensation) take place at one or more times during the evolution of the fluids?

(15) What were the thermal and pressure gradients during mineralization, and how did they change; were the pressures hydrostatic or lithostatic?

(16) How accurate are the P and T estimates from fluid inclusions, and how do they compare with other geobarometers and geothermometers?

(17) What is the relationship of the Cu-depositing fluids to the (various?) fluids yielding the several types of alteration?

When we have answers to these questions (and perhaps a few more), we will be able to say that we have a fair understanding of the conditions of formation of the porphyry Cu (and perhaps Mo and Sn) deposits.

Extrusive-Rock and Volcanic Environments

Future applications of the study of inclusions in extrusive rocks can be considered in terms of either equilibrium processes or rate processes. So far as equilibrium processes are concerned (as in the intrusive-rock and pegmatitic environments), much work is needed to understand the effects of the several potential fluid/fluid immiscibilities. These immiscibilities include silicate/ hydrosaline melt immiscibility in shallow granitic bodies, early silicate/sulfide and late silicate/silicate immiscibility in basalts, and vesiculation by the formation of a low-density H_2O or CO_2 phase. (Rate processes may also enter all these immiscibilities.)

The greatest potential contribution to igneous petrology that may come from inclusion study concerns the volatiles in magmas. The presence of inclusions of a volatile phase, whatever its density, indicates that saturation had been achieved at some time,[4] but the most valuable data are the actual analyses of volatiles such as H_2O and CO_2, but also S and Cl, in silicate-melt inclusions. Such analyses are difficult, and serious questions remain concerning the accuracy of some determinations. The compositions of the gases obtained from melt inclusions should also be compared with those from volcanic vents and with those calculated from thermodynamic data.

Tm and Th determinations on melt inclusions in the various phases in igneous rocks should always be compared with independent evidence of the sequence of crystallization. Definitive experiments also still have to be made on the possibility of volatiles leaking from melt inclusions, both in nature and during Th determinations; until then, all Th determinations on melt inclusions, especially the higher values, should be considered suspect. Chemical analyses of melt inclusions, particularly those from several different host phases, will continue to provide valuable insight into magma mixing and into the differentiation trends of various magmas. The analytical problems, however, are still far from solved. I fully expect that in spite of these difficulties, the study of melt inclusions will become as much an accepted part of igneous petrology as the equivalent study of aqueous inclusions has become for research in ore deposits.

Fluid inclusions from a series of land-based geothermal areas have been studied and have been useful in understanding the circulation patterns. This work needs to be extended, not only because of the direct economic significance, but because such systems are probably the active equivalents of the processes that formed at least some economic hydrothermal mineral deposits. Similarly, study of cores from these land-based systems may help in understanding the more inaccessible deep-sea geothermal areas.

Much less is known about rate processes in inclusions, and many avenues could be explored. Can melt inclusions be used to investigate the conditions leading to oscillatory zoning? What are the diffusion processes and their effects on both sides of the melt inclusion/crystal interface, during and after inclusion trapping? What effects do rate processes have on the composition of the melt trapped in a growing crystal during exsolution of an immiscible phase? Once an inclusion is trapped, what are the rate processes involved in the nucleation and growth of daughter crystals? Also, what are the rates at which H_2 (or O_2) can diffuse out of melt inclusions in various magmatic minerals, once an adequate concentration gradient is established?

----

[4] In this connection, apparently "empty" inclusions in igneous rocks, particularly in phenocrysts, should be checked carefully for gases under pressure.

Upper-Mantle Environments

Unfortunately, the types of inclusion samples available from upper-mantle environments are limited to ultramafic xenoliths ("nodules") and megacrysts from kimberlites and basalts, and diamonds. Diamond would be the ideal host for samples of deep-seated volatiles, but actual fluid inclusions have not been recognized in diamond, and the numerous other possible sources of the gases found in diamond make interpretation of the available analyses difficult. When improved methods of investigating minute inclusions are available, studies of the reported submicroscopic melt inclusions in diamond may be our best chance to learn about the volatiles in the several diamond-forming environments. Analyses of the five noble gases in various diamond samples certainly should help in clarifying some of the more enigmatic aspects of this fascinating mineral and mantle environment. Similar studies of the noble gases in melt inclusions in xenoliths and phenocrysts can be expected to be highly rewarding.

Studies of the CO_2 inclusions in various minerals (mainly olivine) from "nodules" erupted with kimberlites and basalts have yielded much information and offer promise of much more; such nodules are the only samples of solid mantle that are available in quantity. Unfortunately, the inclusions in nodules from kimberlites have been relatively neglected; most of the following suggestions must be based on work done on nodules from basalts, but they are equally applicable to kimberlites and to megacrysts in kimberlites and basalts. The questions begging for answers (or for more complete or precise answers) are numerous:

(1) Which of the several possible trapping mechanisms have yielded most of these CO_2 inclusions?

(2) How much of the obvious shearing in these samples, now outlined by CO_2 inclusions, formed in the mantle, and how much formed during eruption to the surface?

(3) Have the relatively unsheared samples gone through an earlier stage of shearing, and hence developed some of their "primary"-appearing inclusions during subsequent recrystallization?

(4) Why in some olivine samples are the inclusions of CO_2 plus melt sharply faceted negative crystals, yet adjacent simple melt inclusions are subspherical?

(5) What sort of diffusion processes have caused the obvious recrystallization and coarsening of planes of secondary inclusions, and what sort of rates (and hence times) are involved?

(6) What relatively nonvolatile solutes were present in the supercritical CO_2 fluid at the time of trapping?

(7) What was the actual ratio of volatile species in the gas as trapped?

(8) What are the solids that appear to be daughter minerals in some CO_2 inclusions: graphite, spinel, or another mineral?

(9) Why has there been no reaction to form magnesite during cooling?

(10) What sample bias was introduced into the available samples because of the evident decrepitation of the larger inclusions on eruption, and were these larger inclusions cogenetic with the remaining intact inclusions?

(11) Would careful search at high magnification and low temperature reveal small very high density (hence deep-formed) inclusions in olivine?

(12) What role do the exsolved immiscible supercritical CO_2 fluid droplets in the host basalt liquid (i.e., those <u>not</u> trapped in olivine crystals) play in vesiculation on eruption?

(13) What is the possible magnitude of the stretching of high-pressure CO_2 inclusions in olivine during eruption?

(14) What might be learned from a study of the isotopic composition of the He, Ne, Ar, Kr, and Xe, as well as the C, in these CO_2 inclusions?

(15) What are the genetic interrelations of C in solution in the olivine structure, free C and carbonate phases in the nodules, CO_2 in the inclusions, and the C in carbonatites and diamonds?

Extraterrestrial Environments

Study of the lunar samples revealed several apparently new types of melt inclusions (e.g., those containing epitaxially oriented ilmenite daughter crystals and those from immiscible silicate melts), which were later shown to be common on Earth as well. Several other types of lunar inclusions are still enigmatic and need additional studies. Some problems include: (1) The rare examples of apparent gas inclusions; (2) the very high Mg-Fe K_D values for inclusions in olivine; (3) the extremely low-K melt inclusions in ilmenite; (4) the extremely high-SiO_2 inclusions in olivine; (5) the possibilities of a melt-inclusion origin for at least some symplectites in plagioclase or olivine; (6) the large break in available melt-inclusion samples between the early- and the late-crystallizing phases during the crystallization of the lunar basalts; (7) the large apparent difference in cooling rates of inclusions of K-granite melt (some basalt samples contain inclusions of clear glasses, and other basalts of similar grain size (and hence possibly similar cooling rates) have inclusions of coarsely crystalline tridymite and K-feldspar); (8) the wide range in K/Na for immiscible high-silica melts in various basalt samples. The composition and origin of the traces of gas found in the bubbles in tektite glasses are also still open to question and might provide valuable insight into the origin of these strange samples.

Surprisingly few studies have been made of the silicate-melt inclusions in meteorites, and I suspect that far more work will be done with these in the future, as they can reveal much concerning the various environments in which meteorites have formed or to which they have been subsequently subjected. Thus, the distribution of melt inclusions in the olivine of the Murchison carbonaceous chondrite suggests very long annealing periods at high temperature. In addition to the almost ubiquitous silicate-melt inclusions, inclusions of immiscible metal are common and should be examined for metal/silicate equilibria. Less well understood are the "gas" inclusions, the origin of which should be investigated carefully. Even more enigmatic are the <u>liquid</u> inclusions containing moving bubbles in meteorites, particularly in the chondrules of chondrites. Although some of these inclusions may actually be artifacts from sample preparation, others seem valid; if so, they represent an appropriate enigma with which to end both this chapter and this book.

EPILOGUE

The progress since Sorby's time in our understanding of the significance of fluid inclusions has been great, and it is all too easy to sit back, smug and self-satisfied. But there is still much we do not know. We need only look at the status of fluid-inclusion research a century ago, and the concepts held then, to get an inkling of how naive today's work will probably seem a century hence.

NOTES: The more important pages are underlined. Where discussion of the subject is continued on subsequent pages, the first page number is followed by "-on". Some subjects, such as the occurrence of liquid CO_2 and the CO_2/H_2O ratio, and inclusions in a quartz host, are mentioned too frequently to make page listings useful. Some pages are cited on which related material is discussed rather than the subject item itself. Analyses for specific elements are only indexed where they are particularly unusual or significant.

LOCALITY INDEX

NOTES: Transliteration from the Cyrillic has not been consistent among the various sources used. Soviet place names also may be transliterated with different endings added, such as -ia, skii, insk, -koje. The user should check the several possible spellings.

REFERENCES

NOTES:

The numbers in parentheses at the end of each item are the pages in the present text where that item is referenced.

Those references marked with an asterisk (*) are bibliographically lengthy items, mainly from very obscure Russian publications. For brevity, only the author and title are given here, along with the volume and page in the "COFFI" volumes (see Roedder, 1968a) where a full citation and English abstract will be found.

Russian references marked with a dagger (†) are available in English translation in Yermakov, N.P., and others, 1965, Research on the Nature of Mineral-Forming Solutions, with Special Reference to Data from Fluid Inclusions, volume 22 of International Series of Monographs in Earth Sciences: Pergamon Press, Oxford, 743 p. (see Ermakov, 1950). Other translations are indicated where known; recent publications may have been translated subsequently.

As a result of consideration of scientific priority, as well as the reader's convenience, work reported at such events as the Geological Society of America's annual meetings (both verbally and as printed abstracts) is listed and referred to by the year of the meeting, with the year of publication, when different, given in parentheses.

Differences in the transliteration procedures that are used in various Western journals for the cyrillic names of Russian authors have resulted in different spellings of what is probably the same name (e.g., Petersil'e, Petersilie, and Petersilje). To avoid problems in the use of various bibliographic data bases, I have maintained such spellings as they appear in the actual publications.

Academia Sinica (1981) Studies of Fluid Inclusions in Minerals Symp. of the National Meeting on Experimental Studies of the Inclusions in Minerals and the Genesis of Rocks and Minerals: Scientific Publishing House, Beijing, 336 p. (In Chinese). (266)

Ackermann, G., R. Schrader and K. Hoffmann (1964) Untersuchungen an gashaltigen Mineralsalzen II. Teil: Methodik und Ergebnisse der gasanalytischen Untersuchungen. Bergakademie, 16, 676-679. (125, 315)

Adams, L.H. (1931) Equilibrium in binary systems under pressure. I. An experimental and thermodynamic investigation of the system NaCl-H2O. J. Am. Chem. Soc., 53, 3769-3813. (277)

_____ and R.E. Gibson (1930) The melting curve of sodium chloride dihydrate. An experimental study of an incongruent melting at pressures up to twelve thousand atmospheres. J. Am. Chem. Soc., 52, 4252-4264. (233)

Adamson, A.W. (1976) Physical Chemistry of Surfaces, 3rd Ed. Interscience, New York. (38, 61)

Adar, F., M. LeClercq and R.E. Grayzel (1982a) Industrial applications of micro Raman analysis. Am. Laboratory, March 1982, 59-66. (104)

_____, L.D. Ashwal, M.T. Colucci, H. Belkin, E. Roedder, S.C. Bergman, E.K. Gibson, D.J. Henry, R.K. Kotra and J.L. Warner (1982b) A progress report on fluid inclusions in meteorites (abstr.). Meteoritics, 17, 178. (567)

Agar, R.A. (1981) Copper mineralization and magmatic hydrothermal brines in the Rio Pisco section of the Peruvian Coastal Batholith. Econ. Geol. 76, 677-693. (443, 452)

Ahmad, M. (1979) Fluid-Inclusions and Geochemical Studies at the Emperor Gold Mine, Fiji. Ph.D. dissertation, Univ. of Tasmania, Hobart, Tasmania, Australia. 211 p. (427)

Ahmad, S.N. and A.W. Rose (1980) Fluid inclusions in porphyry and skarn ore at Santa Rita, New Mexico. Econ. Geol., 75, 229-250. (52, 443, 446, 447)

Ahrens, L.H., J.B. Dawson, A.R. Duncan and A.J. Erlank, eds. (1975) Physics Chem. Earth 9 (1st Int'l Conf. Kimberlite, 940 p.). (515)

Aines, R.D. and G.R. Rossman (1980) The structural and chemical behavior of hydrous components in minerals at temperatures up to 1000 K (abstr.). Geol. Soc. Am. Abstr. Programs, 12, 377. (84)

_____ and _____ (1984) Water in minerals? A peak in the infrared. J. Geophys. Res. (in press). (36, 84)

_____, S.H. Kirby and G.R. Rossman (1984) Hydrogen speciation in synthetic quartz. Physics Chem. Minerals (in press). (36, 77, 84)

Akhumov, E.I. and B.B. Vasil'ev (1931) [The technological calculations for the equilibrium of the chlorides of potassium, sodium, and magnesium in water at high temperatures]. J. Chem. Ind. (Moscow) 8, (17), 17-23 (in Russian). (245)

Akitzuki, M. (1965a) Electron microscope observations on the crack planes of beryl and quartz. Japan. Assoc. Mineral. Petrol. Econ. Geol. J., 54, 77-86 (In Japanese; English abstr.). (97, 145)

_____ (1965b) Secondary inclusions in fluorite under electron microscope and polarized microscope. Tohoku Univ. Sci. Repts. ser. 3, 9, (2), 347-369. (97, 145)

_____ (1965c) Splitting of liquid inclusions in a potassium alum crystal. Tohoku Univ. Sci. Repts, ser. 3, 9, (2), 371-376. (97, 145)

_____ (1966) Primary liquid inclusions in fluorite under electron microscope. Tohoku Univ. Sci. Repts, ser. 3, 9, (3), 509. (97, 145)

_____ (1967a) Electron microscope observations of galena from Takahi mine and Hosokura mine. Tohoku Univ. Sci. Repts, ser. 3, 10, (1), 41-47. (97, 145)

_____ (1967b) Electron microscope observations of liquid inclusions in fluorite from some Mississippi valley type deposits of America. Tohoku Univ. Sci. Repts, ser. 3, 10, (1), 49-54. (97, 145)

Albee, A.L., J.E. Quick and A.A. Chodos (1977) Source and magnitude of errors in broad-beam analysis (DBA) with the electron probe (abstr.). Lunar Science, VIII, 7-9. (140)

Alderton, D.H.M. and A.H. Rankin (1981) Fluid inclusion geochemistry and its use in mineral exploration (abstr.). Program, Symp. on Current Research on Fluid Inclusions, Univ. Utrecht. Utrecht, The Netherlands, April 22-24, 1981 (unpaginated). (132)

_____, M. Thompson, A.H. Rankin and S.L. Chryssoulis (1982) Developments of the ICP-linked decrepitation technique for the analysis of fluid inclusions in quartz. Chem. Geol. 37, 203-213. (132, 467)

Aldous, R.T.H. and A.H. Rankin (1979) Melt inclusions in olivine from the Palabora carbonatite complex: implications for the origin of phoscorite and copper mineralization at Palabora (abstr.). Program 4th Meeting Geol. Soc. of the British Isles, Univ. of Sheffield, Sheffield, England, Sept. 1979 (unpaginated). (406, 407)

Aleksandrova, E.S., L.A. Bannikova and T.M. Sushchevskaya (1980) Errors in gas analysis by thermal decrepitation of inclusions. Geokhimya, 1970, 1710-1716 (in Russian; translated in Geochem. Int'l, 17, 66-71). (124)

Allman, M. and D.F. Lawrence (1972) Geological Laboratory Techniques. Arco Publ. Co., New York, NY, 335 p. (157)

Allman-Ward, P. and A.H. Rankin (1979) Phase ratio calculations and the prediction of fluid inclusion homogenization temperatures (abstr.). Program 4th Meeting Geol. Soc. of the British Isles, Univ. of Sheffield, Sheffield, England, Sept. 1979 (unpaginated). (102)

Anderson, Jr., A.T. (1967) Possible consequences of composition gradients in basalt glass adjacent to olivine phenocrysts (abstr.). Am. Geophys. Union Trans. 48, 227. (39)

_____ (1973) The before-eruption water content of some high-alumina magmas. Bull. Volcanologique, 37, 530-552. (283, 485)

Anderson, Jr., A.T. (1974a) Evidence for a picritic, volatile-rich magma beneath Mt. Shasta, California. J. Petrol., 15, 243-267. (39, 141, 283, 477)

___ (1974b) Chlorine, sulfur, and water in magmas and oceans. Geol. Soc. Am. Bull., 85, 1485-1492. (485)

___ (1975) Some basaltic and andesitic gases. Rev. Geophys. Space Phys., 13, 37-55. (484)

___ (1976) Magma mixing: Petrological process and volcanological tool. J. Volcan. Geothermal Res., 1, 3-33. (39)

___ and L.P. Greenland (1969) Phosphorus fractionation diagram as a quantitative indicator of crystallization differentiation of basaltic liquids. Geochim. Cosmochim. Acta, 33, 493-505. (551)

___ and J.R. Sans (1975) Volcanic temperature and pressure inferred from inclusions in phenocrysts. Int'l Conf. on Geothermometry and Geobarometry, 5-10 Oct. 1975, Penn. State Univ., Extended Abstr. Penn. State Univ., University Park, PA (unpaginated). (76, 267, 283, 485)

___ and T.L. Wright (1972) Phenocrysts and glass inclusions and their bearing on oxidation and mixing of basaltic magmas, Kilauea volcano, Hawaii. Am. Mineral. 57, 188-216. (484)

Andrawes, F.F. and E.K. Gibson, Jr. (1978) The effect of gaseous additives on the response of the helium ionization detector. Anal. Chem., 50, 1146-1151. (124)

___ and ___ (1979) Release and analysis of gases from geological samples. Am. Mineral., 64, 453-463. (117, 124, 125)

Andreyev, G.V. and M.I. Shvadus (1977) Pressures and temperatures of crystallization of alkaline rocks of Synnyr complex, as inferred from first finds of inclusions of melt. Akad. Nauk SSSR, Dokl. 235, (4), 910-913 (in Russian; translated in Dokl. Acad. Sci. USSR, Earth Sci. Sects., 235, 150-152, 1979). (276, 403)

Angell, C.A. and E.J. Sare (1970) Glass-forming composition regions and glass transition temperatures for aqueous electrolyte solutions. J. Chem. Phys. 52, 1058-1068. (297)

Angus, S., B. Armstrong, K.M. deReuck, V.V. Altunin, O.G. Gadetskii, G.A. Chapela and J.S. Rowlinson (1976) International Thermodynamic Tables of the Fluid State, Vol. 3, Carbon Dioxide: Pergamon Press, Oxford, England, 385 p. (229)

Anonymous (1974) Research Group on the Equilibrium Between Fluids and Minerals. Centre Rech. Petrogr. Geochem., Nancy, 27 p. (in French). (190, 191, 375)

___ (1975) Carbon and its Compounds in Endogenic Processes of Mineral-Formation (Data of Studies of Fluid Inclusions in Minerals). Abstr. Regional Meeting (L'vov. Sept. 1975). COFFI, 8, 7. (6)

Anthony, E.Y., T.J. Reynolds and R.E. Beane (1983) The use of energy dispersive analysis to identify daughter minerals from the Santa Rita porphyry copper deposit, New Mexico (abstr.). Geol. Soc. Am. Abstr. Programs, 15, 516. (145)

Anthony, T.R. and H.E. Cline (1970) The kinetics of droplet migration in solids in an accelerational field. Philos. Mag., 22, (179), 893-901. (66)

Apfel, R.E. (1970) Vapor cavity formation in liquids. Harvard Univ. Acoustics Laboratory, Tech. Memo. No. 62, 207 p. (296)

Arai, Y., G. Kaminishi and S. Saito (1971) The experimental determination of the P-V-T-x relations for the carbon dioxide-nitrogen and the carbon dioxide-methane systems. J. Chem. Eng. Japan 4, 113-122. (243)

Armbruster, T. and F.D. Bloss (1980) Effects of channel H2O and CO2 in cordierite (abstr.): Fortschr. Mineral., 58, 7-8. (363)

Arculus, R.J. and J.W. Delano (1981) Intrinsic oxygen fugacity measurements: techniques and results for spinels from upper mantle peridotites and megacryst assemblages. Geochim. Cosmochim. Acta, 45, 899-913. (505)

Armbrustmacher, T.J. (1979) Replacement and primary magmatic carbonatites from the Wet Mountains area, Fremont and Custer Counties, Colorado. Econ. Geol., 74, 888-901. (407)

Arming, H. and A. Preisinger (1968) Inclusions of gases in minerals. Proc. 5th Int'l Mineral. Assoc. Symp., Cambridge, England 1966, 117-122. (124)

Ashwal, L.D., D.W. Mogk, S.C. Bergman, E.K. Gibson, Jr., D.J. Henry, J.L. Warner and R. Lee-Berman (1981) Liquid-vapor inclusions in achondritic meteorites (abstr.). Meteoritics, 16, 290-291. (566)

___, M.I. Colucci, P. Lambert, D.J. Henry and E.K. Gibson, Jr. (1982) Fluid inclusions in meteorites: direct samples of extraterrestrial volatiles (abstr.). Program Lunar Planetary Institute Conference on Planetary Volatiles, Alexandria, Minnesota, Oct. 9-12, 1982 (unpaginated). (567, 568)

Aspden, J.A. (1980) The mineralogy of primary inclusions in apatite crystals extracted from Alnö ijolite. lithos, 13, 263-268. (404)

Bachinski, S.W. and R.B. Scott (1979) Rare-earth and other trace element contents and the origin of minettes (mica-lamprophyres). Geochim. Cosmochim. Acta, 43, 93-100. (407)

Bailey D.K. (1980) Volatile flux, geotherms, and the generation of the kimberlite-carbonatite-alkaline magma spectrum. Mineral. Mag., 43, 695-699. (407, 504, 505)

Bailey, S.W. (1949) Liquid inclusions in granite thermometry. J. Geol., 57, 304-307. (186, 191)

Bailey, S.W. and E.N. Cameron (1951) Temperatures of mineral formation in bottom-run lead-zinc deposits of the Upper Mississippi Valley, as indicated by liquid inclusions. Econ. Geol., 46, 626-651. (186)

Bain, R.W. (1964) Steam tables 1964. Nat'l. Engineering Lab., Edinburgh, H.M. Stationery Office, 147 p. (500)

Baker, H. (1743) The microscope made easy: R. Dodsley, London. (175, 178)

*Bakhanova, Ye.V., G.B. Levin, B.M. Maydenov and E.Ya. Polyvyannyi (1976) Mineral formation in gold deposits on the basis of isotopes of argon from gas-liquid inclusions (abstr.). COFFI, 10, 10-11. (128)

Batukmenko, I.T. (1964) Determination of refractive indices of inclusions in minerals on the Fedorov stage. In Materialy po Geneticheskoi i Eksperimental'noi Mineralogii, v. 2: Akad. Nauk SSSR Sibirskoye Otelelniye, Inst. Geol. Geofiz. Trudy no. 30, 297-314 (in Russian; English abstr.). (86, 96)

___ (1975) Inclusions in minerals of ultrabasic nodules as indicators of their origin. Inst. Geol. Geophys. Trudy, v. 271 - Deepseated xenoliths and the upper mantle, eds., V.S. Sobolev, N.L. Dobretsov and N.V. Sobolev, 231-235. Novosibirsk, "Nauka," Siberian Branch (in Russian). (515)

___ and Yu.A. Dolgov (1977) Thermobarogeochemical Researches. Geol. Geofiz. 18, (11), 127-135 (in Russian; translated in Soviet Geol. Geophys.. 18, (11), 98-104, 1978). (4)

___, S.S. Kolyago and V.S. Sobelev (1967) Interpretation of thermometric investigations of vitreous inclusions in minerals and preliminary results on artificial inclusions. Akad. Nauk SSSR, Dokl., 175 (5), 1127-1130 (in Russian; translated in Dokl. Acad. Sci. USSR, Earth Sci. Sects., 175, 143-145, 1967). (190)

Balasubramaniam, K.S., V. Panchapakesan and K.C. Sahu (1975) Fluid inclusion geothermometric studies of some Indian fluorites using a fabricated heating stage. J. Geol. Soc. India, 16, 460-464 (in English). (186)

Balce, G.R. (1979) Geology and ore genesis of the porphyry copper deposits in Baguio district, Luzon Island, Philippines. J. Geol. Soc. of the Philippines, 33, (2), 1-43. (443)

Bambauer, H.U. (1961) Spurenelementgehalte und γ-Farbzentren in Quarzen aus Zerklüften der Schweizer Alpen. Schweiz. Mineral. Petrog. Mitt., 41, 335-369. (24, 36)

___, G.O. Brunner and F. Laves (1969) Light scattering of heat-treated quartz in relation to hydrogen-containing defects. Am. Mineral. 54, 718-724. (84)

Barabanov, V.F. (1958) The role of pressure during mineral growth in quartz-wolframite veins. Akad. Nauk SSSR, Dokl., 120, 400-403 (in Russian; translated in Dokl. Acad. Sci. USSR Geol. Sci. Sect., 120, 565-569). (29)

Bardintzeff, J.-M., R. Brousse, R. Clocchiatti and J. Weiss (1980) Evolution of phenocrysts and their melt inclusions in dacite of Stantiaguito Dome, Guatemala. Compt. Rend. Acad. Sci. Paris, 290, 743-746 (in French). (Sic; probably misprint for Bardintzeff). (484)

Bardintzeff, J.-M. and R. Clocchiatti (1980) Melt inclusions in plagioclase phenocrysts of the April 1979 "Nuee Ardente" deposits at Saint-Vincent Soufriere (Antilles): A new approach for a dynamic model. Compt. Rend. Acad. Sci. Paris, 291, 529-532 (in French). (484)

Barker, C.E. (1979) Vitrinite reflectance geothermometry in the Cerro Prieto geothermal system, Baja California, Mexico. M.S. thesis, Univ. California, Riverside, CA, 127 p. (497)

_____ (1983) Influence of time on metamorphism of sedimentary organic matter in liquid-dominated geothermal systems, western North America. Geology, 11, 384-388. (497)

_____ and W.A. Elders (1981) Vitrinite reflectance geothermometry and apparent heating duration in the Cerro Prieto geothermal field. Geothermics, 10, 207-223. (497)

Barker, C.G. (1965a) Mass spectrometric analysis of the gas evolved from some heated natural minerals. Nature, 205, 1001-1002. (123, 302)

_____ (1965b) The use of a mass spectrometer for analyzing the gases liberated by heating or crushing natural minerals. In Proc. 1st Symp. on Application of the A.E.I. MS 10 Mass Spectrometer, Imperial College, London, 13 April 1965. Associated Elec. Industries Ltd., Manchester, 85-95. (118, 123)

_____ (1966) Volatile content of rocks and minerals with special reference to fluid inclusions (abstr.). Geol. Soc. Am. Spec. Paper 101, 10 (pub. 1968). (123)

_____ (1972) Aquathermal pressuring -- Role of temperature in development of abnormal-pressure zones. Am. Assoc. Petrol. Geol. Bull., 56, 2068-2071. (270, 306, 323)

_____ (1978) Pyrolysis of naturally occurring organic matter: A review (abstr.). Geol. Soc. Am. Abstr. Programs, 10, 363. (124)

_____ and M.A. Sommer (1973) Mass spectrometric analysis of the volatiles released by heating or crushing rocks. In Analytical Methods Developed for Application to Lunar Samples Analyses, 56-70. Am. Soc. Testing Materials, Spec. Tech. Pub. 539. (120)

_____ and B.E. Torkelson (1975) Gas adsorption on crushed quartz and basalt. Geochim. Cosmochim. Acta, 39, 212-218. (120, 121)

Barlier, J., J.-P. Ragot and J.-C. Touray (1974) L'évolution des Terres Noires subalpines meridionales d'apres l'analyse minéralogique des argiles et la reflectometrie des particules carbonées: Bur. Rech. Geol. Minières Bull., ser. 2, 2, 533-548. (322)

Barnes, H.L., ed. (1969) Geochemistry of Hydrothermal Ore Deposits. Wiley & Sons, New York, NY, 798 p. (136)

_____ J. Lusk and R.W. Potter II (1969) Composition of fluid inclusions (abstr.). Abstr. 3rd Int'l COFFI Symp. on Fluid Inclusions: Fluid Inclusion Res. --Proc. of COFFI, 2, 13, (39, 190, 435)

Barnes, I., and G.A. McCoy (1979) Possible role of mantle-derived CO2 in causing two "phreatic" explosions in Alaska. Geology, 7, 434-435. (505, 528)

Barnola, J.M., D. Raynaud, A. Neftel and H. Oeschger (1983) Comparison of CO2 measurements by two laboratories on air from bubbles in polar ice. Nature, 303, 410-413. (329)

Baross, J.A. and J.W. Deming (1983) Growth of 'black smoker' bacteria at temperatures of at least 250°C. Nature, 303, 423-426. (495, 513)

_____ M.D. Lilley and L.I. Gordon (1982) Is the CH4, H2 and CO venting from submarine hydrothermal systems produced by thermophilic bacteria? Nature, 298, 366-368. (495, 513)

Barrabé, L. and G. Deicha (1956) Experiences de fusion et de cristallisation magmatique sur des reliquats vitreux des quartz dihexaédriques de la Guadeloupe. Bull. Soc. franc. Mineral. Cristallogr., 79, 146-155. (188, 474)

Barrabé, L. and G. Deicha (1957) Réanimation de magmas et interprétation de quelques particularites de leurs elements de premiere consolidation. Bull. Soc. Geol. Fr., 6th ser., 7, 159-169. (188, 474)

_____ L., P. Collomb and G. Deicha (1957) Utilisation des sphéres polies dan les recherches sur les reliquats magmatiques. Bull. Soc. franc. Mineral. Cristallogr., 80, 450-452. (188, 474)

_____ and _____ (1959) Liberation eruptive de magma reanime sous le microscope de chauffe. Bull. Soc. franc. Mineral. Cristallogr. 82, 163-165. (188)

Barsukov, V.L. and T.M. Sushchevskaya (1973) On the composition evolution of hydrothermal solutions in the process of the formation of tin ore deposits. Geokhimiya, 1973, (4), 491-503 (in Russian; translated in Geochem. Int'l, 10, (2), 363-375, 1974). (266)

Barton, Jr., P.B. (1967) Possible role of organic matter in the precipitation of the Mississippi Valley ores. In Genesis of Stratiform Lead-Zinc-Barite-Fluorite-Deposits: A Symposium. New York, Soc. Econ. Geol. Monograph 3, 371-378. (302, 421)

_____ P.M. Bethke and P. Toulmin, III (1963) Equilibrium in ore deposits. Mineral. Soc. Am. Spec. Paper 1, 171-185. (24, 271)

_____ and M.S. Toulmin (1971) An attempt to determine the vertical component of flow rate of ore-forming solutions in the OH vein, Creede, Colorado. Soc. Mining Geol. Japan, Spec. Issue 2, 132-136 [Proc. IMA-IAGOD Meetings '70, Joint Symp. Vol. 3. (26, 420)

_____ and E. Roedder (1977) Environment of ore deposition in the Creede Mining district, San Juan Mountains, Colorado: Part III. Progress toward interpretation of the chemistry of the ore-forming fluid for the OH vein. Econ. Geol., 72, 1-24. (138, 426)

Basaltic Volcanism Study Project (1981) Basaltic volcanism on the terrestrial planets. Pergamon Press, New York, NY, 1286 p. (482)

Bassett, R.L. and E. Roedder (1981) Water content in Palo Duro salt, Randall and Swisher County cores: Texas Bureau Econ. Geol. Geological Circ. 81-3, 119-121. (319)

Bauer, M. and [J. Spencer (1904) Precious Stones. Charles Griffin & Co., London, 627 p. (507)

Bauman, L., H.J. Blankenburg, O. Leeder and A. Pentzel (1976) Genetic studies of glass sands by means of quartz inclusions. Z. Angewandte Geol., 22, 555-559 (in German). (308)

Bazarov, L.Sh. (1966) A device for freezing inclusions in minerals. In Materialy po genetitscheskoi i eksperimental'noi mineralogii, 4: Novosibirsk, Akad. Nauk SSSR Sibirskoye Otdeleniye Inst. Geol. Geofiz., 231-234 (in Russian; English abstr.). (190)

_____ (1968) A micro thermal chamber for high-temperature studies of inclusions in minerals. Geol. Geofiz. SSSR, 1968, (8), 140-142 (in Russian). (186, 189)

_____ (1975) Inclusions of "melt solutions" in minerals of rare-earth pegmatite. Akad. Nauk. SSSR, Dokl., 215, (4), 940-943 (in Russian; translated in Dokl. Acad. Sci. USSR, Earth Sci. Sects., 215, 102-105; abstr. translated in Int'l Geol. Rev., 16, (9), 1079-1030). (395)

Bazarova, T.Yu. (1965) Mineralothermometric study of inclusions in nepheline rock minerals. Akad. Nauk SSSR, Dokl., 161, (4), 925-928 (in Russian; translated in Dokl. Acad. Sci. USSR, Earth Sci. Sects., 161, 125-128). (89)

_____ and Ya.M. Feigin (1966) Mineralothermometric study of nephelines from the Lovozero massif. Vses. Mineral. Obshch. Zapiskl, 95, (3), 364-366 (in Russian). (89)

_____ and G.A. Kazaryan (1977) Peculiarities of crystallization of leucite phonolite from Azat River-Vedi River area (Armenian SSR). Akad. Sci. SSSR, Dokl. 232 (6), 1418-1420 (in Russian; translated in Dokl. Acad. Sci. USSR, 232, 217-220, pub. 1978). (482)

Bazarova, T.Yu. and A.A. Krasnov (1975) Temperatures and sequence of crystallization of some leucite-bearing basaltoids. Akad. Nauk SSSR, Dokl., 222, (4), 935-938 (in Russian; translated in Dokl. Acad. Sci. USSR, Earth Sci. Sects., 222, 177-180, 1976). (481)

Bazheyev, Ye. D. (1980) Evolution of hydrothermal solutions during formation of tungsten deposits (for the case of the Soktuy ore node). Geokhimiya, 1980 (10), 1461-1467 (in Russian; translated in Geochem. Int'l., 17, 114-119, 1981). (457)

Beales, F.W. and S.A. Jackson (1966) Precipitation of lead-zinc ores in carbonate reservoirs as illustrated by Pine Point ore field, Canada. Inst. Min. Metall. Trans., 75, B278-B285. (417)

Becker, G.F. and A.L. Day (1905) The linear force of growing crystals: Proc. Wash. Acad. Sci., 7, 283-208. (15)

Beckett, J.R., L. Grossman and E. Olsen (1979) Murchison: composition relations between isolated grains and aggregates (abstr.). Meteoritics, 14, 345. (562)

Becquerel, H. and H. Moissan (1890) Étude de la fluorine de Quincié. Compt. Rend. Acad. Sci. Paris, 111 669-672. (117)

Beeunas, M.A. and L.P. Knauth (1983) Isotopic composition of fluid inclusions in Permian halite -- Implications for the isotopic history of sea water (abstr.). Geol. Soc. Am. Abstr. Programs, 15, 524. (317)

Behr, H.J. and E.E. Horn (1982) Fluid inclusion systems in metaplaya deposits and their relationships to mineralization and tectonics. Chem. Geol., 37, 173-189. (349)

Bein, A. and L.S. Land (1982) San Andres carbonates in the Texas Panhandle: Sedimentation and diagenesis associated with magnesium-calcium-chloride brines. Texas Bureau Econ. Geol. Report Invest. 121, 48 p. (319)

Belchic, H.C. (1961) The Winnfield salt dome, Winn Parish, Louisiana, in Interior Salt Domes and Tertiary Stratigraphy of North Louisiana, 1960 Spring Field Trip Guide Book: Shreveport Geol. Soc., Shreveport, LA, 29-47. (315)

Belkin, H.E., B. De Vivo, G. Gianelli and P. Lattanzi (1983) Fluid inclusion reconnaissance study of hydrothermal minerals from geothermal fields of Tuscany (Italy): Program 4th Int'l Symp. on Water-Rock Interaction, Aug. 29-Sept. 3, 1983, Misasa, Japan, 43-47. (501)

_____, E. Roedder and M. Cortini (1984) Fluid inclusion geobarometry from ejected Mt. Somma-Vesuvius nodules. Am. Mineral. (in press). (515, 516, 517, 518, 524, 529, 530)

Bell, P.M., H.K. Mao, E. Roedder and P.W. Weiblen (1975) The problem of the origin of symplectites in olivine-bearing lunar rocks. Proc. Lunar Sci. Conf. 6th, 231-248. (553, 554)

Benhamou, G. and R. Clocchiatti (1976) Glass inclusions in quartz and anorthoclase phenocrysts from Pantelleria peralkaline lavas: a thermometric study. Bull. Soc. franc. Minéral. Cristallogr., 99, 111-116 (in French). (481)

Bennett, J.N. and J.N. Grant (1980) Analysis of fluid inclusions using a pulsed laser microprobe. Mineral. Mag., 43, 945-947. (139)

Benrath, A., F. Gjedebo, B. Schiffers and H. Wunderlich (1937) Über die löslichkeit von Salzen und salzgemischen in Wasser bei Temperaturen oberhalb von 100°. Z. Anorg. Allgem. Chemie, 231, 285-297. (234, 235, 275, 277)

Benson, S.W., C.S. Copeland and D. Pearson (1953) Molal volumes and compressibilities of the system $NaCl-H_2O$ above the critical temperature of water. J. Chem. Physics, 21, 2208-22[12. (253)

Bény, C., N. Guilhaumou and J.-C. Touray (1981) Fluid inclusions in the system $H_2O-NaCl-CO_2-H_2S-S$ in quartz and fluorite from Sierra de Lujar (Granada, Spain): Microthermometry and Raman microprobe data. Compt. Rend. Acad. Sci. Paris, 292, ser. 2, 797-800 (in French). (108, 241)

_____ and _____ (1982) Microthermometry and Raman microprobe fluid inclusions (MOLE analysis) in the $CO_2-H_2S-H_2O-S$ system -- Thermochemical interpretations. Chem. Geol., 37, 113-127. (463)

Berdnikov, N.V. and A.A. Tomilenko (1983) Carbon dioxide-methane inclusions in quartz of granite of the Agusinskii Massif (North Sikhote Alin). Akad. Nauk SSSR, Dokl., 268, (3), 656-659 (in Russian). (243)

Berg, W.F. (1938) Crystal growth from solutions. Proc. Royal Soc. (London), A164, 79-95. (38)

Berglund, L. and J. Touret (1976) Garnet-biotite gneiss in "Systeme du graphite" (Madagascar): petrology and fluid inclusions. Lithos, 9, 139-148. (375)

Bergman, S.C. (1982) Petrogenetic Aspects of the Alkali Basaltic Lavas and Included Megacrysts and Nodules from the Lunar Crater Volcanic Field, Nevada, USA. Ph.D. dissertation, Princeton Univ., Princeton, NJ. (33, 227, 229, 515, 525)

_____ and J. Dubessy (1984) CO_2-CO fluid inclusions in a composite peridotite xenolith: Implications for oxygen barometry. Contrib. Mineral. Petrol. 85, 1-13. (505, 515, 525, 527)

Bergman, W. and H.-J. Blankenburg (1964) Über einige Einschlüsse in Bergkristallen. Silikattechn. 15, (4), 112-116. (115)

Berkeile, C.A. and E. Dowty (1982) Nucleation in laboratory charges of basaltic composition. Am. Mineral., 67, 886-899. (479)

Berkley, J.L., K. Keil and M. Prinz (1980) Comparative petrology and origin of Governador Valadares and other nakhlites. Proc. 11th Lunar Sci. Conf., 1089-1102. (569)

Bernard, A.J. (1973) Metallogenic processes of intra-karstic sedimentation. In G.C. Amstutz and A.J. Bernard (Eds.), Ores in Sediments. Springer, Berlin, 43-57. (419)

Berner, R.A. (1978) Rate control of mineral dissolution under earth surface conditions. Am. J. Sci., 1978, 1235-1252. (39, 59)

Berthelot, M. (1850) Sur quelques phenomenes de dilatation forcée des liquides. Ann. Chim. Physique (Ser. 3) 30, 232-237. (298)

Berzina, A.P., N.A. Shugurova and V.I. Sotnikov (1976) Gaseous composition of mineral-forming solutions at copper-molybdenum deposits. Akad. Nauk SSSR, Dokl., 228, 188-191 (in Russian). (443)

Beskrovnyi, N.S. and B.A. Levedev (1971) Occurrences of sphalerite and other hydrothermal minerals in the Gaurdak sulfur deposit, southeastern Turkmenia. Akad. Nauk SSSR, Dokl., 200, 185-188 (in Russian). (328)

Bethke, P.M. and P.B. Barton, Jr. (1971) Distribution of some minor elements between coexisting sulfide minerals. Econ. Geol., 66, 140-163. (271)

_____ and R.O. Rye (1979) Environment of ore deposition in the Creede mining district, San Juan Mountains, Colorado: Part IV. Source of fluids from oxygen, hydrogen, and carbon isotope studies. Econ. Geol., 74, 1832-1851. (435)

Bibby, D.M. (1978) Trace elements in diamonds of different types. Nature, 276, 379-381. (506, 508)

_____ (1982) Impurities in natural diamond. In Chemistry and Physics of Carbon, Vol. 18, P.A. Thrower, ed. Marcel Dekker, Inc., New York, NY, 1-91. (506, 508, 509)

_____, H.W. Fesq and J.P.F. Sellschop (1978) Trace elements in diamonds of different types. Nature, 276, 379-381. (509)

Bienfait, M. and R. Kern (1965) Établissement de la forme d'équilibre d'un cristal (Méthode de Lemmlein-Klija). Bull. Soc. franc. Minéral. Cristallogr., 87, (4), 604-613. (93)

Biggar, G.M. (1979) Immiscibility in tholeiites. Mineral. Mag., 43, 543-544. (477)

_____ (1983) A reassessment of phase equilibria involving two liquids in the system $K_2O-NaCl-CO_2-FeO-SiO_2$. Contrib. Mineral. Petrol., 82, 259-268. (550)

Bilal, A. and J. Touret (1976) Fluid inclusions in catazonal xenoliths from Bournac (Massif Central, France). Bull. Soc. franc. Minéral. Cristallogr., 99, 134-139 (in French). (515)

_____ (1977) Fluid inclusions in phenocrysts from basaltic lavas of Puy Beaunit (French Massif central). Bull. Soc. franc. Minéral. Cristallogr., 100, 324-328 (in French). (283, 515)

Birch, F. (1966) Compressibility; elastic constants. Geol. Soc. Am. Memoir, 97, 97-173. (70)

Bischoff, J.L. (1980) Geothermal system at 21°N, East Pacific Rise: physical limits on geothermal fluid and role of adiabatic expansion. Science, 207, 1465-1469. (495)

Blacic, J.D. (1975) Plastic-deformation mechanisms in quartz: The effect of water. Tectonophysics, 27, 271-294. (77)

___ (1981) Water diffusion in quartz at high pressure: Tectonic implications. Geophys. Res, Lett., 8, 721-723. (36, 76)

Blank, H., A. El Goresy, J.Janicke, R. Nobiling and V. Traxel (1982) Distribution of trace elements in zoned chromite-ulvöspinel mineral pairs - measured with the proton microprobe (abstr.). Max-Planck-Inst. f. Kernphysik, Jahresbericht 1982, p. 162. (148)

Blattner, P. (1975) Oxygen isotopic composition of fissure-grown quartz, adularia, and calcite from Broadlands geothermal field, New Zealand with an appendix on quartz-K-feldspar-calcite-muscovite oxygen isotope geothermometers. Am. J. Sci., 275, 785-800. (499)

Bloom, M.S. (1981) Chemistry of inclusion fluids: stockwork molybdenum deposits from Questa, New Mexico, and Hudson Bay Mountain and Endako, British Columbia. Econ. Geol., 76, 1906-1920. (443, 452)

*Bobolovich, G.N. (1972) Application of inclusion decrepitation methods for investigation of minerals with excellent cleavage. COFFI, 6, 20-21. (266)

Boctor, N.Z. and H.O.A. Meyer (1977) Oxide and sulfide minerals in kimberlite from Green Mountain, Colorado. In 2nd Int'l Kimberlite Conf. 1977, Extended Abstracts (unpaginated). (511)

___ R.K. Popp and J.D. Frantz (1979) Solubility of hematite in chloride-bearing hydrothermal fluids (abstr.). EOS, Trans. Am. Geophys. Union, 60, 974. (455)

Bodenlos, A.J. (1954) Magnesite deposits in the Serra das Eguas, Brumado, Bahia, Brazil. U.S. Geol. Survey Bull., 975-C, 87-170. (55)

Bodine, Jr., M.W. (1976) Magnesium hydroxychloride: A possible pH buffer in marine evaporite brines? Geology, 4, 76-80. (314)

Bodnar, R.J (1978) Fluid Inclusion Study of the Porphyry Copper Prospect at Red Mountain, Arizona. M.S. thesis, Univ. Arizona, Tucson, AZ. (276)

___ (1981) Use of fluid inclusions in mineral exploration: Comparison of observed features with theoretical and experimental data on ore genesis (abstr.). Geol. Soc. Am. Abstr. Programs, 13, 412. (439, 467, 469)

___ (1982) Fluid inclusions in porphyry-type deposits. Course notes, Mineral Deposits Research Review for Industry, Penn. State Univ., April 6-9, 1982, p. RBI-RB25 (unpub.; ms. in preparation). (287, 288, 289, 448, 449, 450)

___ (1983) A method of calculating fluid-inclusion volumes based on vapor bubble diameters and P-V-T-X properties of inclusion fluids. Econ. Geol., 78, 535-542. (103, 228, 286, 289, 448)

___ and R.E. Beane (1980) Temporal and spatial variations in hydrothermal fluid characteristics during vein filling in pre-ore cover overlying deeply buried porphyry copper-type mineralization at Red Mountain, Arizona. Econ. Geol., 75, 876-893. (23, 167, 227, 443, 452)

___ and P.M. Bethke (1980) Systematics of "stretching" of fluid inclusions (abstr.). EOS, Trans. Am. Geophys. Union, 61, 393. (72, 167, 258, 259, 260)

___ and ___ (1984) Systematics of stretching of fluid inclusions in fluorite and sphalerite as a result of overheating. Econ. Geol. (in press). (71, 72, 167, 258, 259, 260, 277, 285, 428)

___ and J.A.D. Connelly (1983) Fluid pressure as an independent variable in metamorphism obtained from integrated fluid inclusion-mineral phase equilibrium studies (abstr.). Geol. Soc. Am. Abstr. Programs, 15, p. 529. (362)

Bodnar, R.J. and J.A.D. Connelly (1984) A modified Redlich-Kwong equation of state for H2O-CO2 mixtures: Application to fluid inclusion studies. I. Calculated densities, low-temperature phase equilibria, and pressure corrections for H2O-CO2 fluid inclusions. Am. Mineral., (in press). (228)

___ and C.A. Kuehn (1984) Phase relations and interpretation of CO2-bearing fluid inclusions from epithermal gold-silver deposits. Econ. Geol. (in press). (426, 429)

Boettcher, A.L., J.K. Robertson and P.J. Wyllie (1980) Studies in synthetic carbonatite systems: Solidus relationships for CaO-MgO-CO2-H2O to 40 kbar and CaO-MgO-SiO2-CO2-H2O to 10 kbar. J. Geophys. Res., 85, 6937-6943. (410)

Bokii, G.B., G.G. Tsurinov, V.I. Sokol and V.Z. Kolodyazhn'y (1961) Immersion liquids for crystal optical measurements at very low temperatures (from -100°). Zhur. Neorg. Khimii, 6, 1754-1758 (in Russian). (86)

Boland, J.N. and A. Duba (1983) The role of crystalline defects in the distribution of metallic particles in olivine: meteoritic implications (abstr.). Lunar Planet. Sci., XIV, 57-58. (563)

Bonev, I.K. (1969) Liquid inclusions in galena. Compt. Rend. Acad. bulgare Sci., 22, 1289-1292. (72)

___ (1977) Primary fluid inclusions in galena crystals. I. Morphology and origin. Mineralium Deposita, 12, 64-76. (28, 72)

Borak, B. and G.M. Friedman (1981) Textures of sandstones and carbonate rocks in the world's deepest wells (in excess of 30,000 ft. or 9.1 km.): Anadarko Basin, Oklahoma. Sedimentary Geol., 29, 133-151. (335)

Borisenko, A.S. (1977) Cryometric technique applied to studies of the saline composition of solution in gaseous fluid inclusions in minerals. Akad. Nauk SSSR, Sib. Otdel., Geol. i. Geofiz., 1977, (8), 16-27 (in Russian; English abstr.). (249)

Borodzich, E.V., A.N. Eremeev and I.N. Yanitsky (1979) Preliminary results of the study of variations of helium concentration in active seismic zones. Geokhimiya, 1979, (3), 372-377 (in Russian; translated in Geochem. Int'l, 16, 37-41, 1980). (402)

Bostick, N.H. (1973) Time as a factor in thermal metamorphism of phytoclasts (coaly particles). Congr. Int'l de Stratig. et de Géol. du Carbonifère, Septième, Krefeld, Aug. 23-28, 1971, Compte Rendu, 2, 183-193. (330)

Bosworth, W. (1981) Strain-induced preferential dissolution of halite. Tectonophysics, 78, 509-525. (68)

Both, R.A. (1978) Remobilization of mineralization during retrograde metamorphism, Broken Hill, New South Wales, Australia. In Mineralization in Metamorphic Terranes, W.J. Verwoerd, ed. J.L. van Schaik Ltd., Pretoria, South Africa, 481-489. (366)

Bottinga, Y. and P. Richet (1981) High pressure and temperature equation of state and calculation of the thermodynamic properties of gaseous carbon dioxide. Am. J. Sci., 281, 615-660. (227, 228)

___ A. Kudo and D. Weill (1966) Some observations on oscillatory zoning and crystallization of magmatic plagioclase. Am. Mineral, 51, 792-806. (40)

Bouberlova, L. (1977) Analysis of fluid-filled inclusions in minerals. Casopis pro mineralogii a geologii, 22, (2), 151-162 (in English; Czech summary). (125)

Bowen, N.L. (1928) The Evolution of the Igneous Rocks. Princeton Univ. Press, Princeton, NJ. (381, 545)

___ (1937) Recent high-temperature research on silicates and its significance in igneous geology. Am. J. Sci., 33, 1-21. (385)

Bowers, T.S. and H.C. Helgeson (1983a) Calculation of the thermodynamic and geochemical consequences of nonideal mixing in the system H2O-CO2-NaCl on phase relations in geologic systems: Equation of state for H2O-CO2-NaCl fluids at high pressures and temperatures. Geochim. Cosmochim. Acta, 47, 1247-1275. (228, 247, 372)

Bowers, T.S. and H.C. Helgeson (1983b) Calculation of the thermodynamic and geochemical consequences of nonideal mixing in the system H_2O-CO_2-NaCl on phase relations in geologic systems: metamorphic equilibria of high pressures and temperatures. Am. Mineral., 68, 1059-1075. (362, 372)

Boyd, F.R. and J.J. Gurney (1982) Low-calcium garnets: keys to craton structure and diamond crystallization. Carnegie Inst. Washington Year Book 81, 261-267. (507)

*Boyko, S.M. and M.Ye. Markova (1976) Zonal changes of composition of gaseous-liquid inclusions in minerals from Kapcheranginskoe tin ore deposit. COFFI, 10, 35. (137)

Boyle, R. (1672) Essay about the origine and virtues of gems. William Godbid, London. 185 p. (3)

Boyle, R.W. (1954) A decrepitation study of quartz from the Campbell and Negus-Rycon shear zone systems, Yellowknife, Northwest Territories. Canada Geol. Survey Bull. 30, 20 p. (357)
———— (1955) The geochemistry and origin of the gold-bearing quartz veins and lenses of the Yellowknife Greenstone Belt. Econ. Geol. 50, 51-66. (356)

Bozzo, A.T., J.R. Chen and A.J. Barduhn (1973) The properties of hydrates of chlorine and carbon dioxide. In Fourth International Symposium on Fresh Water from the Sea, A. Delyannis and E. Delyannis, eds., 3, 437-451. (241, 373)

Bradbury, J.W. (1975) A reconnaissance study of fluid inclusions from Tertiary intrusives in Colorado. M.S. thesis, Univ. Colorado, Boulder, CO (abstr. in Fluid Inclusion Res. - Proc. COFFI, 8, 28). (444)

Brady, J.D. and J.D. Frantz (1980) A microanalytical technique for determination of aluminum in aqueous solutions. Am. Mineral., 65, 1249-1251. (133)

Brass, G.W. (1980) Stability of brines on Mars. Icarus, 42, 20-28. (249)

Bratus', M.D., Z.V. Stasyuk and R.S. Panchishin (1968) Determination of the composition of gases of individual inclusions in quartz from pegmatite by the omegatron mass-spectrometer. Akad. Nauk SSSR, Dokl., 183, (4), 928-930 (in Russian). (122, 124)
*————, I.M. Svoren' and V.V. Danysh (1975) Inclusions of hydrocarbons in "Marmarosh diamonds" from Carpathians as indicators of migration of oil fluids (abstr.). COFFI, 8, 28. (323)

Bray, C.J. (1980) Mineralization, greisenisation and kaolinisation at Goonbarrow clay pit, Cornwall, U.K. Ph.D. dissertation, Univ. Oxford, Oxford, England. (145)

Brecke, E.A. (1979) A hydrothermal system in the midcontinent region. Econ. Geol., 74, 1327-1335. (424)

Breislak, S. (1818) Institutions géologiques (French trans.) by P.J.L. Campmas). Institutions geologiques, Milan, 1, 468 p. (3)

Brenninkmeijer, C.A.M, P. Kraft and W.G. Mook (1983) Oxygen isotope fractionation between CO_2 and H_2O. Isotope Geoscience, 1, 181-190. (127)

Brewster, D. (1820) On the optical properties and mechanical condition of amber. Edinburgh Philos. J., 2, 332-334. (104, 325)
———— (1823a) On the existence of two new fluids in the cavities of minerals, which are immiscible, and which possess remarkable physical properties. Edinburgh Philos. J. 9, 94-107. (Note: this is a shorter version of Brewster, 1826a.) (3, 86, 90)
———— (1823b) On the existence of two new fluids in the cavities of minerals, which are immiscible, and possess remarkable physical properties. Royal Soc. Edinburgh Trans., 10, 1-41. (52, 82, 86, 94)
———— (1826a) On the refractive power of the two new fluids in minerals, with additional observations on the nature and properties of these substances. Royal Soc. Edinburgh Trans., 10, 407-427. (86)
———— (1826b) On the existence of a group of moveable crystals of carbonate of lime in a fluid cavity of quartz. Edinburgh Philos. J., 9, 268-270. (112, 165)
———— (1827) Notice respecting the existence of the new fluid in a large cavity in a specimen of sapphire. Edinburgh J. Sci., 6, 155-156. (375)
———— (1845) On the existence of crystals with different primitive forms and physical properties in the cavities of minerals; with additional observations on the new fluids in which they occur. Royal Soc. Edinburgh Trans., 16, 11-22 (pub. 1849). (53, 92, 99)
———— (1853) Account of a remarkable fluid cavity in topaz. Philos. Mag., ser. 4, 5, 235-236. (95)
———— (1862) On the pressure cavities in topaz, beryl, and diamond, and their bearing on geological theories. Royal Soc. Edinburgh Trans., 23, 39-44 (pub. 1864). (507)

Bridgeman, P.W. (1912) Water in the liquid and five solid forms, under pressure. Proc. Am. Acad. Arts Sci., 47, 441-558. (228, 299)

Bridgwater, D., J.H. Allaart, J.W. Schopf, C. Klein, M.R. Walter, E.S. Barghoorn, P. Strother, A.H. Knoll and B.E. Gorman (1981) Microfossil-like objects from the Archean of Greenland: a cautionary note. Nature, 289, 51-53. (178)

Bril, H. and C. Ramboz (1982) Tin-wolfram mineralizations from the Brioude-Massiac district and from the south of the Massif Central area (France): A comparative study of mineralogy and associated fluid phases. Compt. Rend. Acad. Sci. (Paris), 294, Ser. II, 387-390 (in French). (457)

Brinson, G. (1966) The nucleation of gases from liquids. J. Australian Inst. Metals, 11, 227-230. (296)

Brock, T.W. (1962) A hot-stage petrographic microscope for glass research. J. Am. Ceramic Soc., 45, 5-7. (188)

Brooks, C. and L. Gélinas (1975) Immiscibility of ancient and modern volcanism. Carnegie Inst. Washington Year Book 74, 240-247. (552)

Brown, Jr., G.E., ed. (1982) The mineralogy of pegmatites. Am. Mineral., 67, 180-189. (388)

Brown, J.S. (1942) Differential density of ground water as a factor in circulation, oxidation and ore deposition. Econ. Geol., 37, 310-317. (289)
————, Ed. (1967) Genesis of Stratiform Lead-Zinc-Barite-Fluorite Deposits (Mississippi Valley-Type Deposits), a Symposium. Econ. Geol. Mon. 3, 443 p. (415)
———— (1970) Mississippi Valley-type lead-zinc ores. Mineralium Deposita, 5, 103-119. (415, 419)

Brown, W.M. (1983) The Genesis of a F-Sn-W Skarn at Mt. Garnet, Queensland - An Example of a Granite-Skarn Hydrothermal System. M.S. thesis, La Trobe Univ., Bundoora, Victoria, Australia, 350 p. (456, 457)

Browne, P.R.L., E. Roedder, and A. Wodzicki (1976) Comparison of past and present geothermal waters, from a study of fluid inclusions, Broadlands field, New Zealand. In J. Cadek, T. Paces, eds., Proceedings International Symposium on Water-Rock Interaction, Czechoslovakia, 1974, Geol. Survey, Prague: 140-149. (203, 275, 499, 500, 501)

Bryan, F.R. and J.C. Neerman (1962) Gas chromatographic identification of major constituents of bubbles in glass. Anal. Chem. 34, (2), 278-280. (124)

Buchanan, D.L. and J. Nolan (1979) Solubility of sulfur and sulfide immiscibility in synthetic tholeiitic melts and their relevance to Bushveld-complex rocks. Canadian Mineral., 17, 483-494. (474)

Buchanan, L.J. (1979) The Las Torres mine, Guanajuato, Mexico. Ore controls of a fossil geothermal system. Ph.D. dissertation, Colorado School of Mines. (426)

Buckley, H.E. (1951) Crystal Growth. Wiley & Sons, New York, NY. (38)

Buerger, M.J. (1932) The negative crystal cavities of certain galena and their brine content: Am. Mineral., 17, 228-233. (114)
———— (1934) The lineage structure of crystals. Z. Kristallogr., 89, 195-220. (15)

Bunsen, R. (1851) Ueber die Processe der vulkanischen Gesteinsbildungen Islands: Annalen Physik u. Chemie, 83, 197-272. Translated in Tyndall, John, and Francis, William, eds., Science Memoirs, Natural Philosophy [New Ser.]: Taylor & Francis, London, 1, pt. 1, 33-98, 1852. (315)

Burlinson, K., J.C. Dubessy, G. Hladky and R.W.T. Wilkins (1983) The use of fluid inclusion decrepitometry to distinguish mineralized and barren quartz veins in the Aberfoyle tin-tungsten mine area, Tasmania. J. Geochem. Explor. (in press). (265)

Burnham, C.W. (1979) Magmas and hydrothermal fluids. In H.L. Barnes, ed., Geochemistry of Hydrothermal Ore Deposits, 2nd ed., J. Wiley & Sons, New York, 71-136. (386)

_____ (1981) Physicochemical constraints on porphyry mineralization. In W.R. Dickinson and W.D. Payne, eds., Relations of Tectonics to Ore Deposits in the Southern Cordillera. Arizona Geol. Soc. Digest, 14, 71-77. (446)

Burnham, C.W. and H. Ohmoto (1980) Late-stage processes of felsic magmatism. Mining Geol. (Japan) Spec. Issue, (8), 1-11. (386, 446)

_____, J.R. Holloway and N.F. Davis (1969) The specific volume of water in the range 1000 to 8900 bars, 20° to 900°C. Am. J. Sci., 267, 70-95. (228, 274, 279)

Burruss, R.C. (1977) Analysis of Fluid Inclusions in Graphitic Metamorphic Rocks from Bryant Pond, Maine, and Khtada Lake, British Columbia: Thermodynamic Basis and Geologic Interpretation of Observed Fluid Compositions and Molar Volumes. Ph.D. dissertation, Princeton University, Princeton, NJ. (194, 195, 205, 207)

_____ (1981a) Analysis of fluid inclusions: Phase equilibria at constant volume. Am. J. Sci., 281, 1104-1126. (104, 233, 243, 244)

_____ (1981b) Hydrocarbon fluid inclusions in studies of sedimentary diagenesis. Mineral. Assoc. Canada Short Course Handbook 6, 138-156. (83, 104, 233, 243, 332)

_____ (1981c) Analysis of phase equilibria in C-O-H-S fluid inclusions. Mineral. Assoc. Canada Short Course Handbook 6, 39-74. (239, 242, 243, 244, 250)

_____ and L.S. Hollister (1979) Evidence from fluid inclusions for a paleo-geothermal gradient at the geothermal test wells sites, Los Alamos, New Mexico. J. Volcan. Geothermal Res., 5, 163-177. (496)

_____, D.J. Toth and R.H. Goldstein (1980) Fluorescence microscopy of hydro-carbon fluid inclusions: Relative timing of hydrocarbon migration events in the Arkoma basin, NW Arkansas (abstr.). EOS, Trans. Am. Geophys. Union, 61, 400. (83, 335)

_____, K.R. Cerone and P.M. Harris (1981) Timing of fracturing and oil migra-tion, Oman fore deep: Evidence from fluid inclusions and burial history analysis (abstr.): Geol. Assoc. Canada - Mineral. Assoc. Canada Abstr., 6, A-7. (335)

_____ and _____ (1983a) Regional distribution of hydrocarbon fluid inclusions in carbonate fracture filling cements: Geohistory analysis and timing of oil migration, Oman foredeep (abstr.). Am. Assoc. Petrol. Geol. Bull., 67, 434. (335)

_____ and _____ (1983b) Fluid inclusion petrography and tec-tonic-burial history of the Al Ali No. 2 well: Evidence for the timing of diagenesis and oil migration, northern Oman Foredeep. Geology, 11, 567-570. (335)

Bursill, L.A. (1983) Small and extended defect structures in gem-quality Type 1 diamonds. Endeavor New Ser., 7, 70-77. (509)

Burt, D.M. (1981) Acidity-salinity diagrams-Application to greisen and porphyry deposits. Econ. Geol., 76, 832-843. (456)

Buseck, P. (1983) Electron microscopy of minerals. Am. Sci., 71, 175-185. (15)

Bussink, R.W. (1981) Fluid inclusion studies of the W-Sn ore deposits of Pana-squeira, Portugal (abstr.). Program, Symposium on Current Research on Fluid Inclusions, Univ. Utrecht, Utrecht, The Netherlands, April 22-24, 1981 (unpaginated). (460)

Butuzov, V.P. and L.V. Bryatov (1956) Growing quartz crystals. Rost Kristallov, Akad. Nauk SSSR Inst. Krist., Dokl. Soveshchaniya 1956, 305-310 (published in 1957); (In Russian; translated in Growth of Crystals, 1st Conf. Repts, Moscow, 1956, Consultants Bureau, New York, NY, 241-244, pub. 1958). (89)

_____, L.N. Khetchikov and A.A. Shaposhnikov (1971) Inclusions in synthetic crystals and their significance in thermobarometry of minerals. COFFI, 4, 109-113. (266)

Byers, C.D., D.W. Muenow and M.O. Garcia (1983) Volatiles in basalts and ande-sites from the Galapagos Spreading Center, 85° to 86°W. Geochim. Cosmochim. Acta, 47, 1551-1558. (486)

Cabri, L.J., H. Blank and A. El Goresy (1983) Quantitative proton microprobe analyses of major sulfides in ore deposits of the Sudbury area (abstr.). Geol. Assoc. Canada/Mineral. Assoc. Canada Abstr., 8, A9. (148)

* Cady, G.H. (1983) Composition of gas hydrates. J. Chem. Educa., 60, 915-918. (241)

Calas, G., A.-Y. Huc and B. Pajot (1976) Utilization of infrared spectrometry for fluid inclusion studies in minerals: feasibility and limits. Bull. Soc. franc. Mineral. Cristallogr., 99, 153-161 (in French). (84)

Cameron, E.N. (1961) Ore Microscopy. Wiley & Sons, New York, NY, 293 p. (157)

_____, R.B. Rowe and P.L. Weis (1953) Fluid inclusions in beryl and quartz from pegmatites of the Middletown district, Connecticut (Part II). Am. Mineral., 38, 218-262. (90, 523)

_____, R.H. Jahns, A.H. McNair and L.R. Page (1949) Internal structure of granitic pegmatites. Econ. Geol. Mon. 2, 115 p. (388)

Candela, P.A. (1983) Halogen trends in apatite as an indicator of magmatic vapor evolution (abstr.). EOS, Trans. Am. Geophys. Union, 64, 343. (390)

Carapezza, M., P.M. Nuccio and M. Valenza (1980) Geochemical precursors of earthquakes. In B. Vodar and P. Marteau, eds., High Pressure Science and Technology. Pergamon Press, Oxford, England, 90-103. (402)

Carignan, J., L. Kheang, A. Brown and L. Gelinas (1979) Étude microthermomon-étrique préliminaire des inclusions fluides associées au gisement volcanogène de Millenbach: Geol. Assoc. Canada, Abstr. Programs, 4, 42. (461)

Carpenter, A.B., M.L. Trout and E.E. Pickett (1974) Preliminary report on the origin and chemical evolution of lead- and zinc-rich oil-field brines in central Mississippi. Econ. Geol., 69, 1191-1206. (418)

Carron, J.-P. (1961) Premieres donnees sur la composition de certains reliquats magmatiques. Compt. Rend. Acad. Sci. Paris, 253, 2016-2018. (98, 140, 485)

Carstens, H. (1968) The lineage structure of quartz crystals. Contrib. Mineral. Petrol. 18, 295-304. (15)

_____ (1969) Arrays of dislocations associated with healed fractures in natural quartz. Norges Geolog. Undersokelse, (258), 368-369. (24, 75)

Carswell, D.A. (1980) Mantle derived lherzolite nodules associated with kimber-lite, carbonatite and basalt magmatism, a review. Lithos, 13, 121-138. (505, 514)

Caruso, L. and G. Simmons (1984) Geological applications of backscattered electron imaging. Proc. Symp. Electron Microscopy in Geology, Tempe, Arizona, 6-11 May, 1981 (in press). (144)

Casadevall, T.J., W. Rose, T. Gerlach, L.P. Greenland, J. Ewert, R. Wunderman and R. Symonds (1983) Gas emissions and the eruptions of Mount St. Helens through 1982. Science, 221, 1383-1385. (489)

Casadevall, T. and H. Ohmoto (1977) Sunnyside mine, Eureka mining district, San Juan County, Colorado: Geochemistry of gold and base metal ore deposi-tion in a volcanic environment. Econ. Geol., 72, 1285-1320. (427, 429)

Castaing, R., H. Bizouard, R. Clocchiatti and M. Havette (1978) Some applica-tions of the electron microprobe and of the ion microanalyser in mineralogy. Bull. Mineral., 101, 245-262 (in French). (148)

Cathelineau, M. (1982) The uranium deposits associated with south-armorican leucogranites and their surrounding rocks: relationships and reactions between mineralization and various structural and geologic settings. Sci. de la Terre, Mem. 42, 375 p. (in French). (462)

Cathles, L.M. (1977) An analysis of the cooling of intrusives by groundwater convection which includes boiling. Econ. Geol., 72, 804-826. (401, 451)
___ (1981) Fluid flow and genesis of hydrothermal ore deposits. Economic Geology 75th Anniversary Volume, 424-457. (451)

Černý, P. (1975) Granitic pegmatites and their minerals: Selected examples of recent progress. Fortschr. Mineral., 52, 225-250. (396)
IV, and R.B. Ferguson (1972) The Tanco pegmatite at Bernic Lake, Manitoba. (396)
Petalite and spodumene relations. Canadian Mineral., 11, 660-678.

Chaigneau, M. (1967) Sur la relation entre l'aspect enfumé des quartz et leur tenure en hydrocarbures. Compt. Rend. Acad. Sci. Paris, ser. D, 265, (20), 1444-1447. (123)

Chamberlin, R.T. (1908) The Gases in Rocks. Carnegie Inst. Washington Pub. 106, 80 p. (119)

Champness, P.E., G. Cliff and G.W. Lorimer (1981) Quantitative analytical electron microscopy. Bull. Mineral., 104, 236-240. (142)

Chen, C.-T.A. (1981) Geothermal system at 21°N. Science, 211, 298. (495)

Chepurov, A.I. (1975) Apparatus for studies of melt inclusions in minerals, chapt. III. In Magmatic Crystallization, as Evidenced by Melt Inclusion Studies (in Russian; translated in Fluid Inclusion Res. -- Proc. of COFFI, 9, 182-195). (186)
___ and N.P. Pokhilenko (1972) Inert medium microchamber used for high-temperature study of inclusions in minerals. Acad. Sci. USSR, Sib. Div., Geol. Geophys., 1972, (6), 139-141 (in Russian). (189)
___, Yu.G. Lavrent'yev, O.S. Pokachalova and Yu.I. Malikov (1974) Melt inclusion composition of minerals of pseudoleucitite, central Aldan. English Geofiz., (Akad. Nauk SSSR, Sib. Otd.), (4), 55-60 (in Russian; English sum.; translated in Sov. Geol. Geophys., 15, (4), 46-50). (484)
___, V.S. Shatskiy, O.S. Pokachalova and Yu.G. Lavrent'yev (1975) Evidence from melt inclusions on the chemistry and crystallization of theralites in some Kuznetsk Alatau intrusions. Geokhimiya, 1975, (4), 595-602 (in Russian; translated in Geochem. Int'l, 12, (2), 242-250). (402, 403, 404)

Chernov, A.A. and D.E. Temkin (1977) Capture of inclusions in crystal growth. In Crystal Growth and Materials. E. Kaldis and H.J. Scheel, eds. North-Holland Pub. Co. 3-77, Amsterdam, The Netherlands. (15, 29)

Chi, G.G. (1981) Symposium on the composition of fluid inclusions in minerals. Geochimica, 1981, 6, (2), 184 (in Chinese). (6)

Chinese Geological Society (1981) Studies of Fluid Inclusions in Minerals, Symp. [1977] Nat'l Meeting on Experimental Studies of the Inclusions in Minerals and the Genesis of Rocks and Minerals. Volume 1, Beijing, Scientific Publishing House (in Chinese), 335 p. (6)

Chivas, A.R. (1976) Magmatic evolution and porphyry copper mineralization of the Koloula igneous complex, Guadalcanal (abstr.). Int'l Geol. Congress, 25th, Abstr., 48. (443)
___ (1981) Geochemical evidence for magmatic fluids in porphyry copper mineralization. Contrib. Mineral. Petrol., 78, 389-403. (446)
___ and R.W.T. Wilkins (1977) Fluid inclusion studies in relation to hydrothermal alteration and mineralization at the Koloula porphyry copper prospect, Guadalcanal. Econ. Geol., 72, 153-169 (276, 443)

Chou, I.-M. (1982) Phase relations in the system NaCl-KCl-H2O. Part I: Differential thermal analysis of the NaCl-KCl liquidus at 1 atmosphere and 500, 1000, 1500, and 2000 bars. Geochim. Cosmochim. Acta, 46, 1957-1962. (58, 245)
___ (1983) Remarks on "migration of brine inclusions in salt." Nuclear Technology, 63, 507-509. (319)

Chou, I.-M. and R.J. Williams (1977) Activity of H_2O in CO_2-H_2O at 600°C and pressure to 8 kilobars (abstr.). Geol. Soc. Am. Abstr. Programs, 9, 928. (239)
___ and ___ (1979) The activity of H_2O in supercritical fluids (abstr.). Lunar Planetary Sci. X, 201-203. (239)

Christie, J.M., D.T. Griggs and N.L. Carter (1964) Experimental evidence of basal slip in quartz. J. Geol., 72, 734-756. (74)

Chryssoulis, S.L. (1983) Study of the effects of feldspar and mica contamination upon the analysis of fluid inclusions by the decrepitation-ICP method. Chem. Geol., 40, 323-335. (132)
___ and N. Wilkinson (1983) High silver content of fluid inclusions in quartz from Guadalcazar granite, San Luis Potosi, Mexico: A contribution to ore-genesis theory. Econ. Geol., 78, 302-318. (143)

Chupina, L.Yu. (1976) Inclusions in minerals of meteorites (abstr.). COFFI, 10, 44. (569)
___ and Yu.A. Dolgov (1976) Investigations of meteorites and inclusions in their transparent minerals. COFFI, 10, 45. (569)

Clabaugh, S.E. (1952) Corundum deposits of Montana. U.S. Geol. Survey Bull. 983, 100 p. (492)

Clark, C.D., E.W.J. Mitchell and B.J. Parsons (1979) Colour centers and optical properties. In J.E. Field, ed., The Properties of Diamond, Academic Press, London, 23-77. (509)

Clark, Jr., S.P., ed. (1966) Handbook of Physical Constants. Geol. Soc. Am. Memoir 97, 587 p. (70)

Clarke, A.R. and M. Cable (1967) A gas chromatograph for analyzing single bubbles in glass. Glass Technology, 8, (3), 82-85. (124)

Clayton, C.J. (1982) Growth history and microstructure of flint (abstr.). Abstrs. 3rd European Regional Mtg. Int'l Assoc. Sedimentologists, 105-107. (306)

Clayton, R.N., L.J.P. Muffler and D.E. White (1968) Oxygen isotope study of calcite and silicates of the River Ranch no. 1 well, Salton Sea geothermal field, California. Am. J. Sci., 266, 968-979. (422)

Clement, C.R. and E.M.W. Skinner (1982) Kimberlite textures I (abstr.). Terra Cognita 2, p. 209. (511)

Clifton, C.G. (1983a) An improved method for locating blind or buried mineralization (abstr.). Geol. Assoc. Canada/Mineral. Assoc. Canada Program Abstr., 8, A13. (122, 467)
___ (1983b) Gas analysis of rocks and minerals: Applications in ore deposit research (abstr.). Geol. Assoc. Canada/Mineral. Assoc. Canada Program Abstr., 8, A-13. (122, 467)

Clocchiatti, R. (1970) Study of glass inclusions and their alteration, a regional example from the Dolomites (Bolzano, Italy). Schweiz. Mineral. Petrog. Mitt. 50, (1), 159-166 (in French). (308)
___ (1971) Composition chimique des inclusions vitreuses des phenocristaux de quartz de quelques laves acides par l'analyse à la microsonde électronique. Compt. Rend. Acad. Sci. Paris, 272, 2045-2047. (485)
___ (1975) Glassy inclusions in crystals of quartz: optical, thermo-optical and chemical studies and geological applications. Soc. Geol. France, Memoires, New Series, 54, (122), 96 p., (in French). (89, 474, 479, 480, 486)
___ (1977a) Melt inclusions in olivine, plagioclase and pyroxene phenocrysts as samples of magmatic liquid during host mineral crystallization. An application to a low-K basalt of emerged mid-oceanic ridge (Asal, T.F.A.I.). Compt. Rend. Acad. Sci. Paris, 284, Ser. D, 2203-2206 (in French). (483)
___ (1977b) Melt inclusions from phenocrysts of olivine and chromite of a "picritic basalt" from the mid-Atlantic ridge: amount of contamination of a basalt with olivine. Compt. Rend. Acad. Sci. Paris, 285, Ser. D, 1155-1158 (in French). (483)

Clocchiatti, R. and H. Bizouard (1979) Mise en évidence de la nature du liquide parental, de son evolution par fractionnement cristallin et d'un mélange de magmas dans les produits de l'éruption fissurale du rift d'Asal (République de Djibouti). Compt. Rend. Acad. Sci. Paris, 289, 647-650. (483)

___ and N. Krasov (1979) Fractional crystallization and immiscibilities in calc-alkaline melts trapped by feldspar phenocrysts of Karimski Volcano lavas (Kamtchatka, USSR). Compt. Rend. Acad. Sci. Paris, 289, Ser. D, 1-4 (in French). (484)

___ and B. Mervoyer (1974) Contribution to the study of quartz from Guadaloupe: VII Conférence géologique des Caraïbes, Point-à-Pitre 1974 (in French; translated in Fluid Inclusion Research -- Proc. of COFFI, 7, 38-39). (468)

___ and ___ (1976) Contribution to the study of quartz crystals from Guadeloupe. Bull. Bureau de recherches géologiques et minières, Fr., ser. 2, sect. 4, (4), 311-324 (in French). (14, 17, 308, 484)

___ and N. Metrich (1977) Comparison between glassy inclusions in cordierite from rhyodacites of Tuscany (San Vincenzo) and those of the rhyodacites of north Tunisia (Aïn ed Deflaia). Compt. Rend. Acad. Sci. Paris, 284, Ser. D., 887-890 (in French). (482, 484)

___, C. Desnoyers, J.-C. Sabroux, H. Tazieff and S. Wilhelm (1976) Relationships between glass inclusion and host anorthoclase megacrysts from the Erebus Volcano. Bull. Soc. franc. Minéral. Cristallogr., 99, 98-110 (in French). (484)

___, A. Havette, J. Weiss and S. Wilhelm (1978) The bytownite megacrysts of Asal Rift. - I. Study of trapped basaltic melts: a new approach to the study of some petrogenetic process. Bull. Minéral., 101, 66-76 (in French). (483)

___ and P. Nativel (1979) Petrogenetic relations between transitional basalts and oceanites from trapped melts in olivine and chromospinel phenocrysts of Piton de la Fournaise (Reunion Island, Indian Ocean). Bull. Mineral., 102, 511-525 (in French). (475, 482, 484)

___ H. Bizouard, A. Havette and R. Brousse (1980) Simultaneous presence of different glasses in the matrix and inclusions of basic plagioclases of calc-alkaline lavas. Immiscibility of silicate melts (abstr.). Int'l Mineral. Assoc., Collected Abstr., 12th General Mtg. 4-6 July 1980, Orleans, France, 127. (476)

___, D. Massare and C. Jehanno (1981) Hydrothermal origin of Zabargad (St. Johns) Red Sea peridot gemstone as proved by their inclusions. Bull. Mineral., 104, 354-360 (in French). (147, 515, 527)

Cloke, P.L. (1963) The geologic role of polysulfides--Part II. The solubility of acanthite and covellite in sodium polysulfide solutions. Geochim. Cosmochim. Acta, 27, 1299-1319. (117)

___ and S.E. Kesler (1979) The halite trend in hydrothermal solutions. Econ. Geol., 74, 1823-1831. (451)

Clynne, M.A. and J.W.J. Wilson (1978) The halite trend in hydrothermal magmatic-hydrothermal mineralization (abstr.). Geol. Soc. Am. Abstr. Programs 10, 381. (246)

Clynne, M.A. and R.W. Potter II (1977) Freezing point depression of synthetic brines (abstr.). Geol. Soc. Am. Abstr. Programs, 9, 930. (250, 289)

___ and J.L. Haas, Jr. (1981) Solubility of NaCl in aqueous electrolyte solutions from 10 to 100°C. J. Chem. Eng. Data 26, 396-398. (250)

Cobb, J.C. (1981) Geology and geochemistry of sphalerite in coal: Ill. Geol. Survey Contract Grant 14-08-0001-G-496 Final Rept. 171 p. (330)

Cohen, L.H. and J.L. Rosenfeld (1979) Diamond: depth of crystallization inferred from compressed included garnet. J. Geol., 87, 333-340. (507)

Cole, D.R. (1983) Time estimates for oxygen isotopic exchange during mineral-fluid interaction in hydrothermal systems. Trans. Geothermal Resources Council, 7, 283-287. (127)

Cole, D.R., H. Ohmoto and A.C. Lasaga (1983) Isotopic exchange in mineral-fluid systems. I. Theoretical evaluation of oxygen isotopic exchange accompanying surface reactions and diffusion: Geochim. Cosmochim. Acta, 47, 1681-1693. (127)

Collins, P.L.F. (1979) Gas hydrates in CO2-bearing fluid inclusions and the use of freezing data for estimation of salinity. Econ. Geol., 74, 1435-1444. (99, 103, 240, 248, 249, 369, 500)

___ (1981) The geology and genesis of the Cleveland tin deposit, western Tasmania: Fluid inclusion and stable isotope studies. Econ. Geol., 76, 365-392. (456, 457)

Coolen, J.J.M.M.M. (1980) Chemical Petrology of the Furua Granulite Complex, Southern Tanzania. Ph.D. dissertation. Free Univ. Amsterdam, Utrecht, The Netherlands, Gua Papers of Geology, Series 1, no. 13-1980, 258 p. (292, 297, 363, 374, 375, 376, 377)

___ (1982) Carbonic fluid inclusions in granulites from Tanzania -- a comparison of geobarometric methods based on fluid density and mineral chemistry. Chem. Geol., 37, 59-77. (375, 376)

Copeland, C.S., J. Silverman and S.W. Benson (1953) The system NaCl-H2O at supercritical temperatures and pressures. J. Chem. Physics, 21, 12-16. (253)

Cornec E. and H. Krombach (1932) Equilibres entre le chlorure de potassium, le chlorure de sodium et l'eau depuis - 23° jusqu'a + 190°. Ann. Chimie, 18, 5-31. (245)

Couture, R.A., M.G. Seitz and M.J. Steindler (1983) Sampling of brine in cores of Precambrian granite from northern Illinois. J. Geophys. Res., 88, 7331-7334. (401)

Coveney, Jr., R.M. (1981) Gold quartz veins and auriferous granite at the Oriental mine, Alleghany district, California. Econ. Geol. 76, 2176-2199. (356)

___ and E.D. Goebel (1983) New fluid inclusion homogenization temperatures for sphalerite from minor occurrences in the Mid-Continent area, in Proc. Oct. 1982 Int'l Conf. on Mississippi Valley-type Lead-Zinc Deposits: G. Kisvarsanyi, et al., eds., Univ. of Missouri Press, Rolla, MO, p. 234-242. (324, 421)

___ and W.C. Kelly (1970) Quartz as a geologic barometer. Michigan Academician 3(2), 45-56. (95, 271)

___ (1971) Dawsonite as a daughter mineral in hydrothermal fluid inclusions. Contrib. Mineral. Petrol., 32, 334-342. (113, 356)

Cox, D.P. (1984) Geology of the Tanama and Helecho porphyry copper deposits and vicinity, Puerto Rico. U.S. Geol. Survey Prof. Paper (in press). (443, 445)

___ I.P. Gonzalez and J.T. Nash (1975) Geology, geochemistry, and fluid-inclusion petrography of the Sapo Alegre porphyry copper prospect and its metavolcanic wallrocks, west-central Puerto Rico. U.S. Geol. Survey J. Res., 3, (3), 313-327. (443)

Craig, J.R., J.F. Light, B.C. Parker, and M.G. Mudrey, Jr. (1975) Identification of hydrohalite. Antarctic J., July/Aug., 1975, 178-179. (199)

Crawford, M.L. (1981) Phase equilibria in aqueous fluid inclusions. Mineral. Assoc. Canada Short Course Handbook 6, 75-100. (249)

___ and V.B. Sisson (1981) CO2-brine immiscibility in high-grade metamorphic rocks (abstr.). Geol. Assoc. Canada/Mineral. Assoc. Canada Abstr. Program, 6, A11. (373)

___ J. Filer and C. Wood (1979a) Saline fluid inclusions associated with retrograde metamorphism. Bull. Mineral., 102, 562-568. (38, 338, 351, 354, 355, 362)

___ D.W. Kraus and L.S. Hollister (1979b) Petrologic and fluid inclusion study of calc-silicate rocks, Prince Rupert, British Columbia. Am. J. Sci., 279, 1135-1159. (38, 351)

Cremer, M., H.N. Elsheimer and E.E. Escher (1972) Microcoulometric measurement of water in minerals. Analytica Chim. Acta, 60, 183-192. (118, 122)

Cl
Cr

Cronn, D.R. and W. Nutmagul (1982) Volcanic gases in the April 1979 Soufriere eruption. Science, 216, 1121-1123. (489)

Cuney, M. (1978) Geologic environment, mineralogy, and fluid inclusions of the Bois Noirs-Limouzat uranium vein, Forez, France. Econ. Geol., 73, 1567-1610. (462)

_____ ([1980]) Preliminary results on the petrology and fluid inclusions of the Rossing uraniferous alaskites. Trans. geol. Soc. S. Africa, 83, 39-45. (462)

_____, M. Pagel and J. Touret (1976) The analysis of gases from inclusions by gas chromatography. Bull. Soc. franc. Minéral. Cristallogr., 99, 169-177 (in French). (125)

Cunningham, Jr., C.G. (1974) Geothermometry and geobarometry of fault plane mineralization -- Ginna project, New York, Appendix C, pp. CIII-CIV2, (17 p.). In Geologic and Geophysical Investigations, Ginna site, Rochester Gas & Electric Corp. Prepared by Dames & Moore Corp. for Rochester Gas & Electric Corp. Submitted in support of proceedings for U.S. Nuclear Regulatory Comm. Docket No. 50-244, April 19, 1974. (283, 355)

_____ (1975) Fluid inclusions as exploration guides in the porphyry environment (abst). Mining Year Book, 1975. Colorado Mining Assoc., 93. (443, 444, 469)

_____ and C. Corollo (1980) Modification of a fluid-inclusion heating/freezing stage. Econ. Geol., 75, 335-337. (194, 195)

_____ and A.V. Heyl (1980) Fluid inclusion homogenization temperatures throughout the sequence of mineral deposition in the Cave-in-Rock area, southern Illinois. Econ. Geol., 75, 1226-1231. (421)

_____, T.A. Stevens, J.D. Rasmussen, and S.B. Romberger (1979) The genesis of the Marysvale, Utah, hydrothermal uranium deposits (abstr.). Geol. Soc. Am. Abstr. Programs, 11, 408. (462)

Cunningham, K.M., M.C. Goldberg and E.R. Weiner (1977) Investigation of detection limits for solutes in water measured by laser Raman spectroscopy. Anal. Chem., 49, 70-75. (104)

Currie, J.B. and S.O. Nwachukwu (1974) Evidence on incipient fracture porosity in reservoir rocks at depth: Bull. Canadian Petrol. Geol., 22, 42-58. (333)

Currie, L.A., ed. (1982) Nuclear and chemical dating techniques. Interpreting the environmental record. Am. Chem. Soc. Symp. Ser. 176, 516 p. (128)

Czamanske, G.K. and P.B. Moore (1977) Composition and phase chemistry of sulfide globules in basalt from the Mid-Atlantic Ridge rift valley near 37°N lat. Geol. Soc. Am. Bull., 88, 587-599. (475)

_____, E. Roedder, F.C. Burns (1963) Neutron activation analysis of fluid inclusions for copper, manganese and zinc. Science, 140, 401-403. (134, 138, 416, 418, 423)

Dahanayake, K. and A.P. Ranasinghe (1981) Source rocks of gem minerals. A case study from Sri Lanka. Mineralium Deposita, 16, 103-111. (375)

Dale, T.N. (1923) The commercial granites of New England. U.S. Geol. Survey Bull., 738, 488 p. (401)

Dalrymple, G.B. and M.A. Lanphere (1969) Potassium-argon dating. W.H. Freeman & Co., San Francisco, CA, 258 p. (128)

Damon, P.E. and J.L. Kulp (1958) Excess helium and argon in beryl and other minerals. Am. Mineral., 43, 433-459. (128)

*Davidenko, N.M. (1968) The practical importance of the study of microinclusions in minerals of western part of Chukotka (abstr.). COFFI, 2, 35-36. (308)

Davis, J.A., N. Rodewald and F. Kurata (1962) Solid-liquid-vapor phase behavior of the methane-carbon dioxide system. Am. Inst. Chem. Eng. J., 8, 537-539. (243)

Davy, Sir Humphry (1822) On the state of water and aeriform matter in cavities found in certain crystals. Royal Soc. London Philos. Trans., 2, 367-376. (3, 91, 115, 212, 241)

Dawson, J.B. and R. Fuge (1980) Halogen content of some African primary carbonatites. Lithos, 13, 139-143. (410)

Dawson, K.M. (1972) Geology of Endako Mine, British Columbia. Ph.D. dissertation, Univ. of British Columbia, Vancouver, B.C., Canada. Diss. Abstr., Inst., 33, (12), Pt. 1, 5912B. (443)

Day, H.W. and P.M. Fenn (1982) Estimating the P-T-X H_2O conditions during crystallization of low calcium granites. J. Geol., 90, 485-507. (386)

Deicha, G. (1950) Essais par écrasement de fragments minéraux pour la mise en evidence d'inclusions de gaz sous pression. Bull. Soc. fran. Minéral. Cristallogr., 73, 439-445. (213)

_____ (1952a) Dispositif experimental pour l'observation directe de la decrepitation des inclusions liquides d'origine hydrothermale. Bull. Soc. franc. Minéral. Cristallogr., 75, 237-245. (115, 219)

_____ (1952b) Les fluides des cristaux de quartz de l'Adrar Mauritanien. Afrique Occidentale Française, Direction mines geol. Bull. 15, 433-436. (82)

_____ (1955) Les Lacunes des Cristaux et Leurs Inclusions Fluides: Signification dans la Genèse des Gites Minéraux et des Roches. Masson et Cie, Paris. 126 p. (5,166)

_____ (1961) Modification des pressions intracristallines et intergranulaires des roches soumises aux variations de température. Bull. Soc. Géol. France, ser. 7, (3), 338-344. (261)

Deines, P. (1980) The carbon isotopic composition of diamonds: relationship to diamond shape, color, occurrence and vapor composition. Geochim. Cosmochim. Acta, 44, 943-961. (506, 507)

Deines, P. (1982) The relationship between inclusion composition and carbon isotopic composition of host diamond (abstr.). Terra Cognita, 2, p. 262. (506, 507)

Dekate, Y.G. (1963) Temperature and geochemical nature of the ore-forming fluids in the Rewat Hill wolfram deposit, Rajasthan, as indicated by liquid inclusions. India Nat'l Inst. Sci. Proc., 29, pt. A, (4), 412-427. (166)

Delaney, J.R. (1982) Generation of high salinity fluids from seawater by two-phase separation (abstr.). EOS, Trans. Am. Geophys. Union, 45, 1135-1136. (495)

_____ and J.L. Karsten (1979) Ion microprobe determination of water in silicate glasses (abstr.). EOS, Trans. Am. Geophys. Union, 60, 966. (486)

_____ and _____ (1981) Ion microprobe studies of water in silicate melts: Concentration-dependent water diffusion in obsidian. Earth Planet. Sci. Lett., 52, 191-202. (148)

_____, and D.S. Kelley (1983) Indirect evidence of hydrothermal temperatures in excess of 400°C from the Mid-Atlantic Ridge (abstr.). Geol. Assoc. Canada/Mineral. Assoc. Canada Program Abstr., 8, A17. (495)

_____, D.W. Muenow and D.G. Graham (1978) Abundance and distribution of water, carbon, and sulfur in the glassy rims of submarine pillow basalts. Geochim. Cosmochim. Acta, 42, 581-594. (486, 488, 489)

_____, D.M. Mogk and B. Cosens (1982) Indirect evidence of boiling hydrothermal systems on the Mid-Atlantic Ridge (abstr.). EOS, Trans. Am. Geophys. Union, 63, 472. (495)

Dele-Dubois, M.-L., P. Dhamelincourt and H.-J. Schubnel (1980) Studies by Raman spectroscopy of inclusions in diamonds, sapphires, and emeralds, parts 1 and 2. Rev. Gemmologie, June 1980, (63), 11-14 (part 1) and September 1980, (64), 13-16 (part 2) (in French). (107, 349)

Delhaye, M. and P. Dhamelincourt (1975) Raman microprobe and microscope with laser excitation. J. Raman Spectros., 3, 33. (105)

_____, J. Barbillat and P. Dhamelincourt (1980) Identification of inclusions and particles by Raman microprobe. In J. Albaigés, ed., Analytical Techniques in Environmental Chemistry, 3. Pergamon Press, Oxford, England, 515-522. (105)

Dellwig, L.F. (1955) Origin of the Salina salt of Michigan. J. Sed. Petrol., 25, 83-110. (309)

Delmas, R., J.M. Ascencio, M. Legrand and D. Raynaud (1979) The atmospheric CO_2 content in the past: Is its estimation possible from polar ice gases? (abstr.). Abstr. XVII General Assembly IUGG, Interdisciplinary Symp., 40. (329)

Deloule, E. and J.F. Eloy (1982) Improvements of laser probe mass spectrometry for the chemical analysis of fluid inclusions in ores. Chem. Geol. 37, 191-202. (139)

Demin, Y. (1970) Structure of the aureoles of evaporation around the ore bodies of some polymetallic ore deposits of Rudni Altai (abstr.), in Collected Abstracts. IMA-IAGOD Meetings. 1970. Tokyo. Science Council of Japan, Tokyo, 256. (266)

Denis, M. (1974) Alteration and associated fluids in the Sierrita porphyry copper (Arizona, USA). Comparison with other deposits of the same type. These de Specialite,CRPG-ENSG-Universite de Nancy, Nancy, France, 146 p. (469)

Denis, M., M. Pichavant, B. Poty and A. Weisbrod (1980) The Sierrita-Esperanza porphyry copper, Arizona, USA. Comparison with other porphyry coppers. Bull. Mineral., 103, 613-622 (in French). (443, 445)

DePaolo, D.J. (1983) Geochemical evolution of the crust and mantle. Rev. Geophys. and Space Phys., 21, 1347-1358. (515, 527)

*Derevyagin, V.S. (1973) Investigations of some microelements in liquid inclusions of rock salt from south of middle Asia (abstr.). COFFI, 7, 46. (314)

Dernov-Pegarov, V.F. and S.D. Malinin (1976) Solubility of calcite in high temperature aqueous solutions of alkali carbonates and the problem of formation of carbonatites. Geokhimiya, 1976, (5), 643-658 (in Russian; translated in Geochem. Int'l., 13, (3), 1-13. (410)

Desnoyers, C. (1980) The Niger (I) carbonaceous chondrite and implications for the origin of aggregates and isolated olivine grains in C2 chondrites. Earth Planet. Sci. Lett., 47, 223-234. (563)

Dewey, C. (1818) Sketch of the mineralogy and geology of the vicinity of Williams' College, Williamstown, Mass. Am. J. Sci., 1, 337-346. (3)

Dhamelincourt, P., J.-M. Beny, J. Dubessy and B. Poty (1979) Analysis of fluid inclusions by the Raman microprobe MOLE. Bull. Mineral., 102, 600-610 (in French). (108)

_____ and H.J. Schubnel (1977) The laser molecular microprobe and its application to mineralogy and gemmology, I. Rev. Gemmologie, 52, 11-14 (in French). (107)

Dimov, V.. V. Breskovska and M. Maleev (1980) Microdiffraction study of the phase composition of precipitates from gas-liquid inclusions in minerals from the Madzharovo ore deposit. Geokhim.Mineral. Petrol., 13, 37-44 (in Russian). (115, 148)

Dix, D.R. and M.P.A. Jackson (1982) Lithology, microstructures, fluid inclusions and geochemistry of rock salt and of the cap-rock contact in Oakwood dome, East Texas: Significance for nuclear waste storage. Texas Bur. Econ. Geol. Rept. Invest. 120, 63 p. (312, 313, 316)

Dmitriyev, S.D. (1970) Reliability of methods of investigation of mineral-forming solutions and pneumatolytic deposition of minerals. Vyssh. Ucheb. Zavedeniy Izvestiya, Geologiya i Razvedka, 1970, (4), 81-86 (in Russian; translated in Int'l Geol. Rev., 13, (5), 1971, 681-684). (266)

Dolgov, Yu.A. (1954) Differentiation by the thermosound method of sedimentary terrigenous quartz contained in Neogene strata of Trans-Carpathia. L'vov. Geol. Obshch. Geol. Sbornik, 1954, 1, 76-87 (in Russian). (308)

_____ (1968) The probable partitioning of gas mixtures during adiabatic expansion of mineral-forming systems. In Mineralogical Thermometry and Barometry, Vol. 1, p. 354-357. "Nauka" Press, Moscow (in Russian). (121)

Dolgov, Yu.A. and L.Sh. Bazarov (1965) A chamber for the investigation of inclusions of mineral-forming solutions and melts at high temperatures. In Mineralogical Thermometry and Barometry, 118-122. "Nauka" Press, Moscow (in Russian). (188, 189)

_____ and N.A. Shugurova (1965) (Ultramicro gas analysis methods). Novosibirsk, Mater. po genetich. i experim. mineralogii [Data on genetic and experimental mineralogy], no. 4 Novosibirsk, 1965 (As quoted by Bazarov (1965); presumably this same reference is quoted by Dolgov and Shugurova (1966a) as Tez. Doklady 2d Vses. Sovesch. po Geotermobarometrii [Thesis reports, 2d All-Union Conference on Geothermometry-Geobarometry]. Further details are unavailable.) (117, 121)

_____ and _____ (1966a) Gases of postmagmatic processes involved in mineral formation. Dokl. Nauk SSSR, Dokl., 170, (6), 1422-1425 (in Russian; translated in Dokl. Acad. Sci. USSR. Earth Sci. Sects., 170, 227-229, 1967). (121)

_____ and _____ (1966b) Research on the compositions of individual gas inclusions. In Materialy po Geneticheskoi i Eksperimental'noi Mineralogii., 4, 173-181. Novosibirsk, Akad. Nauk SSSR Sibirskoe Otdeleniye, Inst. Geologii i Geofiziki (in Russian; English abstr.). (121)

_____ and _____ (1968) Composition of gases in individual inclusions in minerals. In Mineralogical Thermometry and Barometry, 1, Moscow, "Nauka" Press, 290-298 (in Russian). (121)

Dolgov et al., eds., Genetic Studies in Mineralogy. Novosibirsk, Inst. Geol. and Geophy. Sib. Branch Acad. Sci. USSR, 16-21 (in Russian; translated in Fluid Inclusion Res. -- Proc. of COFFI, 10, 56, 1977). (560)

_____ (1976a) Studies of tektites based on inclusions. In Yu.A.

_____ (1976b) Inclusions in the glass spherules and chips from the lunar soil (Luna 16). In Yu.A. Dolgov et al., eds. Genetic Studies in Mineralogy. Novosibirsk, Inst. Geol. and Geophy. Sib. Branch Acad. Sci. USSR, 22-25 (in Russian; translated in Fluid Inclusion Res. -- Proc. of COFFI, 10, 59, 1977). (560)

_____ and V.A. Simonov (1976) Studies of the possibility of diffusion of hydrogen through the walls of inclusions (abstr.). Abstr. 5th All-Union Conf. on Thermobarogeochemistry, Ufa, USSR, 20-23 Sept. 1976. Ufa, Bashkir Section, Acad. Sci. USSR, Inst. of Geol., 167 (in Russian; translated in Fluid Inclusion Res. -- Proc. of COFFI, 10, 60, 1977). (560)

_____ and S.A. Vishnevskiy (1976) High-pressure minerals and inclusions in impactites. In Yu.A. Dolgov et al., eds., Genetic Studies in Mineralogy. Novosibirsk, Inst. Geol. and Geophy. Sib. Branch Acad. Sci. USSR, 12-16 (in Russian; translated in Fluid Inclusion Res. -- Proc. of COFFI, 10, 61, 1977). (560)

_____ and S.A. Vishnevsky (1978) Gas inclusions in glasses and minerals from extraterrestrial samples (abstr.). Proc. 11th Gen. Mtg. Int'l Mineral. Assoc.,3, 21. (569)

_____, V.M. Makagon and V.S. Sobolev (1967) Liquid inclusions in kyanite from metamorphic rocks and pegmatites of the Mamsk region (northeastern Transbaikal]. Akad. Nauk SSSR, Dokl., 175, (2), 444-447 (in Russian; translated in Dokl. Acad. Sci. USSR. Earth Sci. Sects., 175, 164-166, 1967). (89, 362, 374)

_____ and I.T. Bakumenko (1968) Determining the pressure in inclusions by simultaneous use of homogenization and cryometry. COFFI, 2, 37. (190)

_____ Yu. F. Pogrebnyak and N.A. Shugurova (1969a) Composition and pressure of gases in inclusions in tektites. Geokhimiya, (5), 603-609 (in Russian; translated in Geochem. Int'l, 6, (3), 525-531). (560)

_____ N.A. Shugurova, Yu. F. Pogrebnyak (1969b) Gas inclusions in tektites (moldavites). Akad. Nauk SSSR, Dokaldy, 184, (6) 1405-1408. (in Russian). (560)

Drummond, S.E. (1981) Boiling and mixing of hydrothermal fluids: Chemical effects on mineral precipitation. Ph.D. dissertation, Pennsylvania State Univ., University Park, PA, 397 p. (34, 244, 246)

___, and H. Ohmoto (1979) Effects on mineral solubilities in hydrothermal solutions (abstr.). Geol. Soc. Am. Abstr. Programs, 11, 416. (34)

Duba, A.G. and T.J. Shankland (1982) Free carbon and electrical conductivity in the Earth's mantle. Geophys. Res. Lett., 9, 1271-1274. (526)

Dubessy, J., D. Audeoud, R. Wilkins, and C. Kosztolanyi (1982) The use of the Raman microprobe MOLE in the determination of the electrolytes dissolved in the aqueous phase of fluid inclusions. Chem. Geol., 37, 137-150. (107)

___, D. Geisler, C. Kosztolanyi and M. Vernet (1983) The determination of sulfate in fluid inclusions using the M.O.L.E. Raman microprobe. Application to a Keuper halite and geochemical consequences. Geochim. Cosmo. Acta, 47, 1-10. (106)

Dudkin, O.B. (1964) Mineralogy of the Apatite Deposits of the Khibinsk Tundra. "Nauka" Press, Moscow, 235 p. (in Russian). (405)

Duffy, J.R., N.O. Smith and B. Nagy (1961) Solubility of natural gases in aqueous salt solutions. I. Liquidus surfaces in the system $CH_4-H_2O-NaCl$ $CaCl_2$ at room temperatures and at pressures below 1,000 psia. Geochim. Cosmochim. Acta 24, 23-31. (331)

Duke, D.A. and J.D. Stephens (1964) Infrared investigations of the olivine group minerals. Am. Mineral., 49, 1388-1406. (523)

Dumas, J. (1830) Note sur une variété de sel gemme qui décrépite au contact d l'eau. Ann. Chimie Physique, 43, 316-320. (119)

Dunham, K.C. (1970) Mineralization by deep formation waters: A review. Inst. Min. Metall. Trans., Sect. B, 79, B107-B170 and B208-B212. (469)

Dunn, J.R. and D.W. Fisher (1954) Occurrence, properties and paragenesis of anthraxolite in the Mohawk Valley. Am. J. Sci., 252, 489-501. (322)

Durisova, J. (1978) Geothermometry in the minerals from the tin deposits of the eastern Krusne hory Mts. (Czechoslovakia). In Metallization Associated with Acid Magmatism, Vol. 3. Prague, Geol. Survey, 325-335 (in English). (457)

___, B. Charoy and A. Weisbrod (1979) Fluid inclusion studies in minerals from tin and tungsten deposits in the Krusne Hory Mountains (Czechoslovakia). Bull. Mineral., 102, 665-675. (457)

Durney, D.W. (1972) Deformation history of the western Helvetic Nappes, Valais, Switzerland. Ph.D. dissertation, London Univ., London, England, 372 p. (342)

___ (1976) Recent research on fluid inclusions in Australia. Bull. Soc. franc. Mineral. Cristallogr., 99, 128-130 (in French). (186, 193)

___ and J.G. Ramsay (1973) Incremental strains measured by syntectonic crystal growths. In K.A. De Jong and Robert Scholten, eds., Gravity and Tectonics. Wiley & Sons, New York, NY, 67-96. (342, 343)

Dwight, H.E. (1820) Account of the Kaatskill Mountains. Am. J. Sci., 2, 11-29. (3, 98)

*Dzhumailo, V.I., V.N. Vasilenko and V.G. Rylov (1973) Forms of redeposition and conditions of formation of gold in Cu-sulfide ores of the Urup group of deposits, (N. Caucasus). COFFI, 6, 41. (137)

Eadington, P.J. (1974) Microprobe analysis of the non-volatile constituents in fluid inclusions. N. Jahrb. Mineral. Monats., 1974, (11), 518-525. (143)

___ (1978) Composition of fluid inclusions -- non-destructive methods of analysis and the destructive determination of liquid compositions. In Notes for Workshop on Fluid Inclusions: Dept. Geol., La Trobe Univ. Melbourne, Australia (unpaginated, mimeographed). (144)

___ (1983) A fluid inclusion investigation of ore formation in a tin-mineralized granite, New England, New South Wales. Econ. Geol., 78, 1204-1221. (39, 457)

Dolgov, Yu.A., Yu. F. Pogrebnyak and N.A. Shugurova (1971) Gas composition and pressures in inclusions in some tektites and silica glasses. Akad. Nauk SSSR, Dokl. 198, (1), 202-205 (in Russian; translated in Dokl. Acad. Sci. USSR, Earth Sci. Sect., 198, 208-211). (560)

___, A.A. Tomilenko and V.P. Chupin (1976) Inclusions of salt melts-solutions in quartz of deep-seated granites and pegmatites. Akad. Nauk SSSR, Dokl., 226, (4), 938-941 (in Russian; translated in Dokl. Acad. Sci. USSR 226, (4), 206-209). (276, 277)

Dolomanova, E.I. and L.P. Nosik (1977) Possibility of determining the chemical composition of inclusions of hydrothermal solutions in mineral vacuoles by mass spectrometry. Akad. Nauk SSSR, Dokl., 234, 1186-1188 (in Russian; translated in Dokl. Acad. Sci. USSR, 234, 161-163, 1979). (123)

___, V.V. Lider, V.N. Rozhanskii and M.M. Elinson (1966) Composition of solids in some gas-liquid inclusions in motion according to data of X-ray spectral point-analyses [electron microprobe]. Akad. Nauk SSSR, Dokl. 167, 176-179 (in Russian; translated in Dokl. Acad. Sci. USSR, Earth Sci. Sects. 167, 116-119, 1966). (98, 143)

___, ___, and ___ (1968) X-ray spectrographic studies of elemental composition of ultramicrocrystalline phases in gas-liquid inclusions in quartz. In Mineralogical Thermometry and Barometry, Vol. 1, 281-290. "Nauka" Press, Moscow (in Russian). (98, 143)

___, R.V. Boirskala, S.Ye. Borisovskii and L.M. Lupanov (1976) Chemical composition of precipitates in the vacuoles of minerals of tin ore deposits, as indicated by electron microscopy and electron microprobe analysis. COFFI, 10, 63. (113)

Dolomieu, Commandeur Deodat de (1792) Sur de l'Huile de Pétrole dans le Cristal de Roche et les Fluides élastiques tires du Quartz. Observations sur La Physique, 42, 318-319. (3)

Dombrowski, H.J. (1960) Balneobiologische Untersuchungen der nauheimer Quellen. II Mitteilung. Zentralbl. Bakteriologie, Parasitenkunde, Infektionskrankheiten u. Hygiene 178, (1), 83-90. (98)

Dominique, J., J. Lhegu and J.-C. Touray (1973) Evidence of geothermal activity in the Lias of Le Morvan: the Rene-bis vein (La Petite Verrière, Saone-et-Loire, France). Bull. Bureau de recherches geologique et minières, France, 1973, Sect. 2, 389-401 (in French). (496)

Donaldson, C.H. (1975) Calculated diffusion coefficients and the growth rate of olivine in a basalt magma. Lithos, 8, 163-174. (40)

___ and R.W. Brown (1977) Refractory megacrysts and magnesium-rich melt inclusions within spinel in oceanic tholeiites: indicators of magma mixing and parental magma composition. Earth Planet. Sci. Lett., 37, 81-89. (484)

Donnelly, H.G. and D.L. Katz (1954) Phase equilibria in the carbon dioxide-methane system. Indus. Eng. Chem., 46, 511-517. (243)

Dons, J.A. (1966) Barite which decrepitates at room temperatures. Norsk Geol. Tidsskr., 36, (3), 241-248. (116)

*Doroshenko, Yu.P. and N.N. Pavlun (1977) Thermobarogeochemical prospecting-evaluating criteria of tungsten ores (exemplified by Akchatau deposit) (abstr.). COFFI, 10, 65. (457)

___ and ___ (1978) Thermobarogeochemistry of the deposit Akchatau (Central Kazakhstan) (abstr.). COFFI, 12, 45. (457)

Dreibus, G., E. Jagoutz and H. Wanke (1980) Halogens in spinel-lherzolites and their carrier basalts (abstr.). Fortschr. Mineral. 58, pt. 1, 20-21 (in German). (526)

Drier, J.F. (1976) The Geochemical Environment of Ore Deposition in the Pachuca-Real Del Norte District, Hidalgo, Mexico. Ph.D. dissertation, Univ. Arizona, Tucson, AZ, 115 p. (427)

Drozdova, T.V., K.I. Yakubovich and E.F. Konstantinov (1964) On the organic matter from fluorite ores of the Pokrovo-Kireev deposit in the Azov Sea foreland. Geokhimiya, 1964, (6), 573-577 (in Russian; translated in Geochem. Int'l. 1964, (3), 529-531). (178)

Eastoe, C.J. (1976) Fluid inclusion studies of the Panguna and Frieda porphyry coppers, Papua, New Guinea. Abstr., 5th Int'l COFFI Symp. on Fluid Inclusions, Sydney, Australia, Aug. 1976. Fluid Inclusion Res. -- Proc. of COFFI, 8, 53-54. (443)

___ (1978) A fluid inclusion study of the Panguna porphyry copper deposit, Bougainville, Papua New Guinea. Econ. Geol. 73, 721-748. (276, 443, 444, 445, 451)

___ (1979) The formation of the Panguna porphyry copper deposit, Bougainville, Papua New Guinea. Ph.D. disseration, Univ. of Tasmania, Hobart, Tasmania, Australia, 260 p. (246, 451, 452)

___ (1983) Sulfur isotope data and the nature of the hydrothermal systems at the Panguna and Frieda porphyry copper deposits, Papua New Guinea. Econ. Geol. 78, 201-213. (443)

Eadington, P.J. (1982) Problematic fluid inclusions from the Panguna porphyry copper deposit, Bougainville (abstr.). Geol. Soc. Am. Abstr. Programs, 14, 480. (441)

Ebers, M.L. and O.C. Kopp (1979) Cathodoluminescent microstratigraphy in gangue dolomite, the Mascot-Jefferson City district, Tennessee. Econ. Geol. 74, 908-918. (45, 161)

Eggler, D.H., Mysen, T.C. Hoering and J.R. Holloway (1979) The solubility of carbon monoxide in silicate melts at high pressures and its effects on silicate phase relations. Earth Planet. Sci. Lett. 43, 321-330. (505)

___ D.R. Baker and R.F. Wendlandt (1980) FO2 of the assemblage graphite-enstatite-forsterite-magesite: experiment and application to mantle fO2 and diamond formation (abstr.). Geol. Soc. Am. Abstr. Programs, 12, 420. (510)

Einaudi, M.T., L.D. Meinert and R.J. Newberry (1981) Skarn deposits. Economic Geology 75th Anniversary Volume, 317-391. (453, 454)

Elders, W.A. (1980) Hydrothermal minerals; temperature and flow: a model for greenschist metamorphism (abstr.). EOS, Trans. Am. Geophys. Union, 61, 389-390. (500)

Elders, W.A., J.R. Hoagland, E.R. Olson, S.D. McDowell and P. Collier (1978) A comprehensive study of samples from geothermal reservoirs: Petrology and light stable isotope geochemistry of twenty-three wells in the Cerro Prieto geothermal field, Baja California, Mexico. Univ. California, Riverside, Inst. Geophys. and Planet. Phys. Rept. 78/26, 264 p. (498)

Elinson, M.M. (1949) Methods of studying the gas content of rocks. Akad. Nauk SSSR Izv., Otdel. Tekh. Nauk, 269-282 (in Russian). (119)

___ (1956) The study of gases which are occluded in rocks and minerals. Moskov. Geol.-Razvedoch. Inst. Ordzhonikidze Trudy, 29, 195-202 (in Russian). (119)

___ (1968) Method of study of the composition of gases in small samples of minerals and rocks. In Mineralogical Thermometry and Barometry, Vol. 2, 251-255. "Nauka" Press, Moscow (in Russian). (118)

___ and V.S. Polykovskii (1961a) The gases in quartz crystals from Maidantal. Vyssh. Ucheb. Zavedeniy Izv., Geologiya i Razved. 1961, (11), 26-36 (in Russian). (119)

___ and ___ (1961b) Some characteristics of the process of formation of quartz crystal pegmatites as revealed by an investigation of gas inclusions in minerals and rocks. Geokhimiya 1961, (10), 881-890 (in Russian; translated in Geochemistry, (10), 977-987, 1961). (119)

___ and ___ (1963) Gas composition of pneumatolytic-hydrothermal solutions. Geokhimiya 1963, (8), 767-776 (in Russian; translated in Geochemistry, (8), 799-807, 1963). (119)

Ellis, A.J. (1970) Quantitative interpretation of chemical characteristics of hydrothermal systems. Geothermics (1970), Special Issue 2, 516-528. (500)

___ and R.M. Golding (1963) The solubility of carbon dioxide above 100°C in water and in sodium chloride solutions. Am. J. Sci., 261, 47-60. (246, 357, 500)

Ellis, A.J. and W.A.J. Mahon (1977) Chemistry and Geothermal Systems. Academic Press, New York, NY, 392 p. (496, 499, 513)

Ellis, D.E. and P.J. Wyllie (1979) A model of phase relations in the system MgO-SiO2-H2O-CO2 and prediction of the compositions of liquids coexisting with forsterite and enstatite. In F.R. Boyd and H.O.A. Meyer, eds., Kimberlites, Diatremes, and Diamonds: Their Genesis, Petrology, and Geochemistry, Proc. 2nd Int'l Kimberlite Conf., Wash., Am. Geophys. Union, 1, 313-318. (505)

Eloy, J.F. (1984) Geological applications of ionization L.T.E. model in laser mass spectrometry (L.P.M.S. II). Scanning Electron Microscopy (in press). (140)

___ M. Leleu and E. Unsöld (1983) Geological applications of the laser probe mass spectrometer (L.P.M.S. II). Int'l J. Mass Spectrom. Ion Physics, 47, 39-43. (139)

Emons, H.-H., H. Keune and H.-H. Seyfarth (1982) Chemical microscopy. In Comprehensive Analytical Chemistry, Vol. 16, G. Svehla, ed.: Elsevier Sci. Pub. Co., Amsterdam, The Netherlands, 1-328. (156, 207)

Engel, A.E.J. and C.G. Engel (1953) Grenville Series in the northwest Adirondack Mountains, New York. Part I. Geol. Soc. Am. Bull. 64, 1013-1047. (350)

England, P.C. and S.W. Richardson (1977) The influence of erosion upon the mineral facies of rocks from different metamorphic environments. J. Geol. Soc. (London), 134, 201-213. (368)

Entin, M.L., ed. (1968) Abstracts of Reports of Third All-Union Conference on Mineralogical Thermobarometry and Geochemistry of Deep-Seated Mineral-Forming Solutions. Vses. Nauch.-Issl. Inst. Sinteza Mineral. Syr'ya, Moscow (further details unavailable). (5)

Eppler, W.F. (1962) Die diagnostische Bedeutung der Einschlusse in Edelsteinen. Umschau, 15, 472-475. (95)

___ (1981) Inclusions in diamond. J. Gemmology, 8, (1), 1-13. (507)

Erickson, Jr., A.J. (1965) Temperatures of calcite deposition in the Upper Mississippi Valley lead-zinc deposits. Econ. Geol., 60, 506-528. (136, 421)

Ermakov, N.P. (1944) The temperatures of formation of the deposits of optical minerals in central Asia. Sbornik "Sovetskaya Geologiya," 1944, (1), 28-34 (in Russian). (186)

___ (1949) About the primary secondary inclusions in minerals. L'vov. Geol. Obshch. Mineral. Sbornik, 3, 23-27 (in Russian). (23)

___ (1950) Research on the Nature of Mineral-Forming Solutions. University of Kharkov Press, Kharkov, USSR, 460 p. (in Russian) Translated (along with Transactions of the All-Union Research Institute of Piezooptical Mineral Raw Materials (Trudy UNIIP), Vol. 1, part 2, 177 p. and Vol. 2, part 2, 134 p.) in: Yermakov, N.P. et al., 1965, Research on the Nature of Mineral-forming Solutions, with Special Reference to Data from Fluid Inclusions, Vol. 22 of Int'l Ser. of Monographs in Earth Sciences. Pergamon Press, New York, 743 p. (Note - page number references are to those of translation.) (5, 15, 19, 26, 48, 68, 69, 82, 87, 89, 93, 100, 101, 102, 164, 165, 186, 212, 229, 253, 286, 308)

___ (1957) Inclusions of mother liquors in minerals and their significance in theory and practice. Vses. Naucho-Issled. Inst. Pezooptichesk. Mineral. Syr'ya Trudy, 1, (2), 173-175 (in Russian). (52, 468)

___ (1965) The state and activity of the fluids in granitic pegmatites of the chambered type. Int'l Geol. Cong., 22d New Delhi 1964, Dokl. Sovet. Geol. Problem 6, 140-160 (in Russian, with English abstr.; translated in Int'l Geol. Cong., 22d, New Delhi, 1964, Proc., sec. 6, dated 1964, issued 1969), (85, 95, 113, 392)

___ ed. (1966a) Research on Mineral-Forming Solutions (Materials of the First Symposium on Gas-Liquid Inclusions in Minerals, Moscow, May 17-24, 1963). Moscow "Nedra" Press, 264 p. (in Russian; also listed as All-Union Research Inst. Synthetic Mineral Raw Materials, Ministry Geology USSR Trans., 9. (5)

Ermakov, N.P. (1966b) Use of gas-liquid inclusions in prospecting and explora-tion for postmagmatic ore deposits and blind ore bodies. Sovetskaya Geo-logiya, 1966, (9), 77-90 (in Russian; translated in Int'l Geol. Rev., 9, (7), 947-956). [Listed there as "P.P. Ermakov".] (266, 466, 471)

____ (1968a) Mineralogical Thermometry and Barometry. Moscow, "Nauka" Press, 1, 368 p. (in Russian). (5)

____ (1968b) Mineralogical Thermometry and Barometry. Moscow, "Nauka" Press, 2, 320 p. (in Russian). (5)

____ (1969) Geochemical classification of inclusions in minerals. COFFI, 2, 18.(26)

* ____ (1976) Fifth All-Union Meeting on Thermobarogeochemistry (Abstr. of papers), Sept. 20-23, 1976. COFFI, 9, 39. (5)

* ____, ed. (1978) Abstr. of the Sixth All-Union Meeting, Vladivostok, Sept. 15-18, 1978, Volumes 1 and 2, Acad. Sci. USSR, Vladivostok (in Russian). (5)

____ and Yu.A. Dolgov (1979) Thermobarogeochemistry. Nedra Press, Moscow, 271 p. (in Russian). (5)

† ____ and V.A. Kalyuzhnyi (1957) The possibility of determination of real temperatures of mineral-forming solutions. Trudy Vses. Nauch.-Issledovatel. Inst. P'ezooptichesk. Mineral. Syr'ya 1, (2), 41-51 (in Russian; translated in Int'l Geol. Rev., 3, 706-711, 1961). (65, 167)

____ and R.V. Kholmskii (1965) Gas-liquid inclusions as indicators of the age relations between ore and dikes. In Mineralogical Thermometry and Barometry. "Nauka" Press, Moscow, 286-288 (in Russian). (261)

____ and A.G. Kuznetsov (1971) The use of thermo-barogeochemistry methods in the search for hidden ore deposits. Fluid Inclusion Res. -- Proc. of COFFI, 4, 122-125. (266)

† ____ and N.I. Myaz' (1957) Influence of liquid and gaseous inclusions on the weight loss of a mineral by ignition. Vses. Nauchno-Issled. Inst. P'ezooptichesk. Mineral. Syr'ya Trudy, 1 (2), 151-154 (in Russian). (111, 114)

* ____ and A.V. Piznyur (1974) Finding commercial ores by use of thermobaro-geochemical indices of conditions of mineral formation (abstr.). COFFI, 7, 56. (443)

____ and V.N. Trufanov, eds. (1974) Fourth All-Union Conf. on Thermobarogeo-chemistry of Mineral-Forming Processes, Sept. 24-30, 1973, Rostov-on-Don, USSR, Abstracts of Papers: Rostov Univ. Press, Rostov, 351 p. (in Russian). (5)

† ____, V.A. Kalyuzhnyi and N.I. Myaz' (1957) Results of mineral thermometric investigations of some smoky quartz crystals from the Volynia district. Trudy Vses. Nauch.-Issledovatel. Inst. P'ezooptichesk. Mineral. Syr'ya 1, (2), 117-127 (in Russian). (51)

Erwood, R.J., S.E. Kesler and P.L. Cloke (1979) Compositionally distinct, saline hydrothermal solutions, Naica Mine, Chihuahua, Mexico. Econ. Geol., 74, 95-108. (277, 400, 451, 452, 455)

Escobar, R. and A.N. Mariano (1976) On the origin of Colombian emeralds (abstr.): Progr. 2nd Biennial Mineral. Soc. Am. - Friends of Mineralogy Symp., Tucson, Arizona, Feb. 15-17, 1976 (unpaginated). (366)

Etheridge, M.A. and V.J. Wall (1982) High fluid pressures during regional meta-morphism and deformation - implications for mass transport, deformation mechanisms and thermal evolution (abstr.). Geol. Soc. [London] Newsletter, 11, (4) 16. (350, 351)

Etminan, H. (1974) The distribution of fluid inclusions in the porphyry coppers of Sar Cheshmeh and Darreh Zar, Iran. Reunion Annuelle des sciences de la Terre, 2ème, Pont-a-Mousson, April 1974, 167 (in French). (445)

____ (1977) The Porphyry Copper of Sar Cheshmeh (Iran): Role of the Fluid Phases in the Mechanisms of Alteration and Mineralization. Ph.D. disserta-tion, Univ. Nancy, Nancy, France, Sciences de la Terre, Mem. 34, 249 p. (in French; English abstr.). (443, 445)

Eugster, H.P. (1981) Metamorphic solutions and reactions. Phys. Chem. Earth. 13/14, 460-507. (362)

____ (1982) Rock-fluid equilibrium systems. In W. Schreyer, ed., High-Pres-sure Researches in Geoscience. E. Schweizerbart'sche Verlag., Stuttgart, 501-518. (362)

____ and I.-M. Chou (1979) A model for the deposition of Cornwall-type mag-netite deposits. Econ. Geol., 74, 763-774. (455)

____ and G.B. Skippen (1967) Igneous and metamorphic reactions involving gas equilibria. In P.H. Abelson, ed., Researches in Geochemistry, Vol. 2. Wiley & Sons, New York, NY, 492-520. (368)

Evans, A.M., H.W. Haslam and R.P. Shaw (1979) Porphyry style copper-molybdenum mineralization in the Ballachulish igneous complex, Argyllshire, with spe-cial reference to the fluid inclusions. Proc. Geol. Assoc. 91, 47-51. (443)

Evans, Jr., S.H. and W.P. Nash (1979) Diffusion gradients in natural silicic liquids (abstr.). EOS, Trans. Am. Geophys. Union, 60, 402. (40)

Faber, H. (1941) On the salt-solutions in microscopic cavities in granites. Danmarks Geol. Undersøgelse, ser. 2, (67), 45 p. (111)

Fahley, M.P. (1979) Fluid Inclusion Study of the Tonopah District, Nevada. M.S. thesis, Colo. School of Mines, Golden, Colorado. (426, 427)

Fairbairn, H.W. (1943) Gelatin coated slides for refractive index immersion mounts. Am. Mineral., 28, 396-397. (112)

Feininger, T. (1970) Emerald mining in Colombia: History and geology. Mineral. Rec., 1, (4), 142-153. (365)

Feklichev, V.G. (1962) Study of zoned crystals of beryl from cavities of pegma-tites. Akad. Nauk SSSR, Inst. Mineralogii, Geokhimii i Kristallokhimii Redkikh Elementov Trudy, 1962, (8), 166-196 (in Russian). (82, 100, 101)

____ (1963) Chemical composition of minerals of the beryl group, character of isomorphism, and position of principal isomorphous elements in the crys-tal structure. Geokhimiya, 1963, (4), 391-401 (in Russian; translated in Geochemistry, 1963, (4), 410-421). (128)

Feldman, M.D. and J.J. Papike (1981) Metamorphic fluid composition from the Notch Peak aureole, Utah (abstr.). EOS, Trans. Am. Geophys. Union, 62, 435. (455)

Fenn, P.M. and W.C. Luth (1973) Hazards in the interpretation of primary fluid inclusions in magmatic minerals (abstr.). Geol. Soc. Am. Abstr. Programs, 5, 617. (30, 40)

Ferguson, J. and K.L. Currie (1971) Evidence of liquid immiscibility in alkaline ultrabasic dikes at Callander Bay, Ontario. J. Petrol., 12, (3), 561-585. (410)

Ferry, J.M. (1979) A map of chemical potential differences within an outcrop. Am. Mineral., 64, 966-985. (38, 346, 350)

____ (1982) Regional metamorphism of the Vassalboro Formation, South-Central Maine, U.S.A.: A case study of the role of fluid in metamorphic petrogene-sis (abstr.). Geol. Soc. [London] Newsletter, 11, (4), 14-15. (350, 351)

Fesq, H.W., D.M. Bibby, C.S. Erasmus, E.J.D. Kable and J.P.F. Sellschop (1975) A comparative trace element study of diamonds from Premier Finsch and Jagersfontein mines, South Africa. Phys. Chem. Earth, 9, 817-836. (508)

Field, J.E. (1979) The Properties of Diamond. Academic Press, London, 674 p. (506, 508)

Fieni, C., M. Bourote-Denise, P. Pallas and J. Touret (1978) Aqueous liquid inclusions in feldspars and phosphates from Peetz chondrite (abstr.). Meteoritics, 13, 460-461. (2, 534, 566)

Fisher, D.E. (1983) Rare gases from the undepleted mantle? Nature, 305, 298-300. (527)

Fisher, J.R. (1976) The volumetric properties of H_2O--a graphical portrayal. U.S. Geol. Survey J. Res., 4, 189-193. (226, 227)

Flamini, A., A.G. Graziani and O. Grubessi (1975) Inorganic inclusions in amber. Archaeometry, 17, pt. 1, 110-112. (325)

Freund, F. (1982) Volume instabilities in the mantle as a possible cause for kimberlite form. Terra Cognita, 2, 263-265. (117, 505)
___, H. Kathrein, H. Wengeler, R. Knobel and H.J. Heinen (1980) Carbon in solid solution in forsterite--a key to the intractable nature of reduced carbon in terrestrial and cosmogenic rocks. Geochim. Cosmochim. Acta, 44, 1319-1333. (117, 380, 505, 531)

Reil, H. Wengeler, H. Kathrein, R. Knobel, G. Oberheuser, G.C. Matti, D. Reil, U. Knipping and J. Kötz (1983) Hydrogen and carbon derived from dissolved H_2O and CO_2 in minerals and melts. Bull. Mineral., 106, 185-200. (117, 505)

Frey, M, K. Bucher, E. Frank and J. Mullis (1980a) Alpine metamorphism along the Geotraverse Based-Chiasso -- a review. Eclogae geol. Helv., 73, (2), 527-546. (354, 367, 368)
___, M. Teichmüller, R. Teichmüller, J. Mullis, B. Künzi, A. Breitschmid, U. Gruner and B. Schwizer (1980b) Very low-grade metamorphism in external parts of the Central Alps: illite crystallinity, coal rank and fluid inclusion data. Eclogae geol. Helv., 73, (1), 173-203. (347, 353, 498)

Freidman, G.M. (1983) Textures of sandstones and carbonate rocks in the world's deepest wells (in excess of 30,000 ft or 9.1 km): Anadarko basin, Oklahoma - Reply. Sed. Geology, 35, 156-157. (306, 335)
___, S.A. Reeckmann and B. Borak (1981) Carbonate deformation mechanism in the world's deepest wells (-9 km). Tectonophysics, 74, T15-T19. (306, 335)

Friedman, I. and J.R. O'Neil (1977) Compilation of stable isotope fractionation factors of geochemical interest. In M. Fleischer, ed., Data of Geochemistry, U.S. Geol. Survey Prof. Paper 440-KK, 12 p. (498)
___, L.H. Adami, J.D. Gleason and K. Hardcastle (1970) Water, hydrogen, deuterium, carbon, carbon-13 and oxygen-18 content of selected lunar material. Science, 167, 538-540. (540)

Friedrichsen, H. and G. Morteani (1979) Oxygen and hydrogen isotope studies on minerals from Alpine fissures and their gneissic host rocks, Western Tauern Window (Austria). Contrib. Mineral. Petrol., 70, 149-152. (365)

Fronde, C. (1962) The System of Mineralogy. Volume III, Silica Minerals. Wiley & Sons, New York, NY, 334 p. (93)

Frost, B.R. (1979) Mineral equilibria involving mixed-volatiles in a C-O-H fluid phase: The stabilities of graphite and siderite. Am. J. Sci., 279, 1033-1059. (366)

Fuchs, L.H., E. Olsen, and K.J. Jensen (1973) Mineralogy, mineral-chemistry and composition of the Murchison (C2) meteorite. Smithsonian Contrib. Earth Sci. No. 10, 39 p. (562, 563, 564)

Funkhouser, J.G. and J.J. Naughton (1968) Radiogenic helium and argon in ultramafic inclusions from Hawaii. J. Geophys. Res., 73, 4601-4607. (128, 219)
___ and I.L. Barnes (1965) Some problems of dating Hawaiian rocks by the K-Ar method (abstr.). Am. Geophys. Union Trans., 46, 547.

Fuzikawa, K. (1982) Fluid Inclusion and Oxygen Isotope Studies of the Nabarlek Uranium Deposit, N.T., Australia. Ph.D. dissertation, Univ. Adelaide, Adelaide, Australia, 226 p. (147, 203, 219, 462)

Gallagher, D. (1940) Albite and gold. Econ. Geol., 35, 698-736. (359)
Gandolfi, G. (1967) Discussion upon methods to obtain X-ray "powder patterns" from a single crystal. Mineral. Petrogr. Acta, 13, 67-74. (113)
Gardner, F.M. (1972) Hollow apatites in a layered basic intrusion, Norway. Geol. Mag., 103, 285-287. (384)
Gat, J.R. and R. Gonfiantini, eds. (1981) Stable isotope hydrology, deuterium and oxygen-18 in the water cycle: Tech. Repts. Ser. 210, Int'l Atomic Energy Agency, Vienna, 340 p. (127)
Gehrig, M. (1980) Phasengleichgewichte und PVT-Daten ternärer Mischungen aus Wasser, Kohlendioxid und Natriumchlorid bis 3 kbar und 550°C. Ph.D. dissertation, Univ. of Karlsruhe, Karlsruhe, W. Germany. (247, 280, 406)

Floran, R.J., M. Prinz, P.F. Hlava, K. Keil, C.E. Nehru and V.R. Hinthorne (1978) The Chassigny meteorite: A cumulate dunite with hydrous amphibole-bearing melt inclusions. Geochim. Cosmochim. Acta, 42, 1213-1219. (569)

Foley, N.K., P.M. Bethke, and R.O. Rye (1982) A re-interpretation of δDH_2O values of inclusion fluids in quartz from shallow ore bodies (abstr.). Geol. Soc. Am. Abstr. Programs, 15, 489-490. (435)

Folk, R.L. (1955) Note on the significance of "turbid" feldspars. Am. Mineral., 40, 356-357. (111)
___ and R. Assereto (1976) Comparative fabrics of length-slow and length-fast calcite and calcitized aragonite in a Holocene speleotherm, Carlsbad Caverns, New Mexico. J. Sed. Petrol., 46, 486-496. (327)
___ and C.E. Weaver (1952) A study of the texture and composition of chert. Am. J. Sci., 250, 498-510. (97)

Fonteilles, M., B. Guy and P. Soler (1980) Etude du processus de formation des gites de skarns de Salan et Costabonne. In Z. Johan, ed., Minéralisations liées aux Granitoïdes, Mem. Bur. Recherches Geol. Minières, 99, 259-282. (455)

Foo, B.N. (1977) Mineral paragenesis, fluid inclusion studies and geochemistry of the Sungei Lembing tin lodes, West Malaysia (abstr.). Inst. Mining Metall. Trans., Sect. B, 86, B163. (457)

Ford, M.M. and P.G. Feiss (1982) Fluid inclusion studies on barren quartz veins in the Carolina Slate Belt, central and southern North Carolina (abstr.). Geol. Soc. Am. Abstr. Programs, 14, 18. (356)

Forester, F.H. and J. Touret (1980) Granulites, eclogites, peridotites in France. Sciences de la Terre, Nancy, 23-3, 1-42 (Livret guide exc. 114C, 26 Cong. Geol. Int.). (515)

Fouque, F. and A. Michel-Lévy (1879) Minéraux reproduits artificiellement par voie ignée. Bull. Soc. fr. Minéral. 2, 105-113. (474)
___ and ___ (1881) Reproduction des basaltes et mélaphyres labradoriques des diabases et dolérites à structure ophitique I. Bull. Soc. fr. Minéral. 4, 275-279. (474)

Francheteau, J. and 14 others (1979) Massive deep-sea sulfide ore deposits discovered on the East Pacific Rise. Nature, 277, 523-528. (494, 495)

Francis, A.W. (1954) Ternary systems of liquid carbon dioxide. J. Phys. Chem., 58, 1099-1114. (86)

Frantz, J.D., D. Virgo, and B.O. Mysen (1982) Time-lapse spectroscopy for fluorescence radiation rejection. Carnegie Inst. Washington Year Book 81, 437-440. (106)

Freckman, J.T. (1978) Fluid inclusion and oxygen isotope geothermometry of rock samples from Sinclair #4 and Elmore #1 boreholes, Salton Sea geothermal field, Imperial Valley, California, U.S.A. M.S. thesis, Univ. California, Riverside, CA, 155 p. (190)
___ and E.R. Olson (1978) Polythermal history of fracture filling episodes in active geothermal systems (abstr.). In Int'l Assoc. Genesis of Ore Deposits, 5th Symp., Snowbird, Alta, Utah, 1978. Program and Abstr. (Alta, Utah), 91. (498)

Fredriksson, K., A. Noonan and J. Nelen (1973) Meteoritic, lunar and Lonar impact chondrules. The Moon, 7, 475-482. (561)

Freeman, T. (1973) Temporal dolomite-calcite sequence and its environmental implications (abstr.). Am. Assoc. Petrol. Geol. Bull., 57, 780. (321)

Freestone, I.C. (1979) Immiscibility in tholeiites. Mineral. Mag., 43, 544-546. (477)

French, B.M. (1966) Some geological implications of equilibrium between graphite and a C-H-O gas phase at high temperatures and pressures. Rev. Geophys., 4, 223-253. (405, 406)
___ and D.L. Hamilton (1980) The role of liquid immiscibility in the genesis of carbonatites -- an experimental study. Contrib. Mineral. Petrol., 73, 105-117. (410)

Ge
Gr

Gilmour, P. (1982) Grades and tonnages of porphyry copper deposits. In S.R. Titley, Ed., Advances in Geology of the Porphyry Copper Deposits, Southwestern North America. Univ. Arizona Press, Tucson, AZ, 7-35. (439)

Gilson, T.R. and P.J. Hendra (1970) Laser Raman Spectroscopy. Wiley-Interscience, New York, NY. (105)

Giordano, T.H. and H.L. Barnes (1981) Lead transport in Mississippi Valley-type ore solutions. Econ. Geol. 76, 2200-2211. (421)

Gize, A.P. and T.C. Hoering (1980) The organic matter in Mississippi Valley-type deposits. Carnegie Inst. Washington Year Book, 79, 384-388. (421)

Glassley, W.E. (1983a) Deep crustal carbonates as CO₂ fluid sources. Evidence from metasomatic reaction zones. Contrib. Mineral. Petrol, 84, 15-24. (527)

―――― (1983b) The role of CO₂ in the chemical modification of deep continental crust. Geochim. Cosmo. Acta, 47, 597-616. (380)

Godbeer, W.C. and R.W.T. Wilkins (1977) The water content of a synthetic quartz. Am. Mineral. 62, 831-832. (119)

Goguel, R. (1963) Die chemische Zusammensetzung der in den Mineralen einiger Granite und ihrer Pegmatite eingeschlossenen Gase und Flüssigkeiten. Geochim. Cosmochim. Acta, 27, 155-181. (119, 120, 131)

―――― (1964) Untersuchungen zum Chemismus der in Graniten und Metamorphiten eingeschlossenen Gase und Flüssigkeiten (abstr.). Fortschr. Mineral. 41, 190. (119, 131)

Gold, T. (1979) Terrestrial sources of carbon and earthquake outgassing. J. Petrol. Geol., 1, (3), 3-19. (512)

―――― and S. Soter (1980) The deep-earth-gas-hypothesis. Sci. Am. 246, (6), 154-161. (512)

Goldsztaub, S., D. Gérard, J.-P. Deville and B. Lang (1966) Observations au moyen d'électrons de faible énergie d'un cristal de muscovite clivé dans l'ultra-vide. Compt. Rend. Acad. Sci. Paris, 262, ser. B. (26), 1718-1720. (123)

Gorokhov, S.S., B.A. Dorogovin, L.N. Khetchikov (1975) Inclusions of the mineral-forming medium in eclogite and their genetic significance. Akad. Nauk SSSR, Dokl., 225, (2), 412-414 (in Russian; translated in Dokl. Acad. Sci. USSR, 225, 141, published in 1977). (374)

Goss, B.G. and L.M. Cathles (1983) Fluid inclusion study of the porphyry copper deposit at Ely, Nevada. Geol. Soc. Am. Abstr. Programs, 15, 583. (443)

Gould, S.J. (1983) False promise, good science. Natural History, 92, 20-26. (579)

Gow, A.J. and T. Williamson (1975) Gas inclusions in the Antarctic ice sheet and their significance. U.S. Army Cold Regions Research and Eng. Lab., Research Rept. 339, 23 p. (73, 329)

Graber, R., F.J. Sawkins and J. Kowalik (1979) Gas analysis studies at Casapalca, Peru - Implications for ore genesis (abstr.). Geol. Soc. Am. Abstr. Programs, 11, 80. (437)

Graham, J., C.R.M. Butt and R.B.W. Vigers (1983) Charging as a source of error in microprobe analyses. CSIRO (Australia) Div. Mineralogy Research Rev. 1983, 237-238. (141, 483)

Grant, J.N., C. Halls, W. Avila and G. Avila (1977) Igneous geology and the evolution of hydrothermal systems in some subvolcanic tin deposits of Bolivia. Geol. Soc. London Bull., Spec. Issue,7, 117-126. (444)

―――― , S.M.F. Sheppard and W. Avila (1980) Evolution of the porphyry tin deposits of Bolivia. Mining Geol. (Japan) Spec. Issue, (8), 151-173. (44, 443, 444, 445)

Grant, N.K. and M.C. Bliss (1983) Strontium isotope and rare element variations in non-sulfide minerals from the Elmwood-Gordonsville mines, central Tennessee. In G. Kisvarsanyi et al., eds., Proc. Int'l Conf. on Mississippi Valley type lead-zinc deposits. Univ. of Missouri-Rolla, Rolla, MO, 206-210. (324)

Gehrig, M., H. Lentz and E.U. Franck (1979) Thermodynamic properties of water-carbon dioxide-sodium chloride mixtures at high temperatures and pressures. In K.D. Timmerhaus and M.S. Barber, eds., International Conference on High-Pressure Science and Technology (6th, 1977, Univ. of Colorado): Vol. 1, Physical Properties and Material Synthesis. Plenum Press, New York, NY, 539-542. (240, 246, 247)

Gerlach, H. and S. Heller (1966) Concerning artificially produced fluid inclusions in rock salt crystals. Deutsche Gesell. Geol. Wiss.. Reihe B, Mineral, Lagerstattenforsch., 195-214 (in German) (68)

Gerlach, T.M. (1980) Chemical characteristics of the volcanic gases from Nyiragongo lava lake and the generation of CH₄-rich fluid inclusions in alkaline rocks. J. Volcan. Geotherml. Res., 8, 177-189. (489)

―――― (1982) Interpretation of volcanic gas data from tholeiitic and alkaline mafic lavas. Bull. Volcanol., 45, 235-244. (489)

Geronsin, R.L. (1980) Chemical Relationship of the Mississippi-Valley Type Ore Deposits in Missouri, Oklahoma, and Kansas. M.S. thesis, Univ. Missouri-Rolla, Rolla, MO, 158 p. (424)

Giardini, A.A. and C.E. Melton (1975a) The nature of cloud-like inclusions in two Arkansas diamonds. Am. Mineral., 60, 931-933. (507, 508, 509, 510)

―――― and ―――― (1975b) Gases released from natural and synthetic diamonds by crushing under high vacuum at 200°C, and their significance to diamond genesis. Fortschr. Mineral., 52, Spec. Issue, Int'l Mineral. Assoc. -- Papers 9th Mtg, Berlin-Regensburg 1974, 455-464. (509)

―――― and ―――― (1976) The significance of gases released from natural diamonds by crushing and by graphitization (abstr.). Int'l Geol. Cong., 25th, Abstr., 807. (402, 509)

―――― and ―――― (1981) Experimentally-based arguments supporting large crustal accumulations of nonbiogenic petroleum. J. Petrol. Geol., 4, (2), 187-190. (510)

―――― and ―――― (1983) A scientific explanation for the origin and location of petroleum accumulations. J. Petrol. Geol. 6, (2), 127-138. (510) and R.S. Mitchell (1982) The nature of the upper 400 km of the earth and its potential as the source for non-biogenic petroleum. J. Petrol. Geol., 5, 173-190. (510)

―――― , G.V. Subbarayudu and C.E. Melton (1976) The emission of occluded gas from rocks as a function of stress: Its possible use as a tool for predicting earthquakes. Geophys. Res. Lett. 3, 355-358. (402)

―――― , C.E. Melton and R.S. Mitchell (1982) The nature of the upper 400 km of the earth and its potential as the source for nonbiogenic petroleum. J. Petrol. Geol., 5, 173-190. (509)

Gibson, Jr., E.A. (1973) Thermal analysis-mass spectrometer computer system and its application to the evolved gas analysis of Green River shale and lunar soil samples. Thermochim. Acta, 5, 243-255. (122, 123, 124)

―――― and S.M. Johnson (1972) Thermogravimetric-quadrupole mass-spectrometric analysis of geochemical samples. Thermochim. Acta, 4, 1-8. (122, 124)

―――― , R.K. Kotra and J.L. Warner (1982) Direct analysis of trapped vapors and fluids in silicate samples utilizing the laser microprobe-gas chromatograph technique. Lunar Planetary Sci. XIII, 261-262. (140)

Giddings, J.C., M.N. Myers, L. McLaren and R.A. Keller (1968) High pressure gas chromatography of nonvolatile species. Science, 162. 67-73. (124)

Gigashvili, G.M. (1969) Primary solid-gas inclusions in quartz from Volhynian pegmatites. L'vov. Gos. Univ. Mineralog. Sbornik, 23, (4), 398-404 (in Russian; translated in Fluid Inclusion Res. -- Proc. of COFFI, 3, 88-94, 1970). (398)

Giggenbach, W.F. (1982) Carbon-13 exchange between CO₂ and CH₄ under geothermal conditions. Geochim. Cosmochim. Acta, 46, 159-165. (128, 352)

Gilletti, B.J. and R.A. Yund (1984) Oxygen diffusion in quartz. J. Geophys. Res. (in press). (36)

610

†Grushkin, G.G. (1958) Physicochemical factors affecting equilibrium during mineralization of the Aurakhmat fluorite deposit (central Asia). Vses. Nauchno-Issled. Inst. P'ezooptichesk. Mineral. Syr'ya Trudy, 2, (2), 81-92 (in Russian). (114)

_____, and P.L. Prikhid'ko (1952) Changes in chemical composition, concentration, and pH of gaseous-liquid inclusions in successive fluorspar series. Vses. Mineral. Obshch. Zapiski, 81, (2), 120-126 (in Russian; translated in Int'l. Geol. Rev., 1, (12), 66-71, 1959). (114)

Gübelin, E.J. (1953) Inclusions as a Means of Gemstone Identification. Gemological Inst. Am., Los Angeles, CA, 220 p. (4, 29, 365, 375)

_____ (1974) Internal World of Gemstones: ABC Edn Zürich, Zürich 234 p. (In English). (29, 375)

Guha, J., J. Leroy and D. Guha (1979) Significance of fluid phases associated with shear zone Cu-Au mineralization in the Doré lake complex, Chibougamau, Québec. Bull. Minéral., 102, 569-576. (53, 351)

_____, G. Archambault and J. Leroy (1983) A correlation between the evolution of mineralizing fluids and the geomechanical development of a shear zone as illustrated by the Henderson 2 mine, Quebec. Econ. Geol., 78, 1605-1618. (465)

Guilhaumou, N. (1982) Accurate analysis of fluid inclusions by the laser molecular microprobe (MOLE) and by microthermometry. Travaux Lab. Géol. Ecole Normale Supérieure, (14), 68 p. (in French; English abstract). (105, 108, 245, 347, 348)

_____, N.P. Dhamelincourt, J.-C. Touray and J. Barbillat (1978) Raman microprobe analysis of gaseous inclusions in the system N₂-CO₂ (abstr.). Compt. Rend. Acad. Sci. Paris, 287, Series D, 1317-1319 (in French). (108)

_____, and J. Touret (1981) Study of fluid inclusions in the system N₂-CO₂ from dolomite and quartz of northern Tunisia. Results from cryomicroscopy and microRaman analysis. Geochim. Cosmochim. Acta, 45, 657-673 (in French). (104, 272, 347, 348)

Gunter, W.D., I.-M. Chou and S. Girsperger (1983) Phase relations in the system NaCl-KCl-H₂O Part II: Differential thermal analysis of the halite liquidus in the NaCl-H₂O binary above 450°C. Geochim. Cosmochim. Acta, 47, 863-873. (234, 235)

Gurney, J.J. (1979) Inclusions in diamonds from southern Africa. Geokongres 79, 18th Congress of the Geol. Soc. of South Africa, Part 1, 164-175. (507)

Guseva, E.V., F.P. Melnikov, R.Yu. Orlov and M.E. Uspenskaya (1983) Spectroscopy of Raman scattering in the investigation of mineral gas-liquid inclusions. Dokl. Akad. Nauk SSSR, 272, 197-200 (in Russian). (106)

Gustafson, L.B. and J.P. Hunt (1975) The porphyry copper deposit at El Salvador, Chile. Econ. Geol., 70, 857-912. (443)

Gutmann, J.T. (1977) Textures and genesis of phenocrysts and megacrysts in basaltic lavas from the Pinacate volcanic field. Am. J. Sci., 277, 833-861. (3)

Güven, N. and L.L. Carney (1979) The hydrothermal transformation of sepiolite to stevensite and the effect of added chlorides and hydroxides. Clays Clay Minerals, 27, (4), 253-260. (339)

Haapala, I. (1980) Fluid inclusions in the apatite of the Sokli carbonatite, Finland — A preliminary report. Geologi, 32, (7), 83-87. (408, 411)

_____, and K. Kinnunen (1979) Fluid inclusions in cassiterite and beryl in greisen veins in the Eurajoki stock, southwestern Finland. Econ. Geol., 74, 1231-1238. (457)

Haar, L., J. Gallagher, and G. Kell (1979) Thermodynamic properties of fluid water. In J.J. Straub and K.S. Scheffler, eds., Water and Steam, Their Properties and Current Industrial Applications, Proc. of the 9th Int'l Conf. on the Properties of Steam, Sept. 10-14, 1979, Technische Univ. München, F.R.G. Pergamon Press, New York, NY, 69-82. (228)

Haas, J.L. (1970) An equation for the density of vapor-saturated NaCl-H₂O solutions from 75° to 325°C. Amer. J. Sci., 269, 489-493. (234)

Grappin, C., P. Sallot, C. Sabouraud and J.-C. Touray (1979) Variations of the Cl/Br, Na/Br, and K/Br ratios in fluid inclusions of quartz in the Bramans-Termignon evaporites. Vanoise, French Alps. Chem. Geol., 25, 41-52 (in French). (349)

Gratier, J.P. (1982) Experimental and natural deformation of rock by solution-deposition with mass transfer. Bull. Minéral., 105, 291-300 (in French). (71, 74)

_____, and L. Jenatton (1983) Deformation and reequilibration of fluid inclusions depending on temperature, internal pressure and stress (abstr.). European Current Research on Fluid Inclusions, April 6-8, 1983, Orleans, France, 31. (74)

Green, H.W. II (1972) A CO₂-charged asthenosphere. Nature Phys. Sci., 238, 2-5. (380)

_____ (1979) Trace elements in the fluid phase of the Earth's mantle. Nature, 277, 465-467. (525)

_____ and S.V. Radcliffe (1975) Fluid precipitates in rocks from the earth's mantle. Geol. Soc. Am. Bull., 86, 846-852. (2, 36, 380, 505, 522, 531)

Greenwood, H.J. (1973) Thermodynamic properties of gaseous mixtures of H₂O and CO₂ between 450°C and 800°C and 0 to 500 bars. Am. J. Sci., 273, 561-571. (239)

_____ and H.L. Barnes (1966) Binary mixtures of volatile compounds. In Handbook of Physical Constants, ed. S.P. Clark, Jr., Geol. Soc. Am. Mem. 97, revised ed., 385-400. (281)

Greig, J.W. (1927) Immiscibility in silicate melts. Am. J. Sci., 13, 1-44, 133-154. (545)

Grigoriev, D.P. (1944) New observations on the result of gravitational shifting of crystals in veins of Alpine type in the near polar Urals. Akad. Nauk SSSR, Dokl., 5, 198-200 (in Russian). (27)

_____ (1948) On the subject of recognition of primary and secondary liquid inclusions in minerals. Mineral. Sbornik L'vov. Geolog. Obshch., 1948 (2), 75-81 (in Russian). (18, 19)

_____ and A.G. Zhabin (1975) Ontogeny of Minerals. Moscow, Nauka Press. 339 p. (in Russian). (310)

_____ et al., eds. (1973) G.G. Lemmlein, Morphologic and Genetic Crystallography. "Nauka" Press, Moscow, 328 p. (in Russian). (16)

Grishina, S.N. (1979) Microanalysis of gas phase of inclusions in minerals. Zapiski Vses. Mineral. Obsh., 108, (5), 617-621 (in Russian; translated in Fluid Inclusion Res. — Proc. of COFFI, 13, in press). (121)

Grogan, R.M. and J.C. Bradbury (1968) Fluorite-lead-zinc deposits of the Illinois-Kentucky mining district. In J.D. Ridge, ed., Ore Deposits in the United States, 1933-1967, A.I.M.E., New York, NY, 370-399. (416)

Groshenko, A.R. (1968a) Development of technique for studies of inclusions of mineral-forming media by the homogenization method. In Mineralogical Thermometry and Barometry, Vol. 2, "Nauka" Press, Moscow, 95-98 (in Russian). (186)

_____ (1968b) [Notice of new instruments]. Fluid Inclusion Res. — Proc. of COFFI, 2, 4-7. (186)

Grossman, L. (1972) Condensation in the primitive solar nebula. Geochim. Cosmochim. Acta, 36, 597-619. (562)

_____ and E. Olsen (1974) Origin of the high-temperature fraction of C2 chondrites. Geochim. Cosmochim. Acta, 38, 173-187. (562)

_____ and J.M. Lattimer (1979) Silicon in carbonaceous chondrite metal: relict of high-temperature condensation. Science, 206, 449-451. (562)

Grove, T.L. and A.E. Bence (1977) Experimental study of pyroxene-liquid interaction in quartz-normative basalt 15597. Proc. 8th Lunar Sci. Conf., 1549-1579. (554)

Grubenmann, U. (1904) Die Kristallinen Schiefer. Borntråger & Co., Berlin. (353, 367)

Harmon, R.S., T.C. Atkinson and P.L. Smart (1979a) Late Pleistocene palaeoclimate temperatures from fluid inclusion isotope studies in stalagmites (abstr.). Program 4th Mtg Geol. Soc. of the British Isles, Univ. Sheffield, Sept. 1979 (unpaginated). (327)

____ and J.R. O'Neil (1979b) D-H ratios in speleothem fluid inclusions - guide to variations in the isotopic composition of meteoric precipitation. Earth Planet. Sci. Lett., 42, 254-266. (327)

Harrington, B.J. (1905) On an interesting variety of fetid calcite and the cause of its odor. Am. J. Sci., ser. 4, 19, 345-348. (116)

Harris, D.M. (1979a) Preeruption variations of H_2O, S, and Cl in a subduction zone basalt (abstr.). IAVCEI Abstracts and Timetable, Int'l Union Geodesy and Geophys., XVII General Assembly, Canberra, Australia, Dec. 1979. (486, 488)

____ (1979b) Preeruption variations of H_2O, S, and Cl in a subduction zone basalt (abstr.). EOS, Trans. Am. Geophys. Union, 60, 968. (486)

____ (1981a) The Concentrations of H_2O, CO_2, S, and Cl During Pre-Eruption Crystallization of Some Mantle-Derived Magmas: Implications for Magma Genesis and Eruption Mechanisms. Ph.D. dissertation, Univ. Chicago, Chicago, IL, 233 p. (117, 122, 477, 486, 488)

____ (1981b) The concentration of CO_2 in submarine tholeiitic basalts. J. Geol., 89, 689-701. (117, 122, 486, 513)

____ (1981c) The micro-determination of H_2O, CO_2, and SO_2 in glass using a 1280°C microscope vacuum heating stage, cryopumping, and vapor pressure measurements from 77 to 273 K. Geochim. Cosmochim. Acta, 45, 2023-2036. (117, 122, 486)

____ (1981d) Vesiculation and eruption of a subduction zone basalt (abstr.). EOS, Trans. Am. Geophys. Union 62, 1084. (488)

____ and A.T. Anderson, Jr. (1983) Concentrations, sources, and losses of H_2O, CO_2, and S in Kilauean basalt. Geochim. Cosmochim. Acta, 47, 1139-1150. (482, 487, 488)

____ (1984) Volatiles H_2O, CO_2, S, and Cl in a subduction zone basalt. Contrib. Mineral. Petrol. (in press). (482, 488)

Harris, H.J.H, K. Cartwright and T. Torii (1979) Dynamic chemical equilibrium in a polar desert pond: a sensitive index of meteorological cycles. Science, 204, 301-303 (see also correction in Science, 204, 909). (249)

Harris, J.W. (1968) The recognition of diamond inclusions - Pt. 1: Syngenetic mineral inclusions; Pt. 2: Epigenetic mineral inclusions. Ind. Diamond Rev., 28, 402-410 and 458-461. (507)

____ (1979) Physical and chemical constraints on the formation of natural diamond in the upper mantle. Diamond Res., 1979, 2-6. (506)

____ and J.J. Gurney (1979) Inclusions in diamond. In J.E. Field, ed., The Properties of Diamond. Academic Press, London, 555-591. (506, 507, 508)

Hart, R., J. Dymond, L. Hogan and J.G. Schilling (1983) Mantle plume noble gas component in glassy basalts from Reykjanes Ridge. Nature, 305, 403-407. (527)

Hartley, W.N. (1876) On variations in the critical point of carbon dioxide in minerals, and deductions from these and other facts. J. Chem. Soc., 30, 237-250. (82, 239)

____ (1877a) Observations on fluid-cavities. J. Chem. Soc., 31, 241-249. (103)

____ (1877b) On attraction and repulsion of bubbles by heat. Royal Soc. [London], Proc., 26, 137-149. (51, 92)

Hatch, J.R., H.J. Gluskoter and P.C. Lindahl (1976) Sphalerite in coals from the Illinois Basin. Econ. Geol., 72, 613-624. (330)

Hattori, K. and H. Sakai (1979) D/H ratios, origins, and evolution of the ore-forming fluids for the Neogene veins and Kuroko deposits of Japan. Econ. Geol., 74, 535-555. (525)

Haas, Jr., J.L. (1971) The effect of salinity on the maximum thermal gradient of a hydrothermal system at hydrostatic pressure. Econ. Geol., 66, 940-946. (269, 437, 501)

____ (1976) Physical properties of the coexisting phases and the thermochemical properties of the H_2O component in boiling NaCl solutions. U.S. Geol. Surv. Bull., 1421-A, 73 p. (69, 228, 274, 275, 276, 277)

____ (1978) An empirical equation with tables of smoothed solubilities of methane in water and aqueous sodium chloride solutions up to 25 weight percent 360°C, and 138 MPa. U.S. Geol. Surv. Open-File Rept 78-1004, 41 p. (248)

Haggerty, S.E. and B.M. McMahon (1979) Evidence for carbonatite liquids in kimberlites (abstr.). Mineral. Soc. (London) Bull., (45), 3. (408, 511)

Hall, W.E. and I. Friedman (1963) Composition of fluid inclusions, Cave-in-Rock fluorite district, Illinois, and Upper Mississippi Valley zinc-lead district. Econ. Geol., 58, 886-911. (118, 127, 422, 423)

____ and J.T. Nash (1974) Fluid inclusion and light stable isotope study of the Climax molybdenum deposits, Colorado. Econ. Geol., 69, 884-901. (443, 444, 445, 450)

Hallbauer, D.K. (1982) Characterization of pyrite by combined fluid inclusion and trace element studies (abstr.) Program 111th Annual Mtg, AIME, Feb. 14-18, 1982, 84. (145)

____ (1983) Geochemistry and fluid inclusions in detrital minerals as guides to their provenance and distribution. Proc., Int'l Conf. on Applied Mineralogy, Johannesburg, June, 1981 (in press). (144, 145)

____ and E.J.D. Kable (1979) Geochemical and fluid inclusion studies of quartz pebbles in Witwatersrand conglomerates and their relationship to gold mineralization. Geokongres 79, 18th Congress of the Geol. Soc. South Africa, Part 1, 176-186. (308)

____ (1982) Fluid inclusions and trace element content of quartz and pyrite pebbles from Witwatersrand conglomerates: Their significance with respect to the genesis of primary deposits. In G.C. Amstutz et al., eds., Ore Genesis - The State of the Art, Springer-Verlag; Berlin, W. Germany, 742-752. (147, 308)

Halsor, S.P., C.A. Chesner, W.I. Rose, and T.J. Bornhorst (1983) An introduction to fluid inclusions: A color videotape teaching aid for economic geology classes (abstr.). Geol. Soc. Am. Abstr. Programs, 15, 590. (201)

Hamberg, A. (1895) Studien über Meereis und Gletschereis. Kgl. Svenska Vetenskapsakad. Bihang till Handl., 21, pt. 2, (2), 1-13. (116, 118)

Hamilton, D.L., I.C. Freestone, J.B. Dawson and C.H. Donaldson (1979) Origin of carbonatites by liquid immiscibility. Nature, 279, 52. (408, 410)

Hanor, J.S. (1973) The role of in situ densities in the migration of subsurface brines. Geol. Soc. Am. Abstr. Program 5, 651-652. (424)

____ (1979) A mechanism for precipitating lead and zinc from sedimentary brines: Partitioning of CO_2 into a methane gas phase (abstr.). Geol. Soc. Am. Abstr. Programs, 11, 438. (422)

____ (1980) Dissolved methane in sedimentary brines: potential effect on the PVT properties of fluid inclusions. Econ. Geol., 75, 603-609. (248, 324, 330, 333, 421)

____ (1981) Composition of fluids expelled during compaction of Mississippi Delta sediments. Geo-Marine Lett., 1, 169-172. (306)

Hare, P.E., T.C. Hoering, and K. King, eds. (1980) Biogeochemistry of amino acids. J. Wiley & Sons, New York, NY, 558 p. (333)

Harmon, R.S. and T.C. Atkinson (1980) Interpretation of past climates from D/H and $^{18}O/^{16}O$ ratios of speleothem calcite and fluid inclusions. 26th Int'l Geol. Congress, 2, 658. (328)

____ and H.P. Schwarcz (1980) Oxygen-hydrogen isotope relationship in meteoric water: evidence for change during glacial periods (abstr.). Geol. Soc. Am. Abstr. Programs, 12, 442. (328)

Hattori, K. and H. Sakai (1980) Implications of D/H and 18O/16O ratios of ore fluids for the Neogene vein-type and Kuroko mineralization of Japan. In J.D. Ridge, ed., Proc. 5th Quadrennial IAGOD Symp., Vol. I, E. Schweizerbart'sche Verlags., Stuttgart, W. Germany, 297-307. (525)

Havette, A. and J. Weiss (1976) Identification of solid inclusions in lava phenocrysts by ion analyser. Bull. Soc. franc. Mineral. Cristallogr., 99, 165-168 (in French; English abstr.). (148)

Hawes, G.W. (1881) On liquid carbon dioxide in smoky quartz. Am. J. Sci., 3d ser., 21, 203-209. (92, 111)

*Hayakawa, N., M. Nambu and T. Aoshima (1969) Trial manufacture of a heating stage for use under high pressure. COFFI, 2, 45. (186)

__ and __ (1973) Studies on fluid inclusion geothermometers (2nd report) - filling temperature analysis. COFFI, 2, 45. (in Japanese; English abstr.). (186)

Hayashi, M., S. Taguchi and T. Yamasaki (1981) Activity index and thermal history of geothermal systems. Trans. Geothermal Resources Council, 5, 177-180. (497)

Hayba, D.O. (1983) A compilation of fluid inclusion and stable isotope data on selected precious- and base-metal deposits. U.S. Geol. Surv. Open-File Rept. 83-450, 24 p. (426)

Haynes, F.M., S.E. Kesler and M. Taylor (1983) Analytical evidence for a wide range of fluid inclusion compositions. Mascot-Jefferson City zinc district, East Tennessee (abstr.). Geol. Soc. Am. Abstr. Programs, 15, 593. (145, 424)

Haynes, S.J. and M.A. Mostaghel (1979) Formation temperature of fluorite in the Lockport Dolomite in upper New York State as indicated by fluid inclusion studies—with a discussion of heat sources—a discussion. Econ. Geol., 74, 154-159. (324)

Haynes, V. (1959) Compromise growth surfaces on pegmatite minerals. Am. Mineral., 44, 1089-1096. (18)

Heald-Wetlaufer, P., N.K. Foley and D.O. Hayba (1982) Applications of doubly polished sections to the study of ore deposits. In R.D. Hagni, ed., Process Mineralogy II: Applications in Metallurgy, Ceramics, and Geology. Metallurgical Soc. of AIME, New York, 451-468. (154, 157, 158)

__, D.O. Hayba, N.K. Foley and J.A. Goss (1983) Comparative anatomy of epithermal precious- and base-metal districts hosted by volcanic rocks. U.S. Geol. Survey Open-File Rept. 83-710 (abstr. in Geol. Assoc. Canada/Mineral. Assoc. Canada Program with Abstr., 8, A31, 1983). (426)

Hedenquist, J.W. (1982) Fluid Flow in the Waiotapu system, New Zealand: Implications for its potential. Pacific Geothermal Conf., Nov. 1982, Univ. Auckland, New Zealand, Proc., 61-67. (248, 498)

__ (1983) Waiotapu, New Zealand: Geochemical Evolution and Mineralization of an Active Geothermal System. Ph.D. dissertation, Univ. Auckland, Auckland, New Zealand. (498, 501)

__ and R.W. Henley (1984) Effect of CO2 on freezing point depression measurements of fluid inclusions -- evidence from active geothermal systems and application to epithermal studies. Econ. Geol. (in press). (500)

Heinrich, E.W. (1980) The Geology of Carbonatites. Robert Krieger Pub. Co., New York, NY, 585 p. (reprint of 1966 volume). (406)

__ and R.J. Anderson (1965) Carbonatites and alkalic rocks of the Arkansas river area, Fremont County, Colorado. 2. Fetid gas from carbonatite and related rocks. Am. Mineral., 50, 1914-1920. (117, 122, 123)

Heinrich, K.F.J. (1980) Electron Beam X-Ray Microanalysis. Van Nostrand Reinhold, New York, NY, 608 p. (140)

Hekinian, R., M. Chaigneau and J.L. Cheminee (1973) Popping rocks and lava tubes from the Mid-Atlantic Rift Valley at 36°N. Nature, 245, 371-373. (490)

__, M. Fevrier, J.L. Bischoff, P. Picot and W.C. Shanks (1980) Sulfide deposits from the East Pacific Rise near 21°N. Science, 207, 1433-1444. (494)

Helgeson, H.C. and D.H. Kirkham (1974) Theoretical prediction of the thermodynamic behavior of aqueous electrolytes at high pressures and temperatures: I. summary of the thermodynamic/electrostatic properties of the solvent. Am. J. Sci., 274, 1089-1198. (228)

Helzel, M. (1969) Gas chromatographic analysis of gaseous inclusions in glass. Ceramic Bull., 48, 287-290. (124)

Hendel, E.M and L.S. Hollister (1981) An empirical solvus for CO2-H2O -2.6 wt. % salt. Geochim. Cosmochim. Acta, 45, 225-228. (247, 372, 373)

Henley, R.W. and A.J. Ellis (1983) Geothermal systems ancient and modern: a geochemical review. Earth-Sci. Reviews, 19, 1-50. (426, 496)

Henley, R.W. and A. McNabb (1978) Magmatic vapor plumes and ground-water interaction in porphyry copper emplacement. Econ. Geol. 73, 1-20. (446, 450, 451, 495)

__, R.J. Norris and C.J. Paterson (1976) Multistage ore genesis in the New Zealand geosyncline, a history of post-metamorphic lode emplacement. Mineralium Deposita, 11, 180-196. (356)

Henniker, J.C. (1949) The depth of the surface zone of a liquid. Rev. Modern Phys., 21, 322-341. (38)

Henry, D.L. (1978) A Study of Metamorphic Fluid Inclusions in Granulite Facies Rocks of the Eastern Adirondacks. A.B. thesis, Princeton Univ. Princeton, NJ. (196, 206)

Hervig, R.L., J.V. Smith, I.M. Steele, J.J. Gurney, H.O.A. Meyer and J.W. Harris (1980) Diamonds: Minor elements in silicate inclusions; pressure-temperature implications. J. Geophys. Res., 85, 6919-6929. (507)

Herzog, L.F., T.J. Eskew and R.L. Erwin (1962) The analysis of 10^{-14} to 10^{-5} cc STP noble gas samples by mass spectrometry. In 1961 Transactions of Eighth Vacuum Symposium and Second International Congress. Pergamon Press, New York, W., 581-591. (122)

Hewett, D.F. and M. Fleischer (1960) Deposits of the manganese oxides. Econ. Geol., 55, 1-55. (37)

Heyen, G., J. Dubessy and C. Ramboz (1982a) Modelling of phase equilibria in the system CO2-CH4-C2H6 below 50°C and 100 bar. Application to inclusion fluids. Compt. Rend. Acad. Sci. Paris, 294, ser. II, 261-264 (in French). (244)

__, C. Ramboz and J. Dubessy (1982b) Modelling of phase equilibria in the system CO2-CH4 below 50°C and 100 bar. Application to inclusion fluids. Compt. Rend. Acad. Sci. Paris, 294, ser. II, 203-206 (in French). (244)

Hibbard, J.J. (1979) Myrmekite as a marker between preaqueous and post-aqueous phase saturation in granitic systems. Geol. Soc. Am. Bull., Pt. 1, 90, 1047-1062. (387)

Hickling, R. (1965) Nucleation of freezing by cavity collapse and its relation to cavitation damage. Nature, 206, 915-917. (493)

Hidden, W.E. (1882) A phenomenal find of fluid-bearing quartz crystals. New York Acad. Sci. Trans., 1881-1882, 1, 131-136. (2)

Higgins, N.C. (1980) Fluid inclusion evidence for the transport of tungsten by carbonate complexes in hydrothermal solutions. Canadian J. Earth Sci. 17, 823-830. (458)

Hikita, H., S. Asai, H. Ishikawa, M. Seko and H. Kitajima (1979) Diffusivities of carbon dioxide in aqueous mixed electrolyte solutions. Chem. Eng. J., 17, 77-80. (350)

Hilbert, R. (1979) pVT-Daten von Wasser und von wässrigen Natriumchlorid-Lösungen bis 873K, 4000 Bar und 25 Gewichtsprozent NaCl. Ph.D. dissertation, Tech. Hochschule, Karlsruhe, W. Germany, 212 p. (234)

Hildreth, E.W. (1977) The Magma Chamber of the Bishop Tuff: Gradients in Temperature, Pressure, and Composition. Ph.D. dissertation, Univ. California, Berkeley, CA, 328 p. (388)

__ (1979) The Bishop tuff: Evidence for the origin of compositional zonation in silicic magma chambers. Geol. Soc. Am. Spec. Paper 180, 43-75. (389)

Hite, R.J., F.E. Rush, A.H. Balch, J.J. Daniels, J.D. Friedman, R. Watts and H.D. Ackerman (1979) Geologic exploration at Salt Valley, Utah (abstr.). Proc. of the National Waste Terminal Storage Program Information Mtg. Oct. 30-Nov. 1, 1979, Columbus, Ohio: ONWI 62, 73-75. (121)

Hoagland, A.D. (1973) Appalachian zinc-lead and the deposits of middle Tennessee (abstr.). Geol. Soc. Am. Abstr. Program, 5, (5) 404. (417)

Hoagland, L.P. (1951) Moving inclusion. Gemmologist, 20, 128. (92)

Hodgman, C.D. (ed.) (1953) Handbook of Chemistry and Physics, 35th Edition. Chemical Rubber Pub. Co.. Cleveland, Ohio. (230)

Hoefs, J. (1965) Ein Beitrag zur Geochemie des Kohlenstoffs in magmatischen und metamorphen Gesteinen. Geochim. Cosmo. Acta, 29, 399-428. (524)

___ and G. Morteani (1979) The carbon isotopic composition of fluid inclusions in Alpine fissure quartzes from the western Tauern Window (Tyrol, Austria). N. Jahrb. Mineral. Monats., 1979, 123-134. (351, 352, 354)

___ and H.A. Stalder (1977) The carbon isotope composition of CO_2-bearing inclusions in fissure quartz from the central Alps. Schweiz. Mineral. Petrog. Mitt., 57, 329-347. (351, 352, 368)

Hoernes, S. and H. Friedrichsen (1978) Oxygen and hydrogen isotope study of the polymetamorphic area of the Northern Ötztal-Stubai Alps (Tyrol). Contrib. Mineral. Petrol., 67, 305-315. (365)

Holland, R.A.G., C.J. Bray and E.T.C. Spooner (1978) A method for preparing doubly polished thin sections suitable for microthermometric examination of fluid inclusions. Mineral. Mag., 42, 407-408. (157)

Holland, T.J.B. (1979) High water activities in the generation of high pressure kyanite eclogites of the Tauern Window, Austria. J. Geol., 97, 1-27. (374)

___ (1983) Aqueous eclogite facies fluid inclusions (abstr.). Geol. Soc. Newsletter (G. Britain), 12, (2), 13-14. (374)

Hollister, L.S. (1982) Metamorphic evidence for rapid (2 mm/yr) uplift of a portion of the Central Gneiss Complex, Coast Mountains, B.C. Canadian Mineral. 20, 319-332. (371)

___ and R.C. Burruss (1976) Phase equilibria in fluid inclusions from the Khtada Lake metamorphic complex. Geochim. Cosmochim. Acta, 40, 163-175. (35, 62, 75, 76, 202, 241, 242, 243, 249, 282, 344, 362, 368, 370, 371)

___ and M.L. Crawford (eds.) (1981) Fluid Inclusions: Applications to Petrology. Mineral. Assoc. Canada Short Course Handbook, 6, 304 p. (iv, 4, 150, 193, 338)

___ and M. Sherwood (1980) Metamorphic evidence for rapid uplift, Coast Ranges, British Columbia, Canada (abstr.). Int'l Geol. Cong., 26th, Paris, 1980, Abstr., 51. (371)

___, R.C. Burruss, D.L. Henry and E.M. Hendel (1979) Physical conditions during uplift of metamorphic terranes, as recorded by fluid inclusions. Bull. Mineral., 102, 555-561. (53, 75, 370, 371, 375)

___, E. Roedder, R.C. Burruss, E.T.C. Spooner and J. Touret (1981) Practical aspects of microthermometry. In L.S. Hollister and M.L. Crawford, eds., Fluid Inclusions: Applications to Petrology. Mineral. Assoc. Canada Short Course Handbook 6, 278-304. (149, 150, 151, 200, 207, 222, 299, 345)

Holloway, J.R. (1976) Fluids in the evolution of granitic magmas: Consequences of finite CO_2 solubility. Geol. Soc. Am. Bull., 87, 1513-1518. (386)

___ (1981) Compositions and volumes of supercritical fluids in the Earth's crust. Mineral. Assoc. Canada Short Course Handbook, 6, 13-38. (228)

Holser, W.T. (1963) Chemistry of brine inclusions in Permian salt from Hutchinson, Kansas. In Symposium on Salt, Northern Ohio Geol. Soc., Cleveland, Ohio, 86-95. (133, 135, 136, 314, 315)

___ (1979a) Mineralogy of evaporites. In R.G. Burns, ed., Marine Minerals, Mineral. Soc. Am. Short Course Notes, 6, 211-294 (republished as Reviews in Mineralogy, 6). (309)

Holser, W.T. (1979b) Trace elements and isotopes in evaporites. In R.G. Burns, ed., Marine Minerals, Mineral. Soc. Am. Short Course Notes, 6, 295-346 (republished as Reviews in Mineralogy, 6). (311)

Hosking, K.F.G. and D.W.L. Spry (1955) A note concerning the emission of odours when certain sulphides are vigorously rubbed on, or with certain metals. Camborne, England, Camborne School Mines Mag., 55, 8-9; as cited in Mineral. Abstr. 16, (107), 1963. (117)

Hoy, R.B., R.M. Foose and B.J. O'Neill, Jr. (1962) Structure of Winnfield salt dome, Winn Parish, Louisiana. Am. Assoc. Petrol. Geol. Bull., 46, 1444-1459. (See also Geol. Soc. Am. Spec. Paper 88, 409-410, 1968.) (119, 315)

Huang, C.-I. (1976) An Isotopic and Petrologic Study of the Contact Metamorphism and Metasomatism Related to Copper Deposits at Ely, Nevada. Ph.D. dissertation, Pennsylvania State Univ., University Park, PA. (443)

___ (1977) Fluid inclusion study of well cuttings from Magmamax #2 drill-hole, Salton Sea geothermal area, California (abstr.). Econ. Geol. 72, (4) 730. (497)

Huang, W.L., J.K. Robertson and P.J. Wyllie (1973) Melting relations of muscovite to 30 kilobars in the system $KAlSi_3O_8$-Al_2O_3-H_2O. Am. J. Sci, 273, 415-427. (385)

Hughes, T.H. and R.E. Lynch, Jr. (1973) Barite in Alabama. Geol. Surv. Alabama Circ. 85. (186)

Humphreys-Owen, S.P.F. (1949) Crystal growth from solution. Proc. Roy. Soc., A, 197, 218-237. (38)

Hunter, J. and E. Sang (1873) Observations and experiments on the fluid in the cavities of calcareous spar. Royal Soc. Edinburgh Proc., 8, (86), 126-130. (92)

Huntley, H.E. (1955) Radioactivity in quartz inclusions. Nature, 176, 1229-1230. (114)

Ikorskii, S.V. (1962) Some inclusions in nepheline rocks of the Khibina and Lovozero alkalic massives. In I.V. Belkov, ed. Materialy po Mineralogii Kolskogo Poluostrowa, 2. Apatity, Akad. Nauk SSSR Kolskii Filial-Vsesoyuz. Mineral. Obshch., 80-83 (in Russian). (165, 404)

___ (1965) Gas-liquid and gas inclusions in nepheline alkalic rocks of the Khibinsk massif. In Mineralogical Thermometry and Barometry. "Nauka" Press, Moscow, 233-237 (in Russian). (404)

___ (1966) Inclusions of villiaumite and their connection with organic matter in rock-forming minerals of the Khibina massif. Geokhimiya, 1966, (8), 1002-1003 (in Russian; translated in Geochem. Int'l, 3, (4), 79). (96)

___ (1967a) Organic materials in the minerals of igneous rocks, as illustrated by the Khibina alkalic massif: Leningrad, "Nauka" Press, 121 p. (in Russian). (5)

___ (1967b) Bitumens in minerals of igneous rocks (as illustrated by eudialyte of the Khibina alkalic massif). Geol. Geokhim. Goryuch. Iskop.. (9), 22-29 (in Russian). (113)

___ (1968) Inclusions of organic substance in rock-forming minerals of the Khibiny alkalic massif. In Mineralogical Thermometry and Barometry, Vol. 1. "Nauka" Press, Moscow, 142-161 (in Russian). (100, 113, 404)

___ and E.A. Kireeskaya (1975) On CO_2 sorption during gas extraction from rocks and minerals in the vacuum mill. Geokhimiya, 1975, (11), 1712-1719. (120)

___ and A.M. Romanikhin (1964) Forms of hydrocarbon gases in nepheline rocks of the Khibina alkalic massif. Geokhimiya, 1964, (3), 276-281 (in Russian; English abstr. in Geochem. Int'l, (2), 245, 1964). (404)

Imai, H., S. Takenouchi and K. Nagano (1976) Fluid inclusion study of the Mamut porphyry copper deposit, Sabah, Malaysia. Abstr. 5th Int'l Symp. on Fluid Inclusion Research, Sydney, Australia, Aug. 1976, Fluid Inclusion Res. -- Proc. of COFFI, 8, 77-78. (443)

Ingerson, E. (1947) Liquid inclusions in geologic thermometry. Am. Mineral., 32, 375-388. (23, 160)

Ingerson, E. (1954) Nature of the ore-forming fluids at various stages--a suggested approach. Econ. Geol., 49, 727-733. (37)

____ (1965) Discussion [Cedar City fluids]. In Symposium--Problems of Postmagmatic Ore Deposition, Prague, 1963, Vol. 2, p. 457-458. (See discussion, p. 458-459.) Czechoslovakia Geol. Surv., Prague. (37)

Irving, A.J. and M.A. Dungan, eds. (1980) The Jackson Volume. Am. J. Sci. 280-A, 868 p. (515)

____ and E.A. Mathez (1982) The origin of glass in ultramafic xenoliths (abstr.). Terra Cognita, 2, p. 243. (527)

Irwin, W.P. and I. Barnes (1980) Tectonic relations of carbon dioxide discharges and earthquakes. J. Geophys. Res., 85, 3115-3121. (402)

Ishihara, S., ed (1974) Geology of Kuroko Deposits. Mining Geol. (Japan), Spec. Issue 6, 435 p. (425)

Ishkov, Yu.M. and F.G. Reyf (1980) Laser spectral analysis of the liquid in individual inclusions. Geokhimiya, 1980, (9), 1407-1412 (in Russian; translated in Geochem. Int'l, 17, 76-79, 1980). (139, 297)

Ito, E. and A.T. Anderson, Jr. (1983) Submarine metamorphism of gabbros from the Mid-Cayman Rise: Petrographic and mineralogic constraints on hydrothermal processes at slow-spreading ridges. Contrib. Mineral. Petrol., 82, 371-388. (495)

*Ivanova, G.F. (1976) Mineralogy and geochemistry of the tungsten ore mineralization in Mongolia. COFFI, 10, 107-109. (457)

Iwao, S., H. Akabori, M. Koizumi and H. Minato (1953) Electron micrographs of some silicate rocks, with special reference to the micropores with fluid inclusions. Japan. Assoc. Mineral. Petrol.. Econ. Geol. J., 37, (5), 167-178 (in Japanese; English abstr.). (97)

Jackson, E.D. and T.L. Wright (1970) Xenoliths in the Honolulu volcanic series, Hawaii. J. Petrol., 11, 405-430. (514)

Jackson, N.J., J.McH. Moore and A.H. Rankin (1977) Fluid inclusions and mineralization at Cligga Head, Cornwall, England. J. Geol. Soc. London, 134, 343-349. (457)

Jackson, S.A. and F.W. Beales (1967) An aspect of sedimentary basin evolution: The concentration of Mississippi Valley-type ores during late stages of diagenesis. Bull. Canadian Petrol. Geol., 15, 383-433. (417)

Jacobs, G.K. and D.M. Kerrick (1981a) Methane: an equation of state with application to the ternary system H_2O-CO_2-CH_4. Geochim. Cosmochim. Acta, 45, 607-614. (228, 244)

____ and ____ (1981b) Devolatilization equilibria in H_2O-CO_2 and H_2O-CO_2-NaCl fluids: an experimental and thermodynamic evaluation at elevated pressures and temperatures. Am. Mineral., 66, 1135-1153. (363)

Jahns, R.H. (1955) The study of pegmatites. Econ. Geol. 50th Anniversary Volume, Pt. II, 1025-1130. (388)

____ and C.W. Burnham (1969) Experimental studies of pegmatite genesis: I. A model for the derivation and crystallization of granitic pegmatites. Econ. Geol., 64, 843-864. (387, 388)

James, D.E. (1983) Volcanology, geochemistry, and petrology, 1979-1982: EOS, Trans. Am. Geophys. Union, 64, 481. (486, 506)

James, O.B. and J.J. McGee (1979) Consortium breccia 73255: Genesis and history of two coarse-grained "norite" clasts. Proc. 10th Lunar Sci. Conf., 713-743. (554)

Janardhan, A.S., R.C. Newton and J.V. Smith (1979) Ancient crustal metamorphism at low PH_2O: Charnockite formation at Kabbaldurga, south India. Nature, 278, 511-514. (375)

Jarosevich, E. (1971) Chemical analysis of the Murchison meteorite. Meteoritics, 6, 49-52. (562)

Jefferis, R.G. and B. Voight (1981) Fracture analysis near the mid-ocean plate boundary Reykjavik-Hvalfjördur area, Iceland. Tectonophysics, 76, 171-236. (496)

Jeffery, P.G. and P.J. Kipping (1963) The determination of constituents of rocks and minerals by gas chromatography. II. The determination of some gaseous constituents. Analyst, 88, 266-271. (125)

Jeffrey, G.A. (1963) The geometrical approach to the structure of water and the clathrate hydrates. In Proc. Conf. on Desalination Research, Woods Hole, Massachusetts, 19 June-14 July 1961. Nat'l Acad. Sci. - Nat'l Res. Council Pub. 942, 156-172. (99)

Jehl, V. (1975) Le Métamorphisme et les Fluides Associés des Roches Océaniques de l'Atlantique Nord. Thèse de docteur-ingenieur, Univ. Nancy I, Nancy, France. (204, 205, 207, 210, 211, 495)

____, B. Poty and A. Weisbrod (1977) Hydrothermal metamorphism of the oceanic crust in North Atlantic Ocean. Bull. Soc. Geol. Fr., 19, (6), 1213-1221 (in English). (495)

Jenks, G.H. and H.C. Claiborne (1981) Brine migration in salt and its implications in the geologic disposal of nuclear waste. Oak Ridge Nat'l Lab. ORNL 5818, 164 p. (66, 319)

Jessberger, E. and W. Gentner (1972) Mass spectrometric analysis of gas inclusions in Muong Nong glass and Libyan Desert glass. Earth Planet Sci. Lett., 14, 221-225. (560)

Johan, Z., ed. (1980) Porphyres cupriferes dans leur contexte magmatique. In Z. Johan, ed., Minéralisations Liées aux Granitoides, Mem. Bur. Recherches Geol. Minières, 99, Part 1. (445)

____ and L. Le Bel (1978) Origin of chromitite layers in rocks of ophiolitic suite (abstr.). Int'l Mineral. Assoc. XI Gen'l Mtg, Abstr., 1, 51-52. (463)

____ and ____ (1980) Genèse des couches et podes de chromite dans les complexes ophiolitiques (abstr.). Int'l Geol. Congress, 26th, Abstr., 950. (463)

____, J.L. Robert and M. Volfinger (1982) Role of reducing fluids in the origin of chromite deposits from ophiolitic complexes (abstr.). Geol. Assoc./Mineral. Assoc. Canada Program with Abstr., 7, p. 58. (463)

____, H. Dunlop, L. Le Bel, J.L. Robert and M. Volfinger (1983) Origin of chromite deposits in ophiolitic complexes: evidence for a volatile- and sodium-rich reducing fluid phase (abstr.). Fortschr. Mineral, 61, 105-107. (463)

Johannes, W. and W. Schreyer (1981) Experimental introduction of CO_2 and H_2O into Mg-cordierite. Am. J. Sci., 281, 299-317. (363)

Johnson, D.A., ed (1979) H.C. Sorby Centenary Issue. Sheffield Univ. Geol. Soc., J., 7, (4), 181-193. (4)

Johnson, P.W. (1961a) All about emeralds, natural or synthetic. Lapidary J., 15, 118-131. (92, 365)

____ (1961b) The Chivor emerald mine. J. Gemmology, 8, (4), 126-152. (92, 365)

Johnston, D.A. (1978) Volatiles, Magma Mixing, and the Mechanism of Eruption of Augustine Volcano, Alaska. Ph.D. dissertation, Univ. of Washington, 177 p. (485)

Johnston, Jr., W.D. and R.D. Butler (1946) Quartz crystal in Brazil. Geol. Soc. Am. Bull., 57, 601-650. (55)

Kalyuzhnyi, V.A. (1954) Measurement of the index of refraction of free liquids and mother liquors included in minerals by using a Fedorov stage. L'vov. Geol. Obshch. Mineral. Sbornik, 8, 315-344 (in Russian). (86)

____ (1955a) Liquid inclusions in minerals as a geologic barometer. L'vov. Geol. Obshch. Mineral. Sbornik, 9, 64-84 (in Russian; translated in Int'l Geol. Rev., 2, (2), 181-195, 1960). (117)

____ (1955b) Practical verification of the accuracy of the method of measurement of refractive indces of substances enclosed in minerals using the Fedorov stage. L'vov. Gosudarst. Univ. Ivana Franko Uchenye Zapiski, Geol. ser., 35, (8), 187-193 (in Russian). (86)

Kalyuzhnyi, V.A. (1956) New observations on phase transformations in liquid inclusions. L'vov. Geol. Obshch. Mineral. Sbornik, 10, 77-80 (in Russian). (51, 101, 392)

— (1957) Results of pH measurements in solutions from liquid inclusions. Geokhimiya, 1957, (1), 77-79 (in Russian; translated in Geochemistry, (1), 93-96, 1957). (136)

— (1958a) The study of the composition of captive minerals in polyphase inclusions. L'vov. Geol. Obshch. Mineral. Sbornik, 12, 116-128 (in Russian; translated in Int'l Geol. Rev., 4, (2), 127-138, 1962). (95, 96, 100, 113, 392)

— (1958b) Adaptation of a microthermal chamber for analysis of liquid inclusions. Vses. Nauchno-Issled. Inst. P'ezooptichesk. Mineral. Syr'ya Trudy, 2, (2), 43-47 (in Russian). (186)

— (1958c) Homogenization curves in mineralogical thermometry and their plotting. Vses. Nauchno-Issled. Inst. P'ezooptichesk. Mineral. Syr'ya Trudy, 2, (2), 7-18 (in Russian). (51, 87)

— (1960) Methods of Study of Multiphase Inclusions in Minerals. Kiev, Izdatel. Akad. Nauk Ukrainskoy RSR, 169 p. (in Ukrainian). (5, 113, 136, 188)

— (1961) Mineral-forming solutions from inclusions in minerals. Materialy Komis. Mineral. i Geokhim. Mezhdunar. Geol. Kong., Karpato-Balkan. Assots., (1), 159-173 (in Ukrainian): Referativnyi Zhur., Khim., Abs. 15G70 (in Russian) 1962. (113, 136, 474)

— (1965) Optic and thermometric studies of glass inclusions in phenocrysts of hyalodacite of the Transcarpathians. Akad. Nauk SSSR, Dokl., 160, 438-441 (in Russian; translated in Dokl. Acad. Sci. USSR, Earth Sci. Sects., 160, 142-145, 1965). (56, 86, 188, 474)

— (1971) The refilling of liquid inclusions in minerals and its genetic significance. L'vov. Gos. Univ. Mineral. Sbornik, 25, 124-131 (in Russian). (45, 75)

— (1973) New instruments for studies of inclusions of mineral-forming fluids and principles of their use. COFFI, 7, 90. (186, 190)

— (1976) Thermobaric parameters of mineralogenesis based on inclusions of H2O-NaCl type, (abstr.). COFFI, 9, 64. (246, 451)
and G.M. Gigashvili (1976) New cryometric stage with liquid thermostating medium for studies of inclusions of fluids in minerals. Mineral. Sbornik, L'vov. Univ., 30, (1), 34-36 (in Russian; translated in Fluid Inclusion Res.-- Proc. of COFFI, 10, 312-314, 1977). (190)
and L.I. Koltun (1953) Some data on pressures and temperatures during formation of minerals in Nagol'nyy Kryazh, Donets Basin. L'vov. Geol. Obshch. Mineral. Sbornik, 7, 67-74 (in Russian). (278, 279)
and N.A. Mikolaichuk (1968) Determination of the contents of solid phases of salts in the dry residue of solutions of microinclusions by means of electron diffraction. In Mineralogical Thermometry and Barometry, Vol. 2: "Nauka" Press, Moscow, 72-75 (in Russian). (148)
and O.S. Shchiritsya (1962) Physicochemical characteristics of aqueous carbon dioxide solutions in the mineralization media of the Nagol'nyi Ridge (Donets Basin) polymetallic veins. Geol. Zhur., 22, (2), 29-41 (in Ukrainian). (86)
and I.M. Svoren' (1978) Rational method of release and analysis of gas components of inclusions during mechanical destruction of minerals (abstr.). COFFI, 12, 81. (123)
and D.K. Vozryak (1967) Thermodynamic and geochemical characteristics of mineral-forming solutions of pegmatites of the "Zanorysh" type (from liquid inclusions in minerals). L'vov. Geol. Obshch. Mineral. Sbornik, 21, (1), 49-61 (in Russian). (113, 392)

Kalyuzhnyi, V.A., Yu.V. Lyakhov, Z.S. Gryn'kiv, Z.I. Kovalishin and D.K. Voznyak (1966) On age relationships and composition of gas-liquid inclusions in quartz from Volynian pegmatites. In N.P. Ermakov, ed., Research on Mineral-Forming Solutions (Materials of the First Symposium on Gas-Liquid Inclusions in Minerals). "Nedra" Press, Moscow, 112-120. Also listed as All-Union Research Inst. Synthetic Mineral Raw Materials, Ministry Geol. U.S.S.R. Trans.. 9, (in Russian; translated in Geochem. Int'l, 4, (3), 626-633, 1967). (398)

Kamilli, R.J. (1978) The genesis of stockwork molybdenite deposits: Implications from fluid inclusion studies at the Henderson mine. Geol. Soc. Am. Abstr. Programs, 10, 431. (276, 277)
and H. Ohmoto (1977) Paragenesis, zoning, fluid inclusion, and isotopic studies of the Finlandia Vein, Colqui District, Central Peru. Econ. Geol. 72, 950-982. (427)

Kaneoka, T. and N. Takaoka (1980) Rare gas isotopes in Hawaiian ultramafic nodules and volcanic rocks: Constraint on genetic relationships. Science, 208, 1366-1368. (527)

Karasev, V.V. (1958) Gas evolution on cleaving quartz crystals in a high vacuum. Zhur. Eksptl. i Teoret. Fiziki, 34, 1330-1331 (in Russian). (123)

Karpinskii, A.P. (1880) On the occurrence of inclusions of liquid carbon dioxide in mineral substances. Gornyi Zhur., 2, (4-5), 96-117 (in Russian). (117)

Karwowski, L. (1977) Geochemical conditions of greisenization in the Izerskie Mountains foothills (Lower Silesia). Archiwum Mineralogiczne, 33, (2), 83-148 plus 11 plates (in Polish with 7-page English summary and figure captions). (457)
and A. Kozlowski (1972) Thermogravimetric method of determination of decrepitation temperature. Bull. L'Acad. Polonaise Sci., 20, (1), 11-17 (in English; Russian abstr.). (212)

Karzhavin, V.K. (1976) The kinetic characteristics of the degassing of minerals on heating. Geokhimiya 1976, (11), 1701-1713 (in Russian; translated in Geochem. Int'l, 13, (6), 58-69). (125)

Kazahaya, K. (1983) Isotopic study of fluid inclusion in Japanese vein-quartz. M.S. thesis, Tokyo Inst. Technology, 130 p. (573)

Keenan, J.H., F.G. Keyes, P.G. Hill and J.G. Moore (1969) Steam Tables; Thermodynamic Properties of Water, Including Vapor, Liquid and Solid Phases. Wiley & Sons, New York, NY, 162 p. (274, 279, 281)
and (1978) Steam tables: Thermodynamic properties of water, including vapor, liquid, and solid phases (SI units). J. Wiley & Son, New York, NY, 156 p. (228)

Keevil, N.B. (1942) Vapor pressures of aqueous solutions at high temperatures. Am. Chem. Soc. J., 64, 841-850. (234, 235, 275)

Kekulawala, K.R.S.S., M.S. Paterson, and J.N. Boland (1978) Hydrolytic weakening in quartz. Tectonophysics, 46, T1-T6. (74)
and (1981) An experimental study of the role of water in quartz deformation. In Mechanical Behavior of Crustal Rocks, Geophys. Monograph 24, Am. Geophys. Union, 49-60. (74, 343)

Keller, P.C. (1981) Carbonatitic volcanism in the Kaiserstuhl alkaline complex: Evidence for highly fluid carbonatitic melts at the Earth's surface. J. Volcan. Geothermal Res., 9, 423-431. (408)

Keller, W.D. and R.F. Littlefield (1950) Inclusions in the quartz of igneous and metamorphic rocks. J. Sed. Petrol., 20, 74-84. (308)

Kelly, W.C. (1974) Fluid inclusions and other features in hydrothermal quartz from the Pennhurst and Downingtown sites. Appendix F, p. F1-F33, in Geologic Report, Limerick Generating Station, Limerick, Pennsylvania. Prepared by Dames and Moore Corp. for Philadelphia Electric Co., Job 4852-022, July 1974. (355)
(1975) Analysis of fluid inclusions in quartz crystals from Greene County nuclear power plant site: Preliminary Safety Analysis Rept. Appendix 2C, Power Authority of the State of New York Docket No. 50549, Vol. 4, 26 p. (355)

Kelly, W.C. and P.A. Burgio (1983) Cryogenic scanning electron microscopy of fluid inclusions in ore and gangue minerals. Econ. Geol., 78, 1262-1267. (147)

___ and E.N. Goddard (1969) Telluride Ores of Boulder County, Colorado. Geol. Soc. Am. Mem., 109, 237 p. (186, 190, 437)

___ and R.O. Rye (1979) Geologic, fluid inclusion, and stable isotope studies of the tin-tungsten deposits of Panasqueira, Portugal. Econ. Geol., 74, 1721-1819. (457, 459, 460)

___ and F.S. Turneaure (1970) Mineralogy, paragenesis and geothermometry of the tin and tungsten deposits of the Eastern Andes, Bolivia. Econ. Geol., 65, 609-680. (16, 43, 95, 441, 443, 457, 458, 459)

___ and G.A. Wagner (1977) Paleothermometry by combined application of fluid inclusion and fission track methods. N. Jahrb. Mineral. Monats., 1977, 1-15. (461)

___, J.M. Sharp, Jr. and D.E. White (1983a) Penrose Conf. Rept: Hydrodynamics and geochemistry of ore generation in sedimentary environments. Geology, 11, 309-311. (311, 417)

___, E.U. Peterson, F.S. Frederick and W.C. Bigelow (1983b) Cryogenic scanning electron microscopy and energy dispersive analysis of hydrothermal fluid inclusions (abstr.). Geol. Soc. Am. Abstr. Programs, 15, 610. (147)

Kendall, A.C. and P.L. Broughton (1978) Origin of fabrics in speleothems comprised of columnar calcite crystals. J. Sed. Petrol., 48, 519-538. (327)

Kennedy, G.C. (1950a) "Pneumatolysis" and the liquid inclusion method of geologic thermometry. Econ. Geol., 45, 533-547. (76, 265)

___ (1950b) Pressure-volume-temperature relations in water at elevated temperatures and pressures. Am. J. Sci., 248, 540-564. (88)

___ (1954) Pressure-volume-temperature relations in CO_2 at elevated temperature and pressures. Am. J. Sci., 252, 225-241. (229, 230, 231, 279, 280, 283)

Kerkis, T.Yu. and V.P. Kostyuk (1963) Mineral thermometric study of the Botogol nepheline, Eastern Sayan. Akad. Nauk SSSR, Dokl., 150, 1125-1127 (in Russian; translated in Dokl. Acad. Sci. USSR, Earth Sci. Sects., 150, 125-127, 1965). (89)

Kern, R. and G. Mattern (1963) Réalisation d'une platine de microscope -35°C à +100°C à l'aide de thermoéléments à effet Peltier. Bull. Soc. franç. Minéral. Cristallogr. 86, 427-428. (190)

Kerrich, R. (1974) Aspects of Pressure Solution as a Deformation Mechanism. Ph.D. dissertation, Imperial College, London, 255 p. (187, 342)

___ (1976) Some effects of tectonic recrystallization on fluid inclusions in vein quartz. Contrib. Mineral. Petrol., 59, 195-202. (76, 345)

___ (1977) Yellowknife gold mineralization: the product of metamorphic degassing (abstr.). Geol. Soc. Am. Abstr. Programs, 9, 1048-1049. (356)

___ and I. Allison (1978) Vein geometry and hydrostatics during Yellowknife mineralization. Canadian J. Earth Sci., 15, 1653-1660. (356)

___ and W.S. Fyfe (1981) The gold carbonate association: source of CO_2 and CO_2 fixation reactions in Archean lode deposits. Chem. Geol., 33, 265-294. (356)

___, R.D. Beckinsale and N.J. Shackelton (1978) The physical and hydrothermal regime of tectonic vein systems: Evidence from stable isotope and fluid inclusion studies. N. Jahrb. Mineral. Abh., 131, 225-239. (352)

Kerrick, D.M. (1972) Experimental determination of muscovite and quartz stability with $PH_2O < P_{total}$. Am. J. Sci., 272, 946-958. (385)

___ (1977) The genesis of zoned skarns in the Sierra Nevada, California. J. Petrol., 18, 144-181. (455)

___ and G.K. Jacobs (1981) A modified Redlich-Kwong equation for H_2O, CO_2 and H_2O-CO_2 mixtures at elevated pressures and temperatures. Am. J. Sci., 28, 735-767. (228, 238, 239, 240)

Kesler, S.E. (1973) Copper, molybdenum and gold abundances in porphyry copper deposits. Econ. Geol., 68, 106-112. (439)

Kessen, K.M., M.S. Woodruff, and N.K. Grant (1981) Gangue mineral $^{87}Sr/^{86}Sr$ ratios and the origin of Mississippi Valley-type mineralization. Econ. Geol., 76, 913-920. (324, 422)

Khaibullin, I.Kh., B.Ye. Novikov, A.M. Copeilovich and A.M. Besedin (1980) Phase diagrams for steam solutions and caloric properties of two- and three-component systems: $H_2O-NaCl$, $H_2O-Na_2SO_4$, and $H_2O-NaCl-Na_2SO_4$. In J. Straub and K. Scheffler, eds., Water and Steam. Pergamon Press, New York, NY, 641-647. (235, 286)

*Kharlamov, Ye.S. (1973) Construction of an apparatus for fine control of the speed of freezing of preparations in a cryostage by the use of liquid nitrogen. COFFI, 7, 99. (90)

Khetchikov, L.N. and L.A. Samoilovich (1970) The possibilities of the decrepitation method in mineral thermometry. Akad. Nauk SSSR, Izvest., Ser. Geol., 1970, (7), 92-98 (in Russian; translated in Fluid Inclusion Res. -- Proc. of COFFI, 3, 94-100). (265, 266)

___, V.S. Balitskii and N.R. Gasparyants (1966) Possible determination of chemical composition and concentration of mineralizing solutions from the chemical composition of gas-liquid inclusions in minerals. Akad. Nauk SSSR, Dokl., 168, (5), 1179-1182 (in Russian; translated in Dokl. Acad. Sci. USSR, Earth Sci. Sects., 168, 209-212, 1966). (137)

___ and V.F. Dernov-Pegarev (1968) The recrystallization of quartz in hydrothermal solutions of sulfides and fluorides of certain alkali metals. In Mineralogical Thermometry and Barometry, Vol. 1. "Nauka" Press, Moscow, 91-94 (in Russian). (137)

Khitarov, D.N. (1965a) Contemporary investigation of the composition and other properties of fluid inclusions in minerals. In Mineral'nye Mikrovklyucheniya. Akad. Nauk SSSR, Inst. Mineral. Geokhim. i Kristallokhim. Redkikh Elementov, Moscow, 74-263 (in Russian). (5)

___ (1965b) Determinations of microquantities of carbon dioxide in gas-liquid inclusions in minerals with the help of an adapted conductometric method. In Mineralogical Thermometry and Barometry. "Nauka" Press, Moscow, 135-141 (in Russian). (118)

___ (1978) The state of the investigations of the chemical composition of inclusions of mineral-forming fluids (abstr.). In Int'l Assoc. Genesis of Ore Deposits, 5th Symp., Snowbird, Alta, Utah, 1978, Prog. and Abstr., (Alta, Utah), 113. (190)

___ and I.P. Vovk (1963) Improved conductometric method of determination of trace amounts of carbon dioxide in gas-liquid inclusions of minerals. Akad. Nauk SSSR, Inst. Mineral. Geokhim. Kristallkhim. Redkikh Elementov, Trudy, (18), 142-146 (in Russian). (118)

Khitarov, N.I. and E.V. Rengarten (1956) On the subject of geochemistry of carbonic acid in granitic intrusions. Geokhimiya, 1956, (2), 74-77 (in Russian; translated in Geochemistry (2), 198-202, 1956). (119)

___ and N.E. Lebedeva (1958) Chemical composition of liquid inclusions in Iceland spar and genetic problems. Geokhimiya, 1958, (3), 214-221 (in Russian; translated in Geochemistry, (3), 269-278, 1958). (117)

Khitarov, D.M., Y.G. Shmar and Y.V. Reutin (1980) Physical and chemical peculiarities of solutions which formed hydrothermal deposits in the urans [i.e. of uranium]. Sovet. Geologiya, 1980, (5), 90-100. (462)

Khodakov, G.S. (1966) Activated adsorption of inert gases under ordinary conditions on freshly formed solid surfaces prepared by grinding. Akad. Nauk SSSR, Dokl, 168, 158-159 (in Russian; translated in Phys. Chem. Dokl., 168, 312-313). (120)

Khodakovskiy, I.L. (1965) Characterization of hydrothermal solutions according to the data of the study of gas-liquid inclusions in minerals. In Mineralogical Thermometry and Barometry. Moscow, "Nauka" Press, 174-203 (in Russian). (5)

Khollef, M.M. (1975) Genetic significance of fissures and inclusions in detrital quartz of Oligocene sediments at Cairo, Egypt. Chem. Erde, 34, 302-308. (308)

Kobe, K.A. and R.E. Lynn (1953) The critical properties of elements and compounds. Chem. Rev., 52, 117-236. (90)

Kofler, L. (1934) Über einen Mikroschmelzpunktapparat mit am Apparat gezichtem Thermometer. Mikrochemie New Ser., 15, 242-246. (188)

Kogarko, L.N. and B.P. Romanchev (1973) Crystallization temperature of agpaitic magma. Akad. Nauk SSSR, Dokl., 212, (4), 957-960 (in Russian; translated in Dokl. Acad. Sci. USSR, 212, 169-171, 1974). (403)

___ and ___ (1977) Temperature, pressure, redox conditions, and mineral equilibria in agpaitic nepheline syenites and apatite-nepheline rocks. Geokhimiya, 1977, (2), 199-216 (in Russian; translated in Geochem. Int'l, 14, 113-128). (403, 405)

Koivula, J.I. (1981) The hidden beauty of amber: new light on an old subject. Gems and Gemology, Spring 1981, 34-36. (325)

Kokubu, N, T. Mayeda and H.C. Urey (1961) Deuterium content of minerals, rocks, and liquid inclusions from rocks. Geochim. Cosmochim. Acta, 21, 247-256. (119, 126)

Kolomenskii, V.D., L. Yu. Chupina, Yu. G. Laurentev, G.M. Ivanova and L.P. Pospelova (1978) Inclusions in olivine of the Bragin pallasite, Meteoritika, 37, 140-143, (in Russian). (569)

*Kolonin, G.R. and Ya.A. Kosals (1977) Proposed physico-chemical model of rare-metal ore formation exemplified by the Dzhida ore field (abstr.). COFFI, 10, 131. (457)

Konev, P.N. and B.Ya. Chalov (1972) Study of detrital quartz of the Takatin suite in the Kolva-Vishera region for paleogeographic reconstruction. Litologiya i Polez. Isko, 7, (5), 21-25 (in Russian; translated in Lithology and Mineral Resources, 7, (5), 557-561, 1973). (308)

Konnerup-Madsen, J. (1977) Composition and microthermometry of fluid inclusions in the Kleivan Granite, South Norway. Am. J. Sci., 277, 673-696. (204, 247, 344, 383)

___ (1979a) Fluid inclusions in quartz from deep-seated granitic intrusions, south Norway. Lithos, 12, 13-23. (400)

___ (1979b) Fluid inclusions associated with a metamorphosed molybdenite mineralization in Vest-Agder, south Norway. Econ. Geol., 74, 1221-1230. (366)

___ (1980) Fluid inclusions in minerals from igneous rocks belonging to the Precambrian continental Gardar rift province, South Greenland: The alkaline Ilimaussaq intrusion and the alkali acidic igneous complexes. Ph.D. dissertation, Copenhagen Univ., Copenhagen, Denmark, 140 p. (404)

___ and J. Rose-Hansen (1982) Volatiles associated with alkaline igneous rift activity: Fluid inclusions in the Ilimaussaq intrusion and the Gardar granitic complexes (south Greenland). Chem. Geol., 37, 79-93. (404, 405, 406)

___, E. Larsen and J. Rose-Hansen (1979) Hydrocarbon-rich fluid inclusions in minerals from the alkaline Ilimaussaq intrusion, South Greenland. Bull. Mineral., 102, 642-653. (405, 514)

___, J. Rose-Hansen and E. Larsen (1981) hydrocarbon gases associated with alkaline igneous activity: evidence from compositions of fluid inclusions. Rapp. Grønlands geol. Unders., 103, 99-108. (405, 406)

Kormushin, V.A. (1960) New micro heat chamber for the study of gas-liquid inclusions in thin sections. Vses. Nauchno-Issled. Inst. Pezooptichesk Mineral. Syr'ya Trudy, 4, (1), 119-122 (in Russian). (186)

___ (1965) A very simple method for freezing gas-liquid inclusions to obtain data on the concentrations of the solutions. In Mineralogical Thermometry and Barometry. "Nauka" Press, Moscow, 172-173 (in Russian). (190)

___ and A.B. Darbadaev (1978) Apparatus for collecting gas from gas-liquid inclusions by the thermal method (abstr.). COFFI, 12, 94. (119)

*Kornilov, V.F. (1968) Determination of the genesis of glass sands by thermometric methods (abstr.). COFFI, 1, 62. (308)

Kholief, M.M. and H.A. Hamed (1976) Effective method for correlation between some hydrocarbon productive horizons of Miocene age in the Gulf of Suez with the help of inclusions in quartz. Cent. Rech. Pau, Bull., 10, 53-65. (308)

Khoteev, A.D. (1980) Problems of standardization of decrepitation conditions. In V.I. Rekharskiy, ed., Methods and Devices for Studies of Inclusions in Mineral-Forming Media, "Nauka" Pub. House, Moscow, 31-39 (in Russian). (265)

Killingley, J.S. and D.W. Muenow (1975) Volatiles from Hawaiian submarine basalts determined by dynamic high temperature mass spectrometry. Geochim. Cosmochim. Acta, 39, 1467-1473. (486)

___ and ___ (1975b) Thermal stress-induced release of CO_2 inclusions in olivine on cooling from high temperatures. Am. Mineral., 60, 148-151. (261, 265, 489)

Kim, M.Y. (1981) Fluid inclusion studies relating to tungsten-tin-copper mineralization at the Ohtani mine, Japan. J. Geosci., Osaka City Univ., 24, 109-162 (in English). (457)

King, M.B. (1969) Phase Equilibrium in Mixtures. Pergamon Press, New York, NY. (233)

Kinnunen, K.A. (1981) Comparison of fluid inclusion assemblages of Outokumputype ore outcrops and boulders in eastern Finland. Geol. Surv. Finland. Rept. Invest. 51, 40 p. (468)

Kinsland, G.L. (1977) Formation temperature of fluorite in the Lockport Dolomite in upper New York State as indicated by fluid inclusion studies -- with a discussion of heat sources. Econ. Geol., 72, 849-854. (324)

___ (1979) Formation temperature of fluorite in the Lockport Dolomite in upper New York State as indicated by fluid inclusion studies -- with a discussion of heat sources -- a reply. Econ. Geol.74, 159-164. (324)

Kirby, S.H. and H.W. Green, III (1980) Dunite xenoliths from Hualalai volcano: Evidence for mantle diapiric flow beneath the island of Hawaii. Am. J. Sci., 280-A, 550-575. (528)

___ and J.W. McCormick (1979) Creep of hydrolytically weakened synthetic quartz crystals oriented to promote [2110] <0001> slip: a brief summary of work to date. Bull. Mineral., 102, 124-137. (74)

Kirkpatrick, R.J. (1981) Kinetic theory of nucleation in silicate melts (abstr.). EOS, Trans. Am. Geophys. Union, 62, 1064-1065. (479)

Kita, I., S. Matsuo and H. Wakita (1982) H_2 generation by reaction between H_2O and crushed rock: an experimental study on H_2O degassing from the active fault zone. J. Geophys. Res., 87, 10789-10795. (402)

Klatt, E. (1979) The fluid inclusions in the granulites of North Lapland and their host rocks. Fortschr. Mineral., 57, 62-63 (in German). (375)

Klevtsov, P.V. and G.G. Lemmlein (1959a) Determination of the minimum pressure of quartz formation as exemplified by crystals from the Pamir. Vses. Mineralog. Obshch. Zapiski, 88, (6), 661-666 (in Russian; translated in Fluid Inclusion Res. -- Proc. of COFFI, 10, 320-326, 1977). (89, 90, 276, 277)

___ and ___ (1959b) Pressure corrections for the homogenization temperatures of aqueous NaCl solutions. Akad. Nauk SSSR, Dokl., 128, (6), 1250-1253 (in Russian; translated in Dokl. Acad. Sci. USSR, 128, (1-6), 995-997, 1960). (89)

Klosterman, M.J. (1981) Applications of fluid inclusions to burial diagenesis in carbonate rock sequences. Applied Carbonate Research Prog. Tech. Ser. Contrib. 7, 102 p. (Louisiana State Univ., Baton Rouge, LA). (320)

Knauth, L.P. and M.B. Kumar (1978) Trace water content of salt in Louisiana salt domes. Science, 213, 1005-1007. (121, 122)

___ and ___ (1983) Isotopic character and origin of brine leaks in the Avery Island salt mine, south Louisiana, USA. J. Hydrology, 66, 343-350. (319)

*Korobeynikov, A.F. and A.V. Matsyushevskiy (1973) Application of methods of mineral thermometry for prospecting and evaluation of ore necks in endogenic gold deposits. COFFI, 6, 81-82. (465)

Koster van Groos, A.F. and P.J. Wyllie (1969) Melting relationships in the system $NaAlSi_3O_8$-NaCl-H_2O at one kilobar pressure, with petrological applications. J. Geol., 77, 581-605. (400, 446)

____ and ____ (1973) Liquid immiscibility in the join $NaAlSi_3O_8$-$CaAl_2Si_2O_8$-Na_2CO_3-H_2O. Am. J. Sci., 273, 465-487. (410)

Kostyleva, Ye. Ye. (1964) Some methods of study of ore-bearing quartz and an attempt to apply them. Moscow, "Nauka" Press, 98 p. (in Russian). (5)

Kosukhin O.N. (1977) Low-temperature melt inclusions in the quartz of chambered pegmatites. Geol. Geofiz. (Akad. Nauk SSSR, Sib. Otd.), 18, (10), 66-72 (in Russian; translated in Sov. Geol. Geophys., 18, (10), 56-61, 1978). (387, 397)

Kotra, R.K. and E.K. Gibson, Jr. (1982) Direct analysis of fluid and vapor inclusions using laser microprobe-gas chromatography (abstr.). EOS, Trans. Am. Geophys. Union, 63, 450. (117, 124, 140)

____ and ____ (1983) Direct analysis by laser microprobe-gas chromatography of trapped ancient volatiles (abstr.). 7th Int'l Conf. on the Origins of Life, and 4th Mtg Int'l Soc. Study Origin Life, Mainz, Germany, Program (unpaginated). Abstract B-2-10. (117, 124, 140)

Kovalevich, V.M. (1975) Conditions of forming of Verkhnevorotyshchenskiye saline deposits in the region of Stebnik (based on inclusions in halite). Kiev, Inst. Geol. Geochem. Fuels, 44, 42-50 (in Russian). (314, 316)

Kovalishin, Z.I. (1968) Influence of crushing minerals on the liberation of gases from inclusions. In Mineralogical Thermometry and Barometry. Vol. 2, "Nauka" Press, Moscow, 31-33 (in Russian). (120)

Koval'skii, V.V. and N.V. Cherskii (1973) The carbon isotope composition of diamonds. Industrial Diamond Rev., Feb. 1973, 54-56. (506)

Kozlowski, A. (1978) Pneumatolytic and hydrothermal activity in the Karkonosze-Izera block. Acta Geol. Polonica, 28, 171-222. (133, 387, 457)

____ (1981) Melt inclusions in pyroclastic quartz from the Carboniferous deposits of the Holy Cross Mts., and the problem of magmatic corrosion. Acta Geol. Polonica, 31, 273-284. (468)

____ and L. Karwowski (1973) Hydrated salt melt as mineral forming medium of high-temperature mineral associations from Alam Kuh (Iran). COFFI, 7, 113-114. (389)

____ (1974) Chlorine/bromine ratio in fluid inclusions. Econ. Geol., 69, 268-271. (133)

Kramer, J.R. (1965) History of sea water. Constant temperature-pressure equilibrium models compared to liquid inclusion analyses. Geochim. Cosmochim. Acta, 29, 921-945. (118, 133, 136)

Kranz, R. (1966) Organische Fluor-Verbindungen in den Gaseinschlüssen der Wölsendorfer Flussspäte. Naturwissenschaften, 53, 593-600. (117, 123)

____ (1968) Participation of organic compounds in the transport of ore metals in hydrothermal solution. Inst. Mining Metall. Trans., sec. B, 77, B26-B36; abstr.76, 11, 1967. (119, 123)

Krasov, N.F. and R. Clocchiatti (1979) Liquation of silicate melt and its possible petrogenetic role according to the data of melt inclusion study. Doklady Akad. Nauk SSSR, 248, (1), 201-204 (in Russian; translated in Dokl. Acad. Sci. USSR, 248, 92-95, 1981). (476)

Kravtsov, A.I., G.I. Voytov, V.A. Ivanov and O.I. Kropotova (1976) Gases and bitumens in rocks of "Udachnaya" pipe. Akad. Nauk SSSR, Dokl., 228, (5), 1204-1207 (in Russian). (512)

Kravtsov, A.I., G.I. Voytov, V.A. Bobrov, A.P. Akimov, A.N. Ivanov and L.I. Serdyukov (1979a) Gases (chemical and isotopic composition) of the kimberlite pipe "Mir." Akad. Nauk SSSR, Dokl., 245, (4), 950-953 (in Russian). (511)

____, M.I. Kuchev and A.I. Fridman (1979b) Effects of seismicity on natural-gas carbon isotope ratios. Geokhimiya, 1979, (3), 387-390 (in Russian; translated in Geochem. Int'l, 16, 45-47, 1980). (402)

Krendelev, F.P., L.B. Zozulenko and L.M. Orlova (1972) Temperature of homogenization and composition of gases in gas-fluid inclusions in pebbles from ancient conglomerates of sulfide type. Fluid Inclusion Res. -- Proc. of COFFI, 3, 101-107. (307)

____ (1973) Properties of solutions that have metamorphosed metalliferous conglomerate as shown by the study of gas-liquid inclusions in quartz pebbles and regenerated minerals. Akad. Nauk SSSR, Dokl., 212, 713-716 (in Russian; translated in Dokl. Acad. Sci. USSR, Earth Sci. Sects., 212, 160-163, pub. 1974). (307)

Kreulen, R. (1977) CO_2-rich Fluids During Regional Metamorphism on Naxos, a Study on Fluid Inclusions and Stable Isotopes. Ph.D. dissertation, Univ. Utrecht, Utrecht, The Netherlands, 87 p. (in English; Dutch abstr.). (351, 374)

____ (1980) CO_2-rich fluids during regional metamorphism on Naxos (Greece): Carbon isotopes and fluid inclusions. Am. J. Sci., 280, 745-771. (351, 528)

____ and R.D. Schuiling (1982) N_2-CH_4-CO_2 fluids during formation of the Dome de l'Agout, France. Geochim. Cosmochim. Acta, 46, 193-203. (125, 348, 349, 354)

Kronenberg, A.K, S.H. Kirby, R.D. Ames and G.R. Rossman (1983) Hydrogen uptake in hydrothermally annealed quartz: Implications for hydrolytic weakening (abstr.). EOS, Am. Geophys. Union Trans., 64, 839. (36, 76)

Krupka, K.M., H. Ohmoto, and F.E. Wickman (1977) A new technique in neutron activation analysis of Na/K ratios of fluid inclusions and its application to the gold-quartz veins at the O'Brien mine, Quebec, Canada. Canadian J. Earth Sci., 14, 2760-2770. (134)

Kubota, Y. (1979) Hydrothermal rock alteration in the northern Hachimantai geothermal field. J. Japan. Geotherm. Energy Assoc., 16, (4), Ser. No. 63, 15-31 (in Japanese; English abstr.). (497)

Kuhnert-Brandstätter, M. (1982) Thermomicroscopy of organic compounds. In G. Svehla, ed., Comprehensive Analytical Chemistry, Vol. 16, Elsevier Sci. Publ. Co., Amsterdam, The Netherlands, 329-499. (207)

Kul'chitskaya, A.A. (1974) Inclusions of mineral-forming solution in gypsum and their significance. In Ye.K. Lazarenko, ed., Mineralogy of Sedimentary Deposits. Nauk. Dumka, Kiev, 34-38 (in Russian). (316)

Kulikov, I.V. (1982a) The influence of unstable genetic conditions on the properties of fluorite crystals from Tyrmyauz deposit. Int'l Mineral. Assoc., 13th General Mtg, Varna, September 19-25, 1982, Abstr. of Papers, 229, (111, 454, 455)

____ (1982b) Multiphase brine inclusions in fluorite and calcite and their genetic significance. Akad. Nauk SSSR, Dokl., 264, (4), 958-961 (in Russian). (454, 455)

Kunkel, W.B. (1950) The static electrification of dust particles on dispersion into a cloud. J. Appl. Physics, 21, 820-832. (120)

Kuroda, Y., Y. Hariya, T. Suzuoki and S. Matsuo (1982) D/H fractionation between water and melts of quartz, K-feldspar, albite and anorthite at high temperature and pressure. Geochem. J., 16, 73-78, (400, 486)

Kurshev, S.A, and V.N. Trufanov (1965) An attempt to study gas-liquid and solid inclusions in minerals under the electron microscope, in Mineralogical Thermometry and Barometry. "Nauka" Press, Moscow, 112-117 (in Russian). (97, 145)

Kurz, M.D. and W.J. Jenkins (1981) The distribution of helium in oceanic basalt glasses. Earth Planet. Sci. Lett., 53, 41-54. (506, 527)

Kurz, M.D., S.R. Hart and D. Clague (1981) Helium isotopic systematics of oceanic islands and Loihi seamount (abstr.). EOS, Trans. Am. Geophys. Union, 62, 1083. (527)

Kuznetsova, S.V., Zh.V. Kulik and E.E. Lazarenko (1983) Amino acids in the gas-liquid inclusions from Ukrainian black rocks. Dopov. Akad. Nauk Ukr. RSR, Ser. B; Geol., Khim. Biol. Nauki, 1983, (9), 15-17 (in Ukrainian). (132, 333)

Kvenvolden, K.A. and M.A. McMenamin (1980) Hydrates of natural gas: a review of their geologic occurrence. U.S. Geol. Survey Circular, 825, 11 p. (241)

___, K. Weliky, H. Nelson and D.J. Des Marais (1979) Submarine seep of carbon dioxide in Norton Sound, Alaska. Science, 205, 1264-1266. (528)

Kwak, T.A.P. and P.W. Askins (1981) Geology and genesis of the F-Sn-W (-Be-Zn) skarn (wrigglite) at Moina, Tasmania. Econ. Geol., 76, 439-467. (457)

___ and T.M. Tan (1981) The importance of $CaCl_2$ in fluid composition trends--evidence from the King Island (Dolphin) skarn deposit. Econ. Geol., 76, 955-960. (451, 455)

Kweiyang Institute Research Group (1978) Mineralogy and petrology of the Kirin meteorite and its formation and evolution. Scientia Sinica, 21, 805-822 (in English). (566, 569)

Kyser, T.K. and W. Rison (1982) Systematics of rare gas isotopes in basic lavas and ultramafic xenoliths. J. Geophys. Res., 87, 5611-5630. (527)

Lagache, M. and A. Weisbrod (1977) The system: two alkali feldspars-KCl-NaCl-H_2O at moderate to high temperatures and low pressures. Contrib. Mineral. Petrol., 62, 77-101. (389)

Lamar, J.E. and R.S. Shrode (1953) Water soluble salts in limestones and dolomites. Econ. Geol., 48, 97-112. (115)

Lambert, J.B. and J.S. Frye (1982) Carbon functionalities in amber. Science, 217, 55-57. (327)

Land, L.S. and K.L. Milliken (1981) Feldspar diagenesis in the Frio Formation, Brazoria County, Texas Gulf coast. Geology, 9, 314-318. (339, 423)

___ and D.R. Przebindowski (1981) The origin and evolution of saline formation water, Lower Cretaceous carbonates, south-central Texas, USA. J. Hydrology, 54, 51-74. (311, 339)

Landis, G.P. (1972) Geologic, Fluid Inclusion, and Stable Isotope Studies of a Tungsten-Base Metal Ore Deposit: Pasto Bueno, Northern Peru. Ph.D. dissertation, Univ. of Minnesota, Minneapolis, MN, 195 p. (457, 458)

___ and R.O. Rye (1974) Geologic, fluid inclusion, and stable isotope studies of the Pasto Bueno tungsten-base metal ore deposit, Northern Peru. Econ. Geol., 69, 1025-1059. (457, 458)

Langway, Jr., C.C. (1958) Bubble pressures in Greenland glacier ice. Union Géodes. Géophys. Int'l, Assoc. Int'l d'Hydrologie Scientifique, Pub. 47, Symp. Chamonix, 336-349. (11)

Larhidi, N., R. Clocchiatti, A. Havette, N. Metrich and J. Weiss (1980) Chemical and thermometric study of hyperalkaline lavas (pantellerites and comendites) of St. Pietre Island (southern Sardinia, Italy) (abstr.). Abstracts, 1980 Int'l Mineral. Assoc. Meeting, 131-132 (in French). (484)

Larimer, J.W. (1968) Experimental studies on the system Fe-MgO-SiO_2-O_2 and their bearing on the petrology of chondritic meteorites. Geochim. Cosmochim. Acta, 32, 1187-1207. (569)

Larson, L.T., J.D. Miller, J.E. Nadeau and E. Roedder (1973) Two sources of error in low-temperature inclusion homogenization determination and corrections on published temperatures for the East Tennessee and Laisvall deposits. Econ. Geol., 68, 113-116. (167, 185, 258, 259, 416, 428)

Laudise, R.A. (1979) Recent progress in hydrothermal quartz crystallization. In D.D. Timmerhaus and M.S. Barber, eds. High-Pressure Science and Technology, 6th AIRAPT Conf. 1977, Vol.1 Physical Properties and Material Synthesis, Plenum Press, New York, NY, 963-969. (24)

Laul, J.C. (1979) Neutron activation analysis of geological materials. Atomic Energy Rev., 17, (3), 603-695. (134)

Lawler, J.P., and M.L. Crawford (1983) Stretching of fluid inclusions resulting from a low-temperature microthermometeric technique. Econ. Geol. 78, 527-529. (258)

Lazarenko, E.K., D.K. Voznyak, V.I. Pavlishin and V.I. Shelukhin (1976) Typomorphic peculiarities of quartz crystals with inclusions of methane solutions (Donets basin). Akad. Nauk SSSR, Dokl., 231, (6), 1446-1449 (in Russian). (323)

†Laz'ko, E.M. (1957) Some genetic particularities of Kurumkan deposits, determined by thermal analysis. Vses. Nauchno-Issled. Inst. P'esooptichesk. Mineral. Syr'ya Trudy, 1, (2), 129-133 (in Russian). (464)

Laz'ko, Ye.Ye. (1977) Crystalline inclusions in minerals of kimberlitic rocks and their petrogenic significance. Akad. Nauk SSSR Dokl., 234, 918-921 (in Russian; translated in Dokl. Acad. Sci. USSR, 234, 203-207, 1979). (511)

Leach, D.L. (1973) A Study of the Barite-Lead-Zinc Deposits of Central Missouri and Related Mineral Deposits in the Ozark Region. Ph.D. dissertation, Univ. Missouri-Columbia, Columbia, MO, 186 p. (319, 330, 420)

___ (1979) Temperature and salinity of the fluids responsible for minor occurrences of sphalerite in the Ozark region of Missouri. Econ. Geol., 74, 931-937. (324)

___ (1980) Nature of mineralizing fluids in the barite deposits of central and southeast Missouri. Econ. Geol., 75, 1168-1180. (322, 324)

___, L.P. Rowan and J.A. Hedal (1983) Evidence for ore fluid migration in the Bonneterre Formation, southeast Missouri (abstr.). Geol. Soc. Am. Abstr. Programs, 15, 625. (320, 421)

Leach, T.M. (1982) An evaluation of fluid inclusions as a geothermal exploration tool. Proc. Pacific Geothermal Conf., Nov. 1982, Univ. Auckland, Auckland, New Zealand, 475-478. (496, 497)

Leake, B.E., C.M. Farrow and R. Townend (1979) A pre-2,000 Myr old granulite facies metamorphosed evaporite from Caraiba, Brazil? Nature, 277, 49-50. (350, 364)

LeAnderson, P.J. (1981) Calculation of temperature and $X(CO_2)$ values for tremolite-K feldspar-diopside-epidote assemblages. Canadian Mineral., 19, 619-630. (346)

Le Bas, M.J. (1977) Carbonatite-nephelinite volcanism: an African case history. Wiley & Sons, New York, NY, 347 p. (407)

___ and C.D. Handley (1979) Variation in apatite composition in ijolitic and carbonatitic igneous rocks. Nature, 279, 54-56. (408)

Le Bel, L. (1976) Preliminary note on the mineralogy of solid phases in quartz phenocryst inclusions in the porphyry copper from Cerro Verde/Santa Rosa, S Peru. Bull. Soc. Vaudoise Sci. Nat., 73, (350), 201-208 (in French; translation in Fluid Inclusion Res. -- Proc. of COFFI, 9, 167-172). (145, 146, 443, 444, 445)

___ (1980) Caractéristiques de la phase fluide associée à la minéralisation de Cerro Verde-Santa Rosa. In Z. Johan, ed., Minéralisations liées aux Granitoïdes, Mem. Bur. Recherches Geol. Minières, 99, 129-139. (443, 445)

___ and E. Oudin (1982) Fluid inclusion studies of deep-sea hydrothermal sulfide deposits on the East Pacific Rise near 21°N. Chem. Geol., 37, 129-136. (495)

Lemmlein, G.G. (1929) Sekundäre Flüssigkeitseinschlüsse in Mineralien. Z. Kristallogr., 71, 237-256 ("Georg Laemmlein" in original). (59)

___ (1930) Corrosion and regeneration of quartz phenocrysts in quartz porphyries. Akad. Nauk. SSSR, Dokl., ser. A, (13), 341-344 (in Russian). (14, 17)

___ (1937) Crystallographic investigation of quartz from Sura-Iz mountain. Subarctic Ural, Ural Ser. of the SOPS and the Petrographic Inst., 6, 87-92 (in Russian). (342)

Lemmlein, G.G. (1946) On the origin of flat quartzes with "white band": Problems in mineralogy, geochemistry, and petrography. Izdat. Akad. Nauk, SSSR, 98-109 (in Russian; translated into German in Schweizer Strahler, 2, (12), 430-440 (1972)). (342)

___ (1950) Al-Biruni's mineralogical information. Moscow-Leningrad, Sbornik Biruni [Collected papers], 106-127 (in Russian). (As quoted in Lemmlein, 1956b). (3)

___ (1951) The fissure-healing process in crystals and changes in cavity shape in secondary liquid inclusions. Akad. Nauk SSSR, Dokl., 78, 685-688 (in Russian). (60)

___ (1952) Migration of a liquid inclusion in a crystal toward a source of heat. Akad. Nauk SSSR, Dokl., 85, 325-328 (in Russian). (66)

___ (1953) The inverted heated microscope for observations and microphotography at high temperatures. Mineral. Petrogr. Trudy, 2, 157-162 (in Russian). (186)

___ (1956a) Formation of fluid inclusions and their use in geological thermometry. Geokhimiya, 1956, (6), 84-94 (in Russian; translated in Geochemistry, (6), 630-642, 1956). (71, 72, 258)

___ (1956b) Russian literature on liquid inclusions in minerals and on geological thermometer. Supplement to Russian translation of F.G. Smith, Historical Development of Inclusion Thermometry. Inostrannaya Literatura, Moscow, 123-166 (an English translation of the supplement, E. Roedder, ed., is available from the U.S. Geol. Surv. Library, call number 830 qSm5ha). (4)

___ and P.V. Klevtsov (1961) Relations among the principal thermodynamic parameters in a part of the system H_2O-NaCl. Geokhimiya, (2), 133-142 (in Russian; translated in Geochemistry, (2), 148-158). (234, 263, 264)

___ and M.O. Kitya (1952a) New data on the deposition of crystal substance on cavity walls of liquid inclusions. Akad. Nauk SSSR, Dokl., 82 (5), 765-768 (in Russian; translated in Int'l Geol. Rev., 2, (2), 120-124, 1960). (48, 101, 160, 303)

___ and ___ (1952b) Distinctive features of the healing of a crack in a crystal under conditions of declining temperature. Akad. Nauk SSSR, Dokl., 87, (6), 957-160 (in Russian; translated in Int'l Geol. Rev., 2, (2), 125-128, 1960). (59, 67)

___ and I.A. Ostrovskii (1962) The conditions for formation of minerals in pegmatites according to data on primary inclusions in topaz. Akad. Nauk SSSR, Dokl., 142, 81-83 (in Russian; translated in Sov. Phys.-Dokl., 7, (1), 4-6, 1962). (96, 101, 261, 392, 398)

LeRibault, L. (1974) External and internal study of detrital quartz. Rend. Soc. Ital. Mineral. Petrol., 30, 373-416 (in French). (308)

Leroy, J. (1978) The Margnac and Fanay uranium deposits of the La Crouzille district (Western Massif Central, France): Geologic and fluid inclusion studies. Econ. Geol., 73, 1611-1634. (462)

___ (1979) L'étalonnage de la pression interne des inclusions fluides lors de leur décrépitation. Bull. Mineral., 120, 584-593. (70, 71)

Lesnyak (1964) Bases of analysis of physicochemical properties of mineral-forming solutions from the inclusions in minerals. L'vov, L'vov. Univ. Izdatel'stvo, 219 p (in Russian). (5)

Li, Y., Z. Rui and L. Cheng (1981) Fluid inclusions and mineralization of the Yulong porphyry copper (molybdenum) deposit. Acta Geol. Sinica, 8, (3), 216-231 (in Chinese; English abstr.). (443)

Li, Z., F. Sun and M. Li (1981) Study on microtextures in minerals and microtextures of the meteorites. Scientia Sinica, 24, 975-979 (in English). (566)

Lilley, M.D., M.A. de Angelis and L.I. Gordon (1982) CH_4, H_2, CO and MgO in submarine hydrothermal vent waters. Nature, 300, 48-50. (495, 513)

Lindgren, W. and W.L. Whitehead (1914) A deposit of jamesonite near Zimapan, Mexico. Econ. Geol., 9, 435-462. (96)

Linke, W.F. (1958) Solubilities Inorganic and Metal-Organic Compounds, Vol. 1, 4th Edn. D. Van Nostrand Co., Princeton, NJ, 1487 p. (249)

___ (1965) Solubilities, Inorganic and Metal-Organic Compounds, Vol. II, 4th Edn. Amer. Chem. Soc., Washington, D.C., 1914 p. (232)

Lipin, B.R. (1976) The origin of Fra Mauro basalts. Lunar Sci. VII, 495-497. (554)

Lippolt, H.J. and W. Gentner (1963) K-Ar dating of some limestones and fluorites (examples of K-Ar ages with low Ar-concentrations), in Radioactive Dating [Int'l Atomic Energy Agency, Symp., Athens, Nov. 19-23, 1962, Proc.]. Int'l Atomic Energy Agency, Vienna, 239-244. (119, 128)

Little, W.M. (1955) A Study of Inclusions in Cassiterite and Associated Minerals. Ph.D. dissertation, Univ. Toronto, Toronto, Canada, 83 p. (summarized in Little, 1960). (186, 265)

___ (1960) Inclusions in cassiterite and associated minerals. Econ. Geol., 55, 485-509. (265)

Liu, C.-T. and W.T. Lindsay, Jr. (1972) Thermodynamics of sodium chloride solutions at high temperatures. J. Solution Chem., 1, 45-69. (278)

*Lkhamsuren, J. (1976) Fluid inclusion data on the fluorite occurrences in Eastern Mongolia. COFFI, 8, 105. (137)

Llambias, H. (1963) Sobre inclusiones halladas en cristales de inderita, borax y topacio de la Argentina y consideraciones sobre su empleo como termometro geologico. Asoc. Geol. Argentina Rev., 18, 129-138 (in Spanish). (87, 94)

Lofoll, F. (1972) Endoscopie des grains de sable de la Cote Francaise de la Manche. Mem. BRGM, 79, 263-267 (in French; English abstr.). (308)

Lohmann, K.C. (1983) Role of fluid inclusions in diagenesis of metastable marine cements (abstr.). Am. Assoc. Petrol. Geol. Bull 67, (3), p. 505. (320)

Lokerman, A.A. (1962) The possibility of study of the interrelations of dikes and mineralization from inclusions in minerals. L'vov. Geol. Obshch., Mineral. Sbornik, 16, 312-317 (in Russian). (167, 261, 465)

___ (1965) Mineral thermometry in connection with the problem "Ores and Dikes." In Mineralogical Thermometry and Barometry. "Nauka," Moscow, 288-290 (in Russian). (167, 261, 465)

London, D., E.T.C. Spooner and E. Roedder (1982) Fluid-solid inclusions in spodumene from the Tanco pegmatite, Bernic Lake, Manitoba. Carnegie Inst. Washington Year Book, 81, 334-339. (147, 261, 394, 395, 396)

Long, D.T. and E.E. Angino (1982) The mobilization of selected trace metals from shales by aqueous solutions: Effects of temperature and ionic strength. Econ. Geol., 77, 646-652. (424)

Longhi, J. D. Walker and J.F. Hays (1978) The distribution of Fe and Mg between olivine and lunar basaltic liquids. Geochim. Cosmochim. Acta, 42, 1545-1558. (554)

Loomis, T.P. (1983) Metamorphic petrology. Rev. Geophys. Space Phys., 21, 1386-1394. (362)

Lord, H.C., III (1965) Molecular equilibria and condensation in a solar nebula and cool stellar atmospheres. Icarus, 4, 279-288. (562)

Loskutov, A.V. (1955) Electric furnace with water cooled jacket for heating under the microscope. Vses. Mineral. Obshch. Zapiski, 84, 374-376 (in Russian). (186)

___ (1959) Interferometric study of quartz inclusions. Nauchno-Issled. Inst. Geol. Arktiki Trudy, Ministerstva Geol. i Okhrany Nedr. SSSR, 96, 174-180 (in Russian). (160)

___ (1962) Liquid and multiphase inclusions in natrolite. In Belkov, I.V., ed. Materialy po Mineral. Kolskogo Poluostrova, Vol. 2. Apatity, Akad. Nauk SSSR Kolskii Filial-Vses. Mineral. Obshch., 84-95 (in Russian). (115)

Lovell, J.S. (1979) Gas chromatographic determination of sulfur gases, hydrocarbons, carbon dioxide and water in fluid inclusions (abstr.). Program 4th Mtg Geol. Soc. of the British Isles, Univ. Sheffield, Sheffield, England, Sept. 1979 (unpaginated). (126)

Lu
Ma

Lu, H.Z. and M.L. Crawford (1982) The geologic features and fluid inclusion studies of Dongpo tungsten skarn ore deposit (abstr.). EOS, Trans. Am. Geophys. Union 63, 1127. (125, 455)

Luckscheiter, B. and G. Morteani (1980a) Microthermal and chemical studies of fluid inclusions in minerals from Alpine veins from the penninic rocks of the central and western Tauern Window (Austria/Italy). Lithos, 13, 61-77. (363, 374)

___ and ___ (1980b) The fluid phase in eclogites, glaucophane-bearing rocks and amphibolites from the central Tauern Window as deduced from fluid inclusion studies. Tschermaks Mineral. Petrogr. Mitt., 27, 99-111. (374)

___ and ___ (1981) The H contents of quartz from alpine veins from the Penninic rocks of the central and western Tauern Window (Austria/Italy). Tschermaks Mineral. Petrogr. Mitt., 28, 223-228. (36, 84)

Lupton, J.E. (1983a) Helium isotopes and other volatiles as indicators of submarine hydrothermal activity (abstr.). Geol. Assoc. Canada/Mineral. Assoc. Canada Abstr., 8, A43. (506)

___ (1983b) Terrestrial inert gases: Isotope tracer studies and clues to primordial components in the mantle. Ann. Rev. Earth Planet Sci., 11, 371-414. (506)

Lutts, B.G., I.A. Petersil'ye and V.K. Karzhavin (1976) Composition of gaseous substance in rocks of upper mantle of the Earth. Akad. Nauk SSSR, Dokl., 226, (2), 440-443 (in Russian). (511)

Lyakhov, Y.V. (1966) Mineral composition of multiphase inclusions in mortions from Volynian pegmatites. In N.P. Ermakov, ed., Research on Mineral-Forming Solutions (Materials of 1st Symp. on Gas-Liquid Inclusions in Minerals. Moscow, May 17-24, 1963). "Nedra" Press, Moscow, 92-100 (in Russian). Also listed as All-Union Research Inst. Synthetic Mineral Raw Materials, Ministry Geology USSR, Trans., 9; translated in Geochem. Int'l, 4, (3), 618-625 (1967). (113, 392)

___ (1973) Errors in determining pressure of mineralization from gas-liquid inclusions with halite, their causes and ways of eliminating them. Zap. Vses. Mineral. Obshch., 102, (4), 385-393 (in Russian). (277)

Lyubinetskaya, A.V., B.V. Zatsikha, Z.V. Shabo and G.P. Mamchur (1980) Nature and genetic features of organic minerals and matter of the Slaviansk mercury deposit mineralization. L'vov. Geol. Obshch., Mineral. Sbornik 34, (1), 32-39 (in Russian). (463)

McCrone, W.C. (1970) Scientist's notebook: Insight, Oct., 1970. Walter C. McCrone Assoc., Inc., Chicago, Illinois. (572, 573)

___ (1971) Choice of analytical tool: Amer. Laboratory, April, 1971. (572, 573)

McCulloch, D.S. (1959) Vacuole disappearance temperatures of laboratory-grown hopper halite crystals. J. Geophys. Res., 64, 849-854. (76)

Macdonald, A.J. and E.T.C. Spooner (1981) Calibration of a Linkam TH 600 programmable heating-cooling stage for microthermometric examination of fluid inclusions. Econ. Geol., 76, 1248-1258. (196, 203, 204, 205, 206)

MacDonald, G.J. (1983) The many origins of natural gas. J. Petrol. Geol., 5, 341-362. (512)

MacDonald, K.C., K. Becker, F.N. Speiss and R.D. Ballard (1980) Hydrothermal heat flux of the "black smoker" vents on the East Pacific Rise. Earth Planet. Sci. Lett., 48, 1-7. (494)

MacDonald, R.H. (1979) Occurrence and characteristics of polymineralic silicate inclusions in chromite grains, Eastern Bushveld complex (abstr.). Geol. Soc. Am. Abstr. Programs, 11, 470. (463)

MacDougall, J.D. (1979) Refractory spherulites and inclusions in Murchison (abstr.). Meteoritics, 14, 477-478. (562)

Machairas, G. (1963a) Etude quantitative du gaz carbonique des inclusions fluides des minéraux. Compt. Rend. Acad. Sci. Paris, 256, 2883-2884. (114, 137)

___ (1963b) Métallogénie de l'or Guyane française. Bur. Recherches Géol. et Minières Mém., 22, 163 p. (137)

___ (1963c) Les pegmatites de la Guyane française: géologie, pétro-graphie, géochimie. Chronique Mines Rech. Minière, 31, (322), 264-275. (137)

Mack, S. and L.M. Ferrell (1979) Saline water in the foothill suture zone, Sierra Nevada Range, California. Geol. Soc. Am. Bull, Pt. 1, 90, 666-675. (401)

Mackin, J.H. (1971) Rational and empirical methods of investigation in geology. In C.C. Albritton, Jr., ed., The Fabric of Geology. Addison-Wesley Pub. Co., Reading, Massachusetts, 135-163. (80)

McLaren, A.C. and P.P. Phakey (1965a) A transmission electron microscope study of amethyst and citrine. Australian J. Phys., 18, 135-141. (74, 97)

___ and ___ (1965b) Dislocations in quartz observed by transmission electron microscopy. J. Appl. Phys., 36, 3244-3246. (74, 97)

___ and ___ (1966) Transmission electron microscope study of bubbles and dislocations in amethyst and citrine quartz. Australian J. Phys., 19, 19-24. (61, 74, 84, 97, 111, 148)

McLimans, R.K. (1981) Applications of fluid inclusion studies to reservoir diagenesis and petroleum migration: Smackover Formation, U.S. Gulf Coast, and Fateh field, Dubai (abstr.). Am. Assoc. Petrol. Geol. Bull., 65, 957. (320 333)

___, H.L. Barnes and H. Ohmoto (1980) Sphalerite stratigraphy of the Upper Mississippi Valley zinc-lead district, southwest Wisconsin. Econ. Geol., 75, 351-361. (417)

MacQueen, R.W. (1976) Sediments, zinc and lead, Rocky Mountain Belt, Canadian Cordillera: Geosciences Canada, 3, (2), 71-81. (469)

McSween, H.Y., Jr. (1977) On the nature and origin of isolated olivine grains in carbonaceous chondrites. Geochim. Cosmochim. Acta, 41, 411-418. (562, 563, 566)

Magara, K. (1973) Compaction and fluid migration in Cretaceous shales of western Canada: Canada Geol. Survey Paper 72-18, 81 p. (424)

___ (1981) Possible primary migration of oil globules. J. Petrol. Geol., 3, (3), 325-331. (333)

Mahon, W.A.J., G.D. McDowell, and J.B. Finlayson (1980) Carbon dioxide: Its role in geothermal systems. New Zealand J. Sci., 23, 133-148. (501)

Mahtab, M.A. (1982) Geomechanical aspects of gas outbursts in Louisiana salt mines. Bull. Assoc. Eng. Geologists, 19, 389-400. (315)

Maier, S. and E.U. Franck (1966) Die Dichte des Wassers von 200 bis 850°C und von 1,000 bis 6,000 bar. Ber. Bunsengesell. Phys. Chemie, 70, (6), 639-645. (88)

Makogon, Y.F. (1974) Hydrates of Natural Gas. "Nedra" Press, Moscow, 208 p. (in Russian; translated by W.J. Cieslewicz, PennWell Pub. Co., Tulsa, Oklahoma, 237 p., 1981). (241)

Malakhov, V.V. (1977) Investigation of gas-liquid inclusions in minerals by gas chromatography. Geokhimiya, 1977, (8), 1192-1198 (in Russian; translated in Geochem. Int'l, 14, (4), 142-147). (125)

Mal'kov, B.A. and G.N. Bobolovich (1977) Genesis of kimberlite as shown by study of inclusions in calcite and apatite. Akad. Nauk SSSR, Dokl., 234, (2), 436-439 (in Russian; translated in Dokl. Acad. Sci. USSR, Earth Sci. Sects., 234, 181-184, 1979). (511)

622

Melton, C.E. and A.A. Giardini (1976) Experimental evidence that oxygen is the principal impurity in natural diamond. Nature, 263, 309-310. (509)

___ and ___ (1980) The isotopic composition of argon included in an Arkansas diamond and its significance. Geophys. Res. Lett., 7, 461-464. (509, 510)

___ and ___ (1981) The nature and significance of occluded fluids in three Indian diamonds. Am. Mineral., 66, 746-750. (509, 510)

___ and ___ (1982) The evolution of the Earth's atmosphere and oceans. Geophys. Res. Lett., 9, 579-582. (509, 510)

___, C.A. Salotti and A.A. Giardini (1972) The observation of nitrogen, water, carbon dioxide, methane, and argon as impurities in natural diamonds. Am. Mineral., 57, 1518-1523. (509)

Mercado, G.S. (1976) Movement of geothermal fluids and temperature distribution in the Cerro Prieto geothermal field, Baja California, Mexico. Second UN Symp. on the Development and Use of Geothermal Resources, San Francisco, CA, 1975 Vol.1, 492-494. (498)

Mercer, P.D. (1967) Analysis of the gases released on cleaving muscovite mica in ultrahigh vacuum and of gases which remain adsorbed on the freshing cleaved surface. Vacuum, 17, 267-270. (123)

Mercier, J.-C.C. and A. Nicolas (1975) Textures and fabrics of upper-mantle peridotites as illustrated by xenoliths from basalts. J. Petrol., 16, 454-487. (515)

Merkel, G.A., S.E. Haggerty and F.R. Boyd (1976) A unique olivine megacryst from Monastery (abstr.). EOS, Trans. Am. Geophys. Union, 57, 355. (511)

Merlich, B.V. and N.M. Datsenko (1972) The composition of aqueous extracts of liquid inclusions in sulphur and authigenic minerals of sulfur deposits of the Precarpathians. L'vov. Gos. Univ. Mineral. Sbornik, 26, 73-88 (in Russian). (328)

Metrich, N. and R. Clocchiatti (1979) Primary melts trapped in phenocrysts of a pantelleritic lava from Ethiopian Rift: a study of crystallization and setting processes of the flow. Compt. Rend. Acad. Sci. Paris, 289, 57-60 (in French). (482, 484)

Metzger, F.W., B.E. Nesbitt and W.C. Kelly (1975) Scanning electron microscopy of daughter minerals in fluid inclusions (abstr.). Geol. Soc. Am. Abstr. Programs, 7, 1199. (145)

___, W.C. Kelly, B.E. Nesbitt and E.J. Essene (1977) Scanning electron microscopy of daughter minerals in fluid inclusions. Econ. Geol., 72, 141-152. (145, 146, 407, 438)

Meyer, C. (1950) hydrothermal Wall-Rock Alteration at Butte, Montana. Ph.D. dissertation, Harvard Univ., Cambridge, Massachusetts. (160, 186)

Meyer, H.O.A. (1982) Mineral inclusions in natural diamond. In D.M. Eash, ed., Proc. Int'l Gemological Symp., Gemological Inst. Am., Santa Monica, CA, 445-465. (506, 507)

___ and H.-M. Tsai (1979) Inclusions in diamond and the mineral chemistry of the upper mantle. In L.H. Ahrens, ed., Origin and Distribution of the Elements. Pergamon Press, N.Y., 631-644. (507)

Michelsen, H. (1971) In situ measurements of the refractive indices of liquid and gaseous inclusions in minerals. Fortschr. Mineral., 52, Spec. Issue (IMA-Papers 9th Mtg. Berlin-Regensburg 1974), 465-473. (86)

Mikhailov, M.Yu. (1981) Transformation of the form of gas-liquid inclusions in beryls. Geologiya i Geofizika, 22, (9), 127-132 (in Russian; translated in Sov. Geol. Geophys., 22, (9), 111-115). (60)

___ and V.S. Shatskii (1975) Silite heater for high-temperature microthermal chamber. In V.S. Sobolev, ed., Mineralogy of Endogenetic Formations from Inclusions in Minerals, W. Siberian Publishing House (All-Union Mineral. Soc., W. Siberia Branch [ZSOVMO], Trudy, (2), Novosibirsk), 109-110 (in Russian). (189)

*Mikhaylova, G.V., L.L. Kunin and G.B. Naumov (1973) Application of laser for analysis of gaseous inclusions in fluorite (abstr.). COFFI, 6, 102-103. (117, 138)

Manning, D.A.C. and M. Pichavant (1983) The role of fluorine and boron in the generation of granitic melts. In M.P. Atherton and C.D. Gribble, eds., Migmatites, Melting and Metamorphism. Shiva Pub. Ltd., Cheshire, UK, 94-109. (389)

Marshall, W.L., C.E. Hall, and R.E. Mesmer (1981) The system dipotassium hydrogen phosphate-water at high temperatures (100-400°C); liquid-liquid immiscibility and concentrated solutions. J. Inorg. Nuc. Chem., 43, 449-455. (52)

Marutani, M. and S. Takenouchi (1978) Fluid inclusion study of stockwork siliceous orebodies of Kuroko deposits at the Kosaka mine, Akita, Japan. Mining Geol., 28, 349-360. (425, 426)

†Maslova, I.N. (1968) Chemical investigation of fluid inclusions in fluorite by ultramicro methods. Vses. Nauchno-Issled. Inst. P'ezooptichesk. Mineral. Syr'ya Trudy, 2, (2), 119-121 (in Russian). (136)

___ and ___ (1961) Ultramicrochemical investigation of compositions of the liquid and vapor phases in two-phase inclusions in quartz from Volynia. Geokhimiya, 1961, (2), 169-173 (in Russian; translated in Geochemistry, (2), 190-195, 1961). (117, 135)

Mathez, E.A. and J.R. Delaney (1981) The nature and distribution of carbon in submarine basalts and peridotite nodules. Earth Planet. Sci. Lett., 56, 217-232. (526)

___, V.J. Dietrich and A.J. Irving (1982) Abundances of carbon in mantle xenoliths from alkalic basalts (abstr.). Terra Cognita, 2, 199-200. (528)

Matsuhisa, Y., J.R. Goldsmith and R.N. Clayton (1979) Oxygen isotopic fractionation in the system quartz-albite-anorthite-water. Geochim. Cosmochim. Acta, 43, 1131-1140. (127)

Matsui, E., E. Salati and O.J. Marini (1974) D/H and $^{18}O/^{16}O$ ratios in waters contained in geodes from the basaltic province of Rio Grande do Sul, Brazil. Geol. Soc. Am. Bull., 85, 577-580. (325)

Matsuo, S. (1984) Occurrence and chemical form of volatiles in the mantle. Int'l Geol. Congress, Moscow (in press). (504, 525, 528)

Mattey, D.P., C.I. Dillinger and A.E. Fallick (1983) Hydrogen isotopic composition of water in fluid inclusions in the Peetz L6 chondrite (abstr.). (569)

Matthews, A. and R.D. Beckinsale (1979) Oxygen isotope equilibration systematics between quartz and water. Am. Mineral., 64, 232-240. (127)

___, J.R. Goldsmith, and R.N. Clayton (1983) On the mechanisms and kinetics of oxygen isotope exchange in quartz and feldspars at elevated temperatures and pressures. Geol. Soc. Am. Bull., 94, 396-412. (127)

Maugh, T.H., II (1980) Separations by MS speed up, simplify analysis. Science, 209, 675-677. (122)

Maxwell, J.A., S. Abbey and W.H. Champ (1970) Chemical composition of lunar material. Science, 167, 530-531. (541)

Melikhov, I.V., M.A. Prokofiev, Yu.N. Synchev and V.N. Sidorov (1981) Study of liquid inclusions in microcrystals. J. Crystal Growth, 51, 292-298. (84)

Mel'nikov, F.P. (1965) Preliminary results of cryometric investigations of mineral-forming solutions in inclusions. In Mineralogical Thermometry and Barometry. "Nauka" Press, Moscow, 129-134 (in Russian). (190)

___ (1968) Cryometric study of inclusions in samples of a mineral-forming medium. In N.P. Ermakov, ed., Mineralogical Thermometry and Barometry, Vol. 2, New Methods and Results of Investigations of the Conditions of Ore Formation. "Nauka" Press, Moscow, 56-61 (in Russian). (190)

Melson, W.G. (1983) Monitoring the 1980-1982 eruptions of Mount St. Helens: Compositions and abundances of glass. Science, 221, 1387-1391. (482, 484)

Melton, C.E. and A.A. Giardini (1974) The composition and significance of gas released from natural diamonds from Africa and Brazil. Am. Mineral., 58, 775-782. (509, 510)

___ and ___ (1975) Experimental results and a theoretical interpretation of gaseous inclusions found in Arkansas natural diamonds. Am. Mineral., 60, 413-417. (509, 510)

Mori, Y.H. (1978) Configurations of gas-liquid two-phase bubbles in immiscible liquid media. Int'l J. Multiphase Flow 4, 383-396. (53)

Morse, S.A. and L.K. Coachman (1983) Ocean chemistry during glacial time - a comment. Geochem. Cosmochim. Acta, 47, 1539-1540. (328)

Muecke, G.K., ed. (1980) Mineral. Assoc. Canada Short Course in Neutron Activation Analyses in the Geosciences. Mineral. Assoc. Canada, Toronto, 279 p. (134)

Muenow, D.W. (1973) High temperature mass spectrometric gas-release studies of Hawaiian volcanic glass. Pele's Tears. Geochim. Cosmochim. Acta, 37, 1551-1561. (486)
_____, D.G. Graham, N.W.K. Liu and J.R. Delaney (1979) The abundance of volatiles in Hawaiian tholeiitic submarine basalts. Earth Planet. Sci. Lett., 42, 71-76. (122, 486, 487, 488, 489)

Muffler, L.J. P. and C.L. Hofeling (1979) Inventory of drilling activities of the U.S. Geological Survey in the United States, fiscal years 1979-1980. USGS Open-file rept. 79-1567. (497)

Muggli, R.Z. (1979) The laser Raman microprobe. In W.C. McCrone, J.G. Dolly and S.J. Palenik, eds., The Particle Atlas, Edition Two, Volume V, Light Microscopy Atlas and Technique. Ann Arbor Sci. Pub., Ann Arbor, Michigan, 1147-1154. (105)

Müller, O and W. Gentner (1968) Gas content in bubbles of tektites and other natural glasses. Earth Planet. Sci. Lett., 4, 406-410. (560)

Muller, P.H. and J.A. Schufle (1968) Shift in temperature of maximum density of water in capillaries. J. Geophys. Res., 73, 3345-3348. (296)

Mullis, J. (1975) Growth conditions of quartz crystals from Val d'Illiez (Valais, Switzerland). Schweiz. Mineral. Petrogr. Mitt., 55, 419-429. (342, 347, 353)
_____ (1976) The growth environment of quartz crystals from Val d'Illiez, Wallis, Switzerland. Schweiz. Mineral. Petrogr. Mitt., 56, 219-268 (in German). (346, 353)
_____ (1979) The system methane-water as a geologic thermometer and barometer from the external part of the Central Alps. Bull. Mineral., 102, 526-536. (241, 353)

Munasinghe, T. and C.B. Dissanayake (1981) The origin of gemstones of Sri Lanka. Econ. Geol., 76, 1216-1225. (375)

Munoz, J.L. and A. Swenson (1981) Chloride-hydroxyl exchange in biotite and estimation of relative HCl/HF activities in hydrothermal fluids. Econ. Geol., 76, 2212-2221. (390)

Muramatsu, Y. and M. Nambu (1982a) Fluid inclusion study on the contact metamorphic tungsten ore deposits of the Yaguki mine, Fukushima Prefecture, Japan. Mining Geol. (Japan), 32, 107-116 (in Japanese; English abstr.). (455)
_____ and _____ (1982b) Fluid inclusion studies of pyrometasomatic iron-copper ore deposits and igneous rocks at the Kamaishi mining district, Iwate prefecture, Japan (I). Fluid inclusions in igneous rocks. J. Japan Assoc. Mineral. Petrol. Econ. Geol., 77, 7-17 (in Japanese; English abstr.). (455)
_____ and _____ (1982c) Fluid inclusion studies of pyrometasomatic iron-copper ore deposits and igneous rocks at the Kamaishi mining district, Iwate ore deposit. J. Japan Assoc. Mineral. Petrol. Econ. Geol., 77, 181-190 (in Japanese; English abstr.). (455)

Murase, T. and A.R. McBirney (1973) Properties of some common igneous rocks and their melts at high temperatures. Geol. Soc. Am. Bull., 84, 3563-3592. (282)

Murck, B.W., R.C. Burruss and L.S. Hollister (1978) Phase equilibria in fluid inclusions in ultramafic xenoliths. Am. Mineral., 63, 40-46. (515, 525, 527)

Murray, R.C. (1957) Hydrocarbon fluid inclusions in quartz. Am. Assoc. Petrol. Geol. Bull., 41, 950-956. (82, 123, 124, 332)

Milledge, H.J., M.J. Mendlessohn, M. Seal, J.E. Rouse, P.K. Swart and C.T. Pillenger (1983) Carbon isotopic variation in spectral type II diamonds. Nature, 303, 791-792. (507)

Miller, C. and W. Richter (1982) Solid and fluid phases in lherzolite and pyroxenite inclusions from the Hoggar, central Sahara. Geochem. J., 16, 263-277. (515)

Miller, J.D. (1968) Determination of Temperatures of Fluorite Formation by Fluid Inclusion Thermometry, East Tennessee Zinc District. M.S. thesis, Univ. of Tennessee, Knoxville, TN, 41 p. (186, 416)
_____ (1969) Fluid inclusion temperature measurements in the East Tennessee zinc district. Econ. Geol., 64, 109-110. (416)

Miller, K.R. (1980) Petrology, hydrothermal mineralogy, stable isotope geochemistry and fluid inclusion geothermometry of borehole Mesa 31-1, East Mesa geothermal field, Imperial Valley, California. M.S. thesis, Univ. of California, Riverside, CA 113 p. (498)

Miller, M.F. and T.J. Shepherd (1984) The determination of lead in fluid inclusions using voltammetric trace analysis: an exploratory investigation. Chem. Geol., 42, 249-259. (133)

Miller, R.E., D.A. Brobst, and P.C. Beck (1972) Fatty acids as a key to the genesis and economic potential of black bedded barites in Arkansas and Nevada. Geol. Soc. Am. Abstr. Program, 4, (7), 596. (422)

Milovsky, G.A., B.F. Zlenko and A.M. Gubanov (1978) Conditions of formation of scheelite ores in the Chorukh-Dayron mineralized area (as revealed by a study of gas-liquid inclusions). Geokhimiya, 15, (1), 79-86 (in Russian; translated in Geochem. Int'l, 15, (1), 45-52). (276)

* _____ and Ye.V. Frolov (1973a) New possibilities of chromatographic determination of concentration of gas components in mineral-forming solutions. COFFI, 7, 143. (138)

Mogk, D.W. (1979) Contact metamorphism of carbonate rocks at Cave Ridge, Snoqualmie Pass, Washington (abstr.). Geol. Soc. Am. Abstr. Programs, 11, 117. (455)

*Moiseenko, V.G. (1976) Conditions of formation of gold-ore deposits of the Southern part of (Soviet) Far East. COFFI, 10, 180-181. (137)
_____ and V.V. Malakhov (1979) Physico-chemical conditions of endogene ore formation. Moscow, "Nauka" Publishing House, 200 p. (In Russian). (5)

Monier, J.C. and R.J. Hocart (1950) An equipment for the microscopic examination of metals and crystals in polarized light at temperatures from -130° to +35°C. Jour. Sci. Instruments, 27, 302. (191)

Moore, C.H., Y. Druckman, J.A.D. Dickson and A.H. Stueber (1983) Jurassic subsurface calcite cementation, central Gulf of Mexico (abstr.). Am. Assoc. Petrol. Geol. Bull., 67, 518-519. (320)

Moore, F. and D.J. Moore (1979) Fluid-inclusion study of mineralization at St. Michael's Mount, Cornwall. Trans. Inst. Min. Metal., Sect. B, B57-B60. (457)

Moore, J.G. (1979) Vesicularity and CO2 in mid-ocean ridge basalt. Nature, 282, 250-253. (490, 491, 505)
_____ and L. Calk (1971) Sulfide spherules in vesicles of dredged pillow basalt. Am. Mineral., 56, 476-488. (490)
_____ and J.P. Lockwood (1973) Origin of comb layering and orbicular structure, Sierra Nevada batholith, Californica. Geol. Soc. Am. Bull., 84, 1-20. (397)
_____, J.N. Batchelder and C.G. Cunningham (1977) CO2-filled vesicles in mid-ocean basalt. J. Volcan. Geothermal Res., 2, 309-327. (283, 487, 490, 513)

Moore, W.J. and J.T. Nash (1974) Alteration and fluid inclusion studies of the porphyry copper ore body at Bingham, Utah. Econ. Geol., 69, 631-645. (443, 452)

Morey, G.W. and E. Ingerson (1937) Pneumatolytic and hydrothermal alteration and synthesis of the silicates. Econ. Geol., 32, supplement, 607-761. (231)

Myaz', N.I. and B.A. Simkiv (1965) A cooling microchamber for the detection of CO2 in gas-liquid inclusions. In Mineralogical Thermometry and Barometry. "Nauka" Press, Moscow, 171-172 (in Russian). (190)

Mysen, B.O. (1977) The solubility of H2O and CO2 under predicted magma genesis conditions and some petrological and geophysical implications. Rev. Geophys. Space Phys., 15, (3), 351-361. (504)

____ (1981) Rare earth element partitioning between minerals and (CO2 + H2O) vapor as a function of pressure, temperature, and vapor composition. Carnegie Inst. Washington Year Book, 80, 347-349. (526)

Nabelek, P.I., L.A. Taylor and G.E. Lofgren (1978) Nucleation and growth of plagioclase and the development of textures in a high-alumina basaltic melt. Proc. Lunar Sci. Conf., 9th, 725-741. (539)

Nacken, R. (1921) Welche Folgerungen ergeben sich aus dem Auftreten von Flüssigkeitseinschlüssen in Mineralien? Centralbl. Mineral. 1921: 12-20, 35-43. (279)

Nadeau, J.E. (1967) Temperatures of Fluorite Mineralization by Fluid Inclusion Thermometry, Sweetwater Barite District, East Tennessee. Master's thesis, Univ. of Tennessee, Knoxville, TN 36 p. (186)

____ (1968) Temperatures of fluorite mineralization by fluid inclusion thermometry. Sweetwater barite district, East Tennessee (abstr.). Geol. Soc. Am. Spec. Paper 115, 490. (186)

Nagano, K., S. Takenouchi and H. Imai (1977) Fluid inclusion study of the Mamut porphyry copper deposit, Sabah, Malaysia. Mineral. Geol. (Japan), 27, 201-212 (in English). (276)

Nahnybida, C. I. Hutcheon and J. Kirker (1982) Diagenesis of the Nisku formation and the origin of the late-stage cements. Canadian Mineral., 20, 129-140. (320)

Naldrett, A.J., E.L. Hoffman, A.H. Green and S.R. Naldrett (1979) The composition of Ni-sulfide ores: constraints on ore genesis. Bull. Mineral., 102, 455-462. (475)

Nambu, M., T. Sato, N. Hayakawa and Y. Ohmori (1977) On the microanalysis of fluid inclusions with the ion microanalyzer (abstr.). Mining. Geol. (Japan), 27, p. 40 (in Japanese; extended English summary in Fluid Inclusion Res. -- Proc. of COFFI, 10, 326-329). (148)

____ and K. Hara (1978) Fluid inclusion study of Inari copper vein, Ani mine, Akita Perfecture. Mining Geol. (Japan), 28, 25-34 (in Japanese; English abstr.). (186, 190, 457)

Narr, W. and R.C. Burruss (1982) Origin of reservoir fractures in Little Knife Field, North Dakota (abstr.). Am. Assoc. Pet. Geol. Bull., 66, 612. (334)

____ and J.B. Currie (1982) Origin of fracture porosity - example from Altamont field, Utah. Am. Assoc. Petrol. Geol. Bull., 66, 1231-1247. (272, 334)

Nash, J.T. (1972) Fluid-inclusion studies of some gold deposits in Nevada. U.S. Geol. Surv. Prof. Paper, 800-C, C15-C19. (427)

____ (1976) Fluid-inclusion petrology--data from porphyry copper deposits and applications to exploration. U.S. Geol. Survey Prof. Paper, 907D, 16 p. (443, 444, 469)

____ and C.G. Cunningham, Jr. (1973) Fluid-inclusion studies of the fluorspar and gold deposits, Jamestown district, Colorado. Econ. Geol., 68, 1247-1262. (437, 438, 470)

____ and ____ (1974) Fluid-inclusion studies of the porphyry copper deposit at Bagdad, Arizona. U.S. Geol. Survey J. Res., 2, 31-34. (443)

Naslund, H.R. and T.G. Theodore (1971) Ore fluids in the porphyry copper deposit at Copper Canyon, Nevada. Econ. Geol., 66, 385-399. (441, 443)

Naslund, H.R. (1983) The effect of oxygen fugacity on liquid immiscibility in iron-bearing silicate melts. Am. J. Sci., 283, 1034-1059. (477)

Naumov, G.B., O.F. Mironova, V.B. Naumov (1976) Carbon compounds in fluid inclusions of hydrothermal quartz. Geokhimiya, 1976, (8), 1243-1251 (in Russian; translated in Geochem. Int'l, 13, 164-171). (4)

Naumov, V.B. (1976) Results obtained in laboratories of the USSR on thermometric measurements of inclusions in standard samples of quartz. Geokhimiya, 1976, (7), 1109-1112 (in Russian; translated in Fluid Inclusion Res. -- Proc. of COFFI, 8, 208-211). (202)

____ and S.D. Malinin (1968) A new method of determination of pressure by means of gas-liquid inclusions. Geokhimiya, 1968, (4), 432-441 (in Russian; translated in Geochem. Int'l, 5, 382-391, 1968). (279, 280)

____ and G.B. Naumov (1980) Mineral-forming fluids and the physicochemical laws of their evolution. Geokhimiya, 1980, (10), 1450-1460 (in Russian; translated in Geochem. Int'l, 1980, 104-113, 1981). (286)

____ and B.V. Shapenko (1978) Iron concentration in high-temperature chloride solutions (abstr.). COFFI, 12, 129. (445)

____ and V.V. Shapenko (1980) Concentration of iron in high-temperature chloride solutions according to data on fluid inclusions. Geokhimiya, 1980, (2), 231-238 (in Russian). (454)

____, V.S. Balitskiy and L.N. Khetchikov (1966) Correlation of the temperatures of formation, homogenization, and decrepitation of gas-fluid inclusions. Akad. Nauk SSSR, Dokl., 171, (1), 146-148 (in Russian). (70, 71, 277, 280)

Naydenov, B.M., E.Y. Polyvyannyi and V.G. Bogolepov (1978) The present state and perspectives of studies of variations of isotope composition of argon in gas-liquid inclusions in minerals. Geokhimiya, 1978, (12), 1866-1882 (in Russian). (128)

Nedachi, M. (1974) Mineralization of the Kohoku gold, silver and copper ore deposits, Miyagi Prefecture, Japan. Tohoku Univ., Sci. Reports, Ser. 3, 12, 331-394. (138)

____ (1980) Chlorine and fluorine contents of rock-forming minerals of the Neogene granitic rocks in Kyushu, Japan. Mining Geol. (Japan) Spec. Issue, (8), 39-48. (390)

Nelson, J.D. and E. Vey (1968) Relative cleanliness as a measure of lunar soil strength. J. Geophys. Res., 73, 3747-3764. (120)

Nelson, R.C. (1973) Fluid inclusions as a clue to diagenesis of carbonate rocks (abstr.). Geol. Soc. Am. Abstr. Programs, 5, 748. (319)

Nesbitt, B.E. and W.C. Kelly (1975) Fluid and magmatic inclusions in the carbonatite at Magnet Cove, Arkansas (abstr.). Geol. Soc. Am. Abstr. Programs, 7, 1212. (145)

____ and ____ (1977) Magmatic and hydrothermal inclusions in carbonatite of the Magnet Cove Complex, Arkansas. Contrib. Mineral. Petrol., 63, 271-294. (145, 407)

Newhouse, W.H. (1932) The composition of vein solutions as shown by liquid inclusions in minerals. Econ. Geol., 27, 419-436. (23, 114, 136)

____ (1933) The temperature of formation of the Mississippi Valley lead-zinc deposits. Econ. Geol., 28, 744-750. (23)

Newitt, D.M., M.U. Pai and N.R. Kuloor (1956) Carbon dioxide. In F. Din, ed., Thermodynamic Functions of Gases, Vol. 1, Butterworths, London, 102-134. (280, 281)

Newton, R.C., J.V. Smith and B.F. Windley (1980) Carbonic metamorphism, granulites and crustal growth. Nature, 288, 45-50. (380, 505)

Nichols, F.A. (1976) Spheroidization of rod-shaped particles of finite length. J. Mater. Sci., 11, 1077-1082. (59)

____ and M.M. Mullins (1965) Morphological changes of a surface of revolution due to capillarity-induced surface diffusion. J. Appl. Phys., 36, 1826-1835. (63)

Nicol, W. (1828) Observations on the fluids contained in crystallized minerals. Edinburgh New Philos. J., 5, 94-96. (3, 114)

____ (1829) On the cavities containing fluids in rock-salt. Edinburgh New Philos. J., ser. 2, 7, 111-113. (92)

Nielsen, C.H. and H. Sigurdsson (1979) Quantitative methods for electron microprobe analysis of sodium in natural and synthetic glasses. Am. Mineral., 66, 547-552. (141)

Nielsen, J.M. and F.G. Foster (1960) Unusual etch pits in quartz crystals. Am. Mineral., 45, 299-310. (24, 68)

Nooner, D.W., W.S. Updegrove, D.A. Flory, J. Oro', and G. Mueller (1973) Isotopic and chemical data of bitumens associated with hydrothermal veins from Windy Knoll, Derbyshire, England. Chem. Geol., 11, 189-202. (333)

Nordstrom, D.K. (1983) Preliminary data on the geochemical characteristics of groundwater at Stripa. Proc. Workshop Geol. Disposal of Radioactive Waste, Oct. 1982: In Situ Experiments in Granite. Organiz. Econ. Co-op Devel., Stockholm, 143-153. (401)

Norman, D.I. (1978) Analysis of Rb, Sr and Sr isotopes in fluid inclusion waters (abstr.). In Int'l Assoc. of the Genesis of Ore Deposits, 5th Symp., Snowbird, Alta, Utah, Program and Abstracts: (Alta, Utah), 135. (134)

_____ (1981) Gases in the mica minerals: A possible exploration tool for hydrothermal ore deposits (abstr.). Geol. Soc. Am. Abstr. Programs, 13, 520. (122, 467)

_____ (1983) Gases in epithermal Ag-Au ore fluids (abstr.). Geol. Soc. Am. Abstr. Programs, 15, 654. (122)

_____ and C. Bernhardt (1984) Analysis of gases in inclusions from evaporites, Salado Formation, New Mexico. Geochem. Cosmochim. Acta (in press). (315)

_____ and G.P. Landis (1980) Source of mineralizing components in hydrothermal ore fluids as evidenced by $^{87}Sr/^{86}Sr$ and stable isotope data from the Pasto Bueno deposit, Peru (abstr.). Geol. Soc. Am. Abstr. Programs, 12, 493. (134)

_____ and _____ (1983) Source of mineralizing components in hydrothermal ore fluids as evidenced by $^{87}Sr/^{86}/Sr$ and stable isotope data from the Pasto Bueno deposit, Peru. Econ. Geol., 78, 451-465. (457, 458)

Norris, R.J. and R.W. Henley (1976) Dewatering of a metamorphic pile. Geology, 4, 333-336. (339)

Norton, D. (1979) Transport phenomena in hydrothermal systems: the redistribution of chemical components around cooling magmas. Bull. Mineral., 102, 471-486. (401, 451)

_____ and L.M. Cathles (1979) Thermal aspects of ore deposition. In H.L. Barnes, ed., Geochemistry of Hydrothermal Ore Deposits, 2nd Edn. Wiley & Sons, New York, NY, 611-631. (451)

Nutt, D.C. (1961) Significance of gas inclusions in the study of glacier ice. Polarforschung, 5, (1/2), (in English) (pub. 1963). (116)

Oberheuser, G., H. Kathrein, G. Demortier, H. Gonška and P. Freund (1983) Carbon in olivine single crystals analyzed by the $^{12}C(d,p)^{13}C$ nuclear microprobe electron spectrography. Geochim. Cosmochim. Acta, 47, 1117-1129. (505)

O'Hara, P.F. (1980) Intersecting isograds in central Arizona: Evidence for CO_2-H_2O exchange during prograde metamorphism (abstr.). Geol. Soc. Am. Abstr. Programs, 12, 299-300. (354)

Ohle, E.L. (1959) Some considerations in determining the origin of ore deposits of the Mississippi Valley type. Econ. Geol., 54, 769-789. (424)

_____ (1980) Some considerations in determining the origin of ore deposits of the Mississippi Valley type - Part II. Econ. Geol., 75, 161-172. (424)

Ohmoto, H. (1968) The Bluebell Mine, British Columbia, Canada - Part I: Mineralogy, Paragenesis, Fluid Inclusions, and the Isotopes of Lead, Carbon, Oxygen, and Hydrogen; Part II: Chemistry of the Hydrothermal Fluids (Gases and Salts in Fluid Inclusions). Ph.D. dissertation, Princeton Univ., Princeton, NJ, 95 p. (124, 186, 438)

_____ and A.C. Lasaga (1982) Kinetics of reactions between aqueous sulfates and sulfides in hydrothermal systems. Geochim. Cosmochim. Acta, 46, 1727-1745. (421)

_____ and R.O. Rye (1970) The Bluebell mine, British Columbia, I, Mineralogy, paragenesis, fluid inclusions, and the isotopes of hydrogen, oxygen, and carbon. Econ. Geol., 65, 417-437. (186, 438, 439, 461)

Ohmoto, H. and R.O. Rye (1979) Isotopes of sulfur and carbon. In H.L. Barnes, ed., Geochemistry of Hydrothermal Ore Deposits. Wiley & Sons, New York, NY, 509-567. (127)

_____ and B.J. Skinner, eds. (1984) The Kuroko and related volcanogenic massive sulfide deposits. Econ. Geol. Monograph 5 (in press). (425)

O'Keefe, J.A. (1976) Tektites and Their Origin. Amsterdam, Elsevier Sci. Pub. Co., 254 p. (558, 560)

_____ P.D. Lowman, Jr. and K.L. Dunning (1962) Gases in tektite bubbles: Science, 137, 228. (102, 560)

_____ L.S. Walter and F.M. Wood, Jr. (1964) Hydrogen in a tektite vesicle. Science, 153, 39-40. (102, 560)

Olander, D.R., A.J. Machiels and S. Yagnik (1980) Thermal migration of brine inclusions in salt. Office of Nuclear Waste Isolation Tech. Rept. ONWI-208, 94 p. (66)

Olsen, E. and L. Grossman (1974) A scanning electron microscope study of olivine crystal surfaces. Meteoritics, 9, 243-254. (562)

_____ (1978) On the origin of isolated olivine grains in type 2 carbonaceous chondrites. Earth Planet Sci. Lett., 41, 111-127. (562)

O'Neil, J.R. (1971) [$^{13}C/^{12}C$ ratios of CO_2 from Hawaiian lava lake] (abstr.). U.S. Geol. Survey Prof. Paper 750, A-126. (528)

_____ and E.D. Jackson (1969) [Secondary nature of carbonates in Hawaiian basalt] (abstr.). U.S. Geol. Surv. Prof. Paper, 650, A-126. (528)

Oppenheim, V. (1948) The Muzo emerald zone, Columbia, S.A. Econ. Geol., 43, 31-38. (366)

Orr, F.M., Jr. (1983) [CO_2 flooding]. J. Petrol Tech., July 1983, 1285-1286. (245)

_____ A.D. Yu and C.L. Lien (1981) Phase behavior of CO_2 and crude oil in low-temperature reservoirs. Soc. Petrol. Eng. J., 21, 480-492. (245, 333)

Orville, P.M. (1972) Plagioclase cation exchange equilibria with aqueous chloride solution: Results at 700°C and 2000 bars in the presence of quartz. Am. J. Sci., 272, 234-272. (389, 400)

Ostapenko, G.T. and L.N. Khetchikov (1968) Equations for the liquid-vapor phase transition with changing temperature in closed systems at constant volume. Geokhimiya, 1968,(4), 485-488 (in Russian; abstr. translated in Geochem. Int'l, 5, 412, 1968). (87)

Oudin, E.,C. Fouillac and L. Le Bel (1981) Étude minéralogique et géochimique des dépôts sui fures sous-marins actuels de la ride Est-Pacifique (21°N). Documents du BRGM 25, 241 p. (495)

Ozima, M. and S. Zashu (1983a) He, Ar isotopes and K-Ar ages of diamonds (abstr.). Papers presented to the 8th Symp. on Antarctic Meteorites, Tokyo, 17-19 Feb., 1983, Nat'l Inst. Polar Res., Tokyo, 78. (510)

_____ (1983b) Primitive helium in diamonds. Science, 219, 1067-1068. (508)

_____ and O. Nitoh (1983) $^{3}He/^{4}He$ ratio, noble gas abundance and K-Ar dating of diamonds - an attempt to search for the records of early terrestrial history. Geochim. Cosmo. Acta, 47, 2217-2224. (508)

Padovani, E.R., S.B. Shirey and G. Simmons (1982) Characteristics of microcracks in amphibolite and granulite facies grade rocks from southeastern Pennsylvania. J. Geophys. Res., 87, (B10), 8605-8630. (21, 343, 345, 361)

Pagel, M. (1975) Cadre Géologique des Gisements d'Uranium dans la Structure Carswell (Saskatchewan, Canada). "Etude des Phases Fluides." These 3e cycle, Nancy, France, 157 p. (344)

_____ (1977) Microthermometry and chemical analysis of fluid inclusions from the Rabbit Lake uranium deposit, Saskatchewan, Canada (abstr.). Inst. Mining Metall. Trans., 86(B), B157-B158. (321)

_____ and H. Jaffrezic (1977) Chemical analyses of inclusions in quartz and dolomite from the Rabbit Lake uranium deposit (Canada). Methodology and genetic importance. Compt. Rend. Acad. Sci. Paris, 284(D), 113-116 (in French). (462)

Pagel, M., B. Poty and S.M.F. Sheppard (1980) Contribution to some Saskatchewan uranium deposits mainly from fluid inclusion and isotopic data. In J. Ferguson and A.B. Goleby, eds., Proc. Int'l Uranium Symp. on the Pine Creek Geosyncline. Int'l Atomic Energy Agency, Vienna, 639-654. (349, 462)

Palin, J.M. and D.I. Norman (1982) Volatiles in phylosilicates, Copper Flat porphyry deposit, southwest New Mexico—a potential exploration (sic.) tool (abstr.). Geol. Soc. Am. Abstr. Programs, 14, 223. (122, 467)

*Pal'mova, L.G. (1972) Correlation of formation and decrepitation temperatures of synthetic quartz, calcite, and pyrite. COFFI, 5, 81. (266)

Panina, L.I. (1966) Certain data concerning temperature conditions of formation of the Synnyr alkaline massif. Akad. Nauk SSSR, Dokl., 170, (6), 1411-1413 (in Russian; translated in Dokl. Acad. Sci. USSR, Earth Sci. Sects., 170, 173-175, 1967). (89)

Panov, V.V. (1975) Paleo-temperature studies on gas-liquid inclusions in halite. Akad. Nauk BSSR Dokl., 19, 257-260. (316)

*Pashkov, Yu.N. and G.O. Piloyan (1973) On the theory of the decrepitation method. COFFI, 6, 119. (266)

_____, A.V. Timofeev and E.I. Kotov (1968) Analysis of possible errors in using temperatures of homogenization. In Mineralogical Thermometry and Barometry, Vol. 2. "Nauka" Press, Moscow, 236-243 (in Russian). (186)

Pasteris, J.D. (1981a) Occurrence of graphite in serpentinized olivines in kimberlite. Geology, 9, 356-359. (504, 505)

_____ (1981b) Significance of sulfide textures and mineralogy in mantle xenoliths (abstr.). Geol. Soc. Am. Abstr. Program, 7, 526. (504, 505)

_____ (1981c) Kimberlites: Strange bodies? EOS, Trans. Am. Geophys. Union, 62, 713-716. (504, 511)

_____ (1983) Adaptation of SGE-USGS heating-freezing stage for operation down to -196°C. Econ. Geol., 78, 164-169. (197)

_____, R. Patel, S.C. Bergman and F. Adar (1983) Comparative spectroscopy and microthermometry on fluid inclusions in a mantle xenolith (abstr.). EOS, Trans. Am. Geophys. Union, 64, 340. (84)

_____ and K.R.S.S. Kekulawala (1979) The role of water in quartz deformation. Bull. Mineral., 102, 92-98. (74)

Patterson, D.J., H. Ohmoto and M. Solomon (1981) Geologic setting and genesis of cassiterite-sulfide mineralization at Renison Bell, western Tasmania. Econ. Geol., 76, 393-438. (457)

Patty, F.A., ed. (1962) Industrial Hygiene and Toxicology (2nd edition). Interscience Pub., New York, NY, 1548 p. (116)

Peacor, D.R., W.C. Kelly and D.F. Blake (1983) Growth strain, optical anisotropy and fluid entrapment in fluorite (abstr.). Geol. Soc. Am. Abstr. Programs, 15, 660. (24)

Pecher, A. (1981) Experimental decrepitation and reequilibration of fluid inclusions in synthetic quartz. Tectonophysics, 78, 567-583. (61, 71, 74)

_____ and A.M. Boullier (1983) Reequilibration of fluid inclusions under high confining pressure (abstr.). Program Symp. European Current Research on Fluid Inclusions, April 6-8, 1983, Orleans, France, 50. (74)

Peck, J.H. (1982) Fluid inclusion studies for dating fault movement. In O.C. Farquhar, ed., Geotechnology in Massachusetts. Univ. of Massachusetts Graduate School, Amherst, MA, 443-446. (355)

Peck, L.C. and V.C. Smith (1970) Quantitative chemical analysis of lunar samples. Science, 167, 532. (541)

*Pedan, M.V., Yu.L. Gritsay and Kh.U. Koval'chuk (1978) Gas-liquid inclusions as source of salt in water during dressing process of ferruginous quartzites in the Krivoy Rog basin (abstr.). COFFI, 12, 139. (111, 349)

Pedersen, A.K. (1979) Basaltic glass bodies with high-temperature equilibrated immiscible sulfide bodies with native iron from Disko, Central West Greenland: Contrib. Mineral. Petrol., 69, 397-407. (475)

Pering, K.L. (1973) Bitumens associated with lead, zinc and fluorite ore minerals in North Derbyshire, England. Geochim. Cosmochim. Acta, 37, 401-417. (333)

Perry, E.C., Jr. and C.W. Montgomery, eds. (1982) Isotope Studies of Hydrologic Processes. Northern Illinois Univ. Press, De Kalb, IL, 118 p. (27)

Perthuisot, V., N. Guilhaumou and J.-C. Touray (1978) Les inclusions fluids hypersalines et gazeuses des quartz et dolomites du Trias évaporitique Nord-tunisien. Essai d'intérpretation géodynamique. Bull. Soc. Géol. Fr., 20, 145-155. (347, 349)

Pesheck, P.S., L.E. Scriven and H.T. Davis (1981) Cold stage scanning electron microscopy of crude oil and brine in rock. Scanning Electron Micros., 1981, 515-524. (147)

Petersen, E.U., J.W. Valley and E.J. Essene (1981) Fluorphlogopite and fluortremolite in Adirondack marbles: phase equilibria and C-O-H-F-S fluid compositions (abstr.). EOS, Trans. Am. Geophys. Union, 62, 442. (363)

Petersil'e, I.A. (1961) The presence of gas in rocks of the Khibiny tundra. In tundr. Akad. Nauk SSSR, Kol'sk. Filial, Moscow-Leningrad, 89-102 (in Russian). (405)

_____ (1962) Origin of hydrocarbon gases and dispersed bitumens of the Khibina alkalic massif. Geokhimiya, 1962, (1), 15-29 (in Russian; translated in Geochemistry, (1), 14-30, 1962). (405)

_____ (1963) Organic substances in the igneous and metamorphic rocks of the Kola Peninsula. In A.P. Vinogradov, ed., Khimia Zemnol kory. Vol. 1 (Chemistry of the Earth's Crust, Vernadsky Centenary Volume). Akad. Nauk SSSR, Moscow, 48-62 (in Russian; English translation Israel Program Sci. Transl., Jerusalem, 1966). (405)

Petersilie, I.A. and H. Sørensen (1970) Hydrocarbon gases and bituminous substances in rocks from the Ilímaussaq alkaline intrusion, South Greenland. Lithos, 3, 59-76. (125, 405, 528)

Petersilje, I.A. and W.A. Pripachkin (1979) Hydrogen, carbon, nitrogen and helium in gases of igneous rocks. In L.H. Ahrens, ed., Origin and Distribution of the Elements. Proc. 2nd Symp., Paris, May 1977. Pergamon Press, Oxford, 541-545. (405)

Petrichenko, O.I. (1973) Methods of Study of Inclusions in Minerals of Saline Deposits. "Naukova Dumka" Pub. House, Kiev, 92 p. (in Ukrainian; translated in Fluid Inclusion Res. - Proc. of COFFI, 12, 1979). (29, 31, 69, 73, 111, 135, 137, 143, 215, 314, 316)

_____ (1977) Atlas of Microinclusions in the Minerals of Saline Rocks. Izdat. Naukova Dumka, Kiev, 182 p. (in Russian). (23, 309, 313, 316)

_____ (1978) Nature of carbon dioxide inclusions in halite from metamorphosed salt-bearing beds. COFFI, 12, 140-141. (315)

_____ and V.S. Shaidetskaia (1976) Inclusions in halite from veins in diabase. COFFI, 9, 104. (137)

_____ and V.S. Shaydetskaya (1968) Physico-chemical conditions of recrystallization of halite in rock salts. In Mineralogical Thermometry and Barometry, Vol. 1. "Nauka" Press, Moscow, 348-351 (in Russian). (102, 135, 136, 137)

_____ (1973) Determination of Eh and chemical composition of solutions of individual inclusions. COFFI, 6, 121-122. (137, 314)

_____ and E.P. Slivko (1973a) On the conditions of formation of Permian salt deposits of Donbass. COFFI, 7, 167. (137, 314, 316)

_____ (1973b) On the mineral formation conditions during the formation of Permian salt deposits of the Donbas. L'vov. Geol. Obshch. Mineral. Sbornik, 27, (3), 263-274 (in Russian). (137, 314)

_____ (1974) The effect of clayey material on the change in microelement content in salty solutions (abstr.). Abstr. Int'l Symp. on Water-Rock Interaction, Czechoslovakia, Sept. 9-17, 1974, Geol. Survey, Prague, 33 (in English). (315)

Petrichenko, O.I.; V.M. Kovalevich and V.N. Chalyi (1974) Geochemical conditions of salt origin in Tortonian evaporite basin at NW Predkarpat'ye. Geologiya i geokhimiya goryuchikh iskopaemykh, 41, 74-80 (in Russian). (137, 314)

Petrovic, R. (1969) Coalescence of fluid inclusions and their removal from minerals exposed to confining pressure greater than the pressure of the fluid inclusions (abstr.). Fluid Inclusion Res. -- Proc. of COFFI, 2, 20. (66, 67)

Petrovskaya, N.V. (1973) Native Gold. "Nauka" Press, Moscow, 348 p. (in Russian). (28)

_____, M.M. Elinson and L.A. Nikolaeva (1971) Composition and formation conditions of gas inclusions in native gold (abstr.). Int'l Geochemical Congress, Moscow, 1971. Abstr. of Reports, 1, 326-327 (in English). (28, 73, 357)

* _____ and _____ (1973) Composition and formation of gas inclusions in native gold. COFFI, 7, 167. (357)

Pfaff, Fr. (1871) Über den Gehalt der Gesteine an mechanische eingeschlossenem Wasser und Kochsalz. Annalen Physik Chemie (Poggendorffs Annalen), 143, (ser. 5, v. 23), 610-620. (117)

Phillips, W. (1875) The rocks of the mining districts of Cornwall and their relation to metalliferous deposits. Geol. Soc. London Quart. J., 31, 319-345. (3, 186)

Philpotts, A.R. (1967) Origin of certain iron titanium oxide and apatite rocks. Econ. Geol., 62, 303-315. (385)

_____ (1981) Liquid immiscibility in silicate melt inclusions in plagioclase phenocrysts. Bull. Mineral., 104, 317-324, (476)

_____ (1982) Compositions of immiscible liquids in volcanic rocks. Contrib. Mineral. Petrol., 80, 201-218. (477)

Picciotto, E.E. (1950) Distribution de la radioactivité dans les roches éruptives. Soc. belge géol. Bull., 59, 170-198. (114)

Pichavant, M., C. Ramboz and A. Weisbrod (1982) Fluid immiscibility in natural processes: use and misuse. I. Phase equilibria analysis--a theoretical and geometrical approach. Chem. Geol., 37, 1-27. (248, 371)

Pinckney, D.M. and J. Haffty (1970) Content of zinc and copper in some fluid inclusions from the Cave-In-Rock district, southern Illinois. Econ. Geol., 65, 451-458. (423)

_____ and R.O. Rye (1972) Variation of $18O/16O$, $13C/12C$, texture, and mineralogy in altered limestone in the Hill mine, Cave-in-[Rock] District, Illinois. Econ. Geol., 67, 1-18. (422)

Pineau, F. M. Javoy and Y. Bottinga (1976) $13C/12C$ ratios of rocks and inclusions in popping rocks of the Mid-Atlantic Ridge and their bearing on the problem of isotopic composition of deep-seated carbon. Earth Planet. Sci. Lett., 29, 413-421. (490, 528)

_____, F. Behar and J. Touret (1981) The isotopic geochemistry of the Bamble granulite facies (Norway) and the origin of the deep crust carbonic fluids. Bull. Mineral., 104, 630-641 (in French). (528)

Piperov, N.B., N.P. Penchev and I.K. Bonev (1977) Primary fluid inclusions in galena crystals, II. Chemical composition of the liquid and gas phase. Mineralium Deposita, 12, 77-89. (72, 179)

Pisutha-Arnond, V. and H. Ohmoto (1980) Chemical and isotopic compositions of the Kuroko ore-forming fluids (abstr.). Geol. Soc. Am. Abstr. Programs, 12, 500-501. (425, 426)

†Piznyur, A.V. (1957) The genetic connection, in inclusions of minerals, of quartz veins of the Barsukchi deposit with the intrusion of granites. Vses. Nauchno-Issled. Inst. P'ezooptichesk. Mineral. Syr'ya Trudy, 1, (2), 135-143 (in Russian). (111, 464)

_____ (1968) On pressure during the formation of the Zhireken copper-molybdenum deposit (East Transbaikalia). Akad. Nauk SSSR, Dokl., 179, (5), 1186-1188 (in Russian). (276)

Planetary Sciences Unit, Univ. of Cambridge (1982) Mantle methane-fool's gold? Nature, 300, 312-313. (512)

Plank, R. and J. Kuprianoff (1929) Thermal qualities of carbon dioxide in gaseous, liquid, and solid state. Z. tec. Physik, 10, 93-101 (see also Zeit. Gesell. Kalte-Ind. Beihefte 1, 1, 1929). (230)

Plass, G.N. and V.R. Stull (1963) Carbon dioxide absorption for path lengths applicable to the atmosphere of Venus. J. Geophys. Res., 68, 1355-1363. (523)

Plumlee, G.S., C.J. Hackbarth and A.R. Campbell (1983) Applications of infrared microscopy to ore deposit research (abstr.). Geol. Soc. Am. Abstr. Programs, 15, 662. (84, 161)

*Pokhilenko, N.P. and L.V. Usova (1978) Secondary melt inclusions in olivines from kimberlite pipe "Udachnaya" (Yakutia), (abstr.). COFFI, 11, 165. (511)

Poland, E.L. (1982) Stretching of Fluid Inclusions in Fluorite at Confining Pressures up to 1 Kilobar. M.S. thesis, Univ. of California, Berkeley, CA, 90 p. (71, 72, 286)

Poliak, B.G., E.M. Prasolov and V. Čermák (1982) Mantle helium in "juvenile" fluid and the nature of the geothermal anomaly of Krasné Hory (Czechoslovakia). Akad. Nauk SSSR, Dokl., 263, (3), 701-705 (in Russian). (513)

Pomareanu, V. (1959) Apparatus for the determination of the temperature of homogenization of fluid inclusions in minerals. Analele Stiintifice Univ. "Al I. Cuza," Iaşi, new ser., sec.2 (Stiinte nat.), 5, fasc. 1, 119-124 (in Romanian; French summary). (186)

Popivnyak, I.V. and Ye.Ye. Laz'ko (1979) Inclusions of soldified melt in minerals of kimberlitic rocks of western Yakutia. Akad. Nauk SSSR, Dokl., 244, 194-197 (in Russian; translated in Dokl. Acad. Sci. USSR, 244, 86-89, 1981). (511)

Potter, A.P. (1975) Silicate Liquid Inclusions in Olivine Crystals from Kilauea, Hawaii. M. Sc. thesis, Queen's Univ., Kingston, Ontario, Canada. (141, 142)

Potter, II, R.W. (1977) Pressure corrections for fluid-inclusion homogenization temperatures based on the volumetric properties of the system NaCl-H2O. U.S. Geol. Survey J. Res. 5, 603-607. (263, 264, 356, 366)

_____ and D.L. Brown (1975) The volumetric properties of aqueous sodium chloride solutions from 0° to 500°C at pressures up to 2000 bars based on a regression of the available literature data. U.S. Geol. Survey Open-File Rept. No. 75-636, 31 p. (234, 263)

_____ and _____ (1977) The volumetric properties of aqueous sodium chloride solutions from 0° to 500°C and pressures up to 2000 bars based on a regression of available data in the literature. U.S. Geol. Survey Bull. 1421-C, 36 p. (228, 274, 276)

_____ and M.A. Clynne (1978a) Solubility of highly soluble salts in aqueous media--part 1. NaCl, KCl, CaCl2, Na2SO4, and K2SO4 solubilities to 100°C. J. Res. U.S. Geol. Survey, 6, 701-705. (250, 288)

_____ and _____ (1978b) Pressure correction for fluid inclusion homogenization temperatures (abstr.). Int'l Assoc. on the Genesis of Ore Deposits, 5th Symp., Snowbird, Alta, Utah, Program and Abstr., (Alta, Utah), 146. (263, 268, 288)

_____ and _____ (1978c) The solubility of the noble gases He, Ne, Ar, Kr, and Xe in water up to the critical point. J. Solution Chem., 7, 837-844. (340)

_____, D.R. Shaw and J.L. Haas, Jr. (1975) Annotated bibliography of studies on the density and other volumetric properties for major components in geothermal waters, 1928-1974. U.S. Geol. Survey Bull. 1417, 78 p. (289)

_____, R.S. Babcock and D.L. Brown (1977) A new method for determining the solubility of salts in aqueous solutions at elevated temperatures. J. Research, U.S. Geological Survey, 5, (3), 389-395. (235, 250, 286)

_____, M.A. Clynne and D.L. Brown (1978) Freezing point depression of aqueous sodium chloride solutions. Econ. Geol., 73, 284-285. (232, 233)

Prikazchikov, L.A. (1959) On tube canals in morion crystals of pegmatites from Volhynia. Vses. Mineral. Obshch. Zapiskt, 88, 99-102 (in Russian). (2)

—, Yu.G. Sorokin, A.A. Moskalyuk and A.S. Vesel'ev (1964) Giant crystals of quartz from a pegmatite body. Vses. Mineral. Obshch. Zapiski, 93, 212-219 (in Russian). (2, 135, 136)

*Prokhorov, V.G. A.E. Miroshnikov and I.A. Khayretdinov (1968) Certain problems in comprehensive utilization of physico-chemical methods in research, COFFI, 1, 21. (465)

Prone, A. (1981) Utilization des quartz bipyramides et de leurs inclusions comme marqueurs paleogeographiques (exemple du bassin provençal-France) (abstr.). Program, Symposium on Current Research on Fluid Inclusions, Univ. of Utrecht, Utrecht, The Netherlands, April 22-24, 1981 (unpaginated). (308)

Puchner, H.F. and H.D. Holland (1966) Studies in the Providencia area, Mexico; III Neutron activation analyses of fluid inclusions from Noche Buena. Econ. Geol., 61, 1390-1398. (134)

Pulou, R. and C. Baudracco-Gritti (1978) A decrepitometric recorder with electronic filter. Bull. Mineral., 101, 402-405 (in French). (212)

Puzanov, L.S. and A.I. Partsevsky (1978) Genetic type of the Seligdar apatite deposit (Central Aldan). Akad. Nauk SSSR, Dokl., 243, (1), 179-182 (in Russian). (407)

— and Ye.N. Kashintseva (1978) On two varieties of high-temperature anhydrite at the Seligdar apatite deposit. Akad. Nauk SSSR, Dokl., 242, (5), 1170-1172 (in Russian). (407)

Qi, X.L. (1983) High dense CO_2 fluid inclusions in peridotitic nodules in some alkali-basalts from the east of China (abstr.). Program Symp. on European Current Research on Fluid Inclusions, Orleans, France, April 6-8, 1983, p. 55. (515)

Quinn, E.L. and C.L. Jones (1936) Carbon Dioxide. Am. Chem. Soc. Monogr. 72, Rheinhold, New York. (85, 241, 280, 517)

Raade, G. and J. Haug (1980) Rare fluorides from a soda granite in the Oslo region, Norway. Mineral. Record, 11, 83-91. (390)

Radtke, A.S., R.O. Rye and F.W. Dickson (1980a) Geology and stable isotope studies of the Carlin gold deposit. Econ. Geol., 75, 641-672. (426, 427)

—, R.W. Wittkopp and C. Heropoulos (1980b) Genesis of gold-bearing quartz veins of the Alleghany district, Calif. (abstr.). Geol. Soc. Am. Abstr. Program, 12, 148. (356)

Rama, S.N.I., S.R. Hart and E. Roedder (1965) Excess radiogenic argon in fluid inclusions. J. Geophys. Res., 70, 509-511. (119, 128, 340, 430)

Raman, C.V. and K.S. Krishnan (1928) A new type of secondary radiation. Nature, 121, 501-502. (104)

Ramboz, C. (1979) A fluid inclusion study of the copper mineralization in Southwest Tintic District (Utah). Bull. Mineral., 120, 622-632. (443, 444, 445, 452)

—— (1980) Problems related to the analysis of the composition of complex carbonic fluids by microthermometric techniques. Compt. Rend. Acad. Sci. Paris, ser. D, 290, 499-502 (in French). (244)

—, M. Pichavant and A. Weisbrod (1982) Fluid immiscibility in natural processes. Use and misuse of fluid inclusion data. II. Interpretation of fluid inclusion data in terms of immiscibility. Chem. Geol., 37, 29-48. (371)

Rankin, A.H. (1975) Fluid inclusion studies in apatite from carbonatites of the Wasaki area of western Kenya. Lithos, 8, 123-136. (408, 409)

—— (1977) Fluid-inclusion evidence for the formation conditions of apatite from the Tororo carbonatite complex of eastern Uganda. Mineral. Mag., 41, 155-164. (161, 408)

— and D.H.M. Alderton (1983) Fluid inclusion petrography of SW England granites and its potential in mineral exploration. Mineralium Deposita, 18, 335-347. (467)

Pottorf, R.J., J.B. Murowchick and H.L. Barnes (1977) Fluid inclusion analyses, Appendix A, 6 p., prepared for Stone and Webster Engrg. Corp., Fitzpatrick Nuclear Power Plant site, Job. No. E-19214. Submitted in support of proceedings for U.S. Nuclear Regulatory Comm. Docket No. 50333-550/570, Feb. 1978. (355)

Poty, B. (1966) Solid inclusions and "mineralogical plumb-lines": the age of the "La Gardette" quartz vein (Isère). Sci. de la Terre, 11, 41-53 (in French). (27)

—— (1968) La Croissance des Cristaux de Quartz dans les Filons sur l'Exemple du Filon de la Gardette (Bourg d'Oisans) et des Filons du Massif du Mont Blanc. Thesis, Univ. of Nancy I, Nancy, France, Dec. 15, 1967, 209 p. (190)

—— (1969) La croissance des cristaux de quartz dans les filons sur l'exemple du filon de la Gardette (Bourg-d'Oisans) et des filons du massif du Mont-Blanc. Sci. de la Terre Mem., 17, 162 p. (27, 161, 355)

— and H.A. Stalder (1970) Cryometric determination of the salt and gas contents of solutions contained in quartz crystals from fissures in the Swiss Alps. Schweiz. Mineral. Petrogr. Mitt., 50, (1), 141-154 (in German). (99, 349, 354)

— and A. Weisbrod (1974) Fluid inclusion studies in quartz from fissures of Western and Central Alps. Schweiz. Mineral. Petrogr. Mitt., 54, 717-752. (351, 363)

—, J. Leroy and L. Jachimowicz (1976) A new device for measuring temperatures under the microscope: the Chaixmeca microthermometry apparatus. Bull. Soc. franc. Mineral. Cristallogr., 99, 182-186 (in French; translated in Fluid Inclusion Res.---Proc. of COFFI, 9, 173-178). (104, 186, 190, 191, 194, 203, 204, 205, 241)

Powell, R. (1982) Fluids and melting under upper amphibolite facies conditions (abstr.). Geol. Soc. Newsletter (London), 11, (4), 15. (383)

Powers, D.W., S.J. Lambert, S.E. Shaffer, L.R. Hill and W.D. Weart, eds., (1978) Geological characterization report, Waste Isolation Pilot Plant (WIPP) site, southeastern New Mexico. Sandia Laboratories, SAND, 78-1596. (317, 318)

Preece, III, P.R. and R.E. Beane (1982) Contrasting evolutions of hydrothermal alteration in quartz monzonite and quartz diorite wall rocks at the Sierrita porphyry copper deposit, Arizona. Econ. Geol., 77, 1621-1641. (23, 443, 450)

Preisinger, A. and W. Huber (1964) Zur Bestimmung kleinster Gaseinschlüsse in Feldspaten (abstr.). Fortschr. Mineral., 41, 183. (124)

Price, L.C. (1976) Aqueous solubility of petroleum as applied to its origin and primary migration. Am. Assoc. Petrol. Geol. Bull., 60, 213-244. (116, 333, 422)

—— (1979) Aqueous solubility of methane at elevated pressures and temperatures. Am. Assoc. Petrol. Geol. Bull., 63, 1527-1533. (248, 330)

—— (1981a) Primary petroleum migration by molecular solution: consideration of new data. J. Petrol. Geol., 4, 89-101. (333)

—— (1981b) Aqueous solubility of crude oil to 400°C and 2000 bars pressure in the presence of gas. J. Petrol. Geol., 4, 195-223. (116, 333)

—— (1982) Organic geochemistry of core samples from an ultra-deep hot well (300°C, 7 km). Chem. Geol., 37, 215-228. (347)

—, J.L. Clayton and L.L. Rumen (1981) Organic geochemistry of the 9-6 km Bertha Rogers No. 1 well, Oklahoma. Organic Geochem., 3, 59-77. (347)

—, L.M. Wenger, T. Ging and C.W. Blount (1983) Solubility of crude oil in methane as a function of pressure and temperature. Organic Geochem., 4, 201-221. (333)

Price, N.J. (1975) Fluids in the crust of the earth. Sci. Progress, 62, 59-87. (424)

Price, W.F., A.T. Huntingdon and D.K. Bailey (1977) The effect of crushing on the release of volatile components from heated obsidian. Mineral. Mag., 41, 551-553. (120)

Rankin, A.H. and R.T.H. Aldous (1979) Tritolyl phosphate--a suitable immersion oil for fluid inclusion freezing-stage studies. Mineral. Mag., 43, 315-316. (153)

___ and F. Greenaway (1978) Macroscopic inclusions of fluid in British fluorites from the mineral collection of the British Museum (Natural History). Bull. British Mus. Nat. Hist. (Geol.), 30, 295-325. (2, 172, 417)

___ and M.J. Le Bas (1974a) Nahcolite (NaHCO3) in inclusions in apatites from some E. African ijolites and carbonatites. Mineral. Mag., 39, 564-570. (113, 408)

___ and ___ (1974b) Liquid immiscibility between silicate and carbonate melts in naturally occurring ijolite magma. Nature, 250, 206-209. (408, 409, 410)

___, D.H.M. Alderton, M. Thompson and J.E. Goulter (1982) Determination of uranium/carbon ratios in fluid inclusion decrepitates by inductively coupled plasma emission spectroscopy. Mineral. Mag. 46, 179-186. (132, 462)

Rasumny, J. (1960) Detection of free CO2 gas in minerals. Int'l Geol. Congress, 21st, Copenhagen 1960, Reports, pt. 1, 27-29 (in French). (116)

Ravich, M. and F. Borovaya (1949) Phase equilibria in ternary water-salt systems at elevated temperatures. Izvest. Sektora. Fiz.-Khim. Anal., Inst. Obshch. i Neorg. Khim. Akad. Nauk SSSR, 19, 68-81 (in Russian; translated in Fluid Inclusion Res.-- Proc. of COFFI 10, 330-340, 1977). (245, 278)

___ and ___ (1960) Crystallization of melts of chlorides of K and Na in the presence of water vapor. Izvest. Sektora. Fiz.-Khim. Anal. Inst. Obshch. i Neorg. Khim., Akad. Nauk SSSR, 20, 165-183 (in Russian). (245)

Reed, M.H. (1983) Seawater-basalt reaction and the origin of greenstones and related ore deposits. Econ. Geol., 78, 466-485. (496)

Reese, C.L. (1898) Petroleum inclusions in quartz crystals. Am. Chem. Soc. J., 20, 795-797. (83)

Reid, A.M, and T.E. Bunch (1975) The nakhlites - II. Where, when, and how? Meteoritics, 10, 317-324. (569)

Rekharsky, V.I., ed. (1980) Methods and devices for studies of inclusions of mineral-forming media. "Nauka" Pub. House, Moscow, 200 p. (in Russian). (5)

Rex, R.W. (1983) The origin of the brines of the Imperial Valley, California Trans. Geothermal Resources Council, 7, 321-324. (498)

Reyf, F.G. (1973) Inclusions of melt in quartz of postorogenic granites of central Buryatiya and the pressures and temperatures accompanying their formation. Akad. Nauk SSSR, Dokl., 213, (4), 918-921 (in Russian; translated in Dokl. Acad. Sci. USSR, 213, 172-174, 1975; abstr. in Int'l Geol. Rev., 16, (2), 241, 1974). (385)

Reynolds, J.H. (1960) Rare gases in tektites. Geochim. Cosmochim. Acta, 20, 101-114. (120)

Reynolds, T.J. and R.E. Beane (1984) The evolution of hydrothermal fluid characteristics through time at Santa Rita, New Mexico, porphyry copper deposit. Econ. Geol. (in press). (443)

Rhodes, J.M., M.A. Dungan, D.P. Blanchard and P.E. Long (1979) Magma mixing at mid-ocean ridges: Evidence from basalts drilled near 22°N on the Mid-Atlantic Ridge. In J. Francheteau, ed., Processes at Mid-Ocean Ridges, Tectonophysics, 55, 35-61. (482, 484)

Rhodes, R.G. (1950) A low-temperature microscope stage. Rev. Sci. Instruments, 27, 333-334. (191)

Ricci, J.E. (1951) The Phase Rule. D. Van Nostrand Co., New York, NY, 505 p. (231, 233)

Rich, R.A. (1979) Fluid inclusion evidence of Silurian evaporites in southeastern Vermont. Geol. Soc. Am. Bull., 90, Pt. 2, 1628-1643 (on microfiche; summary in Geol. Soc. Am. Bull., Pt. 1, 90, 901-902). (349, 350)

Richardson, S.M. and H.Y. McSween, Jr. (1978) Textural evidence bearing on the origin of isolated olivine crystals in C2 carbonaceous chondrites. Earth Planet Sci. Lett., 37, 485-491. (563)

Richter, D.H. and J.F. Abell (1953) A simple high-temperature microscope heating stage. Am. Mineral., 38, 1269-1271. (186, 187, 206)

Rickard, D.T., M. Willden, Y. Marde and R. Ryhage (1975) Hydrocarbons associated with lead-zinc ores at Laisvall, Sweden. Nature, 255, 131-133. (333)

Riley, J.P. (1975) Tables of physical and chemical constants relevant to marine chemistry. Appendix to Vol. 4, 2nd Ed., J.P. Riley and G. Skirrow, eds. Chemical Oceanography. Academic Press, London. (329)

Ringwood, A.E. (1977) Synthesis of pyrope-knorringite solid solution series. Earth Planet Sci. Lett., 36, 443-448. (507)

Ripley, E.M. and H. Ohmoto (1977a) Mineralogic, sulfur isotope, and fluid inclusion studies of the stratabound copper deposits at the Raul mine, Peru. Econ. Geol., 72, 1017-1041. (461)

___ and ___ (1977b) Oxygen and hydrogen isotopic studies of ore deposition and metamorphism at the Raul mine, Peru. Geochim. Cosmochim. Acta, 43, 1633-1643. (461)

Rison, W. and H. Craig. (1981) Loihi seamount: Mantle volatiles in the basalts (abstr.). EOS, Trans. Am. Geophys. Union, 62, 1083. (495)

Robert, F., L. Merlivat, and M. Javoy (1981) Hydrogen isotopic composition of fluid inclusions in the St.-Severin meteorite (abstr.) Terra Cognita, Spring, 1981, 47-48. (569)

Roberts, N.K. and G. Zundel (1979) IR studies of long-range surface effects -- excess proton mobility in water in quartz pores. Nature, 278, 726-728. (38)

Roberts, S.A. (1973) Pervasive early alteration in the Butte district, Montana (abstr.) Econ. Geol. 68, 909-910. (445)

Robie, R.A., P.M. Bethke, M.S. Toulmin and J.L. Edwards (1966) X-ray crystallographic data, densities, and molar volumes of minerals. In, S.P. Clark, Jr., ed., Handbook of Physical Constants, Geol. Soc. Am. Mem., 97, 27-74. Revised ed.. (278)

Robinson, B.W. (1974) The origin of mineralization at the Tui mine, Te Aroha, New Zealand, in the light of stable isotope studies. Econ. Geol., 69, 910-925. (496)

___ (1983) The GeoSEM: a scanning electron microscope optimized for mineralogy. CSIRO (Australia) Div. Mineralogy Research Rev., 1983, 234-237. (144)

Roedder, E. (1958) Technique for the extraction and partial chemical analysis of fluid-filled inclusions from minerals. Econ. Geol., 53, 235-269. (36, 81, 93, 111, 119, 129, 130, 131, 132, 341)

___ (1960a) Fluid inclusions as samples of the ore-forming fluids. Int'l Geol. Congress, 21st, Copenhagen 1960, Reports, pt. 16, 218-229. (42, 95, 294, 358, 418)

___ (1960b) Primary fluid inclusions in sphalerite crystals from the OH vein, Creede, Colorado (abstr.). Geol. Soc. America Bull., 71, 1958. (436)

___ (1962a) Studies of fluid inclusions I: Low temperature application of a dual-purpose freezing and heating stage. Econ. Geol., 57, 1045-1061. (91, 98, 187, 190, 191, 232, 240, 241, 433, 525)

___ (1962b) Ancient fluids in crystals. Sci. Am., 207, (4) 38-47. (20, 30, 60, 331)

___ (1963) Studies of fluid inclusions II: Freezing data and their interpretation. Econ. Geol., 58, 167-211. (8, 24, 59, 82, 83, 85, 86, 89, 91, 95, 98, 103, 112, 118, 214, 241, 272, 292, 297, 310, 312, 322, 351, 365, 390, 391, 392, 436, 439, 440)

___ (1965a) Liquid CO2 inclusions in olivine-bearing nodules and phenocrysts from basalts. Mineral., 50, 1746-1782. (2, 34, 45, 71, 82, 85, 191, 216, 219, 230, 265, 282, 285, 295, 374, 380, 488, 489, 505, 514, 515, 517, 518, 520, 523, 524, 525, 526, 527)

___ (1965b) Evidence from fluid inclusions as to the nature of the ore-forming fluids. In Symposium-Problems of Postmagmatic Ore Deposition, Prague, 1963. Prague Geol. Survey, 2, 375-384. (20, 25, 43, 44, 45)

Roeder, E. (1965c) Non-Brownian bubble movement in fluid inclusions--a thermal gradient detector of extreme sensitivity and rapid response (abstr.). Geol. Soc. Am. Spec. Paper, 87, 140 (pub. 1966). (92, 93, 168, 178)

―――― (1966) Bouncing bubbles, or who put the pep in Mother Nature's pop (abstr.). Washington Acad. Sci. J., 56, 168-169. (92, 168, 178)

―――― (1967a) Metastable superheated ice in liquid-water inclusions under high negative pressure. Science, 155, 1413-1417. (58, 59, 93, 292, 293, 298, 299, 300, 301)

―――― (1967b) Fluid inclusions as samples of ore fluids. In H.L. Barnes, ed., Geochemistry of Hydrothermal Ore Deposits. 1st edition, Holt, Rinehart & Winston, New York, NY, 515-574. (45, 99, 102, 292, 294, 423, 441)

―――― (1967c) Device for sensing thermal gradients. U.S. Patent No. 3,344,699, granted Oct. 3, 1967. (168, 178)

―――― (1967d) Environment of deposition of stratiform (Mississippi Valley-type) ore deposits, from studies of fluid inclusions. Soc. Econ. Geol. Monograph 3, 349-362. (330, 420, 421)

――――, ed. (1968a) Fluid Inclusion Research -- Proceedings of COFFI (an annual summary of world literature: Vols. 1-6 (1968-1972), privately printed by and available from the editor: vol. 6 (1973) onward, printed by and available from the Univ. of Michigan Press or the editor). (4, 6, 93, 212, 266, 272, 283, 285, 398, 400, 402, 411, 414, 421, 439, 455, 462, 470)

―――― (1968b) Fluid inclusion shape, a transient feature of little diagnostic value (abstr.). Fluid Inclusion Res. — Proc. of COFFI, 1, 4. (168)

―――― (1968c) Temperature, salinity, and origin of the ore-forming fluids at Pine Point, Northwest Territories, Canada, from fluid inclusion studies. Econ. Geol., 63, 439-450. (191, 418, 420)

―――― (1968d) Environment of deposition of the disseminated lead ores at Laisvall, Sweden, as indicated by fluid inclusions. Int'l Geol. Congr. 23rd, Prague 1968, Repts., sec. 7, Endogeneous ore deposits, Proc., 389-401. (259, 416, 420)

―――― (1968e) The noncolloidal origin of "colloform" textures in sphalerite ores. Econ. Geol., 63, 451-471. (420)

―――― (1969) Varvelike banding of possible annual origin in celestite crystals from Clay Center, Ohio, and in other minerals. Am. Mineral., 54, 796-810. (322, 420)

―――― (1970a) Application of an improved crushing microscope stage to studies of the gases in fluid inclusions. Schweiz. Mineral. Petrogr. Mitt., 50, pt. 1, 41-58. (33, 213, 214, 216, 430, 492)

―――― (1970b) Laboratory studies on inclusions in minerals of Ascension Island granitic blocks, and their petrologic significance. In Yu.A. Kuznetsov, ed., Problems of Petrology and Genetic Mineralogy. V.S. Sobolev Memorial Vol. II, "Nauka" Press, Moscow, 247-258 (in Russian; translated in Fluid Inclusion Res. -- Proc. of COFFI, 5, 1972, 129-138). (265, 400)

―――― (1971a) Metastability in fluid inclusions. Soc. Mining Geol. Japan, Spec. Issue 3, 327-334 (Proc. IMA-IAGOD Meetings '70, IAGOD Vol.). (45, 58, 67, 68, 69, 291, 292, 344, 420)

―――― (1971b) Fluid inclusion studies on the porphyry-type ore deposits at Bingham, Utah, Butte, Montana, and Climax, Colorado. Econ. Geol., 66, 98-120. (90, 116, 160, 245, 264, 439, 440, 443, 444, 451, 469)

―――― (1971c) Natural and laboratory crystallization of lunar glasses from Apollo 11. Mineral. Soc. Japan, Spec. Paper 1, 5-12 (Proc. IMA-IAGOD Meeting '70, IM Vol.). (303, 536)

―――― (1971d) Fluid inclusion evidence on the environment of formation of mineral deposits of the southern Appalachian Valley. Econ. Geol., 66, 777-791. (271, 420)

―――― (1972) Composition of Fluid Inclusions. U.S. Geol. Survey Prof. Paper 440JJ, 164 p. (4, 30, 33, 43, 44, 45, 52, 53, 55, 58, 62, 79, 85, 88, 89, 92, 98, 100, 101, 102, 103, 104, 128, 129, 134, 166, 168, 178, 212, 246, 272, 288, 310, 314, 315, 332, 392, 398, 404, 418, 421, 430, 491, 513, 526, 555)

Roeder, E. (1973) Fluid inclusions from the fluorite deposits associated with carbonatite of Amba Dongar, India, and Okorusu, South West Africa. Inst. Mining Metall. Trans., Sect. B, 82, B35-B39. (408, 411)

―――― (1974) Preliminary Mars sample study - effects of sample sterilization on data available from fluid inclusions: p.1-22 in Appendix III of "On the petrological, geochemical, and geophysical characterization of a returned Mars surface sample and the impact of biological sterilization on the analyses." Unnumbered report dated April 1974, NASA Johnson Space Center, Houston. (534)

―――― (1976a) Fluid inclusion evidence on the genesis of ores in sedimentary and volcanic rocks. In K.H. Wolf, ed., Handbook of Strata-bound and Stratiform Ore Deposits, 2, 67-110. Elsevier, Amsterdam, The Netherlands. (43, 150, 181, 205, 271, 330, 415, 416, 418, 420)

―――― (1976b) Petrologic data from experimental studies on crystallized silicate melt and other inclusions in lunar and Hawaiian olivine. Am. Mineral. 61, 684-690. (475, 477, 479, 544, 553)

―――― (1977a) Stable and metastable fluid inclusion data, Browns Canyon fluorspar district, Chaffee Country[sic.], Colorado, and similar epithermal and hot-spring(?) deposits. In Problems of Ore Deposition. 4th IAGOD Symposium, Varna, Bulgaria, 1974, 2, 186-195. (61, 168, 275, 428, 496)

―――― (1977b) Fluid inclusions as tools in mineral exploration. Econ. Geol., 72, 503-525. (212, 266, 285, 413, 451, 464, 465, 466, 467, 469)

―――― (1977c) Changes in ore fluid with time, from fluid inclusion studies at Creede, Colorado. In Problems of Ore Deposition, 4th IAGOD Symposium, Varna, Bulgaria, 1974, 2, 179-185. (284, 432)

―――― (1978a) [A new ultramicrochemical analytical technique for fluid inclusions] (abstr.). U.S. Geol. Survey Prof. Paper 1100, Geol. Surv. Res. 1978, 177. (144)

―――― (1978b) Silicate liquid immiscibility in magmas in the system $K_2O-FeO-Al_2O_3-SiO_2$: an example of serendipity. Geochim. Cosmochim. Acta, G.B., 42, 1597-1617. (33, 545)

―――― (1979a) Origin and significance of magmatic inclusions. Bull. Mineral., 102, 487-510. (13, 28, 33, 57, 141, 150, 282, 283, 382, 474, 476, 477, 478, 483, 485, 535, 561)

―――― (1979b) Fluid inclusions as samples of ore fluids. In H.L. Barnes, ed., Geochemistry of Hydrothermal Ore Deposits, 2nd edition. Wiley & Sons, New York, NY. (43, 70, 273, 416, 423)

―――― (1979c) Fluid inclusion evidence on the environment of sedimentary diagenesis--a review. SEPM Spec. Pub. 89-107. (30, 307, 322, 324, 330, 416, 424)

―――― (1979d) Silicate liquid immiscibility in magmas. In H.S. Yoder, Jr., ed., The Evolution of the Igneous Rocks, 50th Anniv. Perspectives. Princeton Univ. Press, Princeton, New Jersey, NJ, 15-57. (476, 545, 551, 553)

―――― (1979e) Melt inclusions in 75075 and 78505 -- The problem of anomalous low-K inclusions in ilmenite revisited. Proc. 10th Lunar Sci. Conf., 1, 249-257. (546, 555, 556)

―――― (1981a) Use of lunar materials in space construction. Space Solar Power Review, 2, 249-258. (535, 549)

―――― (1981b) Problems in the use of fluid inclusions to investigate fluid-rock interactions in igenous and metamorphic processes. Fortschr. Mineral., 59, 267-302. (388)

―――― (1981c) Are the 3,800-Myr-old Isua objects micro-fossils, limonite-stained fluid inclusions, or neither? Nature, 293, 459-462. (178)

―――― (1981d) Origin of fluid inclusions and changes that occur after trapping. Mineral. Assoc. Canada Short Course Handbook, 6, 101-137. (358)

―――― (1981e) Natural occurrence and significance of fluids indicating high pressure and temperature. Phys. Chem. Earth, 13/14, 9-39. (388, 405, 514)

―――― (1981f) Significance of Ca-Al-rich silicate melt inclusions in olivine crystals from the Murchison type II carbonaceous chondrite. Bull. Mineral., 104, 339-353 (in English). (562, 563, 564, 565)

Roedder, E., and B.J. Skinner (1968) Experimental evidence that fluid inclusions do not leak. Econ. Geol., 63, 715-730. (3, 76, 77, 156, 167, 267, 302, 342, 441)

——— and R.L. Smith (1964) Liquid water in pumice vesicles, a crude but useful dating method (abstr.). Geol. Soc. Am. Spec. Paper 82, Abstr. for 1964, p. 164 (published in 1965). (493)

——— and P.W. Weiblen (1970a) Silicate liquid immiscibility in lunar magmas, evidenced by melt inclusions in Apollo 11 rocks. Science, 167, 641-644. (476, 535, 539, 546, 547, 556)

——— and ——— (1970b) Lunar petrology of silicate melt inclusions, Apollo 11 rocks. Proc. Apollo 11 Lunar Sci. Conf., Geochim. Cosmochim. Acta. Suppl. 1, 801-837. (14, 43, 141, 475, 480, 535, 536, 537, 539, 541, 546, 550, 552, 555, 557, 558)

——— and ——— (1970c) Silicate immiscibility in lunar rocks. Geotimes, 15, (3), 10-13. (535)

——— and ——— (1971) Petrology of silicate melt inclusions, Apollo 11 and Apollo 12 and terrestrial equivalents. Proc. 2nd Lunar Sci. Conf., 507-528. (13, 16, 33, 141, 476, 479, 535, 536, 540, 542, 544, 545, 546, 547, 548, 550, 551, 552, 553, 554, 557)

——— and ——— (1972a) Silicate melt inclusions and glasses in lunar soil fragments from the Luna 16 core sample. Earth Planet. Sci. Lett. 13, 272-285. (141, 535, 546, 549, 552, 553)

——— and ——— (1972b) Petrographic features and petrologic significance of melt inclusions in Apollo 14 and 15 rocks. Proc. 3rd Lunar Sci. Conf., 261-279. (57, 535, 536, 546, 549, 550, 551, 552, 553, 554)

——— and ——— (1973a) Petrology of some lithic fragments from Luna 20. Geochim. Cosmochim. Acta, 37, 1031-1052. (52, 535, 546, 555, 557)

——— and ——— (1973b) Petrology of melt inclusions in Apollo samples 15598 and 62295, and of clasts in 67915 and several lunar soils. Proc. 4th Lunar Sci. Conf., 681-703. (535, 546)

——— and ——— (1975) Anomalous low-K silicate melt inclusions in Ilmenite from Apollo 17 basalts. Proc. 6th Lunar Sci. Conf., 147-164. (141, 555, 556)

——— and ——— (1977a) High-silica glass inclusions in olivine of Luna-24 samples. Geophys. Res. Lett., 4, 485-488. (546, 558)

——— and ——— (1977b) Compositional variation in late-stage differentiates in mare lavas, as indicated by silicate melt inclusions. Proc. 8th Lunar Sci. Conf., 1767-1783. (41, 141, 546, 552)

——— and ——— (1977c) Barred olivine "chondrules" in lunar spinel troctolite. Proc. 8th Lunar Sci. Conf., 2641-2654. (561)

——— and ——— (1978) Melt inclusions in Luna-24 soil fragments. In R.B. Merrill and J.J. Papike, eds., Mare Crisium: the View from Luna-24 ——. Proceedings of the Conference on Luna-24, Houston, Texas, Dec. 1-3, 1977. Pergamon Press, New York, NY, 495-522. (33, 535, 536, 542, 546, 554, 558, 559)

———, B. Ingram and W.E. Hall (1963) Studies of fluid inclusions III: Extraction and quantitative analysis of inclusions in the milligram range. Econ. Geol., 58, 353-374. (106, 114, 118, 121, 125, 126, 131, 134, 422, 423, 471)

———, A.V. Heyl and J.P. Creel (1968) Environment of ore deposition at the Mex-Tex deposits, Hansonburg district, New Mexico, from studies of fluid inclusions. Econ. Geol., 63, 336-348. (16, 23, 43, 44, 45, 65, 422)

Roeder, P.L. (1974) Activity of iron and olivine solubility in basaltic liquids. Earth Planet. Sci. Lett. 23, 397-410. (488)

——— and R.F. Emslie (1970) Olivine-liquid equilibrium. Contrib. Mineral. Petrol., 29, 275-289. (554)

Rogers, P.S.Z. and K.S. Pitzer (1982) Volumetric properties of aqueous sodium chloride solutions. J. Phys. Chem. Ref. Data, 11, 15-81. (234)

Roedder, E. (1982b) Possible Permian diurnal periodicity in NaCl precipitation, Palo Duro Basin, Texas. In T.C. Gustavson et al., eds, Geology and Geohydrology of the Palo Duro Basin, Texas Panhandle, Texas Bur. Econ. Geol. Geological Circ. 82-7, 101-104. (319, 320)

——— (1982c) Int'l Gemological Symp. Proc., 1982, Gemological Inst. Am., Santa Monica, CA, 479-502. (365, 366)

——— (1983a) Discussion of "A re-assessment of phase equilibria involving two liquids in the system K_2O-Al_2O_3-FeO-SiO_2" by G.M. Biggar. Contrib. Mineral. Petrol., 82, 284-290. (142, 550)

——— (1983b) Geobarometry of ultramafic xenoliths from Loihi Seamount, Hawaii, on the basis of CO_2 inclusions in olivine. Earth Planet. Sci. Lett., 66, 369-379. (71, 505, 515, 517, 518, 519, 520, 529)

——— (1984a) The fluids in salt. Am. Mineral. (in press). (18, 66, 308, 319)

——— (1984b) Fluid inclusion evidence bearing on the environments of gold deposition. In Proc. Symp. Gold 82, Geol. Soc. Zimbabwe Spec. Pub. 1, A.A. Balkema, Rotterdam, The Netherlands. (307, 356, 357, 358)

——— and R.L. Bassett (1981) Problems in the determination of the water content of rock-salt samples and its significance in nuclear-waste storage siting. Geology, 9, 525-530. (136, 308, 312, 319)

——— and H.E. Belkin (1978) Fluid inclusions in core samples from ERDA No. 9 borehole, WIPP site, New Mexico: Status Report 2 to Sandia Natl. Laboratories, Jan. 1, 1978 (31 p.); reprinted in part, with authorship ambiguous, in D.W. Powers, S.J. Lambert, S.E. Shaffer, L.R. Hill and W.D. Weart, eds., 1978, Geological characterization report, Waste Isolation Pilot Plant (WIPP) site, southeastern New Mexico Sandia Laboratories, SAND 78-1596, 7-47 through 7-70. (315, 492)

——— and ——— (1979a) Application of studies of fluid inclusions in Permian Salado salt, New Mexico, to problems of siting the Waste Isolation Pilot Plant. In G.J. McCarthy, ed., Scientific Basis for Nuclear Waste Management, Vol. 1, Plenum Press, New York, NY, 313-321. (18, 66, 72, 73, 259, 270, 309, 310, 311, 312, 313, 315, 316, 317)

——— and ——— (1979b) Fluid inclusions in salt from the Rayburn and Vacherie domes, Louisiana. U.S. Geol. Survey Open-File Rept 79-1675, 25 p. (215, 312, 313, 315, 318)

——— and ——— (1980a) Thermal gradient migration of fluid inclusions in single crystals of salt from the Waste Isolation Pilot Plant Site (WIPP). In C.J.M. Northrup, ed., Scientific Basis for Nuclear Waste Management, Vol. 2, Plenum Press, New York, NY, 453-464. (66, 259, 319, 320)

——— and ——— (1980b) Migration of fluid inclusions in polycrystalline salt under thermal gradients in the laboratory and in Salt Block II (abstr.). Proc. 1980 Int'l Waste Terminal Storage Program Info. Mtg, ONWI 212, 361-363. (115, 319)

——— and ——— (1981) Petrographic study of fluid inclusions in salt core samples from Asse mine, Federal Republic of Germany. U.S. Geol. Survey Open-File Rept 81-1128 (32 p.). (215, 216, 308, 312)

——— and R.J. Bodnar (1980) Geologic pressure determinations from fluid inclusion studies. Ann. Rev. Earth Planet. Sci., 8, 263-301. (34, 35, 70, 263, 266, 269, 274, 276, 277, 278, 279, 281, 283, 356, 358, 398, 479)

——— and I.-M. Chou (1982) A critique of "Brine migration in salt and its implications in the geologic disposal of nuclear waste." Oak Ridge National Lab. Rept. 5818, by G.H. Jenks and H.C. Claiborne. U.S. Geol. Survey Open-File Rept 82-1131, 31 p. (66, 319)

——— and D.S. Coombs (1967) Immiscibility in granitic melts, indicated by fluid inclusions in ejected granitic blocks from Ascension Island. J. Petrol., 8, 417-451. (9, 13, 27, 31, 43, 44, 82, 98, 99, 236, 239, 283, 295, 363, 397, 399, 446, 489)

——— and O.C. Kopp (1975) A check on the validity of the pressure correction in inclusion geothermometry, using hydrothermally grown quartz. Fortschr. Mineral., 52, Special Issue, 431-446. (211)

Romanchev, B.P. (1972) Inclusion thermometry and the formation conditions of some carbonatite complexes in East Africa. Geokhimiya, 1979, (2), 172-179 (in Russian; translated in Geochem. Int'l, 9, 115-120, 1972). (407)

_____ (1977) On the reliability of homogenization temperatures of primary inclusions. Geokhimiya, 1977, (5), 726-735 (in Russian; translated in Geochem. Int'l, 14, 65-73). (481)

_____ and S.V. Sokolov (1979) Liquation in the production and geochemistry of the rocks in carbonatite complexes. Geokhimiya, 1979, (2), 229-240 (in Russian; translated in Geochem. Int'l, 16, 126-135, 1980). (407)

Rona, P.A. (1980) TAG hydrothermal field: Mid-Atlantic Ridge crest at latitude 26°N. J. Geol. Soc. London, 137, 385-402. (494)

_____, K. Bostrom and S. Epstein (1980) Hydrothermal quartz from the Mid-Atlantic Ridge. Geology, 8, 569-572. (496)

_____, R.M. Harbison, B.G. Bassinger, R.B. Scott and A.J. Nalwaik (1976) Tectonic fabric and hydrothermal activity of Mid-Atlantic Ridge crest (lat. 26°N). Geol. Soc. Am. Bull, 87, 661-674. (494)

Rosasco, G.J. (1980) Raman microprobe spectroscopy. In R.J.H. Clark and R.E. Hester, eds. Advances in Infrared and Raman Spectroscopy, Vol. 7. Heyden Co. London, 223-282. (105)

_____ and E.S. Etz (1977) The Raman microprobe: a new analytical tool. Research and Development, 28, 20-35. (105)

_____ and E. Roedder (1979) Application of a new Raman microprobe spectrometer to nondestructive analysis of sulfate and other ions in fluid phases in fluid inclusions in minerals. Geochim. Cosmochim. Acta, 43, 1907-1915. (54, 105, 106, 365, 440, 441, 568)

_____, E.S. Etz and W.A. Casalt (1975a) The analysis of discrete fine particles by Raman spectroscopy. Applied Spec., 29, 396. (105)

_____, E. Roedder and J.H. Simmons (1975b) Laser activated Raman spectroscopy for nondestructive partial analysis of individual phases in fluid inclusions in minerals. Science, 190, 557-560. (105, 108, 322, 323, 416, 423)

Rose, H. (1839) Über das Knistersalz von Wieliczka. Ann. Physik Chemie, 48, 353-361. (119)

Rose, Jr., W.I., A.T. Anderson, Jr., L.G. Woodruff and S.B. Bonis (1978) The October 1974 basaltic tephra from Fuego volcano: Description and history of the magma body. J. Volcan. Geotherm. Res., 4, 3-53. (489)

_____, R.E. Stoiber and L.L. Malinconico (1982) Eruptive gas compositions and fluxes of explosive volcanoes: Budget of S and Cl emitted from Fuego volcano, Guatemala. In R.S. Thorpe, ed., Orogenic Andesites and Related Rocks, Wiley & Sons. New York, 669-676. (488)

_____, R.L. Wunderman, M.F. Hoffman and L. Gale (1984) Atmospheric hazards of volcanic activity from a volcanologist's point of view: Fuego and Mount St. Helens. J. Volcan. Geotherm. Res. (in press). (488)

Rosenberg, P.E. and J.W. Mills (1966) A mechanism for the emplacement of magnesite in dolomite. Econ. Geol., 61, 582-586. (55)

Rosenbusch, H. (1887) Mikroskopische Physiographie der Mineralien und Gesteine. Band 2. E. Schweizerbart'sche Verlagshandlung. Stuttgart, Germany. (545)

Rost, R. (1964) Surfaces of and inclusions in moldavites. Geochim. Cosmochim. Acta, 28, 931-936. (115)

Rovetta, M.R. and E.A. Mathez (1982) Magnesite and other minerals in fluid inclusions in a lherzolite xenolith from an alkali basalt (abstr.). Terra Cognita, 2, p. 229. (526)

Rowan, L.P., P.M. Bethke, and R.J. Bodnar (1983) Stretching of fluid inclusions in fluorite at confining pressures up to one kilobar (abstr.). Geol. Soc. Am. Abstr. Programs, 15, 674. (72)

Rowlinson, J.S. (1969) Liquids and Liquid Mixtures, 2nd edition. Plenum Press New York, NY, 371 p. (233)

Rozen, O.M., S.A. Sidorenko and N.N. Kuznetsova (1977) Scapolite and apatite as indicators of composition of volatiles in metamorphism of the granulite complex of the Kola Peninsula. Akad. Nauk SSSR, Dokl., 237, 441-444 (in Russian; translated in Dokl. Acad. Sci. USSR, 237, 211-213, 1980). (363)

Rubin, H. and C. Roth (1979) On the growth of instabilities in groundwater due to temperature and salinity gradients. Advances in Water Resources, 1979, 2, 69-76. (289)

Ruiz, J., S.E. Kesler, L.M. Jones and J.F. Sutter (1980) Geology and geochemistry of the Las Cuevas fluorite deposit, San Luis Potosi, Mexico. Econ. Geol., 75, 1200-1209. (427)

Rumble, III, D. (1982) Oxygen isotope equilibration and permeability enhancement during metamorphism (abstr.). Geol. Soc. (London) Newsletter, 11, 15. (352)

_____, J.M. Ferry, T.C. Hoering, and A.J. Boucot (1982) Fluid flow during metamorphism at the Beaver Brook fossil locality, New Hampshire. Am. J. Sci., 282, 886-919. (352)

Rush, C.A. (1954) Moving bubbles in negative crystals (abstr.). Acta Crystallogr., 7, 672. (92)

Rutherford, M.J. (1963) Geothermometry of Liquid Inclusions in Quartz, Coronation Mine, Flin Flon Area, Saskatchewan. M.S. thesis, Univ. of Saskatchewan, Saskatoon, Sask., Canada, 40 p.; abstr. in Canadian Mining J., 85, (4), 1220, April 1964. (89, 119, 461)

_____, P.C. Hess and G.H. Daniel (1974) Experimental liquid line of descent and liquid immiscibility for basalt 70017. Proc. 5th Lunar Sci. Conf.. 569-583. (30)

Ryabchikov, I.D. and A.L. Boettcher (1980) Experimental evidence at high pressure for potassic metasomatism in the mantle of the Earth. Am. Mineral.. 65, 915-919. (526)

_____, W. Schreyer and K. Abraham (1982) Composition of aqueous fluids in equilibrium with pyroxenes and olivines at mantle pressures and temperatures. Contrib. Mineral. Petrol., 79, 80-84. (526)

Ryan, J.A., J.J. Grossman and W.M. Hansen (1968) Adhesion of silicates cleaned in ultra high vacuum. J. Geophys. Res., 73, 6061-6070. (123)

Rye, D.M. and R.O. Rye (1974) Homestake gold mine, South Dakota: I. Stable isotope studies. Econ. Geol., 69, 293-317. (356)

Rye, R.O. (1965) The Carbon, Hydrogen, and Oxygen Isotopic Composition of the Hydrothermal Fluids Responsible for the Lead-Zinc Deposits at Providencia, Zacatecas, Mexico. Ph.D. dissertation, Princeton Univ., Princeton, NJ, 108 p. (see also Econ. Geol., 61, 1399-1427, 1966). (127)

_____ (1966) The carbon, hydrogen, and oxygen isotope composition of the hydrothermal fluids responsible for the lead-zinc deposits at Providencia, Zacatecas, Mexico. Econ. Geol., 61, 1399-1427. (437)

_____ and J. Haffty (1969) Chemical composition of the hydrothermal fluids responsible for the lead-zinc deposits at Providencia, Zacatecas, Mexico. Econ. Geol., 64, 629-643. (437)

_____ and J.R. O'Neil (1968) The O18 content of water in primary fluid inclusions from Providencia, North-Central Mexico. Econ. Geol., 63, 232-238. (127, 302, 437)

_____ and _____ (1977) Oxygen and hydrogen isotopic studies of ore deposition and metamorphism at the Raul mine, Peru. Geochim. Cosmochim. Acta, 43, 1633-1643. (461)

_____ and F.J. Sawkins (1974) Fluid inclusion and stable isotope studies on the Casapalca Ag-Pb-Zn-Cu deposit, Central Andes, Peru: Econ. Geol. 69, 181-205. (436, 437)

Rykart, R. (1977) Zum Wachstum plattiger Quarze mit "Faden". Schweizer Strahler, 4, 209-221. (342)

Saager, R. (1967) New techniques for polishing ore minerals. Univ. of Witwatersrand Econ. Geol. Res. Unit Info., Circ., 39, 13 p. (157)

Sabourang-Rosset, C. (1973) Rapports Cl/Br des inclusions liquides des cristaux de gypse de divers gisements: Correlations avec les données de la microcryoscopie et interprétations génétiques (abstr.). Paris, Réunion Ann. des Sci. de la Terre, 1973, 375. (315)

Sabouraud-Rosset, C. (1974) Determination par activation neutronique des rapports Cl/Br des inclusions fluides de divers gypses. Correlation avec les données de la microcryoscopie et interprétations génétiques. Sedimentology, 21, 415-431. (315)

_____ (1976) Solid and Liquid Inclusions in Gypsum. Thèse d'Etat, Univ. Paris Sud, Centre d'Orsay, 173 p. (in French). (315)

Saito and H.S. Slavochani (1980) Replacement of solid primary inclusions inside dipyramidal quartz of the Eocene of Northern Tunisia. Bull. Minéral., 103, 54-58 (in French). (75)

_____, J.-C. Macquar and H. Rouvier (1980) The fluid inclusions, evidence and false evidence of the conditions of deposition. Some examples in the Pb-Zn Ba-F mineralization from the southern Massif Central of France. Minerallium Deposita, 15, 211-230. (71, 259, 421)

†Safronov, G.M. (1957) In reference to the morphology of liquid inclusions in crystals of quartz from Pamir. Vses. Nauchno-Issled. Inst. P'ezooptichesk. Mineral. Syr'ya Trudy, 1, (2), 155-159 (in Russian). (92)

Sage, B.H. and W.H. Lacey (1949) Volumetric and Phase Behavior of Hydrocarbons. Gulf Pub. Co., Houston, TX, 299 p. (233)

Saito, T. (1951) On the impurities in fluorites (Studies on the fluorites from Japan, II). Hokkaido Univ. Fac. Sci. J., ser. 4, 7, 383-388. (114)

Sakai, H., T.J. Casadevall, and J.G. Moore (1983) Chemistry and isotope ratios of sulfur in basalts and volcanic gases at Kilauea Volcano, Hawaii. Geochim. Cosmochim. Acta, 46, 729-738. (486)

Sakhibgareyev, R.S. and L.N. Lashkova (1977) Mineral corrosion by petroleum and bitumens. Akad. Nauk SSSR, Dokl., 234, 1452-1455 (in Russian; translated in Dokl. Acad. Sci. USSR, 234, 268-271, 1979). (35)

Sakuyama, M. and I. Kushiro (1979) Vesiculation of hydrous andesitic melt and transport of alkalies by separated vapor phase. Contrib. Mineral. Petrol., 71, 61-66. (389)

Sang, E. (1873) Notice of a singular property exhibited by the fluid enclosed in crystal cavities. Royal Soc. Edinburgh Proc., 8, (86), 87-88. (92)

Sato, M. (1978) Oxygen fugacity of basaltic magmas and the role of gas-forming elements. Geophys. Res. Lett., 5, 447-449. (505)

_____, K.A. McGee, A.J. Sutton and S.S. Schultz (1982) Hydrogen monitoring along seismogenic faults (abstr.). EOS, Trans. Am. Geophys. Union, 63, 1043. (402)

Sato, T. (1972) Behaviors of ore-forming solutions in seawater. Mining Geol. (Japan), 22, 31-42 (in English). (469)

Savel'yeva, N.I. and G.B. Naumov (1979) Methods of analyzing salt composition for liquid inclusions in minerals. Geokhimiya, 1979, (5), 730-736 (in Russian; translated in Geochem. Int'l, 16, (3), 65-70, 1979). (117, 131)

Savkevich, S.S. (1970) Amber. Nedra Press, Leningrad, 190 p. (in Russian). (327)

Sawkins, F.J. (1964) Lead-zinc ore deposition in the light of fluid inclusion studies. Providencia mine, Zacatecas, Mexico. Econ. Geol. 59, 883-919. (437)

_____ (1966) Ore genesis in the North Pennine orefield in the light of fluid inclusion studies. Econ. Geol., 61, 385-401. (190)

_____ (1977) Fluid inclusion studies of the Messina copper deposits, Transvaal, South Africa. Econ. Geol., 72, 619-631. (115)

_____ and D.A. Scherkenbach (1981) High copper content of fluid inclusions in quartz from northern Sonora: Implications for ore-genesis theory. Geology, 9, 37-40. (445)

Saylor, C.P. (1965) A study of errors in the measurement of microscopic spheres. Appl. Optics, 4, 477-486. (103)

Schairer, J.F. and N.L. Bowen (1947) Melting relations in the systems Na2O-Al2O3-SiO2 and K2O-Al2O3-SiO2. Am. J. Sci., 245, 193-204. (549)

Scheffen-Lauenroth, Th., K. Klapper and R.A. Becker (1981) Growth and perfection of organic crystals from undercooled melt. I. Benzil. J. Crystal Growth, 55, 557-570. (75)

Schiffries, C.M. (1982) The petrogenesis of a platiniferous dunite pipe in the Bushveld Complex: Infiltration metasomatism by a chloride solution. Econ. Geol. 77, 1439-1453. (389, 463)

Schlossmacher, K. (1932) Bauer's Edelsteinkunde (3rd edition). Bernard Tauchnitz, Leipzig, 871 p. (507)

Schneider, G.M. (1978) High-pressure phase diagrams and critical properties of fluid mixtures. Chemical Thermodynamics (London), 2, 105-146. (245)

_____, E. Stahl, and G. Wilke, eds. (1980) Extraction with supercritical gases. Verlag Chemie, Weinheim, W. Germany, 189 p. (245)

Scholander, P.F. and D.C. Nutt (1960) Bubble pressure in Greenland icebergs. J. Glaciology, 3, 671-678. (116)

Schreyer, W. (1983) Metamorphism and fluid inclusions in the basement of the Vredefort Dome, South Africa: Guidelines to the origin of the structure. J. Petrol. 24, (1), 26-47. (377)

_____ and O. Medenbach. (1981) CO2-rich fluid inclusions along planar elements of quartz in basement rocks of the Vredefort Dome, South Africa. Contrib. Mineral. Petrol., 77, 93-100. (375, 377, 378)

Schulling R.D. and R. Kreulen (1979) Are thermal domes heated by CO2-rich fluids from the mantle? Earth Planet. Sci. Lett., 43, 298-302. (350, 374, 505)

Schulze, D.J. (1981) Mantle-derived calcite and phlogopite in discrete nodules from a Kentucky kimberlite: Evidence for primary kimberlitic liquids (abstr.) EOS, Trans. Am. Geophys. Union, 62, 414. (512)

_____ (1984) Evidence for primary kimberlitic liquids in discrete nodules from a Kentucky kimberlite. J. Geol. (in press). (512)

Schwarcz, H.P and C.Yonge. (1983) Isotopic composition of palaeowaters as inferred from speleothem and its fluid inclusions. In Paleoclimates and Palewaters: A collection of environmental isotope studies. Int'l Atomic Energy Agency, Vienna, 115-133. (328)

_____ R.S. Harmon, P. Thompson and D.C. Ford (1976) Stable isotope studies of fluid inclusions in speleothems and their paleoclimatic significance. Geochim. Cosmochim. Acta, 40, 657-665. (327)

Scott, D.H. (1948) The decrepitation method applied to minerals with fluid inclusions. Econ. Geol., 43, 637-654. (212)

*Sedletskii, V.I., V.N. Trufanov and Yu.G. Maiskii (1973) Mechanism of mobilization and the nature of highly-mineralized brines in halogenide formations (abstr.). COFFI, 7, 199. (314, 317)

Sekerka, R.F. (1973) Morphological stability. In P. Hartman, ed., Crystal Growth: An Introduction. American Elsevier Pub. Co., New York, NY, 403-443. (14)

Sella, C. and G. Deïcha (1962a) Étude au microscope electronique des pores intergranularies des gangues et des roches. Compt. Rend. Acad. Sci. Paris, 254, 2796-2798. (19, 97, 111, 145)

_____ (1962b) Lacunes de cristallisation et pores intergranulaires du quartz (abstr.). In Sydney Breese, Jr., ed., Electron Microscopy. Repts 5th Int'l Congress Electron Microscopy, no. GG4, 2 p. Academic Press, New York, NY. (19, 97, 111, 145)

_____ (1963) Importance des cavités intra et intercristallines dans l'architecture des minéraux et des roches. J. Microscopie, 2, (2), 283-296. (19, 97, 111, 145)

Sellschop, J.P.F. (1975) Evidence on the environment of diamond genesis from trace element studies of natural diamonds. Diamond Res., 1975, 35-41. (506)

_____ (1979) Nuclear probes in physical and geochemical studies of natural diamonds. In J.E. Field, ed., The Properties of Diamond. Academic Press, London, 107-163. (508)

_____ D.M. Bibby, C.S. Erasmus and D.W. Mingay (1974) Determination of impurities in diamond by nuclear methods. Diamond Res., 1974, 43-50. (508)

_____ H.J. Annegarn, C. Madiba, R.J. Keddy and M.J. Renan (1977) Hydrogen in diamond. Diamond Res., 1977, 2-4. (508)

Sellschop, J.P.F., C.C.P. Madiba, H.J. Annegarn and S. Shongwe (1979) Volatile light elements in diamond. Diamond Res., 1979, 24-30. (508, 509)

Selverstone, J. (1982) Fluid inclusions as petrogenetic indicators in granulite xenoliths, Pali-Aike volcanic field, southern Chile. Contrib. Mineral. Petrol., 79, 1-9. (375, 379 505, 515, 531)

___ and D.P.F. Stern (1983) Petrochemistry and recrystallization history of granulite xenoliths from the Pali-Aike volcanic field, Chile. Am. Mineral., 68, 1102-1112. (515)

___, F. Spear, G. Franz and G. Morteani (1983) P-T-t paths for hornblende + kyanite + staurolite garbenschists: High-pressure metamorphism in the Western Tauern Window, Austria (abstr.). EOS, Trans. Am. Geophys. Union. 64, 351-352. (362)

Serdiuchenko, D.P. (1981) On the volatiles in the structure of pyroxene. Akad. Nauk SSSR, Dokl., 259, 462-466 (in Russian). (504)

Seward, T.M. and E.U. Franck (1981) The system hydrogen-water up to 440°C and 2500 bar pressure. Berichte Bunsen-Gesells. phys. Chemie, 85, 2-7. (231)

*Seyful'-Mulyukov, R.B., A.N. Serengin, B.A. Sokolov, Yu.R. Mazor and F.P. Mel'nikov (1978) Use of thermobarogeochemical studies in oil geology (abstr.). COFFI, 12, 171. (333)

Shaffer, N.R. (1981) Possibility of Mississippi Valley-type mineral emplacement in Indiana. Indiana Dept. Natural Res. Geol. Surv. Spec. Rept. 21, 49 p. (324)

*Sharonov B.N., Yu.V. Lir and A.V. Kozlov (1973) On question of interpretation of decrepigraphs of vein quartz (abstr.). COFFI, 7, 203-204. (266)

*Shatagin, N.N. and B.A. Dorogovin (1978) Reconstruction of the paleoerosion level of a weathering granite massif by fluid inclusion studies in quartz from granite fragments from conglomerates (abstr.). COFFI, 11, 193. (308)

Shaw, H.R. (1974) Diffusion of H2O in granitic liquids: Part I, Experimental data; Part II, Mass transfer in magma chambers. In A.W. Hoffman et al., eds., Geochemical Transport and Kinetics. Carnegie Inst. Washington Pub. 634, 139-170. (389)

___ for chemical variations in zoned magma chambers (abstr.). Geol. Soc. Am. Abstr. Programs 8, 1102. (389)

*Shaydetskaya, V.S. (1975) Inclusions of liquid CO2 in halite from Devonian salts of Dniepr-Donets rift (abstr.). COFFI, 8, 163. (137)

*Shcheka, S.A., N.A. Kurentsova, I.M. Romanenko and V.M. Chubarov (1978) Inclusions of melt in nodules in basalts, related to problem of primary basalt magma (abstr.). COFFI, 11, 194-195. (527)

___, V.V. Malakhov and I.M. Romanenko (1979) The composition of the gaseous-liquid inclusions in the minerals of basic-ultrabasic nodules from volcanic rocks. Akad. Nauk SSSR, Sib. Otd., Geol. i Geofiz., 20, (7), 70-74 (in Russian; translated in Sov. Geol. Geophys. 20, 59-63, 1979). (526)

*Shchepetkin, Yu.V., I.I. Nesterov, N.Kh. Kulakhmetov and A.V. Ryl'kov (1976) Decrepitation in the recognition and correlation of sedimentary beds (abstr.). COFFI, 10, 252. (308)

Shearman, D.J. (1970) Recent halite rock, Baja California, Mexico. Trans. Inst. Mining Metall., 879, 155-162. (311)

___ (1978) Evaporites of coastal sabkha: Ancient sabkha deposits; and Halite in sabkha environments; Section 2. In W.E. Dean and B.C. Schreiber, eds., Marine Evaporites: Lecture Notes for Short Course, 4. Soc. Econ. Paleo. Mineral., Oklahoma City, OK, 6-42. (311)

Sheftal', N.N. (1956) The problem of gaseous-liquid inclusions in quartz for geothermometry and geomanometry. Akad. Nauk SSSR, Inst. Kristallogr. Trudy 1956, (12), 111-118 (in Russian). (100)

Shelton, K.L. (1983) Composition and origin of ore-forming fluids in a carbonate-hosted porphyry copper and skarn deposit: A fluid inclusion and stable isotope study of Mines Gaspé, Quebec. Econ. Geol.,78, 387-421. (443)

___ and P.M. Orville (1980) Formation of synthetic fluid inclusions in natural quartz. Am. Mineral., 65, 1233-1236. (61, 70, 343)

Shepherd, E.S. (1938) The gases in rocks and some related problems.' Am. J. Sci., 5th ser., 35A, 311-351. (120)

Shepherd, T.J. (1977) Fluid inclusion study of the Witwatersrand gold-uranium ores. Phil. Trans. Roy. Soc. Lond., 286A, 549-565. (307)

___ (1981) Temperature-programmable, heating-freezing stage for microthermometric analysis of fluid inclusions. Econ. Geol. 76, 1244-1247. (196)

___ and D.P.F. Darbyshire (1981) Fluid inclusion Rb-Sr isochrons for dating mineral deposits. Nature, 290, 578-579. (134)

Sheppard, S.M.F., R.L. Nielsen and H.P. Taylor (1971) Hydrogen and oxygen isotope ratios in minerals from porphyry copper deposits. Econ. Geol., 66, 515-542. (446)

Shimazu, M. and J. Yajima (1973) Epidote and wairakite in drill cores at the Hachimantai geothermal area, northeastern Japan. Japan Assoc. Mineral. Petrol. Econ. Geol. J., 68, 363-371. (497)

Shirely, S.D., G. Simmons and E.R. Padovani (1979) Angular, oriented microtubes in metamorphic oligoclase (abstr.). EOS, Trans. Am. Geophys. Union, 60, 424. (21)

Shmonov, V.M. and K.I. Shmulovich (1974) Molal volumes and equation of state of CO2 at temperatures from 100 to 1000°C and pressures from 2000 to 10,000 bars. Akad. Nauk SSSR, Dokl., 217, 935-938 (in Russian; translated in Dokl. Acad. Sci. USSR, 217, 206-209, 1975). (227, 229, 231, 283)

Shmulovich, K.I., V.M. Shmonov, V.A. Mazur and A.G. Kalinichev (1980) Relations of P-V-T and activity-concentration in the system H2O-CO2 (homogeneous solutions) Geokhimiya, 1980, (12), 1807-1824 (in Russian). (240)

Shoji, H. and C.C. Langway, Jr. (1982) Air hydrate inclusions in fresh ice core. Nature, 298, 548-550. (329)

Shreve, R.L. (1967) Migration of air bubbles, vapor figures, and brine pockets in ice under a temperature gradient. J. Geophys. Res., 72, 4093-4100. (68)

Shugurova, N.A. (1968) Chemical basis of method of gas analysis of individual inclusions in minerals. In Mineralogical Thermometry and Barometry, Vol. 2. "Nauka" Press, Moscow, 18-23 (in Russian). (121, 128)

___, Yu.A. Dolgov and G.M. Ivanova (1976) Composition of gaseous inclusions in silicate spherules of various origin. In Yu.A. Dolgov et al., eds., Genetic Studies in Mineralogy, Novosibirsk, Inst. Geol. and Geophys., Sib. Branch Acad. Sci. USSR, 3-8 (in Russian). (560)

Shukaylo, L.G. (1979) The composition of microinclusions in apatites of Neogene hypabyssal intrusions in Transcarpathia, (L'vov. Gos. Univ.), Mineral. Sbornik, 33, (2), 53-60 (in Russian). (384)

Shukla, P.N., B.K. Kothari and P.S. Goel (1979) Nitrogen in tektites and natural glasses: Earth Planet Sci. Lett., 46, 138-140. (560)

Sillitoe, R.H., C. Halls and J.N. Grant (1975) Porphyry tin deposits in Bolivia. Econ. Geol., 70, 913-927. (439, 441)

Silvestri, O. (1882) Sulla natura chimica di alcune inclusioni liquide contenute in cristalli naturali di solfo della Sicilia. I Gazzetta Chimica Italiana, 12, 7-9. (328)

Simanovich, I.M. and G.V. Ivensen (1972) Inclusions of minerals and mineral-forming medium in clastic quartz. Litologiya i Polez. Isko., 7, (5), 34-50 (in Russian; translated in Lithol. Mineral. Resources, 7, (5), 568-581, 1973). (308)

Simmons, G. and D. Richter (1976) Microcracks in rocks. In R.G.J. Strens, ed., The Physics and Chemistry of Minerals and Rocks. Wiley & Sons, New York, NY, 105-137. (157, 343, 345)

Sine, F.L. (1925) Antozonite from Monteagle Township, Hastings County, Ontario. toronto Univ. Studies, Geol. Ser., (20), 22-24. (117)

Sisson, V.B. (1979) Petrologic and Fluid Inclusion Studies of the Wissahickon Formation, Southeastern Pennsylvania and the Lessard Formation, Ontario, Canada. Honors thesis, Bryn Mawr College, Bryn Mawr, Pennsylvania, 75 p. (248)

Si
Sp

Snee, L.W., J.F. Sutter and W.C. Kelly (1983) Mineralization history of the Panasqueira, Portugal, tin-tungsten deposit by high-precision $^{40}Ar/^{39}Ar$ age-spectrum dating of muscovites (abstr.). Geol. Soc. Am. Abstr. Programs, 15, 691. (460)

Sobolev, A.V., L.V. Dmitriev, V.L. Barsukov, V.N. Nevsorov and A.B. Slutsky (1980) The formation conditions of the high-magnesium olivines from the monomineralic fraction of Luna 24 regolith. Proc. 11th Lunar Sci. Conf., 105-116. (403, 542, 554, 558)

Sobolev, V.S. and V.P. Kostyuk, eds. (1975) Magmatic Crystallization Based on a Study of Melt Inclusions. "Nauka" Press, Novosibirsk (in Russian; translated in part in Fluid Inclusion Res. -- Proc. of COFFI, 9, 182-253). (Note - page number references are to those of the original.) (17, 18, 186, 189, 190, 384, 400, 402, 403, 474)

Sørensen (1970) A preliminary examination of fluid inclusions in nepheline, sorensenite, tugtupkite and chkalovite from the Ilímaussaq alkaline intrusion, South Greenland. Medd. om Grønland, 181, (11), 32 p. (52, 121)

——— and I.T. Bakumenko (1972) Crystallization temperature and gas phase composition of alkaline effusives as indicated by primary melt inclusions in the phenocrysts. Bull. Volcanologique, 35, pt. 2, 479-496. (481)

——— and V.P. Kostyuk (1974) Inclusions in the minerals of some types of alkaline rocks. In H. Sørenson, ed., The Alkaline Rocks. Wiley & Sons, New York, NY, 389-401. (402, 481)

——— and K. Yagi (1975) Crystallization temperature of wyomingite from Leucite Hills. Contrib. Mineral. Petrol., 49, 301-308. (481)

———, I.T. Bakumenko and V.P. Kostyuk (1976) The possibility of using melt inclusions for petrologic conclusions. Akad. Nauk SSSR, Sib. Otdel., Geol. i Geofiz. 1976, (5), 140-149 (in Russian; translated in Fluid Inclusion Res. -- Proc. of COFFI, 9, 178-181). (403)

Sofer, Z. (1978) The isotopic composition of hydration water in gypsum. Geochim. Cosmochim. Acta, 42, 1141-1149. (318)

Solovova, I.P., I.D. Ryabchikov, V.I. Kovalenko and V.B. Naumov (1982) Inclusions of high-density CO_2 in mantle lherzolites. Dokl. Akad. Nauk SSSR, 263, (1), 179-182 (in Russian). (515)

Sommer, M.A. (1977) Volatiles H_2O, CO_2, and CO in silicate melt inclusions in quartz phenocrysts from the rhyolitic Bandelier air-fall and ashflow tuff, New Mexico. J. Geol., 85, 423-432. (123, 397, 485, 486)

——— and L.S. Schramm (1984) An analysis of the water concentrations in silicate melt inclusions in quartz phenocrysts from the Bandelier Tuff, Jemez Mountains, New Mexico. J. Geol. (in press). (486)

Sorby, H.C. (1858) On the microscopic structure of crystals, indicating the origin of minerals and rocks. Geol. Soc. London Quart. J., 14, pt. 1, 453-500. (3, 4, 6, 50, 86, 94, 114, 119, 164, 186, 252, 263, 266, 375, 474, 571)

Sørenson, H., ed. (1974) The Alkaline Rocks. Wiley & Sons, New York, NY, 622 p. (402)

Sotnikov, V.I. and A.P. Berzina (1975) Peculiarities of metasomatism at the Kal'makyr deposit. In V.S. Sobolev, ed., Materials on Genetic and Experimental Mineralogy, 8. Inst. Geol. Geophys., No. 184: "Nauka" Press, Novosibirsk, Siberian Div., 217-236 (in Russian). (443)

——— and A.A. Proskuryakov (1973) Physicochemical conditions of endogenous processes during subvolcanic ore-formation. COFFI, 7, 217-218. (443)

Sourirajan, S. and G.C. Kennedy (1962) The system H_2O-NaCl at elevated temperatures and pressures. Am. J. Sci., 260, 115-141. (89, 90, 96, 234, 235, 245, 274, 275, 276, 278, 286, 448, 451)

Sparks, R.S.J. (1978) The dynamics of bubble formation and growth in magmas: a review and analysis. J. Volcan. Geotherm. Res., 3, 1-37. (490, 492)

Sisson, V.B., M.L. Crawford and P.H. Thompson (1981) CO_2-brine immiscibility at high temperatures, evidence from calcareous metasedimentary rocks. Contrib. Mineral. Petrol., 78, 371-378. (248, 373)

Sjögren, Jh. (1893) On large fluid inclusions in gypsum from Sicily. Uppsala Univ. Geol. Inst. Bull. 1, 277-281. (328)

Skeaff, J.M., C.W. Bale, A.D. Pelton and W.T. Thompson (1979) Selection of ternary fused chlorides for the electrowinning of lead and zinc based on calculated thermodynamic properties. Canada Centre for Mineral and Energy Tech. Rept., 79-23, 44 p. (58, 100)

Skinner, B.J. (1953) Some considerations regarding liquid inclusions as geologic thermometers. Econ. Geol., 48, 541-550. (3, 76, 186)

——— (1966) Thermal expansion. In S.P. Clark, Jr., ed., Handbook of Physical Constants, Revised edition. Geol. Soc. Am. Mem., 97, 75-96. (69, 278, 282)

——— and D.L. Peck (1969) An immiscible sulfide melt from Hawaii. Econ. Geol. Monogr. 4, 310-322. (540)

———, D.E. White, H.J. Rose and R.E. Mays (1967) Sulfides associated with the Salton Sea geothermal brine. Econ. Geol., 62, 316-330. (498)

Skropyshev, A.V. (1957) Gaseous-liquid inclusions in crystals of Iceland spar. L'vov. Geol. Obshch. Mineral. Sbornik, 11, 303-321 (in Russian; translated in Int'l Geol. Rev., 1, (9), 1-11, 1959). (114, 136)

*Skryabin, V.B. (1976) Melt inclusions in granites of the Voronezh massif (abstr.). COFFI, 10, 265. (385)

Skryabin, V.Yu. (1978) Composition of crystallized inclusions of granite melt. Akad. Nauk. SSSR, Dokl., 242, (2), 416-418 (in Russian; translated in Dokl. Acad. Sci. USSR, 242, 144-147, 1981). (385)

Slack, J.F. (1980) Multistage vein ores of the Lake City district, western San Juan Mountains, Colorado. Econ. Geol., 75, 963-991. (427)

Sliwko, M.M. (1955) Investigations of Tourmalines from Certain Deposits in the U.S.S.R.- L'vov, Izdatel'stvo L'vovskogo Univ., 125 p. (in Russian). (96)

——— (1958) Inclusions of solutions in tourmaline crystals, Vses. Nauchno-Issled. Inst. P'ezooptichesk. Mineral. Syr'ya Trudy, 2, (2), 63-68 (in Russian). (96)

Smalley, R.E. (1982) Mass-selective laser photoionization. J. Chem. Education, 59, 934-939. (574)

Smirnov, V.I., et al., eds. (1965) Mineralogical Thermometry and Barometry. "Nauka" Press, Moscow, 328 p. (in Russian). (5)

Smith, D.L., B. Evans and B.J. Wanamaker (1982) Crack healing in silicates: Experimental observations on quartz and olivine (abstr.). EOS, Trans. Am. Geophys. Union, 63, 437. (63, 530)

Smith, Jr., D.M., T. Albinson and F.J. Sawkins (1982) Geologic and fluid inclusion studies of the Tayoltita silver-gold vein deposit, Durango, Mexico. Econ. Geol., 77, 1120-1145. (203, 426, 427)

Smith, F.G. (1953) Historical Development of Inclusion Thermometry, Univ. Toronto Press, Toronto, Canada, 149 p. (3, 4, 100, 212)

——— and W.M. Little (1953) Sources of error in the decrepitation method of liquid inclusions. Econ. Geol., 48, 233-238. (265)

Smith, F.W. (1973) A simple microscope freezing stage. Mineral. Mag., 39, 366-367. (190)

Smith, J.V., J.S. Delaney, R.L. Hervig and J.B. Dawson (1981) Storage of F and Cl in the upper mantle: geochemical implications. Lithos, 14, 133-147. (506)

Smith, R.L. (1979) Ash-flow magmatism. Geol. Soc. Am. Spec. Paper, 180, 5-27. (389)

Smith, S.P. (1977) Noble gases in igneous rocks and minerals (abstr.). EOS, Trans. Am. Geophys. Union, 58, (6), 536. (506)

——— and B.M. Kennedy (1983) The solubility of noble gases in water and in NaCl brine. Geochim. Cosmo. Acta, 47, 503-515. (310)

Spear, F.S. and J. Selverstone (1983) Water exsolution from quartz: Implications for the generation of retrograde metamorphic fluids. Geology, 11, 82-85. (2, 36)

Spooner, E.T.C. (1980) Cu-pyrite mineralization and seawater convection in oceanic crust - the ophiolitic ore deposits of Cyprus. In D.W. Strangway, ed., The Continental Crust and its Mineral Deposits. Geol. Assoc. Canada Spec. Paper 20, 685-704. (496)

Sprunt, E.S. and A. Nur (1979) Microcracking and healing in granites: New evidence from cathodoluminescence. Science, 205, 495-497. (21, 45, 59, 161)

Stackelberg, M. von and H.R. Müller (1954) Feste Gas-hydrate II Struktur und Raumchemie. Z. Elektrochemie, 58, 25-39. (99)

Stalder, H.A. (1974) Petrographische und mineralogische Untersuchungen im Grimselgebiet. Schweiz. Mineral. Petrogr. Mitt., 44, 187-398. (351)

_____ and J.C. Touray (1970) "Fensterquartz" with methane-bearing inclusions from the western part of the northern sedimentary Swiss Alps. Schweiz. Mineral. Petrogr. Mitt., 50, 109-130 (in German). (342, 347, 353)

Stalkup, F.I. (1978) Carbon dioxide miscible flooding: Past, present and outlook for the future. J. Petrol. Tech. (Aug. 1978), 1102. (245)

Staudacher, T. and C.J. Allègre (1982) Terrestrial xenology. Earth Planet. Sci. Lett., 60, 389-406. (506, 527)

Steele, I.M. (1983) Factors affecting hydrogen determination by ion microprobe (abstr.). Geol. Soc. Am. Abstr. Programs, 15, 696. (148, 486)

Stegmüller, L. (1952) Bestimmung der optischen Natur durchsichtiger Einschlusskörper, entwickelt am Fluorit. Heidelberger Beitr. Mineral. Petrog., 3, 186-192. (86)

Stephenson, T.E. (1952) Sources of error in the decrepitation method of study of liquid inclusions. Econ. Geol., 47, 743-750. (265)

Sternfeld, J.N. (1981) The hydrothermal petrology and stable isotope geochemistry of two wells in the Geysers geothermal field, Sonoma County, California. M.S. thesis, Univ. of California, Riverside, CA, 202 p. (498)

Stettler, A. and P. Bochsler (1979) He, Ne and Ar composition in a neutron activated sea-floor basalt glass. Geochim. Cosmochim. Acta, 43, 157-169. (506)

Steven, T.A., C.G. Cunningham and M.N. Machette (1981) Integrated uranium systems in the Marysvale volcanic field, west-central Utah. In P.C. Goodell and A.C. Waters, eds. Uranium in Volcanic and Volcaniclastic Rocks. Am. Assoc. of Petrol. Geol. Studies in Geology, (13), 111-122. (462)

Stevens, R.E. and M.K. Carron (1948) Simple field test for distinguishing minerals by abrasion pH. Am. Mineral., 33, 31-49. (137)

Stevenson, F.J. (1962) Chemical state of the nitrogen in rocks. Geochim. Cosmochim. Acta, 26, 797-809. (119)

Stewart, D.B. and R.W. Potter, II (1979) Application of physical chemistry of fluids in rock salt at elevated temperature and pressure to repositories for radioactive waste. In G.J. McCarthy, ed., Scientific Basis for Nuclear Waste Management, Vol. 1. Plenum, New York, NY, 297-311. (276, 278)

_____, B.F. Jones, E. Roedder and R.W. Potter (1980) Summary of United States Geological Survey investigations of fluid-rock-waste reactions in evaporite environments under repository conditions. Underground Disposal of Radioactive Wastes, Int'l Atomic Energy Agency, Vienna, IAEA-SM-243/97, 335-344. (319)

Stoessel, R.K. and C.H. Moore (1983) Chemical constraints and origins of four groups of Gulf Coast reservoir fluids. Am. Assoc. Petrol. Geol. Bull., 67, 896-906. (311)

Stoller, G., D. Borcsik and H.D. Holland (1971) Chlorine in intrusives: A possible prospecting tool. Econ. Geol. 66, 361-367. (464)

Stolper, E. (1982a) Water in silicate glasses: An infrared spectroscopic study. Contrib. Mineral. Petrol., 81, 1-17. (84, 386, 486)

_____ (1982b) The speciation of water in silicate melts. Geochim. Cosmochim. Acta, 46, 2609-2620. (84, 386, 486)

Stone, H.W. (1943) Solubility of water in liquid carbon dioxide. Indus. Eng. Chem., 35, 1284-1286. (237, 525)

Stormer, J.C. and I.S.F. Carmichael (1971) Fluorine-hydroxyl exchange in apatite and biotite: A potential igneous geothermometer. Contrib. Mineral. Petrol., 31, 121-131. (390)

Stosch, H.-G. (1982) Rare earth element partitioning between minerals from anhydrous spinel peridotite xenoliths. Geochim. Cosmochim. Acta, 46, 793-811. (526)

Stueber, A.M. and P. Pushkar (1983) Application of strontium isotopes to origin of Smackover brines and diagenetic phases, southern Arkansas (abstr.). Am. Assoc. Petrol. Geol. Abstr. Ann. Convention, Dallas, (1983), 166. (320)

Suess, H.E. (1951) Gas content and age of tektites. Geochim. Cosmochim. Acta, 2, 76-79. (118, 558)

Sugisaki, R. (1981) Deep-seated gas emission induced by the earth tide: A basic observation for geochemical earthquake prediction. Science, 212, 1264-1266. (402)

_____, M. Ido, H. Takeda, Y. Isobe, Y. Hayashi, N. Nakamura, H. Satake and Y. Mizutani (1983) Origin of hydrogen and carbon dioxide in fault gases and its relation to fault activity. J. Geol., 91, 239-258. (402)

Sultanxodjaev, A.N., I.G. Chernov and T. Zakirov (1976) Hydrogeoseismic precursors to the Gasli earthquake. Rept Acad. Sci. Uzbekistan, (7), 51-53 (in Russian). (402)

Surles, T.L. (1978) Chemical and Thermal Variations Accompanying Formation of Garnet Skarns near Patagonia, Arizona. M.S. thesis, Univ. of Arizona, Tucson, AZ, 54 p. (455)

Sushchevskaya, T.M. and B.N. Ryzhenko (1977) Calculation of the composition of tin-bearing hydrothermal solutions. Geokhimiya, 1977, (7), 1091-1095 (in Russian; translated in Geochem. Int'l 14, (4) 88-91). (120)

_____, V.A. Dorofeeva and S.N. Knyazeva (1977) Comparison of two methods of calculating pH from fluid inclusion data. Geokhimiya, 1977, (6), 939-942 (in Russian). (136)

_____, B.N. Ryzhenko, S.N. Knyazeva, V.V. Malakhov and V.L. Barsukov (1978) Redox potential of tin-bearing hydrothermal solutions. Geokhimiya, 1978, (8), 1129-1138 (in Russian). (138)

Sutton, F.M. and A. McNabb (1977) Boiling curves at Broadlands geothermal field, New Zealand. New Zealand J. Sci., 20, 333-337. (500, 501)

Sutton, J.R. (1928) Diamond, a Descriptive Treatise. Thomas Murby & Co., London, 118 p. (507)

Sutton, Jr., R.L. (1964) Bubble crystals and notes on enhydros. Lapidary J., 18, 924-934. (92, 324)

Suzuoki, T., Y. Kuroda and S. Matsuo (1975) Hydrogen extractions from silicate minerals by sodium carbonate fusion and its application to deuterium analyses of fluid inclusions in some olivines. Geochim J., 9, 107-111. (121, 527)

Swallow, J.C. and J. Crease (1965) Hot salty water at the bottom of the Red Sea. Nature, 205, 165-166. (494)

Swanenberg, H.E.C. (1976) Fluid inclusion bubbles in a thermal gradient (abst.). Fluid Inclusion Res.--Proc. of COFFI, 8, 177. (168)

_____ (1979) Phase equilibria in carbonic systems, and their application to freezing studies of fluid inclusions. Contrib. Mineral. Petrol., 68, 303-306. (104, 242, 243, 244, 245, 282)

_____ (1980) Fluid inclusions in high-grade metamorphic rocks from S.W. Norway. Geologica Ultraiectina, Univ. Utrecht., (25), 147 p. (35, 62, 67, 68, 71, 74, 229, 230, 241, 258, 280, 344, 345, 346, 375, 378, 379, 522)

Swanson, S.E. (1979) The effect of CO_2 on phase equilibria and crystal growth in the system $KAlSi_3O_8-NaAlSi_3O_8-CaAl_2Si_2O_8-SiO_2-H_2O-CO_2$ to 8000 bars. Am. J. Sci., 279, 703-720. (386)

Swart, P.K., M.M. Grady, C.T. Pillinger, R.S. Lewis and E. Anders (1983a) Interstellar diamonds in meteorites. Science, 220, 406-410. (507)

_____, C.T. Pillinger, H.J. Milledge and M. Seal (1983b) Carbon isotopic variations within individual diamonds. Nature, 303, 793-795. (507)

Switzer, G.S. (1975) Composition of three glass phases present in an Apollo 15 basalt fragment. Mineral. Sci. Invest., 1972-1973, Smithsonian Contrib. Earth Sci., 14, 25-30. (552)

Taguchi, S. (1981) Underground thermal structure of geothermal fields revealed by the distribution of homogenization temperature of fluid inclusions. J. Geothermal Res., 3, 165-177 (in Japanese; English abstr. and fig. captions). (295, 497)

____ (1982) Underground thermal structure of geothermal fields revealed by the distribution of homogenization temperature of fluid inclusions. J. Geothermal Res. Soc. Japan, 3, (3), 165-177 (in Japanese with English abstr.). (497)

____ (1983) Study on Geothermal Geology of the Kirishima Volcanic Region. Ph.D. dissertation, Kyusha Univ. (in English). (497)

____ and M. Hayashi (1982) Application of the fluid inclusion thermometer to some geothermal fields in Japan. Geothermal Resources Council Trans., 6, 59-62. (497)

____ and ____ (1983) Fluid inclusion study in some geothermal fields of Kyushu, Japan (extended abstr.). Fourth Int'l Symp. Water-Rock Interaction. Aug. 29-Sept. 3, 1983, Misasa, Japan, 459-462. (497)

____, T. Fujino and M. Hayashi (1979) Homogenization temperature measurements of fluid inclusions in quartz and anhydrite from the Hatchobaro geothermal field, Japan, and its applications for geothermal development. Geothermal Resources Council Trans, 3, 705-708. (497)

____, M. Okaguchi and T. Yamasaki (1980) Reduction in the length of fission tracks by geothermal heating and its application to thermal history. Rept. Res. Inst. Industrial Sci., Kyushu Univ. No. 72, 21-26 (in Japanese; English abstr.). (497)

____, Y. Matsumoto, M. Hayashi, T. Fujino and T. Yamasaki (1981) Geothermal structure of Kirishima volcano in southern Kyushu Japan. IAVCEI Symp. on Arc Volcanism, 363-364. (497)

Tait, P.G. and W. Swan (1874) Notes on Mr. Sang's communication of 7th April 1873 on a singular property possessed by the fluid enclosed in crystal cavities in Iceland spar. Royal Soc. Edinburgh Proc., 8,.(87), 247-254. (92)

Takahashi, E. (1978) Partitioning of Ni^{2+}, CO^{2+} Fe^{2+}, Mn^{2+} and Mg^{2+} between olivine and silicate melts: compositional dependence of partition coefficient. Geochim. Cosmochim. Acta, 42, 1829-1844. (554)

Takenouchi, S. (1962) Polyphase inclusions in the quartz from the Taishu mine, Nagasaki Prefecture. Mining Geol. (Japan), 12, (55), 294-297 (in Japanese; English abstr.). (95)

____ (1970) Fluid inclusion study by means of heating-stage and freezing-stage microscope. Mining Geol. (Japan), 20, (103), 31-40 (in Japanese; English abstr.). (190)

____ (1972) Glass inclusions in quartz of volcanic rocks from mining areas (abstr.). Int'l Geol. Congr., 24th, Abstr., Section 10, Geochem., 325. (480)

____ (1980a) Fluid inclusion studies in Japan and their application to mineral exploration. In 4th Joint Mtg MMIJ-AIME, 1980, Tokyo, Tech. Session A-1, 1-16. (425, 455, 457)

____ (1980b) Preliminary studies on fluid inclusions of the Santo Tomas II (Philex) and Tapian (Marcopper) porphyry copper deposits in the Philippines. Mining Geol. (Japan) Spec. Issue (8), 141-150. (443)

____ and H. Imai (1968) On the salinity of liquid inclusions in quartz crystals. COFFI, 2, 73. (190, 425)

____ (1971) Fluid inclusion study of some tungsten-quartz veins in Japan. Soc. Mining Geol. Japan, Spec. Issue 3, 345-350 (Proc. IMA-IAGOD Meetings '70, IAGOD Vol.). (457)

____ and H. Imai (1975) Glass and fluid inclusions in acidic igneous rocks from some mining areas in Japan. Econ. Geol., 70, 750-769. (485)

Takenouchi, S. and T. Katsura (1972) Volcanic glass inclusions in rhyolite and tuff from the Chitose mine, Hokkaido. Mining Geol. (Japan), 22, (5) 383-391. (480)

____ and G.C. Kennedy (1964) The binary system H_2O-CO_2 at high temperatures and pressures. Am. J. Sci., 262, 1055-1074. (51, 103, 237, 238, 288)

____ (1965a) Dissociation pressures of the phase $CO_2 \cdot 5$-$3/4H_2O$. J. Geol., 73, 383-390. (99, 241)

____ (1965b) The solubility of carbon dioxide in NaCl solutions at high temperatures and pressures. Am. J. Sci., 263, 445-454. (103, 246, 281, 282, 369)

Tan, T.H. (1979) The Genesis of the King Island Scheelite (Dolphin) Deposit as Determined from Fluid Inclusion and Mineral Chemical Zoning Studies. Ph.D. dissertation, La Trobe Univ., Bundoora, Victoria, Australia, 367 p. (454, 455)

____ and T.A.P. Kwak (1979) The measurement of the thermal history around the Grassy granodiorite, King Island, Tasmania, by use of fluid inclusion data. J. Geol., 87, 43-54. (455)

Taylor, B.E. (1981) Hydrothermal fluids in the Mother Lode gold deposits of California (abstr.). EOS, Trans. Am. Geophys. Union, 62, 1059. (356)

____ E.E. Foord and H. Friedrichsen (1979) Stable isotope and fluid inclusion studies of gem-bearing granitic pegmatite-aplite dikes, San Diego Co., California. Contrib. Mineral. Petrol., 68, 197-205. (388, 390)

Taylor, Jr., H.P. (1977) Water/rock interactions and the origin of H_2O in granitic batholiths. J. Geol. Soc. London, 133, 509-558. (401)

____ (1979a) Oxygen and hydrogen isotope evidence for meteoric-hydrothermal alteration and ore deposition in the epithermal gold-silver deposits associated with volcanic centers in western Nevada: A case study of the Tonopah deposit (abstr.). Geol. Soc. Am. Abstr. Programs, 11, 526. (127, 427)

____ (1979b) Oxygen and hydrogen isotope relationships in hydrothermal mineral deposits. In H.L. Barnes, ed., Geochemistry of Hydrothermal Ore Deposits. Wiley & Sons, New York, NY, 236-277. (127)

Terashima, S. and S. Ishihara (1980) Anomalous chlorine contents of Miocene granitoids from Tsushima, Japan. Japan Assoc. Mineral. Petrol. Econ. Geol. J., 75, 62-67. (389)

Theodore, T.G. and D.W. Blake (1978) Geology and geochemistry of the west ore body and associated skarns, Copper Canyon porphyry copper deposits, Lander County, Nevada. U.S. Geol. Survey Prof. Paper, 748-C, 85 p. (443, 456)

____ and W.D. Menzie (1984) Fluorine-deficient porphyry molybdenum deposits in the western North American Cordillera. Proc. 1982 IAGOD Symposium, Tbilisi, USSR (in press). (439)

____, W.N. Blair and J.T. Nash (1982) Preliminary report on the geology and gold mineralization of the Gold Basin-Last Basin mining districts, Mohave County, Arizona. U.S. Geol. Survey Open-File Rept. 82-1052, 322 p. (464)

Thisse, Y., E. Oudin and C. Ramboz (1983) Boiling fluids in the Red Sea metalliferous sediments (abstr.). Symp. Program, European Current Research on Fluid Inclusions, April, 1983, Univ. d'Orleans, France, 52. (494)

Thoma, K. and D. Eckart (1964) Untersuchungen an gashaltigen Mineralsalzen I. Teil: Bergmännische Untersuchungen. Bergakademie, 16, 674-676. (315)

Thomas, D.M. (1979) Gases from Hawaiian volcanoes (abstr.). Hawaii Symp. on Intraplate Volcanism and Submarine Volcanism, Hilo, Hawaii, July 16-22, 1979. Int'l Assoc. on Volcanism and Chemistry Earth's Interior, 175. (528)

Thomas, R. (1979) Analysis of Inclusions for the Thermodynamic and Physical Chemical Characterization of Ore-Forming Processes in Magmatic and Post-Magmatic Environments. Dissertation, Bergacademie Freiberg, Freiberg, East Germany, 248 and 83 p. (two vols.; in German). (113, 456, 457)

____ and L. Bauman (1980) Ergebnisse von thermometrischen und kryometrischen Untersuchungen an Kassiteriten des Erzgebirges. Z. geol. Wiss., Berlin 1980, (10), 1281-1299. (456, 457)

Thompson, A.B. (1982) Wet and dry metamorphism and melting (abstr.). Geol. Soc. Newsletter (London), 11, (4), 14. (383)

Thompson, J.M., S.S. Howe and W.E. Hall (1983) Chemical analysis of fluid inclusions by ion chromatography (abstr.). Rocky Mtn. Conf., 25th Annual Mtg., Denver, Colorado, Aug. 14-17, 1983, Abstr. and Mtg Program, 108. (132)

Thompson, M. (1981) Prospects for the use of the inductively coupled plasma (ICP) in fluid inclusion analysis (abstr.). Program, Symp. on Current Research on Fluid Inclusions, Univ. Utrecht, Utrecht, The Netherlands, April 22-24, 1981 (unpaginated). (131, 132)

____, A.H. Rankin, S.J. Walton, C. Halls and B.N. Foo (1980) The analysis of fluid inclusion decrepitate by inductively-coupled plasma atomic emission spectroscopy: an exploratory study. Chem. Geol., 30, 121-133. (131, 132, 467)

Thompson, P., H.P. Schwarz and D.C. Ford (1976) Stable isotope geochemistry, geothermometry, and geochronology of speleothems from West Virginia. Geol. Soc. Am. Bull., 87, 1730-1738. (327)

Thompson, T.B., T. Lyttle, J.R. Pierson, and L.W. Osborne (1979) Genesis of the Bokan Mountain U-Th deposits southeastern Alaska (abstr.). Geol. Soc. Am. Abstr. Programs, 11, 527. (462)

Thoms, R.L. and J.D. Martinez (1980) Blowouts in domal salt. 5th Int'l Symp. on Salt: Northern Ohio Geol. Soc., Cleveland, Ohio, 405-411. (315)

Thomsen, L. (1980) ^{129}Xe on the outgassing of the atmosphere. J. Geophys. Res., 85, (88), 4374-4378. (325, 527)

Thornber, C.R. and J.S. Huebner (1984) Dissolution of olivine in basaltic liquids. I: Experimental observations and applications. Am. Mineral. (in press). (40)

Tillman, J.E. (1983) Exploration for reservoirs with fracture-enchanced permeability. Oil and Gas J., Feb. 1983, 165-180. (333)

____ and H.L. Barnes (1983) Deciphering fracturing and fluid migration histories in the Northern Appalachian Basin. Am. Assoc. Petrol. Geol. Bull., 67, 692-705. (333)

Titley, S.P. (1978) Copper, molybdenum, and gold content of some porphyry copper systems of the southwestern and western Pacific. Econ. Geol., 78, 977-981. (439)

____, ed. (1982) Advances in Geology of the Porphyry Copper Deposits, Southwestern North America. Univ. of Arizona Press, Tucson, AZ, 560 p. (439, 441)

____ and R.E. Beane (1981) Porphyry copper deposits (Parts I & II). Economic Geology 75th Anniversary Volume, 214-269. (441)

Tobin, M.C. (1971) Laser Raman Spectroscopy: Chemical Analysis, Vol. 35, P.J. Elving and I.M. Kolthoff, ser. eds. Wiley-Interscience, New York, NY. (105)

Todd, B.J. (1956) Mass spectrometer analysis of gases in blisters in glass. Soc. Glass Tech. Trans., 40, 32T-38T. (122)

Todheide, K. (1963) Das Zweiphasengebiet und die Kritische Kurve im System Kohlendioxyd-Wasser bis zu Drucken von 3500 Bar. Ph.D. dissertation, George August Univ., Göttingen, W. Germany, 54 p. (238, 281)

____ and E.U. Franck (1963) Das Zweiphasengebiet und die Kritische Kurve im System Kohlendioxid-Wasser bis zu Drucken von 3500 bar. Z. Phys. Chemie, new ser., 37, 387-401. (103, 279, 369, 372, 373)

Tokunaga, M. and H. Honma (1974) Fluid inclusions in the minerals from some Kuroko deposits. Mining Geol. Spec. Issue, 6, 385-388. (425)

Tokunaga, M., T. Miyazawa, H. Honma, and H. Park (1970) Formation temperature of some black ore deposits in Akita Prefecture, Japan (abstr.). Collected Abstr., IMA-IAGOD Meetings 1970: Tokyo. Science Council of Japan, Tokyo, 246. (425)

Tolansky, S. and P.G. Morris (1947) An interferometric survey of the micas. Mineral. Mag., 28, 137-145. (160)

Tomilenko, A.A. and Yu.A. Dolgov (1978) Conditions of formation of the "granulated" quartz from the Borus chain (western Sayan). Akad. Nauk SSSR, Dokl., 242, (5) 1173-1176 (in Russian). (372)

Tomilenko, A.A., V.P. Chupin and Yu.A. Dolgov (1976) Conditions of formation of metamorphic rocks, derived from data of studies of inclusions. In Yu.A. Dolgov et al., eds., Genetic Studies in Mineralogy. Inst. Geol. Geophys., Sib. Branch Acad. Sci. USSR, Novosibirsk, 138-141 (in Russian). (35)

____, N.V. Berdnikov and L.P. Karasakov (1977) Cryometric study on inclusions in rocks of the deep-seated Chogar metamorphic complex, eastern Siberia. Akad. Nauk SSSR, Dokl., 234, (5), 1189-1192 (in Russian; translated in Dokl. Acad. Sci. USSR, Earth Sci. Sects., 234, 163-166, 1979). (375)

Torgerson, T., J.E. Lupton, D.S. Sheppard and W.F. Giggenbach (1982) Helium isotope variations in the thermal areas of New Zealand. J. Volcan. Geothermal Res., 12, 283-298. (513)

Torza, S. and S.G. Mason (1969) Coalescence of two immiscible liquid drops. Science, 163, 813-814. (53, 54)

Toulmin, III, P. and S.P. Clark, Jr. (1967) Thermal aspects of ore formation. In H.L. Barnes, ed., Geochemistry of Hydrothermal Ore Deposits (1st Edition). Hold, Rinehart & Winston, New York, NY, 437-464. (269)

Touray, J.-C. (1968) Recherches geochimiques sur les inclusions à CO_2 liquide. Bull. Soc. franc. Mineral. Cristallogr., 91, 367-382. (124, 211)

____ (1973) Evidence (from fluid inclusions) of fossil geothermal areas (abstr.). Reun. Ann. Sci. Terre (Progr. Resumes) 1973, 398 (in French). (496)

____ (1976) Activation analysis for liquid inclusion studies: a brief review. Bull. Soc. franc. Mineral. Cristallogr., 99, 162-164. (124, 134)

____ and J. Barlier (1975) Liquid and gaseous hydrocarbon inclusions in quartz monocrystals from "Terres Noires" and "Flysch a Helminthoides" (French Alps). Fortschr. Mineral. Spec. Issue to v. 52, 419-426. (322)

____ and M. Guilhaumou (1984) Characterization of H_2S bearing fluid inclusions. Bull. Mineral. (in press). (108, 244, 463)

____ and A. Jauzein (1967) Inclusions à methane dans les quartz des "terres noires" de la Drome. Compt. Rend. Acad. Sci. Paris, 264, (16), ser. D, 1957-1960. (323)

____ and F. Lantelme (1966) Analyse des inclus des minéraux méthode du chauffage progressif. Bull. Soc. franc. Mineral. Cristallogr., 89, 394-398. (122)

____ and C. Sabouraud (1970) Metastable inclusion brines in fluorite from Quezzanne. Econ. Geol., 65, 216-218. (93, 99, 294)

____ and J.-P. Sagon (1967) Inclusions à methane dans les quartz des marnes de la region de Mauléon (Basses-Pyrénées). Compt. Rend. Acad. Sci. Paris, 265, (18), ser. D, 1269-1272. (90, 124, 323)

____ and F. Tona (1974) Geodynamic interpretation of a fluid inclusion study: first example in the Sierra de Lujar (Granda, Spain). Revue Geog. Phys. Geol. Dynam., ser. 2, 16, pt. 1, 71-74 (in French). (355)

____ and A. Ziserman (1981) Observations sur la note "Les inclusions fluides, témoins et faux-témoins des conditions de dépôt. Quelques exemples pris dans les minéralizations de Pb, Zn, Ba, F du Sud du Massif Central Français" de Mineralium Deposita, 211-230, (1980). Mineralium Deposita, 16, 177-179 (see also Sabouraud et al., 181-183). (421)

Touret, J. (1969) Le Socle Précambrien de la Norvège Méridionale. These, Univ. Nancy, Nancy I, France, 3 vol., 609 p. (375)

____ (1971) The granulite facies in Southern Norway. I: The mineral associations; II: The fluid inclusions. Lithos, 4, 239-249 and 423-436 (in French; English abstr.). (362, 375, 376)

____ (1974) Granulite facies and CO_2 fluids. Centenaire de la Soc. Geol. des Domaines Cristallins, Liege, 1974, 267-287 (in French; English abstr.). (375, 377)

____ (1977) The significance of fluid inclusions in metamorphic rocks. In D.G. Fraser, ed., Thermodynamics in Geology. D. Reidel Publ. Co., Dordrecht, The Netherlands, 203-227. (71, 258, 338, 344, 376, 579)

Touret, J. (1981) Fluid inclusions in high grade metamorphic rocks. Mineral. Assoc. Canada Short Course Handbook, 6, 182-208. (344, 362, 363, 380, 383, 384, 515)

_____ (1982) An empirical phase diagram for a part of the N_2-CO_2 system at low temperature. Chem. Geol., 37, 49-58. (245, 383, 384)

_____ and Y. Bottinga (1979) Equation of state of CO_2: application to carbonic inclusions. Bull. Mineral., 102, 577-583 (in French). (227, 228)

_____, D.C. Smith and S.A. Kechid (1982) Fluid inclusions in some eclogites and gneisses from the western gneiss region, Norway (abstr.). Terra Cognita, 2, p. 318. (374)

Travers, M.W. (1898) The origin of gases evolved on heating mineral substances, meteorites, etc. Royal Soc. (London) Proc., 64, 130-142. (120)

Treiman, A.H. and E.J. Essene (1981) Liquid immiscibility in the Oka carbonatite (abstr.). EOS, Trans. Am. Geophys. Union, 62, 415. (408)

Trofimov, V.S. (1974) Amber. "Nedra" Pub. House, Moscow, 183 p. (in Russian).

Trufanov, V.N. (1967) Determination of pH of mineralizing solutions. Geokhimiya, 1967, 694-702 (in Russian; translated in Geochem. Int'l, 4, (3), 567-575, 1967). (137)

_____ (1972) Thermodynamics of post-magmatic processes of mineral formation of northern Caucasus. In V.I. Smirnov, ed. The Ore-Forming Environment as Determined from Inclusions in Minerals. "Nauka" Press, Moscow, 107-115 (in Russian). (52, 186)

_____ and S.A. Kurshev (1968) Determination of the composition of inclusions in the electron microscope. In Mineralogical Thermometry and Barometry, Vol. 2. "Nauka" Press, Moscow, 70-72 (in Russian). (148)

_____ and N.G. Rodzyanko (1973) Essential achievements and perspectives of development of thermobarogeochemistry of ore-forming processes at the Northern Caucasus (abstr.). COFFI, 6, 158. (84, 186)

*

Tsinober, L.I., V.E. Khadzhi, L.A. Gordienko and M.I. Samoilovich (1968) The nature of defects in synthetic α-quartz. In N.N. Sheftal', ed., Growth of Crystals, Vol. 6A, Consultants Bur., New York, 25-36. (24)

Tsui, T.-F. (1976) Laser Microprobe Analysis of Fluid Inclusions in Quartz. Ph.D. dissertation, Harvard Univ., Cambridge, Massachusetts, 156 p. (138)

_____ and H.D. Holland (1979) The analysis of fluid inclusions by laser microprobe. Econ. Geol., 74, 1647-1653. (138, 139, 436)

Tsusue, A., T. Mizuta, M. Watanabe and K.G. Min (1981) Jurassic and Cretaceous granitic rocks in South Korea. Mining Geol. (Japan), 31, (4), 1-76. (390)

Turner, F.J. and J. Verhoogen (1960) Igneous and Metamorphic Petrology. (2nd Edition), McGraw-Hill Book Co, New York, NY. (387, 388)

Tuttle, O.F. (1949) Structural petrology of planes of liquid inclusions. J. Geol., 57, 331-356. (62, 63)

_____ (1952) Origin of the contrasting mineralogy of extrusive and salic rocks. J. Geol., 60, 107-124. (397, 474)

Uchameyshvili, N.E. and N.I. Khitarov (1965) On the chemical composition of solutions of liquid inclusions in barites. In Mineralogical Thermometry and Barometry. "Nauka" Press, Moscow, 227-232 (in Russian). (136)

Umova, M.A., R.I. Glebov and P.N. Shibanov (1957) A study of the chemical composition of gaseous inclusions in quartz from various deposits. Akad. Nauk SSSR, Dokl., 112, 519-521 (in Russian). (119)

_____ and _____ (1960) Apparatus for extraction of gas-liquid inclusions from minerals. Determination of chemical composition of inclusions. Sverdlovsk, Ekaterinburg Gorn. Inst. Trudy. (36), 49-59; Referativnyi Zhur., Geol., Abstr. 9V440, 1961 (in Russian). (119)

Urabe, T. and T. Sato (1978) Kuroko deposits of the Kosaka mine, Northeast Honshu, Japan -- Products of submarine hot springs on Miocene sea floor. Econ. Geol., 73, 161-179. (425)

Urusova, M.A. (1974) Phase equilibria and thermodynamic characteristics of solutions in the systems NaCl-H_2O and NaOH-H_2O at 350-550°C. Geokhimiya, 1974, (9), 1360-1366 (in Russian; translated in Geochem. Int'l, 11, (5), 944-950, 1974). (234)

_____ (1975) Volume properties of aqueous solutions of sodium chloride at elevated temperatures and pressures. Zh. Neorg. Khim., 20, 3103-3110 (in Russian; translated in Russian J. Inorg. Chem., 20, (11), 1717-1721). (234, 274, 275, 276, 277, 278)

_____ and M.I. Ravich (1971) Vapor pressure and solubility in the sodium chloride-water system at 350° and 400°C. Russian J. Inorg. Chem. 16, (10), 1534-1535. (234)

Ushakovskii, V.T. (1966) Vein quartz from one of the rock crystal deposits in the southern Urals. Vses. Mineral. Obshch. Zapiski, 95, 229-232 (in Russian). (111)

Usselman, T.M., G.E. Lofgren, C.H. Donaldson and R.J. Williams (1975) Experimentally reproduced textures and mineral chemistries of high-titanium mare basalts. Proc. Lunar Sci. Conf., 6th, 997-1020. (539)

Vanderslice, T.A. and N.R. Whetten (1962) Cleavage of alkali halide single crystals in high vacua. Analysis of evolved gases. J. Chem. Physics, 37, 535-539. (123)

van Gizel, P. (1979) Manual of the Techniques and Some Geological Applications of Fluorescence Microscopy, a Workshop Sponsored by Assoc. Stratigraphic Palynologists, 12th Annual Mtg Dallas, Texas, 1979. Core Labs., Inc., Dallas, TX, 55 p. (83, 335)

Vanko, D.A. and F.C. Bishop (1982) Occurrence and origin of marialitic scapolite in the Humboldt Lopolith, N.W. Nevada. Contrib. Mineral. Petrol., 81, 277-289. (48, 295, 363)

*Vasilenko, V.N. and S.A. Kurshev (1977) Certain physico-chemical peculiarities of formation of native gold in the NW Caucasus, S regions of Siberia, and in the Far East (abstr.). COFFI, 10, 289. (137)

Vdovykin, G.P., Y.I. Bodunov, A.N. Izosimova, A.I. Botkunov, N.A. Utkina, A.B. Bochkovskaya and I.N. Zueva (1979) Bitumens in kimberlites of the "Mir" pipe. Akad. Nauk SSSR, Dokl., 245, (4), 941-945 (in Russian). (512)

Veitch, J.D. and D.G. McLeroy (1972) Organic mobilization of ore metals in low-temperature carbonate environments (abstr.). Geol. Soc. Am. Abstr. Program, 4, (7), 110-111. (422)

Velchev, V.N. and F.P. Mel'nikov (1965) An attempt to apply the microscope with the Kofler stage to thermo- and cryometric investigations of liquid inclusions in some minerals. In Mineralogical Thermometry and Barometry. "Nauka" Press, Moscow, 123-128 (in Russian). (190)

Vil'denberg, Y.V., I.V. Vysotskiy, G.I. Voytov and R.N. Murogova (1978) Gases in salts from Fore-Caspian depression and adjacent regions. Akad. Nauk SSSR, Dokl., 239, (6), 1430-1433 (in Russian). (315)

Vincent, D., R. Clocchiatti and Y. Langevin (1981) First fission track dating of glass inclusions in quartz phenocrysts from pyroclastic formations. Compt. Rend. Acad. Sci. Paris, 293, Ser. D, 229-234, (in French). (479)

Visser, W. (1982) Maximum diagenetic temperature in a petroleum source-rock from Venezuela by fluid inclusion geothermometry. Chem. Geol., 37, 95-101. (335, 347)

Vochten, R. E. Esmans and W. Vermejrsch (1977) Study of the solid and gaseous inclusions in fluorites from Wösendorf (Bavaria, F.R. of Germany) and Margnac (Haute Vienne, France) by microprobe and mass spectrometry. Chem. Geol., 20, 253-264. (123)

Vogelsang, H. (1867) Philosophie der Geologie und mikroskopische Gesteinsstudien. Cohen and Sohn, Bonn, Germany. (474)

Vogelsang, H. and H. Geissler (1869) Ueber die Natur der Flüssigkeitseinschlüsse in gewissen Mineralien. Ann. Physik Chemie, 137, ser. 5, 17, 56-75. (116, 185)

Vollbrecht, A. (1981) Tectonogenesis of the Münchberg Gneiss Massif (quartz-fabric investigations and microthermometry on fluid inclusions), Göttinger Arb. Geol. Paläont., 24, 122 p. (Thesis, Georg-August-Univ., Göttingen, W. Germany, 1980) [in German; English abstr.). (354)

Volokhov, I.M. (1975) Evaluation of accuracy of thermobarogeochemical reconstructions of formation of magmas and magmatic hearths. Akad. Nauk SSSR, Sib. Otdel., Geol. i Geofiz. no. 1, (181), 12-19 (in Russian; translated in Fluid Inclusion Res. -- Proc. of COFFI, 8, 214-219). (402)

Vorob'ev, E.L. (1978) Exploration for potassium salts on the Siberian Platform according to the fluid inclusions in hydrothermal minerals. Teor. Prakt. Termobarogeokhim., (Dokl. 5th Vses. Soveshch., 1976) N.P. Ermakov, ed.. "Nauka" Press, Moscow, 206-209 (in Russian). (314)
____ and V.I. Lozhkin (1976) The identification of daughter minerals of fluid inclusions by means of X-ray coloration. Year Book Siberian Inst. Geochem., 1975, 281-283 (publ. 1976), (in Russian; translated in Fluid Inclusion Res. -- Proc. of COFFI, 9, 253-254). (98)

Vozniak, D.K., V.K. Kvasnitsa and V.M. Krochuk (1981) Hardened melt inclusions in the baddeleyite from the carbonatites of Sea-of-Azov area. Akad. Nauk SSSR, Dokl., 259, 952-955 (in Russian). (407)

*Voznyak, D.K. (1968) Conditions of formation of topaz in syngenetic minerals in Volyn' pegmatites (abstr.). COFFI, 1, 4). (101, 261, 398)
____ and Yu.A. Galaburda (1977) Stage for cryometric research of inclusions in minerals. L'vov. Gos. Univ., Mineral. Sbornik, 31, (1), 63-65 (in Russian). (190)
____ and V.A. Kalyuzhnyi (1974a) Decrepitated inclusions and their significance for reconstruction of mineral formation (illustrated by quartz from pegmatites from Volhyn). In Ye.K. Lazarenko, ed., Typomorphism of Ukrainian Quartz. Acad. Sci. Ukr. SSR, Kiev, 18-24 (in Russian; translated in Fluid Inclusion Res. -- Proc. of COFFI, 7, 272-275). (71)
____ and ____ (1974b) Change of shape of inclusions in minerals of variable composition and its effect on the composition of the parent solution isolated in vacuoles. Akad. Nauk. SSSR, Dokl., 212, (5), 1192-1195 (in Russian; translated in Dokl. Acad. Sci. USSR, 212, 140-143, 1974). (49, 477)
____ and ____ (1976) Use of decrepitation of inclusions to reconstruct P-T conditions of mineral formation (using quartz from Volyn pegmatites as an example). L'vov Gos. Univ., Mineral. Sbornik, 2, (30), 31-40 (in Russian). (285)
*____, V.N. Kvasnitsa and Yu.A. Galaburda (1974) Typomorphic peculiarities of "Marmarosh diamonds." In Ye.K. Lazarenko, ed., Typomorphism of Ukrainian Quartz. Naukoya Dumka Pub. House, Kiev, 29-82 (in Russian). (323)

Wahler, W. (1956) Über die in Kristallen eingeschlossenen Flüssigkeiten und Gase. Geochim. Cosmochim. Acta, 9, 105-135. (86, 89, 104, 117, 119, 121)

Wakita, H., Y. Nakamura, I. Kita, N. Fujii and K. Notsu (1980) Hydrogen release: New indicator of fault activity. Science, 210, 188-190. (402)

Walenczak, Z. (1977a) Optical ultramicroscopic observations of vesicles in opaque minerals of the Pultusk meteorite (Poland). Bull. Acad. Polon. Sci. Ser. Sci. de la Terr., 25, 15-18 (in English). (566)
____ (1977b) Liquid phase vesicles in the Pultusk meteorite. Bull. Acad. Polon. Sci. Ser. Sci. de la Terre, 25, 19-22 (in English). (566)

Walker, D., R.J. Kirkpatrick, J. Longhi and J.F. Hays (1976) Crystallization history of lunar picritic basalt sample 12002: phase equilibria and cooling-rate studies. Geol. Soc. Am. Bull., 87, 646-656. (554)

Walsh, J.N. and R.A. Howie (1980) An evaluation of the performance of an inductively coupled plasma source spectrometer for the determination of the major and trace constituents of silicate rocks and minerals. Mineral. Mag., 43, 967-974. (132)

Walshe, J.L. and M. Solomon (1981) An investigation into the environment of formation of the volcanic-hosted Mt. Lyell copper deposits using geology, mineralogy, stable isotopes, and a six-component chlorite solid solution model. Econ. Geol., 76, 246-284. (461)

Walther, J. and E. Althaus (1983) Fluid inclusions in ultramafic xenoliths of volcanites from Southern and Western Germany (abstr.). Fortschr. Mineral., 61, p. 217. (527)

Wanamaker, B.J., S.C. Bergman and B. Evans (1982) Crack healing in silicates: Observations on natural lherzolite nodules (abstr.). EOS, Trans. Am. Geophys. Union, 63, 437. (63, 530)

Wand, U., H.-M. Nitzsche, K. Mühle and K. Wetzel (1980) Nitrogen isotope composition in natural diamonds-first results. Chem. Erde, 39, 85-87. (509)

Wardlaw, N.C. and W.M. Schwerdtner (1966) Halite-anhydrite seasonal layers in the Middle Devonian Prairie Evaporite Formation, Saskatchewan, Canada. Geol. Soc. Am. Bull., 77, 331-342. (310, 311)

Ware, R.K. and P.P. Pirooz (1967) Gas adsorption on freshly broken glass surfaces--a source of error in analysis of bubbles in glass. Glass Tech., 8, (3), 86-87. (120)

Harner, J.L., J.D. Ashwal, S.C. Bergman, E.K. Gibson, Jr., D.J. Henry, R. Lee-Berman, E. Roedder and H.E. Belkin (1983) Fluid inclusions in stony meteorites. Proc. 13th Lunar Sci. Conf., Pt. 2, J. Geophys. Res., 88, Suppl., A731-A735. (2, 566, 567, 568, 569)

Watanabe, M. (1981) Reconnaissance study on the fluid inclusions in some Jurassic and Cretaceous granitic rocks in Republic of Korea. Mining Geol. (Japan), 31, (4), 109-115. (398)

Watanabe, S., K. Mishima and S. Matsuo (1983) Content and isotopic ratio of carbonaceous materials incorporated in olivine crystals from the Hualalai volcano, Hawaii - an approach to mantle carbon. Geochem. J., 17, (2), 95-104. (528)

Watkinson, D.H. and P.R. Mainwaring (1981) Sodic solid and fluid inclusions in chromite from stratiform and podiform chromites in Canada (abstr.). Geol. Assoc. Canada/Mineral. Assoc. Canada Abstr., 6, A-60. (463)

Matmuff, G. (1978) Geology and alteration-mineralization zoning in the central portion of the Yandera porphyry copper prospect, Papua New Guinea. Econ. Geol., 73, 829-856. (443)

Watson, E.B. (1976) Glass inclusions as samples of early magmatic liquid: determinative method and application to a South Atlantic basalt. J. Volcan. Geothermal Res., 1, 73-84. (40, 483)
____ (1979) Diffusion of cesium ions in H_2O-saturated granitic melt. Science, 205, 1259-1260. (39)

Weast, R.C. (1980) CRC Handbook of Chemistry and Physics, 60th edition, CRC Press, Boca Raton, Florida. (228)

Weathers, M.S., J.M. Bird, R.F. Cooper and D.L. Kohlstedt (1979) Differential stress determined from deformation-induced microstructures of the Moine thrust zone. J. Geophy. Res., 84, 7495-7505. (75)

Weaver, J.H. and G. Margaritondo (1979) Solid-state photoelectron spectroscopy with synchrotron radiation. Science, 206, 151-156. (113)

Webster, R. (1952) The secrets of the synthetic emerald: The Gemmologist, 21, (1 and 2), 117-121, 140-145. (365)

Weeks, W.F. and S.F. Ackley (1982) The growth, structure, and properties of sea ice. Cold Regions Research and Eng. Lab. Monograph 82-1. (250, 329)

Weiblen, P.W. (1977) Examination of the liquid line of descent of mare basalts in light of data from melt inclusions in olivien. Proc. 8th Lunar Conf., 1751-1765. (483)
____ and E. Roedder (1973) Petrology of melt inclusions in Apollo samples 15598 and 62295, and of clasts in 67915 and several lunar soils. Proc. 4th Lunar Sci. Conf., 681-703, (535, 546, 552)
____ and ____ (1976) Compositional interrelationships of mare basalts from bulk chemical and melt inclusion studies. Proc. 7th Lunar Sci. Conf., 1449-1466. (536, 546)

Weidner, J.R. (1982) Iron-oxide magmas in the system Fe-C-O. Canadian Mineral., 20, 555-566. (385)

Weis, P.L. (1953) Fluid inclusions in minerals from zoned pegmatites of the Black Hills, South Dakota. Am. Mineral., 38, 671-697. (87)

Weisbrod, A. and B. Poty (1975) Thermodynamics and geochemistry of the hydrothermal evolution of the Mayres pegmatite, southeastern Massif Central (France). (Part II). Petrologie, 1, (1), 1-16; Part II, 89-102 (in English). (370, 387)

_____ and J. Touret (1976) Fluid inclusions in geochemistry and petrology: present trends. Bull. Soc. franc. Minéral. Cristallogr., 99, 140-152 (in French). (222, 338)

Weissberg, B.G., P.R.L. Browne and T.M. Seward (1979) Ore metals in active geothermal systems. In H.L. Barnes, ed., Geochemistry of Hydrothermal Ore Deposits, 2nd Edn. Wiley & Sons. New York, NY, 738-780. (358)

Welhan, J.A. and H. Craig (1979) Methane and hydrogen in East Pacific Rise hydrothermal fluids. Geophys. Res. Lett., 6, 829-831. (513)

_____ and _____ (1983) Methane and helium in deep earth gases (abstr.). Am. Chem. Soc. 186th Mtg. Program, Div. of Geochem., Section A, Paper 14 (unpaginated). (513)

Wells, R.A. (1953) Peculiar inclusion found in synthetic emerald. Gems and Gemology, 7, 283. (365)

Wendlandt, R.F. (1981) Influence of CO_2 on melting of model granulite facies assemblages: a model for the genesis of charnockites. Am. Mineral., 66, 1164-1174. (383)

_____ (1982) Sulfide saturation of basalt and andesite melts at high pressures and temperatures. Am. Mineral., 67, 877-885. (475, 505)

Werre, Jr., R.W., R.J. Bodnar, P.M. Bethke and P.B. Barton, Jr. (1979) A novel gas-flow fluid inclusion heating/freezing stage (abstr.). Geol. Soc. Am. Abstr. Programs, 11, 539. (187, 191, 196, 197, 204)

Wetlaufer, P.H., P.M. Bethke, P.B. Barton, Jr. and R.O. Rye (1979) The Creede Ag-Pb-Zn-Cu-Au district, Central San Juan Mountains, Colorado: a fossil geothermal system. In J.D. Ridge, ed., Papers on Mineral Deposits of Western North America, IAGOD 5th Quadrennial Symp. Proc. Vol. 2. Nevada Bur. Mines and Geol. Rept. 33, 159-164. (426, 496)

White, D.E. (1957) Thermal waters of volcanic origin. Geol. Soc. Am. Bull., 68, 1637-1658. (24)

_____ (1967) Mercury and base-metal deposits with associated thermal and mineral waters. In H.L. Barnes (Ed.), Geochemistry of Hydrothermal Ore Deposits, Holt, Rinehart and Winston, New York, NY, 575-631. (418)

_____ (1981) Active geothermal systems and hydrothermal ore deposits. Economic Geology 75th Anniversary Volume, 392-423. (496)

_____, L.J.P. Muffler and A.H. Truesdell (1971) Vapor-dominated hydrothermal systems compared with hot-water systems. Econ. Geol., 66, 75-97. (450)

White, S. (1973) Dislocations and bubbles in vein quartz. Nature, Phys. Sci., 243, 11-14. (35, 36)

White, W.H., A.A. Bookstrom, R.J. Kamilli, M.W. Ganster, R.P. Smith, D.E. Ranta and R.C. Steininger (1981) Character and origin of Climax-type molybdenum deposits. Economic Geology 75th Anniversary Volume, 270-316. (443)

White, W.S. (1971) A paleohydrologic model for mineralization of the White Pine copper deposit, Northern Michigan. Econ. Geol., 66, 1-33. (418)

Whitney, J.A. and J.C. Stormer, Jr. (1983) Igneous sulfides in the Fish Canyon Tuff and the role of sulfur in calc-alkaline magmas. Geology, 11, 99-102. (386, 475)

Whitney, J.A., F.O. Simon, J.J. Hemley and N.F. Davis (1979) Iron concentrations in chloride solutions equilibrated with synthetic quartz monzonite assemblages, Part 1: Sulfur free systems (abstr.). Geol. Soc. Am. Abstr. Programs, 11, 540. (445, 455)

Wiebe, R. and V.L. Gaddy (1940) The solubility of carbon dioxide in water at various temperatures from 12 to 40° and at pressures to 500 atmospheres. Critical phenomena. Am. Chem. Soc. J., 62, 815-817. (237)

Wilcox, R.E. (1983) Refractive index determination using the central focal masking technique with dispersion colors. Am. Mineral., 68, 1226-1236. (176)

Wilkins, R.W.T. (1977) Fluid inclusion assemblages of the stratiform Broken Hill ore deposit, New South Wales, Australia. Science, 198, 185-187. (366)

_____ (1979) Formation, modification and destruction of fluid inclusions. Notes for fluid inclusion workshop. Dept. Geol., La Trobe Univ. La Trobe, Bundoora, Victoria, Australia, Feb. 1979 (mimeographed, 8 p.). (14, 17, 35, 47, 70, 181)

_____ and J.P. Barkas (1978) Fluid inclusions, deformation and recrystallization in granite tectonites. Contrib. Mineral. Petrol., 65, 293-299. (21, 36, 73, 345)

_____ and J.R. Bird (1978) Characterization of healed fracture surfaces in fluorite by etching and proton irradiation. Lithos, 13, 11-18. (19)

_____ and _____ (1980) The use of proton irradiation to reveal growth and deformation features in fluorite. Am. Mineral., 65, 374-380. (19, 161)

_____ and A. Ewald (1982) A reconsideration of the theoretical basis of fluid inclusion decrepitometry. Program. Symp. on Current Research on Fluid Inclusions, Univ. of Utrecht, The Netherlands, April 22-24, 1981 (unpaginated). (265)

_____ and A.C. McLaren (1981) The formation of syngenetic fluid inclusions from etch pits in crystals. N. Jahrb. Mineral. Monats., 1981, 220-224. (17, 24)

_____ and W. Sabine (1973) Water content of some nominally anhydrous silicates. Am. Mineral., 58, 508-516. (84, 122, 127, 364)

_____ and D.A. Sverjensky (1977) The role of fluid inclusions in the exsolution of clinopyroxene in bustamite from Broken Hill, New South Wales, Australia. Am. Mineral., 62, 465-474. (68)

_____, J.R. Bird and M.D. Scott (1978) Decoration of growth and deformation dislocations in fluorite. Proc. 2nd Australian Conf. on Nuclear Techniques of Analysis, May 1978, Australian Inst. of Nuclear Sci. Eng., Lucas Heights, New South Wales, 33-35. (19)

_____ and A. Ewald (1981) Observations on deformation microstructures and fluid inclusions in proton-irradiated halite. N. Jahrb. Mineral. Abb., 141, 240-257. (19, 73)

_____, G. Hladky and K. Burlinson (1983) The rapid distinction of ore veins and barren veins by fluid inclusion decrepitometry. CSIRO (Australia) Div. Mineralogy Research Rev., 1983, 219-221. (265)

Williams, A.E. and W.A. Elders (1981) Oxygen isotope exchange in rocks and minerals from the Cerro Prieto geothermal system: indicators of temperature distribution and fluid flow. Symp. Cerro Prieto Geothermal Field, Baja California, Mexico, 3rd, March 24-26, 1981, San Francisco, California, 149-158. (498)

Wilson, J.C. (1978) Ore-fluid-magma relationships in a vesicular quartz latite porphyry dike at Bingham, Utah. Econ. Geol., 73, 1287-1307. (276)

Wilson, J.W.J., S.E. Kesler, P.L. Cloke and W.C. Kelly (1980) Fluid inclusion geochemistry of the Granisle and Bell porphyry copper deposits, British Columbia. Econ. Geol., 75, 45-61. (441, 443, 445, 451, 452)

Winograd, I.J. and F.N. Robertson (1982) Deep oxygenated ground water: anomaly or common occurrence? Science, 216, 1227-1229. (331)

Wise, D.U. (1964) Microjointing in basement, Middle Rocky Mountains of Montana and Wyoming. Geol. Soc. Am. Bull., 75, 287-306. (62, 67, 344)

_____ (1982) Linesmanship and the practice of Linear Geo-art. Geol. Soc. Am. Bull., 93, 886-888. (201)

Wlotzka, F. (1961) Geochemistry of nitrogen. Geochim. Cosmochim. Acta, 24, 106-154 (in German). (119)

Wodzicki, A. and F.E. Bowen (1979) The petrology of Poor Knights Island: a fossil geothermal field note. N. Zealand J. Geol. Geophys., 22, 751-754. (496)

Wood, D.L. and K. Nassau (1968) The characterization of beryl and emerald by visible and infrared absorption spectroscopy. Am. Mineral., 53, 777-800. (84)

Woods, T.L., P.M. Bethke, R.J. Bodnar and R.W. Werre, Jr. (1981) Supplementary components and operation of the U.S. Geological Survey gas-flow heating/freezing stage. U.S. Geol. Survey Open File Rept 81-954, 12 p. (197, 201)

—— E. Roedder and P.M. Bethke (1982) Fluid-inclusion data on samples from Creede, Colorado, in relation to mineral paragenesis. U.S. Geol. Survey Open File Rept 82-313, 77p. (77, 429, 431, 435)

Wosinski, J.F. and J.R. Kearney (1966) Methods of determining probable sources of gaseous inclusions in glass. Am. Ceramic Soc. Bull., 45, 1001-1004. (122)

Wright, A.W. (1881) On the gaseous substances contained in the smoky quartz of Branchville, Conn. Am. J. Sci., 3rd ser., 21, 209-216. (116, 119)

Wright, T.L. and P.C. Doherty (1970) A linear programming and least squares computer method for solving petrologic mixing programs. Geol. Soc. Am. Bull., 81, 1995-2008. (512)

Wyllie, P.J. (1979a) Kimberlite magmas from the system peridotite-CO_2-H_2O. In F.R. Boyd, and H.O.A. Meyer, eds., Kimberlites, Diatremes, and Diamonds: Their Geology, Petrology, and Geochemistry. Proc. 2nd Int'l Kimberlite Conf., Am. Geophys. Union, Washington, D.C., 1, 319-329. (407, 410, 504)

—— (1979b) Magmas and volatile components. Am. Mineral, 64, 469-500. (407, 410, 504)

—— (1980) The origin of kimberlite. J. Geophs. Res., 85, 6902-6910. (504)

Yakubova, V.V. (1952) An experiment in the study of inclusions in minerals of pegmatites from Murzinka (Urals). Mineral. Muzeya Akad. Nauk SSSR Trudy, 4, 102-121 (in Russian). (100)

—— (1955) An attempt at study of inclusions in minerals of Mursinka (Ural) pegmatites. Mineral. Muzeya Akad. Nauk SSSR Trudy, 7, 132-150 (in Russian; translated in Int'l Geol. Rev., 1, (8), 52-66, 1959). (100)

Yannas, I. (1968) Vitrification temperature of water. Science, 160, 298-299. (223)

Yardley, B.W.D. (1979) Fluid inclusion studies in metamorphic rocks (abstr.). Program 4th Mtg Geol. Soc. of the British Isles, Univ. of Sheffield, Sept., 1979 (unpaginated). (362)

—— (1982) How do fluids move through metamorphosing rock? (abstr.). Geol. Soc. (London) Newsletter, 11, 16-17. (345)

—— T.J. Shepherd and J.P. Barber (1983) Fluid inclusion studies of high-grade rocks from Connemara, Ireland. In M.P. Atherton and C.D. Gribble, eds., Migmatites, Melting, and Metamorphism. Shiva Pub. Ltd., Cheshire, UK, 110-126. (362)

Yasinskaya, A.A. (1967) Inclusions in stony meteorites. Mineral. Sbornik L'vov Gos. Univ., 21, 278-281 (in Russian; translated in Fluid Inclusion Res. -- Proc. of COFFI, 2, 149-153, (1969)). (2, 566)

—— (1978) Genetic classification of inclusions in cosmic minerals (abstr.). COFFI, 11, 232. (569)

Yermakov, N.P. -- See Ermakov, N.P.

Yoder, Jr., H.S. (1955) Role of water in metamorphism. In A. Poldervaart, ed., Crust of the Earth -- A Symposium. Geol. Soc. Am. Spec. Paper, 62, 505-524. (2)

* Yonge, C.J. (1981) Fluid inclusions in speleothems as paleoclimate indicators. In B.F. Beck, ed.. Proc. 8th Int'l Congress of Speleology, Vol. 1, Dept. Geol., Georgia Southwestern College, Americus, GA, 301-304. (328)

York, D., J.A. Hanes, P. Kuybida, C.M. Hall, W.J. Kenyon, A. Masliwec, S.D. Scott and E.T.C. Spooner (1980) The direct dating of ore minerals (abstr.). EOS, Trans. Am. Geophys. Union, 61, 399 (also Int'l Geol. Cong., 26th, Abstr. 1036, 1980). (128)

—— A. Masliwec, C.M. Hall, P. Kuybida, W.J. Kenyon, E.T.C. Spooner and S.D. Scott (1982a) Grant 62, Direct dating of ore minerals. Ontario Geol. Survey Misc. Paper 103, 213-219. (128)

—— P. Kuybida, J.A. Hanes, C.M. Hall, W.J. Kenyon, E.T.C. Spooner and S.D. Scott (1982b) $^{40}Ar/^{39}$ dating of pyrite. Nature, 300, 52-53. (128, 466)

Yoshida, T. (1979) Fluid inclusion study and ore forming process of the Iwani deposit, Shimane prefecture, Japan. Mining Geol. (Japan), 29, 21-31 (in Japanese; English abstr.). (425, 426)

Young, R.A. (1962) Mechanism of the phase transition in quartz. Final Rept. on Contract No. AF 49(638)-624 (Project A-447). Eng. Experiment Sta., Georgia Inst. Tech., Atlanta, GA, (ASTIA Document AD276235). (24)

Ypma, P.J.M. (1963) Rejuvenation of Ore Deposits as Exemplified by the Belledonne Metalliferous Province. Ph.D. dissertation, Univ. Leiden, Leiden, The Netherlands, 213 p. (in English; French and Dutch summaries). (52, 75, 76, 233, 261, 281)

—— (1965) An instrument for geobarometry of fluid inclusions (abstr.). Geol. Soc. Am. Spec. Paper 87, 190 (pub. 1966). (115)

—— (1969) The analysis and significance of volatile constituents in fluid inclusions and their P-T relations (abstr.). Fluid Inclusion Res. -- Proc. of COFFI, 2, 23-24. (76, 120)

—— (1979a) Destructive determination of gas compositions. Notes for fluid inclusion workshop. Dept. Geol., La Trobe Univ., La Trobe, Bundoora, Victoria, Australia, (mimeographed, 21 p.). (124, 125)

—— (1979b) Mineralogical and geological indications for the petroleum potential of the Etosha Basin, Namibia (S.W. Africa). Proc. Koninklijke Nederland. Akad. Wetenschappen, Ser. B, 82, (1), 91-112. (334, 347)

Yu, F.-S. and C.Y. Lin (1978) The heating microscope stage and its application in mineral exploration. K'uang Yeh Chi Shu 1978, 16, (1), 31-45 (in Chinese). (186)

Yurimoto, H. and S. Sueno (1984) Element distribution between olivine and the glass inclusion: an ion microprobe study. Geochem. Cosmochim. Acta, (in press). (148)

Yushkin, N.P. and B.I. Srebrodol'skii (1965) Composition of liquid inclusions in the crystals of native sulfur. L'vov. Gos. Univ. Mineral. Sbornik, 19, 229-236 (in Russian). (136, 328)

Zakharchenko, A.I. (1950) Study of liquid inclusions in quartz. L'vov. Geol. Obshch. Mineral. Sbornik, 4, 167-187 (in Russian). (114, 136)

—— (1955) Mineral-forming solutions and origins of quartz veins in the practice of the study of the quartz veins of Pamir and of the inclusions of solutions in minerals. Gosgeoltekhizdat, Vyp. 6, 106 p. (in Russian). (101)

* —— (1971) On specialization of granites and conditions of formation of related ore deposits (abstr.). Int'l Geochemical Congress, Moscow, 1971, Abstr. Reports, 2, 670-671 (in English). (399)

—— (1973) Pegmatite-forming melt-solutions, on the basis of their inclusions in minerals of granite-chamber pegmatites. COFFI 6, 172-183. (398)

—— (1976) Transition of melts into fluid solutions, the evolution of their composition, nature and metal content (on inclusions in minerals of granites and chambered pegmatites). Abstracts, 5th Int'l COFFI Symp. on Fluid Inclusions. Fluid Inclusion Res. -- Proc. of COFFI, 8, 200-201. (398)

WI
Za

Zakrzhevskaya, N.G. (1964) The problem of the origin of the gases in the rocks of the Khibina apatite deposits. Akad. Nauk SSSR, Dokl., 154, 118-120 (in Russian; translated in Dokl. Acad. Sci. USSR, Earth Sci. Sects., 154, 128-131, 1964). (404)

Zatsikha, B.V., O.I. Petrichenko, B.V. Dolyshny and V.A. Las'kov (1973) Genetic pecularities of mineral formation in the Slaviansk mercury deposit. L'vov. Gos. Univ. Mineral. Sbornik, 27, (4), 326-332 (in Russian). (463)

Zentilli, M. and P.H. Reynolds (1977) Dating of fluid inclusions in ore deposits with the 40Ar/39Ar method (abstr.). Geol. Assoc. Canada/Mineral. Assoc. Canada Programs Abstr. 2, 57. (128, 340)

Zhdanov, A.V., G.A. Satunkin, V.A. Tatarchenko and N.N. Talyanskaya (1980) Cylindrical pores in a growing crystal. J. Crystal Growth, 49, 659-664. (31)

Zhovtula, B.D. (1976) Studies of inclusions in quartz from liparite tuffs in Beregovo hill country. L'vov Gos. Univ., Mineral. Sbornik, 30, (1), 60-64 (in Russian; English abstr.). (186)

Ziegenbein, D. and W. Johannes (1982) Activities of CO2-H2O mixtures, derived from high-pressure mineral equilibrium data. In W. Schreyer, ed., High-Pressure Researches in Geoscience. E. Schweizerbart'sche Verlags., Stuttgart, 493-500. (240)

Zimmermann, J.-L. (1966) Mass spectrometer study of occluded fluids in some quartz samples. Compt. Rend. Acad. Sci. Paris, ser. D, 263, (5), 461-464 (in French). (122)

——— (1981) The liberation of H2O, CO2 and hydrocarbons from cordierites: Kinetics, structural sites and petrogenetic implication. Bull. Minéral., 104, 325-338 (in French). (363)

——— M. Arnold and J.-J. Guillou (1979) Différence entre le chimisme des inclusions visibles et celui de l'ensemble des composés volatils extraits des cristaux de quartz du filon du Gour Nègre (Gard). Compt. Rend. Acad. Sci. Paris, 288, ser. D, 863-866. (124)

Zimmerman, R.V. and S.E. Kesler (1981) Fluid inclusion evidence for solution mixing, Sweetwater (Mississippi Valley-type) district, Tennessee. Econ. Geol., 76, 134-142. (417)

Zindler, A. and E. Jagoutz (1984) Trace element and Nd and Sr isotope systematics of peridotite nodules from Peridotite Mesa, San Carlos, Arizona. Geochim. Cosmo. Acta, (in press). (526)

Zirkel, F. (1866) Über die mikroskopische Zusammensetzung und Structur der diessjärigen Laven von Nea-Kammeni bei Santorin. Neues Jahrb. Mineral., Geol. Paläont., 53, 769-787. (164, 474)
——— (1870) Mikromineralogische Mittheilungen. N. Jahrb. Mineral., Geol. Paläont., 1870, 801-832. (3, 114)
——— (1873) Die Mikroskopische Beschaffenheit der Mineralien und Gesteine. Wilhelm Englemann, Leipzig, Germany, 502 p. (3), 49, 82, 11, 474)
——— (1876) Microscopical petrography. U.S. Geol. Explor. 40th Parallel (King), 6, 297 p. (398)

Zolensky, M.E. and R.J. Bodnar (1982) Identification of fluid inclusion daughter crystals using Gandolfi X-ray techniques. Am. Mineral., 67, 137-141. (112, 113, 394, 444, 445)